Final
8-28 Tues
10:45 - 1:15

Sm
2016

Modern Mathematics

Seventh Edition

Modern Mathematics

Seventh Edition

Ruric E. Wheeler
Samford University

Brooks/Cole Publishing Company
Pacific Grove, California

CONTEMPORARY UNDERGRADUATE MATHEMATICS SERIES
Consulting Editor: Robert J. Wisner

Brooks/Cole Publishing Company
A Division of Wadsworth, Inc.

Printed in the United States of America
10 9 8 7 6 5 4 3 2 1

Library of Congress Cataloging-in-Publication Data
Wheeler, Ruric E., [date]
Modern mathematics.

Includes bibliographies and index.
1. Mathematics—1961- . I. Title.
QA39.2.W46 1988 510 87-25018
ISBN 0-534-08580-6

Sponsoring Editor: Jeremy Hayhurst
Editorial Assistant: Maxine Westby
Production Editor: Joan Marsh
Production Assistant: Dorothy Bell
Manuscript Editor: David Hoyt
Permissions Editor: Carline Haga
Interior Design: John Edeen
Cover Design: Victoria A. Vandeventer
Art Coordinator: Lisa Torri
Interior Illustration: Ayxa
Photo Editor: Carline Haga
Typesetting: Syntax International
Cover Printing: Phoenix Color Corp.
Printing and Binding: R. R. Donnelley & Sons Co., Crawfordsville, Indiana

Photo Credits
Page 12, Figure of Pythagoras courtesy of The New York Public Library, Astor, Lenox, and Tilden Foundations. Page 60, Figure of George Cantor courtesy of the Smith Collection, Rare Book and Manuscript Library, Columbia University. Page 70, Figure of René Descartes courtesy of The New York Public Library, Astor, Lenox, and Tilden Foundations. Page 79, Figure of Leopold Kronecker courtesy of The New York Public Library, Astor, Lenox, and Tilden Foundations. Page 144, Figure of Gottfried von Leibniz courtesy of the Smith Collection, Rare Book and Manuscript Library, Columbia University. Page 170, Figure of Eratosthenes courtesy of Culver Pictures, New York. Page 195, Figure of Evariste Galois courtesy of the Smith Collection, Rare Book and Manuscript Library, Columbia University. Page 362, Figure of Euclid courtesy of The New York Public Library, Astor, Lenox, and Tilden Foundations. Page 426, Photograph of gem tourmaline courtesy of the American Museum of Natural History, New York. Photograph of cadmium sulfide crystals courtesy of General Motors Research Laboratories. Page 427, Figures of The Parthenon and Head of Apollo of Belvedere courtesy of The Bettmann Archive, Inc. Page 428, Photographs of symmetry of butterfly and beetle courtesy of the American Museum of Natural History, New York. Page 429, Photograph of snow crystals courtesy of the National Oceanic and Atmospheric Administration. Photograph of mosaic window courtesy of Bulloz Art Reference Bureau. Page 430, "Study of Regular Division of the Plane with Horseman" by M. C. Escher. Copyright (c) BEELDRECHT, Amsterdam/VAGA, New York Collection Haags Gemeentemuseum. Photograph of geometry of a beehive courtesy of The American Museum of Natural History, New York. Page 431, Photograph of Athenian Vase (*Dipylon Style*) courtesy Museum of Fine Art, Boston. Page 523, Figure of René Descartes courtesy of The New York Public Library, Astor, Lenox, and Tilden Foundations. Page 559, Figure of Girolamo Cardano courtesy of The Bettmann Archive, Inc.

Preface

To the Instructor The seventh edition of this book is written in response to the many suggestions from previous users and in accordance with the recommendations of CUPM of the Mathematical Association of America and the eight basic recommendations of NCTM for the 1980's.

Problem Solving As suggested in the recommendations of NCTM, this book emphasizes problem solving throughout the book. Chapter 1 is devoted to problem solving and critical thinking. Each chapter is introduced with a problem to be solved using Polya″s problem solving suggestions, and each section of the book is introduced with a problem that can be solved using the material of the section. Yet, this book does not depend solely on Polya″s outline. Instead, other procedures of problem solving receive equal attention.

Critical Thinking All recent recommendations from major studies such as the Carnegie Foundation list critical thinking as an essential ingredient in general education for both the liberal arts student and the prospective elementary education teacher. Recent issues of *Focus, National Forum, Educational Leadership, Educational Researcher*, and the *Phi Delta Kappan* carried articles on critical thinking skills. Some states are including critical thinking in the statewide testing program. The missing ingredient in most of this surge of interest in critical thinking is that there is little application of critical thinking in mathematics (or arithmetic), the field best suited for emphasizing this subject. In this revision the student is introduced to critical thinking starting with the very first section of the book. If critical thinking can be developed in this study, then problem solving will automatically improve. Critical thinking, problem solving, and mathematical skills—these concepts are fundamental for a good education in the liberal arts and in elementary school teacher training.

Problems, Problems, Problems The greatest strength of this book has always been and continues to be the problems. Mathematics is not a spectator sport. To learn and appreciate mathematics we must become involved in the solving of problems. There are problems that every student will find easy (marked with an ***R***, routine), problems that require some thinking (marked with a ***T***, requiring thought), and ***C*** (challenge) problems. The Just for Fun problems are both challenging and thought-provoking.

Goals The author has three major goals in the writing of this revision.

(a) To write an elementary textbook that will be not only readable but interesting in order to develop a greater appreciation of mathematics.

(b) At the same time to present solid mathematical procedures in order to improve mathematical skills and mathematical maturity.
(c) To develop critical thinking while at the same time enhance computation skills.

Flexibility A great deal of time and effort have been expended to design a manuscript that provides maximum flexibility for the teacher. Nearly everyone has "his or her own particular arrangement of topics." Topics have been placed in this revision so that the instructor can pick topics and jump around to meet the needs of a class. For example, teach the metric system or integers wherever you wish because both topics are designed as a module to be moved if desired. Some want to teach integers before fractions and some after fractions. You make the choice.

Student Activities Manual (Nothing Like It) Although the Student Activities Manual was written primarily for prospective elementary school teachers, most of the activities would be useful and helpful to anyone. In addition to activities (coordinated with each section of the textbook) the manual contains a calculator corner, computer enrichment, discussion of flowcharts, errors commonly made, suggestions for the solution of Just for Fun problems, and a practice test for each chapter. You will find nothing like it because each section contains problems from elementary school textbooks to show the need for the material of the section.

Competency Examinations With a small amount of supplemental material in the Teacher's Manual (which can be xeroxed by the teacher), this revision satisfies competency requirements in states such as California, Florida, New York, and Texas. The Manual also contains sample competency examinations.

Transformational Geometry New material using transformational geometry provides a simple, easy to understand, unifying approach to congruence, similarity, constructions, and the development of geometric formulas. The simplicity of the new approach facilitates the inclusion of additional geometric concepts.

Logo Three sections on logo provide excellent activities in geometry and elsewhere.

Study Aids Students will find the study aids to be both interesting and helpful: biographical sketches that emphasize the person, historical notes in each "Flashback", points to emphasize in each "Point of Interest", Just for Fun problems that provide a challenge, a very comprehensive chapter summary and review, answers to odd-numbered problems and Just for Fun in the back of the book, an up-to-date bibliography at the end of each chapter, and calculator problems marked with a calculator symbol.

Instructor's Manual This manual, available from the publisher, assists the teacher by providing: detailed outlines for various types of courses; answers for those problems, not included in the textbook; two possible tests (one regular, one multiple choice) for each chapter; supplemental material (with permission from author to xerox) to satisfy competency requirements of many states.

Abbreviated Course outlines

One Semester, Liberal Arts Majors. Chapters: 1 and Appendix; 2, (Sec. 1, 2, 3); 4, (Sec. 6 and 7); 5, (Exclude Sec. 7); 6; 12; 13; and 14.

One Semester, Elementary Education Majors. Chapters; 1, (Exclude Sec. 6 and 7); 2; 3; 4, (Exclude Sec. 4 and 5); 5; 6, (Sec. 1, 3, and 4); 8; 10, (Sec. 3 and 4); 11, (Sec. 1, 2, 3, and 5).

Two Semester, Liberal Arts Major. Chapters; 1, (including Appendix); 2, (Sec. 1, 2, and 3); 4; 5; 6; 7; 8; 9; 10; 11, (Sec. 1–3); 12; 13; 14.

Two Semester Elementary Education Majors. Chapters; 1, (including Appendix); 2; 3; 4; 5; 6, (Sec. 1–4); 7, (Sec. 1, 6–8); 8; 9; 10; 11; 12 (1–3); 13, (Sec 1–3); 14.

More comprehensive outlines are available in the Teacher's Manual.

To the Student

I recognize that many students who use this book begin their study with a great deal of fear and anxiety. I have been very pleased through the years to watch this uncertainty for many change to a positive attitude toward mathematics, as indicated by one letter I received last fall.

"I entered your Ma 111 class this fall with fear and concern. I had been out of school for ten years and was never a strong student of mathematics. After a few weeks, I found I could read the book and work many of the problems. Not only have I learned a great deal of mathematics, I have gained valuable confidence. Thank you very much."

You, too, can have a different attitude toward mathematics, but only if you remember "Mathematics is not a spectator sport." To develop this appreciation of mathematics you must study carefully each section and work a sizeable number of problems. Your maturity in mathematics will come by solving problems.

You will find that at the beginning of each section I have written notes to you. In these notes I hope to accomplish a variety of things. I shall describe briefly in nonmathematical language the content of the section. Often I shall indicate where the material of the section is to be used elsewhere in the book. In general, I will try to place the specifics of what you are learning in the larger context of the goals of the course.

You may wish to refer to the *Student Activities Manual* to test your knowledge at the end of each chapter. (See description on preceding page.)

I wish you a successful semester. Mathematics can be fun if you will approach this study with the right attitude. Good luck!

Acknowledgments

I am most grateful to the following mathematicians who assisted not only in the sixth edition but in previous editions as well. These friends have been most helpful.

Alabama — Lila Acker, Samford University; Angela Hernandez, University of Montevallo; Dwight Holcomb, Snead State Junior College; Mary Hudson, Samford University; Walter G. McQueen, Chattahoochee Valley Community College; Winston Moody, Alabama A&M University.

California — M. A. Fitting, San Jose State University.

Hawaii Richard K. Coburn, Brigham Young University.

Illinois P. J. Blus, National College of Education; Benny F. Tucker, Illinois State University.

Indiana Don S. Balka, Saint Mary's College.

Kansas Helen Kriegsman, Pittsburg State University.

Maine Anthony Soychak, University of Southern Maine.

Maryland John P. Jones, Frostburg State College

Massachusetts Arnold F. Checchi, Salem State College; Joanne Collamore, Anna Maria College; John Tobey, North Shore Community College.

Minnesota Francis T. Hannick, Mankato State University.

Mississippi Reba H. Davis, Northeast Mississippi Junior College.

New Hampshire Richard Evans, Plymouth State College.

North Carolina Grace P. Foster, Beaufort County Community College; Gladys M. Kerr; Wingate College; Emmett S. Sams, Mars Hill College; Joyce E. Speas, Elon College.

Oregon Ron Waite, BMCC

Pennsylvania John T. Held, Gettysburg College.

South Carolina R. A. Lytle, University of South Carolina; Wade H. Sherard, Furman University; Charles M. Walker, University of South Carolina-Union.

South Dakota Verlyn Lindell, Augustana College.

Tennessee Charles E. Hampton, Free Will Baptist College; Wanda Vickery, Dyersburg State Community College.

Texas Robert Netherland, Tareton State University; Sister M. Geralda Schaefer, Pan American University; Donald P. Skow, Pan American University; Evelyn Sowell, University of Texas at Tyler.

Wisconsin Barbara L. Boe, Edgewood College; Harald M. Ness, University of Wisconsin Center.

Wyoming A. Duane Porter, University of Wyoming.

Many mathematics teachers have made suggestions or have responded to a questionnaire concerning changes that should be made in the seventh edition. I am most appreciative of suggestions made by the following people:

Alabama Janelle Elrod

Arkansas Tommy Leavelle

California M. Kurshed Ali, H. U. Coulson, Ann Louis, Douglas B. McLeod, Sherylln Rawls, V. Merriline Smith, W. A. Stannard, Ann Watkins, Norman T. Woo

Colorado Gail Barnhart, Curtiss O. Mallory, C. O. Worm

Florida Vicki Schell

Georgia	Professor Hicks, John Stevens
Illinois	Jerry L. Frang
Iowa	Glen Blume, Cheryl Wangsness
Kentucky	Mary Catherine Brown, Joe K. Smith
Louisiana	Mary T. Benson, Virginia Eddleman, Jane Ellett, Anita Fleming, Larry Foster, Richard Fritsche, James A. Gauthier, Karen Moore, Lee Wilder
Maine	Richard Paul
Massachusetts	Peter K. Gurau, Raymond Hebert, Al Roy
Michigan	Zakaria Musallen
Minnesota	Virginia Christian
Mississippi	Kay Matkins
Montana	Sharon B. Walen
Nebraska	Clinton Ludeman
New Jersey	Joan Guetti, Ilene Katz
New Mexico	Vern Witten
New York	Janet Elmore
North Carolina	A. M. Black, Robert K. Colling, John W. Dunn, Pat Jones, W. P. Love, Nancy Medford, Charles D. Mooney, Harold Williford
Ohio	Mary Coughlin, Janice M. Green, Lowell Leake, Charles Nicewonder, Paula Saintignon
Oregon	Ron Smit, Geraldine Swenson
Pennsylvania	Bernadette Antkoviak, G. Kip Bollinger, Arlene Dowshen, Donald E. Gracey, Robert A. Leume, Joseph L. Rousseau, W. H. Seybold, Ronald L. Shubert, Clark Taylor
South Carolina	Lucy Snead
Utah	Patricia Henry, Kent Kidman, Harry Wickes
Wisconsin	Catherine Kiaie
Texas	Genelle Beck, Harlon Haertig, Marsha May, Patricia Ann Rhodes, Reta Parrish, Samuel R. Thompson, Cindy Trout, Kenneth W. Wunderlich
West Virginia	Kathleen Kleinfelter
Wyoming	Katheryne Earl, Maryanne Marietta, Harry Oxley
Puerto Rico	Marilu Lebron Vazguez
Canada	W. E. Pfaffenberger, E. B. Horne, Jim Vance

I appreciate the suggestions and comments of the reviewers of the first six editions: Marilee Adams, Evelyn D. Bell, Richard W. Billstein, Irving Drooyan, Arnold M. Dunn, Daniel J. Ewy, Hilton G. Falahee, Betty Freemal, Robert Fullerton, Reuben J. Goering, George Henderson, Irving Hollingshead, Alan Holz, John Huber, Lyndal R. Hutcherson, Larry Johnson, M. Bonnie Kelterborn, F. J. Lorenzen, William W. Mitchell, Jr., Linda Oldaker, Cada Parrish, John M. Peterson, Robert L. Poe, Daniel H. Reeves, Jr., Wayne Rich, Sister Geralds Schaefer, Wade H. Sherard III, James Simpson, James R. Smart, Joe K. Smith, Karl J. Smith, V. Merriline Smith, Karl Stromberg, Edward C. Wallace, and Robert J. Wisner.

I am especially appreciative of reviewers of this edition: Bernadette Antkoviak, Harrisburg Area Community College; Keith Craswell, Western Washington University; Richard Fritsche, Northeast Louisiana University; Thomas A. McCready, California State University–Chico; Wayne Rich, Utah State University; and J. K. Smith, Northern Kentucky University.

Thank you, Robert J. Wisner and Craig Barth for your support through the years. The production of this edition is a team effort: the sterling leadership of Jeremy Hayhurst and Joan Marsh, Sharon Kinghan, Maxine Westby, Lisa Torri, Dorothy Bell, and all the others at Brooks/Cole.

Finally, sincere appreciation goes to my wife, Joyce, who has been understanding of a husband who has spent hours in the office working on this book.

Ruric E. Wheeler

Contents

* These sections may be omitted without loss of continuity or comprehension.

CHAPTER 4

Elementary Number Theory and the System of Integers 160

CHAPTER 5

Introduction to the Rational Numbers 217

CHAPTER 6

Real Numbers and a Return to Problem Solving 280

CHAPTER 7

An Introduction to Computers 319

Modern Mathematics

Seventh Edition

CHAPTER 1
Critical Thinking, Problem Solving, and Logic

Three sailors and a monkey are shipwrecked on a deserted island, where they discover that coconuts are the only available food. The men collect coconuts from all the trees, but because the sun is setting, they decide to wait until morning to divide the supply. However, once Bob and Carl are asleep, Al arises, divides the coconuts into three equal shares, gives the remaining coconut to the monkey, hides his share, restacks the others, and returns to sleep. Later, Bob sneaks out of bed, divides the pile of coconuts into three equal shares, has one nut left over, gives it to the monkey, hides his pile, restacks the remaining coconuts, and returns to sleep. Toward morning, Carl, too, carefully moves from his bed, divides the pile into three equal shares, has one coconut left over, gives it to the monkey, hides his pile, restacks the remaining coconuts, and returns to bed. After this busy night, the sailors meet again and divide the pile of coconuts. Again, one coconut is left over for the monkey. What is the least possible number of coconuts in the original pile? (See the solution on p. 53)

Do you want to be able to analyze and comprehend complex descriptions and directions? Do you want to improve your problem-solving ability? Do you want to improve your test scores on true-false, multiple-choice, and standardized tests such as the GRE, the NTE, and the LSAT? This chapter will introduce to you some aids to productive thinking, sometimes called **critical thinking,** that should improve your test-taking techniques and make you a more productive citizen.

There is a new emphasis upon higher-order cognitive skills that lead to improved thinking. This emphasis on critical thinking could be the most important educational trend of our time. Recent issues of *Focus, National Forum,* and *Educational Leadership* have carried articles on the teaching of thinking skills. This long-overlooked but important concept is now receiving special attention. Many colleges today are requiring courses in critical thinking. For example, the California State University system has a course in critical thinking as a graduation requirement.

Many courses in critical thinking are taught apart from mathematics, but the reasoning that should accompany any good mathematics course leads to improved critical thinking. Since some mathematics courses are being taught without emphasis on cognitive skills, a special effort will be made to emphasize these skills in this book.

All sections of Chapter 1 (problem solving and logic) contribute to the development of critical thinking skills, but the first two sections call particular attention to this subject. The thread of critical thinking will extend throughout the book. Additional material on cognitive reasoning and formal logic will be found in the appendix, for those who wish to give added emphasis to this subject. Problem solving will be emphasized throughout the book.

1 Critical thinking: From verbal phrases to mathematics

■ If the third number is less than the first number, circle the second number unless the second number is less than the fourth number. In this case, circle the number that is the difference between the first number and the sixth number.

6 2 3 5 4 1 ■

How does solving simple problems such as this one improve thinking? First let's define *thinking* as any mental activity that helps to solve a problem, make a decision, or assist in understanding. It is a search for answers. In English and reading courses, the emphasis is placed on reading for comprehension. Believe it or not, reading is also important in mathematics. When reading for understanding, you cannot just skim the material or read at speeds of a thousand or more words a minute. Reading for comprehension usually demands slow, thoughtful, analytical reading.

Much attention is given in this section to helping you read instructions and follow directions precisely. Emphasis will be given to a step-to-step analysis of what you read. Creative guidelines will be given to help you develop this type of reading.

Also, we start practicing a very powerful critical thinking tool—**converting verbal statements to mathematical shorthand or terminology.** Converting from statements or words to mathematical expressions is the beginning of algebra and can be taught at all levels. In this section, we consider very elementary concepts. We will build on these concepts as we study the various number systems in the first half of this book.

Reading material that involves mathematics is somewhat different from reading other material. It is best to read at a rate that affords the greatest comprehension; in fact it may be necessary to read some material more than once. Make certain you do not overlook key words, phrases, and ideas. Sometimes it helps to ask the question, "Do I completely understand what I have read?" If questions come to mind while reading, you should mark the place and come back to the questions on a second reading.

We now present four carefully selected guidelines that will help you to understand what you read.

Guidelines for Comprehension

1. Note key words or phrases.
2. Write down what is being asked for.
3. Restate in your own words what you have read.
4. Prepare a visual representation (a picture, a graph, or a chart).

Let's apply these guidelines to the following example.

Example 1 • If the word *happiness* contains more than 7 letters (including repeated letters) and fewer than 3 vowels, circle the second vowel. Otherwise, circle the consonant that precedes the last vowel in the word. (*h* is not a vowel.)

Solution

1. Some key words are *more than, fewer than, and, otherwise,* and *precedes.*
2. We are to circle a letter in *happiness.*
3. (a) Count the letters in *happiness.* Are there more than 7?
   (b) Count the vowels in *happiness.* Are there fewer than 3?
   (c) Are both (a) and (b) satisfied? If so, circle the second vowel.
   (d) If either (a) or (b) is not satisfied, circle the consonant that precedes the last vowel.
4. A pictorial representation may or may not be helpful in this problem.

	Letters	*Vowels*
Number	9	3

There are 9 (more than 7) letters in *happiness.* There are exactly 3 vowels in *happiness,* so "fewer than three" is not satisfied. The statement that both (a) and (b) are true is not correct. Therefore we circle the consonant that precedes the last vowel in *happiⓝess.* •

Example 2 • On a certain day, I ate lunch at Wendy's, went swimming in the campus pool, attended a photography class, and had a doctor's appointment for a checkup. Wendy's is closed on Mondays; there is no swimming in the pool on weekends; the photography class meets only on Monday, Wednesday, and Friday; and my doctor has office hours only on Monday, Tuesday, and Friday. On which day of the week did I do all these things?

Solution

1. A key word is *and,* requiring that several conditions be satisfied.
2. It is a day of the week that is needed.
3. On which day of the week did I eat lunch at Wendy's, go swimming, attend a photography class, and see my doctor?
4. The fourth guideline is most important for this example. Let's use a chart.
   Step 1. Write the days of the week in some order.

   S M T W Th F Sat

Step 2. Wendy's is closed on Monday.

S ~~M~~ T W Th F Sat

Step 3. The pool is closed on weekends.

~~S~~ ~~M~~ T W Th F Sat

(M was crossed out in Step 2.)

Step 4. The photography class meets only on Monday, Wednesday, and Friday. This means it does not meet on Tuesday and Thursday.

~~S~~ ~~M~~ ~~T~~ W ~~Th~~ F ~~Sat~~

(M, S, and Sat are already crossed out.)

Step 5. My doctor has office hours only on Monday, Tuesday, and Friday, so this office is closed on Wednesday.

~~S~~ ~~M~~ ~~T~~ ~~W~~ ~~Th~~ F ~~Sat~~

Step 6. On which day of the week did I do all these things?

Friday •

We return now to the problem at the beginning of this section.

Example 3 • If the third number is less than the first number, circle the second number unless the second number is less than the fourth number. In this case, circle the number that is the difference between the first number and the sixth number.

6 2 3 5 4 1

Solution Some key words are *less than* and *unless.* We are to circle a number in the list. The third number (3) is less than the first number (6), so we would write the second number *unless* the second number (2) is less than the fourth number (5). The "unless" clause is *true,* so we move to the next statement. We circle the number that is the difference between the first number and the sixth number. $6 - 1 = 5$, so we circle the 5.

6 2 3 ⑤ 4 1 •

From our practice in reading, we move now toward expressing verbal statements in mathematical shorthand.

Example 4 • Translate the following into mathematical symbols:
(a) The difference between eighteen and the product of three and four
(b) Four plus the product of six and eight

Solution (a) $18 - 3 \cdot 4$ (b) $4 + 6 \cdot 8$ •

A symbol that may be replaced by any number of a set of elements is called a **variable.** We use $\square$, x, y, n, etc., to represent variables. Ordinarily, variables are associated with the study of algebra, but there is no reason why they should not be a part of all mathematics courses.

Example 5 • Translate the following into mathematical language:
(a) Six added to the product of nine times a certain number
(b) Eight less than the product of three times a number

Solution (a) $9 \cdot x + 6$ (b) $3 \cdot x - 8$ •

In the preceding example, note that $9 \cdot x$ can also be written $9x$, and $3 \cdot x$ as $3x$.

Example 6 • Replace the variable x with the number 3 and evaluate.
(a) $8x + 4 =$ (b) $2(x + 1) - x$

Solution (a) $8 \cdot 3 + 4 = 28$ (b) $2(3 + 1) - 3 = 5$ •

Sentences that can be identified as either true or false are called **statements.** Each of the following is a statement.
(a) $8 + 2 = 2 + 8$ (a true statement of equality)
(b) $7 - 3 = 8 - 4$ (a true statement of equality)
Statements of equality are called **equations.** Statements of inequality may be written using any of the following symbols:

$\neq$: is not equal to

$>$: is greater than

$<$: is less than

Precise definitions for *less than* and *greater than* will be given in Chapter 2. Your intuitive understanding of these symbols will be sufficient for this section.

Example 7 • (a) $6 + 3 > 5 - 2$ (a true statement of inequality)
(b) $5 + 4 \neq 2 + 1$ (a true statement of inequality)
(c) $7 - 3 > 2 + 4$ (a false statement of inequality) •

Many sentences involving variables cannot be classified as being true or as being false. These are not statements but are called **open sentences.**

Example 8 •

Equation	*Inequality*
$x + 4 = 7$	$x + 1 < 10$ •

flashback

The father of critical thinking, Socrates, lived more than 2400 years ago. Confused meanings, inadequate evidence, and self-contradictory beliefs were prevalent throughout Greece. Socrates developed a probing method of questioning that exposed intellectual confusion and challenged accepted ideas. It ultimately led to his execution. However, Socrates' insights have been emphasized through the years by such intellectuals as Voltaire, Newman, John Stuart Mill, and William Graham Sumner.

Just for fun

Five students are sitting in a circle. Two pairs of students have the same color hair. Those with the same color hair are not sitting next to each other. Celeste is on the right side of Teresa and on the left side of Jane. Celeste has the same color hair as Ruth. Jane has the same color as Alice. Identify the positions of the girls in the circle.

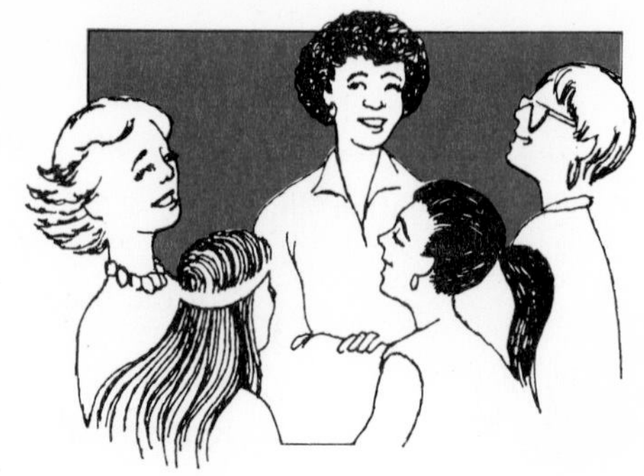

A very important step in the development of critical thinking skills is the ability to write open sentences that represent verbal expressions.

Example 9 • Translate the following expressions into mathematical sentences.
(a) A number added to fifteen is equal to twenty-three.
(b) A number less three is greater than seven.

Solution (a) $x + 15 = 23$
(b) $\square - 3 > 7$ •

Variables such as x in (a) and $\square$ in (b) are actually place-holders for replacements. After a replacement is made, the open sentence becomes a statement because it can be classified as true or as false.

Example 10 • Use the open sentences of the preceding example, replacing x by 8 in (a) and $\square$ by 8 in (b) and classiyfing the statements as true or false.

Solution (a) $8 + 15 = 23$ (true)
(b) $8 - 3 > 7$ (false) •

If a replacement makes an open sentence into a true statement, then the replacement is called a **solution** of an equation or inequality. The possible replacements of a variable of an equation or an inequality make up the **domain** of the variable.

Example 11 • Is there a solution for $x + 3 = 8$ on the domain $\{1, 2, 3, 4, 5, 6\}$?

Solution Yes, x replaced by 5 gives a true statement: $5 + 3 = 8$. So 5 is a solution of $x + 3 = 8$. •

Example 12 • Find the solutions for $2x + 3 < 9$ on the domain $\{1, 2, 3, 4\}$.

Solution For $x = 1$, $2 \cdot 1 + 3 < 9$ (true)
For $x = 2$, $2 \cdot 2 + 3 < 9$ (true)
For $x = 3$, $2 \cdot 3 + 3 < 9$ (false)
For $x = 4$, $2 \cdot 4 + 3 < 9$ (false)
So 1 and 2 are solutions of $2x + 3 < 9$. •

Exercise set 1

R 1. If the word contains more than 7 letters and fewer than 4 vowels, circle the first vowel. Otherwise, circle the consonant that precedes the first vowel.

(a) mathematics (b) surely
(c) deadly (d) lively
(e) frankly (f) summertime

2. In the alphabet, which letter is as far away from T (and in the same direction) as M is from Q?

3. If you are facing southeast and turn 90° left, then do an about-face and then turn 90° right, where are you facing?

4. 1 2 7 3 5 4 8

Add the third number to the seventh number, then divide by 5 and write the quotient, unless it is greater than 2. In this case, find the sum of the first number and the next-to-last number, divide by 5, and write the quotient.

5. Find a number whose distance below 15 is twice as great as the distance that 6 is above 3.

(a) 11 (b) 7 (c) 8 (d) 9 (e) 6

6. If deleting the letter *y* from the word *candy* leaves a 4-letter conjunction, mark out the *y* in the word *candy*. Otherwise, circle the *a* in the word *candy* in the third occurrence of the word in this exercise.

7. Translate each of the following into mathematical symbols.

(a) 7 minus 3 (b) The sum of 16 and 7
(c) 8 divided by 4 (d) 8 exceeds 3 by 5
(e) The product of 3 and 17
(f) 10 is 4 more than twice three.
(g) 16 is equal to the product of 2 and 8.
(h) 9 is less than the sum of 5 and 7.

8. Find the solutions for each sentence.

Sentence	*Replacements*
(a) $n + 3 = 6$	$\{2, 3, 4, 5\}$
(b) $n - 3 = 6$	$\{7, 8, 9, 10\}$
(c) $4n = 20$	$\{5, 6, 7, 8\}$
(d) $n \div 3 = 5$	$\{14, 15, 16, 17\}$
(e) $6 + n = 15$	$\{7, 8, 9, 10\}$
(f) $12 - n = 4$	$\{5, 6, 7, 8\}$

9. Use $\{0, 1, 2, 3, 4, 5\}$ as the domain. Find the solution for each sentence.

(a) $2x = x + 4$ (b) $2w = w + 3$
(c) $21y = 3y + 2y$ (d) $2y = 5 + y$

10. Translate the following verbal phrases into mathematical expressions or sentences.

(a) Four added to a certain number
(b) Six subtracted from a certain number
(c) Fifteen subtracted from a certain number
(d) Eight more than a certain number
(e) Six added to the product of 9 and a certain number
(f) Thirty more than twice a certain number
(g) Sixteen less than the product of 10 and a certain number
(h) Fourteen plus the product of 3 times a certain number
(i) Twenty subtracted from the product of 5 and a certain number
(j) Twenty-five more than the product of 3 and a certain number
(k) Six plus this number is equal to fifteen.
(l) Twenty is less than this number minus six.
(m) The sum of two and this number is greater than 8.

T 11. Translate the following verbal phrases into mathematical expressions or sentences.

(a) The product of x and y decreased by the sum of x and y
(b) The difference between two numbers is 6 and the smaller number is s; what is the larger number?
(c) The quotient of 4 times x and 6 is equal to 8.
(d) The difference of 5 times a number and 7 is equal to 8.
(e) If 20 pens cost x dollars, what is the price of one pen?
(f) If x pens cost 20 dollars, what is the price per pen?
(g) Three less than twice x

12. Find the solution for each open sentence.

Sentence	*Replacements*
(a) $n + 2 > 6$	$\{4, 5, 6, 7\}$
(b) $n - 4 < 5$	$\{6, 7, 8, 9\}$
(c) $2n + n > 9$	$\{0, 1, 2, 3\}$
(d) $2n - 3 > 4$	$\{1, 2, 3, 4\}$
(e) $3 - n < 3$	$\{0, 1, 2, 3\}$

13. The college activity center is open Monday through Saturday. Swimming classes meet daily except Wednesdays. Racketball instruction is available daily except Tuesdays and Saturdays. Daily classes are available in tennis. Basketball instruction is given every other day starting on Mondays. Volleyball starts on Tuesdays and continues daily till the end of the week. On which day does the largest number of activities occur?

14. Express as a mathematical sentence and solve if possible on the domain of counting numbers 1 through 30.
 (a) What number added to 7 equals 10?
 (b) Three times what number equals 21?
 (c) Five subtracted from what number equals 12?
 (d) One number is 6 more than another, and their sum is 20. Find the numbers.
 (e) Three times a number divided by 2 is equal to the sum of the number and six. Find the answer.
 (f) Twice the sum of a number and 4 is equal to 14. Find the number.
 (g) Six more than twice a number is 20. Find the number.
 (h) The sum of two different numbers is 12. The larger is twice the smaller. Find the smaller number.

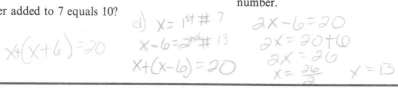

2 Critical thinking and inductive reasoning

■ Leslie has taken a job in which she is to be paid 1 penny for the first day, 2 pennies for the second day, 4 pennies the third day, 8 pennies the fourth day, and so on. How much will she receive for the fifteenth day? ■

The preceding problem is solved by observing patterns. What does the word *pattern* mean to you? We consider a pattern to be any kind of regularity that can be recognized by the mind. You will see many different kinds of patterns in this section. You will first learn to recognize very simple nonmathematical patterns, then you will study some number patterns, and finally you will learn to recognize patterns that will be valuable to you as you gain mathematical maturity and become a problem solver.

Patterns pervade mathematics. Students are introduced to patterns at the most elementary level through addition and multiplication tables and algorithms. We will emphasize the contributions of the search for patterns to critical thinking. Much of our thinking involves generalizations from patterns; such thinking is called **inductive reasoning.**

Begin your study of patterns by looking at two problems that you might have encountered in an elementary school textbook.

Example 13 • Problems on IQ tests as well as problems in elementary textbooks relate patterns to input-output machines. We are told that for the given input we get the following output.

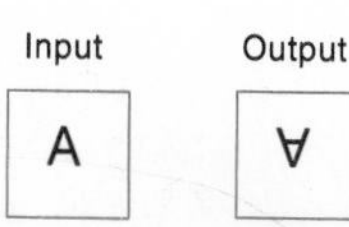

What would the machine give as output for the following input?

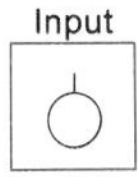

Select your answer from the following.

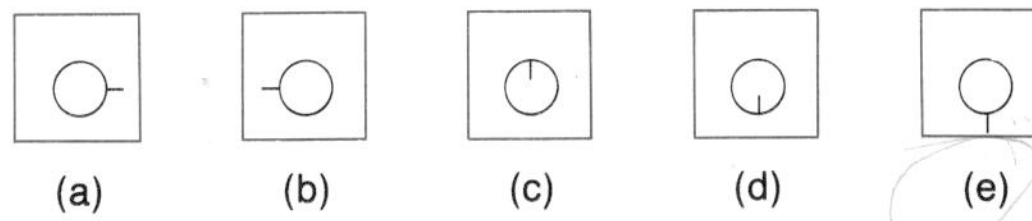

Solution Did you observe that the machine inverted (turned upside down) the input image? Hence the answer is (e). •

Example 14 • The dot printer is also often introduced in a study of patterns. Suppose a dot printer is programmed to print the following designs.

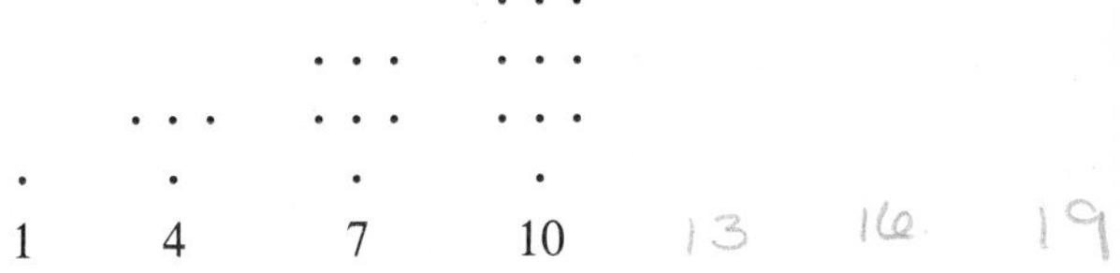

What number will the next pattern represent?

Solution Did you add another row of three dots to the design? The answer is 13. •

Patterns occur in many forms, such as the sequence of letters in the alphabet shown in the following example.

Example 15 • Suppose you were asked to write the next two letters that would appear in the following list.

$$B, A, E, D, H, G, K, J, \_\_\_\_, \_\_\_\_$$

Would your answer be *P*, *O* or *S*, *R* or *N*, *M* or *N*, *O*, or would you say, "I don't know," because you were unable to find an answer?

Solution In order to discover patterns, many students need to develop the ability to concentrate on the meaning of each part of the problem. To emphasize the meaning of each part, let's describe what we discover.

First note that *A*, which follows *B* in the series, is the letter preceding *B* in the alphabet. This is a possible pattern. Are there any more pairs of letters with this pattern as a repetition? Are we in luck? How about *E*, *D*; *H*, *G*; and *K*, *J*? Thus, the two letters we are trying to discover must be such that the second letter precedes the first letter in the alphabet. What are the letters? Maybe we need some additional information.

Since the letters evidently go in pairs, let's look at the first letter of each pair. Study carefully the relationship between *B* and *E*. What do you note? There are two letters in the alphabet between *B* and *E*. Could this be the beginning of a pattern? Yes; there are two letters between *E* and *H*, and two between *H* and *K*. If there are two between K and our first unknown letter, this letter will be N. From our first discovery, the next letter will be M. Thus, from our list of possible answers, N, M was correct. (Note: In this type of problem, the answer is not necessarily unique, as will be discussed later.) •

Example 16 • In an unfamiliar language, *zer mon* means "brown dog," *mil mon miko* means "little brown bird," and *kon miko* means "song bird." What is the word for *little* in this language?

Solution This is a fairly easy problem, but students often miss this type of problem because they do not take time to discover the patterns systematically.

"Bird" is in both "little brown bird" and "song bird." The common word in *mil mon miko* and *kon miko* is *miko,* so *miko* means "bird." "Brown" is in both "brown dog" and "little brown bird." *Mon* is common to *zer mon* and *mil mon miko.* Therefore, *mon* means "brown." The only word left for "little" in *mil mon miko* is *mil;* so *mil* means "little." •

We now look at patterns involving numbers.

Example 17 • The counting numbers are the numbers $\{1, 2, 3, 4, \ldots\}$. Those counting numbers that are divisible by 2, $\{2, 4, 6, \ldots\}$, are called **even numbers,** and those not divisible by 2, $\{1, 3, 5, 7, \ldots\}$, are called **odd numbers.** What can you conjecture about the sum of two odd numbers?

Solution

$$\begin{array}{llll} 1+3=4, & 3+5=8, & 5+7=12, & \ldots \\ 1+5=6, & 3+7=10, & 5+9=14, & \ldots \\ 1+7=8, & 3+9=12, & 5+11=16, & \ldots \\ \vdots & \vdots & \vdots & \end{array}$$

It appears from the examples that the sum of two odd numbers is an even number. •

Example 18 • By looking at several examples, make a guess or a **conjecture** about the sum of two consecutive odd numbers.

Solution

$$\begin{array}{lll} 1+3=4, & 3+5=8, & 5+7=12 \\ 7+9=16, & 9+11=20, & \ldots \end{array}$$

What kind of numbers are the sums? What number divides the sums of consecutive odd numbers? •

Example 19 • Now examine the repeated sum of consecutive odd numbers starting with 1.

Solution

$$1=1$$
$$1+3=4$$
$$1+3+5=9$$
$$1+3+5+7=16$$
$$1+3+5+7+9=25$$

Do you see the pattern? The sum of the first two consecutive odd numbers is $1 + 3 = 2 \cdot 2$; the sum of the first three odd numbers is $1 + 3 + 5 = 3 \cdot 3$; the sum of the first four is $1 + 3 + 5 + 7 = 4 \cdot 4$; and the sum of the first five is $1 + 3 + 5 + 7 + 9 = 5 \cdot 5$. Without addition, what is your guess about $1 + 3 + 5 + 7 + 9 + 11$? Did you guess $6 \cdot 6 = 36$? Now use this pattern to find the sum of the first ten odd counting numbers. (Answer: 100) •

Example 20 • A multiplication table of 9's provides an interesting study of patterns.

$$1 \cdot 9 = 09$$
$$2 \cdot 9 = 18$$
$$3 \cdot 9 = 27$$
$$4 \cdot 9 = 36$$
$$5 \cdot 9 = 45$$
$$6 \cdot 9 = 54$$
$$7 \cdot 9 = 63$$
$$8 \cdot 9 = 72$$
$$9 \cdot 9 = 81$$

What patterns do you notice? Note that the sum of the two digits in the answer $(1 + 8, 2 + 7, 3 + 6, 4 + 5, \ldots)$ equals 9. Notice also that the tens digit (digit on the left) increases by 1 and that the units digit (digit on the right) decreases by 1. What will happen if you continue the table? $11 \cdot 9 = 99$ and $9 + 9 = 18$, which is not 9. However, $1 + 8 = 9$. Try $43 \cdot 9 = 387$. Notice that $3 + 8 + 7 = 18$ and $1 + 8 = 9$. Are you ready to attempt to describe this pattern? Try other multiplications to check your conjecture. •

Among the mathematical patterns studied by the Pythagoreans, ancient Greek mathematicians, was a pattern for a set of numbers called **triangular numbers.** (See Figure 1.) The numbers of blocks in the successive triangles are triangular numbers.

Figure 1

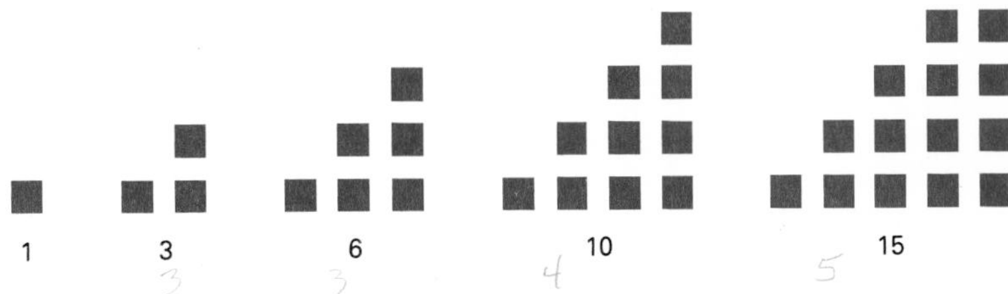

Example 21 • Determine the geometric pattern that governs the triangular numbers and see if you can write down the next two triangular numbers after 15.

Solution It appears that each successive triangle is obtained from the preceding one by drawing a new column on the right. In the fourth triangle, a column of four blocks is added. In the fifth triangle, a column of five blocks is needed. What are the next two triangular numbers? Did you get 21 and 28? Draw these. •

Now that you are thinking geometrically, you are ready for the next example.

BIOGRAPHY

Pythagoras
584(?) B.C.–495(?) B.C.

Tradition, fable, and even mystery surround our knowledge of the life and works of Pythagoras. He was primarily a philosopher and a religious mystic. From his travels throughout the ancient kingdoms of Egypt and Babylon, he learned a strange mixture of magic, science, astrology, and mathematics. He established a cult called the *Order of the Pythagoreans* based on the motto "All is number." The followers of Pythagoras believed that the study of mathematics led to the elevation of the soul, and in this pursuit many worthwhile mathematical discoveries were thereby achieved, all of which were credited to Pythagoras.

Example 22 • The top four diagrams form a series that changes according to some rule. (See Figure 2.) Try to discover the rule and choose from among the alternatives the diagram that should occur next.

Solution Looking at the four rectangles from left to right; notice that some lines have been removed in each step. Two vertical lines are removed from the first rectangle to obtain the second rectangle. What is removed from the second to obtain the third rectangle? (One horizontal line.) What is removed from the third to obtain the

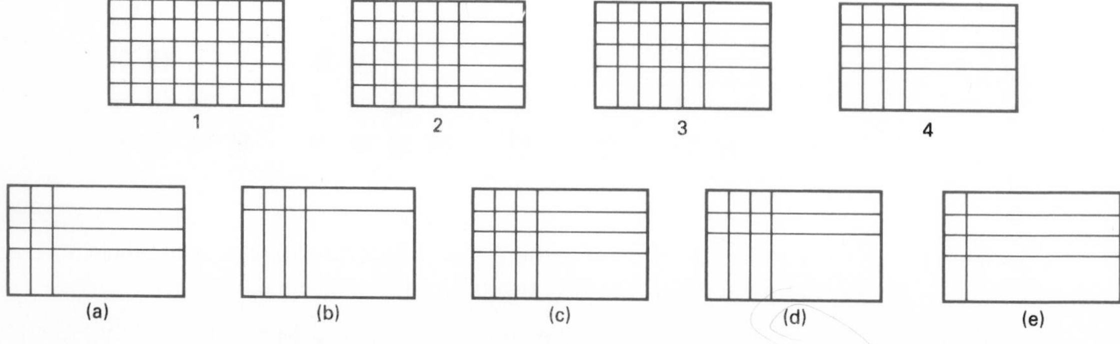

Figure 2

fourth rectangle? (Two vertical lines.) One possible pattern could be to remove two vertical lines and then one horizontal line and repeat the process. Did you get answer (d)? •

Discovering patterns, forming conjectures, and generalizing results are all demonstrated in a study of number sequences.

A number sequence

> A *number sequence* is a collection of numbers arranged in order so that there is a first term, a second term, a third term, and so on. Sequences are arranged left to right, and the numbers are separated by commas.

Example 23 • What pattern can be detected for the sequence

5, 8, 11, 14, _____, _____, _____, . . .?

Find the next three terms.

Solution Each term of the sequence is three more than the preceding term. A reasonable continuation of the pattern would indicate that the next three terms of the sequence are 17, 20, 23. •

Strictly speaking, there is often no unique answer to this type of problem. For example, you would probably guess that the next three terms of

10, 13, 18, 25, _____, _____, _____

are 34, 45, and 58. The reasoning you used to obtain these answers might have involved increasing each term of the sequence by increasing odd numbers (3, 5, 7, . . .):

$$10 + \underline{3} = 13$$

$$13 + \underline{5} = 18$$

$$18 + \underline{7} = 25$$

$$25 + \underline{9} = 34$$

$$34 + \underline{11} = 45$$

$$45 + \underline{13} = 58$$

However, the sequence 10, 13, 18, 25, . . . may come from an entirely different pattern. A very ingenious person might reason thus:

$$10 + 3 = 13$$

$$13 + 5 = 18$$

$$18 + 7 = 25$$

$$25 + 11 = 36$$

$$36 + 13 = 49$$

$$49 + 17 = 66$$

This person's pattern involves increasing each number by consecutive primes (a special kind of number that we will study in Chapter 4). This example illustrates that patterns obtained inductively (defined later in this section) are not necessarily unique.

Consider the sequence

$$3, 7, 11, 15, 19, \ldots$$

Observe that we use the three dots both to indicate that terms are omitted and (in this case) that the sequence continues in the same manner indefinitely. Note also that

$$3 + 4 = 7$$
$$7 + 4 = 11$$
$$11 + 4 = 15$$
$$15 + 4 = 19$$

Thus we suggest that this sequence can be obtained by adding the constant 4 to each term to obtain the next term. Such a sequence is called an **arithmetic sequence,** and 4 is called the **common difference** between terms. $4 = 7 - 3$; $4 = 11 - 7$; etc.

Note that we can also reason as follows:

$$7 = 3 + 1 \cdot 4 \quad \text{or} \quad a + 1 \cdot d$$
$$11 = 3 + 2 \cdot 4 \qquad a + 2 \cdot d$$
$$15 = 3 + 3 \cdot 4 \qquad a + 3 \cdot d$$
$$19 = 3 + 4 \cdot 4 \qquad a + 4 \cdot d$$

So if a is the first term of an arithmetic sequence and d is the common difference, the sequence can be written as

$$a, a + d, a + 2d, a + 3d, \ldots$$

Example 24 • Find a pattern and guess the next three terms of 2, 6, 18, 54, ______, ______, ______.

Solution You undoubtedly discovered that you can obtain any term by multiplying the preceding term by 3. Thus the next three terms are 162, 486, and 1458. •

The sequence in the preceding example is called a **geometric sequence.** In general, it can be described by

$$a, ar, ar^2, ar^3, \ldots .$$

where a is the first term and r is the **common ratio** (a term divided by the preceding term).

Example 25 • Find the next three elements for each of the following patterns.

(a) 1, 4, 1, 7, 1, 10, 1, 13, 1, . . .
(b) 1, 4, 9, 16, 25, . . .

Solution In (a), you probably first noted that every other member of the sequence is 1. The alternate elements increase by 3. The sequence is 1, 4, 1, 4 + 3, 1, 7 + 3, 1, 10 + 3, 1, ... The next three elements are 13 + 3, 1, 16 + 3; or 16, 1, 19. In (b), the members of the sequence are the squares ($x^2 = x \cdot x$ or $3^2 = 3 \cdot 3$) of consecutive counting numbers:

$$1 = 1^2$$

$$4 = 2^2$$

$$9 = 3^2$$

$$16 = 4^2$$

$$25 = 5^2$$

The next three elements are 6^2, 7^2, and 8^2 or 36, 49, and 64. This sequence can also be obtained by adding successive odd numbers to the preceding terms $1 + \underline{3} = 4$; $4 + \underline{5} = 9$; $9 + \underline{7} = 16$; $16 + \underline{9} = 25$; $25 + \underline{11} = 36$; $36 + \underline{13} = 49$; $49 + \underline{15} = 64$; and so on. •

To describe a sequence, we often try to find a pattern that relates the number of a term to the term itself. Let's return to Example 23 where each term was obtained by adding 3 to the preceding term:

$$5, 8, 11, 14, 17, \ldots$$

Note that 5 can be written as 3(1) + 2, where 1 is the number of the first term. Let's see if we can replace 1 by 2 (for the second term) and get the second term, 3(2) + 2 = 6 + 2 = 8 (the second term); 3(3) + 2 = 11 (third term); and 3(4) + 2 = 14 (fourth term). From this pattern, the nth term is $3(n) + 2$, usually written as $3n + 2$.

Number of Term	1	2	3	4	. . .	n
Term	5	8	11	14	. . .	
	3(1) + 2	3(2) + 2	3(3) + 2	3(4) + 2	. . .	$3(n) + 2$

Example 26 • A sequence is built on the pattern $2n + 1$. Find the terms of the sequence that correspond to $n = 1, 2, 3, 4$, and 5.

Solution For $n = 1$, $2(1) + 1 = 3$; for $n = 2$, $2(2) + 1 = 5$; for $n = 3$, $2(3) + 1 = 7$; for $n = 4$, $2(4) + 1 = 9$; for $n = 5$, $2(5) + 1 = 11$. Therefore, the terms of the sequence are 3, 5, 7, 9, 11, ... •

Example 27 • Find an expression for the nth term of the arithmetic sequence

$$a, a + d, a + 2d, a + 3d, \ldots$$

Solution

Number of Term (n)	1	2	3	4 . . .	n
Term	a	$a + d$	$a + 2d$	$a + 3d$	. . .

$$n = 2 \qquad a + (2 - 1)d = a + d$$

$$n = 3 \qquad a + (3 - 1)d = a + 2d$$

$$n = 4 \qquad a + (4 - 1)d = a + 3d$$

Just for fun

Assume that you have six sections of chain, each consisting of four links. If the cost of cutting open one link is 10¢ and welding it together again is 25¢, what is the least it should cost to have the six pieces joined into one chain?

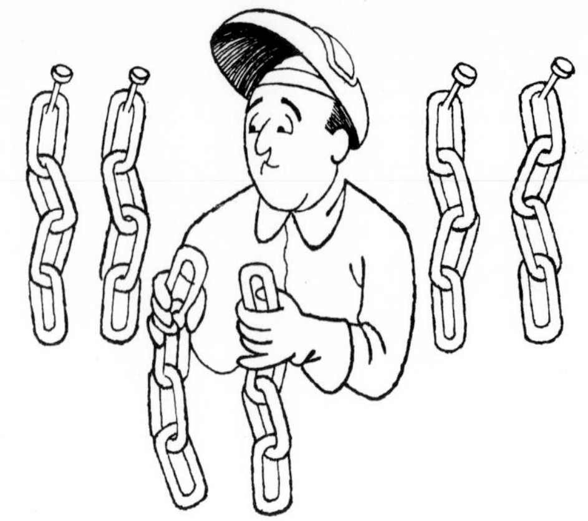

Thus the nth term seems to be $a + (n - 1)d$. See if this yields the first term when $n = 1$. •

Let's return now to the problem in the introduction of this section.

Example 28 • Leslie has taken a job in which she is to be paid 1 penny for the first day, 2 pennies for the second day, 4 pennies for the third day, 8 pennies for the fourth day, and so on. How much will she receive for the fifteenth day?

Solution

Number of Term	1	2	3	4	...	15
Term	1	$1 \cdot 2$	$1 \cdot 2^2$	$1 \cdot 2^3$	...	$1 \cdot 2^{14}$

You recognized this to be a geometric sequence and the fifteenth term is $1 \cdot 2^{14} = 16{,}384$ pennies, or $163.84 •

The type of reasoning used in this section in the prediction of missing terms of a sequence and in the generalization of an nth term is called **inductive reasoning.**

Inductive reasoning

> *Inductive reasoning* is utilized when one arrives at a general conclusion from limited observations.

Many times, a scientist makes observations, discovers regularities, and formulates conclusions. In science this is called experimental research; in mathematics, we say that the scientist is reasoning inductively. Similarly, business people, lawyers, and, in fact, all of us use inductive reasoning regularly. You should note that inductive reasoning in itself does not prove that a unique general pattern exists.

In this book, the reasonableness of a procedure or a formula is emphasized rather than the mathematical development of the formula. Illustrative examples will often be used to motivate general results. Thus, inductive reasoning or an intuitive approach will be used throughout much of this text.

Exercise set 2

1. Which word is different from the collection of other words?

   (a) run (b) ride (c) walk
   (d) jog (e) hike

2. Find a pattern and then complete the blanks.

   (a) △, □, □, △, □, □, ______, ______, . . .
   (b) △, □, ⬠, ______, ______, . . .
   (c) *a, b, a, b, b, a, b, b,* ______, ______, . . .
   (d) $x, x+1, x+2, x+3, x+4,$ ______, ______, . . .
   (e) *A, C, E, G, I, K,* ______, ______, . . .
   (f) 1, 2, 3, 5, 6, 7, 9, 10, 11, ______, ______, . . .

3. Each sequence of letters has a pattern. Find a pattern and write the next three letters.

   (a) *A, Y, X, B, W, V, C, U, T,* —, —, —
   (b) *B, D, B, B, F, B, B, B, H, B, B,* —, —, —
   (c) *M, N, L, P, Q, O, S, T, R,* —, —, —

4. The pattern of which set of letters is different from the other four sets?

   (a) *ADBC* (b) *RUST* (c) *FIGH*
   (d) *UVXY* (e) *LOMN*

5. Find a pattern and write the next two sets of six letters.

   EFGHIJ *JEFGHI* *IJEFGH*

6. Select the letter that should appear next in each of the following sequences.

   (a) *B, A, F, E, J, I,* ______
   (b) *A, C, F, H, K, M, P,* ______
   (c) *A, B, Y, Z, D, E, V, W, G, H,* ______

7. All of these are bleeps.

None of these is a bleep.

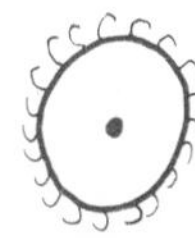

Which of these are bleeps?

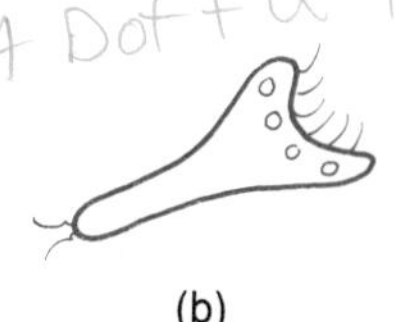
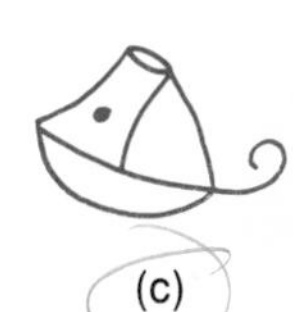

(a) (b) (c)

8. In an unfamiliar language, *enic lod nam* means "nice old man," *enic moor* means "nice room," and *elitle nam* means "little man." How would you say "old room"?

9. In a different language, *lobo strino* means "fat dog," *tropo lobo wasca* means "very fat boy," and *tropo bludo* means "very dumb." How do you say "dumb boy"?

10. Look for a pattern in the first three products and then find the fourth product.

    (a)

1	11	111	1111
1	11	111	1111
1	121	12321	?

    (b)

9	99	999	9999
9	9	9	9
81	891	8991	?

11. Write the next three square numbers.

    1 4 9 16

12. A pentagonal number is the sum of a triangular number and a square number. Write the next three pentagonal numbers.

1 5 12 22

13. With a horizontal line, separate the dots representing triangular numbers and the square numbers in Exercise 12.

14. Using the expression for the nth term of a sequence, find the first five terms of each sequence. Begin with $n = 1$.

(a) $n^2 + 1$ (b) $n^2 + n$
(c) $4n - 1$ (d) $3n^2 + 2$

15. Verbally describe a pattern for each sequence.

(a) 7, 12, 17, 22, . . .
(b) 5, 8, 11, 14, 17, . . .
(c) 3, 9, 27, 81, 243, . . .
(d) 1, 8, 27, 64, 125, . . .

16. Determine a pattern and then find three additional members for each sequence.

(a) 2, 6, 18, 54, 162, . . .
(b) 1, 4, 7, 10, . . .
(c) 3, 5, 9, 15, 23, . . .
(d) 1, 3, 1, 8, 1, 13, 1, . . .
(e) 4, 1, 5, 2, 6, . . .
(f) 7, 10, 6, 9, 5, . . .

17. Complete the blanks so that the following are arithmetic sequences.

(a) 2, 7, 12, ______, ______, . . .
(b) 5, ______, 11, 14, . . .

18. Complete the terms so that the following are geometric sequences.

(a) 1, 3, 9, 27, ______, ______
(b) 4, ______, 36, 108, ______, ______

T 19. Study the given array and then find the following numbers without actually computing the squares.

$$2^2 - 1^2 = 3$$
$$3^2 - 2^2 = 5$$
$$4^2 - 3^2 = 7$$
$$5^2 - 4^2 = 9$$

(a) $6^2 - 5^2$ (b) $7^2 - 6^2$
(c) $8^2 - 7^2$ (d) $9^2 - 8^2$

20. To get a bill passed, the Governor called three legislators on Monday. He asked each one to call three legislators on Tuesday, and each of these was to call three legislators on Wednesday. If this process continued through Sunday and no legislator was called twice, how many were called?

21. The sum of the first n terms of an arithmetic sequence is given by $S = (a + l)n/2$ where $l = a + (n - 1)d$. Use these two formulas to find the sum of the first 8 terms of the sequences in Exercise 17.

22. One famous number sequence was first studied by and named for Leonardo de Pisa, also called Fibonacci. Determine the pattern and the next three elements of the Fibonacci numbers given below.

1, 1, 2, 3, 5, 8, 13, 21, 34, . . .

23. At the beginning of the year, I owned two rabbits, a male and a female. At the end of the first month and each month thereafter, this original pair gave birth to a new pair, male and female. Furthermore, each new pair, in the second month after their birth, gave birth to another pair and continued to do so in subsequent months. So, at the end of the first month, I had 2 pairs, at the end of the second month I had 3 pairs, and at the end of the third month I had 5 pairs. How many did I have after 4 months? 5 months? 6 months? Compare your results with the sequence in Exercise 22.

24. In the study of biology, one learns that a male bee has only one parent, his mother. A female bee has both father and mother. Complete the picture of the family tree of a male bee, where ♂ represents male and ♀ represents female. Do you notice any relationship between Fibonacci numbers of Exercise 22 and

(a) the number of ancestors in each generation?
(b) the number of male ancestors in each generation?
(c) the number of female ancestors in each generation?

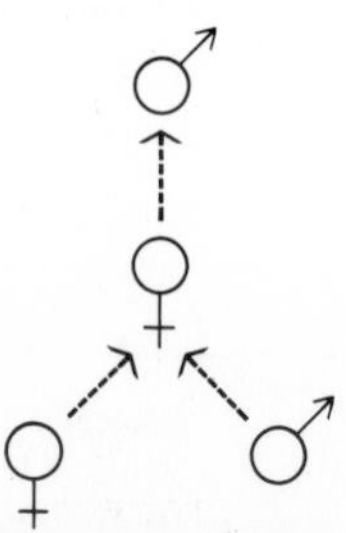

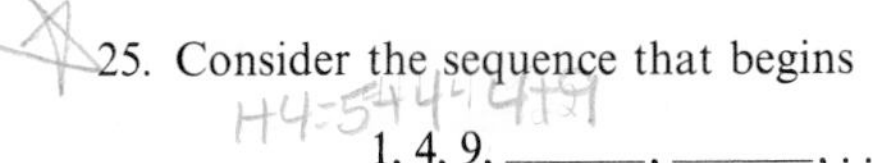

25. Consider the sequence that begins

$$1, 4, 9, \_\_\_\_, \_\_\_\_, \ldots$$

(a) Describe a rule that would yield 16 as the fourth term. What would then be the fifth term?
(b) Describe a rule that would yield 22 as the fourth term. What would then be the fifth term?

26. Find an expression for the nth term of the geometric sequence

$$a, ar, ar^2, ar^3, \ldots$$

C 27. John notices that with daily practice he can increase the number of free throws he makes before missing. On the first day he makes only 1; on the second day 5; on the third day 14; and on the fourth day 30. If this pattern should continue, how many can he expect to make on the fifth day?

28. Find an expression for the nth term of each part of Exercise 15, where n represents a counting number.

29. Find the sum of the first four terms of the Fibonacci sequence; the first five terms; the first six terms.

30. The sum of n terms of a geometric sequence with first term a and ratio r is given by the following formula.

$$S = \frac{a(r^n - 1)}{r - 1}$$

What is the sum of the first six terms of each?

(a) 10, 100, 1000, 10,000, . . .
(b) 4, 16, 64, 256, 1024, 4096, . . .
(c) 4, 12, 36, 108, . . .

31. Consider a chain letter where 10¢ is sent to the originator of the letter by each person who receives a letter and where each recipient of the letter mails 1 copy of the letter to 5 more people. How much money has the originator received after 8 mailings?

3 An introduction to problem solving

■ Dan buys only blue socks and brown socks. He keeps all his socks in one drawer, and in that drawer he has 8 blue socks and 6 brown socks. If he reaches into the drawer without looking, what is the smallest number of socks he must take out to be sure of getting two of the same color? ■

In this section, we introduce suggestions that should help you to answer the preceding question. The first two sections of this chapter on critical thinking have provided techniques for understanding problem solving. In fact, successful problem solving requires creative use of all facets of critical thinking. In this section and the next, we will study four general steps in problem solving. In this section, we will focus on the fact that many times the first step, "understanding the problem," is all that is needed to facilitate a solution.

As we discuss techniques of problem solving, you must be aware that problem solving is more than memorizing the techniques. Rather it is a process whereby you apply your knowledge and skills to understand and to satisfy the demands of unfamiliar situations. Hence, although we shall study problem solving in the context of mathematics, the strategies we develop have application in a wide range of situations. Also, the techniques we are introducing are useful in solving complicated problems throughout the book. However, to make it easier for you to comprehend now, we sometimes apply these techniques to simple "trick problems."

As we have discussed, detecting a pattern is an important step in solving most problems. However, as the problems become more complex, we often encounter obstacles long before we discover a helpful pattern. In this section, we shall examine several steps that may be helpful in solving many problems, whatever the level of difficulty.

In the 1950's, a very successful research mathematician named George Polya began to write a series of insightful articles and books on problem solving. Polya identifies four steps that have characterized problem solving from the ancient Greeks to the present day.

Four Steps in Problem Solving

1. Understanding the problem.
2. Devising a plan.
3. Carrying out the plan.
4. Looking back; seeing if your solution makes sense.

Under the heading "Understanding the Problem," we ordinarily ask a number of questions in the hope that the answers will provide an understanding of the problem. Typically the questions include:

How many?
How much?
What was?
Who is?
Who has?
What is to be found?

Example 29 • On their way back to the University, Joy, Beth and Dill took turns driving. Joy drove 50 miles more than Beth. Beth drove twice as far as Dill. Dill only drove 10 miles. How many miles is the trip back to the university? (List some questions that would be asked in understanding this problem.)

Solution Understanding the problem: Do you know how far anyone drove? (Yes, Dill drove 10 miles.) How many miles more did Joy drive than Beth? (50). What is the relationship between the number of miles driven by Beth and the number driven by Dill? (Beth drove twice as far as Dill.) What are we trying to find? (The distance back to the univerisity.) •

Although asking questions similar to the ones in the preceding example will usually give an understanding of a problem, some additional suggestions may be helpful.

Suggestions for understanding

1. Identify what you are trying to find.
2. Don't impose conditions that do not exist.
3. Strip the problem of irrelevant details.

Once you have carefully summarized the information given in the problem, the suggestion "Identify what you are trying to find" enables you to find the answer easily for simple trick problems such as the following.

1. A farmer has 17 calves. All but 9 die. How many does he have left?
2. Is it legal for a man to marry his widow's sister? (Exercise 1)
3. How many 2¢ stamps are there in a dozen? (Exercise 2)
4. There was a blind beggar who had a brother, but this brother had no brother. What was the relationship between the two? (Exercise 4)

Let's consider the first of these trick problems.

Example 30 • A farmer has 17 calves. All but 9 die. How many does he have left?

Solution Did you get 8? Be careful! What is the unknown? How many calves are living? What are the data? Originally, the farmer had 17 calves (irrelevant!). All but 9 died (important!). Hence, 9 are living. •

This suggestion (Identify the unknown) is used here to solve some fairly easy problems, but it is an essential step in problem solving at all levels of mathematics.

Some problems are difficult because we impose conditions that do not exist. Be careful when you consider the following problems.

1. Five apples are in a basket. How can you divide them among five girls so that each gets an apple, but one apple remains in the basket? (Exercise 8)
2. Two United States coins total 30¢, but one of the coins is not a nickel. Can you explain?

Let us look carefully at the second of these two examples.

Example 31 • Two United States coins total 30¢, but one of the coins is not a nickel. Can you explain?

Solution You are tempted to assume that neither of the coins is a nickel. This assumption is incorrect. One coin is not a nickel, but the other one can be. So the two coins are a nickel and a quarter. •

The need to strip the problem of irrelevant details is very similar to the preceding.

1. Dan buys only blue and brown socks. He keeps all his socks in one drawer; and in that drawer he has 8 blue socks and 6 brown socks. If he reaches into the drawer without looking, what is the smallest number of socks he must take out to be sure of getting two of the same color?
2. Ms. A: How much will 1 cost?
   Salesman: 30¢.
   Ms. A: How much will 15 cost?
   Salesman: 60¢.
   Ms. A: I'll take 615.
   Salesman: That will be 90¢.
   Explain. (Exercise 26)

3. One container holds two quarts of water, and another container holds one quart of red grape juice. One cup (.25 quart) of water is transferred to the container of grape juice and mixed thoroughly. Then one cup of the mixture is transferred back to the container of water. Is there more grape juice in the water or more water in the grape juice? (Exercise 27)

Let's look at the first problem. Students are often misled by the number of blue socks and the number of brown socks. These numbers are not pertinent to solving the problem. To secure the answer, let's analyze what can happen as Dan draws socks one by one from the drawer.

After Dan has drawn two socks, they are either the same color or different colors. If they are the same color, he has a pair in two draws. If they are different colors, the next sock drawn must match one of them. Thus, at most three draws are necessary to get a pair of socks that match.

Polya's second step (Devising a plan) often requires a considerable amount of creativity. However, there are a number of strategies that help us formulate a plan to solve a problem. We will examine three such strategies in this section, seven in the next section, and others from time to time as we move through the text.

Strategies

1. Make a chart or a table.
2. Draw a picture.
3. Eliminate impossible situations.

The first strategy (make a chart or a table) was introduced in Section 1. Sometimes a chart or a table makes obvious the solution to a problem.

Example 32 • The Astros are in first place and the Braves are in fifth. The Dodgers are midway between the two. If the Reds are ahead of the Braves and the Giants are immediately behind the Dodgers, then who is in second place?

Solution By writing down the five positions and placing the information on the chart, we immediately see the answer. In Table 1, place the Astros in first place and the Braves in fifth place. For the Dodgers to be exactly in the middle, they must be in third place. The Giants are immediately behind the Dodgers, so they must be in fourth place. Since the Reds are ahead of the Braves, the Reds must be in second place.

Table 1

First place	*Second place*	*Third place*	*Fourth place*	*Fifth place*
Astros		Dodgers		Braves

•

The strategy of drawing a picture often facilitates solving a problem. Whether drawing a picture leads immediately to a solution or not, it is important for the understanding of a problem. We use this strategy throughout this text. We shall learn to draw arrow diagrams, one-dimensional graphs, two-dimensional graphs, histograms, frequency polygons, circle charts, and many other pictures. In each case, our goal will be to understand the problem and sometimes actually to solve the problem. As an illustration of using pictures to assist in problem solving, consider the following simple examples.

1. A beetle is at the bottom of a bottle that is 6 inches deep and wants out. Each day, the beetle climbs up 1 inch, and each night it slides back $\frac{1}{2}$ inch. How long will it take the beetle to climb out of the bottle?
2. Each hour, an airplane leaves New York for London, England, and another leaves England for New York. The trip across the Atlantic takes 6 full hours. How many planes will a given plane leaving New York meet during the trip to London? (Exercise 30)
3. Volumes I, II, III, and IV of a very boring set of books stand side by side on a shelf. A bookworm has bored its way straight through the set from the very first page of Volume I to the very last page of Volume IV. Each volume is 2 inches thick and each cover is $\frac{1}{8}$ inch thick. How far has the bookworm bored? (Exercise 20)

Let's look at the first of these examples.

Example 33 • A beetle is at the bottom of a bottle that is 6 in. deep and wants out. Each day, the beetle climbs up 1 in., and each night it slides back $\frac{1}{2}$ in. How long will it take the beetle to climb out of the bottle?

Solution To understand this problem, consider the sketch of the bottle in Figure 3. On the first day, the beetle reaches the 1-in. plateau but slides back to $\frac{1}{2}$ in. at night. On the second day, it reaches $1\frac{1}{2}$ in., but again at night the beetle slides back to the 1-in. mark. This process continues until the 11th day. When the beetle reaches the top on the 11th day, it will just crawl out, rather than slide back.

Figure 3

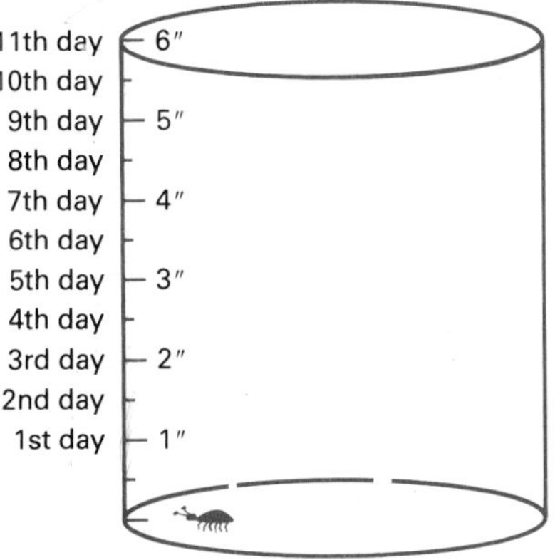

•

The strategy of elimination involves critical thinking and reasoning. Combined with making a table and with the strategy to be introduced in the next section (Guess and Check), this strategy is probably the one most commonly used by people in everyday affairs. We introduce the idea through the construction of a

Table 2

	Leon	*Sarah*	*Russo*	*Sue*	*Sharon*
Affirmative	No	Yes	Yes		
Negative	Yes	No	No		
Hard College	No	Yes		No	
Sloan College	Yes	No		Yes	

table to present an organized method for eliminating incorrect situations, as seen in the next example. For more difficult problems, selecting the clues that lead to elimination is very important.

Example 34 • Five students, Leon, Sarah, Russo, Sue, and Sharon, participate in a debate tournament in which each team must have at least one affirmative debater and one negative debater. Two attend Hard College and three attend Sloan. Three are affirmative debaters, two negative. Leon and Sue attend the same college. Russo and Sharon attend different schools. Sarah and Russo are on the same side. Sue and Sharon are on opposite sides. A negative debater from Hard College was selected outstanding debater. Who was that person?

Solution Leon and Sue attend the same college. Russo and Sharon attend different colleges. What does this tell you? (Leon and Sue, along with either Russo or Sharon attend Sloan.) Sarah must attend Hard. Why? Sarah and Russo are affirmative debaters along with either Sue or Sharon. Leon is a negative debater. Why? The other negative debater is either Sue or Sharon. Sue is not from Hard College. Sharon can be from Hard College. Therefore, the outstanding debater is Sharon. •

Now let us consider one final problem and follow it all the way through Polya's four-step process.

Example 35 • What is the smallest number of pitches that a baseball pitcher can make in the course of pitching a full-length baseball game?

Solution Did you get 27? Have you made the unjustified assumption that the pitcher making the minimum number of pitches would win and thus have to pitch 9 innings?

Let us examine the data. In this case, the data would arise from our knowledge of the game of baseball.

(a) Every full-length game lasts at least $8\frac{1}{2}$ innings.
(b) The winning pitcher (if not relieved) always pitches at least 9 innings.
(c) A losing pitcher for the visiting team may pitch as few as 8 innings.
(d) In each complete inning pitched, a pitcher must face at least 3 batters and throw at least 1 pitch to each batter.

Now do you understand the problem better? Is your answer 24? Not quite! Remember that the pitcher must lose. How about a home run in the eighth inning? Thus, the answer is 25 pitches. •

Polya's last step (Looking back; seeing if your solution makes sense) is very important. Once the solution is found, one is tempted to skip this fourth step. Yet it is in this step that one is forced to examine the thinking that produced the solution. Here are four components of a successful review of a solution.

Just for fun

How can you place 21 balls in 4 boxes so that there is an odd number of balls in each box?

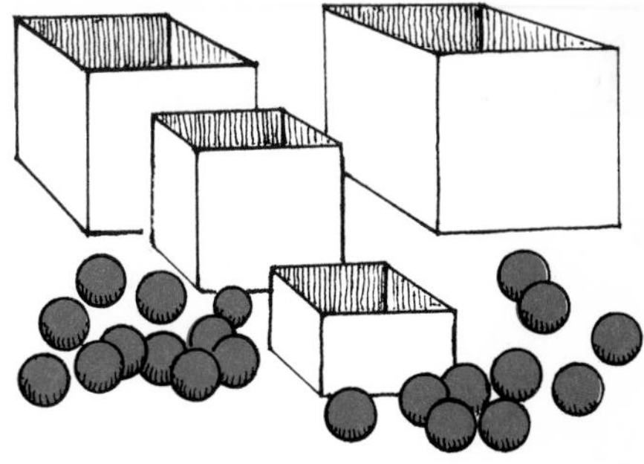

(a) **Check the results.** Let's look at the problem of determining the minimum number of pitches in a baseball game. Suppose in the last of the first inning, one pitch is made to each of three batters, and each is thrown out at first. (Of course there are other ways the three batters could be "retired" on these pitches.) The same happens in innings two through seven. In the last of the eighth inning, the first batter hits a home run, and the next three batters are "retired" as in the preceding innings. At the end of the first half of the ninth inning the pitcher's team has failed to score. The game is over, and the pitcher has thrown the ball 25 times. This illustration is just one of several ways the pitcher could have thrown only 25 pitches.

(b) **Interpret the solution.** For this example this step has already been accomplished, but for other problems it is an important step.

(c) **Is there another way to find the same solution?** The solving of the problem could have used some of the strategies discussed in the next section.

(d) **Is there a second** (and maybe different) **solution?** For our example, the answer is *yes* for procedure and *no* for the 25 pitches. For example, the home run could have been hit in any inning.

In the problems that follow, proceed slowly and make certain you understand the problem by using one or all of the suggestions given in this section. **Don't expect to solve all the problems correctly; we didn't do them all right the first time either.** Do recognize this fact: when correctly approached, seemingly difficult problems may turn out to be easy.

Exercise set 3

Indicate the suggestions for understanding the problem employed in solving Exercises 1 through 12.

1. Is it legal for a man to marry his widow's sister?
2. How many 2¢ stamps are there in a dozen?
3. The number of doughnuts in the storage container doubles every minute. The container became full at 10:30 A.M. When was it one-half full?
4. There was a blind beggar who had a brother, but this brother had no brother. What was the relationship between the two?
5. It takes 1 hour and 20 minutes to drive to the airport, yet the return trip takes only 80 minutes using the same route and driving at what seems to be the same speed. How can this be?

6. A fox is at the bottom of a 16-foot hole. Each day he climbs up 3 feet, and each night he slides back 2 feet. How long will it take him to get out of the hole?

7. How much dirt is there in a hole 2 feet wide, 4 feet long, and 3 feet deep?

8. Five apples are in a basket. How can you divide them among five girls so that each gets an apple, but one apple remains in the basket?

9. A train leaves New York for Chicago, traveling at a rate of 100 miles an hour. Another train leaves Chicago for New York an hour later, traveling at 75 miles an hour. When the two trains meet, which train is nearer to New York?

10. A bottle and a cork cost $1.10, and the bottle costs $1.00 more than the cork. How much does the cork cost?

11. The Browns have six daughters. Each daughter has two brothers. How many children are there in the Brown family?

12. Two ducks before a duck, two ducks behind a duck, and a duck in the middle are how many ducks?

In Exercises 13 through 17, write down a number of questions for each that would assist you in understanding the problem. Then solve the problem.

13. Jodi has 8 coins totaling 57 cents. What are the coins if she does not have a half dollar?

14. The football team is trying to gain weight. Tim, Jim, George, and Rick each check their weight. Tim weighs twice as much as Jim. Jim is fifty pounds lighter than George. Rick is ten pounds heavier than George. If Rick weighs 200 pounds, how much does Tim weigh?

15. Four boxes weigh 60 pounds. Each box is twice as heavy as the next box, and the lightest box weighs 20 pounds less than the two middle ones together. What are the weights of the boxes?

16. Ron is buying old comic books for eight cents, old buttons for four cents, and old toys for sixty cents. He sells each of them for five cents profit. How much does he make if he buys and sells four comic books, four buttons, and two toys?

17. Lou drives a small car, getting 28 miles per gallon in town and 34 miles per gallon on the highway. Her fuel tank holds 15 gallons. Regular gasoline sells for 80 cents a gallon. What is the cost of a full tank of gasoline?

18. In the exercises listed, which facts could be classified as irrelevant?
(a) Exercise 16 (b) Exercise 17

19. If Tom's father is Dick's son, how are Tom and Dick related?

20. Volumes I, II, III, and IV of a very boring set of books stand side by side on a shelf. A bookworm has bored its way straight through the set from the very first page of Volume I to the very last page of Volume IV. Each volume is 2 inches thick and each cover is $\frac{1}{8}$ inch thick. How far has the bookworm bored?

21. A train 1 mile long moving at a rate of 1 mile per hour enters a tunnel 1 mile long. How long does it take the train to pass completely through the tunnel?

T 22. How long will it take to cut a 10-foot log into five 2-foot lengths, allowing 2 minutes per cut?

23. In a machine, a gum ball costs 5¢. There are five different colored gum balls in the machine. How much money do you need to be certain you get three gum balls of the same color?

24. Newlywed Sonya makes a toast in a pan that holds only two slices. After browning one side of a slice, she turns it over. Each side takes 30 seconds. How can she brown both sides of three slices in the shortest period of time?

25. You have 8 sticks: 4 of them are exactly half the length of the other 4. Enclose three equal squares with them, making sure that there is no overlapping.

C 26. Ms. A: How much will 1 cost?
Salesman: 30¢.
Ms. A: How much will 15 cost?
Salesman: 60¢.
Ms. A: I'll take 615.
Salesman: That will be 90¢.
Explain.

27. One container holds two quarts of water, and another holds one quart of red grape juice. One cup (0.25 quart) of water is transferred to the container of grape juice and mixed thoroughly. Then one cup of the mixture is transferred back to the container of water. Is there more grape juice in the water or more water in the grape juice?

28. Form four triangles using six matchsticks so that they touch only at their endpoints.

29. Cut a 3-by-8 rectangle into two pieces with one cut, and form a 2-by-12 rectangle with the pieces.

30. Each hour, an airplane leaves New York for London. The trip across takes 6 hours. At the same times, airplanes leave London for New York. How many planes will a given plane meet on the trip from New York to London?

31. How can you measure 1 mL of medicine using only a 4-mL container and a 7-mL container?

32. How can you obtain 6 gallons of cider from a tank if you have only a 4-gallon container and a 9-gallon container?

4 Strategies for problem solving

■ Fifteen people are in a room, and each shakes hands once and only once with everyone else. How many handshakes are there? ■

In most situations, understanding a problem is not enough to yield a solution. You must use the thinking process to solve many problems. In the previous section we started listing particular strategies that would assist in problem solving. In this section we discuss five additional strategies. Keep in mind that several other strategies will be mentioned as we proceed through the book. Please understand that none of these strategies will magically yield a solution to each problem you attempt, but they often give a helpful starting point in solving a wide range of problems.

Learning to solve problems requires you to stretch and enhance your critical thinking techniques. Therefore, this is an important section in the development of your thinking maturity.

Polya's second major step in problem solving is "Make a plan." The strategies discussed below, alone or in concert, provide the rough outline of plans that will solve a great many problems and use the thinking skills we have studied in the previous three sections.

Strategies for making a plan

1. Make a chart or a table.
2. Draw a picture.
3. Eliminate impossible situations.
4. Look for a pattern.
5. Try a simpler version of the problem.
6. Restate the problem.
7. Guess, test, and revise.
8. Work backward.
9. Design a model.
10. Use an algorithm.

One of the key techniques in problem solving is to look for a pattern. Further, if you examine a simpler version of the problem, you will find it is easier to detect a pattern if one exists. This dual strategy of **looking for a pattern** in **a simple version** of the problem provides a potent weapon in solving problems similar to the following.

1. Without adding all 30 numbers, find the sum of the first 30 odd numbers.
2. How many people can be seated at 10 square (4-person) tables lined up end to end?
3. Fifteen people are in a room, and each shakes hands once and only once with everyone else. How many handshakes are there?

Example 36 • Without adding all 30 numbers, find the sum of the first 30 odd numbers.

Solution To find a pattern, we look at simple versions of the problem.

$$\text{2 odd numbers:} \quad 1 + 3 = 4 = 2 \cdot 2$$

$$\text{3 odd numbers:} \quad 1 + 3 + 5 = 9 = 3 \cdot 3$$

$$\text{4 odd numbers:} \quad 1 + 3 + 5 + 7 = 16 = 4 \cdot 4$$

From this pattern, it seems that the sum of the first 30 odd numbers would be $30 \cdot 30 = 900$. •

Example 37 • How many people can be seated at 10 square (4-person) tables lined up end to end?

Solution Again, let's look at simple versions of the problem. See Figure 4.

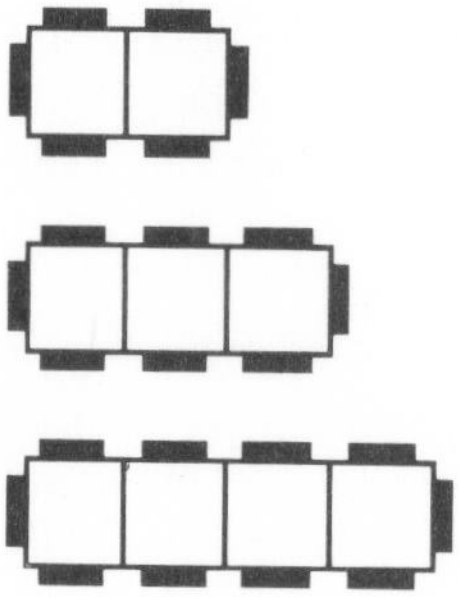

		Number of seats originally − Number lost
	2 tables	$4(2) - 2(1) = 6$
	3 tables	$4(3) - 2(2) = 8$
	4 tables	$4(4) - 2(3) = 10$
	n tables	$4(n) - 2(n - 1)$

Figure 4

Thus for 10 tables ($n = 10$),

$$4(10) - 2(10 - 1) = 22$$

Twenty-two people can be seated. •

Example 38 • Fifteen people are in a room, and each shakes hands once and only once with everyone else. How many handshakes are there?

Solution Let's consider simple versions of this problem to see if we can find a pattern.

Two people: $A \leftrightarrow B$
(1 handshake)

Three people: $A \leftrightarrow B \quad B \leftrightarrow C$
$A \leftrightarrow C$
$2 + 1 = 3$ handshakes

Four people: $A \leftrightarrow B \quad B \leftrightarrow C \quad C \leftrightarrow D$
$A \leftrightarrow C \quad B \leftrightarrow D$
$A \leftrightarrow D$
$3 + 2 + 1 = 6$ handshakes

Five people: $A \leftrightarrow B \quad B \leftrightarrow C \quad C \leftrightarrow D \quad D \leftrightarrow E$
$A \leftrightarrow C \quad B \leftrightarrow D \quad C \leftrightarrow E$
$A \leftrightarrow D \quad B \leftrightarrow E$
$A \leftrightarrow E$
$4 + 3 + 2 + 1 = 10$ handshakes

From the pattern, we infer that 15 people shake hands this many times:

$$14 + 13 + \cdots + 3 + 2 + 1 = 105 \quad \bullet$$

Many times, it is necessary to restate the problem before a solution can be obtained.

1. If 3 hens lay 3 eggs in 3 days, how many eggs will 300 hens lay in 300 days?
2. To select the best two-person debate team in the United States, 117 teams are selected, with the understanding that when a team loses a match, the team is dropped from the tournament. How many games must be played to select a winner? (Exercise 20)

Let us look at the first of these examples.

Example 39 • If 3 hens lay 3 eggs in 3 days, how many eggs will 300 hens lay in 300 days?

Solution In restating the problem, we may wish to strip off unnecessary details. If 3 hens lay 3 eggs in 3 days, we see that 1 hen will lay 1 egg in 3 days. Hence, the problem restated is this:

If a hen lays an egg every 3 days, how many eggs will 300 hens lay in 300 days?

It is seen immediately after the restatement that 1 hen will lay 100 eggs in 300 days. So 300 hens will lay

$$300 \cdot 100 = 30{,}000$$

eggs in 300 days. •

No single plan applies to as many problems as the **guess, test, and revise** cycle. Although the plan is often long and cumbersome to complete and may yield only an approximate answer, it often provides a starting point that inspires more efficient plans.

Example 40 • Jane has 20 coins consisting of nickels and dimes, totaling \$1.20. How many of each does she have?

Solution Suppose we guess that Jane has 2 dimes and 18 nickels. The value of these coins is

$$2(0.10) + 18(0.05) = \$1.10$$

Since the value is less than \$1.20, the actual solution must have more than 2 dimes. Now let's guess 6 dimes and 14 nickels with a value of

$$6(0.10) + 14(0.05) = \$1.30$$

This time, we have too many dimes since the value is more than \$1.20. We make a third guess and we arrange the guesses as shown in Table 3.

Table 3

Dimes	*Nickels*	*Check*
2	18	$2(0.10) + 18(0.05) = 1.10$
6	14	$6(0.10) + 14(0.05) = 1.30$
5	15	$5(0.10) + 15(0.05) = 1.25$

Finally, we try 4 dimes and 16 nickels.

$$4(0.10) + 16(0.05) = \$1.20$$

Thus, the answer is 4 dimes and 16 nickels. •

Example 41 • Ed, Paul, and Aaron are stranded on an island, and none of them can swim. They construct a raft to take them to shore, but they estimate that the raft will sink with the weight of more than 100 kilograms (kg). Ed weighs 80 kg, Paul weighs 50 kg, and Aaron weighs 40 kg. How can the raft be used to get all three to shore?

Solution We shall begin the solution with a guess and then reason whether additional steps satisfy the restrictions placed in the problem.

1. Try Ed and Aaron on the raft. Too heavy.
2. Try Ed and Paul on the raft. Too heavy.
3. Paul and Aaron together weigh less than 100 kg, so they take the raft to shore.
4. Paul takes the raft back to the island.
5. Since Ed and Paul cannot get on the raft together, Ed takes the raft to shore.
6. Aaron takes the raft back to the island.
7. Paul and Aaron take the raft to shore.

Thus the solution is complete. •

A strategy that is sometimes useful in designing a plan involves **working backward** on the problem. This strategy is illustrated in the following example.

Example 42 • *A* and *B* play a game in which they alternate selecting any one of the numbers 1, 2, 3, 4, 5, 6. When a number is selected, it is added to the sum of the numbers already selected. To win the game, the final sum must be 50. If *A* goes first, there is a way for *A* to win every game. What should *A* do to win every game?

Just for fun

Three students, Joan, James, and Edward, agree to be part of an experiment in reasoning. First, the students are shown 5 ribbons—3 blue and 2 red. The students are told that a ribbon will be placed on each of their backs and that each is to try to determine the color of the ribbon. Joan can see the ribbons on James and Edward, James can see the ribbon on Edward, but Edward cannot see the other ribbons. When asked the color of the ribbon on her back, Joan says that she does not know. Similarly, James cannot reason the answer. However, when Edward is finally asked the question, he is able to give a correct answer. Why?

Solution This problem can be solved by working backward. In order to win, A must force B to obtain some total between 44 and 49 on the next-to-last play, so A must have a total of 43. To obtain a 43, A must force B to have some total between 37 and 42, so A must have a 36. We can complete this plan by constructing Table 4.

Table 4

For A to get a score of	*B must have a score from*	*So A wants*
50	44–49	43
43	37–42	36
36	30–35	29
29	23–28	22
22	16–21	15
15	9–14	8
8	2–7	1

A can always choose 1 in the first move. Thus, if A starts first and selects numbers to obtain sums of 1, 8, 15, 22, 29, 36, and 43, there is no way B can win. •

The ninth strategy of designing a model will be used in just about every chapter of this book. You began your first work on writing a mathematical model in Section 1 of this chapter as you wrote verbal expressions in mathematical language, as open sentences.

The most widely used problem-solving strategy in this book is "determine an appropriate algorithm and use it." For our purposes, an algorithm is a sequence of instructions, or a pattern, or a procedure for solving a problem. For example, suppose you are asked to add

$$\frac{1}{7}+\frac{3}{7}$$

A well-known algorithm, to be discussed in Chapter 5, states that you add the numerators and keep the common denominator. That is,

$$\frac{1}{7}+\frac{3}{7}=\frac{4}{7}$$

Practically all your computations involve using well-known algorithms. The key to using algorithms is the ability to choose the correct one for the problem at hand. Problems throughout this book will help you sharpen this skill.

Exercise set 4

R 1. Before solving the problems, look at the following exercises and quickly guess the strategy or strategies that you might use to solve each problem. Then, when you have solved the problem, check to see if you have used the strategy guessed.

(a) Exercise 3 (b) Exercise 4
(c) Exercise 5 (d) Exercise 6
(e) Exercise 17 (f) Exercise 20
(g) Exercise 21 (h) Exercise 28

2. A rectangular house is built so that every wall has a window opening on the south. A bear is seen from one of the windows. What color is the bear?

3. How many different squares are in the following figure?

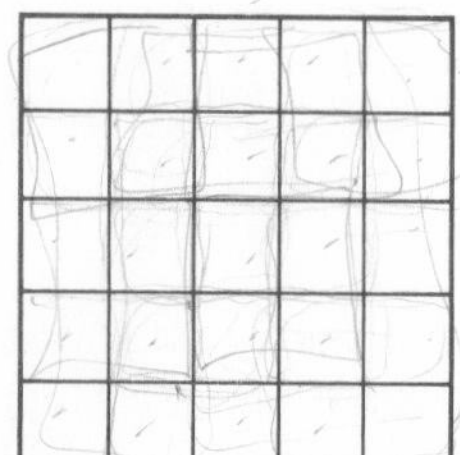

4. There are 10 posts in a row in Brooke's backyard. To decorate for a party, she strings crepe paper between posts. How many streamers will she need?

5. One of a group of 9 dimes is counterfeit and lighter than the others. Explain how you could determine which one is counterfeit with only two weighings on a balance scale.

6. Each member of the Wheeler family gives a gift at Christmas to all other members of the family. Find the total number of gifts given by the ten members of the family.

7. Remove 4 toothpicks to make 4 triangles that are the same size.

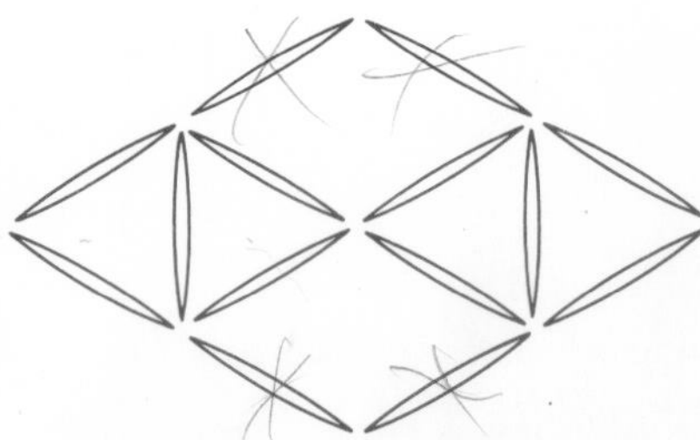

8. (a) Remove 2 toothpicks from the given figure so that you have 2 squares of different sizes.

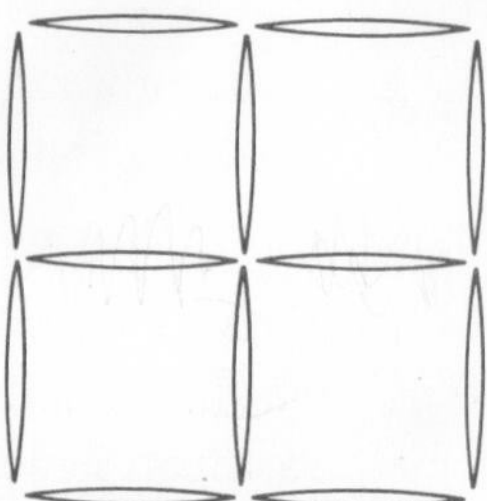

(b) Remove 2 toothpicks so that you have 3 squares of the same size.

9. In the barnyard is an assortment of chickens and pigs. Counting heads, one finds 13; counting legs, one finds 46. How many pigs and how many chickens are there?

10. Charles and Gary earn the same amount of money although one works 5 days more than the other. If Charles earns $28 a day and Gary earns $48 a day, how many days does each work?

11. What is the largest sum of money—all in coins and no silver dollars—that I could have in my pocket without being able to give change for a dollar, half dollar, quarter, dime, or nickel?

12. You have $1 in change. You have at least one of each coin less than a half dollar. You do not have a half dollar. What is the smallest number of coins you can have?

13. Jodi traded dollar bills for dimes and quarters (the same number of each). What is the smallest number of dollars she could have had?

14. How many cubes are there in this figure?

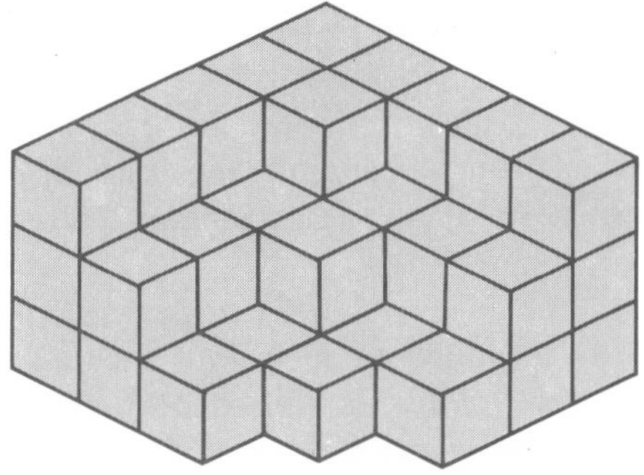

15. Arrange the first six counting numbers in the six circles so that the sum of each side of the triangle totals (a) 9 and (b) 10. (c) Are there other possible common sums?

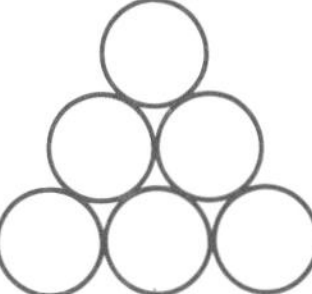

16. Twenty-five coins total $1.00. If one of the coins is a quarter, what are the other coins?

T 17. If a bird and a half lays an egg and a half in a day and a half, how many eggs will 33 birds lay in 11 days?

18. If 3 cats catch 3 rats in 100 minutes, how many cats will catch 100 rats in 100 minutes?

19. A container holds a mixture of 37 peanuts, 14 cashews, and 25 pecans. How many nuts must you take out to be sure of getting at least two of one kind? At least three of one kind?

20. To select the best two-person debate team in the United States, 117 teams are selected, with the understanding that when a team loses a match, the team is dropped from the tournament. How many games must be played to select a winner?

21. Can you determine which of 10 dimes is counterfeit (lighter than others) with only three weighings on a balance scale?

22. These four steps are made of cubes. How many cubes are needed to make 20 steps?

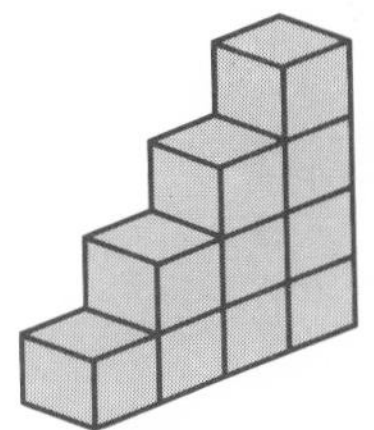

23. It takes 8 days for a truck to cross the desert. The truck can carry gasoline for only 5 days. What is the smallest number of trucks to start the trip in order for one to get across and the others to return?

24. If it takes 12 minutes to cut a log into 3 pieces, how long would it take to cut the log into 4 pieces?

25. Draw a straight line across the face of a clock so that the sums of the numbers on each side of the line are the same.

26. All six faces of a large cube are painted black. Then the cube is cut into 27 smaller cubes of equal size. (a) How many small cubes are painted on three sides? (b) On two sides? (c) On one side? (d) Not painted at all?

C 27. Labels were incorrectly placed on each of the three boxes. Explain how it is possible to determine the correct labels by removing just one ball from one box.

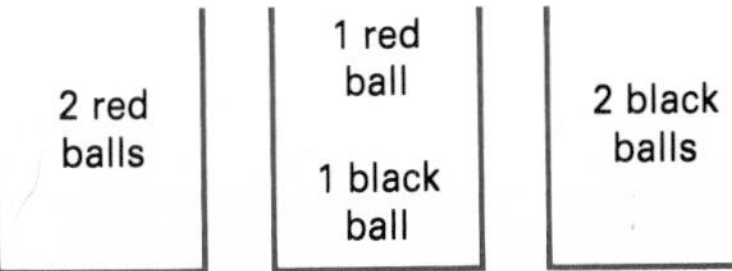

28. If 2 boys can paint 2 houses in 2 weeks and 3 girls can paint 3 houses in 3 weeks, how many houses can 6 boys and 6 girls paint in 12 weeks?

29. At Yute's store, he claims to be able to weigh all whole-number weights up to 40 pounds on a balance scale using only 4 weights. Find the four weights he will need and show how he will do the weighing.

30. Consider a six-room house with furniture arranged as in the accompanying figure. Can you interchange the desk and the bookcase in such a way that there is never more than one piece of furniture in a room at a time? The other three pieces of furniture do not need to return to their original places.

Cabinet		Desk
Television set	Sofa	Bookcase

31. Thirty points are marked on a circle, and straight-line segments are drawn connecting each point with all the other points. How many line segments are drawn?

32. A miniature checkerboard contains 16 small squares. A domino covers 2 of these squares. Can you place dominoes on the checkerboard so as to leave the corners uncovered, as shown in the figure, and cover the remainder of the board?

5 Critical thinking about relationships

■ People who are concerned about the environment support the clean air bill. Senator Dalton supports the clean air bill. Isn't it great that Senator Dalton is concerned about the environment? ■

Critical thinking mistakes, leading to illegitimate conclusions such as the preceding argument, are prevalent in all walks of life. Careful attention to critical thinking will allow us to replace such thinking with appropriate reasoning skills.

Pictures (called *Venn diagrams*) are introduced to help you analyze statements and combine several statements in order to obtain a conclusion. For example, assuming the truth of a number of statements, you determine whether or not you can obtain a true conclusion.

Even the most elementary of reasoning skills can be very useful. We consider in this book several elementary skills, such as:

(a) Drawing inferences from single premises
(b) Standardizing sentences stated in ordinary language
(c) Drawing inferences from multiple premises
(d) Avoiding mistakes with symmetrical and transitive relationships
(e) Working with analogies

Of course, not all of these skills can be studied in this one section. Some of them will be presented in the appendix.

We will begin this section by discussing Venn diagrams, a type of illustration used to represent relationships. Such diagrams, extensively used by the English logician John Venn, will be used not only in this section but also in several sections of Chapter 2 and Chapter 13. Consider the following example.

Example 43 • (a) All D are A. (For instance, all dogs are animals.)
(b) Some P are D. (For instance, some people are Democrats.)
(c) No C is a D. (For instance, no cat is a dog.)

Solution (a) Let's consider the dogs as being interior to one circle and the animals as interior to another circle. Since all dogs are animals, the circle of D's lies entirely within (could coincide with) the circle of A's. (See Figure 5.) If a part of the circle of D's were outside the circle of A's, this would imply that some dogs were not animals. If the circle of A's were entirely within the circle of D's, this would imply that all animals were dogs.

Figure 5

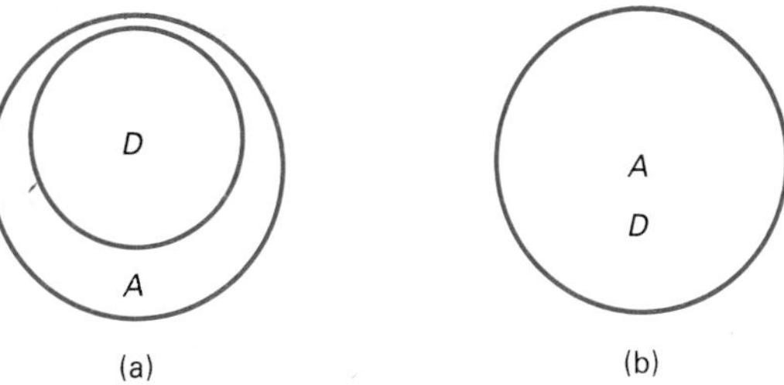

(b) The only thing about which we are certain is that there is a P that is a D. This requirement is met in (a), (b), or (c) of Figure 6.

Figure 6

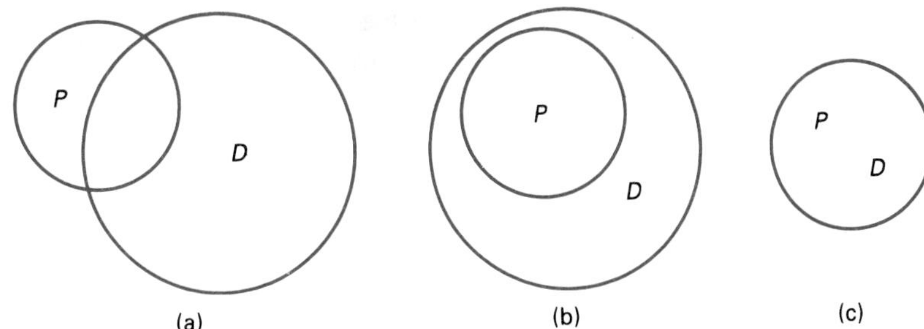

(c) The circle representing cats is disjoint from the circle representing dogs because not one C is a D. (See Figure 7.) •

Figure 7

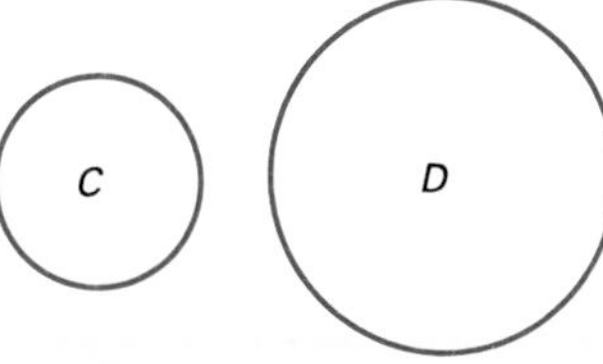

The words *all*, *no*, and *some* are referred to as **quantifiers.** We need to be able to use these and be certain that our conversations are correct from the standpoint of logic. We first use these quantifiers to draw inferences from single premises using Venn diagrams.

Example 44 • (a) "Some bullies are students" can be validly inferred from "Some students are bullies"; however,

(b) "All people are bullies" cannot be validly inferred from "All bullies are people."

This analysis becomes clearer when we examine the Venn diagrams in Figure 8. •

Figure 8

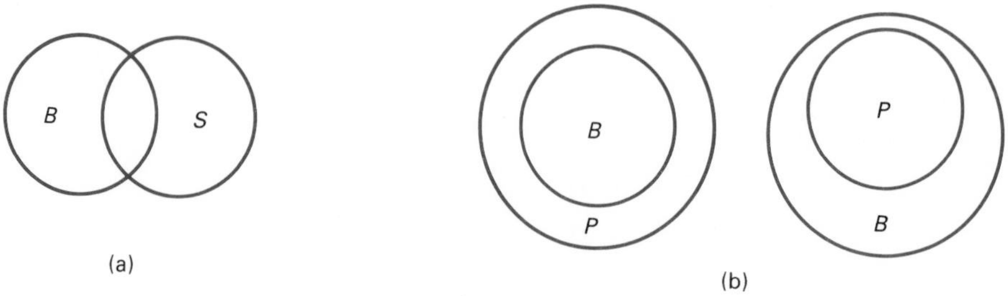

B S

(a)

B P P B

(b)

Example 45 • "No dishonest kids are kids in this class" can be validly inferred from "No kids in this class are dishonest," but

"No teachers are rich" and "Not all teachers are rich" do not logically say the same thing. •

Care must be taken in the use of *only* as a quantifier, in place of *all.* This is seen clearly when we realize that sentences beginning with *only* could begin with *all,* but with the subject and predicate reversed. Let's rewrite the first statement in the following example using *all.*

Example 46 • Only kids in my sorority are friends of mine. All kids in my sorority are friends of mine (incorrect). All friends of mine are kids in my sorority (correct). •

Now let's use our knowledge of Venn diagrams to go from a sequence of statements (assumed to be true and called **premises**) to a conclusion.

Example 47 • All chickens are fowls.
All leghorns are chickens.

Now consider the following assertions made on the basis of these premises.
(a) Can you be certain that all leghorns are fowls?
(b) Can you be certain that all fowls are leghorns?

Figure 9

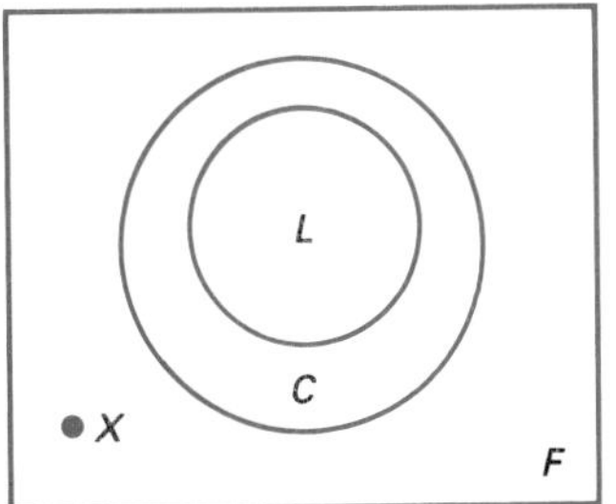

Solution (a) In Figure 9, let the interior of the small circle represent leghorns. According to the second premise, all leghorns (L) are chickens (C), so the circle of leghorns is *within* the circle of chickens. Likewise, the circle of chickens (C) is within the square of fowls (F). Thus anything within the circle of leghorns (L) is automatically within the square of fowls (F). Therefore, all leghorns are fowls.

(b) In Figure 9, note that there is a fowl (at X) that is not a leghorn. Thus the answer to the second question is *no.* •

In (a) of the preceding example, we say that the reasoning is **valid;** in (b), we say that the reasoning is **invalid.** Also, the preceding example illustrates the following truth.

A Venn diagram

> *A Venn diagram* representing a situation in which the conclusion is not true is sufficient to prove the reasoning invalid.

Example 48 • Some college professors are absentminded.
Mr. Smith is a college professor.
Therefore, Mr. Smith is absentminded.

Solution The premises (the first two sentences) of this problem can be illustrated in three ways by Venn diagrams. In Figures (a) and (b), Mr. Smith is a college professor and also absentminded. However, Figure 10(c) demonstrates that Mr. Smith can be a college professor without being absentminded. Thus, the truth of the premises does not require the truth of the conclusion, so Mr. Smith may or may not be absentminded. The argument contains a conclusion that is not an inescapable result of the premises. Thus, the reasoning is not valid, or it is invalid. •

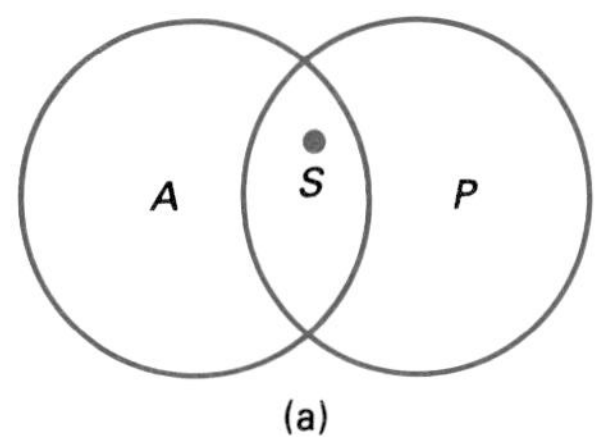

(a)

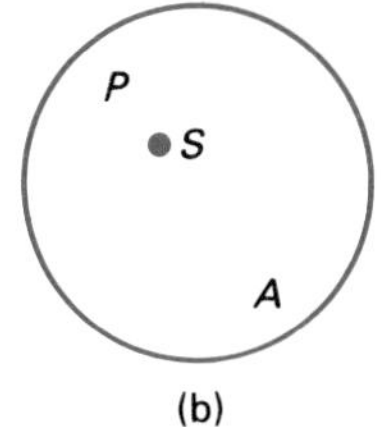

(b)

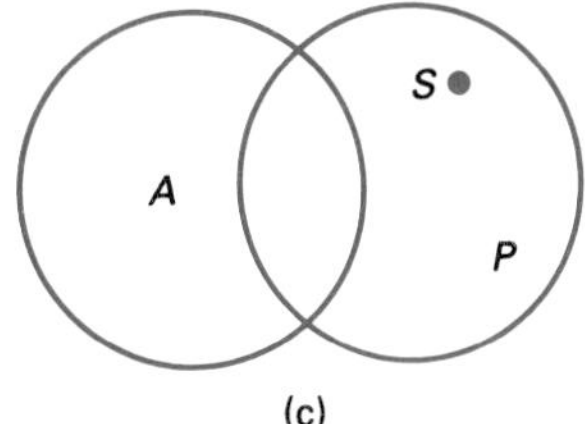

(c)

A: Absentminded people
P: Professors
S: Mr. Smith

Figure 10

Let's turn now to the problem in the introduction.

Example 49 • People who are concerned about the environment support the clean air bill. Senator Dalton supports the clean air bill.
Therefore, Senator Dalton is concerned about the environment.

Solution Let E represent people who are concerned about the environment. All these support the clean air bill (C). In Figure 11 Senator Dalton (D) supports the clean air bill (C) but is not necessarily inside E. Therefore, the argument is invalid. •

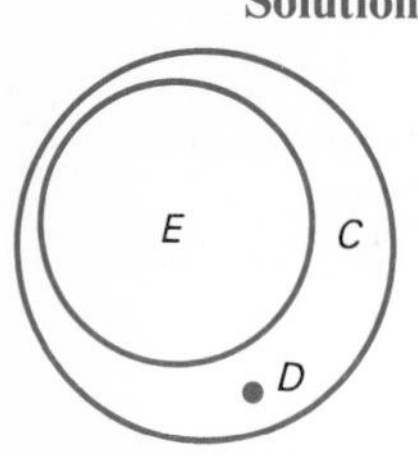

Figure 11

We need to remember that there is a difference between a valid argument and a true conclusion. A conclusion is true if the inference or argument is valid and all the premises are true.

Example 50 • All college professors are brilliant. Dr. Wheeler is a college professor. Therefore, Dr. Wheeler is brilliant.

Solution The argument is valid, but there may be some question about the truth of the first premise. Therefore, there is some question about the truth of the conclusion. •

We conclude this section with a short discussion of analogies. An **analogy** is the recognition of a resemblance between two relationships. Let's look at a simple analogy.

A bat is related to baseball as a paddle is related to Ping-Pong.

To understand this analogy, one must show that the relationship between a bat and baseball is the same as that between a paddle and Ping-Pong. We need to find what is called a **relationship sentence.**

Relationship sentence: *A ______ is used to hit the ball when playing ______.*
*For example, a **bat** is used to hit the ball when playing **baseball** and a **paddle** is used to hit the ball when playing **Ping-Pong**.*

As you begin to work with analogies, it is better to write the two relationships given in the analogy before attempting to write a relationship sentence.

Example 51 • Write the two facts given in the following analogy as preliminary steps for writing a relationship sentence.

Ice cream is to milk as bread is to flour.

Solution **Statement 1.** Ice cream is made from milk.
Statement 2. Bread is made from flour.
Relationship sentence: ______ *is made from* ______. •

Example 52 • Bedspread is to bed as carpet is to floor.

Just for fun

Three couples were on a hike when it became necessary to cross a river in a small boat having a maximum capacity of two. Because the men were extremely jealous, no woman could be left with a man unless her date was present. How did they manage to cross the river?

Solution **Statement 1.** Bedspread covers a bed.
Statement 2. Carpet covers a floor.
Relationship sentence: ______ *covers* ______. •

In the next examples, words are placed in blanks to form the best analogy. A relationship sentence will assist you in obtaining the best answer.

Example 53 • Cow is to grass as ______ is to ______.

(a) cow; beef (b) farm; pig (c) frog; flies
(d) saddle; stallion (e) horse; barn

Solution One possible relationship sentence is

______ *eats* ______.

Cow does not eat beef; farm does not eat pig; saddle does not eat stallion; and horse does not eat barn. However, a frog eats flies. So the answer is (c) (frog; flies). •

Exercise set 5

R 1. Make a Venn diagram showing each of the following.
(a) All im's are elm's.
(b) Some men are fat.
(c) No cams are dills.
(d) All men are persons and some men are lazy.

2. Make a Venn diagram showing
(a) the relationship among cats, dogs, and animals.
(b) the relationship among boys, girls, and blond people.
(c) some x's are y's; some x's are z's; no y's are z's.
(d) some y's are x's; no z's are x's; some z's are y's.
(e) all x's are y's; some y's are z's; no x's are z's.

3. Write a relationship sentence for each of the following:
(a) Ham is to pig as steak is to cow.
(b) Ears are to hear as nose is to smell.
(c) Soap is to hands as shampoo is to hair.
(d) Carpenter is to hammer as accountant is to calculator.
(e) Fingernail is to finger as toenail is to toe.
(f) Writing is to pen as sewing is to needle.
(g) "Pigging out" is to obese as dieting is to skinny.
(h) Cry is to baby as bark is to dog.

Write a relationship sentence and determine which pair of words fits best the blanks in Exercise 4 through 8,

4. Cabinet is to dishes as ______ is to ______.

   (a) Food; ice (b) Closet; clothes
   (c) Vacuum cleaner; rug (d) Knife; fork

5. Water is to liquid as ______ is to ______.

   (a) Orange; juice (b) Coffee; cream
   (c) Milk; cheese (d) Ice; solid

6. Fat is to skinny as ______ is to ______.

   (a) Pretty; beautiful (b) Wrinkles; old
   (c) Loud; quiet (d) Disbelief; heretic

7. Food is to a cook as ______ is to ______.

   (a) Clothes; tailor (b) Book; library
   (c) Water; dry (d) Teeth; clean

8. Apple is to fruit as ______ is to ______.

   (a) Cabbage; vegetable (b) Painting; art
   (c) Juice; orange (d) Berry; wine

9. Write what you feel is the best possible relationship sentence for each of the following:

   (a) 5 is to 25 as 8 is to 40.
   (b) 8 is to 20 as 40 is to 52.
   (c) 10 is to 11 as 6 is to 11.

For the next few problems, assume that the first two statements are correct. Use a Venn diagram to answer the questions.

10. All cats are elephants. Some elephants are red. Can you be certain that some cats are red?

11. Some cows are birds. All birds have two legs. Can you be certain that some cows have two legs? Can you be certain that all birds are cows? Can you be certain that all cows have two legs?

12. Some clothes are old. No old things are good. Can you be certain that some clothes are good? Can you be certain that some clothes are not good?

13. Some beans are green. All green things are edible. Can you be certain that all beans are edible? Can you be certain that some beans are not edible?

Using Venn diagrams, determine whether the third statement is correct if it is assumed that the first two statements are true in Exercises 14 through 20.

T 14. All fast runners are athletes. No people with long legs are athletes. Therefore, no people with long legs are fast runners.

15. All spheres are round. All circles are round. Therefore, all circles are spheres.

16. Carelessness leads to accidents. Mrs. Yeager had an accident. Therefore, Mrs. Yeager was careless.

17. All cats are intelligent animals. Fido is an intelligent animal. Therefore, Fido is a cat.

18. Some em's are red. All im's are em's. Therefore, some im's are red.

19. All athletes are fast runners. Some people with short legs are not fast runners. Therefore, no athletes have short legs.

20. All wise boys are interested in girls. George is interested in girls. Therefore, George is wise.

21. Write a valid conclusion for the following premises. All sales must come from production, and consumption must come from sales.

In Exercises 22 through 27, select what you think is the best answer. Check yourself by writing a relationship sentence.

22. State is to governor as ______ is to ______.

    (a) Mayor; city
    (b) President; country
    (c) Government; business
    (d) Senate; congress
    (e) Business; manager

23. ______ is to pipes as dentist is to ______.

    (a) Tobacco; teeth
    (b) Nurse; dental assistant
    (c) Bagpipes; root canal
    (d) Doctor; oral hygiene
    (e) Plumber; teeth

24. ______ is to sewing as an artist is to ______.

    (a) Homemaker; picture
    (b) Dress; paint brush
    (c) Making; painter
    (d) Cloth; color
    (e) Seamstress; painting

25. ______ are to doctors as ______ are to veterinarians.

    (a) Nurses; dogs (b) People; animals
    (c) Hospitals; cats (d) Medicine; shots
    (e) Dogs; pigs

26. ______ is to reporter as ______ is to author.

    (a) Newspaper; pencil (b) Paper; ink
    (c) Pages; ideas (d) Newspaper; book
    (e) Editor; plot

27. Snow is to ______ as ice is to ______.

(a) Pads; helmet (b) Clouds; cold
(c) Skiing; skating (d) Feet; head
(e) Jacket; shoes

C 28. (a) What conclusion can be drawn about Dr. X?

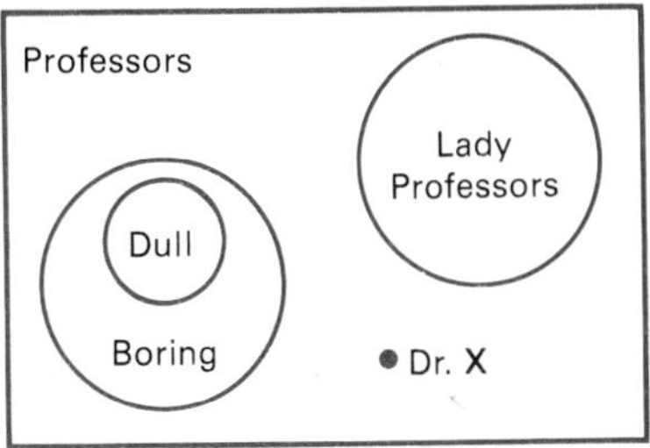

(b) I represents integers; R, rational numbers; Ir, irrational numbers; Re, real numbers; Im, imaginary numbers; and C, complex numbers. What conclusion can be drawn about x? (Although it will not affect the answer, to be precise one needs to note that the region of Re outside of both R and Ir contains no elements.)

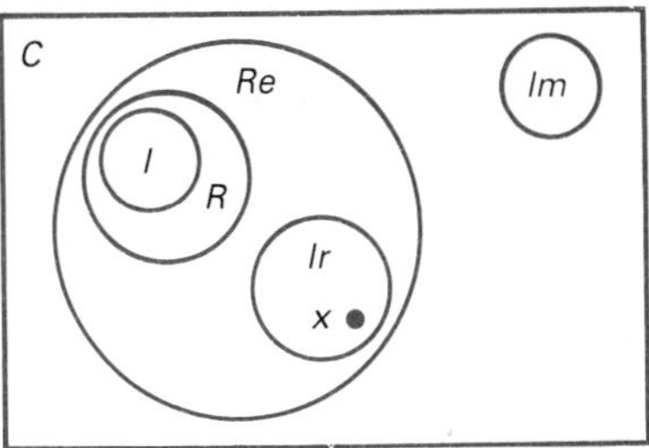

(c) What conclusions can be drawn concerning the members of set R?

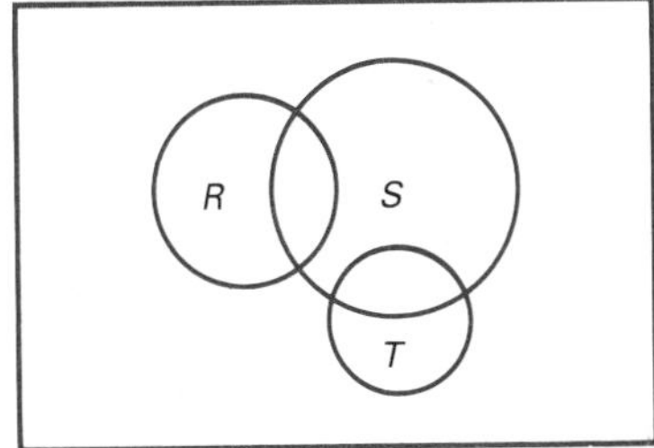

29. Using Venn diagrams and assuming the first three statements are true, determine which of the possibilities are correct.
All x's are y's.
Some x's are z's.
Some x's are w's.
Therefore,

(a) all w's are y's.
(b) all z's are y's.
(c) some y's are w's.
(d) some x's are not y's.

*6 Critical thinking and an introduction to logic

■ I will pass this mathematics course or I will be sad. This statement can be true in three different ways. Can you enumerate them? The answer to this question will be obvious as we study this section. ■

Now that you are familiar with some critical thinking skills and have an understanding of some of the strategies of problem solving, you are ready to take a giant step in the development of your cognitive skills. One important ingredient in all thinking is the use of symbolic logic. The discussion in this section introduces the terminology and some of the procedures of logic.

Also in this section, you will learn to use truth tables. One way to prove that something is true is to validate its truth for all situations. This we shall do with truth tables.

One important ingredient both in critical thinking and in mathematical reasoning is logic. The following discussion introduces the terminology and some of the procedures of logic. In mathematics, as well as in ordinary communication, **assertions** similar to the ones below are often encountered.

1. Fred ordered hamburgers, and Leslie selected hot dogs.
2. Income tax rates will be increased, or yearly appropriations will be decreased.
3. It is not true that George failed statistics.

Such examples illustrate sentences involving **connectives** like "and," "or," and "not." These and other connectives play an important role in one of the basic problems of elementary logic—determining whether complicated statements are true or false from the truth or falsity of their parts. This type of logic in everyday language is called *common sense,* whether it is demonstrated by a politician, a sociologist, an educator, or a student on a true-false test.

We learned in an earlier section that a **statement is a declarative sentence that can be assigned a truth value;** that is, it can be classified as true or as false, but not as both.

Since some sentences are neither true nor false, not all sentences are statements. Consider the following examples, which are not statements.

Example 54 • (a) Keep your eyes on the road. (A command)
(b) What time is it? (A question)
(c) Oh, what a beautiful sunset! (An exclamation)
(d) I am lying to you. (A paradox) •

When a statement is true, its truth value is denoted by T. When a statement is false, its truth value is denoted by F. No statement is both true and false. Every statement has either truth value T or truth value F. Notice that the following sentences assert something and have truth values in the sense that they can be classified as true or as false.

Example 55 • (a) President Lincoln was born in Texas. (F)
(b) $2 + 3 = 5$ (T)
(c) 8 is less than 7. (F) •

"He is president of the Owl's Club" and "$x + y = 9$" are statements if we know to whom *he* refers and what numbers x and y represent. Otherwise they are not statements.

Ordinarily, simple statements express only one thought, and compound statements are formed by simple statements joined by connectives. The three sentences previously given as examples of statements are all simple statements.

The following statements are compound statements:

1. The birth rate in Freedom County has increased, or there are more young married couples in the county.
2. The number of majors in sociology at Graduatum College has increased, and the number of majors in physics has decreased.

Compound statement

> A statement is a **compound statement** if it contains a connective; otherwise, the statement is a **simple statement.**

Let's look now at the difference in opposite and negation of a statement. The opposite of white is commonly accepted to be black, but the negation of white is anything not white. The opposite of "turn right" is "turn left," but the negation of "turn right" includes turn left, go straight ahead, or even go sideways.

Negation of a statement

> The *negation of statement* p is the statement "It is not true that p" (denoted by $\sim p$).

The negation of "New York is the capital of the United States" is "It is not true that New York is the capital of the United States." The use of the grammatical form "It is not true that p" sometimes results in an awkward sentence structure. For many statements the negation may be more simply expressed by negating the predicate. For instance, the negation of the preceding statement could be written as "New York is not the capital of the United States." By definition, the negation of a statement is false when the statement is true and true when the statement is false. This fact is summarized as follows.

Truth value of $\sim p$

> If p is true, then $\sim p$ is false; and if p is false, then $\sim p$ is true.

Table 5

p	$\sim p$
T	F
F	T

Table 5 summarizes the two possibilities for p and $\sim p$. Such a table, called a *truth table,* emphasizes the fact that p and $\sim p$ cannot be simultaneously true or simultaneously false.

Consider the statement "Fred ordered hamburgers and Leslie selected hot dogs." We accept this statement as true, provided that on the day in question Fred actually requested hamburgers and Leslie actually helped herself to hot dogs. On the other hand, we consider the statement false if Fred did not order hamburgers, regardless of Leslie's preference.

A conjunction

> Given any statements p and q connected by "and," the statement "p and q" (denoted by $p \wedge q$) is called a *conjunction.*

Since p is either true or false and since q is either true or false, there are four possibilities for the combined truth values of p and q. Table 6 illustrates these possibilities. In the third column of the table, we note the truth value of $p \wedge q$ for

Table 6

p	q	$p \wedge q$	*Examples*
T	T	T	$2 + 3 = 5$ and 1 is less than 2.
T	F	F	$2 + 3 = 5$ and 4 is less than 3.
F	T	F	December has 34 days and Christmas is on December 25.
F	F	F	December has 32 days and January has 22 days.

each of the four possibilities. When p is true and q is true, then $p \wedge q$ is true. When p is true and q is false, $p \wedge q$ is false. When p is false and q is true, $p \wedge q$ is false. Finally, when p is false and q is false, $p \wedge q$ is false.

Truth value of a conjunction

The *conjunction,* "p and q," is true, if both p and q are true; otherwise "p and q" is false.

If you had a true-false question in political science—for instance, "The Democrats have control of the House of Representatives or the Republicans control the Senate"—would the answer be true or false?

A disjunction

Given any statements p and q connected by "or," the statement "p or q" (denoted by $p \vee q$) is called a *disjunction.*

Statements using the connective "or" are false only when both parts are false. Table 7 illustrates the truth values for the statement "p or q" for all possible combinations of truth values of p and of q.

Table 7

p	q	$p \vee q$	*Examples*
T	T	T	Sugar is sweet or December has 31 days.
T	F	T	Sugar is sweet or the earth is flat.
F	T	T	$2 + 3 = 7$ or 2 is less than 7.
F	F	F	$2 + 3 = 7$ or 2 is greater than 5.

Truth value of a disjunction

The *disjunction,* p or q, is true if p is true, if q is true, or if both p and q are true.

Just for fun

Mr. Kojak, assistant principal at Local High School, discovered that a student had thrown a rock and broken a window. Quickly remembering that he had seen five students in the area after school, he called all five into his office. There each student made two true statements and one false statement. In recording the statements, Kojak realized who the guilty student was. Dude said, "I did not break the window. I never even learned to throw rocks—Butch did it." Butch replied, "Little Joe is the one who threw the rock. Dude lied when he said I was guilty. I am innocent." Little Joe angrily shouted, "I did not do it. Tanner and I are old pals. Bushy is the guilty person." Bushy denied this charge with "I did not break the window. Dude does not know who threw the rock. Everyone else is just trying to pass the buck." Tanner added "Butch is guilty. I don't even like Little Joe. I did not throw the rock." Which student was suspended?

When working with truth tables, one needs to be careful to list all possible truth values. For example, in statements involving p, q, and r, p can be T or F, q can be T or F, and r can be T or F. Thus, eight combinations are needed to consider all possibilities.

Example 56 • Make a truth table for $p \wedge (q \vee r)$.

Solution

Table 8

p	q	r	$q \vee r$	$p \wedge (q \vee r)$
T	T	T	T	T
T	T	F	T	T
T	F	T	T	T
T	F	F	F	F
F	T	T	T	F
F	T	F	T	F
F	F	T	T	F
F	F	F	F	F

•

Exercise set 6

R 1. Determine which of the following are statements and classify each statement as either true or false.

(a) George Washington was born in Alabama.
(b) Good morning.
(c) $5 + 4 = 9$
(d) $x + 5 = 8$
(e) Close the door.
(f) $2 \cdot 3 = 7$
(g) $3x = 6$
(h) Help stop inflation.

2. If p is T and q is F, find the truth values. $\sim(\ )$ represents the negation of everything inside the parentheses.

(a) $\sim p \wedge q$ (b) $\sim p \vee \sim q$
(c) $\sim(p \vee q)$ (d) $\sim(p \wedge q)$
(e) $\sim(\sim p \wedge q)$ (f) $\sim(p \vee \sim q)$

3. Find the truth values of the statements in Exercise 2 if p is T and q is T.

4. Find the truth values of the statements in Exercise 2 if p is F and q is F.

5. In each of the parts below, construct two statements, one using "or" and the other using "and," and find the truth values of each.

(a) November has 30 days. Thanksgiving is always on November 25.
(b) The smallest counting number is 2. 10 is not a multiple of 5.
(c) $2 + 3 = 4 + 1$ $8 \cdot 6 = 4 \cdot 12$
(d) Triangles are not squares. 3 is smaller than 5.

6. Translate the following statements into symbolic form, using $A, B, C, D, \wedge, \vee$, and $\sim$, where A, B, C, and D denote the following statements.

A, it is snowing.
B, the roofs are white.
C, the streets are not slick.
D, the trees are beautiful.

(a) It is snowing and the trees are beautiful.
(b) The trees are beautiful or it is snowing.
(c) The streets are not slick and the roofs are not white.
(d) The trees are not beautiful and it is not snowing.

7. Using the statements of Exercise 6, translate the symbolic statements below into English sentences.

(a) $A \wedge \sim B$ (b) $\sim B \vee \sim C$
(c) $A \wedge (B \vee C)$ (d) $(A \vee \sim C) \wedge D$
(e) $\sim(A \wedge \sim D)$ (f) $\sim(A \wedge \sim C)$

8. If "it is snowing" and "the trees are beautiful" are true and "the roofs are white" and "the streets are slick" are false, classify the following statements as true or as false.

(a) It is snowing and the trees are beautiful.
(b) The trees are beautiful or it is snowing.
(c) It is not snowing and the roofs are white.
(d) It is not snowing or the streets are slick.
(e) The streets are not slick and the roofs are not white.
(f) The trees are not beautiful and it is not snowing.

T 9. Construct truth tables to illustrate the truth values for

(a) $p \vee \sim q$ (b) $p \wedge (q \vee r)$
(c) $p \wedge \sim p$ (d) $(p \vee \sim p) \wedge q$
(e) $(p \vee \sim q) \wedge r$ (f) $\sim(p \vee q) \wedge r$

10. Let p represent "Jane is clever" and q represent "Charles is clever." Use "slow" as the negative of clever. Write the following using the symbols of logic.

(a) Charles is slow and Jane is clever.
(b) It is not true that Jane is clever and Charles is slow.
(c) Either Charles is slow or Jane is slow.
(d) Both Charles and Jane are slow.

11. In the construction of a truth table, how many different combinations of T's and F's do you need for statements involving

(a) p only? (b) p and q?
(c) p, q, and r? (d) p, q, r, and s?

12. Let p represent "The stock market is bullish." Let q represent "The Dow average is increasing." Finally, let r represent "The price of utilities is decreasing." Translate each of the following into words.

(a) $\sim p \wedge q$ (b) $\sim p \vee \sim r$
(c) $(p \vee r) \wedge q$ (d) $(\sim r \vee \sim p) \wedge q$

13. Translate the following sentences into symbols.
    (a) Either I invest my money in stocks, or I put my money in a savings account.
    (b) I do not invest in stocks or deposit my money in a savings account but buy gold.
    (c) I do not buy gold, and I do not deposit my money in a savings account.

14. Let p represent the statement "The price of automobiles will rise," and q the statement "The inflation rate will increase." Translate each of the following sentences into symbols.
    (a) The price of automobiles will rise, and the inflation rate will increase.
    (b) The price of automobiles will rise, or the inflation rate will increase.
    (c) The price of automobiles will not rise, and the inflation rate will increase.
    (d) The price of automobiles will not rise, or the inflation rate will not increase.

C 15. Find the truth value of each of the following when q is true and p and r are both false.
    (a) $(p \vee q) \wedge \sim(\sim p \vee r)$ (b) $(p \wedge \sim q) \vee (r \wedge \sim q)$
    (c) $\sim(r \wedge p) \wedge \sim(q \vee r)$ (d) $(r \wedge p) \vee (\sim q \wedge r)$

*7 Critical thinking and conditionals

■ If Lee takes Sue to the prom, then I'll be a monkey's uncle. This is a true statement because Lee and Sue are not speaking these days. ■

How do we know the above is a true statement? The answer comes from the truth of conditionals.

Sometimes, instead of making an outright assertion or statement, one may wish to restrict the statement by means of making it a **conditional statement.** For example, instead of saying "I carry my umbrella," one may wish to say "If it rains, I carry my umbrella." This statement contains a condition concerning the carrying of an umbrella. "If the sun shines, we go on a picnic," "If I get a bonus, then I take the family on a vacation," and "If a number is not divisible by 2, then it is odd" are examples of conditional statements.

This section focuses on the various forms that conditionals can assume and on ways to determine when they are logically true.

To begin our discussion we consider the following definition.

A conditional

> A *conditional* (sometimes called an implication) is a statement of the form "If p, then q," where p and q are statements. It is denoted by $p \rightarrow q$.

Let p represent "It snows" and q represent "The streets will be slick." "If it snows, then the streets will be slick" is denoted by "If p, then q," where p is called the **antecedent** (or hypothesis) and q the **consequent** (or conclusion). This same conditional may be worded in various ways, several of which are listed for comparison.

Example 57 • (a) If it snows, then the streets will be slick.
(b) If it snows, the streets will be slick.
(c) The streets will be slick if it snows.

Using the letters p and q, the same conditionals may be written as follows.
(a) If p, then q.
(b) If p, q.
(c) q if p. •

In practice, we use "if-then" statements in many different ways. Sometimes the statements indicate a decision on the part of the speaker. For instance, "If it rains, I shall carry my umbrella today" indicates a decision. We occasionally use "if-then" to connect two statements that are actually unrelated. "If Lee takes Sue to the prom, then I'll be a monkey's uncle" involves two statements, both of which are assumed false. Other "if-then" statements indicate logical relationships; "If a number is divisible by 4, then the number is divisible by 2." We want to formulate a procedure, which will apply in all these cases, for classifying a conditional as true or as false.

In order to decide when to classify a conditional as true and when to classify it as false, we shall think of "if p, then q" as a promise. I promise my wife that "If I get a bonus, we will take a vacation." The promise will be broken (that is, the conditional is false) if I get a bonus and we do not take a vacation (that is, if p is true and q is false). If I do not get a bonus, we can either take or not take a vacation, and the conditional is true because the promise was made relative to what would happen if the bonus were granted.

Truth value of a conditional

When the **antecedent of a conditional is false,** the conditional is defined to be true regardless of the truth or falsity of the consequent. A conditional is false only when the antecedent is true and the consequent is false.

This fact is illustrated in Table 9.

Table 9

p	q	$p \to q$	*Examples*
T	T	T	If there is water pollution in Lake Erie, then fish are dying.
T	F	F	If $2 + 3 = 5$, then $2 + 4 = 7$.
F	T	T	If $4 + 1 = 7$, then $2 \cdot 3 = 6$.
F	F	T	If a triangle is a square, then each of its angles has a measurement of 90°.

Closely related to a conditional are its *converse,* its *inverse,* and its *contrapositive.*

Converse of a conditional

The *converse* of the conditional "If p, then q" is the conditional "If q, then p" (that is, the converse of $p \to q$ is $q \to p$).

"If it is cold, then it is snowing" is the converse of "If it is snowing, then it is cold." This example shows that the truth of a conditional in no way ensures the truth of its converse. "If it is snowing, then it is cold" is a true conditional, but the fact that it is cold does not imply that it is snowing. Assuming that the truth of a conditional ensures the truth of its converse is a common fallacy, often subtly used in political and advertising campaigns. For instance, "If a man is a communist, he is a socialist" is a true conditional. Yet its converse has incorrectly been asserted at times to be true: "If a man is a socialist, then he is a communist."

Similarly, the truth of a conditional certainly does not require that its converse be false. "If $2 + 2 + 2 = 6$, then $3 \cdot 2 = 6$" and its converse, "If $3 \cdot 2 = 6$, then $2 + 2 + 2 = 6$," are both true.

Inverse of a conditional

The *inverse* of a given conditional is the conditional that results when p and q are replaced by their negations (that is, the inverse of $p \to q$ is $\sim p \to \sim q$).

The inverse of "If a polygon is a rectangle, then it is a parallelogram" is "If a polygon is not a rectangle, then it is not a parallelogram."

Although the truth of a conditional does not ensure the truth of the inverse of the conditional, much advertising assumes that you will make this mistake.

Example 58 • "If you use Stay-White toothpaste, you will be sexy." The advertisers hope you will assume that the inverse is true: "If you do not use Stay-White toothpaste, you will not be sexy." •

Contrapositive of a conditional

The *contrapositive* of the conditional "If p, then q" is "If $\sim q$, then $\sim p$" (that is, the contrapositive of $p \to q$ is $\sim q \to \sim p$).

The contrapositive of "If it is snowing, then it is cold" is "If it is not cold, then it is not snowing." Common sense dictates that the two statements say the same thing. The fact that the truth of the contrapositive ensures the truth of the associated conditional will be shown later in this section.

To summarize the definitions of converse, inverse, and contrapositive, consider the following example.

Example 59 •
(a) *Conditional.* If Mr. Jones buys a new automobile, he will select a red compact.
(b) *Converse.* If Mr. Jones selects a red compact, he is buying a new automobile.
(c) *Inverse.* If Mr. Jones does not buy a new automobile, he will not select a red compact.
(d) *Contrapositive.* If Mr. Jones does not select a red compact, he is not buying a new automobile. •

Closely associated with the conditional statement is the **biconditional.**

Biconditional

"p if and only if q," called a **biconditional**, is denoted by $p \leftrightarrow q$.

For example, let p be the simple statement "I am tired," and let q be the statement "I exercise." The biconditional $p \leftrightarrow q$ is "I am tired if and only if I exercise." $p \leftrightarrow q$ can be explained as

$$(p \to q) \wedge (q \to p)$$

Thus, from the truth table in Table 10, it is seen that for $p \leftrightarrow q$ to be true, it is necessary that when p is true q must be true, and when p is false q must be false. Note that the truth values for $p \leftrightarrow q$ and $(p \to q) \wedge (q \to p)$ are the same.

Table 10

p	q	$p \leftrightarrow q$	$p \to q$	$q \to p$	$(p \to q) \wedge (q \to p)$
T	T	T	T	T	T
T	F	F	F	T	F
F	T	F	T	F	F
F	F	T	T	T	T

Sometimes, instead of writing "p if and only if q," we write "p is necessary and sufficient for q." "If p, then q" has the same meaning as "p is sufficient for q"; that is, in order to conclude q, it is sufficient to know p. Writing "p is necessary for q" is equivalent to writing "if q, then p" and to writing "q only if p." We may write "Two triangles are congruent if and only if corresponding sides are congruent" as "A necessary and sufficient condition for two triangles to be congruent is that corresponding sides be congruent."

Example 60 • Construct a truth table to determine when $\sim(p \wedge q) \to \sim r$ is true.

Solution The eight possible combinations of the truth values of p, q, and r are listed in columns A, B, and C of Table 11. In D, the truth values of $p \wedge q$ are listed; in E, the negations of the truth values of D are listed; the truth values for $\sim r$ are found in F; and the truth values for the given implication are tabulated in G. •

Table 11

A	*B*	*C*	*D*	*E*	*F*	*G*
p	q	r	$p \wedge q$	$\sim(p \wedge q)$	$\sim r$	$\sim(p \wedge q) \to \sim r$
T	T	T	T	F	F	T
T	T	F	T	F	T	T
T	F	T	F	T	F	F
T	F	F	F	T	T	T
F	T	T	F	T	F	F
F	T	F	F	T	T	T
F	F	T	F	T	F	F
F	F	F	F	T	T	T

Just for fun

The mobster Big Maf's body was found floating in his swimming pool. When Detective Columbus was called to investigate the murder, he determined a number of facts involving the prime suspects. Either the butler, James, was at home or Big Maf's girlfriend, Blondie B, was out. If the private movie projector was on, Blondie B was at home. But if it was off, Big Maf's first assistant. Godchild, murdered Big Maf. On the other hand, if Godchild did not commit the murder, the butler did it. However, James was not at home at the time of the murder; a witness verified that he was at the local curb market. Detective Columbus quickly made his arrest. Whom did he arrest? What reasoning did he follow?

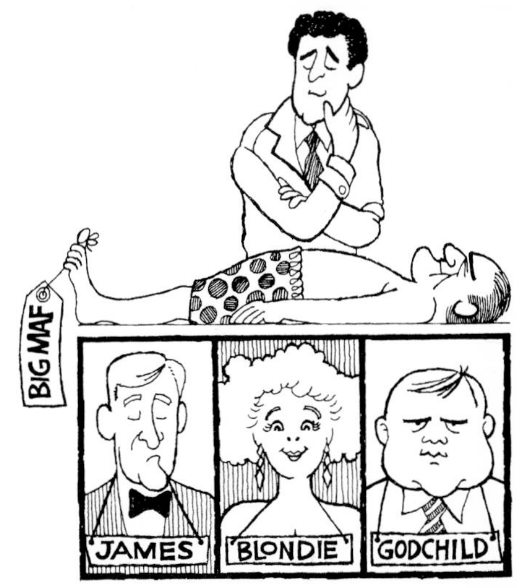

When two statements p and q have the same truth values, we say that p is *logically equivalent* to q. •

Logically equivalent

> Two statements are said to be *logically equivalent* if they have the truth values.

A conditional and its contrapositive are logically equivalent, as the following example demonstrates.

Example 61 • "If you wash your hair with *Head* shampoo, your dandruff will disappear" is a conditional whose contrapositive is "If your dandruff does not disappear, then you do not wash your hair with *Head* shampoo."

Solution Table 12 verifies that the two statements are logically equivalent; here p represents "you wash your hair with Head shampoo" and q represents "your dandruff will disappear."

Table 12

p	q	$\sim p$	$\sim q$	$p \to q$	$\sim q \to \sim p$
T	T	F	F	T	T
T	F	F	T	F	F
F	T	T	F	T	T
F	F	T	T	T	T

We observe that, for all possible values of p and q, the truth values for $p \to q$ and $\sim q \to \sim p$ are exactly the same. Thus, a conditional and its contrapositive are logically equivalent. •

Example 62 • Using a truth table, show that the converse of a conditional is not logically equivalent to the conditional.

Solution In Table 13, note that the truth values for $p \to q$ and $q \to p$ are not the same for given values of p and of q. Therefore, a conditional and its converse are not logically equivalent.

Table 13

p	q	$p \to q$	$q \to p$
T	T	T	T
T	F	F	T
F	T	T	F
F	F	T	T

•

Exercise set 7

R 1. Let A represent "It is snowing"; B, "The roofs are white"; C, "The streets are slick"; and D, "The trees are beautiful." Write the following in symbolic notation.

(a) If it is snowing, then the trees are beautiful.
(b) If it is not snowing, then the roofs are not white.
(c) If the streets are not slick, then it is not snowing.
(d) If the streets are slick, then the trees are not beautiful.

2. In Exercise 1, assume A is true, B and C are both false, and D is true. Classify the conditional statements as either true or false.

3. State the converse, the inverse, and the contrapositive of each of the following conditional statements.

(a) If a triangle is a right triangle, then one angle has a measure of 90 degrees.
(b) If a number is a prime, then it is odd.
(c) If two lines are parallel, then alternate interior angles are equal.
(d) If Joyce is smiling, then she is happy.
(e) If x is divisible by 10, then x is divisible by 5.
(f) If Tom is John's father, then John is Tom's son.

4. Write each of the following statements in the form "If . . . then."

(a) Triangles are not squares.
(b) Birds of a feather flock together.
(c) Honest politicians do not accept bribes.
(d) A rolling stone gathers no moss.

5. Write the converse and contrapositive of each part of Exercise 4.

6. Using the notation of Exercise 1, translate the following into sentences.

(a) $\sim A \to B$
(b) $\sim C \to \sim B$
(c) $(\sim B \wedge \sim C) \to A$
(d) $(A \vee B) \to \sim C$
(e) $\sim(A \to \sim B)$
(f) $(\sim C \vee \sim A) \to \sim B$

7. Using the truth values of Exercise 2, classify the statements in Exercise 6 as either true or false.

8. Use the following statements to form a conditional, its converse, its inverse, and its contrapositive.

(a) I teach third grade. I do not teach in high school.
(b) I own a car. I have a driver's license.

9. Commercials and advertisements are often based on the assumption that naive audiences assume that the converse, inverse, contrapositive, and original statement are all true. Write the converse, inverse, and contrapositive of each statement below. Do they have the same truth values?

(a) If you brush your teeth with White-as-Snow, then you have fewer cavities.
(b) If you like this book, then you love mathematics.
(c) If you want to be strong, then you eat Barlies for breakfast.
(d) If you use Wave, then your clothes are bright and colorful.
(e) If you are tall, dark, and sexy, then you use Velvet after-shave lotion.

T 10. For the conditional statement "If I use leaded gasoline, I pollute the atmosphere," find the following.

(a) Contrapositive of the converse.
(b) Converse of the contrapositive.

11. Construct truth tables for the following.

(a) $\sim p \rightarrow \sim q$ (b) $(p \wedge q) \rightarrow r$
(c) $p \rightarrow (q \wedge r)$ (d) $(p \vee q) \rightarrow r$
(e) $(p \vee q) \rightarrow (r \wedge p)$ (f) $(p \vee \sim q) \rightarrow r$
(g) $(p \vee q) \leftrightarrow (q \vee p)$ (h) $(p \wedge q) \leftrightarrow (q \wedge p)$

C 12. Answer true or false, where p represents "A car is a Chevrolet" and q represents "A car is a General Motors product." Assume $p \rightarrow q$ is true.

(a) $q \rightarrow p$
(b) p is necessary and sufficient for q
(c) p if and only if q
(d) If q, then p
(e) p is a sufficient condition for q

13. Construct a truth table to show that $p \rightarrow q$ and $\sim(p \wedge \sim q)$ have the same truth values.

14. Using truth tables, determine whether the following pairs of statements are logically equivalent.

(a) $\sim(\sim p)$; p (b) $\sim(p \wedge q)$; $\sim p \vee \sim q$
(c) $\sim(p \vee q)$; $\sim p \wedge \sim q$ (d) $\sim(p \rightarrow q)$; $p \wedge \sim q$
(e) $p \rightarrow q$; $\sim p \vee q$ (f) $\sim(p \vee q)$; $\sim p \vee \sim q$
(g) $p \vee \sim q$; $\sim(\sim p \vee q)$ (h) $p \vee \sim q$; $\sim(\sim p \wedge q)$
(i) $r \vee (s \wedge t)$; $(r \vee s) \wedge (r \vee t)$

Solution to introductory problem

Understanding the problem. Let us restate the problem. Three times the following procedure is executed.
(a) The remaining pile is divided into three equal piles with one extra coconut.
(b) One of the three piles is hidden, while the other two equal piles are combined and left.
Finally, the remnant after these three incursions is divided into three equal piles with one coconut remaining.

Devise a plan. Work the problem backward. Note that the sum of the last three piles plus 1 must be an even number because it is the sum of two equal previous piles. This situation must be true for the total number of coconuts at all times except at the beginning. This fact allows us to rule out possibilities as we look at various cases.

Carry out the plan. Let n be the number of coconuts received by each sailor at the end. Now n cannot be even, since $n + n + n + 1$ would be odd, which is a contradiction. If $n = 1$, then

$1 + 1 + 1 + 1 = 4$ (Carl then divides into piles of 2.)

$2 + 2 + 2 + 1 = 7$ (odd)

which is a contradiction. If $n = 3$, then

$3 + 3 + 3 + 1 = 10$ (Carl then divides into piles of 5.)

$5 + 5 + 5 + 1 = 16$ (Bob then divides into piles of 8.)

$8 + 8 + 8 + 1 = 25$ (odd)

which is a contradiction. If $n = 5$,

$5 + 5 + 5 + 1 = 16$ (Carl then divides into piles of 8.)

$8 + 8 + 8 + 1 = 25$ (odd)

which is a contradiction. If $n = 7$,

$7 + 7 + 7 + 1 = 22$ (Carl then divides into piles of 11.)

$11 + 11 + 11 + 1 = 34$ (Bob divides into piles of 17.)

$17 + 17 + 17 + 1 = 52$ (Al divides into piles of 26.)

$26 + 26 + 26 + 1 = 79$

The least number of coconuts in the pile is 79.

Looking back. Al divides the pile into three stacks of 26, hides his 26, and gives the extra nut to the monkey. This leaves 52 nuts in the pile. Bob divides the pile into three stacks of 17, hides his 17, and gives the extra nut to the monkey. This leaves 34 nuts in the pile. Carl divides the pile into stacks of 11, hides his 11, and gives the extra one to the monkey. This leaves 22 nuts in the pile. In the morning they divide the pile, giving each sailor 7 nuts and the monkey 1.

Summary and review

Guidelines for comprehension

1. Note key words and phrases.
2. Write down what is being asked for.
3. Restate key phrases in your own words.
4. Where appropriate, prepare a visual representation.

Translating from verbal statements using open sentences

1. Obtain an intuitive understanding of $=$, $<$, and $>$.
2. Distinguish between an open sentence and a statement.
3. A replacement makes an open sentence into a statement. If the statement is true, the replacement is called a *solution*. The list of possible replacements is called the *domain*.

Number sequences

1. An arithmetic sequence is characterized by a common difference between terms.
2. In a geometric sequence, a term divided by a preceding term gives a common ratio.

Problem solving

1. Polya's four steps
   (a) Understand the problem.
   (b) Devise a plan.
   (c) Carry out the plan.
   (d) Look back.
2. Understanding a problem
   (a) Identify what you are trying to find.
   (b) Don't impose conditions that do not exist.
   (c) Strip the problem of irrelevant details.
3. Strategies for problem solving.
   (a) Make a chart or a table.
   (b) Draw a picture.
   (c) Eliminate impossible situations.
   (d) Look for a pattern.
   (e) Try a simpler version of the problem.
   (f) Restate the problem.
   (g) Guess, test, and revise.
   (h) Work backward.
   (i) Design a model.
   (j) Use an algorithm.

Venn diagrams

1. Venn diagrams can be used to deduce conclusions from statements involving quantifiers such as "all" and "some."

Logic

1. Understand
   (a) negation, $\sim p$
   (b) conjunction, $p \wedge q$
   (c) disjunction, $p \vee q$
   (d) conditionals, $p \rightarrow q$
   (e) biconditional, $p \leftrightarrow q$
2. Associated with every conditional "if p then q" is its
   (a) converse, if q then p.
   (b) contrapositive, if $\sim q$ then $\sim p$.
   (c) inverse, if $\sim p$ then $\sim q$.
3. Two statements are logically equivalent if they have the same truth values.
4. The contrapositive is logically equivalent to the original proposition.

Analogies

1. Recognizing the relationship sentence is the key to understanding analogies.

Review exercise set 8

1. Let p represent "Logic is easy," q "Algebra is hard," and r "Arithmetic is useful." Write each symbolic statement in words.

   (a) $r \to (p \wedge q)$ (b) $\sim(p \vee q)$
   (c) $\sim r \wedge \sim q$ (d) $p \wedge (q \vee \sim r)$

2. Consider the pattern below.

   (a) Is the 31st square shaded?
   (b) Is the 501st square shaded?

3. If the word *animal* has more than three consonants and fewer than four vowels, cross out the entire word. Otherwise, place an asterisk over the middle consonant.

4. For the set of numbers

   8, 4, 1, 5, 7, 3, 6

   subtract the fourth number from the first number. Multiply your answer by 2. If your answer is less than 6, write the answer. Otherwise, multiply your answer by 3 and write the result. What did you write?

5. Express the following verbal expressions as mathematical expressions:

   (a) Two less than the product of three times a number.
   (b) Three less than a number equals seven.
   (c) The product of four and a number, minus five equals seven.
   (d) The product of four times the sum of a number and three equals eleven.

6. Using the replacements 1, 2, 3, 4, 5, solve

   (a) $x + 3 = 7$ (b) $2x = 8$
   (c) $2x - 1 = 5$ (d) $3x + 1 = 13$
   (e) $2(x + 1) = 8$ (f) $3(x - 1) = 6$

7. Write the converse and the contrapositive of each statement.

   (a) If it does not rain, we shall play tennis.
   (b) If I study diligently, I shall pass the course.

T 8. Complete the following truth table.

p	q	$p \to q$	$p \wedge q$	$p \vee q$
				F
T		F		
			T	

9. Three mothers—Mary, Eve, and Jean—have between them a total of 14 children. Mary has 2 boys and Eve has the same number of girls. Eve has no more children than Mary, who has 3 children. Jean has 6 more girls than boys and the same number of boys as Mary has girls. How many girls each do Jean and Mary have?

10. There were 16 members of Jim's soccer team. At halftime all but 9 of them were hurt. How many were able to play the second half?

11. Study two rows of Pascal's triangle at a time to determine patterns of regularity. Note that the sum of two consecutive numbers in a row gives the number between them in the next row. For example, the sum of 1 and 3 in the fourth row equals 4 (circled) in the fifth row, and the sum of 3 and 3 is 6 (circled). Complete the sixth and seventh rows of the array.

    1
    1 1
    1 2 1
    1 3 3 1
    1 ④ ⑥ 4 1

12. Find the pattern.

    $1 =$
    $3 + 5 =$
    $7 + 9 + 11 =$
    $13 + 15 + 17 + 19 =$
    $21 + 23 + 25 + 27 + 29 =$

    Predict the sum of $31 + 33 + 35 + 37 + 39 + 41$.

13. Find the next three terms of each sequence.
    (a) $A, Z, C, X, E, V, G, \ldots$
    (b) $C, B, A, G, F, E, K, J, I, \ldots$
    (c) $1, 5, 9, 13, \ldots$
    (d) $3, 6, 9, 18, 21, 42, 45, \ldots$

14. Use Venn diagrams to test whether the conclusion is a necessary result of the given facts.
    (a) All students who study hard make A's.
    All math students study hard.
    Fred is a math student.
    Therefore, Fred makes an A.
    (b) Some business majors find jobs.
    All marketing majors are business majors.
    Therefore, some marketing majors find jobs.

15. Find the truth value of each of the following when q is false and p and r are both true.
    (a) $(p \vee q) \wedge \sim(p \vee r)$
    (b) $(p \wedge \sim q) \vee (r \wedge \sim q)$
    (c) $\sim(r \wedge p) \wedge \sim(q \vee r)$
    (d) $(r \wedge p) \vee (\sim q \wedge r)$
    (e) $(\sim r \wedge p) \rightarrow q$
    (f) $r \rightarrow \sim(p \vee q)$

16. Can you determine a pattern?

$$1 + 2 = 3$$

$$4 + 5 + 6 = 7 + 8$$

$$9 + 10 + 11 + 12 = 13 + 14 + 15$$

Find the next two lines. Show why this pattern always works.

17. Suppose you have only an 11-minute and a 7-minute hourglass. How would you time the cooking of a vegetable that demands exactly 15 minutes of cooking?

Find valid conclusions, where possible, for the set of premises given in Exercises 18 through 22.

18. All college students are clever.
    Paul is clever.

19. All triangles are polygons.
    All polygons are plane figures.

20. All right angles are equal.
    Angles A and B are right angles.

21. Not all students are radicals.
    Some students are Republicans.
    Henry is a student.

22. All A's are C's.
    Some D's are A's.
    Some D's are not C's.

Using Venn diagrams, select valid conclusions from the given possibilities in Exercises 23 and 24.

C 23. All whole numbers are integers.
    All integers are rationals.
    No irrationals are rational.
    All rationals are reals.
    All irrationals are reals.
    x is an integer.
    Therefore,
    (a) x is a whole number.
    (b) x is a real number.
    (c) x is an irrational number.
    (d) x is not a rational number.

24. All authors are intelligent.
    Some men fron New York are authors.
    Some men from Philadelphia are not intelligent.
    Therefore,
    (a) no man from Philadelphia is an author.
    (b) no authors are from Philadelphia.
    (c) no authors are from New York.
    (d) some New York men are intelligent.

Bibliography

1. Aichele, Douglas B. "'Pica-Centro'—A Game of Logic." *Arithmetic Teacher* (hereafter abbreviated *AT*) (May 1972), 359–361.
2. Bruni, James V. "Problem Solving for the Primary Grades." *AT* (February 1982), 10–15.
3. Campbell, Patricia. "Using a Problem-Solving Approach in the Primary Grades." *AT* (December 1984) 11–14.
4. Carmen, R. A., & M. J. Carmen. "Number Patterns." *AT* (December 1970), 637–639.

5. Charles, Randall I., Robert P. Mason, & Dianne Garner. *Problem-Solving Experiences in Mathematics.* Reading, Mass.: Addison-Wesley, 1985.

6. DeVault, M. Vere. "Doing Mathematics Is Problem Solving." *AT* (April 1981), 40–43.

7. Edmonds, G. F. "An Intuitive Approach to Square Numbers." *Mathematics Teacher* (hereafter abbreviated *MT*) (February 1970), 113–117.

8. Eisenberg, Theodore A., & John S. Van Beynen. "Mathematics Through Visual Problems." *AT* (February 1973), 85–90.

9. Feeman, George F. "Reading and Mathematics." *AT* (November 1973), 523–529.

10. Gardner, M. "The Multiple Fascinations of the Fibonacci Sequence." *Scientific American* (March 1969), 116–200.

11. Gardner, Martin. "Mathematical Games Department." *Scientific American* (January 1976), 118–122.

12. Greenes, Carole E., & Linda Schulman. "Developing Problem-Solving Ability with Multiple-Condition Problems." *AT* (October 1982), 18–21.

13. Hater, Mary Ann, Robert B. Kane, & Mary Ann Byrne. "Building Reading Skills in the Mathematics Class." *AT* (December 1974), 662–668.

14. Hendrichs, Margaret, & Tom Sisson. "Mathematics and the Reading Process: A Practical Application of Theory." *MT* (April 1980), 253–257.

15. Hervey, M. A., & B. H. Litevieller. "Polygonal Numbers: A Study of Patterns." *AT* (January 1970), 33–38.

16. Jensen, Rosalie, & David R. O'Neill. "Classical Problems for All Ages." *AT* (January 1982), 8–12.

17. Johnson, Martin L. "An Observation on Patterns." *AT* (March 1974), 244–245.

18. Knifong, J. Dan, & Grace M. Barton. "Understanding Word Problems." *AT* (January 1985), 13–17.

19. Kraitchik, Maurice. *Mathematical Recreations.* New York: Dover Publications, 1953.

20. Krulik, Stephen. "Problem Solving: Some Considerations." *AT* (December 1977), 51–52.

21. Krulik, Stephen. "To Read or Not to Read, That is the Question!" *MT* (April 1980), 248–252.

22. Krulik, Stephen, & Ann M. Wilderman. "Mathematics Class + Strategy Games = Problem Solving." *School Science and Mathematics* (March 1976), 221–225.

23. Krulik, Stephen, & Jesse R. Rudnick. *Problem Solving: A Handbook for Teachers.* Boston, Mass.: Allyn & Bacon, 1980.

24. Lee, Kil S. "Guiding Young Children in Successful Problem Solving." *AT* (January 1982), 15–17.

25. McGinty, Robert L., & Lawrence N. Meyerson. "Problem Solving: Look Beyond the Right Answer." *MT* (October 1980), 501–503.

26. Mercer, Gene B., & John R. Kolb. "Three Dimensional Tic-Tac-Toe." *MT* (February 1971), 119–122.

27. Pace, Angela. "Understanding and the Ability to Solve Problems." *AT* (May 1961), 226–233.

28. Spencer, Patricia J., & Frank K. Lester. "Second Graders Can Be Problem Solvers!" *AT* (September 1981), 15–17.

29. Stiff, Lee V. "Understanding Word Problems." *AT* (March 1986), 163–165.

30. Thornton, Carol A. "A Glance at the Power of Patterns." *AT* (February 1977), 154–157.

31. Warman, Michele. "Fun with Logical Reasoning." *AT* (May 1982) 26–30.

32. Wheatley, Charlotte L., & Grayson H. Wheatley. "Problem-Solving in the Primary Grades." *AT* (April 1984) 22–25.

33. Whimbey, Arthur, & Jack Lochhead. *Problem Solving and Comprehension: A Short Course in Analytical Reasoning.* Philadelphia, Pa.: Franklin Institute Press, 1980.

34. Whitin, David. "Patterns with Square Numbers." *AT* (December 1979), 38–39.

35. Zur, Mordecai, & Frederick L. Silverman. "Problem Solving for Teachers." *AT* (October 1980), 48–50.

CHAPTER 2

Tools for Problem Solving: Sets and Numbers

In a certain apartment complex live 20 men: 8 are married and own automobiles; 12 own automobiles and have television sets; and 11 are married and have television sets. What is the largest possible number of married men?

In this chapter, you will acquire a useful tool. Sets are valuable in understanding the properties of whole numbers, in defining events in probability, and in geometric definitions. In fact, sets help us to understand many topics of mathematics that would be difficult to understand otherwise.

After a quick introduction to sets you will be introduced to the first of several number systems that we shall study—the whole numbers. Addition, multiplication, subtraction, and division will be defined on this system and properties will be discussed. Make certain you understand the associative and commutative properties of addition and multiplication, the distributive property of multiplication over addition, and the existence of additive and multiplicative identities. We shall continue to use these properties as we advance from the whole numbers to the integers and then to the rational numbers and finally to the real numbers. The number line is introduced to give a visual representation of the operations and to illustrate the concept of order.

In this chapter we continue our emphasis on critical thinking with a large number of word problems.

1 An introduction to sets

■ Place eight tennis balls in three baskets of different sizes so that there is an odd number of balls in each basket. ■

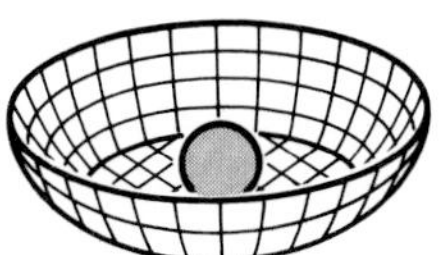

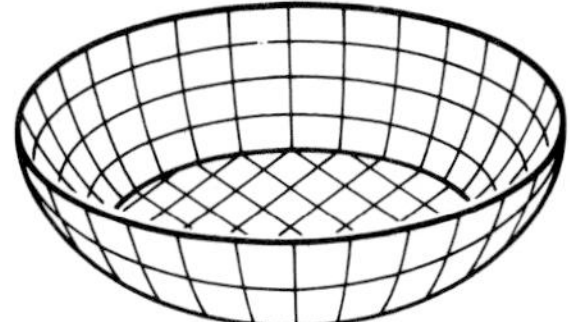

The preceding is a "trick problem" involving subsets. This and other properties of sets will be introduced in this section.

For many of you, the material in this section will be a review. However, if you have never studied sets, don't worry; all the beginning definitions and concepts are presented. You will find Venn diagrams, introduced in Chapter 1, to be especially useful when working with intersection and union of sets. Problem-solving techniques may be used to discover the number of subsets in a set with n elements. See the "Just for Fun" on p. 66.

Although this book will give little emphasis to what is called **set theory,** some familiarity with the notation and language of sets is useful and important. During the latter part of the 19th century, while working with mathematical entities called *infinite series,* Georg Cantor found it helpful to borrow a word from common usage to describe a mathematical idea. The word he borrowed was *set.*

Our purposes will be served if we describe a set as follows.

A set

> A *set* is a *collection* of objects or symbols possessing a property that enables one to determine whether a given object is in the collection.

We sometimes say a set is a *well-defined* collection, meaning that, given an object and a set, we are able to determine whether or not the object is in the set. The individual objects in a set are called **elements** of the set. They are said *to belong to* or *to be members of* or *to be in* the set. The relationship between objects of the set and the set itself is expressed in the form *is an element of* or *is a member of.*

Element of a set

> $x \in A$ means x is an *element* of set A. $x \notin A$ means x is *not an element* of set A.

Often it is possible to specify a set by listing its members within braces. This method of describing a set is called *the* **tabulation** *method* (sometimes called *the* *roster method*). The set of counting numbers less than 10 can be written

$$\{1, 2, 3, 4, 5, 6, 7, 8, 9\}$$

A set remains the same regardless of the order of tabulating the elements. For example, $\{1, 2, 3\}$ is the same set as $\{2, 1, 3\}$, $\{3, 2, 1\}$, $\{3, 1, 2\}$, $\{1, 3, 2\}$, or $\{2, 3, 1\}$. In fact, two sets are said to be **equal** if they contain exactly the same elements. If $A = B$, then A and B have exactly the same elements.

Example 1 • Glenda, Cathie, and Martia are the counselors in the admissions office. They constitute the set $A = \{\text{Glenda, Cathie, Martia}\}$. Cathie $\in A$; Linda $\notin A$. Can you identify other elements of set A? •

Sometimes sets have so many elements that it is tedious or difficult or even impossible to tabulate them. Sets of this nature may be indicated by a descriptive statement or a rule. The following sets are well specified without a tabulation of members: the counting numbers less than 10, the even numbers less than 1000, the past presidents of the United States, and the football teams in Pennsylvania.

The difficulty of tabulating sets is minimized by using **set-builder notation** that encloses within braces a letter, or symbol, representing an element of the set followed by a qualifying description of the element. For example, let A represent the set of counting numbers less than 10; then

$$A = \{n \mid n \text{ is a counting number less than } 10\}$$

This notation is read *the set of all elements n such that each n is a counting number less than* 10.

BIOGRAPHY

Georg Cantor

1845–1918

Born in Russia, Cantor moved to Germany at age 11 and lived there the rest of his life. He is known today as the originator of set theory. He received ridicule and abuse from his contemporaries for his concept that if two infinite sets could be put in one-to-one correspondence, then they had the same cardinality. Perhaps this lack of acceptance contributed to the fact that he later underwent a series of mental breakdowns and died in a mental hospital.

Example 2 • Use set-builder notation to denote the set of current United States senators.

Solution $\{x \mid x \text{ is a U.S. senator}\}$. The set is read *the set of all x such that each x is a U.S. senator.* •

Frequently, three dots (called an *ellipsis*) are used to indicate the omission of terms. The set of even counting numbers less than 100 may be written as $\{2, 4, 6, \ldots, 98\}$. This notation saves time in tabulating elements of large sets, but it can be ambiguous unless the set has been specified completely by another description. For example, $\{2, 4, \ldots, 16\}$ could be $\{2, 4, 8, 16\}$ or it could be $\{2, 4, 6, 8, 10, 12, 14, 16\}$.

An ellipsis is also used to indicate that a sequence of elements continues indefinitely. For example, consider the set of **natural,** or **counting, numbers.**

$$\{1, 2, 3, 4, 5, \ldots\}$$

The set of natural numbers is an example of an **infinite set,** described informally as one that contains an unlimited number of elements. In contrast, a **finite set** contains zero elements or a natural number of elements.

An empty set

> A set that contains no elements is called the *empty* or *null set* and is denoted by either $\varnothing$ or $\{\ \}$.

The relationship between two sets such as $A = \{1, 3, 5, 7\}$ and $B = \{1, 2, 3, 4, 5, 6, 7, 8\}$ is described by the term *subset.*

A subset

> Set A is said to be a *subset* of set B, denoted by $A \subseteq B$, if and only if each element of A is an element of B.

Example 3 • If $A = \{x, y\}$ and $B = \{w, x, y, z\}$, then $A \subseteq B$ since each element of A is an element of B. •

Example 4 • If $P = \{1, 4, 7\}$ and $Q = \{4, 7, 1\}$, then $P \subseteq Q$ since each element of P is an element of Q. •

To show that A is not a subset of B (denoted by $A \nsubseteq B$), one must find at least one element in A that is not in B. Let's use this idea to examine whether $\varnothing$ is a subset of some set B. Using our problem-solving techniques, let's list all possibilities.

1. Either $\varnothing$ is a subset of B
2. Or $\varnothing$ is not a subset of B

In possibility 2, if $\varnothing$ is not a subset of B, then there is an element of $\varnothing$ not in B. But $\varnothing$ has no elements. Consequently, possibility 2 cannot be true. The only alternative is for possibility 1 to be true.

Subset $\varnothing$

> For any set B, $\varnothing \subseteq B$.

Since all dogs are animals, the set of dogs is a subset of the set of animals. Moreover, the set of dogs is a proper subset of the set of animals, since there are animals that are not dogs.

A proper subset

> Set A is said to be a *proper subset* of set B, denoted by $A \subset B$, if and only if each element of A is an element of B and there is at least one element of B that is not an element of A.

If a discussion is limited to a fixed set of objects and if all elements to be discussed are contained in this set, then this "overall" set is called the **universal set,** or simply the **universe.**

A Venn diagram, such as that in Figure 1, was introduced in Chapter 1 and is very useful in representing the relationship between a set and the universe. The universal set can be regarded as the region bounded by the rectangle and the set under consideration as the region bounded by the circle (or some other closed figure within the rectangle). $x \in A$ means that x is a point in the circular region.

Figure 1

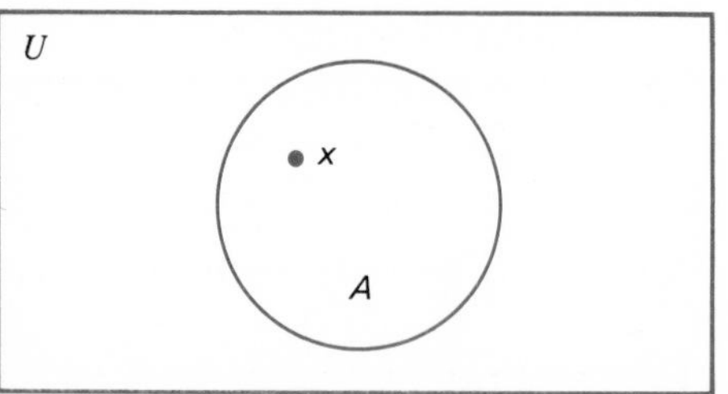

The region outside set A and inside the universe represents the *complement* of A, denoted by $\bar{A}$ (sometimes by A' or $\sim A$) and read "A bar" or "bar A."

Complement of a set

> The *complement* of set A is the set of elements in the universe that are not in set A.

If A is a subset of the universe U, then the complement of A can be written in set-builder notation as $\bar{A} = \{x | x \in U \text{ and } x \notin A\}$.

Example 5 • If $U = \{a, b, c, d\}$ and $A = \{b, c\}$, find $\bar{A}$.

Solution $\bar{A} = \{a, d\}$. •

Example 6 • If the universe is "all college students" and if A is "the set of college students who have made all A's," then "all college students who have made at least one grade lower than an A" is the complement of A and is represented by the shaded region of Figure 2.

Figure 2

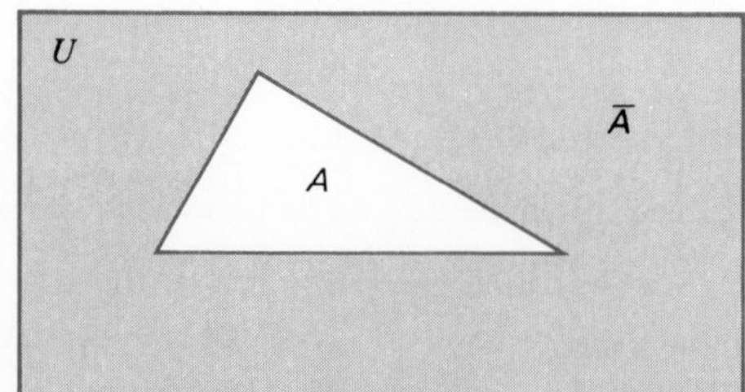

•

Special notations are used for discussing the relationship among members of two or more sets.

Intersection of sets

> The *intersection* of two sets A and B (denoted by $A \cap B$) is the set of all elements common to both A and B.

Using set-builder notation, $A \cap B = \{x \mid x \in A \text{ and } x \in B\}$, and $A \cap B$ is read "*A intersection B*."

Union of sets

> If A and B are any two sets, the *union* of A and B, denoted by $A \cup B$, is the set consisting of all the elements in set A or in set B or in both A and B.

Using set-builder notation, $A \cup B = \{x \mid x \in A \text{ or } x \in B\}$, and $A \cup B$ is read "*A union B*."

The shaded regions in Figure 3 compare intersection and union under different situations for sets A and B. Note in (a) that A and B overlap or have elements in common. In (b), A is a proper subset of B. In (c), A and B have no elements in common or $A \cap B = \varnothing$. In (c) A and B are said to be disjoint.

Figure 3

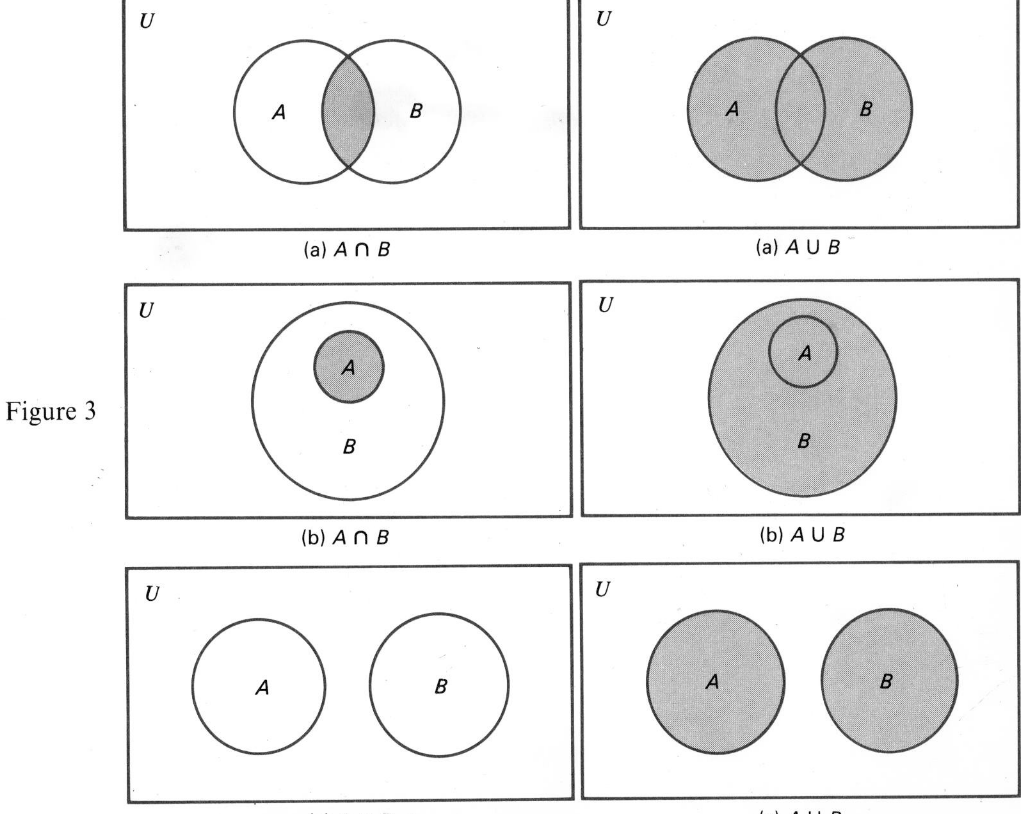

Disjoint sets

> Two sets A and B are said to be *disjoint* if and only if $A \cap B = \varnothing$—that is, if the two sets have no elements in common.

Example 7 • Let A represent a committee of {Joe, Dave, Sue, John, Jack} and B represent a second committee of {Sue, Edward, Cecil, John}. The intersection of these two sets, $A \cap B$, is {Sue, John}. •

Example 8 • If $A = \{1, 2, 3\}$ and $B = \{6, 7\}$, find $A \cup B$.

Solution $A \cup B = \{1, 2, 3, 6, 7\}$ •

Recall that $A \cup B$ is the set of all elements that belong to A or B or to both A and B. If there are elements common to both sets, they are listed only once in the union. Given

$$A = \{a, b, c, d, e\} \quad \text{and} \quad B = \{c, d, e, f, g\}$$

then

$$A \cup B = \{a, b, c, d, e, f, g\}$$

Example 9 • Let set A represent the secretaries in the Order Department of Jemison Stores, {Kaye, Aster, David, Brenda}, and let set B represent the secretaries associated with billing, {Doris, Gayle, Brenda}; then $A \cup B$ = {Kaye, Aster, David, Brenda, Doris, Gayle} because each of these secretaries belongs to either or A or B. •

Example 10 • If $A = \{1, 2\}$, then $A \cup A = \{1, 2\}$. •

Example 11 • If $U = \{x \mid x \text{ is a counting number less than } 10\}$, $A = \{2, 4, 6\}$, $B = \{1, 2, 3, 4, 5\}$, and $C = \{3, 5, 7\}$, find $A \cap B$, $A \cap C$, $A \cup C$, and $\bar{B}$.

Solution

$A \cap B = \{2, 4\}$	elements in common
$A \cap C = \varnothing$	no elements in common
$A \cup C = \{2, 3, 4, 5, 6, 7\}$	elements in either A or C or both
$\bar{B} = \{6, 7, 8, 9\}$	elements in the universe, not in B •

The following properties of intersection and union of sets can be illustrated using Venn diagrams. If desired, they can be proved using truth tables, as seen in Exercise set 1.

Properties of intersection and union

For all sets *A*, *B*, and *C*

1. Commutative properties:

$$A \cup B = B \cup A; \qquad A \cap B = B \cap A$$

2. Associative properties:

$$A \cup (B \cup C) = (A \cup B) \cup C; \qquad A \cap (B \cap C) = (A \cap B) \cap C$$

3. Identity properties:

$$A \cup \varnothing = A; \qquad A \cap U = A$$

Diagram (a) of Figure 4 represents the formation of $(A \cap B) \cap C$. Similarly, (b) represents the formation of $A \cap (B \cap C)$. It can be seen that the double-shaded region representing $(A \cap B) \cap C$ is the same as the region representing $A \cap (B \cap C)$, illustrating the associative property, $(A \cap B) \cap C = A \cap (B \cap C)$.

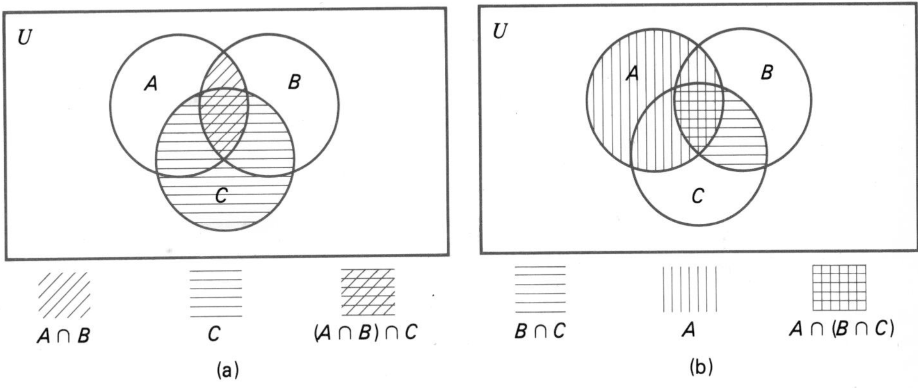

Figure 4

Example 12 • Let $A = \{x|x \text{ is a state whose name begins with } A\}$ and $B = \{x|x \text{ is a southeastern state}\}$. To satisfy both properties, $A \cap B = \{\text{Alabama}\}$ and $B \cap A = \{\text{Alabama}\}$; therefore, $A \cap B = B \cap A$, illustrating the commutative property of intersection. •

The **commutative property of union** is easily understood, as noted in the next example.

Example 13 • Consider $A = \{1, 3\}$ and $B = \{3, 5, 7\}$. Then $A \cup B = \{1, 3, 5, 7\}$ and $B \cup A = \{1, 3, 5, 7\}$; thus, $A \cup B = B \cup A$. •

Example 14 • If $A = \{1, 2, 3, 4\}$ and $B = \varnothing$, then

$$A \cup B = \{1, 2, 3, 4\} \cup \varnothing = \{1, 2, 3, 4\} = A \quad \bullet$$

Just for fun

Use problem-solving techniques to discover the number of subsets in a set with n elements. Hint: Use the strategy of trying simple versions of the problem; then look for a pattern.

Example 15 • In Figure 5, (a) and (b) represent the formation of $(A \cup B) \cup C$, whereas (c) and (d) represent $A \cup (B \cup C)$. Note in (b) and (d) that $(A \cup B) \cup C$ and $A \cup (B \cup C)$ are represented by the same shaded region. This demonstrates that the **associative property** holds for union. •

Figure 5

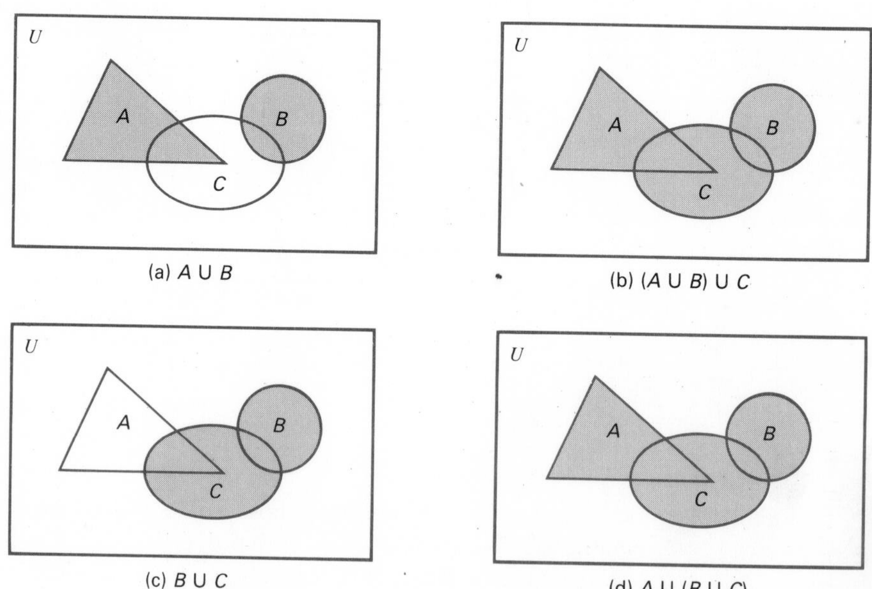

Example 16 • The introductory problem asks that we place eight tennis balls in three baskets of different sizes so that there is an odd number in each basket.

Solution The problem did not state that the baskets be disjoint. In fact, the hint suggested subsets. So place one basket in another. One solution is given in Figure 6.

Figure 6

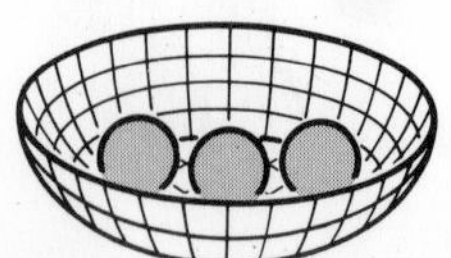 •

Exercise set 1

In this exercise set we shall use the counting numbers $\{1, 2, 3, \ldots\}$, *the odd counting numbers* $\{1, 3, 5, 7, \ldots\}$, *and the even counting numbers* $\{2, 4, 6, \ldots\}$.

R 1. Let A be the set of all counting numbers less than 16. Which of the following statements are true and which are false? Justify your answer.

(a) $11 \in A$ (b) $81 \in A$
(c) $\{5\} \in A$ (d) $\{1, 2, 3, \ldots, 15\} \in A$
(e) $14 \in A$ (f) $0 \in A$

2. List within braces the members of the sets below.

(a) The counting numbers less than or equal to 16.
(b) The set of even counting numbers.
(c) The set of women presidents of the United States.

3. Express the sets in Exercise 2 using set-builder notation.

4. For each set in the left column, choose the sets from the right column that are subsets of it.

(a) $\{a, b, c, d\}$ (a) $\{\ \}$
(b) $\{o, p, k\}$ (b) $\{1, 4, 8, 9\}$
(c) Set of letters in the word *book* (c) $\{o, k\}$
(d) $\{2, 4, 6, 8, 10, 12\}$ (d) $\{12\}$
(e) $\{b\}$

5. Classify each statement as true or as false.

(a) $\{z, r\} \subseteq \{x, y, z, r\}$
(b) {Brenda, Sharon, Glenda} $\subseteq$ {Brenda}
(c) $\{7, 2, 6\} \subseteq \{2, 6, 7\}$
(d) $6 \subseteq \{4, 5, 6, 7\}$
(e) $\{p, q, r\} \in \{p, q, r, s\}$

6. Which of the following sets are well defined?

(a) The set of great baseball players
(b) The set of beautiful horses
(c) The set of students in this class
(d) The set of counting numbers smaller than a million

7. Insert the appropriate symbol $\{\in, \notin, \subset$ or $\subseteq\}$ to make the following statements true.

(a) 3 ______ $\{1, 2, 3\}$
(b) $\{2\}$ ______ $\{1, 2, 3\}$
(c) $\{1, 2, 3\}$ ______ $\{1, 2, 3\}$
(d) 0 ______ $\{1, 2, 3\}$

8. Form the union and the intersection of the following pairs of sets.

(a) $R = \{5, 10, 15\}$ $T = \{15, 20\}$
(b) $M = \{1, 2, 3\}$ $N = \{101, 102, 103, 104\}$
(c) $A = \{0, 10, 100, 1000\}$ $B = \{10, 100\}$
(d) $G = \{$odd counting numbers less than 100$\}$
$H = \{$even counting numbers between 1 and 31$\}$
(e) $A = \{x, y, z, t\}$ $B = \{x, y, r, s\}$

9. Describe in words using sets A and B the elements in regions (a), (b), (c), and (d).

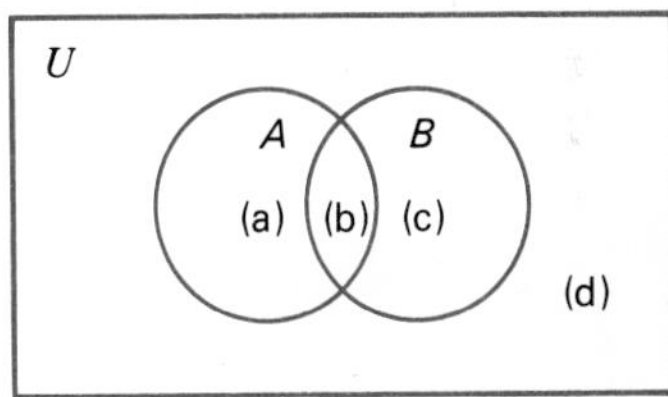

10. If $U = \{a, b, c, d, e, f, g, h\}$, $A = \{b, d, f, h\}$, and $B = \{a, b, e, f, g, h\}$, find the following.

(a) $\bar{A}$ (b) $\bar{B}$ (c) $A \cap B$
(d) $\overline{A \cap B}$ (e) $\bar{A} \cap \bar{B}$ (f) $\overline{A \cup B}$

11. Let U be the set of students at Myschool; let A be the set of female students, B the set of male students, C the set of students who ride the bus, and D the set of members of athletic teams. Denote each of the following symbolically.

(a) The set of female athletes
(b) The set of male athletes who ride the bus
(c) The set of all males who neither ride the bus nor are athletes
(d) The set of females who either ride the bus or are not athletes

12. Shade the portion of a similar diagram that will illustrate each of the sets below.

(a) $A \cup B$ (b) $\bar{A} \cap \bar{C}$
(c) $\overline{A \cup B}$ (d) $\bar{A} \cup \bar{B}$
(e) $A \cap (B \cap C)$ (f) $A \cup (B \cup C)$

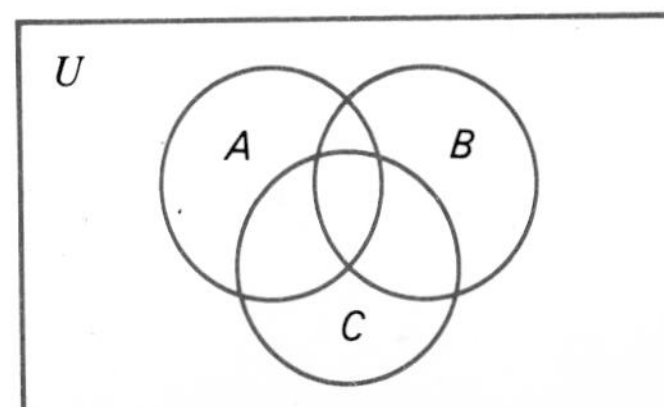

T 13. Draw a Venn diagram that illustrates each situation below.

(a) In Ourtown, no elementary teacher teaches in high school, but some high school teachers teach in college. Let U be all teachers in Ourtown.

(b) At Ourtown High, all mathematics teachers have a chalkboard. Some have an overhead projector and some have a cassette recorder, but no mathematics teacher has all three. Let U be all teachers at Ourtown High.

14. In the accompanying figure, the sets of elements of closed regions are indicated by (a), (b), . . . , (h). Express the following in terms of these sets.

(a) $\overline{A}$	(b) $A \cap B$
(c) $A \cap B \cap C$	(d) $\overline{A} \cup \overline{B}$
(e) $A \cup B \cup C$	(f) $A \cap \overline{B}$
(g) $\overline{A \cup B}$	(h) $\overline{A \cup B \cup C}$

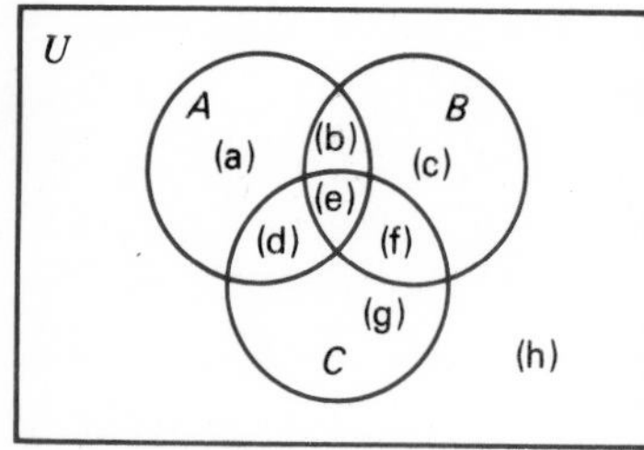

15. (a) List all the subsets of $\{a, b\}$.
(b) List all the subsets of $\{a, b, c\}$.
(c) List all proper subsets of $\{a, b, d\}$.

16. The complement of A relative to B is the set of elements in B that are not in A. It is written as $B - A$ or

$$\{x \mid x \in B \text{ and } x \notin A\}$$

Let $A = \{3, 4, 5, 6\}$ and $B = \{5, 6, 7, 8\}$. Find

(a) $B - A$ (b) $A - B$

C 17. In a Venn diagram, a single set partitions the universe into two distinct regions. Two sets partition the universe into a maximum of four regions.

(a) What is the maximum number of regions into which three sets will partition the universe?
(b) Four sets?
(c) Five sets?
(d) Use problem-solving techniques to conjecture an answer for n sets.

18. Using A, B, and/or C, describe the elements of the closed regions denoted by (a), (b), (c), (d), (e), (f), (g), and (h).

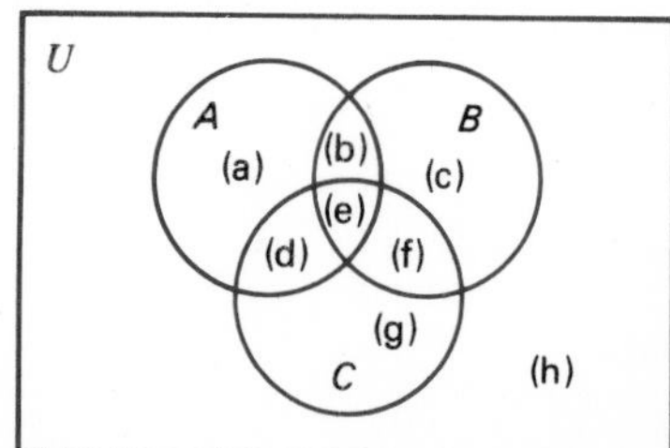

19. Use Venn diagrams to illustrate *De Morgan's laws*.

(a) $\overline{A \cap B} = \overline{A} \cup \overline{B}$ (b) $\overline{A \cup B} = \overline{A} \cap \overline{B}$

20. To prove that $(A \cap B) \cap C = A \cap (B \cap C)$ using truth tables, the truth values of $(A \cap B) \cap C$ must be the same as the truth values of $A \cap (B \cap C)$ for all 8 possibilities of A, B, and C. $x \in A$ is indicated by T and $x \notin A$ by F. Complete the table and determine if $(A \cap B) \cap C = A \cap (B \cap C)$.

$x \in A$	$x \in B$	$x \in C$	$x \in (A \cap B)$	$x \in (A \cap B) \cap C$	$x \in (B \cap C)$	$x \in A \cap (B \cap C)$
T	T	T	T	T	T	T
T	T	F	T	F	—	—
T	F	T	—	—	F	F
T	F	F	F	—	F	—
F	T	T	—	—	—	—
F	T	F	—	—	—	—
F	F	T	—	—	—	—
F	F	F	—	—	—	—

21. Using the same truth values as given in Exercise 20, prove the following.

(a) $A \cup (B \cup C) = (A \cup B) \cup C$
(b) $A \cup (B \cap C) = (A \cup B) \cap (A \cup C)$

22. Demonstrate with Venn diagrams the following properties.

(a) $A \cup (B \cup C) = (A \cup B) \cup C$
(b) $A \cup (B \cap C) = (A \cup B) \cap (A \cup C)$

Set theory can be applied to analyze the power of voting coalitions. A winning coalition consists of any set of voters who can carry a proposal. Answer the questions about voting coalitions in Exercises 23 and 24.

23. A committee has five members: A, B, C, D, E. For a proposal to be passed, it must have at least three votes. List all the possible winning coalitions.

    EXAMPLE. $\{A, B, D\}$ is a winning coalition; $\{B, D\}$ is not a winning coalition.

24. A town council consists of five councilmen whose votes are weighted according to the number of citizens in their districts. Councilman A has 6 votes; councilman B, 5 votes; C, 4 votes; D and E, 1 vote apiece. Nine or more votes are required to carry an issue.

    (a) List all winning coalitions.

    EXAMPLE. $\{A, B\}$ and $\{A, C, D\}$ are winning coalitions.

    (b) Review the list in part (a) and pick all winning coalitions having the property that if any voter were removed from this coalition, the coalition would fail to win.

    EXAMPLE. $\{A, B\}$ has this property, but if D were removed from $\{A, C, D\}$, the resulting coalition $\{A, C\}$ would still win.

    (c) Do councilmen D and E appear on any coalitions that you listed in part (b)?

    (d) What does this say about the power of D and E?

    Note: Similar arguments have been used in the courts to force modification of certain voting schemes.

2 Properties of sets

■ What is the relationship between open sentences (introduced in Chapter 1) such as

$$x + y = 4, \qquad x > y, \qquad 3x = 2y$$

and relations? ■

To answer this question and to define a relation, we first need definitions of two terms: *ordered pair* and *Cartesian product*. Cartesian products will be defined in terms of ordered pairs, and relations will be defined in terms of Cartesian products. Cartesian products will be used later in this chapter to give an alternate definition of multiplication.

Let us now turn our attention to **Cartesian product.** When the union or intersection of two sets is formed, the result is a set containing some or all of the same elements. The Cartesian product is not in the same form as either set but instead uses ordered pairs of elements.

The concept of an *ordered pair* can be found frequently in real life. Consider, for instance, the following actions: put on socks, put on shoes; fry chicken, eat chicken; start mower, cut grass. Do you perform these actions in the order given? Does the order in which they are performed make a difference? In an ordered pair, the order of the elements is important.

Ordered pair

An *ordered pair* is an element (x, y) formed by taking x from a set and y from a set in such a way that x is designated as the *first* element and y as the *second* element.

If x and y are elements of any two sets (or the same set), they can be used to form an ordered pair denoted by (x, y). This pair is ordered in the sense that (x, y) and (y, x) are not equal unless $x = y$. Equality for ordered pairs is defined as follows.

Equality of ordered pairs

$(a, b) = (c, d)$ if and only if $a = c$ and $b = d$.

Example 17 • A traveler can go from Chicago to Miami by auto, plane, train, or bus and from Miami to Nassau by plane or ship. The different ways of traveling from Chicago to Nassau through Miami can be described in terms of a set of ordered pairs.

$$\begin{aligned}\{&(\text{auto, plane}), && (\text{auto, ship}),\\ &(\text{plane, plane}), && (\text{plane, ship}),\\ &(\text{train, plane}), && (\text{train, ship}),\\ &(\text{bus, plane}), && (\text{bus, ship})\}\end{aligned}$$

If A represents {auto, plane, train, bus} and B represents {plane, ship}, then the set of ordered pairs representing the different means of travel is denoted by $A \times B$ and is called the *Cartesian product* of the sets A and B. •

The Cartesian product

The *Cartesian product*, denoted by $A \times B$, of two sets A and B is the set of all ordered pairs (a, b) with the first element chosen from A and the second element chosen from B. In set-builder notation,

$$A \times B = \{(a, b) \mid a \in A \text{ and } b \in B\}$$

BIOGRAPHY

René Descartes

1596–1650

René Descartes, a French mathematician, was a man with many hats. Most noted as a philosopher, he was also a soldier, a scientist, and a mathematician. As a philosopher, he ran into conflict with the Roman Catholic Church. Many of his works were banned, and he was once brought to court on charges of atheism and vagrancy. As a mathematician, he wrote only one formal treatise attached as an appendix to a longer scientific work. However, some have classified his invention of the Cartesian plane as "one of the great steps in the progress of the exact sciences."

Example 18 • If $A = \{r, s, t\}$ and $B = \{w, u, v\}$, find $A \times B$.

Solution $A \times B = \{(r, w), (r, u), (r, v), (s, w), (s, u), (s, v), (t, w), (t, u), (t, v)\}$ •

Note that the Cartesian product includes every pair that can be constructed by choosing a left partner from A and a right partner from B. $A \times B$ is sometimes called the *cross product* of A and B and is read "*A cross B*." The definition does not specify that the sets used in forming the cross product must be distinct. Examples illustrate that they may be the same or different.

Example 19 • (a) If $A = \{1, 2, 3\}$ and $B = \{4, 5\}$, then

$$A \times B = \{(1, 4), (1, 5), (2, 4), (2, 5), (3, 4), (3, 5)\}$$

(b) If $A = \{a, b\}$, form $A \times A$. By definition, the ordered pairs $(x, y) \in A \times A$ must be such that $x \in A$ and $y \in A$. Thus,

$$A \times A = \{(a, a), (a, b), (b, a), (b, b)\}$$ •

We will now consider *relations* as a subset of a Cartesian product set. Everyone is familiar with such common nonmathematical relations as "Glenda is engaged to Jerry" and "Leigh is the daughter of Richard." The expressions *is engaged to* and *is the daughter of* are connectives that define relations on the set of all people. From the Biblical narrative, the connective "is the father of" gives a relation between Adam and Abel but not between Cain and Abel.

Relations can often be better understood by using a visual representation. Let R represent Richard; T, Tom; S, Sanders; L, Leigh; G, Glenda; and M, Maxine. From Figure 7, can you pick out who is the daughter of whom?

Leigh is the daughter of Richard.
Maxine is the daughter of Sanders.
Glenda is the daughter of Tom.

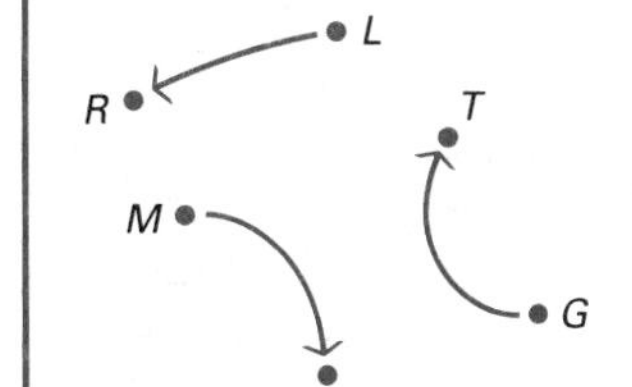

Figure 7

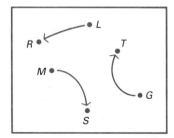

Example 20 • The following sets of ordered pairs suggest various relations.

(a) (Hawthorne, *The Scarlet Letter*), (Crane, *The Red Badge of Courage*), (Steinbeck, *Of Mice and Men*), (Twain, *Huckleberry Finn*) are included in the relation "x is the author of the book y."

(b) (Clark Gable, *Gone with the Wind*), (Julie Andrews, *The Sound of Music*), (Charlton Heston, *Ben Hur*), (Omar Sharif, *Dr. Zhivago*) are included in the relation "x is a star in the movie y."

The following example illustrates intuitively why the mathematical definition of a relation involves the concept of a subset of a Cartesian product. •

Example 21 • Consider the Cartesian product of the set $A = \{$Atlanta, New York, Chicago, Buffalo$\}$ and the set $B = \{$Georgia, New York$\}$. $A \times B$ is $\{$(Atlanta, Georgia), (New York, Georgia), (Chicago, Georgia), (Buffalo, Georgia), (Atlanta, New York), (New York, New York), (Chicago, New York), (Buffalo, New York)$\}$. If x is an element of A and y is an element of B, the relation "x is a city in y" would be defined by the set $\{$(Atlanta, Georgia), (New York, New York), (Buffalo, New York)$\}$. •

Notice that this new set is a subset of $A \times B$. Indeed, all relations on sets can be understood in this way.

A relation

> A relation R from set A to set B is a set of ordered pairs whose first elements are in A and whose second elements are in B. Notice $R \subseteq A \times B$.

Example 22 • Let A represent all men and B represent all women. Then R, the set of all married couples, is a subset of $A \times B$, and so $(c, d) \in R$ means that c is the husband of d. •

As a special case of this definition, a relation R from set A to set A is a subset of $A \times A$. In this case, we say that R is a relation on A.

Example 23 • The relation *less than* on the set $A = \{1, 2, 3, 4\}$ is defined by $\{(1, 2), (1, 3), (1, 4), (2, 3), (2, 4), (3, 4)\}$, which is a subset of $A \times A$. •

We now classify relations as being *reflexive, symmetric,* and/or *transitive.*

Reflexive relation

> A relation R on a set A is *reflexive* if and only if $(a, a) \in R$ for every $a \in A$.

For instance, the relation *is the same species as* is reflexive on the set of species of animals. A dog is the same species as a dog. In mathematics, equality is a reflexive relation on the set of counting numbers because $1 = 1$, $2 = 2$, and, in general, $a = a$.

In terms of a visual representation, there is a loop at every point with a reflexive relation. If D represents dog; C, cat; and H, horse; we have the loops in Figure 8(a) for *is the same species as.*

The relation *is the mother of* is not reflexive because one is not the mother of himself or herself. Although geometric *congruence* is reflexive, *less than* and *greater than* are not.

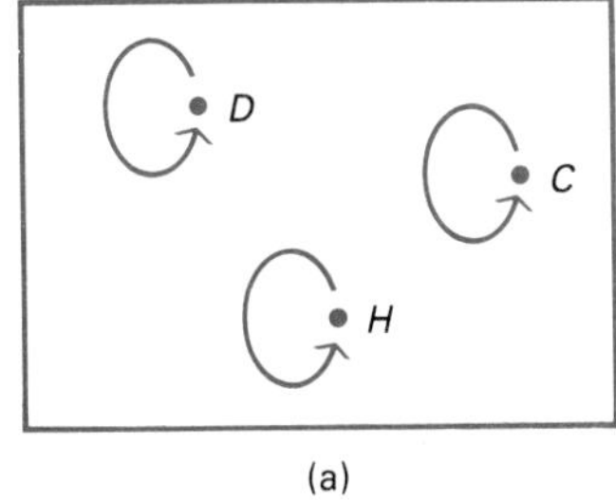

(a)

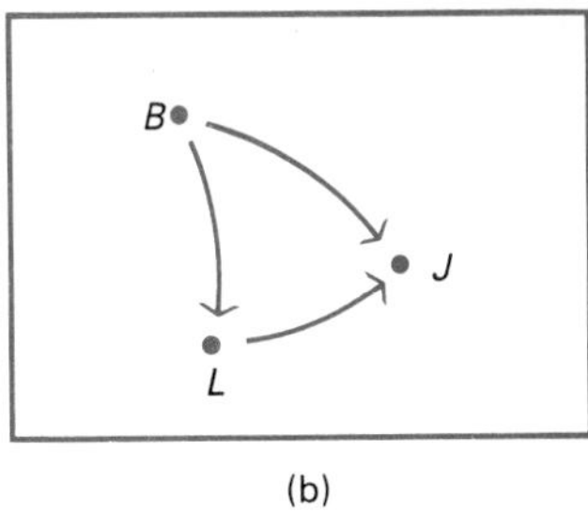

(b)

Figure 8

Symmetric relation

A relation R is said to be *symmetric* if and only if

$$(a, b) \in R \quad \text{implies that} \quad (b, a) \in R$$

For example, the relation *is the brother of* is a symmetric relation for the set of all males. If Bob is the brother of Louis, then Louis is also the brother of Bob.

In terms of a visual representation, every correspondence must have an arrow pointed in both directions. *Is a brother of* is not symmetric on the set of all people, as seen using B (Bob), L (Louis), and J (Janice). In Figure 8(b), the arrow on the connection between Janice and the boys goes in only one direction.

In a symmetric relation, whenever a is related to b, b is related to a. Perpendicularity on the set of lines in a plane is symmetric even though it is not reflexive. However, *is the mother of* is not symmetric because if Amy is the mother of Patty, Patty cannot be the mother of Amy. Similarly, *is less than* and *is greater than* are not symmetric.

Transitive relation

A relation R is *transitive* if and only if

$$(a, b) \in R \quad \text{and} \quad (b, c) \in R \quad \text{implies that} \quad (a, c) \in R$$

Consider the relation *is the same age as,* which is transitive. If Burt is the same age as Rick and Rick is the same age as Bob, then Burt is the same age as Bob. This transitive relation is shown in Figure 9. Interpret Figure 9 for Gail, Susan, and Janice.

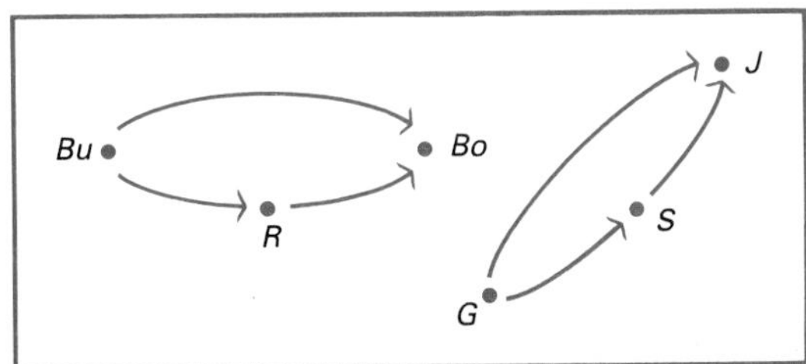

Figure 9

The transitive property for relations states that whenever a is related to b and b is related to c, then a is related to c. Note that we are not excluding the possibility that a and c can be the same. *Is equal to, is greater than,* and *is less than* are good examples of relations for which the transitive property holds. However, a relation like *is the mother of* is not transitive. Elsie is the mother of Amy and Amy is the mother of Patty; but Elsie is not the mother of Patty. Perpendicularity is also not a transitive relation.

The transitive property of equality can be generalized to involve any finite number of equalities; for instance, if

$$a = b, \quad b = c, \quad c = d, \quad d = e, \quad \text{and} \quad e = f$$

then

$$a = f$$

Equivalence relation

> A relation that is reflexive, symmetric, and transitive is called an *equivalence relation.*

Is the same age as and *is equal to* are examples of equivalence relations. *Is greater than* is not an equivalence relation. Why not?

Example 24 • Determine if the relation $R = \{(1, 1), (1, 2), (1, 3), (2, 1), (2, 2), (2, 3), (3, 1), (3, 2), (3, 3)\}$ is an equivalence relation on $A \times A$ where $A = \{1, 2, 3\}$.

Solution The relation is reflexive because all three elements (1, 1), (2, 2), and (3, 3) are members of the relation.

The relation is symmetric because for each $(a, b) \in R$, then (b, a) is also in R. For instance, (1, 2) and (2, 1), (1, 3) and (3, 1), (2, 3) and (3, 2) are in the set.

The relation is transitive. For (1, 2) and (2, 1), the element $(1, 1) \in R$; for (1, 3) and (3, 2), the element $(1, 2) \in R$; for (2, 3) and (3, 1), the element $(2, 1) \in R$. For every (a, b) and (b, c) in the relation, the element (a, c) is in the relation.

Therefore, the relation is an equivalence relation. •

An interesting feature of an equivalence relation on $A \times A$ is that the relation partitions, or separates, the set A into disjoint subsets known as *equivalence classes.* The equivalence relation *is the same sex as* would partition the set {Leo, Mary, William, Sue, Jane} into two equivalence classes {Leo, William} and {Mary, Sue, Jane}. Note that these two sets are disjoint and that every member of the equivalence class is related to every other member in the class.

We conclude this section with properties of sets that are necessary to develop the properties of addition and multiplication of whole numbers from set properties.

Distributive properties

> For all sets A, B, and C,
> (a) $C \times (A \cup B) = (C \times A) \cup (C \times B)$
> (b) $(A \cup B) \times C = (A \times C) \cup (B \times C)$

Example 25 • Consider $A = \{1, 2\}$, $B = \{2, 3\}$, and $C = \{1, 5\}$. Show that $(A \cup B) \times C = (A \times C) \cup (B \times C)$ and $C \times (A \cup B) = (C \times A) \cup (C \times B)$.

Solution

$$\begin{aligned}(A \cup B) \times C &= \{1, 2, 3\} \times \{1, 5\} \\ &= \{(1, 1), (2, 1), (3, 1), (1, 5), (2, 5), (3, 5)\}\end{aligned}$$

Now

$$A \times C = \{1, 2\} \times \{1, 5\} = \{(1, 1), (2, 1), (1, 5), (2, 5)\}$$

and

$$B \times C = \{2, 3\} \times \{1, 5\} = \{(2, 1), (3, 1), (2, 5), (3, 5)\}$$

so

$$\begin{aligned}(A \times C) \cup (B \times C) &= \{(1, 1), (2, 1), (3, 1), (1, 5), (2, 5), (3, 5)\} \\ &= (A \cup B) \times C\end{aligned}$$

Similarly,

$$\begin{aligned}C \times (A \cup B) &= \{1, 5\} \times \{1, 2, 3\} \\ &= \{(1, 1), (1, 2), (1, 3), (5, 1), (5, 2), (5, 3)\}\end{aligned}$$

$$C \times A = \{1, 5\} \times \{1, 2\} = \{(1, 1), (1, 2), (5, 1), (5, 2)\}$$

and

$$C \times B = \{1, 5\} \times \{2, 3\} = \{(1, 2), (1, 3), (5, 2), (5, 3)\}$$

Hence

$$\begin{aligned}(C \times A) \cup (C \times B) &= \{(1, 1), (1, 2), (1, 3), (5, 1), (5, 2), (5, 3)\} \\ &= C \times (A \cup B)\end{aligned}$$ •

This example illustrates the distributive property of the Cartesian product over union.

The Cartesian product fails to be either associative or commutative, as illustrated in the following example. However, $A \times C$ has the same number of elements as $C \times A$, and $(A \times B) \times C$ has the same number of elements as $A \times (B \times C)$.

Example 26 • If $A = \{1, 2\}$, $B = \{4, 5\}$, and $C = \{6, 7, 8\}$, then

$$B \times C = \{(4, 6), (4, 7), (4, 8), (5, 6), (5, 7), (5, 8)\}$$

$$C \times B = \{(6, 4), (6, 5), (7, 4), (7, 5), (8, 4), (8, 5)\}$$

Just for fun

You lost your football schedule of the Northeast Conference, but you found the picks of three sportswriters for this weekend.

A: Reds, Blues, Bulldogs, Tigers
B: Blues, Panthers, Reds, Rockets
C: Rockets, Bulldogs, Jets, Reds

(Notice that no one picked the Angels.) Use this information to construct a schedule.

Since $B \times C \neq C \times B$ for this example, the Cartesian product fails to be commutative. To form $A \times (B \times C)$, each element of A is paired with an ordered pair of $B \times C$ to obtain

$$\{[1, (4, 6)], [1, (4, 7)], [1, (4, 8)], [1, (5, 6)], [1, (5, 7)], [1, (5, 8)], [2, (4, 6)], [2, (4, 7)], [2, (4, 8)], [2, (5, 6)], [2, (5, 7)], [2, (5, 8)]\}$$

Moreover,

$$\begin{aligned}(A \times B) \times C = \{&[(1, 4), 6], [(1, 5), 6], [(2, 4), 6], [(2, 5), 6],\\ &[(1, 4), 7], [(1, 5), 7], [(2, 4), 7], [(2, 5), 7],\\ &[(1, 4), 8], [(1, 5), 8], [(2, 4), 8], [(2, 5), 8]\}\end{aligned}$$

Since $(A \times B) \times C \neq A \times (B \times C)$ for this example, the operation of cross product for sets is not associative. However, note that $(A \times B) \times C$ and $A \times (B \times C)$ have the same number of elements in this example and that each element of one set can be associated in a natural way with an element of the other. For example, the element $[1, (4, 6)]$ corresponds in a natural way with $[(1, 4), 6]$. •

Consider a Cartesian product involving the null set, such as $A \times \varnothing$. $A \times \varnothing$ consists of ordered pairs of elements in which the first element is from A and the second element is from $\varnothing$. Since $\varnothing$ has no elements, it is impossible to form any ordered pairs of elements for $A \times \varnothing$. Thus,

$$A \times \varnothing = \varnothing$$

In like manner, it is easy to see that $\varnothing \times A = \varnothing$.

Let's return now to the problem with which we introduced this section. Consider the equation, $x + y = 4$. When $x = 1$ and $y = 3$, $1 + 3 = 4$. So the ordered pair $(1, 3)$ is said to satisfy the equation. $(4, 0)$, $(2, 2)$, $(0, 4)$, and $(3, 1)$ are also solutions. Likewise $(5, 2)$ and $(2, 3)$ are ordered pairs that satisfy $x > y$ and $3x = 2y$, respectively. The set of ordered pairs that satisfy an open sentence involving x and y is called a *relation in x and y*.

Exercise set 2

R 1. Can you determine a relation between the elements of the ordered pairs described in the sets below?

(a) {(washing machine, Maytag), (book, Brooks/Cole), (iron, Sunbeam), (sewing machine, Singer), (tractor, John Deere)}
(b) {(bicycle, 2), (car, 4), (tricycle, 3), (unicycle, 1), (boat, 0), (stagecoach, 4)}

2. Draw a visual representation and then examine each of the following relations to determine if each is reflexive, symmetric, or transitive.

(a) Is heavier than
(b) Is the uncle of
(c) Lives in the same city as
(d) Runs faster than
(e) Works at the same time as
(f) Is the same color as
(g) Is a parent of
(h) Is married to

3. Classify each of the following mathematical relationships as reflexive, symmetric, or transitive. Determine whether each is an equivalence relation.

(a) Is not equal to
(b) $\subseteq$
(c) Is less than
(d) Has the same area as
(e) Is greater than
(f) $\subset$
(g) Is the same size as
(h) Intersects (lines)
(i) Is perpendicular to
(j) Is less than or equal to

4. Determine whether each of the following relations on set A is reflexive, symmetric, or transitive.

(a) $\{(1, 1), (1, 2), (2, 1), (2, 2)\}$ on $A = \{1, 2\}$
(b) $\{(3, 3), (3, 4), (4, 3)\}$ on $A = \{3, 4\}$
(c) $\{(c, c), (c, d), (c, e), (d, d), (d, e), (e, e)\}$ on $A = \{c, d, e\}$

5. If $A = \{a, b, c\}$ and $B = \{r, s, t\}$, tabulate the elements of the indicated Cartesian products.

(a) $A \times B$ (b) $A \times A$
(c) $B \times A$ (d) $B \times B$

6. Let $A = \{a, b, c\}$, $B = \{c, d, e\}$, and $C = \{a, c, x\}$. Tabulate the following Cartesian product sets.

(a) $A \times (B \cap C)$ (b) $A \times (B \cup C)$
(c) $(A \cap B) \times C$ (d) $(A \cup C) \times B$

7. Find $B \times C$ for each of the pairs of sets below.

(a) $B = \{3\}$; $C = \{0\}$ (b) $B = \{3\}$; $C = \{\ \}$

8. The Cartesian product $B \times C$ is given by the sets shown. Find a B and C for each.

(a) $\{(1, 1), (1, 2), (1, 3), (4, 1), (4, 2), (4, 3)\}$
(b) $\{(1, 4), (1, 5), (0, 4), (0, 5)\}$
(c) $\{(6, 6), (6, 7), (6, 8)\}$
(d) $\{(x, x), (x, y), (x, z), (y, x), (y, y), (y, z), (z, x), (z, y), (z, z)\}$

9. Three professors at the University of Mystate have student assistants. If $P = \{$Mardis, Boyd, Kiefer$\}$ and $S = \{$Janet, Judy, Marion$\}$, find all the possible combinations of professors and assistants by listing the elements of $P \times S$.

T 10. (a) Count the number of elements in $A \times B$ and $A \times A$ of Exercise 5.
(b) If A has four elements and B has three elements, how many elements are in $A \times B$? In $B \times B$? In $B \times A$? In $A \times A$?
(c) Let A be a set with r elements and B a set with s elements. How many elements are in $A \times A$? In $A \times B$? In $B \times B$?

11. If $A = \{1, 2, 3, 4\}$ and $B = \{2, 5\}$, tabulate the elements (x, y) of $A \times B$ so that (a) x is less than y. (b) x is unequal to y. (c) x is equal to y. (d) x is greater than or equal to y.

12. Which of the relations defined on A in Exercise 11 are equivalence relations?

13. If $A = \{1, 2\}$, $B = \{3, 4, 5\}$, and $C = \{4, 5\}$, tabulate the elements of the following sets.

(a) $B \times (C \times A)$ (b) $A \times (B \times C)$

14. Give an example of a relation that is reflexive and symmetric but not transitive.

15. Relation R is defined on the set $\{1, 3, 5, 7\}$. Give an R that satisfies the following conditions.

(a) R is transitive but not reflexive and symmetric.
(b) R is transitive, reflexive, and symmetric.
(c) R is reflexive and transitive but not symmetric.

16. Consider the set of names {Asa, Jim, Allen, Ben, John, Betty, Bob, Joe, Jack, Alto, Alisa, Amy, Bill,

Jill}. Identify the equivalence classes for the relations (a) has the same number of letters in name; (b) name begins with the same letter.

C 17. Verify that $(A \cup B) \times C = (A \times C) \cup (B \times C)$ with a truth table by showing that the truth values are the same for $(x, y) \in (A \cup B) \times C$ and $(x, y) \in (A \times C) \cup (B \times C)$.

18. Using truth tables, verify that $C \times (A \cup B) = (C \times A) \cup (C \times B)$.

3 The number of elements in a set

■ A pollster found that in a group of 100 people, 40 use brand *A*, 25 use both brand *A* and brand *B*, and everyone uses either brand *A* or brand *B*. How many people use brand *B*? ■

By associating a number with the elements in a set, using Venn diagrams, and utilizing problem-solving techniques, you should be able to solve the preceding problem.

The number of elements in a set provides the basis for counting. Is the number of this page a cardinal number, ordinal number, or nominal number? You will learn in this section the distinctions among such numbers.

Determining the number of elements in a set seems very simple. However, in probability theory (Chapter 13), we encounter some challenging problems involving the number of elements in sets. The material of this section provides the background needed to solve these problems.

The number concept held by mathematicians today is the result of many centuries of study. Leopold Kronecker, a German mathematician of the 19th century, is supposed to have said, "God created the natural numbers; all else is the work of man." His statement has been interpreted by some as meaning that man's mind is naturally endowed with the power to comprehend the concept of the counting numbers, whereas other numbers are a result of the inventiveness of man.

Before discussing the concept of a number, we need an understanding of one-to-one correspondence and equivalence.

One-to-one correspondence

A one-to-one correspondence between sets *A* and *B* is a pairing of the elements of *A* and *B* such that for every element *a* of *A* there corresponds exactly one element *b* of *B* and for every element *b* of *B* there corresponds exactly one element *a* of *A*.

BIOGRAPHY

Leopold Kronecker
1823–1891

Born in Prussia of prosperous Jewish parents, Kronecker received his early education from private tutors and from his father, whose love for philosophy was passed on to his son. Although he was wealthy, Kronecker did not forsake his first love, mathematics. Most of his works have a strong arithmetical flavor; he attempted to explain everything in terms of whole numbers and in a finite number of steps.

Example 27 • Consider the sets $A = \{1, 2, 3\}$ and $B = \{\text{John, Jane, Jill}\}$. One way of placing these two sets in a one-to-one correspondence is to use double-headed arrows, as shown here.

$$\begin{array}{ccc} 1 & 2 & 3 \\ \updownarrow & \updownarrow & \updownarrow \\ \text{John} & \text{Jane} & \text{Jill} \end{array}$$

Of course, these two sets could be placed in a one-to-one correspondence in several other ways. For instance, $1 \leftrightarrow$ Jill, $2 \leftrightarrow$ John, and $3 \leftrightarrow$ Jane. Try to establish a one-to-one correspondence between

$$A = \{1, 2, 3, 4\} \quad \text{and} \quad B = \{\text{John, Jane, Jill}\}$$

If we let John correspond to 1, Jane to 2, and Jill to 3, no element is left to correspond to 4. Thus, a one-to-one correspondence cannot be set up between these two sets. •

A number of sets can be placed in one-to-one correspondence with a given set. For example, $\{a, b, c, d\}$ can be placed in one-to-one correspondence with $\{1, 2, 3, 4\}$, and so can $\{\square, \triangle, \square, \bigcirc\}$, $\{x, y, z, w\}$, and $\{140, 180, 190, 200\}$. This characteristic leads to the following definition.

Equivalent sets

> If a one-to-one correspondence exists between two sets, the sets are said to be *equivalent* (or *matched*).

Example 28 • If a room contains 70 seats, one person is sitting in each of the seats, and no one is standing, then the set of seats and the set of people are equivalent. •

The term *equivalent* should not be confused with *equality*. If two sets are equal, then each element of one set must equal a corresponding element of the other, and conversely. However, sets are equivalent if a one-to-one correspondence exists between elements. Thus, equal sets are always equivalent, but equivalent sets are not always equal.

Example 29 • $\{a, b, c, d\}$ and $\{b, c, a, d\}$ are equal sets; however, $\{a, b, c, d\}$ is equivalent to but not equal to $\{x, y, z, w\}$. •

Note that the relation *is equivalent to* on a collection of sets is an equivalence relation on that collection.

Consider four sets: {1, 2, 3}, {*, #, %}, $\{a, b, c\}$, and {○, △, □}. What do these sets have in common? The answer is that they are equivalent. Intuitively, another property is probably obvious to you. A property that equivalent sets have in common is called their **cardinal number.**

Cardinal number

> The property that is possessed by each set of a class of equivalent sets is the *cardinal number* of each set. Basically, the cardinal number of a set is the number of elements in the set.

The symbol $n(A)$ represents the number of elements in set A. It will be read *the number of A*. This number is the cardinal number of all sets equivalent to A. Suppose that there are 30 students in a classroom. If S stands for the set of students, then $n(S) = 30$. It should be noted that the number of a set is another property of a set that does not depend in any way on the kind of items represented as elements or on the order in which the elements are listed. The use of counting numbers to tell how many objects are in a set is called **cardinal** usage. "There are 24 hours in a day" uses the counting number 24 in a cardinal sense.

In the phrase "the fifth section of the second chapter of this book," numbers are used in an **ordinal** sense. *Fifth* and *second* refer to position or order. The following example illustrates the cardinal and ordinal use of counting numbers.

Example 30 • The administration building is five stories high (cardinal). The president's office is on the second floor (ordinal). •

Counting numbers may also be used in a **nominal sense** for naming things, as in social security numbers, bank account numbers, and zip codes.

Let's return now to the problem in the introduction of the section.

Example 31 • A pollster found that in a group of 100 people, 40 use brand A, 25 use both brand A and brand B, and everyone uses either brand A or brand B. How many people use brand B?

Can you help find the answer?

Solution To solve this problem, we use the outline given in Chapter 1 for solving problems.

Understanding the Problem. Let's draw a Venn diagram (Figure 10) to visualize the given information. Now $n(A \cup B) = 100$. Why? The region bounded by the circle and designated as A represents the people who use brand A, or $n(A) = 40$. The region bounded by the circle and designated as B represents the people who use brand B. What does the shaded region represent? Is $n(A \cap B) = 25$?

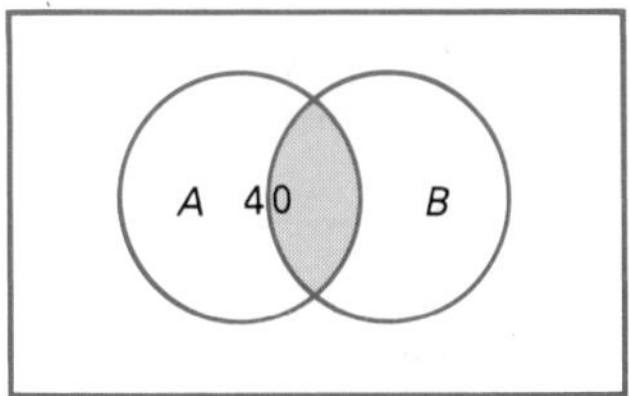

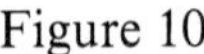
Figure 10

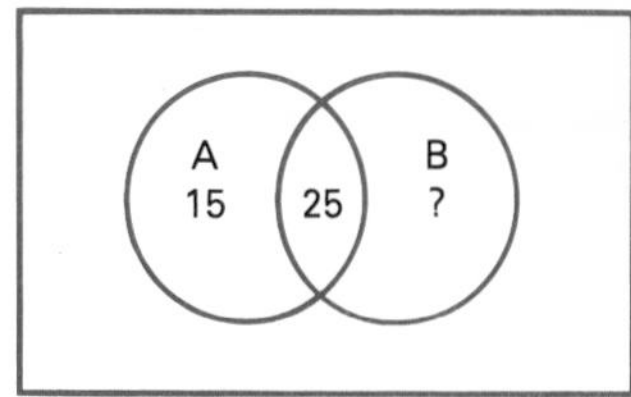

Figure 11

Devising a plan. We can easily solve the problem by placing numbers in each closed region of Figure 11. Since 25 of the 40 people who use brand A also use brand B, 15 use brand A only. Now let's see if we can determine how many use brand B only.

Carrying out the plan. Since 100 people were surveyed and since everyone contacted uses either brand A or brand B, the total number in the three regions bounded by the circles of Figure 11 is $n(A \cup B) = 100$, or

$$15 + 25 + ? = 100$$

Thus, the number using brand B only is 60. Therefore, the total number using brand B is

$$n(B) = 25 + 60 = 85$$

Looking back. Looking at Figure 11 (replacing ? by 60), does the total in all regions equal 100? (Yes, $15 + 25 + 60 = 100$.) Do 40 people use brand A? (Yes, $15 + 25 = 40$.) Do 25 people use brand A and brand B? (Yes, $n(A \cap B) = 25$.) •

The following development demonstrates the theory of the preceding example.

Find the number of elements that are in either set A or set B denoted by $n(A \cup B)$. The answer would not necessarily be

$$n(A) + n(B)$$

as we discover in Figure 12. In this figure, the number of elements in the closed regions of the Venn diagram are denoted by (a), (b), (c), and (d). Thus,

$$n(A \cup B) = (a) + (b) + (c)$$

$$n(A) = (a) + (b)$$

$$n(B) = (b) + (c)$$

$$n(A) + n(B) = (a) + (b) + (b) + (c)$$

Figure 12

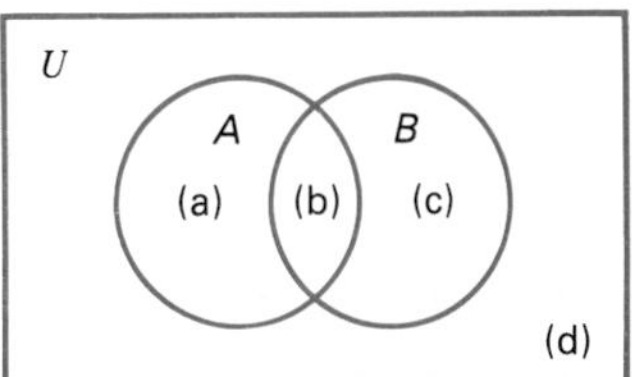

So $n(A) + n(B)$ contains (b) more elements than $n(A \cup B)$. Since (b) $= n(A \cap B)$, we have the following.

Number of elements in $A \cup B$

> The number of elements that are in either A or B is
>
> $$n(A \cup B) = n(A) + n(B) - n(A \cap B)$$
>
> where $A \cap B$ consists of elements common to A and B.

Example 32 • Out of 1500 freshmen, 13 students failed English; 12 students failed mathematics; and 7 students failed both English and mathematics. How many students failed English or mathematics?

Solution

$$n(E) = 13$$

$$n(M) = 12$$

$$n(E \cap M) = 7$$

To find $n(E \cup M)$,

$$\begin{aligned} n(E \cup M) &= n(E) + n(M) - n(E \cap M) \\ &= 13 + 12 - 7 \\ &= 18 \quad \bullet \end{aligned}$$

Example 33 • Out of 100 freshmen at Hard College, 60 are taking English, 50 are taking history, 30 are taking mathematics, 30 are taking both English and history, 16 are taking both English and mathematics, 10 are taking both history and mathematics, and 6 are taking all these courses. How many of the freshmen are enrolled in either English, history, or mathematics?

Solution Again we use problem-solving techniques to find a solution.

Understanding the problem. Let E represent those enrolled in English, H in history, and M in mathematics. The example states that

$$n(E) = 60 \qquad n(E \cap H) = 30$$

$$n(H) = 50 \qquad n(E \cap M) = 16$$

$$n(M) = 30 \qquad n(H \cap M) = 10$$

$$n(E \cap H \cap M) = 6$$

We are to find $n(E \cup H \cup M)$.

Devising a plan. First we draw a Venn diagram, letting the circular regions representing E, H, and M overlap with $n(E \cap H \cap M) = 6$ (Figure 13(a)). By finding the numbers within each closed region, we can then find $n(E \cup H \cup M)$.

Figure 13

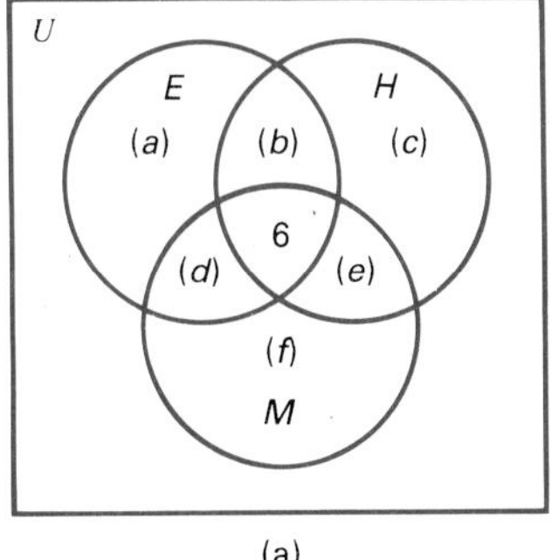

(a)

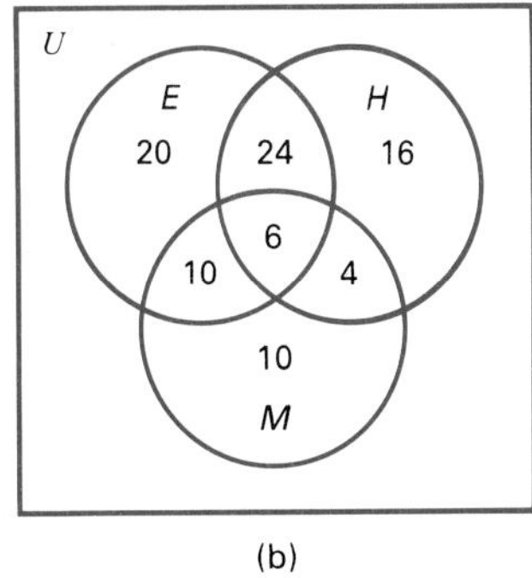

(b)

Carrying out the plan. Let's denote the number of elements in closed regions by (a), (b), (c), (d), etc., as indicated in Figure 13(a).

$$(b) + 6 = n(E \cap H) = 30 \quad \text{so} \quad (b) = 24$$

$$(e) + 6 = n(H \cap M) = 10 \quad \text{so} \quad (e) = 4$$

$$(d) + 6 = n(E \cap M) = 16 \quad \text{so} \quad (d) = 10$$

$$(a) + (b) + (d) + 6 = n(E) = 60 \quad \text{or} \quad (a) + 24 + 10 + 6 = 60$$

so
$$(a) = 20$$

$$(b) + (c) + (e) + 6 = n(H) = 50 \quad \text{or} \quad 24 + (c) + 4 + 6 = 50$$

so
$$(c) = 16$$

$$(f) + (e) + 6 + (d) = n(M) = 30 \quad \text{or} \quad (f) + 4 + 6 + 10 = 30$$

so
$$(f) = 10$$

From Figure 13(b),

$$N(E \cup H \cup M) = 20 + 24 + 16 + 10 + 4 + 10 + 6 = 90 \quad \bullet$$

The preceding example outlines a procedure that could be used to prove the following theorem.

Number of elements in $A \cup B \cup C$

For any three sets A, B, and C,

$$n(A \cup B \cup C) = n(A) + n(B) + n(C) - n(A \cap B) - n(A \cap C) - n(B \cap C) + n(A \cap B \cap C)$$

Just for fun *Dan has 16 pieces of candy. Nine contain caramel; 10 contain coconut; and 6 contain both caramel and coconut. Use problem-solving techniques along with sets to find how many of the candies contain neither caramel nor coconut.*

Let's verify the answer for the preceding example using this theorem.

$$n(E \cup H \cup M) = n(E) + n(H) + n(M) - n(E \cap H) - n(E \cap M) - n(H \cap M) + n(E \cap H \cap M)$$

$$n(E \cup H \cup M) = 60 + 50 + 30 - 30 - 16 - 10 + 6 = 90$$

This is the same answer obtained previously.

Exercise set 3

R 1. Place a C by the phrases involving cardinal number concepts and an O by those involving ordinal number concepts.

(a) 16 sheep (b) Page 9
(c) 82nd Congress (d) Third grade
(e) Five people (f) Fifth session
(g) 7 days in a week (h) 10th round
(i) 12 months in a year (j) \$15
(k) A dozen eggs (l) Second period

2. Let set $A = \{1, 2, 3, 4, 5, 6\}$. Find all the sets in the following listing that are equivalent to set A.

(a) $\{a, b, c, d, e, f\}$
(b) $\{1, 2, 3, 4, 5, 6, \ldots\}$
(c) $\{4, 5, \ldots, 9\}$
(d) $\varnothing$
(e) {Jane, Mary, Sue, Louise}
(f) $\{8, 9, 10, 11, 12, 13\}$

3. Exhibit a one-to-one correspondence between one of the sets $\{1\}$, $\{1, 2\}$, $\{1, 2, 3\}$, $\{1, 2, 3, 4\}$, or $\{1, 2, 3, 4, 5\}$, and the following sets.

(a) $\{b\}$ (b) $\{\varnothing, \lambda\}$
(c) $\{a, b, c, d, e\}$ (d) $\{4, 6, 8\}$
(e) $\{10, 40, 30, 50, 70\}$ (f) $\{2\}$

4. In the accompanying figure, the numbers indicate the cardinal number of the closed region. Find the following.

(a) $n(A)$ (b) $n(B)$
(c) $n(A \cap B)$ (d) $n(A \cup B)$
(e) $n(\bar{A})$ (f) $n(\bar{B})$
(g) $n(U)$ (h) $n(\bar{A} \cap \bar{B})$
(i) $n(\bar{A} \cup \bar{B})$ (j) $n(\bar{A} \cap B)$

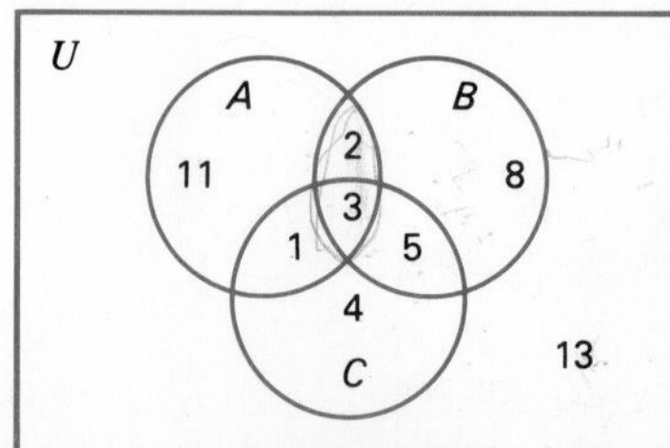

5. Indicate which statements below are true and which are false.

   (a) Sets $A = \{\bigcirc, \triangle, \text{⌂}, \square, \text{▭}\}$, $B = \{1, 6, 8, 3, 0\}$, and $C = \{\text{Rick, Steve, Allan, Paul, Ed}\}$ are equivalent.
   (b) Sets A and B in part (a) have the same cardinal number because they are equal.
   (c) Although sets A, B, and C in part (a) are not equal, they still have the same cardinal number.
   (d) The smallest element in the set of natural numbers is 0.
   (e) A subset of a finite set may be infinite.
   (f) A subset of an infinite set may be finite.
   (g) A subset of an infinite set may be infinite.
   (h) $n\{0\} = 0$
   (i) $n(\varnothing) = 0$
   (j) $n\{\ \} = 1$
   (k) $n\{0, 1, 2\} = 3$

6. If $A = \{a, b, c, d\}$ and $R = \{r, s, t, v\}$, explain why (a) A and R are not equal sets. (b) A and R are equivalent sets.

7. Define $A = \{1, 2, 3\}$, $B = \{1\}$, $C = \varnothing$, and $D = \{a, b, c, d\}$. Determine whether the statements shown are true or false.

   (a) $n(A \times B) = n(B \times A)$ (b) $A \times B = B \times A$
   (c) $n(A \times B) = n(A)$ (d) $n(A \times D) = 12$
   (e) $A \times C = C \times D$
   (f) $n[(A \times B) \times C] = n[A \times (B \times C)]$

T 8. For $A = \{2, 3, \ldots, 6\}$ and $B = \{6, 7, \ldots, 10\}$, show that

   (a) $n(A \cup B) = n(A) + n(B) - n(A \cap B)$
   (b) $n(A \times B) = n(A) \cdot n(B)$
   (c) $n(A \times A) = n(A) \cdot n(A)$

9. Indicate whether the statements below are true or false for all sets A and B. If a statement is false, give an example to demonstrate that it is false.

   (a) $n(A \cup B) = n(A) + n(B)$
   (b) $n(A \cap B) = n(A) - n(B)$
   (c) If $n(A) = n(B)$, then $A = B$
   (d) If $A = B$, then $n(A) = n(B)$
   (e) $n(A \times B) = n(A) \cdot n(B)$
   (f) $n(A) + n(B) = n(A \cup B) - n(A \cap B)$
   (g) $n(\bar{A} \cup \bar{B}) = n(\overline{A \cup B})$
   (h) $n(A \cap B) = n(A \cup B) - n(A \cap \bar{B}) - n(\bar{A} \cap B)$

10. In Atlanta, 600,000 people read newspaper A, 450,000 read newspaper B, and 160,000 read both newspapers. How many read either newspaper A or newspaper B?

C 11. Out of 1000 first-year women in a certain university, 300 pledged sororities, 200 live off campus, and 150 pledged sororities and live off campus.

   (a) Draw a Venn diagram labeling and counting the four relevant parts of the diagram, where S represents sorority pledges and O represents living off campus.
   (b) How many pledged sororities or live off campus?
   (c) How many did not pledge sororities or live off campus?
   (d) How many live off campus and did not pledge sororities?
   (e) How many pledged sororities and do not live off campus?
   (f) How many pledged sororities or do not live off campus?

12. A poll was taken of 100 students at a nonresidential college to find out how they got to campus. The results were as follows: 28 mentioned car pools; 31 used buses; and 42 said that they drove to school alone. In addition, 9 used both car pools and buses; 10 used both a car pool and sometimes their own cars; only 6 used buses as well as their own cars. Of the 100 respondents, only 4 used all three methods of arriving on campus.

   (a) Draw a Venn diagram labeling each of the *eight* relevant pieces.
   (b) How many used none of the three methods of transportation discussed above?
   (c) How many used car pools exclusively to arrive on campus?
   (d) How many used buses exclusively to get on campus?

13. A discount store ran a special sale on hair spray, paper cups, and boys' pants. One salesperson reported that 50 people took advantage of the sale. Thirty people purchased hair spray; 25 purchased paper cups; 35 purchased boys' pants; 14 purchased both hair spray and paper cups; 15 purchased both paper cups and boy's pants; 22 purchased both hair spray and boys' pants; and 10 purchased all three. Was the salesperson's report accurate? Explain.

14. In how many ways can $\{a, b, c, d\}$ be placed in a one-to-one correspondence with itself? Use problem-solving techniques to determine in how many

ways a set with 7 elements can be placed in a one-to-one correspondence with itself.

15. A pollster on a college campus interviewed the women students as to the shampoo they used. Of the 7900 on campus, she reported that 4500 used Shiny Shampoo, 4850 used Dandruff Doo, and 4750 used Glow and Gleam. Of these women, 3000 used both Shiny Shampoo and Dandruff Doo, 2750 used Dandruff Doo and Glow and Gleam, 2250 used Shiny Shampoo and Glow and Gleam, and 1750 used all three. Was this poll accurate? Why or why not?

4 Whole number addition

■ Pam adds down.

```
 6    6 + 3 =  9
 3    9 + 4 = 13
 4   13 + 2 = 15
 2
--
15
```

Roy adds up.

```
 6    9 + 6 = 15
 3    6 + 3 =  9
 4    2 + 4 =  6
 2
--
15
```

Sue adds every other term.

```
 6
 3    6 + 4 = 10
 4    3 + 2 =  5     10 + 5 = 15
 2
--
15
```

All get the same answer. Why? ■

We have been competent arithmeticians for so long that we use the properties of addition without any thought. We change at our convenience the order of addition ($3 + 2 = 2 + 3$) and regroup addition $(2 + 3) + 4 = 2 + (3 + 4)$, scarcely conscious that we are using powerful properties that have impact throughout much of mathematics. In this section, you will see a definition of addition in terms of sets and then observe how the properties of addition are developed from the properties of sets.

We now consider a new set consisting of the natural numbers and the number 0.

Whole numbers

The set composed of the *natural numbers and zero* is called the set of *whole numbers.*

Consider the number zero for a moment. The familiar statement "Zero is nothing" is false. Zero is defined to be the cardinal number of the null set, $n(\varnothing) = 0$, and it answers such questions as "How many elephants are in your class?" The next step is to examine the operations on the set of whole numbers.

Suppose you have |||| Delicious apples and ||| Jonathans. (See Figure 14(a).) How many apples do you have? Note that |||| represents the number of Delicious apples and that ||| gives the number of Jonathan apples. Now place the Delicious apples and the Jonathan apples in a fruit bowl. By placing together the abstract number representations of the number of Delicious apples and the number of Jonathans, ||||||| now represents the number of apples, Figure 14(b). This idea of putting together (or taking the union) leads to the definition of addition. Let's look at some examples.

Figure 14

Example 34 • Consider the two sets $A = \{a, b, c\}$ and $B = \{\triangle, \lambda, \bigcirc, \square\}$.

$$n(A) = 3 \quad \text{and} \quad n(B) = 4$$

How many elements are in $A \cup B$?

Solution If the elements of A and B are placed together in $A \cup B$ and counted, $n(A \cup B) = 7$. Thus,

$$n(A) + n(B) = n(A \cup B)$$

or

$$3 + 4 = 7 \quad \bullet$$

Example 35 • Let $A = \{a, b, c\}$ and $B = \{c, d, e, h\}$.

$$n(A) + n(B) = 3 + 4 = 7$$

But $A \cup B = \{a, b, c, d, e, h\}$, and so $n(A \cup B) = 6$. In this example,

$$n(A) + n(B) \neq n(A \cup B) \quad \bullet$$

What is the distinction between these two examples? In the first one the sets are disjoint, whereas in the second the letter c is common to both sets. Combining these ideas, we arrive at a precise definition of addition.

Addition of whole numbers

Let a and b represent any two whole numbers and choose A and B to be disjoint finite sets so that $n(A) = a$ and $n(B) = b$. Then $a + b = n(A \cup B)$.

Note that the operation of addition is defined in terms of two numbers, a and b; that is; addition is a **binary operation.**

If $A = \{r, s, c, d, e\}$ and $B = \{x, y, z, w\}$, then $n(A) = 5$ and $n(B) = 4$. Note that A and B are disjoint since $A \cap B = \varnothing$. Next,

$$A \cup B = \{r, s, c, d, e, x, y, z, w\}$$

and $n(A \cup B) = 9$. It follows that

$$n(A \cup B) = n(A) + n(B)$$

or

$$9 = 5 + 4$$

The 5 and 4 are called **addends,** and 9 is called the **sum.**

Let's look now at the representation of whole numbers on a number line and see how we can visually represent addition of whole numbers. A **number line** is a representation of a geometric line extending endlessly in two opposite directions. The line is marked with two fundamental points, one representing 0 and one representing 1, as in Figure 15(a). These points mark the ends of what is called a **line segment** of length 1 (a **unit segment**). Once the length of a unit segment is established, other points are marked to the right of 0 on the number line, as in Figure 15(b).

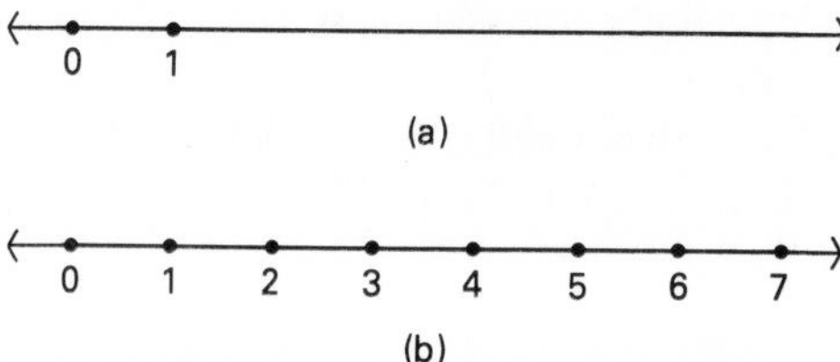

Figure 15

Thus, to the right of 0 a number line is divided into equal divisions by points, each point representing a whole number. No matter how large a whole number may be, it can be matched with a point on a number line.

We use parts of lines with arrows on the end, called **arrow diagrams** or **directed line segments**, to represent numbers involved in operations on the whole numbers. For example, let's consider the sum of 2 and 3. Starting at 0 on the number line in Figure 16(a), draw an arrow two units long. At the end of the first arrow, draw another arrow to the right three units long. Note that the point of the second arrow is at 5. Therefore, $2 + 3 = 5$. Note also in Figure 16(b) the representation of $3 + 4$, and $4 + 3$.

If $n(A) = a$ and $n(B) = b$ where a and b are whole numbers, then A and B are finite sets, and it is intuitively evident that $A \cup B$ exists and is a finite set. Thus,

flashback

The Hindus introduced the number zero before A.D. 800. Not only did this new number provide an enlargement for the set of counting numbers, but zero was a necessity in the Hindu-Arabic numeration system, which we will study in the next chapter.

Figure 16

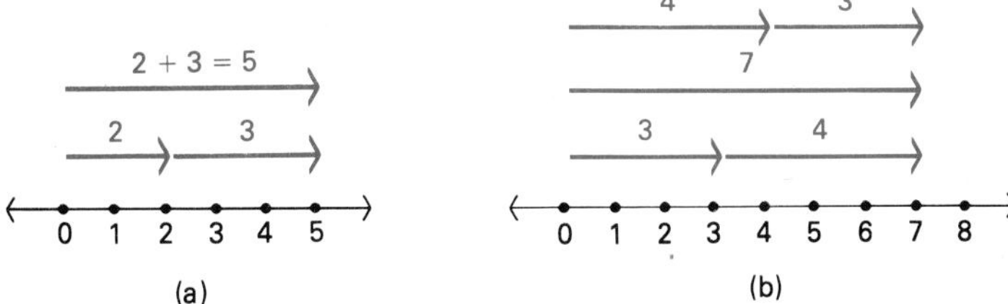

from the definition $a + b = n(A \cup B)$, we see that $a + b$ exists uniquely as a whole number.

Closure property for addition

> The *closure property for addition* for the whole numbers states that for any two whole numbers a and b there exists a unique whole number $a + b$.

Note that closure has two parts: uniqueness and existence. Existence stipulates that an answer exists in the set of whole numbers for the addition of any two whole numbers. Uniqueness stipulates that only one answer exists; that is, when 2 and 7 are added, the unique answer 9 is obtained.

The closure property for the addition of whole numbers can be extended to any finite number of whole numbers; that is, $4 + 5 + 0 + 1 + 3 = 13$, a whole number.

We should note that not every set of numbers is closed under the operation of addition. For example, the set of odd numbers is not closed under addition. Why not?

Example 36 • Which of these sets are closed under addition?

(a) $\{0, 1, 2, 3\}$

(b) $\{x \mid x \text{ is an even whole number}\}$

Solution (a) $2 + 3 = 5$, and 5 is not in the set. Therefore, the set is not closed under addition.

(b) The sum of any two even whole numbers is an even whole number. For example, $2 + 6 = 8$ (which is even). Therefore, the set is closed under addition. •

Suppose you add $3 + 4 = 7$ and then add $4 + 3 = 7$. Are you surprised that the answers are the same? This can be seen visually in Figure 16(b). Note that if you add the arrow diagram for 4 at the end of that for 3, you terminate at the same point (namely 7) where you terminate by adding the arrow diagram for 3 at the end of 4. This property for the addition of whole numbers is called *commutativity*.

Commutative property of addition

> For all whole numbers a and b,
>
> $$a + b = b + a$$

Although the commutative property of addition is obvious from the visual representation in Figure 16(b), let's see if this result can be obtained from the definition of addition in terms of the union of sets.

Now let A and B be any two sets such that $a = n(A)$ and $b = n(B)$, with $A \cap B = \varnothing$. Since $A \cup B = B \cup A$, then $A \cup B$ and $B \cup A$ are equivalent, or $n(A \cup B) = n(B \cup A)$. Thus,

$$a + b = n(A \cup B) = n(B \cup A) = b + a$$

When adding three or more numbers, we must group the numbers to specify the order in which the operations are to be performed. Suppose that we wish to add 2, 3, and 7. Should we add 2 and 3 and then 7, or should we add the sum of 3 and 7 to 2? Usually we indicate grouping of additions by the use of parentheses ( ) or brackets [ ]. $(2 + 3) + 7$ means that 2 and 3 are added and 7 is added to the answer. $2 + (3 + 7)$ means that the sum of 3 and 7 is added to 2. The parentheses serve as mathematical punctuation. Perform these additions and see if you find the same answer. If you do, then you have verified for this example another important property of addition, the *associative* property.

Associative property of addition

For all whole numbers a, b, and c,

$$(a + b) + c = a + (b + c)$$

Example 37 •

$$3 + 4 + 7 = (3 + 4) + 7 = 7 + 7 = 14$$

or

$$3 + 4 + 7 = 3 + (4 + 7) = 3 + 11 = 14$$ •

When four numbers are to be added, the numbers may be associated or grouped in several ways.

$$(3 + 2) + (5 + 6) = 5 + 11 = 16$$
$$3 + [(2 + 5) + 6] = 3 + [7 + 6] = 3 + 13 = 16$$
$$[3 + (2 + 5)] + 6 = [3 + 7] + 6 = 10 + 6 = 16$$

That they yield the same number results from a generalization or extension of the associative property of addition.

In simple proofs such as in the next example we often use the transitive property of equalities "If $a = b$ and $b = c$, then $a = c$." In the next example, we show that $(2 + 3) + 5 = 5 + (2 + 3)$ (or $a = b$). Then we show that $5 + (2 + 3) = (5 + 2) + 3$ (or $b = c$). Therefore, $(2 + 3) + 5 = (5 + 2) + 3$ (or $a = c$). Notice also that because of the closure property of addition we are considering $(2 + 3)$ as one term.

Example 38 • Using the commutative and associative properties for addition, show that $(2 + 3) + 5 = (5 + 2) + 3$.

Solution

$(2 + 3) + 5 = 5 + (2 + 3)$ Commutative property of addition
$5 + (2 + 3) = (5 + 2) + 3$ Associative property of addition

Therefore,

$(2 + 3) + 5 = (5 + 2) + 3$ Transitive property of equalities •

Just for fun

In a magic square, the sum of the numbers in every row, column, and diagonal is the same. Rearrange the numbers shown to make a magic square with a magic sum of 12.

Consider the addition $4 + 0$. We may associate with 4 the set $A = \{a, b, c, d\}$ and with 0 the null set $\varnothing$. (See Figure 17.) The union of A and $\varnothing$ is A, and $A \cap \varnothing = \varnothing$. Thus, $n(A) + n(\varnothing) = n(A \cup \varnothing) = n(A)$. Hence, $4 + 0 = 4$. We call 0 *the additive identity for addition.*

Figure 17

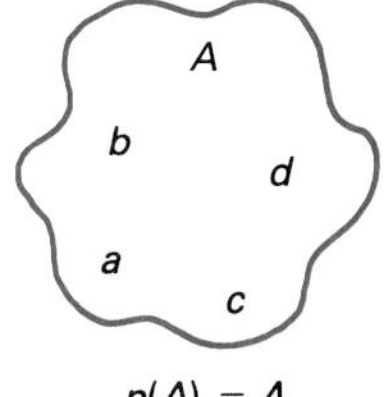

$n(A) = 4$

$n(\varnothing) = 0$

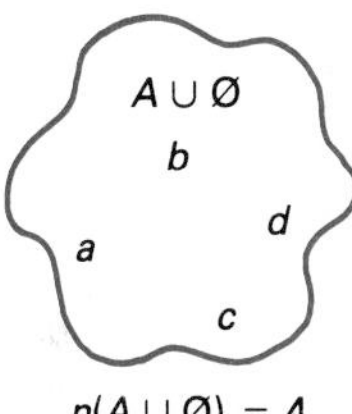

$n(A \cup \varnothing) = 4$

Additive identity for addition

> 0 is called *the additive identity* since, for any whole number a,
>
> $$a + 0 = a \quad \text{and} \quad 0 + a = a$$

Example 39 •

$$4 + 0 = 4$$

$$0 + 3 = 3$$

$$0 + 0 = 0 \quad \bullet$$

In Chapter 1, we introduced the finding of solutions of open sentences. The values that satisfy an open sentence are the elements of the **solution set** of the open sentence. In this section, we will use the set of whole numbers as the domain.

Example 40 • Find the solution set for

(a) $x + 7 = 11$

(b) $x + 6 = 4$

using the set of whole numbers as the domain.

Solution (a) $x = 4$ or $\{4\}$ is the solution set, since $4 + 7 = 11$
(b) $\varnothing$ is the solution set, as there is no whole number that yields 4 when added to 6. •

Exercise set 4

R 1. For the given pairs of sets, state whether $n(R) + n(S) = n(R \cup S)$.

(a) $R = \{3, 2, 7, 1\}$, $S = \{4, 6, 5\}$
(b) $R = \{3, 2, 4, 7\}$, $S = \{1, 4, 6, 5\}$
(c) $R = \{x|x$ is a whole number greater than 1 and less than $6\}$
$S = \{x|x$ is a whole number greater than 4 and less than $9\}$
(d) $R = \{x|x$ is a whole number greater than 5 and less than $8\}$
$S = \{x|x$ is a whole number greater than 6 and less than $7\}$
(e) $R = \varnothing$, $S = \{0\}$

2. The problems below are applications of properties of addition described in this section. State which properties are being applied in each example.

(a) $2 + 3 = 3 + 2$
(b) $4 + (2 + 3) = (4 + 2) + 3$
(c) $6 + 0 = 6$
(d) $(a + c) + d = a + (c + d)$
(e) $[2 + (4 + 6)] + 8 = (2 + 4) + (6 + 8)$
(f) $2 + (5 + 7) = 2 + (7 + 5)$
(g) $5 + 0 = 0 + 5$
(h) $(8 + 2) + 3 = 3 + (8 + 2)$
(i) $(4 + 2) + 3 = 3 + (4 + 2)$
(j) $(2 + 5) + 3 = (5 + 2) + 3$

3. Addition can be simplified using the associative property of addition.

(a) $27 + 46 = (27 + 3) + 43 = 30 + 43 = 73$
(b) $49 + 58 =$ ______ $+ 50 =$
(c) $56 + 78 =$ ______ $+ 70 =$
(d) $263 + 85 = 260 +$ ______ $=$

4. Frames provide an easy transition to the study of algebra as well as providing a clear way of stating problems. Fill in the frames to make the following number statements true.

(a) $3 + (5 + 7) = (5 + \square) + 3$
(b) $5 + 0 = \square$
(c) $7 + \square = 9 + 7$
(d) $5 + (\square + 6) = (6 + 5) + 9$

5. Show the following additions on a number line.

(a) $3 + 5$
(b) $5 + 3$
(c) What do (a) and (b) demonstrate?
(d) Show on a number line that $(2 + 3) + 4 = 2 + (3 + 4)$, the associative property of addition.

6. Determine what property of addition is illustrated by filling in the frames.

(a) $\square + (\triangle + \bigcirc) = (\square + \triangle) + \bigcirc$
(b) $\square + 0 = \square$
(c) $\square + \bigcirc = \bigcirc + \square$
(d) $\square + (\triangle + \bigcirc) = \square + (\bigcirc + \triangle)$

7. Which of the following sets have closure with respect to addition?

(a) $\{1, 2, 3, 4, 5, \ldots\}$
(b) $\{2, 4, 6, 8, 10, \ldots\}$
(c) $\{1, 3, 5, 7, \ldots\}$
(d) $\{0, 1\}$
(e) $\{x|x$ is a whole number greater than $15\}$
(f) $\{x|x$ is a natural number less than $10\}$
(g) $\{0, 3, 6, 9, 12, \ldots\}$
(h) $\{0\}$

8. What is wrong with the following?

(a) {Glenda, Joyce, Kathy} + {Linda, June} = 5
(b) $4 \cup 7 = 11$
(c) n\{Robert, Ed, Bill\} + n\{Tom, Bill\} = n\{Robert, Ed, Bill, Tom\}
(d) $n(A) + n(B) = n(A + B)$
(e) $n(A \cup B) = n(A) \cup n(B)$
(f) $6 + \varnothing = 6$
(g) $5 \cup \varnothing = 5$
(h) $A \cup 0 = A$
(i) $a + \varnothing = a$

9. Often, tables are made for operations. These tables make verification of the properties of the operation easier. For example, consider the addition table for whole numbers where the numbers on the left are added to the numbers on top to obtain answers inside. In the sum $a + b$, let the first number a represent the number on the left and the second number b, the number on top.

+	0	1	2	3	4	5	6	⋯
0	0	1	2	3	4	5	6	⋯
1	1	2	3	4	5	6	7	⋯
2	2	3	4	5	6	7	8	⋯
3	3	4	5	6	7	8	9	⋯
4	4	5	6	7	8	9	10	⋯
5	5	6	7	8	9	10	11	⋯
6	6	7	8	9	10	11	12	⋯
⋮	⋮	⋮	⋮	⋮	⋮	⋮	⋮	

(a) Look for a pattern relative to the diagonal from upper left to lower right.
(b) Is $3 + 4 = 4 + 3$?
(c) Is there an additive identity in the table? What pattern shows this?
(d) Consider the diagonal from upper left to lower right. The fact that numbers at equal distances from the diagonal are equal is a result of which property?
(e) Since every whole number on the left added to every whole number at the top produces a whole number inside, what property of the addition of whole numbers is demonstrated?

T 10. Let us invent a new binary operation on the whole numbers. The new sum is defined to be the first number added to twice the second.

(a) Is this operation commutative?
(b) Is it associative?
(c) Is there an additive identity?

11. Consider those sets in Exercise 7 that have closure with respect to addition. Determine whether the associative and commutative properties of addition hold and whether they have an additive identity.

12. Sophomore Sandy had a difficult time understanding this section on addition. After solving two exercises, she looked in the back of the book only to discover that she had the wrong answers! Study Sandy's proof below and find her mistakes.

$(6+7)+(0+1)=(6+7)+0$	Additive identity
$(6+7)+0=6+7$	Commutative property
$6+7=13$	Addition
Therefore,	
$(6+7)+(0+1)=13$	Transitive property of equalities

13. List the elements in the solution set for each of the following, using the set of whole numbers as the domain.

(a) $x + 4 = 9$ (b) $6 + x = 8$
(c) $8 = 5 + x$ (d) $x + 6 = 8$
(e) $x + 5 = x + 4$ (f) $3 + x = x + 3$

14. We often add a column of numbers by adding down.

$$\begin{array}{r} 6 \\ 3 \\ 4 \\ 2 \\ \hline \end{array}$$

If we then add up (as a check), what property or properties of addition are we demonstrating?

Set up an equation to represent each of the next two exercises. Then solve the problem.

15. John hiked 6 km more than Xan. Together they hiked 22 km. How far did each hike?

16. If Les had 4 more tickets to the football game, he would have 9 tickets. How many tickets does he have?

C 17. Working from left to right, verify the following addition facts. State the reason for each step you use.

(a) $(2 + 4) + 3 = (3 + 2) + 4$
(b) $(2 + 0) + 1 = 1 + 2$
(c) $8 + (5 + 2) = 2 + (8 + 5)$
(d) $6 + (9 + 1) = 9 + (6 + 1)$
(e) $(a + b) + (c + d) = (b + d) + (a + c)$
(f) $(a + d) + (b + c) = (b + d) + (a + c)$

18. Verify that each of the following is equal to $a + (b + c)$.

(a) $a + (c + b)$ (b) $(a + b) + c$
(c) $(b + c) + a$ (d) $(a + c) + b$
(e) $c + (a + b)$ (f) $(c + b) + a$

19. The given table is similar to the ordinary addition table in Exercise 9. This table defines a binary operation $\oplus$ for a set $S = \{1, 2, 3\}$. Study the patterns of the table and answer the following.

$\oplus$	1	2	3
1	2	3	1
2	3	1	2
3	1	2	3

(a) Is S closed under the operation $\oplus$?

(b) Is the operation ⊕ commutative?
(c) Is the operation associative? (Test several examples.)
(d) Is there an identity element for ⊕ in S? If so, name it.

20. Using the fact that $n[(A \cup B) \cup C] = n[A \cup (B \cup C)]$, verify the associative property of addition for whole numbers.

21. Demonstrate the associative and commutative properties of addition by separating the set $\{l, m, n, o, p, q, r, s, t, u\}$ into disjoint sets, the union of which is the given set, and by adding the cardinal numbers of each disjoint set.
(a) Commutative (b) Associative

5 Whole number multiplication

■ The ten-person cheerleading squad at Hard College is authorized to purchase three sets each of matching uniforms. The head cheerleader reasoned that they would need to purchase

$$3 \cdot 4 \text{ (male)} + 3 \cdot 6 \text{ (female) uniforms}$$

The faculty advisor submitted a purchase order for $3 \cdot 10$ uniforms. Who was correct? ■

Since $12 + 18 = 30$ and $3 \cdot 10 = 30$, both are obviously correct. When we prove that this happens in general, we have the distributive property of multiplication over addition. However, we must first define multiplication. This we will do in terms of repeated addition. An alternate definition will be given in terms of Cartesian product.

Do you remember the ordeal of memorizing all your multiplication tables? Would you have been appropriately grateful for the commutative property of multiplication that reduced your work by nearly half? This and other properties of whole number multiplication will be discussed in this section. The associative property of multiplication can be used to help you to do mental arithmetic. The distributive property of multiplication is the tool that underlies many algebraic manipulations. These properties provide the basis on which are built our algorithms for arithmetic on large numbers in the next chapter.

Now that the properties of the addition of whole numbers have been defined and developed, we focus on **multiplication,** the operation of finding the product of two numbers. The first approach to multiplication of whole numbers is to consider multiplication as a shorthand for addition. That is,

$$2 + 2 + 2 = 3 \cdot 2$$

By combining three sets of two elements each (as in Figure 18), we have a pictorial representation of $3 \cdot 2$.

A second concept of multiplication can be illustrated with a display (Figure 19).

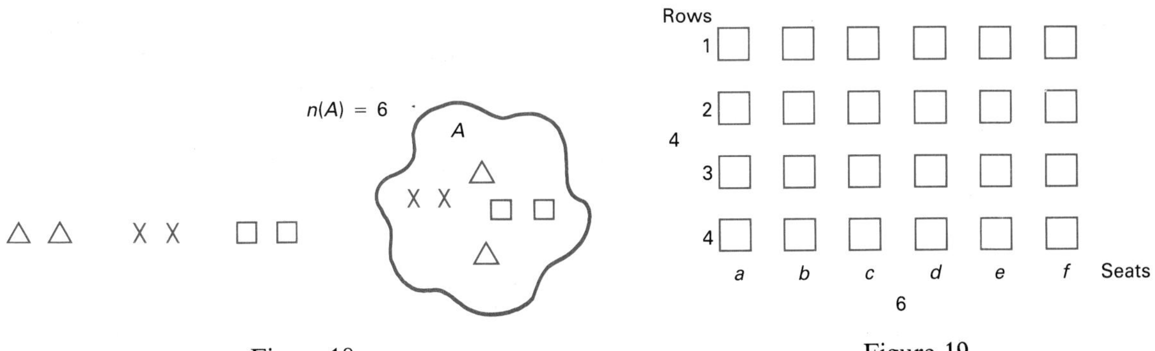

Figure 18

Figure 19

Example 41 • If a classroom has four rows of desks with six desks in each row, how many desks are in the room?

Solution Designate the desks in each row *a*, *b*, *c*, *d*, *e*, *f* and number the rows 1, 2, 3, 4.

There are obviously four rows, each with six desks. Thus the number of desks in the room (using addition) is

$$4 \cdot 6 = 6 + 6 + 6 + 6 = 24$$

There are $4 \cdot 6 = 24$ desks in the room. •

These examples illustrate the following interpretation of multiplication.

Multiplication as repeated addition

> If a is a natural number and b is a whole number, then $ab = b + b + b + \cdots + b$, where b occurs a times.

Example 42 •

$$5 \cdot 4 = 4 + 4 + 4 + 4 + 4 = 20$$

$$2 \cdot 5 = 5 + 5 = 10$$

$$7 \cdot 0 = 0 + 0 + 0 + 0 + 0 + 0 + 0 = 0$$ •

Example 43 • Demonstrate multiplication on a number line as repeated addition. (See Figure 20.)

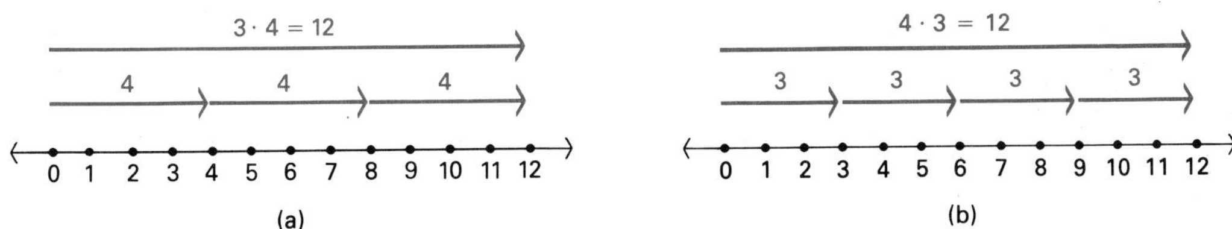

Figure 20 •

An alternate definition of multiplication can be given in terms of Cartesian product. Suppose a football stadium has six lettered exits and four numbered gates.

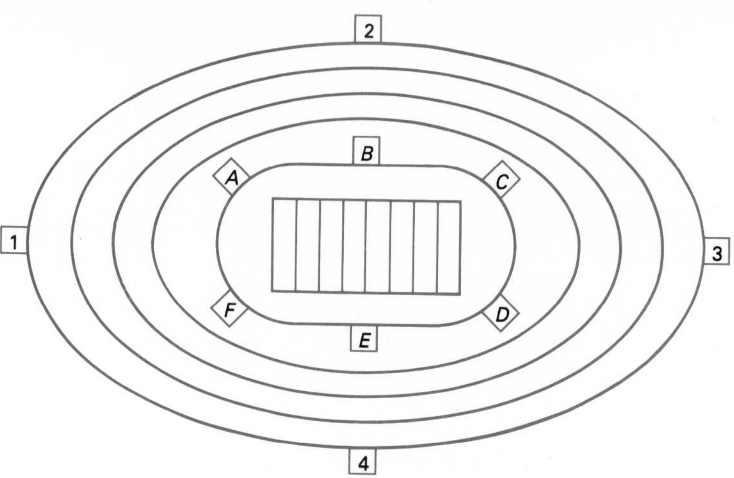

Figure 21

(See Figure 21.) If you are inside the park, in how many different ways can you leave through a lettered exit and a numbered gate?

Suppose you list the possibilities as

$$\{(A, 1), (A, 2), (A, 3), (A, 4),\\ (B, 1), (B, 2), (B, 3), (B, 4),\\ (C, 1), (C, 2), (C, 3), (C, 4),\\ (D, 1), (D, 2), (D, 3), (D, 4),\\ (E, 1), (E, 2), (E, 3), (E, 4),\\ (F, 1), (F, 2), (F, 3), (F, 4)\}$$

Does the preceding tabulation look familiar? It is the Cartesian product

$$\{A, B, C, D, E, F\} \times \{1, 2, 3, 4\}$$

The number of ways you can leave the stadium is the cardinal number of the Cartesian product of the set of lettered exits and the set of numbered gates:

$$n(\{A, B, C, D, E, F\} \times \{1, 2, 3, 4\})$$

Example 44 • Set $A = \{a, b, c\}$ has the cardinal number 3 and set $B = \{k, l\}$ has the cardinal number 2. The cardinal number of the set

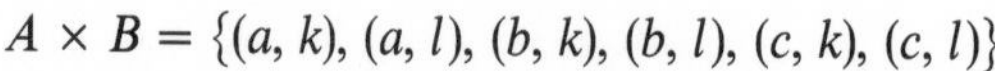

$$A \times B = \{(a, k), (a, l), (b, k), (b, l), (c, k), (c, l)\}$$

is the number of squares in Figure 22 representing the elements of $A \times B$ or 6. Note that $3 \cdot 2 = n(A) \cdot n(B) = n(A \times B) = 6$. •

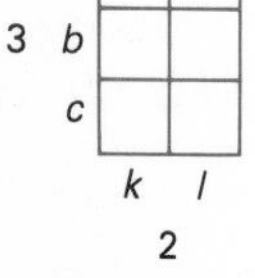

Figure 22

This discussion leads to the following definition.

Multiplication of whole numbers

> Let a and b represent any two whole numbers. Select finite sets A and B such that $n(A) = a$ and $n(B) = b$. Then
>
> $$\mathbf{a \cdot b = n(A \times B)}$$

This definition states that, to multiply two numbers a and b, we count the number of terms in the Cartesian product $A \times B$. In an expression indicating multiplication, such as $4 \cdot 3 = 12$, the 3 and 4 are called **factors** and 12 is the **product.** (4 is sometimes called the **multiplier** and 3 the **multiplicand.**) For example,

$$\underbrace{a \cdot b}_{\text{factors}} = \underset{\text{product}}{c}$$

We also write $a \cdot b$ as ab, as $(a)(b)$, as $a(b)$, as $(a)b$, and sometimes as $a \times b$. However, the notation $a \times b$ will not be used in this book because it is easily confused with the Cartesian product.

If a and b are whole numbers, then A and B are finite sets, and it is intuitively evident that $A \times B$ exists and is a finite set. Thus, from the definition $ab = n(A \times B)$, ab exists and is a unique whole number.

Closure property for multiplication

The *closure property for multiplication* of whole numbers states that for any two whole numbers, a and b, there exists a unique whole number $a \cdot b$.

Example 45 • If 3 students each check out 4 books from the library, then the number of books borrowed is 12. If 4 students each take 3 books, the total number of books borrowed is still 12. This situation is represented by the arrays in Figure 23(a) and (b). Also, in Figure 20, note that $3 \cdot 4$ in (a) produces the same answer as $4 \cdot 3$ in (b).

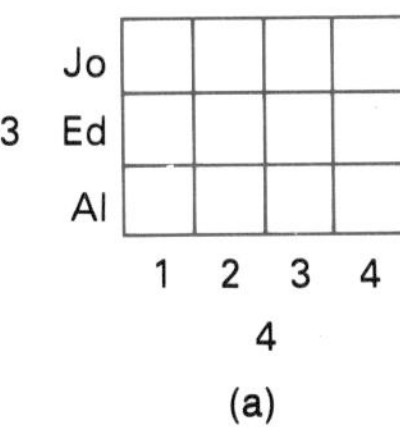

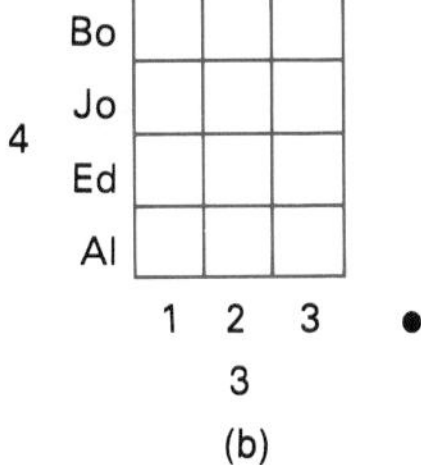

Figure 23 •

Sets of blocks (called *flats*) are also effective for demonstrating properties of multiplication. These blocks may be arranged in either of the ways shown in Figure 24.

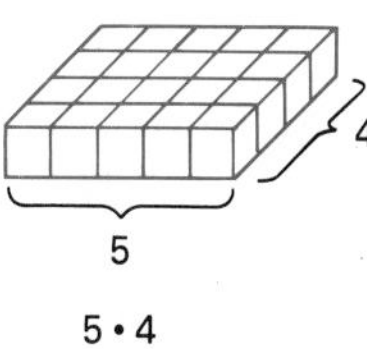

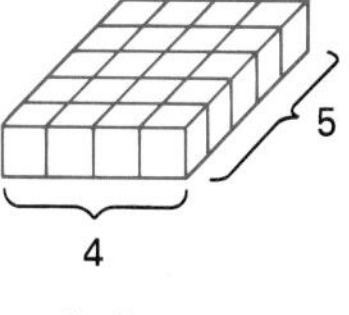

Figure 24

$5 \cdot 4 = 4 \cdot 5$

The preceding examples illustrate the *commutative property of multiplication.*

Commutative property of multiplication

For all whole numbers a and b,

$$a \cdot b = b \cdot a$$

To prove this property, choose sets A and B so that $n(A) = a$ and $n(B) = b$. Now $A \times B$ is equivalent to $B \times A$, so

$$n(A \times B) = n(B \times A)$$

or

$$a \cdot b = n(A \times B) = n(B \times A) = b \cdot a$$

In multiplying $2 \cdot 3 \cdot 5$, note that the same answer is obtained by different groupings.

$$(2 \cdot 3) \cdot 5 = 6 \cdot 5 = 30$$

$$2 \cdot (3 \cdot 5) = 2 \cdot 15 = 30$$

By the transitive property of equality,

$$(2 \cdot 3) \cdot 5 = 2 \cdot (3 \cdot 5)$$

This example illustrates the *associative property of multiplication of whole numbers.*

Associative property of multiplication

For all whole numbers a, b, and c,

$$(ab)c = a(bc)$$

Visually the associative property of multiplication is illustrated in Figure 25.

Example 46 • Use the properties of multiplication to show that $b \cdot (4 \cdot 3) = 3 \cdot (4 \cdot b)$.

Solution

$b \cdot (4 \cdot 3) = (4 \cdot 3) \cdot b$	Commutative property of multiplication
$(4 \cdot 3) \cdot b = (3 \cdot 4) \cdot b$	Commutative property of multiplication
$(3 \cdot 4) \cdot b = 3 \cdot (4 \cdot b)$	Associative property of multiplication

Therefore,

$b \cdot (4 \cdot 3) = 3 \cdot (4 \cdot b)$ Transitive property of equalities •

The associative property can be used to give meaning to the product of any number of factors. Consider the multiplication of four factors, such as $abcd$. Using a generalization or repetition of the associative property, this product can be regarded as $(ab) \cdot (cd)$, or $a[b(cd)]$, or $[a(bc)]d$, all of which give the same answer.

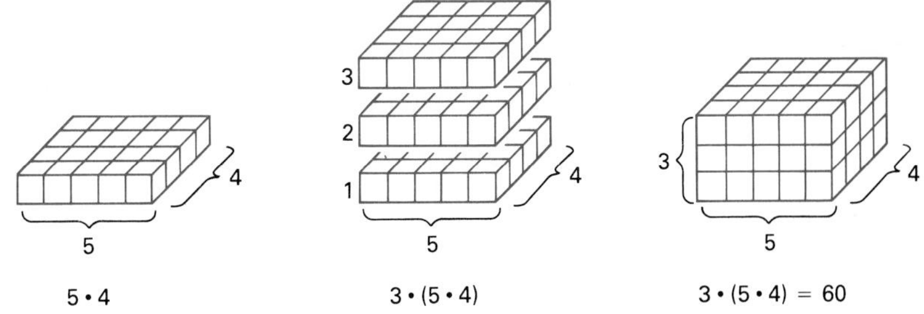

Figure 25

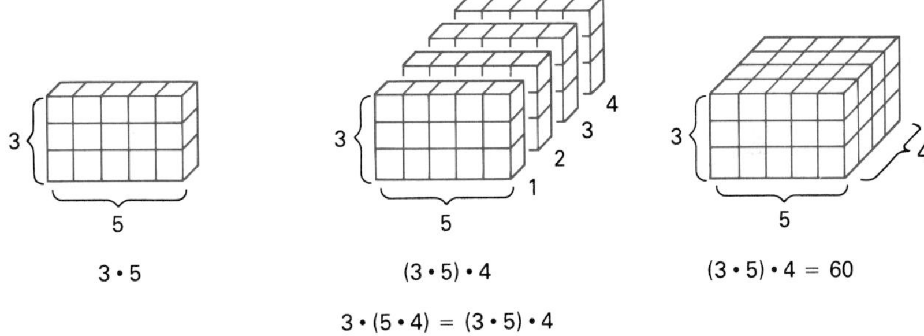

3 • (5 • 4) = (3 • 5) • 4

Example 47 • Compute $3 \cdot 5 \cdot 2 \cdot 4$ in several different ways.

Solution

$$(3 \cdot 5) \cdot (2 \cdot 4) = 15 \cdot 8 = 120$$

$$3[5(2 \cdot 4)] = 3[5 \cdot 8] = 3 \cdot 40 = 120$$

$$[3(5 \cdot 2)] \cdot 4 = [3 \cdot 10] \cdot 4 = 30 \cdot 4 = 120 \quad \bullet$$

Example 48 • Consider the multiplication $3 \cdot 1$. If $Q = \{a, b, c\}$ and $R = \{a\}$, then $n(Q) = 3$ and $n(R) = 1$.

$$Q \times R = \{(a, a), (b, a), (c, a)\}$$

By counting, we see that the number of elements in $Q \times R$ is the same as the number of elements in Q. Therefore, $n(Q \times R) = n(Q) = 3$. $R \times Q = \{(a, a), (a, b), (a, c)\}$, and $n(R \times Q) = n(Q) = 3$. So

$$3 \cdot 1 = 1 \cdot 3 = 3$$

The preceding discussion illustrates the following.

Multiplicative identity for whole numbers

> 1 is the multiplicative identity since, for any whole number a,
>
> $$1 \cdot a = a \quad \text{and} \quad a \cdot 1 = a$$

What is the product $a \cdot b$ if $b = 0$? For example, consider $3 \cdot 0$.

$$3 \cdot 0 = 0 + 0 + 0 = 0$$

In general, we would expect $a \cdot 0$ to be 0.

The preceding example and the commutative property of multiplication illustrate the following **multiplicative property of zero.**

Multiplicative property of zero

> If a is any whole number, then
>
> $$0 \cdot a = 0 \quad \text{and} \quad a \cdot 0 = 0$$

So far, the problems in this chapter have involved either addition or multiplication, but not both. Now we combine these two operations in a single expression, such as $2(6 + 4)$. The parentheses indicate that the addition of 6 and 4 should be performed first; then the result should be multiplied by 2 to obtain $2 \cdot 10 = 20$. However,

$$(2 \cdot 6) + (2 \cdot 4) = 12 + 8 = 20$$

so

$$2(6 + 4) = (2 \cdot 6) + (2 \cdot 4)$$

To visualize this property, look at the 3-by-8 array in Figure 26(a). This array shows that $3 \cdot 8 = 24$. Since $8 = 6 + 2$, then

$$3 \cdot 8 = (3 \cdot 6) + (3 \cdot 2) = 18 + 6 = 24$$

as shown by Figure 26(b).

Figure 26

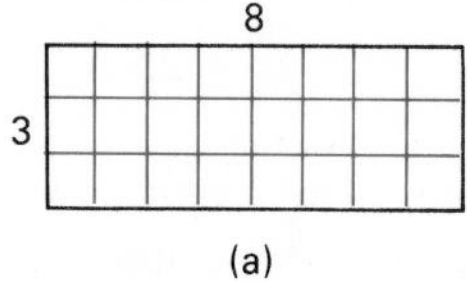

(a)

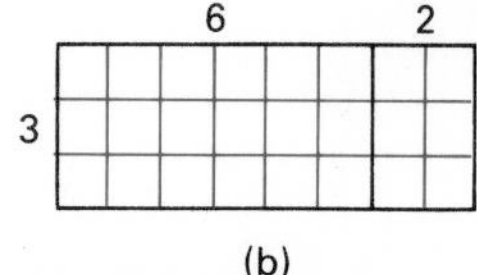

(b)

These examples illustrate another important property of whole numbers, called the *distributive property of multiplication over addition.*

Distributive property of multiplication over addition

> For all whole numbers a, b, and c,
>
> $$a(b + c) = ab + ac \quad \text{and} \quad (b + c)a = ba + ca$$

Example 49 •

$$(5 + 4)(8) = 9 \cdot 8 = 72$$

$$(5 + 4)(8) = (5 \cdot 8) + (4 \cdot 8) = 40 + 32 = 72$$ •

The distributive property is very important in our study of place value in the next chapter.

Example 50 •

$$\begin{aligned} 30 + 40 &= 3 \cdot 10 + 4 \cdot 10 \\ &= (3 + 4) \cdot 10 \\ &= 7 \cdot 10 \\ &= 70 \end{aligned}$$ •

Just for fun

The equation below needs some operation signs inserted to make it a true statement. Can you make the sides balance?

$$6 \quad 6 \quad 6 \quad 6 = 5$$

Example 51 • The distributive property is useful in the process of multiplying mentally. For instance, (34)(21) could be solved without pencil and paper if considered as 34(20 + 1). The distributive property gives

$$(34)(21) = (34)(20) + (34)(1) = 680 + 34 = 714 \quad \bullet$$

Example 52 •

$$\begin{aligned} 63 \cdot 4 &= (60 + 3) \cdot 4 \\ &= 60 \cdot 4 + 3 \cdot 4 \\ &= 240 + 12 \\ &= 252 \quad \bullet \end{aligned}$$

The distributive property of multiplication over addition is fundamental to many manipulations that are an important part of algebra.

Example 53 •

$$3(x + 2y) = 3x + 6y$$

$$(2pq + 3q) = (2p + 3)q$$

$$a[b + c + d] = a[(b + c) + d] = a(b + c) + ad = ab + ac + ad \quad \bullet$$

Sometimes additions and multiplications are contained in the same problem without parentheses to indicate the order of performing operations. In such problems, perform the multiplication and then the addition.

Example 54 •

$$\begin{aligned} 3 + 5 \cdot 4 &= 3 + 20 \\ &= 23 \quad \bullet \end{aligned}$$

We conclude this section with the following important reminder. The properties of this and the preceding section are the rules of arithmetic, and, except by chance, an answer is wrong if these properties are not followed. They are not something merely to memorize and illustrate but to follow as you do computations. As we move from the whole numbers to more inclusive number systems, we shall be careful to develop the number systems to be consistent with these properties.

Exercise set 5

R 1. Classify each of the following statements either as true or as false.

(a) $8 \cdot 0 = 8$ (b) $(16 + 1) + 0 = 17$
(c) $7 + 0 = 0$ (d) $(ab)(0) = 0$
(e) $(16 + 1)0 = 17$ (f) $(1 + 1)1 = 1$

2. Determine what property of multiplication is illustrated by filling in the frames.

(a) $\square \cdot \bigcirc = \bigcirc \cdot \square$
(b) $1 \cdot \square = \square$
(c) $0 \cdot \square = 0$
(d) $\bigcirc \cdot (\square + \triangle) = \bigcirc \cdot \square + \bigcirc \cdot \triangle$
(e) $\bigcirc \cdot (\square \cdot \triangle) = (\bigcirc \cdot \square) \cdot \triangle$

3. Using the associative and commutative properties of multiplication, complete the following statements as illustrated in the example.

EXAMPLE. $16 \cdot 3 = (8 \cdot 2) \cdot 3 = 8 \cdot (2 \cdot 3) = 8 \cdot 6$

(a) $18 \cdot 5 = 9 \cdot$ _____ (b) $24 \cdot 9 = 8 \cdot$ _____
(c) $8 \cdot 36 =$ _____ $\cdot 9$ (d) $35 \cdot 14 =$ _____ $\cdot 70$

4. Rename the following numbers using the distributive property.

(a) $2(3 + 4)$ (b) $5a + 3a$
(c) $(4 \cdot 2) + (4 \cdot 3)$ (d) $ax + a$
(e) $4xay + 2xz$ (f) $ax + bx + cx$
(g) $ax + 5x$ (h) $a(a + 3) + 2(a + 3)$
(i) $2(x + 4) + 3(x + 4)$

5. Name the property that is illustrated.

(a) $(6 + 5) \cdot (4 + 8) = (5 + 6) \cdot (8 + 4)$
(b) $(5 \cdot 4) \cdot 7 = 5 \cdot (4 \cdot 7)$
(c) $(6 + 5) \cdot (4 + 8) = (4 + 8) \cdot (6 + 5)$
(d) $(5 \cdot 4) \cdot 7 = (4 \cdot 5) \cdot 7$
(e) $4 + 3 = (4 + 3) + 0$
(f) $(7 \cdot 3) \cdot 2 = 7 \cdot (3 \cdot 2)$
(g) $1 \cdot (4 + 2) = (2 + 4)$
(h) $(a)(c) + b = (c)(a) + b$
(i) $c(a + b) = ca + cb$
(j) $(a + c)d + (a + c)e = (a + c)(d + e)$
(k) $5 + (2 + 7) = (2 + 7) + 5$
(l) $6(5 \cdot 4) = (5 \cdot 4) \cdot 6$
(m) $6(5 \cdot 8) = 6(8 \cdot 5)$

6. Group 2, 3, 5, and 6 for multiplication in at least three different ways; for example, one way is $2 \cdot [3(5 \cdot 6)]$. Perform the multiplications to show that all the products are the same.

7. The distributive property of multiplication over addition can be generalized as follows:

$$a(b + c + d + \cdots) = ab + ac + ad + \cdots$$

Compute in two ways:

(a) $5(3 + 7 + 5 + 6)$ (b) $11(5 + 8 + 10 + 4)$

8. Find the solution set on the domain of whole numbers.

(a) $2x + 2 = 6$ (b) $3a + 6 = 9$
(c) $5b + 4 = 4$ (d) $4b + 4 = 12$

In the Exercises 9 through 12, write an equation that serves as a model or represents the given problem. Then find the solution set to get the answer for the problem.

9. Aaron's soccer team averaged 3 goals per game. How many points were scored in 8 games?

10. The average of five numbers is 100. Four of the numbers are 1, 2, 3, and 4. Find the fifth number.

11. Smackos cost 20¢ each. Kate has 80¢ to spend. How many Smackos can she purchase?

12. John won the long jump by jumping 2 m farther than his next competitor. If John jumped 7 m, how far did his competitor jump?

13. In the accompanying multiplication table for whole numbers, numbers on the left are multiplied by numbers on top to get answers inside.

(a) Look for a pattern relative to the diagonal from upper left to lower right.
(b) Is $3 \cdot 4 = 4 \cdot 3$? What does this example demonstrate?
(c) Is there a multiplicative identity? What pattern shows this?
(d) The fact that numbers at equal distances from the diagonal from upper left to lower right are equal is a result of which property?
(e) Since every whole number on the left multiplied by every whole number on top seems to produce a whole number inside, what property of multiplication is demonstrated?
(f) How is the multiplicative property of zero indicated in the table?

(g) Discover the patterns about multiplication by 2 and by 5.

·	*0*	*1*	*2*	*3*	*4*	*5*	...
0	0	0	0	0	0	0	...
1	0	1	2	3	4	5	...
2	0	2	4	6	8	10	...
3	0	3	6	9	12	15	...
4	0	4	8	12	16	20	...
5	0	5	10	15	20	25	...
⋮	⋮	⋮	⋮	⋮	⋮	⋮	

T 14. Discuss whether the following statements are correct or incorrect.

(a) Addition is distributive over addition.

$$a + (b + c) = (a + b) + (a + c)$$

(b) Addition is distributive over multiplication.

$$a + (bc) = (a + b) \cdot (a + c)$$

(c) Multiplication is distributive over multiplication.

$$a(b \cdot c) = (ab) \cdot (ac)$$

15. Determine whether the following sets of numbers are closed with respect to multiplication. Does the set contain an identity element for multiplication?

(a) $\{1, 2, 3, 4, 5\}$
(b) $\{1\}$
(c) The even whole numbers
(d) The union of $\{0\}$ and the odd whole numbers
(e) $\{x \mid x$ is a whole number greater than $10\}$
(f) $\{x \mid x$ is a whole number less than $2\}$
(g) $\{1, 4, 7, 10, 13, 16, \ldots\}$
(h) $\{0, 1\}$
(i) $\{0\}$

16. Use the distributive property to write each indicated product as a sum.

(a) $4(x + 3)$
(b) $a(a + 7)$
(c) $x(x + a)$
(d) $4(a + 3b + 2c)$
(e) $4y(y + x + 2z)$

17. Use the distributive property to complete the following calculations.

(a) $3(12) = 3(10 + 2) =$
(b) $3(242) = 3(200 + 40 + 2) =$
(c) $3(132) =$
(d) $2(3, 243) =$

18. Use your calculator to verify that multiplication is repeated addition with the following problems:

(a) $8 \cdot 31$
(b) $7 \cdot 63$

C 19. Use the distributive property to write each indicated product as a sum.

(a) $(x + 3)(x + 2)$
(b) $(a + 4)(a + 1)$
(c) $(x + y)(x + y)$
(d) $(a + 2b)(3a + b)$

20. Use the distributive property to write each sum as a product.

EXAMPLE. $4x + 12 = 4(x + 3)$

(a) $8x + 16$
(b) $4ab + b^2 + 3b$
(c) $6x^2 + 18x$
(d) $ax^2 + bx$
(e) $3(x + 2) + x(x + 2)$
(f) $a(x + 3) + 6(x + 3)$

21. Working from left to right, give explicit reasons for each step in verifying that the following statements are true.

(a) $(2 \cdot 3)4 = (4 \cdot 2)3$
(b) $(8 + 2)7 = 7(2 + 8)$
(c) $(6 + 1)5 = (1)(5) + (5)(6)$
(d) $3(4 + 7) = (4)(3) + (3)(7)$
(e) $2(3 \cdot 1) = (1 \cdot 2)3$
(f) $(ab)c = (ca)b$

22. The given table is similar to an ordinary multiplication table. The binary operation ⊙ is defined on $S = \{1, 2, 3\}$ by the table.

(a) Is S closed under the operation ⊙?
(b) Is the operation ⊙ commutative?
(c) Is the operation associative?
(d) Is there an identity in S for ⊙? If so, name it.

⊙	*1*	*2*	*3*
1	1	2	3
2	2	1	3
3	3	3	3

23. Consider the operation given in the table and answer the questions in Exercise 22. A search for patterns should be helpful.

*	x	y	z
x	x	y	z
y	y	y	z
z	z	z	z

24. Define a new operation # such that, for any whole numbers a and b, $a \# b = 3a + 4b$. Find $a \# b$ for the following values of a and b.

(a) $a = 2, b = 0$ (b) $a = 10, b = 13$
(c) $a = 6, b = 23$ (d) $a = 2, b = 2$
(e) $a = 11, b = 1$ (f) $a = 7, b = 17$

6 Whole number subtraction, division, and order

■ "Aaron started reading on page 3. He stopped on page 9. How many pages did he read?" "Jodi wants her dad to purchase a horse. She learns that a horse needs 6 bales of hay per month. She has money to purchase 24 bales. For how long will the horse have hay from this purchase?" Such simple problems lead to operations that we classify as inverse operations. ■

Frequently, you do something and then undo it. You put on your coat, then take it off; open the door, then close it; pull down the window shade, then raise it. Such pairs of opposites are also encountered in mathematics. These opposites are called **inverses**—inverse operations, inverse elements, and inverse functions.

In this section, we shall discuss the inverse operations of addition and multiplication. You may have already guessed that we are describing the operations of *subtraction* and *division.* We shall see that these operations can also be viewed in several different ways.

To continue our emphasis on critical thinking, we will use inverse operations to solve simple algebraic sentences. Through the centuries, algebra has been one of the most important tools for achieving and recording shortcuts in thinking.

We shall first examine subtraction, the inverse of addition. Consider the sum

$$8 + 3 = 11$$

Observe that

$$11 - 3 = 8 \quad \text{and} \quad 11 - 8 = 3$$

The addition problem gives rise to two subtraction problems. In general, when b is subtracted from a, the result is the number c (called the **difference**). The difference c has the property that when b is added to c, the result is a.

Subtraction

The difference $a - b$ in the *subtraction* of whole number b from whole number a is equal to whole number c if and only if $c + b = a$.

Example 55 •

$$16 - 5 = 11 \quad \text{because} \quad 11 + 5 = 16$$
$$8 - 3 = 5 \quad \text{because} \quad 5 + 3 = 8$$
$$10 - 0 = 10 \quad \text{because} \quad 10 + 0 = 10$$
•

This difference, $a - b$, may be read as "a minus b," "a subtract b," or "a take away b." The number a is called the **minuend,** and b is called the **subtrahend.**

Next consider subtraction on a number line. Let's illustrate 8 minus 3 in two different ways in Figure 27. First, since subtraction is the inverse of addition, we shall consider the arrow diagram representing the number to be subtracted to point toward the left instead of the right. Place the arrow diagram for 3 at the end of the arrow diagram for 8, but pointing toward the left in Figure 27(a). In Figure 27(b), we ask the question, "What whole number added to 3 gives an answer of 8?" In both cases, note that $8 - 3 = 5$.

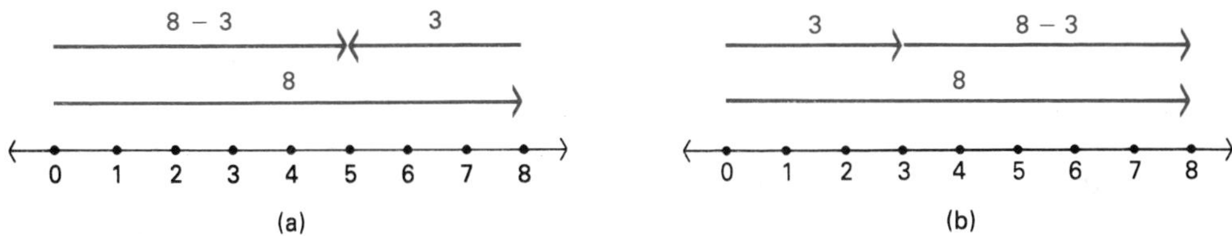

Figure 27

The definition does not imply that the whole number c exists for all pairs a and b. For instance, subtract 5 from 3 ($3 - 5 = ?$). What whole number added to 5 will give an answer of 3? There is no such number in the system of whole numbers. Thus, the set of whole numbers is *not closed* under the operation of subtraction.

Previously, we have solved such equations as $x + 3 = 8$ by the method of "guess, test, and revise." We now present the first formal procedure for solving equations. By the definition of subtraction, $8 - 3 = x$ if and only if $8 = x + 3$. Since $8 - 3 = 5$, the solution for $x + 3 = 8$ is $x = 5$.

Example 56 • Solve $x + 7 = 9$.

Solution By the definition of subtraction,

$$x = 9 - 7 \quad \text{or} \quad x = 2 \quad \bullet$$

Example 57 • Solve $n + 10 = 21$.

Solution $n = 21 - 10 = 11$ •

In Section 4, we defined addition in terms of the union of two disjoint sets. Now, let's consider subtraction in terms of taking away a subset from a given set to obtain a new set.

Example 58 • Five *A*, *B*, *C* blocks are in the baby's playpen. The baby shoves 2 blocks out of the playpen and onto the floor. How many blocks remain in the playpen?

Solution In this example of the subtraction $5 - 2$, two blocks are isolated or are taken away from the 5 blocks, leaving 3. (See Figure 28.) •

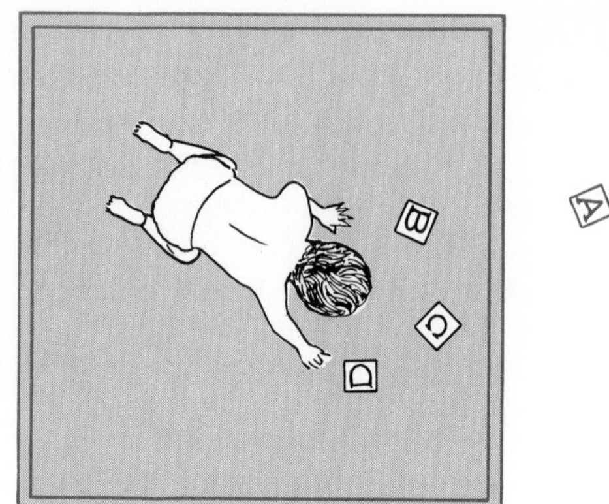

Figure 28

Multiplication also has an inverse operation: *division.* Just as subtraction "undoes" addition, division "undoes" multiplication. For instance, since $2 \cdot 8 = 16$, then 16 divided by 8 equals 2, and 16 divided by 2 equals 8. Division for whole numbers is defined in the following manner.

Division of whole numbers

> If x is a whole number and y is a natural number, then x divided by y is equal to a whole number z if and only if $z \cdot y = x$.

Symbolically, the operation of division is commonly denoted in any of four ways.

$$x \div y = z, \qquad x/y = z, \qquad \frac{x}{y} = z, \quad \text{or} \quad y\overline{)x}^{\,z}$$

In all these cases, x is called the **dividend,** y is the **divisor,** and z is the **quotient.**

Example 59 • Use a number line to illustrate $12 \div 3$.

Solution On the number line in Figure 29, we are seeking 3 arrow diagrams of equal length whose sum is 12. The diagram shows that

$$12 \div 3 = 4$$

Figure 29

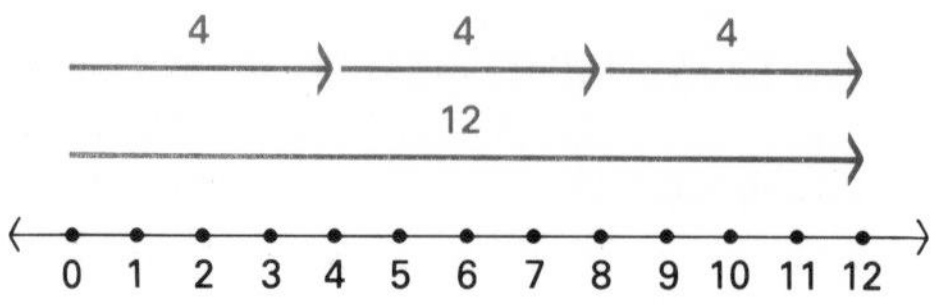

•

Example 60 • (a) $12 \div 3 = 4$ because $4 \cdot 3 = 12$
(b) $45 \div 9 = 5$ because $5 \cdot 9 = 45$
(c) $16 \div 5 = \boxed{?}$ because $\boxed{?} \cdot 5 = 16$ •

There is no whole number that when multiplied by 5 gives 16. Thus, division is not closed on the set of whole numbers. In the next chapter, we shall introduce a concept called a *remainder*, such that

$$16 \div 5 = 3 \qquad \text{with a remainder of } 1$$

Division can also be illustrated by partitioning a set into equivalent subsets. Joyce has 12 pieces of candy. She wishes to share her candy with her two friends, Ann and June. In Figure 30, the set of 12 pieces of candy has been partitioned into 3 equivalent sets, and each girl has 4 pieces of candy, or $12 \div 3 = 4$.

Figure 30

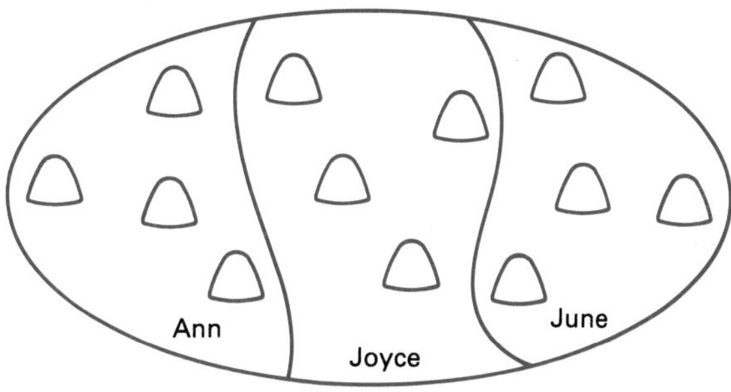

Example 61 • In a cleanup of the OZT fraternity house, 48 empty cola bottles were collected. These were placed in cases holding 12 bottles each. How many cases were needed to collect the empty bottles?

Solution The freshman pledge of OZT who was assigned the task was not proficient in division, so he computed the number of cases as follows.

$$\begin{array}{rl} 48 & \text{bottles} \\ -12 & \text{bottles in 1 case} \\ \hline 36 & \\ -12 & \text{bottles in 1 case} \\ \hline 24 & \\ -12 & \text{bottles in 1 case} \\ \hline 12 & \\ -12 & \text{bottles in 1 case} \\ \hline 0 & \end{array}$$

By counting, he noted that he had used 4 cases for the empty bottles. •

In the preceding example, we illustrated that just as multiplication can be considered as repeated addition, division can be considered as repeated subtraction.

The number 0 plays a very special role in division. Consider the problem $0 \div 9 = ?$. We need to know what number multiplied by 9 is 0. The answer is clearly 0. Thus, $0 \div 9 = 0$. We may generalize and say that 0 divided by any whole number other than zero is always 0.

Next, consider the problem $4 \div 0 = ?$. As a multiplication problem, it can be written as $4 = 0(?)$. There is no solution to this problem, since 0 multiplied by any number is 0 and not 4. Also, consider $0 \div 0 = ?$. This division can be written as

$0 = 0(?)$. Note that any number substituted for the question mark would make this statement true, and thus $0 \div 0$ has no unique answer. We avoid these situations by **excluding all division by 0. We say that division by 0 is undefined.**

Division involving zero

> If $a \neq 0$, then $0 \div a = 0$ because $a \cdot 0 = 0$. If $a \neq 0$, then $a \div 0$ is undefined because $a = 0 \cdot d$ is false for any d. $0 \div 0$ is undefined because there is no unique quotient.

Using a combination of subtraction and division, one can solve open sentences of the form $2x + 3 = 9$.

Example 62 • Find the solution set for $2x + 3 = 9$.

Solution By the definition of subtraction, $2x = 9 - 3$, or $2x = 6$. By the definition of division, $x = 6 \div 2$, or $x = 3$. •

In studying the basic operations for whole numbers, we have unconsciously been working with order relations. If a and b are different whole numbers, then there is a natural number n so that either $a + n = b$ or $b + n = a$. If we consider 2 and 6, then 2 is less than 6 because $2 + 4 = 6$; but 6 is greater than 5 because $6 = 5 + 1$.

Less Than and Greater Than

1. If a and b are any whole numbers, then ***a* is said to be less than *b*,** denoted by $a < b$, if and only if there exists a nonzero whole number n such that $a + n = b$.
2. If a and b are any whole numbers, then ***a* is said to be greater than *b*,** written $a > b$, if and only if there exists a nonzero whole number d such that $a = b + d$.

Example 63 • (a) $6 < 14$ because there exists a nonzero whole number 8 such that $6 + 8 = 14$.
(b) $10 > 4$ because there exists a nonzero whole number 6 such that $10 = 4 + 6$. •

If a is greater than b, then b is less than a; and if a is less than b, then b is greater than a. Thus the expressions $x < y$ and $y > x$ express the same idea. Sometimes the equality relation is combined with these *inequalities* to give a **less than or equal to** relation. The symbol $a \leq b$ means that either $a < b$ or $a = b$.

By using a number line, we can easily discuss the *greater than* and *less than* relations. When comparing two numbers on a number line, the number on the right is greater than the one on the left, and the one on the left is less than the one on the right. Some obvious relations are $1 < 2$, $2 < 3$, $3 < 4$, and so forth, as seen in Figure 31.

Figure 31

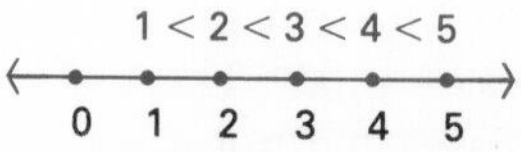

$1 < 2 < 3 < 4 < 5$

0 1 2 3 4 5

Just for fun

Can you write the numbers 0 through 9, using exactly four 4's? For instance,

$$5 = \frac{4 \cdot 4 + 4}{4}$$

Open sentences such as $x + 4 < 6$ can be solved by substituting whole numbers for x.

$$x = 0 \text{ is a solution, since } 0 + 4 < 6$$

$$x = 1 \text{ is a solution, since } 1 + 4 < 6$$

$$x = 2 \text{ is not a solution, since } 2 + 4 \nless 6$$

The solution set on the domain of whole numbers is $\{0, 1\}$.

In Chapter 1, we considered statements involving "and" and "or." You probably remember that an *or* statement is false only if both parts are false, and an *and* statement is true only if both parts are true. Use this information to solve the following open sentences.

Example 64 • Find the solution set of each on the domain of whole numbers.
(a) $x + 2 = 5$ and $x + 1 = 6$
(b) $2x + 3 \leq 7$
(c) $x + 3 = 5$ or $x + 1 = 6$

Solution (a) The solution for $x + 2 = 5$ is $x = 3$; the solution for $x + 1 = 6$ is $x = 5$. There is no solution that will satisfy both. Therefore, the solution set is $\varnothing$.
(b) $2x + 3 \leq 7$ means that $2x + 3 < 7$ or $2x + 3 = 7$. The solution set is $\{0, 1, 2\}$.
(c) 2 satisfies the first equation and 5 the second equation. The solution set is $\{2, 5\}$. •

Exercise set 6

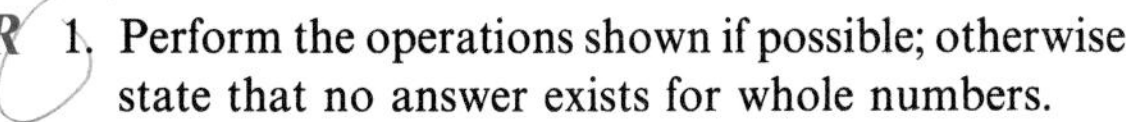

R 1. Perform the operations shown if possible; otherwise state that no answer exists for whole numbers.

(a) $6 - 4$ (b) $2 - 7$
(c) $4 - (3 + 2)$ (d) $8 + 4 - 7$
(e) $6 - (8 - 5)$ (f) $6 - (12 - 3)$

2. Illustrate on the number line the following computations.

(a) $7 - 5 = 2$ (b) $14 - 6 = 8$
(c) $5 + 7 = 12$ (d) $6 + 8 = 14$
(e) $4 \cdot 3 = 12$ (f) $2 \cdot 5 = 10$
(g) $12 - 7 = 5$ (h) $10 - 4 = 6$

3. Using the definition of less than, classify the inequalities as either true or false.

(a) $6 < 9$ (b) $8 < 8$
(c) $1 < 1$ (d) $5 < 7$
(e) $7 + 3 \geq 10$ (f) $8 + 9 > 13$
(g) $3 + 3 \leq 6$ (h) $4(6 + 2) \leq 27$

4. For each of the following statements, use the definition of subtraction to write two other statements that can be derived from the given statement.

(a) $5 + 16 = 21$ (b) $(7 + 4) - 3 = 8$
(c) $a + c = f$ (d) $2(8 + 5) - 4 = 22$

5. Use a counterexample to prove that for the operation of subtraction on the whole numbers (a) closure does not hold; (b) commutativity does not hold; (c) associativity does not hold.

6. Compute where possible.

(a) $72 \div 9$ (b) $14 \div 0$
(c) $0 \div 16$ (d) $3(5 + 1) \div 2$
(e) $8(3) \div 4$ (f) $(3 \cdot 2) \div 6$

7. Consider the following subtraction. By counting the number of times 8 must be subtracted from 24 to obtain 0, determine the quotient $24 \div 8$.

$$\begin{array}{r} 24 \\ -\ 8 \\ \hline 16 \\ -\ 8 \\ \hline 8 \\ -\ 8 \\ \hline 0 \end{array}$$

8. Use the technique of Exercise 7 to divide (a) $36 \div 9$. (b) $126 \div 21$.

9. The distributive property of multiplication over subtraction is stated as

$$a(b - c) = ab - ac$$

Illustrate this property by working the following problems in two ways.

(a) $6(8 - 5)$ (b) $8(5 - 3)$ (c) $17(10 - 4)$

10. Label as either true or false.

(a) $(48 \div 12) \div 2 = 48 \div (12 \div 2)$
(b) $(12 + 6) \div 3 = (12 \div 3) + (6 \div 3)$
(c) $18 \div (3 + 3) = (18 \div 3) + (18 \div 3)$
(d) $3 + (9 \cdot 4) = 3 \cdot (9 + 4)$

11. Is the relation *less than* an equivalence relation? Illustrate each requirement with selected whole numbers.

12. Find the solution set for each of the following sentences, with the whole numbers as the domain.

(a) $x + 3 < 7$ (b) $2x < 6$
(c) $3x < 9$ (d) $5x < 11$

13. Each of the following statements is false, in general. Find an example in each case that illustrates this fact. Then find a special example for each statement for which the statement is true.

(a) $x - y = y - x$
(b) $(x - y) - z = x - (y - z)$
(c) $x - 0 = 0 - x = x$
(d) $(x + y) \div z = x + y \div z$

For Exercises 14 through 17, set up an equation representing each problem. Then solve the equation.

14. Marlyn is 5 years older than Jean, and the sum of their ages is 27. Find Jean's age.

T 15. Leroy bought \$4 football tickets from his savings of \$11. How many did he buy and how much of his savings remain? Solve as an inequality and find the largest solution.

16. The width of a rectangle is 2 cm less than its length. If the perimeter (distance around) is 32 cm, find the length of the rectangle.

17. Sarah picked 60 apples each from 6 trees and packed them in a basket holding 8 apples each. How many baskets did she fill?

18. Illustrate each of these properties with an example.

(a) Distributive property of division over addition for the set of whole numbers
(b) Distributive property of division over subtraction for the set of whole numbers

19. (a) For what whole number values of a and b does $a \div b = b \div a$?
(b) What is the value of $24 \div (4 \div 2)$?
(c) What is the value of $(24 \div 4) \div 2$?
(d) What can you conclude from parts (b) and (c) about the associativity of division for whole numbers?

20. What conditions must we place on the values of a, b, and c so that the following are defined as whole numbers?

(a) $a - b$
(b) $(a - b) - c$
(c) $(a - c) - b$
(d) When $(a - b) - c$ and $(a - c) - b$ are both defined, are they necessarily equal?

21. Find the solution set on the domain of whole numbers.

(a) $2x + 1 = 7$	(b) $9 = 3x + 3$
(c) $2x + 1 = 3$	(d) $6x - 1 < 24$
(e) $3x + 1 < 5$	(f) $14 > 2x + 9$

22. Perform the following operations on a calculator.
(a) $420 \div 3 - 50 \cdot 2$
(b) $210 \cdot 2 \div 3 - 4$
(c) $410 - 3 \cdot 7 - 4$
(d) $(21 - 6)(17 - 4)$

C 23. Find the truth set on the domain of whole numbers.
(a) $x > 5$ or $x < 7$
(b) $x \geq 5$ and $x \leq 7$
(c) $x > 7$ and $x > 5$
(d) $x > 0$ or $x > 5$
(e) $x + 4 = 8$ or $x + 2 = 4$
(f) $x + 4 = 8$ and $x + 2 = 4$

24. If $a < b$, use the definition of $<$ to show that $a + c < b + c$.

Solution to introductory problem

In a certain apartment complex live 20 men: 8 are married and own automobiles; 12 own automobiles and have television sets; and 11 are married and have television sets. What is the largest possible number of married men?

Understand the problem. Let M represent married men; A, those who own automobiles; and T, those who own television sets. We are given

$$n(M \cap A) = 8$$

$$n(M \cap T) = 11$$

$$n(A \cap T) = 12$$

$$n(U) = 20$$

We wish to find the largest possible value for $n(M)$.

Draw a picture. The Venn diagrams in Figure 32 assist in finding a solution.

Strategy. The strategy employed is to reason backward to a solution. $n(M)$ will be a maximum when $n(\bar{M})$ is a minimum. $n(\bar{M})$ will be a minimum when (f), (c), (d), and 12 – (a) hold the smallest values the conditions of the problem will allow. No conditions are imposed on (f), (c), and (d) so suppose they each are 0. 12 – (a) is smallest when (a) is largest, and the largest value (a) can assume is 8. One sees that in order for the total number of men to be 20

$$(e) + 8 + 3 + 4 = 20$$

$$(e) = 5$$

In this case

$$n(M) = 5 + 8 + 3 = 16$$

Therefore, the maximum number of married men is 16.

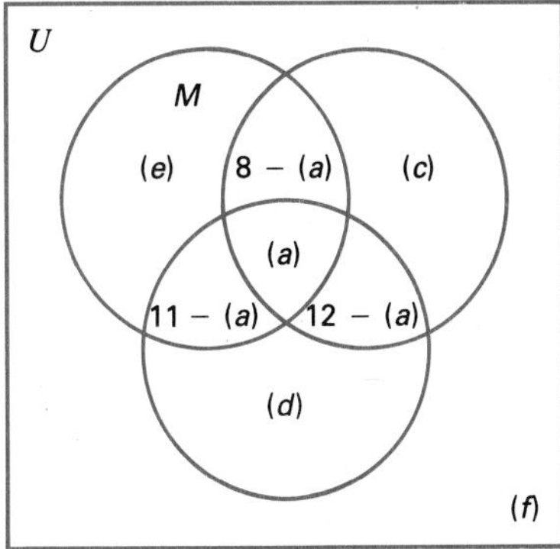

Figure 32

Summary and review

Sets

1. Notation:
   (a) $x \in A$ means x is an element of A.
   (b) $\{x \mid x$ has property $Q\}$ is read "the set of all x such that x has property Q."
   (c) $\varnothing$ or $\{\ \}$ represents the *empty set*.
   (d) $A \subseteq B$ means that A is a *subset* of B.
   (e) $A \subset B$ means that A is a *proper subset* of B ($A \subseteq B$ but $A \neq B$).
   (f) $\bar{A}$ is the *complement* of set A.
   (g) $A \cap B$ is the *intersection* of sets A and B.
   (h) $A \cup B$ is the *union* of sets A and B.
   (i) $A - B$ is the complement of B relative to A.
   (j) $A \times B$ is the *Cartesian product* of A and B.
   (k) $n(A)$ is the number of elements in A.
2. Operations on sets:
   (a) The complement of A is the set of elements in the universe that are not in A.
   (b) The union of sets A and B is the set of all elements in either A or B or both.
   (c) The intersection of sets A and B is the set of elements common to both A and B.
   (d) The difference $A - B$ is the set of all elements of A that are not in B.
   (e) The Cartesian product is the set of all ordered pairs formed by taking the first element from A and the second element from B.
3. Properties of set operations:
   (a) *Identity properties.* An identity property indicates that there is a special set so that the operation with that set and set A gives set A.
      (i) $A \cup \varnothing = A$
      (ii) $A \cap U = A$
   (b) *Commutative properties.* A commutative property indicates that the order in which the operation is performed does not affect the results.
      (i) $A \cup B = B \cup A$
      (ii) $A \cap B = B \cap A$
   (c) *Associative properties.* An associative property indicates that the manner in which objects are grouped for multiple application of an operation does not affect the results.
      (i) $A \cup (B \cup C) = (A \cup B) \cup C$
      (ii) $A \cap (B \cap C) = (A \cap B) \cap C$
   (d) *Distributive properties.* Distributive properties indicate that two operations interact according to a special pattern.
      (i) $A \cup (B \cap C) = (A \cup B) \cap (A \cup C)$ is called the *distributive property of union over intersection.*
      (ii) $A \cap (B \cup C) = (A \cap B) \cup (A \cap C)$ is called the *distributive property of intersection over union.*
      (iii) $A \times (B \cup C) = (A \times B) \cup (A \times C)$ and $(A \cup B) \times C = (A \times C) \cup (B \times C)$ are *distributive properties of Cartesian product over union.*
   (e) Observe that the operation of Cartesian product
      (i) does not have an identity set.
      (ii) is not commutative.
      (iii) is not associative.
4. If there exists a one-to-one correspondence between sets A and B, the sets are equivalent.
5. $n(A \cup B) = n(A) + n(B) - n(A \cap B)$

Numbers

1. Natural numbers (counting numbers): $\{1, 2, 3, 4, 5, \ldots\}$
2. Number usage
   (a) cardinal usage: describes the number of elements in a set
   (b) ordinal usage: describes position or order
   (c) nominal usage: names things

Relations

1. A relation from A to B is a subset of $A \times B$.
2. A relation on A is a relation from A to A. A relation R on A is
   (a) reflexive if $(a, a) \in R$ for all $a \in A$.
   (b) symmetric if $(a, b) \in R$ implies that $(b, a) \in R$.
   (c) transitive if $(a, b) \in R$ and $(b, c) \in R$ implies that $(a, c) \in R$.
3. A relation that is reflexive, symmetric, and transitive is called an *equivalence relation.* Equivalence relations partition the set A into disjoint sets called *equivalence classes.*

Whole numbers

1. The numbers $\{0, 1, 2, 3, 4, 5, \ldots\}$ are called *whole numbers.*

2. If A and B are disjoint sets with $n(A) = a$ and $n(B) = b$, then $a + b$ is defined by $n(A \cup B)$.
3. Properties of addition of whole numbers
   (a) If a and b are whole numbers, there is a unique whole number c with $a + b = c$: *Closure property of addition*
   (b) $a + b = b + a$: *Commutative property of addition*
   (c) $(a + b) + c = a + (b + c)$: *Associative property of addition*
   (d) $a + 0 = a$: 0 is the *identity for addition*
4. $a < b$ if there is a nonzero whole number n with $a + n = b$.
5. For whole numbers a and b, subtraction is defined by $a - b = c$ if there is a whole number c with $a = b + c$. Thus, subtraction is the *inverse* of addition.
6. $a \cdot b$ is defined to be b added a times.
7. Properties of multiplication of whole numbers
   (a) If a and b are whole numbers, there is a unique whole number c with $a \cdot b = c$: *Closure property of multiplication*
   (b) $a \cdot b = b \cdot a$: *Commutative property of multiplication*
   (c) $(a \cdot b) \cdot c = a \cdot (b \cdot c)$: *Associative property of multiplication*
   (d) $a \cdot (b + c) = (a \cdot b) + (a \cdot c)$: *Distributive property of multiplication over addition*
   (e) $a \cdot 1 = a$: 1 is the *identity for multiplication*
   (f) $a \cdot 0 = 0$: The *zero property of multiplication*
8. If a and b are whole numbers, division is defined by $a \div b = c$ if $b \cdot c = a$. Division is the inverse operation of multiplication.

Review exercise set 7

R 1. Select the computations that equal 0.

   (a) $5 - 5$ (b) $0 \div 4$ (c) $4 \div (3 - 3)$
   (d) $4(2 - 2)$ (e) $(5 - 5) \div 2$ (f) $(2 \cdot 3) \div 0$
   (g) $(7 - 7) \div (4 - 4)$

2. Select the true statements from the list below.
   (a) If $n + 6 = 11$, then $n \geq 5$.
   (b) If $n + 89 = 91$, then $n < 2$.
   (c) If $n + 13 = 12$, then n is not a whole number.
   (d) If $n + 16 = 14$, then $1 < n < 2$.
   (e) If $0 - 6 = n$, then n is a whole number.
   (f) If $0 \div 6 = n$, then n is a whole number.
   (g) If a is a natural number, then $(a - 0)/a = 1$.
   (h) If a is a natural number, then $(a - a)/a = 1$.
3. Write definitions for the following sets in set-builder notation.

   (a) $A \cap B$ (b) $A \cup B$ (c) $A \times B$

4. Given sets $A = \{1, 2, \ldots, 5\}$ and $B = \{4, 5, \ldots, 10\}$ find each of the following.
   (a) $A \cup B$ (b) $A \cap B$
   (c) A subset of B that can be put into a one-to-one correspondence with A
   (d) $A \cap (B \cup A)$
5. Indicate whether the following statements are true or false. If false, explain why they are false.
   (a) The transitive property holds true for the relation

   $$\{(a, b), (b, c), (c, d), (a, c), (a, d)\}.$$

   (b) If set A is reflexive, then it is also symmetric.
   (c) The set $\{1, 2, 3, \ldots, 1{,}000{,}000\}$ is an infinite set.
   (d) The commutative, associative, and identity properties hold for both union and intersection.
   (e) The commutative and associative properties hold for the Cartesian product.
   (f) If two sets are equivalent, they are equal.
   (g) If sets A and B are equal, they are equivalent.
6. Complete the following.

   (a) $A \cap \varnothing = ?$ (b) $A \cup \varnothing = ?$
   (c) $A \cap U = ?$ (d) $A \cup U = ?$
   (e) $A \cap \bar{A} = ?$ (f) $A \cup \bar{A} = ?$
   (g) $U \cap \varnothing = ?$ (h) $U \cup \varnothing = ?$

7. If $A = \{a, b, c\}$, $B = \{b, c, d, e\}$, and $C = \{c, e, g\}$, demonstrate the following properties of sets.
   (a) Commutative property of intersection (use A and B)
   (b) Commutative property of union (use B and C)
   (c) Associative property of intersection
   (d) Associative property of union

(e) Distributive property of intersection over union
(f) Distributive property of union over intersection

8. What property of operations on whole numbers is used in each of the following?

(a) $(4 + 5) + 7 = 4 + (5 + 7)$
(b) $3(5 \cdot 10 + 6) = 3(5 \cdot 10) + 3 \cdot 6$
(c) $xy + yx = xy + xy$
(d) $b = 1 \cdot b$

9. $10 - 7 = 3$ because ______.
$30 \div 5 = 6$ because ______.
$7 > 5$ because ______.
$3 < 14$ because ______.

10. (a) What sum is represented by this arrow diagram?
(b) What product is represented by this arrow diagram?

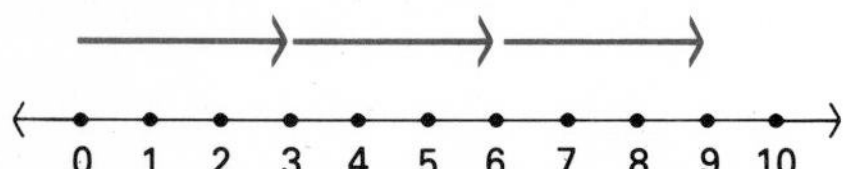

11. (a) What is the smallest whole number?
(b) What is the largest whole number?

12. Considering set $C = \{6, 7, 8, \ldots, 15\}$ and A and B as defined in Exercise 4, find each of the following.

(a) $A \cap (B \cup C)$ (b) $A \cap (B \cap C)$
(c) $C \cup (B \cup A)$ (d) $B \cup (A \cap C)$

13. If $A = \{0, 3, 6, 9\}$, classify each statement as true or false.

(a) $\varnothing \subset A$ (b) $\varnothing \in A$
(c) $\{0\} \subset A$ (d) $\{\varnothing\} \subset A$

14. Classify as reflexive, symmetric, or transitive. Determine if either is an equivalence relation.

(a) Is shorter than
(b) Has the same color hair as

T 15. Assume that $\boxdot$ and $\oplus$ represent operations defined on set S, when x, y, and z are elements of S. Describe what the following statements mean.

(a) $\boxdot$ is commutative.
(b) $\oplus$ is commutative.
(c) $\boxdot$ is associative.
(d) $\oplus$ is associative.
(e) $\boxdot$ is distributive over $\oplus$.

16. Find the solution set on the domain of whole numbers.

(a) $3x + 2 = 8$
(b) $3x + 2 < 8$
(c) $0 < x - 4$
(d) $2x - 7 = 3$
(e) $2x + 3 = 7$ or $x + 3 < 5$
(f) $2x + 3 = 7$ and $x + 3 < 5$

17. In the given Venn diagram, shade each set listed below.

(a) $(A \cap B) \cup (D \cup E)$
(b) $(E \cap C) \cap B$

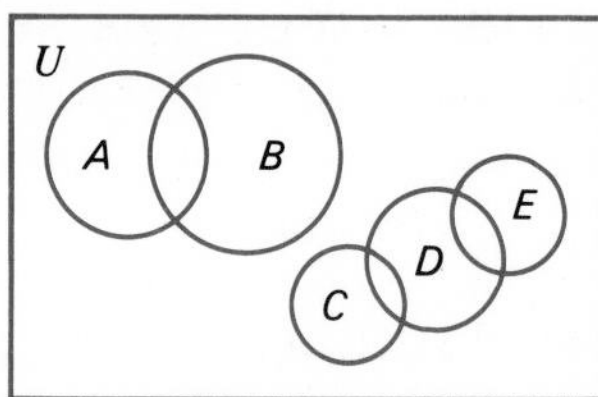

18. Using symbolic notation, describe the sets indicated by the shaded part of each diagram.

(a)

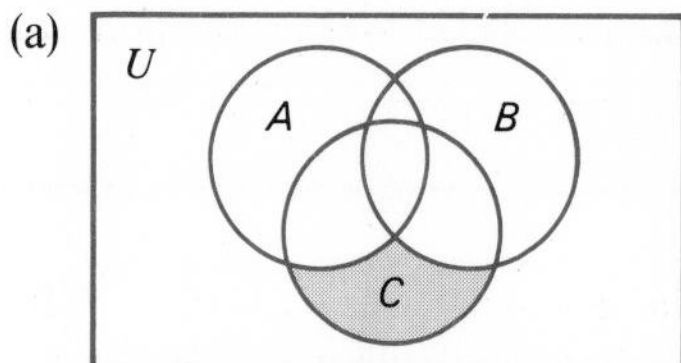

(b)

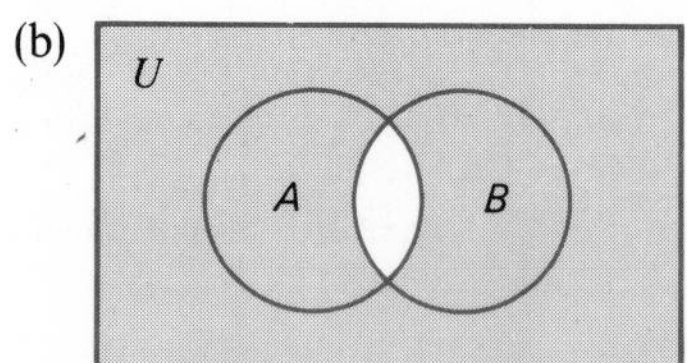

(c)

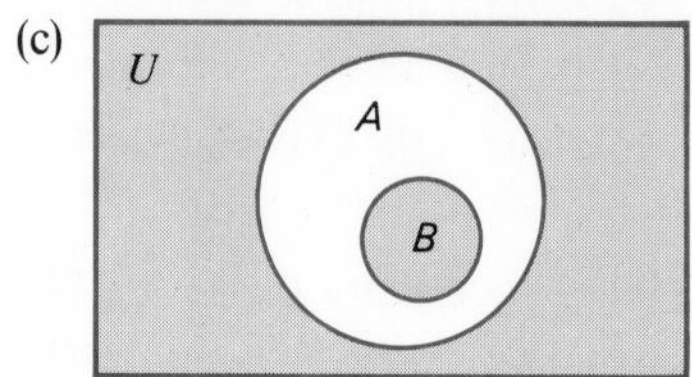

(d)

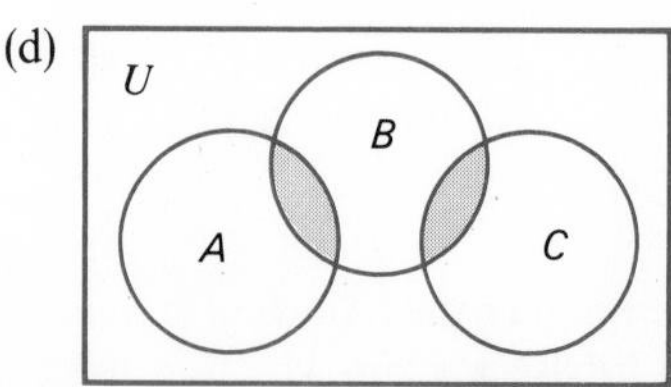

C 19. Show that the following are true. Give the reason for each step of the discussion.

(a) $(110)(9) - (9)(6) = 9(110 - 6)$
(b) $3(a + 0) = (a)(3)$
(c) $4(23 \cdot 25) = 25(23 \cdot 4)$
(d) $(108)(6) - (6)(108) = 0$
(e) $c(d \cdot e) = e(d \cdot c)$
(f) $a + (b + c) = c + (b + a)$

20. Given set $A = \{1, 2, 3, 4\}$ and set $B = \{5, 6, 7, 8\}$, tabulate the relation R on A where (x, y) is contained in R if and only if $x + y \in B$ and $x \neq y$.

21. In surveying a small group of preschool children, a psychologist determined that 14 of the children liked the television program "At Home," 17 liked the program "Back and Forth," and 11 liked the program "Catch It!" Of these children, 11 liked both *A* ("At Home") and *B* ("Back and Forth"), 6 liked both *B* and *C* ("Catch It!"), and 5 liked both *A* and *C*. Only 2 of the children liked all three programs. Determine from this information each of the following.

(a) $n(A \cup B)$
(b) $n(B \cup C)$
(c) $n(A \cup C)$
(d) $n(A \cup B \cup C)$

22. An employment agency has 50 prospective employees to place in jobs. It is known that 7 are poor employment risks, since 5 of the 7 are lazy and 4 of the 7 have poor health.

(a) How many employees are both lazy and have poor health?
(b) How many are lazy or have poor health?
(c) How many are not lazy and have poor health?
(d) How many are lazy and do not have poor health?
(e) How many are not lazy and do not have poor health?
(f) How many are not lazy or do not have poor health?

Bibliography

1. Bell, Max S. "What Does 'Everyman' Really Need from School Mathematics?" *MT* (March 1974), 196–202.
2. Feinstein, I. "Sets: Pandora's Box?" *School Science and Mathematics* (June 1973), 495–500.
3. Freeman, William, & Diana Kroll. "A Mathematics Fair in the Lower Grades." *AT* (November 1974), 624–628.
4. Gardner, Martin. *Aha.* New York: W. H. Freeman, 1978.
5. Geddes & Lipsey. "The Hazards of Sets." *MT* (October 1969), 454.
6. Murray, Peter J. "Addition Practice Through Partitioning Sets of Numbers." *AT* (October 1976), 430–431.
7. Nelson, Rebecca S. "Multiplication Games that Every Child Can Play." *AT* (October 1979), 34–35.
8. Scheuer, Donald W., Jr., & David E. Williams. "Ideas." *AT* (May 1980), 22–24.
9. Schoen, Harold L. "Some Difficulty in the Language of Sets." *AT* (March 1974), 236–237.
10. Smith, Douglas, & William R. Topp. *An Activity Approach to Elementary Concepts of Mathematics.* Reading, Mass.: Addison-Wesley Pub., 1981.
11. Tabler, M. Bernadine, & Marilyn H. Jacobson. "Ideas." *AT* (May 1981), 19–24.
12. Vilenkin, N. *Stories About Sets.* New York: Academic Press, 1969.
13. Wheatley, Grayson H., & Alan Van Duinen. "Mathematical Road Maps: A Teaching Technique." *AT* (January 1976), 18–20.

CHAPTER 3
Numeration Systems

Each letter represents a different digit in these additions. Further, both additions give the same sum. What are X, A, B, G, and H?

In the development of the language of mathematics, one of the first essential steps for any ancient civilization was the creation of a numeration system to express quantitative discoveries. The Egyptians made marks on papyri (reed paper) and the Babylonians formed shapes in clay.

Our present numeration system, the Hindu-Arabic system, had its origin about the third century B.C. The symbols for this system seemed to originate in India. The date and origin of the invention of zero is unknown; it was being used by the Hindus in India in the sixth century A.D. In the eighth century A.D., this positional system with symbol zero was translated into Arabic and found its way into Europe. The battle between those favoring the Roman system of numeration and this new system lasted more than 400 years. Even the church got in the act. Those not using Roman numerals were chastized for using "heathen" numerals.

It will be our privilege in this chapter to examine several ancient numeration systems. A short historical account of the development of numeration should help you understand our base ten numeration system. The simplicity of the base ten numeration system leads to very efficient algorithms or computational schemes. Finally, numeration systems in bases other than ten are introduced to help you to understand the base ten system better.

1 History of numeration systems

■ While demolishing an old building last year, the construction crew was surprised to find the numeral

M C M X C

on the cornerstone. Why were they surprised? ■

In this section, we will examine some ancient numeration systems, including the Roman system. We will also take a close look at our own base ten numeration system and marvel at its simplicity and usefulness. The creation of a numeration system using place value ranks as one of the great accomplishments in human history.

It is hard for us to realize the importance of the development of a simple, efficient system of representing numbers. Modern science and modern commerce would be centuries behind their present development if they had had to struggle with any of the many numeration systems that have been tried and discarded.

One of the earliest systems of numeration was the tally system, which used marks or strokes to represent numbers. For instance, |||||| could have represented a herd of six animals. Even today, tally marks are used to represent numerals like 1, 2, 3, . . . when keeping score in certain games. As an illustration, 5 and 𝍸 are symbols for the same number.

In general, early systems of numeration are characterized as additive systems, multiplicative systems, and place-value systems. Additive systems rely primarily on the addition of numbers represented by symbols; that is, a number represented by a set of symbols is the sum of the numbers represented by the individual symbols.

The **Egyptian system** is one of the earliest additive systems of which we have a record. The Egyptians used hieroglyphics, or picture symbols, along with tally marks to represent numbers (see Table 1).

Table 1 Egyptian Numerals

1	𓏺	Stroke
10	𓎆	Heelbone
100	𓍢	Scroll
1000	𓆼	Flower
10,000	𓂭	Pointed finger
100,000	𓆐	Burbot fish
1,000,000	𓁨	Astonished person

Example 1 • (a) 𓍢𓍢𓎆𓎆𓎆𓏺𓏺𓏺𓏺 $= 234$

(b) 𓎆𓏺𓏺𓏺𓏺𓏺 $= 15$

(c) 𓂭𓆼𓆼𓍢𓍢𓍢𓎆𓏺𓏺𓏺𓏺𓏺𓏺 $= 12{,}316$

(d) 𓆼𓍢𓍢𓍢𓍢𓍢𓍢𓎆𓎆𓎆𓎆𓏺𓏺𓏺𓏺 $= 1644$ •

Since the Egyptian system is an additive system, symbols could be placed in any order to represent a number.

𓎆𓎆𓎆𓎆𓏺𓏺𓏺 $= 10 + 10 + 10 + 10 + 3 = 43$

𓎆𓏺𓎆𓏺𓎆𓏺𓎆 $= 10 + 1 + 10 + 1 + 10 + 1 + 10 = 43$

There was no "place value" in this system (whereas in our system 916 is different from 619).

About 3000 B.C., the great **Babylonian** civilization flourished in what is today the Middle East. In this period, people wrote numerals on clay tablets with pieces of wood. A vertical wedge ▼ represented 1 and the symbol ❮ represented 10 (see Table 2). Symbols for numbers from 1 to 59 were formed by additive use of these symbols. For example, ❮❮▼▼▼ represented 23. The Babylonian numeration system used the idea of place value with the places representing different multiples of 60. The various places are indicated by a wider space between the characters or by inverting the order of ▼ and ❮; that is, if a ▼ appears to the left of ❮, it lies in a new place.

Table 2 Babylonian Symbols

▼	represented 1
❮	represented 10

Example 2 •

▼ ❮❮▼ or $60 + 21 = 81$

❮▼ ❮▼▼▼ or $11(60) + 13 = 673$

▼▼ ❮❮ ❮❮❮ or $2(60)(60) + 20(60) + 30 = 8430$

▼▼ ❮ or $2(60) + 10 = 130$ •

flashback

The system of numeration used today evolved gradually over the centuries, proceeding through many developmental stages as various cultures tested, modified, and described different patterns of numeration. It is not an exaggeration to say that the present system is probably one of the greatest inventions of the human mind.

Because many early efforts to represent numbers are lost in the dimness of time, only part of the development of numerals is described here. Certain interesting facts about the numeration systems of several civilizations have been discovered by archaeologists, and examining the patterns of a few of these early systems will help us appreciate our own.

Despite the clever use of place value, the numeration system was flawed by the absence of a symbol for zero. This meant that whether ▾< represented 1(60) + 10 or 1(60)(60) + 10 could only be determined by the context in which the numeral was used.

The early **Chinese-Japanese** numeration system was a base ten system similar to the system we use today. However, the symbols were different, and the numerals were written in vertical columns with the powers of 10 placed in descending order from top to bottom. Further, a symbol for a power of 10 (such as 百 for 100) was placed below each digit to signify the digit's place value. The numerals are listed in Table 3. (Modern Chinese and Japanese symbols are written horizontally.)

Table 3 Chinese-Japanese Numerals

一	二	三	四	五	六
1	2	3	4	5	6
七	八	九	十	百	千
7	8	9	10	100	1000

Example 3 •

七	7		
十	10	or	7(10) + 8 = 78 •
八	8		

Example 4 •

四	4		
千	1000		
三	3	or	4(1000) + 3(10) + 5 •
十	10		
五	5		

At the peak of its civilization, around A.D. 100, the Roman Empire needed an elaborate system of numeration for record-keeping and accounting, a situation created by the collection of taxes and by commerce in the vast empire. Essentially, the **Roman system** was an **additive** system with *subtractive* and *multiplicative* features. If symbols decreased in value from left to right, their values were to be added; however, if a symbol had a smaller value than a symbol on its right, it was to be subtracted. For example, CX = 100 + 10 = 110, but XC = 100 − 10 = 90. The position of a letter was important, since XC and CX stood for different numbers, but the Romans did not use place value as we know it today.

Roman numerals are still in use today. You recognize that the letters I, V, X, L, C, D, and M represent 1, 5, 10, 50, 100, 500, and 1000, respectively. Not more than three identical symbols are ordinarily used in succession. For example, IV is usually used for 4 instead of IIII. Likewise, there are never more than two symbols involved in the subtractive feature. The only subtractive symbols allowed are IV, IX, XL, XC, CD, and CM.

$$\text{LXXVI} = 76, \qquad \text{XLIV} = 44$$

To write large numbers using Roman numerals, one utilizes a multiplicative feature involving bars above the symbols. For example, $\overline{\text{V}}$ indicates 5 multiplied by 1000, or 5000. $\overline{\overline{\text{V}}}$ represents 5,000,000. Thus, a symbol with a bar above it indicates a number represented by the symbol multiplied by 1000; a double bar means multiplication by 1,000,000.

Example 5 • (a) $\overline{\text{IV}}\text{DCXLVII} = 4{,}647$
(b) $\overline{\overline{\text{L}}}\text{MDXXI} = 50{,}001{,}521$ •

Example 6 • While demolishing an old building last year, the construction crew was surprised to find the numeral

M C M X C

on the cornerstone. Why were they surprised?

Solution M C M X C represents 1990. Surely there was a mistake on the cornerstone. •

Another interesting numeration system comes from the early civilizations of Guatemala and Honduras. Although evidence suggests that the numeration system was developed before the advent of the Mayan Indians, most people still describe this system as the Mayan numeration system. At one time it seems to have been based on 5, but in later forms it was based on 20. Instead of figures to represent a number, dots and horizontal lines were used. Each dot was a unit and each line represented 5. The largest number that was recorded in one place with these symbols was 19, three bars and four dots: ⩸. Numbers greater than 19 were written in terms of base 20, making use of a symbol for zero that looked somewhat like a football, ⬭.

Their system used 18(20) instead of $(20)^2$, $18(20)^2$ instead of $(20)^3$, but in spite of this discrepancy, their place-value system of numeration was ingenious. The Mayans (as well as the Chinese and Japanese) wrote their numerals in vertical form.

Example 7 •

(a)

·	1(20) = 20
⬭	0 = 0
	20

(b)

··	2(20) = 40
····	4 = 4
	44

(c)

·	1(18)(20) = 360
⬭	0(20) = 0
——	5 = 5
	365

(d)

· over ——	6(18)(20) = 2160
· over ——	6(20) = 120
⬭	0 = 0
	2280

(e)

··	$2(18)(20)^2$ = 14,400
· over ——	6(18)(20) = 2,160
⬭	0(20) = 0
··· over ——	8 = 8
	16,568

•

The numerals we use today were invented in India by the Hindus and brought to Europe by Arabs. Consequently, our numeration system is called *Hindu-Arabic system*. Its important characteristics are listed below.

Characteristics of Hindu-Arabic System

1. Each symbol represents a cardinal number.
2. The position of the symbol in the numeral has a place meaning.
3. All numbers are constructed from ten basic symbols or digits.
4. There is a representation for zero.

Since the Hindu-Arabic system uses groups of ten, it is called the **decimal system.** By counting in sets of ten, we need ten symbols in our numeration system. These symbols, called *digits,* are written 0, 1, 2, 3, 4, 5, 6, 7, 8, and 9. With these ten digits, we can represent every whole number. (The word **digit** means "a finger or toe," and it is natural to assume that our number system is based on ten digits *because* we have ten fingers.)

The decimal system uses the idea of **place value** to represent the size of a group in a grouping process. The size of the group represented by a digit depends on the position of the digit in the numeral. The digit also tells how many there are of a particular group. In the numeral 234, the 2 represents two groups of 100, (200), the 3 represents three groups of 10, (30), and the 4 represents four 1's, (4). The idea of place value makes the decimal system very convenient in many ways.

Example 8 • Table 4 shows the place value of the digits in 243, 104,679, and 13,580.

Solution

Table 4

	millions	*hundred thousands*	*ten thousands*	*thousands*	*hundreds*	*tens*	*ones*
243					2	4	3
104,679		1	0	4	6	7	9
13,580			1	3	5	8	0

•

Example 9 • The digit 4 is found in each of the numerals 654, 456, and 546. Because of its place, it is multiplied by different values. In 654, the 4 represents four 1's; in 456, it represents four 100's; and in 546, it represents four 10's. •

In a base ten numeral, each successive place to the left represents a group 10 times the size of the preceding group. Beginning at the right and moving toward the left, the first place digit tells how many groups of 1; the second place indicates the number of groups of 10; the third place tells how many groups of 10 times 10; the next place gives the number of groups of 10 times 10 times 10; and so on.

Just for fun *Consider the symbols that appeared on a recently unearthed Babylonian tablet. Translate these symbols (numbers) and determine exactly what purpose the tablet served.*

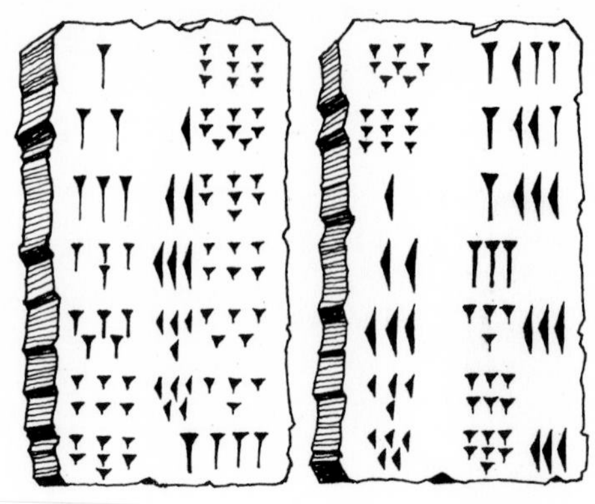

Thus, 2346 is an abbreviation for

$$2(10 \cdot 10 \cdot 10) + 3(10 \cdot 10) + 4(10) + 6(1)$$

or

$$2(1000) + 3(100) + 4(10) + 6$$

which is called the **expanded notation** for 2346.

Example 10 • Write the following numerals in expanded notation.
(a) 7642 (b) 80,002 (c) 36,200

Solution (a) 7642 = 7(1000) + 6(100) + 4(10) + 2
(b) 80,002 = 8(10,000) + 2
(c) 36,200 = 3(10,000) + 6(1000) + 2(100) •

Sometimes (b) is written as

$$80{,}002 = 8(10{,}000) + 0(1000) + 0(100) + 0(10) + 2(1)$$

and (c) as

$$36{,}200 = 3(10{,}000) + 6(1000) + 2(100) + 0(10) + 0(1)$$

The 7642, the 80,002, and the 36,200 are said to be written in **standard** base ten notation.

Exercise set 1

R 1. Write the following numerals in expanded notation.

EXAMPLE. 4321 = 4(1000) + 3(100) + 2(10) + 1

(a) 43,736 (b) 354,555 (c) 14,136
(d) 315,161 (e) 3,111,005 (f) 20,004

2. Write each in standard base ten notation.

(a) 3(10,000) + 4(1000) + 7(100) + 8(10) + 6
(b) 4(1000) + 2(10) + 9
(c) 7(100) + 4(10)
(d) 5(10,000) + 6

3. State the place value of the digit 6 in each number below.

(a) 6374 (b) 6
(c) 376 (d) 456,987,243

4. In the numeral 34,576, what place values do each of the following represent?

   (a) 4 (b) 7 (c) 3 (d) 5

5. In the numeral 12,435, which is greater,

   (a) the value represented by 2 or the value represented by 3?
   (b) the value represented by 1 or the value represented by 4?

6. Write the following as Roman numerals.

   (a) 76 (b) 101 (c) 189
   (d) 44 (e) 148 (f) 948

7. Suppose that you are given the digits, 3, 5, and 7.

   (a) What is the largest number that can be represented in base ten by these three digits?
   (b) What is the smallest number that can be represented in base ten by these three digits?

8. As a basis of comparison for the various systems, complete the table. If one system does not use a single symbol, use a combination of symbols.

Hindu-Arabic	0	1	5	10	50	100
Egyptian						
Roman						
Babylonian						
Chinese-Japanese						
Mayan						

9. Work Exercise 7 for the digits 0, 1, 2, 3, 4. (0 cannot be the first digit in the five-digit numbers.)

10. What numbers are represented by the following Roman symbols?

    (a) XXIX (b) MCMLXXVI
    (c) MDCCLXXVI (d) $\overline{\text{XI}}$XI
    (e) $\overline{\overline{\text{X}}}$DCXLIX (f) $\overline{\overline{\text{III}}}$DCXLIV

11. What numbers are represented by these Egyptian symbols?

    (a) |||| (b) ∩∩||
    (c) ∩∩∩|||| (d) 𓆼∩∩𓁨
    (e) 𓍢𓍢𓍢∩∩𓆐𓆐 (f) 𓂭𓂭𓆼𓆼𓆼

12. Write the Hindu-Arabic symbols that represent the same numbers as the Babylonian numerals shown.

    (a) ▼▼ (b) ‹‹
    (c) ‹▼ (d) ‹ ‹ ▼▼
    (e) ▼ ‹‹▼ ▼▼ (f) ‹▼ ‹‹ ‹

13. What Hindu-Arabic numerals are represented by the following Chinese-Japanese symbols?

    (a) 一百四十二 (b) 二百五十九 (c) 三千四百一十

14. What numbers are represented by the following Mayan symbols?

    (a) $\dot{—}$ (b) $\dot{=}$ (c) $\ddot{—}$ (d) $\dot{\equiv}$
    (e) $\overset{....}{—}$ (f) $\overset{...}{\equiv}$ (g) ·· / $\dot{—}$ / ··· (h) — / ⊜ / ⊜

15. Write the following numerals using Egyptian symbols.

    (a) 14 (b) 76 (c) 184 (d) 1728

16. Write the numerals in Exercise 15 using Babylonian symbols.

17. Write the numerals in Exercise 15 using Chinese-Japanese symbols.

18. Write the numerals in Exercise 15 using Mayan symbols.

T 19. Arrange the following from largest to smallest.

    (a) 𓍢𓍢𓍢∩∩𓆐𓆐 ‹ ‹ ▼ 一百四十二

    (b) 二百五十七 ∩∩∩|||| ▼ ‹‹▼

20. Write the numeral for the number that precedes and the number that follows each of the following.

    (a) MLXXXIX

(b) ‹‹‹‹‹▼▼▼▼▼▼▼▼▼

(c) ∩|∩|∩|∩|∩|∩|∩|∩|∩|

C 21. Find the sum and the difference of each pair of numbers.

(a) LXIV and XIV
(b) MCMLXXVI and MDCCLXXVI
(c) ‹▼ ‹‹▼▼ and ▼▼▼ ‹▼▼
(d) 99∩∩||| and 9∩∩∩∩||

22. Kathryn had a dream in which she was selling roses in an international market. She started out with 143 roses. Then her first customer, an Egyptian, bought ∩||||| of them. Soon afterward a Babylonian asked to buy ‹‹‹▼▼▼▼▼▼▼▼ roses, and later XIV roses were bought by a Roman. At the end of the day an Oriental asked Kathryn to sell him all the roses that remained. How many roses did he buy? Write the answer in Chinese-Japanese numerals.

23. You are a farmer in ancient times who produced a crop of 150 bushels of barley, which is 40 more bushels of barley than you produced last year and twice as many bushels of corn as you produced this year. You hope to increase production of both corn and barley next year by at least 15 bushels each.
(a) Using Egyptian symbols, record the number of bushels of barley produced this year and the number that you hope to produce next year.
(b) Using Babylonian symbols, record this year's production of barley and corn.

2 Using exponentials in addition and subtraction algorithms

■ Use your problem-solving techniques to place each of the digits 0 to 9 in one of the blocks to make an addition problem. A guess has been made about placement of two of the digits to start you on your way and to reduce the number of different possible answers.

```
    □ □ 9
 +  3 □ □
 ---------
  □ □ □ □
```
■

In Chapter 1, we reminded you that one of the most important tools in problem solving is our knowledge of a variety of common algorithms. In the next two sections, we shall discuss the very important algorithms we use daily. Whenever we perform arithmetic operations with single-digit numerals, we write the answer from memory. However, when operations involve larger numbers, it is necessary to have a pattern or procedure to follow. Such a pattern or procedure is called an **algorithm.**

The algorithms we use today are refined versions of ones that have been passed down from generation to generation. Many good patterns for performing the various operations exist. We shall study these algorithms in this section and the next, as well as in the laboratory projects in the student activities manual.

In order to discuss carefully our algorithms for arithmetic, we need to express numbers in expanded notation. As discussed in Section 1, this required cumbersome use of 10, 100, 1000, 10,000, and so on. To simplify our discussion of algorithms, we shall develop exponential notation. Problem-solving strategies will be used to obtain the necessary patterns that govern the use of this notation.

An exponential is a type of abbreviation in mathematics. For instance, $5 \cdot 5 \cdot 5 \cdot 5 \cdot 5 \cdot 5$ is written in exponential form as 5^6. The symbol a^2 means $a \cdot a$; the symbol a^3 means $a \cdot a \cdot a$. These ideas lead to the following definition.

Exponents

> If n is a counting number, then a^n is the product obtained by using a as a factor n times, or
>
> $$a^n = \underbrace{a \cdot a \cdot a \cdot a \cdot \cdots \cdot a}_{n \text{ factors}}$$
>
> and a^1 is defined to be a.

Each part of the exponential expression a^n has a name. The superscript n is called the *exponent;* the number a is called the *base;* and the complete symbol is read "the nth power of a" or "a to the nth power." A number expressed in the form a^n is said to be written in exponential form.

Example 11 • In 6^3, 6 is the base and 3 is the exponent; moreover, 6^3 is read "the third power of 6," "6 to the third power," "the cube of 6," or "6 cubed," and it means "$6 \cdot 6 \cdot 6$." •

We now use two of the strategies of problem solving to find expressions for $a^m \cdot a^n$, $(ab)^m$, and $(a^m)^n$. We shall consider simple examples and then look for a pattern. Write

$$(a) \cdot (a^2) = (a) \cdot (a \cdot a) = a \cdot a \cdot a = a^3 \qquad = a^{1+2}$$

$$(a) \cdot (a^3) = (a) \cdot (a \cdot a \cdot a) = a \cdot a \cdot a \cdot a = a^4 \qquad = a^{1+3}$$

$$(a) \cdot (a^4) = (a) \cdot (a \cdot a \cdot a \cdot a) = a \cdot a \cdot a \cdot a \cdot a = a^5 \qquad = a^{1+4}$$

Thus,

$$(a) \cdot (a^n) = (a) \cdot \underbrace{(a \cdot a \cdot a \cdot \cdots \cdot a)}_{n \text{ factors}} \qquad = a^{1+n}$$

Now

$$(a^2) \cdot (a^4) = (a \cdot a) \cdot (a \cdot a \cdot a \cdot a) = a \cdot a \cdot a \cdot a \cdot a \cdot a = a^6 \qquad = a^{2+4}$$

$$(a^3) \cdot (a^4) = (a \cdot a \cdot a) \cdot (a \cdot a \cdot a \cdot a) = a \cdot a \cdot a \cdot a \cdot a \cdot a \cdot a = a^7 = a^{3+4}$$

flashback

In 1795, Joseph Louis Lagrange, a French mathematician, said, "Arithmetic and geometry are the wings of mathematics. For, whenever we have reached a result, in order to make use of it, it is requisite that it be translated into numbers or lines." Lagrange followed this statement with a lecture on algorithms. From the time of the development of decimal notation, mathematicians have worked to improve addition and subtraction algorithms.

More generally, if m and n are natural numbers,

$$a^m \cdot a^n = \underbrace{(a \cdot a \cdot a \cdot \cdots \cdot a)}_{m \text{ factors}} \cdot \underbrace{(a \cdot a \cdot a \cdot \cdots \cdot a)}_{n \text{ factors}} = a^{m+n}$$

That is, when multiplying like bases, add the exponents.

Note that

$$(2^3)^2 = 2^3 \cdot 2^3 = 2^{3+3} = 2^6 = 2^{3 \cdot 2}$$

and

$$(a^2)^4 = a^2 \cdot a^2 \cdot a^2 \cdot a^2 = a^{2+2+2+2} = a^8 = a^{2 \cdot 4}$$

In general, $(a^m)^n = a^{m \cdot n}$. That is, when raising a power to a power, multiply exponents.

Now

$$(ab)^2 = (ab) \cdot (ab) = a \cdot a \cdot b \cdot b = a^2 \cdot b^2$$

What properties of multiplication are used?

$$(ab)^3 = (ab) \cdot (ab) \cdot (ab) = a \cdot a \cdot a \cdot b \cdot b \cdot b = a^3 \cdot b^3$$

What properties of multiplication are used? In general,

$$(ab)^m = \underbrace{ab \cdot ab \cdot ab \cdot \cdots \cdot ab}_{m \text{ factors}} = a^m b^m$$

In later chapters, we shall have the opportunity to discuss powers with exponents other than counting numbers. In the interim, let us observe that, for the sake of consistency with these later results, a^0 must be defined to be 1. Therefore, $\boldsymbol{a^0 = 1}$ **for all** $\boldsymbol{a \neq 0}$.

Example 12 • (a) $10^0 = 1$ (b) $10^3 \cdot 10^2 = 10^{3+2} = 10^5$
(c) $(10x)^3 = 10^3 \cdot x^3$ (d) $x^4 \cdot x^9 = x^{13}$ •

The preceding properties of exponents were demonstrated for counting number exponents. These same properties hold for all exponents that will be used in this book. A summary of these properties follows.

Exponential properties

If a and b are any numbers unequal to 0 and m and n are any exponents, then the following properties hold.

(a) $a^m \cdot a^n = a^{m+n}$ (b) $a^0 = 1$ (c) $a^1 = a$
(d) $(a^m)^n = a^{m \cdot n}$ (e) $(ab)^m = a^m b^m$

Since 10 is the base of our numeration system, let us examine successive powers of 10.

$$10^0 = 1$$
$$10^1 = 10$$
$$10^2 = (10)(10) = 100$$
$$10^3 = (10)(10)(10) = 1000$$
$$10^4 = (10)(10)(10)(10) = 10{,}000$$
$$10^5 = (10)(10)(10)(10)(10) = 100{,}000$$

Therefore, 2346 can be written in terms of decreasing powers of 10 as

$$2(10)^3 + 3(10)^2 + 4(10)^1 + 6(10)^0$$

Note that the digits represent or "count" powers of 10 in a strictly decreasing order.

$$4321 = 4(10)^3 + 3(10)^2 + 2(10)^1 + 1(10)^0$$
$$15 = 1(10)^1 + 5(10)^0$$
$$12{,}000 = 1(10)^4 + 2(10)^3 + 0(10)^2 + 0(10)^1 + 0(10)^0$$

These numbers are said to be written in **expanded form in powers of the base.**

Example 13 • $3(10)^4 + 4(10) + 2 \neq 342$ because the terms involving $(10)^3$ and $(10)^2$ are missing, implying that the terms are $0(10)^3$ and $0(10)^2$. Thus

$$\begin{aligned} 3(10)^4 + 4(10) + 2 &= 3(10)^4 + 0(10)^3 + 0(10)^2 + 4(10) + 2 \\ &= 30{,}042 \end{aligned}$$ •

Exponents are very helpful as we examine the validity of our algorithms for addition and subtraction.

Example 14 • Addition: 421 + 176.

Solution

$$\begin{array}{l} 400 + 20 + 1 \\ 100 + 70 + 6 \\ \hline 500 + 90 + 7 = 597 \end{array} \quad \text{or} \quad \begin{array}{l} 4(10)^2 + 2(10) + 1(1) \\ 1(10)^2 + 7(10) + 6(1) \\ \hline 5(10)^2 + 9(10) + 7(1) \end{array} \quad \text{or} \quad \begin{array}{r} 421 \\ 176 \\ \hline 597 \end{array}$$ •

Example 15 • Subtraction: 87 − 35.

Solution

$$\begin{array}{l} 80 + 7 \\ -(30 + 5) \\ \hline 50 + 2 = 52 \end{array} \quad \text{or} \quad \begin{array}{l} 8(10) + 7(1) \\ 3(10) + 5(1) \\ \hline 5(10) + 2(1) \end{array} \quad \text{or} \quad \begin{array}{r} 87 \\ 35 \\ \hline 52 \end{array}$$ •

In order to verify the validity of our usual addition algorithm in which we rename or regroup numerals, we need to use the associative and commutative properties and the distributive property of multiplication over addition, as well as closure.

Example 16 • Rewrite $4(10)^2 + 36(10)$ so that each coefficient of a power of 10 is less than 10. Give reasons for each step. Then write the result as a number in standard form.

Solution

$4(10)^2 + 36(10) = 4(10)^2 + [3(10) + 6] \cdot (10)$	Renaming 36 in expanded form
$= 4(10)^2 + [3(10)^2 + 6(10)]$	Distributive property of multiplication over addition and the multiplication of exponentials
$= [4(10)^2 + 3(10)^2] + 6(10)$	Associative property of addition
$= (4 + 3)(10)^2 + 6(10)$	Distributive property of multiplication over addition
$= 7(10)^2 + 6(10)$	Addition
$= 760$	Decimal system of numeration •

This idea of renaming or regrouping numbers greater than 10 plays an integral part in various algorithms. The example below verifies the familiar addition algorithm.

Example 17 • Compute $17 + 8$.

Solution

$17 + 8 = (10 + 7) + 8$	Renaming 17
$= 10 + (7 + 8)$	Associative property of addition
$= 10 + 15$	Addition
$= 10 + (10 + 5)$	Renaming 15
$= (10 + 10) + 5$	Associative property of addition
$= (1 \cdot 10 + 1 \cdot 10) + 5$	Multiplicative identity
$= (1 + 1)10 + 5$	Distributive property of multiplication over addition
$= 2(10) + 5$	Addition
$= 25$	Decimal system of numeration •

The preceding example illustrates the mathematical validity of the usual procedure for adding 17 and 8. The following organization may make even clearer the connection between the properties of whole number addition and our usual algorithm for addition.

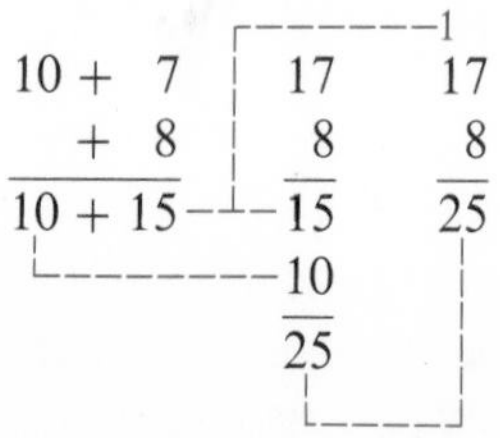

```
              ┌-------1
 10 +  7      |  17    17
    +  8      |   8     8
 ---------    |  ---   ---
 10 + 15------┴--15    25
 └-------------- 10     |
                ---     |
                 25     |
                 └------┘
```

The words **rename** and **regroup** are used rather than the terms *borrow* and *carry* because they describe more accurately the computational procedures involved.

Example 18 • Addition: 267 + 315.

Solution Again, let's write this sum so as to demonstrate our usual algorithm.

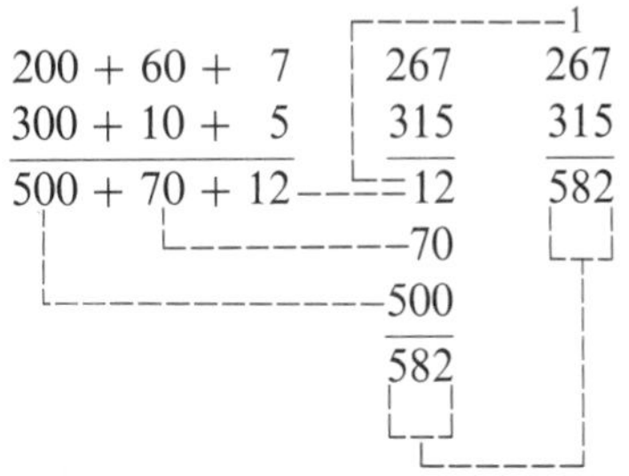

```
                    ┌-------1
200 + 60 +  7       | 267     267
300 + 10 +  5       | 315     315
---------------     └---      ---
500 + 70 + 12-----==12        582
 |     |                       |_|
 |     └----------70            |
 └----------------500           |
                  ---           |
                  582           |
                  |_|           |
                   └------------┘
```

In order to perform some subtractions, we must rename or regroup in the opposite direction.

Example 19 • Perform the following subtraction.

$$\begin{array}{r} 843 \\ -267 \\ \hline \end{array}$$

Solution In expanded form, this problem becomes

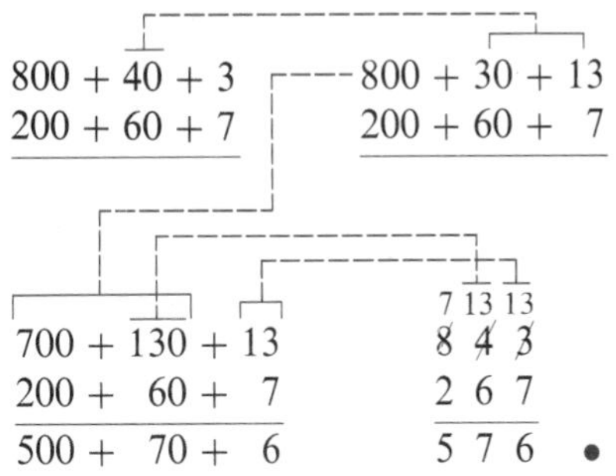

```
800 + 40 + 3     800 + 30 + 13
200 + 60 + 7     200 + 60 +  7
------------     -------------

                    7 13 13
700 + 130 + 13      8  4  3
200 +  60 +  7      2  6  7
--------------      -------
500 +  70 +  6      5  7  6
```

The fact that these manipulations are valid is directly dependent on the properties of arithmetic we have discussed previously.

Now that we understand the algorithms for addition and subtraction, let us work a simple example, writing down the thought process that governs the computation.

Example 20 • Find 715 − 348 and discuss the thinking involved during the computation.

Solution (a) We need more ones. Rename to show 10 more ones. 1 ten + 5 ones = 0 tens + 15 ones. Subtract the ones.

$$\begin{array}{rrr} & {\scriptstyle 0} & {\scriptstyle 15} \\ 7 & \cancel{1} & \cancel{5} \\ -3 & 4 & 8 \\ \hline & & 7 \end{array}$$

(b) We need more tens. Rename to show 10 more tens. 7 hundreds + 0 tens = 6 hundreds + 10 tens. Subtract the tens.

$$\begin{array}{rrr} {\scriptstyle 6} & {\scriptstyle 10} & {\scriptstyle 15} \\ 7 & \cancel{1} & \cancel{5} \\ -3 & 4 & 8 \\ \hline & 6 & 7 \end{array}$$

(c) Subtract the hundreds.

$$\begin{array}{rrr} {\scriptstyle 6} & {\scriptstyle 10} & {\scriptstyle 15} \\ \cancel{7} & \cancel{1} & \cancel{5} \\ -3 & 4 & 8 \\ \hline 3 & 6 & 7 \end{array}$$ •

The algorithms we have presented are commonly used and are fairly efficient. However, there are other excellent algorithms; we give a brief discussion of five of these. A technique called *subtraction by equal additions* achieved great popularity in the European schools of the fifteenth and sixteenth centuries. This algorithm is based on the fact that the difference in two numbers is unchanged if one adds the same amount to both numbers. In the subtraction

$$\begin{array}{r} 25 \\ -8 \\ \hline \end{array}$$

one can add 10 in the units digit of 25 and 1 in the tens digit of 8 to obtain

$$\begin{array}{rr} & {\scriptstyle 1} \\ 2 & 5 \\ -1 & 8 \\ \hline 1 & 7 \end{array}$$

Example 21 •

$$\begin{array}{r} 327 \\ -148 \\ \hline \end{array}$$

Solution

Step 1
$$\begin{array}{rr} & {\scriptstyle 1} \\ 32 & 7 \\ 15 & 8 \\ \hline & 9 \end{array}$$

Step 2
$$\begin{array}{rrr} & {\scriptstyle 1} & {\scriptstyle 1} \\ 3 & 2 & 7 \\ 2 & 5 & 8 \\ \hline & 7 & 9 \end{array}$$

Step 3
$$\begin{array}{rrr} & {\scriptstyle 1} & {\scriptstyle 1} \\ 3 & 2 & 7 \\ 2 & 5 & 8 \\ \hline 1 & 7 & 9 \end{array}$$ •

Example 22 • The "scratch method" of addition is very useful when adding several numbers. For example, add 67 + 56 + 48.

$$\begin{array}{rcl} & {\scriptstyle 2} & \\ & 67 & \\ {\scriptstyle 3} & \cancel{5}\cancel{6} & {\scriptstyle 3} \\ & 4\cancel{8} & {\scriptstyle 1} \\ \hline & 171 & \end{array}$$

$7 + 6 = 13$. Scratch out the 6 and replace with a 3. $3 + 8 = 11$. Scratch out the 8 and replace with a 1. The number of scratches in the first column is 2, so add this number to the tens column. There is one scratch in the second column, so add 1 to the third column. •

Example 23 • In the "scratch" method of subtraction, we begin on the left. In

$$\begin{array}{r} {}^{13} \\ 7\,3 \\ -4\,7 \\ \hline \not{3} \\ 2\,6 \end{array}$$

we subtract 4 from 7, getting 3. Then "scratch through" the 3, changing it to 2, and then subtract 7 from 13. •

Example 24 • The Austrian method of subtraction consists of the following steps. Write $764 - 348$ as

$$\begin{aligned} (700 + 60 + 4) - (300 + 40 + 8) &= \\ (700 + 60 + 10 + 4) - (300 + 40 + 10 + 8) &= \quad \text{(Add 10 in each set of (parenthesis)} \\ (700 + 600 + 14) - (300 + 50 + 8) &= \quad \text{(Group 10 appropriately)} \\ (700 - 300) + (60 - 50) + 14 - 8 &= \\ 400 + 10 + 6 &= 416 \end{aligned}$$

•

Example 25 • "Subtraction by taking complements" consists of the adding to both numbers the same sum. This sum is selected so that the number to subtract is a multiple of 10.

$$\begin{array}{r} 1734 \\ -468 \\ \hline \end{array} \qquad \begin{array}{r} 1734 \\ +532 \\ -532 \\ -468 \end{array} \qquad \begin{array}{r} 2266 \\ -1000 \\ \hline 1266 \end{array}$$

•

Example 26 • A second method of subtraction using complements defines the complement of a digit a to be $9 - a$. To perform a subtraction, replace the digits of the number being subtracted by their complements. Then, instead of subtracting, add. Then subtract 1 from the left digit and add 1 to the right digit. This gives the answer for the difference.

$$\begin{array}{r} 5316 \\ -3927 \\ \hline \end{array} \qquad \begin{array}{r} 5316 \\ +6072 \\ \hline 11388 \\ -1\ \ +1 \\ \hline 1389 \end{array}$$

•

Just for fun

Jack discovered, in climbing his giant beanstalk, that the giant had a numeration system based on "fee, fie, foe, fum." When the giant counted his golden eggs, Jack heard him count "fee, fie, foe, fum, fot, feefot, fiefot, foefot, fumfot, fotfot, feefotfot, ..." Jack believes that the giant has 20 eggs. Can you guess the other nine numerals that the giant used to finish the counting?

Let's return now to the problem at the beginning of this section. Place the digits 0 to 9 in one of the blocks to make an addition problem. Note that we have made a guess about the placement of 3 and 9.

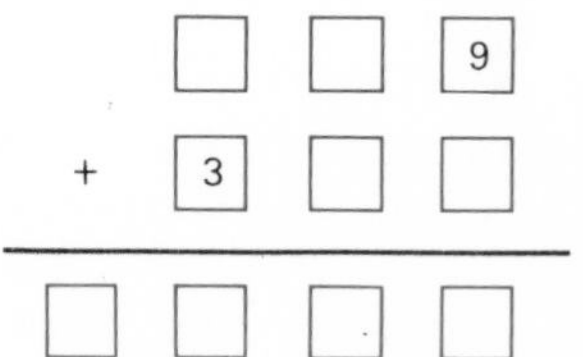

Let us first consider the leftmost digit of the answer. Since the block will be filled as the consequence of a regrouping process, it must be 1 (the biggest number we ever regroup in two-number addition results from $9 + 9 = 18$). In order to get the 1 we just placed, the number to be added to 3 is either 6, 7, or 8. Let's start with 8. Since 1 has been used, the second blank in the answer is 2. ($8 + 3 + 1 = 12$.) Therefore, the sum of the two numbers in the second column must exceed 10. Trials of 5 and 6 and 6 and 7 lead to complications. Then we try 4 and 5; $(4 + 5) + 1 = 10$. This leaves 7 and 6 for the last two blocks. One answer is

$$\begin{array}{r} 859 \\ +347 \\ \hline 1206 \end{array}$$

Can you find another answer?

Exercise set 2

R 1. Complete the following table.

10^9	1,000,000,000	billion
______	100,000,000	hundred million
10^7	______	ten million
10^6	______	______
______	100,000	______
10^4	______	______
10^3	______	______
______	______	hundred
______	______	ten
10^0	______	______

2. Using exponents, write the following numerals in simpler form.

(a) (2)(2)(2)(2)(2) (b) (4)(4)(4)(3)(3)
(c) $7 \cdot 7 \cdot 7 \cdot 5 \cdot 5$ (d) $3 \cdot 3 \cdot 3 \cdot 2 \cdot 2 \cdot 2 \cdot 2$
(e) (3)(3)(3)(3)(3)(3)(3) (f) $x \cdot x \cdot x \cdot y \cdot y$

3. Perform the indicated operations.

(a) 3^4 (b) $(x^4)(x^7)$
(c) 9^0 (d) $(2^3)(2^0)$

4. Write the answer in standard form.

(a) $(3 \cdot 10^4)(4 \cdot 10^3)$ (b) $(7 \cdot 10^4)(8 \cdot 10^6)$

5. Write the following in standard form.

(a) $4(10)^3 + 5(10)^2 + 7(10)^1 + 8(10)^0$
(b) $9(10)^4 + 7(10)^2 + 3(10)^0$
(c) $8(10)^3 + 3(10)^1 + 5(10)^0$
(d) $3(10)^4 + 5(10)^3 + 7(10)^1$
(e) $4(10)^2 + 5(10)^0$
(f) $2(10)^0$
(g) $3(10)^2 + 4(10)^4 + 6(10)^3 + 7$

6. Write the following numerals in "expanded form," using powers of 10.

(a) 3403 (b) 6,741,007
(c) 34,001,620 (d) 4001
(e) 4,000,000 (f) 20,001,100

7. Multiply and write the answer as a decimal numeral.

(a) $10^4[4(10)^2 + 5(10) + 6(10)^0]$
(b) $10^3[2(10)^4 + 7(10)]$
(c) Study the patterns produced in (a) and (b), and conjecture what happens when a number is multiplied by a power of 10.

8. Perform the following additions and subtractions in two ways. (i) Write each number in expanded notation. (ii) Write each number in expanded notation using powers of 10.

(a) $\begin{array}{r} 364 \\ +423 \\ \hline \end{array}$ (b) $\begin{array}{r} 476 \\ -324 \\ \hline \end{array}$ (c) $\begin{array}{r} 426 \\ -14 \\ \hline \end{array}$

(d) $\begin{array}{r} 1758 \\ +241 \\ \hline \end{array}$ (e) $\begin{array}{r} 1894 \\ -562 \\ \hline \end{array}$ (f) $\begin{array}{r} 8547 \\ -2413 \\ \hline \end{array}$

9. Rewrite each number so that each coefficient of a power of 10 is less than 10. [In $4(10)^2$, 4 is the coefficient of 10^2.] Give reasons for each step.

(a) $5(10)^2 + 13(10)$
(b) $7(10)^3 + 25(10)^2 + 5(10) + 17$
(c) $16(10) + 17$
(d) $1(10)^3 + 39(10) + 9$

10. Complete the blanks in the following calculations.

(a) $\begin{array}{r} 768 \\ +574 \\ \hline \end{array}$ or $\begin{array}{r} 700 + \_\_ + 8 \\ 500 + 70 + 4 \\ \hline \_\_ + \_\_ + \_\_ = \_\_ \end{array}$

(b) $\begin{array}{r} 865 \\ -378 \\ \hline \end{array}$ or

$\begin{array}{r} 800 + \_\_ + \_\_ \\ -(300 + 70 + 8) \\ \hline \end{array}$ or

$\begin{array}{r} 800 + \_\_ + \_\_ \\ -(300 + 70 + 8) \\ \hline \_\_ + \_\_ + \_\_ \end{array}$ or

$\begin{array}{r} \_\_ + \_\_ + 15 \\ -(300 + 70 + 8) \\ \hline \_\_ + \_\_ + \_\_ = \_\_ \end{array}$

11. Perform the following computations using the expanded notation, as illustrated in Exercise 10.

(a) 46 + 75 (b) 248 + 574
(c) 60 − 37 (d) 136 − 29

T 12. The Red River Ecology Club had a paper drive and collected 3 truckloads of paper. Each truck carried 3 large boxes of paper; each box contained 3 bales of paper; and each bale contained 3 stacks of paper. How many stacks of paper were there in all?

Express the answer in exponential form and then simplify.

13. Place digits 1, 2, 3, 4, 5, 6, 7, 8 in the boxes.

$$\begin{array}{r} \square\ \square\ \square\ \square \\ +\ \square\ \square\ \square\ \square \\ \hline \end{array}$$

to obtain

(a) the greatest sum
(b) the least sum

14. In Exercise 13, place the digits to obtain

(a) the greatest difference
(b) the least difference

15. (a) Order the following exponential numbers from smallest to largest (increasing order): 10^2, 7^3, 4^0, 2^{10}, 3^7, 9^4, 4^9.
(b) Order the following exponential numbers from largest to smallest (decreasing order): 2^5, 10^2, 3^4, 9^1, 136^0, 5^4, 8^2.

16. One way to combine exponential factors is to change the bases so that they are identical. Simplify the following problems by obtaining a common base [such as 2 in (a)].

(a) $2^3 \cdot 8^4$ (b) $5^5 \cdot 125$ (c) $16^2 \cdot 2^2$
(d) $3^4 \cdot 27^2$ (e) $16^6 \cdot 4$ (f) $9^2 \cdot 27^1$

17. Find a value for n to make each of the following a true statement.

(a) $2^n = 16$ (b) $n^2 = 64$
(c) $3^4 = 9^n$ (d) $(4^2)^n = 2^8$
(e) $(a^2)(a^3)(a^n) = a^8$ (f) $(a^6)^n = a^6$
(g) $n^n = 1$ (h) $n^n = 2n$
(i) $2^n = n^2$

18. Is it true that $(a^x)(a^y) = (a^y)(a^x)$, where a, x, and y are counting numbers? Why?

19. Several times in this section, we rewrote a problem using expanded notation and then drew lines to show as clearly as possible the connection between the properties of arithmetic and numeration and our usual algorithms.

EXAMPLE

$$\begin{array}{r} 26 \\ +46 \\ \hline \end{array} \qquad \begin{array}{r} 20 + 6 \\ 40 + 6 \\ \hline 60 + 12 \end{array} \qquad \begin{array}{r} 26 \\ 46 \\ \hline 12 \\ 60 \\ \hline 72 \end{array} \qquad \begin{array}{r} {}^{1} \\ 26 \\ 46 \\ \hline 72 \end{array}$$

Rewrite the following problems in this way.

(a) $\begin{array}{r} 34 \\ +27 \\ \hline \end{array}$ (b) $\begin{array}{r} 63 \\ -36 \\ \hline \end{array}$ (c) $\begin{array}{r} 167 \\ +245 \\ \hline \end{array}$ (d) $\begin{array}{r} 378 \\ -196 \\ \hline \end{array}$

20. If you have a calculator with a y^x key, find:

(a) 2^5
(b) $(2^5)^4$
(c) 2^{20}
(d) What have you verified?

C 21. Consider the following justification of the algorithm used in the addition of 28 and 9. List the properties of whole numbers that are applied in each step.

(a) $28 + 9 = (2 \cdot 10 + 8) + 9$
(b) $= 2 \cdot 10 + (8 + 9)$
(c) $= 2 \cdot 10 + 17$
(d) $= 2 \cdot 10 + (1 \cdot 10 + 7)$
(e) $= (2 \cdot 10 + 1 \cdot 10) + 7$
(f) $= (2 + 1) \cdot 10 + 7$
(g) $= 3 \cdot 10 + 7$
(h) $= 37$

22. (a) In Exercise 21, which steps validate the following procedure?

$$\begin{array}{r} {}^{1} \\ 28 \\ +9 \\ \hline 7 \end{array}$$

(b) Which steps validate the addition of 1 + 2 and the final answer of 37?

$$\begin{array}{r} {}^{1} \\ 28 \\ +9 \\ \hline 37 \end{array}$$

23. (a) Study the two examples to understand this particular algorithm for adding column by column. Explain why and how this algorithm works.

$$\begin{array}{r} 623 \\ +397 \\ \hline 10 \\ 11 \\ 9 \\ \hline 1020 \end{array} \qquad \begin{array}{r} 223 \\ 997 \\ +275 \\ \hline 15 \\ 18 \\ 13 \\ \hline 1495 \end{array}$$

(b) Add 499 + 276 using this algorithm.
(c) Add 745 + 628 + 211 using this algorithm.

24. Use Austrian subtraction to subtract

(a) $\begin{array}{r} 5706 \\ -3407 \\ \hline \end{array}$ (b) $\begin{array}{r} 329 \\ -146 \\ \hline \end{array}$

(c) 1634
 −985

(d) 1275
 −968

25. Compute the answers in Exercise 24 using complements.

26. Work Exercise 24 using the second method of taking complements.

27. Work Exercise 24 using the "scratch method."

3 Multiplication and division algorithms

■ Consider this algorithm for multiplying (256)(324):

	2	5	6	
0	0 / 6	1 / 5	1 / 8	3
8	0 / 4	1 / 0	1 / 2	2
2	0 / 8	2 / 0	2 / 4	4
	9	4	4	

Notice that the answer, 82,944, is read from the top of the left margin, down to the bottom, and then to the right of the bottom margin. ■

There are several other unusual procedures or algorithms for multiplication. Before we introduce these, you must first understand why the algorithm you have used for many years works so effectively.

Recall that we emphasized in Chapter 2 that division was not possible for every pair of whole numbers. Now we observe that division is always possible when the divisor is not zero. However, the division we are discussing gives a quotient and a remainder; it is not the exact division discussed earlier. Furthermore, although exact division was defined as the inverse of multiplication, our usual algorithm for division is best understood in terms of repeated subtraction.

To introduce multiplication algorithms, we use a diagram to represent a product as partial products. First, we show how this diagram relates to the distributive property of multiplication over addition, and then we use the partial products to demonstrate a vertical format.

Suppose we want to multiply $3 \cdot 16$. This can be written as $3(10 + 6)$, and by the distributive property of multiplication over addition

$$3(10 + 6) = 3 \cdot 10 + 3 \cdot 6$$

A visual representation of this multiplication is shown in Figure 1.

Figure 1

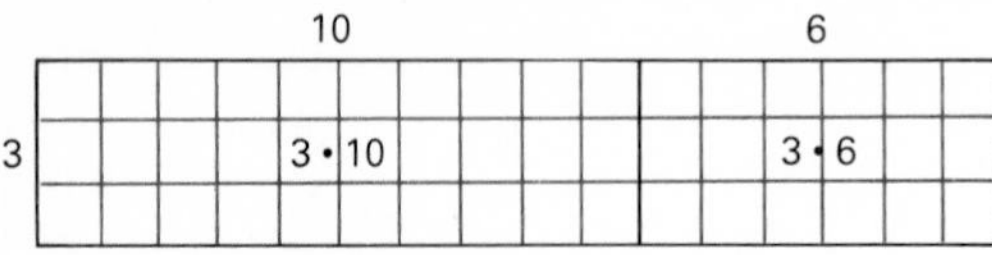

Example 27 • Multiply $3 \cdot 16$ using a horizontal format.

Solution

$3 \cdot 16 = 3(10 + 6)$	Definition of addition
$= 3 \cdot 10 + 3 \cdot 6$	Distribution property of multiplication over addition
$= 30 + 18$	Multiplication
$= 48$	Addition •

Example 28 • Compute $3 \cdot 16$ using a vertical format (partial products) and tie in with our usual format for multiplication.

Solution From Figure 1, we have:

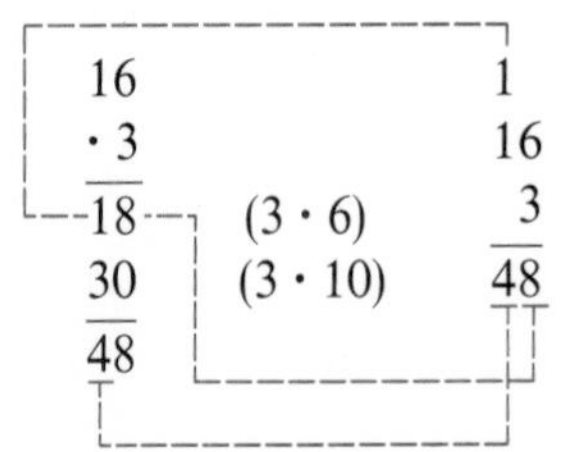

•

We now show the multiplication of $13 \cdot 14$ as four partial products (Figure 2). This represents two applications of the distributive property of multiplication over addition, as seen in the next example.

Example 29 • Show $13 \cdot 14$ in horizontal format.

Solution

$$\begin{aligned} 13 \cdot 14 &= 13(10 + 4) \\ &= 13 \cdot 10 + 13 \cdot 4 \\ &= (10 + 3) \cdot 10 + (10 + 3) \cdot 4 \\ &= 10 \cdot 10 + 3 \cdot 10 + 10 \cdot 4 + 3 \cdot 4 \end{aligned}$$

(See Figure 2.) •

Example 30 • Compute $13 \cdot 14$ using a vertical format (with partial products) and tie to the usual format for multiplication.

flashback

When we look at our usual algorithm for multiplication, it is difficult for us to realize that there have been no significant developments in our numeration system for more than 500 years. Maybe you can develop a new and fascinating numeration system or a simpler algorithm for computation.

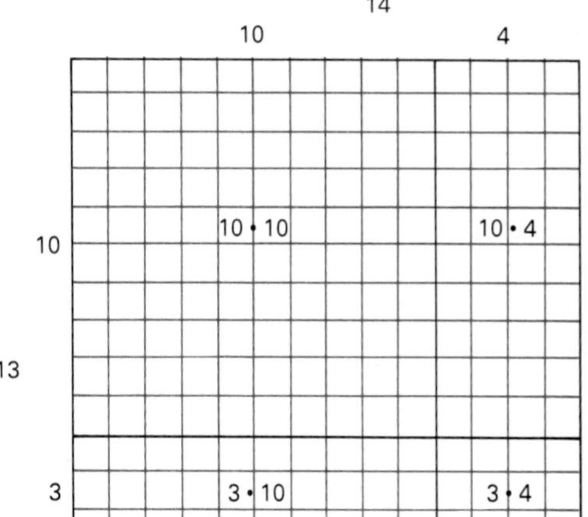

Figure 2

Solution From Figure 2,

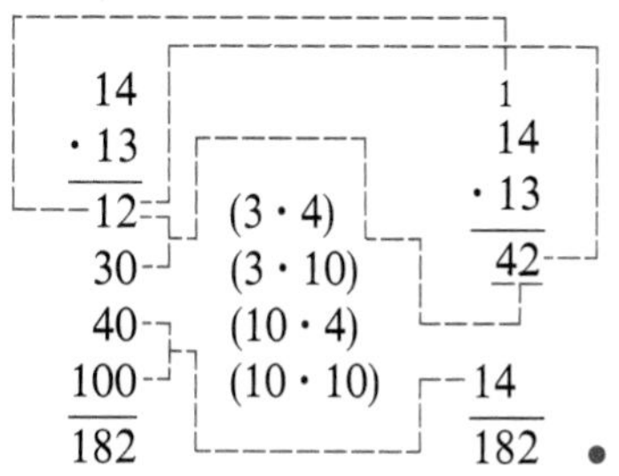

Example 31 • Multiplication: 47 · 53.

Solution Note in this example that the distributive property of multiplication over addition must be applied twice.

(a) Horizontal format

$$\begin{aligned} 47 \cdot 53 &= 47(50 + 3) \\ &= 47 \cdot 50 + 47 \cdot 3 \qquad \text{(Why?)} \\ &= (40 + 7) \cdot 50 + (40 + 7) \cdot 3 \\ &= 40 \cdot 50 + 7 \cdot 50 + 40 \cdot 3 + 7 \cdot 3 \qquad \text{(Why?)} \\ &= 2000 + 350 + 120 + 21 \\ &= 2491 \end{aligned}$$

(b) Vertical format

$$\begin{array}{rl} 53 & \\ \underline{47} & \\ 21 & (7 \cdot 3) \\ 350 & (7 \cdot 50) \\ 120 & (40 \cdot 3) \\ \underline{2000} & (40 \cdot 50) \\ 2491 & \end{array}$$

(c) Vertical format, powers of 10

$$\begin{array}{r} 5(10) + \ \ 3(1) \\ 4(10) + \ \ 7(1) \\ \hline 35(10) + 21(1) \\ 20(10)^2 + 12(10) \qquad\qquad \\ \hline \end{array} \qquad \text{or} \qquad \begin{array}{r} 5(10) + 3(1) \\ 4(10) + 7(1) \\ \hline 3(10)^2 + 7(10) + 1 \\ 2(10)^3 + 1(10)^2 + 2(10) \qquad \\ \hline 2(10)^3 + 4(10)^2 + 9(10) + 1 = 2491 \end{array}$$

(d) Usual format

$$\begin{array}{r} 53 \\ 47 \\ \hline 371 \\ 212 \\ \hline 2491 \end{array} \quad \bullet$$

Example 32 • Multiply 247 by 1000.

Solution $1000 = 10^3$ and $247 = 2(10)^2 + 4(10) + 7(1)$

$$\begin{aligned} 1000(247) &= 10^3[2(10)^2 + 4(10) + 7(1)] \\ &= 2(10)^5 + 4(10)^4 + 7(10)^3 \\ &= 247{,}000 \quad \bullet \end{aligned}$$

Note the pattern in the preceding example. To multiply a number N by 10^k, one would annex k zeros to the right of the digits of N.

There are several other interesting algorithms for multiplication. For sake of brevity, we consider only two: Russian peasant multiplication and lattice multiplication. Russian peasant multiplication, used in medieval Europe, involves repeatedly doubling one of the numbers to be multiplied, and halving the other. Any time a remainder occurs in the halving process, it is dropped. We then cross out all lines in which the entry in the halving column is an even number. The entries in the doubling column are added to get the answer.

Example 33 • Multiply $37 \cdot 168$.

Solution

Halving	*Doubling*
37	168
~~18~~	~~336~~
9	672
~~4~~	~~1344~~
~~2~~	~~2688~~
~~1~~	~~5376~~
	6216

$37 \cdot 168 = 6216$ •

Example 34 • Multiply $476 \cdot 753$ by the lattice method of multiplication.

Solution The problem is to find the product of 476 and 753. Since the two numbers to be multiplied are represented by numerals with three digits each, a square containing nine small squares of equal size is drawn (Figure 3). The numerals 4, 7, 6 are written across the top of the square, and 7, 5, 3 are written along the right side, from top to bottom. Diagonals (from lower left to upper right) are drawn in each of the squares to form a *lattice* design. Products of pairs of digits taken from the top and the right side of the rectangle are now entered in the squares. The tens digit of the product is written above the diagonal, and the units digit below the diagonal. Thus, in the first square, $6 \cdot 7 = 42$, so 4 is above the diagonal and 2 is below it. Now add the elements between adjacent diagonals, beginning at the lower right corner. There is only one element below the first diagonal. It is an 8, so write down an 8 in the space at the bottom of this diagonal. The sum of the elements between the next two diagonals is $1 + 1 + 0$, and the answer 2 is recorded below the area between diagonals. If the sum is more than 9, record the units digit as before and regroup the tens digit to be added to the elements between the next two adjacent diagonals. When all diagonal elements are totaled, the answer is read down the left side and along the bottom. The answer is 358,428.

Figure 3

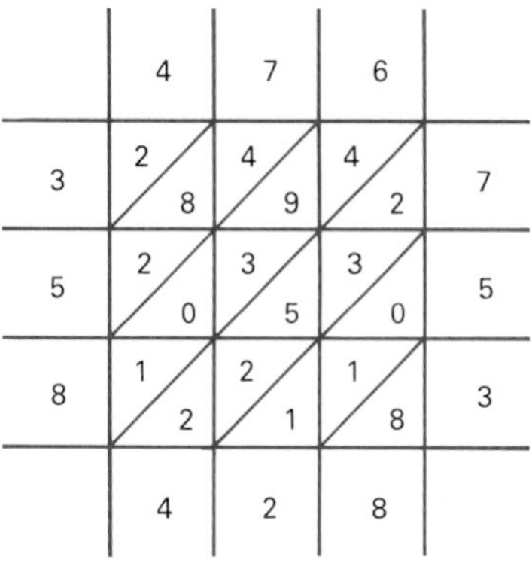

In the preceding example, the two factors contained the same number of digits. If the two numbers to be multiplied contain a different number of digits, a rectangle rather than a square is drawn. Like the square, the rectangle is then divided into small squares of equal size, and the lattice is formed by drawing the diagonals. Multiplication of the numbers is performed as in the example.

In Section 6 of Chapter 2, division was defined as the inverse of multiplication. If a and b ($b \neq 0$) are any two whole numbers, then $a \div b$ is some whole number c (if it exists) such that $a = bc$. It was emphasized that division is not possible for every pair of whole numbers. For example, it is not possible to compute $8 \div 5$ in the system of whole numbers. It is possible, however, in every division problem involving whole numbers (where the divisor is unequal to zero) to find an answer consisting of a quotient and a remainder, both of which are whole numbers.

The Division Algorithm

If a and b ($b \neq 0$) are whole numbers, then there exist unique whole numbers q and r such that $a = bq + r$, where $0 \leq r < b$.

This theorem states that for any two whole numbers a and b ($b \neq 0$), a can be divided by b to obtain two numbers, q (the quotient) and r (the remainder) such that $a = bq + r$, where $0 \le r < b$. If we call b the divisor and a the dividend, then

dividend = divisor · quotient + remainder

Example 35 • If $a = 47$ and $b = 7$, write a in the form $bq + r$, where $0 \le r < b$.

Solution $47 = 7(6) + 5; \quad q = 6$ and $r = 5 \quad 0 \le 5 < 7$ •

Example 36 • If $a = 54$ and $b = 9$, write a in the form $bq + r$, where $0 \le r < b$.

Solution $54 = 9(6) + 0; \quad q = 6$ and $r = 0$ •

Many procedures are known for performing the operation of division. Generally, they are algorithms given in such a way that the properties of whole numbers relative to the operations involved are completely obscured. Most people use a process of guessing, multiplying, subtracting, and then guessing again. If an initial guess proves wrong, it is replaced by a new guess. This process is called *long division.* The Division Algorithm states that the answer exists, so after a sufficient number of guesses one should always find the correct answer.

However, long division becomes more understandable when viewed as repeated subtraction. For instance, the division $16 \div 5$ asks the question, "How many 5's are contained in 16?" We find the answer by repeated subtractions.

$$\begin{array}{r} 16 \\ -5 \\ \hline 11 \\ -5 \\ \hline 6 \\ -5 \\ \hline 1 \end{array}$$

Thus, there are three 5's in 16, with a remainder of 1. This format (sometimes called *scaffolding*) is used in the next two examples.

Example 37 • Find $20 \div 6$.

Solution Note in (b) that instead of subtracting 6 three times, 18 or $(3 \cdot 6)$ is subtracted once.

(a)
$$\begin{array}{r|l} 6\,)\ 20 & \\ 6 & 1\ (6) \\ \hline 14 & \\ 6 & 1\ (6) \\ \hline 8 & \\ 6 & 1\ (6) \\ \hline 2 & 3 \end{array}$$

(b)
$$\begin{array}{r|l} 6\,)\ 20 & \\ 18 & 3\ (6\text{'s}) \\ \hline 2 & \end{array}$$

The answer is $20 \div 6 = 3$ with a remainder of 2. •

A pair of slight modifications in format brings us to our usual algorithm for division. Observe that the "guessing" portion of long division is really the question, "How large a multiple of the divisor can I subtract and have a remainder that is a whole number?"

Just for fun

In the computations below, a given letter stands for the same digit no matter where it occurs. Can you unravel the puzzle and find the digit represented by each letter?

```
           X V U                 P Q R
    2 X 5 ) U 7 X 9 Z            · Q Q
            Y X Z              S R Q Q
            X Z T 9          S R Q Q
            X 5 W 5          P M P R Q
                T 4 Z
                T 4 Z
```

Example 38 • Compute 7410 ÷ 23.

Solution The largest multiple of the form (power of 10) times 23 that we can subtract from 7410 is (100)(23). After (100)(23) is subtracted three times, a (10)(23) can be subtracted twice. Finally (1)(23) is subtracted twice, leaving a remainder of 4.

(a)

23) 7410	
2300	100 (23's)
5110	
2300	100 (23's)
2810	
2300	100 (23's)
510	
230	10 (23's)
280	
230	10 (23's)
50	
23	1 (23)
27	
23	1 (23)
4	322

(b)

23) 7410	
6900	300 (23's)
510	
460	20 (23's)
50	
46	2 (23's)
4	322

The answer is 7410 ÷ 23 = 322 with a remainder of 4.

Notice that in part (b) of this example we subtracted (300)(23) instead of subtracting (100)(23) three times and (20)(23) instead of subtracting (10)(23) twice. This shortcut will be employed in future use of the scaffold format in this text. •

Instead of writing the parts of the quotient "down the scaffold" as in the preceding example, the answer can be written above the division sign in what we shall call the **full notation format.** Alternately, we can write the answer using the algorithm we have used for many years, designated here as **usual format.**

Example 39 • Compute $8134 \div 38$ by first using the scaffold format, then the full notation format, and finally the usual format.

Solution (a) Scaffold format (b) Full notation format (c) Usual format

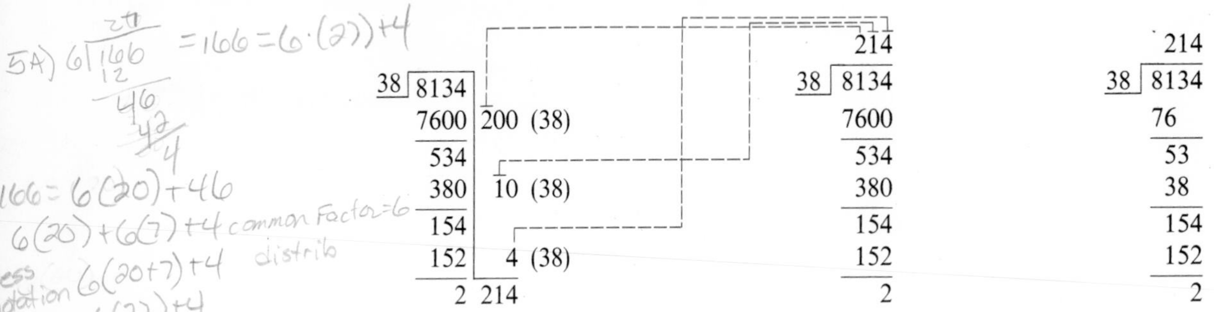

$8134 \div 38 = 214$ with a remainder of 2 or $38 \cdot 214 + 2 = 8134$ •

Exercise set 3

R 1. Draw an array for each product that illustrates the distributive property of multiplication over addition.

(a) $6 \cdot 28$ (b) $8 \cdot 34$
(c) 3(106) (d) 4(67)

2. Using Exercise 1, perform the following computations in the three ways demonstrated in the examples: (i) horizontal format, (ii) vertical format (partial products), and (iii) usual format.

(a) $6 \cdot 28$ (b) $8 \cdot 34$
(c) 3(106) (d) 4(67)

3. For each part of Exercise 2, draw lines connecting equivalent parts of (ii) and (iii).

4. For the following pairs of numbers, let a be the first number of the pair and b the second number. Find whole numbers q and r for each pair such that $a = bq + r$, where $0 \leq r < b$.

(a) 72, 11 (b) 16, 9 (c) 11, 18
(d) 106, 13 (e) 51, 14 (f) 25, 39
(g) 54, 9 (h) 176, 21

5. Perform the following divisions in three ways as demonstrated in the last two examples in this section: (i) scaffold format, (ii) full notation format, (iii) usual format.

(a) $166 \div 6$ (b) $324 \div 7$ (c) $1425 \div 8$

6. Draw lines connecting the equivalent parts of (i) and (ii) for each part of Exercise 5.

7. Fill in the missing numerals:

(a)
```
        4  7
      · 8  4
   1  __  8
 3 __  6
 3 __ __  8
```

(b)
```
          2  4  6
        · 5  7  8
     1 __ __  8
   1 __ __  2
 1 __ __  0
 1 __ __ __ __  8
```

(c)
```
          2
 21 | 6  7  8
      6 __
        __  8
         4  2
        ____
```

8. Follow the directions of Exercise 2 or 5 and perform the following computations.

(a) $32 \cdot 26$ (b) $84 \cdot 76$
(c) $92 \cdot 74$ (d) $1075 \div 27$
(e) $35{,}304 \div 314$ (f) $35{,}714 \div 87$

9. Draw lines connecting the equivalent parts of (ii) and (iii) for multiplication and (i) and (ii) for division of Exercise 8.

10. Study the pattern in the multiplication algorithm below and use this algorithm to do the computations.

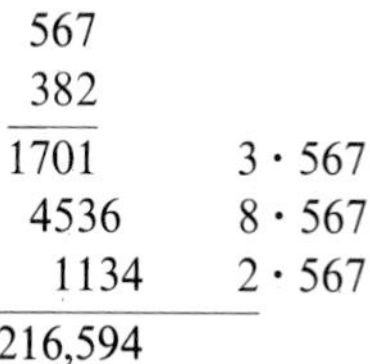

567	
382	
1701	$3 \cdot 567$
4536	$8 \cdot 567$
1134	$2 \cdot 567$
216,594	

(a) $745 \cdot 74$ (b) $365 \cdot 947$ (c) $848 \cdot 1724$

11. Use the incomplete example

$$\begin{array}{r|l} & 4 \\ \hline 33 & 1605 \\ & 1320 \\ \hline & 285 \end{array}$$

as a basis to answer (a), (b), and (c).

(a) Discuss the meaning of 4 in the quotient.
(b) Discuss the reason for placing the 4 above the 0.
(c) Some people omit the 0 in 1320 and say "Bring down the 5." Discuss the meaning of this procedure.

12. Perform the following divisions:

(a) $16{,}048 \div 21$ (b) $16{,}044 \div 21$

Explain how you can tell if the division is exact or if there is a remainder. Find the remainder.

13. Perform the following division by repeated subtraction.

$$2473 \div 23$$

Can you check your answer, using the calculator to perform the operation another way?

14. Multiply 15,873 by 7, 14, and 21. From the pattern, guess the answer for 49(15,873). Check your answer.

15. Multiply $11 \cdot 11$, $111 \cdot 111$, and $1111 \cdot 1111$. Now guess the answer for $11111 \cdot 11111$. Check your guess.

C 16. Multiply using Russian peasant multiplication.

(a) $14 \cdot 36$ (b) $54 \cdot 17$
(c) $23 \cdot 102$ (d) $51 \cdot 318$

17. Compute using the lattice method of multiplication.

(a) $34 \cdot 176$ (b) $56 \cdot 742$
(c) $374 \cdot 1728$ (d) $748 \cdot 964$

18. (a) Compute the q and r of the Division Algorithm for the division of 438 by 100, 10, and 1.
(b) Compute the q and r of the division of 8104 by 1000, 100, 10, and 1.
(c) Can you generalize your work to produce a "rule" for division by 10 or a power of 10?

*4 Patterns for nondecimal bases

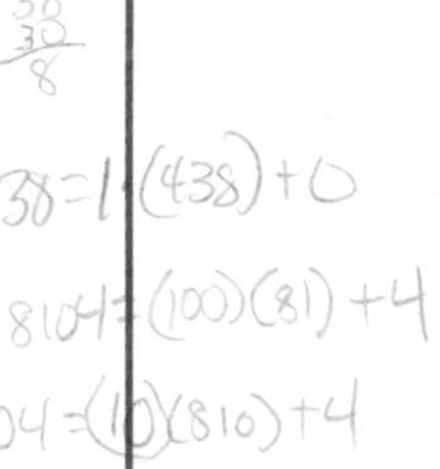

■ Linda learned that numbers may often be expressed in a form different from base ten numeration. She was surprised when her teacher asked her, "Which expressions name the same number 18?"

(a) 3 half dozens (b) one dozen and four
(c) 33_{five} (d) $2(7) + 4$ ■

A study of numeration systems with bases other than ten helps to provide a more complete understanding of the decimal numeration system. In this section, we introduce different number bases, with particular emphasis on base seven and base two numeration systems.

A 17th-century German mathematician, Gottfried Wilhelm von Leibniz, is reported to have invented and used the binary numeration system, which has two for a base. An electric switch needs only two digits (0 and 1) to represent *off* and *on*, so a switch position can be represented by a binary symbol. Pictures taken in outer space are transmitted back to earth in binary notation so that computers can convert the knowledge into pictures.

The use of binary numbers can help in solving problems involving certain number tricks. See the introductory problem for Section 5.

BIOGRAPHY

Gottfried Wilhelm von Leibniz

1646–1716

Gottfried Wilhelm von Leibniz was a philosopher, diplomat, lawyer, historian, and geologist as well as a well-known mathematician. His most noteworthy contribution to mathematics was his invention of the calculus (simultaneously with but independent of Isaac Newton). Despite the scope of his investigations, Leibniz was not a cloistered academician. He was heavily involved in the political maneuvering of Europe and worked tirelessly for reconciliation between the Catholic and Protestant faiths. The last years of Leibniz's life were shadowed by a quarrel with Newton over who first invented the calculus. The resulting feud almost broke relations between European and British mathematicians and embittered and isolated Leibniz. It is said that only a faithful secretary attended his funeral.

In this section, we study numeration systems with bases other than base ten. Base ten has ten digits (0, 1, 2, 3, 4, 5, 6, 7, 8, 9). Base seven has only seven (0, 1, 2, 3, 4, 5, 6), and base two has only two (0, 1).

Counting in base seven begins 1, 2, 3, 4, 5, 6, 10, ... However, this is not the "10" that we are accustomed to seeing. In base seven, 10 represents one group of 7 and no 1's. The next counting number, 11, represents one 7 and one 1. The counting process continues with 12, 13, 14, 15, 16, and then 20; 20 represents two 7's and no 1's. If you continue counting in this numeration system, you will soon reach the numeral 66. What is the next numeral? Be sure you understand that it is 100.

Examine the x's listed in groups of 7 in Figure 4. In part (a), the numeral representing the x's could be written 12_{seven}, representing one group of 7 and 2 ones. If the x's correspond to days of the week, then the 12_{seven} would represent 1 week and 2 days. The numeral representing the x's in part (b) could be written as 35_{seven} to mean three groups of 7 and 5 ones. We write the subscript "seven" to show that the numeral is not the "35" we usually mean in base ten.

Figure 4

xxxxxxx	xx	$= 1(7) + 2(1)$ $= 12_{\text{seven}}$

(a)

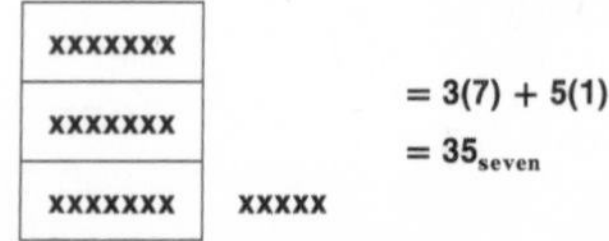

(b)

Consider now a base seven numeral written in expanded notation. Remember that $(10)^2$ represents base squared and 10 represents one base.

$$243_{\text{seven}} = [2(10)^2 + 4(10) + 3]_{\text{seven}}$$

The base squared, $(10)^2$, is equivalent to 7^2 in base ten; and the (10) is equivalent to 7 in base ten. Thus, in base ten, we have

$$243_{\text{seven}} = [2(7)^2 + 4(7) + 3]_{\text{ten}}$$

This example provides an easy method for changing 243_{seven} to a base ten numeral. Thus,

$$243_{\text{seven}} = (98 + 28 + 3)_{\text{ten}} = 129_{\text{ten}}$$

Example 40 •

$$\begin{aligned} 3462_{\text{seven}} &= [3(7)^3 + 4(7)^2 + 6(7) + 2]_{\text{ten}} \\ &= (1029 + 196 + 42 + 2)_{\text{ten}} \\ &= 1269_{\text{ten}} \end{aligned}$$ •

Binary numerals are based on groups of two, just as the preceding arithmetic was based on groups of seven and the decimal system is based on groups of ten. The groups in Figure 5 are set up for counting in the binary system. Figure 5(a) represents one group of $(\text{two})^2$ elements, one group of two elements, and one element, or 111_{two}. In part (b), there are no groups of two elements, so the symbol becomes 1101_{two}. The first eleven counting symbols in the base two system are 1, 10, 11, 100, 101, 110, 111, 1000, 1001, 1010, 1011. What is the next symbol?

Figure 5

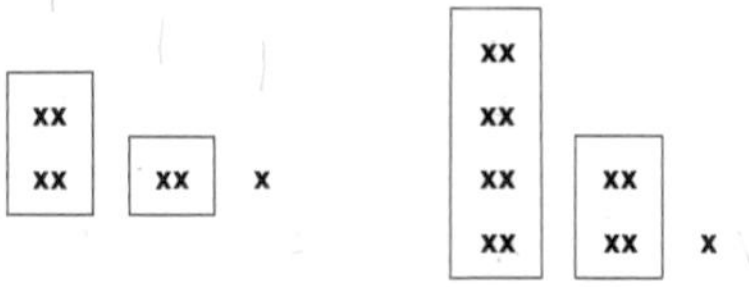

(a)	(b)
[xx xx] [xx] x	[xx xx xx xx] [xx xx] x
$= 1(2)^2 + 1(2) + 1(1)$	$= 1(2)^3 + 1(2)^3 + 0(2) + 1$
$= 111_{\text{two}}$	$= 1101_{\text{two}}$

It is easy to convert a binary numeral to a base ten numeral, as the following examples illustrate. Remember that $10_{\text{two}} = 2_{\text{ten}}$; $(10)^2_{\text{two}} = 2^2_{\text{ten}}$; and $(10)^3_{\text{two}} = 2^3_{\text{ten}}$.

Example 41 • What decimal numeral is equivalent to 1001_{two}?

Solution

$$\begin{aligned} 1001_{\text{two}} &= [1(10)^3 + 0(10)^2 + 0(10) + 1]_{\text{two}} \\ &= [1(2)^3 + 0(2)^2 + 0(2) + 1]_{\text{ten}} \\ &= 9_{\text{ten}} \end{aligned}$$ •

Example 42 • Change 10111_{two} to a base ten numeral.

Solution

$$\begin{aligned} 10111_{\text{two}} &= [1(2)^4 + 0(2)^3 + 1(2)^2 + 1(2) + 1]_{\text{ten}} \\ &= 23_{\text{ten}} \quad \bullet \end{aligned}$$

Suppose that we wish to change 23 in base ten to a base seven numeral. What is the largest power of 7 contained in 23? Is there a 7^2 contained in 23? No, $7^2 = 49$ is larger than 23. Is there a 7 contained in 23? Yes, $3(7) = 21$ is contained in 23, with a remainder of 2. Hence, $23_{\text{ten}} = [3(7) + 2]_{\text{ten}} = [3(10) + 2]_{\text{seven}} = 32_{\text{seven}}$. Do you see that the pattern or rule for changing from base ten to base seven requires division by powers of 7?

Example 43 • Express 59 as a numeral in base two.

Solution Since the numeral is to be expressed in base two, remember that in base ten the powers of 2 are 1, 2, 4, 8, 16, 32, 64, and so on. Since 59 is less than 64, it is first necessary to find the number of 32's in 59.

```
32 | 59   | 1
     32
     --
     27
```

The remainder is 27. The next step is to determine how many 16's are in 27.

```
32 | 59   | 1
     32
16 | 27   | 1
     16
 8 | 11   | 1     How many 8's are in 11?
      8
 4 |  3   | 0     How many 4's are in 3?
      0
 2 |  3   | 1     How many 2's are in 3?
      2
 1 |  1   | 1     How many 1's are in 1?
      1
      -
      0
```

Hence,

$$\begin{aligned} 59_{\text{ten}} &= [1(32) + 1(16) + 1(8) + 0(4) + 1(2) + 1(1)]_{\text{ten}} \\ &= [1(2)^5 + 1(2)^4 + 1(2)^3 + 0(2)^2 + 1(2) + 1]_{\text{ten}} \\ &= [1(10)^5 + 1(10)^4 + 1(10)^3 + 0(10)^2 + 1(10) + 1]_{\text{two}} \end{aligned}$$

Thus,

59 is expressed in base two as 111011_{two} •

The theory illustrated for bases two and seven may be applied to other bases as indicated in the following example.

Example 44 • (a) Give a visual representation of 234_{five}.
(b) Find a decimal numeral equal to 2201_{three}.
(c) Change 4695 to a base eight numeral.

Solution (a)

Figure 6

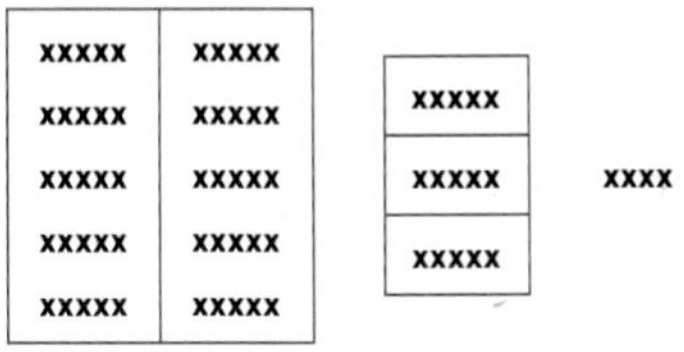

xxxxx xxxxx
xxxxx xxxxx
xxxxx xxxxx
xxxxx xxxxx
xxxxx xxxxx

xxxxx
xxxxx
xxxxx

xxxx

(b)
$$\begin{aligned} 2201_{\text{three}} &= [2(10)^3 + 2(10)^2 + 0(10) + 1]_{\text{three}} \\ &= [2(3)^3 + 2(3)^2 + 1]_{\text{ten}} \\ &= 54 + 18 + 1 \\ &= 73 \end{aligned}$$

(c) Now $8^2 = 64$; $8^3 = 512$; and $8^4 = 4096$.

4096 | 4695 | 1
4096
512 | 599 | 1
512
64 | 87 | 1
64
8 | 23 | 2
16
7

$$\begin{aligned} 4695 &= [1(8)^4 + 1(8)^3 + 1(8)^2 + 2(8) + 7]_{\text{ten}} \\ &= [1(10)^4 + 1(10)^3 + 1(10)^2 + 2(10) + 7]_{\text{eight}} \\ &= 11127_{\text{eight}} \end{aligned}$$ •

A system that has received wide attention and that has even been suggested as a replacement for the decimal system is the base twelve **duodecimal system.** An argument for having twelve as a base is that twelve has more divisors than ten; the only counting numbers that divide 10 are 1, 2, 5, 10, whereas the divisors of 12 are 1, 2 3, 4, 6, 12. Georges Buffon, a French naturalist, suggested over 200 years ago that the base twelve system be universally adopted, a suggestion that has been carried into this century. Even today, the Duodecimal Society of America advocates the base twelve system as a replacement for our present system.

Just for fun *Select a number between 1 and 15. Tell on which card or cards, C, A, R, and/or D, the number appears. We can tell you the number that you selected. Suppose that you tell us that the number is on cards C and D. We tell you, without looking, that the number is 9. How does the trick work? Hint: Write the numbers 1 to 15 in binary notation.*

C	A	R	D
1	2	4	8
3	3	5	9
5	6	6	10
7	7	7	11
9	10	12	12
11	11	13	13
13	14	14	14
15	15	15	15

Figure 7 illustrates how we group by dozens. The *x*'s in part (a) are denoted by 13_{twelve} because there is one group of a dozen and three ones. In Figure 7(b), there are two groups of a dozen and nine ones, written 29_{twelve}.

Figure 7

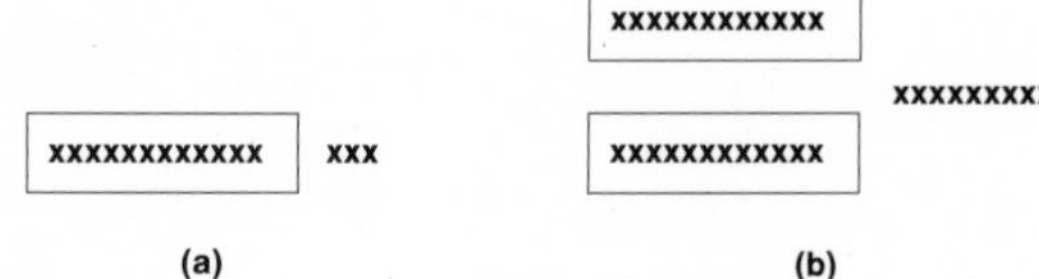

To group and count by twelves requires that two new symbols be introduced. T and E are the symbols commonly used for what we usually call ten and eleven. The base twelve numerals are as follows: 1, 2, 3, . . . , 8, 9, T, E, 10, 11, 12, . . . , 19, 1T, 1E, 20, 21, ... You recognize that T_{twelve} is 10_{ten}; $E_{\text{twelve}} = 11_{\text{ten}}$; and $10_{\text{twelve}} = 12_{\text{ten}}$.

Exercise set 4

R 1. Write the first 20 counting numbers, using each of the following as a base.

(a) Seven (b) Two

2. Group 15 dots to find the corresponding numeral in each of the following bases.

(a) Base ten (b) Base seven
(c) Base two

3. Repeat Exercise 2 for 54 dots.

4. Write in the indicated base the preceding counting number and the next consecutive counting number for each of the following.

(a) 16_{seven} (b) 666_{seven}
(c) 111_{two} (d) 1011_{two}

5. Write each of the given numerals as a base ten numeral.

(a) 601_{seven} (b) 660_{seven}
(c) 1110_{two} (d) 101101_{two}

6. Change each of the base ten numerals below to a base seven numeral.

(a) 9 (b) 35 (c) 55
(d) 285 (e) 1000 (f) 5280

7. Repeat Exercise 6 by changing each of the numerals to a base two numeral.

8. How many different digits are needed for base five? Base eight? Base sixteen?

9. Find the missing digit for each of the following.

(a) $1\_\_011_{two} = 19_{ten}$ (b) $\_\_34_{seven} = 221_{ten}$
(c) $3\_\_4_{seven} = 186_{ten}$ (d) $11\_\_11_{two} = 27_{ten}$

10. Write the first 50 counting numbers in the duodecimal system.

11. Work Exercise 10 for base five.

12. Work Exercise 3 for the following bases.

(a) Twelve (b) Five (c) Eight

13. Write in the indicated base the counting numbers that precede and that follow each of the following.

(a) EEE_{twelve} (b) 607_{eight} (c) 222_{three}

14. Write each of the following as a base ten numeral.

(a) 157_{nine} (b) $E6_{twelve}$ (c) 504_{six}
(d) 2010_{four} (e) 430_{eight} (f) $T0E2_{twelve}$

15. Work Exercise 6 by changing each expression to base twelve.

16. Work Exercise 6 by changing each expression to base five.

17. Choose the largest number from each of the following lists.

(a) 5_{six}, 11_{four}, 101_{three}
(b) 122_{three}, 112_{four}, 76_{eight}
(c) ET_{twelve}, 101_{eight}, 11110_{two}
(d) 325_{twelve}, 523_{six}, 10122_{three}

T 18. Find the base indicated by the letter *b*.

(a) $67_{ten} = 61_b$
(b) $12_{ten} = 1100_b$
(c) $234_{ten} = 176_b$

19. A bookstore ordered 8 gross, 7 dozen, and 5 erasers. Express the number of erasers as a base ten numeral.

20. Hubert decided to fill out an application for employment as a freight-car loader in base six. He listed his age as 25, his height as 145 inches, and his weight as 302 pounds. The supervisor, glancing over the application, thought Hubert a giant and decided to hire him. What were Hubert's actual statistics?

21. (a) Write the largest three-digit number found in base four, base nine, and base twelve.
(b) Write the smallest four-digit number in each of these bases.
(c) What base ten numbers equal these six numbers?

22. Change the following numerals to the base indicated.

(a) 231_{four} to base twelve
(b) $27TE_{twelve}$ to base six

23. A time machine carried an adventurer far into the future. Although welcomed by the natives of Septuland, he began to question their honesty when they sent him 7 wives rather the 10 that had been promised him and only 19 free movie passes when they had promised 25. In what base did these natives do their mathematics?

24. If the adventurer in Exercise 23 was promised the following number of articles by the natives, determine the number (from the base ten adventurer's point of view) that he would actually receive.

(a) 15 suits (b) 225 bananas
(c) 12 milkshakes (d) 24 rugs
(e) 113 records (f) 66 neckties

C 25. Can you characterize even and odd numbers merely by looking at the units digit of a given number when it is expressed in base two? In base three? In base four? In base five?

26. Write the Biblical phrase "three score and ten" as an expression in base twenty, using the symbol T for ten. Do the same for "four score and seven" from the Gettysbury Address. What numbers do these expressions represent in base ten?

27. You have two quarters, four nickels, and two pennies. Use base five to write the number indicating your financial status.

*5 Computations in different bases

■ The Acme Potato Chip Company received 6 freight cars supposedly full of potatoes in 100-lb. bags. It was learned that the automatic weighing machine was broken for a while and that some of the freight cars were full of 90-lb. bags of potatoes. Mr. Pate said, "Let's load some sacks of potatoes from each freight car into our truck. Then in one weighing we shall locate the freight cars containing sacks with the short weight." How can he accomplish this feat? ■

Hint: This problem can be solved using ideas from the binary number system. In addition, computations in bases other than ten provide excellent practice and increase understanding of the algorithms in base ten. In fact, the same algorithms are used in other bases as are used in base ten.

In this section, we shall do many of the examples in two ways. We shall write the problems in expanded notation in base seven (or base two) and then do computations in that base. Alternatively, we shall do the computations in base ten and then think backward to the base in question. If you learn to understand the difference in these two procedures, then you have not only mastered computation in different bases; you have also attained proficiency in the algorithms given in Sections 2 and 3 of this chapter.

To perform operations in any base, one needs to have available addition and multiplication tables. To avoid having to memorize tables in base two and base seven, these tables are provided in this section. Verify the additions in Tables 5 and 6.

Table 5
Addition Table, Base Two

+	0	1
0	0	1
1	1	10

Table 6
Addition Table, Base Seven

+	0	1	2	3	4	5	6 ↓
0	0	1	2	3	4	5	6
1	1	2	3	4	5	6	10
2	2	3	4	5	6	10	11
3	3	4	5	6	10	11	12
4	4	5	6	10	11	12	13
→ 5	5	6	10	11	12	13	14
6	6	10	11	12	13	14	15

Example 45 • Find $5_{\text{seven}} + 6_{\text{seven}}$.

Solution Looking at the row and column designated by arrows in Table 6, $5_{\text{seven}} + 6_{\text{seven}} = 14_{\text{seven}}$. •

It is easy to see that the algorithm for the addition of numbers in base ten is applicable to the addition of numbers expressed in other bases.

Example 46 • Compute $26_{\text{seven}} + 34_{\text{seven}}$.

Solution In the computation below, we use Table 6 to see that $6 + 4 = 13$ in base seven.

$$\begin{aligned} 26_{\text{seven}} &= \left[\begin{array}{l} 2(10) + 6 \\ 3(10) + 4 \\ \hline 5(10) + 13 \\ 5(10) + [1(10) + 3] \\ 6(10) + 3 \end{array}\right]_{\text{seven}} \\ +34_{\text{seven}} & \\ &= 63_{\text{seven}} \end{aligned} \qquad \text{or} \qquad \left[\begin{array}{l} 2(7) + 6(1) \\ 3(7) + 4(1) \\ \hline 5(7) + 10(1) \\ 5(7) + 1(7) + 3(1) \\ 6(7) + 3(1) \end{array}\right]_{\text{ten}} = 63_{\text{seven}}$$ •

Some people are more successful in working this type of problem when they "think" base ten (that is, they perform all operations in base ten) and write the answers in terms of the given base. The solution on the right in the preceding example illustrates this procedure.

The renaming process is performed in the next example without writing the numerals in expanded form. If you have difficulty understanding the computations, write each numeral in expanded form.

Example 47 • Find $111_{\text{two}} + 101_{\text{two}}$.

Solution

$$\left[\begin{array}{r} 111 \\ 101 \\ \hline 1100 \end{array}\right]_{\text{two}}$$ •

Example 48 • What is $11_{\text{seven}} - 2_{\text{seven}}$?

Solution Looking in Table 6, we find the answer to be 6_{seven}. (*Hint:* What number must be added to 2_{seven} to obtain 11_{seven}?) •

Example 49 • Compute $624_{\text{seven}} - 246_{\text{seven}}$.

Solution

$$\begin{array}{r} 624_{\text{seven}} \\ -246_{\text{seven}} \\ \hline \end{array} \qquad \begin{array}{r} 6(10)^2 + 2(10) + 4 \\ -[2(10)^2 + 4(10) + 6] \\ \hline \end{array} \qquad \begin{array}{r} 5(10)^2 + 11(10) + 14 \\ -[2(10)^2 + 4(10) + 6] \\ \hline 3(10)^2 + 4(10) + 5 \end{array} = 345_{\text{seven}}$$

We work the same problem by performing the operations in base ten.

$$\begin{array}{r} 624_{\text{seven}} \\ -246_{\text{seven}} \\ \hline \end{array} \qquad \begin{array}{r} 6(7)^2 + 2(7) + 4 \\ -[2(7)^2 + 4(7) + 6] \\ \hline \end{array} \qquad \begin{array}{r} 5(7)^2 + 8(7) + 11 \\ -[2(7)^2 + 4(7) + 6] \\ \hline 3(7)^2 + 4(7) + 5 \end{array} = 345_{\text{seven}}$$ •

To perform multiplications and divisions, we need Tables 7 and 8.

Table 7
Multiplication Table, Base Two

·	0	1
0	0	0
1	0	1

Table 8
Multiplication Table, Base Seven

·	0	1	2	3	4	5	6
0	0	0	0	0	0	0	0
1	0	1	2	3	4	5	6
2	0	2	4	6	11	13	15
3	0	3	6	12	15	21	24
4	0	4	11	15	22	26	33
5	0	5	13	21	26	34	42
6	0	6	15	24	33	42	51

Example 50 • Multiply 3 and 5 in base seven.

Solution From Table 8, in the row with label 3 and in the column with 5 at the top, we find 21. Thus

$$[3 \cdot 5 = 21]_{\text{seven}} \quad \bullet$$

Example 51 • Find the product of 216_{seven} and 14_{seven}.

Solution

$$\left[\begin{array}{r} 2(10)^2 + 1(10) + \ \ 6 \\ 1(10) + \ \ 4 \\ \hline 11(10)^2 + 4(10) + 33 \\ 2(10)^3 + \ \ 1(10)^2 + 6(10) \qquad \\ \hline \end{array}\right]_{\text{seven}}$$

or

$$\left[\begin{array}{r} 2(10)^2 + \qquad 1(10) + 6 \\ 1(10) + 4 \\ \hline 1(10)^3 + 1(10)^2 + (3 + 4)(10) + 3 \\ 2(10)^3 + 1(10)^2 + \qquad 6(10) \qquad \\ \hline \end{array}\right]_{\text{seven}}$$

or

$$\left[\begin{array}{r} 2(10)^2 + 1(10) + 6 \\ 1(10) + 4 \\ \hline 1(10)^3 + 2(10)^2 \qquad\quad + 3 \\ 2(10)^3 + 1(10)^2 + 6(10) \qquad \\ \hline 3(10)^3 + 3(10)^2 + 6(10) + 3 \end{array}\right]_{\text{seven}} \quad \text{or} \quad \left[\begin{array}{r} 216 \\ 14 \\ \hline 1203 \\ 216\ \ \\ \hline 3363 \end{array}\right]_{\text{seven}}$$

The same problem can also be worked by performing all operations in base ten and writing the answers in terms of base seven.

$$\left[\begin{array}{r} 2(7)^2 + 1(7) + 6 \\ 1(7) + 4 \\ \hline 8(7)^2 + 4(7) + 24 \\ 2(7)^3 + 1(7)^2 + 6(7) \quad\quad \\ \hline \end{array}\right]_{\text{ten}} \quad \text{or} \quad \left[\begin{array}{r} 2(7)^2 + 1(7) + 6 \\ 1(7) + 4 \\ \hline 1(7)^3 + 2(7)^2 + 0(7) + 3 \\ 2(7)^3 + 1(7)^2 + 6(7) \quad\quad \\ \hline 3(7)^3 + 3(7)^2 + 6(7) + 3 \end{array}\right]_{\text{ten}}$$

$$216_{\text{seven}} \cdot 14_{\text{seven}} = 3363_{\text{seven}}$$

Consider the following calculation and the suggested helps in base ten.

$$\left[\begin{array}{r} 216 \\ 14 \\ \hline 1203 \\ 216\quad \\ \hline 3363\quad \end{array}\right]_{\text{seven}} \left[\begin{array}{l} 4 \cdot 6 = 24 = 3(7) + 3 \\ 4 \cdot 1 + 3 = 7 = 1(7) + 0 \\ 4 \cdot 2 + 1 = 9 = 1(7) + 2 \\ \text{first line } 1203 \\ 1 \cdot 6 = 6,\ 1 \cdot 1 = 1, \text{ and } 1 \cdot 2 = 2 \\ \text{second line } 216 \end{array}\right]_{\text{ten}}$$ •

Example 52 • Multiply 1011 and 101 in base two.

Solution

$$\left[\begin{array}{r} 1011 \\ 101 \\ \hline 1011 \\ 1011\quad \\ \hline 110111 \end{array}\right]_{\text{two}}$$ •

Division will be illustrated with the scaffold format and in expanded form involving powers of the base.

Example 53 • Divide 11010_{two} by 1011_{two}.

Solution

$$\begin{array}{r|r|l} 1011 & 11010 & \\ & 10110 & 10 \\ \cline{2-2} & 100 & \\ & 0 & 0 \\ \cline{2-2} & 100 & \end{array}$$

The answer is 10_{two} with a remainder of 100_{two}. •

Example 54 • Divide 1662_{seven} by 24_{seven}.

$$\begin{array}{r|r} & 5(10) + 2 \\ \hline 2(10) + 4 & 1(10)^3 + 6(10)^2 + 6(10) + 2 \\ & 1(10)^3 + 5(10)^2 + 6(10) \quad\quad \\ \cline{2-2} & 1(10)^2 + 0(10) + 2 \\ & 5(10) + 1 \\ \cline{2-2} & 2(10) + 1 \end{array}$$

The answer is 52_{seven} with a remainder of 21_{seven}. •

Just for fun

Scientists use the binary system in the coded messages sent into space with the hope that some intelligent being might decipher the message. Use the guide below, where each space is labeled either 1 or 0 according to the message below, and darken the spaces marked by 0. What is this message that has been sent into space?

0	1	0	1	0	0	1	0	1	0
0	1	0	1	0	1	1	0	1	0
0	0	1	1	0	0	1	1	0	1
0	1	0	1	0	1	1	1	0	1
0	1	0	1	0	0	1	1	0	1
1	1	1	1	1	1	1	1	1	1
0	0	0	1	1	1	1	1	1	1
0	0	0	0	0	0	0	0	0	0
0	0	0	1	1	1	1	0	1	0
1	1	1	1	1	1	1	0	1	0

Let's return now to the problem in the introduction to this section.

Understand the problem. There are 6 freight cars. Some are filled with 90-lb. and some with 100-lb. bags of potatoes. We are to load one or more sacks of potatoes from each freight car onto a truck. Then in one weighing we want to find the freight cars with the 90-lb. bags of potatoes.

Devise a plan. Line up the six freight cars in a row and label them

$$1, 2, 3, 4, 5, 6$$

Let's take 1 bag of potatoes from the first freight car, 2 from the second, 4 from the third, 8 from the fourth, 16 from the fifth, and 32 from the sixth. So we have $1 + 2 + 4 + 8 + 16 + 32 = 63$ bags of potatoes. Place all of these in a truck and find the total weight. If all sacks weighed 100 lb., the correct answer should be 6300 lb. more than the truck. Suppose the answer is 5870 lb. more than the truck (that is, the potatoes in the truck are 430 lb. too light). Since each light sack differs from 100 lb. by 10 lb., then there are 43 light sacks in the truck.

Since the number of sacks taken from each freight car is a power of two, we need to write 43 as a sum of powers of two (binary number).

$$43 = 1(32) + 0(16) + 1(8) + 0(4) + 1(2) + 1$$

So the 90-lb. bags are found in the freight cars with labels 6, 4, 2, and 1, as 32 sacks come from freight car 6, 8 from car 4, 2 from car 2, and 1 from car 1.

Exercise set 5

R 1. Perform the following additions.

(a) $304_{\text{seven}} + 366_{\text{seven}}$ (b) $1111_{\text{two}} + 101_{\text{two}}$
(c) $1011_{\text{two}} + 1001_{\text{two}}$ (d) $562_{\text{seven}} + 453_{\text{seven}}$

2. Perform the following subtractions.

(a) $1100_{\text{two}} - 11_{\text{two}}$
(b) $100101_{\text{two}} - 10011_{\text{two}}$
(c) $1111_{\text{seven}} - 555_{\text{seven}}$
(d) $404_{\text{seven}} - 65_{\text{seven}}$

3. Perform the following multiplications.

(a) $(1101_{\text{two}})(11_{\text{two}})$ (b) $(366_{\text{seven}})(34_{\text{seven}})$
(c) $(454_{\text{seven}})(205_{\text{seven}})$ (d) $(1101_{\text{two}})(101_{\text{two}})$

T 4. Make an addition table for the following bases.

(a) Three (b) Five (c) Twelve

5. Perform the following operations.

(a) $35_{\text{twelve}} + 15_{\text{twelve}}$
(b) $2304_{\text{five}} + 121_{\text{five}}$
(c) $12_{\text{twelve}} - 9_{\text{twelve}}$
(d) $121_{\text{three}} + 22_{\text{three}}$
(e) $8T2_{\text{twelve}} + 26E_{\text{twelve}}$
(f) $321_{\text{five}} - 143_{\text{five}}$
(g) $503_{\text{twelve}} - 2TE_{\text{twelve}}$
(h) $714_{\text{twelve}} - ET_{\text{twelve}}$
(i) $320_{\text{five}} - 43_{\text{five}}$
(j) $3E8_{\text{twelve}} + 3TT_{\text{twelve}}$

6. Make a multiplication table for the given bases.

(a) Three (b) Five (c) Twelve

7. Perform the indicated operations.

(a) $(44_{\text{five}})(4_{\text{five}})$ (b) $(22_{\text{twelve}})(5_{\text{twelve}})$
(c) $(323_{\text{five}})(43_{\text{five}})$ (d) $(40E_{\text{twelve}})(3T_{\text{twelve}})$
(e) $(2012_{\text{three}})(21_{\text{three}})$ (f) $(2222_{\text{three}})(22_{\text{three}})$

8. For each of the bases given, name the largest and the smallest number represented by a three-digit numeral of the form *abc*, where $a \neq 0$. Find the difference of these two numbers.

(a) Five (b) Eleven
(c) Three (d) Six

9. Our time system is a base 60 system. Add and subtract the following pairs of times.

(a) 6 hours 40 minutes 31 seconds
3 hours 51 minutes 37 seconds
(b) 8 hours
6 hours 10 minutes 15 seconds

10. Perform the following divisions.
(a) $11101_{\text{two}} \div 11_{\text{two}}$ (b) $460_{\text{seven}} \div 23_{\text{seven}}$
(c) $1666_{\text{seven}} \div 102_{\text{seven}}$

C 11. (a) If $243 + 461 = 724$, what base is being used?
(b) If $576 + 288 = 842$, what base is being used?
(c) If $4203 + 434 = 10142$, what base is being used?
(d) If $4203 + 434 = 5041$, what base is being used?
(e) If $4203 + 434 = 4640$, what base is being used?

12. For what possible bases is the following addition correct?

$$\begin{array}{r} 4203 \\ 434 \\ \hline 4637 \end{array}$$

13. For what base is each of the following true?

(a) $10 - 4 = 4$ (b) $2 \cdot 3 = 12$ (c) $11 \div 3 = 4$

14. Check the given problem to show that it could have been performed in base three or base four. Could it have been performed in any other bases?

$$\begin{array}{r} 101 \\ 12\,\overline{)\,1212} \\ 12 \\ \hline 01 \\ 0 \\ \hline 12 \\ 12 \\ \hline \end{array}$$

15. Determine the error in each calculation below and then correct each mistake.

(a)
$$\begin{array}{r} 11010010_{\text{two}} \\ -11111_{\text{two}} \\ \hline 10998899_{\text{two}} \end{array}$$

(b)
$$\begin{array}{r} 2132_{\text{nine}} \\ -2004_{\text{nine}} \\ \hline 128_{\text{nine}} \end{array}$$

(c)
$$\begin{array}{r} 47E_{\text{twelve}} \\ +145_{\text{twelve}} \\ \hline 606_{\text{twelve}} \end{array}$$

(d)
$$\begin{array}{r} 435_{\text{seven}} \\ 313_{\text{seven}} \\ +111_{\text{seven}} \\ \hline 1060_{\text{seven}} \end{array}$$

16. Perform the following operations. Leave the answer in terms of the base of the numeral on the left.

(a) $563_{\text{seven}} - ET_{\text{twelve}}$
(b) $11011_{\text{two}} + 323_{\text{four}} + TE_{\text{twelve}}$
(c) $10011_{\text{two}} + 423_{\text{five}}$
(d) $E4T_{\text{twelve}} - 11111_{\text{two}}$

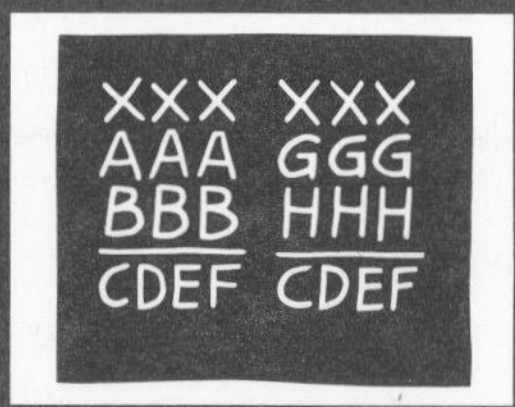

Solution to introductory problem

Understand the problem. Each letter represents a different digit in these additions. Further, both additions give the same sum. What are X, A, B, G, and H?

Devise a plan. By guess and revise, and by your knowledge of numeration systems, you should obtain the answer below.

$$\begin{array}{rcl} XXX &=& 888 \\ AAA &=& 555 \\ BBB &=& 666 \\ \hline CDEF &=& 2109 \end{array} \qquad \begin{array}{rcl} XXX &=& 888 \\ GGG &=& 444 \\ HHH &=& 777 \\ \hline &=& 2109 \end{array}$$

Summary and review

Exponential notation

1. For whole numbers n, a^n is defined by
   (a) $a^n = \underbrace{a \cdot a \cdot a \cdot \cdots \cdot a}_{n \text{ factors}}$ if n is a counting number
   (b) $a^0 = 1$
2. Operations involving exponents
   (a) $a^x \cdot a^y = a^{x+y}$ (b) $(a^x)^y = a^{xy}$
   (c) $(ab)^x = a^x \cdot b^x$
3. In the symbol b^a, b is called the *base* and a is called the *exponent*, b^a is called the *a*th power of b.

Numeration systems

4. If the numerals are written as sums of products of powers of b, we call the system a base b numeration system.
   (a) The binary numeration system is a base two system.
   (b) The decimal numeration system is a base ten system.
5. Historical numeration systems
   (a) The Egyptian system was an additive numeration system with neither a concept of place value nor zero. The Egyptian numerals with their modern equivalents are (1, 𓏺), (10, 𓎆), (100, 𓍢), (1000, 𓆼), (10,000, 𓂭), (100,000, 𓆐), (1,000,000, 𓁨).
   (b) The Babylonian system was an additive system that included a base sixty place-value feature but had no zero. The symbols used were 𒌋 for ten and 𒁹 for one. The numbers 1 to 59 were formed additively, and the place-value system was used to form larger numbers.
   (c) The Chinese-Japanese system was a base ten place-value system. However, the symbols for the numbers were written vertically, and the places were named. The numerals are (1, 一), (2, 二), (3, 三), (4, 四), (5, 五), (6, 六) (7, 七), (8, 八), (9, 九). The place-value names are (10, 十), (100, 百), (1000, 千).
   (d) The Roman system was an additive system with one peculiar place-value principle: If a symbol were preceded by a symbol of lesser value, the lesser value was subtracted. The Roman numerals with their modern equivalents are (1, I), (5, V), (10, X), (50, L), (100, C), (500, D), (1000, M). A bar over a numeral multiplied the value of the numeral by 1000.
   (e) The Mayan numeration system was a place-value system with a base of 20. A dot represented 1; a line represented 5; and 𝋠 represented zero. The numerals were written in vertical form. One eccentricity of this base twenty system is the fact that the third place represented (18)(20) instead of $(20)^2$, and the fourth place represented $(18)(20)^2$ instead of $(20)^3$.
6. (a) When a base ten number is written as a sum of multiples of 1, 10, 100, 1000, 10,000, . . . , we say that the number is written in expanded notation.
   (b) When a base b number is written as a sum of multiples of powers of b, such as b^1, b^2, b^3, $b^4, \ldots$, we say that the number is written in expanded notation in terms of its base.

Algorithms for arithmetic

7. We can more clearly understand our usual algorithms for addition and subtraction if we write the numerals in either expanded notation or expanded notation in terms of powers of 10.

8. We can more clearly understand our usual algorithm for multiplying whole numbers if we understand multiplication using a horizontal format, using a vertical format, and using a vertical format with powers of 10.

9. The Division Algorithm: If a and b are whole numbers with $b \neq 0$, then there exist whole numbers q and r with $0 \leq r < b$ such that $a = bq + r$. Here q is called the *quotient* of the division $a \div b$, and r is the *remainder*.

10. We can more clearly understand our usual algorithm for division using a scaffold format, using a vertical format with full notation and using our usual format.

Review exercise set 6

R 1. Simplify.

(a) $(3)(3)(3)^4$ (b) $(8)^2 \cdot (2)^5$
(c) $(9)^0(4)$ (d) $[(a)^3(a)^5]^4$
(e) $(2)^4(2)^3$ (f) $2^0 \cdot 3^0$

2. Determine the next counting number in the original base for each of the following and write the answer as a base ten numeral.

(a) 10110_{two} (b) 46_{seven}
(c) 466_{seven} (d) 101011_{two}

3. The numeral 333 is a three-digit numeral. If you were permitted to change one of its digits from 3 to 4, which digit should be changed to

(a) alter the size of the number the least?
(b) alter the size of the number the most?

4. How much smaller will the number 476 be if

(a) 3 is subtracted from the second digit?
(b) 4 is subtracted from the units digit?
(c) 1 is subtracted from the hundreds digit?

5. Write the following numbers in base seven and in base two.

(a) 47 (b) 106 (c) 421 (d) 621

6. Write the following as base ten numerals.

(a) 543_{seven} (b) 406_{seven}
(c) 1101_{two} (d) 10001_{two}

7. Perform the following additions and subtractions in two ways. (i) Write each number in expanded notation. (ii) Write each number in expanded notation using powers of 10.

(a) $\begin{array}{r} 436 \\ 243 \\ \hline \end{array}$ (b) $\begin{array}{r} 524 \\ +361 \\ \hline \end{array}$

(c) $\begin{array}{r} 476 \\ -243 \\ \hline \end{array}$ (d) $\begin{array}{r} 584 \\ -361 \\ \hline \end{array}$

8. Draw a visual representation of the following multiplications (distributive property), show this multiplication in horizontal format, show this multiplication in expanded vertical format, and then show multiplication in usual format.

(a) $3 \cdot 276$ (b) $4 \cdot 265$

9. Find the value of k to make the following statements true.

(a) $x^3 \cdot x^4 \cdot x^k = x^{10}$
(b) $(x^2)^k = x^4$
(c) $3^{12} = 9^k$
(d) $2^k = 32$
(e) $5^k \cdot 125 = 5^7$
(f) $3^k \cdot 27 = 3^8$

T 10. Use expanded notation, vertical format, and then our usual algorithm for addition (or subtraction) in each problem.

(a) $\begin{array}{r} 754 \\ +647 \\ \hline \end{array}$ (b) $\begin{array}{r} 847 \\ +686 \\ \hline \end{array}$

(c) $\begin{array}{r} 1431 \\ -\ 572 \\ \hline \end{array}$ (d) $\begin{array}{r} 2436 \\ -\ 568 \\ \hline \end{array}$

11. Work the problems in Exercise 10 using powers of 10.

12. Perform the following divisions in three ways: (i) scaffold format, (ii) full notation format, (iii) usual format.

    (a) $1728 \div 32$ (b) $2301 \div 47$
    (c) $23{,}410 \div 37$ (d) $2435 \div 867$

13. A time machine took an adventurer back to visit many ancient civilizations. The adventurer found that his work in elementary mathematics helped him understand these civilizations. What answers did he find to the following problems? (Write the answers in base ten.)

    (a) CXLIV + DCLXVI
    (b) ▼▼▼▼▼ ‹ + ‹ ▼▼▼▼
    (c) 𓆼 9199 ∩|∩∩ − 919∩919|99|

14. Perform the indicated operations.

    (a) $3123_{\text{five}} + 2314_{\text{five}}$
    (b) $930E_{\text{twelve}} - 3T4_{\text{twelve}}$
    (c) $(T2_{\text{twelve}})(58_{\text{twelve}})$
    (d) $(234_{\text{six}})(4_{\text{six}})$
    (e) $(455_{\text{six}})(24_{\text{six}})$
    (f) $(58E_{\text{twelve}})(E_{\text{twelve}})$

15. Express 109 in

    (a) Egyptian numerals.
    (b) Babylonian numerals.
    (c) Chinese-Japanese numerals.
    (d) Roman numerals.

C 16. Perform the following divisions, leaving the answer in the same base as the a in $a \div b$.

    (a) $(4T0E_{\text{twelve}}) \div (1101_{\text{two}})$
    (b) $(1204_{\text{five}}) \div (212_{\text{three}})$

17. Name the base used in the following problems.

    (a) $54 \div 8 = 8$
    (b) $403 - 134 = 214$
    (c) $452 - 263 = 156$
    (d) $1111 + 1000 = 10111$
    (e) $(48)(32) = 1294$
    (f) $(604)(35) = 31406$

Bibliography

1. Baroody, Arthur J. "The Value of Informal Approaches to Mathematics Instruction and Remediation." *AT* (January 1986), 14–18.
2. Bates, Tom, & Leo Rousseau. "Will the Real Division Algorithm Please Stand Up?" *AT* (March 1986), 42–46.
3. Beattie, Ian D. "Modeling Operations and Algorithms." *AT* (February 1986), 23–28.
4. Bennett, Albert B., Jr., & Leonard T. Nelson. *Mathematics: an Activity Approach.* Boston, Mass.: Allyn & Bacon, 1979.
5. Boykin, Wilfred E. "The Russian-Peasant Algorithm: Rediscovery and Extension." *AT* (January 1973), 29–32.
6. Bradford, John W. "Methods and Materials for Learning Subtraction." *AT* (February 1978), 18–20.
7. Brown, Stephen I. "A New Multiplication Algorithm: On the Complexity of Simplicity." *AT* (November 1975), 546–554.
8. Cacha, Frances B. "Subtraction: Regrouping with Flexibility." *AT* (May 1975), 402–404.
9. Cacha, Frances B. "Understanding Multiplication and Division of Multidigit Numbers." *AT* (May 1972), 349–354.
10. Greenwood, May. "Critique on the Chisanbop Finger Calculating Math." *AT* (March 1979), 18–21.
11. Hall, William D. "Division with Base-Ten Blocks." *AT* (November 1983), 21–25.
12. Hardin, Daniel D. "Teaching Base Three? In a Pig's Ear!" *AT* (September 1979), 48–49.
13. Ikeda, Hitoshi, & Masere Ando. "A New Algorithm for Subtraction?" *AT* (December 1974), 716–719.
14. Killian, Lawrence, Edna Cahill, Carolann Ryan, Deborah Sutherland, & Diane Taccetta. "Errors That Are Common in Multiplication." *AT* (January 1980), 22–25.
15. King, Irv. "Giving Meaning to the Addition Algorithm." *AT* (May 1972), 345–348.
16. Leutzinger, Larry P., & Glenn Nelson. "Let's Do It: With Powers of Ten." *AT* (February 1980), 8–13.

17. Panker, Andy. "Pattern Discovery with Binary Trees." *MT* (May 1979), 337–341.

18. Peera, Zehra. "Number Patterns and Bases." *AT* (October 1982), 52–53.

19. Reardin, C. Richard, Jr. "Understanding the Russian Peasant." *AT* (January 1973), 33–35.

20. Shaw, Jean M. "Let's Do It: A-Plus for Counters." *AT* (September 1983), 10–14.

21. Smith, Douglas B., & William R. Topp. *An Activity Approach to Elementary Concepts of Mathematics.* Reading, Mass.: Addison-Wesley, 1961.

22. Smith, Karl J. "Inventing a Numeration System." *AT* (November 1973), 550–553.

23. Wheatley, Charlotte L., & Grayson H. Wheatley. "How Shall We Teach Column Addition? Some Evidence." *AT* (January 1978), 18–19.

CHAPTER 4

Elementary Number Theory and the System of Integers

A trainer in an athletic department was asked to arrange the towels in the locker room in stacks of equal size. When he separated the towels into sets of 4, one was left over. When he tried stacks of 5, one was left over. The same was true for stacks of 6. However, he was successful in arranging the towels in stacks of 7 each. What is the smallest possible number of towels in the locker room?

The great mathematician Karl Friedrich Gauss (1777–1855) said, "Mathematics is the queen of the sciences, and arithmetic is the queen of mathematics." By *arithmetic,* Gauss meant the subject of this chapter—number theory.

As an introduction to elementary number theory, we shall examine a number and ask which natural numbers divide it. For instance,

271,352,468,925,721,358,270

is a large number. Is it divisible by 6? By 7? By 9? In this chapter, we shall formulate procedures to test for divisibility.

Also, we shall define greatest common divisor and least common multiple, both of which are very useful in solving a variety of problems. We later use the least common multiple in Chapter 5, because the least common multiple of two denominators is the least common denominator.

If Thanksgiving is on November 26, on what day of the week does Christmas fall, or if July 4th is on Sunday, what will be the date of Labor Day? Problems of this nature are easily solved using what are called *congruences.* You will find this material easy because we introduce this subject of congruences with the arithmetic of a clock.

In this chapter, you will be exposed to a number of unsolved problems. Maybe you can find a solution to one of these.

1 Divisibility

■ While discussing a new machine for packing packages of potato chips, Mr. Pate explained that the previous hour's production of 190,236 large sacks had been packed in boxes of 8 each. One of his employees questioned Mr. Pate's production figures. Who was correct? ■

A study of the divisibility properties will not only provide an immediate answer to the preceding problem but will also serve as the beginning of a study of **number theory.** Number theory is concerned primarily with relationships between natural numbers. This study has involved some of the great minds through the ages: Pythagoras (550 B.C.), Euclid (330 B.C.), Fermat (1601–1665), and Euler (1707–1783). Moré recently, the dėvelopment of the computer has sparked new and widespread interest in number theory.

In this section we develop important properties of divisibility. Then we formulate rules for determining divisibility by several natural numbers. You will be surprised by some of these properties and will learn much about our number system as you come to understand these properties.

We shall first formulate some rules that determine divisibility by several small natural numbers. Let us begin by defining the expression "*a* divides *b*."

Divides If a is any natural number and b is any whole number, then a divides b (denoted by $a|b$) if and only if there exists a whole number c such that $b = ac$. Alternatively, we say that b is divisible by a.

Example 1 •

$$24 \div 8 = 3 \quad \text{or} \quad 24 = 8 \cdot 3$$

so

8 divides 24
24 is divisible by 8
$8|24$ •

If a divides b, then it can be said, equivalently, that

a is a **divisor** of b
a is a **factor** of b
b is a **multiple** of a

Example 2 • $2|16$ (read as *2 divides 16*) because $2 \cdot 8 = 16$. Thus, 2 is a factor of 16, and 16 is a multiple of 2. •

Example 3 • Classify the following either as true or as false, where a is a natural number.

(a) $1|a$ (b) $a|a$ (c) $3|42$

(d) $4|0$ (e) $0|5$ (f) $0|0$

Solution (a) $1|a$ is true because $a = 1 \cdot a$.

(b) $a|a$ is true because $a = a \cdot 1$.

(c) $3|42$ is true because $42 = 3 \cdot 14$.

(d) $4|0$ is true because $0 = 4 \cdot 0$.

The definition of *divides* does not include division by 0 for the following reasons.

(e) $0|5$ is false because there is no whole number x such that $5 = 0 \cdot x$.

(f) $0|0$ is false because there is not a *unique* whole number x for which $0 = x \cdot 0$. •

Be careful to note the distinction between the statement $a|b$ (a divides b) and the symbol a/b, an alternate notation for the operation $a \div b$. But note that $a|b$ does mean that b/a is a whole number. We use $a \nmid b$ to indicate that a does not divide b; that is, $3 \nmid 8$ because there is no whole number x such that $8 = 3x$.

Example 4 • Note that $4|20$, since $20 = 4 \cdot 5$. Furthermore, $20|460$, since $460 = 20 \cdot 23$. Replacing 20 by $4 \cdot 5$ gives

$$460 = (4 \cdot 5) \cdot 23 = 4(5 \cdot 23)$$

Thus, $460 = 4 \cdot 115$, and so $4|460$. In general, if $x|y$ and $y|z$, then $x|z$. •

Example 5 • $4|24$ and $4|36$, since $24 = 4 \cdot 6$ and $36 = 4 \cdot 9$. Therefore,

$$24 + 36 = (4 \cdot 6) + (4 \cdot 9) = 4(6 + 9) = 4 \cdot 15$$

So $4|(24 + 36)$. In general, if $x|y$ and $x|z$, then $x|(y + z)$. •

Example 6 • Use the fact that $5|10$ to show that $5|60$.

Solution Now $10 = 2 \cdot 5$, so $5|10$. To see that $5|(10 \cdot 6)$, write 10 as $2 \cdot 5$.

$$\begin{aligned} 10 \cdot 6 &= (2 \cdot 5) \cdot 6 \\ &= (5 \cdot 2) \cdot 6 \qquad \text{Why?} \\ &= 5(2 \cdot 6) \qquad \text{Why?} \\ &= 5(12) \end{aligned}$$

So $5|(10 \cdot 6)$. In general, if $x|y$ or $x|z$, then $x|yz$; that is, if x divides y or x divides z, then x divides the product of y and z. •

The preceding discussion suggests some of the following properties of divisibility.

Properties of divisibility

> Let x be a natural number and y and z be whole numbers.
>
> **1.** $1|y$ and $x|x$.
> **2.** Transitive property. If $x|y$ and $y|z$ where $y \neq 0$, then $x|z$.
> **3.** If $x|y$ and $x|z$, then $x|(y + z)$ and $x|(y - z)$.
> **4.** If $x|y$ and $x|(y + z)$ or $x|(y - z)$, then $x|z$.
> **5.** If $x|y$ or $x|z$, then $x|yz$.
> **6.** If $x|y$ and $x \nmid z$, then $x \nmid (y + z)$ and $x \nmid (y - z)$.

Example 7 • (a) $1|7$ and $7|7$ illustrate Property 1.

(b) $3|6$ and $6|24$; therefore, $3|24$. This example illustrates the transitive property of divisibility given in Property 2.

(c) $7|49$ and $7|84$; hence, $7|133$ because $133 = 49 + 84$. $6|30$ and $6|18$, so $6|12$, since $12 = 30 - 18$. These examples illustrate the property of divisibility in Property 3.

(d) Now $7|21$ and $7|84$; thus, by Property 4, $7|63$ since $84 = 63 + 21$.

(e) Since $3|18$, then $3|(18 \cdot 20)$ by Property 5, even though 3 does not divide 20.

(f) $3|12$ and $3 \nmid 5$; therefore, $3 \nmid (12 + 5)$ or $3 \nmid 17$, by Property 6. •

Now let's use good problem-solving strategies to verify some of these properties.

Example 8 • Show that if $x|y$ and $x|z$, then $x|(y + z)$. (See Property 3.)

Solution *What information is given?* We are given two facts, $x|y$ and $x|z$. *Can we restate this data in another way?* Yes. Since $x|y$, the definition of "divides" asserts that there is a whole number k with $y = kx$. Similarly, there is a whole number l with $z = lx$.

What is unknown? What are we trying to show? We are trying to show that $x|(y + z)$. Look at the definition of "divides" on p. 161 to see what this means. We need to find some whole number m such that $y + z = mx$.

What is our plan? We shall try to find the connection between the information we know and the desired conclusion.

$$y + z = kx + lx$$

$$y + z = (k + l)x \qquad \text{Why}$$

Since $k + l$ is a whole number, it can be the m desired. We have $y + z = mx$. The definition of divisibility tells us that $x|(y + z)$. •

Now, can this be turned around? Is the converse true? If $x|(y + z)$, must $x|y$ and $x|z$?

Example 9 • If $x|(y + z)$, x may or may not divide either y or z. For instance, $3|24$; $24 = 11 + 13$, so $3|(11 + 13)$, but $3 \nmid 11$ and $3 \nmid 13$. •

Now suppose $x|y$ and $y|z$. Property 2 says that $x|z$. Can you prove this property?

Example 10 • If $x|y$ and $y|z$, then $x|z$.

Solution We are given the facts that $x|y$ and $y|z$. Restated, this says there are whole numbers k and l with $y = kx$ and $z = ly$. To connect what we know with what we wish to show, multiply both sides of the equation $y = kx$ by l to obtain

$$y = kx$$
$$ly = l(kx)$$
$$ly = (lk)x$$

Since $ly = z$, then $z = (lk)x$. Since lk is a whole number, by definition, $x|z$. •

Property 3 can be generalized to include the sum of any finite number n of whole numbers. If

$$a|b_1, \quad a|b_2, \quad a|b_3, \quad \ldots, \quad \text{and} \quad a|b_n$$

then

$$a|(b_1 + b_2 + b_3 + \cdots + b_n)$$

Similarly, Property 4 may be generalized to include the sum of any finite number n of whole numbers. If

$$a|b_1, a|b_2, \ldots, a|b_{n-1}$$

and if

$$a|(b_1 + b_2 + \cdots + b_{n-1} + b_n)$$

then $a|b_n$.

The process of determining whether a number is divisible by another number often becomes a matter of inspection. Let's examine why this statement is true, using several examples. To understand the criterion for divisibility by 2, consider whether $2|54{,}236$. Recall that 54,236 can be written as

$$54{,}236 = 5(10{,}000) + 4(1000) + 2(100) + 3(10) + 6$$

Now $2|10{,}000$, so $2|5(10{,}000)$ by Property 5. Likewise, $2|4(1000)$, $2|2(100)$, and $2|3(10)$. Now if $2|6$ (and it does), then by the generalization of Property 3, $2|54{,}236$. So we can tell if a number is divisible by 2 just by looking to see whether 2 divides its last digit.

In general, any number N can be written as

$$N = a_k\, 10^k + a_{k-1}\, 10^{k-1} + \cdots + a_2\, 10^2 + a_1\, 10 + a_0$$

where each of the a's is a digit. Now $2|10$, $2|100$, $2|1000$, and, in general, $2|10^k$ for any natural number k; thus, if $2|a_0$, then

$$2|(a_k\, 10^k + \cdots + a_2\, 10^2 + a_1\, 10 + a_0)$$

To test divisibility by 4, we must examine the last two digits of a numeral as shown in the following example.

Example 11 • Let us show that 54,236 is divisible by 4. Again

$$54{,}236 = 5(10)^4 + 4(10)^3 + 2(10)^2 + 3(10) + 6$$

Now $4|10^n$ for $n = 2, 3, 4$. So $4|5(10)^4$, $4|4(10)^3$, and $4|2(10)^2$ by Property 5. Hence if $4|[3(10) + 6]$ or $4|36$ (and it does), then by the generalization of Property 3, $4|54{,}236$. Hence the number is divisible by 4, since the last two digits of the numeral for the number are divisible by 4. •

You cannot tell whether a number is divisible by 9 or 3 by looking at its last digits. The next example demonstrates this divisibility rule.

Example 12 • Let us demonstrate that 4236 is divisible by 3. Write

$$4236 = 4(1000) + 2(100) + 3(10) + 6$$

Note that $3|999$, $3|99$, and $3|9$. Thus, write

$$\begin{aligned} 4236 &= 4(999 + 1) + 2(99 + 1) + 3(9 + 1) + 6 \\ &= 4(999) + 2(99) + 3(9) + 4 + 2 + 3 + 6 \qquad \text{Why?} \end{aligned}$$

Now $3|4(999)$, $3|2(99)$, and $3|3(9)$ by Property 5. Hence if $3|(4 + 2 + 3 + 6)$ (and it does), then by the generalization of Property 3, $3|4236$. •

In general, write a number N as

$$N = a_k\,10^k + \cdots + a_3\,10^3 + a_2\,10^2 + a_1\,10 + a_0$$

Since $10^k - 1 + 1 = 10^k$, the general form of a natural number N can be written as

$$\begin{aligned} N = a_k(10^k - 1 + 1) &+ \cdots + a_3(10^3 - 1 + 1) \\ &+ a_2(10^2 - 1 + 1) + a_1(10 - 1 + 1) + a_0 \end{aligned}$$

Rearrange as

$$\begin{aligned} a_k(10^k - 1) + \cdots + a_3(10^3 - 1) + a_2(10^2 - 1) + a_1(10 - 1) \\ + (a_k + \cdots + a_3 + a_2 + a_1 + a_0) \end{aligned}$$

Now $3|(10 - 1)$, $3|(10^2 - 1)$, $3|(10^3 - 1)$, and, in general, $3|(10^k - 1)$, where k is any natural number. Thus, $3|N$ if and only if

$$3|(a_k + \cdots + a_3 + a_2 + a_1 + a_0)$$

This expression is easily recognized as the sum of the digits of the number. Thus a number is divisible by 3 if and only if the sum of the digits of its base ten numeral is divisible by 3.

Example 13 • Is 4972 divisible by 11? To study this divisibility criterion, note that $11|99$ but $11 \nmid 999$. Instead, $11|1001$. This suggests the following manipulation.

$$\begin{aligned} 4972 &= 4(1000) + 9(100) + 7(10) + 2 \\ &= 4(1000 + 1 - 1) + 9(100 - 1 + 1) + 7(10 + 1 - 1) + 2 \\ &= 4(1000 + 1) - 4 + 9(100 - 1) + 9 + 7(10 + 1) - 7 + 2 \\ &= 4(1001) + 9(99) + 7(11) - 4 + 9 - 7 + 2 \end{aligned}$$

Now $11|4(1001)$, $11|9(99)$, and $11|7(11)$, so $11|4972$ if $11|(9 + 2 - 4 - 7)$. Now $9 + 2 - 4 - 7 = 0$. Since $11|0$, then $11|4972$. •

The preceding examples illustrate the plausibility of the following divisibility criteria.

Divisibility criteria

A whole number x is **divisible**

1. by 2 (or 5) if and only if the number named by the last digit of x is divisible by 2 (or 5).
2. by 3 (or 9) if and only if the number named by the sum of its digits is divisible by 3 (or 9).
3. by 4 if and only if the last two digits of x represent a number divisible by 4; by 8 if and only if the last three digits of x represent a number divisible by 8.
4. by 6 if and only if x is divisible by both 2 and 3.
5. by 7 if and only if it satisfies the following property. Subtract the double of the right-hand digit from the number represented by the remaining digits. If the difference is divisible by 7, then the original number is divisible by 7.
6. by 10 if and only if x ends in 0.
7. by 11 if and only if the difference (larger minus smaller) in the sum of the odd-numbered digits and the sum of the even-numbered digits is divisible by 11.

Example 14 •
(a) 756 is divisible by 3 because $7 + 5 + 6 = 18$, and $3|18$.
(b) $4|536$ because $4|36$.
(c) 678,342,570 and 417,235 are both divisible by 5, but 736 is not divisible by 5.
(d) 6294 is divisible by 6 because 4 is divisible by 2, and $6 + 2 + 9 + 4 = 21$, and $3|21$.
(e) 362,789,576 is divisible by 8 because $8|576$.
(f) 31,383 is divisible by 11 because $(3 + 3 + 3) - (8 + 1) = 9 - 9 = 0$, and 0 is divisible by 11.
(g) To determine if $7|6055$, repeat the operation described above until the difference is small enough that divisibility by 7 is obvious. $7|6055$ because

$$605 - 2(5) = 595$$

and because

$$59 - 2(5) = 49$$

and 7 divides 49. •

Example 15 • Test the divisibility of 30,492 by

(a) 9 (b) 7 (c) 11

Just for fun *Find the one composite number among the eight numbers. A composite number is divisible by some number other than itself and 1.*

31
331
3331
33331
333331
3333331
33333331
333333331

Solution (a) Since $3 + 0 + 4 + 9 + 2 = 18$ and $9|18$, then $9|30{,}492$.
(b) Now

$$3049 - 4 = 3045$$
$$304 - 10 = 294$$
$$29 - 8 = 21$$

and $7|21$; so $7|30{,}492$.

(c)

$$3 + 4 + 2 = 9$$
$$0 + 9 = 9$$
$$9 - 9 = 0$$

and $11|0$; so $11|30{,}492$. •

Let us return now to the problem with which we introduced this section.

Example 16 • While discussing a new machine for packing packages of potato chips, Mr. Pate explained that the previous hour's production of 190,236 large sacks had been packed in boxes of 8 each. One of his employees questioned Mr. Pate's production figures. Who was correct?

Solution Of course, the employee was correct. Evidently he understood divisibility. $8 \nmid 190{,}236$ because $8 \nmid 236$. Therefore, 190,236 sacks would not pack evenly into boxes containing 8 each. •

Exercise set 1

R 1. List three distinct natural numbers that are divisors of each of the following.
(a) 35 (b) 216 (c) 1001 (d) 299

2. Classify as true or false. Explain your answers that are false.
(a) 39 is a multiple of 9.

(b) 11 is a factor of 111.
(c) 4 is a multiple of 8.
(d) 9 is a factor of 12.
(e) 8 is divisible by 24.
(f) If every digit of a number is divisible by 3, the number is divisible by 3.
(g) 12 divides 6.
(h) $5|25$ and $5\nmid 4$; therefore $5\nmid 25 \cdot 4$.
(i) If $a|b$, then $a|b^2$ $(a \neq 0)$.
(j) $5\nmid 11$, $5\nmid 4$; therefore, $5\nmid (11 + 4)$.

3. Test each for divisibility by 4 and 8.

(a) 8116 (b) 9320 (c) 26,428

4. Which of the following numbers are divisible by 11?

(a) 704 (b) 7051 (c) 81,928

5. Which of the following numbers are divisible by 7?

(a) 6468 (b) 1848 (c) 8162

6. Which of the following numbers are divisible by 5? By 10?

(a) 7280 (b) 625 (c) 89,001

7. Classify the following as true or false. If false, give a counter-example. Let a, b, c, and d be natural numbers.

(a) If $a|(b + c)$, then $a|b$ and $a|c$.
(b) If $a|b$ and $a|c$, then $(a + b)|c$.
(c) If $a\nmid b$ and $a\nmid c$, then $a\nmid bc$.
(d) If $ab|c$, then $a|c$ and $b|c$.
(e) If $a|bc$, then $a|b$ and $a|c$.
(f) If $a|b$, $a|c$, $a|d$, and $a|e$, then $a|(b + c + d + e)$.
(g) If $a\nmid b$ and $a\nmid c$, then $a\nmid (b + c)$.
(h) $0|b$
(i) $a|0$ $a \neq 0$
(j) $1|b$
(k) $a|1$

8. State the property that justifies each of the following:

(a) $3|3$.
(b) $3|6$, $6|18$; therefore $3|18$.
(c) $4|12$, $4\nmid 3$; therefore $4|36$.
(d) $4|8$; $4\nmid 9$; therefore $4\nmid (8 + 9)$.
(e) $3|6$, $3|9$, $3|12$, and $3|15$; therefore, $3|42$.

9. Without actually dividing, test each of these numbers for divisibility by 5, by 6, by 7, by 8, by 9, and by 11.

(a) 6944 (b) 81,432 (c) 1,076,770
(d) 50,177 (e) 1,161,914 (f) 162,122
(g) 57,293 (h) 214,890 (i) 482,144
(j) 150,024 (k) 23,606 (l) 63,372

10. Answer "yes" or "no" and explain your answer.

(a) Is $2^3 \cdot 3^2 \cdot 5^7$ a factor of $2^4 \cdot 3^9 \cdot 5^7$?
(b) Is $3^7 \cdot 5 \cdot 17^4$ a factor of $3^7 \cdot 5^2 \cdot 17^3$?
(c) Does $7^2 \cdot 3^5$ divide $7^5 \cdot 3^6$?
(d) Is $7^4 \cdot 11^5 \cdot 5^4$ a multiple of $7^5 \cdot 11^3 \cdot 5$?

T 11. Suppose 3 new states were added to the United States, making a total of 53. Could stars of our new flag be arranged in equal rows and columns? Design an arrangement for 53 stars.

12. (a) Without zeros, write a five-digit number divisible by 9 and by 5.
(b) Without zeros, write a seven-digit number divisible by 11 and by 6.
(c) Without zeros, write a six-digit number divisible by 7 and by 8.
(d) Without zeros, write an eight-digit number divisible by 4 and by 3.

13. If a leap year is one whose date is divisible by 4, which of the following are leap years? Years that end in two zeros must be divisible by 400 in order to be leap years.

(a) 348 (b) 1492 (c) 1984 (d) 1776
(e) 1940 (f) 1000 (g) 500 (h) 1024

14. The 135 county bus drivers are having a meeting.

(a) Jim's job is to place chairs in the auditorium. He wants to have exactly one chair for each driver; yet he also wants each row to contain the same number of chairs. The space available could hold 10, 11, 12, 13, 14, or 15 rows. Testing for divisibility, how many rows should Jim have?
(b) At the meeting, the superintendent plans to divide the 135 drivers into study groups. If he wants equal numbers in each group with no more than 10 to a group, what are the possible group sizes?

15. Could Mr. Jones put the 2142 papers on his desk

(a) into 6 equal piles? (b) into 18 equal piles?
(c) into 7 equal piles? (d) into 8 equal piles?

16. John has 2144 pieces of candy. Can he divide them evenly among 8 friends? Why?

17. (a) Give an example of a three-digit number divisible by 2 and not by 4.
(b) Give an example of a four-digit number divisible by 4 and not by 8.

(c) Give an example of a number divisible by 4 and not by 2.

18. Write 167,532 in expanded notation and verify that $4 \mid 167{,}532$ because $4 \mid 32$.

19. Write 726,664 in expanded notation and verify that $8 \mid 726{,}664$ because $8 \mid 664$.

20. The symbol $n!$ is defined to be

$$n(n-1)(n-2)\cdot\cdots\cdot 3\cdot 2\cdot 1$$

that is, $4! = 4\cdot 3\cdot 2\cdot 1 = 24$. Using the definition "$a \mid b$, if $b = ac$," find c for the following.

(a) $7 \mid 7!$ (b) $5 \mid 6!$ (c) $6 \mid 5!$

C 21. Write N as $a_k10^k + \cdots + a_3 10^3 + a_2 10^2 + a_1 10^1 + a_0$ and verify the divisibility properties for the following.

(a) 4 (b) 5 (c) 8

22. True or false? Give reasons.

(a) $5 \nmid (5! + 1)$ (b) $n \nmid (n! + 1)$

23. Write down two digits and repeat them two more times to form a six-digit number. (For example, one such number would be 232,323.)

(a) Test several such numbers for division by 3 and formulate a conjecture.

(b) Repeat (a) for 7.

(c) Repeat (a) for 13.

(d) Show that (a), (b), and (c) are always true.

24. Write a four-digit number and repeat it to form an eight-digit number.

(a) Divide by 73 and find the remainder.

(b) Repeat the experiment with another four-digit number.

(c) Show that the result is always true.

25. What is the smallest natural number divisible by all the natural numbers 1 through 9?

26. Make up a three-digit number that is divisible by the product of its digits. Do not use 111, 112, or 115.

27. (a) Make up an example to demonstrate that the test for divisibility by 3 in base eight is the same as the test for divisibility by 11 in base ten. Show this fact in general.

(b) Show that the tests for divisibility by 2, 4, and 8 hold for base twelve numerals.

28. (a) Devise some divisibility tests in base seven.

(b) Devise some divisibility tests in base six.

2 Primes, composites, and factorization

■ Can a teacher arrange 61 students as a rectangle (that is, an equal number of rows with an equal number in each row)? ■

You will have the answer to the preceding problem when you learn what we mean by a prime number and how to factor numbers as a product of primes. Prime numbers are elements of a subset of natural numbers. In fact, we can partition the natural numbers into three sets: primes, composites, and the number 1. We conclude this section by writing a given number as a product of primes. For this factorization, we state what is called the **Fundamental Theorem of Arithmetic.**

We introduce primes by a procedure called the *Sieve of Eratosthenes.* Make an array of the first 100 natural numbers (see Table 1). Cross out the 1. Then cross out all the numbers greater than 2 that are divisible by 2 (every even number except 2). Next, cross out all numbers divisible by 3, but do not cross out 3. Notice that 4, the next consecutive number, has already been crossed out. Cross out all numbers divisible by 5 except 5 itself. Continue this process for every natural

BIOGRAPHY

Eratosthenes

270–190 B.C.

Over 2000 years ago, a Greek mathematician named Eratosthenes developed a procedure known as the *Sieve of Eratosthenes* for finding primes. Eratosthenes was a very talented mathematician (also a distinguished athlete, geographer, and poet) who not only knew that the earth was spherical but computed its diameter to within 50 miles of the actual value. It's interesting to note that the Europeans were still arguing this subject 1600 years later. His knowledge of the earth and the sun enabled him to find the distance between cities with remarkable accuracy.

number up to 100 that has not been crossed out. When this task is completed, circle the numbers not crossed out; the array will resemble Table 1.

Table 1

~~1~~	2	3	~~4~~	5	~~6~~	7	~~8~~	~~9~~	~~10~~
11	~~12~~	13	~~14~~	~~15~~	~~16~~	17	~~18~~	19	~~20~~
~~21~~	~~22~~	23	~~24~~	~~25~~	~~26~~	~~27~~	~~28~~	29	~~30~~
31	~~32~~	~~33~~	~~34~~	~~35~~	~~36~~	37	~~38~~	~~39~~	~~40~~
41	~~42~~	43	~~44~~	~~45~~	~~46~~	47	~~48~~	~~49~~	~~50~~
~~51~~	~~52~~	53	~~54~~	~~55~~	~~56~~	~~57~~	~~58~~	59	~~60~~
61	~~62~~	~~63~~	~~64~~	~~65~~	66	67	~~68~~	~~69~~	~~70~~
71	~~72~~	73	~~74~~	~~75~~	~~76~~	~~77~~	~~78~~	79	~~80~~
~~81~~	~~82~~	83	~~84~~	~~85~~	~~86~~	~~87~~	~~88~~	89	~~90~~
~~91~~	92	~~93~~	~~94~~	~~95~~	~~96~~	97	~~98~~	~~99~~	100

The numbers circled in this array are the **prime numbers** less than or equal to 100. Notice that none of these circled numbers is a multiple of any natural number that precedes it (other than 1).

A prime number

A natural number p, greater than 1, is *prime* if and only if the only natural number divisors of p are 1 and p.

Notice that the smallest prime is 2 and that 2 is the only prime that is even. The first ten primes are

$$\{2, 3, 5, 7, 11, 13, 17, 19, 23, 29, \ldots\}$$

The dots at the end indicate that there are still more primes. How many more primes exist? Is there a largest prime? Can you find a prime greater than 100? Can you find a prime greater than 100,000? About 300 B.C., Euclid proved that there are infinitely many primes. (See Exercise 24 in Exercise set 4.)

To test whether a number N is prime, we could divide it by all primes less than N. For instance, to determine whether 91 is prime, we could test all primes less than 91. To reduce this work, let's return to the Sieve of Eratosthenes and make an observation as we cross out numbers.

Prime	*Observation*
2	The first number (not crossed out) that 2 divides is $4 = 2^2$.
3	The first number (not crossed out) that 3 divides is $9 = 3^2$.
5	The first number (not crossed out) that 5 divides is $25 = 5^2$.
7	The first number (not crossed out) that 7 divides is $49 = 7^2$.

So if a number is less than the square of a prime and is composite, it is divisible by a number smaller than the prime. If it is not divisible by such a number, it is prime.

Finding a prime

When **testing** to see if a number is **prime,** we need to try only primes whose squares are less than or equal to the number being tested.

In the Sieve of Eratosthenes (Table 1), the numbers other than 1 that are crossed out are *composite* numbers.

A composite number

A natural number greater than 1 is said to be **composite** if it has a natural number divisor other than itself and 1.

The first ten composite numbers are

$$\{4, 6, 8, 9, 10, 12, 14, 15, 16, 18, \ldots\}$$

Once again, the dots indicate that there are many more composite numbers. Notice that 1 is neither prime nor composite. 1 is called a **unit.** Also remember that a prime number has only two divisors—itself and 1. A composite number has more than two divisors.

When a composite number is written as a product of all its prime divisors, we say that we have a **prime factorization.** Two methods are commonly employed to find all the prime factors of a composite number. The first method consists of repeated division starting with the smallest prime (2) and dividing by it as long as it divides the last quotient; then going to the next larger prime; and continuing until all prime factors have been obtained.

Example 17 • What are the prime factors of 72?

Solution

	Compact Form
$72 \div 2 = 36$	$2\,\overline{)\,72}$
$36 \div 2 = 18$	$2\,\overline{)\,36}$
$18 \div 2 = 9$	$2\,\overline{)\,18}$
$9 \div 3 = 3$	$3\,\overline{)\,9}$
$3 \div 3 = 1$	$3\,\overline{)\,3}$
$72 = 2 \cdot 2 \cdot 2 \cdot 3 \cdot 3 = 2^3 \cdot 3^2$	1

The second method involves factoring the number into any two easily recognized factors and then factoring the factors.

$$72 = (12)(6) = (4 \cdot 3)(3 \cdot 2) = (2 \cdot 2 \cdot 3) \cdot (3 \cdot 2)$$
$$= 2 \cdot 2 \cdot 2 \cdot 3 \cdot 3 \quad \bullet$$

Some elementary textbooks employ **factor trees** to illustrate this procedure for factoring a number into its prime factors. (See Figure 1.) Can you draw still another factor tree for 210?

Figure 1

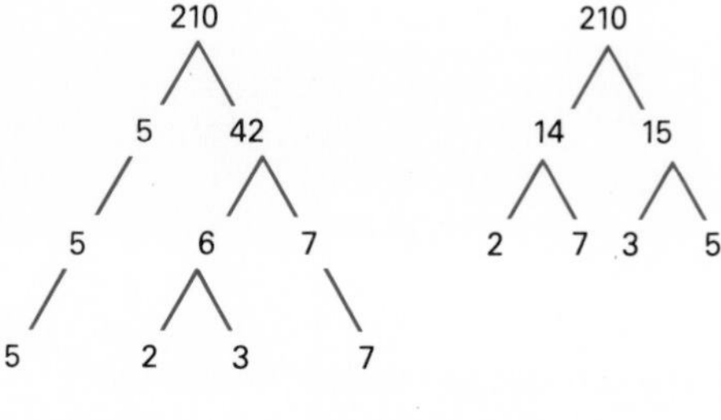

$210 = 2 \cdot 3 \cdot 5 \cdot 7$

Interestingly, the factorization of a composite number into prime factors is unique.

The Fundamental Theorem of Arithmetic

> Every composite natural number can be factored uniquely (except for the order of the factors) into a product of primes.

You may have wondered why 1 was not defined to be a prime. If 1 were a prime, this theorem would not be true. For example,

$$6 = 2 \cdot 3 \quad \text{and} \quad 6 = 1 \cdot 2 \cdot 3$$

So the factorization into primes would not be unique.

The Fundamental Theorem of Arithmetic asserts that if x is any composite number, then it can be written as $x = p_1 p_2 \cdots p_n$, where each p_i is a prime and the primes are unique except for the order in which they are written. The factors of 70 (2, 5, and 7) are unique, but they may be written as $2 \cdot 7 \cdot 5$ or as $5 \cdot 7 \cdot 2$,

Just for fun

Construct a 3 × 3 magic square using only primes. (The answer will be given for the one with the smallest total.)

and so on. Note that it is possible for the p's to be equal; that is, a factor may be used more than once in the Fundamental Theorem: $24 = 2 \cdot 2 \cdot 2 \cdot 3$.

Example 18 • To illustrate the uniqueness of the factors in the prime factorization of a composite number, we factor 360 in two ways, as shown in Figure 2. In both cases, $360 = 2 \cdot 2 \cdot 2 \cdot 3 \cdot 3 \cdot 5$.

Figure 2

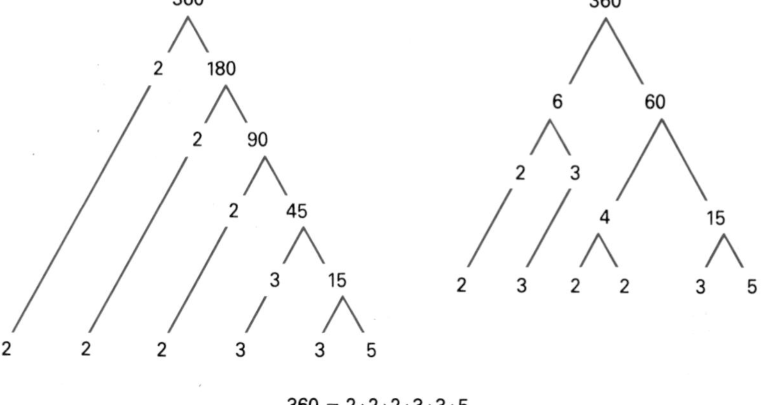

The Fundamental Theorem of Arithmetic holds for any composite number in the whole number system. Let us look at a number system for which the Fundamental Theorem does not hold.

Example 19 • Consider the system of numbers of the form $\{3x + 1 | x = 0, 1, 2, 3, \ldots\}$. This set of numbers can be written as $\{1, 4, 7, 10, 13, 16, \ldots, 25, \ldots\}$. Note that 4, 10, and 25 are all "primes" in this system, since they have no factors greater than 1 in the system.

Notice also the two prime factorizations of 100: $100 = 10 \cdot 10$ and $100 = 4 \cdot 25$. 4, 10, and 25 are all primes. Since 100 can be expressed as a product of primes in two ways, the Fundamental Theorem of Arithmetic is not true for this system. •

Exercise set 2

R 1. Factor into prime factors.

(a) 144 (b) 592 (c) 612 (d) 162

2. Which of the following are primes?

(a) 149 (b) 89 (c) 87 (d) 43
(e) 737 (f) 411 (g) 91 (h) 1003

3. Find the sum of the prime factors of each composite given below.

(a) 24 (b) 33 (c) 57
(d) 2000 (e) 108 (f) 299

4. Express each number as the product of its prime factors.

(a) 54 (b) 36 (c) 120
(d) 51 (e) 141 (f) 76
(g) 144 (h) 178 (i) 256

5. Apply the Sieve of Eratosthenes to find all primes less than 250.

6. What is the intersection of the set of prime numbers less than 25 and the set of odd natural numbers less than 25? What is the union?

7. (a) How many primes are less than 50? Less than 100? Less than 150?
   (b) How many composites are less than 50? Less than 100? Less than 150?

8. Use a factor tree to find the prime factorization of each of the following.

(a) 1512 (b) 810 (c) 1836

9. Use the compact form of the division technique to write the numbers in Exercise 8 as a product of primes.

10. What is the largest prime you need to use (a) in checking whether 689 is a prime? (b) In testing 7001?

T 11. Find the smallest natural number that has exactly the following number of distinct (that is, different) prime divisors, and describe the procedure used to determine the answer.

(a) Two (b) Three (c) Four
(d) Five (e) Six (f) Seven

12. Although there is a unique set of prime factors for any given number, many numbers can be expressed as a product in other ways. Express each of the following as (i) the product of two equal numbers, (ii) the product of a composite and a prime, and (iii) the product of two composites.

(a) 3^6 (b) 64 (c) 100 (d) 144

13. (a) List the divisors of 3^4.
    (b) How many of these divisors are primes?

14. $4!$ represents $4 \cdot 3 \cdot 2 \cdot 1!$; $3! = 3 \cdot 2 \cdot 1$. Can you show that $4! + 2$, $4! + 3$, $4! + 4$ are all composite numbers?

15. (a) Show that $5! + 2$, $5! + 3$, $5! + 4$, $5! + 5$ is a sequence of four consecutive composite numbers.
    (b) Will $7! + 2$, $7! + 3$, $7! + 4$, $7! + 5$, $7! + 6$, $7! + 7$ be a sequence of six consecutive composite numbers?

16. (a) Write a sequence of 10 consecutive composite numbers.
    (b) Write a sequence of n consecutive composite numbers.

C 17. Will $n! + 2, n! + 3, \ldots, n! + n$ for $n \geq 2$ always be a sequence of $n - 1$ composite numbers? Why?

18. (a) Determine the first 20 numbers in the number system discussed in the last example of this section.
    (b) How many of these numbers are primes?
    (c) Is this system closed under multiplication?
    (d) Write 484 as a product of primes in two ways.

19. Consider the subset of numbers defined by

$$\{2x + 1 \mid x = 0, 1, 2, \ldots\}$$

(a) Define a prime on this set of numbers; list the first ten primes.
(b) Does a Fundamental Theorem of Arithmetic hold?

3 Greatest common divisor and least common multiple

■ The cheerleaders for the Hard College Bulldogs want to buy the same number of red pompons and black pompons. What is the least number of each they can buy if red are shipped in boxes of 8 and black are shipped in boxes of 6? ■

In this section, we introduce the greatest common divisor and the least common multiple. Which of these two concepts do you feel can be used to solve the preceding problem?

As you may have guessed, the greatest common divisor of two or more natural numbers is the greatest natural number that will divide the numbers. Likewise, the least common multiple of two natural numbers is the least natural number that both the numbers will divide. You may be surprised at the many applications of these two concepts. Probably the most common usage of the least common multiple involves the least common denominator of fractions. A least common denominator is the least common multiple of two or more denominators.

If a natural number d is a divisor of each of two natural numbers b and c, then d is called a **common divisor** of b and c.

Example 20 • The divisors of 18 are {1, 2, 3, 6, 9, 18}.
The divisors of 30 are {1, 2, 3, 5, 6, 10, 15, 30}.
The set of common divisors of 18 and 30 are {1, 2, 3, 6}. •

In the set of common divisors of any two numbers, there is a greatest number, called the *greatest common divisor* (or *highest common factor*). The greatest common divisor of 18 and 30 is 6.

Greatest common divisor

The *greatest common divisor* (abbreviated g.c.d.) of two natural numbers, a and b, is the largest natural number d such that $d|a$ and $d|b$.

Thus, the greatest common divisor of two natural numbers a and b is the largest natural number d that is a divisor of both numbers. It is denoted by

$$d = \text{g.c.d}\ (a, b)$$

We saw previously that g.c.d. (18, 30) = 6.

Finding the greatest common divisor of two natural numbers is easy when the numbers are small. In this case, we write the numbers as products of primes. We then note the powers of the primes that are common to both numbers. The g.c.d. is the product of these common powers of primes. We illustrate this method by the following example.

Example 21 • Find g.c.d. (70, 90).

Solution

$$70 = 2 \cdot 5 \cdot 7 \qquad \text{and} \qquad 90 = 2 \cdot 3 \cdot 3 \cdot 5$$

Note that 2 and 5 are divisors of both 70 and 90 and that 2 and 5 are the only prime factors that 70 and 90 have in common. Thus,

$$\text{g.c.d. } (70, 90) = 2 \cdot 5 = 10 \quad \bullet$$

Example 22 • Find the g.c.d. (144, 180).

Solution

$$144 = 2 \cdot 2 \cdot 2 \cdot 2 \cdot 3 \cdot 3 = 2^4 \cdot 3^2$$

$$180 = 2 \cdot 2 \cdot 3 \cdot 3 \cdot 5 = 2^2 \cdot 3^2 \cdot 5$$

Note that two 2's and two 3's are divisors of both 144 and 180. So

$$\text{g.c.d. } (144, 180) = 2^2 \cdot 3^2 = 36 \quad \bullet$$

The g.c.d. of three natural numbers a, b, and c can be found by pairing. First, find the g.c.d. $(a, b) = e$. Then

$$\text{g.c.d. } (a, b, c) = \text{g.c.d. } (e, c)$$

However, the easiest way is again to write each number as a product of primes and find those primes common to all numbers.

Example 23 • Find g.c.d. (54, 72, 84).

Solution

$$54 = 2 \cdot 3^3$$

$$72 = 2^3 \cdot 3^2$$

$$84 = 2^2 \cdot 3 \cdot 7$$

$$\text{g.c.d. } (54, 72, 84) = 2 \cdot 3 = 6 \quad \bullet$$

Example 24 • A teacher decides to use discussion groups in four classes. There are 24 students in the first class, 32 in the second, 28 in the third, and 40 in the fourth class. If the teacher has the same number of groups in every class and wants every group in a given class to have an equal number of students, the largest number of groups that can be used is g.c.d. (24, 32, 28, 40).

$$24 = 2^3 \cdot 3$$

$$32 = 2^5$$

$$28 = 2^2 \cdot 7$$

$$40 = 2^3 \cdot 5$$

$$\text{g.c.d. } (24, 32, 28, 40) = 2^2 = 4$$

So each class will have 4 groups. •

In some books, the greatest common divisor is classified as greatest common factor and abbreviated as g.c.f. Of course, the largest common factor (or the largest number that is a factor of two given numbers) is the largest number that will divide the two numbers. Sometimes natural numbers have no common factors except 1.

Relatively prime

If the greatest common divisor of two natural numbers a and b is 1, then a and b are *relatively prime.*

Example 25 • (a) 5 and 7 are relatively prime because g.c.d. (5, 7) = 1.
(b) 23 and 124 are relatively prime because g.c.d. (23, 124) = 1.
(c) 111 and 6 are not relatively prime because g.c.d. (111, 6) = 3. •

Note that all the natural numbers less than a prime number are relatively prime to that number. For instance, each of the numbers 1, 2, 3, 4, 5, and 6 are relatively prime to the prime number 7.

The writing of a natural number as a product of prime factors can be quite cumbersome when the number is large. Consequently, we need a more practical method for finding the g.c.d. for large numbers. The method considered at this time is based on what we earlier called the **Division Algorithm.** This theorem states that, for natural numbers a and b, $a \geq b$, a can always be written as

$$a = bq + r$$

where q and r are whole numbers with $0 \leq r < b$. This representation is accomplished by division. If $a = 7$ and $b = 2$, then

$$\begin{array}{r} 3 \\ 2\overline{)7} \\ 6 \\ \hline 1 \end{array}$$

can be written as $7 = 2(3) + 1$, where $q = 3$ and $r = 1$ and obviously $0 \leq 1 < 2$.

The following property facilitates the finding of the g.c.d. of two natural numbers using the Division Algorithm.

Euclidean Algorithm

If a and b are two natural numbers with $a \geq b$, and r is the remainder when a is divided by b, then

$$\text{g.c.d. } (a, b) = \text{g.c.d. } (b, r)$$

The preceding property enables you to find the g.c.d. of two numbers by dividing several times, each time letting the divisor become the dividend and the remainder become the divisor. The last nonzero remainder is the g.c.d. (a, b).

Example 26 • Find the g.c.d. (180, 144).

Solution

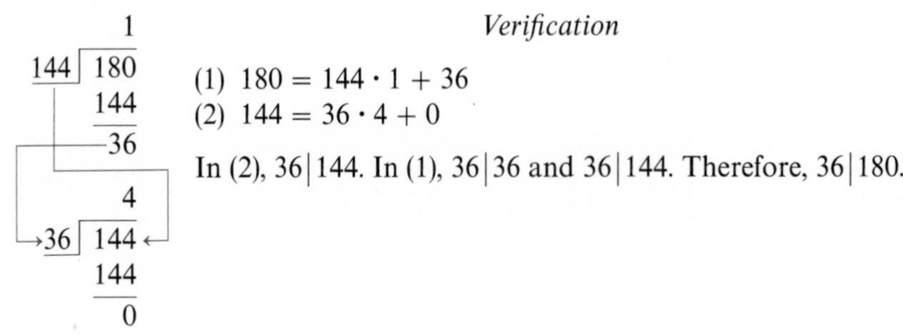

Verification

(1) $180 = 144 \cdot 1 + 36$
(2) $144 = 36 \cdot 4 + 0$

In (2), $36 \mid 144$. In (1), $36 \mid 36$ and $36 \mid 144$. Therefore, $36 \mid 180$.

The last nonzero remainder is 36. Therefore,

$$\text{g.c.d. } (180, 144) = 36 \quad \bullet$$

Example 27 • Find g.c.d. (3360, 4576).

Solution

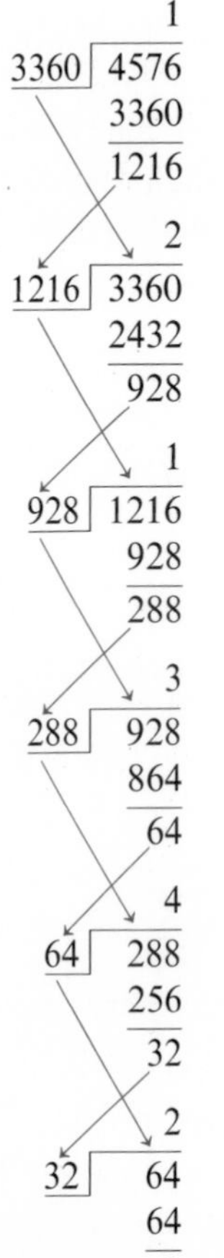

Verification

(1) $4576 = 3360 \cdot 1 + 1216$
(2) $3360 = 1216 \cdot 2 + 928$
(3) $1216 = 928 \cdot 1 + 288$
(4) $928 = 288 \cdot 3 + 64$
(5) $288 = 64 \cdot 4 + 32$
(6) $64 = 32 \cdot 2$

In (6), $32 \mid 64$. In (5), $32 \mid 32$ and $32 \mid 64$; therefore, $32 \mid 288$. In (4), $32 \mid 64$ and $32 \mid 288$; therefore, $32 \mid 928$. In (3), $32 \mid 288$ and $32 \mid 928$; therefore, $32 \mid 1216$. In (2), $32 \mid 928$ and $32 \mid 1216$; therefore, $32 \mid 3360$. In (1), $32 \mid 1216$ and $32 \mid 3360$; therefore, $32 \mid 4576$.

The last nonzero remainder is 32. Therefore,

$$\text{g.c.d.}\ (3360, 4576) = 32 \quad \bullet$$

Example 28 • Find g.c.d. (144, 104).

Solution

$$144 = (104)1 + 40$$
$$104 = (40)2 + 24$$
$$40 = (24)1 + 16$$
$$24 = (16)1 + 8$$
$$16 = (8)2 + 0$$

Since the last nonzero remainder is 8,

$$\text{g.c.d.}\ (144, 104) = 8 \quad \bullet$$

We are not only interested in the common divisors of two or more numbers, but also in common multiples of numbers. Remember that if $b|c$, then c is said to be a **multiple** of b. *A number m is a common multiple of two natural numbers b and c if it is a multiple of both b and c.* How many common multiples can two natural numbers have? Do you agree that there are infinitely many?

Example 29 • Multiples of 5 are

$$5, 10, 15, 20, 25, \ldots$$

Multiples of 2 are

$$2, 4, 6, 8, 10, 12, 14, 16, 18, 20, \ldots$$

Common multiples of 2 and 5 are

$$10, 20, 30, \ldots$$

The smallest of these numbers is 10. So the least common multiple is 10. •

Example 30 • Examine the multiples of 6 and 9. The multiples of 6 are $\{6, 12, 18, 24, 30, 36, \ldots\}$, and the multiples of 9 are $\{9, 18, 27, 36, \ldots\}$. The common multiples of 6 and 9 are $\{18, 36, 54, 72, \ldots\}$, and 18 is the least of the common multiples; so 18 is the least common multiple. •

Least common multiple

A natural number m is the *least common multiple* (abbreviated l.c.m.) of two natural numbers b and c if m is the least natural number divisible by both b and c. We write

$$m = \text{l.c.m.}\ (b, c)$$

The least common multiple of two natural numbers is therefore the least natural number that is divisible by both of the numbers. It is never less than the larger number involved. Thus, l.c.m. (3, 4) = 12; l.c.m. (4, 5) = 20; l.c.m. (6, 15) = 30; and l.c.m. (7, 35) = 35.

Because finding the l.c.m. by finding the set of common multiples and then selecting the least is a rather cumbersome process, additional methods must be considered. The first procedure uses prime factorization. **The l.c.m. of two natural numbers is the product of the highest powers of all the different prime factors that occur in factoring the numbers.**

Example 31 • (a) Find l.c.m. (8, 12)
(b) Find l.c.m. (120, 180).

Solution (a) The set of multiples of 8 is $\{8, 16, 24, 32, 40, \ldots\}$. The set of multiples of 12 is $\{12, 24, 36, 48, \ldots\}$. By inspection, we see that 24 is the least of the common multiples. Let's examine 8, 12, and 24 in factored form.

$$\begin{array}{r|l} 2 & 24 \\ 2 & 12 \\ 2 & 6 \\ 3 & 3 \\ & 1 \end{array} \qquad \begin{array}{r|l} 2 & 8 \\ 2 & 4 \\ 2 & 2 \\ & 1 \end{array} \qquad \begin{array}{r|l} 2 & 12 \\ 2 & 6 \\ 3 & 3 \\ & 1 \end{array}$$

$$24 = 2 \cdot 2 \cdot 2 \cdot 3 \qquad 8 = 2 \cdot 2 \cdot 2 \qquad 12 = 2 \cdot 2 \cdot 3$$

Note that 24 contains 2 the largest number of times it occurs in either 8 or 12. 24 contains 3, which also occurs once in 12. So

$$\text{l.c.m. } (8, 12) = 2^3 \cdot 3 = 24$$

Let's use this procedure to find l.c.m. (120, 180).

(b) $120 = 2^3 \cdot 3 \cdot 5$ and $180 = 2^2 \cdot 3^2 \cdot 5$. Thus, l.c.m. (120, 180) $= 2^3 \cdot 3^2 \cdot 5 = 360$. •

Example 32 • Find l.c.m. (24, 15, 20, 6).

Solution

$$24 = 2^3 \cdot 3$$
$$15 = 3 \cdot 5$$
$$20 = 2^2 \cdot 5$$
$$6 = 2 \cdot 3$$

l.c.m. (24, 15, 20, 6) $= 2^3 \cdot 3 \cdot 5 = 120$ •

If the l.c.m. of several numbers is to be found, a procedure involving division by primes is sometimes quicker.

Example 33 • Find the l.c.m. (12, 30).

Solution Divide both 12 and 30 by 2, finding 6 and 15; divide 6 and 15 by 3, finding 2 and 5. Since 2 and 5 are primes, the process is complete. In order to obtain the l.c.m. by this procedure, the division process is continued until the row of answers (after division) consists of relatively prime numbers.

$$\begin{array}{r|rr} 2 & 12 & 30 \\ \hline 3 & 6 & 15 \\ \hline & 2 & 5 \end{array} \qquad \text{l.c.m. } (12, 30) = 2 \cdot 3 \cdot 2 \cdot 5 = 60$$

Suppose that we change the order of the division and divide by 2, 2, and then 3.

$$\begin{array}{r|rr} 2 & 12 & 30 \\ \hline 2 & 6 & 15 \\ \hline 3 & 3 & 15 \\ \hline & 1 & 5 \end{array} \qquad \text{l.c.m. } (12, 30) = 2 \cdot 2 \cdot 3 \cdot 1 \cdot 5 = 60$$

In the second line of this procedure, notice that 2 does not divide 15; thus, we simply bring down the 15. •

Example 34 • Find l.c.m. (24, 15, 20, 6).

Solution

$$\begin{array}{r|rrrr} 2 & 24 & 15 & 20 & 6 \\ \hline 3 & 12 & 15 & 10 & 3 \\ \hline 5 & 4 & 5 & 10 & 1 \\ \hline 2 & 4 & 1 & 2 & 1 \\ \hline & 2 & 1 & 1 & 1 \end{array}$$

l.c.m. $(24, 15, 20, 6) = 2 \cdot 3 \cdot 5 \cdot 2 \cdot 2 \cdot 1 \cdot 1 \cdot 1 = 120$ •

To keep from confusing the processes of finding the g.c.d. and the l.c.m., it is useful to understand the relationship between the two.

Formula for least common multiple

> The **least common multiple** of two natural numbers a and b can be found from the greatest common divisor of a and b by computing the quotient
>
> $$\text{l.c.m. } (a, b) = \frac{ab}{\text{g.c.d. } (a, b)}$$

Example 35 • Find the l.c.m. of 144 and 180.

Solution

$$144 = 2^4 \cdot 3^2$$

$$180 = 2^2 \cdot 3^2 \cdot 5$$

$$\text{g.c.d. } (144, 180) = 2^2 \cdot 3^2$$

$$\begin{aligned} \text{l.c.m. } (144, 180) &= \frac{(144)(180)}{\text{g.c.d. } (144, 180)} \\ &= \frac{2^4 \cdot 3^2 \cdot 2^2 \cdot 3^2 \cdot 5}{2^2 \cdot 3^2} \\ &= 2^4 \cdot 3^2 \cdot 5 \\ &= 720 \end{aligned}$$ •

Just for fun

Three neighborhood dogs barked consistently last night. Spot, Patches, and Lady began with a simultaneous bark at 11:00 P.M. Then Spot barked every 4 minutes, Patches every 2 minutes, and Lady every 5 minutes. When Mr. Jones suddenly awakened at 1:00 A.M., did he do so because all was quiet at the moment or because all three dogs barked at once?

The solution to the introductory problem is given as follows:

Example 36 • The cheerleaders of the Hard College Bulldogs want to buy the same number of red and black pompons. What is the least number of each they can buy if red pompons are shipped in boxes of 8 and black pompons are shipped in boxes of 6?

Solution **Understand the problem.** The number of red pompons must be a multiple of 8, and the number of black pompons purchased must be a multiple of 6. Since they are purchasing the same number of red and black, this number must be a multiple of both 6 and 8.

Devise a plan. We want the least common multiple or

$$\text{l.c.m.}\ (6, 8) = 24$$

Carry out the plan. Therefore, the cheerleaders will purchase 3 boxes of red pompons (or $3 \cdot 8 = 24$ red pompons) and 4 boxes of black pompons (or $4 \cdot 6 = 24$ black pompons). •

Exercise set 3

R 1. Find the greatest common divisor and the least common multiple for each pair of numbers.

(a) 105 and 30 (b) 15 and 21
(c) 66 and 90 (d) 60 and 108
(e) 16 and 42 (f) 57 and 90
(g) 10 and 9 (h) 252 and 96

2. Find the greatest common divisor and the least common multiple for each of the following.

(a) 24, 30, 42 (b) 26, 36, 39
(c) 4600, 224, 228 (d) 45, 36, 24
(e) 15, 39, 30, 21, 70 (f) 42, 96, 104, 18

3. (a) Express g.c.d. (a, b) as a product of primes.

$$a = 3^3 \cdot 5^2 \cdot 2$$

$$b = 3 \cdot 5 \cdot 2^3$$

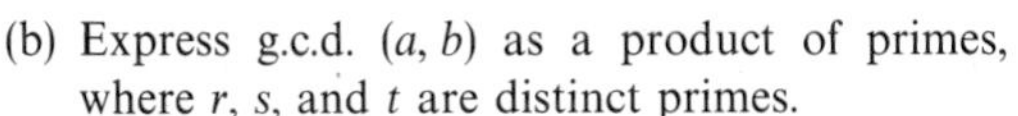

(b) Express g.c.d. (a, b) as a product of primes, where r, s, and t are distinct primes.

$$a = r \cdot s^3 \cdot t^2$$

$$b = r^4 \cdot s^2 \cdot t^5$$

(c) Express l.c.m. (a, b) in part (a) as a product of primes.

(d) Express l.c.m. (a, b) in part (b) as a product of primes.

4. (a) When does l.c.m. $(a, b) = ab$?
   (b) When does l.c.m. $(a, b) = a$?
   (c) When does g.c.d. $(a, b) = ab$?
   (d) When does g.c.d. $(a, b) = a$?

5. Find the least common multiple of the following by two different processes.

   (a) 44 and 92 (b) 45 and 72
   (c) 146 and 124 (d) 252 and 74
   (e) 840 and 1800 (f) 8125 and 1980

T 6. Using the Euclidean Algorithm, find the greatest common divisor.

   (a) 1122 and 105 (b) 4652 and 232
   (c) 2244 and 418 (d) 735 and 850
   (e) 220 and 315 (f) 486 and 522
   (g) 912 and 19,656 (h) 7286 and 1684

7. (a) What are the g.c.d. and l.c.m. of a and b if a and b are distinct primes?
   (b) If $a \mid b$, compute both g.c.d. (a, b) and l.c.m. (a, b).
   (c) If a is any natural number, compute g.c.d. (a, a) and l.c.m. (a, a).
   (d) If a and b are composite, under what conditions will l.c.m. $(a, b) = ab$?
   (e) What is the relationship between a and b if l.c.m. $(a, b) = a$?
   (f) What are the l.c.m. and g.c.d. of 1 and 4, 1 and 101, 1 and a?
   (g) If g.c.d. $(a, b) = 1$, find l.c.m. (a, b).
   (h) Find g.c.d. (a^2, a) and l.c.m. (a^2, a).

8. The l.c.m. of two numbers is $2^2 \cdot 3^4 \cdot 7 \cdot 11 \cdot 13$. The g.c.d. of the same two numbers is $2 \cdot 3 \cdot 7$. One of the numbers is $2^2 \cdot 3 \cdot 7 \cdot 11$. What is the other number?

9. Two joggers are running around a track. One jogger does a mile in 4 minutes, but the second one takes 10 minutes. If they start at the same time and same place, how long will it take them to be at this place together if they continue to run?

10. When Jill counts the pennies in her bank by 5's, there are 3 pennies left over. When she counts them by 3's, there are 2 left over. What is the least number of pennies in the bank?

11. An elementary art teacher has three art classes with 21, 35, and 28 students, respectively. The teacher wants to order some equipment that can be used by groups in each class. What is the largest number of students in a group in each class so that each group has the same number of students?

12. Before checking with the caterer, the cook cuts a cake into 35 equal pieces and an identical cake into 42 equal pieces. The caterer, however, insists that the cakes be cut exactly alike. Into how many pieces must each cake now be cut?

13. Event A happens every 15 minutes and event B happens every 18 minutes. If events A and B happened together at 6:00 P.M., when is the next time that they will happen together?

14. Three cars are warming up at a circular race track. Car A goes around the track in 5 minutes, car B in 8 minutes, and car C in 12 minutes. If all three cars begin at the same moment, how many minutes will elapse before they are all again at the starting point at the same time?

15. Jane is planning a party at which she expects 16, 24, or 12 guests. Since she is unsure of the number of guests but wants each guest to have available an equal number of hors d'oeuvres with none left over, how many should she order?

C 16. Is l.c.m. $(a, b, c) = (a \cdot b \cdot c)/\text{g.c.d. } (a, b, c)$?

17. Suppose we consider finding the g.c.d. of two numbers as an operation. Let *aGb represent finding g.c.d.* (a, b). Find example, 2G3 = 1, 2G4 = 2, 8G12 = 4, and so on. Which of the following properties hold for this operation on the set of natural numbers?

   (a) closure (b) commutativity
   (c) associativity (d) an identity
   (e) an inverse

18. Let *aLb represent finding l.c.m.* (a, b). Answer the questions in Exercise 17.

19. Consider the operations defined in Exercises 17 and 18.

   (a) Does L distribute over G?
   (b) Does G distribute over L?

20. For any natural number x, prove that

   (a) g.c.d. $(xa, xb) = x \cdot \text{g.c.d. } (a, b)$.
   (b) l.c.m. $(xa, xb) = x \cdot \text{l.c.m. } (a, b)$.

*4 Some unsolved problems in number theory

■ The divisors of 6 (excluding 6 itself) are 1, 2, and 3. Note that the sum of these divisors 1, 2, and 3 equals 6. Is 6 the only number for which this is true? Is 6 the only number so that the sum of its divisors (excluding the number itself) equals the number? ■

The answer to the preceding question is "No," and we will define and find such numbers in this section.

Many people believe that all mathematical problems have been solved, that mathematics today is simply a matter of learning what has already been done. However, the opposite is true. More mathematics has been developed since 1900 than in all recorded history to this date. The more mathematics that is developed, the more problems that arise. Each day there are numerous new problems to be solved.

Because there are so many easily stated and famous problems that have defied solution through the years, number theory is fascinating to many people. In this section, we shall present some of these unsolved problems. We hope they provide a new challenge that will encourage you to continue your study of mathematics. To consider some of these unsolved problems, we introduce you to some new terms: perfect, deficient, abundant, and amicable.

Let's begin this discussion with some numbers with which you are familiar. Even and odd numbers have been used previously in this book, but a formal definition will now be given.

Even and odd numbers

(a) A whole number n is an **even number** if and only if $2|n$.
(b) A whole number n is an **odd number** if it is not an even number.

The set of even numbers can be listed as

$$\{0, 2, 4, 6, \ldots\} \quad \text{or} \quad \{x | x = 2n, n \in W\}$$

The set of odd numbers can be tabulated as

$$\{1, 3, 5, 7, \ldots\} \quad \text{or} \quad \{x | x = 2n + 1, n \in W\}$$

Let's prove a simple theorem involving even numbers.

Example 37 • The set of even numbers is closed under addition.

Solution Let $2x$ and $2y$ represent two even numbers. Then the sum of the two even numbers

$$2x + 2y = 2(x + y)$$

But $(x + y) \in W$. Thus,

$$2x + 2y = 2z \qquad \text{where } z \in W$$

So the sum of any two even numbers is an even number. •

Now let us consider the sum of all the **proper divisors** of a number. (The proper divisors of a number are all divisors except the number itself.) The sum of the proper divisors of a number is sometimes equal to, sometimes less than, and sometimes greater than the number. This observation leads to the following definitions.

Perfect, deficient, abundant

If the sum of the proper divisors of a number is equal to the number, the number is called a *perfect* number; if this sum is less than the number, it is called a *deficient* number; if it is greater than the number, the number is an *abundant* number.

Example 38 • From Table 2, it is obvious that 6 is a perfect number; 4, 8, 9, 10, 14, and 15 are deficient numbers; and 12 is an abundant number. •

Table 2

Number	*Sum of proper divisors*
4	$1 + 2 = 3$
6	$1 + 2 + 3 = 6$
8	$1 + 2 + 4 = 7$
9	$1 + 3 = 4$
10	$1 + 2 + 5 = 8$
12	$1 + 2 + 3 + 4 + 6 = 16$
14	$1 + 2 + 7 = 10$
15	$1 + 3 + 5 = 9$

Perfect numbers seem to be quite rare. The next one after 6 is 28 $(1 + 2 + 4 + 7 + 14 = 28)$; the next to appear is 496; the fourth is 8128; and the fifth is 33,550,336. Clearly, finding perfect numbers is a rather difficult task. We might ask: How many perfect numbers are there? The answer is "No one knows." Perhaps there are infinitely many, but can you prove it? Can you answer the question: Are there any odd perfect numbers? All the perfect numbers that have been found at this time are even numbers.

point of interest

Much of the recent interest in number theory relates to using computers to find larger and larger primes. Most of these are of the form $2^p - 1$, where p is a prime, and are called **Mersenne primes** after the French mathematician Marin Mersenne (1588–1648).

In 1963 at the University of Illinois, $2^{11213} - 1$ was discovered to be a prime. In 1971, Dr. Bryant Tuckerman of IBM found that $2^{19937} - 1$ was a prime. That $2^{21701} - 1$ (a number with 6533 digits) is a prime was discovered in 1978 by two high school students, Laura Nickel and Curt Noll. In 1979, Curt Noll proved that $2^{23209} - 1$ is prime, and, also in 1979, Harry L. Nelson and David Slowinski found $2^{44497} - 1$ to be prime. It has 13,395 digits. By the time this edition is published, larger primes will undoubtedly have been discovered.

It has been proved (by Euclid) that if $2^n - 1$ is a prime, then $N = 2^{n-1}(2^n - 1)$ is perfect, and every even perfect number is of this form.

Example 39 • (a) $2^2 - 1 = 3$ is a prime, so $N = 2(3) = 6$ is perfect.
(b) $2^3 - 1 = 7$ is a prime, so $N = 2^2(2^3 - 1) = 28$ is perfect.
(c) $2^5 - 1 = 31$ is a prime, so $N = 2^4(2^5 - 1) = 496$ is perfect because

$$1 + 2 + 4 + 8 + 16 + 31 + 62 + 124 + 248 = 496$$

(d) $2^7 - 1 = 127$ is a prime, so $N = 2^6(2^7 - 1) = 8128$ is perfect. •

Some additional unsolved problems involve amicable numbers. From Greek to medieval times, amicable numbers were believed to have a mystical relationship to human friendship.

Amicable

> Two numbers are said to be *amicable* if each is the sum of the proper divisors of the other.

One example of an amicable number pair is 220 and 284. The proper divisors of 220 are 1, 2, 4, 5, 10, 11, 20, 22, 44, 55, 110, whose sum is 284, and the sum of the proper divisors of 284 is $1 + 2 + 4 + 71 + 142 = 220$. The fact that 1184 and 1210 are amicable was discovered by a 16-year-old boy. 17,296 and 18,416 are also amicable. Perhaps you can discover a pair. An unanswered question is: How many amicable pairs are there? Are there infinitely many? No one knows.

Another famous unsolved problem deals with twin primes. **Twin primes** are primes that differ by 2. There are many examples of twin primes, such as 3 and 5, 5 and 7, 11 and 13, 17 and 19, and 29 and 31. Yet an unanswered question is: Are there infinitely many such pairs? Again, no one knows. The largest pair of twin primes known at present is 140,737,488,353,699 and 140,737,488,353,701.

The last unsolved problems to be considered here are called **Goldbach's Conjectures.** Goldbach, an 18th-century mathematician, advanced two conjectures.

1. Every even number greater than 4 is the sum of two odd primes.

Example 40 •

$$6 = 3 + 3$$
$$8 = 3 + 5$$
$$10 = 3 + 7 = 5 + 5$$
$$12 = 5 + 7$$
$$14 = 3 + 11 = 7 + 7$$
$$16 = 3 + 13 = 5 + 11$$
$$18 = 5 + 13 = 7 + 11$$
$$20 = 3 + 17 = 7 + 13$$
$$22 = 3 + 19 = 5 + 17 = 11 + 11$$

•

Just for fun

Professor Abstract's car license plate consists of a three-digit number. He challenges the class to find this number from the following clues. If you add 1 to the number, it is divisible by 7; if you add 4, it is divisible by 8; and if you add 7, it is divisible by 9.

2. Every odd number greater than 7 is the sum of three odd primes.

Example 41 •

$$9 = 3 + 3 + 3$$
$$11 = 3 + 3 + 5$$
$$13 = 3 + 3 + 7 = 3 + 5 + 5$$
$$15 = 3 + 5 + 7 = 5 + 5 + 5$$
$$17 = 3 + 3 + 11 = 3 + 7 + 7 = 5 + 5 + 7$$
$$19 = 3 + 3 + 13 = 3 + 5 + 11 = 5 + 7 + 7$$
$$21 = 3 + 5 + 13 = 3 + 7 + 11 = 5 + 5 + 11 = 7 + 7 + 7$$ •

We hope that this section has shown that there are many unsolved problems relating to number theory. The same is true in all branches of mathematics.

Exercise set 4

R 1. Classify the following numbers as perfect (P), deficient (D), or abundant (A).

(a) 12 (b) 14 (c) 28
(d) 30 (e) 50 (f) 72
(g) 226 (h) 300 (i) 496

2. (a) What is the intersection of the set of odd counting numbers and the primes?
   (b) What is the intersection of the set of even counting numbers and the primes?

3. Which of the following are true statements? Use a counterexample to verify that each false statement is false.
   (a) If the product of two natural numbers is odd, then both numbers are odd.
   (b) If the product of two natural numbers is even, then both numbers are even.

   (c) If the product of two natural numbers is even, then at least one of the numbers is even.

4. Can a prime number ever be perfect? Deficient? Abundant?

5. (a) Prove that every prime is deficient.
   (b) Prove that if p is a prime, then p^n is deficient for every natural number exponent n.

   This exercise shows that there are infinitely many deficient numbers. Why?

6. Someone conjectured that $n^2 - n + 41$ generated only prime numbers.
   (a) Show this is true for $n = 1, 2, 3, 4, 5, \ldots, 10$.
   (b) Does $n = 40$ produce a prime?
   (c) Does $n = 41$ produce a prime?
   (d) Find a composite number different from 41^2 that is of form $n^2 - n + 41$.

7. Find a pair of twin primes that are between

(a) 35 and 45 (b) 55 and 65 (c) 65 and 75

8. Pierre Fermat (1601–1665) thought that the formula $P = 2^{2^n} + 1$ would give only prime numbers. Does this formula give a prime for

(a) $n = 0$? (b) $n = 1$? (c) $n = 2$?
(d) $n = 3$? (e) $n = 5$?

(For $n = 5$, try dividing by 641.)

9. Show that 1184 and 1210 are amicable numbers.

10. For each of the following, give an example.

(a) A composite number of the form $4n + 1$
(b) A composite number of the form $3^p + 1$, where p is a prime
(c) A prime number of the form $n^2 + n$, when n is a natural number
(d) A prime number of the form $4n + 1$, when n is a natural number
(e) A prime number of the form $8n + 1$, where n is a natural number

11. Verify that the odd numbers are or are not closed under

(a) addition (b) multiplication

12. Someone has conjectured that $n^2 + n + 17$ generates prime numbers.

(a) Test for $n = 1, 2, 3, 4, 5, 6, 7, \ldots, 17$.
(b) How many primes did you discover?

T 13. Find all twin primes between 100 and 200.

14. The great mathematician Fermat proved the following theorem: "If p is prime and a is any natural number, the natural number given by $a^p - a$ is divisible by p." Verify this theorem when

(a) $a = 6, p = 5$ (b) $a = 4, p = 7$
(c) $a = 11, p = 3$

15. An 18th-century English mathematician named Wilson proved the following theorem. "The number $N = [1 \cdot 2 \cdot 3 \cdot \cdots \cdot (p - 1)] + 1$, where p is a prime, is divisible by p." Verify that it is true when

(a) $p = 5$ (b) $p = 7$
(c) $p = 11$ (d) $p = 13$

16. A number n is termed *multiperfect* if there exists a natural number x such that the sum of the proper divisors of n equals $x \cdot n$. Determine whether the following numbers are multiperfect.

(a) 496 (b) 28 (c) 120 (d) 132

17. Part 1 of Goldbach's Conjectures is that every even number greater than 4 is the sum of two odd primes. Write each of the following as such a sum.

(a) 24 (b) 78 (c) 44

18. Part 2 of Goldbach's Conjectures is that every odd number greater than 7 can be expressed as the sum of three odd primes. Write each of the following as such a sum.

(a) 25 (b) 77 (c) 43

19. Prove that if the first of Goldbach's Conjectures is true, then the second will necessarily be true.

20. (a) Prove that the sum of any two odd numbers is an even number.
(b) Prove that the sum of any even number and any odd number is an odd number.
(c) Prove that the product of any even number and odd number is an even number.

21. Prime numbers like 3, 5, 7 are called *prime triplets.* These three numbers are consecutive primes that differ by 2. Can you find another set of prime triplets? Make a conjecture about prime triplets. Prove your conjecture.

C 22. Fermat's Last Theorem states that there are no natural numbers a, b, c, n such that

$$c^n = a^n + b^n$$

for $n > 2$. Let $n = 3$ and try all numbers up to 10 for a, b, and c to illustrate this theorem.

23. Prove that if A is an abundant number, then nA is abundant for every natural number n. Since 12 is abundant, this exercise shows that there are infinitely many abundant numbers. Why?

24. The following proof is given that there are infinitely many primes.

Proof: Assume that there are but n primes, where n is a natural number; that is, assume there is only a finite number of primes. We may list all these primes as $2, 3, 5, 7, \ldots, p$. Now form a number R by multiplying together all these n primes; then add 1 to the result.

$$R = (2 \cdot 3 \cdot 5 \cdot 7 \cdot \cdots \cdot p) + 1$$

R is either a prime number or a composite number, for it is clearly greater than 1. If R is a prime number, R is not one of the primes $2, 3, 5, 7, \ldots, p$ since R is larger than each of these primes.

But if R is not one of the primes listed, we have a contradiction, for we assumed that we had listed all the primes.

The other possibility is that R is a composite number. If R is a composite number, some prime p_i will divide R. Now

$$R = (2 \cdot 3 \cdot 5 \cdot \cdots \cdot p_i \cdot \cdots \cdot p) + 1$$

Finish the proof.

25. Show that $2^6(2^7 - 1)$ is perfect.

Skip

5 Modular arithmetic

■ If July 4 is on Friday this year and next year is a leap year, on what day will July 4 be next year? If Thanksgiving is on November 28, Christmas will be on what day of the week? If Valentine's Day is 47 days before Easter, on what day of the week will it fall? ■

In this section, we study mathematical systems with a finite number of elements. More specifically, we consider finite mathematical systems using the theory of congruences. This theory will be easy for you to understand, since it is introduced as the arithmetic of a 12-hour clock.

Figure 3

To understand what characterizes a mathematical system, let us consider a normal 12-hour clock, which has only 12 numbers on it. The clock in Figure 3 is such a clock that has been modified by removing the minute hand and replacing the 12 by 0.

When considering future time, we add a fixed number of hours to the present time. The time is now 9:00 A.M. and you have an appointment in 5 hours. Five hours after 9:00 A.M. is 2:00 P.M. See Figure 4. Thus, 9:00 A.M. ⊕ 5 hours = 2:00 P.M., where ⊕ is used as the symbol for clock addition.

Figure 4

Figure 5

In the same way, past events are denoted by subtraction. We had breakfast 4 hours ago. The time is now 9:00 A.M. At what time did we have breakfast? Study Figure 5. Letting the symbol ⊖ denote clock subtraction, 9 A.M. ⊖ 4 hours = 5:00 A.M. Thus, we had breakfast at 5:00 A.M.

If we were unfamiliar with time as given by a 12-hour clock, this new type of arithmetic might be difficult. However, we can easily verify that each of the following examples is correct.

Example 42 • Add 9 hours to 6:00 P.M.

Solution $6 \oplus 9 = 3$ A.M. Check with your watch. •

Example 43 • Add 27 hours to 7:00 A.M.

Solution $7 \oplus 27 = 10$. The clock time is 10:00 A.M. Did you get this answer? •

Example 44 • It takes 7 hours to drive to your house. You wish to arrive by 3:00 P.M. When should you start?

Solution $3 \ominus 7 = 8$, or you should start by 8:00 A.M. •

Next, consider the numbers 0 through 11 as the elements of a set, and consider addition to be based on counting in a clockwise direction on the face of a clock. Forgetting any A.M. or P.M. designations, a table of addition facts for a 12-hour clock is shown in Table 3. Note that no matter where we start on the clock, we are always at the starting position 12 hours later. Thus, for any element b, $b + 12 = b$.

Clock multiplication is defined in terms of repeated addition.

Table 3

$\oplus$	*0*	*1*	*2*	*3*	*4*	*5*	*6*	*7*	*8*	*9*	*10*	*11*
0	0	1	2	3	4	5	6	7	8	9	10	11
1	1	2	3	4	5	6	7	8	9	10	11	0
2	2	3	4	5	6	7	8	9	10	11	0	1
3	3	4	5	6	7	8	9	10	11	0	1	2
4	4	5	6	7	8	9	10	11	0	1	2	3
5	5	6	7	8	9	10	11	0	1	2	3	4
6	6	7	8	9	10	11	0	1	2	3	4	5
7	7	8	9	10	11	0	1	2	3	4	5	6
8	8	9	10	11	0	1	2	3	4	5	6	7
9	9	10	11	0	1	2	3	4	5	6	7	8
10	10	11	0	1	2	3	4	5	6	7	8	9
11	11	0	1	2	3	4	5	6	7	8	9	10

Example 45 • (a) On a clock find $3 \odot 2$, and (b) $3 \odot 5$.

Solution (a) Start at any time and add 2 hours three times. It is easy to see that this is equivalent to adding 6 hours.
(b) $3 \odot 5 = 3$ (Do you see why?) •

Clock division is defined in terms of multiplication. The quotient may or may not exist. 3 ⨸ 5 equals a quotient x such that $3 = 5 \odot x$. From (b) of the preceding example, $x = 3$. The easiest computation involves interpreting division as repeated subtraction. How many times do we need to subtract 5 from 3 in order to get 0? Verify that the answer is 3.

Example 46 • Find (a) 4 ⨸ 8 and (b) 7 ⨸ 2 if they exist.

Solution (a) $4 \oslash 8 = 2$, because one would need to subtract two 8's to get 0.
(b) $7 \oslash 2 = x$ where $7 = 2 \odot x$. No x exists, because adding any 2 x's would give an even answer, not a 7. •

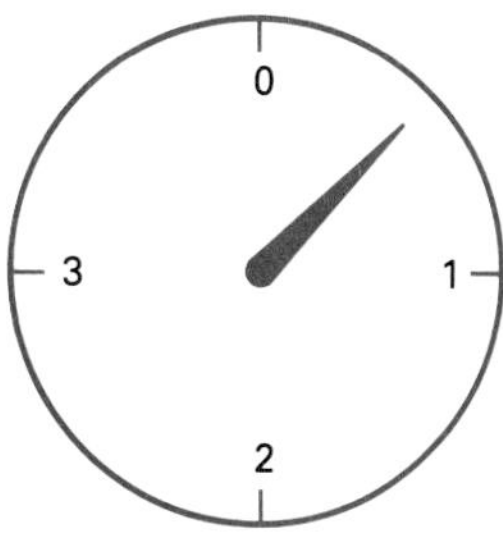

Figure 6

The construction of a finite mathematical system can also be illustrated by a 4-minute clock. This type of clock might be used to time the rounds and intermissions in a boxing match. The sketch in Figure 6 represents the face of such a clock. Note that 0 is used in place of 4 or that, at the end of 4 minutes, the clock starts over.

This mathematical system will contain only four numbers—0, 1, 2, and 3. Addition in this system (denoted by $\oplus$) is defined and interpreted to mean that the hand of the clock moves the same number of positions as the number to be added. If the hand is at 1 and moves for 2 minutes, then it will be at 3; so $1 \oplus 2 = 3$. In a similar manner, if the hand is at 2 and it moves for 2 minutes, then it will be at 0; so $2 \oplus 2 = 0$. Finally, suppose that the hand is at 3 and it moves for 2 minutes. Then it will be at 1; so $3 \oplus 2 = 1$.

The operation of addition for this system can be displayed by a table. Check the entries in Table 4 by visualizing the movement of the hand of the 4-minute clock.

Table 4

$\oplus$	*0*	*1*	*2*	*3*
0	0	1	2	3
1	1	2	3	0
2	2	3	0	1
3	3	0	1	2

Example 47 • (a) $2 \oplus 3 = 1$, from Table 4.
(b) $2 \ominus 3 = 3$, since in Table 4 we see that $3 \oplus 3 = 2$. •

The 12-hour clock and the 4-minute clock are illustrations of **modular arithmetics.** This new type of number system may seem strange to you. Yet it is the kind of mathematics that is used for many machines with dials or controls.

In this system, emphasis is placed on the fact that two distinct whole numbers differ by a multiple of some natural number. For instance, 3 minutes and 17 minutes register the same way on a 7-minute clock, since 17–3 is a multiple of 7. Numbers that differ by multiples of a given natural number are said to be congruent modulo the given natural number.

Congruent modulo m

> Two whole numbers a and b are *congruent modulo m* (where m is a natural number called the **modulus**) if and only if m divides the difference of a and b.

The relationship is denoted by $a \equiv b \pmod{m}$ and is read *a is congruent to b (mod m)*. Thus two numbers are congruent modulo m if their difference is divisible by m.

Example 48 •

$$17 \equiv 3 \pmod 7 \quad \text{because} \quad 7 \mid (17 - 3)$$
$$14 \equiv 8 \pmod 6 \quad \text{because} \quad 6 \mid (14 - 8)$$
$$21 \equiv 5 \pmod 4 \quad \text{because} \quad 4 \mid (21 - 5)$$
$$10^2 \equiv 2 \pmod 7 \quad \text{because} \quad 7(10^2 - 2)$$
$$6 \not\equiv 4 \pmod 3 \quad \text{because} \quad 3 \nmid (6 - 4)$$
•

Another criterion for determining if natural numbers a and b are congruent is given by the following.

Requirement for congruence

> a and b are **congruent modulo *m*** if their remainders are equal when divided by m.

Example 49 • (a) $17 \equiv 35 \pmod 3$ because 17 divided by 3 leaves a remainder of 2, and 35 divided by 3 also produces a remainder of 2.
(b) $701 \equiv 7001 \pmod 7$ because when the numbers are divided by 7, the remainder is 1 in each case. •

In particular, $a \equiv 0 \pmod m$ means that $m \mid a$; conversely, if $m \mid a$, then $a \equiv 0 \pmod m$.

Addition in a system modulo m is clearly the same as addition of whole numbers except when the sum is greater than or equal to m. When the sum is greater than or equal to m, we divide the sum by m and use the remainder in place of the ordinary sum. For instance, Table 4 is a mod 4 addition table. Table 5 is a mod 5 addition table. The following examples use the table for addition mod 5. Check the work by using the table.

Example 50 •

$$2 + 1 \equiv 3 \pmod 5 \qquad 2 + 3 \equiv 0 \pmod 5$$
$$2 + 2 \equiv 4 \pmod 5 \qquad 2 + 4 \equiv 1 \pmod 5$$
•

Table 5

+	*Addition mod 5*				
	0	*2*	*3*	*4*	
0	0	1	2	3	4
1	1	2	3	4	0
2	2	3	4	0	1
3	3	4	0	1	2
4	4	0	1	2	3

Table 6

·	*Multiplication mod 5*				
	0	*1*	*2*	*3*	*4*
0	0	0	0	0	0
1	0	1	2	3	4
2	0	2	4	1	3
3	0	3	1	4	2
4	0	4	3	2	1

Multiplication in a system modulo m is the same as multiplication of whole numbers except when the product is greater than or equal to m; then it is reduced by the congruence relationship $a \equiv b \pmod m$ to a number less than m. Table 6 illustrates multiplication mod 5.

Example 51 • (a) $3 \cdot 3 \equiv 4 \pmod 5$
(b) $2 \cdot 3 \equiv 1 \pmod 5$
(c) $2 \div 3 \equiv 4$ since $3 \cdot 4 \equiv 2 \pmod 5$
(d) $4 \div 3 \equiv 3$ since $3 \cdot 3 \equiv 4 \pmod 5$ •

According to Table 6, division by 3 (mod 5) is always possible because every element occurs in the fourth row. Similarly, division by 1, 2, and 4 are always possible in mod 5 arithmetic. You will discover that division is not always possible in other finite systems. For example, make a mod 4 multiplication table and look at the division by 2.

Many properties of equality carry over to congruence.

Properties of congruence

If $a \equiv b \pmod m$ and $c \equiv d \pmod m$, then

1. $a + c \equiv b + d \pmod m$.
2. $a - c \equiv b - d \pmod m$, $a \geq c$ and $b \geq d$.
3. $ka \equiv kb \pmod m$, when k is any natural number.
4. $ac \equiv bd \pmod m$.

Example 52 • It is easy to verify that $16 \equiv 2 \pmod 7$ and $38 \equiv 17 \pmod 7$. Now without using the definition of (mod 7), argue that

(a) $54 \equiv 19 \pmod 7$ (b) $22 \equiv 15 \pmod 7$ (c) $608 \equiv 34 \pmod 7$

Of course, each of these is true by one of the Properties 1, 2, or 4 just noted. •

Example 53 • In each problem below, find a whole number answer less than the modulus.

(a) $5 + 2 \pmod 3$ (b) $18 + 5 \pmod 7$ (c) $4 - 8 \pmod 5$

Solution (a) $5 + 2 \equiv 1 \pmod 3$ since $5 + 2 = 7 = 2(3) + 1$
(b) $18 + 5 \equiv 2 \pmod 7$ since $18 + 5 = 23 = 3(7) + 2$
(c) $4 - 8 \equiv 1 \pmod 5$ since $4 + 5 - 8 = 1$ (Note that the modulus, 5, is added to make subtraction possible.) •

We can apply the concepts of congruence mod 7 to solve calendar problems like the ones with which we began this section. Note that the calendar for November is given in Figure 7.

Figure 7

November						
S	M	T	W	T	F	S
				1	2	3
4	5	6	7	8	9	10
11	12	13	14	15	16	17
18	19	20	21	22	23	24
25	26	27	28	29	30	

BIOGRAPHY

Evariste Galois

1811–1832

The term *group* was first used by the brilliant young French mathematician Galois, whose genius was not recognized during his lifetime. At school, he did so much work in his head that his teachers called him a show-off. At 16, he sent several mathematical discoveries to the French Academy, but they were lost. At age 20, he had an affair with a young woman and subsequently participated in a duel over her "honor." He spent the entire night before the duel writing down his ideas. The next morning, he was killed. Think of the contributions he might have made to mathematics if he had lived beyond age 20.

Notice that the five Thursdays have dates 1, 8, 15, 22, and 29; they differ by multiples of 7. The same is true for any other day of the week. Two weeks after Friday the 9th will be Friday the

$$9 + 2 \cdot 7 = 23\text{rd}$$

Example 54 • What day of the week will be November 1 one year beyond the calendar of Figure 7? (Exclude leap year.)

Solution A normal year has 365 days. Since $365 = (52)7 + 1$, $365 \equiv 1 \pmod 7$. Dates that differ by a multiple of 7 fall on the same day of the week; 52 weeks after a Thursday is a Thursday. One day later is a Friday. Therefore, November 1 a year later will be on a Friday. •

Example 55 • Christmas is on Wednesday this year. On what day will it be next year if next year is a leap year?

Solution There are 366 days in a leap year.

$$366 \equiv 2 \pmod 7$$

Consequently, Christmas will be 2 days after Wednesday, on Friday. •

Now return to the problems at the beginning of the section and see if you have mastered the concepts necessary to solve them.

Some of the finite mathematical systems we have been studying are examples of a group. Consider the definition of a group.

Just for fun

Can you write a schedule for 12 basketball teams so that each team will play every other team only once and no team will be idle? (Hint: Try mod 1 arithmetic to see if it is possible. How many games will each team play?)

A group

> A group is a set of elements, together with a binary operation $\otimes$, with the following properties:
>
> **1.** There is closure with respect to $\otimes$.
> **2.** The operation is associative.
> **3.** There is an identity element i in the set such that
>
> $$i \otimes a = a \otimes i = a$$
>
> where a is any element of the set.
> **4.** Every element a has an inverse for $\otimes$.

If the operation is commutative, the group is called an *Abelian group.*

We have identified the set of elements in the mod 4 finite system as $\{0, 1, 2, 3\}$ and have defined an operation $\oplus$ on this system in Table 4. We next examine Table 4 carefully to determine if the set with this operation is a group.

Example 56 • Show that the set $\{0, 1, 2, 3\}$ with operation $\oplus$ forms an Abelian group.

Solution It is obvious in examining the addition table that the closure property for addition is satisfied, since for each ordered pair of elements (a, b), with a and b from the set, there exists a unique element $a \oplus b$ in the set.

The associative property also holds for addition. The truth of the associative property may be illustrated by testing several cases at random to see whether $(a \oplus b) \oplus c = a \oplus (b \oplus c)$. To prove this fact, we would need to test all cases.

An identity element for addition exists and is the number 0 (zero). For every element a contained in the set, $a \oplus 0 = 0 \oplus a = a$.

Each element in the set has an additive inverse, since for each element a, there exists an element b such that $a \oplus b = 0$. This fact is clear, since the identity element 0 appears in every row and every column of the addition table. The inverse of 0 is 0, since $0 \oplus 0 = 0$. The inverse of 1 is 3, since $1 \oplus 3 = 0$. Similarly, the additive inverse of 2 is 2, and the inverse of 3 is 1. Note that both 0 and 2 are their own additive inverses, and 1 and 3 are inverses of each other.

The commutative property is satisfied, since for each pair of elements a and b in the set,

$$a \oplus b = b \oplus a$$

This fact can be determined without testing all possible sums in the table. The table is symmetric about the diagonal extending from the upper left corner to the lower right corner. Since each sum $a \oplus b$ is reflected across the diagonal to $b \oplus a$, symmetry indicates that $a \oplus b = b \oplus a$. Therefore, addition modulo 4 on $\{0, 1, 2, 3\}$ forms an Abelian group. •

Exercise set 5

R 1. Perform, if possible, each of the following operations on a 12-hour clock.

(a) $5 \oplus 8$ (b) $3 \ominus 7$ (c) $7 \ominus 11$
(d) $9 \oplus 6$ (e) $4 \odot 3$ (f) $7 ⨸ 3$
(g) $9 ⨸ 2$ (h) $8 ⨸ 5$

2. Perform, if possible, each of the following operations on a 4-minute clock.

(a) $2 \oplus 3$ (b) $2 \oplus 5$ (c) $2 \odot 3$
(d) $3 \odot 1$ (e) $2 ⨸ 3$ (f) $3 ⨸ 2$
(g) $1 ⨸ 3$ (h) $3 \odot 3$

3. (a) If Mr. Smith goes to work at 8:00 A.M. and works 8 hours, at what time does his work day end?
(b) The Smiths are leaving on a car trip at 5:00 A.M. tomorrow. Assuming that the trip is made in exactly 17 hours, at what time will they arrive at their destination?
(c) Mr. Smith, exhausted from the long trip, fell into bed 1 hour after their arrival and slept for 14 hours. At what time did he awaken?

4. Find t for the following, where the answer is some numeral on a 12-hour clock.

(a) $t = 8 + 7$ (b) $t = 5 - 8$
(c) $t = 9 - 11$ (d) $t = (4)(7)$
(e) $t = (3)(9)$ (f) $t = 2 - 7$

5. Work Exercise 4 for a 24-hour clock.

6. The fourth of July holiday occurs on Tuesday this year. On what day of the week will it occur next year if next year is not a leap year? Is a leap year?

7. Christmas is on Friday this year. In how many years will it be on Sunday if no leap years are involved? If next year is a leap year?

8. Labor Day falls on Monday. This year, it was on Monday, September 4. On what date will it be next year if next year is not a leap year? If next year is a leap year?

9. (a) Is $4 + 7 \equiv 15 + 4 \pmod 8$?
(b) Is $6 + 2 \equiv 31 + 19 \pmod{10}$?
(c) In what modulus is $3 + 8 \equiv 5$?
(d) For what modulus is $5 + 9 \equiv 6$?

10. Label each as either true or false.

(a) $6 \cdot 5 \equiv 4 \pmod 6$ (b) $18 - 5 \equiv 1 \pmod{12}$
(c) $6 \equiv 5 \pmod 3$ (d) $79 \equiv 17 \pmod 4$
(e) $7 + 4 \equiv 5 \pmod 6$ (f) $6 \equiv 3 \pmod 8$
(g) $8 \equiv 7 + 1 \pmod 5$ (h) $34 \equiv 7 \pmod{13}$

11. Without tables, perform the following operations and in each case reduce the answer to a whole number less than the modulus.

(a) $2 + 3 \pmod 4$ (b) $2 \cdot 4 \pmod 5$
(c) $1 - 5 \pmod 3$ (d) $15 - 2 \pmod 7$
(e) $3 + (2 + 5) \pmod 6$ (f) $1 - 13 \pmod 4$
(g) $3(12) \pmod 5$ (h) $(12)(5) \pmod 7$
(i) $15 - 4 \pmod{11}$ (j) $2(4 + 11) \pmod 7$
(k) $5^3 \pmod 9$ (l) $4^3 \pmod{12}$
(m) $(22)(46) \pmod 7$ (n) $(624)(589) \pmod 9$

T 12. By inspection, find a number x that will make each of these statements true.

(a) $x \equiv 5 \pmod 3$ (b) $x + 10 \equiv 9 \pmod 4$
(c) $x + 5 \equiv 7 \pmod 9$ (d) $x + 20 \equiv 13 \pmod 3$
(e) $x + 4 \equiv 6 \pmod 7$ (f) $x + 5 \equiv 7 \pmod 5$
(g) $x + 3 \equiv 9 \pmod{11}$ (h) $x + 4 \equiv 3 \pmod 7$
(i) $x^2 \equiv 1 \pmod 3$ (j) $x^2 \equiv 0 \pmod 4$

13. Consider a three-way light switch with four positions in the order off, dim, bright, brightest. If Sam starts with the position named first in the following, at which position will he be when he completes the number of turns given?

(a) off, 4 (b) brightest, 5 (c) dim, 6
(d) bright, 3 (e) off, 7 (f) dim, 23

14. Suppose we agree to order the numbers in a 4-hour clock as

$$0 < 1 < 2 < 3$$

(a) Does the transitive property for $<$ hold?
(b) If $a < b$, is $a + c < b + c$?

15. Use your knowledge of the number of days in each month to answer the following questions.

(a) If April 1 is on Saturday, on what day is May 30?

(b) If Pat's birthday is exactly 6 weeks before Christmas, what is its date?
(c) If Christmas is on Monday, on what day will Diane's birthday (January 29) be?
(d) Do the 125th and 256th days of the year fall on the same day of the week? Why or why not?

16. An operation ⊗ is defined by each of the following tables. Determine whether each is a group. An Abelian group.

(a)

⊗	x	y
x	x	y
y	y	x

(b)

⊗	x	y
x	x	y
y	y	y

(c)

⊗	x	y	z
x	y	z	x
y	z	x	y
z	x	y	z

C 17. Construct addition and multiplication tables for a 5-minute clock (modulus 5).

(a) Is the set a group under addition modulo 5?
(b) Is the set a group under multiplication modulo 5?
(c) Is either an Abelian group?

18. Construct addition and multiplication tables for a 7-minute clock (modulus 7).

(a) Is the set a group under addition modulo 7?
(b) Is the set a group under multiplication modulo 7?
(c) Is either an Abelian group?

6 The system of integers

■ The temperature in Chicago at noon was 0°C. Before bedtime, the temperature had dropped 9°. What was the new temperature? ■

The everyday need for negative numbers should be evident to you. Such numbers provide the most convenient method for distinguishing between 10° above zero and 10°C below zero, a gain of 6 yards or a loss of 6 yards on the football field, and "I owe \$50" or "I have \$50."

In this section, you will review the operations of addition and subtraction for integers. Do you remember from high school that $^{-}8 + 6 = {}^{-}2$? You will be pleased to note that the extension or enlargement of whole numbers to the integers is made in such a way that all previous properties of operations are maintained. Thus, the whole numbers are contained as a subset of the integers, called the *nonnegative integers*.

Number theory properties are often expressed in terms of positive integers and sometimes in terms of integers. We have chosen instead the procedure of presenting number theory properties in terms of natural numbers. Now we introduce the integers with the understanding that all number theory properties discussed hold for positive integers.

In order to explain our definition of negative integers, consider the problem of subtraction from 0. What does $0 - a$ equal, or what is x such that $a + x = 0$? Can we, for example, find numbers x, y, and z such that $1 + x = 0$, $2 + y = 0$, or $5 + z = 0$? If by numbers we mean the set of whole numbers 0, 1, 2, . . . , the answer is "no." However, let us see if we can invent some new numbers that will satisfy such requirements. If you earn \$10 and then spend the \$10, your net holding is \$0. If the money spent is symbolized as $^-$\$10, then we can say that

$$\$10 + (^-\$10) = \$0$$

Similarly, if you begin a walk to the drugstore, travel three blocks, and then remember that your money is at home and return the three blocks, your distance from home is zero. $3 + (^-3) = 0$. Let $^-1$ be a number such that $1 + {^-1} = 0$; $^-2$ a number such that $2 + {^-2} = 0$; and $^-5$ a number such that $5 + {^-5} = 0$. In general, ^-n will be a number such that $n + {^-n} = 0$.

Additive inverse

We form a new set of numbers in which each number n has a unique *additive inverse* ^-n.

$$n + {^-n} = {^-n} + n = 0$$

Thus, $^-1$ is the only number that when added to 1 gives 0; $^-20$ is the only number that when added to 20 gives 0; and, in general, ^-n is the only number that when added to n gives 0. If ^-n is the additive inverse of n, then n is the additive inverse of ^-n. For example, 6 is the additive inverse of $^-6$ because

$$6 + (^-6) = 0$$

What is the additive inverse of 0? It is a number $^-0$ such that $0 + {^-0} = 0$. But 0 itself has the property that $0 + 0 = 0$. In order that the additive inverse be unique, we therefore insist that $^-0$ be the same as 0.

Set of integers

The *set of integers* I is the set

$$I = W \cup {^-N}$$

where ^-N represents the additive inverses of the natural numbers or

$$I = \{\ldots {^-3}, {^-2}, {^-1}, 0, 1, 2, 3, \ldots\}$$

The natural numbers, as part of the integers, are often called *positive integers.* The set of positive integers is sometimes written as $\{+1, +2, +3, \ldots\}$ or as $\{^+1, {^+2}, {^+3}, \ldots\}$, but in this book the plus signs will be omitted. The positive integers and 0 are called the **nonnegative integers.** The set

$$\{\ldots {^-4}, {^-3}, {^-2}, {^-1}\}$$

is called the set of **negative integers.**

point of interest

^-x should not read as "negative x," since x can be either a positive or negative number; that is, ^-x can be positive if x is negative. For instance, if $x = {^-5}$, then ^-x is 5. Read ^-x as "the additive inverse of" or "the opposite of" x, and you'll save yourself a lot of confusion.

A number line can be constructed for the integers just as one was constructed for the whole numbers. Mark an arbitrarily chosen point as 0, called the *origin.* Then measure equal segments to the right to determine points labeled 1, 2, 3, 4, . . . and to the left for points labeled $^-1$, $^-2$, $^-3$, $^-4$, . . . , as illustrated in Figure 8. By this process, we are setting up a one-to-one correspondence between a subset of points on the line and the integers. The positive integers extend to the right of zero, and the negative integers extend to the left of zero. Pairs of additive inverses, such as $^-5$ and 5, are represented by points at equal distances on each side of zero.

Figure 8

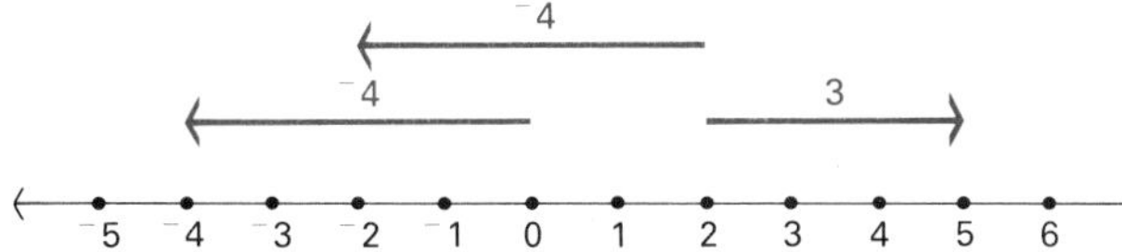

Notice that integers may also be represented by arrow diagrams, as in Figure 8. The integer 3 may be represented by an arrow diagram directed toward the right and of the same length as the line segment from 0 to 3 on the number line; for example, 3 can be represented by the line segment from 2 to 5 on the number line. Thus an arrow diagram may be *translated* or moved along the number line, as shown in Figure 8. In the same figure, integer $^-4$ is represented by two arrow diagrams of length 4; note that both diagrams are directed toward the left.

In Chapter 2, we illustrated the addition of whole numbers using arrow diagrams. This procedure is reviewed in Figure 9(a) for the sum $5 + 3$. This concept is then extended in Figure 9(b) for the addition of $^-5 + {}^-3$.

It is obvious that

$$^-5 + {}^-3 = {}^-8$$

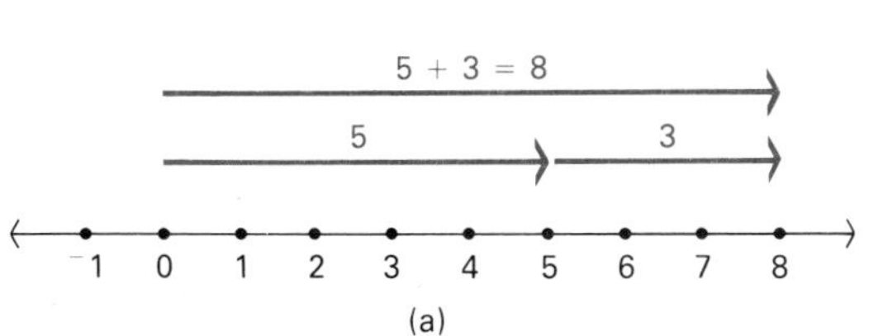

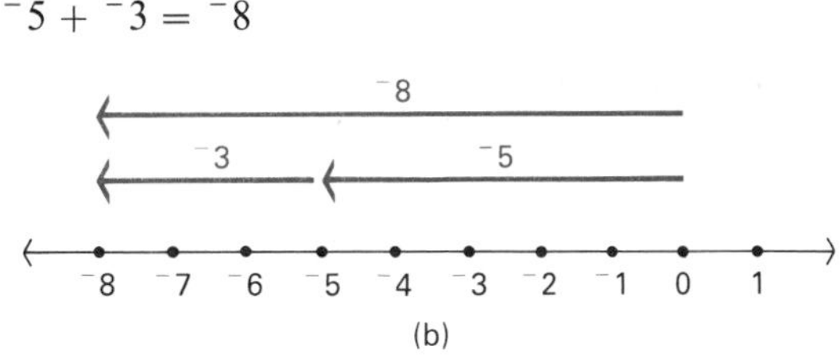

Figure 9

flashback

Negative numbers were used by Hindu mathematicians for the first time between 400 and 800 A.D. However, Western mathematicians resisted them for hundreds of years. During the Medieval and Renaissance European periods, practical mathematics found no place for negative numbers. The concept of a negative number was called absurd by the famous mathematician Diophantus of Alexandria. In 1225, the mathematician Fibonacci stated that the equation $a + x = b$, with a greater than b, had no solution unless it expressed that a person was in debt. As late as the 17th century, negative numbers were referred to as "false" or "fictitious." Not until the publication of Descartes' *La Géométrie* in 1637 did the negative numbers achieve legitimacy in the language of mathematics.

Using the pattern from the preceding number line, the answers can be obtained in the following example. (If you have difficulty, draw a number line).

Example 57 • Find
(a) $^{-}4 + {}^{-}3$ (b) $^{-}2 + {}^{-}7$

Solution (a) $^{-}7$ (b) $^{-}9$ •

Now see if you agree with Part 2 of the definition on page 203.

Example 58 • On a number line, find $6 + {}^{-}2$.

Solution

Figure 10

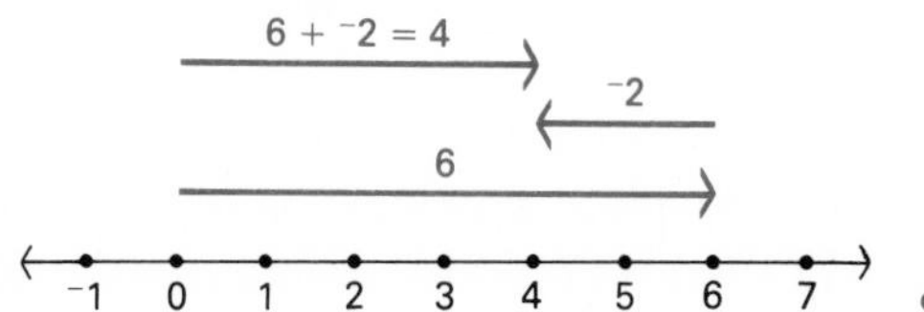

•

Example 59 • From the pattern established in the preceding example, find
(a) $5 + {}^{-}3$ (b) $^{-}2 + 6$ (c) $7 + {}^{-}2$
(If you have difficulty, draw a number line).

Solution (a) 2 (b) 4 (c) 5 •

Example 60 • On a number line, find the sum $^{-}8 + 5$.

Solution

Figure 11

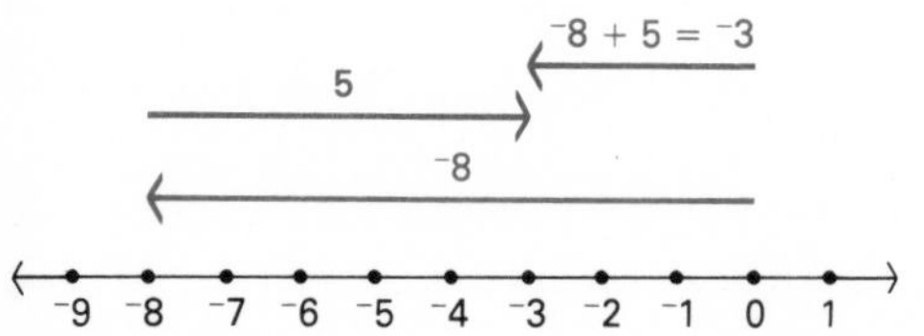

•

Example 61 • From the pattern established in the preceding example, find
(a) $^{-}7 + 2$ (b) $4 + {}^{-}9$ (c) $^{-}5 + 3$
(Again, if you have difficulty, draw a number line.)

Solution (a) $^{-}5$ (b) $^{-}5$ (c) $^{-}2$ •

Do the preceding answers agree with Parts 3 and 4 of the definition on page 203? We now validate the reasonableness of this definition algebraically. Recall that the system of whole numbers consists of the set $W = \{0, 1, 2, 3, \ldots\}$ and two operations $+$ and $\cdot$ that satisfy several properties: closure, associative, commutative, and distributive properties and existence of additive and multiplicative identities. Extending the whole numbers to the system of integers is done in such a way that the two operations $+$ and $\cdot$ still have these properties and have one additional property, the existence of a unique additive inverse for each element. The operations of $+$ and $\cdot$ are defined so that the operations involving whole numbers within the system of integers are the same as within the system of whole numbers.

System of integers

The *system of integers* consists of the set I of integers $\{\ldots, ^-4, ^-3, ^-2, ^-1, 0, 1, 2, 3, 4, \ldots\}$ and two binary operations, addition, $+$, and multiplication, $\cdot$, with the following properties for any integers a, b, and c.

Closure Properties

1. $a + b$ is a unique integer.
2. ab is a unique integer.

Commutative Properties

3. $a + b = b + a$
4. $ab = ba$

Associative Properties

5. $(a + b) + c = a + (b + c)$
6. $(ab)c = a(bc)$

Distributive Property of Multiplication over Addition

7. $a(b + c) = ab + ac$
8. $(b + c)a = ba + ca$

Identity Elements

9. There is a unique integer 0, such that

$$0 + a = a + 0 = a$$

10. There is a unique integer 1, such that

$$1 \cdot a = a \cdot 1 = a$$

Additive Inverses

11. For each a in I, there exists a unique additive inverse ^-a, such that

$$a + {}^-a = {}^-a + a = 0$$

Let us now discuss the addition of two integers, using the properties listed in the preceding definition. We consider first the case where both integers are negative. For example, consider $^-5 + {}^-3$.

$({}^-5 + {}^-3) + (5 + 3) = ({}^-5 + {}^-3) + (3 + 5)$	Commutative property of addition
$= {}^-5 + [{}^-3 + (3 + 5)]$	Associative property of addition
$= {}^-5 + [({}^-3 + 3) + 5]$	Associative property of addition
$= {}^-5 + [0 + 5]$	Additive inverse
$= {}^-5 + 5$	Additive identity
$= 0$	Additive inverse

Therefore,

$$(^-5 + {}^-3) + (5 + 3) = 0 \qquad \text{Transitive property of equalities}$$

Thus the number $^-5 + {}^-3$ is the additive inverse of $5 + 3$. But, by definition, the unique additive inverse of $5 + 3$ is $^-(5 + 3) = {}^-8$. Hence,

$$^-5 + {}^-3 = {}^-8$$

We can see that the sum above is reasonable. Now let us consider a situation where one integer is positive and the other integer is negative.

Example 62 • How should the sum $6 + {}^-2$ be defined so as to satisfy the properties of addition?

Solution

$$\begin{aligned} 6 + {}^-2 &= (4 + 2) + {}^-2 && \text{Renaming 6} \\ &= 4 + (2 + {}^-2) && \text{Associative property of addition} \\ &= 4 + 0 && \text{Additive inverse} \\ &= 4 && \text{Additive identity} \end{aligned}$$

Thus, $6 + {}^-2 = 4$. But $6 - 2 = 4$, so

$$6 + {}^-2 = 6 - 2 \quad \bullet$$

Notice that since the addition of integers is commutative, $a + {}^-b = {}^-b + a$, or, in this case, $^-2 + 6 = 6 + {}^-2 = 4$.

Now let's consider another example of one integer positive and the other negative. See if you can discover the difference.

Example 63 • Consider the sum $^-8 + 5$.

Solution From a previous example, $^-8 = {}^-5 + {}^-3$. Thus,

$$\begin{aligned} {}^-8 + 5 &= (^-5 + {}^-3) + 5 && \text{Renaming } {}^-8 \\ &= (^-3 + {}^-5) + 5 && \text{Commutative property of addition} \\ &= {}^-3 + (^-5 + 5) && \text{Associative property of addition} \\ &= {}^-3 + 0 && \text{Additive inverse} \\ &= {}^-3 && \text{Additive identity} \end{aligned}$$

Therefore,

$$^-8 + 5 = {}^-3 \qquad \text{Transitive property of equalities}$$

But $^-(8 - 5) = {}^-(3) = {}^-3$, so

$$^-8 + 5 = {}^-(8 - 5) \quad \bullet$$

To satisfy the conditions discussed on page 201 and illustrated by the preceding examples, the definition of addition of integers is formulated as follows.

Addition of integers

> The **sum of two integers** (denoted by +) is defined by the following cases, where a and b are whole numbers.
>
> 1. $a + b = n(A \cup B)$ where $a = n(A)$, $b = n(B)$, and $A \cap B = \varnothing$ for sets A and B
> 2. $^-a + ^-b = ^-(a + b)$
> 3. $a + ^-b = ^-b + a = (a - b)$ if $a > b$
> 4. $a + ^-b = ^-b + a = 0$ if $a = b$
> 5. $a + ^-b = ^-b + a = ^-(b - a)$ if $a < b$

Verify the following examples.

Example 64 •

$$2 + (^-5) = ^-(5 - 2) = ^-3$$
$$7 + (^-4) = 7 - 4 = 3$$
$$^-3 + (^-5) = ^-(3 + 5) = ^-8$$
$$^-7 + 9 = 9 + (^-7) = 9 - 7 = 2$$
$$^-9 + 5 = 5 + (^-9) = ^-(9 - 5) = ^-4$$
$$^-16 + (^-17) = ^-(16 + 17) = ^-33$$ •

For whole numbers x, y, and z, $x - y = z$ if and only if $x = y + z$. Subtraction for integers will be defined in a way that is consistent with this definition of subtraction for whole numbers.

Subtraction of integers

> If x and y are integers, the **difference** in the subtraction of y from x (denoted by $x - y$) is the integer z if and only if $x = y + z$.

Example 65 •

$6 - 4 = 2$ because $6 = 4 + 2$

$^-2 - 5 = ^-7$ because $^-2 = 5 + ^-7$

Now $^-4 - 2 = ^-6$, since $^-4 = 2 + ^-6$. Likewise, $^-4 + ^-2 = ^-6$. So $^-4 - 2 = ^-4 + ^-2$. This example illustrates that subtraction is equivalent to the addition of the additive inverse of the number being subtracted. •

In general, for integers x, y, and z, $x - y = z$ if $x = y + z$.

$$\begin{aligned} x + ^-y &= (y + z) + ^-y && \text{Why?} \\ &= ^-y + (y + z) && \text{Why?} \\ &= (^-y + y) + z && \text{Why?} \\ &= 0 + z && \text{Why?} \\ &= z && \text{Why?} \end{aligned}$$

Thus

$$x - y = x + {}^{-}y$$

Subtraction property

> If x and y are integers, then there always exists a unique integer z such that $x - y = z$. z can be written as $z = x + {}^{-}y$.

To subtract in the set of integers, change the sign of the number being subtracted and then proceed as in addition. This process makes it possible to reduce subtractions involving positive and negative integers to calculations involving addition only.

Example 66 •

$$8 - 6 = 8 + {}^{-}6 = 2$$

$$^{-}12 - {}^{-}10 = {}^{-}12 + 10 = {}^{-}2 \quad \bullet$$

Let's look at the definition of subtraction on the number line and then apply the preceding property of subtraction.

Example 67 • Find $^{-}3 - 4$.

Solution By definition, the answer is some number that when added to 4 gives $^{-}3$. Take a look at such a number in Figure 12(a).

$$^{-}3 - 4 = {}^{-}7 \quad \bullet$$

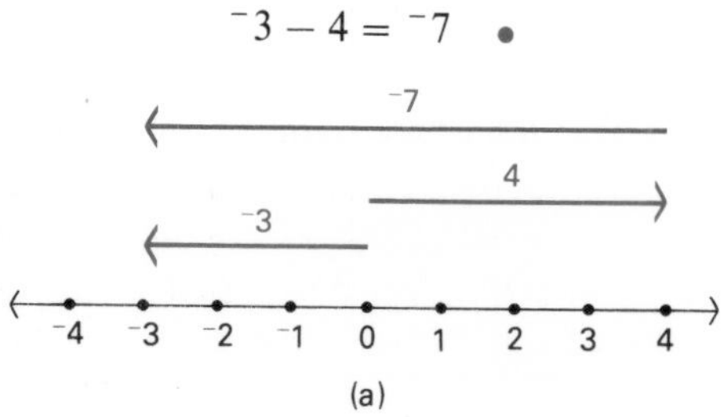

(a)

Figure 12

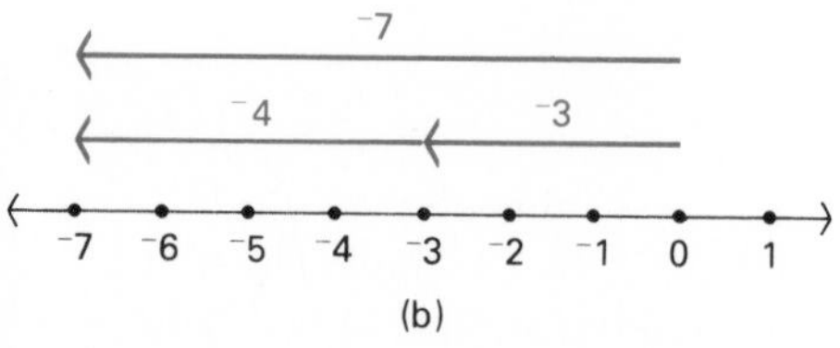

(b)

Now let's consider subtraction as changing direction of the arrow and then adding (see Chapter 2) in Figure 12(b). To subtract 4, the arrow diagram will be the same as that for $^{-}4$.

$$^{-}3 - 4 = {}^{-}3 + {}^{-}4 = {}^{-}7$$

Example 68 • Find the solution set for each open sentence on the domain of integers, and graph on a number line.

(a) $x + {}^{-}4 = {}^{-}6$ (b) $x + {}^{-}4 < {}^{-}3$

Just for fun

A boy and a girl ran a 100-meter race. The girl crossed the finish line when the boy had gone 95 meters, so she won the race by 5 meters. When they raced a second time, the girl wanted to make the contest more even so she started at ${}^-5$ meters. If the two ran at the same constant speed as before, who won the race? (If you think it was a tie, you had better start using your problem-solving techniques.)

Solution (a) If $x + {}^-4$ equals ${}^-6$, then, by the definition of subtraction, $x = {}^-6 - {}^-4 = {}^-6 + 4 = {}^-2$. This solution is represented by the colored dot in Figure 13(a).

(b) From the definition of subtraction, $x + {}^-4 = x - 4$, so $x - 4 < {}^-3$. x can be $0, {}^-1, {}^-2, {}^-3, {}^-4, \ldots$, as shown in Figure 13(b). •

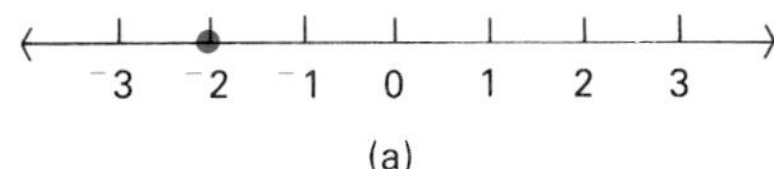

(a)

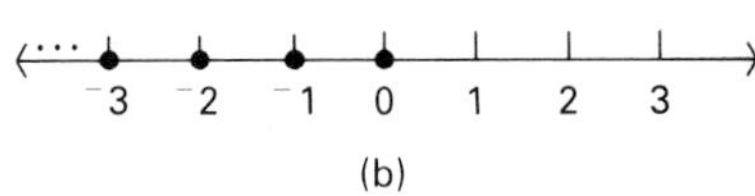

(b)

Figure 13

Since the operation of subtracting 5 is equivalent to adding ${}^-5$, most books use the same symbol for the sign of a number and the operation of subtraction. That is, ${}^-5$ is indicated by -5, the same symbol used in subtracting 5 from 8, $8 - 5$. Starting with the next chapter, this book will use this standard notation. We conclude this section with a definition of absolute value. Absolute value of a number is always positive or zero.

Absolute value

> The undirected distance of an integer from 0 is called the *absolute value* of the integer. The absolute value of an integer n is written $|n|$. Thus
>
> $$|n| = n \text{ if } n > 0$$
>
> $$|n| = 0 \text{ if } n = 0$$
>
> $$|n| = \text{the additive inverse of } n \text{ if } n < 0$$

Example 69 • $|{}^-6| = 6$; $|3| = 3$; $|{}^-4| = 4$; $|0| = 0$

You can see that the absolute value of any integer is always a nonnegative integer. •

Exercise set 6

R 1. What are the additive inverses of the following integers?

(a) 5 (b) $^{-}3$ (c) 0
(d) $^{-}8$ (e) a (f) ^{-}a

2. Draw arrow diagrams on a number line to represent the integers below. Start each arrow diagram at 0.

(a) $^{-}2$ (b) $^{-}4$ (c) $^{-}3$
(d) 8 (e) 6 (f) $^{-}5$

3. Draw arrow diagrams starting at $^{-}3$ for each integer in Exercise 2.

4. From the following list, select the pairs that are additive inverses.

(a) $^{-}2$, 2 (b) $^{-}3$, $(^{-}3)$
(c) 7, $^{-}8$ (d) $(2 \cdot 3)$, $^{-}(3 + 3)$
(e) $(2 + 3)$, $^{-}(2 + 3)$ (f) 0, 0
(g) 2, 2 (h) 1, $^{-}1$

5. Perform the indicated operations.

(a) $2 - 5$ (b) $3 - {}^{-}1$
(c) $^{-}6 - 2$ (d) $4 - 13$
(e) $^{-}10 + {}^{-}3$ (f) $6 - {}^{-}2$
(g) $^{-}8 + {}^{-}3$ (h) $13 - {}^{-}6$
(i) $^{-}9 - 7$ (j) $500 - {}^{-}5$
(k) $(6 - 4) - {}^{-}2$ (l) $^{-}7 - (6 - 4)$
(m) $(^{-}4 + 8) + {}^{-}6$ (n) $(6 + {}^{-}6) + {}^{-}2$
(o) $(^{-}3 + {}^{-}2) + 5$ (p) $(^{-}8 + {}^{-}7) + 8$
(q) $(a + b) + {}^{-}b$ (r) $(x + y) + {}^{-}(x + y)$

6. Find the following sums by using the number line.

(a) $^{-}6 + {}^{-}5$ (b) $6 + {}^{-}3$
(c) $^{-}5 + {}^{-}3$ (d) $8 + {}^{-}1$
(e) $^{-}6 + {}^{-}21$ (f) $0 + {}^{-}7$
(g) $(8 + {}^{-}2) + {}^{-}7$ (h) $(^{-}9 + {}^{-}3) + 4$
(i) $2 + {}^{-}6$

7. For each of the properties shown, state whether the property holds for subtraction of integers. If a property does not hold, then disprove it by a counterexample. If a property is valid, illustrate it with an example.

(a) Associative (b) Commutative

8. If the domain is the set of integers, find the solution set for each and graph the solution set on a number line.

(a) $x + {}^{-}3 = 2$, (b) $x - {}^{-}5 = 4$,
(c) $x + {}^{-}5 < {}^{-}2$, (d) $x - {}^{-}6 < {}^{-}3$

9. Find the sum by grouping the positive terms together and the negative terms together, and then perform the operation.

(a) $^{-}3 + 4 + 7 + {}^{-}6 + {}^{-}8$
(b) $^{-}6 - 7 + 8 - {}^{-}6 + {}^{-}5$

10. Suppose that you have \$89 in the bank. If you write one check for \$79 and then deposit \$69, how much money will you have in the bank? How much do you have in the bank if you write a second check for \$100?

11. The temperature was recorded last night as $^{-}16$°C. At noon, the temperature had increased by 8°. What was the temperature at noon?

12. Determine whether Fred or Ed has more points. Fred has earned 10 points, lost 6, lost 2, earned 5, lost 7, lost 3, and gained 8. Ed has earned 6 points, earned 2, earned 4, lost 7, and lost 5.

13. Dennis has a checking account. At the first of the month, he deposited \$448 to a previous balance of \$121. He later wrote six checks for \$135, \$225, \$12, \$20, \$139, and \$80 and then made another deposit of \$250. If the service charge was \$10 for the overdrawn account, what balance did Dennis have at the end of the month?

14. The University of Mystate football team, the Wild Geese, made the following plays in their first game. In each case, tell whether the team made the 10 yards for a first down and give the number of yards they made over the necessary 10 or the number of yards they lacked to make the first down.

(a) The Geese received the ball on the 4-yard line and gained 15 yards.
(b) The Geese completed a pass for 8 yards and then were pushed back 4 yards. After one futile attempt to move the ball, they ran around the left end for 5 yards.
(c) The Geese marched down the field for 7 yards. Then the quarterback fell down while trying to complete a pass and lost 3 yards. They made 4 yards in each of the next two plays.
(d) The Geese received a 5-yard penalty on the first play for being off sides. They completed a 7-yard pass and a 9-yard pass and twice tested the middle for 3 yards each try.

15. Express these changes as positive or negative integers:
    (a) The change in elevation from an airplane 10,000 feet above sea level to a submarine 500 feet below sea level.
    (b) The change in position of the ball if the football team first loses 24 yards and then gains 8 yards.
    (c) The change in elevation from a desert 100 feet below sea level to the top of a mountain 3400 feet above sea level.

16. Classify either as true or as false.
    (a) Every integer is a whole number.
    (b) The set of integers is the same as the set of additive inverses of whole numbers.
    (c) Every counting number is an integer.
    (d) Every integer has an additive inverse.
    (e) The intersection of the positive integers and the negative integers is the null set.
    (f) The union of the negative integers and the counting numbers is the set of integers.

17. (a) With a signed number, describe the position of a vein of coal 150 meters under ground level.
    (b) Aristotle was born in 384 B.C. and Euclid was born in 365 B.C. Who was born first?
    (c) Which temperature is higher, $^-22$°C or $^-12$°C?
    (d) Explain each of the above with a diagram.

18. Evaluate.

    (a) $|^-7|$ (b) $|0|$ (c) $|12|$
    (d) $|^-6| + |^-5|$ (e) $|2| + |^-3|$ (f) $|^-8| - |^-6|$

19. Find the additive inverse and the absolute value of each of the following.

    (a) $^-(21 + 5)$ (b) $^-(a - b)$ (c) ^-c
    (d) $|x|$ (e) $^-|5|$ (f) $|x - y|$

T 20. (a) What is the smallest nonnegative integer?
    (b) What is the smallest positive integer?
    (c) What is the largest positive integer?
    (d) What is the smallest negative integer?
    (e) What is the largest negative integer?
    (f) Is $+0$ different from -0?

21. Find the following sums.

    (a) $^-2x + 3x + {}^-7x - 4x$
    (b) $^-4a - 6a - (^-2a) - 5a$
    (c) $^-10y^2 + (^-3y^2) - 4y^2 - (^-6y^2)$
    (d) $^-8k^2 - (^-3k^2) - 2k^2 + (^-4k^2)$

22. If W stands for the whole numbers, I the integers, ^-I the negative integers and ^+I the positive integers, find the following.

    (a) $^-I \cap {}^+I$, (b) $^-I \cap W$,
    (c) $^-I \cup {}^+I$, (d) $^-I \cup W$,
    (e) $^+I \cap W$, (f) $W \cup {}^+I$

23. Working from left to right, identify the property or properties that make the following true.

    (a) $(8 + 4) + {}^-2 = (8 + {}^-2) + 4$
    (b) $(6 + {}^-6) + 0 = (0 + 6) + {}^-6$
    (c) $(^-4 + 6) + {}^-2 = (^-2 + 6) + {}^-4$
    (d) $(^-6 + {}^-4) + 2 = (^-6 + 2) + {}^-4$

C 24. For the domain of integers, graph the solution set on a number line.

    (a) $x < 4$ and $x > {}^-3$
    (b) $x < 2$ and $x < {}^-3$
    (c) $x < 2$ or $x < {}^-3$
    (d) $x > {}^-1$ or $x < 5$

7 Integer multiplication and division

■ The football team at Mystate U had some difficulty in their last game. On each of four sets of downs in the first quarter, the team lost 6 yards. How many yards did the team lose during the first quarter? ■

In this section, you will be introduced to the operations of multiplication and division for integers. Again, the extension from whole numbers to integers is made in such a way that the properties of multiplication and division for whole numbers hold for integers.

A study of this section demands care. Most mistakes made in the multiplication and division of integers are careless ones. Don't let the notation of this section cause you difficulty.

Read again the problem presented in the introduction of this section. Of course, this example can be considered as a multiplication problem: $4 \cdot {}^{-}6 = ?$ To solve such problems, we define multiplication for the integers so that the closure, commutative, associative, and identity properties of multiplication, the multiplicative property of 0, and the distributive property of multiplication over addition, as discussed for the whole numbers, hold for the set of integers.

Example 70 • How should $4 \cdot {}^{-}6$ be defined so as to be in agreement with the properties of multiplication of whole numbers?

Solution In order to be consistent with the idea that multiplication by a natural number is repeated addition,

$$4 \cdot {}^{-}6 = {}^{-}6 + {}^{-}6 + {}^{-}6 + {}^{-}6 = {}^{-}24$$

Thus, the football team lost 24 yards in the first quarter. •

Using multiplication as repeated addition, note that $4({}^{-}6) = {}^{-}24$ on the number line in Figure 14.

Figure 14

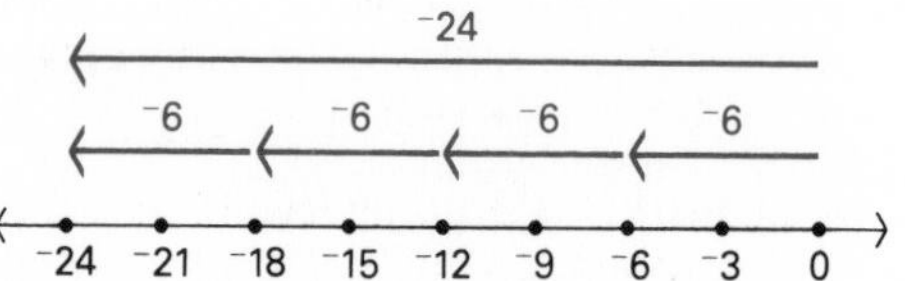

Since multiplication is commutative,

$${}^{-}6 \cdot 4 = 4 \cdot {}^{-}6 = {}^{-}24$$

This idea of multiplication is further emphasized by looking at the following pattern.

$$3 \cdot 4 = 12$$
$$2 \cdot 4 = 8$$
$$1 \cdot 4 = 4$$
$$0 \cdot 4 = 0$$
$${}^{-}1 \cdot 4 = \_\_\_$$
$${}^{-}2 \cdot 4 = \_\_\_$$
$${}^{-}3 \cdot 4 = \_\_\_$$
$${}^{-}4 \cdot 4 = \_\_\_$$
$${}^{-}5 \cdot 4 = \_\_\_$$
$${}^{-}6 \cdot 4 = {}^{-}24$$

Notice that the answer is decreased by 4 each time the multiplier is decreased by 1. Can you complete the blanks from this pattern?

Example 71 • How should we define $^-3 \cdot {}^-4$ if the properties of multiplication of whole numbers are to hold for integers?

Solution Consider $^-3 \cdot {}^-4 + {}^-3 \cdot 4 = {}^-3({}^-4 + 4) = {}^-3(0) = 0$. Thus, $^-3 \cdot {}^-4$ is the additive inverse of $^-3 \cdot 4$, and $^-3 \cdot 4 = {}^-12$. But the unique additive inverse of $^-12$ is 12, so $^-3 \cdot {}^-4 = 12$. •

The preceding product can also be inferred from the following pattern.

$$\begin{aligned} 3 \cdot {}^-4 &= {}^-12 \\ 2 \cdot {}^-4 &= {}^-8 \\ 1 \cdot {}^-4 &= {}^-4 \\ 0 \cdot {}^-4 &= 0 \\ {}^-1 \cdot {}^-4 &= \underline{\quad} \\ {}^-2 \cdot {}^-4 &= \underline{\quad} \\ {}^-3 \cdot {}^-4 &= \underline{\quad} \end{aligned}$$

Notice that as the multipliers decrease by 1, the answers increase by 4. Complete the blanks and see if your answers agree with the preceding example.

Such examples illustrate the following definition of multiplication of integers.

Multiplication of integers

> The **product of two integers** is defined by the following cases, where a and b are whole numbers.
>
> **1.** $a \cdot b = n(A \times B)$, where $a = n(A)$ and $b = n(B)$ for sets A and B.
> **2.** $^-a \cdot {}^-b = ab$
> **3.** $^-a \cdot b = b \cdot {}^-a = {}^-(ab)$

Example 72 •

$$\begin{aligned} 5({}^-3) &= {}^-15 \\ ({}^-7)(6) &= {}^-42 \\ ({}^-6)({}^-17) &= 6 \cdot 17 = 102 \\ ({}^-2)({}^-5)({}^-3) &= ({}^-2)[({}^-5)({}^-3)] \\ &= ({}^-2)[5 \cdot 3] \\ &= ({}^-2) \cdot 15 \\ &= {}^-30 \end{aligned}$$ •

Multiplication has an inverse operation, division. For instance, since (2)(8) = 16, then 16 divided by 8 equals 2, and 16 divided by 2 equals 8. Division for integers is defined in the following manner.

Division of integers

> If x and y are integers with $y \neq 0$, then x divided by y is equal to an integer z if and only if $x = y \cdot z$.

Just for fun

Insert a pair of parentheses to make the following a true sentence.

$$1 - 2 + 3 - 4 + 5 - 6 = {}^-1$$

To determine whether the answer is positive or negative in an exact division, we refer to the definition of multiplication. Since $x \div y = z$ if and only if $x = y \cdot z$, the sign of z will be fixed so that the multiplication equals x. We may summarize as follows.

Sign of a quotient

> The quotient, if it exists, of two positive or two negative integers is a positive integer; the quotient, if it exists, of a negative and a positive integer, in either order, is a negative integer.

Example 73 •

$${}^-12 \div 3 = {}^-4 \quad \text{since} \quad 3 \cdot {}^-4 = {}^-12$$

$$45 \div {}^-5 = {}^-9 \quad \text{since} \quad {}^-5 \cdot {}^-9 = 45$$

$${}^-24 \div {}^-6 = 4 \quad \text{since} \quad {}^-6 \cdot 4 = {}^-24$$

$${}^-16 \div 5 = ?$$ •

Division of integers is not closed, since ${}^-16 \div 5$ has no meaning in the set of integers. There is no integer such that when multiplied by 5 the result is ${}^-16$. The number system will be extended in Chapter 5 to alleviate this deficiency.

Remember that zero plays a very special role in division.

Properties of zero

> Let a be any integer. If $a \neq 0$, then $0 \div a = 0$, because $a \cdot 0 = 0$. If $a \neq 0$, then $a \div 0$ is undefined, because $a = 0 \cdot d$ is false for any d. $0 \div 0$ is undefined, because there is no unique quotient.

Exercise set 7

R 1. Write the following products and quotients as integers.

(a) $5({}^-9)$ (b) ${}^-7({}^-4)$
(c) $0 \div {}^-9$ (d) ${}^-24 \div {}^-3$
(e) ${}^-7(46)$ (f) ${}^-5(42)$
(g) ${}^-x({}^-y)$ (h) ${}^-h(g)$
(i) $(3 \cdot {}^-2) \div {}^-3$ (j) $8({}^-3) \div {}^-4$
(k) ${}^-9({}^-5 \cdot {}^-3)$ (l) ${}^-3({}^-2 \cdot 4)$

2. Perform the following computations.

   (a) $^-3(^-4 \cdot 6)$ (b) $(^-2 \cdot {}^-5) \cdot 7$
   (c) $(^-6 \cdot {}^-4) \cdot (8 \cdot {}^-2)$ (d) $(^-6 \cdot {}^-4) \cdot (0 \cdot 3)$
   (e) $(23 - 5) \div {}^-6$ (f) $(^-6 + {}^-18) \div {}^-4$
   (g) $(^-27 + {}^-5) \div {}^-4$ (h) $^-3(5 + {}^-1) \div 6$
   (i) $(^-4 + {}^-8) \cdot (5 + {}^-7)$ (j) $(^-6 + {}^-1)(^-3 + {}^-2)$
   (k) $(^-x \cdot {}^-y) \cdot {}^-z$ (l) $(^-4 + 4)(3 + {}^-6)$

3. Compute each part in two different ways, using the distributive property.

   (a) $^-1(3 + {}^-5)$ (b) $(3 + {}^-2)(4)$

4. Generalize the distributive property for integers and expand each of the following into the sum of three terms.

   (a) $^-5(a + b + c)$ (b) $3(2x + y + z)$
   (c) $^-y(3 + {}^-2x + 4z)$ (d) $^-x(2 + 3z + y)$

5. Under what conditions will the product of two factors be

   (a) greater than 0? (b) equal to 0?
   (c) less than 0?

T 6. (a) For what integral values of a and b does $a \div b = b \div a$?
   (b) What is the value of $^-24 \div (4 \div {}^-2)$?
   (c) What is the value of $(^-24 \div 4) \div {}^-2$?
   (d) What can you conclude from parts (b) and (c) about the associativity of division for integers?

7. Solve on the domain of integers.

   (a) $^-2x = 4$ (b) $^-3x = {}^-9$
   (c) $^-2x - 3 = 5$ (d) $^-3x + 4 = {}^-5$

In Exercises 8 through 15, first find an equation to represent each problem. Then solve the equation.

8. Signs are placed at regular intervals along a mountain road to indicate changes in altitude. At Greenbriar Lodge, the sign indicates an elevation of 876 meters. Seventeen kilometers down the road, another sign indicates that the elevation is 672 meters. What is the average elevation change per kilometer traveled from the lodge?

9. A plane flying at an altitude of 10,000 meters descends at a rate of 800 meters per minute for 10 minutes. What is the altitude of the plane relative to a mountain of height 2000 meters?

10. Senator Snodgrass received an illegal contribution of $255,000 from an anonymous donor. How many $1500 donations would it take to exceed this generous gift?

11. In northeastern Alaska, the temperature has fallen 4°C per day for a week. If the temperature is now $^-45$°C, what was the temperature 5 days ago?

12. (a) In Exercise 11, what was the temperature 2 days ago?
    (b) If the pattern continues, what will be the temperature 2 days from now?

13. The temperature now at the North Pole is $^-40$°C. The temperature is expected to drop 6 degrees per day for the next few days. Write an open sentence for the temperature t days from now.

14. In Exercise 13, what will be the temperature 4 days from now?

15. The highest temperature during the day in Atlanta, Georgia is 94°F. The high temperature is expected to decrease 2°F per day for several days. When will the highest temperature be 88°F?

16. Let

$$W = \{0, 1, 2, 3, \ldots\}$$

$$M = \{0, {}^-1, {}^-2, {}^-3, \ldots\}$$

$$I = \{\ldots {}^-3, {}^-2, {}^-1, 0, 1, 2, 3, \ldots\}$$

   Which sets are closed under

   (a) addition? (b) subtraction?
   (c) multiplication? (d) division?

17. For integers x and y, under what condition will $x = {}^-y$?

18. $a(b - c) = a(b + {}^-c)$ Why?
    $= ab + {}^-ac$ Why?
    $= ab - ac$

    This verifies the ________ property of multiplication over subtraction.

19. Use Exercise 18 to simplify:

    (a) $2(3 - 5)$ (b) $^-3(4 - 8)$

20. (a) Tom's five test grades in his mathematics class are 90, 74, 46, 67, and 53. What is his average grade?
    (b) Subtract the average grade from each of his five test grades. Add the five numbers obtained. Are you surprised at the answer?

C 21. $(a + b) \cdot (a - b) = (a + b) \cdot (a + {}^-b)$
$= (a + b) \cdot a + (a + b) \cdot {}^-b$
$= a^2 + b \cdot a + a \cdot {}^-b + b \cdot {}^-b$
$= a^2 + ba - ba - b^2$
$= a^2 - b^2$

Use this to simplify:

(a) $(x-2)(x+2)$ (b) $(3x+5)(3x-5)$
(c) $(2x-y)(2x+y)$ (d) $(3x+4y)(3x-4y)$

22. Use Exercise 21 to write each of the following as a product.

(a) x^2-9 (b) $4y^2-25$
(c) $9x^2-y^2$ (d) $4x^2-9y^2$

23. Solve each of the following on the domain of integers.

(a) $2x < {}^-6$ (b) $3x < {}^-9$
(c) $2x+4>2$ (d) $3x+5>8$

Solution to introductory problem

Understanding the problem. When the trainer arranges the towels in 4 stacks, there is one left over; that is, if the number of towels is divided by 4, there is a remainder of 1. The same is true for division by 5 and by 6. Upon arranging in 7 stacks, there are no extra towels. Therefore, the number of towels is divisible by 7. We are to find the smallest possible number of towels.

Devising a plan. Plan 1: Write the multiples of 7 and divide each by 4, 5, and 6. Find the smallest multiple so that each of the three divisions produces a remainder of 1. This may be time consuming, though it will certainly produce the answer.

Plan 2: Let's use the fact that if $n \div 4$ leaves a remainder of 1, then $4|(n-1)$. Likewise, $5|(n-1)$ and $6|(n-1)$. Since $7|n$, then $n=7k$.

$$4|(7k-1)$$

$$5|(7k-1)$$

$$6|(7k-1)$$

Now the smallest number that 4, 5, and 6 will divide is

$$\text{l.c.m.}(4, 5, 6) = 60$$

Since every common multiple is a multiple of the least common multiple,

$$60l = 7k-1 \quad \text{or} \quad 7k = 60l+1 \quad \text{or} \quad 7|(60l+1)$$

Carry out the plan. Let l take on values 1, 2, 3, . . . until a $(60l+1)$ is found that is divisible by 7.

$$7\nmid 61 \quad 7\nmid 121 \quad 7\nmid 181 \quad 7\nmid 241 \quad \text{but} \quad 7|301$$

Looking back. Check to see if 301 towels is a valid answer.

Summary and review

Divisibility

1. If b is a natural number and a is any whole number, then b divides a (denoted $b|a$) if and only if there is a unique whole number c such that $a=bc$.
2. The following are basic divisibility properties for natural numbers x, y, and z.
   (a) $1|x$ and $x|x$.
   (b) If $x|y$ and $y|z$, then $x|z$.
   (c) If $x|y$ and $x|z$, then $x|(y+z)$ and $x|(y-z)$.
   (d) If $x|y$ and $x\nmid z$, then $x\nmid(y+z)$ and $x\nmid(y-z)$.
   (e) If $x|y$ and $x|(y+z)$ [or $x|(y-z)$], then $x|z$.
   (f) If $x|y$ and b is any natural number, then $x|by$.
   (g) If $x|y$ or $x|z$, then $x|yz$.
3. Generalization of divisibility properties

   If a divides the sum of n whole numbers and also divides each one of $n-1$ of these whole numbers, then a divides the remaining whole number.

Divisibility tests (base ten)

1. A whole number is divisible by 2 or 5 if and only if the units digit of the number is divisible by 2 or 5.

2. A whole number is divisible by 4 if and only if the last two digits of the number represent a number divisible by 4.
3. A whole number is divisible by 8 if and only if the last three digits of the number represent a number divisible by 8.
4. A whole number is divisible by 3 or 9 if and only if the sum of its digits is divisible by 3 or 9.
5. A whole number is divisible by 6 if and only if it is divisible by both 2 and 3.
6. A whole number is divisible by 7 if and only if it satisfies the following property. Subtract the double of the right-hand digit from the number represented by the remaining digits. If the difference is divisible by 7, then the original number is divisible by 7. Repeat the process until divisibility is obvious.
7. A whole number is divisible by 10 if and only if it ends in 0.
8. A whole number is divisible by 11 if and only if the difference in the sum of the digits in the odd-numbered positions and the sum of the digits in the even-numbered positions is divisible by 11.

Primes and composite numbers

1. A whole number p, greater than 1, is *prime* if and only if the only divisors of p are 1 and p.
2. A whole number is said to be *composite* if and only if it has a natural number divisor other than itself and 1.
3. If n is not divisible by any prime p such that $p^2 \leq n$, then n is prime.
4. When a composite number is written as a product of all of its prime divisors, the product is called a *prime factorization* of the number.
5. The Fundamental Theorem of Arithmetic. Every composite whole number has a unique prime factorization (except for the order of the factors).

Greatest common divisor (g.c.d.) *and least common multiple* (l.c.m.)

1. The *greatest common divisor* of two whole numbers a and b is the largest whole number d such that $d|a$ and $d|b$.
2. If the greatest common divisor of two whole numbers a and b is 1, then a and b are *relatively prime.*
3. If a and b are two whole numbers with $a \geq b$, and r is the remainder when a is divided by b, then g.c.d. $(a, b) =$ g.c.d. (b, r).
4. Euclidean Algorithm. In the repeated application of the preceding property until a remainder of zero is obtained, the last nonzero remainder is g.c.d. (a, b).
5. A whole number m is the *least common multiple* of two whole numbers a and b if m is the least nonzero whole number divisible by both a and b.
6. $\text{l.c.m.}\ (a, b) = \dfrac{ab}{\text{g.c.d.}\ (a, b)}$

Additional definitions of number theory

1. Primes of the form $2^p - 1$, where p is a prime, are called *Mersenne primes.*
2. The proper divisors of a whole number are all the whole number divisors except the number itself.
3. Perfect, deficient, abundant. If the sum of the proper divisors of a number is equal to the number, the number is called a *perfect* number; if this sum is less than the number, it is called a *deficient* number; if it is greater than the number, the number is an *abundant* number.
4. Two numbers are said to be *amicable* if each is the sum of the proper divisors of the other.

Goldbach's conjectures

1. Every even number greater than 4 is the sum of two odd primes.
2. Every odd number greater than 7 is the sum of three odd primes.

Clock and modular arithmetic

1. Two whole numbers a and b are congruent modulo m if and only if $m|(a - b)$.
2. a and b are congruent modulo m if their remainders are equal when divided by m.
3. If $a \equiv b \pmod{m}$ and $c \equiv d \pmod{m}$, then
   (a) $a + c \equiv b + d \pmod{m}$
   (b) $a - c \equiv b - d \pmod{m}$
   (c) $ka \equiv kb \pmod{m}$, where k is any natural number
   (d) $ac \equiv bd \pmod{m}$
4. A *group* is a set of elements, together with a binary operation $\otimes$, with the following properties.
   (a) There is closure with respect to $\otimes$.
   (b) The operation is associative.
   (c) There is an identity element i in the set such that
   $$i \otimes a = a \otimes i = a$$
   where a is any element of the set.
   (d) Every element a has an inverse for $\otimes$.

The integers

1. The set of integers I consists of $\{\ldots {}^{-}3, {}^{-}2, {}^{-}1, 0, 1, 2, 3, \ldots\}$.

2. $I = \{\text{positive integers}\} \cup \{\text{negative integers}\} \cup \{0\}$.

3. For each integer a, there exists a unique *additive inverse* ${}^{-}a$ such that $a + {}^{-}a = {}^{-}a + a = 0$.

4. Addition is defined by the following cases where a and b are whole numbers.

   (a) $a + b = n(A \cup B)$ where $a = n(A)$, $b = n(B)$, and $A \cap B = \varnothing$ for sets A and B.
   (b) ${}^{-}a + {}^{-}b = {}^{-}(a + b)$
   (c) $a + {}^{-}b = {}^{-}b + a = (a - b)$ if $a > b$
   (d) $a + {}^{-}b = {}^{-}b + a = 0$ if $a = b$
   (e) $a + {}^{-}b = {}^{-}b + a = {}^{-}(b - a)$ if $a < b$

5. The subtraction of y from x is defined as $x - y = z$ if and only if $x = y + z$.

$$x - y = x + {}^{-}y$$

6. Multiplication is defined by the following cases where a and b are whole numbers.

   (a) $a \cdot b = n(A \times B)$, where $a = n(A)$ and $b = n(B)$ for sets A and B
   (b) ${}^{-}a \cdot {}^{-}b = ab$
   (c) ${}^{-}a \cdot b = b \cdot {}^{-}a = {}^{-}ab$

7. The division $a \div b = c$ if and only if $a = bc$.

   (a) If $a \neq 0$, then $0 \div a = 0$, because $a \cdot 0 = 0$.
   (b) If $a \neq 0$, $a \div 0$ is undefined, because $a = 0 \cdot d$ is false for any d.
   (c) $0 \div 0$ is undefined, because there is no unique quotient.

Review exercise set 8

R 1. Determine if the following numbers are prime.

   (a) 1 (b) 157 (c) 83 (d) 391

2. Factor the following numbers into primes.

   (a) 126 (b) 525 (c) 2475 (d) 252

3. Use the factorizations from Exercise 2 to compute

   (a) g.c.d. (126, 525). (b) l.c.m. (525, 2475).
   (c) g.c.d. (126, 252, 525). (d) l.c.m. (126, 252).

4. Compute on a 4-minute clock.

   (a) $3 + 3$ (b) $2 - 3$ (c) $3 - 2$ (d) $2 \cdot 3$

5. Perform the following computations without a table. In each case, reduce the answer to a whole number less than the modulus.

   (a) $4 + 5 \pmod 6$ (b) $8 - 11 \pmod 5$
   (c) $6 \cdot 3 \pmod 7$ (d) $14 \cdot 5 \pmod 8$

6. Show without actual division that 12 divides 936. Give reasons for your answer.

7. What is the greatest common divisor of 5734 and 12,862? What is the least common multiple?

8. If 1 is the greatest common divisor of two numbers, what can you say about their least common multiple?

9. Suppose that the least common multiple of two numbers is the same as their greatest common divisor. What can you say about the numbers?

10. Classify each of the following as true or as false.

   (a) If a number is divisible by 3, it must be divisible by 9.
   (b) $12 \mid 6$.
   (c) Every odd number is a prime number.
   (d) If a number is divisible by 8, it must be divisible by 4.
   (e) $7 \mid 7$.
   (f) If a number is divisible by 2 and 10, it must be divisible by 20.

11. Perform the indicated calculations.

   (a) $36 - ({}^{-}7) - 18$
   (b) ${}^{-}12 - 15 + ({}^{-}7) - ({}^{-}26)$
   (c) $50 - 64 - 13 + 10$
   (d) $0 - [2 - ({}^{-}3)]$
   (e) ${}^{-}6 - ({}^{-}6)$
   (f) ${}^{-}1 - ({}^{-}3) - ({}^{-}4) - 17$
   (g) ${}^{-}6(3 - {}^{-}5) - ({}^{-}7)$
   (h) ${}^{-}x + (2x + {}^{-}3x)$
   (i) ${}^{-}x({}^{-}z + y)$
   (j) $x(w + {}^{-}y) + xw$

(k) $(^{-}30 \div 6) \cdot {}^{-}8$
(l) $^{-}4 + (^{-}16 \div {}^{-}8)$

12. In the given tables, insert the appropriate sign. Put a question mark if there is insufficient information to determine the sign.

(a)

Add	+	−
+		
−		

(b)

Subtract	+	−
+		
−		

(c)

·	+	−
+		
−		

(d)

÷	+	−
+		
−		

13. Show that each of the following holds. Justify each step.
(a) $4(^{-}3) + (^{-}4)(2) = 4(^{-}3 + {}^{-}2)$
(b) $(4 \cdot {}^{-}3) = 3 \cdot {}^{-}4$
(c) $^{-}4(^{-}8 \cdot 2) = (8 \cdot {}^{-}2) \cdot {}^{-}4$

14. Show the following operations on a number line.
(a) $^{-}2 + 3$ (b) $^{-}2 + {}^{-}3$ (c) $4 \cdot {}^{-}2$
(d) $3 + {}^{-}3$ (e) $^{-}5 + 2$ (f) $4 + {}^{-}3$

T 15. The operation $\otimes$ is defined to be the smaller of two numbers if they are unequal and one of the numbers if they are equal.
(a) Make a table defining the operation for the set $\{1, 2, 3, 4, 5\}$
(b) Is the operation commutative?
(c) Is there an identity element?
(d) Is the operation associative?
(e) Do all the elements of the set have an inverse with respect to the operation?

16. Classify as true or as false. If false, demonstrate with a counterexample.
(a) If $c|(a + b)$, then $c|a$ and $c|b$.
(b) If g.c.d. $(a, b) = 1$, then a and b cannot both be even.
(c) If $c|(a + b)$, then $c|a$ or $c|b$.
(d) If g.c.d. $(a, b) = 2$, then both a and b are even.
(e) $1|a$ for all whole numbers a.
(f) $a|0$ for all natural numbers a.
(g) If $c|a$ and $c|b$, then $c|ab$.
(h) l.c.m. $(a, b)|$g.c.d. (a, b).
(i) If $c|ab$, then $c|a$ or $c|b$.

17. If November 30 falls on Wednesday, on what day will it fall next year if next year is not a leap year? Is a leap year?

18. If May 19 is on Thursday, what day is the Fourth of July of the same year?

19. (a) Find five consecutive composite numbers all less than 30.
(b) Explain why $(25 \cdot 24 \cdot \cdots \cdot 3 \cdot 2) + 2$ and $(25 \cdot 24 \cdot \cdots \cdot 3 \cdot 2) + 3$ are composite.
(c) How many consecutive numbers of the form $(25 \cdot 24 \cdot 23 \cdot \cdots \cdot 3 \cdot 2) + k$ $(k = 2, 3, 4, \ldots)$ are composite?

20. Find the smallest natural number divisible by 2, 3, 4, 5, and 6.

21. Is it possible to have exactly six consecutive composite numbers between two primes? Justify your answer by finding an example.

22. Work Exercise 21 for seven consecutive composite numbers.

23. Solve on the domain of integers.
(a) $^{-}2x = {}^{-}4$ (b) $x + {}^{-}4 = 8$
(c) $^{-}2x - 5 = {}^{-}3$ (d) $^{-}3x - 4 = {}^{-}5$
(e) $x - 4 < {}^{-}2$

C 24. For each natural number n,
(a) show that $3|(n^3 - n)$.
(b) show that $5|(n^5 - n)$.
(c) does $9|(n^9 - n)$?

25. Find the smallest natural number divisible by 2, 3, 4, . . . , 20.

26. Which of the following systems are groups? Abelian groups?
(a) System in Exercise 15
(b) System defined by a 24-hour clock under addition

Bibliography

1. Battista, Michael T. "A Complete Model for Operations on Integers." *AT* (May 1983), 26–31.

2. Bezuszka, Stanley J. "A Test for Divisibility by Primes." *AT* (October 1985), 36–38.

3. Brown, G. W. "Searching for Patterns of Divisors." *AT* (December 1984), 32–34.

4. Brown, Stephen I. *Some Prime Comparisons.* National Council of Teachers of Mathematics, 1978.

5. Burton, Grace M., & J. Dan Knifong, "Definitions for Prime Numbers." *AT* (February 1980), 44–47.

6. Chang, Lisa. "Multiple Methods of Teaching the Addition and Subtraction of Integers." *AT* (December 1985), 14–19.

7. Dubisch, Roy. "Generalizing a Property of Prime Numbers." *AT* (February 1974), 93–94.

8. Francis, Richard L. "A Note on Perfect Numbers." *MT* (November 1975), 606–607.

9. Gardella, Francis J. "Divisibility—Another Route." *AT* (March 1984), 55–56.

10. Grady, M. B. Tim. "A Manipulative Aid for Adding and Subtracting Integers." *AT* (November 1978), 40.

11. Hall, Wayne H. "These Three: *Minus, Negative, Opposite.*" *AT* (December 1974), 712–713.

12. Henry, Loren L. "Another Look at Least Common Multiple and Greatest Common Factor." *AT* (March 1978), 52–53.

13. Kindle, E. Glenn. "Droopy, The Number Line, and Multiplication of Integers." *AT* (December 1976), 647–650.

14. Lamb, Charles E., & Lyndal R. Hutcherson. "Greatest Common Factor and Least Common Multiple." *AT* (April 1984), 43–44.

15. Millema, Wilbur. "Exploring Primes," *AT* (October 1982).

16. Miller, Ann. "Divisor Lattices and the Teaching of GCD and LCM." *MT* (September 1977), 510–513.

17. Miller, G. Frank, & Maurice Jeffrey. "'Zero': An Addition Game with Integers." *MT* (January 1977), 52–53.

18. Neuner, Albert R. "A Modulo Line." *AT* (March 1973), 214.

19. Niemann, Christopher E. "Let's Play Mod 7." *AT* (May 1976), 348–350.

20. Pasquali, Giorgio. "Discovering a Formula that Generates Even Perfect Numbers." *MT* (October 1976), 469–470.

21. Prielipp, Robert W. "Perfect Numbers, Abundant Numbers, and Deficient Numbers." *MT* (December 1970), 692–696.

22. Schaefer, Sister M. Geralda. "Motivational Activities in Elementary Mathematics." *AT* (May 1981), 17–18.

23. Schatz, Mary Christine. "Comments on Euclid's Proof of the Infinitude of Primes." *MT* (December 1975), 676–677.

24. Scherzer, Laurence. "A Simplified Presentation for Finding the LCM and the GCF." *AT* (May 1974), 415–416.

25. Singer, Richard. "Modular Arithmetic and Divisibility Criteria." *MT* (December 1970). 653–656.

26. Troutman, Andria P., & Betty K. Lichtenberg. *Mathematics: A Good Beginning, Strategies for Teaching Children* (3rd ed.). Monterey, Calif.: Brooks/Cole, 1987.

CHAPTER 5
Introduction to the Rational Numbers

A used car dealer was asked how many cars he had sold during the week. He replied, "Monday I sold one-half of my cars plus half a car. Tuesday I sold a third of what remained plus one-third of a car. Wednesday I sold a fourth of what remained plus one-fourth of a car. Thursday I sold a fifth of what remained plus one-fifth of a car. I have not sold any today, Friday. I now have 11 cars left on the lot." Of course one cannot divide a car, so how many cars did the salesman have on the lot at the beginning of the week?

In the system of integers, it is impossible to divide 6 candy bars equally among 4 youngsters or to express 20 laps as part of a 100-lap race. Also you will recall that the system of integers does not have closure for division. In this chapter, we shall introduce some new numbers, called *fractions,* by extending the system of integers to a new system of numbers called *rational numbers.*

Recall that the system of whole numbers was extended to the system of integers in order to provide closure for subtraction. We now extend the system of integers to the system of **rational numbers** to provide closure for division.

In addition, we will represent rational numbers as decimals and review the algorithms for working with decimals.

Thus in this chapter, not only will you acquire a better understanding of fractions and decimals, but you will improve your dexterity in using both of these.

1 The set of rational numbers

■ You recall that you divided the numerator and the denominator of a fraction by 5 and then by 3 and got $\frac{2}{5}$. You have lost the original fraction. Can you find it? ■

In this section, we will study the Fundamental Law of Fractions, which will enable you to solve the preceding problem. We begin by learning that a fraction can be interpreted in three ways—as a division problem, as a part of a whole, and as an element in a mathematical system. Although you have used fractions for many years, you may not have full appreciation of this concept. Has it ever occurred to you that several fractions may represent the same number concept, but only one whole number represents a given number concept? This property of fractions should help you to understand rational numbers.

Rational numbers can be represented by numerals of the form $\frac{2}{3}$, $\frac{4}{7}$, $\frac{15}{9}$, called **fractions.** You are already familiar with numerals of this type. The symbol $\frac{2}{3}$ involves the pair of integers 2 and 3, where 2 is the numerator and 3 is the denominator. Rational numbers, represented by fractions, are used in the following ways.

1. A **division problem** or a solution to a multiplication problem.
2. A **part** of a whole.
3. An **element** of a mathematical system.

Consider the division $2 \div 3$. Obviously, there is no answer on the set of integers. However, let us define a new number, denoted by the fraction $\frac{2}{3}$ (or 2/3), such that

$$3 \cdot \left(\frac{2}{3}\right) = 2$$

In general, $a \div b$ where $b \neq 0$ has an answer denoted by a/b such that

$$b\left(\frac{a}{b}\right) = a$$

Not only does this new number provide closure for division, it also enables us to express parts of a whole; that is, $\frac{3}{4}$ means three of four equal parts.

Example 1 • Consider a set of "like" regions, as shown in each part of Figure 1. Certain regions in each part are shaded. By pairing the number of shaded regions with the total number of regions, write the result. Figure 1(a) shows that $\frac{5}{9}$ of the square is shaded; part (b) shows that $\frac{2}{4}$ of the circle is shaded; and part (c) shows that $\frac{1}{3}$ of the triangle is shaded. Intuitively, these fractions represent numbers (called **rational numbers**) that express five out of nine equal parts (or $5 \div 9$), two out of four equal parts (or $2 \div 4$), and one out of three equal parts (or $1 \div 3$). •

Figure 1

(a)

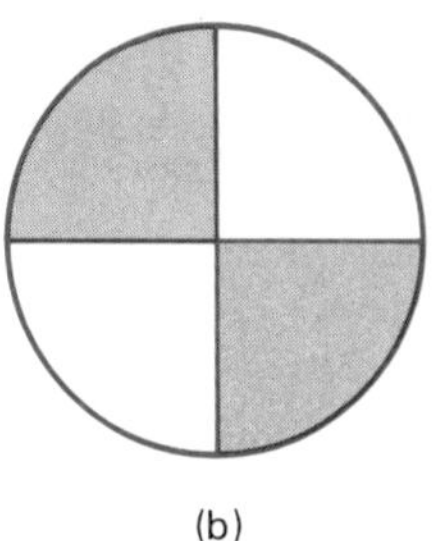

(b)

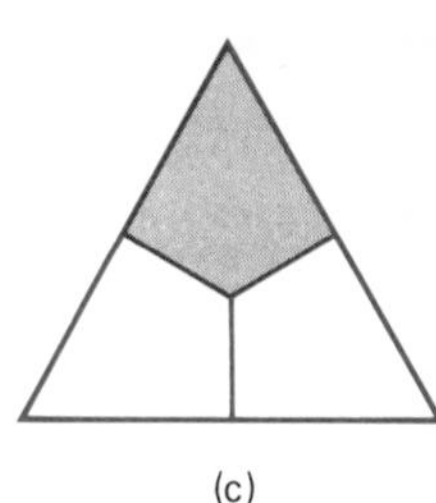

(c)

A rational number

> A *rational number* is a quotient, $x \div y$, of two integers x and y, where $y \neq 0$. This number may be represented by a fraction
>
> $$\frac{x}{y} \quad \text{or} \quad x/y$$

Actually, a rational number can also be represented as a terminating or repeating decimal, as an ordered pair (x, y), or as a division $x \div y$, but it may always be represented as a fraction.

Let's represent now a point on a number line by a fraction. (See Figure 2.) Consider the line segment from 0 to 3 on a number line. As best you can, divide this segment into 4 equal parts. (One procedure would be to cut a strip of paper the length of the segment and fold twice.) Each part of the division would be of length $\frac{3}{4}$. Thus a point can be located on a number line that is represented by $\frac{3}{4}$.

Figure 2

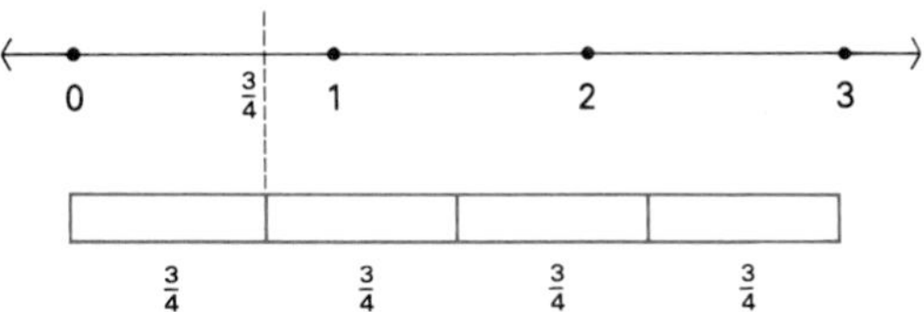

In discussing the concept of a fraction using a number line, we divide each unit interval into a given natural number of divisions of equal length. For example, if we divide each unit into 3 divisions as in Figure 3, our fractions will have denominators of 3. The numerators will count the number of subintervals to the left and right of 0.

Figure 3

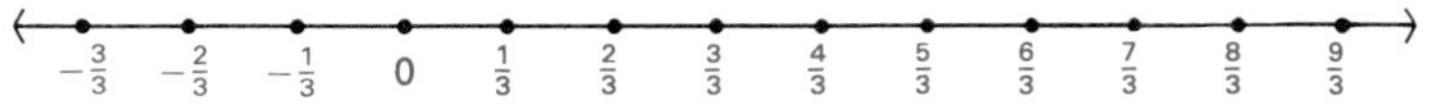

point of interest

The word *fraction* is derived from the Latin word *fractio,* meaning "to break into parts." In the fraction a/b, a is called the **numerator,** and b is called the **denominator.** If a pie is cut into b equal parts and a of them are eaten, then a/b represents the fraction of the pie that is eaten.

In general, any fraction a/b may be represented on a number line by dividing each unit interval into b subintervals of equal length. Then a/b represents the point that is a subintervals to the right or left of the point 0.

To discover relationships among fractions, we examine the division of the rectangle in Figure 4 into 3 equal parts, 6 equal parts, and 12 equal parts. The shaded region of Figure 4 is 1 part out of the 3 parts (represented by $\frac{1}{3}$) or 2 parts out of 6 (represented by $\frac{2}{6}$) or 4 parts out of 12 (represented by $\frac{4}{12}$). Notice that $\frac{4}{12}$, $\frac{2}{6}$, and $\frac{1}{3}$ all represent the same shaded region of Figure 4.

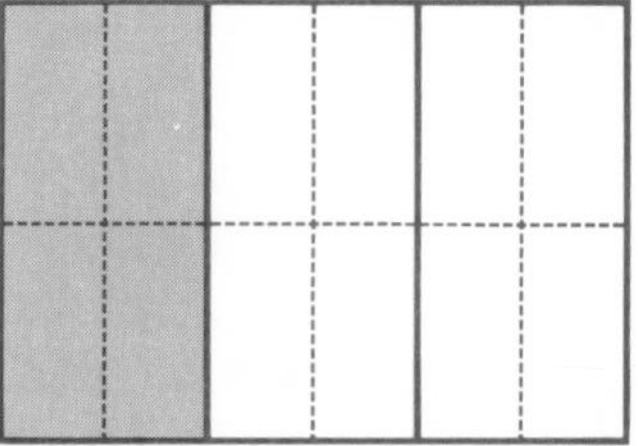

Figure 4

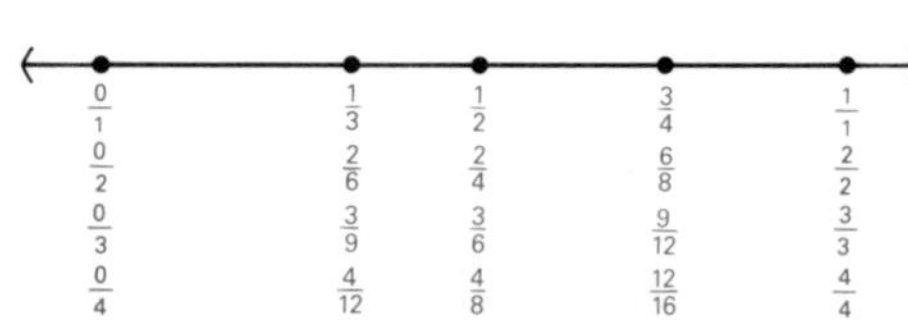

Figure 5

Similarly, we note in Figure 5 that a rational point on the number line is represented by many (in fact, infinitely many) different fractions. For example, $\frac{1}{3}$, $\frac{2}{6}$, $\frac{3}{9}$, and $\frac{4}{12}$ all represent the same point.

Such fractions that represent the same rational number point on the number line are called **equivalent fractions.** Each of these fractions is said to be *equivalent* to the other fractions, and a set such as

$$\left\{\ldots, \frac{-4}{-12}, \frac{-3}{-9}, \frac{-2}{-6}, \frac{-1}{-3}, \frac{1}{3}, \frac{2}{6}, \frac{3}{9}, \frac{4}{12}, \ldots\right\}$$

is called an **equivalence class** of fractions.

Equivalent fractions are sometimes denoted by $\cong$, but we shall use the equality sign to state that

$$\frac{1}{3} = \frac{2}{6} = \frac{3}{9}$$

flashback

The earliest civilizations had a very incomplete knowledge of fractions. Even the illustrious Egyptian civilization could use only unit fractions such as $\frac{1}{2}$, $\frac{1}{3}$, $\frac{1}{4}$, and $\frac{1}{10}$. A fraction was indicated by placing a special symbol over the numeral for a counting number. (The exception was two-thirds.) We show that here using a dot:

$$\dot{2} = \frac{1}{2} \qquad \dot{10} = \frac{1}{10} \qquad \dot{3} = \frac{1}{3} \qquad \ddot{3} = \frac{2}{3}$$

How would you divide 6 loaves of bread among 10 men?

Each man would receive $\dot{2}\ \dot{10}$.

That is, $6 \div 10 = \frac{1}{2} + \frac{1}{10}$. However, about 2400 B.C., the civilization that flourished between the Tigris and the Euphrates rivers (presently Iraq) developed a system of fractions that could be considered as the "great-great-grandfather" of the system we use today.

The Fundamental Law of Fractions will be used throughout this chapter to obtain equivalent fractions (or equal rational numbers).

Fundamental Law of Fractions

For any fraction a/b and any natural number c,

$$\frac{a}{b} = \frac{ac}{bc} \quad \text{or} \quad \frac{ca}{cb} = \frac{a}{b}$$

Example 2 • 9/15 can be written as $(3 \cdot 3)/(5 \cdot 3)$, for $9 = 3 \cdot 3$ and $15 = 5 \cdot 3$. Thus, by the Fundamental Law of Fractions,

$$\frac{9}{15} = \frac{3 \cdot 3}{5 \cdot 3} = \frac{3}{5} \quad \bullet$$

Example 3 •

$$\frac{r^2}{r^5} = \frac{1 \cdot r^2}{r^3 \cdot r^2} = \frac{1}{r^3} \qquad r \neq 0$$

because $r^2 = 1 \cdot r^2$ and $r^5 = r^3 \cdot r^2$. •

In the next to last example, 3/5 is said to be in **lowest terms, simplest form,** or **reduced form.**

Reduced fractions

A fraction a/b is in its **simplest form** when b is positive and the largest positive integer that will divide both a and b is 1.

Using the Fundamental Law of Fractions, we can place a fraction in its simplest form by repeatedly dividing out the factors that are common to both the numerator and denominator. For instance,

$$\frac{120}{255} = \frac{5 \cdot 24}{5 \cdot 51} = \frac{24}{51} = \frac{3 \cdot 8}{3 \cdot 17} = \frac{8}{17}$$

Example 4 • Reduce $\frac{1020}{1380}$ to lowest terms.

Solution Since g.c.d. (1020, 1380) = 60, write

$$\frac{1020}{1380} = \frac{60 \cdot 17}{60 \cdot 23} = \frac{17}{23} \quad \bullet$$

The Fundamental Law of Fractions may also be used to show that two fractions are equivalent.

Example 5 • Show that $\frac{35}{42}$ and $\frac{40}{48}$ are equivalent.

Solution Two procedures are used: (a) writing each fraction in simplest form and (b) writing each fraction with a denominator that is the product of the two denominators.

(a) $\dfrac{35}{42} = \dfrac{5 \cdot 7}{6 \cdot 7} = \dfrac{5}{6}$ and $\dfrac{40}{48} = \dfrac{5 \cdot 8}{6 \cdot 8} = \dfrac{5}{6}$

(b) $\dfrac{35}{42} = \dfrac{35 \cdot 48}{42 \cdot 48} = \dfrac{1680}{42 \cdot 48}$ and $\dfrac{40}{48} = \dfrac{40 \cdot 42}{48 \cdot 42} = \dfrac{1680}{42 \cdot 48}$ •

By carefully studying the second procedure of the preceding example, we see an interesting relationship. Since $35 \cdot 48 = 1680 = 40 \cdot 42$, the fractions $\frac{35}{42}$ and $\frac{40}{48}$ are equivalent. This fact can be generalized as follows.

Equivalent fractions

> The fractions a/b and c/d ($b, d \neq 0$) are *equivalent* (or the two rational numbers represented by a/b and c/d are equal) if and only if $ad = bc$.

Example 6 • $\dfrac{2}{5} = \dfrac{6}{15}$ because $2(15) = 5(6) = 30$ •

Example 7 • Are the fractions $\frac{3}{8}$ and $\frac{6}{24}$ equivalent?

Solution $3 \cdot 24 = 72$ and $6 \cdot 8 = 48$. Since $72 \neq 48$, $\frac{3}{8} \neq \frac{6}{24}$. •

The Fundamental Law of Fractions holds for fractions involving letters as well as numbers.

Example 8 • Write x^2y/xy^4 in simplest form, $x \neq 0$ and $y \neq 0$.

Solution

$$\frac{x^2y}{xy^4} = \frac{\not{x} \cdot x \cdot \not{y}}{\not{x} \cdot y \cdot y \cdot y \cdot \not{y}} = \frac{x}{y^3} \quad \bullet$$

Example 9 • Let's return now to the introductory problem. You recall that you divided the numerator and the denominator of a fraction by 5 and then by 3 and got $\frac{2}{5}$. You lost the original fraction. Can you find it?

Solution Now that you understand the Fundamental Law of Fractions, the solution is easy. The opposite of dividing is multiplying, so the original fraction was

$$\frac{2}{5} = \frac{2 \cdot 5 \cdot 3}{5 \cdot 5 \cdot 3} = \frac{30}{75} \quad \bullet$$

We conclude this section with some important remarks. You probably have noted that we made a distinction between a fraction and a rational number. A fraction is a numeral representing a number; the number it represents is a rational number. Two fractions that represent the same rational number are said to be *equivalent*, whereas the corresponding rational numbers are said to be *equal*. However, once the distinction between rational numbers and fractions is made, there is no need to belabor it; in general usage, the terms are used interchangeably. This is why we have used the statement that the two fractions $\frac{3}{7}$ and $\frac{18}{42}$ are equal.

Just for fun *If you double $\frac{1}{6}$ of a fraction and then multiply by the fraction, the answer is $\frac{1}{27}$. What is the fraction?*

Exercise set 1

R 1. What fraction does the shaded portion of each diagram illustrate?

(a)

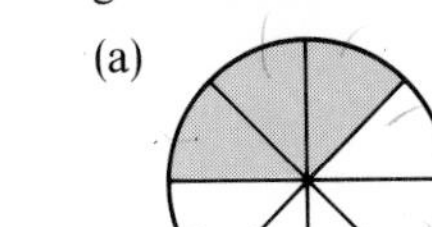

(b)

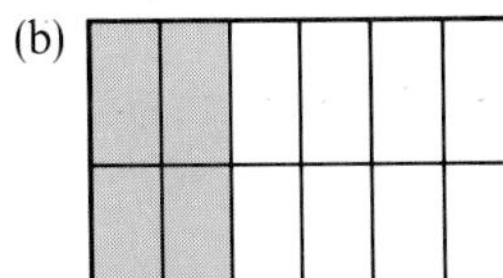

(c)

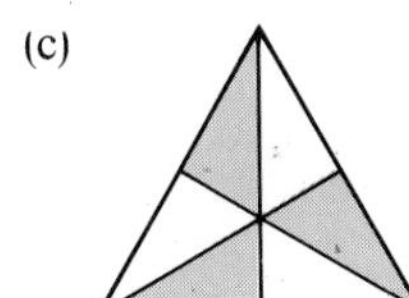

(d)

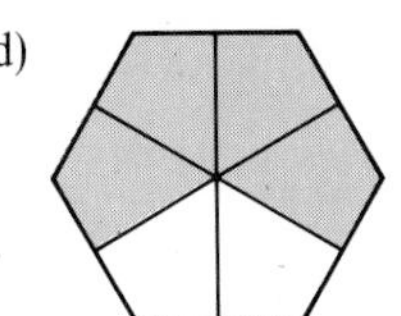

(e)

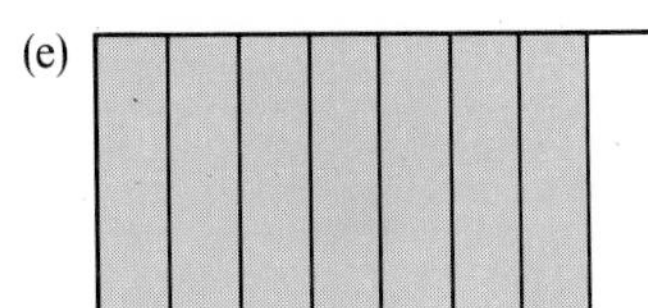

2. Write five fractions equivalent to each of the following fractions.

(a) $\frac{3}{4}$ (b) $\frac{2}{3}$ (c) $\frac{0}{8}$

3. Which of the following are pairs of equivalent fractions?

(a) $\frac{4}{1}$ and $\frac{8}{2}$ (b) $\frac{6}{1}$ and $\frac{18}{2}$

(c) $\frac{0}{1}$ and $\frac{0}{2}$ (d) $\frac{7}{20}$ and $\frac{21}{60}$

(e) $\frac{3a+b}{3c}$ and $\frac{a+b}{c}$, where a, b, and c are any whole numbers with $b \neq 0$ and $c \neq 0$.

(f) $\frac{7a}{7b+c}$ and $\frac{a}{b+c}$, where $7b + c \neq 0$ and $b + c \neq 0$.

4. Write whole numbers that correspond to the following fractions.

(a) $\frac{15}{5}$ (b) $\frac{16}{2}$ (c) $\frac{27}{9}$

5. Find the values of x such that the following fractions are equivalent.

(a) $\frac{1}{4} = \frac{2x}{24}$ (b) $\frac{7x}{10} = \frac{42}{20}$ (c) $\frac{0}{6} = \frac{x}{4}$

6. Represent with a fraction

(a) the part of the collection of dots inside the rectangle.
(b) the part of the collection of dots outside the rectangle.
(c) the part of the collection of dots in the triangle.
(d) the part of the collection of dots in the intersection of the triangle and the rectangle.

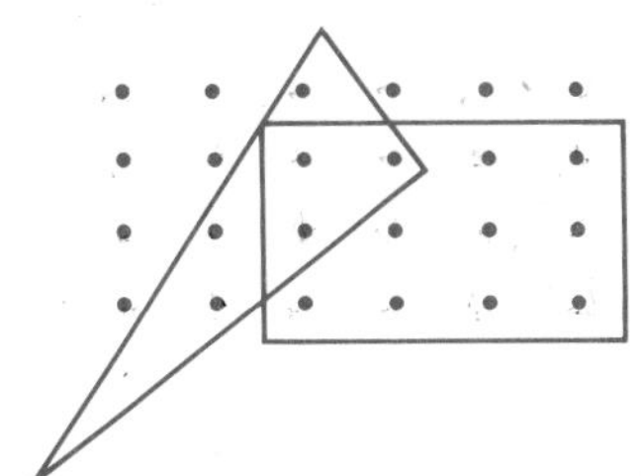

7. Write each of the following in simplest form. In (c) and (i), use the g.c.d. of numerator and denominator.

(a) $\frac{162}{88}$ (b) $\frac{252}{210}$ (c) $\frac{308}{418}$

(d) $\frac{14}{261}$ (e) $\frac{x^2y}{yz^2}$ (f) $\frac{r^2s^3t^4}{r^2s^4t}$

(g) $\frac{x^3yz}{x^2y^2z^4}$ (h) $\frac{xy + xz}{x^2y + x^2z}$ (i) $\frac{-416}{512}$

8. Find the fraction a/b, with specified numerator or denominator that equals the given fraction. Make sure that both a and b are whole numbers; observe that in some problems this is impossible.

(a) $\frac{7}{3}$, $b = 21$ (b) $\frac{2}{3}$, $b = 27$ (c) $\frac{4}{9}$, $b = 14$

(d) $\frac{5}{3}$, $b = 10$ (e) $\frac{3}{4}$, $a = 30$ (f) $\frac{9}{5}$, $a = 27$

(g) $\frac{5}{9}$, $a = 8$ (h) $\frac{7}{16}$, $a = 32$ (i) $\frac{5}{12}$, $a = 10$

(j) $\frac{7}{3}$, $a = 14$ (k) $\frac{3}{5}$, $b = 10$ (l) $\frac{7}{3}$, $b = -12$

9. Write five members of the set described, where a and b are whole numbers and $b \neq 0$.

(a) $\{x \mid x = a/b \text{ and } a + b = 5\}$
(b) $\{x \mid x = a/b \text{ and } a + b < 7\}$
(c) $\{x \mid x = a/b \text{ and } a - b < 2\}$
(d) $\{x \mid x = a/b \text{ and } a - b < 4\}$
(e) $\{x \mid x = a/b \text{ and } a = b\}$

10. A set of fractions equivalent to a/b can be written as

$$\left\{\frac{ca}{cb} \,\middle|\, a \text{ and } b \text{ are whole numbers with } b \neq 0 \text{ and } c \text{ is any natural number}\right\}$$

Write a set of fractions equivalent to each of the following fractions.

(a) $\frac{3}{5}$ (b) $\frac{7}{9}$ (c) x/y

11. Which of the following fractions are in simplest form?

(a) $\frac{9}{17}$ (b) $\frac{2}{4}$ (c) $\frac{7}{14}$

(d) $\frac{20}{49}$ (e) $\frac{3}{17}$ (f) $\frac{9}{14}$

12. Find a fraction equivalent to $\frac{1}{2}$ such that the product of the numerator and the denominator is 72.

T 13. Determine if the following fractions are equal by writing each in simplest form.

(a) $\frac{750}{2000}, \frac{39}{104}$ (b) $\frac{42}{172}, \frac{51}{215}$

14. Find a replacement value for x.

(a) $x/9 = \frac{5}{3}$ (b) $\frac{4}{6} = x/-2$
(c) $x/72 = \frac{5}{8}$ (d) $3/x = \frac{4}{16}$

15. If W is the set of whole numbers, I is the set of integers, and Q is the set of rational numbers, indicate which of the following are true and which are false.

(a) $W \subset Q$ (b) $W \subset (Q \cup I)$
(c) $W \subseteq I$ (d) $I \cap Q = W$
(e) $W \cap Q = W \cap I$ (f) $W \cup Q = I \cup W$

16. (a) If $a = c$, is a/d equal to c/d? Why?
(b) If $b = d$, is a/b equal to c/d? Why?
(c) If $a/b = c/d$ and $b = d$, what is true about c and a?

17. Place an appropriate restriction on the unknowns and write each of the following in simplest form.

(a) $\frac{2 \cdot 3^2 \cdot 5^3 \cdot 7}{2^2 \cdot 3 \cdot 5 \cdot 11}$ (b) $\frac{4a^6b^5}{8a^4b^2}$

(c) $\frac{4x^3y^7}{8xy^9}$ (d) $\frac{3c + 6d}{9e}$

(e) $\frac{a(x + y)}{b^2(x + y)}$ (f) $\frac{a}{2a + ba}$

(g) $\frac{a}{2a + b}$ (h) $\frac{ab}{a + b}$ (i) $\frac{ab(a + b)}{a - b}$

18. Write three fractions representing each of the shaded regions of the following figures.

(a) 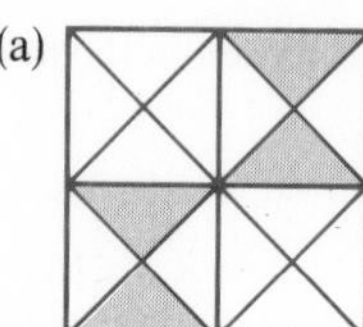

(b) 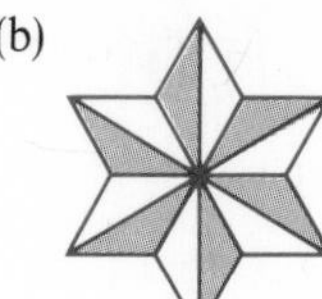

19. Determine whether or not the two fractions are equal by using the fact that $a/b = c/d$ if and only if $ad = bc$.

(a) $\frac{9}{15}, \frac{1200}{2000}$ (b) $\frac{6}{24}, \frac{41}{164}$ (c) $\frac{33}{43}, \frac{62}{189}$

20. Two third-grade teachers at Ourtown Elementary gave their classes the same test. Of Mr. Brown's 27 students, 20 passed the test, and 22 of Ms. Gray's 29 also passed. What part of each class passed the test?

21. (a) Show that the relationship "is equivalent to" is an equivalence relation on the set of fractions.
(b) Find some equivalence classes for this equivalence relation.

22. Write in simplest form.

(a) $\dfrac{2^{10} - 2^{9}}{2^{11} - 2^{10}}$ (b) $\dfrac{3^{12} - 3^{9}}{3^{15} - 3^{12}}$

2 Addition and subtraction of rational numbers

■ Use four numbers, 2, 4, 5, and 8, two as numerators and two as denominators of two fractions, such that each fraction is less than 1 and the difference in the two fractions is a maximum. ■

From the preceding problem you would expect that in this section you will be introduced to addition and subtraction of rational numbers. Two procedures will be introduced for adding rational numbers. The first method uses the product of the two denominators and the Fundamental Law of Fractions. The most common procedure for adding rational numbers makes use of a least common denominator (l.c.m. of two denominators). We shall make the extension from the integers to rational numbers in such a way that all the properties for addition of integers will be preserved for addition of rational numbers.

Since the integers are a subset of the rational numbers, the operations on the system of rational numbers must be consistent with the operations on the integers. Keep this point in mind as we define addition for rational numbers. To help visualize the addition of two rational numbers, let's consider the number line.

Example 9 • We already know that $3 + 4 = 7$, or, using the corresponding rational numbers,

$$\frac{3}{1} + \frac{4}{1} = \frac{7}{1}$$

as seen in Figure 6(a). •

Figure 6

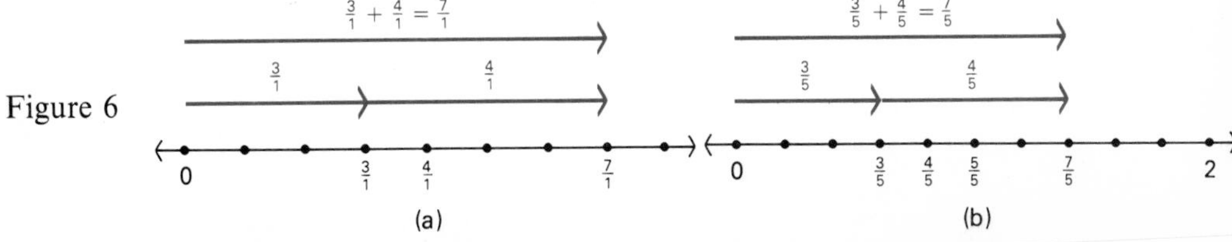

In general, we define the addition of rational numbers with denominators of 1 by

$$\left(\frac{a}{1}\right) + \left(\frac{b}{1}\right) = \frac{a+b}{1}$$

Now consider the case where the denominators are equal but are not 1.

Example 10 • On the number line in Figure 6(b), observe the sum of $\frac{3}{5}$ and $\frac{4}{5}$. We define $(\frac{3}{5}) + (\frac{4}{5}) = (3 + 4)/5 = \frac{7}{5}$. •

Example 11 • Let's illustrate the same concept for $\frac{3}{10}$ and $\frac{4}{10}$ by using a rectangle divided into ten parts as illustrated in Figure 7. Three parts out of ten are represented by horizontal lines and four parts out of ten by vertical lines. It is easy to see that seven parts out of ten have been shaded.

$$\frac{3}{10} + \frac{4}{10} = \frac{3+4}{10} = \frac{7}{10} \quad \bullet$$

Figure 7

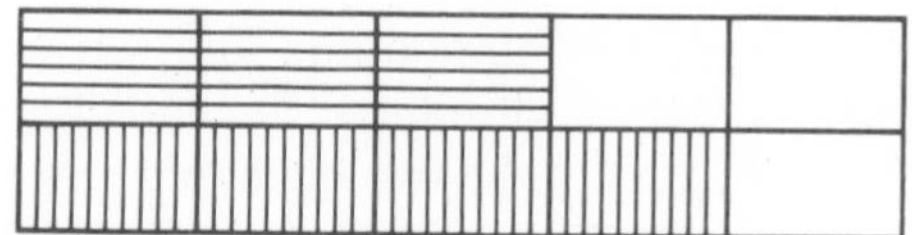

In general,

$$\left(\frac{a}{c}\right) + \left(\frac{b}{c}\right) = \frac{a+b}{c}$$

Now consider the case in which the denominators are unequal.

Example 12 • Add $\frac{2}{3}$ and $\frac{4}{5}$.

Solution Using the fact that a/b and ac/bc are equal,

$$\frac{2}{3} = \frac{2 \cdot 5}{3 \cdot 5} \quad \text{and} \quad \frac{4}{5} = \frac{3 \cdot 4}{3 \cdot 5}$$

The denominators are now the same, and we add according to the preceding discussion. Thus,

$$\frac{2}{3} + \frac{4}{5} = \frac{(2 \cdot 5) + (3 \cdot 4)}{3 \cdot 5} \quad \bullet$$

In general, let a/b and c/d represent rational numbers. The denominators can be made identical by applying the Fundamental Law of Fractions. To get equivalent fractions with denominators, bd,

$$\frac{a}{b} = \frac{ad}{bd} \quad \text{and} \quad \frac{c}{d} = \frac{bc}{bd}$$

Therefore,

$$\frac{a}{b}+\frac{c}{d}=\frac{ad}{bd}+\frac{bc}{bd}=\frac{ad+bc}{bd}$$

Because of these considerations, we define addition of rational numbers in the following manner.

Addition of rational numbers

(a) If a/b and c/b are *rational numbers* (note equal denominators), then

$$\frac{a}{b}+\frac{c}{b}=\frac{a+c}{b}$$

(b) If a/b and c/d are *any two rational numbers*, then

$$\frac{a}{b}+\frac{c}{d}=\frac{ad+bc}{bd}$$

(Note that (a) is a special case of (b).)

The equivalence of this two-part definition is demonstrated by adding $\frac{3}{7}+\frac{2}{7}$. By (a),

$$\frac{3}{7}+\frac{2}{7}=\frac{3+2}{7}=\frac{5}{7}$$

By (b),

$$\frac{3}{7}+\frac{2}{7}=\frac{3\cdot 7+7\cdot 2}{7\cdot 7}=\frac{35}{49}$$

However,

$$\frac{5}{7}=\frac{35}{49}$$

Example 13 • (a) $\frac{2}{3}+\frac{4}{7}=\frac{2(7)+3(4)}{3\cdot 7}=\frac{14+12}{21}=\frac{26}{21}$

(b) $\frac{5}{7}+\frac{-2}{4}=\frac{5(4)+7(-2)}{7\cdot 4}=\frac{20+-14}{28}=\frac{6}{28}=\frac{3}{14}$ •

In our definition of addition of fractions, we find a common denominator for the pair of fractions, but we do not find the least common denominator. The procedure of finding the least common denominator of the fractions to be added has the advantage of working with smaller numbers. In this procedure, if the denominators are not the same, the Fundamental Law of Fractions is applied as before to each fraction to make the denominators identical. However, in order to make the operation as simple as possible, we look for what is called the *least common denominator*.

Least common denominator

> The *least common denominator* is the l.c.m. of the denominators of the fractions to be added.

Example 14 • (a) $\frac{5}{6}+\frac{7}{15}=\frac{5\cdot 5}{6\cdot 5}+\frac{7\cdot 2}{15\cdot 2}=\frac{25}{30}+\frac{14}{30}=\frac{39}{30}=\frac{13}{10}$

(b) $\frac{5}{24}+\frac{11}{36}=\frac{5}{2^3\cdot 3}+\frac{11}{2^2\cdot 3^2}=\frac{5\cdot 3}{2^3\cdot 3^2}+\frac{11\cdot 2}{2^3\cdot 3^2}=\frac{15+22}{2^3\cdot 3^2}=\frac{37}{72}$

(c) $\frac{3}{16}+\frac{1}{4}+\frac{5}{8}=\frac{3}{16}+\frac{1\cdot 4}{4\cdot 4}+\frac{5\cdot 2}{8\cdot 2}=\frac{3}{16}+\frac{4}{16}+\frac{10}{16}=\frac{17}{16}$ •

Example 15 • Find $\frac{5}{48}+\frac{7}{36}+\frac{1}{54}$.

Solution Factor into prime factors each denominator. Then the least common denominator can be obtained easily.

$$\frac{5}{2\cdot 2\cdot 2\cdot 2\cdot 3}+\frac{7}{2\cdot 2\cdot 3\cdot 3}+\frac{1}{2\cdot 3\cdot 3\cdot 3}$$

The least common denominator is $2\cdot 2\cdot 2\cdot 2\cdot 3\cdot 3\cdot 3=432$

$$\frac{5}{48}+\frac{7}{36}+\frac{1}{54}=\frac{5(3\cdot 3)}{432}+\frac{7(2\cdot 2\cdot 3)}{432}+\frac{1(2\cdot 2\cdot 2)}{432}$$

$$=\frac{45+84+8}{432}$$

$$=\frac{137}{432}$$ •

Sometimes we use mixed numbers to represent the addition of a fraction that corresponds to a whole number and a fraction that does not correspond to a whole number without performing the operation of addition. For instance, consider $\frac{4}{1}+\frac{2}{3}$. The addition $4+\frac{2}{3}$ is often denoted by $4\frac{2}{3}$, which is called a **mixed number.** Thus, $7\frac{3}{8}$ means $7+\frac{3}{8}$, or $\frac{7}{1}+\frac{3}{8}=\frac{59}{8}$. Fractions like $\frac{59}{8}$ are called **improper fractions.** (*Caution:* A common mistake is to write $7\frac{3}{8}$ as $7\cdot\frac{3}{8}$.)

Improper fraction

> An *improper fraction* is one in which the absolute value of the numerator is larger than or equal to the absolute value of the denominator. A fraction in which this is not the case is called a *proper* fraction.

Example 16 • (a) Change $5\frac{3}{4}$ to an improper fraction.

$$5\frac{3}{4}=5+\frac{3}{4}=\frac{5}{1}+\frac{3}{4}=\frac{20+3}{1\cdot 4}=\frac{23}{4}$$

(b) Change $\frac{17}{3}$ to a mixed fraction.

$$\frac{17}{3}=\frac{15+2}{3}=\frac{15}{3}+\frac{2}{3}=\frac{5}{1}+\frac{2}{3}=5+\frac{2}{3}=5\frac{2}{3}$$

or

$$3\overline{)17}\ \text{with quotient } 5,\quad \frac{-15}{2} \qquad \text{and} \qquad \frac{17}{3}=5+\frac{2}{3}=5\frac{2}{3}$$

(c) Add $4\frac{1}{3}$ and $6\frac{1}{2}$.

$$4\frac{1}{3}+6\frac{1}{2}=\left(4+\frac{1}{3}\right)+\left(6+\frac{1}{2}\right)=(4+6)+\left(\frac{1}{3}+\frac{1}{2}\right)$$

$$=10+\frac{5}{6}=10\frac{5}{6}$$

or

$$\begin{array}{rcl} 4\frac{1}{3} & = & 4\frac{2}{6} \\ +6\frac{1}{2} & = & 6\frac{3}{6} \\ \hline & & 10\frac{5}{6} \end{array}$$

(d) Change $-3\frac{5}{8}$ to an improper fraction.

$$-3\frac{5}{8}=-\left(3+\frac{5}{8}\right)=-3+\frac{-5}{8}=\frac{-24}{8}+\frac{-5}{8}=\frac{-29}{8}$$ •

The definition of subtraction for the rational numbers is much like the definition of subtraction for whole numbers and integers; that is, subtraction is the inverse operation of addition. For rational numbers x, y, and z,

$$x-y=z \quad \textbf{if and only if} \quad x=y+z$$

or

$$\frac{a}{b}-\frac{c}{d}=\frac{e}{f} \quad \textbf{if and only if} \quad \frac{a}{b}=\frac{c}{d}+\frac{e}{f}$$

Example 17 • (a) $\frac{7}{3}-\frac{5}{3}=\frac{2}{3}$ because $\frac{7}{3}=\frac{5}{3}+\frac{2}{3}$ •

To find the sum of two fractions, we found a common denominator, rewrote both fractions using the common denominator, and then added numerators. To subtract fractions, we again write them with a common denominator and then subtract numerators.

Subtraction of rational numbers

For rational numbers a/b and c/d,

$$\frac{a}{b} - \frac{c}{d} = \frac{ad - bc}{bd}$$

Example 18 • $$\frac{4}{5} - \frac{2}{3} = \frac{4(3) - 5(2)}{15} = \frac{2}{15}$$ •

For subtraction of integers, it was shown that subtracting an integer was equivalent to adding its additive inverse. The same is true for rational numbers. For rational numbers a/d and b/e,

$$\frac{a}{d} - \frac{b}{e} = \frac{a}{d} + \frac{-b}{e}$$

Example 19 • (a) $\frac{-4}{5} - \frac{2}{3} = \frac{-4}{5} + \frac{-2}{3} = \frac{-12 + -10}{15} = \frac{-22}{15}$

(b) $\frac{5}{4} - \frac{-1}{6} = \frac{5}{4} + \frac{1}{6} = \frac{15 + 2}{12} = \frac{17}{12}$ •

Just as in addition, the computations will be simpler if we use a least common denominator instead of the definition for subtraction.

Example 20 • Complete the subtraction.

$$\frac{17}{12} - \frac{8}{9}$$

Solution The least common denominator of 9 and 12 is 36.

$$\frac{17}{12} = \frac{51}{36} \quad \text{and} \quad \frac{8}{9} = \frac{32}{36}$$

$$\frac{17}{12} - \frac{8}{9} = \frac{51}{36} - \frac{32}{36} = \frac{19}{36}$$ •

Example 21 • Find $4\frac{1}{3} - 2\frac{5}{9}$

Solution

$$4\frac{1}{3} - 2\frac{5}{9} = 4\frac{3}{9} - 2\frac{5}{9}$$

$$= 3 + \left(1 + \frac{3}{9}\right) - 2\frac{5}{9}$$

$$= 3 + \frac{12}{9} - 2\frac{5}{9}$$

$$= 1\frac{7}{9}$$ •

Just for fun *Complete the magic square.*

Let us return now to the problem with which we introduced this section.

Example 22 • Use four numbers, 2, 4, 5, and 8, two as numerators and two as denominators of two fractions, such that each fraction is less than 1 and the difference in the two fractions is a maximum.

Solution First we want the subtrahend to be as small as possible. This will occur with the largest possible denominator and the smallest numerator or $\frac{2}{8}$. The second fraction then becomes $\frac{4}{5}$, or

$$\frac{4}{5} - \frac{2}{8} = \frac{22}{40} = \frac{11}{20} \quad \bullet$$

Exercise set 2

R 1. Perform the following computations, using the definition of addition. Leave answers in simplest form.

(a) $\frac{3}{4} + 0$ (b) $\frac{7}{8} + \frac{2}{3}$ (c) $\frac{5}{9} + \frac{1}{4}$

(d) $\frac{2}{3} + \frac{3}{5}$ (e) $\frac{4}{x} + \frac{5}{2x}$ (f) $\frac{7}{xy} + \frac{4}{y}$

2. In Exercise 1, change each addition sign to a subtraction sign and perform the computation.

3. Find the least common denominator of the following fractions and then add the fractions.

(a) $\frac{5}{8}, \frac{3}{2}, \frac{1}{4}$ (b) $\frac{5}{32}, \frac{5}{4}, \frac{7}{8}$

(c) $\frac{7}{54}, \frac{4}{27}, \frac{5}{6}$ (d) $\frac{3}{4}, \frac{5}{3}, \frac{7}{5}$

(e) $\frac{5}{11}, \frac{11}{5}, \frac{3}{7}$ (f) $\frac{3}{3}, \frac{4}{3}, \frac{7}{9}$

4. Change the following improper fractions to mixed numbers.

(a) $\frac{58}{3}$ (b) $\frac{19}{4}$ (c) $\frac{49}{8}$ (d) $\frac{27}{4}$

5. Change the following mixed numbers to improper fractions.

(a) $3\frac{5}{6}$ (b) $4\frac{3}{4}$ (c) $16\frac{1}{3}$ (d) $50\frac{1}{8}$

6. The sum of two fractions is given in the following. Determine the two fractions, reducing each to lowest terms. For example,

$$\frac{15 + 24}{36} = \frac{15}{36} + \frac{24}{36} = \frac{5}{12} + \frac{2}{3}$$

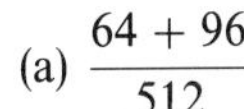

(a) $\frac{64 + 96}{512}$ (b) $\frac{4(5) + 12(9)}{9(5)}$

(c) $\frac{15 + 19}{4}$ (d) $\frac{4(4) + 3(5)}{5(4)}$

7. Find the sum and difference (first minus second) for the following mixed numbers. (Write your answer as a mixed number.)

   (a) $8\frac{4}{21}$, $2\frac{3}{7}$ (b) $89\frac{1}{2}$, $1\frac{12}{13}$
   (c) $23\frac{3}{4}$, $7\frac{7}{9}$ (d) $17\frac{1}{2}$, $8\frac{9}{10}$

8. Compute.

   (a) $\frac{7}{8} + \left(\frac{2}{3} + \frac{1}{2}\right)$ (b) $\left(\frac{7}{8} + \frac{2}{3}\right) + \frac{1}{2}$
   (c) $\frac{7}{8} - \left(\frac{2}{3} - \frac{1}{2}\right)$ (d) $\left(\frac{7}{8} - \frac{2}{3}\right) - \frac{1}{2}$
   (e) What do (a) and (b) demonstrate?
   (f) What do (c) and (d) demonstrate?

9. Carry out the indicated operations.

   (a) $5\frac{1}{2} + 4\frac{3}{4} + 1\frac{1}{3}$ (b) $4\frac{1}{2} - 2\frac{3}{4}$
   (c) $3\frac{5}{9} + 2\frac{2}{3} + 5\frac{1}{2}$ (d) $2\frac{5}{9} + 4\frac{2}{3} + 3\frac{1}{4}$
   (e) $\frac{5}{12} + 7\frac{2}{5}$ (f) $17\frac{7}{12} - 7\frac{2}{5}$
   (g) $17\frac{4}{5} - 6\frac{3}{8}$ (h) $24\frac{5}{8} + 17\frac{4}{9}$
   (i) $80\frac{17}{20} + 19\frac{1}{15}$ (j) $4\frac{1}{6} + 19\frac{1}{10}$

10. Using the fractions $\frac{3}{4}$ and $\frac{2}{5}$, demonstrate for rational numbers

   (a) the commutative property of addition.
   (b) with $\frac{2}{3}$ the associative property of addition.
   (c) that 0/1 is the additive identity.

11. Use a number line to indicate the following additions:

   (a) $\frac{2}{5} + \frac{1}{5}$ (b) $\frac{1}{6} + \frac{3}{6}$
   (c) $\frac{1}{3} + \frac{1}{4}$ (d) $\frac{1}{6} + \frac{1}{4}$

12. Does each of the following properties hold for subtraction of rational numbers? Demonstrate with examples or counterexamples.

   (a) Closure (b) Commutative
   (c) Associative (d) Inverse
   (e) Identity

13. Solve the following equations.

   (a) $x - \frac{2}{3} = \frac{1}{4}$ (b) $x + 2\frac{1}{2} = 3\frac{1}{3}$
   (c) $3x - \frac{4}{3} = 2$ (d) $2x + \frac{1}{3} = 5$

T 14. Using the fact that $-(a/b + c/d) = -a/b + -c/d$, express each of the following in the form a/b, where b is positive.

   (a) $-\left(\frac{3}{5} + \frac{7}{8}\right)$ (b) $-\left(\frac{-7}{5} - \frac{-3}{7}\right)$
   (c) $-\left(\frac{6}{5} + \frac{-7}{9}\right)$ (d) $-\left(\frac{7}{3} - \frac{6}{5}\right)$

15. Find the sum and difference (first minus second) of each of the following.

   (a) $\frac{7}{39}, \frac{2}{91}$ (b) $\frac{13}{243}, \frac{5}{162}$
   (c) $\frac{7}{2^3 \cdot 3^2}, \frac{5}{2^3 \cdot 3^4}$
   (d) $\frac{5}{2^2 \cdot 3^{10} \cdot 7^4}, \frac{8}{3^6 \cdot 5^2 \cdot 7^3}$
   (e) $41\frac{17}{23}, 16\frac{5}{11}$ (f) $74\frac{3}{40}, 8\frac{5}{36}$

16. Find the sum and difference (first minus second) for each pair of the following rational numbers.

   (a) $\frac{3yz}{xy^4z^2}, \frac{3x}{x^2y}$ (b) $\frac{6a^2b}{3xba}, \frac{5x^3b}{14ba^3}$
   (c) $\frac{3}{ab}, \frac{5}{a}$ (d) $\frac{d}{a^3bc^2}, \frac{a}{ab^3c}$
   (e) $\frac{11}{mn^2}, \frac{mnx}{12m^2}$ (f) $\frac{n^2}{16}, \frac{2mn}{9m^3n^2}$

17. For the following problems, formulate an open sentence that serves as a model for the problem.

   (a) Tricia purchased $\frac{1}{2}$ pound of cheese, a $6\frac{1}{4}$-pound ham, a $2\frac{3}{4}$-pound roast, a $12\frac{1}{8}$-pound turkey, and 1 pound of butter.
      (i) What was the total weight of this purchase?
      (ii) If Tricia returned the roast, what was the weight of the remaining purchase?
   (b) About $\frac{3}{10}$ of the cars in the United States are less than 3 years old. About $\frac{1}{4}$ of the cars in the United States are at least 3 years old and less than 6 years old. What fraction of the cars in the United States are less than 6 years old?
   (c) Paul owns $\frac{1}{3}$ of the stock of the company of which he is president. His wife owns $\frac{1}{3}$ of the stock. What fractional part of the stock will his daughter, Brooke, need to purchase in order for the family to own $\frac{3}{4}$ of the stock?
   (d) An Alabama farmer uses $61\frac{1}{3}$ acres of his 100-acre farm for cattle grazing. He has $22\frac{4}{5}$ acres planted in pine trees. How many acres are left for growing hay?
   (e) On a recent cross country trip in an airplane, Lane spent $4\frac{5}{8}$ hours on the first flight and $2\frac{1}{4}$ hours on the second flight. If there was a $1\frac{1}{3}$-hour wait between flights, how long did it take Lane to make the trip?

18. Find the solutions for the equations in Exercise 17 and thus the answers to the questions.

19. What is wrong with each argument that follows?

(a) $\frac{(2+3)}{2} = \frac{2}{2} + 3 = 1 + 3 = 4$

(b) $1 = \frac{8}{8} = \frac{8}{4+4} = \frac{8}{4} + \frac{8}{4} = 2 + 2 = 4$

(c) $\frac{3}{4} + \frac{7}{9} = \frac{27+28}{36+36} = \frac{55}{72}$

20. In order to use calculators efficiently to add fractions, unusual formulas are sometimes useful. Make up examples and illustrate the following formulas.

(a) $\frac{x}{y} + \frac{w}{z} = \frac{\frac{xz}{y} + w}{z}$

(b) $\frac{x}{y} + \frac{w}{z} + \frac{u}{v} = \frac{\frac{\left(\frac{xz}{y} + w\right)v}{z} + u}{v}$

C 21. Suppose addition for rational numbers were defined as

$$\frac{x}{y} \oplus \frac{w}{z} = \frac{x+w}{y+z}$$

(a) Is there closure for $\oplus$?
(b) Does it agree with arithmetic for integers?
(c) Is $\oplus$ commutative?
(d) Is $\oplus$ associative?
(e) Is there an additive identity?
(f) Is there an additive inverse?

22. The definition $a/b + c/d = (ad + bc)/bd$ holds also in nondecimal bases. Perform the following computations.

(a) $\left(\frac{1}{4} + \frac{3}{5}\right)_{\text{seven}}$ (b) $\left(\frac{1}{3} + \frac{1}{2} + \frac{1}{4}\right)_{\text{five}}$

(c) $\left(14\frac{2}{3} + 25\frac{3}{4}\right)_{\text{six}}$ (d) $\left(1\text{E}\frac{3}{\text{T}} + 78\frac{5}{\text{E}}\right)_{\text{twelve}}$

3 Multiplication and division of rational numbers

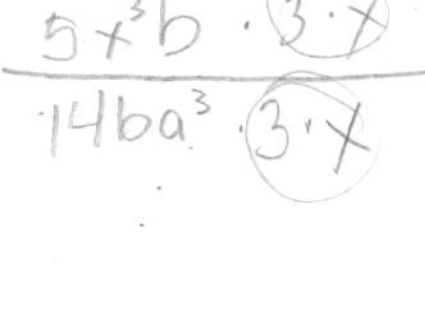

■ John gave all of his apples to his two brothers. To Leo he gave half of his apples and half of an apple. To Ned he gave half of what he had left plus half of an apple. He then had no apples left. How many apples did John have to begin with? (It is interesting to note that at no time were the apples cut, split, or divided.) ■

If you set up this problem as an open sentence, you will need a knowledge of multiplying and dividing rational numbers to solve the problem. Undoubtedly, you have been multiplying and dividing rational numbers for many years. From this section, we trust that you will better understand these operations and will become more proficient in your computations. To help you understand the multiplication of rational numbers, we shall begin our study by interpreting $(\frac{1}{3}) \cdot (\frac{2}{5})$ as $\frac{1}{3}$ of $\frac{2}{5}$. We can then geometrically explain the answer. You probably remember that dividing by a fraction is the same as multiplying by its reciprocal. Why is this true? You should understand why when you finish this section.

Now we extend integer multiplication to multiplication of rational numbers. Let's first consider multiplication of rational numbers that correspond to integers.

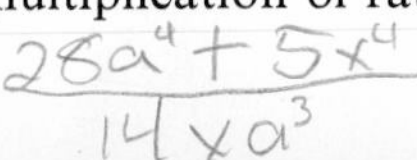

Example 23 • $2 \cdot 3 = 6$ or $\frac{2}{1} \cdot \frac{3}{1} = \frac{6}{1}$ •

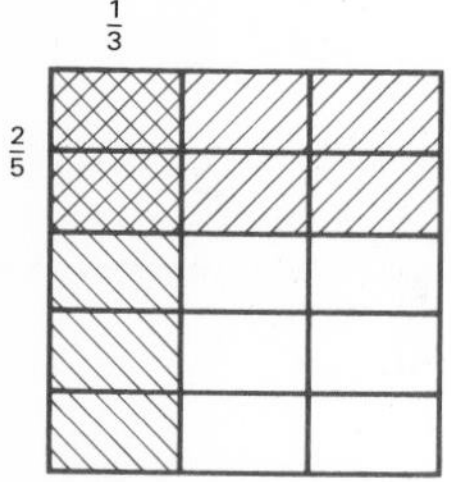

Figure 8

The answer we have given to $(\frac{2}{1}) \cdot (\frac{3}{1})$ can be obtained by multiplying numerators and denominators. The multiplication of $(\frac{1}{3}) \cdot (\frac{2}{5})$ is often interpreted as $\frac{1}{3}$ of $\frac{2}{5}$. This interpretation can be illustrated by dividing a square into 15 like regions with three columns and five rows (Figure 8). Shade any one of the three columns and then shade any two of the five rows. By observation, we see that two parts have been shaded twice, so $(\frac{1}{3}) \cdot (\frac{2}{5})$ is $\frac{2}{15}$. Again, this result can be obtained by multiplying the numerators and multiplying the denominators of the two fractions.

These examples illustrate a concept that is expressed in the following definition.

Multiplication of rational numbers

> If a/b and c/d are any rational numbers, then
> $$\frac{a}{b} \cdot \frac{c}{d} = \frac{ac}{bd}$$

Example 24 • (a) $\frac{4}{5} \cdot \frac{1}{2} = \frac{4 \cdot 1}{5 \cdot 2} = \frac{4}{10}$ (b) $\frac{2}{3} \cdot \frac{7}{11} = \frac{2 \cdot 7}{3 \cdot 11} = \frac{14}{33}$ •

Leaving the numerator and denominator in factored form sometimes simplifies the process of reducing the fraction.

Example 25 • (a) $\frac{7}{8} \cdot \frac{8}{15} = \frac{7 \cdot \overset{1}{\cancel{8}}}{\underset{1}{\cancel{8}} \cdot 15} = \frac{7}{15}$

The 8 is divided out by the Fundamental Law of Fractions.

(b) $\frac{2}{3} \cdot \frac{21}{4} = \frac{\overset{1}{\cancel{2}} \cdot \overset{7}{\cancel{21}}}{\underset{1}{\cancel{3}} \cdot \underset{2}{\cancel{4}}} = \frac{1 \cdot 7}{1 \cdot 2} = \frac{7}{2}$

Using the Fundamental Law of Fractions, the 3 is divided out of 21 in the numerator and 3 in the denominator, and the 2 is divided out of 2 in the numerator and 4 in the denominator. •

A word of caution should be inserted here. Instead of using the Fundamental Law of Fractions, students sometimes randomly cancel numbers and thus make errors. For example,

$$\frac{41 \cdot 5}{3 \cdot 8} \neq \frac{\overset{1}{\cancel{4}}1 \cdot 5}{3 \cdot \underset{2}{\cancel{8}}} \quad \text{Why not?} \qquad \frac{5 + \overset{1}{\cancel{2}}}{\underset{2}{\cancel{4}}} \neq \frac{6}{2} = 3 \quad \text{Why not?}$$

$$\frac{34+3}{9\cdot 2} \neq \frac{\overset{17}{\cancel{34}}+\overset{1}{\cancel{3}}}{\underset{3}{\cancel{9}}\cdot\cancel{2}} \qquad \text{Why not?} \qquad \frac{1\cancel{7}}{\cancel{7}4} \neq \frac{1}{4} \qquad \text{Why not?}$$

Let us now multiply two fractions when one is a whole number. To illustrate, multiply $2(\frac{1}{8})$. The 2 may be written as $\frac{2}{1}$; so the multiplication becomes $(\frac{2}{1})(\frac{1}{8}) = (\frac{2}{8})$. Note that this result is the same as multiplying the whole number times the numerator of the fraction.

Example 26 •

$$3\cdot\frac{7}{40} = \frac{3}{1}\cdot\frac{7}{40} = \frac{21}{40} \quad \bullet$$

1/1 represents the multiplicative identity for rational numbers. It is easy to verify this fact, since

$$\frac{c}{d}\cdot\frac{1}{1} = \frac{c\cdot 1}{d\cdot 1} = \frac{c}{d}$$

Also, a/a, where $a \neq 0$, represents the multiplicative identity, since $\frac{1}{1} = a/a$.

The product of certain rational numbers always produces the multiplicative identity.

Example 27 •

$$\left(\frac{1}{3}\right)\cdot\left(\frac{3}{1}\right) = 1$$

$$\left(\frac{5}{6}\right)\cdot\left(\frac{6}{5}\right) = 1$$

$$\left(\frac{-6}{3}\right)\cdot\left(\frac{3}{-6}\right) = 1 \quad \bullet$$

In general, observe that for $x \neq 0$ and $y \neq 0$, $(x/y)\cdot(y/x) = xy/yx = 1$.

Multiplicative inverse

> In the system of rational numbers, there is a *multiplicative inverse* y/x for every rational number x/y if $x \neq 0$. However, the multiplicative inverse for x/y where $x = 0$ does not exist. The multiplicative inverse of a rational number is called its **reciprocal.**

Multiplication involving mixed numbers may be performed in either of two ways. The mixed numbers may be changed to improper fractions, or the distributive property of multiplication over addition may be used. Both methods are illustrated in the following example.

Example 28 • Multiply $2\frac{3}{4}\cdot 7\frac{1}{8}$.

Solution

$$2\frac{3}{4}\cdot 7\frac{1}{8} = \frac{11}{4}\cdot\frac{57}{8} = \frac{627}{32} = 19\frac{19}{32}$$

Using the distributive property,

$$2\frac{3}{4}\cdot 7\frac{1}{8} = \left(2+\frac{3}{4}\right)\left(7+\frac{1}{8}\right)$$

$$= 2\left(7+\frac{1}{8}\right)+\frac{3}{4}\left(7+\frac{1}{8}\right)$$

$$= 2(7) + 2\left(\frac{1}{8}\right) + \left(\frac{3}{4}\right)(7) + \left(\frac{3}{4}\right)\left(\frac{1}{8}\right)$$

$$= 14 + \frac{2}{8} + \left(5+\frac{1}{4}\right) + \frac{3}{32}$$

$$= 19 + \left(\frac{2}{8}+\frac{1}{4}+\frac{3}{32}\right)$$

$$= 19\frac{19}{32} \quad \bullet$$

Division is defined for rational numbers in a manner similar to division for integers.

Division of rational numbers

> If a/b and c/d are any rational numbers with $c/d \neq 0$, then $a/b \div c/d$ is equal to a rational number e/f where
>
> $$\frac{a}{b} = \frac{c}{d}\cdot\frac{e}{f}$$

Example 29 •

$$\frac{2}{3}\div\frac{4}{5} = \frac{5}{6} \quad \text{because} \quad \frac{4}{5}\cdot\frac{5}{6} = \frac{20}{30} = \frac{2}{3} \quad \bullet$$

In order to determine a convenient technique for dividing one fraction by another, consider the following examples.

Example 30 • Find $\frac{2}{3} \div \frac{4}{5}$.

Solution Let $\frac{2}{3} \div \frac{4}{5} = x$. By the definition of division, $\frac{2}{3} = \frac{4}{5}\cdot x$. Now multiply both sides of the equation by the reciprocal of $\frac{4}{5}$, namely $\frac{5}{4}$.

$$\frac{5}{4}\cdot\frac{2}{3} = \frac{5}{4}\left(\frac{4}{5}\cdot x\right)$$

$$= \left(\frac{5}{4}\cdot\frac{4}{5}\right)x$$

$$= x$$

Therefore, $\frac{2}{3} \div \frac{4}{5} = x = \frac{5}{4}\cdot\frac{2}{3} = \frac{2}{3}\cdot\frac{5}{4}$. •

This example suggests the following property of division.

Division as a multiplication

> If d/c exists, then
>
> $$\frac{a}{b} \div \frac{c}{d} = \frac{a}{b} \cdot \frac{d}{c}$$
>
> where d/c is the *multiplicative inverse* or *reciprocal* of c/d.

Example 31 • (a) $\frac{4}{7} \div \frac{3}{8} = \frac{4}{7} \cdot \frac{8}{3} = \frac{32}{21}$

(b) $\frac{-1}{4} \div \frac{-1}{3} = \frac{-1}{4} \cdot \frac{3}{-1} = \frac{-3}{-4} = \frac{3}{4}$

(c) $\frac{8}{15} \div \frac{4}{9} = \frac{\overset{2}{\cancel{8}}}{\underset{5}{\cancel{15}}} \cdot \frac{\overset{3}{\cancel{9}}}{\underset{1}{\cancel{4}}} = \frac{2 \cdot 3}{5 \cdot 1} = \frac{6}{5}$ •

The preceding property of division of rational numbers is easy to verify from the definition. $a/b \div c/d$ has an answer e/f if and only if

$$\frac{a}{b} = \frac{c}{d} \cdot \frac{e}{f}$$

Now substitute for e/f, $e/f = a/b \cdot d/c$.

$$= \frac{c}{d}\left(\frac{a}{b} \cdot \frac{d}{c}\right) \qquad \text{substitution for } \frac{e}{f}$$

$$= \left(\frac{a}{b} \cdot \frac{d}{c}\right) \cdot \frac{c}{d} \qquad \text{Why?}$$

$$= \frac{a}{b}\left(\frac{d}{c} \cdot \frac{c}{d}\right) \qquad \text{Why?}$$

$$= \frac{a}{b} \cdot 1 \qquad \text{Why?}$$

$$= \frac{a}{b}$$

So $a/b \cdot d/c$ is indeed an answer for $a/b \div c/d$.

Since division is the inverse of multiplication, and since every rational number unequal to 0 has a multiplicative inverse, division of rational numbers is always possible except when the divisor is 0. Moreover, the answer is unique in that it is one of a number of equivalent fractions.

Once again, since 0 plays such a special role in division, here is a word of caution. **Division of a fraction by 0** is not defined since

$$\frac{a}{b} \div 0 \qquad (a \neq 0)$$

cannot equal an answer e/f, because

$$\frac{e}{f} \cdot 0 = 0 \neq \frac{a}{b}$$

Consider the quotient

$$x \div y = \frac{x}{1} \div \frac{y}{1} = \frac{x}{1} \cdot \frac{1}{y} = \frac{x}{y} \qquad (y \neq 0)$$

Thus as noted at the beginning of this chapter, the symbol x/y can be used to represent the division, $x \div y$. For example, 3/4 can be considered as $3 \div 4$. We use this notation to express the quotient of two rational numbers: $a/b \div c/d$ can be written as

$$\frac{\frac{a}{b}}{\frac{c}{d}} \quad \text{or} \quad \frac{\frac{a}{b}}{\frac{c}{d}} = \frac{a}{b} \cdot \frac{d}{c}$$

Example 32 •

$$\frac{\frac{3}{8}}{\frac{2}{3}} = \frac{3}{8} \cdot \frac{3}{2} = \frac{9}{16} \quad \bullet$$

Example 33 •

$$\frac{2 + \frac{3}{4}}{1 + \frac{1}{4}} = \frac{\frac{2}{1} + \frac{3}{4}}{\frac{1}{1} + \frac{1}{4}} = \frac{\frac{8+3}{4}}{\frac{4+1}{4}}$$

$$= \frac{\frac{11}{4}}{\frac{5}{4}} = \frac{11}{\cancel{4}_{1}} \cdot \frac{\cancel{4}^{1}}{5} = \frac{11}{5} \quad \bullet$$

Example 34 • Find $\frac{3}{16}$ of 3376.

Solution As indicated earlier, the phrase "$\frac{3}{16}$ of" is used to indicate multiplication by $\frac{3}{16}$. Thus,

$$\frac{3}{16} \text{ of } 3376 = \frac{3}{16} \cdot 3376 = 633 \quad \bullet$$

Just for fun

In Ahab's will, $\frac{1}{2}$ of his camels were to go to his oldest child, $\frac{1}{3}$ to his second child, and $\frac{1}{9}$ to his youngest child. At his death, there were 17 camels. A clever lawyer borrowed a camel to make 18.

$$\frac{1}{2} \text{ of } 18 = 9$$

$$\frac{1}{3} \text{ of } 18 = 6$$

$$\frac{1}{9} \text{ of } 18 = 2$$

$$\text{Total } \overline{17}$$

Then the lawyer returned the borrowed camel and everybody was happy thereafter. Are you? Why not?

We now use the material of this section to solve the introductory problem.

Example 35 • John gave all of his apples to his brothers. To Leo he gave half of his apples and half an apple. To Ned he gave half of what he had left plus half an apple. He then had no apples left. How many apples did John have to begin with?

Solution Let x be the number of apples at the beginning.

Leo received $\left(\frac{1}{2} \text{ of } x\right) + \frac{1}{2}$ apple.

x less $\left[\left(\frac{1}{2} \text{ of } x\right) + \frac{1}{2}\right] = \frac{x}{2} - \frac{1}{2}$; this is what remains after the gift to Leo.

John received $\left(\frac{1}{2}\right)\left(\frac{x}{2} - \frac{1}{2}\right) + \frac{1}{2} = \frac{x}{4} + \frac{1}{4}$ apples.

The sum of the apples Leo received and John received equals the total number of apples (x).

$$\frac{x}{2} + \frac{1}{2} + \frac{x}{4} + \frac{1}{4} = x$$

$$\frac{3x}{4} + \frac{3}{4} = x$$

$x = 3$ satisfies this equation.

Therefore, John had 3 apples. Look back at the problem and see if this solution makes sense. •

Exercise set 3

R 1. Use rectangular regions such as Figure 8 to show

(a) $\frac{1}{3}\cdot\frac{1}{3}=\frac{1}{9}$ (b) $\frac{1}{5}\cdot\frac{2}{3}=\frac{2}{15}$

2. Write the multiplicative inverse of each of the following, where possible, and perform the operations necessary to verify your answers.

(a) $\frac{7}{1}$ (b) $\frac{3}{5}$ (c) $\frac{0}{8}$

(d) $\frac{6}{11}$ (e) $\frac{w}{z}$ (f) $\frac{z}{w+y}$

3. Perform the following multiplications and divisions and write your answer as a reduced fraction.

(a) $\frac{3}{4}\left(\frac{1}{2}\cdot\frac{2}{3}\right)$ (b) $\frac{4}{3}\left(\frac{1}{3}\cdot\frac{1}{2}\right)$

(c) $\frac{4}{7}\left(\frac{5}{8}\cdot\frac{2}{3}\right)$ (d) $\left(\frac{2}{3}\cdot\frac{0}{1}\right)\left(\frac{13}{2}\right)$

(e) $4\div\frac{1}{9}$ (f) $8\div\frac{1}{4}$

(g) $\left(\frac{2}{3}+\frac{1}{5}\right)\div\frac{7}{10}$ (h) $\left(8\div\frac{5}{11}\right)\left(\frac{2}{3}\div 9\right)$

(i) $\left(\frac{4}{3}\div\frac{7}{8}\right)\cdot\left(\frac{1}{2}\div\frac{3}{4}\right)$ (j) $\left(\frac{3}{2}+\frac{7}{12}\right)\div\left(\frac{2}{5}+\frac{3}{8}\right)$

(k) $\frac{16}{45}\cdot\frac{15}{72}$ (l) $\frac{7}{9}\div\frac{42}{5}$

(m) $\frac{9}{14}\cdot\frac{8}{30}\cdot\frac{7}{27}$ (n) $\frac{16}{27}\cdot\frac{3}{4}\cdot\frac{9}{32}$

4. Simplify each of the following statements, using the same concepts as when simplifying numbers represented by numerals.

(a) $\frac{a}{b}\cdot\frac{abc}{ac}$ (b) $\frac{x+y}{x}\cdot\frac{x}{y}$ (c) $\frac{ab}{c}\left(\frac{a}{e}+\frac{b}{d}\right)$

(d) $2x\cdot\frac{3}{xy}\cdot\frac{y}{z}$ (e) $\frac{y^2}{x}\cdot\frac{x^3}{y^4}$ (f) $\frac{r^2s^3}{t}\cdot\frac{t^2}{rs^2}$

5. Perform the following operations involving mixed numbers.

(a) $7\frac{1}{8}\cdot 6\frac{1}{4}$ (b) $8\frac{1}{7}\cdot 3\frac{1}{2}$ (c) $2\frac{1}{10}\cdot 8\frac{1}{3}$

(d) $16\frac{2}{3}\cdot 4\frac{1}{4}$ (e) $7\frac{1}{2}\div 4\frac{1}{3}$ (f) $\frac{16}{3}\div 2\frac{1}{6}$

(g) $4\frac{7}{8}\div 1\frac{3}{4}$ (h) $6\frac{1}{4}\div\frac{4}{11}$

6. Express each of the following as a fraction in simplest form.

(a) $\dfrac{\frac{3}{8}}{\frac{4}{16}}$ (b) $\dfrac{\frac{1}{7}}{\frac{8}{21}}$ (c) $\dfrac{\frac{1}{3}+\frac{2}{5}}{\frac{5}{6}-\frac{3}{5}}$

(d) $\dfrac{\frac{1}{3}+\frac{1}{4}}{\frac{7}{12}}$ (e) $\dfrac{\frac{2}{3}+\frac{6}{7}}{\frac{13}{27}}$ (f) $\dfrac{1}{\frac{1}{2}+\frac{3}{4}}$

(g) $\dfrac{\frac{3}{4}+\frac{2}{7}}{\frac{3}{4}-\frac{1}{2}}$ (h) $\dfrac{\frac{x-y}{x}}{\frac{x+y}{x}}$ (i) $\dfrac{\frac{z+y}{z-y}}{\frac{z^3}{z-y}}$

7. (a) Make up two examples to show that division for rational numbers is neither commutative nor associative.

(b) Do rational numbers have a division identity? Give reasons to support your answer.

(c) Does the distributive property of division over addition hold true in the system of rational numbers? Demonstrate your answer with an example.

(d) Does the distributive property of multiplication over subtraction hold true for rational numbers? Demonstrate your answer with an example.

T 8. Solve the following equations on the domain of rational numbers.

(a) $\frac{1}{4}x=\frac{5}{8}$ (b) $\frac{3}{4}x=\frac{6}{5}$

(c) $\frac{1}{4} = \frac{7}{5}x$ (d) $x \div \frac{5}{4} = \frac{3}{5}$

(e) $\frac{3}{4} \div x = \frac{1}{4}$ (f) $\frac{5}{8} + \frac{1}{4}x = \frac{3}{4}$

9. For the following problems, set up an open sentence that is a model or represents each written statement.
   (a) Sandy's mathematics class meets two days a week for $\frac{5}{4}$ hours each day. How many hours is that a week?
   (b) Jeff is $6\frac{1}{2}$ feet tall. Ron is $5\frac{3}{4}$ feet tall. How many inches taller than Ron is Jeff?
   (c) A parking lot holds 64 cars. The parking lot is $\frac{7}{8}$ filled. How many spaces remain in the lot?
   (d) A tailor uses $3\frac{1}{2}$ m of material to make a suit. A bolt contains 60 m of material. How many suits can be made from a bolt of material?
   (e) A machine can desalinate $12\frac{1}{3}$ liters of water per hour. How long will it take to desalinate 100 liters?
   (f) John can mow the yard in $\frac{6}{7}$ of an hour. Joel can do the same job in $\frac{5}{6}$ of an hour. How long will it take them together to mow the yard?
   (g) Two-thirds of the students like math. One-half of the students who like math like science. One-fourth of the students who like math and science like spelling. What part of the students like all three subjects?
   (h) Hard College reduced its faculty by $\frac{1}{8}$ one year and by $\frac{1}{5}$ the next year. If there were 168 faculty members left, how many were there originally?
   (i) At Joe's Pizza Parlor a slice of pizza is $\frac{1}{12}$ of a whole pie. Joe noted that there was $\frac{2}{3}$ of a pie left. A customer asked for 9 slices. Was Joe able to serve the customer without baking a new pie?
   (j) Aaron owns $\frac{3}{7}$ of the stock in a company. His sister Jodi owns $\frac{2}{3}$ as much as he owns. What part of the stock does the family own?
   (k) The boy scouts collected $2\frac{1}{2}$ times as many pounds of paper as the girl scouts and the soccer league did together. The soccer league collected 50 pounds. If all three collected 490 pounds of paper, how much did the girl scouts and boy scouts collect?
   (l) Together, Shirley and Linda weigh 190 pounds. Shirley weighs $\frac{2}{3}$ as much as Linda. How much does each weigh?
   (m) Mary bought $2\frac{1}{3}$ yards of ribbon. How many bows can Mary make if each ribbon uses $\frac{3}{8}$ feet of ribbon? How many whole bows does she have? What part of a bow is left over?

10. Solve each open sentence in Exercise 9 and then use the procedure of "Looking Back" to see if your answer makes sense for the problem.

11. Use the formula $d = rt$ (distance equals rate times time) to solve these problems.
   (a) Dennis has been bicycling at an average speed of $2\frac{1}{2}$ kilometers per hour for $3\frac{1}{3}$ hours. How many kilometers has he traveled?
   (b) Jerry has driven $20\frac{3}{4}$ hours to cover $1724\frac{1}{2}$ kilometers. What has been his average speed per hour?

12. (a) John had $\frac{3}{4}$ of a pound of candy. He gave his friend Joel $\frac{1}{3}$ of the candy by weight. What fraction of a pound of the candy did he give Joel?
   (b) Katherine has eaten $\frac{5}{6}$ of a box of candy. If she gives Mary $\frac{1}{2}$ of the remainder, how much does she have left?

13. The following is a recipe for pancakes. (mL represents milliliter.)

325 mL flour	3 eggs
6 mL of salt	60 mL butter
45 mL sugar	375 mL milk
8 mL baking powder	

   (a) Rewrite this recipe to make 4 times as many pancakes.
   (b) Rewrite to make $\frac{1}{3}$ as many pancakes.

14. Seven-twentieths of the cars are red. One-half of the red cars are convertibles. One-third of the red convertibles have CB sets. What part of the cars are red convertibles with CB sets?

15. Express each as a simple fraction.
   (a) $\frac{a}{b} + \frac{c}{d} + \frac{e}{f}$ (b) $\frac{a}{b} \cdot \frac{c}{d} + \frac{e}{f} \cdot \frac{g}{h}$
   (c) $\dfrac{\frac{a}{b} + \frac{c}{d}}{\frac{e}{f} + \frac{g}{h}}$ (d) $\dfrac{\frac{a}{b} \div \frac{c}{d}}{\frac{e}{f} \div \frac{g}{h}}$

C 16. Multiply
   (a) $\left(\frac{1}{4} \cdot \frac{3}{5} \cdot \frac{5}{6}\right)_{\text{seven}}$ (b) $\left(4\frac{9}{T} \cdot 2E\frac{7}{8}\right)_{\text{twelve}}$

4 The system of rational numbers

■ Joyce joined the local weight-reducing club. She weighed $128\frac{1}{2}$ pounds when she joined. During the first two weeks, she lost $7\frac{1}{4}$ pounds. For the next four weeks, she lost weight at a more moderate rate of $1\frac{1}{3}$ pounds per week. If she continues to lose weight at this rate, how long will it take for her to attain the goal of weighing less than 110 pounds? ■

From this problem, you expect that we will be introducing inequalities in this section. In fact, we will study procedures for solving both equations and inequalities. In this section we will also emphasize some of the very important properties of rational numbers and point out that a select number of these properties are called *field properties.*

Learn well how to solve equations and you will have a useful tool for life. Many applied problems can be solved using either arithmetic or algebra (in this case, equations). The use of equations provides a problem-solving strategy that can make seemingly difficult problems quite easy. First, you will be introduced to properties of equations and inequalities that enable you to use this tool.

Let us summarize now the characteristics of rational numbers. A comparison of the three number systems—whole numbers, integers, and rational numbers—should start with the fact that integers have all the properties of whole numbers plus one additional property: For every integer, there is a unique additive inverse. Similarly, the system of rational numbers has all the properties of integers with the additional property that every rational number except $\frac{0}{1}$ (or 0) has a multiplicative inverse. Let's enumerate some of these properties.

The additive identity for rational numbers may be written as $0/b$, since

$$\frac{a}{b}+\frac{0}{b}=\frac{0}{b}+\frac{a}{b}=\frac{a}{b}$$

Since $0/b = \frac{0}{1}$, $\frac{0}{1}$ or the corresponding integer 0 is usually used to represent the additive identity.

For each rational number a/b, define $-(a/b)$ to be the additive inverse such that $a/b + -(a/b) = 0/b = \frac{0}{1} = 0$. Observe that

$$\frac{a}{b}+\frac{-a}{b}=\frac{(ab)+(b\cdot -a)}{b^2}=\frac{ab+-ab}{b^2}=\frac{0}{b^2}=0$$

Thus, $-a/b$ is also an additive inverse of a/b. In like manner,

$$\frac{a}{b}+\frac{a}{-b}=\frac{-ab+ba}{-b^2}=\frac{0}{-b^2}=0$$

and $a/-b$ is an additive inverse of a/b.

Since $-(a/b)$, $-a/b$, and $a/-b$ are all additive inverses for a/b and since the additive inverse of a/b is unique, these fractions all represent the same rational number.

Example 36 • The additive inverse of $\frac{2}{3}$ is $\frac{-2}{3}$ because

$$\frac{2}{3}+\frac{-2}{3}=\frac{2+-2}{3}=\frac{0}{3}=0 \quad \bullet$$

The existence of a multiplicative inverse y/x for each $(x/y) \neq 0$ *almost* gives closure for the division of rationals. The one exception is that $\frac{0}{1}$ (or 0) has no multiplicative inverse.

Before summarizing the properties of rational numbers, let's emphasize that the integers are a subset of the rational numbers through the following correspondence.

$$\begin{array}{cccccccccccc} \ldots & -4 & -3 & -2 & -1 & 0 & 1 & 2 & 3 & 4 & \ldots \\ & \updownarrow & \updownarrow & \updownarrow & \updownarrow & \updownarrow & \updownarrow & \updownarrow & \updownarrow & \updownarrow & \\ \ldots & \frac{-4}{1} & \frac{-3}{1} & \frac{-2}{1} & \frac{-1}{1} & \frac{0}{1} & \frac{1}{1} & \frac{2}{1} & \frac{3}{1} & \frac{4}{1} & \ldots \end{array}$$

The properties that we now enumerate for the rational number system are called **field properties.** The rational number system is the first field that we have developed in this text. The real number system to be discussed in the next chapter is also a field.

System of rational numbers

The system of rational numbers consists of the set Q of all rational numbers and two binary operations, addition (+) and multiplication (·), with the following properties for any rational numbers a/b, c/d, and e/f.

Closure Properties

1. $\frac{a}{b}+\frac{c}{d}=\frac{ad+bc}{bd}$ is a unique rational number in Q.

2. $\frac{a}{b}\cdot\frac{c}{d}=\frac{ac}{bd}$ is a unique rational number in Q.

Commutative Properties

3. $\frac{a}{b}+\frac{c}{d}=\frac{c}{d}+\frac{a}{b}$

4. $\frac{a}{b}\cdot\frac{c}{d}=\frac{c}{d}\cdot\frac{a}{b}$

Associative Properties

5. $\left(\frac{a}{b}+\frac{c}{d}\right)+\frac{e}{f}=\frac{a}{b}+\left(\frac{c}{d}+\frac{e}{f}\right)$

6. $\left(\frac{a}{b}\cdot\frac{c}{d}\right)\cdot\frac{e}{f}=\frac{a}{b}\cdot\left(\frac{c}{d}\cdot\frac{e}{f}\right)$

Distributive Property of Multiplication over Addition

7. $\frac{a}{b}\cdot\left(\frac{c}{d}+\frac{e}{f}\right)=\frac{a}{b}\cdot\frac{c}{d}+\frac{a}{b}\cdot\frac{e}{f}$

8. $\left(\frac{c}{d}+\frac{e}{f}\right)\cdot\frac{a}{b}=\frac{c}{d}\cdot\frac{a}{b}+\frac{e}{f}\cdot\frac{a}{b}$

Identity Elements

9. There exists a unique rational number, $\frac{0}{1}$, such that

$$\frac{a}{b}+\frac{0}{1}=\frac{0}{1}+\frac{a}{b}=\frac{a}{b}$$

10. There exists a unique rational number, $\frac{1}{1}$, such that

$$\frac{a}{b}\cdot\frac{1}{1}=\frac{1}{1}\cdot\frac{a}{b}=\frac{a}{b}$$

Inverse Elements

11. For each a/b there exists a unique additive inverse $-a/b$ such that

$$\frac{a}{b}+\frac{-a}{b}=\frac{-a}{b}+\frac{a}{b}=\frac{0}{1}$$

12. For each $a/b \neq 0$, there exists a unique multiplicative inverse b/a such that

$$\frac{a}{b}\cdot\frac{b}{a}=\frac{b}{a}\cdot\frac{a}{b}=\frac{1}{1}$$

Some interesting properties of rational numbers are introduced through the study of order (or relative size). The number line serves as a good visual aid for interpreting order for rational numbers.

Less than

A rational number a/b is less than c/d if a/b is located to the left of c/d on the number line.

Example 37 • $-\frac{5}{4}$ is less than $\frac{1}{4}$ because it lies to the left of $\frac{1}{4}$ on the number line in Figure 9. •

Figure 9

"Less than" can be defined for rational numbers so that the definition holds on both the integers and the whole numbers:

$$\frac{p}{q} < \frac{r}{s}$$

if and only if there exists some positive rational number e/f such that

$$\frac{p}{q} + \frac{e}{f} = \frac{r}{s}$$

However, such a definition does not provide an easy way to determine whether, for example, $\frac{7}{11}$ is less than $\frac{5}{6}$.

Perhaps the easiest approach to the definition of an appropriate order relationship between two rational numbers is to refer to the definition of equality of fractions. Recall that $p/q = r/s$ if and only if $ps = qr$. This fact suggests a way in which order can be checked for rational numbers.

Less than for rational numbers

> If p/q and r/s are rational numbers expressed with positive denominators, then
>
> $$p/q < r/s \quad \text{if and only if} \quad ps < qr$$

This property can be obtained from the definition of "less than." Suppose that $p/q < r/s$. Then, by definition, there is a rational number $e/f > 0$ with

$$\frac{r}{s} = \frac{p}{q} + \frac{e}{f}$$

Adding $-p/q$ to both sides yields

$$\frac{r}{s} + \frac{-p}{q} = \left(\frac{p}{q} + \frac{e}{f}\right) + \frac{-p}{q}$$

$$= \frac{-p}{q} + \left(\frac{p}{q} + \frac{e}{f}\right) \qquad \text{Why?}$$

$$= \left(\frac{-p}{q} + \frac{p}{q}\right) + \frac{e}{f} \qquad \text{Why?}$$

$$= 0 + \frac{e}{f}$$

So

$$\frac{r}{s} + \frac{-p}{q} = \frac{e}{f} > 0 \qquad \text{or} \qquad \frac{qr - ps}{sq} > 0$$

Since both s and q are positive, $sq > 0$. Therefore,

$$qr - ps > 0 \qquad \text{or} \qquad ps < qr$$

Now work backward to see if you can show that if $ps < qr$, then

$$\frac{p}{q} < \frac{r}{s}$$

Example 38 • (a) $\frac{2}{3} < \frac{5}{4}$ because $2 \cdot 4 < 3 \cdot 5$

(b) $\frac{-4}{5} < \frac{1}{4}$ because $-4 \cdot 4 < 5 \cdot 1$ •

Now that we have the concept of order, we can introduce two additional properties of rational numbers. First, given any two rational numbers, either they are equal or one is less than the other.

Trichotomy property for rational numbers

If a/b and c/d are any rational numbers, then exactly one of the following is true.

$$\frac{a}{b} < \frac{c}{d} \qquad \frac{a}{b} = \frac{c}{d} \qquad \frac{a}{b} > \frac{c}{d}$$

The average of two rational numbers a/b and c/d is $(\frac{1}{2})(a/b + c/d)$, a number that is between a/b and c/d. For example, if a/b is $\frac{2}{3}$ and c/d is $\frac{3}{4}$, then $(\frac{1}{2})(\frac{2}{3} + \frac{3}{4}) = \frac{17}{24}$. It is easy to verify that

$$\frac{2}{3} < \frac{17}{24} < \frac{3}{4}$$

Thus, there is always another rational number between any two rational numbers.

Density property of rational numbers

For any two rational numbers, it is always possible to find another rational number between them.

This property indicates that there are infinitely many rational numbers between any two rational numbers on the number line, so there is no "next" rational number in the sense that 3 is the next positive integer after 2.

Since integers make up a subset of the rational numbers, the following properties hold also for integers. One way to prove these characteristics is to prove them for integers and then for rational numbers.

Properties of inequalities

Let p/q, r/s, and t/v be rational numbers such that $p/q < r/s$. Then

1. $\frac{p}{q} + \frac{t}{v} < \frac{r}{s} + \frac{t}{v}$
2. $\frac{p}{q} \cdot \frac{t}{v} < \frac{r}{s} \cdot \frac{t}{v}$ if $\frac{t}{v} > 0$
3. $\frac{p}{q} \cdot \frac{t}{v} > \frac{r}{s} \cdot \frac{t}{v}$ if $\frac{t}{v} < 0$

Example 39 • Consider the fact that $\frac{2}{3} < \frac{7}{8}$; then

(a) $\frac{2}{3} + \frac{1}{4} < \frac{7}{8} + \frac{1}{4}$ or $\frac{11}{12} < \frac{9}{8}$

(b) $\frac{2}{3} \cdot \frac{1}{4} < \frac{7}{8} \cdot \frac{1}{4}$ or $\frac{1}{6} < \frac{7}{32}$

(c) $\frac{2}{3} \cdot \frac{-1}{4} > \frac{7}{8} \cdot \frac{-1}{4}$ or $\frac{-1}{6} > \frac{-7}{32}$ •

Of course, similar properties exist for equations.

Properties of equalities

Let p/q, r/s, and t/v be rational numbers with $p/q = r/s$. Then

1. $\frac{p}{q} + \frac{t}{v} = \frac{r}{s} + \frac{t}{v}$
2. $\frac{p}{q} \cdot \frac{t}{v} = \frac{r}{s} \cdot \frac{t}{v}$

Let us now use these properties of equalities and inequalities to find solution sets for equations and inequalities.

Example 40 • Solve $x + 4 = 17$ when the domain is the set of rational numbers. To solve this equation, we wish to isolate x on one side of the equation. To do this, we must remove the 4 from the left-hand side. Hence, we add the additive inverse of 4, namely -4, to both sides.

Solution

$x + 4 = 17$	Given
$(x + 4) + -4 = 17 + -4$	If $a/b = c/d$, then $a/b + e/f = c/d + e/f$.
$(x + 4) + -4 = 13$	Addition
$x + (4 + -4) = 13$	Associative property of addition
$x + 0 = 13$	Additive inverse
$x = 13$	Additive identity

Thus, 13 is the solution of the equation. We say that the solution set is $\{13\}$. A solution can be checked by substituting the solution into the original equation:

$$13 + 4 = 17 \quad \bullet$$

Example 41 • Solve $3x + 4 = -5$ on the domain of rational numbers. Again, we wish to isolate x on one side. To accomplish this, we add -4 to both sides. Since $3x$ remains, we multiply both sides of the equation by $\frac{1}{3}$ to undo multiplication by 3.

Solution

$3x + 4 = -5$	Given
$(3x + 4) + -4 = -5 + -4$	If $a/b = c/d$, then $a/b + e/f = c/d + e/f$.
$(3x + 4) + -4 = -9$	Addition
$3x + (4 + -4) = -9$	Associative property of addition
$3x + 0 = -9$	Additive inverse
$3x = -9$	Additive identity
$\left(\frac{1}{3}\right)(3x) = \left(\frac{1}{3}\right)(-9)$	If $a/b = c/d$, then $(a/b)(e/f) = (c/d)(e/f)$.
$x = -3$	Multiplication

The solution set is $\{-3\}$. •

The steps for solving inequalities are very similar to the steps for solving equations.

Example 42 • Solve $x + 3 < -5$ when the domain is rational numbers. Again, we wish to isolate x on one side. To accomplish this, we add -3 to both sides of the inequality.

Solution

$x + 3 < -5$	Given
$(x + 3) + -3 < -5 + -3$	If $a/b < c/d$, then $a/b + e/f < c/d + e/f$.
$(x + 3) + -3 < -8$	Addition
$x + (3 + -3) < -8$	Associative property of addition
$x + 0 < -8$	Additive inverse
$x < -8$	Additive identity

The solution set is $\{x \mid x \text{ is a rational number less than } -8\}$. •

Example 43 • Solve $-2x + -3 \leq 5$ on the rational numbers. Again, we wish to isolate x on one side. To accomplish this, we add 3 to both sides of the inequality. Since $-2x$ remains, we multiply by $-\frac{1}{2}$ to undo multiplication by -2. Remember that multiplying by a negative number reverses the inequality.

Solution

$-2x + -3 \le 5$	Given
$(-2x + -3) + 3 \le 5 + 3$	If $a/b < c/d$, then $a/b + e/f < c/d + e/f$.
$-2x + (-3 + 3) \le 5 + 3$	Associative property of addition
$-2x \le 8$	Addition
$\left(-\frac{1}{2}\right)(-2x) \ge \left(-\frac{1}{2}\right)(8)$	If $a/b < c/d$, $(a/b)(-e/f) > (c/d)(-e/f)$.
$x \ge -4$	Associative property of multiplication and multiplication

$\{x \mid x \text{ is a rational number } \ge -4\}$ •

If the variable in an equation or inequality appears on both sides, we must first use the properties discussed earlier to get all terms involving the variable on the same side.

Example 44 • Solve $5x - 2 = -4x + 7$.

Solution

$$5x - 2 = -4x + 7$$
$$4x + (5x - 2) = 4x + (-4x + 7)$$
$$9x - 2 = 7$$
$$(9x - 2) + 2 = 7 + 2$$
$$9x = 9$$
$$x = 1 \quad \bullet$$

Example 45 • On the rational numbers, solve for x in $\frac{-5}{2}x + \frac{1}{4} < \frac{-7}{8}$.

Solution

$$\left(\frac{-5}{2}x + \frac{1}{4}\right) + \frac{-1}{4} < \frac{-7}{8} + \frac{-1}{4}$$
$$\frac{-5}{2}x < \frac{-9}{8}$$
$$\frac{-2}{5}\left(\frac{-5}{2}x\right) > \frac{-2}{5} \cdot \frac{-9}{8}$$
$$x > \frac{9}{20}$$

$\{x \mid x \text{ is a rational number } > \frac{9}{20}\}$ •

Formulas are equations in which letters are used to represent numbers that satisfy the relationship dictated by the formula. Since the letters in the formula

Just for fun *Professor Abstract gave his report on the grades of the last test in the following manner. One-fifth made A's; one-fourth made B's; one-third made C's; one-sixth made D's; and 12 made F's. How many are in this class?*

represent numbers, the techniques of the section can be used to solve for one letter in terms of the others.

Example 46 • Solve for b in $y = mx + b$.

Solution

$y = mx + b$	Given
$mx + b = y$	Symmetric property of equality
$b + mx = y$	Commutative property of addition
$(b + mx) + (-mx) = y + (-mx)$	If $a/b = c/d$, then $a/b + e/f = c/d + e/f$.
$b + (mx + -mx) = y + (-mx)$	Associative property of addition
$b = y + (-mx)$	Addition
$b = y - mx$	Definition of subtraction •

Let's solve now the introductory problem.

Example 47 • Joyce joined the local weight-watching club. She weighed $128\frac{1}{2}$ pounds when she joined. During the first two weeks she lost $7\frac{1}{4}$ pounds. For the next four weeks she lost weight at a more moderate rate of $1\frac{1}{3}$ pounds per week. If she continues to lose weight at this rate, how long will it take for her to attain the goal of weighing less than 110 pounds?

Solution At the start, Joyce weighs $128\frac{1}{2}$ pounds. She loses $7\frac{1}{4} + 4(1\frac{1}{3}) = 12\frac{7}{12}$ pounds. Her weight at the end of four weeks is

$$128\frac{1}{2} - 12\frac{7}{12} = 115\frac{11}{12} \text{ pounds}$$

She continues to lose weight for x additional weeks, with a new weight of

$$115\frac{11}{12} - x\left(1\frac{1}{3}\right)$$

This needs to be less than 110.

$$115\frac{11}{12} - x\left(\frac{4}{3}\right) < 110$$

$$x\left(\frac{4}{3}\right) > \frac{71}{12}$$

$$x > \frac{71}{12}\cdot\frac{3}{4}$$

$$x > \frac{71}{16} \quad \text{or} \quad 4\frac{7}{16} \text{ weeks} \quad \bullet$$

Exercise set 4

R 1. Insert >, =, or < to express the correct relationship between the following pairs of numbers.

(a) $\frac{9}{10}, \frac{7}{11}$ (b) $\frac{33}{9}, \frac{21}{7}$ (c) $\frac{5}{7}, \frac{2}{5}$

(d) $\frac{29}{3}, \frac{16}{2}$ (e) $\frac{17}{51}, \frac{6}{7}$ (f) $\frac{22}{73}, \frac{25}{71}$

2. Arrange the following sets of rational numbers in increasing order, such as $\frac{1}{2} < 2 < \frac{7}{2}$.

(a) $\frac{71}{100}, \frac{3}{2}, \frac{23}{30}$ (b) $\frac{9}{2}, 4, \frac{165}{41}$

(c) $\frac{2}{3}, \frac{11}{18}, \frac{16}{27}, \frac{67}{100}$ (d) $\frac{25}{28}, \frac{27}{20}, \frac{14}{16}, \frac{79}{90}$

(e) $\frac{22}{7}, \frac{7}{2}, \frac{10}{3}, \frac{156}{50}$ (f) $\frac{51}{95}, \frac{19}{36}, \frac{17}{30}, \frac{14}{29}$

3. (a) For what integral values of x will $x/8$ be less than $\frac{1}{4}$?
   (b) For what integral values of x will $x/9$ be greater than $\frac{1}{3}$?

4. Classify as true or as false.
   (a) Every rational number is an integer.
   (b) Every whole number is a rational number.
   (c) The rationals are a subset of the integers.
   (d) Some whole numbers are not rational numbers.

5. Let W be the whole numbers, F be the nonnegative rational numbers, Q be all the rationals, I be the integers, and N be the negative integers. Which of these sets have the following properties?
   (a) -2 is a member of the set.
   (b) $3/-7$ is a member of the set.
   (c) The set has an additive inverse for each element.
   (d) The set has a multiplicative inverse for each element.
   (e) The set is closed under addition.
   (f) The set is closed under multiplication.
   (g) The set is closed under subtraction.
   (h) The set is closed under division.

6. State the property that is used in each of (a)–(g).

(a) $\frac{3}{7} + \frac{9}{14} < \frac{4}{7} + \frac{9}{14}$

(b) $\frac{3}{7} + \left(\frac{3}{8} + \frac{1}{4}\right) = \frac{3}{7} + \left(\frac{1}{4} + \frac{3}{8}\right)$

(c) $\frac{5}{13}\cdot\frac{4}{9} < \frac{6}{13}\cdot\frac{4}{9}$

(d) $\frac{1}{18} + \left(\frac{3}{11} + \frac{5}{9}\right) = \left(\frac{1}{18} + \frac{3}{11}\right) + \frac{5}{9}$

(e) $\frac{5}{9}\cdot\frac{-3}{4} > \frac{6}{9}\cdot\frac{-3}{4}$

(f) $\frac{2}{3}\cdot\frac{1}{4} + \frac{3}{11}\cdot\frac{1}{4} = \left(\frac{2}{3} + \frac{3}{11}\right)\cdot\frac{1}{4}$

(g) $\frac{5}{9}\cdot\left(\frac{1}{9}\cdot\frac{1}{7}\right) = \frac{5}{9}\left(\frac{1}{7}\cdot\frac{1}{9}\right)$

7. Given the first relationship in each item that follows, is the conclusion necessarily true?

(a) If $\frac{6}{7} + \frac{1}{3} < \frac{29}{23}$, then $\frac{6}{7} < \frac{29}{23} - \frac{1}{3}$.

(b) If $\frac{11}{12} > \frac{2}{5}$, then $\frac{-11}{12} > \frac{-2}{5}$.

(c) If $\frac{-63}{87} < \frac{1}{3}$, and $\frac{1}{3} \leq \frac{23}{69}$, then $\frac{-63}{87} < \frac{23}{69}$.

(d) If $\frac{a}{b} < 0$ and $\frac{c}{d} > 0$, then $\frac{a}{b} \cdot \frac{c}{d} > 0$.

(e) If $\frac{a}{b} < \frac{c}{d}$ and $x < 0$, then $\left(\frac{a}{b}\right)x < \left(\frac{c}{d}\right)x$.

8. Using the average, or midpoint, find three numbers between $\frac{1}{6}$ and $\frac{3}{4}$.

9. If a/b and c/d are unequal rational numbers, is the rational number $(a + c)/(b + d)$ between them?

10. Find the solution set in each case. The domain for the variable is the set of rational numbers.

(a) $2y - 3 = 11$ (b) $5y + 4 < -6$
(c) $3z + (-5) \geq 7$ (d) $2x + -2 = 20$
(e) $6x = 4x + 4$ (f) $2m + 3 = m + 7$
(g) $3x + 5 \geq 7 + x$
(h) $-3m + 4 > -m + 2$
(i) $4x + 3 = -3 + x$ (j) $3x + 3 < x - 1$
(k) $3x - 7 = 4x + (-2)$
(l) $7y - 6 = 6y + (-2)$
(m) $-2x - 3 < 5$ (n) $-3x + (-5) < -2$

11. Solve each of the following equations on the domain of rational numbers and give reasons for performing each step.

(a) $x + 8 = 15$ (b) $t - 15 = 6$
(c) $2x = 12$ (d) $2x + 4 = 6$
(e) $y + 7 = 8$ (f) $3x + 2 = 7$
(g) $5x + 7 = 12$ (h) $2r + 3 = 5$

12. Solve each of the sets of inequalities on the domain of rational numbers and give reasons for performing each step.

(a) $x + 2 < 6$ (b) $-x + 1 \geq 3$
(c) $-2x - 4 < -6$ (d) $3x + 5 \leq -1$

13. Find all rational numbers x in the solution set of each of the following.

(a) $x + \frac{-1}{2} < \frac{5}{2}$ (b) $3x + \frac{3}{10} < \frac{9}{10}$

(c) $2x + \frac{-5}{6} < \frac{-7}{4}$ (d) $2x + 5 < \frac{3}{10}$

(e) $\frac{3x}{2} + -5 < 7$ (f) $\frac{x}{4} + \frac{-1}{2} < -7$

(g) $\frac{x}{3} + (-2) < -5$ (h) $4x + (-3) < -7$

(i) $\frac{2x}{3} - \frac{1}{4} \leq \frac{1}{6}$ (j) $\frac{-3x}{4} + \frac{x}{7} \leq 2$

For Exercises 14 through 18, set up an equation as a model for the problem. Then solve the equation.

T 14. A man leaves $\frac{1}{2}$ of his estate to his wife, $\frac{1}{8}$ plus \$1000 to each of three children, and the remainder to his pet dog. If he has a \$100,000 estate, how much does the dog get?

15. Mrs. Smith's electric bill was six times Mrs. Jones' bill. The two bills totaled \$84. What was the cost of each woman's bill?

16. The difference between two integers is 21. The larger integer is equal to twice the smaller integer plus 20. What are the two integers?

17. Three times Ralph's weight added to 54 kg is equal to 300 kg. How much does Ralph weigh?

18. Jerry's age is four times the age Diane will be in 2 years. The sum of their ages is 78. How old are Jerry and Diane?

19. Which field properties of rational numbers fail if you remove the following from the rational numbers?

(a) 0 (b) 1 (c) -5 (d) $\frac{1}{4}$

C 20. Prove that the positive rational numbers with operations + and · do not form a field.

21. Which of these systems are fields?

(a)

+	0	1
0	0	1
1	1	0

·	0	1
0	0	0
1	0	1

(b)

+	0	1	2
0	0	1	2
1	1	2	0
2	2	0	1

·	0	1	2
0	0	0	0
1	0	1	2
2	0	2	1

22. Which of the following form Abelian groups?

(a) The nonnegative rational numbers under addition
(b) The positive rational numbers under multiplication
(c) The rational numbers, excluding 0, under multiplication
(d) The negative rational numbers under addition

(e) The negative rational numbers under multiplication

23. Solve for the variable indicated.

(a) $A + C = P + R$; R
(b) $A = P + Prt$; r
(c) $A = P + Prt$; t
(d) $y = mx + b$; m
(e) $l = a + (n - 1)d$; d
(f) $ax = bx + 7$; x
(g) $y = mx + b$; x

24. Prove that if $p/q < r/s$, then

(a) $\frac{p}{q} + \frac{t}{v} < \frac{r}{s} + \frac{t}{v}$.

(b) $\frac{p}{q} \cdot \frac{t}{v} < \frac{r}{s} \cdot \frac{t}{v}$ if $\frac{t}{v} > 0$.

(c) $\frac{p}{q} \cdot \frac{t}{v} > \frac{r}{s} \cdot \frac{t}{v}$ if $\frac{t}{v} < 0$.

25. (a) Prove that if $p/q + t/v < r/s + t/v$, then $p/q < r/s$.
(b) If $p/q \cdot t/v < r/s \cdot t/v$ where $t/v > 0$, then $p/q < r/s$.

5 An introduction to decimals

■ Two trains are 53.6 kilometers apart. One train travels 47.3 kilometers away from the other train. The second train travels 61.4 kilometers toward the first train. How far apart are the two trains? ■

You note that the preceding problem involves decimals. Why study decimals? Although considerably more useful than whole numbers, fractions are not always adequate and often lead to awkward manipulative procedures. For example, suppose you were asked to add

$$\frac{4621}{7839} + \frac{5641}{8441}$$

With a hand calculator, these can be easily converted to decimals and the answer given to the accuracy of the calculator. You will learn in Section 6 that all fractions can be expressed as decimals, but in this section you will express a subset of fractions as terminating decimals. A knowledge of this section is especially important in Chapters 6 and 13 (consumer mathematics and probability).

Decimal fractions are constructed by placing a dot, called a **decimal point,** after the units digit and letting the digits to the right of the dot denote, in turn, the number of tenths, hundredths, thousandths, and so forth. If there is no whole number part in a given numeral, a 0 is usually placed before the decimal point (for example, 0.24).

Example 48 • $\frac{3}{10} = 0.3 \quad \frac{19}{10} = 1.9 \quad \frac{165}{10} = 16.5$ •

Note that if the denominator is 10, there is one digit to the right of the decimal point. If the denominator is 100, there are two digits to the right of the decimal point. If the denominator is 1000, there are three, and so on.

Example 49 • $\frac{31}{100} = 0.31 \qquad \frac{191}{100} = 1.91 \qquad \frac{1654}{100} = 16.54$

$$\frac{426}{1000} = 0.426 \qquad \frac{1833}{1000} = 1.833 \qquad \frac{17{,}927}{1000} = 17.927$$ •

Decimals may be illustrated graphically by shading parts of a block consisting of 100 parts. For example, Figure 10 represents 0.26.

Figure 10

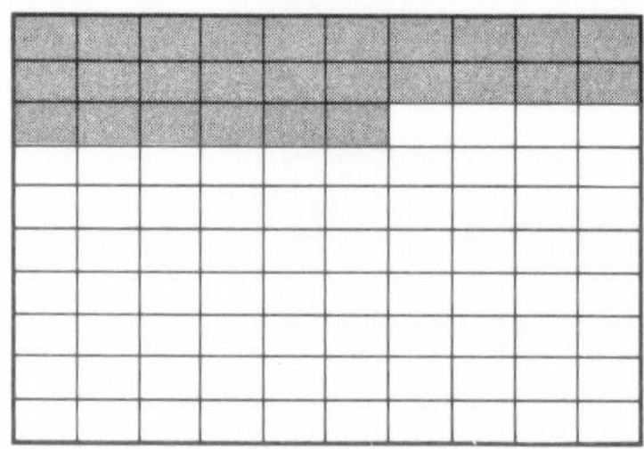

Since decimals involve a power of 10 in the denominator of a fraction, we need to extend to fractions the discussion of exponents (found in Chapter 3).

Example 50 •

$$\frac{6^5}{6^3} = \frac{6 \cdot 6 \cdot 6 \cdot 6 \cdot 6}{6 \cdot 6 \cdot 6} = 6^{5-3} = \frac{1}{6^{3-5}}$$

$$\frac{x^2}{x^7} = \frac{\not{x} \cdot \not{x}}{\not{x} \cdot \not{x} \cdot x \cdot x \cdot x \cdot x \cdot x} = \frac{1}{x^{7-2}} = \frac{1}{x^5}$$ •

In general,

$$\frac{x^m}{x^n} = x^{m-n} \qquad \text{or} \qquad \frac{x^m}{x^n} = \frac{1}{x^{n-m}}$$

Remember that

$$(x^m)^n = x^{mn} \qquad \text{and} \qquad \left(\frac{x}{y}\right)^m = \frac{x^m}{y^m} \qquad y \neq 0$$

Again, since any nonzero number divided by itself is 1, $a^3/a^3 = 1$ $(a \neq 0)$. But by the pattern we observed earlier,

$$\frac{a^3}{a^3} = a^{3-3} = a^0$$

Therefore, a^0 is defined to be 1.

Example 51 • $10^0 = 1$; $\quad 4^0 = 1$; $\quad n^0 = 1$ when $n \neq 0$ •

Similarly, the same pattern can be followed to infer that $1/3^4 = 3^0/3^4$ should equal $3^{0-4} = 3^{-4}$. Thus,

$$3^{-4} = \frac{1}{3^4}$$

In general, a^{-m} is defined to be $1/a^m$, when $a \neq 0$.

Example 52 • (a) $10^{-3} = \frac{1}{10^3}$

(b) $\frac{1}{4^2} = 4^{-2}$

(c) $10^8 \cdot 10^{-5} = 10^8 \cdot \frac{1}{10^5} = 10^{8-5} = 10^3$

(d) $10^{-6} \cdot 10^{-3} = \frac{1}{10^6} \cdot \frac{1}{10^3} = \frac{1}{10^6 \cdot 10^3} = \frac{1}{10^9} = 10^{-9}$ •

Properties of exponents

If a and b are any numbers unequal to 0 and m and n are any exponents, then the following properties hold.

$$a^m \cdot a^n = a^{m+n}$$

$$\frac{a^m}{a^n} = a^{m-n} = \frac{1}{a^{n-m}}$$

$$a^0 = 1$$

$$a^{-m} = \frac{1}{a^m}$$

$$(a^m)^n = a^{m \cdot n}$$

$$(ab)^m = a^m b^m$$

flashback

Decimals were introduced by Simon Stevin, a Flemish engineer, in the 16th century. In his book, *The Tenth*, published in 1585, he advocated the use of decimal fractions and showed how to compute with them. Stevin's notation for the decimal fraction 4.6183 was

4 ⓪ 6 ① 1 ② 8 ③ 3 ④

Stevin did not use a decimal point. Later, in England, the decimal point was written 4·6183, and in many parts of Europe, the comma was (and still is) used as a decimal indicator as in 4,6183.

Example 53 •

$$\left(\frac{2}{3}\right)^2 = \frac{2^2}{3^2} = \frac{4}{9}$$

$$\left(\frac{1}{10}\right)^2 = \frac{1^2}{10^2} = \frac{1}{100}$$

$$\left(\frac{1}{10}\right)^3 = \frac{1^3}{10^3} = \frac{1}{1000} \quad \bullet$$

The following example illustrates important relationships between fractions and decimals.

Example 54 •

$$0.46 = \frac{46}{100} = \frac{4(10) + 6(1)}{100}$$

$$= 4\left(\frac{10}{100}\right) + 6\left(\frac{1}{100}\right) = 4\left(\frac{1}{10}\right) + 6\left(\frac{1}{10}\right)^2 = 4(10)^{-1} + 6(10)^{-2}$$

$$46.7 = \frac{467}{10} = \frac{4(100) + 6(10) + 7(1)}{10} = 4\left(\frac{100}{10}\right) + 6\left(\frac{10}{10}\right) + 7\left(\frac{1}{10}\right)$$

$$= 4(10) + 6 + 7\left(\frac{1}{10}\right) = 4(10) + 6 + 7(10)^{-1}$$

$$6.895 = \frac{6895}{1000} = \frac{6(1000)}{1000} + \frac{8(100)}{1000} + \frac{9(10)}{1000} + \frac{5(1)}{1000}$$

$$= 6 + 8\left(\frac{1}{10}\right) + 9\left(\frac{1}{100}\right) + 5\left(\frac{1}{1000}\right)$$

$$= 6 + 8\left(\frac{1}{10}\right) + 9\left(\frac{1}{10}\right)^2 + 5\left(\frac{1}{10}\right)^3$$

$$= 6 + 8(10)^{-1} + 9(10)^{-2} + 5(10)^{-3} \quad \bullet$$

The decimals we discussed at the beginning of this section are called **terminating decimals.** You note in all our examples of terminating decimals that it is possible to write the decimal as a fraction with a denominator as a power of 10. Also note that $10 = 2 \cdot 5$; $100 = 10^2 = 2^2 \cdot 5^2$; $1000 = 10^3 = 2^3 \cdot 5^3$ and, in general, $10^n = 2^n \cdot 5^n$. This suggests the following property.

Terminating decimals

> The only reduced fractions that can be written as terminating decimals are those in which the denominators can be factored into powers of 2 and 5.

Example 55 • Change $\frac{3}{20}$ to a decimal.

Solution To change to a decimal, we need a power of 10 in the denominator. Multiply the numerator and denominator by 5, getting

$$\frac{3 \cdot 5}{20 \cdot 5} = \frac{15}{100} = 0.15 \quad \bullet$$

Since all terminating decimals can be expressed as fractions with denominators of powers of 10, interesting properties can be developed for multiplying and dividing a decimal by powers of 10.

Example 56 •

$$14.37(10) = \left[1(10) + 4 + 3\left(\frac{1}{10}\right) + 7\left(\frac{1}{10}\right)^2\right] \cdot 10$$

$$= 1(10)^2 + 4(10) + 3 + 7\left(\frac{1}{10}\right)$$

$$= 143.7$$

$$2.615(100) = \left[2 + 6\left(\frac{1}{10}\right) + 1\left(\frac{1}{10}\right)^2 + 5\left(\frac{1}{10}\right)^3\right] \cdot 10^2$$

$$= 2(10)^2 + 6(10) + 1 + 5\left(\frac{1}{10}\right)$$

$$= 261.5$$ •

Do you see a pattern? Now let's divide by powers of 10.

Example 57 • Find $4.2 \div 10$; $4.2 \div 100$; $4.2 \div 1000$.

Solution (a) $\frac{4.2}{10} = \frac{(4.2)(10)}{10(10)} = \frac{42}{100} = 0.42$

(b) $\frac{4.2}{100} = \frac{(4.2)(10)}{(100)(10)} = \frac{42}{1000} = 0.042$

(c) $\frac{4.2}{1000} = \frac{(4.2)(10)}{(1000)(10)} = \frac{42}{10{,}000} = 0.0042$ •

The preceding discussion suggests the following statement.

Powers of 10

> To **divide** a number N by 10^k, where k is a whole number, move the decimal point (in N) k places to the left. In a like manner, to **multiply** by 10^k, move the decimal point k places to the right.

Example 58 • (a) $176.2 \div 10^3 = 0.1762$

(b) $0.163 \cdot 10^4 = 1630.$ •

Example 59 • Perform the addition $0.253 + 0.14$ in two ways.

Solution Now

$$0.253 = \frac{253}{1000} \quad \text{and} \quad 0.14 = \frac{14}{100}$$

$$\frac{253}{1000} + \frac{14}{100} = \frac{253}{1000} + \frac{140}{1000} = \frac{393}{1000} = 0.393$$

Note also that

$$0.253 = \frac{2}{10} + \frac{5}{100} + \frac{3}{1000} \quad \text{and} \quad 0.14 = \frac{1}{10} + \frac{4}{100}$$

Now add the two fractions in this expanded form.

$$\begin{array}{rl}
0.253 = & 2\left(\frac{1}{10}\right) + 5\left(\frac{1}{100}\right) + 3\left(\frac{1}{1000}\right) \\
0.14 = & 1\left(\frac{1}{10}\right) + 4\left(\frac{1}{100}\right) + 0\left(\frac{1}{1000}\right) \\
\hline
 & (2+1)\left(\frac{1}{10}\right) + (5+4)\left(\frac{1}{100}\right) + (3+0)\left(\frac{1}{1000}\right) \\
 & = 3\left(\frac{1}{10}\right) + 9\left(\frac{1}{100}\right) + 3\left(\frac{1}{1000}\right) \\
 & = 0.393
\end{array}$$
•

The preceding addition was accomplished by grouping and adding the coefficients of tenths, hundredths, and thousandths. This same grouping can be accomplished by lining up the decimal points in a vertical column and using the algorithm shown in the next example.

Example 60 •

$$\begin{array}{lcl}
0.253 = \frac{253}{1000} = \frac{253}{1000} \\
0.14 \;\; = \frac{14}{100} = \frac{140}{1000} \\
\hline
0.393 \qquad\qquad \frac{393}{1000} = 0.393
\end{array}$$
•

Example 61 • Find 3.716 + 23.4

Solution

$$\begin{array}{r}
3.716 \\
23.4 \\
\hline
27.116
\end{array}$$

Note that the solution in expanded notation agrees with the algorithm.

$$\begin{array}{rl}
 & \qquad\qquad\quad 1(10)^0 \\
3.716 = & \qquad\qquad\quad 3(10)^0 + 7(10)^{-1} + 1(10)^{-2} + 6(10)^{-3} \\
23.4 \;\; = & 2(10)^1 + 3(10)^0 + 4(10)^{-1} \\
\hline
 = & 2(10)^1 + 7(10)^0 + 1(10)^{-1} + 1(10)^{-2} + 6(10)^{-3} \\
 = & 27.116
\end{array}$$
•

Of course, the preceding algorithm applies to any number of addends.

Example 62 •

$$\begin{array}{r} 3.461 \\ 47.03 \\ 641.007 \\ 34.165 \\ \hline 725.663 \end{array}$$ •

The algorithm for subtraction is similar to the algorithm for addition, except that we subtract instead of adding.

Example 63 •

$$\begin{array}{rcl} 32.4 = \dfrac{324}{10} = & \dfrac{3240}{100} & \\ -6.73 = \dfrac{673}{100} = & \dfrac{673}{100} & \\ \hline 25.67 \qquad\qquad & \dfrac{2567}{100} & = 25.67 \end{array}$$ •

Example 64 • Find 16.14 – 0.237.

Solution

$$\begin{array}{r} 16.140 \\ -0.237 \\ \hline 15.903 \end{array}$$ •

In the preceding example, did it bother you to annex the 0 on 16.14 to get 16.140 before beginning your subtraction? The placing of the 0 at the end of 16.14 does not change the problem.

$$\begin{aligned} 16.14 &= 1(10) + 6 + 1\left(\frac{1}{10}\right) + 4\left(\frac{1}{10}\right)^2 \\ &= 1(10) + 6 + 1\left(\frac{1}{10}\right) + 4\left(\frac{1}{10}\right)^2 + 0\left(\frac{1}{10}\right)^3 \end{aligned}$$

In fact, you can annex as many zeros as you wish and the value does not change.

The algorithm we have used for addition and subtraction in base ten can also be used with other bases.

Example 65 •

Base seven	*Base two*
$\begin{array}{r} 46.143 \\ +\ 24.35 \\ \hline 103.523 \end{array}$	$\begin{array}{r} 111.0101 \\ +\ \ 11.1011 \\ \hline 1011.0000 \end{array}$ •

What is a bit more tricky is changing such expressions to base ten.

Just for fun *Complete the subtraction magic square. (In a subtraction magic square, you sum the numbers on the end and subtract the middle.)*

$$0.06 + 0.18 - 0.09 = 0.15$$

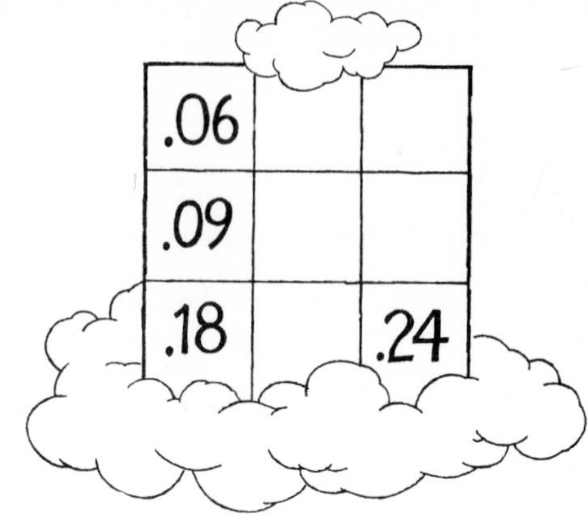

Example 66 • Change 42.14_{five} to base ten.

$$42.14_{\text{five}} = \left[4(10) + 2 + 1\left(\frac{1}{10}\right) + 4\left(\frac{1}{10}\right)^2\right]_{\text{five}}$$
$$= \left[4(5) + 2 + 1\left(\frac{1}{5}\right) + 4\left(\frac{1}{5}\right)^2\right]_{\text{ten}}$$
$$= 20 + 2 + 0.2 + 0.16$$
$$= 22.36 \quad \bullet$$

As you can guess from the preceding procedure, only those decimals in base two and five can be changed to a terminating decimal in base ten.

Exercise set 5

R 1. Write each of the following in expanded form.
(a) 14.72 (b) 4.0016 (c) 0.146

2. Rewrite the following as decimals.
(a) $4(10) + 3(10)^0 + 2(\frac{1}{10}) + 6(\frac{1}{10})^2$
(b) $4(\frac{1}{10})^2 + 5(\frac{1}{10})^3$
(c) $7(10)^0 + 4(\frac{1}{10})^3 + 2(\frac{1}{10})^4$

3. Express each of the following decimal fractions as a fraction.
(a) 0.085 (b) 3.25 (c) 12.74
(d) 0.84 (e) 2.75 (f) 0.00025

4. Express the following fractions as decimal fractions.
(a) $\frac{316}{10}$ (b) $\frac{602}{10{,}000{,}000}$ (c) $\frac{18}{15}$
(d) $\frac{63}{70}$ (e) $\frac{651}{120}$ (f) $\frac{91}{140}$
(g) $\frac{3}{8}$ (h) $\frac{5}{16}$ (i) $\frac{126}{105}$

5. Which of these fractions can be expressed as terminating decimal fractions?
(a) $\frac{36}{15}$ (b) $\frac{21}{70}$ (c) $\frac{48}{36}$
(d) $\frac{27}{60}$ (e) $\frac{10}{30}$ (f) $\frac{21}{36}$

6. Where possible, write the fractions in Exercise 5 as terminating decimals.

7. Perform the following operations.
(a) 28.32 + 7.521 (b) 0.56 + 0.006

(c) $0.3 + 5.00311$ (d) $2.04 + 4.1$
(e) $354.51 - 38.64$ (f) $389.27 - 63.99$
(g) $691.7 - 8.526$ (h) $84.5 - 9.72$

8. Change each decimal to a fraction in Exercise 7, perform the operation, then change each answer back to a decimal. Do you get the same answer?

9. Write each decimal in Exercise 7 in expanded form, using negative exponents, and perform the operation. Then change each answer back to a decimal and check the answer in Exercise 7.

10. Multiply each of the following by 10; by 100; by 10^3.

(a) 41.76 (b) 4.007 (c) 0.00021

T 11. Find the value of the following.

(a) $\dfrac{10^{-1}4^0}{10^{-2}}$ (b) $\dfrac{10^0 4^{-1}}{4^{-3}}$

(c) $(2^{-1})^0$ (d) $(3^{-1})^{-2}$

For Exercises 12 through 15 set up an equation to represent the problem, solve the equation, and then check the answer with data given in the problem.

12. Three students in the mathematics club participated in a bike-a-thon to raise money. John rode 16.2 miles, Jim rode 13.5 miles, and Sue rode enough so that they collected on 40 miles. How many miles did Sue ride?

13. Chang has \$142.36 in his checking account. He writes two checks for \$25 each and one check for \$13.42. How much money is in his account?

14. Sue's gasoline tank on her new car holds 20 gallons. She fills the car with 14.6 gallons. How much gasoline was in the tank already?

15. Robert has agreed with his parents that he will pay his long-distance calls out of his allowance. Last month, Robert made all the long-distance calls; the total phone bill for the family was \$30, excluding tax. How much will Robert pay if the monthly cost of the telephone is \$13.75?

16. (Introductory problem) Two trains are 53.6 km apart. One train travels 47.3 km away from the other train. The second train travels 61.4 km toward the first train. How far apart are the two trains?

C 17. Complete.

(a) $(\frac{1}{3})_{ten} = (\_\_\_\_)_{three} = (0.\_\_\_\_)_{three}$
(b) $(\frac{1}{3})_{ten} = (\_\_\_\_)_{six} = (0.\_\_\_\_)_{six}$
(c) $(\frac{1}{3})_{ten} = (\_\_\_\_)_{nine} = (0.\_\_\_\_)_{nine}$

18. (a) In what base does $\frac{2}{7}$ terminate?
(b) Express 0.34_{six} as a fraction base ten.

19. Discuss the meaning of each of the following.

(a) 101.11_{two} (b) 0.004_{five}
(c) $6T.71_{twelve}$ (d) 0.0645_{seven}

20. Perform the following operations in the indicated bases.

(a) $11.001_{two} + 1.110_{two}$
(b) $0.043_{seven} + 6.45_{seven}$
(c) $12.21_{three} - 2.222_{three}$

6 The arithmetic of decimals

■ Bill has the highest batting average on the Samford Bulldogs, 0.400 after 30 times at bat. In a slump, Bill gets only 2 hits in the next 8 times at bat. What is his new batting average? ■

To solve this problem, you need to understand both multiplication and division of decimals. Therefore, in this section you will review these basic operations. The algorithms will be presented so you will understand the general rules for placing decimals. From the algorithm for division of decimals, a procedure will be developed for expressing fractions as either terminating or repeating decimals. You may be surprised to learn that every fraction can be expressed as either a terminating decimal or a repeating decimal.

The algorithms of Section 5 allow us to add and subtract decimals in much the same manner as we add and subtract whole numbers. We now need algorithms for multiplication and division similar to those for whole numbers.

The multiplication algorithm for decimals is easily discovered by writing the decimals as fractions. Let's multiply 32.4 by 0.12.

$$(32.4) \cdot (0.12) = \frac{324}{10} \cdot \frac{12}{100} = \frac{3888}{1000} = 3.888$$

Note that the answer is obtained by multiplying the numbers 324 and 12 without decimals and dividing the result by 1000 or 10^3.

Note that there is one digit to the right of the decimal point in 32.4, leading to a 10 in the denominator of the equivalent fraction; there are two digits to the right of the decimal point in 0.12, leading to a 10^2 or 100 in the denominator of the equivalent fraction. Since $10 \cdot 10^2 = 10^3$ in the denominator of the answer, the answer as a decimal has three digits to the right of the decimal point. Do you see the pattern? Try your pattern on the following example.

Example 67 • How many digits are there to the right of the decimal point in the answer for $2.001 \cdot 0.02$?

Solution From your pattern, did you get an answer of 5? Let's verify your solution. 2.001 has three digits to the right of the decimal point. The equivalent fraction has a denominator of 10^3. The two digits to the right of the decimal point in 0.02 produce a 10^2 in the denominator of the fraction. The product of the two denominators is 10^5, so the decimal answer has five digits to the right of the decimal point. •

The preceding discussion leads to the following generalization.

Decimal places in multiplication

> If there are r digits to the right of the decimal point in one factor and s digits to the right of the decimal point in a second factor, multiply the two factors, ignoring decimal points, and place the decimal point so that there are $r + s$ digits to the right of the decimal point in the answer.

$$\begin{array}{rl}
32.4 & \text{(1 digit to the right of the decimal point)} \\
\cdot\ 0.12 & \text{(2 digits to the right of the decimal point)} \\
\hline
6\,48 & \\
32\,4 & \\
\hline
3.8\,88 & \text{(3 digits to the right of the decimal point)}
\end{array}$$

This algorithm is further illustrated by expressing the multiplication in expanded form.

Example 68 •

$$\begin{array}{r} 3(10)^1 + 2(10)^0 \quad + 4\left(\frac{1}{10}\right) \\ 1\left(\frac{1}{10}\right) + 2\left(\frac{1}{10}\right)^2 \\ \hline 6\left(\frac{1}{10}\right) + 4\left(\frac{1}{10}\right)^2 + 8\left(\frac{1}{10}\right)^3 \\ 3(10)^0 + 2\left(\frac{1}{10}\right) + 4\left(\frac{1}{10}\right)^2 \qquad\qquad \\ \hline 3(10)^0 + 8\left(\frac{1}{10}\right) + 8\left(\frac{1}{10}\right)^2 + 8\left(\frac{1}{10}\right)^3 \end{array} = 3.888$$ •

The algorithm for multiplication of decimals in base ten applies in other bases.

Example 69 • Multiply $3.41_{\text{five}} \cdot 2.3_{\text{five}}$

Solution

$$\begin{array}{rl} 3.41 & \text{(2 places)} \\ 2.3 & \text{(1 place)} \\ \hline 2123 & \\ 1232 & \\ \hline 14.443 & \text{(3 places)} \end{array}$$ •

Some people have difficulty ordering numbers involving decimals. With whole numbers, a two-digit number was always greater than a one-digit number. The same is not true for decimals. One suggestion for comparing the size of two decimal numbers is to place one just above the other with the decimal points lined up vertically. If the whole parts are equal, then one scans both decimals from left to right until one notes a place where corresponding digits differ. 2.325176 < 2.3255, because for

2.325176
2.3255

1 is less than 5 in the fourth decimal place.

Division by decimals is difficult for many people. Let's consider 4.26 ÷ 2. Write 4.26 as $\frac{426}{100}$ and then

$$\frac{426}{100} \div 2 = \frac{\overset{213}{\cancel{426}}}{100} \cdot \frac{1}{\underset{1}{\cancel{2}}} = \frac{213}{100} = 2.13$$

Think of this division as

$$\begin{array}{r} \text{quotient} \\ \text{divisor}\overline{)\text{dividend}} \end{array} \quad \text{or} \quad \begin{array}{r} 2.13 \\ 2\overline{)4.26} \end{array}$$

Division of decimals

> When a decimal is divided by a whole number, the decimal point in the quotient is placed directly above the decimal point in the dividend.

Any division of decimals may be changed to a division where the divisor is a whole number.

Example 70 • Divide 106.08 by 1.7.

Solution $106.08 \div 1.7$ can be expressed as

$$\frac{106.08}{1.7}$$

Then, by the Fundamental Law of Fractions,

$$\frac{106.08}{1.7} = \frac{(106.08)(10)}{(1.7)(10)} = \frac{1060.8}{17}$$

```
      62.4
17 ) 1060.8
     102
      40
      34
       68
       68
```
•

In the preceding example, to convert $106.08 \div 1.7$ to division by a whole number, we multiplied both 106.08 and 1.7 by 10. Observe that this is equivalent to moving the decimal point one place to the right in both the dividend and the divisor. **In general, if there are *r* decimal places in the divisor, we move the decimal point *r* places to the right in the dividend and divide by the divisor as a whole number.**

Example 71 • Find $0.7371 \div 0.21$.

Solution Using the Fundamental Law of Fractions

$$\frac{0.7371 \cdot 100}{0.21 \cdot 100} = \frac{73.71}{21}$$

```
         3.51
0.21 ) 0.7371     (2 places)
         63
         107
         105
           21
           21
```
•

Now let us turn to the problem of finding the decimal representation of a fraction.

Example 72 • Change to decimals:

(a) $\frac{3}{8}$ (b) $\frac{1}{3}$ (c) $\frac{3}{11}$

Solution

(a)
$$\begin{array}{r} 0.375 \\ 8\overline{)\,3.000} \\ \underline{24} \\ 60 \\ \underline{56} \\ 40 \\ \underline{40} \\ 0 \end{array}$$

(b)
$$\begin{array}{r} 0.333 \\ 3\overline{)\,1.000} \\ \underline{9} \\ 10 \\ \underline{9} \\ 10 \\ \underline{9} \\ 1 \end{array}$$

(c)
$$\begin{array}{r} 0.2727 \\ 11\overline{)\,3.0000} \\ \underline{22} \\ 80 \\ \underline{77} \\ 30 \\ \underline{22} \\ 80 \\ \underline{77} \\ 3 \end{array}$$ •

Notice in part (a) of the previous example that a remainder of 0 is obtained, and division is no longer possible, so the decimal is said to **terminate.** In part (b), since a remainder of 1 repeats, and in part (c), since remainders of 3 and 8 repeat, it should be clear that a remainder of zero will never be attained. These decimal fractions are called **nonterminating.** However, these nonterminating decimals have interesting properties. In the quotient of part (b), the 3 repeats, whereas in the quotient of part (c), the pair of digits 27 repeats. Such decimals are called **repeating decimals.**

Repeating decimal

> A repeating decimal is a nonterminating decimal in which (after a certain point) the same pattern of numbers repeats endlessly.

Example 73 •

0.36 is a terminating decimal.
0.846333 is a terminating decimal.
0.242424 . . . is a repeating decimal.
0.4271271 . . . is a repeating decimal.
0.398210321032103 . . . is a repeating decimal. •

Notice that each of the last three decimal representations in the preceding example has repeating blocks of digits. The fact that a block of digits is repeated indefinitely is denoted by a bar above the digits being repeated.

Example 74 •

(a) $\dfrac{2}{3} = 0.6666\ldots = 0.\overline{6}$

(b) $\dfrac{1}{7} = 0.142857142857\ldots = 0.\overline{142857}$

(c) $\dfrac{19}{444} = 0.04279279\ldots = 0.04\overline{279}$ •

If the division process does not yield a terminating decimal, it will yield a repeating decimal. For example perform the long division of 1.0000000 by 7. The first remainder after subtraction is 3, then 6, then 4, then 5, then 1, then 3. Notice there are only six digits that can occur before a repetition. For a/b, in simplest or reduced form, the largest number of possible remainders without a repetition is $b - 1$. Beyond this the answer is a repetition of previous remainders. Therefore, if the division process does not yield a terminating decimal, it will yield a repeating decimal.

Decimal representation

> Every fraction can be represented by a decimal that either terminates or eventually repeats; conversely, every decimal that either terminates or eventually repeats represents a fraction.

Example 75 • Find a fraction representation for $0.\overline{27}$.

Solution Let $N = 0.272727\ldots$. We shall form a second equation (obtained from the first) so that when we subtract the first equation from the second, the repeating decimal part will become zeros. The block of repeating digits consists of "27." Since there are two digits in the block, we multiply by 100 to accomplish our goal in subtraction. (If there were three repeating digits, we would multiply by 1000, and so on.) Multiplying by 100 shifts the decimal point two places to the right, and we obtain

$$\begin{aligned} 100N &= 27\ .272727\ldots \\ -N &= -0.272727\ldots \\ \hline 99N &= 27 \end{aligned}$$

or

$$N = \frac{27}{99} = \frac{3}{11} \quad \bullet$$

It is interesting to note that every terminating decimal can be written as a repeating decimal. Subtract 1 from the last digit of the number and annex an infinite number of 9's.

Example 76 •

$$26.4 = 26.3\overline{9}$$

$$0.005 = 0.004\overline{9} \quad \bullet$$

To show that the preceding manipulation is correct, verify that $1 = 0.\overline{9}$. Let

$$\begin{aligned} N &= \ \ 0.9999\ldots \\ 10N &= \ \ 9.9999\ldots \\ -N &= -0.9999\ldots \\ \hline 9N &= \ \ 9 \\ N &= \ \ 1 \end{aligned}$$

So

$$1 = 0.\overline{9}$$

Example 77 • (Introductory Problem) Bill has the highest batting average on the Samford Bulldogs, 0.400 after 30 times at bat. In a slump, Bill gets only 2 hits in the next 8 times at bat. What is his new batting average?

Just for fun *Using decimal points and the four fundamental operations, arrange six 4's to obtain an answer of 10.*

Solution Out of 30 times at bat, Bill has

$$0.400(30) = 12 \text{ hits}$$

He now has $12 + 2 = 14$ hits out of

$$30 + 8 = 38 \text{ times at bat}$$

$14 \div 38 = 0.368$ (Bill's new average) •

Exercise set 6

R 1. Perform the following operations using the algorithms of this section.

(a) $7.62 \cdot 0.021$ (b) $4002 \cdot 7.61$
(c) $32.736 \div 12.4$ (d) $1.504 \div 0.32$

2. In Exercise 1, change each decimal to a fraction, perform the computation, and change the answer back to a decimal. Then check your answers with those in Exercise 1.

3. Perform the divisions by first writing each division as a fraction having a whole number denominator.

(a) $4.1\overline{)66.83}$ (b) $0.42\overline{)4.116}$
(c) $0.31\overline{)5.301}$ (d) $2.6\overline{)445.64}$

4. Determine which of the following division problems will have the same quotients without actually dividing. How do you know which quotients will be identical?

(a) $3.2\overline{)36}$ (b) $32\overline{)360}$
(c) $0.320\overline{)3.06}$ (d) $0.032\overline{)3.06}$
(e) $0.32\overline{)3.6}$ (f) $0.032\overline{)0.36}$

5. Perform the following operations.

(a) $(7.05)(0.006)$ (b) $(0.04)(6.011)$
(c) $(2.04)(3.25)$ (d) $(21.7)(74.65)$

6. Place one of the symbols $<$, $>$, or $=$ in each of the □ to make a true statement.

(a) 31.216 □ 31.20
(b) 0.6 □ 0.60
(c) 9.301 □ 9.30
(d) $2\frac{1}{4}$ □ 2.251
(e) 641.7125 □ 641.7131

7. Arrange in order from smallest to largest:

(a) 16.3, 16.259, 16, 16.2591
(b) 0.01365, 0.013, 0.0136, 0.013654
(c) 0.603, 0.063, 0.306, 0.036
(d) 2.402, 2.204, 2.420, 2.024

8. Write the following as repeating decimals.

(a) $\frac{1}{7}$ (b) $\frac{2}{11}$ (c) $\frac{3}{13}$

9. Write each decimal expression as a reduced fraction.

(a) 0.346 (b) $0.\overline{9}$ (c) $0.\overline{18}$
(d) $3.\overline{845}$ (e) $16.45\overline{9}$ (f) 146.1
(g) $1.2\overline{54}$ (h) $0.01\overline{79}$ (i) $6.\overline{8}$
(j) $3.\overline{25}$ (k) $6.\overline{01}$ (l) $4.\overline{247}$
(m) $0.0\overline{09}$ (n) $0.0\overline{132}$

T 10. (a) Write $\frac{2}{3}$ as a repeating decimal.
(b) Write $\frac{4}{6}$ as a repeating decimal.

(c) Conjecture a fact about repeating decimal representations of equivalent fractions.

11. (a) Show that $7.9\overline{9}$ is 8.
    (b) Show that $0.043\overline{9}$ is 0.044.
    (c) Starting with $10/3 = 3.33\overline{3}$, multiply both sides by 3. What is $9.99\overline{9}$?

In Exercises 12 through 15, first find an equation, then solve the equation, and then check your answer by "looking back" at the problem.

12. 7.05 gallons of a certain liquid weighs 40 pounds. How much does one gallon weigh?

13. Kelley's book satchel with books weighs 5.46 kg. It contains 2 books weighing 1.03 kg each and 3 notebooks weighing 0.72 kg each. How much does the satchel weigh?

14. The peel of an orange usually weighs $\frac{1}{20}$ of the total weight of the orange. You buy 50 pounds of oranges for \$19.60. How much do you pay for the peeling?

15. You travel 120 miles in 2 hours and 20 minutes, stop 30 minutes for lunch, and then travel 180 miles in 3 hours and 10 minutes. What is your average speed for the trip?

16. A seamstress purchased 2.40 meters of \$1.90-a-meter material; 5.40 meters at \$3.90 per meter; 7.5 meters at \$1 per meter; 25.20 meters of cording at 10¢ per meter; two spools of thread at 70¢ per spool; and 2.2 meters of pleater tape at 60¢ per meter.

    (a) What was the total purchase price for these goods?
    (b) With 5¢ per dollar sales tax, what was her total bill?

17. At the candy counter, Gertrude weighed all the candy and nuts she had in stock. She found that she had 6.25 kilograms of divinity at 50¢ per $\frac{1}{4}$ kilogram; 4.3 kilograms of fudge at 80¢ per $\frac{1}{4}$ kilogram; 1.8 kilograms of chocolate-covered peanuts at 90¢ per $\frac{1}{2}$ kilogram; and 5.27 kilograms of mixed nuts at 130¢ per $\frac{1}{2}$ kilogram.

    (a) What was the total weight of the candy and nuts?
    (b) What was the monetary value of the supply of each type of candy?
    (c) What was the monetary value of all the candy?

18. Diane needs $2\frac{3}{4}$ m of material at \$1.98 a meter, $4\frac{3}{4}$ m of material at \$2.59 a meter, 5 spools of thread at 70¢ each, and 2 zippers at 75¢ each. Will a 20-dollar bill cover the purchase?

19. Solve:

    (a) $2.16x + 7.3 = 9.05$
    (b) $0.02x - 17.6 = 41.5$

C 20. Perform the following operations.

    (a) $4.2_{\text{five}} \cdot 3.1_{\text{five}}$
    (b) $7.\text{ET}_{\text{twelve}} \cdot 3.\text{T}_{\text{twelve}}$
    (c) $514.6_{\text{seven}} \cdot 2.43_{\text{seven}}$
    (d) $10.101_{\text{two}} \cdot 0.011_{\text{two}}$
    (e) $1.01_{\text{two}} \cdot 1.11_{\text{two}}$
    (f) $102.2_{\text{three}} \cdot 0.21_{\text{three}}$

21. Find a fraction to represent each of the following repeating decimals.

    (a) $0.\overline{x}$
    (b) $0.\overline{xy}$
    (c) $0.\overline{xyz}$
    (d) $0.\overline{xyzw}$

7 Calculators, accuracy, and estimation

■ Myra has an annual salary of \$21,614. In your head, estimate her monthly salary. ■

Someone has said that we live in the age of the calculator or the computer. Throughout this book, we have inserted exercises to be worked using a calculator. In addition, special instructions as to the use of a calculator are found in the Student Activity Manual.

Since the calculator, and sometimes the computer, play such an important role in our lives, we need to be able to estimate whether the answers that appear on the screen are in any way reasonable. We learn in this section to estimate answers. At the same time, we look at the accuracy of answers when working with approximate numbers.

Consider an inexpensive hand-held calculator that can be bought for under $15. It will have keys for each of the arithmetical operations: $+$, $-$, $\cdot$ (or $\times$), and $\div$, and it will output its computations with a floating-point decimal display that is accurate from 6 to 12 places. Since there are many varieties of these machines, no attempt will be made to describe how any one of them is to be operated.

There are several rounding techniques built into the various computers and hand calculators. We briefly discuss two of them. One rounding process simply chops off all digits beyond the number to be displayed by the machine. For instance, if the machine displays six digits and the computation yields 1276.4374, the machine chops off the last two digits and displays 1276.43. Likewise, 2.000345 is displayed as 2.00034.

Example 78 • If 2431.62 is multiplied by 274.654 by hand, the answer is 667,854.15948. A machine that has eight display spaces and that rounds by chopping displays the answer as 667,854.15. •

Many machines round off to the nearest digit. This procedure is more precisely described by the following rules.

Rounding numbers

1. Drop all digits to the right of the place to which you are rounding.
2. If the first (left-most) digit to be dropped is 5, 6, 7, 8, or 9, increase the preceding digit by one.
3. If the first (left-most) digit to be dropped is 0, 1, 2, 3, or 4, leave the preceding digit as it is.

Example 79 • Round off the following numbers to two decimal places.

$$3.1671 \approx 3.17 \qquad 5.315 \approx 5.32$$

$$8.134 \approx 8.13 \qquad 7.245 \approx 7.25$$

$$7.1451 \approx 7.15 \qquad 3.1416 \approx 3.14$$

•

To round to the nearest *one,* look at the tenths digit; to the nearest *tenth,* the hundredths digit, to the nearest *hundredth,* the thousandths digit.

Example 80 • Round off number to the nearest 10.

$$146 \approx 150$$

$$946.3 \approx 950$$

$$24.61 \approx 20$$

•

It is clear from the preceding rules that if the approximation 5.84 is a rounded-off representation of the exact number N, then

$$5.835 \leq N < 5.845$$

One-half of the difference between 5.835 and 5.845 is called the **greatest possible error** and abbreviated as g.p.e. of 5.84.

Example 81 • Find the greatest possible error of the following:
(a) 5.84 (b) 2.1 (c) 2.111

Solution The greatest possible error is
(a) 0.005 (b) 0.05 (c) 0.0005 •

Scientists use exponential notation to express extremely large numbers and extremely small numbers, such as those involved in the velocity of light, the velocity of sound, the distance of a light year, or the mass of an electron. The use of exponents in *scientific notation* allows us to present large and small numbers in a shortcut form that can be written, read, and understood easily.

Scientific notation

> In scientific notation, a number N is written in the form
> $$N = A \cdot 10^n$$
> where $1 \le A < 10$ (such as 2.61) and n is an integral exponent.

To write a number in scientific notation we first move the decimal to place the number between 1 and 10. Then, by inspection we can note the power of 10 that makes equal the two expressions.

Example 82 • Change 432.7 to scientific notation.

Solution
$$1 \le 4.327 < 10$$
$$432.7 = 4.327(10)^x$$

We must multiply 4.327 by 100 to get 432.7. Thus
$$432.7 = 4.327(10)^2 \quad \bullet$$

To change a number from scientific notation to standard form, multiply by the power of 10.

Example 83 • Change $2.174(10)^{-3}$ to standard form.

Solution
$$2.174(10)^{-3} = \frac{2.174}{1000} = 0.002174 \quad \bullet$$

Example 84 • Change the following numbers from scientific notation to standard notation, using only your knowledge of the pattern of multiplying by powers of 10, not direct multiplication.
(a) $2.67 \cdot 10^4$ (b) $2.6705 \cdot 10^3$ (c) $9.2347 \cdot 10^1$
(d) $3.457 \cdot 10^{-7}$ (e) $5.55555 \cdot 10^5$ (f) $6.233 \cdot 10^{-2}$

Solution (a) 26,700 (b) 2670.5 (c) 92.347
(d) 0.0000003457 (e) 555,555 (f) 0.06233 •

Example 85 • Express 0.00012, 467,000, and $453(10)^{-2}$ in scientific notation.

Solution (a) $0.00012 = 1.2 \cdot (10)^{-4}$ (b) $467{,}000 = 4.67 \cdot (10)^5$
(c) $453(10)^{-2} = 4.53(10)^0$ •

The digits in A in the definition of scientific notation are called *significant digits* of the number. In the preceding example, the number 0.0012 has two significant digits.

Example 86 • $1.40 \cdot (10)^5$ has three significant digits.
$1.4 \cdot (10)^2$ has two significant digits.
$1.041 \cdot (10)^7$ has four significant digits. •

The following procedure uses the number of significant digits in the parts of a product (or quotient) to provide a system for rounding the answer.

Product of approximate numbers

When *approximate numbers are multiplied* (or divided), their product (or quotient) should be rounded off to have the same number of significant digits as the factor with the smallest number of significant digits.

Example 87 • (a) $(31.45) \cdot (56.2) = 1767.49$ rounds to 1770 with three significant digits, since 56.2 has only three significant digits.
(b) $(0.023) \cdot (0.45) = 0.01035$ rounds to 0.010 with two significant digits, since 0.45 has only two significant digits.
(c) $635 \div 0.27 = 2351.85$ rounds to 2400 with two significant digits, since 0.27 has only two significant digits. •

Additions and subtractions with approximate numbers can be no more accurate than the least accurate number in the computation. This result may be stated as follows.

Addition of approximate numbers

When *approximate numbers are added* (or subtracted), their sum (or difference) should be rounded to the same decimal place as the term with the greatest possible error.

Example 88 • When we add $3.15 + 0.732$ on a calculator, we obtain 3.882. Since 3.15 has the greatest possible error, we round 3.882 to 3.88.

$$23.4341 - 6.112 = 17.322$$

Why? •

It is inevitable that, on occasion, the user will input the wrong information into a calculator. Such a mistake will often yield a large error in the display. For this reason, an important skill for the calculator user is the ability to estimate the answer mentally in order to recognize large mistakes.

This idea is sometimes formalized by rounding off the larger numbers to one or two significant digits and ignoring those numbers that are small in comparison with the others. The computation can then be performed mentally, or quickly on paper, to check whether the order of magnitude of the answer displayed on the machine is reasonable.

Example 89 •

Subtract	*Check by rounding off to the nearest* 100
4137.6	4100
861.74	900
3275.86	3200

The answer seems reasonable. •

Example 90 •

Add	*Checking by rounding off to the nearest* 1000
4373	4000
79	00
5	0
7580	8000
7037	12,000

Undoubtedly, we have made a mistake. Can you find it? •

Example 91 • Multiply 12.47 and 0.623 and check by rounding each number to one significant digit.

Solution On a calculator, 12.47 ⊡ 0.623 = 7.76881. For the check: $10 \cdot 0.6 = 6$. The answer seems reasonable. •

Example 92 •

Divide	*Check*
4324 ⊟ 2.12 = 196.742	$4000 \div 2 = 2000$

The answer obtained from the calculator is evidently incorrect. •

Example 93 • Myra has an annual salary of $21,614. In your head, estimate her monthly salary.

Solution Since we can easily divide by 10, we approximate the number of months in a year as 10. A salary of

$$\$22,000/10 = \$2200 \text{ per month}$$ •

One good way of placing a decimal point in both multiplication and division is to round off to whole numbers and then estimate the answer.

Just for fun

Select an unusual number (say a decimal such as 26.2347). Insert this number in your calculator. Multiply the number by 0.625; add 1.375; multiply by 3.68; divide by 2.3; add 2.8; subtract the number you started with. Your final result will be 5. Can you explain this trick?

Example 94 • Use estimation to place the decimal point in the answer.

8.99 · 3.2 has an answer with digits 28768

Solution 9 · 3 = 27. Thus, the answer should be close to 27. Therefore, the answer is 28.768. •

Exercise set 7

Estimate each product in Exercises 1 through 4. Then perform the operation and see if your calculator answer makes sense relative to your estimate.

1. 8,316 · 7.53
2. 91.26 · 0.904
3. 428.6 · 4.21
4. 6534 · 0.00391

Estimate the sums or differences in Exercises 5 through 10. Then perform the operation and see if your calculator answer makes sense relative to your estimate.

5. 3,909 + 161.79
6. 5986 − 2653
7. 0.08469 − 0.00104
8. 631,428 − 241,586
9. 425,631 + 27,408
10. 726,410 − 14,190

Estimate each quotient in Exercises 11 through 16. Then perform the operation and see if your answer makes sense relative to your estimate.

11. 84,967 ÷ 25
12. 0.0012 ÷ 468
13. 0.016 ÷ 74
14. 95,640 ÷ 0.018
15. 14625 ÷ 0.0031
16. 0.0321 ÷ 0.0016.
17. Use the technique discussed in this section to classify the answers as probably correct or incorrect without actually performing the computations.
    (a) (7764) · (1172) = 9,099,408
    (b) 76 ÷ 0.011 = 2.2727273
    (c) (41) · (4444) = 18,204
    (d) 77.2 + 489.32 + 66.7 + 3.09 = 421.11
    (e) 777 ÷ 12 = 8.333333
    (f) (42.167) · (21.233) = 8.4231681
    (g) 743.111 − 6.222 = 736.889
18. Estimate the approximate size of each of the following computations and then find the answer by using a calculator.
    (a) 22.17 + 44.274 (b) 26.742 − 21.41
    (c) 2.1233 − 0.002422 (d) 1.0023 − 4.011
    (e) Add: 1763.4212
    112.43
    2.044
19. Estimate the approximate size of each of the following computations and then find the answer by using a calculator.
    (a) 7.223 · 4.2 (b) 215.26 ÷ 24.6
    (c) 0.66 ÷ 0.0022 (d) 8.1 · 0.0012
20. Write the following numbers in scientific notation.
    (a) 11 million (b) 33 billion
    (c) 0.0000033 (d) 67,300,000,000,000
    (e) 0.036754 (f) 3891
21. Determine the number of significant digits and the greatest possible error for each of the following.
    (a) 23.41 (b) $2.7 \cdot 10^{-3}$ (c) $0.301(10)^4$
    (d) 300.4 (e) $168(10)^3$ (f) 0.0012

22. Change these numbers from scientific notation to standard notation that does not use powers of 10.

(a) $4.6 \cdot 10^{-1}$ (b) $3.321 \cdot 10^{6}$
(c) $6 \cdot 10^{-6}$ (d) $7.12 \cdot 10^{-2}$
(e) $2.89 \cdot 10^{10}$
(f) $2.99776 \cdot 10^{10}$ cm per sec (velocity of light)

23. Round off the numerals shown to the nearest hundredth. Then round off the given numbers to the nearest tenth and finally to three significant digits.

(a) 91.4833 (b) 1.6651 (c) 0.8535
(d) 1400.176 (e) 273.871 (f) 436.436

24. For each of the following rounded numbers, what is the greatest possible original number? The least possible original number?

(a) 4000 (b) 4700 (c) 2.1 (d) 3.25

25. Place the decimal point in each by estimating.

(a) (15.2)(3.8) = 5776
(b) (7.32)(4.94) = 361608
(c) (.085)(.07) = 595
(d) (13.4)(0.003) = 402
(e) 339.48 ÷ 0.36 = 943
(f) 1.28 ÷ 3.2 = 4

T 26. Change each factor to scientific notation and then multiply or divide the approximate numbers.

(a) $0.00023 \cdot 526{,}000$ (b) $682{,}000 \div 0.00024$
(c) $2.71 \cdot (10)^{-3} \div 0.0002$
(d) $271 \cdot (10)^{-3} \cdot 1.6$

27. Perform the following additions and subtractions of approximate numbers, rounding off answers to the appropriate number of decimal places.

(a) $2.036 + 2.21 - 0.0072$
(b) $40.672 - 0.05 + 31.6$
(c) $17.6041 + 2.6 \cdot 10^{-2} - 0.014$

28. One liter (L) is approximately 1.06 qt. One kilogram (kg) is approximately 2.2 lb. Perform the following conversions on a calculator, expressing each answer to the correct number of significant digits.

(a) 8.21 lb = ________ kg
(b) 7.867 kg = ________ lb.
(c) 4.2 L = ________ qt.
(d) 7.2 qt. = ________ L

29. Use the calculator procedure of Exercise 20 (p. 233) to find the following.

$$\frac{a}{b} + \frac{c}{d} = \frac{\dfrac{ad}{b} + c}{d}$$

(a) $\frac{2}{3} + \frac{7}{19}$ (b) $\frac{5}{13} + \frac{2}{5}$

(c) $\frac{2}{9} + \frac{2}{11} + \frac{2}{15}$ (d) $\frac{3}{10} + \frac{3}{11} + \frac{3}{13}$

C 30. Devise a scheme to increase the accuracy of multiplication on a calculator. Then find (2,743,215,421) · (5,621,843), accurate to 15 significant digits.

Solution to introductory problem

Understand the problem. The number N of cars that the salesman had at the beginning of the week must be such that the cars sold each day (or the cars remaining on the lot at the end of each day) is a whole number. Secondly, N must be such that the number of cars in the lot on Friday is 11.

Devise a plan. The plan for solving this problem is to express the number of cars on the lot at the end of each day in terms of N. In the process, we shall look for a pattern to save time in computation. Then we shall set the expression for the number of cars on the lot Friday evening equal to 11 and solve the equation.

Carry out the plan. Number sold on Monday:

$$\frac{N}{2} + \frac{1}{2}$$

Number in lot at end of day Monday:

$$N - \left(\frac{N}{2} + \frac{1}{2}\right)$$

which reduces to

$$\frac{N}{2} - \frac{1}{2}$$

Number sold on Tuesday:

$$\frac{1}{3}\cdot\left(\frac{N}{2}-\frac{1}{2}\right)+\frac{1}{3}$$

Number in lot at end of day Tuesday:

$$\frac{N}{2}-\frac{1}{2}-\frac{1}{3}\left(\frac{N}{2}-\frac{1}{2}\right)-\frac{1}{3}$$

which reduces to

$$\frac{N}{3}-\frac{2}{3}$$

Do you suppose we have discovered a pattern? Check to see if the number in the lot at end of Wednesday is

$$\frac{N}{4}-\frac{3}{4}$$

If this is true, on Friday

$$\frac{N}{5}-\frac{4}{5}=11 \quad\text{or}\quad \frac{N}{5}-\frac{4}{5}=\frac{55}{5} \quad\text{or}\quad N=59$$

Looking back. Is

$$\frac{N}{2}-\frac{1}{2}$$

a whole number of cars? Yes,

$$\frac{59}{2}-\frac{1}{2}=29$$

Is

$$\frac{N}{3}-\frac{2}{3}$$

a whole number of cars? Yes,

$$\frac{59}{3}-\frac{2}{3}=19$$

Is

$$\frac{N}{4}-\frac{3}{4}$$

a whole number of cars? Yes,

$$\frac{59}{4}-\frac{3}{4}=14$$

Finally,

$$14-\left(\frac{1}{5}\text{ of }14+\frac{1}{5}\right)=11$$

Summary and review

The set of rational numbers

1. Fractions such as a/b are used to represent the following:
   (a) a division problem or a solution to a multiplication problem
   (b) a part of a whole
   (c) an element of a mathematical system
2. Fractions represent numbers called rational numbers.
3. Two fractions are equivalent if they represent the same (or equal) rational number(s).
4. Fundamental Law of Fractions. a/b is equivalent to ac/bc (or rational numbers a/b and ac/bc are equal).
5. The fraction a/b is in simplest form when $b > 0$ and the largest natural number that will divide both a and b is 1.
6. The fractions a/b and c/d ($b, d \neq 0$) are equivalent (or the rational numbers represented by a/b and c/d are equal) if and only if $ad = bc$.

Addition and subtraction of rational numbers

1. If a/b and c/b are rational numbers (note equal denominators), then $a/b + c/b = (a + c)/b$.
2. If a/b and c/d are any two rational numbers, then

$$\frac{a}{b}+\frac{c}{d}=\frac{ad+bc}{bd}$$

3. The least common denominator of two or more denominators of fractions is the l.c.m. of the denominators.
4. An improper fraction is one in which the numerator is numerically larger than or equal to the denominator. A fraction in which the numerator is numerically smaller than the denominator is called a proper

fraction. Improper fractions can be converted to whole numbers or to mixed numbers such as $3\frac{1}{4}$.

5. Subtraction of rational numbers.

$$\frac{a}{d} - \frac{b}{e} = \frac{ae - bd}{de}$$

Multiplication and division of rational numbers

1. $\frac{a}{b} \cdot \frac{c}{d} = \frac{ac}{bd}$
2. Every rational number x/y ($x \neq 0$, $y \neq 0$) has a multiplicative inverse y/x.
3. $(p/q) \div (r/s) = (p/q) \cdot (s/r)$ where s/r is the multiplicative inverse of r/s.

Properties of rational numbers with operations + and · (called field properties)

Closure properties

1. $\frac{a}{b} + \frac{c}{d} = \frac{ad + bc}{bd}$ is a unique rational number.
2. $\frac{a}{b} \cdot \frac{c}{d} = \frac{ac}{bd}$ is a unique rational number.

Commutative properties

3. $\frac{a}{b} + \frac{c}{d} = \frac{c}{d} + \frac{a}{b}$
4. $\frac{a}{b} \cdot \frac{c}{d} = \frac{c}{d} \cdot \frac{a}{b}$

Associative properties

5. $\left(\frac{a}{b} + \frac{c}{d}\right) + \frac{e}{f} = \frac{a}{b} + \left(\frac{c}{d} + \frac{e}{f}\right)$
6. $\left(\frac{a}{b} \cdot \frac{c}{d}\right) \cdot \frac{e}{f} = \frac{a}{b} \cdot \left(\frac{c}{d} \cdot \frac{e}{f}\right)$

Distributive property of multiplication over addition

7. $\frac{a}{b} \cdot \left(\frac{c}{d} + \frac{e}{f}\right) = \frac{a}{b} \cdot \frac{c}{d} + \frac{a}{b} \cdot \frac{e}{f}$
8. $\left(\frac{c}{d} + \frac{e}{f}\right) \cdot \frac{a}{b} = \frac{c}{d} \cdot \frac{a}{b} + \frac{e}{f} \cdot \frac{a}{b}$

Identity elements

9. There exists a unique rational number, $\frac{0}{1}$ (or 0), such that

$$\frac{a}{b} + \frac{0}{1} = \frac{0}{1} + \frac{a}{b} = \frac{a}{b}$$

10. There exists a unique rational number, $\frac{1}{1}$ (or 1), such that

$$\frac{a}{b} \cdot \frac{1}{1} = \frac{1}{1} \cdot \frac{a}{b} = \frac{a}{b}$$

Inverse elements

11. For each a/b, there exists a unique additive inverse $-a/b$ such that

$$\frac{a}{b} + \frac{-a}{b} = \frac{-a}{b} + \frac{a}{b} = \frac{0}{1}$$

and a unique multiplicative inverse b/a ($a/b \neq 0$) such that

$$\frac{a}{b} \cdot \frac{b}{a} = 1$$

Properties of decimals

1. To divide a number N by 10^k, where k is a natural number, move the decimal point (in N) k places to the left. In like manner, to multiply by 10^k, move the decimal point k places to the right.
2. An algorithm for adding (or subtracting) decimals consists of lining up the decimal points of the numbers in a vertical column, adding without reference to the decimal point, and then placing the decimal point in the answer directly underneath the given decimal points.
3. Multiplication of decimals. If there are r decimal places in one factor and s decimal places in a second factor, then there will be $r + s$ decimal places in the product.
4. When a decimal is divided by a whole number, the decimal point in the quotient is placed directly above the decimal point in the dividend.
5. A terminating decimal is one that repeats zeros after a specific number of decimal places.
6. A repeating decimal is a nonterminating decimal in which (after a certain point) the same pattern of digits repeats endlessly.
7. Every rational number can be represented by a decimal fraction that either terminates or eventually repeats; conversely, every decimal that either terminates or eventually repeats represents a rational number.
8. A rational number a/b (in reduced form) can be expressed as a terminating decimal fraction if and only if the prime factorization of b is of the form

$$b = 2^x \cdot 5^y$$

where x and y are whole numbers.

Rounding numbers

1. If the digit to the right of the last digit to be retained is less than 5, leave the last digit to be retained unchanged.
2. If the digit to the right of the last digit to be retained is 5 or greater, increase the last digit to be retained by 1.

Review exercise set 8

R 1. Write three fractions equivalent to 3/8.

2. Reduce each fraction to simplest form.

   (a) $\frac{24}{36}$ (b) $\frac{ax^2}{axy}$ (c) $\frac{45}{162}$

3. Compute $x + y$, $x - y$, $x \cdot y$ and $x \div y$ for the following pairs of fractions.

   (a) $x = \frac{4}{7}$ and $y = \frac{2}{5}$

   (b) $x = \frac{11}{9}$ and $y = \frac{4}{5}$

   (c) $x = \frac{9}{10}$ and $y = \frac{3}{4}$

   (d) $x = \frac{4}{3}$ and $y = \frac{5}{6}$

4. Perform each of the following computations.

   (a) $\frac{3}{16} \div \frac{4}{6}$ (b) $4\frac{1}{3} \cdot 5\frac{2}{5}$
   (c) $(6\frac{1}{4} - 2\frac{1}{3}) \cdot 7\frac{1}{4}$ (d) $0.216 \cdot 7.4$
   (e) $4.0001 - 1.64$ (f) $8.757 - 2.1$

5. Find the multiplicative inverse of each of the following.

   (a) $\frac{5}{7}$ (b) 15 (c) 0 (d) $\frac{1}{3} - \frac{1}{4}$

6. Using a least common denominator, find

$$\frac{3}{8} - \frac{1}{6} + \frac{11}{18} - \frac{7}{24}$$

7. Put in simplest form a/b.

   (a) $\dfrac{4\frac{1}{2} - 1\frac{3}{4}}{6\frac{5}{6} - 1\frac{7}{8}}$ (b) $\dfrac{4\frac{3}{4} \cdot 2\frac{5}{6}}{8\frac{1}{2}}$

8. Find a terminating or repeating decimal for the following.

   (a) $\frac{3}{8}$ (b) $\frac{5}{18}$ (c) $\frac{15}{24}$

9. Determine the fraction a/b that corresponds to each of the following repeating decimals.

   (a) $14.\overline{23}$ (b) $0.072\overline{01}$ (c) $30.0\overline{12}$
   (d) $0.002\overline{46}$ (e) $0.01\overline{79}$ (f) $3.3\overline{51}$
   (g) $6.5\overline{0}$ (h) $3.66\overline{0}$ (i) $0.24\overline{9}$
   (j) $0.\overline{9}$

10. Perform the following computations and round off answers to the appropriate number of decimal places.

    (a) $3.015 + 2.2 - 0.0074$
    (b) $0.33 \cdot 0.0005$
    (c) $26.5 \div 0.0022$
    (d) $14.671 - 3.1 + 0.0007$

11. Solve the following equations on the domain of rational numbers.

    (a) $4x + 2 = -10$ (b) $5y + -3 = 12$

12. Solve the following inequalities on the domain of rational numbers.

    (a) $5x + 6 < -4$
    (b) $-3x + 2 > -x + 4$

T 13. If $x = 2$, determine the reduced form for each of these rational numbers.

    (a) $\frac{6}{x^2 + 4}$ (b) $\frac{a - 1}{2a - x}$

    (c) $\frac{a + 1}{3(x + 2)}$ (d) $\frac{6x}{3 + (x - 2)}$

    (e) $\frac{2x^2 - 3x - 2}{3 + x^2}$

14. Replace x with a whole number so that each statement is true.

    (a) $\frac{0}{3} + \frac{2}{5} = \frac{2}{x}$ (b) $\frac{1}{8} + \frac{3}{5} = \frac{29}{x}$

    (c) $\frac{3}{5} + \frac{x}{4} = \frac{17}{20}$ (d) $\frac{x}{17} - \frac{2}{3} = \frac{242}{51}$

    (e) $\frac{4}{3} + \frac{x}{4} = \frac{25}{12}$ (f) $\frac{5}{8} + \frac{7}{3} = \frac{71}{x}$

15. Find the rational number solution for each of the following.

(a) $\frac{x}{3} = \frac{-18}{54}$ (b) $2x + -4 = 16$

(c) $-3x - 7 = 41$ (d) $4x - 3 = 5$

(e) $\frac{x}{100} = \frac{480}{54}$ (f) $-3x + 7 = -5$

(g) $\frac{-3x}{2} + 6 = \frac{1}{2}x - \frac{1}{3}$

(h) $\frac{-8}{5} - \frac{3x}{10} = \frac{-3}{5}x + \frac{7}{10}$

16. A will provides that an estate is to be divided among a wife and three children. The wife is to receive \$4000, and the remainder is to be divided four ways, with the wife receiving twice each child's share. If the estate is valued at \$24,000, how much does each receive?

17. Pam is holding 55¢ in her hand. If she has twice as many nickels as quarters and one more dime than quarters, how many coins does she hold? How many coins of each type does she have?

18. To produce a new record, Shoe String Recording Enterprise must make an initial investment of \$600. Then each record produced costs an additional \$2 for manufacturing, sales, and royalties. How many copies of a record can be manufactured if Shoe String Recording Enterprise has available \$1500 of capital?

19. Canoeing enthusiasts find that they can paddle upstream at 4 km per hr and downstream at 12 km per hr. How long can they paddle upstream if they must return to the same spot 4 hours after departing?

20. Two backpackers start toward each other from 2 points on the Appalachian Trail. Initially they are 12 km apart. If one walks 4 km per hr and the other walks 2 km per hr., how long will it be until they meet?

C 21. Find the value in base ten.

(a) 1.0101_{two} (b) 6.54_{seven}
(c) 3.02_{four} (d) $\text{TE}.9_{\text{twelve}}$

22. Express as a decimal in base seven.

(a) $\frac{1}{2}$ (b) $\frac{1}{3}$ (c) $\frac{1}{4}$

23. (a) How do you determine, without division, when a fraction is a terminating decimal in base ten?
(b) Can you find a similar rule in base six?
(c) Can you find a similar rule in base eight?

Bibliography

1. Allison, Joe Frank. "A Picture of the Rational Numbers: Dense but Not Complete." *MT* (January 1972), 87–89.

2. Ballew, Hunter. "Of Fractions, Fractional Numerals, and Fractional Numbers." *AT* (May 1974), 442–444.

3. Bell, Kenneth M., & Donald D. Rucker. "An Algorithm for Reducing Fractions." *AT* (April 1974), 299–300.

4. Chiosi, Lou. "Fractions Revisited." *AT* (April 1984), 46–47.

5. Dana, Marcia E., & Mary Montgomery Lindquist. "Let's Do It: From Halves to Hundredths." *AT* (November 1978), 4–8.

6. Ettline, J. Fred. "A Uniform Approach to Fractions." *AT* (March 1985), 42–44.

7. Gardner, Martin. *Aha!* New York, N.Y.: W. H. Freeman, 1978.

8. Green, George F., Jr. "A Model for Teaching Multiplication of Fractional Numbers." *AT* (January 1973), 5–9.

9. Harris, V. C. "On Proofs of the Irrationality of $\sqrt{2}$." *MT* (January 1971), 19–21.

10. Mielke, Paul T. "Rational Points on the Number Line." *MT* (October 1970), 475–479.

11. Moulton, J. Paul. "A Working Model for Rational Numbers." *AT* (April 1975), 328–332.

12. Ness, Harald M., Jr. "Another Look at Fractions." *AT* (January 1973), 10–12.

13. Prielipp, Robert W. "Decimals." *AT* (April 1976), 285–288.

14. Skypek, Dora Helen. "Special Characteristics of Rational Numbers." *AT* (February 1984), 10–12.

15. Sweetland, Robert D. "Understanding Multiplication of Fractions." *AT* (September 1984), 48–52.

16. Trafton, Paul R., & Judith S. Zawojewski. "Teaching Rational Number Division: A Special Problem." *AT* (February 1984), 20–22.

17. Van de Walle, John, & Charles S. Thompson. "Fractions with Fraction Strips." *AT* (December 1984) 4–9.

CHAPTER 6

Real Numbers and a Return to Problem Solving

Three men attending a convention rented a room at a nearby hotel for \$30, so they paid \$10 each. However, the desk clerk realized that the room only cost \$25, so he sent the bellboy upstairs with \$5 to return to the three men. The bellboy, rewarding himself with a \$2 tip, returned only \$3 to the men. Yet if each man received \$1, then each man paid \$9 for the room. But 3(\$9) + \$2 = \$27 + \$2 = \$29. What happened to the extra dollar?

Mathematics has been a part of the language of the marketplace ever since the first nomad tried to decide whether to trade two goats for three woolen robes or three goats for four woolen robes. The abacus and other simple computing devices were inspired by the needs of problem solving in the marketplace. As a citizen of the modern world, you have replaced the abacus with a hand calculator or a computer, but you must still master the language of the marketplace—discount, percent, markup, finance charge, simple interest, and so on. We introduce you to these concepts in this chapter. All of your computations in solving equations and in working problems will be performed on the real number system. Therefore, we introduce this chapter with a discussion of the real number system.

1 The real number system

■ A certain number, approximated to four decimal places, is 2.3135. Can you tell whether the original number is a rational number? ■

This problem will be given as Exercise 5 in the exercise set at the end of this section. The answer will be easy to obtain after you study this section.

Do you know of decimals that neither terminate nor repeat? If you do, then you are aware of numbers that are not rational numbers. In fact, you have discovered what we call *irrational numbers.* The rational numbers and the irrational numbers make up what we call the **real number system.**

In this section, you will learn about irrational numbers. Then you will study some interesting properties of real numbers.

The fact that every rational number (expressed as a fraction) can be represented by a terminating or a repeating decimal was shown in Chapter 5; conversely, every terminating decimal or repeating decimal represents a rational number. This situation implies that if a number in decimal form neither terminates nor repeats, then it must represent a number other than a rational number. A decimal of this type represents what is called an **irrational number.**

If we construct an infinite decimal with no repeating cycle of digits, we shall have an irrational number. There are many. Consider

$$0.737737773777737777773\ldots$$

where the number of 7's between successive 3's increases by 1 each time. This number is irrational because it is a nonterminating, nonrepeating decimal. It is nonrepeating because each pair of consecutive 3's has one more 7 between them than the preceding pair. Probably the most famous irrational number is pi (π). The number π is not $3\frac{1}{7}$ or 3.1416; π is an unending, nonrepeating decimal whose first few digits are given by 3.14159265358979323846.

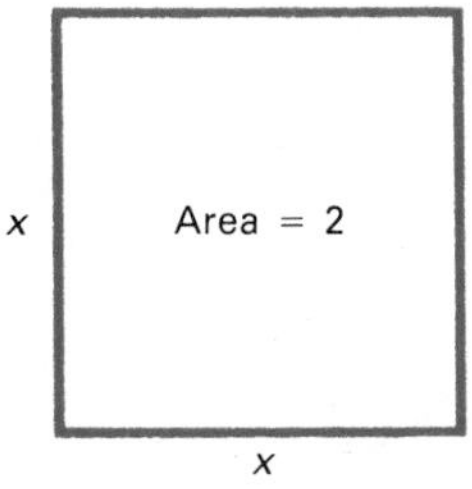

x | Area = 2

x

Figure 1

The ancient Greeks spent a long time debating whether numbers other than rational numbers existed. In fact, for many years the Pythagoreans, a group of religious mystics and mathematicians, asserted that indeed there were no numbers that were not rational. But then one day they began to ponder the question: How long is the side of a square whose area is 2? (See Figure 1.)

The area of a square is given by multiplying the length of a side times itself; so $x \cdot x = 2$. The question is: What number multiplied by itself is equal to 2 (or what is the square root of 2, denoted by $\sqrt{2}$)? They finally proved that $\sqrt{2}$ is not a rational number (as recorded in Euclid's *Elements*).

Example 1 • Prove that $\sqrt{2}$ is an irrational number.

Solution We shall assume that $\sqrt{2}$ is a rational number and show that this leads to a contradiction. Thus, $\sqrt{2}$ must be irrational. Assume that $\sqrt{2}$ is a rational number. Then it can be written as a quotient of two integers a/b.

If

$$\frac{a}{b} = \sqrt{2}$$

then

$$\left(\frac{a}{b}\right)^2 = 2 \qquad \text{and} \qquad a^2 = 2b^2$$

Factor a into prime factors. One of these prime factors will be 2, since $a^2 = 2b^2$. Since each factor of a occurs twice in a^2, a^2 contains an even number of 2's as factors. At the same time, $2b^2$ must contain an odd number of 2's as factors. But the Fundamental Theorem of Arithmetic states that the factorization must be unique. Since we have reached an obvious contradiction, our hypothesis that $\sqrt{2}$ is a rational number must be false. Thus, $\sqrt{2}$ is irrational. In fact, it can be proved that square roots of all positive integers except the perfect squares $\{1, 4, 9, 16, \ldots\}$ are irrational. •

We have stated that every rational number can be represented by either a terminating decimal or a repeating decimal and that an irrational number can be represented by a nonrepeating decimal. The union of these two sets forms the real number system.

A real number

A real number is any number that is either a rational number or an irrational number.

$$\text{Real numbers} = \{\text{rational numbers}\} \cup \{\text{irrational numbers}\}$$

We have previously shown that the rational numbers are dense and that there are infinitely many rational numbers between two given rational numbers. Thus, it seems that the rational numbers may "fill up" the number line. But this is not correct. In fact, we demonstrate now that the irrational numbers are also represented by points on the number line.

Place one side of a unit square on the number line as indicated in Figure 2. In Chapter 10, we shall learn that the sum of the squares of two sides of the right triangle is equal to the square of the side opposite the right angle. Therefore, the diagonal of the unit square is of length $\sqrt{1^2 + 1^2} = \sqrt{2}$. Consider the diagonal with one end at 0, and rotate this diagonal clockwise about the point 0 until it lies on the number line; the end will mark a point on the number line that is a

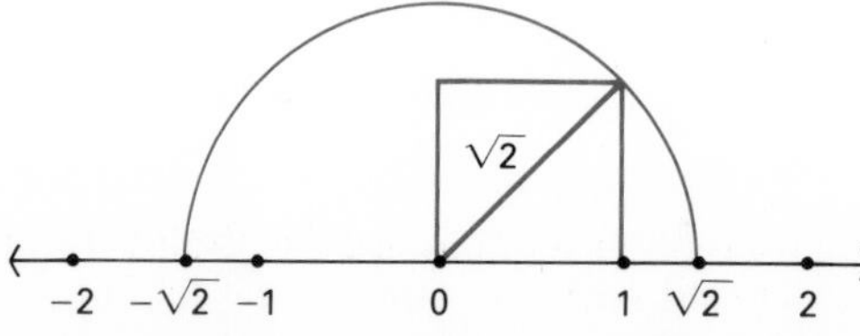

Figure 2

distance of $\sqrt{2}$ from the origin, a point that represents the irrational number $\sqrt{2}$. Similarly, if we rotate the diagonal counterclockwise about the origin, we can mark the point $-\sqrt{2}$. So there are points on the number line ($\sqrt{2}$ and $-\sqrt{2}$, for example) that cannot be represented by rational numbers.

Let us translate our unit square so that its lower left corner is at $\frac{3}{2}$ (Figure 3). Next, rotate the diagonal in both directions until it coincides with the number line. The points located at the ends of the diagonal correspond to $\frac{3}{2}+\sqrt{2}$ and $\frac{3}{2}-\sqrt{2}$. The question arises: Is $\frac{3}{2}+\sqrt{2}$ a rational number or is it an irrational number? If we assume it to be a rational number p/q, then $\frac{3}{2}+\sqrt{2}=p/q$ or $\sqrt{2}=p/q-\frac{3}{2}$. Since the rational numbers are closed under the operations of addition and subtraction, $p/q-\frac{3}{2}$ is a rational number, which implies that $\sqrt{2}$ is a rational number. This, of course, is a contradiction; consequently, $\frac{3}{2}+\sqrt{2}$ is an irrational number. This example illustrates that the sum of an irrational number and a rational number is an irrational number.

Figure 3

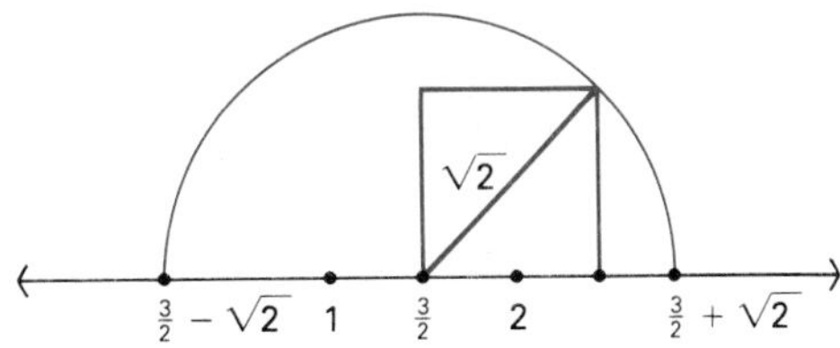

A similar argument can be used to show that the product of a nonzero rational number ($\neq 0$) and an irrational number is irrational. Consider the product of 5 and $\sqrt{2}$. Assume this product is rational, which means it can be written as

$$5\sqrt{2}=\frac{a}{b}$$

$$\frac{1}{5}(5\sqrt{2})=\frac{1}{5}\left(\frac{a}{b}\right)$$

$$\sqrt{2}=\frac{a}{5b}$$

Now $a/5b$ is a rational number, which means that $\sqrt{2}$ is rational. This is a contradiction. Thus, the assumption is incorrect. Therefore, the product of a nonzero rational number and an irrational number is irrational.

Example 2 • Show that $3/\sqrt{2}$ is irrational.

Solution

$$\frac{3}{\sqrt{2}}=\frac{3\sqrt{2}}{\sqrt{2}\cdot\sqrt{2}}=\frac{3\sqrt{2}}{2}=\left(\frac{3}{2}\right)\sqrt{2}$$

which is irrational since the product of a nonzero rational and an irrational number is irrational. •

Just for fun

A rubber ball is known to rebound half the height it drops. If the ball is dropped from a height of 100 feet, how far will it have traveled by the time it hits the ground for

1. *the first time?*
2. *the second time?*
3. *the third time?*
4. *the fourth time?*
5. *the fifth time?*
6. *Look for a pattern.*

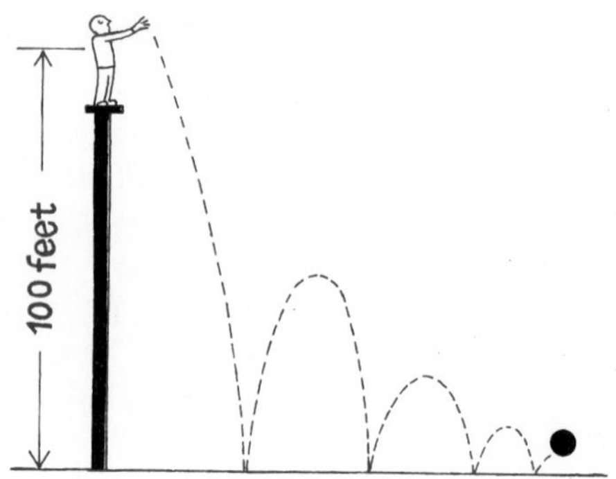

The geometric process of placing the left corner of the unit square at a rational number a/b and then finding $a/b + \sqrt{2}$ and $a/b - \sqrt{2}$ always yields an irrational number. So it is seen that there are as many irrational numbers of the form $a/b + \sqrt{2}$ or $a/b - \sqrt{2}$ as there are rational numbers a/b. The same argument could be used to show that $a/b + \pi$, $a/b + \sqrt{3}$, $a/b + \sqrt{5}$, and so on are irrational numbers and that there are as many in each class as there are rational numbers. What does this mean in terms of the number of irrational numbers?

The real numbers can be placed in one-to-one correspondence with all points on the number line; that is, every point on the number line represents a real number and every real number can be represented by a point on the number line. The rational numbers do not have this characteristic, since there are points on the number line that cannot be represented by rational numbers. Since every point on the number line represents a real number, we shall use the real number system to measure distances in Chapter 8.

In our extension to the system of real number, we demonstrated that there are numbers that cannot be represented as fractions. These numbers, called *irrational numbers,* obey the same rules for the operations of addition, subtraction, multiplication, and division as rational numbers because both were combined into the system of real numbers through the use of unending decimals. This presentation, along with geometric plausibility, illustrates the fact that the real number system satisfies the following properties.

Real number system

1. It is closed under addition.
2. It is closed under subtraction.
3. It is closed under multiplication.
4. It is closed under division—division by zero being excluded.
5. The elements of the set satisfy the commutative and associative properties of addition and multiplication.
6. The elements of the set satisfy the distributive property of multiplication over addition.

7. There exist an additive identity and additive inverses, and a multiplicative identity and multiplicative inverses, except for 0.
8. It is an ordered set.
9. It is dense.

In Figure 4, you note that every negative integer is also an integer, a rational number, and a real number. Fractions that are not integers make up a subset of the rational numbers, which is a subset of the real numbers. Unless stated otherwise, all of the computations in the remainder of the book will be on the system of real numbers.

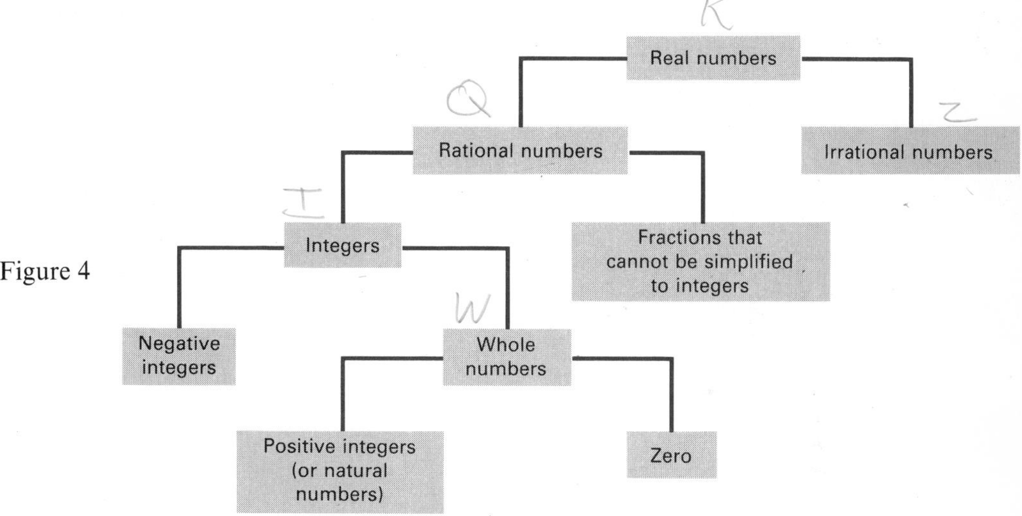

Figure 4

Exercise set 1

R 1. Classify the following numbers as rational or irrational.

(a) $\sqrt{3} - 1$ (b) $6 \cdot \sqrt{3}$ (c) $\dfrac{3}{\sqrt{2}}$

(d) $\dfrac{\sqrt{2}}{4\sqrt{2}}$ (e) $2 \div \sqrt{2}$ (f) $\sqrt{2} + 7$

(g) $\dfrac{\sqrt{2}}{\sqrt{3}}$ (h) $2 - \sqrt{3}$ (i) $\dfrac{57}{1001}$

(j) 0.253125312531 . . .
(k) 26.1311311131111 . . .
(l) 5π (m) 0
(n) 3.1416 (o) 22/7

2. If $R = \{\text{real numbers}\}$, $I = \{\text{integers}\}$, $Q = \{\text{rationals}\}$, $W = \{\text{whole numbers}\}$, and $Z = \{\text{irrationals}\}$, classify the following as true or as false.

(a) $R \cap I = R$ (b) $R \cup Q = Q$
(c) $I \cap Q = Q$ (d) $Q \cap W = W$
(e) $Q \cap Z = \varnothing$ (f) $Q \subset R$
(g) $Q \subset I$ (h) $I \subset Q$
(i) $R \subset Z$ (j) $I \subset (Q \cap R)$
(k) $I \cap (Q \cup Z) = I$
(l) A repeating, nonterminating decimal is a rational number.
(m) A nonrepeating, nonterminating decimal is an irrational number.
(n) $Q \cup Z = R$

(o) Z is closed under addition.
(p) Z is closed under multiplication.
(q) R is closed under both addition and multiplication.

3. Complete the following chart.

Number	Whole Number	Integer	Rational Number	Irrational Number	Real Number
-5	No	Yes	Yes	No	Yes
$2.\overline{71}$	No	No	Yes	No	Yes
0.717117111 . . .					
0					
$5\sqrt{2}$					
3π					
$0.\overline{45}$					
4					

4. (a) Is there a greatest whole number less than 8?
(b) Is there a greatest integer less than 8?
(c) Is there a greatest rational number less than 8?
(d) Is there a greatest irrational number less than 8?
(e) Is there a greatest real number less than 8?
(f) What property is involved in these questions?

5. A certain number, approximated to four decimal places, is 2.3135. Can you tell whether the original number was rational?

6. Show with examples each of the following.
(a) The product of irrational numbers may be rational or irrational.
(b) The sum of irrational numbers may be rational or irrational.
(c) The difference of two irrational numbers may be rational or irrational.
(d) The quotient of two irrational numbers may be rational or irrational.

7. Demonstrate with an example that the sum of an irrational number and a rational number is irrational.

8. Demonstrate with an example that the product of an irrational number and a nonzero rational number is irrational.

9. (a) How many integers are between -2 and 3?
(b) How many rational numbers are between -2 and 3? Name three of them.
(c) How many irrational numbers are between -2 and 3? Name three of them.

10. (a) How many counting numbers are between 0 and 10?
(b) How many integers are between 0 and 10?
(c) How many rational numbers are between 0 and 10?
(d) How many irrational numbers are between 0 and 10?
(e) How many real numbers are between 0 and 10?

T 11. Classify as true or false.
(a) $\sqrt{x} = -8$ for some real x.
(b) $\sqrt{17} = 4.12\overline{3}$
(c) $\pi = 22/7$
(d) $\sqrt{-x} = 4$ for some real x

12. Find two irrational numbers between 0.24 and 0.25.

13. Find two irrational numbers between $0.\overline{46}$ and $0.\overline{47}$.

14. Let $a = 0.565\overline{2}$ and $b = 0.5652020020002 \ldots$.
(a) Find a rational number between a and b.
(b) Find an irrational number between a and b.

15. To eight decimal places, $\pi = 3.14159265 \ldots$. In what decimal place does the decimal representation for each fraction first differ from π?

(a) $\dfrac{22}{7}$ (b) $\dfrac{333}{106}$ (c) $\dfrac{355}{113}$

(d) $\dfrac{31{,}412}{9999}$ (e) $\dfrac{1046}{333}$

16. Arrange the following real numbers from smallest to largest.

0.45, $0.\overline{4}$, 0.44, 0.454454445 . . . , $0.\overline{45}$, 0.455

C 17. If x and y are positive integers such that $x > y$, which one of the following is true: $1/x > 1/y$, $1/x = 1/y$, $1/x < 1/y$? Why?

18. Demonstrate with several positive real numbers x that $x + 1/x \geq 2$.

19. Prove that each of the following is an irrational number.

(a) $3\sqrt{2}$ (b) $3 + \sqrt{2}$ (c) $\dfrac{3}{\sqrt{2}}$

(d) $\sqrt{2} - 3$ (e) $3 - 2\sqrt{2}$ (f) $\dfrac{3}{2} + \sqrt{2}$

(g) $2 + \sqrt{3}$ (h) $\sqrt[3]{2}$

20. Geometrically construct a representation of $\sqrt{3}$ on the number line.

21. Geometrically construct a representation of $\sqrt{4}$ on the number line and note that it is 2.

22. Use base three notation to prove that $\sqrt{2}$ is not rational. (*Hint:* The definition of a rational number still holds in base three. Now look at the units digit of the square of any integer in base three.)

*2 Working with irrational numbers

■ Write an irrational number that is greater than $\frac{2}{3}$ but less than $\frac{3}{4}$. ■

The number of irrational numbers is unlimited. In this section, we will practice locating irrational numbers between rational numbers. We will also discover that some properties of rational numbers do not hold for irrational numbers.

As noted, the square roots of natural numbers can be either rational or irrational. The **principal** or **positive** square root of a number n is the positive number that, when multiplied by itself, gives n.

Example • (a) $\sqrt{16} = 4$ because $4^2 = 16$.
(b) $\sqrt{100} = 10$ because $10^2 = 100$. •

16 and 100 in the preceding example are called **perfect squares.** Any number that is the square of a natural number is said to be a perfect square.

The number inside the **radical sign,** $\sqrt{\ }$, is called the **radicand.** If a number is multiplied by the radical, as in $3\sqrt{5}$, this number is called the **coefficient of the radical.** Three is the coefficient in $3\sqrt{5}$. 5 is the radicand.

Note that

$$x^{1/2} \cdot x^{1/2} = x^{1/2+1/2} = x^1 \qquad \text{and} \qquad (x^{1/2})^2 = x^{(1/2)(2)} = x^1$$

But $\sqrt{x} \cdot \sqrt{x}$ or $(\sqrt{x})^2$ was defined to be x. Thus, $x^{1/2} = \sqrt{x}$.

The preceding discussion leads to the following definition.

Exponentials

For any real number x and natural number n,

$$x^{1/n} = \sqrt[n]{x}$$

whenever $\sqrt[n]{x}$ is meaningful on the set of reals.

Let us apply the rules already developed,

$$x^m \cdot x^n = x^{m+n}$$

$$(x^m)^n = x^{mn}$$

$$(xy)^m = x^m y^m$$

$$\left(\frac{x}{y}\right)^m = \frac{x^m}{y^m}$$

to the real numbers in the preceding definition.

Radicals

Since $(xy)^{1/n} = x^{1/n}y^{1/n}$, then $\sqrt[n]{xy} = \sqrt[n]{x}\sqrt[n]{y}$.

Since $\left(\frac{x}{y}\right)^{1/n} = \frac{x^{1/n}}{y^{1/n}}$, then $\sqrt[n]{\frac{x}{y}} = \frac{\sqrt[n]{x}}{\sqrt[n]{y}}$.

Since $(x^{1/n})^m = x^{m/n}$, then $(\sqrt[n]{x})^m = \sqrt[n]{x^m}$.

Example 3 • Find the value of
(a) $8^{1/3}$
(b) $25^{-1/2}$

Solution (a) $8^{1/3} = \sqrt[3]{8} = 2$

(b) $25^{-1/2} = \dfrac{1}{25^{1/2}} = \dfrac{1}{\sqrt{25}} = \dfrac{1}{5}$ •

An irrational number is said to be in **simplest form** if everything possible has been removed from the radicand.

Example 4 • Write in simplest form (a) $\sqrt{300}$ and (b) $\sqrt{128}$.

Solution (a) $\sqrt{300} = \sqrt{100 \cdot 3} = \sqrt{100} \cdot \sqrt{3} = 10\sqrt{3}$

(b) $\sqrt{128} = \sqrt{64 \cdot 2} = \sqrt{64} \cdot \sqrt{2} = 8\sqrt{2}$

Note in each case that we wrote the number under the radical as a product of a perfect square times another number. •

Example 5 • Simplify $\sqrt[3]{54}$.

Solution

$$\begin{aligned}\sqrt[3]{54} &= \sqrt[3]{27 \cdot 2} \qquad \text{(Why did we select 27?)}\\ &= \sqrt[3]{27} \cdot \sqrt[3]{2}\\ &= 3\sqrt[3]{2}\end{aligned}$$ •

A fraction involving radicals is said to be in **simplest form** if the denominator contains no radical expression. When we remove radicals from denominators, we say that the denominator is **rationalized.**

Example 6 • Rationalize the denominator of $5/\sqrt{2}$.

Solution We need to multiply the numerator and denominator by a number that will make the denominator a perfect square.

$$\frac{5}{\sqrt{2}} = \frac{5\sqrt{2}}{\sqrt{2}\sqrt{2}} = \frac{5\sqrt{2}}{\sqrt{4}} = \frac{5\sqrt{2}}{2} \quad \bullet$$

Example 7 • Rationalize the denominator of $7/\sqrt{8}$.

Solution

$$\frac{7}{\sqrt{8}} = \frac{7\sqrt{2}}{\sqrt{8}\sqrt{2}} = \frac{7\sqrt{2}}{\sqrt{16}} = \frac{7\sqrt{2}}{4} \quad \bullet$$

The arithmetic operations of addition, subtraction, multiplication, and division can be performed with irrational numbers using the properties listed in this section. Using the distributive properties of multiplication over addition, we can add (or subtract) numbers with the same radical and radicand by adding their coefficients.

Example 8 • Add $3\sqrt{5} + 2\sqrt{5}$.

Solution

$$3\sqrt{5} + 2\sqrt{5} = (3 + 2)\sqrt{5} = 5\sqrt{5} \quad \bullet$$

Example 9 • Subtract $7\sqrt{3} - 2\sqrt{12}$.

Solution

$$\begin{aligned} 7\sqrt{3} - 2\sqrt{12} &= 7\sqrt{3} - 2\sqrt{4 \cdot 3} \\ &= 7\sqrt{3} - 4\sqrt{3}. \\ &= (7 - 4)\sqrt{3} \\ &= 3\sqrt{3} \quad \bullet \end{aligned}$$

Of course, we use $\sqrt[n]{a}\sqrt[n]{b} = \sqrt[n]{ab}$ to multiply irrational numbers expressed as radicals and $\sqrt[n]{\frac{a}{b}} = \frac{\sqrt[n]{a}}{\sqrt[n]{b}}$ to divide.

Example 10 • Simply $\sqrt{8} \cdot \sqrt{5}$.

Solution

$$\begin{aligned} \sqrt{8} \cdot \sqrt{5} &= \sqrt{8 \cdot 5} \\ &= \sqrt{40} \\ &= \sqrt{4 \cdot 10} \\ &= 2\sqrt{10} \quad \bullet \end{aligned}$$

Example 11 • Simplify $\dfrac{\sqrt{96}}{\sqrt{3}}$.

Solution

$$\begin{aligned} \frac{\sqrt{96}}{\sqrt{3}} &= \sqrt{\frac{96}{3}} \\ &= \sqrt{32} \\ &= \sqrt{16 \cdot 2} \\ &= \sqrt{16}\sqrt{2} \\ &= 4\sqrt{2} \quad \bullet \end{aligned}$$

In the introduction, we discussed placing irrational numbers between rational numbers. We illustrate two procedures.

Example 12 • Write an irrational number that is greater than $\frac{2}{3}$ and less than $\frac{3}{4}$.

Solution First, we write the two rational numbers as decimals.

$$\frac{2}{3} = 0.66666\overline{6}$$

$$\frac{3}{4} = 0.75$$

The irrational number 0.6676776777 . . . satisfies the requirement.

Now, $\frac{3}{4} - \frac{2}{3} = \frac{1}{12}$, and $\sqrt{2}/100$ is obviously less than $\frac{1}{12}$. Therefore, $\frac{2}{3} + \sqrt{2}/100$ also satisfies our requirement. •

In the preceding example, if we wished to place 10 irrational numbers between $\frac{2}{3}$ and $\frac{3}{4}$, we could write them as

$$\frac{2}{3} + \frac{\sqrt{2}}{1000}$$

$$\frac{2}{3} + \frac{2\sqrt{2}}{1000}$$

$$\frac{2}{3} + \frac{3\sqrt{2}}{1000}$$

$$\vdots$$

$$\frac{2}{3} + \frac{10\sqrt{2}}{1000}$$

Sometimes we need a rational approximation for the square root of a number. One procedure for finding a square root is to use the following algorithm.

Square root

> The steps in an algorithm for finding square roots to any desired accuracy are as follows.
> (i) Let G be a guess for the square root of X.
> (ii) Let $H = X/G$.
> (iii) Let $A = \frac{1}{2}(G + H)$.
> (iv) A is the approximate square root of X. If it is not as accurate as desired, let G be A in Step (i) and repeat the process.

Example 13 • Find the $\sqrt{39}$ by performing the suggested computation three times.

Just for fun *Write in a column any two natural numbers. Continue your vertical column with eight more numbers, each of these numbers being the sum of the preceding two. Find the sum of all ten numbers. Multiply the seventh number from the top by 11. Are the product and the sum the same?*

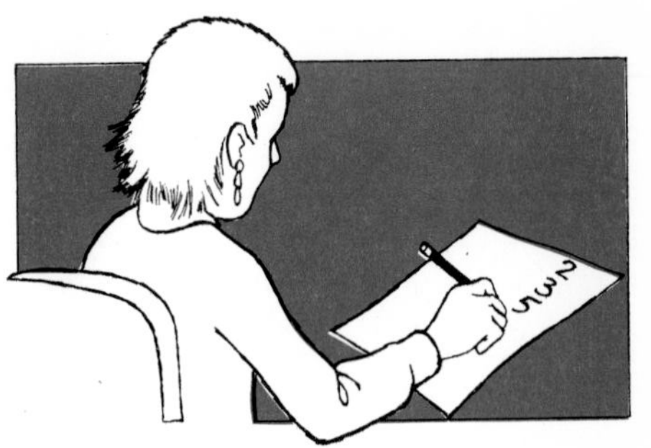

Solution Let's guess that the answer G is 6.

$$H = \frac{39}{6} = 6.5$$

$$A = \left(\frac{1}{2}\right)(6 + 6.5) = 6.25 = G'$$

$$H' = \frac{39}{6.25} = 6.24$$

$$A' = \left(\frac{1}{2}\right)(6.25 + 6.24) = 6.245 = G''$$

$$H'' = \frac{39}{6.245} = 6.24499$$

Therefore, $\sqrt{39}$ is approximately 6.24499. •

Exercise set 2

R 1. Evaluate the following:

(a) $\sqrt{49}$ (b) $\sqrt{121}$
(c) $\sqrt[3]{27}$ (d) $\sqrt[4]{16}$
(e) $-\sqrt{81}$ (f) $-\sqrt{196}$

2. Simplify the following radicals.

(a) $\sqrt{32}$ (b) $\sqrt{75}$ (c) $\sqrt{125}$ (d) $\sqrt{162}$

3. Perform the indicated operation.

(a) $\sqrt{5} + 2\sqrt{5}$ (b) $\sqrt{7} - 4\sqrt{7}$
(c) $3\sqrt{18} - 5\sqrt{2}$ (d) $4\sqrt{6} - \sqrt{24}$
(e) $4\sqrt{12} + 3\sqrt{48}$ (f) $\sqrt{20} + \sqrt{125}$

4. Write each of the following in simplest form.

(a) $\sqrt{50}$ (b) $\sqrt{384}$ (c) $\sqrt{\frac{147}{36}}$ (d) $\sqrt{\frac{450}{169}}$

5. Write without radicals.

(a) $\sqrt[3]{64}$ (b) $\sqrt[4]{81}$ (c) $\sqrt[5]{32}$ (d) $\sqrt{225}$

6. Perform the indicated operation and simplify.

(a) $\sqrt{5} \cdot \sqrt{2}$ (b) $\sqrt{6} \cdot \sqrt{3}$
(c) $\sqrt{18} \cdot \sqrt{12}$ (d) $\sqrt{5} \cdot \sqrt{20}$
(e) $\frac{\sqrt{9}}{\sqrt{3}}$ (f) $\frac{\sqrt{72}}{\sqrt{8}}$

(g) $\dfrac{\sqrt{240}}{\sqrt{6}}$ (h) $\dfrac{\sqrt{144}}{\sqrt{3}}$

7. Rationalize the denominator of each.

(a) $\dfrac{5}{\sqrt{2}}$ (b) $\dfrac{7}{\sqrt{5}}$

(c) $\dfrac{\sqrt{7}}{\sqrt{2}}$ (d) $\dfrac{\sqrt{14}}{\sqrt{6}}$

(e) $\dfrac{\sqrt{15}}{\sqrt{7}}$ (f) $\dfrac{\sqrt{10}}{\sqrt{6}}$

(g) $\dfrac{\sqrt{3}}{\sqrt{8}}$ (h) $\dfrac{\sqrt{5}}{\sqrt{18}}$

8. Simplify as much as possible.

(a) $\dfrac{4^{-2} \cdot 5^{-1}}{2}$ (b) $(10^{1/3})^{-6}$

(c) $\dfrac{32^{-3/5}}{16^{-1/4}}$ (d) $\dfrac{16^{-1/2}}{4^{-3}}$

T 9. Simplify the following:

(a) $\sqrt[3]{-81}$ (b) $\sqrt[4]{64}$

(c) $\sqrt[3]{-24}$ (d) $\sqrt[5]{-32}$

(e) $-\sqrt[3]{-8}$ (f) $\sqrt{\dfrac{-16}{-2}}$

C 10. Solve each of the following and locate approximately the solution on a number line.

(a) $4x + \sqrt{2} = 8$ (b) $\sqrt{2}x - 3 = 5$

(c) $2y + 3 = 4\sqrt{3}$ (d) $\sqrt{3}y + 2 = 7$

11. Place 4 irrational numbers between 0.1 and 0.2.

12. Place 10 irrational numbers between $\frac{1}{4}$ and $\frac{1}{3}$.

13. Using the square root algorithm three times, find the following.

(a) $\sqrt{37}$ (b) $\sqrt{85}$ (c) $\sqrt{5.2}$

(d) $\sqrt{127}$ (e) $\sqrt{245}$ (f) $\sqrt{1007}$

3 Ratio and proportion

■ While wandering down the aisles of my local supermarket, I come to the shelf holding Clean and Preen, my laundry detergent. The supereconomy-size box (6 lb, 12 oz for \$2.28) dwarfs the giant-size box (2 lb, 12 oz for 93¢). Suspiciously, I attempt to determine which is the best buy. What is my decision? ■

We have discussed the various number systems from the whole numbers through the real numbers. Although an additional extension can be made to the complex numbers, the real number system is sufficient for our needs not only in this course but in everyday experiences. Most of the time, in fact, the rational numbers are sufficient for everyday experiences.

In this section, we give renewed emphasis to critical thinking and problem solving. We shift our attention now to the kind of thinking involved in everyday experiences. One of the fundamental skills of the marketplace is the ability to compare, and the simplest mathematical instrument for comparison is the ratio. When the salesperson tells you that three out of every four people who purchase a Limo automobile average 22 miles per gallon, he or she is using two ratios: 3 to 4 and 22 to 1. In this section, you should find ratios as well as proportions to be simple but powerful tools.

The concepts of ratio and proportion are closely related to the rational numbers. Let's begin with ratio. **A ratio expresses a relation in size between two sets.**

Example 14 • Suppose there are 10 boys and 15 girls in our class. The ratio of boys to girls is 10 to 15, or 10:15, or $\frac{10}{15}$. •

Example 15 • On a cold day, the teacher notices that the ratio of gloves to scarfs is 5 to 2, written 5:2 or $\frac{5}{2}$. •

Ratio

> *Ratio* is another name for the quotient of two numbers.

In the first example, the ratio could be written $\frac{10}{15}$.

Example 16 • Express each as a fraction in simplest form: (a) 3.1:32.71, (b) $2\frac{1}{3}:3\frac{1}{4}$.

Solution (a) $\dfrac{3.1}{32.71} = \dfrac{3.1(100)}{32.71(100)} = \dfrac{310}{3271}$

(b) $\dfrac{2\frac{1}{3}}{3\frac{1}{4}} = \dfrac{\frac{7}{3}}{\frac{13}{4}} = \dfrac{7}{3} \cdot \dfrac{4}{13} = \dfrac{28}{39}$ •

Example 17 • A 120-g can of Lovely Mushrooms costs 84¢. A 160-g can of Earthy Mushrooms costs $1.28. Which is the better buy?

Solution We compute the ratio, cost to quantity, for each brand. This is called *unit pricing*.

	Lovely	*Earthy*
Cost	84¢	128¢
Number of grams	120 g	160 g
Ratio	$\dfrac{84}{120} = 0.7$	$\dfrac{128}{160} = 0.8$

Since Lovely Mushrooms have the lower cost to quantity ratio, this is the better buy. •

Two rational numbers a/b and c/d are equal if $ad = bc$. The equality for two ratios is determined in this same manner.

Equal ratios

> If a, b, c, and d are integers with $b \neq 0$ and $d \neq 0$, two ratios a/b and c/d are equal if and only if $ad = bc$.

Example 18 • In mixing fuel for a 2-cycle motor, one must mix 2 parts oil with each 8 parts of gasoline. How much oil is needed to mix with 24 liters (L) of gasoline?

Solution We need to identify the unknown carefully. Let $x =$ the number of liters of oil necessary to mix with 24 L of gasoline. Let us summarize our information in a table.

	Standard mixture	*New mixture*
Oil	2 parts	x L
Gasoline	8 parts	24 L

Since the ratio 2/8 must equal the ratio $x/24$,

$$\frac{2}{8} = \frac{x}{24}$$

$$2 \cdot 24 = 8 \cdot x$$

$$48 = 8x$$

$$6 = x$$

Six liters of oil are needed. •

This discussion leads us to the idea of proportions. Ratios that are equal form proportions.

A proportion

> For integers a, b, c, and d ($b \neq 0$ and $d \neq 0$), the statement that the two ratios a/b and c/d are equal is a *proportion*. Thus, $a/b = c/d$ is a proportion.

Example 19 • Part of a recipe calls for 3 eggs, 1.5 cups of sugar, and 2 cups of flour. If we have only 2 eggs, how should the amounts of sugar and flour be changed to keep the proportions the same?

Solution Suppose $x =$ cups of sugar and $y =$ cups of flour now needed. Again, let's place the given information in a table.

flashback

The fact that musical chords are based on ratios was one of the number facts that fascinated the Pythagoreans. In one demonstration, a taut string was divided into 12 parts. When shortened to 6 parts, the tone was raised an octave. When the string was shortened to 8 or 9, the tone was raised, respectively, a fourth or a fifth. Hence, ratios of $\frac{2}{1}$ (that is, $\frac{12}{6}$), $\frac{3}{2}$, and $\frac{4}{3}$ produced important parts of the musical scale. Legend relates that on his deathbed, Pythagoras urged his followers to practice on the "monochord."

Number of eggs	3	2
Number of cups of sugar	1.5	x
Number of cups of flour	2	y

The ratio of eggs to cups of sugar should be the same in both columns.

$$\frac{3}{1.5} = \frac{2}{x}$$

$$3x = (1.5)(2) = 3$$

$$x = 1$$

Also the ratio of eggs to cups of flour should be the same.

$$\frac{3}{2} = \frac{2}{y}$$

$$3y = 4$$

$$y = \frac{4}{3}$$

The new recipe needs 1 cup of sugar and 4/3 cups of flour. •

Example 20 • A mixture of dog food is composed of 5 parts of meat and 3 parts of cereal. How much cereal is included in 20 kilograms (kg) of dog food?

Solution Each 8 units of dog food contains 3 units of cereal. Therefore, the ratio of cereal to dog food is 3 to 8 or $\frac{3}{8}$. Let x represent the amount of cereal in 20 kg of dog food.

Number of kilograms of cereal	3	x
Number of kilograms of dog food	8	20

$$\frac{3}{8} = \frac{x}{20}$$

$$8x = 60$$

$$x = 7\tfrac{1}{2} \text{ kg of cereal}$$ •

Example 21 • A box contains 6 red balls and 3 green balls. If 4 red balls are added, how many green balls must be added in order to keep the ratio of red balls to green balls the same?

Solution x is the number of green balls to be added.

Number of red balls	6	10
Number of green balls	3	$3 + x$

$$\frac{6}{3} = \frac{10}{3 + x}$$

$$6(3 + x) = 3(10)$$

$$18 + 6x = 30$$

$$6x = 12$$

$$x = 2$$

Two additional green balls are needed. •

Example 22 • In the city of Ft. Thomas, property is taxed at a rate of \$19.75 per \$1000 of assessed value. This fact can be represented by the following proportion.

$$\frac{19.75}{1000} = \frac{\text{tax to be paid}}{\text{assessed value of property}}$$

Hence, if a home is assessed at \$32,000, the tax is computed by solving the following proportion for x.

$$\frac{19.75}{1000} = \frac{x}{32{,}000}$$

$$(19.75)(32{,}000) = 1000x$$

$$x = \$632 \quad \bullet$$

Ratio and proportion are very useful in making scale drawings. For example, 1 centimeter (cm) on a scale drawing may represent 10 m in actual length. Then the ratio is $\frac{1}{10}$, which in this case is centimeters to meters.

Example 23 • In a particular scale drawing, $\frac{1}{2}$ centimeter (cm) represents 1 meter (m).
(a) What length is represented by 2 cm?
(b) 40 m is represented on the drawing by how many centimeters?

Solution (a) The ratio of the scale drawing to the actual length is

$$\frac{\frac{1}{2}}{1} = \frac{1}{2}$$

Just for fun *Look at the ratios of numbers in the circles on the opposite ends of the line segments. All ratios are proportional with one exception. Can you find the exception?*

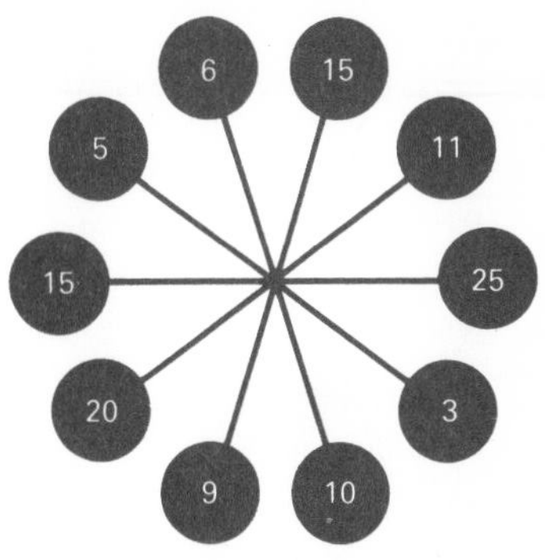

This ratio is centimeters to meters. Let x m correspond to 2 cm. Then

$$\frac{2}{x} = \frac{1}{2} \qquad \text{or} \qquad x = 4 \text{ m}$$

(b) Let y cm correspond to 40 m.

$$\frac{y}{40} = \frac{1}{2}$$

$$2y = 40$$

$$y = 20 \text{ cm} \quad \bullet$$

Exercise set 3

R 1. Express each ratio as a fraction in simplest terms.

(a) 5 to 3 (b) 7:11 (c) 2:7
(d) 2.1:7.3 (e) 0.01:2.3

2. Express in simplest form.

(a) 3:27 (b) $4\frac{1}{3}:5\frac{2}{3}$ (c) $8\frac{1}{2}:11\frac{1}{4}$

3. Solve for x in each proportion below.

(a) $\frac{6}{x} = \frac{9}{12}$ (b) $\frac{x}{21} = \frac{3}{7}$

(c) $\frac{23}{27} = \frac{13}{x}$ (d) $\frac{16}{x} = \frac{31}{4}$

(e) $\frac{25}{5} = \frac{x}{8}$ (f) $\frac{17}{21} = \frac{x}{13}$

4. A mathematics class consists of 16 women and 24 men.

(a) What is the ratio of women to men?
(b) What is the ratio of men to women?
(c) What is the ratio of women to all students in the class?

5. Write the reduced fractions that represent the following ratios.

(a) The mile-per-gallon ratio for a trip of 387 miles on 18 gallons.
(b) The student-to-faculty ratio at a college with 7527 students and a faculty of 215.
(c) The administrator to employee ratio if there are 15 administrators and 85 employees.

(d) The profit-per-lot ratio if 18 lots are sold for a total profit of $32,864.

6. Find the value of x in each of the following proportions.

(a) $x:4 = 6:20$ (b) $4:1 = 8:x$
(c) $9:14 = x:8$ (d) $9:x = 56:30$

7. If the ratio of boys to girls in a class is 3:7, find the following.

(a) The ratio of girls to boys.
(b) The ratio of boys to the total class.
(c) The number of boys in the class if the number in the class is 40.

8. Determine which item in the following pairs is the better buy. Make a table showing unit pricing.

(a) A 900-g can of peaches for 49¢ or an 850-g can for 46¢.
(b) A giant-size box (1500 g) of detergent for $1.51 or a king-size box (2400 g) for $2.47.

9. Determine which item in the following pairs is the better buy.

(a) A 29-ounce can of beans for 49¢ or a 27-ounce can for 46¢.
(b) A giant-size (49 oz) box of corn flakes $1.51, or a king-size box (84 oz) for $2.47.
(c) A 3-pound canned ham for $5.49 or a 5-pound canned ham for $9.49.
(d) A 16-ounce can of corn for 29¢ or a 14.5-ounce can for 27¢.
(e) A 7-ounce box of spaghetti for 29¢ or an 8-ounce box for 30¢.

10. Work the problem given in the introduction of this section.

11. A recipe calls, in part, for 2 eggs, 1 teaspoon of vanilla, 3 cups of flour, and 1.75 cups of sugar. If one ingredient is changed to the amount given below, how must the other ingredients be changed to keep the proportions the same?

(a) 1 egg (b) 2 cups of flour
(c) 3.5 cups of sugar (d) 3 eggs

12. One centimeter (cm) on a map represents 50 kilometers. What is the distance between two towns that are $2\frac{1}{2}$ cm apart on the map?

13. The quantity of poison required to kill a mammal is approximately proportional to the body weight. If 0.5 gram (g) will kill a squirrel weighing 400 g, how much poison is needed to kill a woodchuck weighing 1.5 kilograms?

14. The ratio of boys to girls in a class is 8:7. If the class has 60 students, how many are girls?

15. A mixture of paint requires one part green, one part yellow, and four parts white. How much of each is needed to paint a house requiring 60 liters (L) of paint?

16. If it takes 10 minutes to saw a log into 3 pieces, how long does it take to saw it into 4 pieces? (Be careful!)

17. The ratio of boys to girls in two classes is 3:4 and 4:5. If the two classes have the same number of students, compare the number of boys in the two classes.

18. The electrical resistance of a wire is proportional to its length. The resistance of 4 feet of a certain wire is 3.5 ohms. What resistance should be found in a 14-foot piece of the same wire?

19. The capacities of buckets having the same shapes are proportional to the cubes of the diameters of their tops. A bucket that is 10 inches across the top holds 2 gallons. A second bucket shaped similarly has a diameter of 30 inches. How many gallons will it hold?

20. The strength of a cable is proportional to the square of its diameter. If a $\frac{1}{2}$-inch cable will support 1200 pounds, how much will a $\frac{5}{8}$-inch cable support?

21. Oxygen and hydrogen combine in a fixed ratio to form water. The ratio is about 8 parts of oxygen to 1 part of hydrogen, by weight. How many grams of oxygen are needed to form 32 grams of water?

22. If a, b, c, and d are unequal to zero, then

$$\frac{a}{b} = \frac{c}{d} \quad \text{if and only if} \quad \frac{b}{a} = \frac{d}{c}$$

and

$$\frac{a}{b} = \frac{c}{d} \quad \text{if and only if} \quad \frac{a}{c} = \frac{b}{d}$$

(a) Make up three examples to demonstrate the truth of these statements.
(b) Use properties of rational numbers to prove the preceding statements.

4 The language of percent

■ M-Mart plans a 4th of July sale in which they will mark down garden tools by 18%. What will be the sale price of a $30 wheelbarrow? ■

Percent is a familiar concept, yet many people have failed to master completely this elusive topic. Since percent is used frequently in critical thinking and in decision making, and since it is a fundamental word in the language of the marketplace, further study of percent is in order. Percents are used to compute taxes, report unemployment and inflation rates, determine sale prices and interest on loans, and perform a multitude of other critical tasks.

You will be presented some new terms in this section: commission, discount, markup, and depreciation. Yet you will have no difficulty with this material if you can distinguish between these questions:

1. What percent is A of B?
2. What is p percent of B?
3. If A is p percent of B, then what is B?

Percent is derived from the Latin phrase *per centum* and means *by the hundred.* Thus 46% could be correctly stated *46 hundredths.*

Percent

Percent is a ratio giving the number of parts per hundred, or $n\% = n/100$.

From this definition, it is easy to see that percent is a fraction with a denominator of 100. Thus, to express $\frac{2}{5}$ as a percent, we are asked to find the numerator of a fraction that is equivalent to $\frac{2}{5}$ and has a denominator of 100:

$$\frac{2}{5} = \frac{x}{100} \quad \text{or} \quad \frac{2}{5} = \frac{40}{100}$$

so $\frac{2}{5}$ as a percent is 40%.

To convert a decimal fraction such as 0.016 to a percent, we are asking that

$$0.016 = \frac{16}{1000}$$

be written as a fraction with a denominator of 100.

$$0.016 = \frac{16}{1000} = \frac{1.6}{100}$$

Thus 0.016 can be written as 1.6%.

Example 24 • Express (a) 1.7% and (b) 215% as decimals.

Solution (a) $\frac{1.7}{100} = \frac{1.7(10)}{100(10)} = \frac{17}{1000} = .017$

(b) $\frac{215}{100} = 2.15$ •

Example 25 • Convert the following to decimals: 2.12%, 0.012%, 324%.

Solution

$$2.12\% = 0.0212$$
$$0.012\% = 0.00012$$
$$324\% = 3.24$$ •

Note in the preceding examples that to write a decimal fraction as a percent merely requires that one move the decimal point two places to the right and write the symbol %.

Example 26 • Convert to percents: 0.047, 0.769, 3.56, and 0.00071.

Solution

$$0.047 = 4.7\% \qquad 0.769 = 76.9\%$$
$$3.56 = 356\% \qquad 0.00071 = 0.071\%$$ •

All problems that involve percent are variations of the problem "Some percent of one number equals another number." Note that only three quantities are involved:

1. The percent (sometimes called rate).
2. The first number called the *base*.
3. The second number called the *percentage*.

Percent formula

$$\text{Percentage} = (\text{percent}) \cdot (\text{base})$$

or

$$p = rb$$

Example 27 • Raymond, currently making $12,500 a year, receives a 5.5% raise. The rate is 0.055, and the base amount is $12,500. To compute the amount of the raise (the percentage), use the formula $p = rb$.

Solution

$$\begin{aligned} p &= r \cdot b \\ &= (0.055)(12{,}500) \\ &= \$687.50 \end{aligned}$$

Thus his new salary is

$$\$12{,}500 + \$687.50 = \$13{,}187.50$$ •

In solving problems such as the preceding example, some people like to use the equivalent fraction approach.

Dollars	$x	$12,500
Percent	5.5%	100%

$$\frac{x}{\$12{,}500} = \frac{5.5}{100}$$

$$x = \frac{5.5(\$12{,}500)}{100}$$

$$= 0.055(\$12{,}500)$$

You see that this is the same result obtained by using the formula $p = rb$.

Other types of problems using percent involve solving a simple equation.

Example 28 • Tom weighed 65 kg in January and lost 13 kg by April. What percent of his body weight had he lost?

Solution

$$p = r \cdot b$$

$$13 = r \cdot 65$$

$$r = \frac{13}{65} = 0.20$$

He had lost 20% of his body weight. Using the equivalent fraction approach,

	Loss	*Total*
Weight	13	65
Percent	x%	100%

$$\frac{13}{65} = \frac{x\%}{100\%}$$

$$x = \frac{1300}{65} = 20\% \quad \bullet$$

Example 29 • Eighteen percent of the freshmen at Star College failed English. If 396 freshmen failed English, how many freshmen are enrolled in Star College?

Solution Using $p = r \cdot b$,

$$396 = 0.18b$$

$$b = \frac{396}{0.18} = 2200 \text{ freshmen}$$

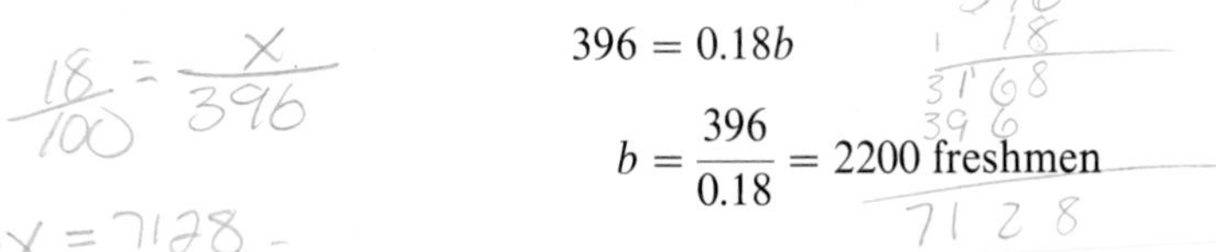

Using the equivalent fraction approach,

	Failed	*Total*
Number of students	396	x
Percent	18%	100%

$$\frac{18\%}{100\%} = \frac{396}{x}$$

$$x = \frac{396(100)}{18} = 2200 \quad \bullet$$

Example 30 • 92 is what percent of 400?

Solution

$$p = r \cdot b$$

$$92 = r(400)$$

$$r = \frac{92}{400} = 0.23 = 23\%$$

So 92 is 23% of 400. Again, using the equivalent fraction approach,

Number	92	400
Percent	x%	100%

$$\frac{92}{400} = \frac{x\%}{100\%}$$

$$x = \frac{9200}{400} = 23\% \quad \bullet$$

In everyday experience, reference is frequently made to a change in percent.

Example 31 • What percent less than 48 is 36?

Solution First, find the change.

$$\text{Change} = 48 - 36 = 12$$

$$p = r \cdot b$$

$$12 = r \cdot 48$$

$$r = \frac{12}{48} = 0.25$$

Thus, 36 is 25% less than 48. Also,

Number	12	48
Percent	$r\%$	100%

$$\frac{12}{48} = \frac{r\%}{100\%}$$

$$r = \frac{1200}{48} = 25\% \quad \bullet$$

Many terms used in everyday business procedures (such as *commission, discount, markup,* and *depreciation*) are computed using percents.

Example 32 • A real estate agent receives as commission 6% of the sales price of the property she sells. If a house is sold for \$38,000, what is the agent's commission?

Solution $p = rb$ where b represents sales price and p represents the commission.

$$\text{Commission} = (\text{rate})(\text{sales price})$$

$$\text{Commission} = (0.06)(38{,}000) = 2280$$

The commission is \$2280. The seller of the house receives \$38,000 − \$2280 = \$35,720. •

Example 33 • The Korner Drugstore marks up its prescription drugs by 42%. If a certain drug costs \$3 from the wholesale distributor, what will be the selling price at the store?

Solution $p = rb$ where b is the cost and p is the markup.

$$\text{Markup} = (0.42)(3)$$

$$\text{Markup} = \$1.26$$

$$\text{Retail price} = \$3.00 + \$1.26 = \$4.26 \quad \bullet$$

Here is the solution to the section introductory problem.

Example 34 • M-Mart plans a 4th of July sale in which they will mark down garden tools by 18%. What will be the sale price of a \$30 wheelbarrow?

Solution $p = rb$ where b is the regular sales price and p is the markdown.

$$\text{Markdown} = (0.18)(30) = \$5.40$$

$$\text{Sale price} = \$30.00 - \$5.40 = \$24.60 \quad \bullet$$

Example 35 • The selling price of a suit of clothes is \$200. If the cost of the article is \$150, what is the percent of markup based on selling price?

Just for fun *The oil boom is on in the isolated frontier territory. Last year, 25% of the women in the territory got married, whereas 2.4% of the 1250 men married these local ladies. Assuming no bigamy occurred, how many women are in the territory?*

Solution

$$\text{Profit} = \text{selling price} - \text{cost}$$
$$= \$200 - \$150$$
$$= \$50$$

$$\text{Rate of markup} = \frac{50}{200} = 0.25$$

The percent of markup based on selling price is 25%. •

Exercise set 4

R 1. Convert to decimals.

(a) 5.5% (b) 0.012% (c) 0.31%
(d) 426% (e) 43.6% (f) 18.4%

2. Convert to percents.

(a) 0.147 (b) 0.4472 (c) 0.005
(d) 0.086 (e) 8.61 (f) 23.14

3. Convert to percents.

(a) $\frac{18}{15}$ (b) $\frac{33}{150}$ (c) $\frac{6}{15}$
(d) $\frac{18}{4}$ (e) $\frac{11}{40}$ (f) $\frac{143}{80}$

4. Using the fact that percent is a ratio, find:

(a) 17% of 200 (b) 23% of $1200
(c) 15% of 823 (d) 125% of 820
(e) 0.06% of 1100 (f) 6% of 88

5. Solve each part of Exercise 4 using the relationship $p = r \cdot b$.

6. A saleswoman makes a commission of 9% on all sales. If her monthly sales amount to $7500, what is her commission that month?

7. Find the markup and retail price on the following goods if the percent of the markup is to be 30%.

Item	*Wholesale cost*
Paring knife	$1.20
Electric clock	$2.40
Can opener	$1.70
Waste can	$3.60

8. (a) What percent of 48 is 12?
(b) 24 is what percent of 96?
(c) Find 16% of 200.
(d) Find 0.01% of 1.632.
(e) What is 78% of 16?
(f) What percent of 150 is 6?
(g) What percent of 18 is 54?
(h) What number is 130% of 96?

9. Fill in the missing entries.

	Cost	*% Markup based on cost*	*Selling price*
(a)	$1400	24	—
(b)	$50	—	$62.50
(c)	—	20	$84

T 10. In some areas of business, the percent of markup is applied to the selling price rather than the cost. In Exercise 9, compute the percent markup based on selling price for each article.

11. There are 23 girls and 19 boys in a freshman history class at Simmons College. Seven people failed the first test. What percent of the class failed the first test?

12. If your automobile payments are \$200 a month, what percent of your \$950-per-month salary must be set aside to pay for your automobile?

13. In Exercise 12, you receive a 4% raise. What is your new monthly salary? What percent of your salary now must be set aside to pay for your automobile?

14. Of the 80,000 football seats in the stadium, 48,640 were filled. What percent of the stadium was filled?

15. If you spend 65% of your leisure time reading, how much leisure time do you have if you spend 13 hours a week reading?

16. A merchant sold a suit for \$87, thereby gaining 30% of the cost. What was the cost of the suit?

17. Mrs. Taylor considers buying a dress that costs \$76, but she decides to have the dress made if she can save at least 25% of the cost. The materials for the dress cost \$37. She finds out the dressmaker's fee and decides to purchase the dress. What is the minimum amount of the fee?

18. Joan Makenny received a 12% raise on her salary of \$15,500. What is her new salary?

19. U.S. oil imports from one OPEC nation increased from 22 million barrels to 29 million barrels in a 2-year period. What percent does this increase represent?

20. A newspaper reports that the Amalgamated Sweepers Union negotiated a raise of \$0.40 per hour. If they were previously making \$4.85 an hour, what is the percent of increase?

21. One serving of Wheats cereal contains 4% of the recommended daily allowance of protein. If one serving of Wheats contains 8 grams of protein, what is the recommended daily allowance for protein?

22. In a presidential election, only 54% of the registered voters voted. If 42 million went to the polls, how many registered voters were there?

23. The average white-collar government employee earned \$4900 in 1969 and \$20,300 in 1982. What was the percent of increase?

24. (a) An airline ticket costs \$84. A federal tax of 12% of this amount is added. What is the total amount of the ticket?
    (b) If the total cost of an airline ticket is \$120.96, what is the amount of federal tax (the rate is 12%)?

25. Sara Friedman makes \$700 a month plus 3% of all contracts she negotiates. If in a given month she generates \$6600 of business, what is her pay that month?

26. Fred Moss wants to sell shirts for \$7.25. If he marks up his items 35% over wholesale cost, what is the greatest amount he can pay for the shirts?

27. Thomas Sporting Goods buys boots for \$69.95 and sells them for \$110.50. What is the percent of markup based on cost?

28. What percent of markdown was used in each of the sale prices given below?

C 29. What was the presale price on the lighting fixture advertised below?

30. Thomas Sporting Goods buys a tennis racket at a wholesale price of \$17.40. The racket is marked up 45%, and then during the March sale it is marked down by 30%.
    (a) What is the sale price?
    (b) What percent of the markup on the wholesale cost does the sale price represent?

31. George Moore was hired at a yearly salary of \$18,500. He received a raise of 11% and then received a raise of 6%. What is his current annual salary?

32. Sales at the Acme Agency increased 30% in 1979, decreased 20% in 1980, and increased 10% in 1981. The sales in 1981 were what percent of sales in 1979?

33. A cotton picking machine picks 70% of the cotton each time it goes over a field. If a farmer runs his machine over a field twice, what percentage of the cotton will be harvested?

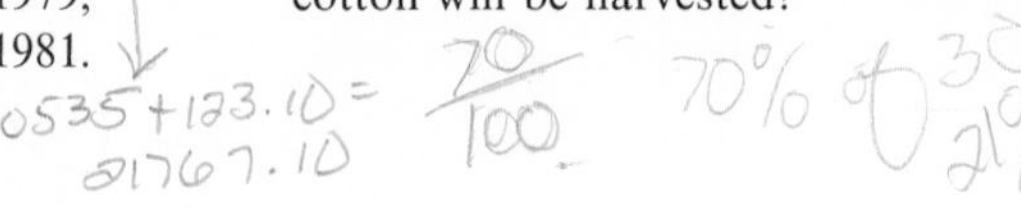

5 Some comparisons of interest rates

■ Claire deposited \$100 in a savings account paying 6% per year compounded quarterly. At the end of the year, her statement indicated that she had \$106 in her account. Claire felt that a mistake had been made in her account. Is she right or wrong? ■

The phrase *compounded quarterly* indicates that this problem involves compound interest. A study of compound interest is very time-consuming without tables or without a hand calculator. In this chapter, you should use a hand calculator.

Almost everyone encounters interest through loans, installment purchases, or savings. The calculations are elementary, but a discussion of the concepts is important for consumers.

Simple interest is interest charged or interest earned on the amount borrowed or saved. Simple interest is computed as a constant percentage of the money borrowed for a specified time, usually a single year or less, and is paid at the end of the specified time.

The sum borrowed is called the *principal, P*, or sometimes the present value; r denotes the rate of interest, usually expressed in percent per year; and t is time expressed in years or fractions of years. By definition, simple interest, I, equals principal multiplied by the interest rate multiplied by the time in years.

$$I = Prt$$

Example 36 • Four hundred dollars is borrowed at 12% simple interest for 3 years. What is the interest?

Solution

$$I = Prt$$

$$I = (400)(0.12)(3) = \$144 \quad \bullet$$

At the end of the term of the loan, the borrower must pay not only the interest but also the sum that was originally borrowed, the principal. Hence, the amount due at the end of the term is given by

$$\textit{Amount} = \textit{principal} + \textit{interest}$$

$$A = P + I = P + Prt$$

Using the distributive property, we can write this formula as

$$A = P(1 + rt)$$

Simple interest

$$I = Prt$$

$$A = P(1 + rt)$$

P is the principal; r, the annual interest rate; t, the time in number of years; I, the interest; and A, the amount owed.

Example 37 • A loan of \$1000 is made for 6 months at a simple interest rate of 8%. How much does the borrower owe at the end of 6 months?

Solution

$$A = P(1 + rt) = \$1000\left[1 + (0.08)\left(\frac{1}{2}\right)\right]$$

$$= \$1000(1 + 0.04)$$

$$= \$1040$$

The borrower owes \$1040. •

To find the present value of an amount at simple interest r, solve the equation $A = P(1 + rt)$ for P to obtain

$$P = \frac{A}{1 + rt}$$

Example 38 • Compute the present value of \$1000 due in 3 months with interest at 6% annually.

Solution

$$P = \frac{\$1000}{[1 + (0.06)(\frac{3}{12})]} = \frac{\$1000}{1.015} = \$985.22$$

The present value is \$985.22 •

Example 39 • One thousand dollars is borrowed for 2 years. At the end of that time, \$1120 is repaid. What percent of simple interest was charged?

Solution

$$A = P(1 + rt)$$

$$1120 = 1000(1 + 2r)$$

$$1120 = 1000 + 2000r$$

$$120 = 2000r$$

$$r = \frac{120}{2000} = 0.06 = 6\% \text{ simple interest}$$ •

Simple interest is not always as simple as it appears. Consider the story of Jill. Jill anticipated an IRS tax refund of \$300 in early May. Being short on cash, she decided on January 19 to borrow \$300 until May 10. Her banker agreed to give her the loan at 7.5% simple interest. First the banker calculated the number of days she would have the money.

Days remaining in January	31 − 19 = 12
Days in February	28
Days in March	31
Days in April	30
Days in May	10
	111

She was comfortable with this computation, but she was quite surprised when the banker computed her interest:

$$I = Prt$$

$$I = (300)(0.075)\left(\frac{111}{360}\right)$$

$$I = \$6.94$$

The banker pointed out that the fraction $\frac{111}{360}$ indicates the portion of the year that she would have the money. Bankers often compute interest on the basis of a 360-day year. In the days prior to widespread use of calculators, this convention made computations easier; **Jill noted that it also made more money for the bank.**

We have learned that simple interest I is found by using the formula $I = Prt$, where P represents the principal, r the rate, and t the time. When interest is computed by this formula, the principal always remains the same. If the interest is added to the principal at the end of each interest period, so that the principal is increased, the interest is said to be **compounded.** The sum of the original principal and all the interest is called the *compound amount,* and the difference between the compound amount and the original principal is the *compound interest.* A comparison of simple interest and compound interest is given in the following example.

Example 40 • Find the simple interest on \$1000 for 3 years at 6%. Then find the compound interest on \$1000 for 3 years at 6% compounded annually.

To emphasize the difference between simple and compound interest, both will be computed year by year.

Solution

	Simple interest	*Compound interest*
For the first year,	$I = \$1000(0.06) = \$60$	$I = \$1000.00(0.06) = \$60$
For the second year,	$I = \$1000(0.06) = \$60$	$I = \$1060.00(0.06) = \$63.60$
For the third year,	$I = \$1000(0.06) = \$60$	$I = \$1123.60(0.06) = \$67.4160$ $= \$67.42$

Thus, the simple interest for 3 years totals \$180, whereas the compound interest totals \$191.02. The compound amount is \$1191.02. Notice that the principal changes each year when interest is compounded, but it remains the same when simple interest is used. Let's compute the compound amount at the end of each year in another way. At the end of the first year, the compound amount is

$$\$1000 + \$1000(0.06) = \$1000(1 + 0.06) = \$1000(1.06)$$

During the second year, the principal is \$1000(1.06). Thus, at the end of the second year the compound amount is

$$\begin{aligned} \$1000(1.06) + \$1000(1.06)(0.06) &= \$1000(1.06)(1 + 0.06) \\ &= \$1000(1.06)(1.06) \\ &= \$1000(1.06)^2 \quad \bullet \end{aligned}$$

This pattern extends easily to the most general case. Suppose that P dollars are deposited at an interest rate of i per period for n periods. Then Table 1 computes the amount that has accrued at the end of that time.

Table 1

Period	*Principal*	*+ Interest*	*= Amount (at end of period)*
1	P	$+ Pi$	$= P(1 + i)$
2	$P(1 + i)$	$+ P(1 + i)i$	$= P(1 + i)^2$
3	$P(1 + i)^2$	$+ P(1 + i)^2 i$	$= P(1 + i)^3$
⋮			
n	$P(1 + i)^{n-1}$	$+ P(1 + i)^{n-1} i$	$= P(1 + i)^n$

This table suggests that the compound amount can be found by multiplying the principal by $(1 + i)^n$ where i is the interest rate per period and n is the number of periods.

Compound interest

The compound amount A that results from a principal P invested at a rate of i per period for n periods is given by the formula

$$A = P(1 + i)^n$$

Example 41 • Find the compound amount that results from investing \$500 at 12% compounded annually for 4 years.

Solution Substituting $P = \$500$, $i = 0.12$, and $n = 4$ into the formula

$$A = P(1 + i)^n$$

$$A = \$500(1 + 0.12)^4 = \$500(1.12)^4$$

To complete this problem, it is necessary to compute $500(1.12)^4$. Although this computation is lengthy when performed by hand, $500(1.12)^4$ can be obtained easily on a calculator. Enter 500 and then multiply by 1.12 four times, or

$$500(1.12)^4 = 786.76$$

So

$$A = \$786.76 \quad \bullet$$

Interest may be compounded for any period of time—annually, semiannually, quarterly, monthly, daily, and so on. When the rate of compound interest is given, it is usually specified as an annual rate called the **nominal rate.** Consequently, if the interest is to be compounded semiannually, this rate must be divided by 2; if the interest is to be compounded quarterly, the nominal rate must be divided by 4, and so forth.

Example 42 • Find the compound amount obtained from an investment of \$2000 at 6% compounded quarterly for 5 years.

Solution Six percent compounded quarterly is $0.06/4 = 0.015$ each quarter. Since there are four quarters in a year, the number of interest periods is $n = 4 \cdot 5 = 20$. Thus, the compound amount is

$$\begin{aligned} A &= P(1 + i)^n \\ &= \$2000(1 + 0.015)^{20} \\ &= \$2000(1.015)^{20} \end{aligned}$$

Multiplying 2000 by 1.015 twenty times on a minicalculator, we obtain $A =$ \$2693.71 •

Once we have established the relationship $A = P(1 + i)^n$, it is rather easy to change the direction of our thinking. Consider the question "How much principal must we invest now at 8% per year compounded quarterly in order to have in 4 years the \$6000 we need to buy a new car?" When the question is asked in this way, the principal for which we are searching is called the *present value.* Thus we are looking for the present value of the compound amount \$6000 due in 4 years at 8% per year compounded quarterly.

$$A = P(1 + i)^n$$

$$6000 = P(1 + 0.02)^{16}$$

$$P = \frac{6000}{(1.02)^{16}}$$

Using a hand calculator, we find that $6000/(1.02)^{16}$ is \$4370.67. Thus the present value of \$6000 due in 4 years at 8% compounded quarterly is \$4370.67.

Example 43 • How much money should be deposited in a Savings and Loan Association paying 6% compounded quarterly in order to have \$3000 in 5 years?

Solution

$$A = P(1 + i)^n$$

$$3000 = P(1.015)^{20}$$

$$P = \frac{3000}{(1.015)^{20}}$$

Using a hand calculator, we find

$$P = \$2227.41$$

Thus \$2227.41 must be deposited now in a Savings and Loan Association in order to have \$3000 in 5 years.

Just for fun

A man goes into a store and says to a salesperson, "Give me as much money as I have with me and I will spend \$10 with you." It is done. The operation is repeated in a second and a third store, after which he has no money left. How much did he have originally?

Exercise set 5

R 1. Compute the simple interest when the principal, rate, and time of the loan are given.

(a) $P = \$500$, $r = 0.08$, $t = 2$ years
(b) $P = \$300$, $r = 0.13$, $t = 4$ years
(c) $P = \$500$, $r = 0.10$, $t = 5$ years

2. What is the amount to be repaid in (a), (b), and (c) of Exercise 1?

3. Find the interest and the amount of a loan of \$3000 for 2 years at 16% simple interest.

4. Find the interest on \$2000 for 8 months at 10% simple interest.

5. Find the amount of a 90-day, \$1500 loan at 12% simple interest.

6. How much interest is earned on \$740 left on deposit for 1 year in a savings account with an annual interest rate of 8.17%?

Find the compound interest and compound amount for the investments in Exercises 7 through 9.

7. (a) \$5000 at 12% compounded annually for 3 years
(b) \$5000 at 12% compounded semiannually for 3 years
(c) \$5000 at 12% compounded quarterly for 3 years

8. (a) \$2000 at 10% compounded annually for 4 years
(b) \$2000 at 10% compounded semiannually for 4 years
(c) \$2000 at 10% compounded quarterly for 4 years

9. (a) \$3000 at 6% compounded annually for 6 years
(b) \$3000 at 6% compounded semiannually for 6 years
(c) \$3000 at 6% compounded quarterly for 6 years

10. Compute the amount to be repaid when the principal, rate, and time of the loan are given.

(a) $P = \$4000$, $r = 0.11$, $t = 2$ years
(b) $P = \$3000$, $r = 0.12$, $t = 6$ months
(c) $P = \$100$, $r = 0.08$, $t = 3$ years

11. Find the simple interest rate on the loan when the principal, the amount repaid, and the term of the loan are given.

(a) Principal \$3500, amount repaid \$4130 in 2 years.
(b) Principal \$1000, amount repaid \$1046.67 in 120 days.
(c) Principal \$500, amount repaid \$510 in 45 days.

Find the present value of the money in Exercises 12 through 15.

12. \$5000 due in 5 years if money is worth 16% compounded annually

13. \$6000 due in 5 years if money is worth 8% compounded semiannually

14. \$7000 due in 4 years if money is worth 12% compounded semiannually

15. \$8000 due in 6 years if money is worth 7% compounded annually

T 16. How much interest will one owe on a \$1000 loan from March 3 to August 7 at 10% simple interest?

17. What is the interest on a $1500 loan from June 15 to September 11 at a simple interest rate of 8%?

18. Fred Jones borrowed $1800 from his father and agreed to pay an interest rate of 4% compounded semiannually. What did he owe when he repaid the loan after 5 years?

19. A loan is negotiated for 3 years at 12% compounded quarterly. A second loan is negotiated for 3 years at 12% simple interest. Compare the amounts of interest paid on the two loans.

20. How much should a father invest for his daughter at 6% compounded annually in order to have
    (a) $5000 at the end of 20 years?
    (b) $7000 at the end of 20 years?

21. Frank Lane borrows $400 from his father to be repaid in 90 days. He is to repay $412. What simple interest rate is he paying?

22. What is the principal of a loan if the interest after 120 days at $6\frac{1}{4}$% is $100?

23. Jane Sims borrows $700 to tide her over until school is out and she has summer employment. She borrows the money on March 2 and is to repay it on July 31. If the simple interest rate is 12%, how much will she repay?

24. Some lending institutions have a minimum amount of interest they must collect on any loan. An institution with a minimum charge of $5 will expect to receive $5 interest on the loan, even though actual interest charges are only $3. How long must $1000 be borrowed at 8% interest to reach the minimum service charge of $5? (Use a banker's year of 360 days in the computation.)

The effects of inflation can be obtained using the formula for compound interest where i becomes the inflation rate per period.

C 25. Find the cost of the following items in 10 years at an annual inflation rate of 6%.
    (a) A $40,000 house
    (b) A 90¢ hamburger
    (c) A $3.50 movie ticket

26. Work Exercise 25 for an 8% annual inflation rate.

27. Work Exercise 25 for a 10% annual inflation rate.

*6 Annual percentage rate

■ Jim's old station wagon finally quit after many years of faithful service. After some tight-fisted bargaining with his car dealer, he found that he still needed $3000 to purchase the car he had chosen as a replacement. A visit to his banker revealed that he could borrow the $3000 to be repaid over a period of three years with monthly payments of $98.78. This would repay the $3000 principal as well as $556 in finance charges. The banker explained that the finance charges consist of $540 of interest computed as 6% add-on interest together with an acquisition fee of $16. The banker closed the discussion by noting that the annual percentage rate would be 11.5%. With his head swimming Jim went home to ponder this information. ■

Many of the credit needs of the consumer on the marketplace are met with loans such as the one Jim is considering. Rather than repaying the loan in one lump sum at the end of the term, the consumer often repays it in a series of equal monthly payments. Such installment credit situations are often encountered when one borrows money to buy a car, furniture, appliances, or arranges a personal loan through a bank or loan agency. In arranging such loans, one encounters a whole new vocabulary—*finance charges, add-on interest, monthly payments, and annual percentage rates.* It is the purpose of this section to clarify meanings of some of these terms and to investigate the computational devices that are used to determine them.

The ***finance charge*** *is the amount that the consumer must pay for his or her credit.* It mainly involves interest, but its legal definition also includes things such as loan fees, acquisition fees, and insurance, which are required to receive the loan. *The* ***monthly payments*** *that the borrower makes to repay the loan are computed by dividing the finance charges plus principal by the number of months the payments are to be made.* For instance, in Jim's loan, the finance charges consisted of

$540 interest
$ 16 acquisition fee
$556 finance charges

and each of his 36 monthly payments equals

$$\frac{\$3556}{36} = \$98.78$$

Example 44 • An automobile is bought for $5200. A down payment of $1200 is made, and the remaining $4000 is to be financed. A loan is arranged in which $780 is to be paid in interest and, in addition, a credit investigation fee and a fire, theft, and casualty insurance policy are required. The credit investigation fee is $10, and the cost of the insurance for the term of the loan is $220.

(a) Find the finance charges. Since the credit investigation fee and the fire and theft insurance are required for the loan, the finance charges total

$ 780
$ 10
$ 220
$1010

(b) Find the monthly payment if the loan is to extend for 3 years (36 months).

$$\text{Monthly payment} = \frac{\$4000 + \$1010}{36} = \$139.17$$ •

Example 45 • Madly in love and without a cent, Fred rushes to the Nice and Easy Finance Company to borrow $300 for a ring. He quickly agrres to pay N.A.E. $29.50 a month for 12 months. What are his finance charges?

$$\text{Finance charges} = (12)(29.50) - 300 = \$54$$

In many instances in which loans with a monthly payment schedule are obtained, it is not clear how the interest portion of the finance charge is computed. However, in some cases, particularly in bank loans, the interest is computed as *add-on interest.* In other words, the interest is computed as if it is simple interest and then added on to the principal as part of the finance charge. Hence, a 6% add-on rate applied to a 2-year loan of $1500 yields interest that equals (0.06)(1500)(2) = $180. •

Example 46 • A bank loan of \$2000 for 18 months is to be repaid in 18 equal installments. The interest is computed as 7.5% add-on interest. An additional \$15 credit check fee is required.

(a) How much interest is to be paid?

$$\text{Interest} = (0.075)(2000)\left(\frac{18}{12}\right)$$

$$= \$225$$

(b) What are the finance charges?

$$\$225 + \$15 = \$240$$

(c) What is the monthly payment?

$$\frac{\$2000 + \$240}{18} = \$124.44 \quad \bullet$$

With a little thought, we see that even though add-on interest is computed as if it were simple interest, in fact this is not the case. With simple interest, the borrower can use the entire principal for the entire term of the loan. In a loan with periodic payments, the borrower starts returning the money immediately. In the later months, he or she has the use of very little of the money. Hence, a 6% add-on interest rate corresponds to a considerably higher interest rate when considered from this perspective.

Because of the many different ways of computing interest on installment loans and because of the misleading aspects of quotation of add-on interest rates, the Federal Truth in Lending Act was passed in 1969. It requires that with each loan, the lending institution quote an Annual Percentage Rate (APR), regardless of how the interest is computed. The APR is a uniform interest rate that can be used for comparative purposes when shopping for a loan and that gives an accurate representation of how much rent one is paying for the loan.

To find the APR, most loan officers rely on tables compiled by the Federal Reserve System. Sample pages from these tables are given in Table A-1 in the appendix. The following steps describe in detail how the table is used.

1. Compute (finance charge · 100) ÷ (amount financed).
2. Looking in the row labeled by the number of payments to be made, find the entry closest to the number found in step 1.
3. Find the percentage rate at the top of the column in which this entry is found. This is the APR rounded to the nearest one-quarter of a percent.

Example 47 • Find the APR for a loan of \$200 to be repaid with six monthly payments of \$36.40.

Solution

$$\text{Finance charges} = 6(36.40) - (200) = \$18.40$$

1. $\dfrac{(\$18.40)(100)}{\$200} = \$9.20$
2. Find the row in Table A-1 for six payments. Find the entry in that row nearest 9.20. The nearest entry is 9.23.

Just for fun

Deposit \$100 in the bank and then make withdrawals in the following manner.

	Withdraw		*Leaving a balance of*
	\$40,		*\$60.*
	\$30,		*\$30.*
	\$18,		*\$12.*
	\$12,		*\$ 0.*
Adding, we have	*\$100*	*and*	*\$102*

The total withdrawal is \$100, whereas the total of the balances is \$102. Can you now go to the bank to demand your extra two dollars?

---

3. The percentage rate at the top of the column is 31.00%. This is the APR (rounded to the nearest one-quarter of a percent). •

Example 48 • A car loan of \$2500 is to be financed for 3 years at $6\frac{1}{2}\%$ add-on interest with no additional finance charges. What is the APR?

The finance charge consists entirely of interest = (2500)(0.065)(3) = \$487.50.

1. $\dfrac{(487.50) \cdot 100}{2500} = \$19.50$

2. Find the row in Table A-1 for 36 payments. In this row find \$19.57.

3. This column is labeled as 12.00%. Twelve percent is the APR, accurate to one-quarter of a percent. •

Exercise set 6

R 1. Compute the interest paid for the following loans at the given add-on interest rate.

(a) \$1000 at 7% add-on interest for 2 years
(b) \$3500 at $6\frac{1}{2}\%$ add-on interest for 4 years
(c) \$5000 at $5\frac{1}{2}\%$ add-on interest for 3 years

2. Compute the finance charges for each of the following loans.

(a) A \$4000 loan for 2 years with \$480 interest, a required \$5 loan investigation fee, and a mandatory \$300 insurance coverage.
(b) A \$3500 loan with interest computed at 6% add-on for 3 years with a \$15 registration fee required.
(c) A \$500 loan to be repaid in 18 payments of \$36 each.
(d) A \$700 loan to be repaid in 24 payments of \$35.50 each.

3. Compute the monthly payments in Exercises 2(a) and 2(b).

4. Riverside Ford advertises that it will sell a new Pinto for \$3036. Furthermore, if the buyer will provide a down payment of \$200, Riverside will arrange financing for the remaining \$2836 with finance

charges of \$800. What is the monthly payment if the loan is for 48 months?

5. N.A.E. Finance Company advertises that \$500 can be repaid in 33 monthly payments of \$21 each. What is the finance charge in this case?

6. Wanda must have \$650 to buy a motorcycle. At N.A.E. Loan Company, she learns that the total interest paid will be \$30 and a registration fee of \$10 is also required. What would be the amount of her six monthly payments?

7. Raymond is buying a new car that costs \$7000. He trades in his old car as a down payment of \$2000. He has two options for financing the loan. He can obtain a 3-year loan with interest computed as 7% add-on, or he can obtain a 4-year loan with the interest computed as $6\frac{1}{2}\%$ add-on. In either case, he will have to pay an additional \$100 in finance charges.

   (a) In which option will he pay the largest dollar amount of finance charges?
   (b) In which option will he pay the largest monthly payment?

T 8. CIP Finance Company offers to loan Bill \$508 for 18 months, to be repaid in 18 monthly payments of \$31.

   (a) What is the finance charge?
   (b) What is the APR (rounded to the nearest $\frac{1}{4}\%$)?

9. Blueport National Bank offers to loan Bill \$500 to be repaid in 18 monthly payments of \$30.83.

   (a) What is the finance charge?
   (b) What is the APR (rounded to the nearest $\frac{1}{4}\%$)?

10. Bellevue National Bank will loan \$3000 for 3 years with interest to be computed as 6% add-on. A \$12 recording fee is required.

   (a) What is the finance charge?
   (b) What are the monthly payments?
   (c) What is the APR (rounded)?

11. First National Bank will finance \$2000 for 24 months toward the purchase of a car. The interest is computed as $6\frac{1}{2}\%$ add-on, with an additional \$11.50 acquisition fee. What is the APR (rounded)?

12. Doris needs to borrow \$1000 for one year. She can receive proceeds of \$1000 from a discounted note with interest of \$98.90, or she can borrow the \$1000 and repay it in monthly installments of \$93. On which loan would she pay the smallest dollar amount of interest?

C 13. Susan investigated three different ways to pay for a new car on which she owed \$3000. Which of these three plans for financing the car gave the best APR?

   (a) The car company offered to let her pay \$100.00 per month for 36 months.
   (b) Her bank offered to let her borrow the money for $6\frac{1}{2}\%$ add-on interest for $2\frac{1}{2}$ years with equal monthly payments.
   (c) Another bank offered her 6% add-on interest for 3 years with equal monthly payments.

Solution to introductory problem

Understand the problem. We first summarize the facts. It was thought that rent for the hotel room was \$30. Each of the three men paid \$10. The rent for the room was only \$25. The bellboy returned \$3 to the three men and kept a \$2 tip.

Make a plan. As good consumers, we must find exactly how the books must balance. Specifically, the amount received by the motel (\$25) and the bellboy (\$2) must equal the amount paid by the guests.

Carry out the plan.

Amount received by motel and bellboy		*Amount paid by guests*
\$25 + \$2	=	3(\$9)

There is no missing dollar. There is no reason why 30 must equal 3(9) + 2. 3(9) + 2 is the sum of the amount paid by the guests and the amount received by the bellboy—an irrelevant sum.

Summary and review

The real numbers

1. A number in decimal form which neither terminates nor repeats is an irrational number.
2. The real number system consists of rational numbers and irrational numbers.
3. In addition to the field properties, the real number system

   (a) is an ordered set (b) is dense.

4. Irrational numbers in exponential or radical form satisfy the following properties:

   (a) $x^{1/n} = \sqrt[n]{x}$

   (b) $(xy)^{1/n} = x^{1/n}y^{1/n}$ or $\sqrt[n]{xy} = \sqrt[n]{x}\sqrt[n]{y}$

   (c) $(x/y)^{1/n} = x^{1/n}/y^{1/n}$ or $\sqrt[n]{\dfrac{x}{y}} = \dfrac{\sqrt[n]{x}}{\sqrt[n]{y}}$

   (d) $(x^{1/n})^m = x^{m/n}$ or $\sqrt[n]{x^m} = (\sqrt[n]{x})^m$

Ratio and proportion

1. *Ratio* is a name used for the quotient of two numbers when this quotient is used to compare the sizes of two quantities.
2. The ratio of cost to quantity is called the *unit price ratio* and is used in comparison shopping.
3. Two ratios a/b and c/d are equal if and only if $ad = bc$.
4. A statement that two ratios are equal is called a *proportion.*

Percent

1. Percent is a ratio giving the number of parts per hundred.
2. Percentage = (percent)(base)

Interest

1. Simple interest is computed by $I = Prt$, where P is the principal, r is the simple interest rate, and time t is expressed in terms of years.
2. Amount $A = P + I$ or $A = P(1 + rt)$
3. For compound interest $A = P(1 + i)^n$, where i is the interest rate per period and n is the number of interest periods.
4. The APR rate is a uniform interest rate that can be used for comparative purposes. We obtain the APR from a table in the appendix.

Review exercise set 7

R 1. Solve for x in each proportion.

(a) $3:4 = x:7$ (b) $4/7 = 12/x$

2. Compute

   (a) 0.5% of \$660 (b) 105% of 77
   (c) 9/8 as a percent (d) 1.007 as a percent

3. Find the simple interest for a loan of \$1000 for 75 days at 12% simple interest.

4. After 2 years, a \$6000 loan is repaid with the amount \$6780. What simple interest rate was paid?

5. (a) An \$89 coat is sold for \$65. What is the percent of markdown?
   (b) John earns a commission of 4% on all sales in excess of \$10,000. What is his commission in a month in which he sells \$28,000 of merchandise?
   (c) A hat has a retail price of \$57.75. What was the wholesale price if the hat was marked up 65% based on cost?

6. Indicate whether each statement is true or false. If false, explain why.

   (a) 4 is an irrational number.
   (b) Every rational number is a real number.
   (c) Every point on a number line represents a real number.
   (d) Every real number is a rational number.
   (e) The irrational numbers are closed under multiplication.

(f) Every repeating decimal represents a real number.
(g) Every irrational number can be represented by a repeating decimal.
(h) $a/0$ is a rational number.
(i) If a and b are rational numbers with $ab = 0$ and $b = 0$, then $a = 0$.
(j) Division is always possible in the rational number system.
(k) 2/3 is the largest rational number less than 3.
(l) 1 is the smallest positive integer as well as the smallest positive rational number.
(m) 8 is the largest integer less than 9.
(n) Zero is a positive rational number.
(o) The set of rational numbers between 2 and 3 is dense.
(p) If a is any rational number, $a^2 > 0$.
(q) -1 is the largest negative rational number.

7. Simplify.

(a) $\sqrt{192}$ (b) $\sqrt{162}$ (c) $\sqrt{\frac{72}{25}}$ (d) $\sqrt{\frac{50}{36}}$

(e) $\sqrt[3]{27}$ (f) $\sqrt[3]{64}$ (g) $\frac{16^{-1/2}}{4^{-2}}$ (h) $\frac{27^{-1/3}}{3^{-2}}$

8. Write two irrational numbers greater than 0.853 and less than 0.854.

9. Find the compound interest and compound amount for an investment of \$4000 at 6% compounded semiannually for 4 years.

10. Find the present value of \$4000 due in 5 years if money is worth 6% compounded semiannually.

11. If the \$180 you spend each month on room and board represents 60% of your monthly income, what is your monthly income?

12. Brooke said "I have 20% less than Jodi," but Jodi replied "I have 25% more than Brooke." If Brooke has \$40 and Jodi has \$50, who is correct?

13. What is the selling price of an article that cost \$8.00 if the markup is 20% based on cost? What is the cost of an article that sells for \$2.40?

T 14. According to the National Safety Council, about 60% of all automobile-accident fatalities for a year occurred during daylight hours. If, during the year, 23,000 automobile-accident fatalities occurred during daylight hours, how many automobile-accident fatalities occurred that year?

15. If an article costs \$76.98 and sells for \$132.98, what is the percentage of markup based on the cost? On the selling price?

16. On which of the following loans will the most interest (dollar amount) be paid?

(a) A loan with principal of \$500 at 12% simple interest for 6 months
(b) A loan of \$500 for 6 months with interest at 12% compounded monthly.

Bibliography

1. Brown, Gerald W., & Lucien B. Kinney. "Let's Teach Them about Ratio." *MT* (April 1973), 352–355.
2. Cole, Blaine L., & Henry S. Weissenfluh. "An Analysis of Teaching Percentages." *AT* (March 1974), 226–228.
3. Dewar, Jacqueline M. "Another Look at the Teaching of Percent." *AT* (March 1984), 48–49.
4. Fennell, Francis (Skip). "The Newspaper: A Source for Applications in Mathematics." *AT* (October 1982), 22–26.
5. Friesen, Charles D. "Check Your Calendar Computations." *AT* (December 1976), 660.
6. Glatzer, David J. "Teaching Percentage: Ideas and Suggestions." *AT* (February 1984), 24–26.
7. Hawthorne, Frank S. "Hand-Held Calculators: Help or Hindrance?" *AT* (December 1973), 671–672.
8. Herron, J. D., & G. H. Wheatley. "A Unit Factor Method for Solving Proportion Problems." *MT* (January 1978), 18–21.
9. Reys, Robert E. "Estimation." *AT* (February 1985), 37–41.
10. Schmalz, R. "The Teaching of Percent." *MT* (April 1977), 340–343.

CHAPTER 7

An Introduction to Computers

Write a computer program to find all primes less than 1000.

No single technological innovation of the last half of the twentieth century can compare with the computer in its influence on all levels of society. Yet for most people the computer remains a mysterious black box, capable of solving great problems and at times perpetrating a bit of mischief (when bills are mailed). The purpose of this chapter is to remove some of the mystery and replace it with a modicum of understanding.

The first point to understand is that computers are not intelligent! They can perform only the basic arithmetic operations ($+$, $-$, $\cdot$, $\div$) and make simple decisions on the basis of the relations $<$, $=$, $>$ and the logical connectives *and* and *or*. The outstanding feature of the computer is that it can perform these operations with astounding speed and, when teamed with a clever human mind, can solve unbelievably complex problems. Our task will be to examine how this team, human and machine, works together to solve problems.

1 An introduction to computer literacy

■ You can "know about" the computer in at least two different ways. You can know about the machine, its components, and how it works internally. On the other hand, you can know how to use the machine and how to make it perform the steps you wish it to perform. ■

In this section, we become familiar with the computer, its potential, and terminology of the various parts of a computer. In Section 2, we turn our attention to programming a computer to make it work for us.

As you have observed, the computer has slipped into almost every aspect of our lives. Sometimes it is conspicuously present, as when we use the automated teller at the bank or when a computer-directed voice calls out prices in the check-out line of the grocery. Other applications are less conspicuous, as when a computer under the hood of a car directs the fine tuning of the motor's performance and informs us of the rate at which we are consuming gasoline. What is the secret of this machine that touches our lives in so many ways? What are its powers?

The everyday, ordinary computer can do essentially five things.

1. It can do four-function arithmetic; that is, it can add, subtract, multiply, and divide.
2. It can compare two numbers; that is, it can determine whether two numbers are equal, and if not, which is larger.
3. It can store and recall values of numbers and characters.
4. It accepts numbers and characters as input data.
5. It will print numbers and characters as output.

Given this fairly modest list of accomplishments, we might ask how the computer has affected our lives in so many ways. The answer lies in these facts: The computer can do these steps in any order prescribed by the human user, it can do them very, very fast, and it can do them very accurately. Because of its speed, it can do seemingly impossible tasks, such as sketching pictures, writing music, and even checking spelling.

The mechanical and electronic components of the computer are called **hardware.** The hardware of all computer systems includes an input unit, a central processing unit, a secondary storage unit, and an output unit. By means of the input unit, the human user communicates with the machine. On small desktop computers and cathode ray terminals (CRTs), the input unit is often a keyboard. The computer provides information to the user through the output unit. On desktop computers, the output can come either to a screen or to an attached printer. Information and instructions to be stored over a long period of time are stored using the secondary storage unit. On desktop computers, this is most often a floppy-disk drive; in computer systems, the secondary storage medium is either a large disk drive or magnetic tape.

The **central processing unit** is where the most important work of the computer is done. Within the central processing unit, there is an accumulator or arithmetic unit, where the intricacies of electronic circuitry enable the machine to add, subtract, multiply, and divide. There is a core memory, where the current lists of directions are stored, as well as any values that are being manipulated by these directions. Finally, there is a control unit, which directs traffic in the machine. It requests information from memory, requests computations from the arithmetic unit, and makes decisions on the basis of comparisons.

Of much more interest to us than hardware are the lists of instructions that can make the computer perform a specific task. These lists of instructions, prepared by human user called a *programmer,* are called **programs.** Programs and collections of programs are called **software.** Skillfully written programs can make the machine perform tasks as varied as handling the inventory of a business, making the machine into a word processor for the writing and editing of documents, or doing long and arduous mathematical computations.

The main purpose of this chapter will be to introduce you to the fundamentals of writing a program to solve a specific problem. In the process, we will learn to use two pieces of software: a **BASIC** interpreter and **Logo.**

Programs are communicated to the machine using computer languages. In the short history of the computer, a variety of languages have been written to serve this purpose, among them such familiar names as FORTRAN, COBOL, Pascal, and BASIC. BASIC, Beginner's All-purpose Symbolic Information Code, was written to allow users to learn quickly the fundamentals of communicating programs to the computer. It is available on virtually all desktop computers. BASIC will be a good starting point for your learning experience.

Logo was developed by Seymour Papert, a researcher at the Massachusetts Institute of Technology. Influenced by the ideas of Jean Piaget, a noted Swiss psychologist, Papert wanted to create a learning tool to help children and others to learn to think and solve problems. Initially, Papert created a turtle robot that children at a control panel could instruct to crawl across a floor covered with butcher paper. As it crawled it sketched a geometric design with a pen that it dragged along the floor. The challenge for the learners was to think of sequences of instructions that would cause the turtle to draw increasingly complex designs. Since most children do not have available a robot turtle to teach to draw designs, Papert helped design the Logo software that provided a turtle for every computer screen and hence a learning tool for every child with access to a computer. You will find that practice with Logo will increase your problem solving skills and your enjoyment of this material.

2 An introduction to the "BASIC" language

■ Write a BASIC program to obtain the pattern shown.

```
X        XXXXX   X         X   XXXXXXXX
X        X   X    X       X    X
X        X   X     X     X     XXXX
X        X   X      X   X      X
X    X   X   X       X X       X
XXXXXX   XXXXX        X        XXXXXXXX
```
■

A program is a sequence of directions that instructs the computer to perform a list of operations in a specified order. In the previous section, we observed that a computer can do arithmetic, store values, and produce output. In this section, we will learn how to instruct the computer to do these things using the BASIC language.

In a computer program, we can instruct the machine to store values in memory locations and retrieve values from memory locations. Variables in BASIC are names of memory locations in the computer in which numbers can be stored. Hence, when we use a variable name in a BASIC statement, we are in fact referring to the number stored in that location. To help us understand this, we can visualize the memory of a computer as a collection of boxes, each having a specific variable name. In a computer program, when we say A = 3, we mean "place the number 3 in the box named A." Similarly, if we say PRINT C, we mean "print the number stored in box C." In the circumstances shown in Figure 1, a 5 would be printed. Note that in BASIC the set of capital letters can be used as variable names.

Figure 1

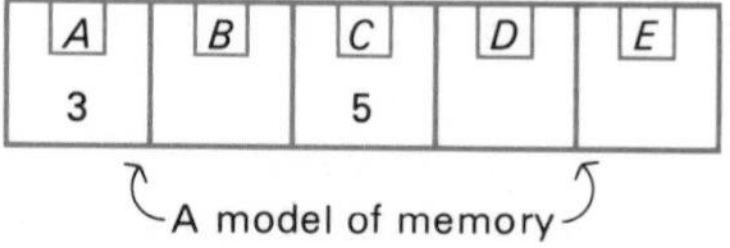

When we wish to assign a value to a memory location in BASIC, we use an assignment command of the form

Variable

Variable = value to be assigned
or LET variable name = value to be assigned.

In many versions of BASIC, the use of the word LET in assignment commands is optional. We will omit it in our discussion. Study the next example to see the various ways that an assignment command can be used.

Example 1 •

Statement	*Action*
10 A = 3	The value 3 is assigned to memory location A.
20 B = 7	The value 7 is assigned to memory location B.
30 C = A + B	The value in memory location C is 10.
40 A = A + 1	The value in A is now 4.

•

Understand that the = sign in an assignment statement does not indicate mathematical equality. Instead, statement 4Ø of the previous example, "A = A + 1," can be interpreted to mean "Let the old value in memory location A be replaced by a new value that is 1 larger." Observe also that in an assignment statement of the form [1]=[2], a single variable always goes in box [1] and a number, an arithmetic expression, or a variable that has previously received a value goes in box 2. Hence, 4 = A is an incorrect statement.

Did you notice the numbers preceding the statements in the preceding example? Since a BASIC program consists of a series of steps in a definite order, each instruction or program statement is numbered (called **line numbers**). Unless instructed otherwise, the machine executes the instructions in the order in which the statements are numbered.

Since the computer executes the program by line numbers rather than the order in which the instructions are entered, it is good practice to leave space between the line numbers of consecutive instructions so that other instructions can be added at a later time if desired. For instance, many people number instructions as 1Ø, 2Ø, 3Ø ,4Ø, and so on, allowing space for as many as nine later instructions (provided they do not change the meaning of the program) between each pair of steps.

Example 2 •

```
10  A = 3
20  Z = A^2
30  PRINT Z
40  END
```

(The statements in this program will be discussed later.) •

We observed that the right side of an assignment command can consist of an arithmetic expression to be evaluated and stored in a memory location. For this reason, we need to discuss arithmetic operations in BASIC. In BASIC, the set of symbols that represent the four arithmetic operations are very similar to those we use daily. In addition, the up arrow ↑ or caret ^ is used in BASIC to represent exponentiation. A special function SQR ( ) is available to compute square roots. Table 1 summarizes these operational symbols and their meanings.

Table 1

BASIC notation	*Operation*	*Example*	*BASIC notation*	*Result*
+	Addition	$5 + 7$	5 + 7	12
−	Subtraction	$9 - 4$	9 − 4	5
*	Multiplication	$3 \cdot 1.2$	3 * 1.2	3.6
/	Division	$15 \div 3$	15/3	5
↑ or ^	Exponentiation	2^3	2^3	8
SQR( )	Square root	$\sqrt{25}$	SQR(25)	5

Example 3 •

BASIC notation	*Arithmetic expression*	*Value*
3^2	3^2	9
(2^2) * 3	$(2^2) \cdot 3$	12
(2^3) − 24	$2^3 - 24$	−16
SQR(16)	$\sqrt{16}$	4
111/37	$111 \div 37$	3

•

To avoid errors, it is important to be aware of the order in which the computer will perform the various operations. Suppose that you typed the computation 3 * 2 + 7 into the machine. Would it compute

$$(3 \cdot 2) + 7 = 6 + 7 = 13$$

or

$$3(2 + 7) = 3 \cdot 9 = 27$$

The computer always follows the following rules of precedence of operations.

Order of operation

1. First it performs all exponentiation (raising to powers).
2. Then it performs all multiplications and divisions from left to right.
3. Next, it performs all additions and subtractions from left to right.
4. The order specified above can be modified by using parentheses. Expressions enclosed in parentheses are evaluated first.

Using these rules, then, 3 * 2 + 7 will be computed in BASIC as 13 and not 27.

Example 4 •

BASIC notation	*Arithmetic expression*	*Value*
5 − 2^2	$5 - 4$	1
2 * 3^2	$2 \cdot 3^2$	18
8/2/2	$(8 \div 2) \div 2$	2
2^3^2	$(2^3)^2$	64
4 + 2^3 − 14	$4 + 2^3 - 14$	−2
8/4 * 2	$\frac{8}{4} \cdot 2$	4
8/(4 * 2)	$\frac{8}{4 \cdot 2}$	1
SQR(9)/(5 − 2)^2	$\sqrt{9} \div (5 - 2)^2$	1/3

•

If you should forget the rules of precedence for operations or are uncertain as to their use, use parentheses. The computer always performs the operations inside parentheses first.

Example 5 •

BASIC notation	*Arithmetic expression*	*Value*
3 * 2 + 7	$3(2) + 7$	13
3 * (2 + 7)	$3(2 + 7)$	27
(3 + 4)^2 − 6	$(3 + 4)^2 - 6$	43
4^(5 − 3)	$4^{(5-3)}$	16
(10 + 5)/(8 − 11)	$\frac{10 + 5}{8 - 11}$	−5
(10 + 5)/3 − 5	$\frac{10 + 5}{3} - 5$	0

•

In writing algebraic expressions with BASIC, multiplication symbols as well as the grouping in quotients must be made explicit.

Example 6 •

Algebraic expression	*BASIC expression*
$2x^2 + 3x + 7$	2 * X^2 + 3 * X + 7
$\dfrac{x^2 + 1}{x - 1}$	(X^2 + 1)/(X − 1)
$b^2 - 4ac$	B^2 − 4 * A * C
$\sqrt{a^2 + b^2}$	SQR(A^2 + B^2)

•

In BASIC language, rational numbers are written as decimals or decimal approximations. For very large and very small numbers, BASIC uses a form of scientific notation. The power of 10 is expressed as E ± exponent. E + Ø2 means multiply by 10^2 or move the decimal point two places to the right. E − Ø3 means multiply by 10^{-3} or move the decimal point three places to the left.

Example 7 • (Assume that our BASIC numbers are rounded to six digits.)

Ordinary form	*BASIC form*
−5	−5
1/8	0.125
2/3	0.666667
π	3.14159
1,000,000,000	1E + Ø9
0.00172	1.72E − Ø3
7325.6	7.3256E + Ø3

•

Of course, it does no good to ask the machine to do an arithmetic computation unless we can instruct it to give us the results of the computation. In BASIC, the output is obtained using the PRINT command.

The PRINT command can be used for four different purposes.

1. To print the value currently stored in a variable.
2. To print the results of a computation.
3. To print the messages and labels.
4. To provide a blank line in output.

Study the following examples to see how this is done.

Example 8 •

Statement	*Action*
1Ø A = 2 2Ø B = 4 3Ø PRINT A, B 4Ø PRINT 2 * A + B^2 5Ø End	In line 3Ø the values of A and B are printed, and in line 4Ø the number $2(2) + 4^2$ is computed and printed. The output is given below.
2 4	(Line 3Ø printed)
2Ø	(Line 4Ø printed)

•

Ordinarily, the result of a PRINT command is printed on one line and the result of the next PRINT command is printed on the next line.

Example 9 •

```
10  PRINT 2 * 3
20  PRINT 3 + 9
30  PRINT 2 * 3, 3 + 9, 8/4, 11 - 4
40  END
```

The answers would be printed as follows.

```
6
12
6    12   2   7
```
•

The PRINT command without a variable, number, or message can be used to obtain a blank line of output.

Example 10 •

```
10  PRINT 2 * 3
20  PRINT 3 + 9
30  PRINT
35  PRINT 7^2
```

The computer response will be as follows.

```
6
12

49
```
•

A PRINT statement followed by words or symbols enclosed in quotation marks will output the statement enclosed in quotation marks exactly as entered.

Example 11 •

```
10  A = 3
20  B = 4
30  LET C = A + B
40  PRINT "THE SUM OF A AND B IS   " C
50  END
```

The computer will print the following.

```
THE SUM OF A AND B IS 7
```
•

Now let us return to the example with which we began this section.

Example 12 • Write a BASIC program to obtain the pattern shown.

```
X         XXXXX   X          X   XXXXXXXX
X         X   X    X        X    X
X         X   X     X      X     XXXX
X         X   X      X    X      X
X    X    X   X       X  X       X
XXXXXX    XXXXX        X         XXXXXXXX
```

Just for fun

In a casino, a customer places quarters on numbers and the house places silver dollars on numbers. The first one to put money on squares with a total of 15 wins what the other has on the board.

EXAMPLE. Customer, 25¢ on 7, the house, \$1 on 8; customer, 25¢ on 2, and the house blocks with \$1 on 6; customer blocks with 25¢ on 1 and the house places \$1 on 4; customer blocks with 25¢ on 5 but the house places \$1 on 3 and wins 4 · 25¢ = \$1 since 8 + 4 + 3 = 15.

The algorithms for computer programs often apply to many other problems. Show that the game above is similar to tic-tac-toe and a 3-by-3 magic square with totals of 15. Can the house always win? [Gardner, 1978]

Solution The key to the solution hinges on the fact that the computer will print any symbols enclosed in quotes exactly as those symbols are typed. Hence, we may print this word with six PRINT commands. The first PRINT command is

```
10  PRINT "X                   XXXXX    X               X    XXXXXXXX"
```

Can you supply the next five PRINT commands? (See Exercise 14.) •

A semicolon or comma at the end of a PRINT line indicates that the output from the next PRINT instruction should be printed on the same line.

Example 13 •

```
30  PRINT "2 * X^2 =  ";
40  PRINT 2 * 4^2
50  END
```

The output will be printed on one line as follows.

```
2 * X^2 = 32
```

•

Notice that the programs in the two previous examples are concluded by an END statement. The last statement to be executed in any BASIC program is an END statement.

Exercise set 1

R 1. Evaluate the given BASIC expressions.

(a) 3^2 (b) 5 − 2
(c) 8/5 (d) (3 * 2)^2
(e) 3 * 2^2 (f) 3 * (2^2)
(g) (1/5) + 2 (h) 1/5 + 2
(i) 1/(5 + 2) (j) 2^4 − 3
(k) 2^(4 − 3) (l) (2^4) − 3
(m) (3 * 7) − 4 (n) 3 * 7 − 4
(o) 3 * (7 − 4) (p) 24/3/2/2
(q) 2^2^2 (r) 8/(2/2)
(s) 1.0E + 03 − 10^2 (t) 6.1E − 02
(u) 2.07E + 04

2. Write a BASIC expression (without parentheses if possible) to evaluate each of the following.

(a) The product of 6 and the cube of 5
(b) The division of the square root of 104 by 16
(c) The product of ten 16's
(d) The square of the sum of 9 and 12
(e) The fourth power of the sum of 16.1 and 17.3

3. Write a BASIC expression for each of the following.

(a) $4x + 7$ (b) $5x^2 - 3x$

(c) $2x^2 + 7x + 4$ (d) $\dfrac{4x^2 - 3}{5x + 7}$

4. Write each of the following BASIC expressions, using ordinary algebraic notation.

(a) 4 * X^2 − 3 * X + 2
(b) (2 * X^2 + 1) * (4 * X + 3)^2
(c) (X − 5)^2 + 4 * (X − 5)^3
(d) (3 * X^2 − 4)^3

5. What will the computer output for each of the following programs?

(a)
```
10  A = 2
20  B = 5
30  PRINT A * B
40  END
```
(b)
```
10  X = 2
20  A = 2 * X^2 + 7 * X − 1
30  PRINT A
40  END
```
(c)
```
10  X = 1
20  LET A = X^2/4 + X^3^2
30  PRINT "THE SOLUTION IS'  "A
40  END
```
(d)
```
10  X = 5
20  PRINT "B^2 − 4 * A * C =  ";
30  PRINT X^2 − 4 * X + 3
40  END
```
(e)
```
10  A = 5
20  PRINT A + 6
30  PRINT 5 * 6, (A + 6)^2
40  END
```
(f)
```
10  X = −2
20  Y = X^3
30  PRINT Y − X
40  END
```

6. Determine which of the following program statements are incorrect, and correct them.

(a) 15 PRINT C * D
(b) 40 PRINT I LOVE YOU
(c) B + C = A
(d) 40 C = A^2
(e) PRINT 2 * A

7. What is the output of this program?

```
10  PRINT "ROW  ";
20  PRINT "ROW  ";
30  PRINT "ROW  ";
40  PRINT "YOUR BOAT"
50  PRINT "GENTLY  ";
60  PRINT "DOWN THE STREAM"
70  END
```

8. Write without the E notation.

(a) 7.63E − 04 (b) 1.3333E − 07
(c) 7.156E + 02 (d) −1.71625E + 01

9. Write each, using E notation.

(a) 4176.2 (b) 714,000 (c) 0.000013
(d) −0.03217 (e) 1,000,000 (f) 0.0000001

T 10. What is wrong with each of the following BASIC programs?

(a)
```
10  10 = C
20  20 = B
30  D = B + C
35  PRINT D
40  END
```
(b)
```
10  C = 7
20  D = 14
30  E = CD
35  PRINT E
40  END
```

11. Use a calculator to find the value of each BASIC expression.

(a) (14.1 − 7.3)^2

(b) 18.6 − 7.1/5
(c) 100.6/5.2/4
(d) 7.1^2 − 5
(e) (14.6/8.5) * (4.1^2 − 6.1)
(f) 14.6/8.5 * (4.1^2 − 6.1)
(g) 14.6/8.5 * 4.1^2 − 6.1
(h) 17.6/5.2/.8/.4

C 12. Perform the following computations and express the answers in BASIC scientific notation.

(a) $(2.02E + 11) \cdot (4.11E - 12)$
(b) $(7.83E + 14) \cdot (4.11E - 11)$
(c) $43 \div 10^{17}$
(d) $(200)^6$

13. Run the following programs on a computer. Try to determine the purpose of the BASIC commands with which you are not familiar.

(a)
```
10 INPUT A, B
20 C = A^2 + B^2
30 D = SQR(C)
40 PRINT "THE SQUARE ROOT IS" D
50 END
```

(b)
```
 5 PRINT "LEG 1", "LEG 2",
 6 PRINT "HYPOTENUSE"
10 READ A, B
15 IF B = 0 THEN 40
20 C = SQR (A^2 + B^2)
25 PRINT A, B, C
30 GOTO 10
35 DATA 3, 4, 7, 8, 1, 3, 0, 0
40 END
```

14. Write a program for the problem in the introduction.

3 Some "BASIC" statements

■ Electric bills in the town of Cary are based on a two-tiered billing system. If usage is less than 1000 kilowatt hours, the bill is \$0.04 per kilowatt hour. For kilowatt hours exceeding 1000, the user pays \$0.025 per kilowatt hour. The Cary Utilities Board needs a program that accepts usage as input and produces the amount of the electric bill as output. Can you produce such a program? ■

In Section 2, we learned how to assign values to memory locations, how to do arithmetic, and how to obtain output. In this section, we will learn how to provide the computer with the values we want it to use in executing our programs. More importantly, we will learn how to instruct the computer to compare two numbers and choose between two different paths of action on the basis of the comparison.

In most programs, we need to provide the computer with the values that we want manipulated in the program. In many programs, we need to do this while the program is running. The INPUT command allows the programmer to place a value in a memory location at any point in the program.

INPUT

The general form of the INPUT statement is

INPUT variable, variable, . . . , variable

Example 14 •

```
INPUT A, B, C
```
•

When the machine executes an INPUT statement, it prints a question mark. The user of the program is expected at that time to type in a numerical value for each variable in the INPUT statement, placing commas between each of the numbers.

Example 15 •

```
10  INPUT A, B
20  PRINT A + B
30  END

?   21, 36
57
```

The computer displays the ? and the user types 21, 36. Note that the output is 57. •

Example 16 • Earlier in the text, we learned that to compute the simple interest due on a loan, we use the formula

$$I = Prt$$

where P is the principal of the loan, r is the annual interest rate, and t is the time of the loan. Write a program to input the principal, rate, and time and compute the interest due on the loan.

Solution

```
10  PRINT "ENTER PRINCIPAL, RATE, AND TIME"
20  INPUT P, R, T
30  I = P * R * T
40  PRINT "THE INTEREST IS" I
50  END
```

When lines 1∅ and 2∅ of this program are executed, the computer will output

ENTER PRINCIPAL, RATE, AND TIME

If we enter the values 200, .12, 2, the computer will output

THE INTEREST IS 48 •

In line 1∅ of the program of the previous example, we printed a prompt that ensures that the computer user knows what to enter from the keyboard. In line 4∅, we printed a label for the value that is output that ensures that the user understands the meaning of this value. It is good programming practice to print a message before every INPUT command and to print a label for all output.

One of the most important capabilities of the computer is its ability to compare two values and then choose a path of action on the basis of this comparison. The BASIC commands that can be used to instruct the computer to choose between two different paths of action are the IF . . . , THEN and GOTO commands. The GOTO command gives the computer instructions to proceed directly to the given line of the program and perform the operation described there.

GOTO statements

The general form of the GOTO statement is

GOTO (line number)

and the computer goes to that line number for the next instruction.

Example 17 •

Statement	*Result*
50 GOTO 30	The computer goes to line 3∅ for the next operation.

•

Example 18 •

```
10 INPUT A, B, C
20 PRINT B^2 - 4 * A * C
30 GOTO 10
40 END
```

In this program, the control of the computer returns to line 1∅ immediately after executing line 2∅. •

The IF . . . THEN command instructs the computer to alter numerical sequence if a certain algebraic condition is true.

IF . . . THEN statement

The general form of the IF . . . THEN statement is

IF (algebraic condition) THEN (line number)

When the algebraic statement is true, the next instruction is found at the line indicated; if false, the computer continues to the next line.

Example 19 •

Statement	*Result*
40 IF X < Y THEN 20	If the value of x is less than y, then the next instruction will be found at line 2∅. If x is not less than y, the program continues with the next line after 4∅.

•

Now consider this BASIC program.

```
10 INPUT C, D
20 S = C + D
30 IF S > 7 THEN 50
40 GOTO 60
50 PRINT S
60 END
```

If the values 2 and 3 are entered for C and D in line 1∅, the value assigned to S in line 2∅ is 5. Since the algebraic expression in line 3∅ ($S > 7$) is false in this case, the machine next executes line 4∅. In line 4∅, the instruction is to execute line 6∅, which is END.

If the values 5 and 6 are entered for C and D in line 1∅, the value assigned to S is 11. Since the expression ($S > 7$) is true in line 3∅, the machine next executes line 5∅ and prints the value of S. Then the machine executes line 6∅.

Now let us return to the problem with which we introduced this section.

Example 20 • Electric bills in the town of Cary are based on a two-tiered billing system. If usage is less than 1000 kilowatt hours, the bill is \$0.04 per kilowatt hour. For kilowatt hours exceeding 1000, the user pays \$0.025 per kilowatt hour.

Solution Before solving this problem on the computer, we must be sure we understand it. Look at these examples.

Usage (K)	*Bill (B)*
700 kilowatt-hours	(0.04)(700) = \$28.00
1200 kilowatt-hours	(0.04)(1000) + (0.025)(200) = \$45.00

After careful thought, we might conclude that a good plan for the solution of this problem might be described in this way.

Enter the usage K
If the usage $K > 1000$
  then the bill $B = .04(1000) + .025(K - 1000)$
  otherwise the bill $B = .04K$
Print the bill B

This design can be translated into a BASIC program in the following way, using IF . . . THEN and GOTO.

Program

```
10 PRINT "ENTER THE USAGE"
20 INPUT K
30 IF K > 1000 THEN 60
40 B = .04 * K
50 GOTO 70
60 B = .04 * 1000 + .025 * (K - 1000)
70 PRINT "THE BILL IS", B
80 END
```
•

As a final note, observe that we have used the relations $>$ and $<$ in the algebraic condition of IF . . . THEN statements. In fact, any of the relations shown in Table 2 may be used.

Table 2

Mathematics symbol	*BASIC symbol*
$=$	`=`
$<$	`<`
$\leq$	`<=`
$>$	`>`
$\geq$	`>=`
$\neq$	`<>` or `><`

Example 21 •

```
40 IF B <= C THEN 80
80 IF N <> 4 THEN 60
```
•

Just for fun

Write a BASIC program for obtaining this drawing.

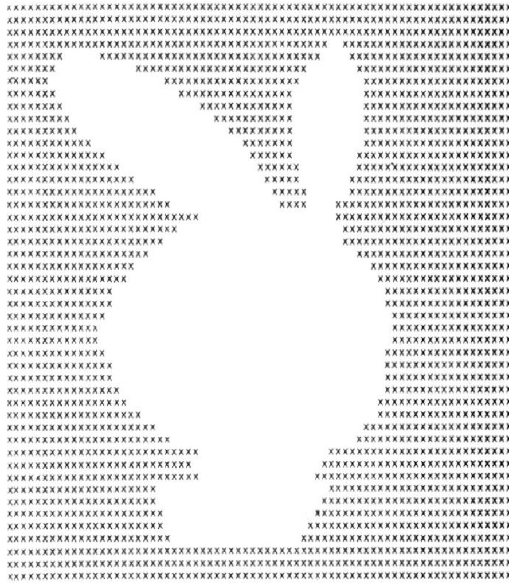

Exercise set 2

R 1. What is the output of the following programs?

(a)
```
10  INPUT A, B
20  X = A^B
30  PRINT A, B, X
40  END
? 3, 4
```

(b)
```
10  INPUT X
20  A = X^2/4 + X^3
30  PRINT "THE VALUE IS " A
40  END
? 1
```

(c)
```
10  PRINT "ENTER THE RATE AND BASE"
20  INPUT R, B
30  P = R * B
40  IF P > 100 THEN 60
50  GOTO 70
60  PRINT "THE PERCENTAGE IS " P
70  END
? .04, 1500
```

(d)
```
10  N = 1
20  IF N >= 10 THEN 60
30  PRINT N * N
40  LET N = N + 1
50  GOTO 20
60  END
```

2. After thinking about a problem, a computer programmer will often write a careful description of her ideas before writing the program. This careful description is sometimes called a **design.** Write a BASIC program for the following designs.

(a) Enter the principal P, rate R, and time T
   Compute the amount $A = P + PRT$
      Print the amount A

(b) Enter the rate R and the base B
   Compute the percentage $P = RB$
      Print the percentage P

(c) Enter the rate R and the base B
   Compute the percentage $P = RB$
   If the percentage $P < 100$
      then print the message,
      "TOO SMALL"
      otherwise print the message,
      "TOO LARGE"

3. Find any errors in the following programs and correct them.

(a)
```
INPUT A, B
C = A^B
PRINT C
END
```

(b)
```
10  INPUT 10, 20 30
20  A = 10 + 20
30  PRINT A
40  END
```

(c)
```
10  INPUT X
20  IF X < 10 THEN X = X + 10
30  GOTO 50
40  PRINT X
50  END
```

4. Charlie has worked hard on the problem of converting measurements in meters to either centimeters or hectometers. He has come up with the following design for a solution to the problem.

   Enter the measurement in meters M
   If $M < 1$
   then convert M to centimeters C using
   $C = 100M$
   Print C
   otherwise convert M to hectometers H using
   $H = M/100$
   Print H

   (a) What is output if 0.872 is entered for M?
   (b) What is output if 241 is entered for M?
   (c) Write a BASIC program for this design.

T 5. Celsius temperatures can be converted to Fahrenheit temperatures by multiplying the Celsius temperatures by $\frac{9}{5}$ and then adding 32. Write a program to input a Celsius temperature and output the corresponding Fahrenheit temperature. Run the program for these temperatures.

   (a) 30°C (b) −10°C (c) 0°C

6. Modify the program from Exercise 5 to output the Fahrenheit temperature only when it falls below 32°.

7. Write a program to compute and print the output for the compound amount A for the following values of P, i, and n, respectively.

   (a) \$100, 0.04, 3 (b) \$2000, 0.02, 4

8. Write a program to obtain the present value or the principal in compound interest where the inputs are A, i, and n. What does the computer print when

   (a) $A = \$1666$, $i = 0.08$, $n = 10$?
   (b) $A = \$3650$, $i = 0.10$, $n = 14$?

9. John makes \$4 an hour for his first 40 hours work each week, and he makes \$5.50 an hour for all hours over 40 hours. Write a program to input the number of hours he works in a given week and output his pay for the week.

C 10. Water in the town of Apex is billed on a two-tiered system. For usage below 8000 gallons, the customer must pay \$1.50 per thousand gallons. If usage exceeds 8000 gallons, the customer must pay \$1.25 for each thousand gallons over 8000. Write a program that inputs usage and outputs the cost.

4 Solving problems using the computer

■ Find the sum of the first 1000 integers. First determine how to do it yourself. Then produce a computer program that will execute your solution. ■

To this point, we have been concerned primarily with the details of the BASIC language. However, when solving a problem using a computer, writing the program is often both the easiest and the last step of the process. More precisely, the fundamental steps in solving a problem on the computer are as follows:

1. Determine how to solve the problem yourself.
2. Break down your solution into small computer-sized steps (input, output, arithmetic, comparison, and storage of values).
3. Write your solution, using the BASIC language or some other language.

In solving a problem on the computer, the most difficult steps of the solution process are often the steps of solving the problem yourself and then breaking the solution into "computer-sized" pieces. To complete these steps, you can bring to bear all the problem-solving skills that you learned in earlier chapters. One good strategy is to summarize your thinking in outline form as you go. Although you may need to revise your outline several times, when completed it will be very helpful in writing the program. We will call such an outline a **design** of your solution.

Example 22 • To be a member of the security force at the M-Mart department store chain, an applicant must exceed 70 inches in height and weigh more than 150 pounds. Write a program that will enable the personnel office to determine if a candidate meets the minimum qualifications.

Solution After some thought, we might arrive at an outline like this:

Enter the height H and the weight W of the applicant
If the height $H <= 70$
  then print the message "Too small"
  otherwise we need to check the weight W

After a little more thought, we might revise the design to look like this:

If the height $H <= 70$
  then print the message "Too small"
  otherwise if the weight $W <= 150$
    then print the message "Too small"
    otherwise print "Satisfactory size"

We are now ready to write a program. We must remember that the consequence to be followed if the algebraic condition is false (indicated by "otherwise" in our design) is specified on the line immediately following the appropriate IF . . . THEN statement.

```
10  PRINT "ENTER THE HEIGHT AND WEIGHT"
20  INPUT H, W
30  IF H <= 70 THEN 70
40  IF W <= 150 THEN 70
50  PRINT "SATISFACTORY SIZE"
60  GOTO 80
70  PRINT "TOO SMALL"
80  END
```
•

Example 23 • Write a program that will instruct the computer to print the squares of the first 25 positive integers. In other words, we wish to print

$$1, 4, 9, 16, 25, 36, 49, \ldots, 625$$

Solution We can obtain the positive integers by initializing a memory location N at 1 and then adding 1 to N repeatedly. N would then hold the values 1, 2, 3, 4, ... After initializing N at 1, we will need to repeat this sequence of steps 25 times.

Compute the term $T = N^\wedge 2$
Print the term T
Let $N = N + 1$

The arrow indicates that we do these three steps repeatedly. We must face the question of how to stop the process after only 25 times. The following refinement should do the job nicely.

Let $N = 1$
If $N > 25$
  then the process should stop
  otherwise the term $T = N\hat{\ }2$
Print T
Let $N = N + 1$

Now, by translating the arrow with a GOTO statement, the BASIC program can be completed as follows.

```
10  N = 1
20  IF N > 25 THEN 70
30  T = N^2
40  PRINT T
50  N = N + 1
60  GOTO 20
70  END  •
```

In the previous example, we repeated a sequence of instructions several (25) times. In this circumstance, the portion of the program that is repeated is called a **loop.** Most computer programs are built around one or more loops. Let us use a loop to solve the problem we posed at the beginning of this section.

Example 24 • Write a program that computes the sum of the first 1000 integers.

Solution Remember that we must have a clear idea about how to do the problem ourselves. Imagine the numbers from 1 to 1000 written in a row across a page. Now start with a sum of 0 and begin to add the integers, one at a time.

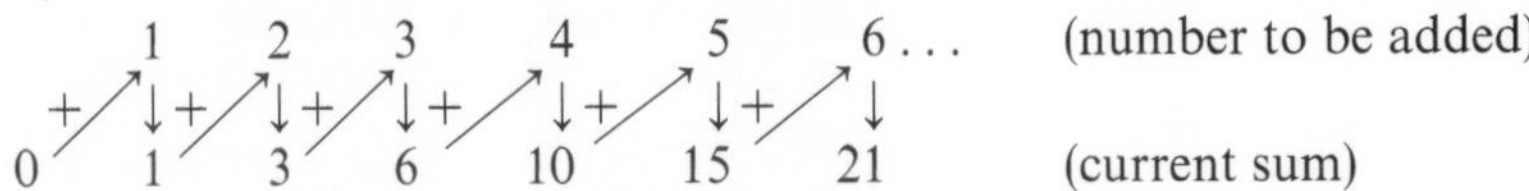

In the program for this problem, we must have a variable S for the current sum and a variable N for the number to be added. It appears that the following steps are performed repeatedly and hence should form the body of a loop.

Increase the current sum S by the value of N, (let $S = S + N$)
Increase N by 1 (let $N = N + 1$)

We need to initialize N at 1 and S at 0, and we need to include a decision statement to stop the loop. A final design might look like this:

Let $S = 0$
Let $N = 1$
If $N > 1000$
If this condition is not satisfied
  Let $S = S + N$
  Let $N = N + 1$
If this condition is satisfied, print the sum S

Again translating the arrow with a GOTO, we get the following program.

```
10  S = 0
20  N = 1
30  IF N > 1000 THEN 70
40  S = S + N
50  N = N + 1
60  GOTO 30
70  PRINT "THE SUM IS " S
80  END
```
•

In the previous two examples, we have used a variable N that "counted" the number of times a loop was completed. That is, N held the value 1 the first time the loop was executed, then incremented to 2, then to 3, and so on. Computer programs use such a "counter" in a variety of ways.

Example 25 • To say that the inflation rate is 6% means that at the end of every year an item costs 1.06 times what it cost the previous year. Assume that we have a constant inflation rate of 6% for a number of years, and suppose a new car now costs $9000. Compute and print the projected price of the car each successive year. Continue until the price of the car has doubled.

Solution After some careful thought, we determine that the following design will solve this problem.

Initialize the cost C at 9000
Initialize the counter N at 0
If the cost C has doubled ($C > 18000$)
  then print "The cost doubled after N years"
Otherwise print the cost C and the counter N
Increment the counter N by 1
Compute cost $C = 1.06C$

Can you write a BASIC program for this solution?

```
10  C = 9000
20  N = 0
30  IF C > 18000 THEN 80
40  PRINT "AFTER" N "YEARS THE COST IS" C
50  N = N + 1
60  C = 1.06 * C
70  GOTO 30
80  PRINT "THE COST DOUBLED AFTER" N "YEARS"
90  END
```
•

Before running a program on the computer, it is useful to check for errors by tracing the program by hand. Label columns for each of the variables used in the program and go through the program line by line, recording the current values held by each variable.

Just for fun

What is the output of this computer program?

```
 10  N = 1
 20  PRINT "ROW ";
 30  IF N = 3 THEN 90
 40  N = N + 1
 50  GOTO 20
 60  PRINT "MINNOW";
 70  PRINT "THE STREAM"
 80  GOTO 150
 90  PRINT "YOUR BOAT ";
100  GOTO 130
110  GOTO 110
120  PRINT "GREEN WATERS";
130  PRINT "GENTLY DOWN ";
140  GOTO 70
150  END
```

Example 26 • Trace this program.

Solution

Program				*Trace*
10 LET F = 1	*F*	*N*	*H*	
20 PRINT F	1	1		Lines 10, 30
			1	Line 60
30 LET N = 1		2		Line 70
40 IF N > 20 THEN 100	1			Line 80
50 PRINT N	2	3	2	Lines 60, 70, 80
60 LET H = N	3	5	3	Lines 60, 70, 80
70 LET N = N + F	5	8	5	Lines 60, 70, 80
80 LET F = H	8	13	8	Lines 60, 70, 80
90 GOTO 40	13	21	13	Lines 60, 70, 80
100 END				

What famous sequence will this program print? •

Exercise set 3

R 1. Give the output for each program.

(a)
```
10  N = 1
20  T = N^3
40  PRINT T
50  IF N = 6 THEN 80
60  N = N + 1
70  GOTO 20
80  END
```

(b)
```
10  T = 1
20  PRINT T
30  IF T > 20 THEN 60
40  T = T + 3
50  GOTO 20
60  END
```

(c)
```
10  INPUT A
20  B = A^2 + 2 * A
40  PRINT A, B
45  IF A = 4 THEN 60
50  GOTO 10
60  END
?1, 2, 3, 4
```

(d)
```
5   S = 0
10  K = 1
15  IF K > 5 THEN 40
20  T = 2 * K
25  S = S + T
30  K = K + 1
35  GOTO 15
40  PRINT S
45  END
```

(e)
```
3   X = 0
6   PRINT "X =  " X, "X^2 =  " X^2
9   X = X + 2
12  IF X > 8 THEN 18
15  GOTO 6
18  END
```

(f)
```
5   N = 1
10  IF N > 5 THEN 40
20  PRINT "H";
25  PRINT "I"
30  N = N + 1
35  GOTO 10
40  END
```

2. Determine which of the following instructions are incorrect and discuss what is wrong.

(a) `GOTO 70`
(b) `20  GO to 40`
(c) `10  GOTO END`
(d) `80  IF X > 10 THEN 80`
(e) `60  IF X >< THEN 40`
(f) `70  IF X = 4 THEN X^2 = 16`

3. Determine the incorrect lines of the following program, discuss what is wrong, and correct the mistakes. Tracing the program may help find the error.

```
10  N = 2
20  S = 0
30  IF N > 10 THEN 40
40  S = S + 2N
45  N = N + 1
50  GOTO 10
60  PRINT S
70  END
```

4. One program discussed in this section prints the squares of the first 25 positive integers. Modify that program to compute the squares of the first 50 positive integers.

5. (a) Write a program to add the first 500 counting numbers.
   (b) Modify this program to input Q and add the first Q counting numbers.
   (c) Trace the program from (b) by hand, using $Q = 7$.
   (d) Execute this program for $Q = 35$.

6. Write a BASIC program to input Q and compute $(S * (S + 1))/2$ where S is the sum of Q counting numbers. Run this program with $Q = 7$ and $Q = 35$.

7. Write a program for this design:

Initialize the counter N at 1
Let the term T start at 1
  If $N > 10$
    then END
    Otherwise print T
      Let $T = T + 3$
      Let $N = N + 1$

T 8. Taxes in the state of West Carolina are paid on the sliding scale shown below. Write a program that will input income and output tax.

Income	*Tax rate*
\$0–\$8000	\$0.02
\$8000–\$16,000	\$160 + 4% of excess over \$8000
\$16,000–\$20,000	\$480 + 6% of excess over \$16,000
\$20,000 and up	\$1320 + 10% of excess over \$20,000

9. Write a program that will accept a measurement in centimeters. If the measurement exceeds 100, it should output the corresponding measurement in meters. If the measurement is less than 1, it should output the corresponding measurement in millimeters. Otherwise, it should output the same measure as was input. Correct labels should be printed with all output.

10. To be a policeman in the town of Postem, you must be between 160 cm and 190 cm tall and weigh between 70 kg and 100 kg. Write a program to determine whether a candidate qualifies to be a policeman.

11. Write a program to compute
    (a) the sum of the squares of the first 100 positive integers

    $$(1 + 4 + 9 + 16 + \cdots)$$

    (b) the sum of the squares of the first Q positive integers where a value for Q is input in the program.
12. Write a program that
    (a) allows you to enter an inflation rate and initial cost of an item.
    (b) computes and prints the projected price of that item at the end of each year for 10 years.

C 13. Write a program to output a table like the following for the first 25 positive integers.

Number	*Square*	*Cube*	*Square root*
1	1	1	1
2	4	8	1.414
3	9	27	1.732

14. Now that computers have been placed in elementary classrooms, a favorite form of punishment may soon be outmoded. Write a program that will produce 500 copies of the sentence, "I will not talk during the reading lesson."
15. A fat lazy frog falls into a well that is 30 feet deep. Each day he climbs 3 feet, only to slip back 2 feet while he sleeps at night. Write a computer program that will compute how long it will take him to climb out of the well.
16. Professor I. M. Lazy wishes to program his computer to compute semester grades. I. M. gives 3 unit tests and a final exam that is weighted double. He averages these scores and assigns grades on the following scale:

$90 \leq$ average	A
$80 \leq$ average < 90	B
$70 \leq$ average < 80	C
average < 70	U

EXAMPLE. John made 65, 85, 75 on 3 unit tests and 80 on the final.

$$\text{Average} = \frac{65 + 85 + 75 + 80 + 80}{5} = 77; \text{ grade} = \text{C}$$

Write a program to input test grades and output both the numerical average and the letter grade.

5 More "BASIC" tools for problem solving

While reading a manual in the computer laboratory, Bill discovered the INT function that is built into BASIC. He learned that INT(number) is the largest integer less than or equal to that number. For example, INT(10.52) = 10, and INT(4.3) = 4. Experimenting a bit, Bill discovered that

INT(10/2) = 5   INT(10/3) = 3   INT(10/4) = 2

Said Bill, "I can use this function to test numbers for divisibility."

■ Determine if you can use INT( ) to write a BASIC program that will accept a number as input and determine if it is divisible by 3. ■

At this point in your study, you have learned enough BASIC to understand the fundamental ideas in programming. In this section, you will learn BASIC commands that make both data entry and the programming of loops more efficient.

We have found the INPUT command to be useful for data entry. This command allows the programmer to enter data at any point in the program and to enter different data each time the program is run. There are times, however, when this is not the best way to input data. When large amounts of data are to be entered, the fast machine must wait for the slow human each time data is needed. The READ . . . DATA statements allow data entry to occur at the time the program is written.

The READ and DATA statements go together as a pair. The general form of the READ statement is "READ variable, variable, . . . , variable," and the general form of the DATA statement is "DATA number, number, . . . , number." When the program is typed into the computer, the computer combines all the numbers in the DATA statements into a long list. Then, each time a READ statement is executed, the computer assigns the next number in the list to the appropriate variable. Immediately after being assigned, the number is dropped from the list.

Example 27 •

Statement	*Action*
10 READ A, B 20 DATA 1, 4, 6, 8 30 READ C, D	The variable A is assigned the value 1; the variable B is assigned the value 4. In line 3∅, the variable C is given value 6, and D is assigned the value 8.

Example 28 •

```
10 DATA 1, 3, 7
20 READ X
30 PRINT X^2
40 IF X = 6 THEN 70
50 GOTO 20
60 DATA 6
70 END
```

(a) The data list formed in the machine is 1, 3, 7, 6.
(b) X is first assigned the value 1; then X receives consecutively the values 3, 7, and 6.
(c) The output is

1 9 49 36

(d) The DATA statements can occur anywhere in the program as long as they precede the END statement. •

Example 29 • The payroll of a large factory is compiled using a computer. Each week, a card is punched for each worker giving an I.D. number, the wage per hour, and the number of hours worked. This information is then input to the computer. Write a program to compute the gross pay of each worker and output it along with the I.D. number.

Solution This problem is solved by multiplying (hours) (wage per hour) and then printing the result. Hence the solution will be built around a loop that includes these instructions.

Get the I.D. number I, the wages W, and the hours H
Compute the pay $P = WH$
Print the I.D. number I and the pay P

A program fragment for this portion of the design is given below. Observe that the program includes enough DATA statements to enter the triple, I, W, H, for each worker in the factory.

```
 10  DATA 101, 4.50, 40, 102, 4.75, 36, . . .
 11  DATA . . .
 12  DATA . . .
  :
 90  READ I, W, H
100  P = W * H
110  PRINT "I.D. NUMBER " I, "GROSS PAY " P
120  GOTO 90
130  END  •
```

The pair of statements, FOR . . . NEXT, gives a more efficient way of writing loops in BASIC. The general form of these two statements is given below.

FOR . . . NEXT statement

FOR (variable = first number) TO (second number) NEXT (variable)

The FOR statement causes the variable tc act as a counter by allowing a computation to be performed as we count from the first number to the second number. The variable in NEXT is the same as in FOR and instructs the computer to return to FOR for instructions.

Example 30 •

Statement	*Result*
30 FOR Y = 1 TO 50 : 70 NEXT Y	Y takes on the value 1. When the program gets to line 70, the next instruction is at line 3∅ and Y has the value 2, continuing in this manner until Y assumes all counting numbers up to and including 50. After the loop has been executed with Y = 5∅, the computer goes to the next line after 7∅.

•

Example 31 • Write a program to compute the sum of the first 10 positive integers. Write the program first without FOR . . . NEXT, then with FOR . . . NEXT.

Solution In Section 4, we first examined the problem of adding lists of numbers. Return to that material to get clearly in mind how the problem is solved. Then compare these two programs.

Just for fun

What is the temperature?

0
B.S.
M.S.
Ph.D.

(Look for a trick.)

```
10  S = 0
20  N = 1
30  If N > 10 THEN 70
40  S = S + N
50  N = N + 1
60  GOTO 30
70  PRINT "THE SUM IS " S
90  END
```

```
10  S = 0
20  FOR N = 1 TO 10
30  S = S + N
40  NEXT N
50  PRINT "THE SUM IS " S
60  END
```
•

The BASIC language has within it a library of mathematical functions that greatly increase the power of BASIC for problem solving. We have already mentioned the SQR ( ) function that computes the square root of a number. Also important is the INT ( ) function that extracts the integer part of a number.

Example 32 •

INT (2.5) = 2
INT (6.1) = 6
INT (4) = 4 •

One nice application of the INT ( ) function is that it can be used to check whether A is divisible by B. A is divisible by B if and only if INT $(A/B) = A/B$.

Example 33 •

$3|6$ since INT $(6/3) = 2 = 6/3$

$3\nmid 7$ since INT $(7/3) = 2 \neq 7/3 = 2.33333$ •

Now we are ready to solve the problem with which we began this section.

Example 34 • Write a program to input an integer N and determine if it is divisible by 3.

```
10  PRINT "INPUT AN INTEGER"
20  INPUT N
30  IF INT (N/3) = N/3 THEN 60
40  PRINT N, "IS NOT DIVISIBLE BY 3"
50  GOTO 70
60  PRINT N, "IS DIVISIBLE BY 3"
70  END
```
•

Exercise set 4

R 1. Give the output of the following programs.

(a)
```
2  S = 0
4  A = 3
6  FOR P = A TO 6
9  S = S + P
10 NEXT P
11 PRINT S
12 END
```

(b)
```
10 READ A, B, C
20 DATA -2, 0, 3
30 PRINT A^2 + 3 * B - C, A, B
40 END
```

(c)
```
10 READ X, Y
20 DATA 4, 1
30 PRINT "X^2 - Y^2 =  ", X^2 - Y^2
40 END
```

(d)
```
6  FOR X = 1 TO 7
10 Y = X^2 + X
16 PRINT X, Y
20 NEXT X
30 END
```

(e)
```
10 PRINT "BASE", "HEIGHT";
20 PRINT "AREA"
30 PRINT
40 DATA 4, 3, 2, 1, 3, 6, -1, -1
50 READ A, B
60 IF B <= 0 THEN 90
70 PRINT A, B, (1/2) * A * B
80 GOTO 50
90 END
```

2. Determine which of the following program statements are incorrect, then correct them.

(a) `15 4 = B`
(b) `20 READ 7, 6, 8`
(c) `40 PRINT "I LOVE YOU"`
(d) `60 PRINT A, B`
(e) `40 C = 2A`
(f) `70 PRINT 2 * A^2`

3. Trace this program to determine the incorrect lines. Correct the mistakes.

```
10 READ A, B
20 DATA 5, 7, 6, 4, 10, 11, 13, 15
30 A^2 - B
35 IF B < 0 THEN 50
40 GOTO 30
50 END
```

4. Write BASIC programs using READ ... DATA statements to assign to A the value of 13 and to B the value of 24; then provide the following output.

(a)
```
THE SUM IS 37
THE PRODUCT IS 312
```

(b)
```
SUM    PRODUCT
37     312
```

(c)
```
SUM    PRODUCT


37     312
```

5. Write a program to input a positive integer and determine if it is divisible by 17.

T 6. Modify the payroll program in this section to include an overtime feature. If the workers work more than 40 hours, they are paid 1.5 times their usual wage per hour for extra hours.

7. Write a program to compute the sum of the first n positive even integers. Use FOR ... NEXT.

8. Write a program using FOR ... NEXT for finding the sum.

$$2^2 + 4^2 + 6^2 + 8^2 + \cdots + (2n)^2$$

9. In the design below, the computer is given a list of measurements in kilometers and changes those measurements to meters.

> Get the measurement in kilometers K
> If K is the Flag ($K < 0$) then END
> otherwise compute $M = 1000K$
> Print M

(a) Write a BASIC program for this design. Include column headings KILOMETERS and METERS. Use READ ... DATA statements for input.

(b) Run the program for this data list: 2, 2.13, 4.634, 2.11141, 3, 3.67. Be sure to include an appropriate flag at the end of this list.

10. Write a program to
    (a) obtain column headings RADIUS, CIRCUMFERENCE, AREA.
    (b) input a radius.
    (c) compute circumference and area.
    (d) output radius, circumference, and area.
    (e) return and do the same thing for another piece of data.
    (f) end the program when a flag is reached. *Hint:* Circumference of a circle is equal to 2π (radius); area of a circle is equal to $\pi \cdot (\text{radius})^2$. Use 3.14159 for π.

C 11. (a) Write a program that will accept a measurement in centimeters, and express it in terms of millimeters, decimeters, meters, dekameters, hectometers, and kilometers. For example, when 21 is input, the output should be the following.

```
21 CM IS EQUIVALENT TO
210 MM
2.1 DM
0.21 M
0.021 DAM
0.0021 HM
0.00021 KM
```

    (b) Run this program for several examples.

12. Write a program that will sum a list of numbers and then compute the average of the list. The program should use a flag placed at the end of the list to determine when to stop the summing loop.

6 An introduction to Logo

■ Teach a turtle to write the word LOVE.
Note: If you think that this author has lost his mind, read on a bit further. ■

Children learn by building complex knowledge out of simple components.

Children learn by manipulating concrete objects, "objects to learn with."

Motivated by these two principles, Seymour Papert of the Massachusetts Institute of Technology developed the Logo computer language, a language intended to help children learn. In this section, we will become acquainted with some of the fundamental concepts of Logo.

Central to the computer screen of a computer running Logo is a small triangular pointer, the **turtle.** A few very simple Logo commands allow a student to direct a turtle to move across a computer screen. As the turtle moves, it draws simple geometric designs. By combining the same simple set of instructions in slightly more complex patterns, the turtle can be used to draw even more complex and elegant designs. Logo provides a laboratory for building complex knowledge from simple components; the turtle becomes an object with which children can learn.

When the Logo software has been loaded into the computer, the turtle is made visible by typing the command **SHOWTURTLE.** Note that the small triangular pointer representing the turtle has both a **location** and a **direction.** Initially, the turtle's position is located near the center of the screen, and its direction points straight up.

The fundamental commands of Logo change either the location or the direction of the turtle. To issue a command, one types the command and then strikes the RETURN or ENTER key on the computer. The command FORWARD 50 (followed by RETURN or ENTER) will move the turtle forward 50 turtle steps, tracing its path as it moves. The command RIGHT 45 will turn the turtle 45 degrees to the right (45 degrees in a clockwise direction). Figure 2 shows the diagram produced by the turtle when these commands are typed and entered.

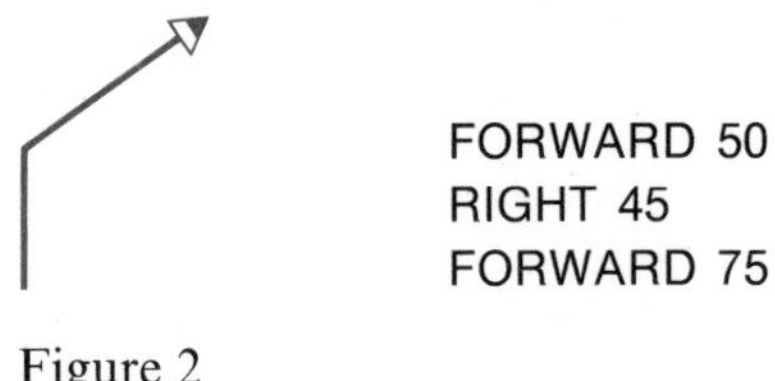

```
FORWARD 50
RIGHT 45
FORWARD 75
```

Figure 2

Without too much difficulty, we can direct the turtle to sketch several familiar geometric shapes. For instance, the following Logo program produces the square shown in Figure 3.

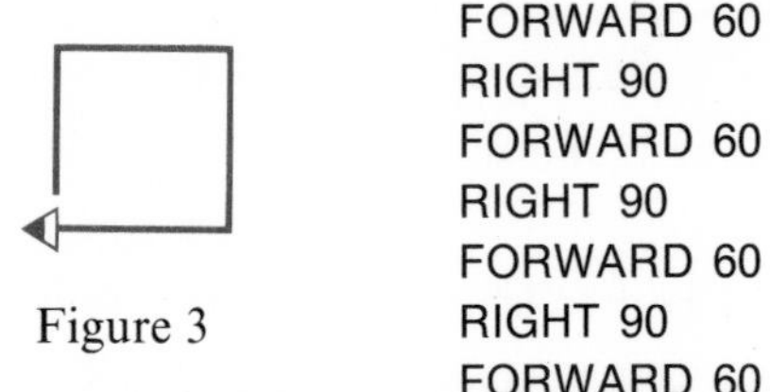

```
FORWARD 60
RIGHT 90
FORWARD 60
RIGHT 90
FORWARD 60
RIGHT 90
FORWARD 60
```

Figure 3

You can see that, with a little imagination, a large variety of figures can be drawn with just the FORWARD and RIGHT commands. There are several other very simple commands that make the process of producing simple designs even simpler. The command BACK 30 causes the turtle to move backwards 30 turtle steps, and the command LEFT 90 produces a turn of 90 degrees in a counterclockwise direction. When the user wishes to draw a picture with disconnected line segments, the command PENUP is very important. After the command PENUP is issued, the path the turtle follows on the screen is not drawn. The command PENDOWN causes the path to be drawn again.

Example 35 • This Logo program produces the design shown in Figure 4.

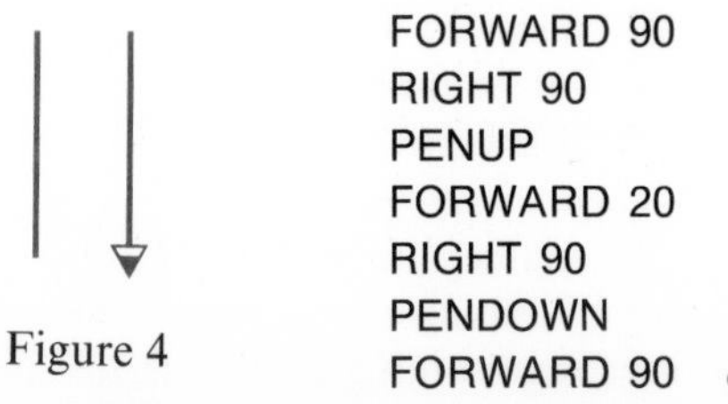

```
FORWARD 90
RIGHT 90
PENUP
FORWARD 20
RIGHT 90
PENDOWN
FORWARD 90
```

Figure 4 •

Just for fun *Write a sequence of Logo instructions that will produce this drawing.*

In long programs, it is awkward to type each LOGO command in full, so the Logo software recognizes two letter abbreviations of each of its commands. In Table 3, you will find the abbreviations of several useful Logo commands.

Table 3

Command	*Abbreviation*
FORWARD	FD
BACK	BK
RIGHT	RT
LEFT	LT
PENUP	PU
PENDOWN	PD
SHOWTURTLE	ST
HIDETURTLE	HT
CLEARSCREEN	CS

Example 36 • The Logo program of the stairsteps in Figure 5 is given below.

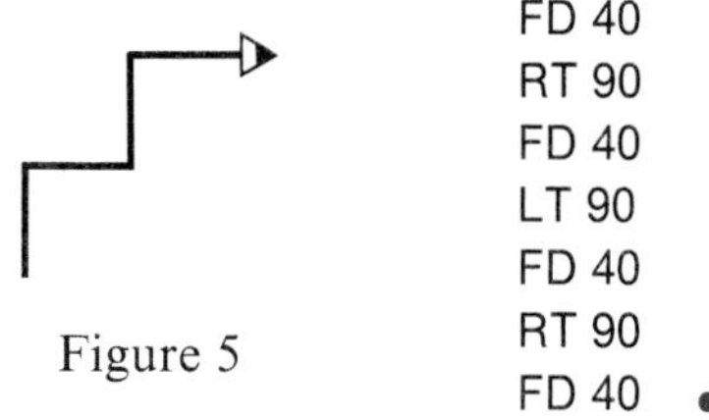

Figure 5

```
FD 40
RT 90
FD 40
LT 90
FD 40
RT 90
FD 40
```
•

If you want the computer to execute each step, one at a time, you use the vertical format. If you write the complete program horizontally as

```
FD 40  RT 90  FD 40  LT 90  FD 40  RT 90  FD 40
```

and then hit the RETURN key the complete drawing will be made.

Exercise set 5

R 1. Use graph paper and sketch the figure that the turtle would sketch as a result of each of these sets of commands. You may wish to regard each unit length on your graph paper as representing 10 turtle steps.

(a)
```
CLEARSCREEN
LEFT 90
FORWARD 50
RIGHT 90
FORWARD 70
RIGHT 90
FORWARD 50
```

(b)
```
CLEARSCREEN
FORWARD 10
PENUP
FORWARD 10
PENDOWN
FORWARD 10
PENUP
FORWARD 10
PENDOWN
FORWARD 10
```

(c)
```
CLEARSCREEN
FORWARD 20
RIGHT 90
FORWARD 20
LEFT 90
FORWARD 20
RIGHT 90
FORWARD 20
LEFT 90
FORWARD 20
```

(d)
```
CLEARSCREEN
RIGHT 30
FORWARD 90
BACK 90
RIGHT 60
PENUP
FORWARD 50
LEFT 120
PENDOWN
FORWARD 90
HIDETURTLE
```

2. Rewrite each of the programs in Exercise 1, using the abbreviations shown in Table 3.

3. Determine which letter of the alphabet would be produced by these Logo programs.

(a)
```
FD 50
LT 30
BK 60
RT 30
FD 50
HT
```

(b)
```
FD 50
RT 90
FD 20
BK 40
HT
```

(c)
```
FD 50
RT 90
FD 20
BK 20
LT 90
BK 20
RT 90
FD 10
HT
```

(d)
```
RT 45
FD 30
BK 60
FD 30
RT 90
FD 30
BK 60
HT
```

4. Experiment using Logo until you can write a Logo program that draws each of these geometric figures.

(a) 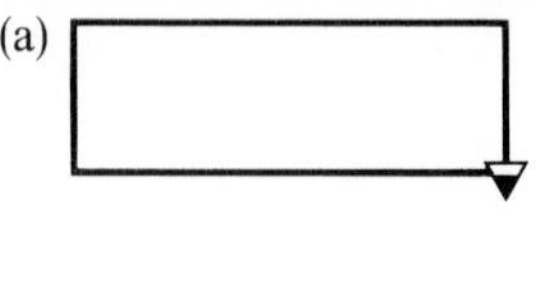

(b)

T 5. Use graph paper and sketch the figure that the turtle would draw if presented with these sets of Logo instructions.

(a)
```
CS FD 50 RT 90 FD 20 LT 90 FD 10
LT 90 FD 50 LT 90 FD 10 LT 90 FD 20
RT 90 FD 50 LT 90 FD 10 HT
```

(b)
```
PU FD 20 PD FD 20 RT 90 FD 20
PU FD 20 PD FD 20 RT 90 FD 20
PU FD 20 PD FD 20 RT 90 FD 20
PU FD 20 PD FD 20 RT 90 FD 20
```

6. Experiment using Logo until you can write a Logo program that will instruct the turtle to draw these letters and numbers.

(a) (b) 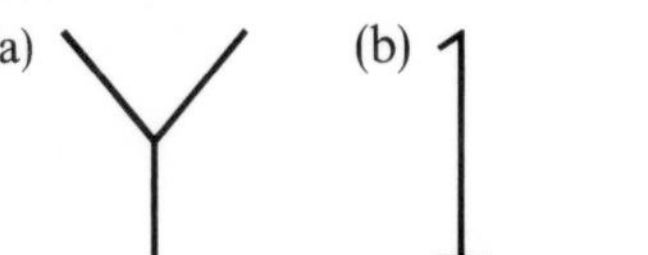

(c) 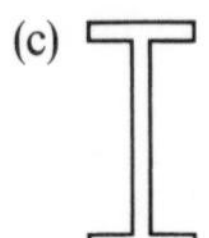

C 7. Experiment with Logo until you can write Logo programs that produce geometric designs similar to these.

(a) (b) 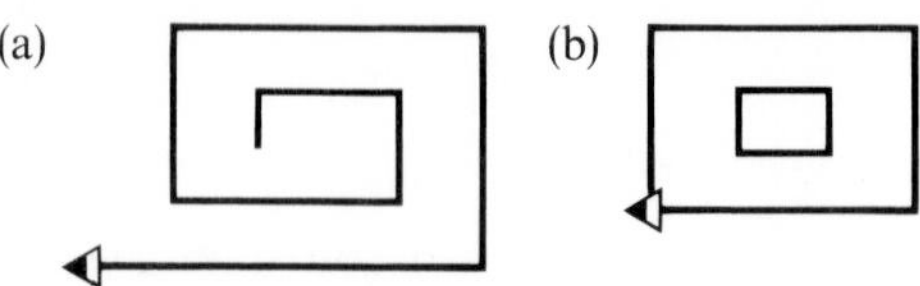

8. Run these Logo programs. Try to guess the function of the Logo commands with which you are not familiar.

(a) `REPEAT 4[FD 50 RT 90]`
(b) `REPEAT 36 [REPEAT 4[FD 50 RT 90] RT 10]`

7 Looping in Logo—the REPEAT command

■ Write Logo commands that will draw this figure.

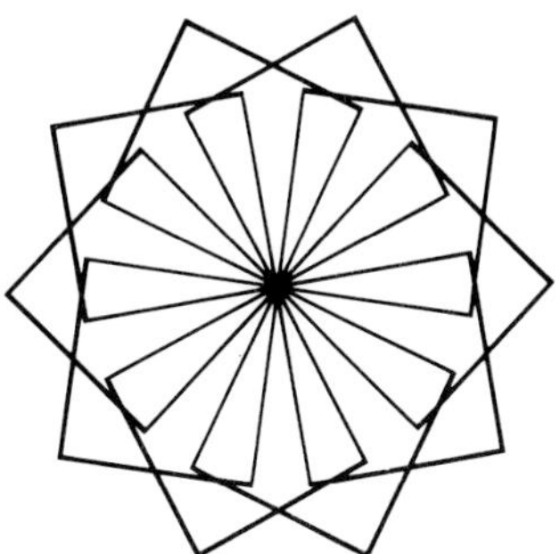

■

The power of the computer is most clearly seen when it is used to perform repetitive tasks. In Logo, the REPEAT command allows the programmer to instruct the computer to repeat one or more tasks a fixed number of times. Since Logo is intended as a laboratory for learning to solve complex problems by packaging simple commands, the REPEAT command is of central importance.

Figure 6

In the previous section, we learned that the following program would enable the turtle to draw the square in Figure 6.

```
FD 60 RT 90 FD 60 RT 90
FD 60 RT 90 FD 60 RT 90
```

Observe that the square is drawn by repeating the two commands FD 60 and RT 90 exactly four times. The Logo command

```
REPEAT 4 [FD 60 RT 90]
```

will also draw the square.

The REPEAT command

The standard form of the REPEAT command is REPEAT (number of repetitions) [commands to be repeated].

The key to using REPEAT successfully is to visualize a complex design in terms of simpler components.

Example 37 • Use REPEAT to draw the stair steps shown in Figure 7.

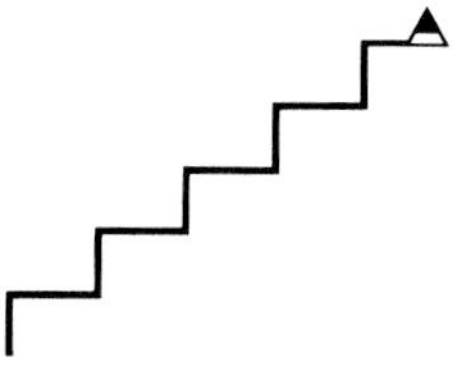

We must first see that the set of stair steps are composed of multiple recurrences of the single unit,

which can be drawn using the commands

Figure 7

```
FD 20 RT 90 FD 20 LT 90
```

Do you see that the single command

```
REPEAT 5 [FD 20 RT 90 FD 20 LT 90]
```

would produce the design in Figure 7? •

The Logo turtle can help us discover in a very concrete way some fundamental facts about geometry. From the following sequences of turn commands where the degrees of turn total 360°,

```
RT 90   RT 90  RT 90  RT 90
LT 120  LT 90  LT 30  LT 60  LT 60
RT 60   RT 60  RT 60  RT 60  RT 60  RT 60
```

we discover the truth of what is sometimes called the **total turtle trip theorem.**

Total turtle trip

> The turtle must turn 360 degrees (or some multiple of 360 degrees) to return to the direction in which it started.

In other words, a full rotation measures 360 degrees. We can use this fact to help draw polygons.

The square in Figure 6 is an example of a *regular polygon.* Other examples of regular polygons are shown in Figure 8. Note that in each case the sides and angles of the figure are of equal measure. When we absorb this fact, it is particularly easy to write a Logo program to draw a regular polygon. To draw a square, the turtle must turn 90 degrees at each of the four corners to complete the square. To draw a regular triangle, the turtle must turn $360 \div 3 = 120$ degrees at each corner to complete the triangle. Similarly, to draw a regular five-sided polygon (a pentagon), the turtle must turn $360 \div 5 = 72$ degrees at each corner. Thus, a command to draw a regular pentagon would look like this:

```
REPEAT 5 [FD 30 RT 72]
```

Can you write commands to draw a regular triangle (equilateral triangle)? Can you write a command to draw a regular hexagon (six-sided polygon)?

Figure 8

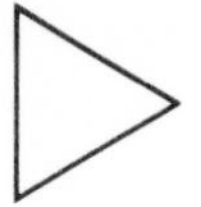
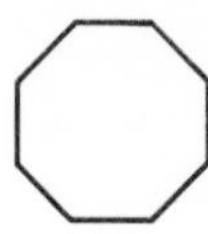
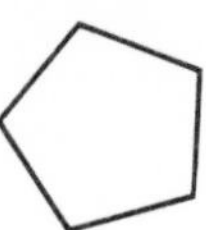

The true power of the REPEAT command is realized when one begins to nest REPEAT commands: that is, when one REPEATs a REPEAT. We have seen that the command

```
REPEAT 4 [FD 40 RT 90]
```

draws a square. The command

```
REPEAT 10 [REPEAT 4 [FD 60 RT 90] RT 36]
```

draws 10 squares, each rotated 36 degrees beyond the previous square. See Figure 9 for the result. Observe that this solves the problem posed at the beginning of this section.

Just for fun *Write a one-line Logo command that will draw this figure.*

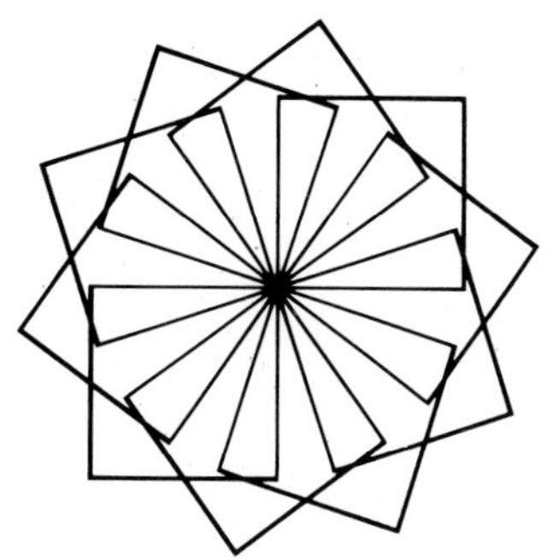

Figure 9

Exercise set 6

R 1. Rewrite the following programs, using the REPEAT command.

(a)
```
FD 30
RT 120
FD 30
RT 120
FD 30
RT 120
```

(b)
```
PU FD 20 PD FD 20 RT 90 FD 20
PU FD 20 PD FD 20 RT 90 FD 20
PU FD 20 PD FD 20 RT 90 FD 20
PU FD 20 PD FD 20 RT 90 FD 20
```

2. Use graph paper and sketch the figure that the turtle would sketch as a result of each of these sets of commands.

(a) REPEAT 4 [FD 10 PU FD 20 PD]
(b) REPEAT 8 [FD 20 BK 20 RT 45]

3. What is wrong with each of the following Logo commands?

(a) FD100
(b) PU 50
(c) REPEAT [FD 50 LT 20]
(d) REPEAT 36 (FD 10 RT 10)

4. Experiment with Logo until you can write a Logo program using REPEAT that draws each of these figures.

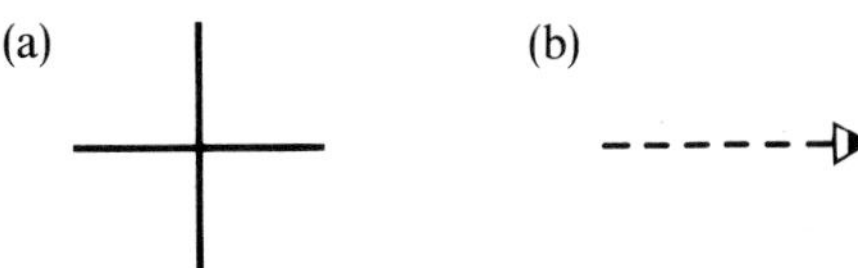

5. Write a Logo command that will cause the turtle to draw a regular triangle. A regular hexagon.

6. The Logo language has the capacity to perform arithmetic operations. For instance 360/10 will be evaluated as $360 \div 10 = 36$. Thus,

```
REPEAT 7 [FD 40 RT 360/7]
```

will draw a regular seven-sided polygon. Use Logo to draw

(a) a 8-sided polygon.
(b) a 10-sided polygon.
(c) an 11-sided polygon.

7. As we have seen, the Logo turtle can be used to draw figures constructed from straight lines. To

draw a circle, we must use the fact that a many-sided regular polygon looks very much like a circle. Use Logo to draw a 36-sided regular polygon. Observe that it is a good approximation of a circle.

T 8. Experiment with Logo until you can write Logo programs that produce geometric designs similar to these.

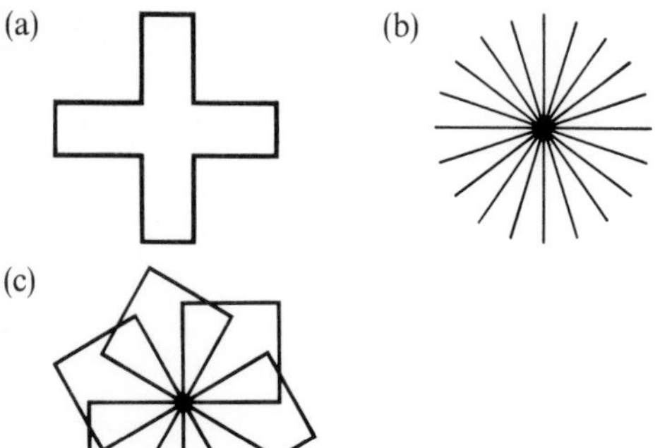

C 9. Interesting designs can be produced when you nest REPEAT commands several levels deep. For instance, the command

```
REPEAT 8 [REPEAT 4 [FD 30 RT 90] LT 45]
```

produces the design below.

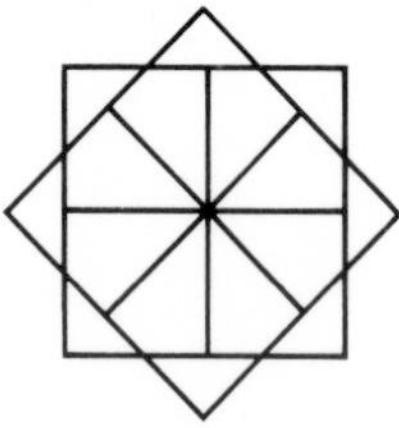

Use nested REPEAT commands to produce designs like these.

(a)

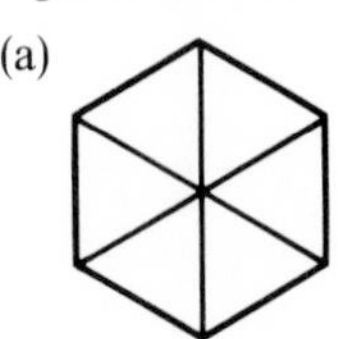

(b)

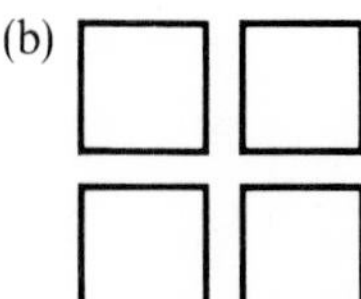

(c)

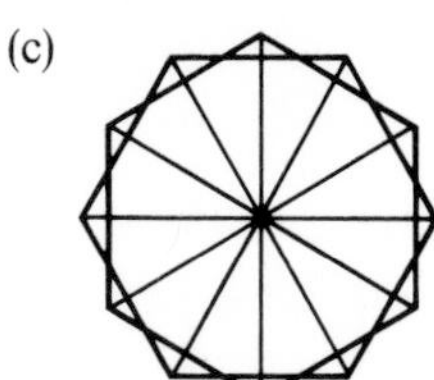

(d)

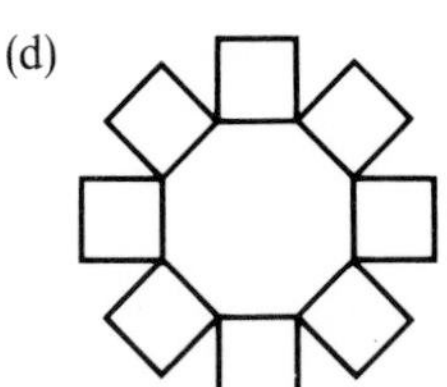

10. One can sketch 5 regular pentagons (a polygon with 5 equal sides and 5 equal angles) with this command

```
REPEAT 5 [REPEAT 5 [FD 30 RT 360/5] RT 360/5]
```

Modify this example to sketch

(a) Six regular hexagons (6 equal sides)
(b) Eight regular octagons (8 equal sides)
(c) Seven polygons with 7 sides
(d) Twelve polygons with 12 sides

8 Building new Logo commands

■ Use Logo to draw a design like this:

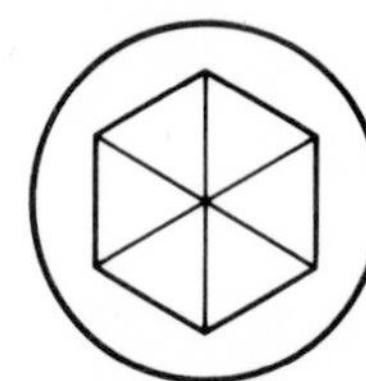

■

In this section, you will learn how to write your own Logo commands. The Logo feature that allows you to do this is called the *procedure*.

Logo was developed to give the user skills in building complex structures out of simple structures. Toward this end, you can group one or more simple Logo commands under a new name to accomplish a more complex task. In this way, you can create new Logo commands. The mechanism for doing this is the **procedure.**

Procedure

> The form of a procedure is
>
> TO ⟨procedure name⟩
> ⟨instructions⟩
> END

For example, the simple procedure LINE described below will draw a line segment 50 turtle steps long.

```
TO LINE
  FD 50
  HT
  END
```

You execute a procedure by merely entering the name of the procedure. Hence, once the entire procedure LINE has been entered, we can draw a line merely by typing LINE. Once a procedure has been entered during a session on the computer, you can use it many times. For the duration of that session, the procedure name LINE has become a new Logo command. Moreover, that procedure can be used to define other procedures.

Example 38 • Since a square is constructed of 4 lines, use the procedure to write a procedure to draw a square.

Solution

```
TO SQUARE
  REPEAT 4 [LINE  RT 90]
  END
```
•

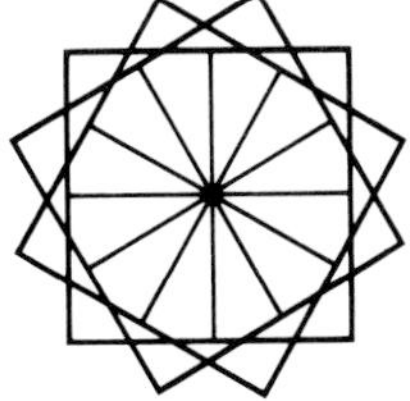

Figure 10

To draw an elaborate design like the one in Figure 10, we need only note that the design is constructed by 12 squares, each rotated $360 \div 12 = 30$ degrees from the previous square. Since the procedure SQUARE is already defined, we can define a procedure to draw this design in the following way:

```
TO FANCYDESIGN
  REPEAT 12 [SQUARE  RT 30]
  END
```

Example 39 • Let us draw the design presented in the problem with which this chapter was introduced (see Figure 11).

Solution The figure appears to be built from triangles and a circle. In Exercise 7 of the previous section, we saw that the following procedure would give a good approximation of a circle.

```
TO CIRCLE
  REPEAT 36 [FD 8 RT 10]
  END
```

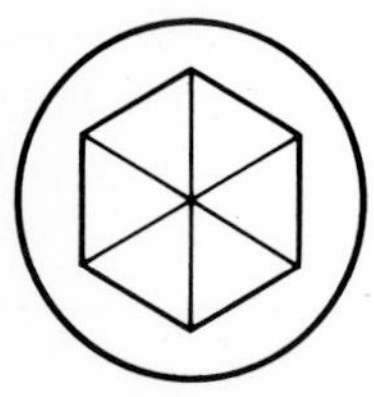

Figure 11

Similarly, a triangle will be drawn by this procedure:

```
TO TRIANGLE
  REPEAT 3 [FD 30  RT 120]
  END
```

The design inside the circle can be drawn using this procedure:

```
TO INNERDESIGN
  REPEAT 6 [TRIANGLE  RT 60]
  END
```

Finally, we try to put it all together with this master procedure:

```
TO BIGDESIGN
  CIRCLE
  INNERDESIGN
  END
```

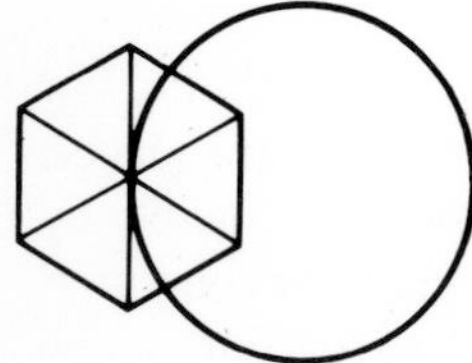

Figure 12

However, when we run this program, we get the result in Figure 12. The difficulty arises from the fact that the inner design is not placed in the center of the circle. With a little experimentation, we find that we can position the turtle in the center of the circle with the commands

```
PU RT 90 FD 46 LT 90 FD 3 PD
```

Hence the master procedure should read

```
TO BIGDESIGN
  CIRCLE
  PU  RT 90  FD 46  LT 90  FD 3  PD
  INNERDESIGN
  END
```
•

By using Logo variables, one can make even more powerful procedures. A Logo variable is a colon followed by a word; numerical values can be stored in the variable name. The procedure LINE defined below has a variable: LENGTH. Following the command LINE with a number causes the turtle to draw a line of the length specified by the number.

```
TO LINE :LENGTH
  FD :LENGTH
  HT
  END
```

For instance, the command LINE 40 would produce a line of length 40 turtle steps; the command LINE 60 would cause the turtle to draw a line 60 steps long.

Example 40 • Consider the procedure TRIANGLE with variable: SIDE.

```
TO TRIANGLE :SIDE
  REPEAT 3 [FD :SIDE  RT 120]
  HT
  END
```

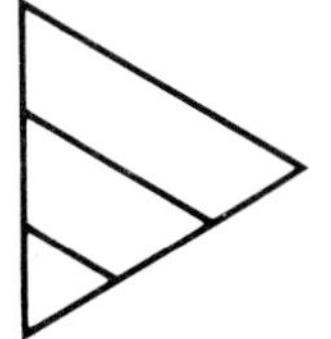

Figure 13

After entering this procedure, enter the commands

```
TRIANGLE 30
TRIANGLE 60
TRIANGLE 90
```

Did you get the results shown in Figure 13? •

Exercise set 7

R 1. Write a Logo procedure that will draw a

(a) rectangle
(b) pentagon (5-sided polygon)
(c) hexagon (6-sided polygon)
(d) octagon (8-sided polygon)

2. What is wrong with each of the following Logo procedures?

(a)
```
TO TRIAL
   REPEAT 4 [FD 50 RT 90]
   REPEAT 5 [FD 60 RT 72]
```

(b)
```
STAR
   REPEAT 10 [FD 60 BK 60 RT 36]
   END
```

(c)
```
TO BALL :SIZE
   REPEAT 36 [FD SIZE RT 10]
   END
```

(d)
```
TO SAMPLE
   FD :LENGTH
   RT :LENGTH
   FD :LENGTH
   END
```

T 3. Consider the procedures TRIANGLE and SQUARE defined below. Suppose that these procedures have already been entered into the machine.

```
TO TRIANGLE
  REPEAT 3 [FD 40 RT 120]
  HT
  END

TO SQUARE
  REPEAT 4 [FD 40 RT 90]
  HT
  END
```

Use graph paper to show what would be drawn if the following commands and procedures were executed.

(a)
```
TRIANGLE
```

(b)
```
TRIANGLE
SQUARE
```

(c)
```
TRIANGLE
PU LT 90 FD 80 RT 90 PD
TRIANGLE
```

(d)
```
TO INSIDE
   SQUARE
   RT 90 FD 20 LT 120
   TRIANGLE
   END
```

4. Use the procedures SQUARE and TRIANGLE from Exercise 3 and write procedures that will draw these figures.

(a)

(b)

(c)

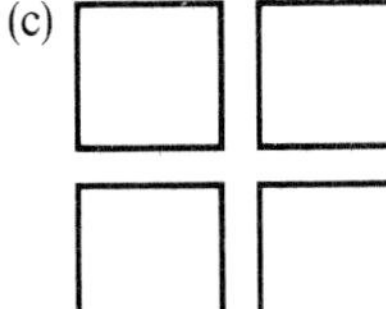

(d)

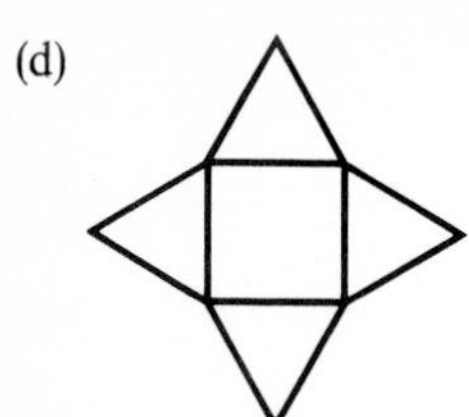

5. Write a procedure to draw each of the following letters.

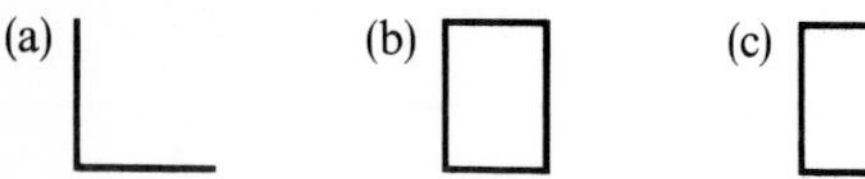

C 6. Use the procedures from Exercise 5 to write a master procedure to write the word LOGO.

7. Correct the following Logo instructions so that they will create the design shown.

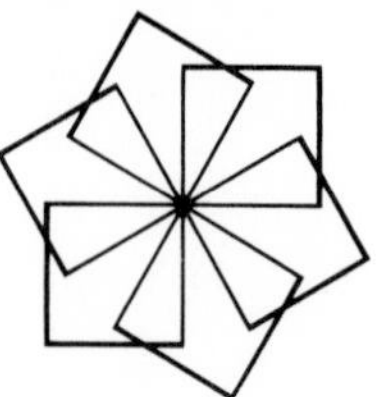

(b) REPEAT 12 [REPEAT 3 [FD 20 RT 60] LT 30]

Solution to introductory problem

Understand the problem. Before finding all primes less than 1000, let us look at a simpler problem. Given a number N, can we tell whether N is a prime? If N is 2, N is certainly prime. If $N > 2$, we can test for divisibility using all integers K with $2 \leq K \leq \sqrt{N}$. If any K divides N, N is not prime; if no K divides N, N is prime.

Divise a plan. To determine if N is prime, we must write a loop to test whether N has divisors. To find all primes less than 1000, we need to place this loop inside a loop that tests all values of N.

Execute the plan.

```
10  PRINT "2 IS A PRIME"
20  FOR N = 3 TO 999
30  LET K = 2
40  IF K > SQR (N) THEN 100
50  IF INT (N/K) = N/K THEN 80
60  LET K = K + 1
70  GOTO 40
80  PRINT N " IS NOT A PRIME"
90  GOTO 110
100 PRINT N " IS A PRIME"
110 NEXT N
120 END
```

Summary and review

Computer literacy

1. Hardware and software
2. Storage units
3. Central processing unit
4. Five things a computer can do

BASIC language

1. Use of a variable
2. Use of +, −, *, /, ^, and SQR ( )
3. The PRINT command
4. The INPUT statement

5. IF . . . THEN commands
6. GOTO commands
7. Use of symbols: =, <, < =, >, > =, < >, or > <
8. READ . . . DATA statements
9. FOR . . . NEXT statements
10. The INT statement

Logo

1. Use of the turtle
2. Commands: FORWARD (FD), BACK (BK), RIGHT (RT), LEFT (LT), PENUP (PU), PENDOWN (PD), SHOWTURTLE (ST), HIDETURTLE (HT), and CLEARSCREEN (CS)
3. The REPEAT command
4. The total turtle trip theorem
5. The procedure:

```
TO ⟨procedure name⟩
    ⟨instructions⟩
END
```

Review exercise set 8

R 1. Write each of the following in BASIC language.

(a) $6(7 + 3)$ (b) $\frac{7}{8} - 14$

(c) $\frac{2}{3} + 8$ (d) $9 - 6^5$

(e) $4^3 - (2 \cdot 3^2)^4$ (f) $\frac{7^4}{2} + 5^3$

2. What are the printouts for the following programs?

(a)
```
10  X = 5
20  Y = 12
30  PRINT X, Y, X * Y
40  X = X + 1
41  Y = Y + 1
42  IF X * Y < 100 THEN 30
43  END
```

(b)
```
10  READ A, B
20  DATA 5, 7, 6, 4, 10, 11, 13, 15
30  PRINT A^2 - B
35  IF B = 15 THEN 50
40  GOTO 10
50  END
```

3. Determine which of the following program statements are incorrect, and correct these statements.

(a) `15  3 = A`
(b) `    B = 4`
(c) `40  PRINT I LOVE YOU`
(d) `60  A = 3C`
(e) `20  PRINT A B`
(f) `40  C = A^2`
(g) `PRINT 2 * A`
(h) `10  GOTO END`
(i) `80  IF X > 10 THEN 80`
(j) `60  IF X >< THEN 40`
(k) `70  IF X = 4 THEN X^2 = 16`
(l) `20  READ 7, 6, 8`

4. Write a BASIC expression for each of the following.

(a) $(4x^2 + 7)/(x - 1)$
(b) $(5x^2 - 3x)^3/(3x^2 + x)$
(c) $(2x^2 + 7x + 4)^3$
(d) $(2/x^2) + 5(x + 7)^3$

5. Find the value of A in each of the following.

(a)
```
20  .X = 2
30  A = 2 * X^2 + 7 * X - 1
40  END
```

(b)
```
20  X = 1
40  A = X^2/3 + X^3^2
60  END
```

6. Write a program to compute and print the first 100 terms of the sequence 1, 4, 7, 10, 13, 16, ...

7. Use a piece of paper to sketch the figure that the Logo turtle would draw if given these commands.

(a)
```
FD 50 RT 135 FD 80 RT 45
FD 50 RT 135 FD 80
```

(b)
```
REPEAT 3 [FD 50 BK 50 RT 90 PU
FD 30 PD LT 90] PU LT 90 FD 90
RT 90 FD 25 LT 90 PD FD 30
```

T 8. Write a series of Logo commands that will draw each letter and then displace the turtle as shown below.

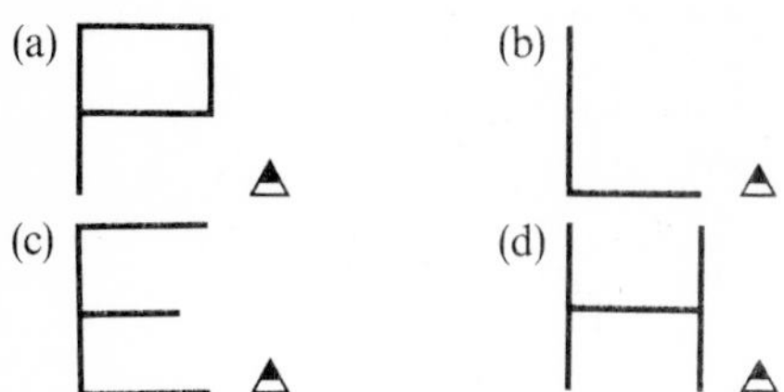

9. Write a program to do the following.
   (a) Print column headings: FEET, INCHES.
   (b) Input feet.
   (c) Compute the equivalent number of inches.
   (d) Output feet and inches in columns.
   (e) Return and repeat for another piece of data.
   (f) End the program when a flag is reached at the end of the data.

10. Write a program to find the sum of the first Q positive even integers.

11. Write a BASIC program to compute $3y^3 - 17y^2 + 4y + 3$ and to print only positive answers.

12. Write a BASIC program to print all the numbers between 100 and 200 that are divisible by 12.

13. Write a program that prints the table below complete with column headings.

Principal	*Rate*	*Time*	*Amount*
100	0.06	3	118
200	0.06	3	236
300	0.06	3	354
400	0.06	3	472

14. Write each of the programs from Exercise 8 as a procedure. Then write a master procedure that prints the word HELP.

15. (a) Write a Logo procedure STEP to draw this figure.

   (b) Write a procedure FOURSTEP that uses STEP to draw this picture.

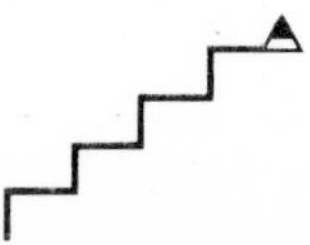

   (c) Write a procedure MANYSTEP that uses FOURSTEP to draw this figure.

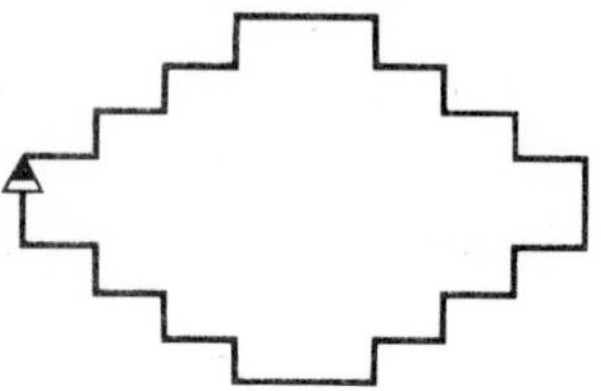

16. (a) Write a Logo procedure that will instruct the computer to draw a single line and displace the turtle 20 turtle steps to the right of the line, as shown below.

   (b) Write a Logo procedure that will draw 8 parallel lines, each 20 steps apart.

C 17. (a) Write a program to find the smallest number in a list of numbers.
   (b) Trace this program by hand.

18. Write a program to input a list of integers and find their sum. The integers should be processed one at a time in a loop. A flag should signal the computer when the list has been completely processed.

19. In the town of Slowdown a driver arrested for exceeding the speed limit of 35 has two choices. He may go directly to the police department and pay a fine of \$5.00 for each mile per hour over 35. His second choice is to wait and appear in traffic court that evening. If he is found guilty, he must pay \$6.00 for each mile per hour in excess of the speed limit. If the fine is not paid, he gets a jail sentence. The data after 1 day's work by the traffic department are as follows.

Ticket number	*SPEED*	*Disposition code*
111	50	1
112	43	2
113	37	3
114	80	4
115	46	1

The disposition codes mean

(1) paid fine immediately.
(2) found guilty and paid fine.
(3) found not guilty.
(4) jailed.

Write a program that will input the data and output the ticket number, speed, and fine (if a fine was paid). If the disposition code was item (3), print NOT GUILTY. If the code was item (4), print JAILED.

20. Write Logo programs that will produce each of these figures.

(a)

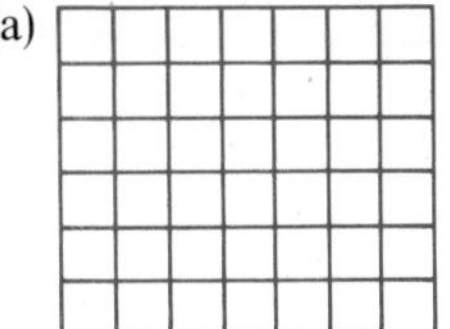

(*Hint:* You might like to use your work from Review Exercise 16.)

(b)

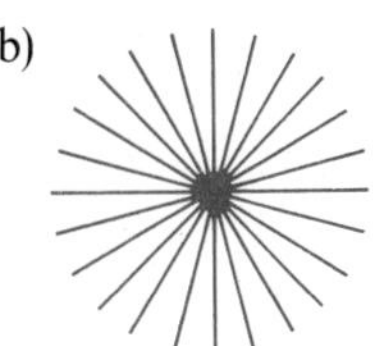

Bibliography

1. Billings, Karen. "Developing Mathematical Concepts with Microcomputer Activities." *AT* (February 1983), 18–19, 57–58.
2. Clemens, Stanley R. "The Gifted and the Micro—Will You Let It Happen?" *AT* (February 1983), 26.
3. Cooper, Richard P. "What Do You Do with the New Computer?" *AT* (October 1982), 46–47.
4. Ducharme, Robert. "Computers (and Students!) Need Explicit Notation." *MT* (May 1978), 448–451.
5. Filliman, Paula K. "Guidelines for Introducing Microcomputers in the Schools." *AT* (February 1983), 16–17, 56.
6. Gardner, Martin, *Aha*! New York. W. H. Freeman, 1978.
7. Hampel, Paul J. "Computer Corner." *AT* (March 1985), 50–51.
8. Hatfield, Larry L. "Teaching Mathematics with Microcomputers: Junior High School." *AT* (February 1983), 44–45, 68–69.
9. Heck, William. "Teaching Mathematics with Microcomputers: Primary Grades." *AT* (February 1983), 27, 63–66.
10. Hill, Shirley A. "The Microcomputer in the Instructional Program." *AT* (February 1983), 14–15, 54–55.
11. Johnson, David C., Ronald E. Anderson, Thomas P. Hansen, & Daniel L. Klassen. "Computer Literacy—What Is It?" *MT* (February 1980), 91–96.
12. Johnson, Jerry. "Education's Lexicon of Computerese." *AT* (February 1983), 46–49.
13. Kansky, Robert. "The C.P.U. and You: Mastering the Microcomputer." *AT* (February 1983), 12–13, 52–54.
14. Kantowski, Mary Grace. "The Microcomputer and Problem Solving." *AT* (February 1983), 20–21, 58–59.
15. Mason, Margie. "Computer Corner." *AT* (February 1986), 58–59.
16. Sadowski, Barbara A. "A Model for Preparing Teachers to Teach with the Microcomputer." *AT* (February 1983), 24–25, 62–63.
17. Thomas, Eleanor M., & Rex A. Thomas. "Exploring Geometry with Logo." *AT* (September 1984) 16–18.
18. Thompson, Charles S., & John Van de Walle. "Patterns and Geometry with Logo." *AT* (March 1985), 6–13.
19. Tobin, Catherine D. "Developing Computer Literacy." *AT* (February 1983), 22–23, 60.
20. Wiebe, James H. "BASIC Programming for Gifted Elementary Students." *AT* (March 1981), 42–44.
21. Winter, Mary Jean. "Teaching Mathematics with Microcomputers: Middle Grades." *AT* (February 1983), 28–29, 66–67.

CHAPTER 8

An Introduction to Geometry

Cutting a circular cake produces interesting problems in geometry. If n concurrent vertical cuts (that is, n vertical cuts through the same point on top) are made, it is easy to see that $2n$ pieces of the cake are produced. But suppose the vertical cuts are not concurrent. The figures at the beginning of this section show the maximum number of pieces. When $n = 1, 2, 3,$ and 4, the number of pieces are 2, 4, 7, and 11, respectively. Use your problem-solving techniques to conjecture the maximum number of pieces with 5 vertical cuts. Then draw a figure and verify your conjecture. What is the maximum number of pieces produced by 10 cuts? (Gardner, 1978)

The next five chapters of this book will be devoted to geometry. We will follow a transformation approach in the development of geometric concepts, for the following reasons.

1. It simplifies and unifies the elementary topics presented.
2. It simplifies the mathematical development (for example, definitions for congruence and similarity cover all figures).
3. It provides a background for additional work in mathematics.

To understand that geometry can be presented in many different ways, let's take a look at the history of geometry. For over 4000 years, people have been studying geometry and its applications to the real world. The name *geo-metry*, literally meaning *earth measurement*, reveals that the beginnings of geometry were in practical applications. The geometry of the Babylonians, for instance, dealt with the computation of areas and volumes. The Egyptians used geometry in building pyramids and surveying fields. Both civilizations applied geometry to astronomy and hence to the development of a calendar. The Greeks were responsible for an important change in direction. They introduced logic into the presentation of geometric ideas, thereby starting the consideration of geometry as an abstract mathematical system.

Geometry is just as important today as it was for the Greeks, Babylonians, and Egyptians. Various geometric concepts are used in navigation, surveying, engineering, architecture, art, and so forth. Geometric figures such as triangles and circles have many utilitarian applications. They also play an important role in the development of theories concerning the basic structure of the universe.

In this chapter, you will experience an intuitive approach to geometry. This intuitive approach includes a description of undefined terms and a discussion of definitions and axioms with many examples as well as statements and discussions of the most important theorems of geometry. Proofs will play an insignificant role; understanding will be emphasized. Although you will find this approach to geometry to be fairly easy, you are advised to learn well the terms of this chapter as background for four additional chapters on geometry.

1 Some basic ideas of geometry

■ Fold a piece of paper. What geometric figure is represented by the fold? Fold the piece of paper again, so that the creases are not parallel. Do the two creases intersect? In how many points? What geometric truth have you demonstrated? What geometric figure does a flashlight beam represent? ■

If your answers to the introductory questions were "line (or line segment), two intersecting lines intersect in a unique point, and a ray," then you probably already understand much of this section. However, we must present the terminology of geometry before considering transformations in the next chapter.

Geometry is the study of sets of points. Therefore, we can use the terminology of Chapter 2 on sets. Someone might say that geometry is a study of figures. This is true also, because we consider a figure to consist of a set of points.

In previous chapters of this book, terms were defined and then used. In this section, we use another approach. For example, points, lines, and planes are fundamental in nature and in man-made figures. We present these as undefined terms and build our definitions of more complicated figures on these concepts.

In the study of geometry, we are concerned with sets of elements called **points.** All geometric figures are sets of points. A point is a unique, exact location in space. It has no dimensions. Physical representations of points in nature include a star seen with the naked eye or the tip of a porcupine's quill. However, we shall consider *point* as an undefined term. In a diagram, a point is usually named by a capital letter, such as A, B, C, or D. (See Figure 1.)

Now think of **space as the set of all points.** Thus, the physical universe may be visualized as completely filled, or dense, with points. The geometric figures that we shall study consist of subsets of these points in space.

Figure 1

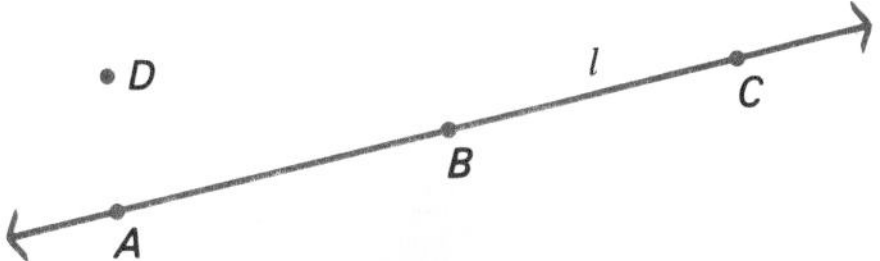

BIOGRAPHY

Euclid

Circa 365 B.C.–275 B.C.

The Greek author of the world's oldest scientific textbook bequeathed a rich heritage to all of us. We first learn of Euclid at the time the Egyptian ruler Ptolemy invited him to form a mathematics department at the University at Alexandria. It is recorded that Euclid was a kind and patient teacher. One student, laboring over his geometry, asked Euclid what reward would come for such study. Euclid sent a servant to give the youth a penny, "since he must make a gain from what he learns."

Euclid's reputation rests on his textbook, *Elements,* written in 13 volumes over 2000 years ago when Euclid was about 40 years old.

A **line** is a particular subset of points in space. (We use the word *line* to mean *straight line* in this book.) The edge of a desk, a sideline of a football field, and the edge of a box are physical models of parts of lines (called *line segments*). Yet these are only physical models of lines, for the idea of a line is an abstraction that we use without definition (an undefined term). Figure 1 represents a line. The arrow on each end indicates that the line extends indefinitely in each direction. A line can be named either by a lowercase letter (line l) or by a pair of capital letters representing any two points on the line, such as line $\overleftrightarrow{AB}$.

A **plane** is another subset of points in space. Intuitively, we describe a plane as being "flat," like a sheet of paper, and extending indefinitely; so a plane contains infinitely many points and infinitely many lines.

When visualizing a plane, we usually think of a set of points lying on some flat surface. Tops of tables and the surfaces of floors and walls are common models of parts of planes. Figure 2 represents a plane. Notice that it is named by α (the lowercase letter *alpha* from the Greek alphabet). Greek letters such as α, β (beta), and γ (gamma) are commonly used to name planes.

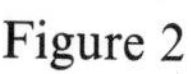
Figure 2

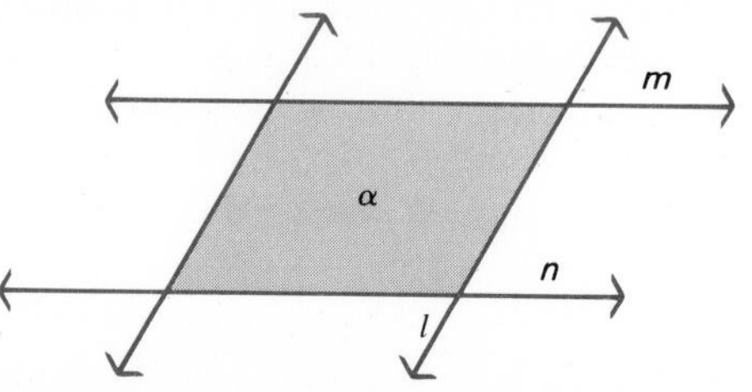

In statements relating points, lines, and planes, other undefined terms, such as *is on, contains, is between,* and *separates,* are used. If the members of a set of points are on a line, they are said to be **collinear,** and the line is said to **pass through** them or to contain them. In Figure 1, A, B, and C are collinear; however, A, B, and D are not. If three points are collinear, then one of the points is **between** the other two. Which one of the points in Figure 1 is between the other two? In Figure 1, point A is on line l and line l is said to contain point A. In set terminology, A is an element of l, or $A \in l$.

As suggested in the introduction, when we fold a piece of paper (waxed paper is best), the resulting crease represents a piece of a line. Fold the paper again so that the creases are not parallel. The two creases intersect in how many points? Your answer demonstrates the following property. **If two distinct lines intersect, then the intersection is exactly one point.**

Consider a point in space. How many lines can contain, or pass through, this point? It should be easy to visualize that infinitely many lines can contain a single point. For example, consider more and more lines drawn through A in Figure 3(a). Two or more lines are said to be **concurrent** if there is exactly one point common to all the lines. For instance, in Figure 3(a), the three lines are concurrent, but in Figure 3(b) they are not. All the lines that can be drawn through a point is sometimes called a **pencil of lines.**

Figure 3

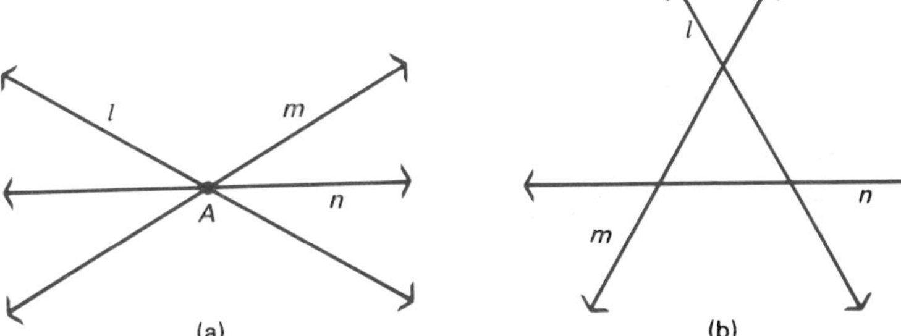

How many lines can contain the same pair of distinct points? Mark two points on a piece of paper. Draw a line containing the two points. (See Figure 4(a).) Can you draw another line through them that is distinct from the first line? Take any two points in a room and stretch a string between them. Any other string stretched between these two points would occupy the same position as the first string. These physical models illustrate the fact that lines are uniquely determined by two fixed points. In general, **there is exactly one line that contains two distinct points.**

Figure 4

If two distinct lines in a *plane* do not intersect, the lines are said to be *parallel.*

Parallel lines

> Two lines in a plane are *parallel* if and only if there is no point common to both lines.

Straight railroad tracks suggest the idea of parallel lines. In Figure 4(b), lines r and s are parallel. When we say that two rays (defined on the next page) or two line segments are parallel, we mean that lines containing these rays or segments are parallel.

The edges of your book illustrate parallel segments. If l and m are parallel, we write $l \| m$, where $\|$ denotes *is parallel to*.

Can you imagine two lines that do not meet anywhere but that are still not parallel? This situation can happen if the two lines are not in the same plane. Notice in Figure 5(a) that r and s will not intersect no matter how far they are extended. Similarly, lines t and s and t and r do not intersect. Note also that they are not parallel, since they do not lie in the same plane. Such lines are called **skew lines.**

Figure 5

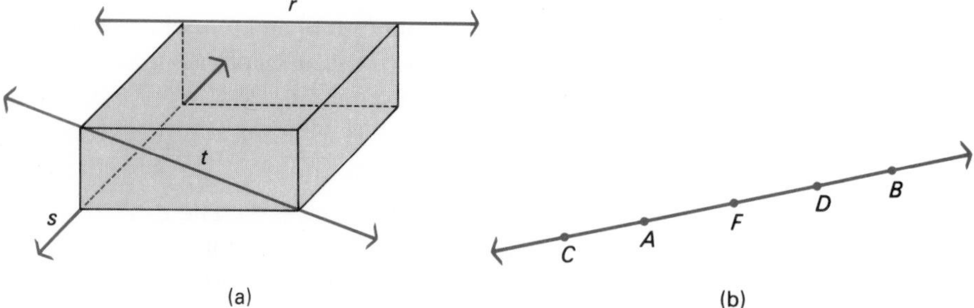

Any point on a line separates the line into three parts: the point itself and two half lines. In Figure 5(b) all the points on the same side of F as D (but not including F) constitute a half line on the D side of F. All the points on the C side of F constitute another half line. If the point of separation is included with one of the half lines, we have the following definition.

Ray

> A ray is the union of a point and one of the half lines determined by the point.

The first letter used to denote a ray names the endpoint; the second letter names some point on the half line and indicates direction. A is the endpoint of ray $\overrightarrow{AB}$. If D is a point on $\overrightarrow{AB}$ other than A, then $\overrightarrow{AB}$ and $\overrightarrow{AD}$ name the same ray. In Figure 5(b), point A determines two rays, $\overrightarrow{AB}$ and $\overrightarrow{AC}$, which are called *opposite rays.* Note that

$$\overrightarrow{AB} \cap \overrightarrow{AC} = \{A\}$$

Line segment

> A *line segment* is a subset of a line consisting of two points on the line and the set of points between them.

Consider the points A and B in Figure 5(b). These two endpoints and all the points between them are called *segment* $\overline{AB}$ and are denoted by either $\overline{AB}$ or

Just for fun

Two points determine one line. Three noncollinear points determine three lines. Without counting (using your problem-solving techniques), find how many lines are determined by 12 points, no three of which are collinear.

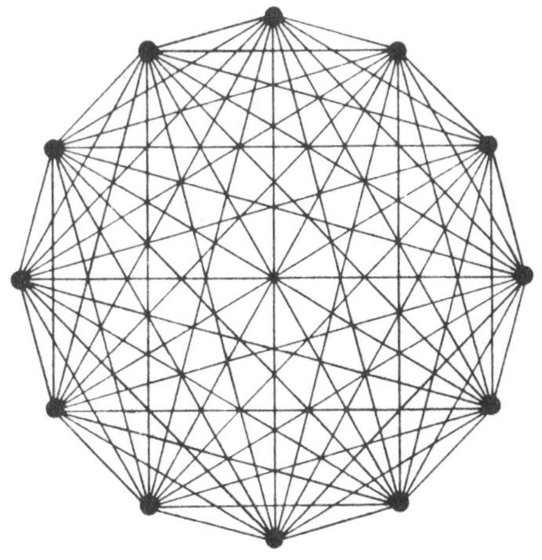

$\overrightarrow{BA}$. Thus, $\overline{AB}$ consists of the points that $\overrightarrow{AB}$ and $\overrightarrow{BA}$ have in common, or $\overline{AB} = \overrightarrow{AB} \cap \overrightarrow{BA}$.

Example 1 • (a) Find $\overline{CD} \cap \overline{AB}$ in Figure 5(b). The answer is $\overline{AD}$, because all the points in $\overline{AD}$ are common to both $\overline{CD}$ and $\overline{AB}$.
(b) Find $\overrightarrow{CA} \cap \overrightarrow{AC}$. The answer is $\overline{AC}$.
(c) Find $\overline{CA} \cap \overline{DB}$. The answer is $\varnothing$.
(d) Find $\overline{AF} \cap \overline{DF}$. The answer is $\{F\}$. •

Example 2 • Using Figure 6, name
(a) $\overleftrightarrow{AB} \cap \overleftrightarrow{AD}$
(b) $\overleftrightarrow{AB} \cap \overleftrightarrow{CD}$
(c) $\overrightarrow{DC} \cap \overrightarrow{CD}$

Figure 6

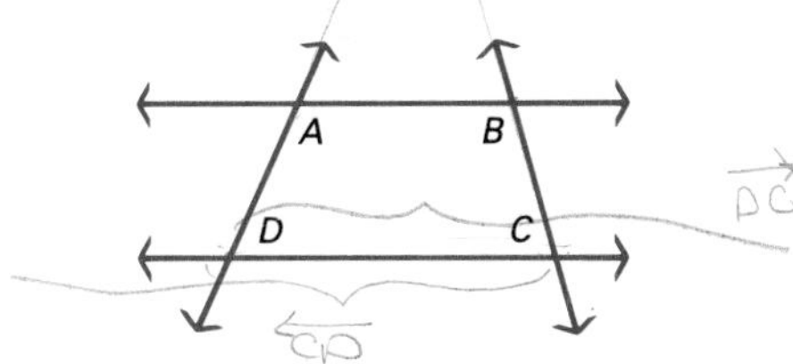

Solution (a) $\{A\}$ (b) $\varnothing$ (c) $\overline{DC}$ •

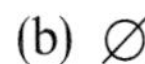

Exercise set 1

R 1. Which geometric concepts are suggested by these physical situations?
(a) A piece of rubber band stretched between two points in space
(b) A pair of opened scissors
(c) The wall of your bedroom
(d) A straightened paper clip
(e) The tip of a needle

(f) The blackboard
(g) The spokes of a bicycle wheel

2. Draw sketches showing the following intersections.
   (a) A line and a ray that intersect in one point
   (b) A line and a line segment that intersect in the line segment itself
   (c) A plane and a line segment that intersect in one point
   (d) Two lines that neither intersect nor are parallel
   (e) A line segment and a ray that intersect in an endpoint

3. Suppose there are four points A, B, C, and D, no three of which are collinear. How many lines do they determine if they are coplanar (lying in the same plane)?

4. Consider the points and lines given in the following figure.

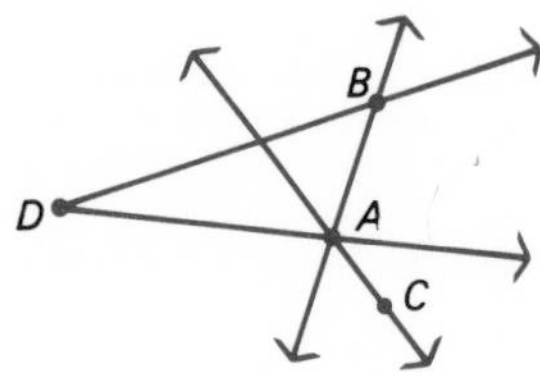

Find each of the following.
   (a) A segment containing point B
   (b) A ray with point A as the endpoint
   (c) Two rays with point D as the endpoint
   (d) A ray with point B as the endpoint
   (e) Two lines
   (f) Three segments containing point A

5. In the following figure classify the pairs of lines as parallel, skew, or intersecting.

(a) $\overleftrightarrow{AB}$ and $\overleftrightarrow{CD}$	(b) $\overleftrightarrow{FH}$ and $\overleftrightarrow{AB}$
(c) $\overleftrightarrow{AB}$ and $\overleftrightarrow{FB}$	(d) $\overleftrightarrow{AE}$ and $\overleftrightarrow{DH}$
(e) $\overleftrightarrow{AE}$ and $\overleftrightarrow{BD}$	(f) $\overleftrightarrow{EG}$ and $\overleftrightarrow{HG}$

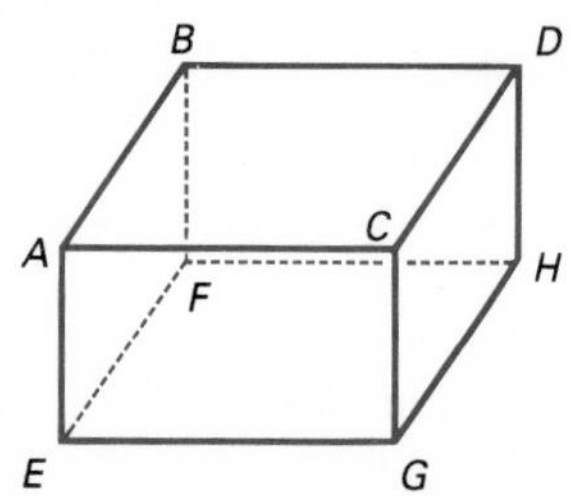

6. Label each statement as true or as false. Rewrite the false statements as true statements.
   (a) A line has an infinite number of different line segments that are subsets of it.
   (b) Space is the set of all points.
   (c) A ray has two endpoints.
   (d) A line segment contains infinitely many points.
   (e) $\overrightarrow{AB}$ is the same set of points as $\overrightarrow{BA}$.
   (f) If A, B, and C are collinear with B between A and C, then $\overrightarrow{AB} \cap \overrightarrow{BC} = \overrightarrow{BC}$.
   (g) In part (f), $\overline{AB} \subset \overline{BC}$.
   (h) A ray separates a plane.
   (i) Two distinct lines either intersect in a point or are parallel.
   (j) A half line separates a plane.
   (k) The endpoints of $\overleftrightarrow{BA}$ are B and A.
   (l) $\overline{AB}$ is the same line segment as $\overline{BA}$.
   (m) A line has one endpoint.
   (n) *If* A, B, and C are collinear points, where $\overline{AB} \cap C = \varnothing$, then $\overrightarrow{AB} \subset \overrightarrow{CA}$.
   (o) For any points A and B, $\overleftrightarrow{AB} = \overleftrightarrow{BA}$.
   (p) If two lines are parallel to a third line, they are parallel to each other.

7. Consider three collinear points D, E, and F with E between D and F.
   (a) Do $\overrightarrow{DE}$ and $\overrightarrow{ED}$ name the same ray?
   (b) Do $\overline{DE}$ and $\overline{ED}$ name the same line segment?
   (c) Do $\overrightarrow{DE}$ and $\overrightarrow{EF}$ have points in common?
   (d) Is each point on $\overline{EF}$ also on $\overrightarrow{ED}$?
   (e) Do $\overleftrightarrow{DE}$ and $\overleftrightarrow{EF}$ name the same line?

8. Consider l, m, and n as three distinct lines in a plane. These lines may be related to each other, as shown in the figure.

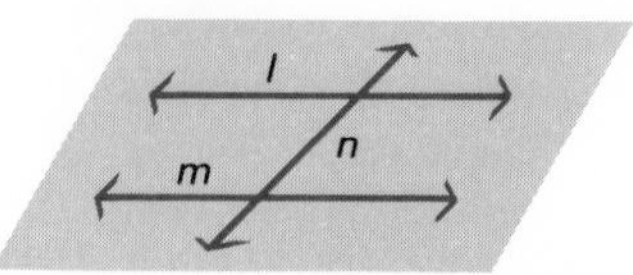

By sketching a figure, show other ways they may be related.

9. Using the line given, determine each of the following and tell which are empty.

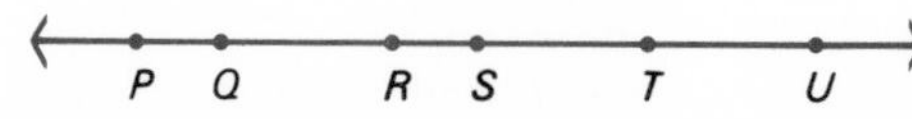

(a) $\overline{PR} \cap \overline{QT}$ (b) $\overline{RS} \cap \overline{ST}$ (c) $\overline{QS} \cap \overrightarrow{RU}$
(d) $\overline{PQ} \cap \overline{ST}$ (e) $\overline{RS} \cap \overrightarrow{QU}$ (f) $\overrightarrow{QS} \cap \overrightarrow{RQ}$
(g) $\overrightarrow{RP} \cap \overrightarrow{RU}$ (h) $\overrightarrow{RP} \cap \overrightarrow{RQ}$ (i) $\overline{RS} \cup \overline{QT}$

10. (a) Draw two segments $\overline{UV}$ and $\overline{XW}$ so that $\overline{UV} \cap \overline{XW}$ is empty but for which $\overleftrightarrow{UV} \cup \overleftrightarrow{XW}$ is one line.
    (b) Draw two segments $\overline{AB}$ and $\overline{CD}$ for which $\overline{AB} \cap \overline{CD}$ is not empty but for which $\overleftrightarrow{AB}$ is not the same as $\overleftrightarrow{CD}$.

11. In a plane, draw three lines and then four lines with the following number of intersections. Do you find that any of the drawings are impossible?
    (a) 0 (b) 1 (c) 2 (d) 3 (e) 4 (f) 5

T 12. Line l contains the four points A, B, C, and D.

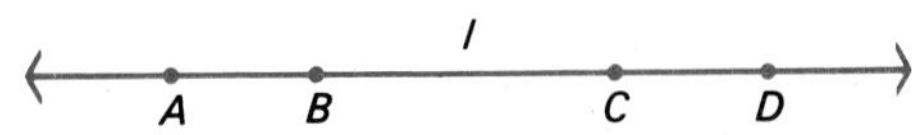

Name two rays on l with the following characteristics.
    (a) Their intersection is a ray.
    (b) Their intersection is a point.
    (c) Their intersection is a segment.
    (d) Their intersection is empty.

13. In Exercise 12, name two line segments such that
    (a) their intersection is one of the segments.
    (b) their intersection is B.
    (c) their intersection is $\overline{BC}$.
    (d) their intersection is empty.

14. Draw a figure to demonstrate each of your answers to the following.
    (a) Is it possible for a ray to be a proper subset of another ray?
    (b) Is it possible for the union of two rays to be a segment?

15. Using the figure, name
    (a) $\overleftrightarrow{AB} \cap \overline{AB}$ (b) $\overleftrightarrow{BC} \cap \overleftrightarrow{CD}$ (c) $\overrightarrow{CD} \cap \overleftrightarrow{BC}$
    (d) $\overrightarrow{CD} \cap \overrightarrow{BC}$ (e) $\overline{AD} \cap \overline{DC}$ (f) $\overrightarrow{CD} \cap \overline{AB}$
    (g) $\overleftrightarrow{AD} \cup \overline{AD}$ (h) $\overrightarrow{BC} \cap \overleftrightarrow{BC}$ (i) $\overrightarrow{AF} \cup \overline{ED}$

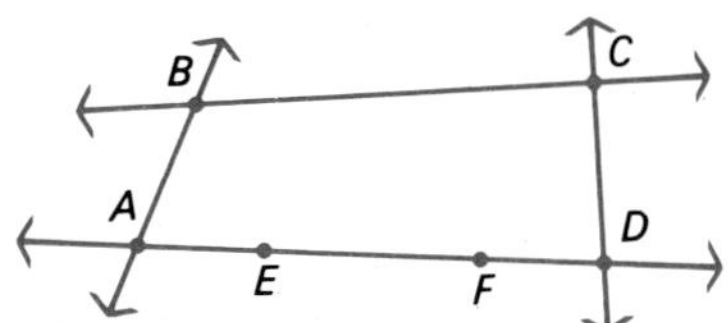

C 16. If all points are in one plane and no three points are collinear, state how many lines are determined by
    (a) 2 points (b) 3 points
    (c) 4 points (d) 5 points
    (e) 6 points (f) n points

17. How many rays are determined by
    (a) three collinear points?
    (b) four collinear points?
    (c) five collinear points?
    (d) n collinear points?

18. Let P be a point not on line m. How many lines may be drawn through P intersecting m? How many planes contain P and m?

19. If all the lines are in the same plane, each line intersects each of the other lines, and no point is common to three lines, how many points are determined by
    (a) 2 lines (b) 3 lines
    (c) 4 lines (d) 5 lines
    (e) 6 lines (f) n lines

20. Using terms such as *segments* and *half lines,* name the following sets, where x represents points on a number line. ($\leq$ means less than or equal to.)
    (a) $\{x \mid x > 1\}$
    (b) $\{x \mid x \leq -2\}$
    (c) $\{x \mid x \geq 3\} \cap \{x \mid x \leq 5\}$
    (d) $\{x \mid x \leq 3\} \cup \{x \mid x \geq 2\}$

21. Two intersecting lines partition a plane into at most four disjoint regions, while three lines (no two of which are parallel) partition the plane into at most seven regions. Determine a pattern and complete the table.

Number of Lines	1 2 3 4 5 . . . 10
Number of Regions	2 4 7

22. Use the fact that there is exactly one line that contains two different points to prove that two distinct nonparallel lines in a plane intersect in exactly one point.

2 Lines, planes, and space

■ Complete each statement using *always, sometimes,* or *never.*

1. A line and a point not on the line are ___________ coplanar.
2. Two intersecting lines are ___________ contained in one plane.
3. Three non-collinear points are ___________ contained in one plane.
4. Two parallel lines are ___________ coplanar. ■

If your answer to each of the preceding questions was *always* then you understand already some of the concepts of this section. An intuitive understanding of the conditions that uniquely determine a plane and the relationship between lines and planes will be helpful as you pursue your study of geometry.

In the previous section, we considered how two lines might be related to each other. Let us consider now how a line and a plane might be related. There are basically three possibilities for the relationship between a line and a plane in space.

Line and a plane

1. A line and a plane may intersect in exactly one point. (Line l intersects plane α in point A in Figure 7(a).)
2. The entire line lies in the plane. (Plane γ contains line m in Figure 7(b).)
3. The line and the plane do not intersect, in which case we say that the line is parallel to the plane. (Line n is parallel to plane α in Figure 7(c).)

Figure 7

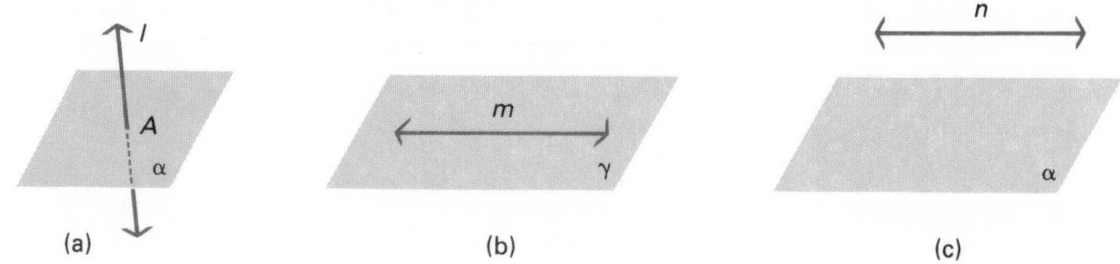

Place two tacks in a bulletin board and stretch a rubber band between them. Notice how the stretched rubber band seems to cling to the bulletin board. Intuitively, this result suggests that **if a line contains two different points of a plane, the line lies in the plane** (Figure 8).

Figure 8

Points that lie in the same plane are said to be **coplanar.** Points A, B, and C in Figure 9(a) are coplanar. However, points A, B, C, and D are not coplanar. Lines that lie in the same plane are also called *coplanar.*

Figure 9

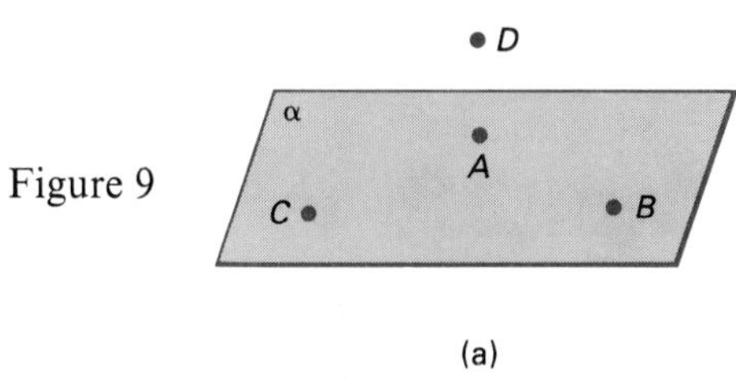

(a)

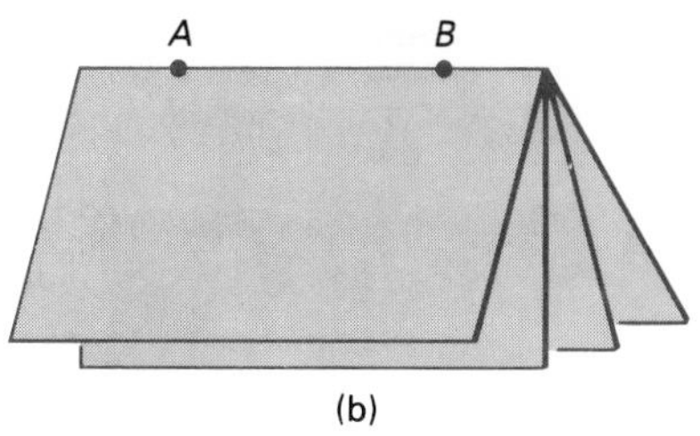

(b)

How many points are needed to determine a plane? Is a plane uniquely determined by two points? Fold several pieces of paper together and put a paper clip along the fold so that they will stay together. Let A and B be two points on the fold. Each half page of paper is part of a plane containing points A and B (see Figure 9(b)). Thus there can be infinitely many planes through two distinct points, and so two distinct points do not determine a plane.

Attempt to balance a piece of cardboard on the tip of a pencil. Does the cardboard tend to tilt in some direction? How many planes can contain one point? Now hold the tips of two pencils under the cardboard. Does the cardboard tilt? How many planes can contain two points? Finally, hold three pencils under the cardboard. Try to tilt the cardboard. What do you find? In like manner, if you sit on a three-legged stool whose legs are not the same length, all three legs will touch the floor. However, try sitting in a four-legged chair having one leg shorter than the other three legs. How many legs will touch the floor? Did you answer three? This discussion suggests the following property of planes. **If A, B, and C are three distinct noncollinear points, then there is one and only one plane containing A, B, and C,** Figure 9(a).

Take a pencil representing a line, a sheet of paper representing a plane, and a point in space. How many ways can you arrange the paper to contain the pencil and the point? The fact that the arrangement was unique suggests that a **line and a point not on the line uniquely determine a plane.**

Take two pencils and arrange them to represent intersecting lines. How many possible ways can you arrange your piece of paper to contain both pencils? This experiment suggests that **two intersecting lines uniquely determine a plane.**

Now arrange your pencils so that they seem to be parallel. Again try to arrange your sheet of paper in different ways to contain the pencils. Do you agree that **two parallel lines uniquely determine a plane?**

Planes

A plane can be determined by

1. three noncollinear points.
2. a line and a point not on that line.
3. two intersecting lines.
4. two parallel lines.

All of the preceding properties relate to the fact that three noncollinear points determine a plane.

Example 3 • (a) Points A, B, and C determine plane α in Figure 10(a).

(b) Line $\overleftrightarrow{AB}$ and C determine plane α in Figure 10(b).

(c) Intersecting lines $\overleftrightarrow{AB}$ and $\overleftrightarrow{CB}$ determine plane α in Figure 10(c).

(d) Lines $\overleftrightarrow{AB}$ and $\overleftrightarrow{CD}$ with $\overleftrightarrow{AB}$ parallel to $\overleftrightarrow{CD}$ determine plane α in Figure 10(d).

(a) (b)

Figure 10

(c) (d) •

Examine the walls of a room. Let the walls represent parts of planes. The intersection of two walls (parts of planes) is clearly a line segment. We generalize this idea by **if the intersection of two distinct planes is not empty, then the intersection is a line,** Figure 11(a).

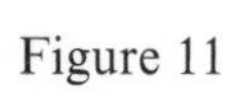

Figure 11

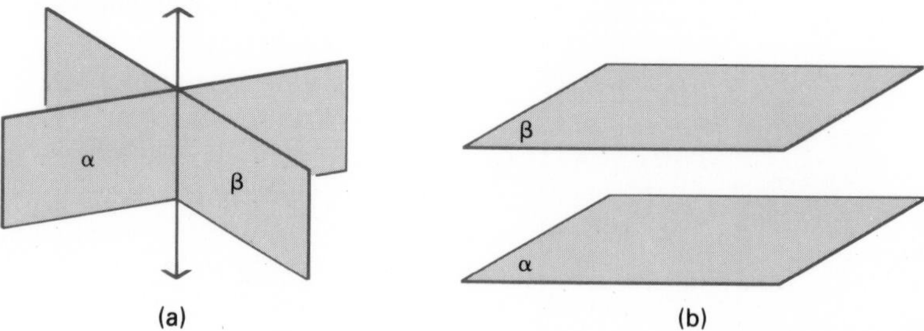

(a) (b)

Planes do not always intersect. Consider the opposite walls of a room as parts of planes. If these walls are accurately constructed, they will not intersect no matter how far they are extended. Such planes are said to be *parallel,* Figure 11(b). The top and bottom of a box represent parts of parallel planes.

Parallel planes

> If the intersection of two distinct planes is empty, then the planes are said to be *parallel.*

Just for fun

Without raising your pencil, draw four line segments that contain all nine dots.

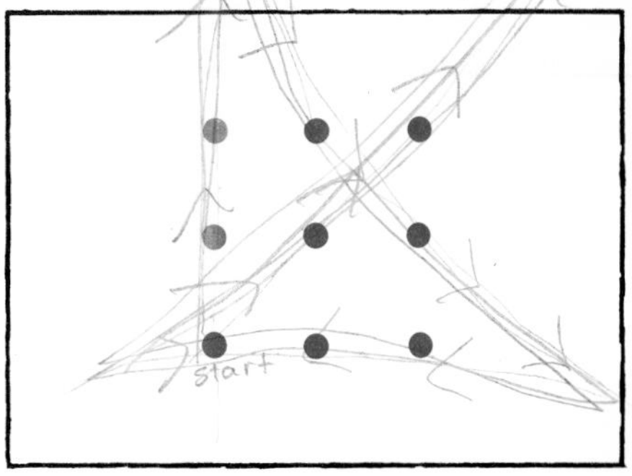

Planes α and β and planes α and γ in Figure 12(a) intersect along the dotted lines. However, plane β does not intersect plane γ because the two planes are parallel.

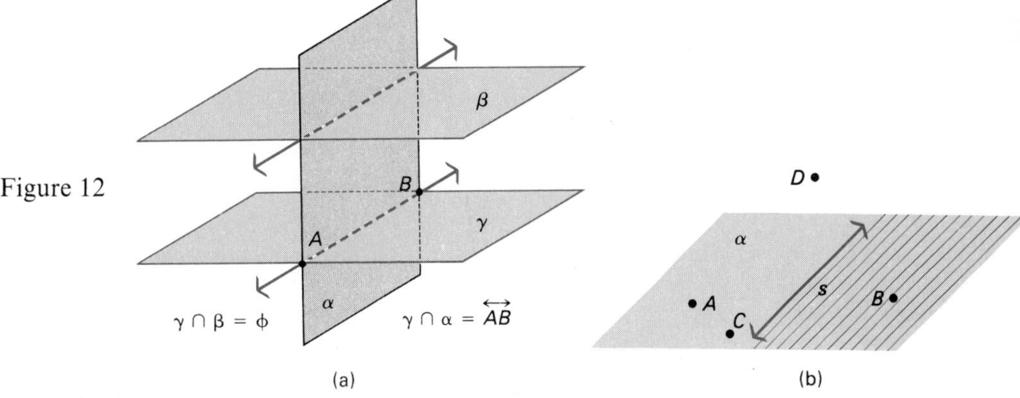

Figure 12

Just as a point separates a line, a line separates a plane into three parts: the line and two half planes. The points on the line do not belong to either half plane. However, the line is called the *edge* or boundary of both half planes. For example, in Figure 12(b), the shaded region represents one half plane and the unshaded region represents the other half plane. The portion of plane α that contains point A is called the A side of line s, and the other half plane is called the B side of line s. If A and B were on the same side of line s, then segment $\overline{AB}$ would lie entirely in one half plane. Does $\overline{AB}$ in Figure 12(b) lie in one half plane? Why or why not? How about $\overline{AC}$? What happens to the segment and the line when two points lie in opposite half planes?

Throughout this section, we have discussed points, lines, and planes in space. **Space** is the set of all points in three dimensions. Recall that a point separates a line into three disjoint sets: the point and two half lines on each side of the point. A line separates a plane into three disjoint sets: the line and the two half planes determined by the line. In a like manner, a plane separates space into three disjoint sets: the plane and the two half spaces determined by the plane.

Exercise set 2

R 1. Carefully fold a piece of paper in half. Hold the paper so that the fold is on a tabletop but neither side of the paper lies on the tabletop. Consider the three planes—the plane of the tabletop and the two planes of the two parts of the folded paper. What is the intersection of these three planes? Now stand the folded paper on its end. What is the intersection of the three planes?

2. Classify the following statements as always true, sometimes true, or never true.
   (a) Two distinct planes intersect in a line.
   (b) The intersection of two distinct lines determines a unique plane.
   (c) If two parallel planes are cut by a third plane, the lines of intersection are parallel.
   (d) If a plane contains one and only one of two parallel lines, it is parallel to the other line.
   (e) Given a plane and a point not on the plane, there is exactly one plane through the given point parallel to the given plane.
   (f) A line parallel to each of two intersecting planes is parallel to the line of intersection of the planes.

3. (a) Are any two lines in space always coplanar?
   (b) Two intersecting lines?
   (c) Two parallel lines?
   (d) Two skew lines?

4. Find the following from the accompanying figure. Label a plane, using three points.

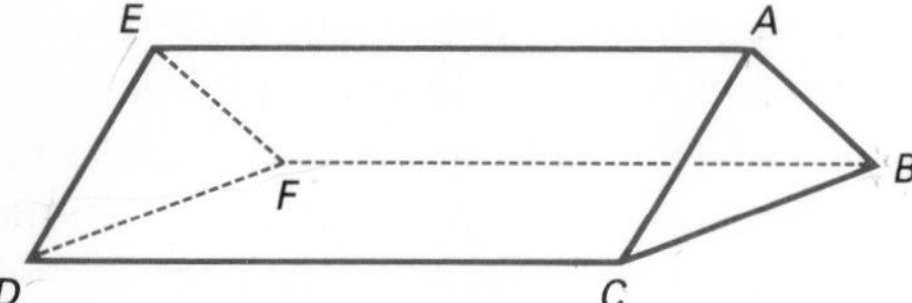

   (a) A pair of parallel planes
   (b) Two intersecting planes
   (c) Three planes that intersect in a point
   (d) A line parallel to a plane
   (e) A line that intersects a plane in one point
   (f) Two skew lines

5. Each of two intersecting lines is parallel to a given plane. Is the plane determined by these lines parallel to the given plane?

6. Planes α and β are parallel and β and γ are parallel. Is α necessarily parallel to γ? Illustrate with a drawing.

7. Do a point and a line necessarily determine a plane regardless of where the point is located?

8. (a) Planes α and β have points A, B, and C in common. Suppose that A, B, and C are noncollinear. What conclusion can you draw about α and β?
   (b) If α and β are distinct planes having points A, B, and C in common, what conclusion can you draw about points A, B, and C?

9. If l is a line and A and B are points not on l, how many planes are there that contain the points A and B and at least one point on line l?

10. If l is a line on plane α and m is a line on plane β and if α and β are parallel, are l and m necessarily parallel?

11. Suppose four points do not lie in the same plane. How many planes are determined by these points?

T 12. Each of two planes intersects a third plane in parallel lines. Are the two planes necessarily parallel? Illustrate with a drawing.

13. Complete the blank with *no, exactly one,* or *at least one.*
   (a) Given any line l and point P not on l, there is ______ plane that is parallel to l and contains P.
   (b) Given any plane α and point P that is not on α, there is ______ that contains P and intersects α.
   (c) Line l is parallel to plane α. There is ______ line in α skew to l.
   (d) Given any plane α and any line l that is parallel to α. There is ______ plane that contains l and intersects α.
   (e) Given any plane α and point P not on α, there is ______ line that contains P and is parallel to α.

14. Two parallel planes intersect a third plane in lines l and m. Are l and m necessarily parallel? Show with a drawing.

15. α, β, and γ are three distinct planes. Show with drawings the various ways the three planes can be related.

C 16. Prove the statement in Exercise 12.

17. Verify that the last three ways of determining a plane (as given on p. 369) are a result of "three noncollinear points determine a plane."

3 Angles and their measures

■ Draw a pair of intersecting lines. Use a protractor to measure the angles formed. What can you say about the angles in the drawing? ■

To produce a physical model of an angle, bend a straight piece of wire in the middle. Angles can also be demonstrated by taking two straws, joining them together with a pipe cleaner, and then bending the resulting object. Thus, angles are subsets of pairs of intersecting lines. In this section, we shall define the measure of an angle. This definition will be useful as we discuss geometric figures in this chapter.

As we have learned, through any point can be drawn an infinite number of rays that lie in a plane. These can be called a **pencil of rays.** (See Figure 13.) The union of any two of these rays is an angle.

Angle

An *angle* is the union of two rays that have the same endpoint.

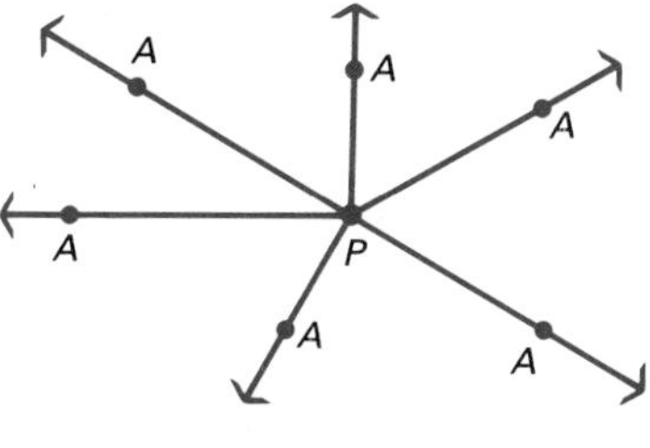

Figure 13

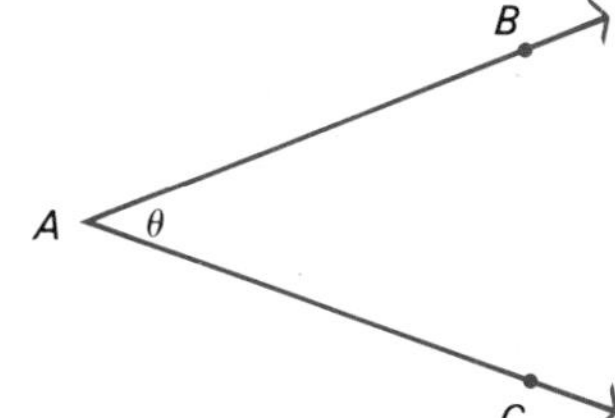

Figure 14

In Figure 14, the rays $\overrightarrow{AB}$ and $\overrightarrow{AC}$ are called the **sides** of the angle; the common endpoint A is called the **vertex.** Although an angle is an abstract idea and thus cannot actually be drawn, a figure like the one in Figure 14 is a visual representation of an angle. An angle may be named in a number of ways.

1. By using the symbol $\angle$ with the vertex, such as $\angle A$.
2. By using the symbol $\angle$ and three points A, B, C with B and C on different sides, such as $\angle BAC$ or $\angle CAB$. The vertex point is indicated by the middle letter.
3. By using the symbol $\angle$ and a letter of the alphabet, such as x or y, or a Greek letter, such as θ, as in $\angle x$ or $\angle \theta$. (In a drawing, the letter is placed inside the angle near the vertex, as seen in Figure 14.)

In Figure 14, if $\overrightarrow{AB}$ were rotated about point A until $\overrightarrow{AB}$ coincides with $\overrightarrow{AC}$, we consider $\overrightarrow{AB}$ to be rotated through angle θ. We now need a unit of measure for such an angle.

Properties of angle measure

To each angle $\angle BAC$ there corresponds a unique real number greater than or equal to zero and less than or equal to 180, called the **degree measure** of the angle and denoted by $m(\angle BAC)$, such that each of the following statements holds.

(a) If both sides of $\angle BAC$ are on the same ray, the measure of the angle is $0°$. (See Figure 15(a).)

(b) If the sides of $\angle BAC$ are opposite rays, as shown in Figure 15(b), then the measure of the angle is $180°$.

(c) If $\overrightarrow{AD}$ is located as shown in Figure 15(c), then

$$m(\angle CAD) + m(\angle DAB) = m(\angle CAB)$$

(d) In each half plane of $\overleftrightarrow{AB}$, there is exactly one ray whose union with $\overleftrightarrow{AB}$ is a given angle whose measure is between $0°$ and $180°$. (See Figure 15(d).)

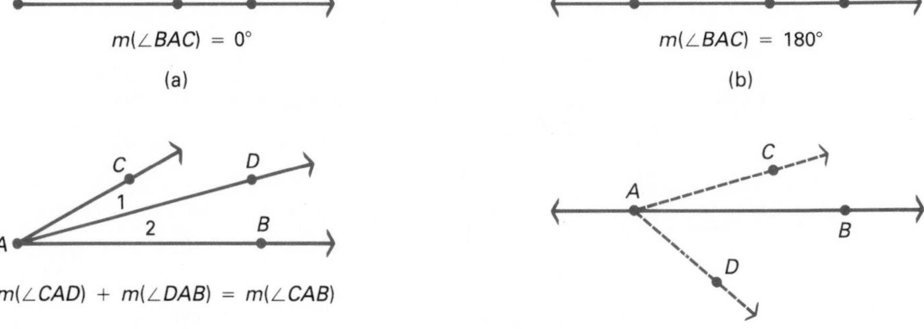

Figure 15

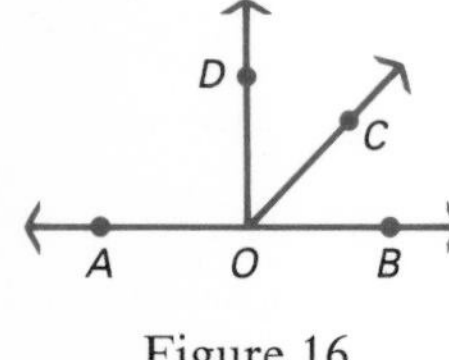

Figure 16

In the preceding discussion, the unit of measurement is the degree, which is indicated by a small circular superscript. For instance, 20 degrees is written $20°$. To understand the size of a unit angle with a measure of one degree, recall that in (a) a straight angle (that is, the rays lie on a line and point in opposite directions) has a measure of $180°$. So in Figure 16, $m(\angle AOB) = 180°$. Consequently, the other angles such as $\angle COB$ possess a measure between $0°$ and $180°$ (from (d).) For ex-

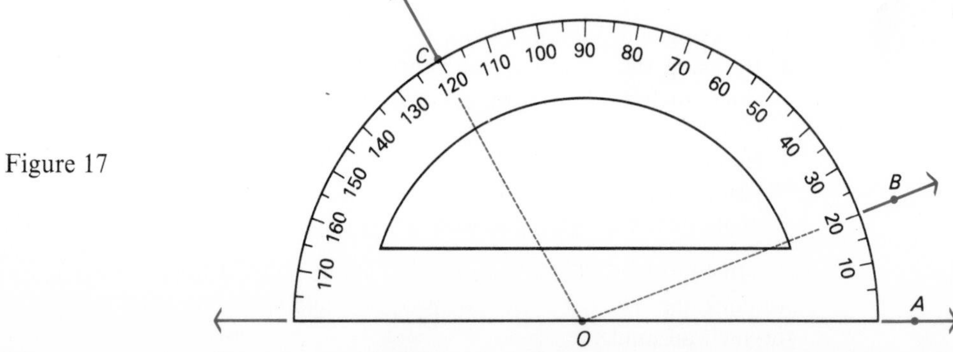

Figure 17

ample, $\angle DOB$ which seems to have half of the measure of $\angle AOB$, has a measure of 90° and is called a **right angle.**

A protractor is an instrument commonly used for measuring angles. In Figure 17, each hash mark indicates an angle of 5°. Find the arrow or dot in the middle of the bottom of the protractor (which is usually indicated by an O). Place that dot on the vertex of the angle to be measured with the 0° line of the protractor on one side of the angle. The other side crosses the protractor at some line, indicating the measure of the given angle. $\angle AOB$ in Figure 17 has a measure of 20°; $\angle AOC$ has a measure of 120°.

Example 4 • To get an intuitive understanding of the sizes of angles, guess the size of each angle in Figure 18, then check by enlarging the rays of the angle and using a protractor to measure each angle.

Figure 18

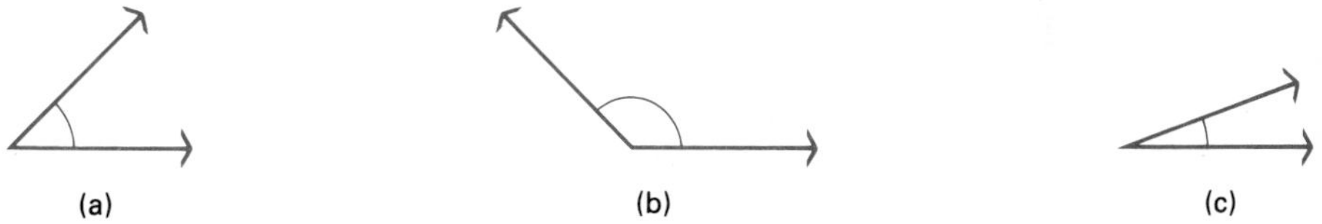

(a) (b) (c)

Solution (a) 45° (b) 135° (c) 20° •

Angles are commonly classified by properties of their measures.

Classification of angles

1. A **right angle** has a measure of 90°.
2. A **straight angle** has a measure of 180° (a line).
3. An **acute angle** has a measure between 0° and 90°.
4. An **obtuse angle** has a measure between 90° and 180°.
5. Two angles are **supplementary** if the sum of their measures is 180°.
6. Two angles are **complementary** if the sum of their measures is 90°.

Example 5 • An angle of 47° is an acute angle (Figure 19(a)), but a 154° angle, such as in (d), is obtuse. Angles with measures of 60° and 30° are complementary angles in (f); and angles with measures of 145° and 35° are supplementary angles in (e).

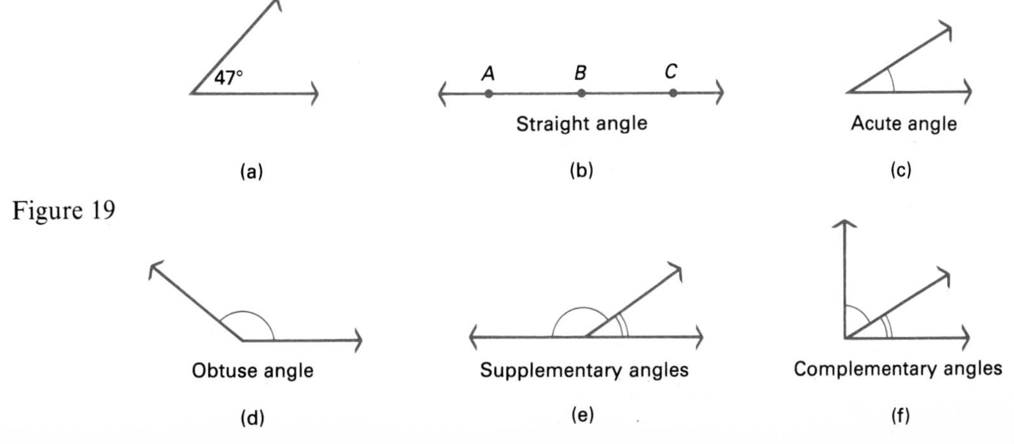

Figure 19

•

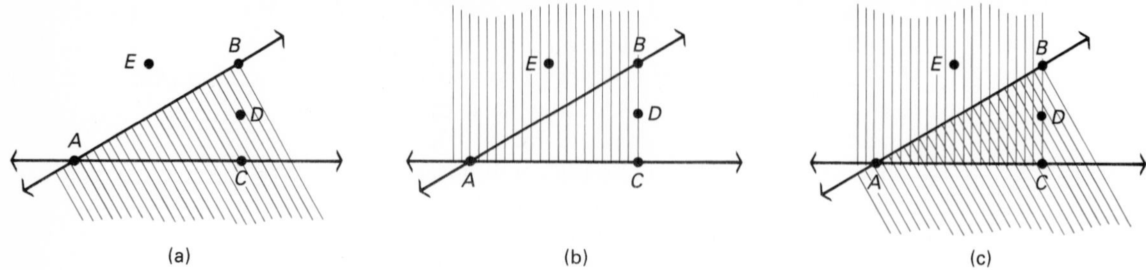

Figure 20

For some definitions, we need the concept of interior of an angle. The interior of $\angle BAC$ (Figure 20(c)) is the intersection of the half plane on the C side of $\overleftrightarrow{AB}$ in (a), and the half plane on the B side of $\overleftrightarrow{AC}$ in (b). D is in the **interior** of $\angle BAC$, and E is in the **exterior.**

Figure 21(a) shows the intersection of two lines. Observe that four angles, $\angle AOB$, $\angle AOC$, $\angle COD$, and $\angle DOB$, are formed by the intersection of these two lines. Since angles AOB and AOC have a common ray $\overrightarrow{OA}$ and a common vertex O, we call them **adjacent angles.**

Figure 21

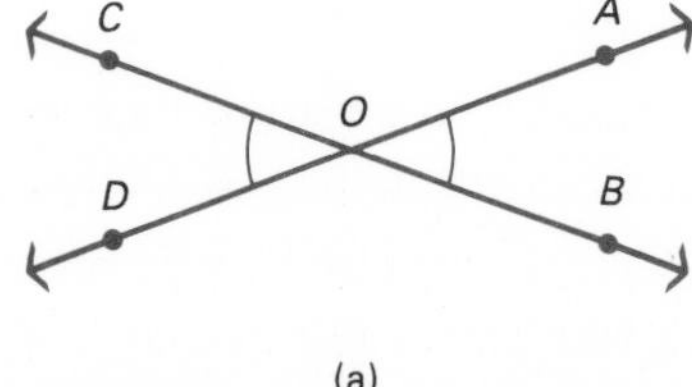

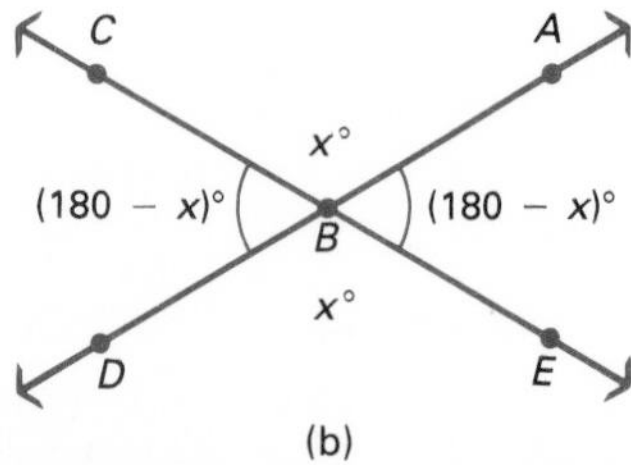

Adjacent angles

> Two coplanar angles that have a common vertex and a common side are called *adjacent angles* if the ray of one angle is not in the interior of the other.

In Figure 21(a), angles COA and DOB are not adjacent angles. The nonadjacent angles formed by the intersection of two lines are called **vertical angles.** Angles COD and AOB are vertical angles.

Vertical angles

> Two angles are vertical angles if and only if the sides of one are rays opposite the sides of the other.

In general, there are always two pairs of angles with the same measure when two lines intersect. In Figure 21(b), let $m(\angle CBA) = x°$. Then measure $m(\angle ABE) = 180° - x°$.

$$m(\angle DBE) + m(\angle ABE) = 180°$$

$$m(\angle DBE) + 180° - x° = 180°$$

$$m(\angle DBE) = x° = m(\angle CBA)$$

Notice that these angles having the same measure are vertical angles.

Measure of vertical angles

> Vertical angles have equal measures.

Suppose two lines intersect and form a right angle, as seen in Figure 22(a). Now $\angle PQR$ and $\angle PQS$ are supplementary angles. So $m(\angle PQS) = 90°$; likewise, $m(\angle RQT)$ and $m(\angle TQS)$ equal 90°. So all four angles are right angles. The two lines are called **perpendicular lines.**

Figure 22

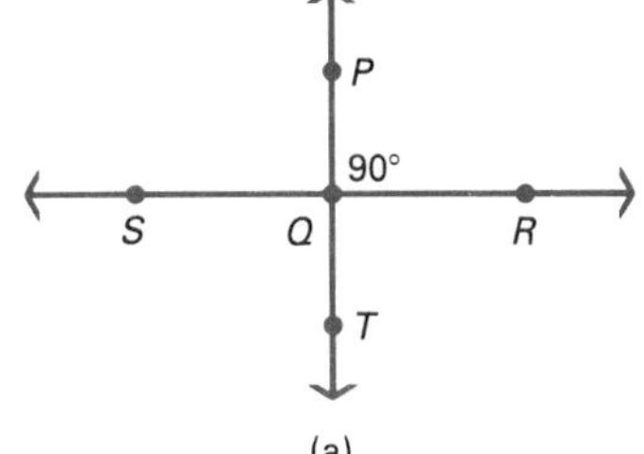

(a)

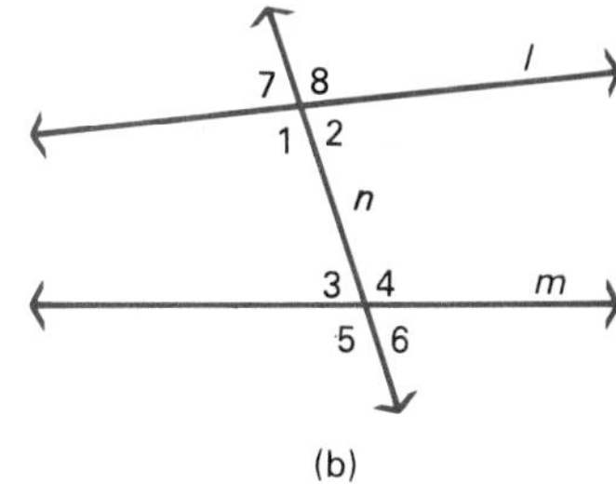

(b)

Perpendicular

> Two intersecting lines, rays, or segments are said to be **perpendicular** if and only if the lines (containing rays and segments) form a right angle. Perpendicular lines m and n are denoted by $m \perp n$.

Any line that intersects a pair of lines in exactly two points is called a **transversal** for those lines. In Figure 22(b), line n is a transversal of lines l and m. For this transversal, angles 1, 2, 3, and 4 are called *interior* angles, whereas angles 5, 6, 7, and 8 are called *exterior* angles. $\angle 1$ and $\angle 4$ and $\angle 2$ and $\angle 3$ are pairs of **alternate interior angles.** $\angle 5$ and $\angle 8$ and $\angle 6$ and $\angle 7$ are pairs of **alternate exterior angles.** $\angle 6$ and $\angle 2$, $\angle 4$ and $\angle 8$, $\angle 5$ and $\angle 1$, and $\angle 3$ and $\angle 7$ are called **corresponding angles.** Recall that angles 3 and 6, 4 and 5, 1 and 8, and 7 and 2 are called **vertical angles.**

The following two properties state important relationships between parallel lines and the angles formed by their transversals.

Angles and transversals

> 1. If lines l and m in a plane are intersected by a transversal in such a way that alternate interior angles are of equal measure, then l is parallel to m.
> 2. Alternate interior angles formed by two parallel lines and a transversal are of equal measure.

Using the fact that pairs of vertical angles are of equal measure, it is clear that several angles formed by a transversal of parallel lines are of equal measure. In Figure 22(b), if l is parallel to m, then $\angle 2$ and $\angle 7$ have equal measures, since they are vertical angles. Likewise, $\angle 2$ and $\angle 3$ have the same measure, since they

Just for fun

At four o'clock, what is the measure of the angle formed by the hands of the clock? What is the measure at three o'clock?

are alternate interior angles. Therefore, from the transitive property of equality, $\angle 7$ and $\angle 3$ have the same measure. In fact, **corresponding angles formed by a transversal between parallel lines always have the same measure.**

Exercise set 3

R 1. Give three names other than $\angle A$ for the given angle. Use a protractor to find the measure of this angle.

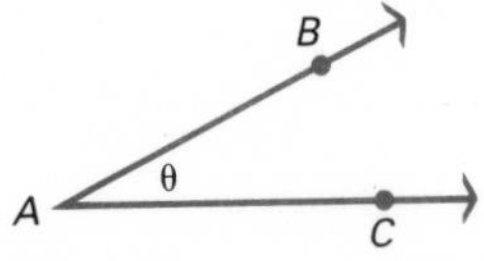

2. Classify each of the following statements as either true or as false.
   (a) Adjacent angles are complementary.
   (b) Vertical angles have equal measures.
   (c) Adjacent angles must have a common vertex.
   (d) Any two intersecting lines form adjacent angles.
   (e) Any two right angles have equal measures.
   (f) Two obtuse angles cannot be adjacent.

3. Are angles x and y adjacent in the following figures? If not, explain.

(a)

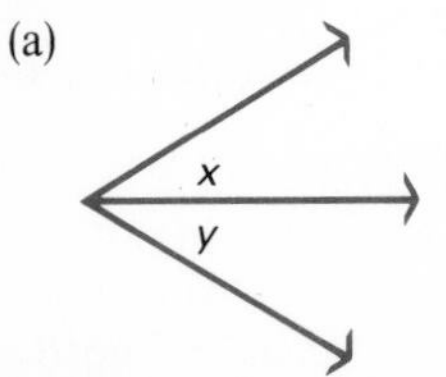

(b)

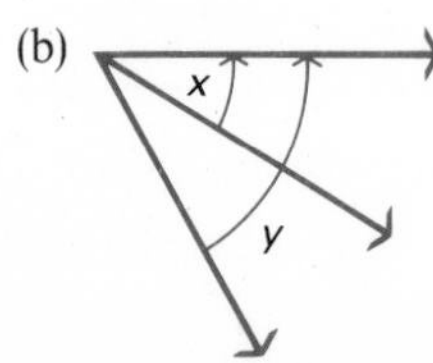

(c)

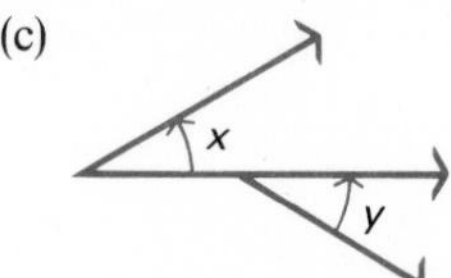

(d)

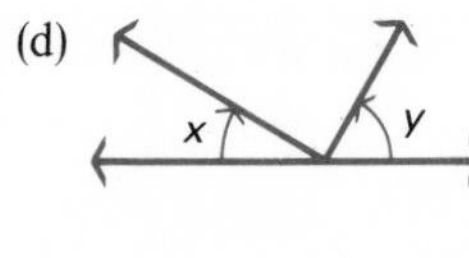

4. For each of the following sets of angles, tell which pairs of the marked angles are adjacent and which are vertical.

(a)

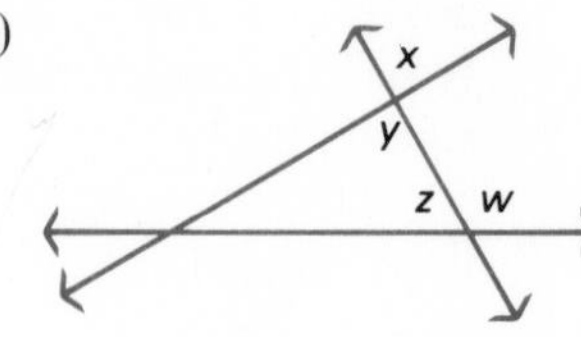

(b)

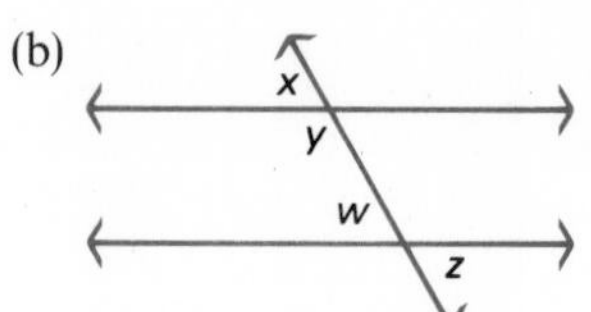

5. Make sketches that show the intersection of two angles as
   (a) a segment. (b) empty.

(c) a ray. (d) exactly two points.
(e) one point. (f) exactly four points.

6. Find two adjacent angles and a vertical angle for each of the following angles in the diagram.

(a) $\angle AOC$ (b) $\angle BOE$ (c) $\angle COD$

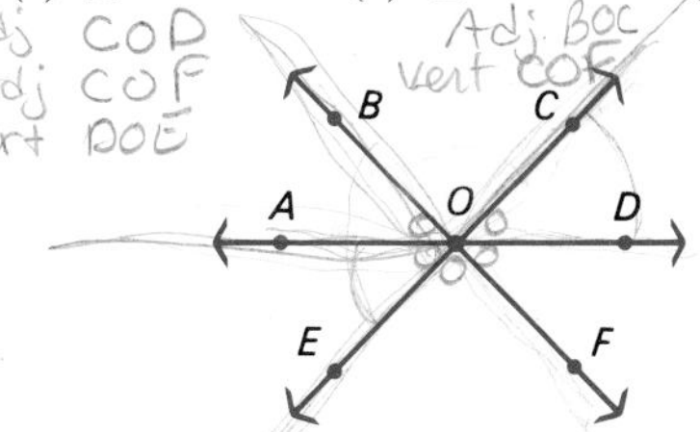

7. (a) If an angle has measure x, then its complements have measure ________.
(b) If an angle has measure y, then its supplements have measure ________.
(c) All supplements of an angle have the same ________.
(d) If two angles have the same measure and are supplementary, then the measure of each angle is ________.

8. Given the figure below, which consists of three parallel lines cut by a transversal, list the angles congruent to $\angle 9$ and explain why each is congruent.

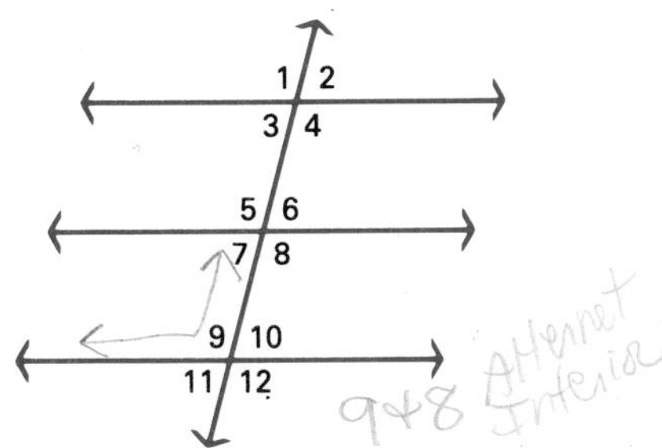

9. Identify each angle given as acute, obtuse, or right.

(a) $\angle FAB$ (b) $\angle ABC$ (c) $\angle CDE$
(d) $\angle EBC$ (e) $\angle FEB$ (f) $\angle AFD$

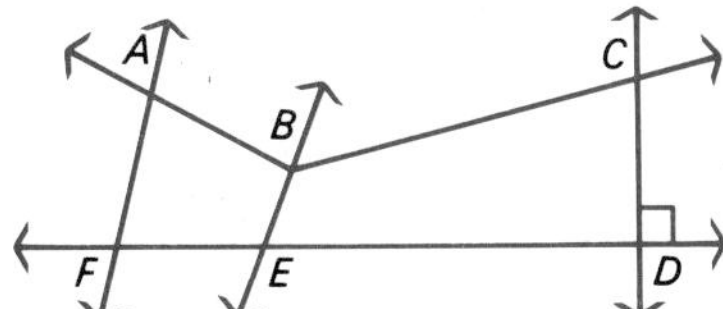

10. If $m(\angle COB) = 40°$ in the figure, then find

(a) $m(\angle AOB)$ (b) $m(\angle AOD)$ (c) $m(\angle COD)$

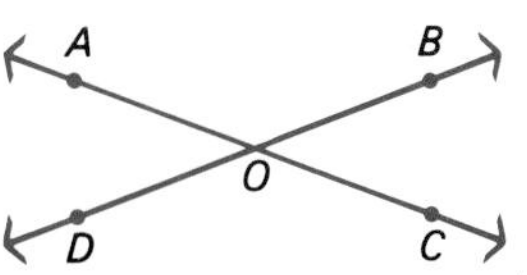

11. In the accompanying sketch, $m(\angle COB) = 35°$ and $\overleftrightarrow{EF}$ is perpendicular to $\overleftrightarrow{AC}$. Find the measure of each of the following angles.

(a) $\angle AOB$ (b) $\angle AOE$ (c) $\angle COD$
(d) $\angle DOF$ (e) $\angle AOD$ (f) $\angle FOC$
(g) $\angle EOB$ (h) $\angle FOB$

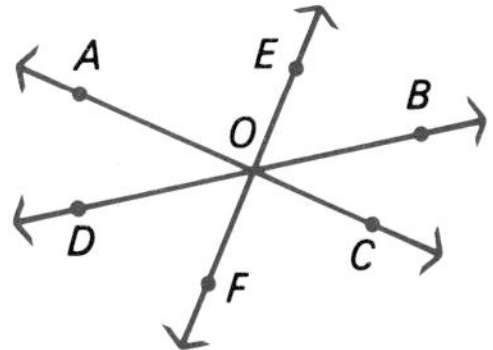

12.

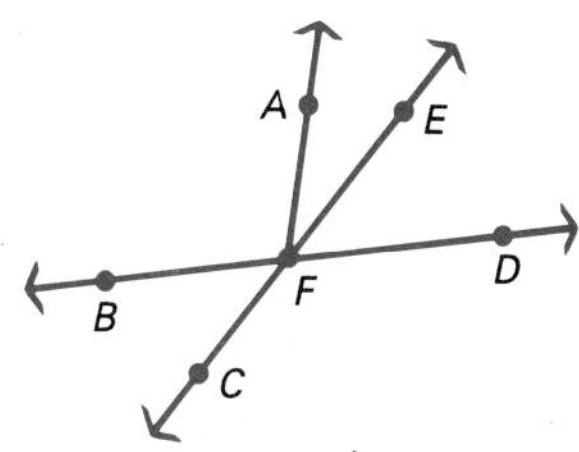

(a) Two angles adjacent to $\angle AFE$ are ________ and ________.
(b) ________ and ________ are a pair of vertical angles.
(c) $\angle BFC$ and $\angle CFD$ are ________ and ________ angles.
(d) If $m\angle BFC = 40°$, then $m\angle CFD =$ ________.
(e) If $m\angle BFC = 40°$, then $m\angle EFD =$ ________.
(f) If $m\angle CFD = 120°$, then $m\angle BFE =$ ________.
(g) Suppose $\overleftrightarrow{AF} \perp \overleftrightarrow{BD}$. Then $m\angle AFB =$ ________.
(h) Suppose $\overleftrightarrow{AF} \perp \overleftrightarrow{BD}$ and $m\angle AFE = 30°$. Then $m\angle EFD =$ ________.

T 13. Consider the intersection of three lines in space.

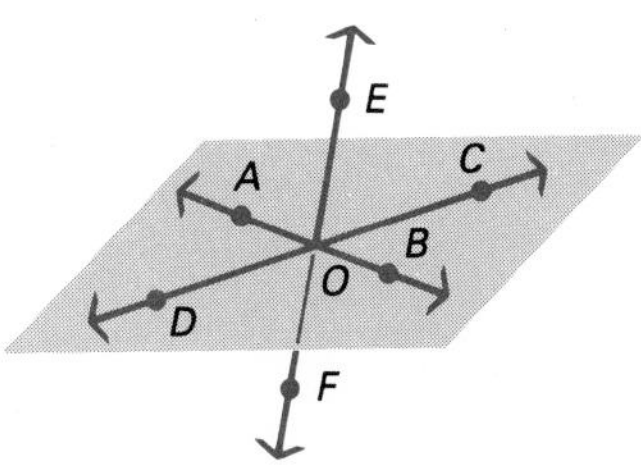

Two of these lines, $\overleftrightarrow{AB}$ and $\overleftrightarrow{CD}$, lie in the same plane; the third line, $\overleftrightarrow{EF}$, does not lie in their plane.

(a) How many angles are formed? Name them.
(b) Name six pairs of adjacent angles.
(c) Which angles are vertical? Name them.

14. In the figure, assume that $m(\angle 4) = m(\angle 8)$. Prove that all the other pairs of corresponding angles are equal.

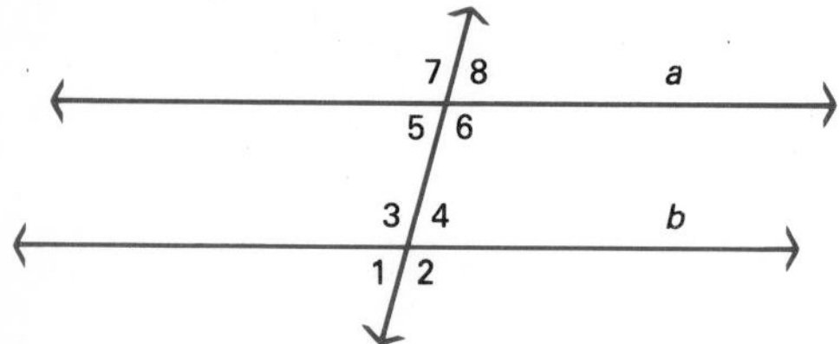

C 15. Suppose that you are given four points in space.
(a) What is the greatest number of lines determined by them? The greatest number of angles?
(b) Let the four points be in one plane. How many lines can be determined by them? Illustrate.

16. (a) Draw three noncollinear points on your paper and label them *A*, *B*, and *C*. Connect the three with line segments. How many segments do you have?
(b) Measure $\angle ABC$, $\angle BCA$, and $\angle CAB$. What is the sum of the measure of these three angles?
(c) Repeat steps (a) and (b) for three other points.
(d) Make a conjecture as to the sum of the measures of the angles in the interior of a triangle.

4 Linear measure and measurement

■ Albert was helping his father to do repairs on their house. Albert was asked to saw a plank to cover a space extending to a window. All Albert could find was a torn tape where the smallest number was 15. If the other end of the tape was at 24, how long should Albert cut the plank? ■

Of course, you know the answer to be 9. In the preceding problem, you probably assumed the 15 and the 24 to be feet, but the answer could have been in centimeters. The need for a standard unit of measurement is clear.

Before introducing linear units of measurement, we need to define what we mean by *measure*. Intuitively, we introduce measure through the use of a compass. Two tools used by ancient geometers to explore geometric concepts through construction were the straight edge and the compass (Figure 23). We introduce the

Figure 23

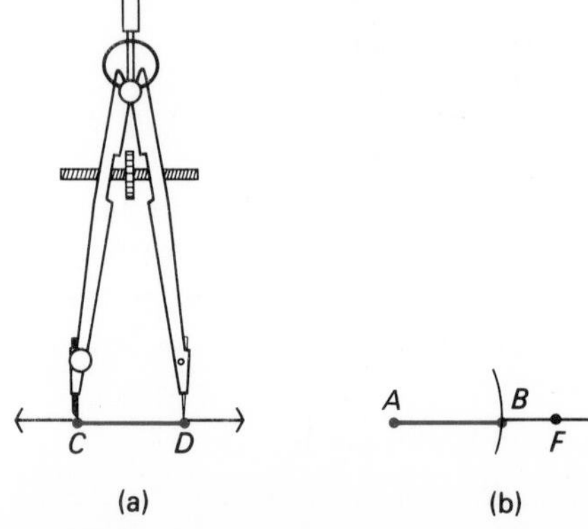

idea of construction by finding a line segment that intuitively has the same length or measure as a given line segment. Note that once we set a compass (Figure 23(a)), spanning a given line segment of a given length, we assume that the points on the compass always produce two points the same distance apart as the given distance.

Construction 1 • Construct a line segment on $\overrightarrow{AF}$ that has the same length as line segment $\overline{CD}$ in Figure 23(a).

(a) Set your compass points at C and at D in Figure 23(a).
(b) Without changing the setting of your compass, place one end at A and draw a portion of a circle (called an arc), cutting the ray at B in part (b).
(c) $\overline{AB}$ and $\overline{CD}$ have the same length. •

One of the important properties of a line segment is that it can be subdivided into a pair of segments of the same length.

Construction 2 • In Figure 24, find a point C between A and B such that $\overline{AC}$ and $\overline{BC}$ have the same length.

Figure 24

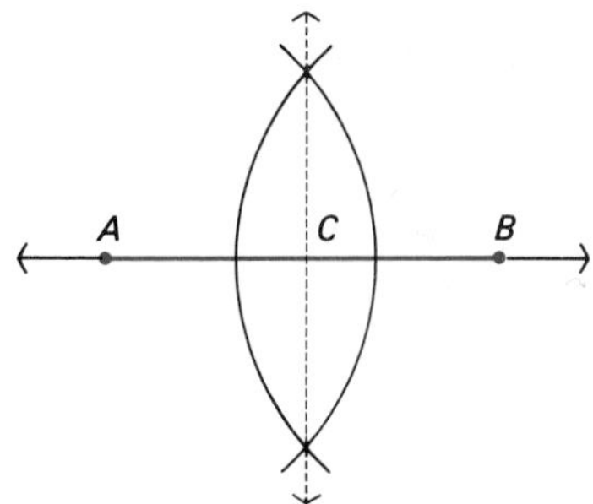

(a) With A as a center of a circle, draw an arc with radius greater than half of $\overline{AB}$.
(b) Similarly, with B as the center, draw an arc with the same radius as in (a).
(c) Use a straightedge to construct a line through the intersections of these two arcs.
(d) The point of intersection of this line and $\overline{AB}$, denoted by C, divides $\overline{AB}$ into two segments $\overline{AC}$ and $\overline{BC}$ that have the same length. C is said to be at the **midpoint** of $\overline{AB}$. •

To measure length, it is necessary to select a line segment that will serve as a unit segment. In Chapter 6, we considered number lines such as that found in Figure 25. On the number line are marked segments of a given length, which we use as unit segments. We reason that there exists a one-to-one correspondence between the set of real numbers and points on the number line. The following summarizes the properties of measure in terms of this correspondence.

Figure 25

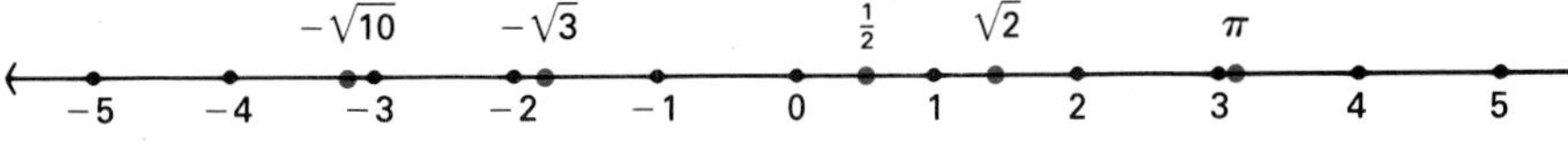

Properties of linear measure

The points on a number line can be placed in correspondence with real numbers so that

(a) to every point on the number line there corresponds exactly one real number called its **coordinate;**

(b) to every real number there corresponds exactly one point called the **graph** of the number on the line;

(c) to each pair of points on the number line there corresponds a unique real number called the **distance** between the points;

(d) for any two distinct points A and B of a line, the line can be coordinatized in such a way that A corresponds to zero and B corresponds to a positive number.

Figure 26

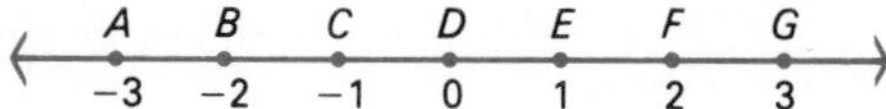

If we calculate distance by the normal process of counting space between two points, then in Figure 26 the distance between E and G is 2. Note that this is $3 - 1 = 2$, where 3 is the coordinate of G and 1 is the coordinate of E. This suggests the following definition, using Figure 27.

Figure 27

Distance

If A and B are points on a number line with coordinates (real numbers) a and b, respectively, then the distance between A and B is given by $|b - a|$. The distance between A and B is sometimes called the measure of AB and is denoted by $m(AB) = m(BA) = |b - a|$.

Example 6 • Using the preceding definition, one can obtain the following measures from Figure 26.

$$m(\overline{EG}) = 3 - 1 = 2 \qquad m(\overline{DG}) = 3 - 0 = 3$$

$$m(\overline{FC}) = 2 - (-1) = 3 \qquad m(\overline{BD}) = 0 - (-2) = 2 \quad \bullet$$

Note three important properties of distance (or measure).

(a) Distance is never negative.

(b) The distance between two points is zero only when the points are identical.

(c) The distance between points is unaffected by the order in which they are considered. That is, $m(\overline{AB}) = m(\overline{BA})$. $m(\overline{AB})$ is also written as AB in the remainder of this book.

Now that we have a procedure for measurement, we need a unit of measurement, as indicated in the introductory problem. The English were the first to have a standardized system of measurement, but residual problems from antiquity contributed to the awkwardness of their system. For instance, a "foot" was originally measured by the length of a king's foot, a "yard" was measured from the tip of a person's nose to fingertip, and an "inch" was measured by three barleycorns laid end to end. You are familiar with the standard English system of measurement as listed in Table 1.

Table 1

12 inches (in.) = 1 foot (ft.)
3 feet (36 in.) = 1 yard (yd.)
$5\frac{1}{2}$ yards ($16\frac{1}{2}$ ft.) = 1 rod
320 rods (1760 yds.) = 1 mile

One big disadvantage of the English system is that there is no simple scheme for remembering the relationship between units. Without constant use, the average person forgets the number of feet in a rod, the number of square yards in an acre, and the number of gallons in a barrel. In addition, computations are often messy.

The rapid growth of science demanded a more convenient and systematic system of measurement. The French called a meeting of scientists from many countries at the end of the 18th century in an effort to establish an international standard of measures. This group developed the metric system, of which the meter is the basic unit of length.

It was intended that the meter be one ten-millionth of the distance from the Equator to the North Pole. Unfortunately, this distance was originally measured incorrectly, and it cannot be measured accurately anyway. As a result, several redefinitions of the meter have been necessary. At present, this international standard of length is based on the orange radiation of the krypton isotope, which has an atomic weight of 86 (Kr_{86}). The conditions under which this radiation must be measured were very rigidly specified by the Eleventh General Conference of Weights and Measures in 1960, which developed the Système International d'Unités, abbreviated SI and known in English as the International System of Units.

A distance between two marks on a platinum bar (when held at a temperature of 0° Celsius) on deposit in the U.S. National Bureau of Standards represents a copy of the international standard for a meter. All other lengths used in the United States are defined in terms of the standard meter. For example, a yard is exactly 0.9144 meter.

The metric system is important not only because of its worldwide adoption but also because computations in the metric system are much easier than computations in the English system, since all conversions are based on powers of 10. The entire SI (metric) system is based on common prefixes attached to root units. These prefixes and their meanings are listed in Table 2. These are used in expressing SI units for length, area, volume, mass, and capacity.

The SI units for measuring length are based on the **meter.** Using the information in Table 2, verify the results in Table 3, which gives the SI length measurements based on the meter. Note the abbreviations or symbols for the measurements. These symbols are generally used with numerals.

Table 2

Prefix	*Meaning*	*Decimal meaning*
mega	one million	1,000,000 or 10^6
kilo	one thousand	1000 or 10^3
hecto	one hundred	100 or 10^2
deka	one ten	10
deci	one-tenth	1/10 or 0.1
centi	one-hundredth	1/100 or 0.01
milli	one-thousandth	1/1000 or 0.001
micro	one-millionth	1/1,000,000 or 0.000001

Table 3

1 megameter (Mm) = 10^6 meters (m)	1 meter = 0.000001 or 10^{-6} megameter
1 kilometer (km) = 1000 or 10^3 meters	1 meter = 0.001 or 10^{-3} kilometer
1 hectometer (hm) = 100 or 10^2 meters	1 meter = 0.01 or 10^{-2} hectometer
1 dekameter (dam) = 10 meters	1 meter = 0.1 or 10^{-1} dekameter
1 decimeter (dm) = 0.1 or 10^{-1} meter	1 meter = 10 decimeters
1 centimeter (cm) = 0.01 or 10^{-2} meter	1 meter = 10^2 centimeters
1 millimeter (mm) = 0.001 or 10^{-3} meter	1 meter = 10^3 millimeters
1 micrometer (μm) = 0.000001 or 10^{-6} meter	1 meter = 10^6 micrometers

The kilometer, meter, centimeter, and millimeter are the most commonly used measures of length. A kilometer is about nine football fields (end zones included) laid end to end. A meter is approximately the distance from a doorknob to the floor, so most doorways are a little over 2 meters high. A dime is about 1 millimeter thick; a nickel is about 2 centimeters wide; a piece of chalk is about 1 cm in diameter; a baseball bat is about 1 m long; and the stitches on a baseball are about 1 mm apart. Figure 28 shows a 10-cm ruler on which a segment $\overline{AB}$ that is 1 cm long has been marked.

Figure 28

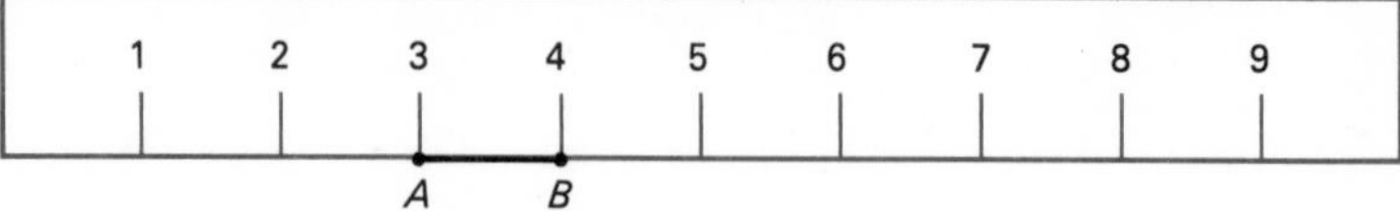

A first step in working with metric units is to develop the ability to estimate the approximate length of objects expressed in metric units. Practice on the problems in the following example.

Example 7 • Select the answer that best estimates the length in each problem.

(a) Distance from pitcher's mound to home plate on a baseball field: 180 cm 18 m 1.8 km

(b) Length of a paper clip: 100 mm 3 cm 0.3 m

(c) Distance walked in 20 minutes: 100 m 5000 cm 1.5 km

(d) Length of a dinner knife: 20 cm 2 m 0.002 km

(e) Width of little finger: 0.1 m	100 mm	1 cm
(f) Width of your hand: 10 cm	0.01 m	10 mm

Solution

(a) 18 m	(b) 3 cm	(c) 1.5 km
(d) 20 cm	(e) 1 cm	(f) 10 cm •

To convert metric units, multiply or divide by a power of 10. To change from a larger unit to a smaller unit, multiply. There will be more smaller units. To change from a smaller unit to a larger unit, divide. There will be fewer larger units.

Example 8 • Express 13 km in terms of meters.

Solution

$$1 \text{ km} = 1000 \text{ m}$$

$$13 \text{ km} = 13{,}000 \text{ m} \quad \bullet$$

Example 9 • Convert (a) 3 m, (b) 43 mm, (c) 41 km, and (d) $2 \cdot 10^8$ km to centimeters.

Solution (a) 1 m = 100 cm
3 m = 300 cm
(b) 10 mm = 1 cm or $1 \text{ mm} = \frac{1}{10} \text{ cm}$
$43 \text{ mm} = \frac{43}{10} = 4.3 \text{ cm}$
(c) 1 km = 100,000 cm
41 km = 4,100,000 cm
(d) $1 \text{ km} = 10^5 \text{ cm}$
$2(10)^8 \text{ km} = 2(10)^8(10)^5 = 2(10)^{13} \text{ cm}$ •

Some people like to move the decimal point in conversions. In going from a smaller unit (on the right) in Figure 29 to a larger unit (on the left), move the decimal point one place to the left for each unit crossed. For example, to change from millimeters to kilometers, move the decimal point 6 places to the left. Alternately, to change from kilometers to millimeters, move the decimal six places to the right, adding appropriate zeros.

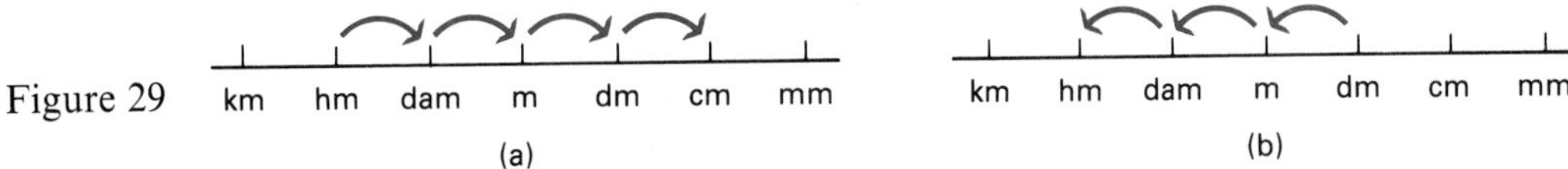

Figure 29

To convert from hectometers to centimeters, we would move four units to the right on the scale. (See Figure 29(a).) Hence we would multiply by 10^4 or move the decimal point 4 places to the right. To convert from decimeters to hectometers, we would move three units to the left on the scale. Hence we would multiply by 10^{-3} or move the decimal point 3 places to the left. (See Figure 29(b).)

Example 10 • Convert the following by moving the decimal point.
(a) 55 mm to m
(b) 170 km to cm
(c) 860 cm to km

Just for fun

The chef baked a cake that measured 4 dm (decimeters) on each side. He wanted to cut the cake so that he would have eight pieces of full thickness, each with each side measuring 2 dm. How did he cut the cake?

Solution (a) 55 mm = .055. m

(b) 170 km = 170.00000. cm

(c) 860 cm = .00860. km •

Example 11 • Scientists use the micrometer (μm) for exceedingly precise scientific measurements. If the maximum error tolerable in grinding a lens is 4 μm, this 4-μm tolerable error equals how many millimeters?

Solution

$$1 \text{ m} = 10^6 \ \mu\text{m}$$

$$1 \text{ m} = 10^3 \text{ mm}$$

$$10^6 \ \mu\text{m} = 10^3 \text{ mm}$$

$$1 \ \mu\text{m} = 10^3/10^6 \text{ mm}$$

$$1 \ \mu\text{m} = 1 \cdot 10^{-3} \text{ mm}$$

$$4 \ \mu\text{m} = 4 \cdot 10^{-3} \text{ mm} \quad \bullet$$

Example 12 • Susie has 6 m of material that she wishes to cut into pieces of length 30 cm. How many pieces will she have?

Solution

$$1 \text{ m} = 100 \text{ cm}$$

$$6 \text{ m} = 600 \text{ cm}$$

600 divided by 30 gives 20 pieces. •

Exercise set 4

R 1. Convert the following measurements to meters.

(a) 37,000 cm (b) 389,000 mm
(c) 12,000 mm (d) $21 \cdot 10^8$ cm
(e) 35,470 km (f) 38,902 km

2. Convert the following measurements to millimeters.

(a) 3789 cm (b) 389,850 km
(c) 1284 m (d) $21 \cdot 10^8$ cm

(e) 35,478 km (f) 38,902 m
(g) 16 cm (h) $412 \cdot 10^4$ m

3. Convert the following measurements to kilometers.

(a) 1700 cm (b) 120,000 mm (c) 1600 m

4. Without using a ruler, draw segments that you think represent the lengths given and then measure them with a ruler to check your estimates.

(a) 12 cm (b) 5 cm (c) 7 cm
(d) 12 mm (e) 24 mm (f) 4 mm

5. Fill in the following blanks.

(a) 32 mm = ______ m
(b) 160 cm = ______ km
(c) 3 km = ______ mm
(d) 216 mm = ______ km
(e) 1,340,000 μm = ______ cm
(f) 1689 km = ______ cm
(g) 48 m = ______ μm

6. Consider the line segments determined by the points A, B, C, D, E, F, and G. Find each of the following measures, where $m(\overline{BC}) = \frac{1}{2}$ and $m(\overline{DE}) = m(\overline{EF}) = \frac{3}{4}$.

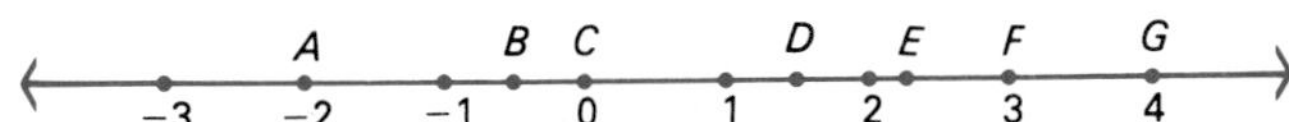

(a) $m(\overline{AB})$ (b) $m(\overline{CE})$ (c) $m(\overline{DB})$
(d) $m(\overline{GE})$ (e) $m(\overline{GA})$ (f) $m(\overline{CD})$

7. Would you use millimeters (mm), centimeters (cm), meters (m), or kilometers (km) to measure the

(a) length of a room?
(b) thickness of a button?
(c) distance from Miami to Louisville?
(d) length of a car?
(e) width of a book?
(f) width of hand?
(g) width of paper clip?
(h) length of football field?
(i) length of a photograph?
(j) diameter of a screw?
(k) altitude of an airplane?

8. Order from shortest to longest.

(a) 10 m, 100 cm, 100,000 mm
(b) 10 cm, 1000 mm, 0.01 m
(c) 0.1 cm, 0.01 m, 0.001 km
(d) 0.1 cm, 0.001 km, 100 mm

9. Measurements are usually approximations. For instance, if I measure a pencil and approximate the answer to be 12.3 cm long, I really mean that its length is between 12.25 cm and 12.35 cm. Complete this table for approximate measurements.

	Shortest possible length	*Measurement*	*Longest possible length*
(a)	______	21.6 cm	______
(b)	______	14.16 km	______
(c)	______	2 mm	______

T 10. Add or subtract.

(a) 3.4 m + 689 cm + 46 cm
(b) 4932 mm − 2.78 m
(c) 7 m − 243 cm
(d) 6.2 m + 541 cm + 4628 mm

11. Select the best measurement for each item.

(a) A man's height: (i) 185 cm (ii) 4 m (iii) 2000 mm
(b) School zone speed limit: (i) 20 mm per hr. (ii) 2400 m per hr. (iii) 25 km per hr.
(c) The width of a Venetian blind slat: (i) 5 m (ii) 10 cm (iii) 50 mm
(d) The length of an average book: (i) 2 mm (ii) 24 cm (iii) 24 m
(e) The width of an average door: (i) 0.01 km (ii) 100 cm (iii) 100 mm

12. Formulate rules for changing the following measurements.

(a) Meters to kilometers
(b) Centimeters to meters

13. (a) How many pieces of string can be cut from a ball of twine 9 m long if each piece is 4.2 cm long?
(b) How many lengths of string 0.5 mm long could be cut from the ball of twine in (a)?
(c) A map has a scale in which the ratio of map distance to actual distance is 1 cm:0.5 km. If the distance between two lakes on the map appears to be 11.2 cm, what is the actual distance in kilometers?
(d) You want to draw a map with a scale of 1 cm: 2 km. What would be the distance on the map separating two towns that are actually 59.2 km apart?

14. In the triplets of numbers shown, find a coordinate x so that $\overline{AB}$ is congruent to $\overline{BC}$.

(a) A at $\frac{1}{2}$, B at 2, C at x
(b) A at $\frac{3}{4}$, B at x, C at 2
(c) A at -1, B at 3, C at x
(d) A at x, B at 1, C at 4

15. Find two possible coordinates for B so that $\overline{AB}$ is a unit segment, where A has the following coordinates.

(a) $\frac{1}{2}$ (b) -3 (c) -4
(d) $\frac{7}{8}$ (e) -1 (f) 3

16. Jane's rectangular swimming pool measures 10 m by 8 m. What is the distance around it in meters? In centimeters?

17. Choose the most sensible answer.

(a) The length of a tennis racket:
68 mm 68 cm 68 m 68 km
(b) The height of an adult:
165 cm 250 cm 315 cm 500 cm
(c) The length of a canoe:
4 mm 4 cm 4 m 4 km
(d) The distance from Denver to New Orleans:
2130 mm 2130 cm 2130 m 2130 km
(e) The width of a light cord:
4 mm 30 mm 60 mm 6 cm

18. Estimate the following in the units given.

(a) The length of an arm in centimeters
(b) The length of a football field in meters
(c) The length of a middle finger in centimeters
(d) The width of a thumb in centimeters
(e) The thickness of a cardboard box in millimeters
(f) The distance between New York and Chicago in kilometers
(g) The length of a tennis court in meters
(h) The height of the tallest mountain in the United States in kilometers
(i) The height of the tallest basketball player on your team in centimeters

C 19. Simon wants to buy a fence to enclose a plot of land whose sides measure 1 hm by 65 dam by 125 m by 5 hm. What length of fencing material should he buy in terms of meters?

5 Simple closed curves

Many interesting figures can be displayed using a geoboard. A geoboard is a board with rows of evenly spaced nails as shown in the following figure.

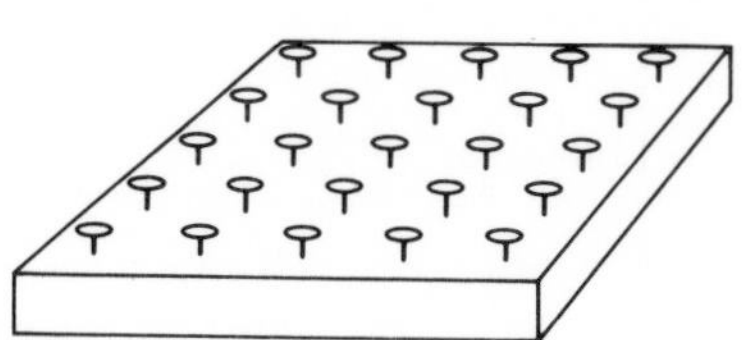

■ By stretching a rubber band around nails on the geoboard, see if you can form a rhombus; try an equilateral triangle. ■

It will be instructive for you to see how many figures introduced in this section can be illustrated by stretching a rubber band around nails in a geoboard. We shall discuss a number of familiar plane figures, such as the triangle and the square, and we shall discuss the angles associated with each. When we use phrases like "angle of a triangle" or "angle of a square," we mean, of course, the portion of the angle formed by the vertex and the lines containing the two sides of the figure. As you study these figures, observe that they bound regions in the plane.

Figure 30

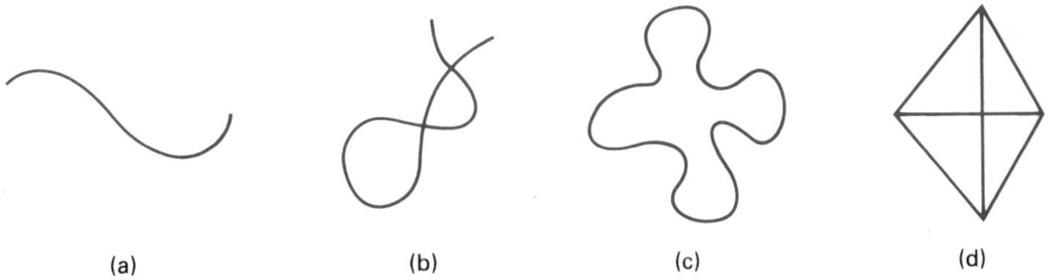

(a) (b) (c) (d)

Think of a piece of thread lying on a tabletop and the many configurations into which it can be arranged. This image should serve as a model for a discussion of plane curves. A **plane curve** can be considered as a set of points that can be drawn without lifting the pencil. This means that lines, rays, line segments, and angles are all plane curves. So are the drawings in Figure 30.

A curve is **simple,** if without lifting the pencil, it can be drawn without retracing any of its points (with the possible exception of its endpoints. (a) and (c) in Figure 30 are simple; (b) and (d) are not. A **closed curve** is drawn by starting and ending at the same point.

In Figure 31(a), the curve is simple but not closed; diagram (b) represents a simple closed curve; diagrams (c) and (d) are not simple closed curves. Why not?

Figure 31

Simple closed curves have a common property. Can you discover it? No, it has nothing to do with size or shape. In fact, the property is so fundamental that you may have overlooked it. Every simple closed curve has an inside and an outside. Actually, a simple closed curve separates the points of the plane into three sets: the set of points on the curve and two sets called the *interior* and the *exterior* of the curve. Although this property may seem obvious, a well-known theorem that states this result is quite difficult to prove.

Jordan curve theorem

Any simple closed curve c separates a plane into three disjoint sets of points called the *interior,* the *exterior,* and the *curve.*

Thus, for any simple closed curve c in the plane, the plane is the union of three sets, no two of which intersect. The three sets are the set c, the interior of c, and the exterior of c. The union of a simple closed curve with its interior is called a **region.**

Regions of a plane are classified as *convex* or *nonconvex.*

Convex region

A region (or set of points) is *convex* if for any two points in the region the line segment joining them lies completely in the region.

A region (or set of points) is **nonconvex** (sometimes called **concave**) if there is at least one segment joining points of the region that is not completely in the region. The region bounded by the closed curve in Figure 32(a) is convex, whereas the region in (b) is nonconvex. Curves bounding convex regions are called **convex curves.** In this section we speak of convex polygons and concave polygons.

Figure 32

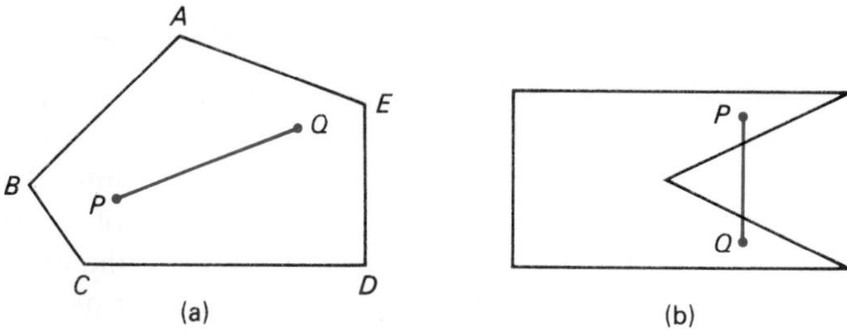

An example of a simple closed curve whose interior is a convex region is a circle.

Circle

> The set of all points in a plane at an equal distance from a fixed point in the plane is a *circle.*

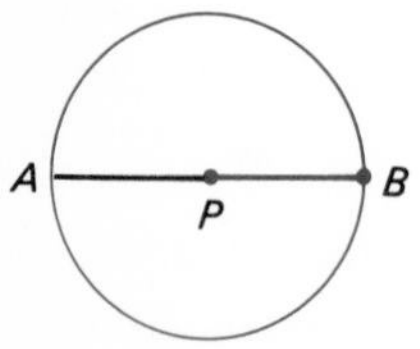

Figure 33

In this definition of a circle, the fixed point is called the **center** of the circle, and any line segment from the fixed point to a point on the circle is called a **radius** of the circle. In Figure 33, a line is drawn through the center of the circle, intersecting the circle in points A and B. The segment $\overline{AB}$ is called a **diameter** of the circle, and $\overline{PB}$ is called a *radius* ($\overline{PA}$ is also a radius).

Other plane curves such as a polygon may be either convex or concave.

Polygon

> A *polygon* is a closed curve that is the union of three or more line segments, $\overline{AB}, \overline{BC}, CD, \ldots$ such that the points $A, B, C, D, \ldots$ are coplanar and distinct, and no three consecutively named points are collinear.

If the segments described in the definition intersect in a point or points in addition to their endpoints, the polygon is not a simple polygon (Figure 34(b)). In this book, we are interested in simple polygons (Figure 34(a)). (a) is convex; (c) is concave.

For the polygon given in Figure 34(a), $\overline{AB}$, $\overline{BC}$, $\overline{CD}$, $\overline{DE}$, $\overline{EF}$, and $\overline{FA}$ are called the **sides** of the polygon $ABCDEF$. The points A, B, C, D, E, and F are called the **vertices** of the polygon. Adjacent vertices are endpoints of the same side, and the diagonals of a polygon bounding a convex region are the line segments joining nonadjacent vertices (such as $\overline{AE}$ in Figure 34(a)).

Adjacent pairs of sides of a polygon are parts of angles of the polygon. When we refer to the angles of a polygon, we generally mean the interior angles. For

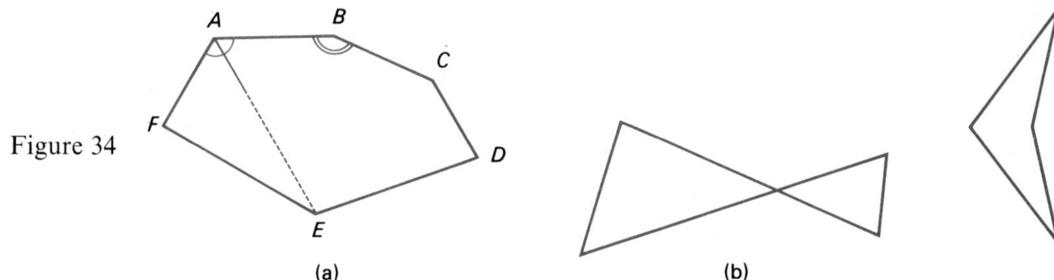

Figure 34

example, $\angle FAB$ and $\angle ABC$ are two interior angles of the polygon in Figure 34(a). A polygon has the same number of angles in its interior as it has sides.

A polygon is named according to the number of sides it has. Table 4 gives the name, number of sides, and number of interior angles of common polygons. Why is three the least number of sides a polygon can have?

Table 4

Number of sides	*Number of interior angles*	*Name of polygon*
3	3	Triangle
4	4	Quadrilateral
5	5	Pentagon
6	6	Hexagon
7	7	Heptagon
8	8	Octagon
9	9	Nonagon
10	10	Decagon
12	12	Dodecagon

Several types of triangles occur often enough to be given special names. (See Figure 35.) Special types of quadrilaterals are listed below.

Quadrilaterals

1. A **parallelogram** is a quadrilateral having both pairs of opposite sides parallel. (See Figure 36(a).)
2. A **trapezoid** is a quadrilateral having at least one pair of opposite sides parallel. (Figure 36(b).)
3. A **rectangle** is a parallelogram with a right angle. (One right angle implies that all four angles are right angles.) (Figure 36(c).)
4. A **square** is a rectangle with two adjacent sides of equal measure. (Figure 36(e).)
5. A **rhombus** is a parallelogram with two adjacent sides of equal measure. (That is, all sides are of equal measure.) (Figure 36(d).)

Just for fun *The unshaded set is a curve. Is it a simple curve?*

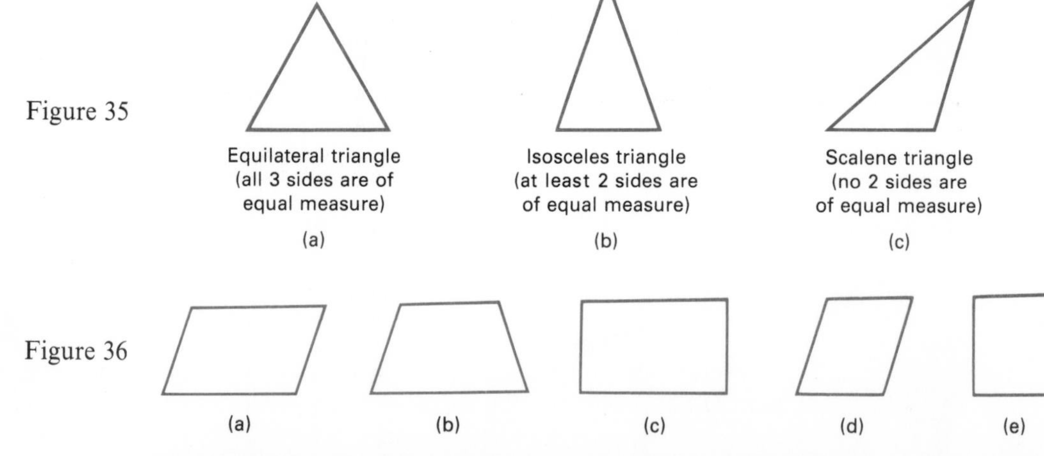

Figure 35

Figure 36

Regular polygon

> A polygon with all sides with equal measure and all angles of equal measure is called a regular polygon. Such polygons are said to be equilateral and equiangular.

Exercise set 5

R 1. Identify the figures that are plane curves and classify these as simple or not simple.

(a) 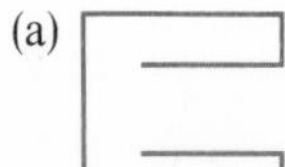

(b) 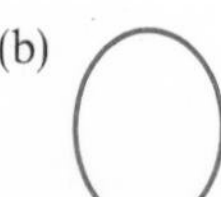

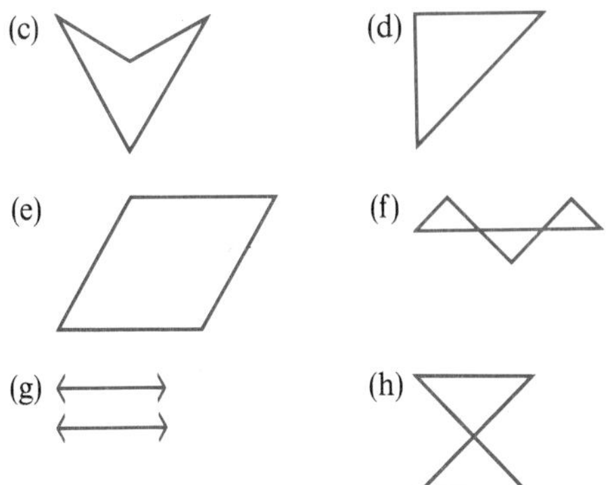

2. In Exercise 1, classify the simple curves as closed or not closed.
3. Which of these figures are simple polygons?

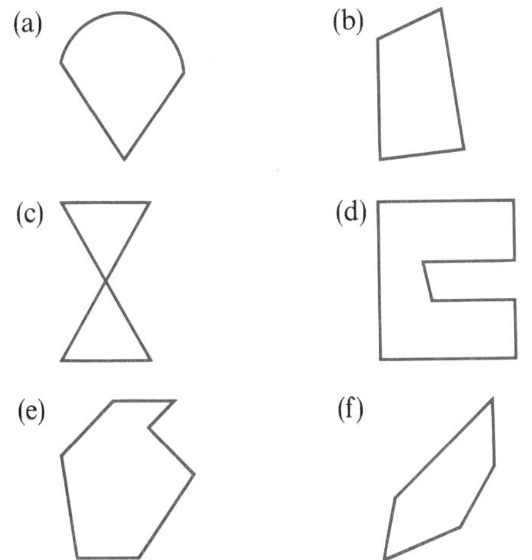

4. Which of the regions in Exercise 3 are convex?
5. Draw if possible a polygonal region with the following characteristics.
   (a) Convex with four sides
   (b) Nonconvex with four sides
   (c) Convex with three sides
   (d) Nonconvex with three sides
   (e) Convex with two sides
6. Sketch a curve that is the union of line segments such that the curve is
   (a) simple but not closed.
   (b) simple and closed.
   (c) a polygon of four sides bounding a convex region.
   (d) a polygon of five sides bounding a nonconvex region.
7. Sketch a line that contains exactly one point of some simple closed curve; two points; three points; four points.
8. Draw two triangles whose intersection is
   (a) one point. (b) two points.
   (c) three points. (d) four points.
   (e) five points. (f) six points.
   Are there other possible nonempty intersections?
9. Classify each of the following as true or as false.
   (a) Every closed curve is simple.
   (b) A square is a rhombus.
   (c) The union of two segments cannot be a simple closed curve.
   (d) A circle always bounds a convex region.
   (e) A parallelogram is a rectangle with a right angle.
   (f) An equilateral triangle has exactly two sides of equal measure.
   (g) A square is a rectangle.
   (h) An isosceles triangle can have a right angle.
   (i) Every polygon has more than two sides.
   (j) An angle is a simple closed curve.
   (k) A rectangle is a regular polygon.
10. Choose from the given characteristics those that apply to the following figures.
    (i) Convex polygon
    (ii) Nonconvex or concave polygon
    (iii) Regular polygon
    (iv) Plane curve
    (v) Simple curve
    (vi) Closed curve

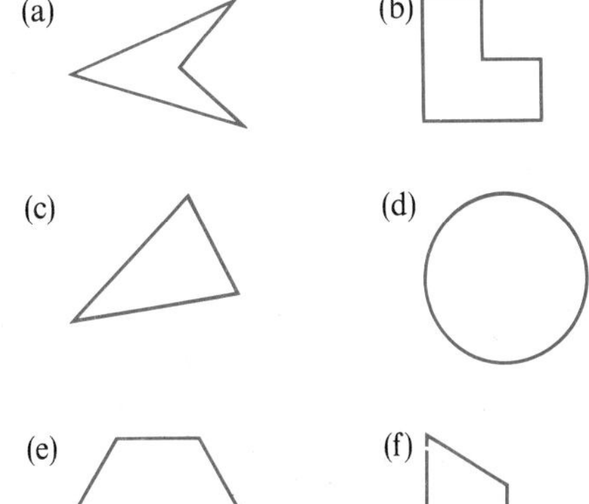

11. Consider these polygons.

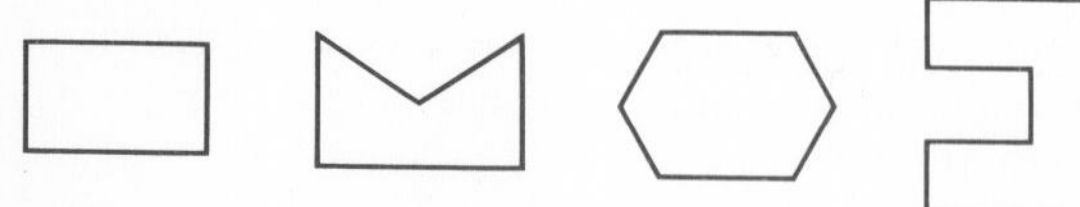

(a) Which are convex?
(b) Which have a diagonal that intersects their exterior?
(c) Can we say, "A concave polygon is a polygon in which at least one diagonal intersects its exterior"?

T 12. Draw an equilateral hexagon that is not equiangular.

13. Draw an equiangular hexagon that is not equilateral.

14. How many diagonals do each of the following have?

(a) Triangle (b) Hexagon (c) Octagon
(d) Can you generalize for an n-sided polygon?

C 15. Let T be the union of $ABCD$ with its interior (the shaded part of the figure). Decide which of the following are true..

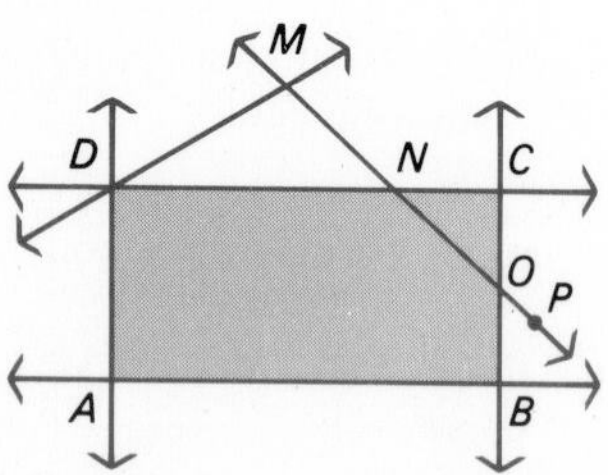

(a) $\triangle NOC \cap ABCD = \{N, O\}$
(b) $\overrightarrow{MP} \cap T = \overline{NO}$
(c) $\overline{MO} \subset T$
(d) $\triangle NOC \subset T$
(e) $\angle DAB \subset T$
(f) D is in the interior of $\angle ABC$.
(g) $M \in \overrightarrow{NO} \cup \overrightarrow{AD}$
(h) $\triangle MDN \cap T = \varnothing$

16. (a) Define the interior of $\triangle ABC$ in terms of the intersection of half planes.
(b) Define the exterior of $\triangle ABC$ in terms of the union of half planes.

17. For the given figure, describe each of these sets of points.

(a) $\triangle DBF \cap \overline{AC}$
(b) (Exterior of $\triangle DBF) \cap \overline{CE}$
(c) (Interior of $\triangle DBF) \cap \overleftrightarrow{CE}$
(d) (Interior of $\triangle DBF) \cap$ (interior of $\triangle CDE$)
(e) (Interior of $\triangle DBF) \cup$ (interior of $\triangle CDE$)
(f) (Interior of $\angle EAF) \cap$ (interior of $\triangle CDE$)

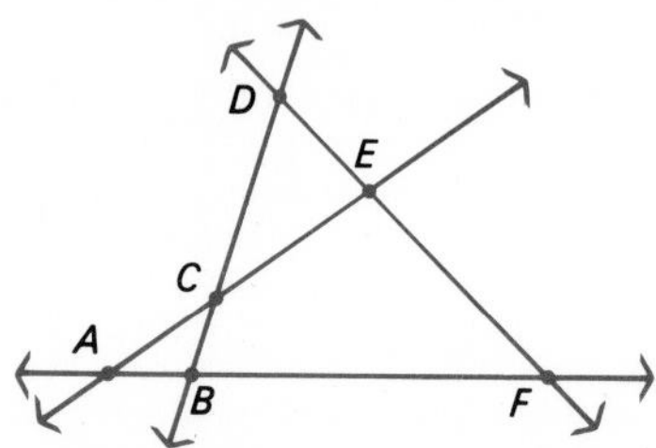

18. Draw a figure that is not a polygon but does satisfy the following: It is the union of three or more coplanar segments such that:

(a) no two segments with a common end point are collinear, and
(b) each segment intersects exactly two other segments at the endpoints.

19. Draw a large picture of each of the following polygons and use a protractor to measure the angles. What is the sum of these measures? Determine a pattern and give the sum of the measures of the interior angles of a n-sided polygon.

(a) Triangle (b) Quadrilateral
(c) Pentagon (d) Hexagon
(e) Heptagon (f) Octagon
(g) n-sided polygon

20. Draw some figures and make conjectures about the truth or falsity of the following statements.

(a) In any quadrilateral, the sum of the opposite angles is 180°.
(b) A line segment from a vertex of a triangle to the midpoint of the opposite side is called a *median*. The three medians of a triangle meet in a point.
(c) The diagonals of a quadrilateral bisect each other.
(d) If the midpoints of the sides of any quadrilateral are connected, they form a parallelogram.
(e) The three altitudes of a triangle meet in a point.
(f) If the midpoints of the sides of any triangle are connected, they form an equilateral triangle.

*6 Two-dimensional measure and the metric system

■ I just purchased a summer home with an 8-hectare lot. Do you have any idea of the size of my lot? You will after you study this section. ■

Life would be simple if all our measurements were one-dimensional. Yet some of our most important measures are two-dimensional. If you buy a farm, you want to know its area. The amount of paint needed to paint a house depends on the surface area.

In this section, we define two-dimensional measure and discuss units of measurement. You will find this section to be a very practical unit of study. The asterisk on the section number means this section may be taught elsewhere, such as before the third or fourth section of Chapter 10.

The measuring process for two-dimensional regions is similar to that for one-dimensional figures: To find a measure, we determine how many nonoverlapping "units" fill the figure. Consider the area bounded by a rectangle with sides that are of measure 6 units and 3 units. (See Figure 37.)

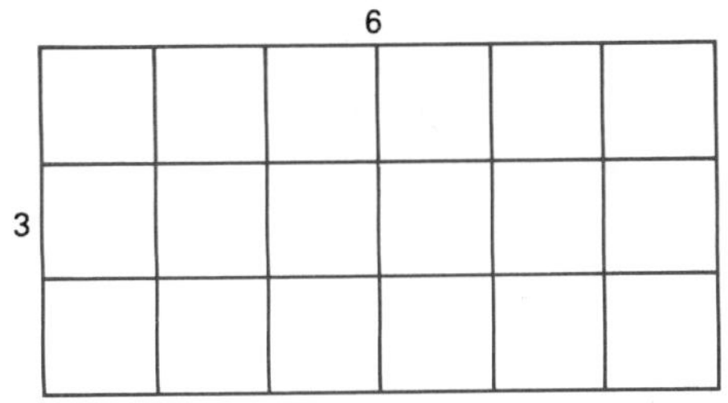

6

3

Figure 37

Six unit squares (the sides of which measure one unit) placed in a nonoverlapping position cover the entire length of the rectangle. Three unit squares cover the width of the rectangle. A total of 18 unit squares completely cover the region bounded by the rectangle. This discussion suggests that the unit square is quite reasonably designated as the unit for two-dimensional measure, called **area.**

Are there other possibilities that could be chosen as the unit for two-dimensional area? How about a unit circle or maybe a unit triangle? Can you see why the unit square is better than these?

The geoboard provides a grid of unit squares that can be used to demonstrate two-dimensional area.

Example 13 • Find the area of the regions bounded by the figures on the geoboards in Figure 38.

Solution
(a) Count the 12 square units in the region bounded by the rectangle in Figure 38(a). Note that the rectangle is 4 units long and 3 units wide and that $12 = 4 \cdot 3$.
(b) Note that there are 7 full unit squares and 10 half units. The area is again 12 square units.
(c) 4 full units + 8 half units = 8 square units

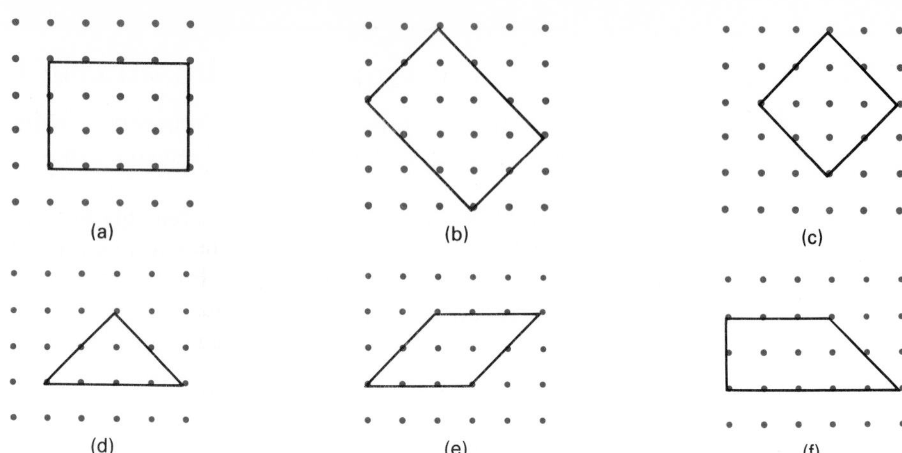

Figure 38

(d) The area of the region bounded by the triangle is 2 full units + 4 half units, or 4 square units.
(e) The area of the region bounded by the parallelogram is 4 full units + 4 half units, or 6 square units.
(f) The area of the region bounded by the trapezoid is 7 full units + 2 half units, or 8 square units. •

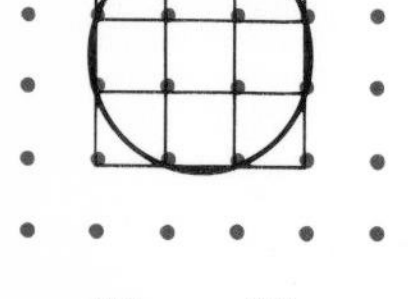

Figure 39

Of course, unit squares cannot always be placed within many nonpolygonal regions to obtain the area visually. You note the difficulty in finding the area of a circle. (See Figure 39.) We shall learn later how to determine the area of a circle in square units.

From your study of the areas bounded by figures on the geoboard, you can see that area has the following properties.

Properties of area

1. *Area is additive;* that is, the area of the whole is equal to the sum of the areas of its nonoverlapping parts. If a region is decomposed into a finite number of regions such that any two regions have in common only line segments, curves, or points, then the area of the whole region is the sum of the areas of the regions into which it is decomposed.
2. *If two figures A and B have the same size and shape, then the area bounded by A equals the area bounded by B.* Theoretically figure A can be picked up and put down on figure B so that they fit perfectly. Thus, they have the same area.
3. *If a region is cut into parts and reassembled in a nonoverlapping manner to form another region, then the two regions have the same area.* In Figure 40, the area of the rectangular region consisting of A and B in part (a) is the same as the area of the triangular region composed of A and B in part (b).
4. *If l and w (called length and width) are the linear measures of two consecutive sides of a rectangle forming the boundary of a rectangular region, then the area of the rectangular region is the product of the length and width, denoted by $A = lw$.*

Figure 40

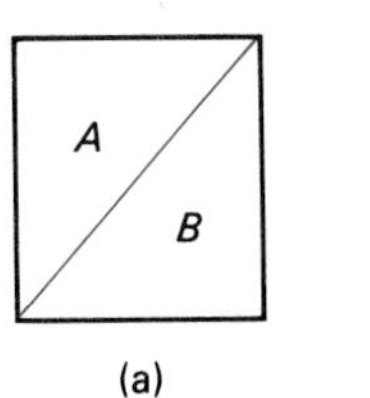

(a)

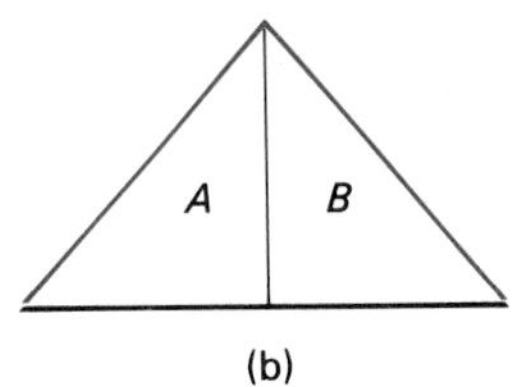

(b)

In Property 2, it is noted that regions with the same size and shape have the same area. The converse is not true. In Figure 40(a), the rectangle has the same area as the triangle in (b), but they obviously do not have the same size and shape.

Using Property 4, we can find the area of the region bounded by any rectangle.

Example 14 • Find the area of the region bounded by a rectangle that has a length of $6\frac{1}{2}$ and a width of 4.

Solution Area $= lw = (6\frac{1}{2})(4) = 26$ •

If the length and width of the rectangle in the preceding example were given in meters, then the area would be 26 square meters, written 26 m^2. Remember that area is measured in square units. Figure 41(a) illustrates a square centimeter (cm^2), which is about the size of one face of a sugar cube.

Figure 41

1 square centimeter

6 square centimeters

1 square centimeter

(a) (b)

Example 15 • Find the area of the region in Figure 41(b).

Solution We count the number of square centimeters in the region to get the area of the region. The area of the region is 8 square centimeters. We denote the answer as 8 cm^2. •

Let us emphasize from the preceding example that an area of 8 cm^2 means that there are 8 squares (1 cm by 1 cm) that fill up the region. The cm^2 means that the unit of measurement is the unit square (1 cm by 1 cm).

Table 5 gives in detail the relationships among area measurements for the metric system. Notice, for instance, that since

$$1 \text{ km} = 10^3 \text{ m}$$

then

$$\begin{aligned} 1 \text{ km}^2 &= (10^3 \text{ m})(10^3 \text{ m}) \\ &= 10^6 \text{ m}^2 \end{aligned}$$

Table 5

$1\ km^2 = 10^6\ m^2$, since $1\ km = 10^3\ m$
$1\ hm^2 = 10^4\ m^2$, since $1\ hm = 10^2\ m$
$1\ dam^2 = 10^2\ m^2$, since $1\ dam = 10\ m$
$1\ dm^2 = 10^{-2}\ m^2$, since $1\ dm = 10^{-1}\ m$
$1\ cm^2 = 10^{-4}\ m^2$, since $1\ cm = 10^{-2}\ m$
$1\ mm^2 = 10^{-6}\ m^2$, since $1\ mm = 10^{-3}\ m$
$1\ \mu m^2 = 10^{-12}\ m^2$, since $1\ \mu m = 10^{-6}\ m$

The square kilometer, the square meter, the square centimeter, and the square millimeter are the most commonly used area measurements, along with hectare (to be defined in this section). The other measurements are seldom used.

Of course, either Table 5 or Table 3 (after squaring) can be used for changing metric units squared; however, many people prefer the metric scale in Figure 42. Note the two bars between units. Recall from Section 1 that for each bar as you move to the right, you move the decimal point that many places to the right. For example, if you were changing from square centimeters to square meters (this time moving to the left), you would move the decimal point four places to the left.

Figure 42

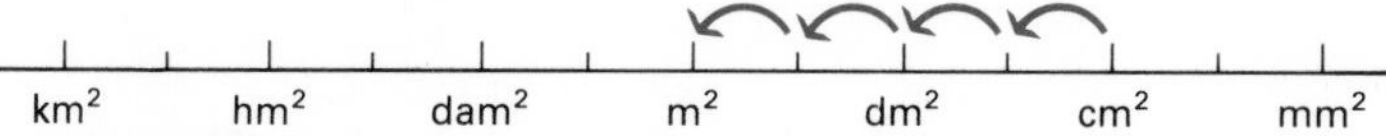

Example 16 • Change 30,000 cm^2 to square meters.

Solution

$$1\ cm^2 = 0.0001\ m^2$$

$$30{,}000\ cm^2 = 30{,}000(0.0001) = 3\ m^2$$ •

Example 17 • Change 4 km^2 to square meters.

Solution

$$1\ km^2 = 1{,}000{,}000\ m^2$$

$$4\ km^2 = 4(1{,}000{,}000) = 4{,}000{,}000\ m^2$$ •

To get a feel for the approximate size of areas expressed in SI units, consider the following.

1. The area of a postage stamp—500 mm^2
2. The area of a dollar bill—100 cm^2
3. The area of a wallet photo—60 cm^2
4. The area of the kitchen floor—20 m^2
5. The area of a small garden—50 m^2

The English system uses special names such as the **acre** for frequently used areas of land measure. Metric measure also has such a unit in the **hectare (ha).** One hectare is 1 hm^2, which is $10^4\ m^2$ or 10,000 m^2.

Example 18 • For a housing project, a residential property developer discovers what he considers to be a perfect tract of land that is rectangular and measures 500 m by 1000 m. How many hectares are contained in this tract?

Just for fun *These figures are made by 16 sticks of equal length. Move 3 sticks so that each becomes four squares of equal area.*

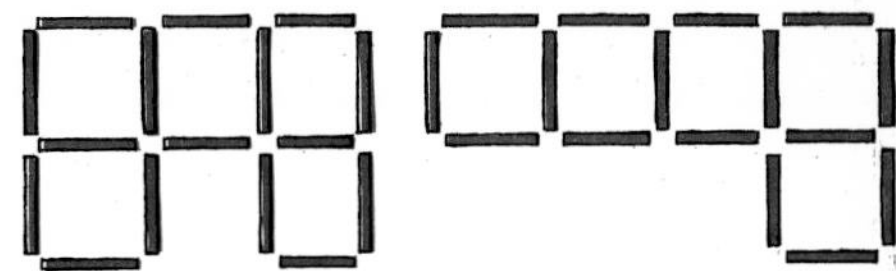

Solution

$$A \text{ (of rectangle)} = l \cdot w$$
$$= 1000 \text{ m} \cdot 500 \text{ m}$$
$$= 500{,}000 \text{ m}^2$$
$$= \frac{500{,}000}{10{,}000} = 50 \text{ ha} \quad \bullet$$

Example 19 • Industrial property covers approximately 24% of a city that has an area of 234 km^2. How much land in hectares is used for industrial purposes?

Solution $234(0.24) = 56.16 \text{ km}^2 = 56.16(10)^2 \text{ ha} = 5616 \text{ ha}$ •

Exercise set 6

R 1. Find the area of the following rectangle by counting; by the formula. Do they agree?

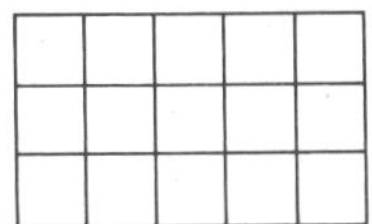

2. Complete the table for the area of the rectangle in Exercise 1 as the width changes by 1 unit. What is the pattern?

Length	*Width*	*Area (square units)*
5	2	
5	3	15
5	4	20
5	5	
5	6	
5	7	
5	8	
5	9	

3. Complete the table for the rectangle in Exercise 1 as the length is increased by 1 unit. What is the pattern?

Length	*Width*	*Area (square units)*
3	3	
4	3	
5	3	15
6	3	18
7	3	
8	3	
9	3	
10	3	

4. Give the area in square units.

(a) 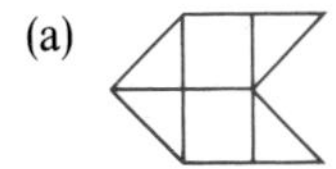

(b) 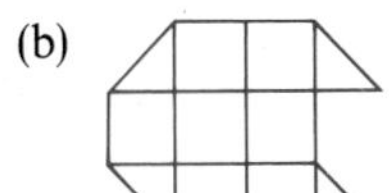

(c)

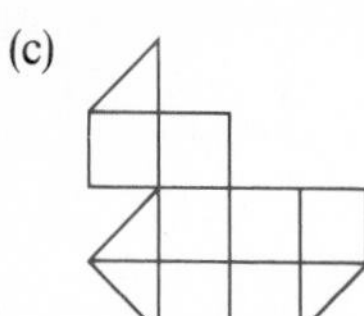

5. Find the nearest whole numbers greater than and less than the areas of the following. Use the fact that if only part of a square is shaded, the shaded area is less than 1 square unit and greater than 0 units.

(a)

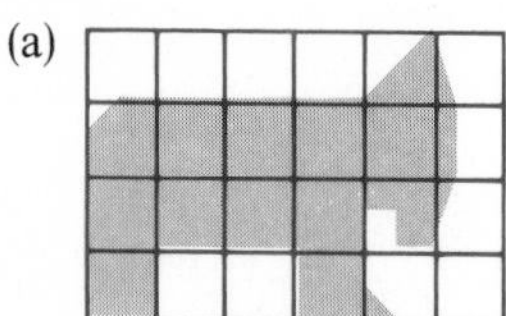

(b)

(c)

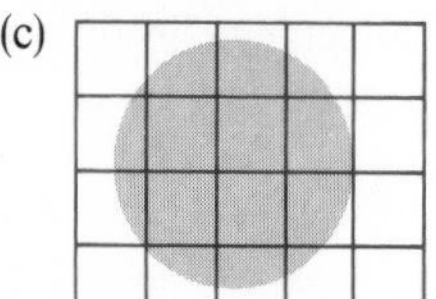

(d)

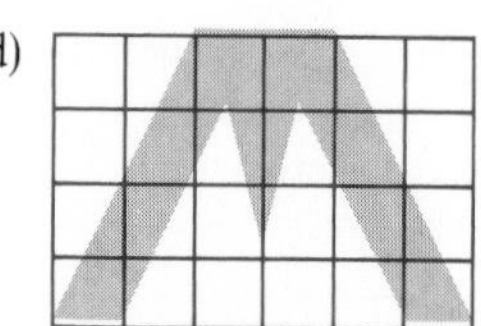

6. Improve or sharpen the answers in Exercise 5 by determining whether a partial shaded square has an area of a half unit, less than a half unit, or more than a half unit.

7. The given grid is made up of squares of 1 cm on a side. Trace the grid and use it to estimate the area of the regions below.

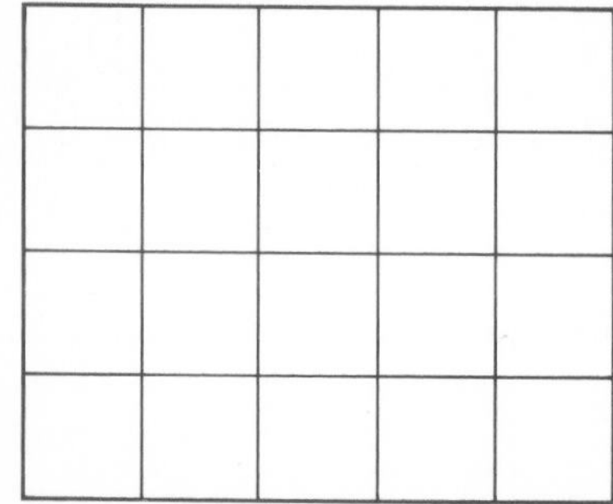

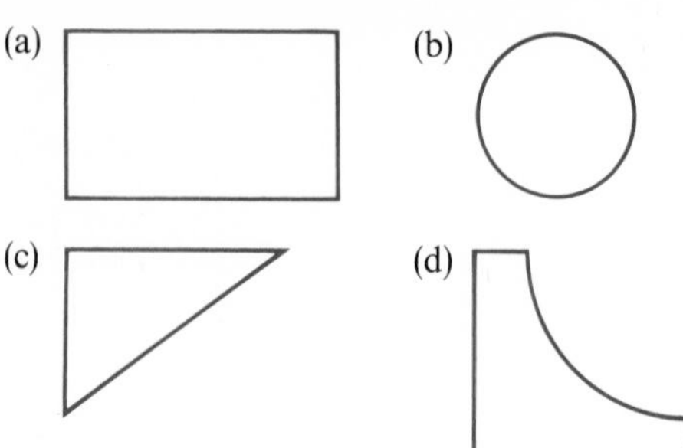

8. These figures have been drawn on geoboards. For each figure, count the square units bounded by the figure and thus find the area.

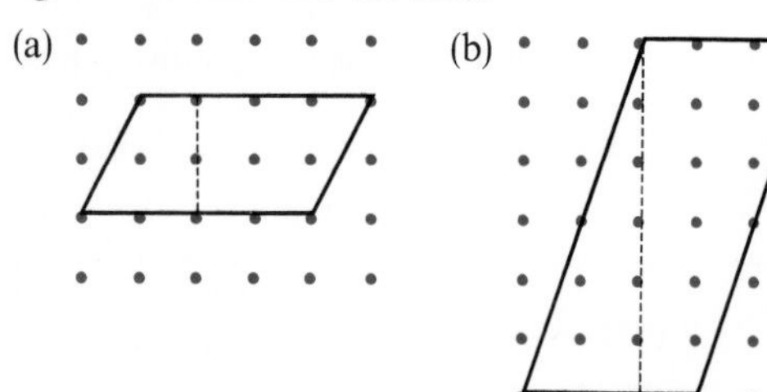

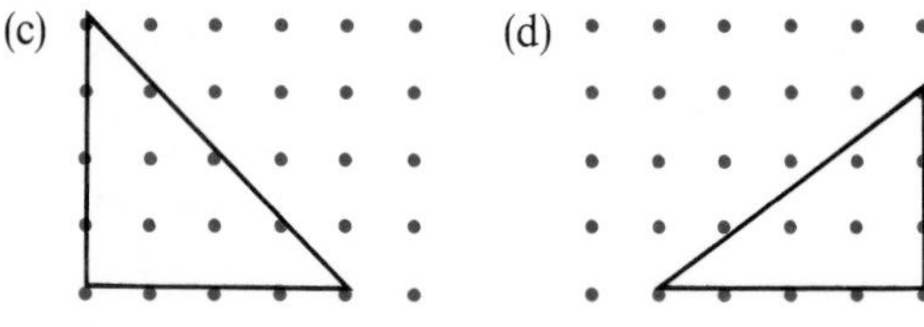

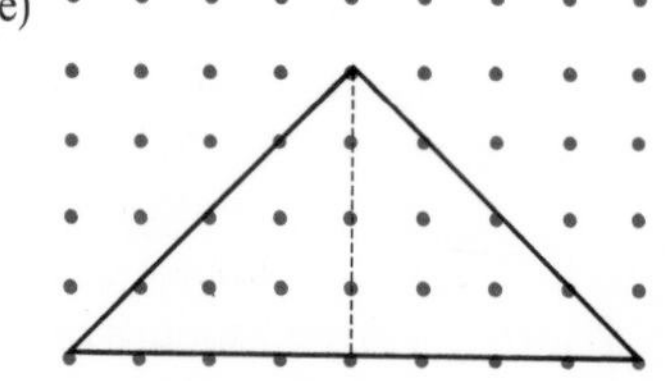

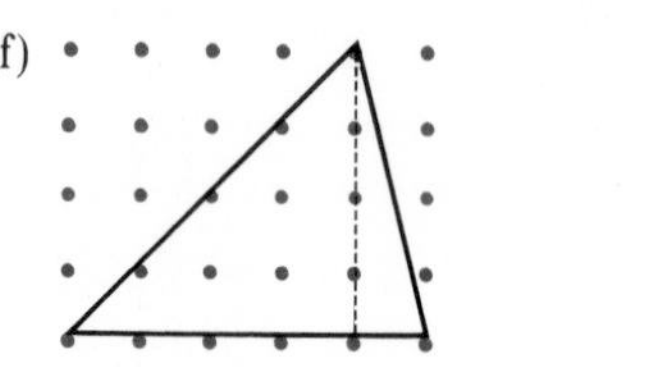

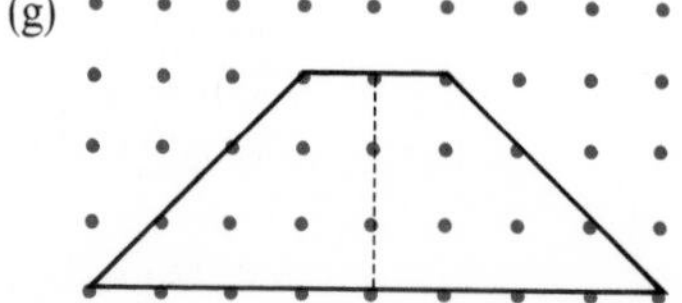

(h)

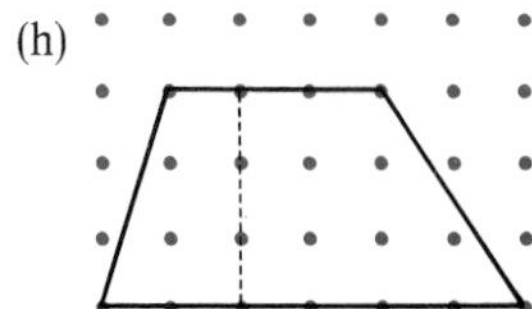

9. Each dashed line in Exercise 8 is called an *altitude*, and the line to which it is perpendicular is called a *base*. Multiply the length of the base times the length of the altitude in (a) and (b) and see if you get the same answers as in 8(a) and 8(b). Conjecture a formula for the area of a parallelogram.
10. Multiply the length of the two legs of the right triangle in (c) and (d) of Exercise 8. Take one-half of your answer. Check with the answers in Exercise 8. Conjecture a formula for the area of a right triangle.
11. For (e) and (f) of Exercise 8, multiply $(\frac{1}{2})$. (altitude)(base). Do you get the same answers as in Exercise 8? Conjecture a formula for the area of any triangle.
12. In (g) and (h) of Exercise 8, multiply the length of the altitude times the average of the lengths of the two bases. Do you get the same answer? Conjecture a formula for the area of a trapezoid.
13. A farm is square and measures 100 meters on a side. What is the area of this farm in hectares?
14. Which of the two areas given below is larger?

    (a) 34 km^2 or 129,000 m^2 (b) 1234 μm^2 or 3 cm^2
    (c) 4 ha or 456,000 m^2 (d) 916 mm^2 or 1 cm^2
15. Convert each measurement below to square centimeters.

    (a) 34 m^2 (b) 328 mm^2 (c) 3100 mm^2
    (d) 41 km^2 (e) 42 km^2 (f) 863 m^2

T 16. Formulate rules for changing

    (a) square meters to hectares.
    (b) square kilometers to hectares.

17. Select the best approximation of each area.

    (a) The front of this book: (i) 4 m^2 (ii) 400 cm^2 (iii) 40 cm^2
    (b) A football field: (i) 8 km^2 (ii) 8 m^2 (iii) 800 m^2

C 18. It seems most natural to measure area in square units. However, what would happen to measurements if the units were not square but rectangular? For parts (a), (b), (c), and (d) below, use the unit U described and determine the number of units contained in a 12-cm-by-8-cm rectangle.

    (a) $U = 1$ cm by 2 cm
    (b) $U = 1$ cm by 4 cm
    (c) $U = 3$ cm by 2 cm
    (d) $U = 3$ cm by 4 cm

Solution to introductory problem

Chords	*Maximum number of regions*	
1	2	
2	4	2
3	7	3
4	11	4

Understand the problem. Let us restate the problem. Cutting the cake is equivalent to using line segments (called *chords*) to divide a circle into regions. One chord divides the circle into 2 regions; two chords into a maximum of 4 regions; three chords into a maximum of 7; and four chords into 11.

Strategy. From the table above, we attempt to determine a pattern. Note that the number of regions formed by 4 chords is the number formed by 3 plus the 4. The number formed by 3 is the number formed by 2 plus

3. The number for 2 is the number formed by 1 plus 2. Let's work backward.

$$n(4) = n(3) + 4$$
$$n(4) = [n(2) + 3] + 4$$
$$n(4) = n(2) + (3 + 4)$$
$$n(4) = [n(1) + 2] + (3 + 4)$$
$$n(4) = 2 + 2 + 3 + 4$$
$$\vdots \qquad \vdots$$
$$n(k) = 2 + 2 + 3 + 4 + \cdots + k$$

Therefore,

$$n(10) = 2 + 2 + 3 + 4 + 5 + 6 + 7 + 8 + 9 + 10 = 56$$

Looking back. Looking back, we see that we can obtain our solution another way. Let us suppose k line segments have been drawn, creating $n(k)$ regions. Let us start drawing the $k + 1$ segments at point A. Immediately, we divide the region adjacent to A, creating a new region. Then each time we intersect one of the first k line segments, we enter a new region and divide it. Hence, with the $(k + 1)$st segment we create $k + 1$ new regions. Since with 1 segment we have 2 regions, with 2 segments we have $2 + 2$ regions. Since with 2 segments we have $2 + 2$ regions, with 3 segments we have $2 + 2 + 3$ regions, and so on. Thus, with k segments we have $2 + 2 + 3 + 4 + \cdots + k$ regions.

Summary and review

1. *Point, line, plane,* and *between* are undefined terms.
   (a) Points on a line are said to be *collinear.*
   (b) Lines containing a common point are *concurrent.*
   (c) If three distinct points are collinear, then one of the points is between the other two.
   (d) Points and/or lines that lie in the same plane are said to be *coplanar.*
2. A *plane* is determined by
   (a) Three noncollinear points
   (b) Two intersecting lines
   (c) A line and a point not on the line
   (d) Two parallel lines
3. A unique line is determined by two distinct points.
4. Two lines in a plane are parallel if and only if there is no point on both lines.
5. If lines r and s are not in the same plane, they are said to be *skew* lines.
6. A *ray* is the union of a point on a line and one of the half lines determined by the point.
7. A line segment is a subset of a line consisting of two points on the line and the set of points between them.
8. Characteristics of planes
   (a) If the intersection of two distinct planes is not empty, then the intersection is a line.
   (b) If a line contains two distinct points of a plane, then the line lies in the plane.
   (c) Planes are parallel if their intersection is empty.
9. If A and B are points on a number line with coordinates a and b, respectively, then $m(\overline{AB}) = m(\overline{BA}) = |b - a| = |a - b|$.
10. An *angle* is the union of two rays that have the same endpoint.
    (a) A *right* angle has a measure of 90°.
    (b) A *straight* angle has a measure of 180°.
    (c) An *acute* angle has a measure between 0° and 90°.
    (d) An *obtuse* angle has a measure between 90° and 180°.
    (e) Two angles are *supplementary* if the sum of their measures is 180°, and *complementary* if the sum of their measures is 90°.
    (f) Two coplanar angles that have a common vertex and a common side are called *adjacent* angles if the ray of one angle is not inside the other.

11. Metric prefixes

Prefix	*Meaning*	*Decimal meaning*
mega	one million	1,000,000 *or* 10^6
kilo	one thousand	1000 *or* 10^3
hecto	one hundred	100 *or* 10^2
deka	one ten	10
deci	one-tenth	1/10 *or* 0.1
centi	one-hundredth	1/100 *or* 0.01
milli	one-thousandth	1/1000 *or* 0.001
micro	one-millionth	1/1,000,000 *or* 0.000001

12. Plane curves

(a) A curve is *simple* if, without lifting the pencil, the curve can be drawn without retracing any of its points (with the possible exception of its endpoints).
(b) A *closed* curve is drawn by starting and ending at the same point.
(c) A simple closed curve separates a plane into three disjoint sets of points: the *interior,* the *exterior,* and the curve.
(d) A region is *convex* if, for any two points in the region, the line segment joining them lies completely in the region.
(e) A *polygon* is a closed curve that is the union of three or more line segments $\overline{AB}$, $\overline{BC}$, $\overline{CD}$, . . . such that A, B, C, D, . . . are coplanar and distinct, and so three consecutively named points are collinear.
(f) Polygons can be classified as triangles (3 sides), quadrilaterals (4), pentagons (5), hexagons (6), heptagons (7), octagons (8), ...
(g) A *quadrilateral* can be classified as a parallelogram, a trapezoid, a rectangle, a square, or a rhombus.

Review exercise set 7

R 1. Classify each as true or as false. If a statement is false, explain why.

(a) A line separates space.
(b) A point separates a line.
(c) For any two points A and B, $\overrightarrow{AB} = \overrightarrow{BA}$.
(d) For any two points A and B, $\overleftrightarrow{AB} = \overleftrightarrow{BA}$.
(e) If $\overrightarrow{AB} = \overrightarrow{AD}$, then $B = D$.
(f) Every hexagon is a polygon.
(g) Every square is a parallelogram.
(h) The sides of any regular polygon are equal.
(i) The three angles of any isosceles triangle are equal.
(j) If the sides of a polygon are equal, the polygon is a regular polygon.
(k) If 4 points are collinear, then they must be coplanar.
(l) If a line segment has the center of a circle for one endpoint and a point on the circle for the other endpoint, it is called a *diameter.*
(m) A circle is a polygon.
(n) A line could be called an infinite set of collinear points.
(o) The union of two half lines is a line.
(p) A segment separates a plane.
(q) Closed curves are simple curves.
(r) All straight lines are curves.
(s) A polygon has more than three sides.
(t) A plane region is a simple closed curve.
(u) Three points, not all on the same line, determine a plane.

2. Can you take the "wobble" out of a table with 4 legs by shortening or lengthening 1 leg only? Explain.

3. Line l lies in plane α and line m in plane β. If l is parallel to m, describe the planes.

4. Classify the following as Always (A), Sometimes (S), or Never (N) true.

(a) For any two points A and B, $AB = BA$.
(b) If $AB = AD$, then $B = D$.
(c) Point P is on line m. There is only one line containing P perpendicular to m.
(d) Two adjacent angles are complementary. Their noncommon rays form a line.
(e) A segment can be a bisector of a line.
(f) An equilateral triangle is a right triangle.
(g) All rectangles are parallelograms.

(h) The radii of a circle are line segments of equal measure.
(i) If two angles are complements then they are adjacent.

5. Draw an angle and a circle that intersect in exactly

(a) no points. (b) one point.
(c) two points. (d) three points.
(e) four points. (f) five points.
(g) more than five points.

6. How many *lines* are determined by four points? How many *line segments* are determined by four points?

7. Make sketches to indicate the intersection of two parallelograms such that the intersection is

(a) an empty set. (b) exactly one point.
(c) exactly two points. (d) exactly four points.
(e) exactly five points. (f) exactly six points.

8. Sketch

(a) a plane parallel to line r and intersecting line s.
(b) skew lines r and s.
(c) a plane containing line r, where r and s are skew lines.

9. Perform these conversions.

(a) 5001 cm = ________ km
(b) 52 mm = ________ cm
(c) 81 km = ________ m
(d) 17 ha = ________ m^2
(e) 2.61 km = ________ cm
(f) 421 dm = ________ mm,

10. Which unit—the kilometer, the meter, the centimeter or the millimeter—would you most likely use to measure the following:

(a) Length of a pencil
(b) Distance from Boston to New York
(c) Thickness of glasses
(d) Width of a postcard
(e) Height of your classroom

T 11. Describe each of the following sets of points with reference to the given figure.

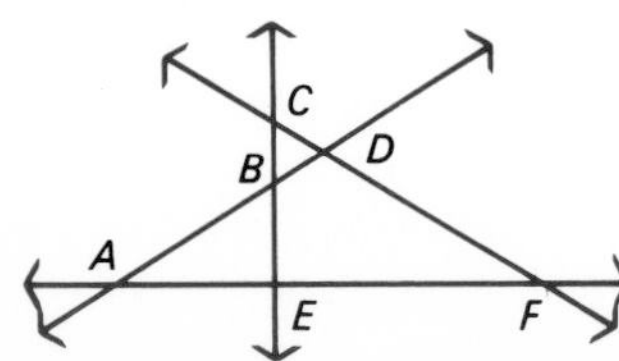

(a) $\overrightarrow{AF} \cap \overline{AE}$
(b) $\overline{CE} \cap \triangle ADF$
(c) (interior $\triangle ADF$) $\cap \overrightarrow{CB}$
(d) $\overline{AE} \cup \overrightarrow{EF}$
(e) $\triangle BAE \cap \triangle ADF$
(f) (interior $\triangle ABE$) $\cup$ (interior $\triangle ADF$)
(g) $\overline{CD} \cap \overleftrightarrow{AD}$
(h) $\overline{EF} \cup \overline{AE}$

12. Can two planes have one and only one point in common?

13. List four ways to determine a plane.

14. Construct a perpendicular to a line from a point not on the line, (a) by paper folding, and (b) by compass and ruler.

15. Repeat Exercise 14 when the point is on the line.

16. (a) Draw a closed curve that is not simple.
(b) Draw a five-sided convex polygon.
(c) Draw a five-sided concave polygon.
(d) Draw a concave parallelogram.

17. Use paper folding to find the center of a circle. Explain your answer.

18. Sketch

(a) a convex hexagon.
(b) a nonconvex quadrilateral.
(c) a nonconvex triangle.
(d) a nonconvex hexagon.
(e) a nonconvex circle.

19. Consider the four points below.

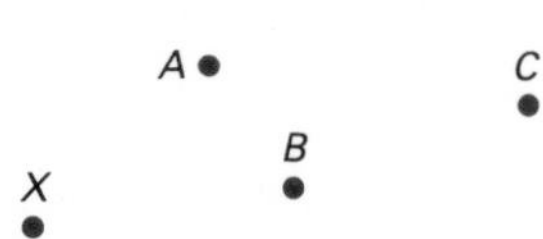

Sketch a closed figure beginning and ending at X containing A, B, and C such that

(a) the figure is convex.
(b) the figure is not a polygon.
(c) the figure is a convex polygon.

20. Consider A and B as being any two distinct points. Classify as true or false.

(a) $\overrightarrow{AB} \subseteq \overline{AB}$ (b) $\overline{AB} \subseteq \overrightarrow{AB}$
(c) $\overline{AB} \subseteq \overleftrightarrow{AB}$ (d) $\overline{AB} \subseteq \overline{AB}$
(e) $\overline{BA} \subseteq \overrightarrow{AB}$ (f) $\overrightarrow{BA} \subseteq \overleftrightarrow{AB}$

21. Can a pair of angles be both complementary and supplementary?

22. Can $\overline{AC} \cap \overline{BD} = \emptyset$ and $\overleftrightarrow{AC} \cap \overleftrightarrow{BD}$ be a point? A line? A segment? Describe, in general, the various possible relationships that can exist between $\overline{AC} \cap \overline{BD}$ and $\overleftrightarrow{AC} \cap \overleftrightarrow{BD}$.

23. One acute angle of a triangle measures 40°. The difference in the other two angles is 20°. Find the three angles.

24. (a) Under what conditions do $\overleftrightarrow{AB}$ and $\overleftrightarrow{CD}$ name the same line?
    (b) Under what conditions do $\overrightarrow{AB}$ and $\overrightarrow{CD}$ name the same ray?

25. Determine whether x is in the interior or the exterior of the closed curve.

Bibliography

1. Bentley, W. A., & W. J. Humphrey, *Snow Crystals.* New York: McGraw-Hill, 1931.
2. Bruni, James V., & Helene J. Silverman. "Making Patterns with a Square." *AT* (April 1977), 265–272.
3. Gardner, Martin. *Aha!* New York: W. H. Freeman, 1978.
4. Horak, Virginia M., & Willis J. Horak. "Using Geometry Tiles as a Manipulative for Developing Basic Concepts." *AT* (April 1983), 8–15.
5. Immerzeel, George. "Geometric Activities for Early Childhood Education." *AT* (October 1973), 438–443.
6. Meggison, G. W. "Rays and Angles." *AT* (May 1974), 433–435.
7. Nelson, Glenn, & Larry P. Leutzinger. "Let's Take a Geometry Walk." *AT* (November 1979), 2–4.
8. Pottinger, Barbara. "Measuring, Discovering and Estimating the Metric Way." *AT* (May 1975), 372–377.
9. Russell, Dorothy S., & Elaine M. Bologna. "Teaching Geometry with Tangrams." *AT* (October 1982), 34–38.
10. Shaw, Jean M. "Student-made Measuring Tools." *AT* (November 1983), 12–15.
11. Smith, Joe K. "Consistent Classification of Geometric Figures." *MT* (November 1976), 574–576.

CHAPTER 9

Using Transformation in Geometry

Jack found a message inside a bottle that had been swept onto the beach. Help him decipher the message. What techniques can you use?

The idea of motion or movement is an important concept of geometry. We introduce the subject casually through a study of flips, slides, and turns. These are mathematically defined concepts and are shown to be distance-preserving transformations.

Transformation theory is used to define congruence and similarity. The congruence concept will be used in the next chapter to derive well-known geometric formulas.

You will undoubtedly find this brief and elementary encounter with motion geometry to be both interesting and accessible.

1 An introduction to flips or reflections

Flips or reflections form a basis for studying various properties of geometric figures. The idea of a reflection is very familiar to you. Every time you look in a mirror, you see a reflected image of yourself.

■ To get an idea of how a reflection works, write a large letter T on a piece of paper. Then fold the paper so that the T is covered and trace the T on the fold. Unfold the paper and hold it to the light so that you can see the T you traced. Your trace should be like the two "traced" T's in Figure 1 that are shown in color.

Figure 1

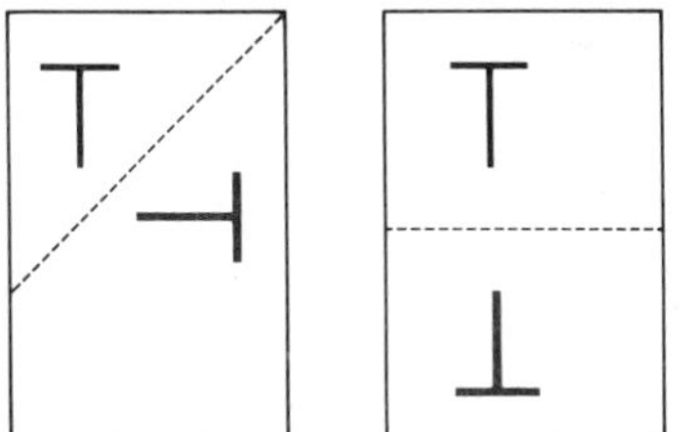

■

The tracing is called the **image** of the reflection, and the line of the fold (shown as a dashed line) is called the **line of reflection.** Often we say reflection *about* the line, or reflection *over* the line.

Example 1 • Find the point A' (called the image of A) corresponding to point A by folding a piece of paper along line m in (a), (b) and (c) of Figure 2.

Figure 2

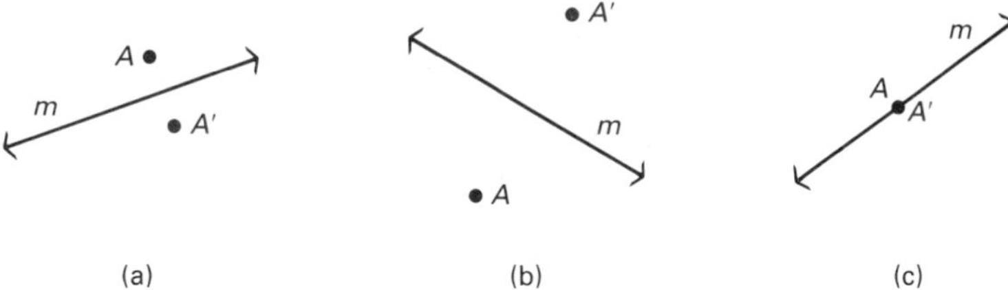

Solution In (c), where A happens to be on the line, A and A' are the same point. •

Now suppose you have printed the word MATH on a piece of paper. Before the ink is dry, the paper is folded, covering the print. When the paper is unfolded you get an image as shown in Figure 3(a).

The preceding example is an illustration of a reflection about a line (sometimes called a *flip*).

A *mira,* made of plexiglass, reflects like a mirror although it is transparent. Thus, in Figure 3(b), A' is the flip image of A where $\overleftrightarrow{PQ}$ is the line of reflection.

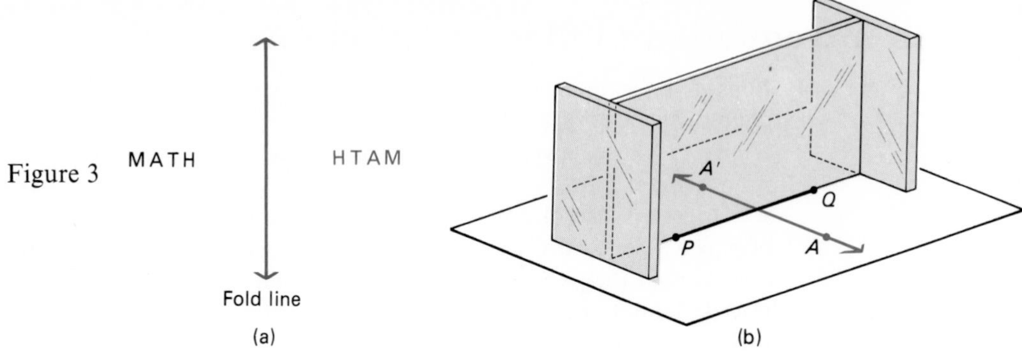

Figure 3

Reflection images or flip images are easy to obtain using paper folding, a mirror, a mira, or just common sense, as indicated in the following example.

Example 2 • Find the flip image of $\triangle ABC$ relative to line m in Figure 4.

Solution The flip image is shown in color. •

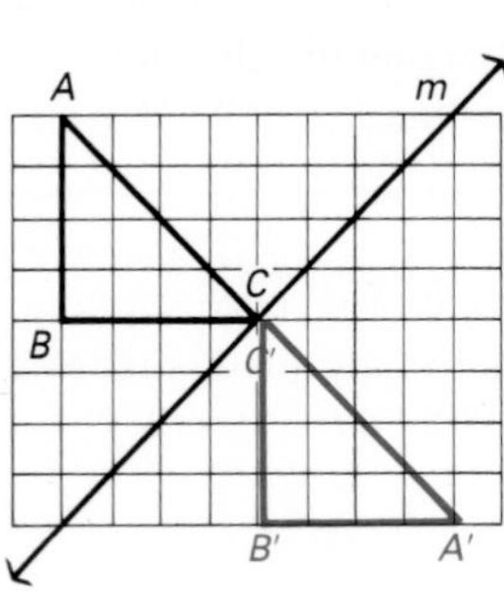

Figure 4

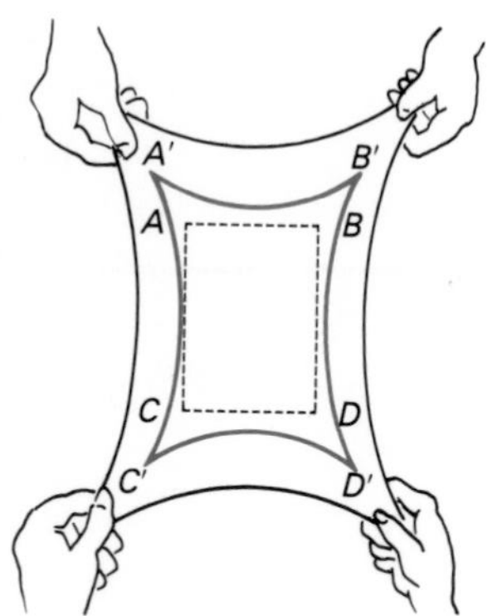

Figure 5

The concept of a reflection is a special case of a **transformation.** Suppose a rectangle, $ABCD$, is drawn on a transparent sheet of rubber, which is stretched as shown in Figure 5. The vertices after the stretching are marked as A', B', C', and D'. Through this stretching, we are associating a unique point with each given point.

$$A \leftrightarrow A'$$

$$B \leftrightarrow B'$$

$$C \leftrightarrow C'$$

$$D \leftrightarrow D'$$

We call such a correspondence a **transformation.**

Transformation

> A transformation of a plane is a one-to-one correspondence between two sets of points of a plane.

A transformation is also called a **mapping.** In the preceding example, A is mapped onto or into A', B onto B', and so on. A' is again called the image of A.

We now introduce the geometric terms to describe mathematically the concept of a reflection. In Chapter 8, we considered the measure of a line segment. The midpoint of a line segment is the point that divides it into two segments with equal measures. Any line, segment, or ray (not coinciding with the given segment) that intersects the given segment at this midpoint is called a **bisector** of the segment. Of particular interest is the **perpendicular bisector** of a segment. This is the bisector (line, segment, or ray) that intersects the given segment in a right angle (90°).

In Figure 3(b), a reflection maps A onto A' about line $\overleftrightarrow{PQ}$ because $\overleftrightarrow{PQ}$ is the perpendicular bisector of $\overline{AA}$. In Figure 2, if you were to place a protractor along $\overline{AA'}$, you would note that line m is perpendicular to $\overline{AA'}$. In fact, it is the perpendicular bisector of $\overline{AA'}$.

Reflection

> A reflection (or flip) about line m is a transformation mapping each point P onto point P' as follows:
> (a) If P is on m, then $P = P'$.
> (b) If P is not on m, then m is the perpendicular bisector of $\overline{PP'}$.

To understand this concept better, we take time to review goemetric constructions.

Construction 1

• Construct a perpendicular to a line from a given point not on the line. Given line r and a point P not on r (see Figure 6(a)):

(a) With P as a center, draw an arc that intersects r in two points, A and B (Figure 6(b)).

(b) With A and B as centers, draw arcs with the same radii as $PA = PB$, intersecting at P and C.

(c) Draw line $\overleftrightarrow{PC}$.

(d) $\overleftrightarrow{PC}$ is perpendicular to $\overleftrightarrow{AB}$.

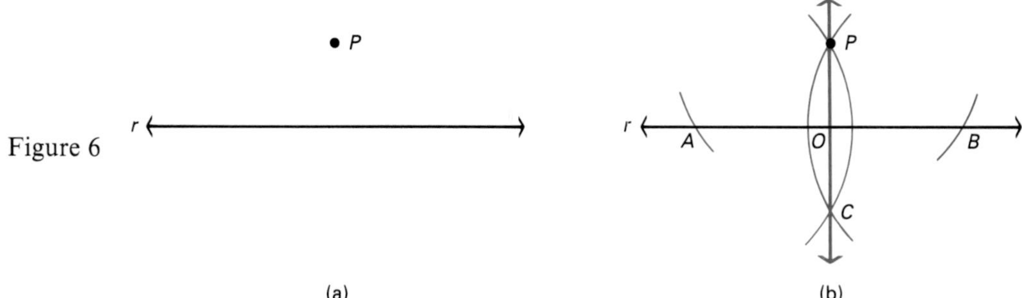

Figure 6

(e) $PO = PC$
(f) r is the perpendicular bisector of $\overline{PC}$. •

Construction 2 • Given a line m and a point A, using geometric constructions, locate a point A' that is the reflection image of A.

(a) Draw an arc, using point A as the center, cutting line m in two points B and C (see Figure 7(b)).
(b) With B as the center, construct an arc containing A.
(c) With C as the center, construct an arc containing A.
(d) The two arcs in (b) and (c) intersect in a second point, which we denote A'. (See Figure 7(c).)
(e) We have constructed $\overline{AA'}$ perpendicular to m such that m is the perpendicular bisector of $\overline{AA'}$. Therefore, A' is the reflection image of A. •

Figure 7

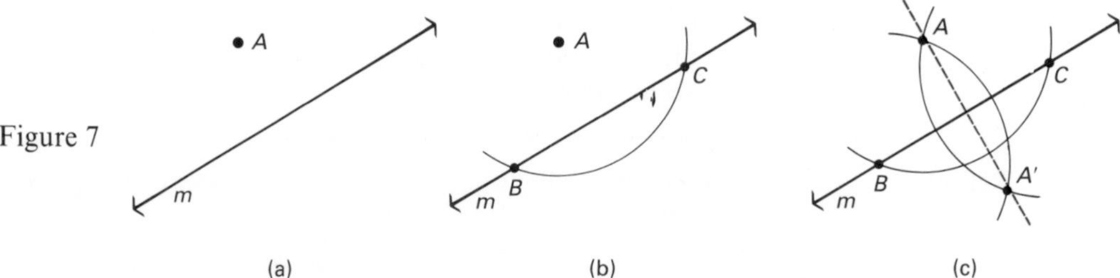

The preceding discussion of constructions assists in the understanding of the reflection of figures about a line.

Example 3 • Taking line segment $\overline{PQ}$, as in Figure 8, reflect or flip it about line l.

Solution Fold your paper along flip line l. Now trace the flip image, $\overline{P'Q'}$. Or again, fold your paper along flip line l and, with a pin, mark points Q' and P' corresponding to Q and P. With a straightedge, draw $\overline{P'Q'}$, which is the flip image of $\overline{PQ}$. •

This same result can be obtained geometrically by the following construction. From point P, construct a line perpendicular to l, calling the point of intersection R. To determine a point P' that corresponds to P, construct a segment $\overline{RP'}$ on $\overleftrightarrow{PR}$ such that $PR = RP'$. Do the same for point Q, where $QS = SQ'$. The flip images of points on $\overline{PQ}$ are on the segment $\overline{P'Q'}$, a reflection of $\overline{PQ}$ about line l.

In Figure 8, set your compass, on P and Q; with the same setting, measure $\overline{P'Q'}$. Are PQ and $P'Q'$ equal? This suggests that a reflection is a distance-preserving transformation.

With constructions, you can also see that if G is a point on $\overline{PQ}$, its image G' will be on $\overline{P'Q'}$. Thus, a reflection preserves the collinearity of points. This same study will demonstrate that if G is a point between P and Q, then G' will be between P' and Q'. Thus, reflection preserves betweenness of points.

To discuss whether reflections preserve angle measure, we first look at a construction of angles with the same measure.

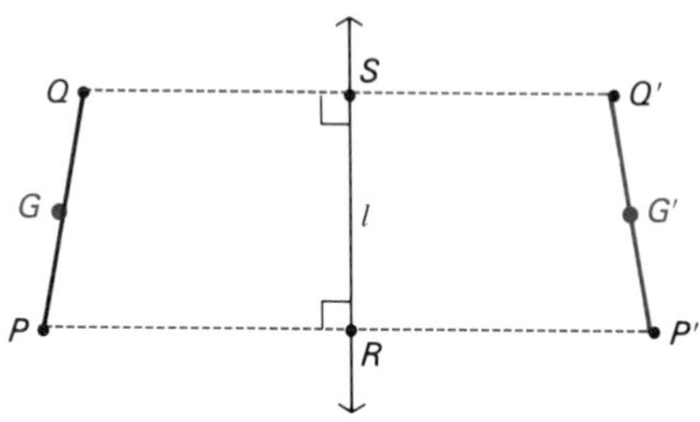

Figure 8

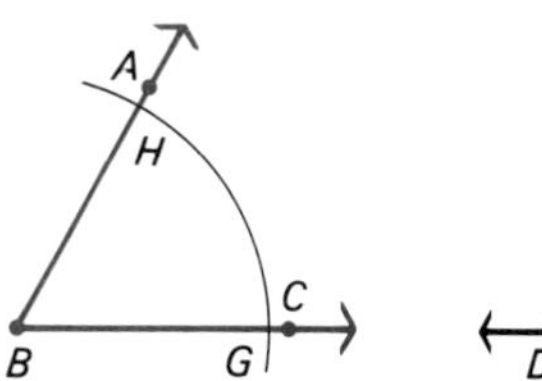

Figure 9

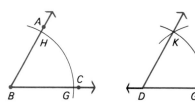

Construction 3 • Construct an angle with the same measure as $\angle ABC$ in Figure 9.

(a) Draw $\overrightarrow{DE}$ as one side of the angle.
(b) With B as the center, draw an arc of any radius, cutting $\overrightarrow{BA}$ at H and $\overrightarrow{BC}$ at G.
(c) With D as the center, draw an arc of the same radius as in part (b), cutting $\overrightarrow{DE}$. Denote the intersection of this arc and $\overrightarrow{DE}$ by G'.
(d) With G' as center, draw an arc of radius HG.
(e) Denote the intersection of these two arcs as K.
(f) Draw $\overrightarrow{DK}$.
(g) $\angle KDE$ has the same measure as $\angle ABC$. •

To demonstrate that reflections preserve angle measure, we examine the following construction.

Construction 4 • Given: $\angle ABC$ and line m in Figure 10(a). We want to reflect $\angle ABC$ about line m in (b).

(a) Construct $\overline{BB'}$ so that m is a perpendicular bisector of $\overline{BB'}$.
(b) Construct $\overline{AA'}$ so that m is a perpendicular bisector of $\overline{AA'}$.
(c) Construct $\overline{CC'}$ so that m is a perpendicular bisector of $\overline{CC'}$.
(d) Draw rays $\overrightarrow{B'A'}$ and $\overrightarrow{B'C'}$.
(e) Now use Construction 3 to construct an angle on $\overrightarrow{B'A'}$ that is the same measure as $\angle ABC$. D' seems to be on $\overrightarrow{B'C'}$, so $\angle ABC$ and $\angle A'B'C'$ have the same measure. •

Figure 10

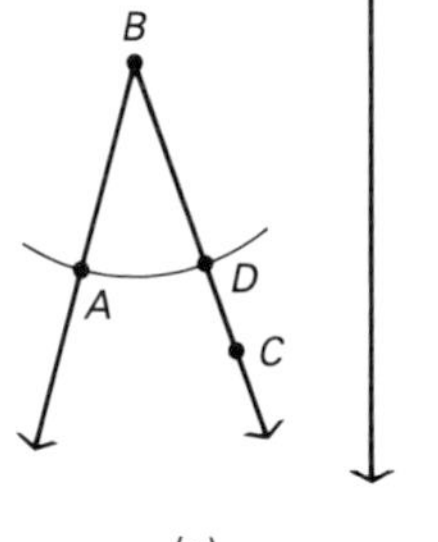

(b)

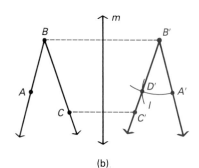

Just for fun *Is the distance around the unshaded region greater than, equal to, or less than the distance around the shaded area?*

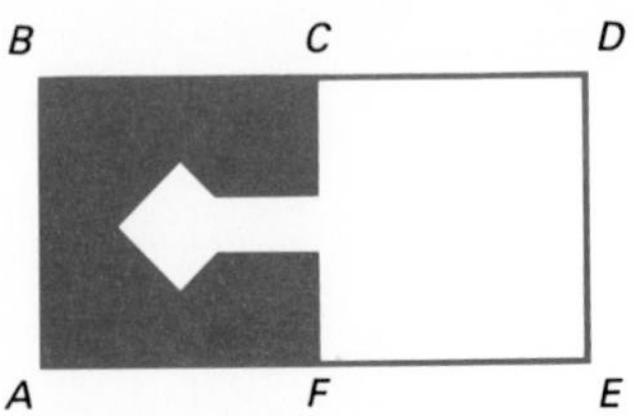

Properties of reflections

Reflection is a transformation that preserves
(a) distance
(b) collinearity of points
(c) betweenness of points
(d) angle measure

Figure 11 shows a quadrilateral $ABCD$, where the vertices are read in a clockwise direction. If we reflect the quadrilateral about line m, the vertices of $A'B'C'D'$ are now read in a counterclockwise direction. Therefore, the reflection has reversed the orientation of the points of vertices. Thus, reflections do not preserve the orientation of points or vertices.

Figure 11

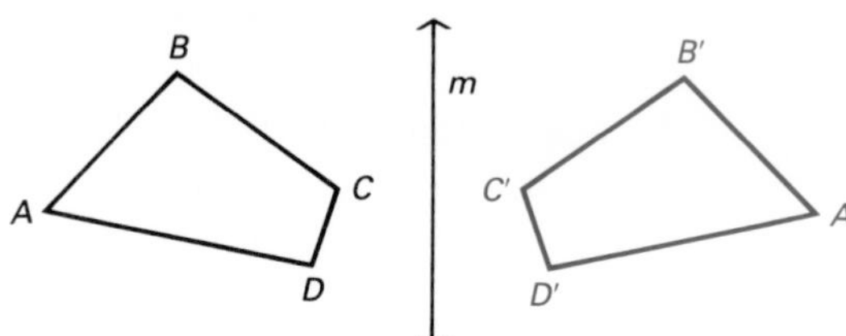

Exercise set 1

Trace the drawing onto thin paper. Then fold the paper and find the reflection image of each given point with respect to the given line m.

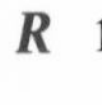

1.

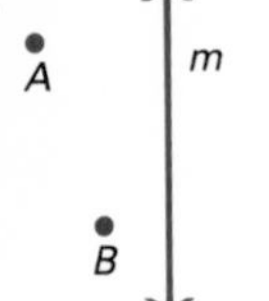

2.

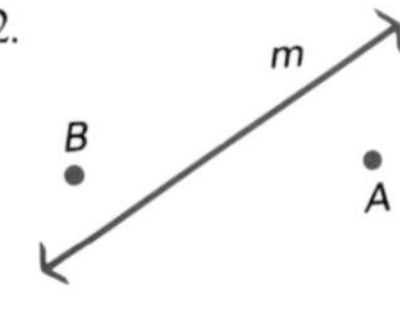

3.

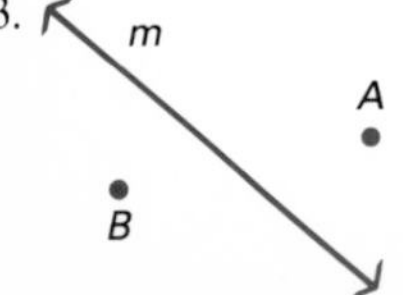

4.

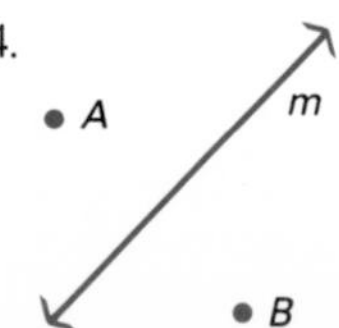

5. 6.

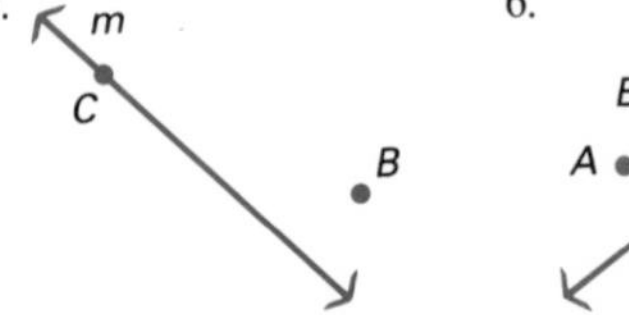

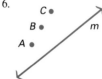

7. (a) In Exercise 4, how are points A and B related?
   (b) In Exercise 5, what is the reflection image of C?
   (c) How are A', B', and C' related in Exercise 6?
8. If R is the reflection image of S over line m, how are m and $\overline{RS}$ related?

9. Trace the following figures on your paper. Then flip each figure about the given line. Draw the flip image.

(a)

(b)

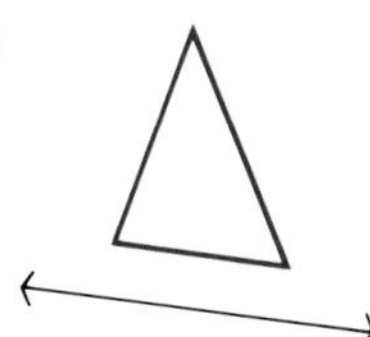

(c)

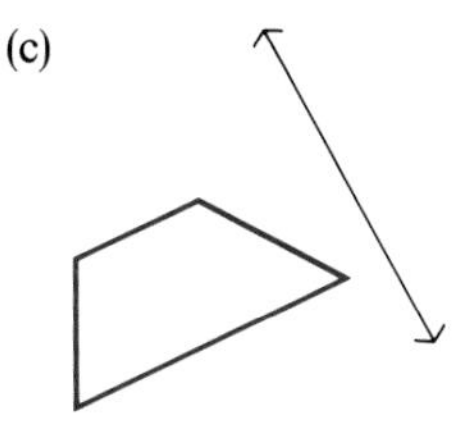

(d)

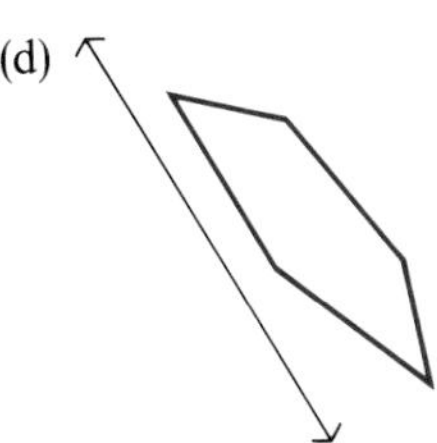

10. Classify as true or false.
    (a) A segment and its reflection image have different lengths.
    (b) A point and its reflection image are the same distance from the reflecting line.
    (c) A reflection is a one-to-one correspondence.
    (d) A point and its image coincide if the point is on the reflecting line.
    (e) The reflection image of an acute angle is always acute.
    (f) The reflection image of a ray is always a ray.
    (g) When two nonintersecting lines are reflected over the same line, their images may intersect.

Copy the drawing. Then reflect $\angle ABC$ over line m.

11.

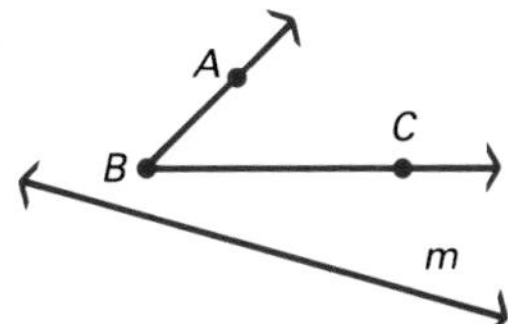

12.

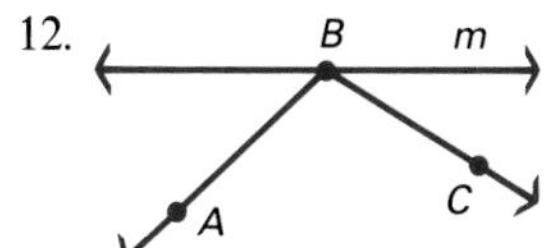

13.

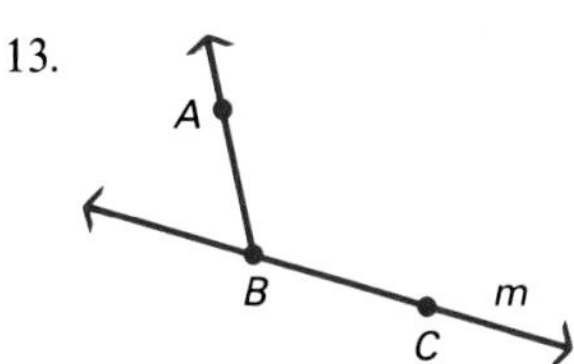

14.

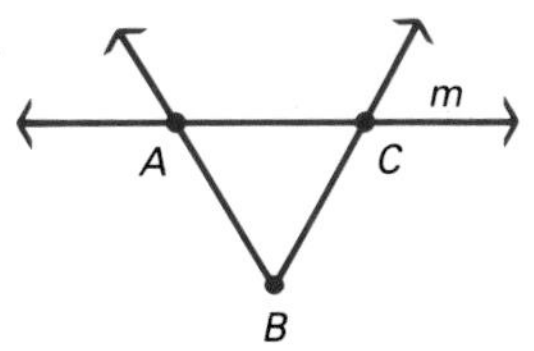

T 15.

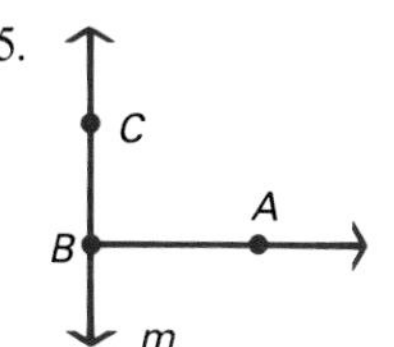

16.

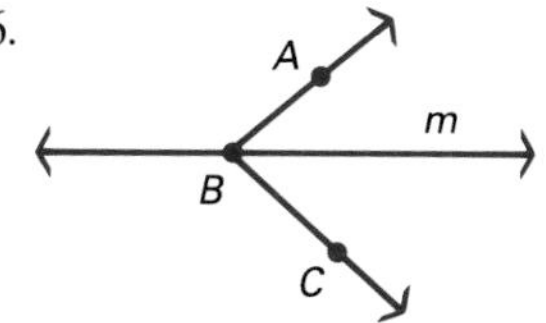

17. Flip the following figures about line *l* and show the flip image.

(a)

(b)

(c)

(d)

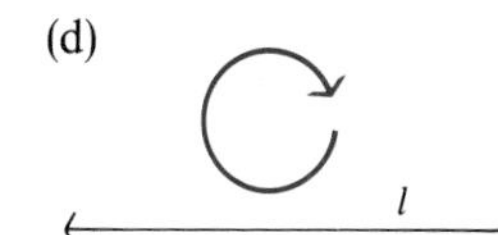

(e)

(f)

18.

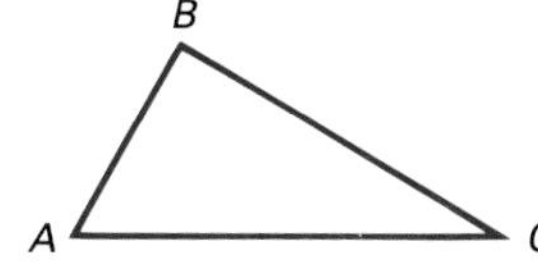

Find these reflecting lines.
(a) *l* so that *B* is the image of *A*.
(b) *m* so that *B* is the image of *C*.
(c) *n* so that *A* is the image of *C*.

19. Draw a reflection of the figure about line l and label the corresponding points.

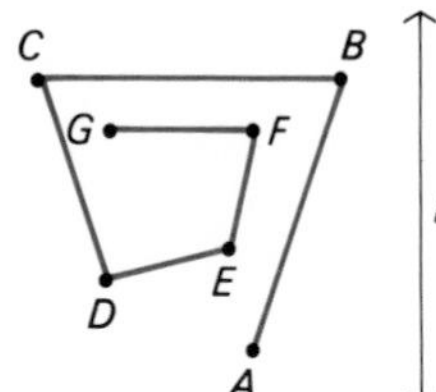

20. When do a point and its image coincide under a reflection?
21. Name four properties preserved by reflections.
22. Could an angle coincide with its reflection image? Explain.
23. Using each figure below, draw a reflection about line l.

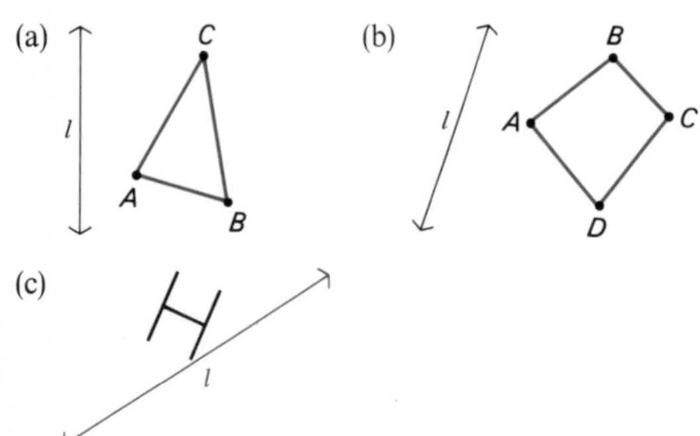

24. Sketch the image of the figure under the reflection about line l.

25. Draw an image for the letter H with the following transformations.
    (a) Reflect H about the leg on which P does not lie.
    (b) Reflect H about the middle bar.

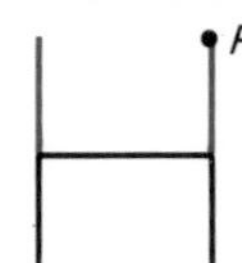

26. Draw a triangle CDE and a line l on your paper. Show how you can draw the flip or reflection image of $\triangle CDE$, using
    (a) a tracing
    (b) paper folding
    (c) ruler and compass construction

C 27. Draw two parallel lines. Then draw line m, making a 45° angle with the parallel lines. Now reflect the lines about m. Are the lines parallel in the image?

28. Find the reflection of the triangle about line m.

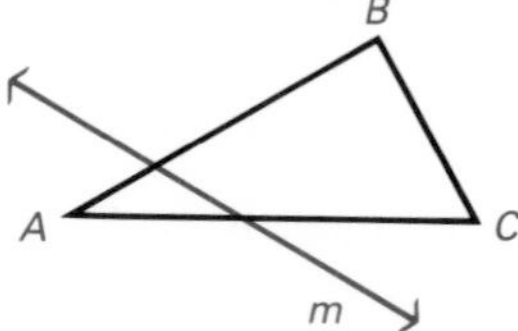

2 Slides or translations

■ Trace the preceding figure on thin paper. Fold on line n and show that the middle car is a reflection image of the car on the left. Fold on line m and note that the car on the right is a reflection image of the middle car about line m. Now mark corresponding points on the car on the left and the car on the right, and measure the distances between these points with a compass. Are all the distances the same? You have discovered a new transformation, called a *slide*. This new transformation is a result of two reflections about parallel lines. ■

The only transformation we have defined so far is a reflection. Other transformations often result from successive reflections, as the following example illustrates.

Example 4 • In Figure 12, $\triangle A'B'C'$ is the reflection image of $\triangle ABC$ over line m. $\triangle A''B''C''$ is the reflection image of $\triangle A'B'C'$ over line n. We are given that m is parallel to n. Measure the distance from A to A'', from B to B'', and from C to C'' and make a conjecture about these distances.

Figure 12

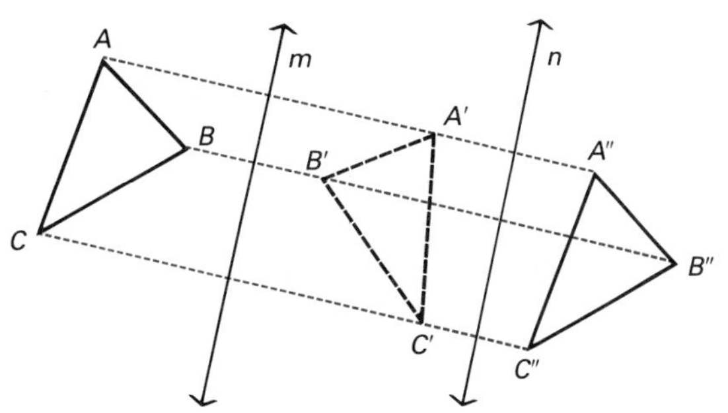

•

In the preceding example, it seems that $\triangle ABC$ has been moved a given distance in a given direction. This leads to the concept of a slide or, more formally, a **translation.**

Example 5 • Slide or translate the car in Figure 13 up the hill in the direction and for the distance indicated by the darkened arrow.

Solution Trace the car and the darkened arrow. Slide your tracing paper along line l until the tail of the arrow on the tracing paper is at the point of the arrow on the original paper. The trace of the figure in the new position coincides with the second figure in Figure 13, with P and Q corresponding to P' and Q'. The second diagram in Figure 13 is the *slide image* of the original figure. Keep this example in mind as we continue the discussion of a slide. •

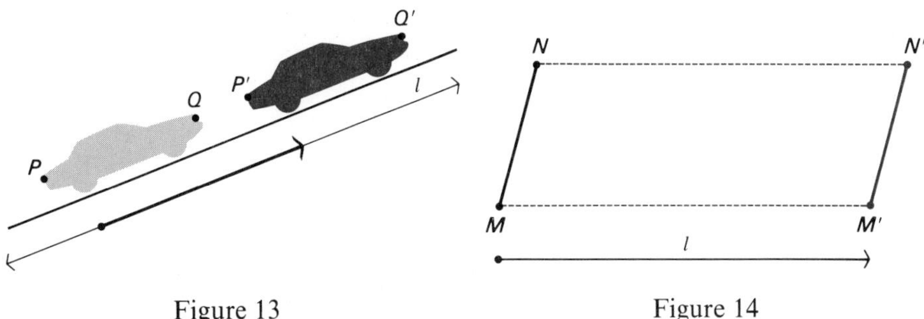

Figure 13

Figure 14

Example 6 • Slide a given line segment $\overline{MN}$, as in Figure 14, 5 centimeters in the direction of ray l. To begin, determine a point M' 5 centimeters from point M in the direction determined by ray l. In fact, slide each point on $\overline{MN}$ 5 centimeters from its original position, parallel to l. The point N' is 5 centimeters from N, and $\overline{NN'}$ is parallel to l. Notice that segments like $\overline{MM'}$ and $\overline{NN'}$ are of length 5 cm and are parallel to l. The side image of segment $\overline{MN}$ is the segment $\overline{M'N'}$, and $M'N' = MN$. •

Example 5 showed how a figure was translated to a new position by sliding it for a given distance in a given direction. In the next example, individual points were translated for a given distance in a given direction, and the resulting slide image was a line segment. The intuitive understanding of slides obtained from these examples provides a valuable background for understanding the idea of translations.

Translation

> A translation (or slide) determined by a ray l and a specified distance is a transformation that maps point M onto point M' so that $\overline{MM'}$ is parallel to and in the same direction as l and MM' equals the specified distance.

Example 7 • In Figure 15(a), translate (or slide) the word MATH two blocks down. In (b), translate the triangle three blocks to the right.

Figure 15

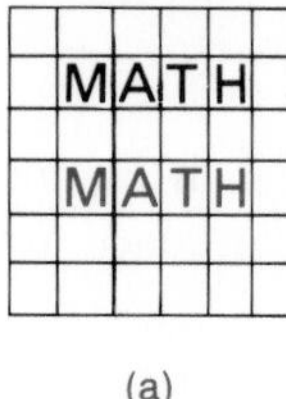

(a)

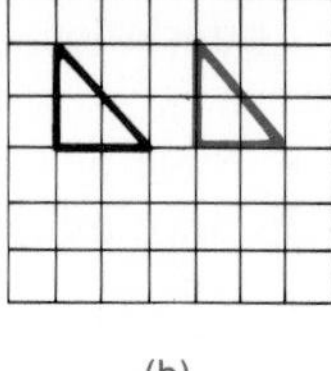
(b)

Solution The translation images are shown in color. •

The ideas associated with the first example in this section are explained by the following definition.

Composite

> Let r_m and r_n represent two reflections about lines m and n, respectively. Then the transformation "apply r_m and then apply r_n to the image obtained by r_m" is called the composite of r_m and r_n and is denoted by composite (r_m, r_n).

If the image of P under r_m is P', and the image of P' under r_n is P'', then the image of P under the composite (r_m, r_n) is P''.

Translation from reflections

> A transformation is a translation (T), or slide, if and only if T = composite (r_m, r_n) where r_m is a reflection over line m, r_n is a reflection over line n, and m is parallel to n.

Just for fun

Can you form four equilateral triangles with six matches of same length?

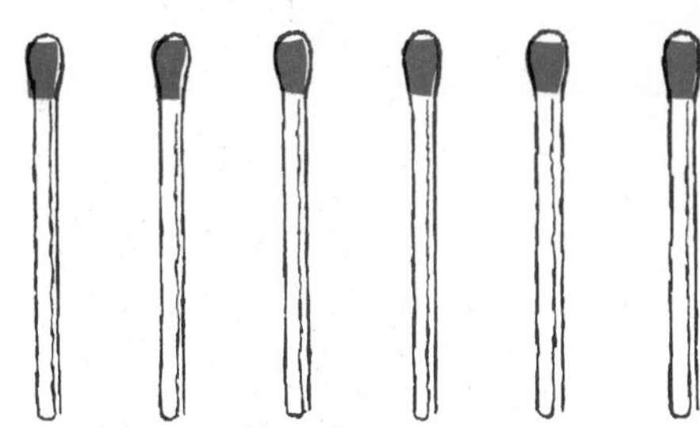

Example 8 • In Figure 16, rectangle $WXYZ$ is reflected over line m and the image is reflected over line n to give rectangle $W'X'Y'Z'$. Note that if $WXYZ$ is translated in the direction of $\overrightarrow{WW'}$ a distance of WW', the same image $W'X'Y'Z'$ is obtained. •

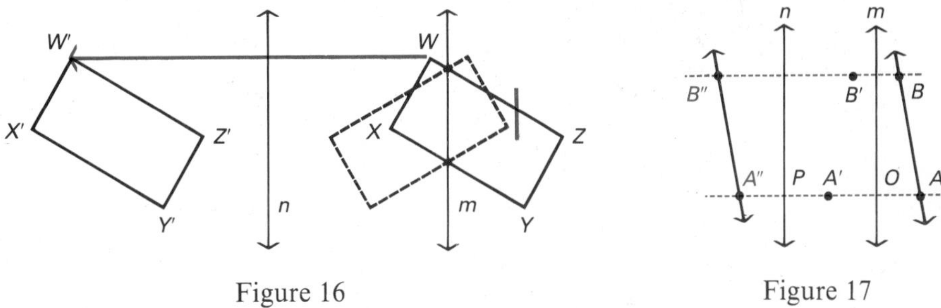

Figure 16

Figure 17

Since a translation can be defined as a composite of two reflections, it is a mapping that preserves collinearity, betweenness, distance, and angle measure.

In Figure 17, A' is the image of A in a reflection over m, and A'' is a reflection of A' over n. Note that

$$AO = A'O \qquad \text{and} \qquad A'P = A''P$$

Now the distance between parallel lines is

$$A'O + A'P$$

The distance of the translation from A to A'' is

$$\begin{aligned} AO + A'O + A'P + A''P &= 2A'O + 2A'P \\ &= 2(A'O + A'P) \end{aligned}$$

Thus, the distance of the translation for this example is twice the distance between parallel lines.

By the definition of a reflection, $\overline{A''A}$ in Figure 17 is perpendicular to the parallel lines.

Translation and parallel lines

If a translation maps A onto A'', then $\overline{AA''}$ is perpendicular to each of the parallel lines defining the translation in terms of reflections and AA'' is twice the distance between the parallel lines.

See if you can establish from Figure 17 that $\overleftrightarrow{BA}$ is parallel to $\overleftrightarrow{B''A''}$, leading to the following result.

Parallel image | Under a translation, a line and its image are parallel.

Exercise set 2

R 1. Complete the blanks.

(a) r_m represents a reflection about ________.

(b) If $m \| n$ and r_m reflects about m and r_n about n, then composite (r_m, r_n) is a transformation called a ________.

(c) A translation preserves ________.

(d) A translation is a composite of ________.

(e) Under a translation, a line is parallel to ________.

(f) If S' and T' are translation images of S and T, then $S'T'$ = ________.

(g) When is a point its own image under a translation? ________.

(h) A reflection acts like a flip; a translation acts like a ________.

2. Trace each given figure on your paper. Then slide each one as indicated. Draw the slide image.

(a) (b)

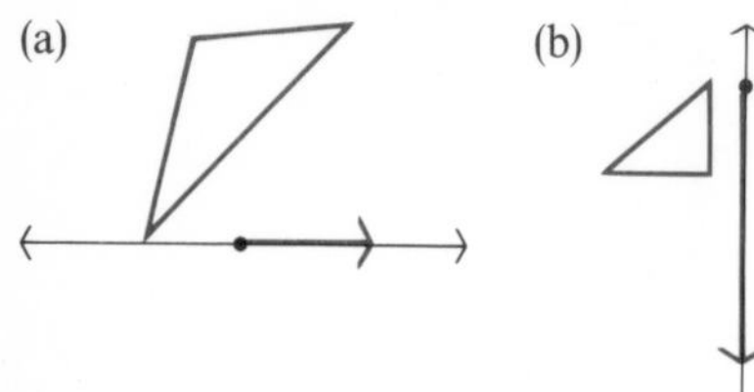

(c)

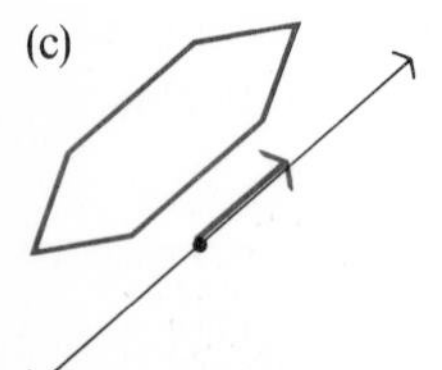

(d)

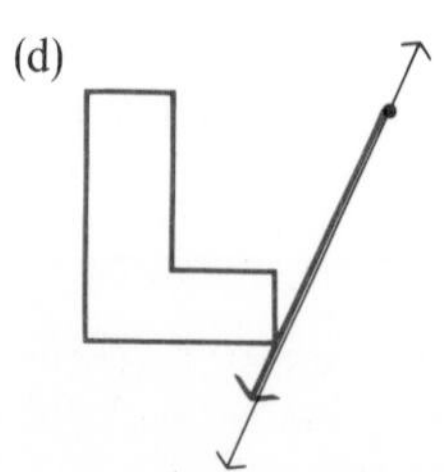

3. Slide each figure 2 centimeters in the direction of $\overrightarrow{PA}$.

(a) (b)

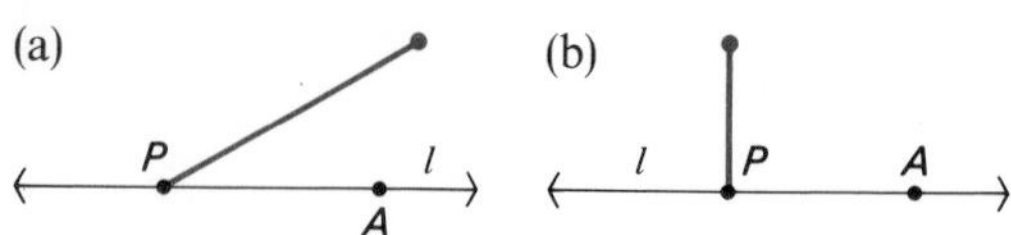

(c) (d)

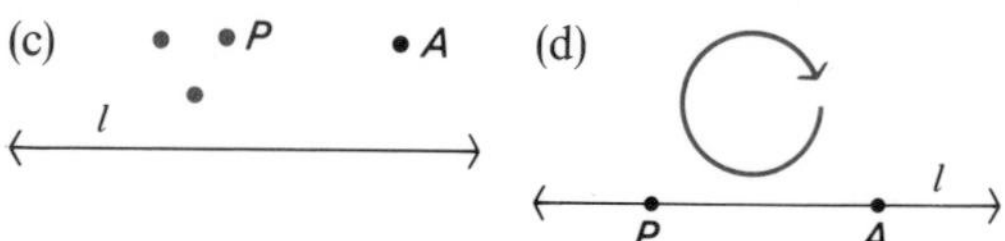

(e) (f)

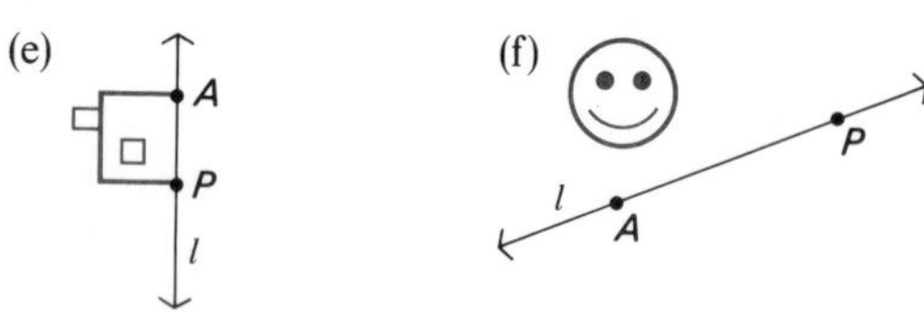

4. Trace each figure below on your own paper and then translate each traced figure 2 centimeters to the right.

(a) (b)

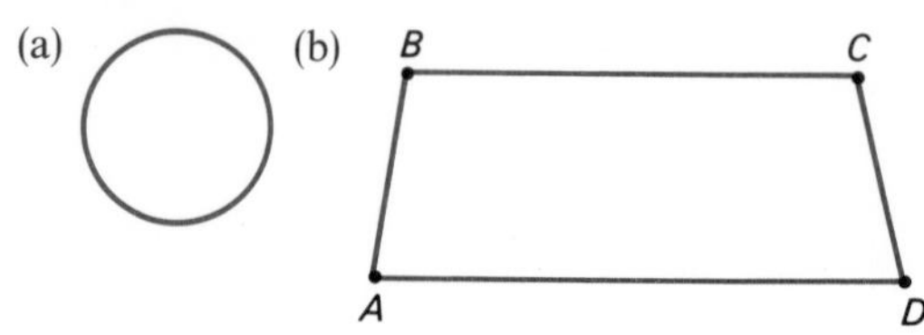

(c)

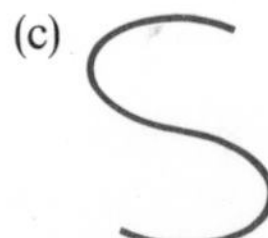

5. Draw an image for the letter H with the following transformation. Translate H until P is at the bottom of the leg.

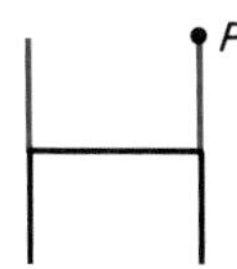

6. Line l is parallel to $\overleftrightarrow{AB}$. Use line l and your compass to find the slide or translation image of point C after a slide of a distance AB in a direction $\overrightarrow{AB}$.

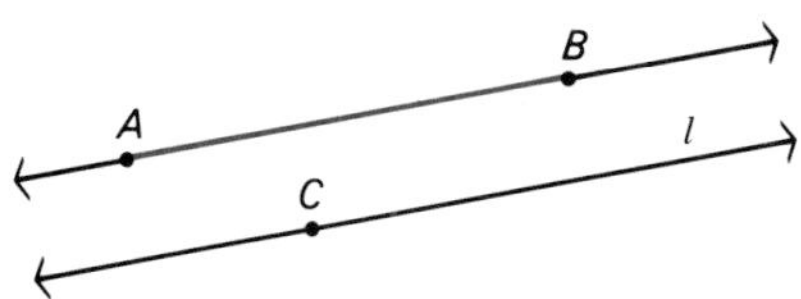

7. Draw two parallel lines in a vertical direction. Then perform the following transformation. Translate the parallel lines 4 cm in the direction of a ray making an angle of 45° counterclockwise from horizontal. Are the lines still parallel after the translation? Do you feel that parallelism is preserved under translations?

8. In the figure below, a translation maps A onto C. Name the image of:

(a) E (b) B (c) K
(d) $\overline{AE}$ (e) $\angle EFK$ (f) $BCGF$

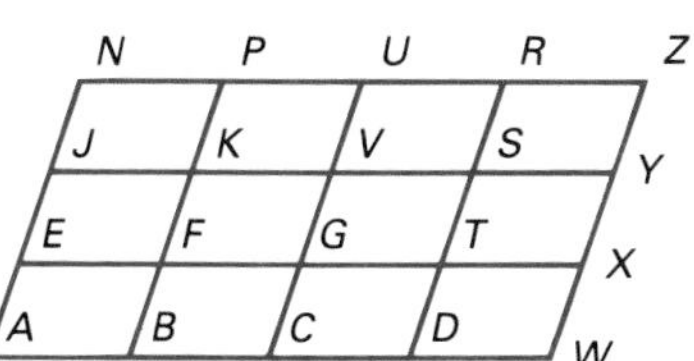

T 9. In Exercise 8, find the images if A is mapped onto F. If A is mapped onto G.

10. Draw a 30° angle with a protractor. Translate the angle 4 cm to the right. Now measure the angle. Do you believe that angles and their measures are preserved under translations?

11. Devise a method for determining the specified distance of the translation in the given figure. Does your method apply to any translation?

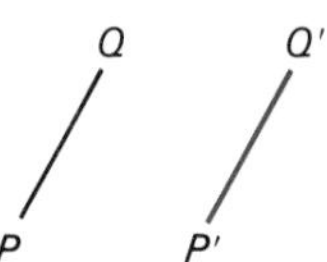

C 12. A rectangular coordinate system involves a number line in two directions, one called x and one called y. The coordinates of a point are given by (x, y). Thus, the coordinates of A are $(-1, 3)$, and the coordinates of A' are $(2, 3)$.

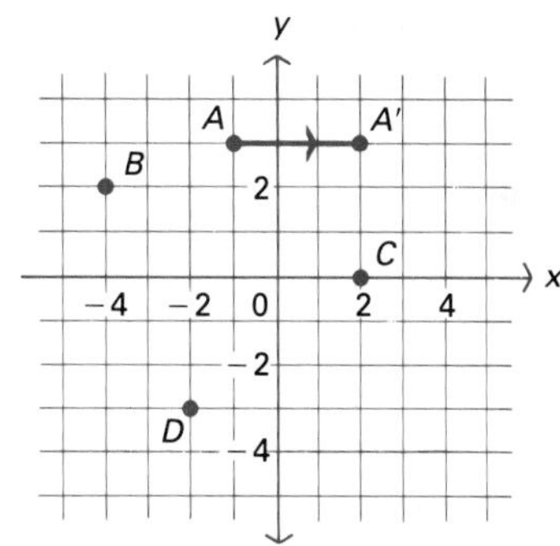

In the figure above, a translation maps A onto A'. Name the coordinate of the translation image of each of the following.

(a) B (b) C (c) D

13. In the figure below a translation maps A onto A'. Name the image of

(a) C (b) B (c) D (d) E

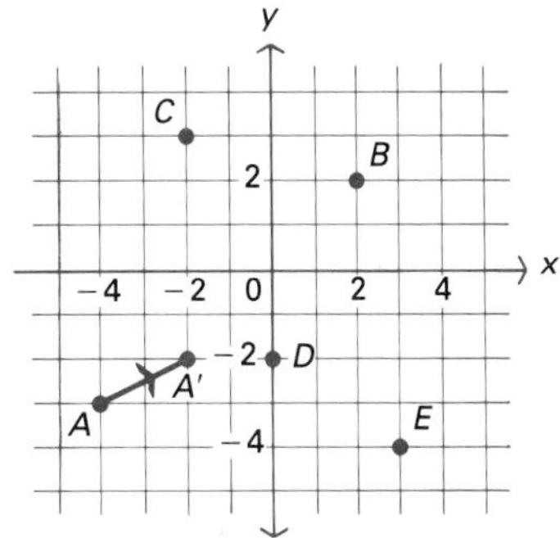

14. Draw 2 points (A and A') 4 cm apart. If A' is the translation image of A, show 2 parallel lines, m and n, so that the translation is composite (r_m, r_n)

15. Find a pair of parallel lines (m and n) such that the translation from $\triangle ABC$ to $\triangle A'B'C'$ is composite (r_m, r_n).

3 Rotations and successive motions

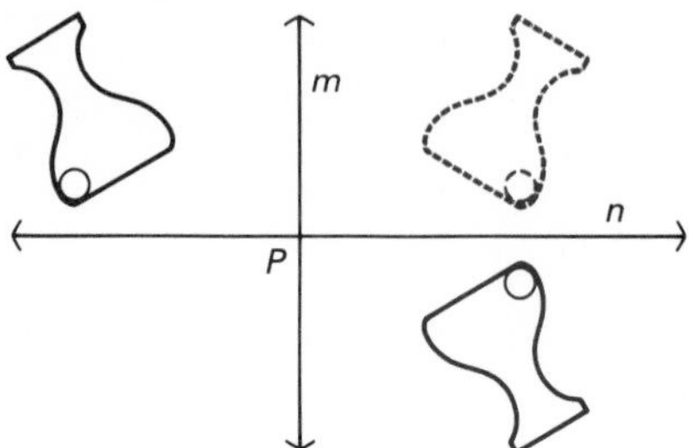

■ Copy the figure and lines m and n. Show that the dashed-line figure is the reflection of the first figure about line m. Then show that the second figure is the reflection of the image (of the first figure) about line n. Now put a pen at point P and see if you can rotate until the first figure coincides with the second. This suggests that turns or rotations may be a composite of reflections. Thus, rotations will have all the properties of reflections. ■

It is usually not too difficult to visualize one motion or transformation. However, when two or more motions have been performed, it can be difficult to visualize what has happened. Do you get the same result if you perform a rotation followed by a reflection as you get by the same reflection followed by the rotation? You will answer this question and others in this section. In the process, you may stumble upon a bit of a surprise.

The composite of two reflections over parallel lines is a translation. Figure 18 shows a composite of two reflections over intersecting lines m and n.

Let r_m reflect figure ABC about line m onto figure $A'B'C'$. Then let r_n reflect figure $A'B'C'$ about line n onto figure $A''B''C''$. In Figure 18, this could have been accomplished by rotating $\overline{OA}$ through an angle θ until it coincides with $\overline{OA''}$. This example illustrates that when two reflecting lines intersect, the composite of the two reflections results in a transformation called a **rotation.**

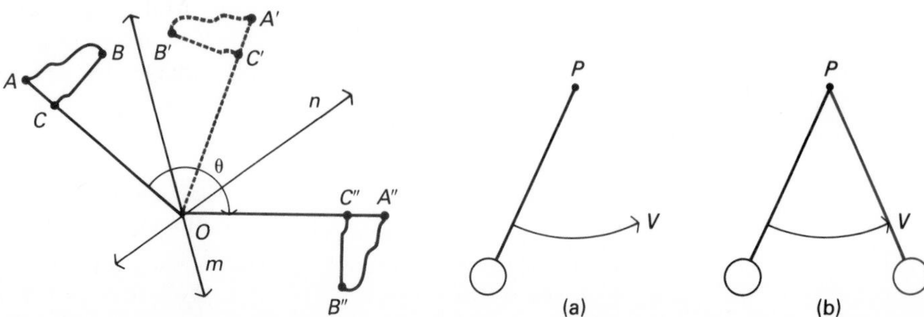

Figure 18

Figure 19

Rotation

> A transformation is a rotation (R) or turn about point P if and only if $R =$ composite (r_m, r_n), where m and n intersect at P (called the center of rotation).

A rotation may be defined in terms of the center of rotation P and the turn angle θ. The center of turn is often denoted by a dot, and the angle of turn is usually given by a curved arrow or by a given number of degrees. The following example illustrates an informal development of a turn, using a "turn arrow" rather than a measured angle.

Example 9 • The clock pendulum in Figure 19 is to swing in the direction and through the distance indicated by the turn arrow. Note the center of turn, point P, on which the pendulum swings.

Solution Trace the pendulum. Keeping the tracing paper fixed with a pin at point P, turn it until the tail of the arrow on the tracing paper is at point V. If possible, trace the pendulum in the new position. The colored pendulum in Figure 19 illustrates the image. •

Example 10 • Rotate the given segment $\overline{XY}$, shown in Figure 20(a), 45° in a counterclockwise direction about X. To perform the rotation, simply establish X as the center of turn and draw a 45° angle with $\overrightarrow{XY}$ as one side of the angle. On the other side, mark point Y' so that $XY = XY'$. •

Figure 20

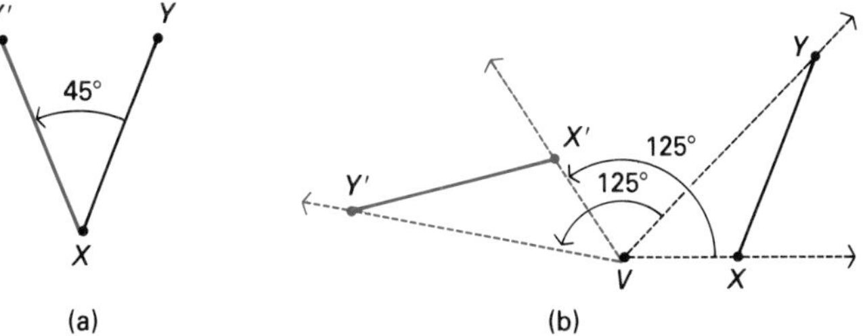

Example 11 • Rotate $\overline{XY}$ 125° counterclockwise about V in Figure 20(b). The rotation in (b) is more complex than in (a), since $\overline{XV}$ and $\overline{YV}$ are on different lines. Rotate segment $\overline{XV}$ counterclockwise 125° about V. That is, with V as center of turn and $\overrightarrow{VX}$ as one ray, draw an angle of 125°. Mark X' on the new ray so that $VX = VX'$. Similarly, rotate $\overrightarrow{VY}$ 125° about V, marking point Y' such that $VY = VY'$. $X'Y'$ is equal to XY. Thus, $\overline{X'Y'}$ is the image of a 125° rotation of $\overline{XY}$ about V. •

Rotation

> A rotation (or turn), determined by a point P and a given angle, maps point R into point R' in such a way that $\angle RPR'$ is of the given measure and $RP = R'P$. The direction (clockwise or counterclockwise) of the turn must be specified.

One goal in Section 5 is to define congruence in terms of transformations. However, to accomplish this goal, we must allow the possibility of successive transformations. As indicated previously, images are sometimes produced as a result of several consecutive transformations.

Example 12 • The E on the left in Figure 21 is to be mapped onto the E on the right by successive transformations. First, let's translate the figure 5 units to the right. Then let's translate the figure 2 units up. Finally, we rotate the figure 90° counterclockwise about P'. •

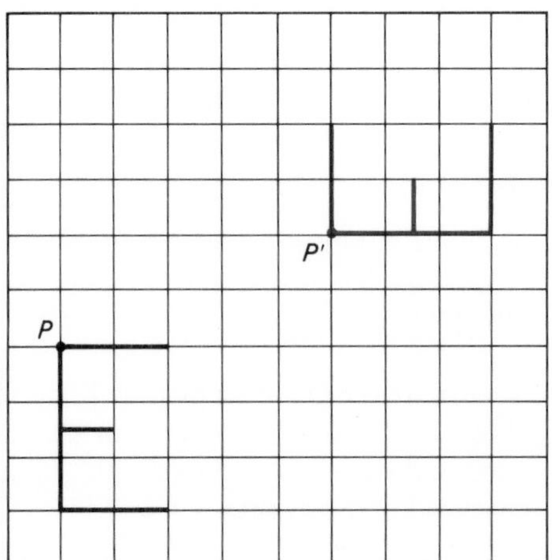

Figure 21

Note that a translation of 5 cm to the right and a translation of 4 cm to the right could be accomplished by a translation of 9 cm to the right. Likewise, a rotation of 135° counterclockwise about some point P followed by a rotation of 60° clockwise about P could be accomplished by a rotation of 75° counterclockwise about P. This suggests that a translation followed by a translation is always a translation, and that a rotation followed by a rotation is a rotation. Is the same true of reflections?

Example 13 • In Figure 22, a translation of 6 units up, followed by a rotation of 90° clockwise about the image of B, followed by a translation of 1 unit to the right pro-

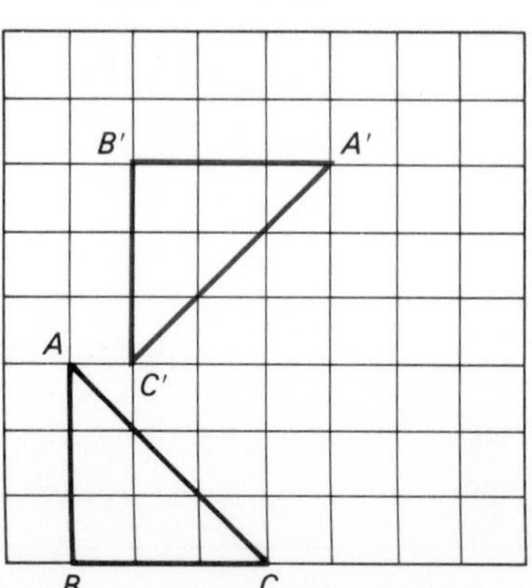

Figure 22

Just for fun *Can this figure be obtained by rotations of a triangle? Explain.*

duces image $\triangle A'B'C'$ from $\triangle ABC$. Notice that the same image $A'B'C'$ could have been obtained from ABC by other sequences of transformations. •

Example 14 • In Figure 23(a), reflect the triangle about line l so as to produce the dotted figure. Then translate it in the direction and distance indicated on line m in order to get the shaded figure. •

Example 15 • What motions have been performed in Figure 23(b) to produce the shaded figure? (Note the dotted figure given to help you determine the intermediate motion.)

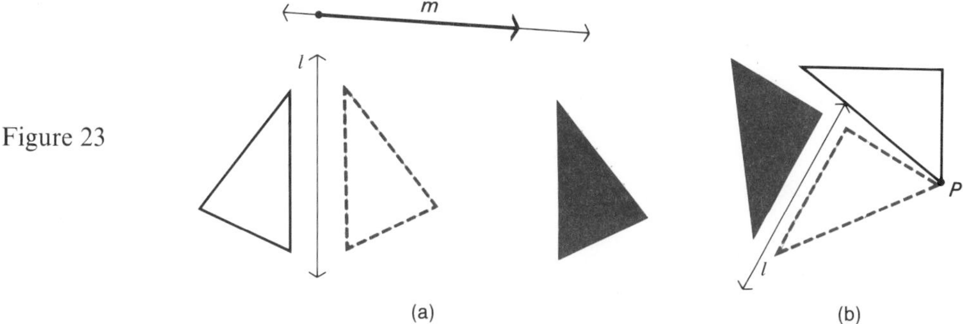

Figure 23

Solution After studying the figure, you should realize that a rotation of 60° counterclockwise about the point P and then a reflection about line l have occurred. •

Example 16 • Study the transformations that map the word MATH into the image shown in Figure 24. Do you see that this can be accomplished by a translation of 3 units to the right followed by a reflection about line l? •

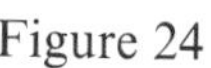

Figure 24

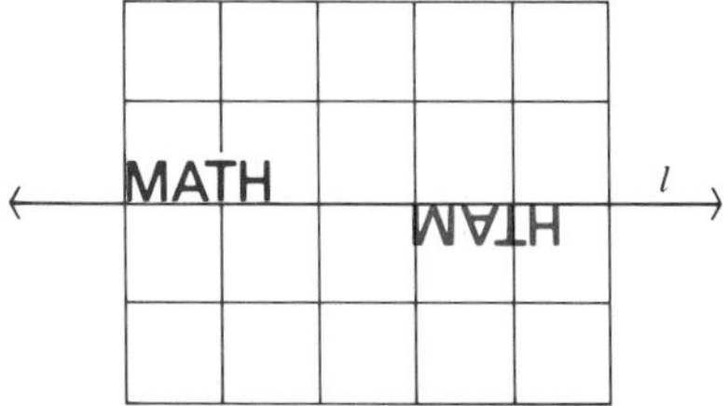

The preceding example introduces the following definition.

Glide reflection

A translation followed by a reflection about a line parallel to the translation is called a *glide reflection.*

Exercise set 3

R 1. Trace the following figures on your paper. Then turn each figure about the given point through the rotation indicated by the turn angle.

(a) 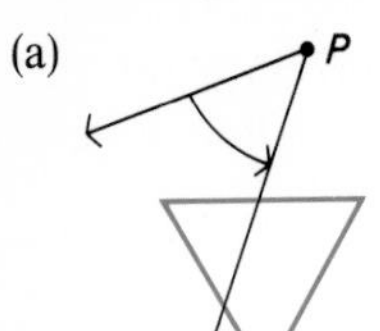

(b) 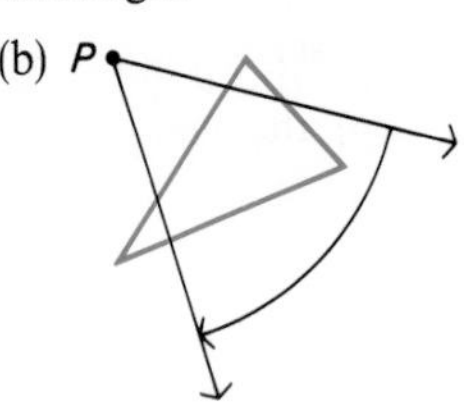

(c) 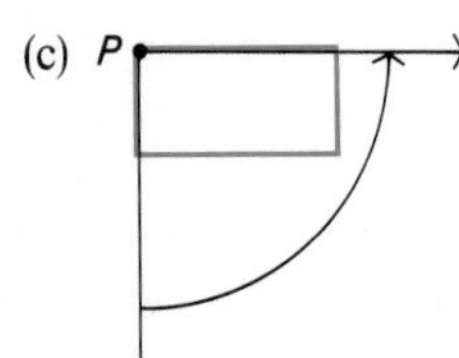

(d) 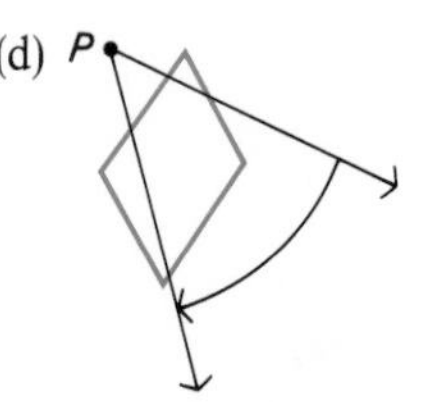

2. Rotate each figure 120° counterclockwise about P.

(a) 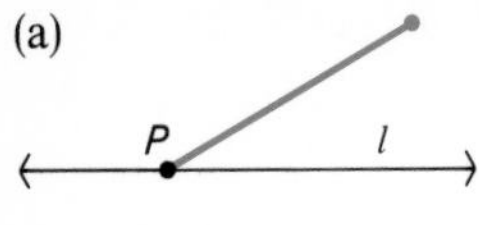

(b) 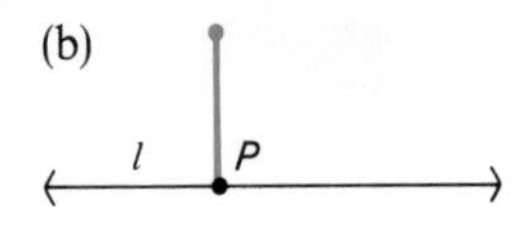

(c) 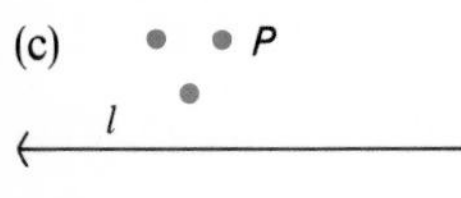

(d) 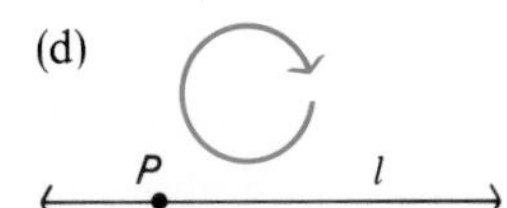

(e) 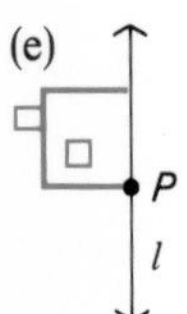

(f) 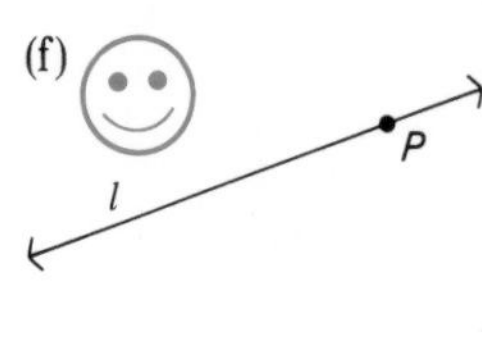

3. Sketch the image of the figure under the rotation shown.

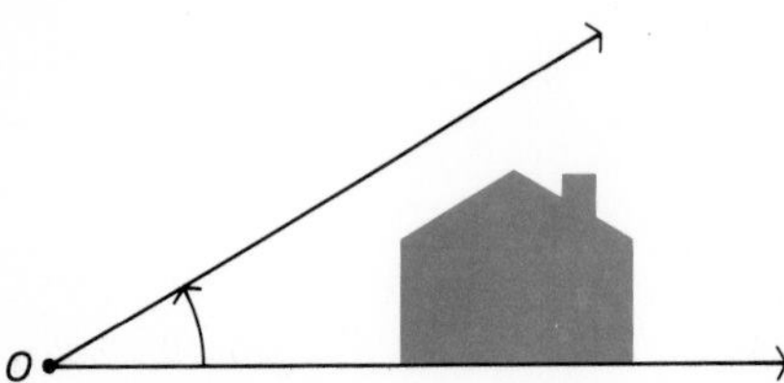

4. Rotate each of the following figures 120° counterclockwise about point A.

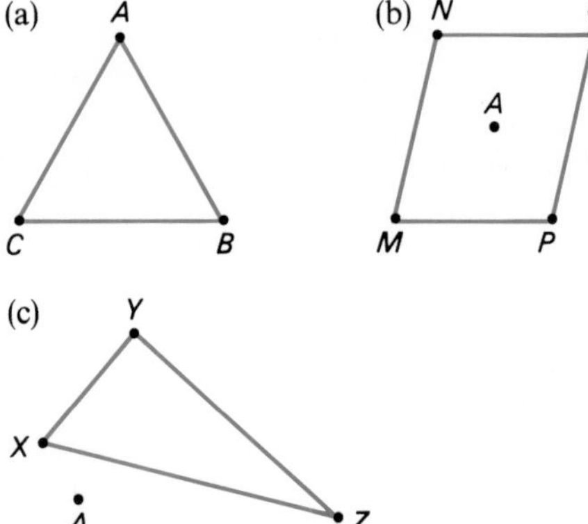

5. Draw an image for the letter H with the following transformation. Rotate H 180° clockwise about P.

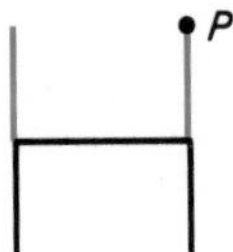

6. In each part, perform in succession the two slides in the direction and distance indicated by arrow diagrams. Does the order of making a slide make a difference?

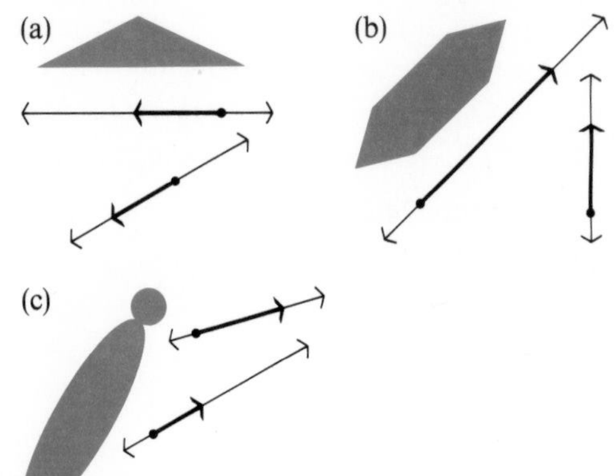

7. Perform in succession the two turns (order not important) for each figure shown, both of which occur about the same vertex V.

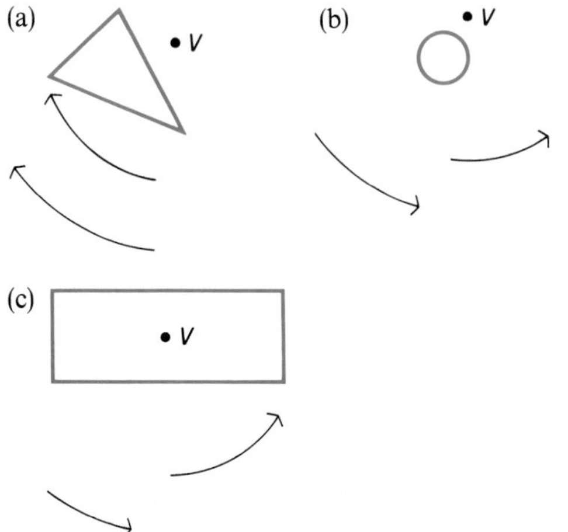

8. Perform two consecutive flips about the line of reflection given in each figure.

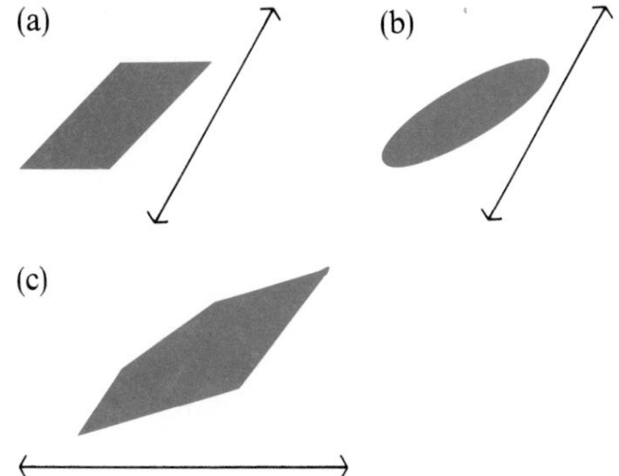

T 9. In each pair of figures shown, the shaded figure is the image of the other under a translation, reflection, or rotation. Identify the motion that has occurred.

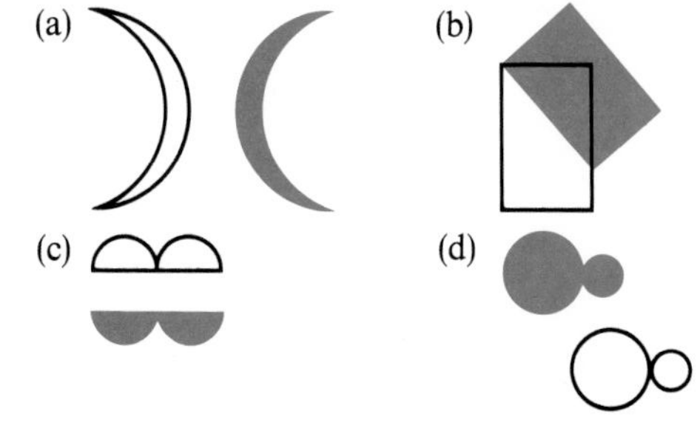

10. Determine whether a translation, reflection, or rotation has been performed in the following figures.

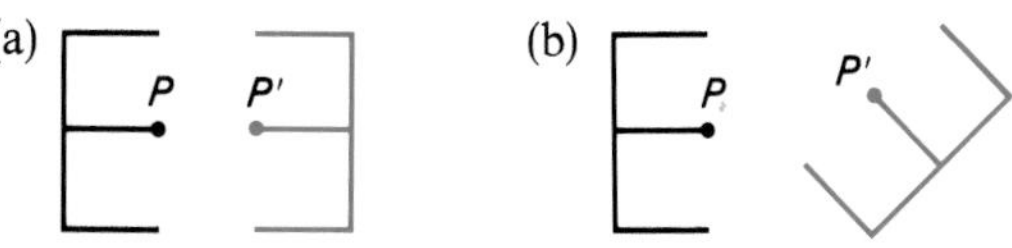

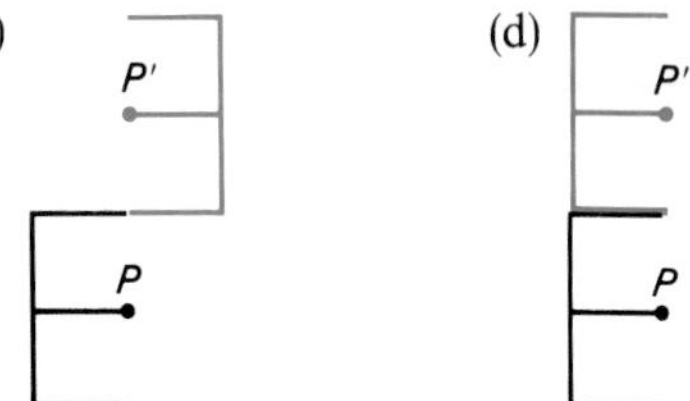

11. Describe the motion that will produce
    (a) the image on the right from the one on the left.
    (b) the image on the left from the one on the right.

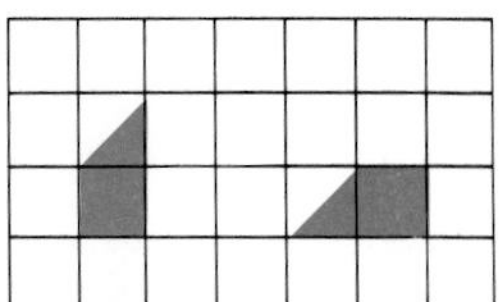

12. Draw two parallel lines in a vertical direction. Rotate the parallel lines 60° counterclockwise about a point 2 cm to the right of the right line. Are the images parallel?

C 13. The quadrilaterals in the figure are images under two reflections about lines *m* and *n*. Find a rotation that will accomplish the same result. What is the point of rotation?

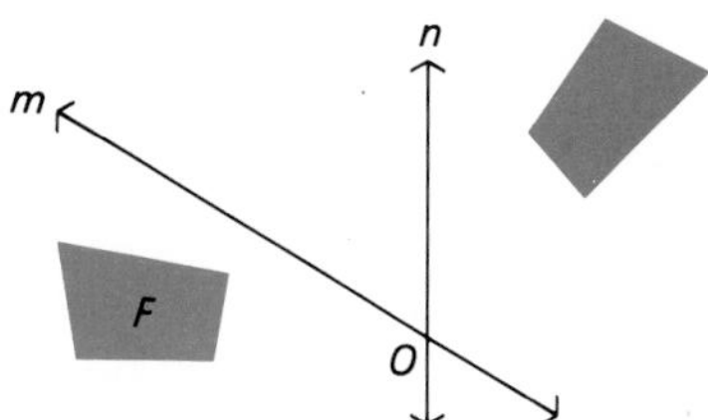

14. Draw the in-between image to show a reflection followed by a rotation to give image $\triangle A''B''C''$ from $\triangle ABC$.

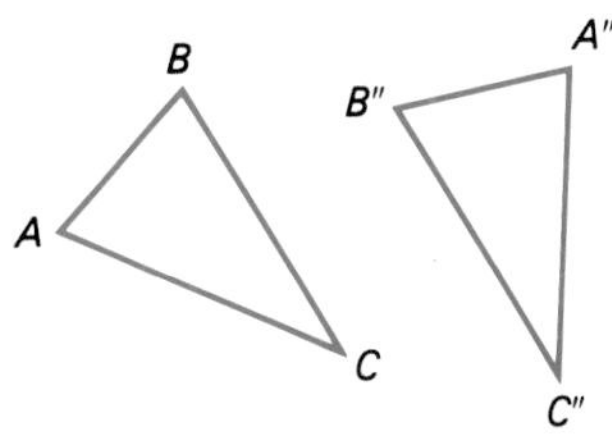

15. Draw a 30° angle counterclockwise from the vertical. Rotate the angle 60° clockwise about a point 2 cm to the right of the vertex of the angle. Do you believe that angles and their measures are preserved under rotations?

*4 Patterns in nature and art

■ Have you noticed geometric figures as they occur in nature? In this section, we shall call attention to a few of many examples. Triangles are found on a cross section of the gem tourmaline [Figure 25(a)], and regular polygons are seen in a cross section of cadmium sulfide crystals [Figure 25(b)]. ■

Figure 25

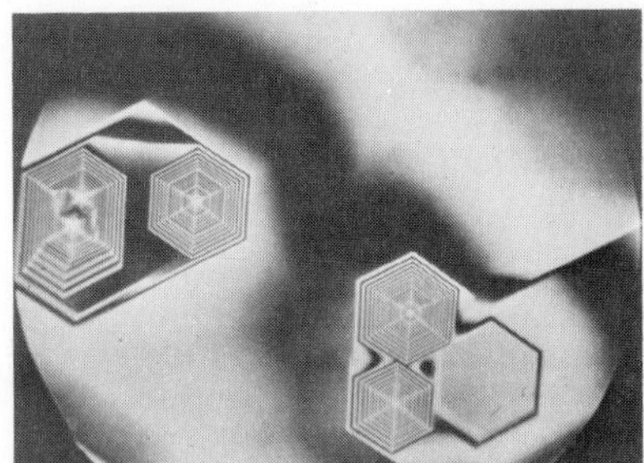

Likewise, geometry makes a contribution to art. The relationship between geometry and art usually appears in three ways—proportion, symmetry, and subject matter. All of these will be considered in this section.

Consider the five rectangles in Figure 26. Which of these seems most pleasing to your eye? If you were an early Greek, you would choose rectangle (c) because the ratio of length to width is equal to the so-called **golden ratio** (approximately 1.6). Greek artists felt that artworks employing the golden ratio were pleasing to the eye and conveyed a sense of serenity and harmony. Among the Greek artworks in which the golden ratio was used are the Parthenon in Athens [Figure 27(a)] and the famous statue of Apollo of Belvedere [Figure 27(b)].

Figure 26

Think back to your primary-school days of cutting paper hearts. Did you have difficulty making the two sides match? Finally you resorted to folding your paper, drawing one side of the heart, cutting both sides of paper along the design, and opening the fold to have a perfect heart. The fold line is the **line of symmetry** of this heart (see Figure 28). Formally, this type of symmetry is called **reflectional symmetry.** A figure is a reflection of another figure about a line l if corresponding to each point A of one there is a point A' of the other such that $\overline{AA'}$ is perpendicular to the line l at some point M and $AM = A'M$. (See Figure 28.)

This intuitive approach to symmetry can now be given meaning with the following mathematical definition.

(a) The Parthenon (448 – 432 B.C.)

(b) Statue of Apollo of Belvedere

Figure 27

Symmetry

> A *symmetry* of a figure is a transformation of the figure for which the image is the original figure.

Figure 28

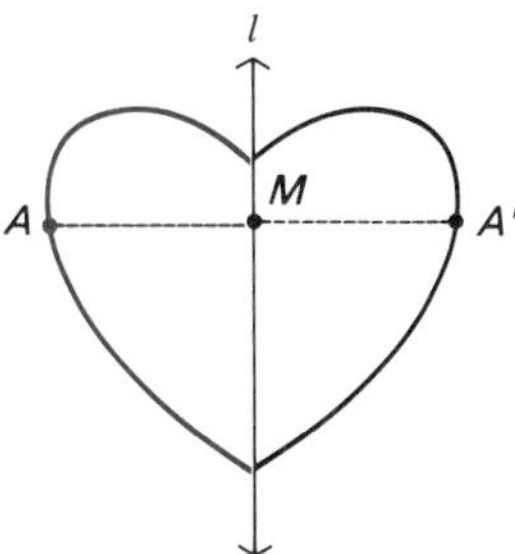

Some figures have more than one line of symmetry. To produce a figure with two lines of symmetry, fold a paper and then fold the crease back over itself, cut out the desired figure, and then unfold. The two perpendicular creases are lines of symmetry or reflection. (See Figure 29.)

Figure 29

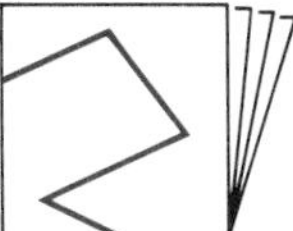

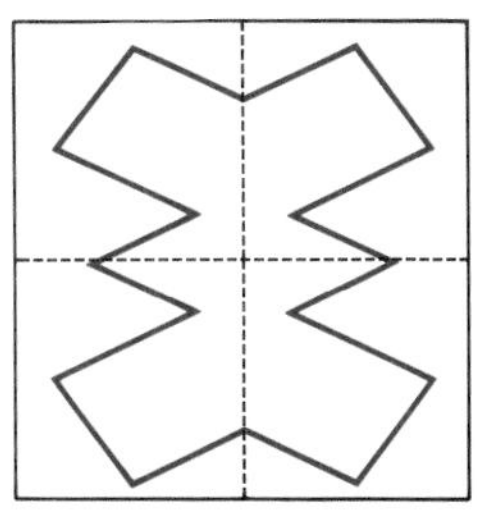

There are many examples of lines of symmetry in nature. Do you see a line of symmetry for the butterfly and the beetle in Figure 30?

Figure 30

Example 17 • Line l is a line of symmetry for Figure 31. The figure on one side of the line is a reflection of the figure on the other side of the line. •

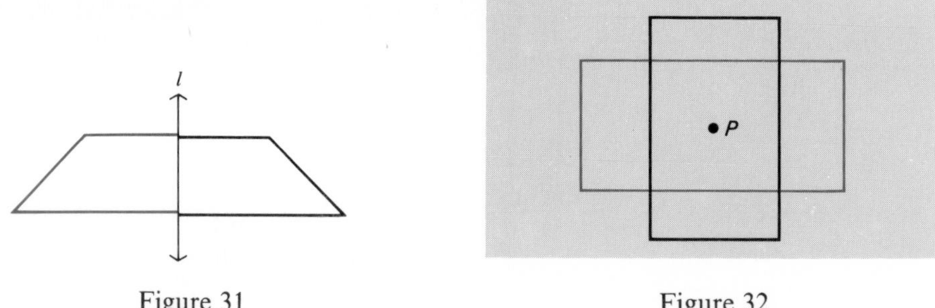

Figure 31

Figure 32

Figures may also possess a second type of symmetry called **rotational symmetry.** Rotational symmetry involves a central point of symmetry, which might be called a *point of balance.* A figure has a rational symmetry if there is a rotation (of less than 360°) about the point of symmetry that moves the figure onto itself. The following example uses the tracing method to help determine a point of symmetry.

Example 18 • Trace the rectangle in Figure 32 on another sheet. Determine the central point and call it P. Test P as a point of symmetry by rotating the tracing of the rectangle about the point P. If the tracing can be turned so that the original and the tracing coincide in less than a full rotation, then the point P is a point of symmetry.

Solution Notice that the original rectangle and the tracing do not coincide in Figure 32 when the tracing of the rectangle has been rotated 90°. However, the rectangle and its tracing do coincide after a rotation of 180°. Thus this rectangle has 180° rotational symmetry about point P. •

There are many examples in nature of rotational symmetry. In *Snow Crystals* [Bentley & Humphrey, 1931] are found more than 2000 pictures of snowflakes that display examples of both rotational and reflectional symmetry. Find some 60° rotational symmetries in the picture of snowflakes in Figure 33 and the portion of the mosaic window in Figure 34.

Figure 33

Figure 34

Example 19 • Determine whether the two figures in Figure 35 have rotational symmetry. If so, how much rotation is involved in each?

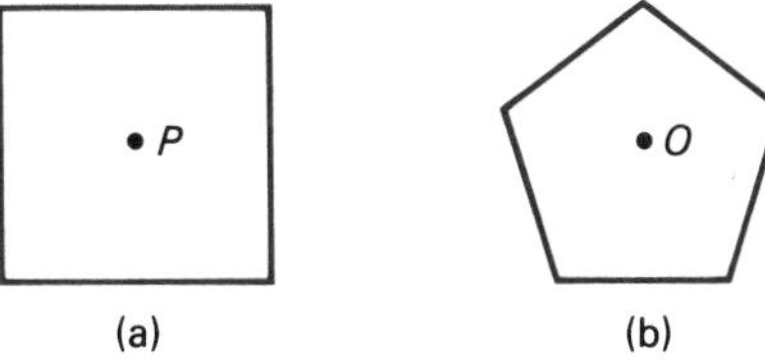

Figure 35

Solution The square has 90° rotational symmetry, and the pentagon has 72° rotational symmetry. •

The geometric concepts of symmetry and proportion are always important in artistic design, but there are times when geometric patterns become the content of the art. For instance, paintings by artists such as M. C. Escher have intrigued mathematicians because of their emphasis on geometric form. Escher made use of tessellations in Figure 36. Note the repetition of the drawing of a man on a horse.

Just for fun

This pretty Swiss miss is extremely clever at working geometrical cutting puzzles. She has discovered a way of cutting the piece of wallpaper in her right hand into two pieces that will fit together to form the Swiss flag held in her left. The white cross in the center of the flag is actually a hole in the paper. The cutting must follow the lines ruled on the paper. Can you discover her secret?

Tessellation

A *tessellation* is a pattern of shapes (obtained by reflections, translations, and rotations) that covers the plane without overlapping the component parts and without leaving gaps.

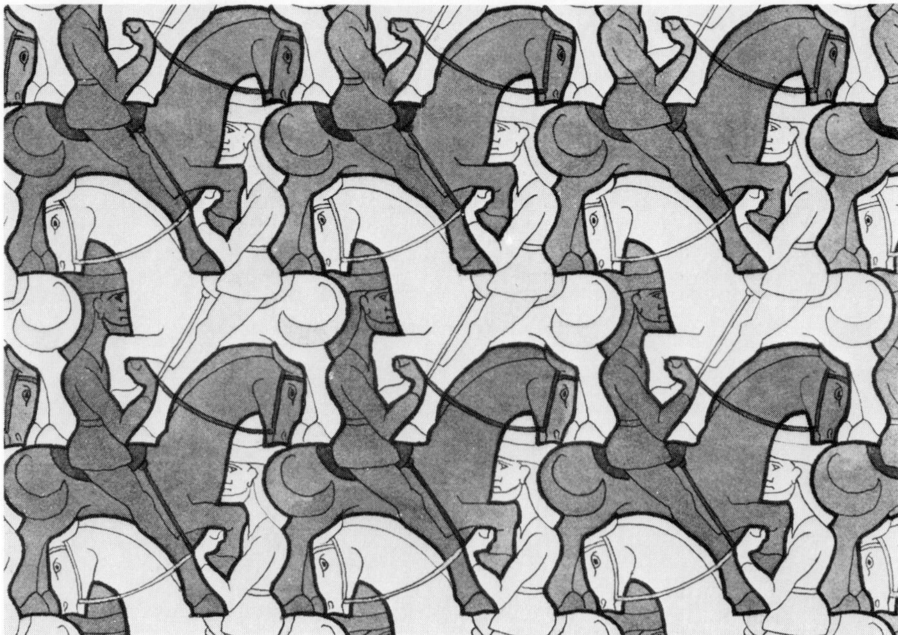

Figure 36

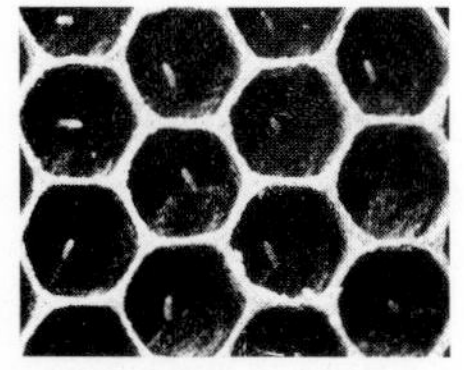

Figure 37

An excellent example can be found in the geometry of a beehive. The picture of a cross section of a honeycomb in Figure 37 is a form of a tessellation.

Tessellations have been used as patterns for rugs, paintings, fabrics, pottery, and architecture since ancient times, as illustrated in Figure 38. The Moors of Spain were quite proficient in decorating walls and floors with tessellations using colored tiles.

Figure 38

Example 20 • The tessellation in Figure 39(a) is a pattern from a floor tile. The basic element is a square tile.

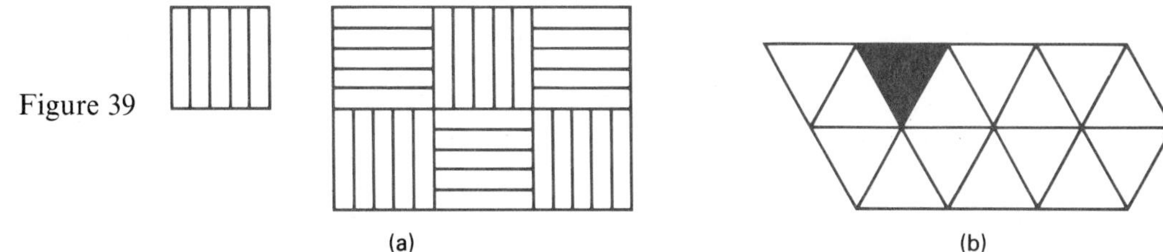

Figure 39

(a) (b)

The tessellation in Figure 39(b) is another example of a repetition of a polygonal figure. In this figure, the repeated element is an equilateral triangle, one of which is shaded. The tessellation is formed by a 60° rotation of the triangle about one of its vertices. •

Exercise set 4

R 1. In the given sketches, find those cases in which l appears to be a line of symmetry.

(a)

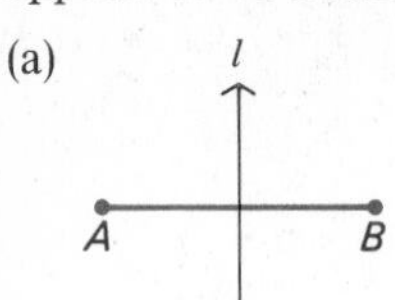

(b)

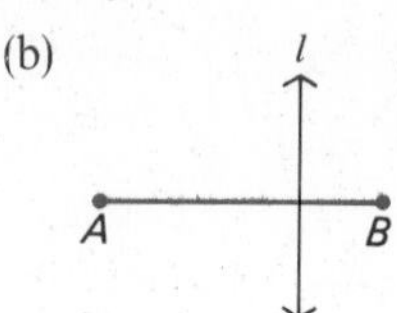

(c)

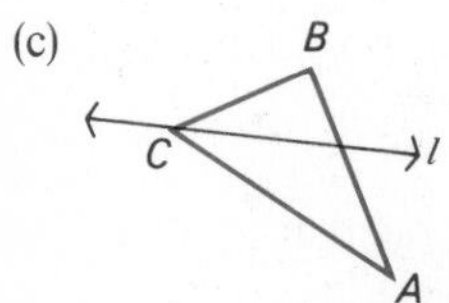

(d)

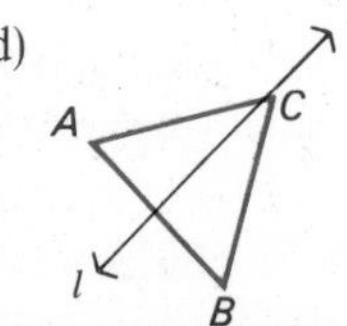

2. In the given sketches, decide which figures have a line of symmetry. Draw all lines of symmetry for these figures.

(a)

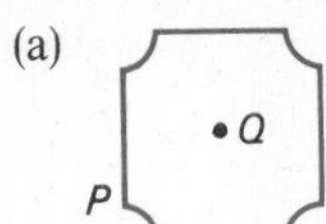

(b)

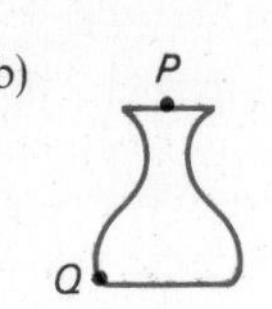

(c)

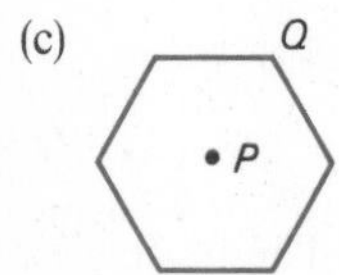

(d)

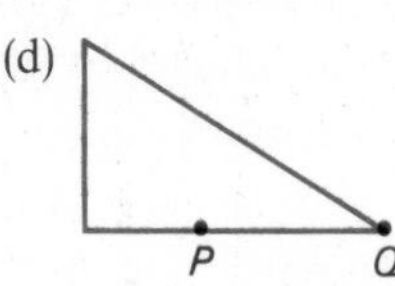

3. Determine which figures in Exercise 2 have a rotational symmetry about P. About Q.

4. How many lines of symmetry are there for each of the following regular polygons? How many degrees are there in the smallest rotational symmetry?

(a) Triangle (b) Square (c) Pentagon
(d) Hexagon (e) Heptagon (f) Octagon

5. (a) Which of the letters that follow have two lines of reflection?
(b) Which have rotational symmetry?
(c) Which have two lines of reflection but no rotational symmetry?

A B H I M N O P T X Z

T 6. With which of the following polygons can one tessellate the plane? Draw a portion of the plane to show the tessellations.

(a)

Parallelogram

(b)

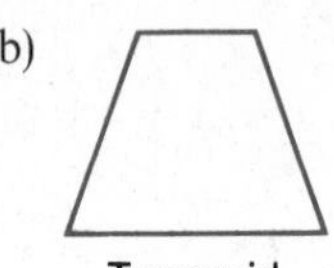

Trapezoid

(c)

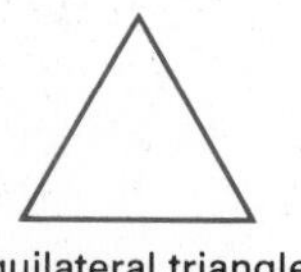

Equilateral triangle

(d)

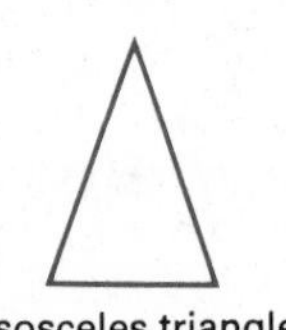

Isosceles triangle

(e)

Pentagon

(f)

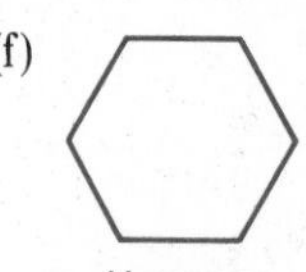

Hexagon

(g)

Kite

(h)

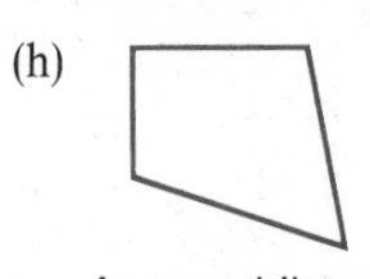

Any quadrilateral

(i)

Scalene triangle

(j) (k)

Nonconvex quadrilateral Heptagon

7. A tessellation is a *regular tessellation* if it is constructed from regular polygons of one size and shape such that each vertex figure (the polygon formed by connecting midpoints of edges emanating from a given vertex) is a regular polygon.

(a) What is the vertex figure of a tessellation of squares?

(b) What is the vertex figure of a tessellation of equilateral triangles?

(c) What is the vertex figure of a tessellation of hexagons?

8. An equilateral triangle is called a *reptile* (abbreviation for repeating tile) because four equilateral triangles can be used to form a larger equilateral triangle. Which of the following is a reptile? Demonstrate your answer.

(a)  (b)  (c) 

(d)  (e) 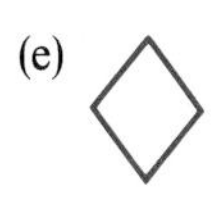 (f) 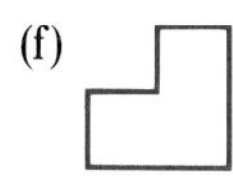

9. How many lines of symmetry do you note in the following figures? What is the smallest rotational symmetry?

(a)

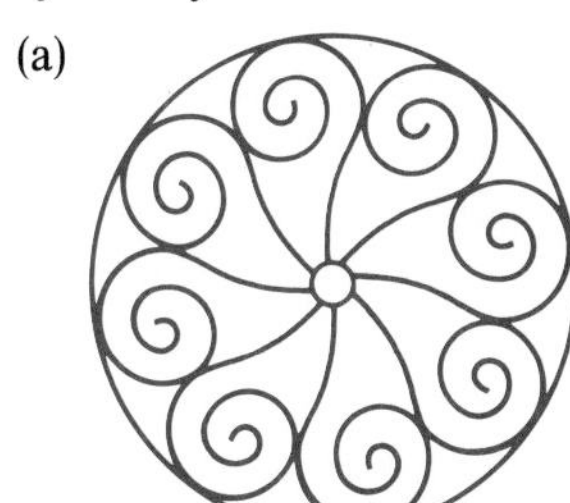

(b)

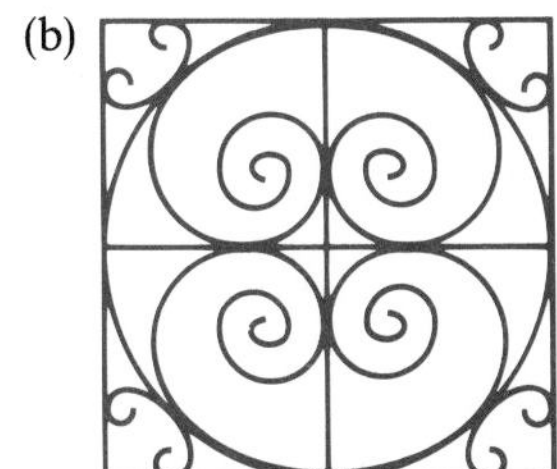

(c)

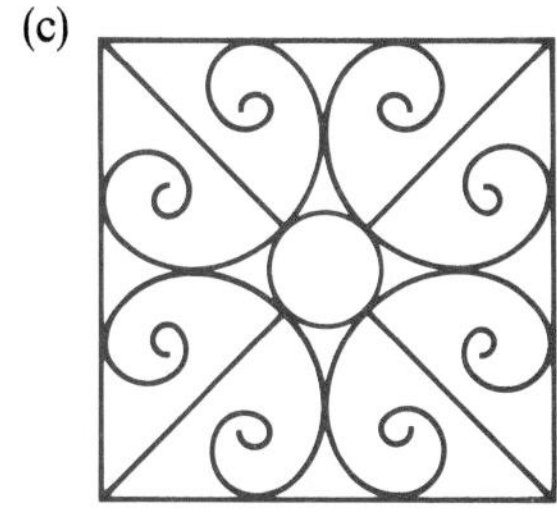

10. Suppose that the following design covers the entire plane. Find at least four transformations that will map the design onto itself.

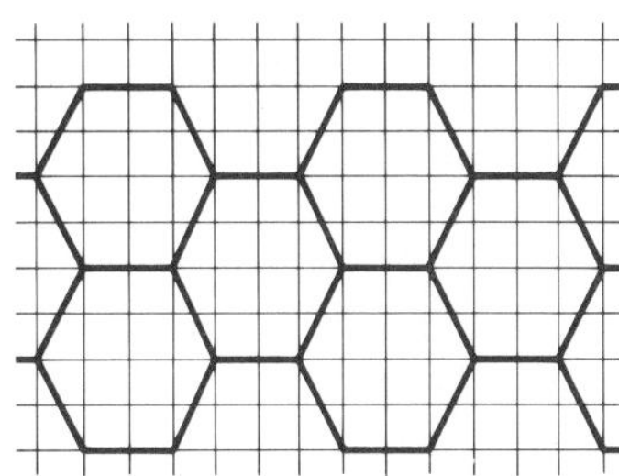

11. Sketch figures with the following qualities.

(a) Exactly one line of symmetry
(b) Exactly two lines of symmetry
(c) Exactly three lines of symmetry
(d) More than three lines of symmetry
(e) Rotational but not reflectional symmetry
(f) Reflectional but not rotational symmetry

C 12. Another way to form tessellations is to construct circles with a compass such as those given. Join with the shortest line segments the points where the circles intersect.

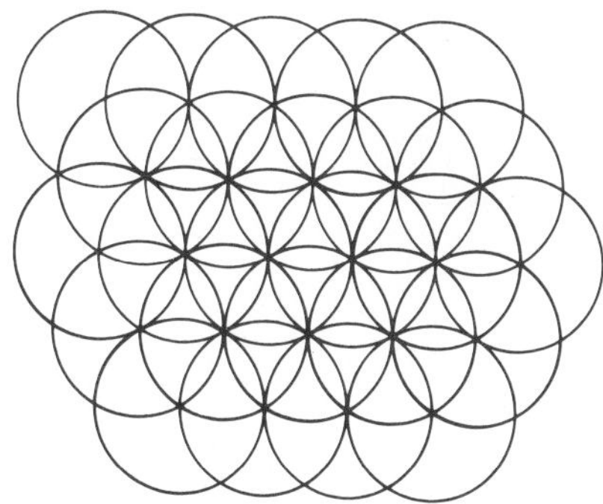

Do you see a tessellation of

(a) triangles? (b) squares?
(c) pentagons? (d) hexagons?
(e) heptagons? (f) octagons?

13. Which of the following regular polygons will form a tessellation by a rotation about a vertex?

(a) Triangle (b) Square (c) Pentagon
(d) Hexagon (e) Heptagon (f) Octagon
(g) Nonagon (h) Decagon (i) Dodecagon

5 Transformations and congruence

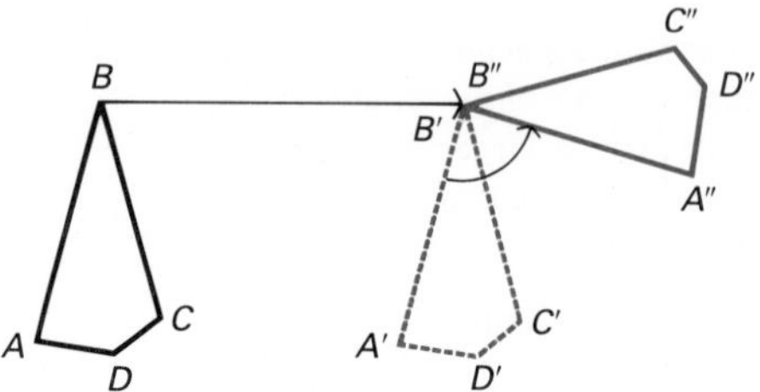

■ Polygon $A'B'C'D'$ has been obtained from polygon $ABCD$ by a translation of length BB'. Polygon $A''B''C''D''$ has been obtained from polygon $A'B'C'D'$ by a rotation counterclockwise through $\angle A'B'A''$ about B'. You can demonstrate with a compass that the sides of the polygon $A''B''C''D''$ have the same measure as found in $ABCD$. With a protractor, you can demonstrate that angle measurements are the same in $A''B''C''D''$ as in $ABCD$. This experiment provides an introduction to the concept of congruence. ■

To sharpen your intuition, consider the tracing of a given figure. *A* **tracing** *is made by placing a thin sheet of paper over a figure and moving a pencil over every line of the figure so that an exact copy of the figure is made on the thin paper.*

Briefly study the stick figure in Figure 40(a). Take a thin sheet of paper and place it on top of the figure. Trace the figure on the new sheet of paper by going over every line. Notice that, as the figure is traced, the "copy" of the figure is exactly the same size and shape as the original. Every line and point matches. The stick figure, Figure 40(b), can be thought of as a trace of the given figure. Tracing figures provides the groundwork for the study of congruent figures.

Figure 40

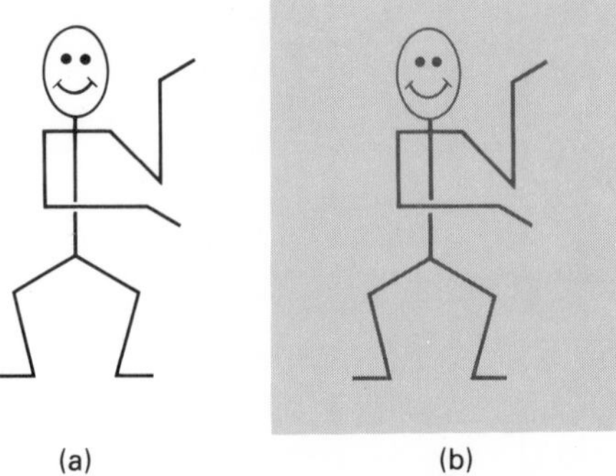

(a) (b)

Is it now intuitively obvious that tracing a figure and moving the tracing is nothing more in most cases than translating the figure? Sometimes the tracing must be rotated, and sometimes the tracing must be turned over (reflected), but the movement of the tracing can be described by either a translation, a reflection, or a rotation.

Congruent figures

> Two figures A and B are said to be congruent if and only if there is a composite of reflections, translations, or rotations that maps figure A onto figure B or figure B onto figure A.

Since reflections, translations, and rotations preserve linear measure, angle measure, collinearity, and betweenness, we say that two figures, or two collections of points, are congruent if and only if they have the same size and shape.

Example 21 • $\overline{AC} \cong \overline{A'C'}$ in Figure 41(a), because $\overline{A'C'}$ is the image of $\overline{AC}$ under a rotation about point C or a reflection about line m. Intuitively, if a trace were made of $\overline{AC}$, you could place this trace on $\overline{A'C'}$ and it would coincide point by point. Likewise, with a compass you can demonstrate that $AC = A'C'$ •

Figure 41

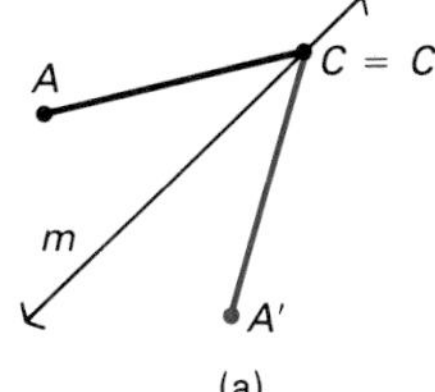

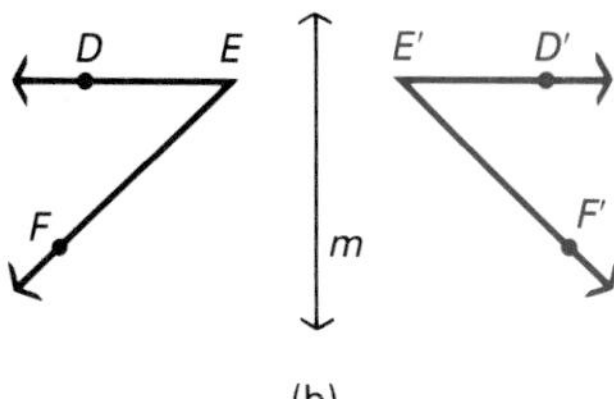

Example 22 • $\angle DEF \cong \angle D'E'F'$ in Figure 41(b), because $\angle D'E'F'$ is the image of $\angle DEF$ under a reflection. Again, you can intuitively show this congruent relationship with a tracing. •

In Euclid's *Elements,* if two triangles are congruent, they can be made to coincide. That is, one can be placed exactly on the other. A special case of the above definition is the modern way of saying the same thing as Euclid said.

Congruent triangles

> Two triangles are congruent if and only if there is a composite of reflections, translations, or rotations that maps one onto the other.

Example 23 • In Figure 42, $\triangle ABC \cong \triangle A'B'C'$ because $\triangle A'B'C'$ can be obtained from $\triangle ABC$ by a composite of a reflection and a rotation. •

Figure 42

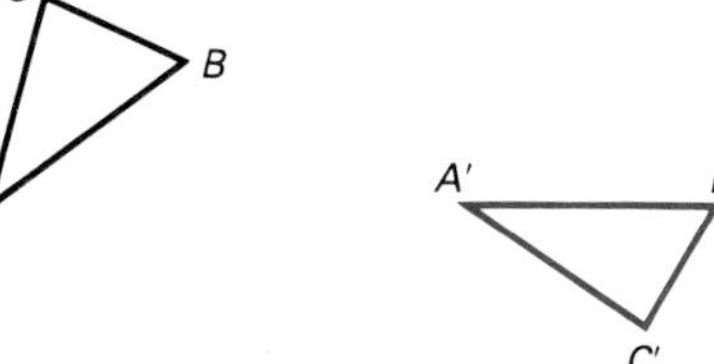

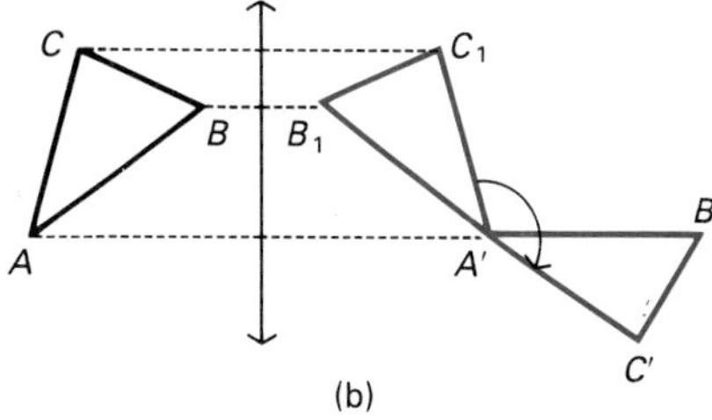

Just for fun *How many squares can you find in the figure? How many are congruent to the one containing four small squares?*

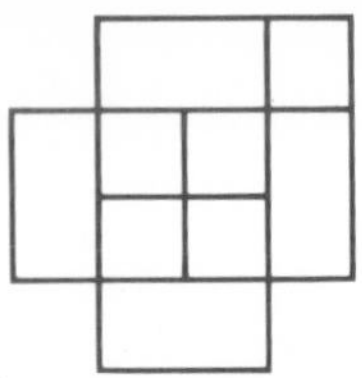

Characteristics of congruent figures, which we will use in the next chapter, are listed as follows.

Congruent figures

(a) Congruent segments have the same length.
(b) Congruent angles have the same measure.
(c) Corresponding sides of congruent figures are congruent.
(d) Corresponding angles of congruent figures are congruent.

Recall that rotations and translations are a composite of two reflections. A glide reflection can be considered as a composite of three reflections (two for the translation and one for the reflection). A composite of two or more translations or two or more rotations can be written as one translation and one rotation. This leads to the following summary.

Given any two congruent figures in a plane, one can be obtained from the other by a composite of at most three reflections.

Exercise set 5

Use your compass on Exercises 1 through 2.

R 1. Draw line segment $\overline{AB}$. Is $\overline{AB} \cong \overline{AB}$? Is $\overline{BA} \cong \overline{AB}$?

2. Consider the line segments $\overline{AB}$, $\overline{CD}$, and $\overline{EF}$ as given.

Is $\overline{AB} \cong \overline{CD}$? Is $\overline{CD} \cong \overline{EF}$? Is $\overline{AB} \cong \overline{EF}$?

3. Which pairs of figures are congruent?

(a)

(b)

(c)

(d)

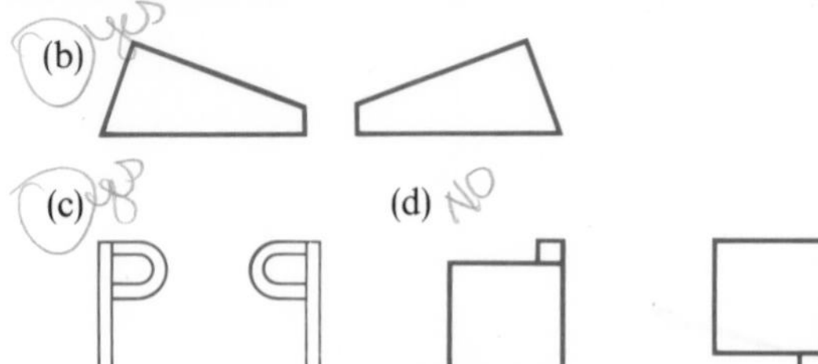

4. Classify each of the following statements either as true or as false.

(a) If $\overline{AB}$ is congruent to $\overline{AD}$, then B and D name the same point.
(b) If $\overline{AB} \cong \overline{BC}$ and $\overline{BC} \cong \overline{CD}$, then $\overline{AB} \cong \overline{CD}$.
(c) If $\overline{AC} \cong \overline{AD}$, then C and D name the same point.
(d) The union of two different segments may be congruent to one of the segments.

(e) Two congruent figures have the same shape.
(f) If segments $\overline{AB}$ and $\overline{BC}$ are congruent, then they are of equal length.

5. For the given figure, complete each blank.

(a) $\overline{CD} \cong$ ________
(b) $\angle ADC \cong$ ________
(c) $\triangle ABC \cong$ ________
(d) $BD =$ ________

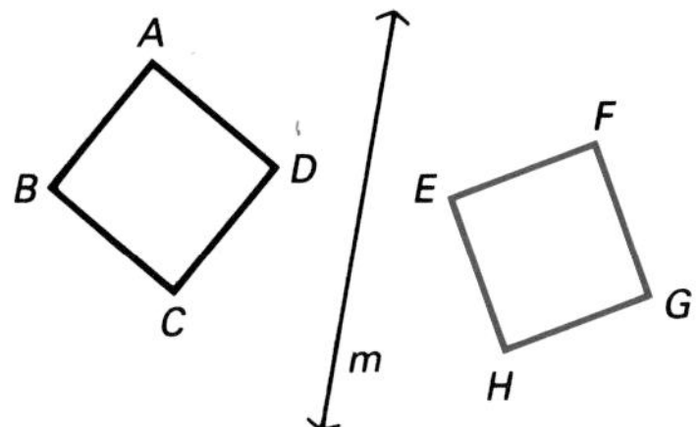

6. Which of the following statements involve incorrect notation? Explain.

(a) $m(\angle A) \cong m(\angle B)$
(b) $\angle C > \angle D$
(c) $\angle E \cong \angle F$
(d) $m(\angle E) = m(\angle F)$

7. Describe one composite of reflections, translations, or rotations that maps each of these designs onto itself.

(a)

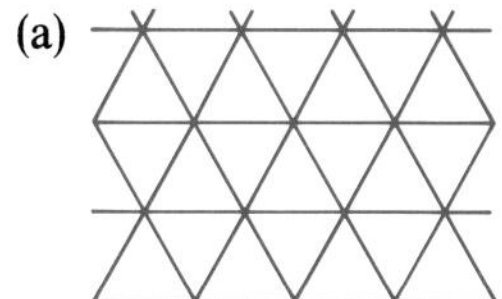

(b)

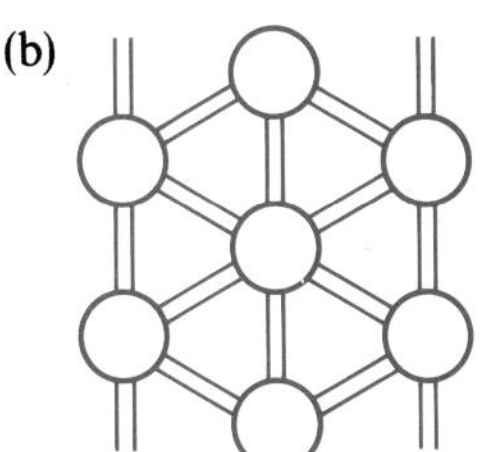

(c)

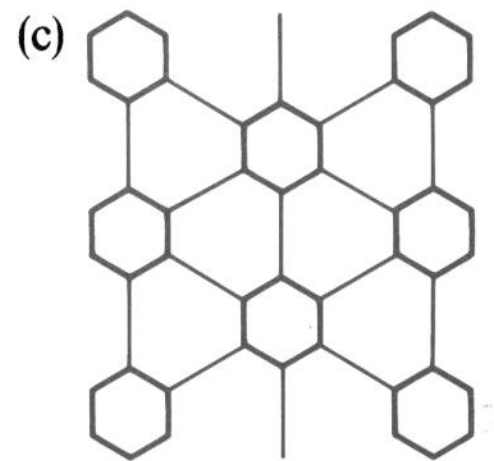

8. Given $\triangle ABC \cong \triangle DEF$, find one composite of reflections, translations, or rotations that will map $\triangle ABC$ onto $\triangle DEF$.

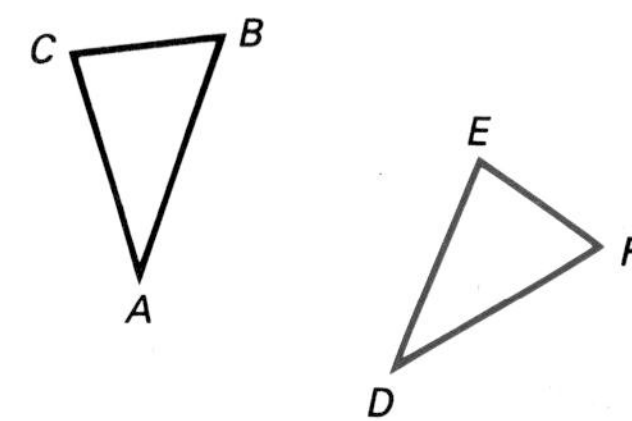

9. For each pair of congruent figures, name one composite of reflections, translations, or rotations that will map one figure onto the other.

(a)

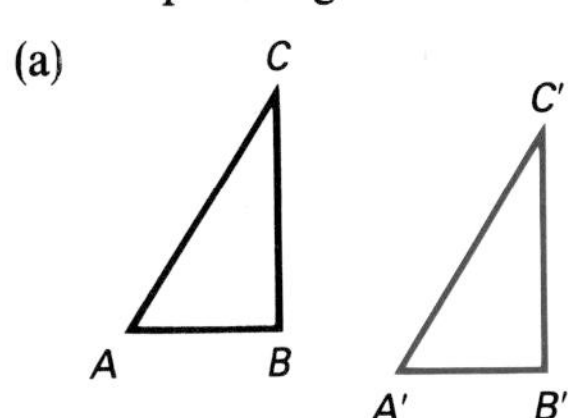

(b)

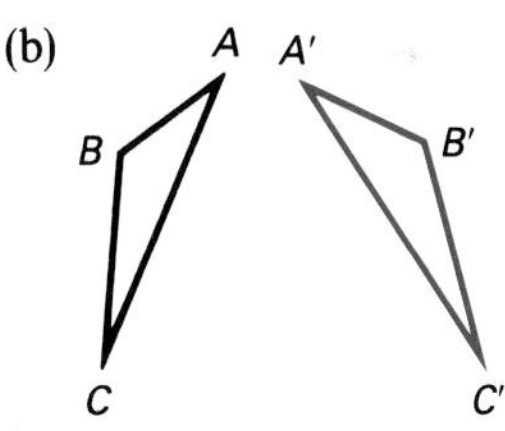

(c)

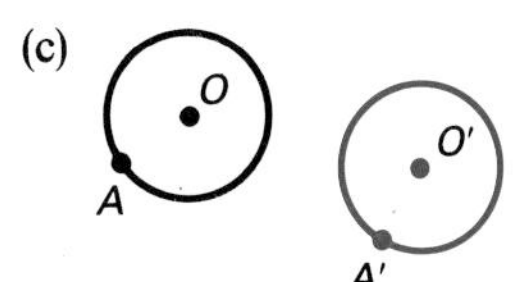

(d)

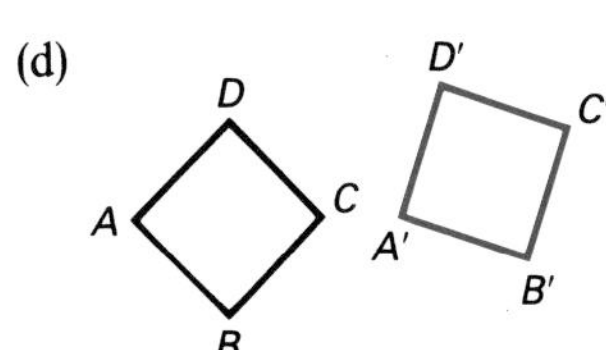

(e)

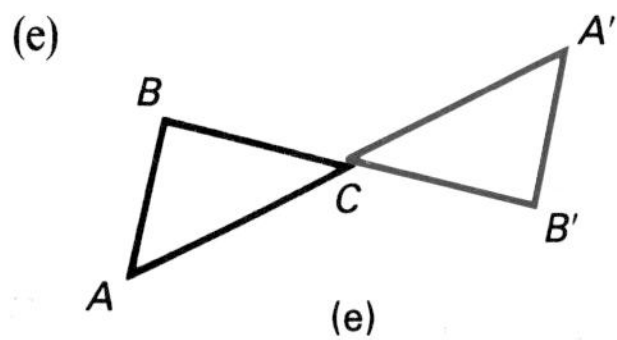

T 10. Draw two congruent triangles, $\triangle ABC$ and $\triangle A'B'C'$, such that $\triangle ABC$ can be mapped on $\triangle A'B'C'$ by a

(a) reflection (b) rotation
(c) translation (d) glide reflection

11. Draw two congruent circles, C_1 and C_2, such that one can be mapped on the other by a

(a) reflection (b) rotation
(c) translation (d) glide reflection

12. Suppose that a composite of reflections, translations, or rotations maps circle C_1 onto C_2. Prove that the radii are equal.

Prove the conclusion using an appropriate composite of reflections, translations, or rotations in Exercises 13 and 14.

C 13.

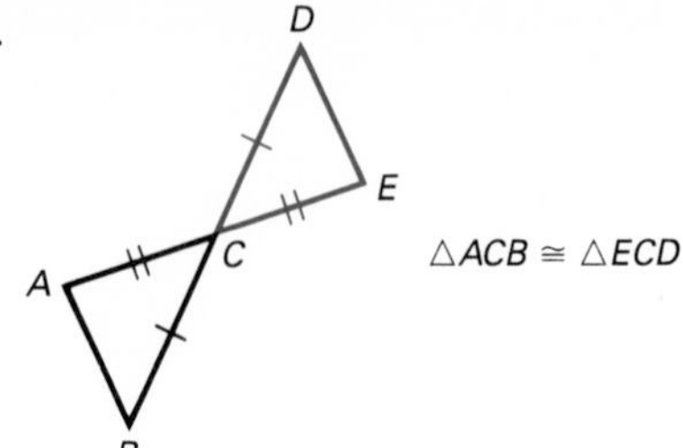

14.

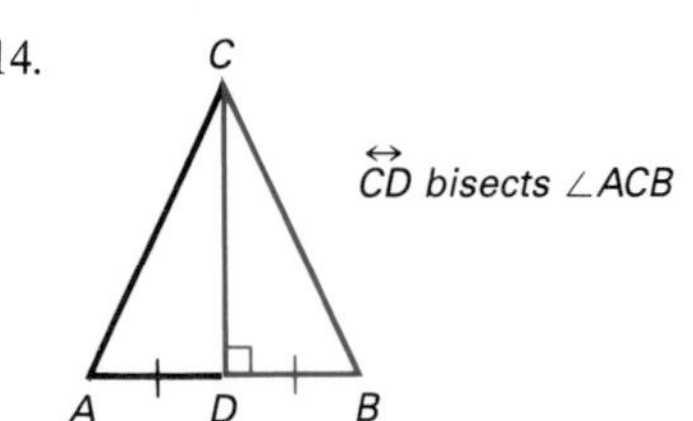

6 Similarity transformations

■ Take a pen flashlight and hold it in front of a rectangular piece of cardboard with the light ray perpendicular to the center of the cardboard. Note the shadow of the cardboard on the wall. The shadow is the same shape as the cardboard but is not the same size. We intuitively describe the two figures as being **similar.** ■

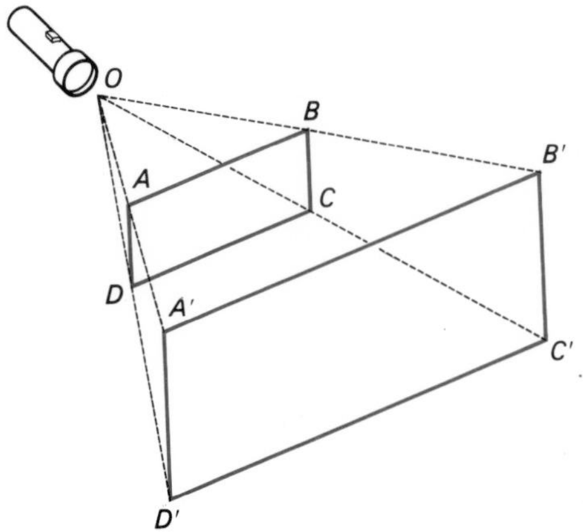

Similar figures are used repeatedly in photography, engineering, map-making, architecture, and interior design to make scale drawings. Actually, similar figures occur in the use of a microscope, a telescope, or simply a pair of reading glasses. You will study many interesting applications of similarity in this section.

The three transformations of reflection, translation, and rotation all preserve congruence; that is, they map figures into congruent figures. Let's consider a transformation that preserves shape but not size. Such a transformation is called a **similarity transformation.** We intuitively think of a similarity transformation as one that may change the size but not the shape of a figure.

Similarity transformation

A similarity is a transformation of a plane that changes distances by a constant factor k, the **scale factor** of the similarity. If A' and B' are the images of A and B, then $A'B' = k(AB)$.

The following properties hold for similarity transformations.

Properties of similarity transformations

A given similarity transformation with scale factor k
(a) maps each segment onto a segment k times as long.
(b) maps each angle onto a congruent angle.

Similarity transformations enable us to consider all kinds of similar figures.

Similar figures

Geometric figures are similar if (denoted by $\sim$) and only if there is a similarity transformation that maps one onto the other.

In many similarity transformations, we are not concerned with the numerical value of the scale factor, but simply with the fact that it exists. In Figure 43, if the two triangles are similar, we can show a correspondence or ratio between sides, determined by points A and A', B and B', and C and C'.

$$A'B' = kAB \quad \text{or} \quad \frac{A'B'}{AB} = k$$

$$B'C' = kBC \quad \text{or} \quad \frac{B'C'}{BC} = k$$

$$C'A' = kCA \quad \text{or} \quad \frac{C'A'}{CA} = k$$

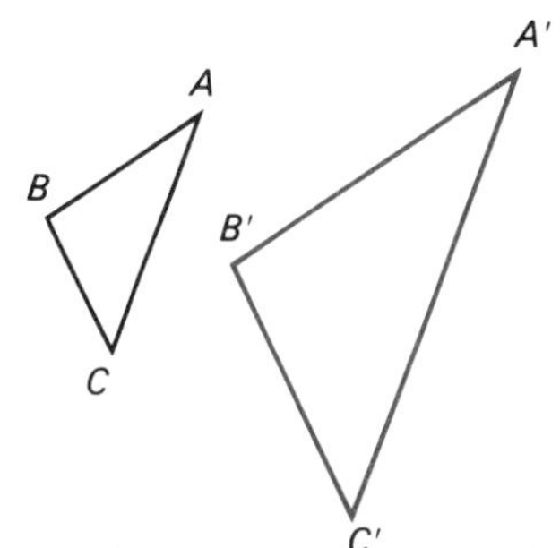

Figure 43

Therefore,

$$\frac{A'B'}{AB} = \frac{B'C'}{BC} = \frac{C'A'}{CA}$$

Example 24 • In Figure 44, the two triangles are similar with a scale factor of 2. This is shown by the fact that each segment in $\triangle ABC$ is double the corresponding segment in $\triangle A'B'C'$. •

Figure 44

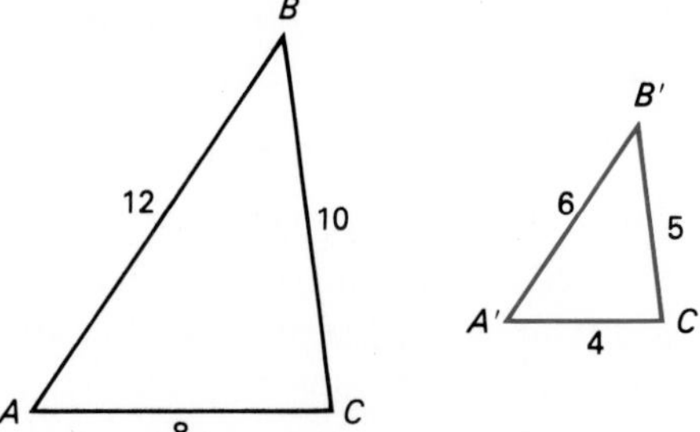

As with congruent triangles, there are special properties to use as shortcuts in showing two triangles to be similar.

Similar triangles

1. Two triangles are **similar** if and only if corresponding angles are congruent in Figure 45,

$$\triangle ABC \sim \triangle DEF \qquad \text{if and only if}$$

$$\angle A \cong \angle D, \qquad \angle B \cong \angle E, \qquad \angle C \cong \angle F$$

Figure 45

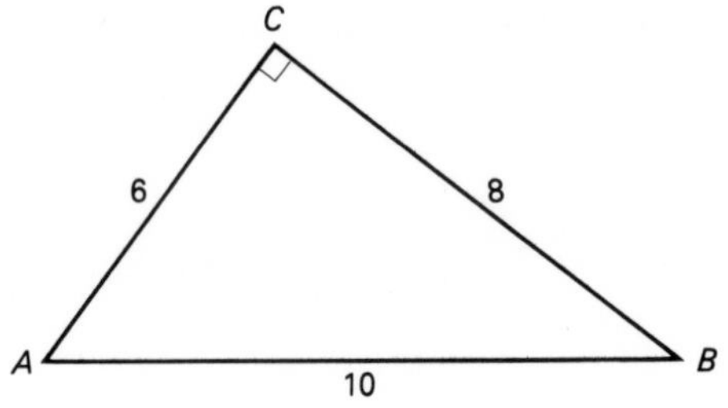

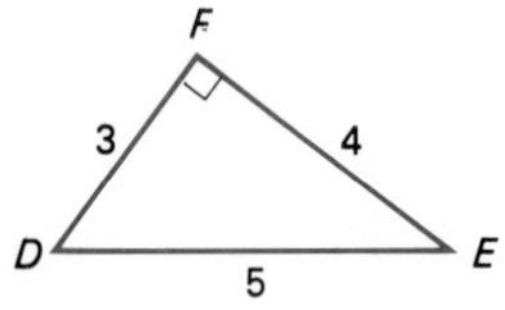

2. Two triangles are **similar** if and only if the ratios of corresponding sides are equal. In Figure 44,

$$\frac{AB}{A'B'} = \frac{BC}{B'C'} = \frac{CA}{C'A'}$$

Example 25 • Compare the similar triangles given in Figure 45. $\overline{AC}$ corresponds to $\overline{DF}$, $\overline{AB}$ corresponds to $\overline{DE}$, and $\overline{CB}$ corresponds to $\overline{FE}$. Notice that

$$\frac{AC}{DF} = \frac{AB}{DE} = \frac{CB}{FE}$$

or

$$\frac{6}{3} = \frac{10}{5} = \frac{8}{4} \quad \bullet$$

Example 26 • Suppose that $\triangle ABC \sim \triangle FGH$. (See Figure 46.)

(a) If $m(\angle A) = 40^\circ$ and $m(\angle H) = 30^\circ$, find $m(\angle B)$ and $m(\angle C)$.

(b) If $AB = 3$, $BC = 5$, $FH = 12$, and $GH = 10$, find FG and AC.

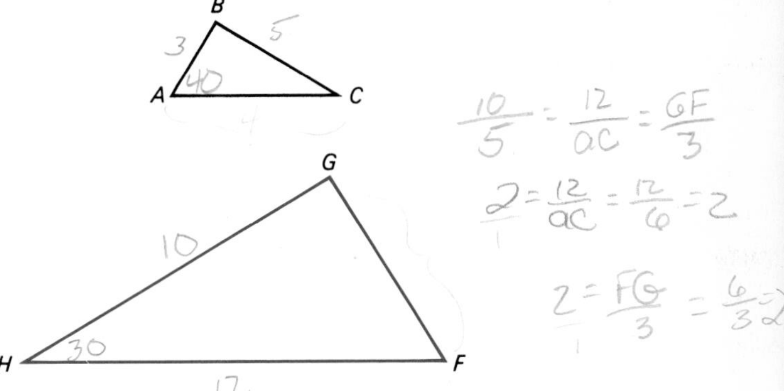

Figure 46

Solution (a) Since $\triangle ABC \sim \triangle FGH$, $\angle C \cong \angle H$. Thus $m(\angle C) = 30^\circ$. Now

$$m(\angle B) + m(\angle C) + m(\angle A) = 180^\circ$$

or

$$m(\angle B) + 30^\circ + 40^\circ = 180^\circ$$

or

$$m(\angle B) = 110^\circ$$

(b) Since $\triangle ABC \sim \triangle FGH$,

$$\frac{AB}{FG} = \frac{BC}{GH}$$

or

$$\frac{3}{FG} = \frac{5}{10}$$

so

$$FG = 6$$

Similarly,

$$\frac{AC}{FH} = \frac{BC}{GH}$$

or

$$\frac{AC}{12} = \frac{5}{10}$$

so

$$AC = 6 \quad \bullet$$

Example 27 • How high is a flagpole that casts a shadow of 24 m if a nearby post 3 m high casts a shadow of 6 m? Since the sun rays make the same angles with the horizontal for the triangles involving the flagpole and the post, and since both the flagpole and the post are assumed to make right angles with the horizontal, the two triangles are similar, as indicated in Figure 47. Thus,

$$\frac{x}{24} = \frac{3}{6}$$

or

$$6x = 72$$

or

$$x = 12$$

And so the flagpole is 12 m high. •

Figure 47

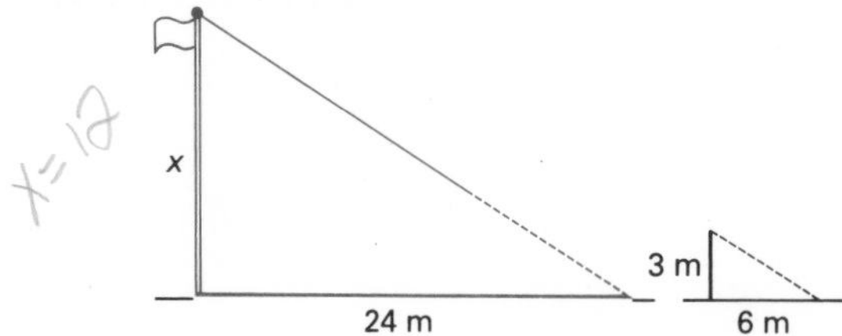

The following similarity properties of polygons come from the general definition of similarity transformations.

Similar polygons

> Two *polygons* are similar if and only if there exists a one-to-one correspondence between the vertices of the polygons such that
>
> **1.** corresponding angles have equal measures.
> **2.** ratios of the lengths of corresponding segments are equal.

Similarity of polygons is the mathematical basis of most *scale drawings.* For instance, a floor plan of a house may be drawn on a small piece of paper and yet

Just for fun

Because of optical illusions, the comparative size of different figures cannot always be visually determined. In part (a), can you determine which segment is longer by looking at them? In part (b), can you guess which set of parallel lines is longer? In part (c), which curve is longer?

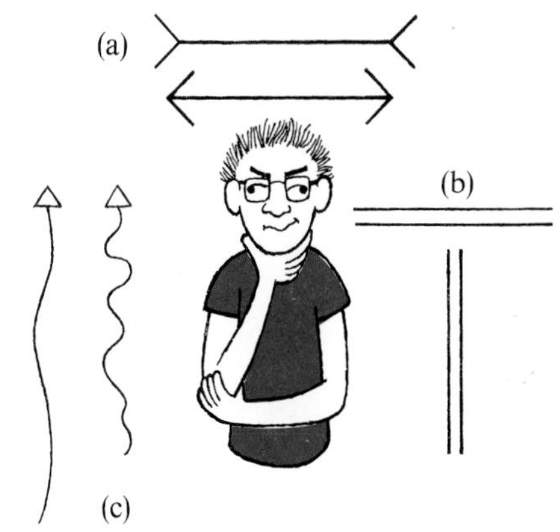

represent the shape of the house exactly. For such drawings, a scale is usually indicated; 1 cm on a scale drawing may represent 1 m in the actual measure of the house. In this case,

$$\frac{m(\text{scale drawing of a segment})}{m(\text{segment represented by scale drawing})} = \frac{1}{100}$$

The measure x of the scale drawing of a room 10 m long is

$$10 \cdot \frac{1}{100} = \frac{10}{100} = 0.1 \text{ m} = 10 \text{ cm}$$

Exercise set 6

R 1. Assume that the following pairs of figures are similar, and find x.

(a)

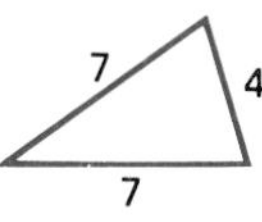

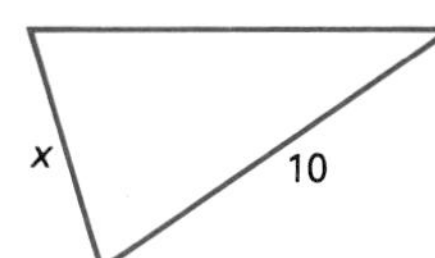

(b)

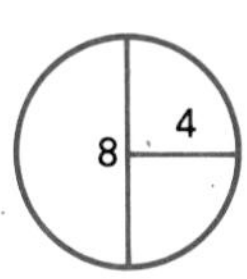

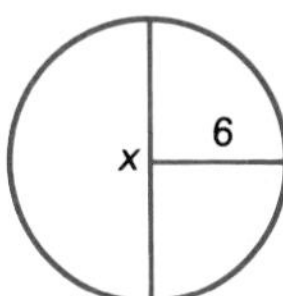

(c)

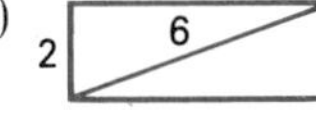

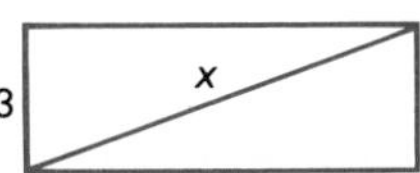

(d)

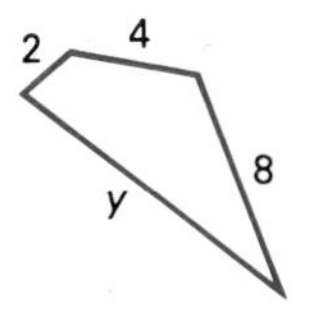

(e) In (d), find y.

2. Assume that the three triangles shown are similar. Find the measure of the unknown sides.

(a)

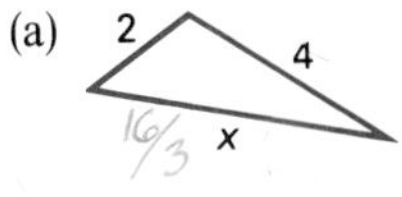

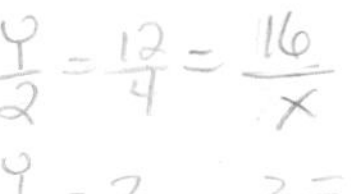

(b)

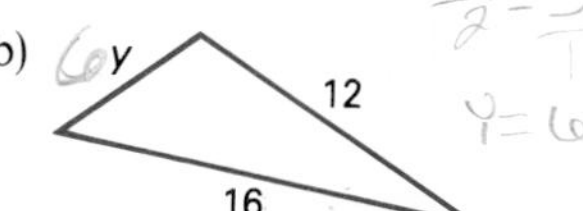

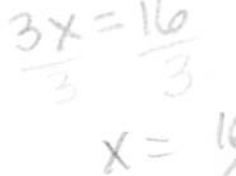

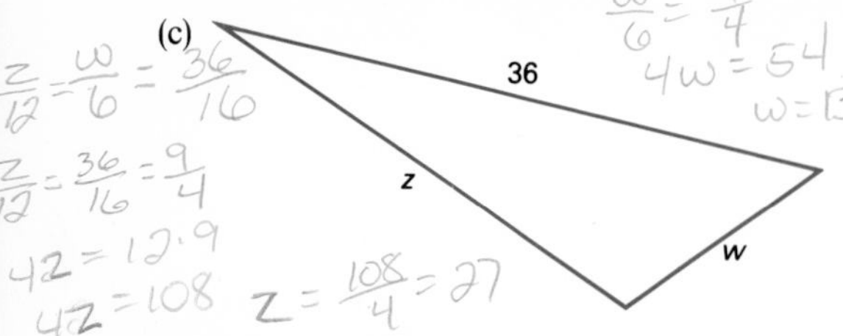

3. (a) Given $\triangle ABC \sim \triangle DEF$. If $m(\angle A) = 70°$ and $m(\angle E) = 40°$, find $m(\angle C)$ and $m(\angle F)$.
   (b) Given $\triangle GHI \sim \triangle JKL$. If $GH = 5$, $GI = 7$, $JK = 8$, and $KL = 9\frac{3}{5}$, find HI and JL.
   (c) Given $\triangle STU \sim \triangle VWX$, $TU = 8$, $SU = 8$, $VW = 4$, and $m(\angle TSU) = 60°$, find ST, WX, and $m(\angle WVX)$.

4. A snapshot is 4 cm wide and 6 cm long. It is enlarged so that it is 14 cm wide. How long is the enlarged picture? What is its perimeter?

5. At the same time that a yardstick held vertically casts a 4-ft. shadow, a vertical flagpole casts a 24-ft. shadow. How high is the flagpole?

6. On a certain map, 1.5 cm is given as representing 60 km. If the distance between two cities on the map is 4 cm, what is the distance in kilometers between the two cities?

7. Find x in the following, assuming in each part that the triangles are similar.

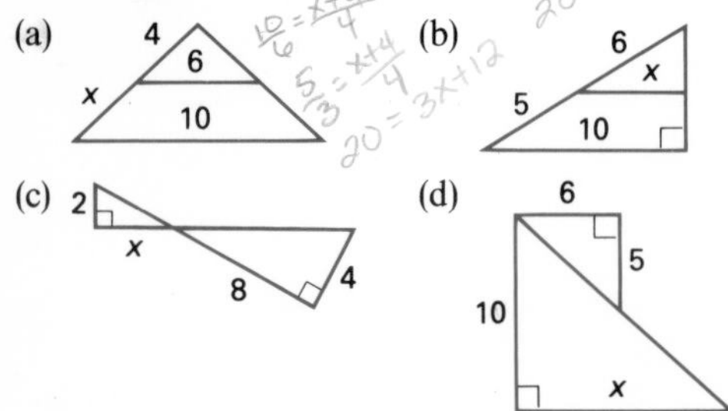

8. Prepare a scale drawing to represent a floor plan 8 m by 5 m, using a scale of 1 cm to 2 m.

9. Find the measurements (approximately) of the actual objects, using the scale drawings as given.

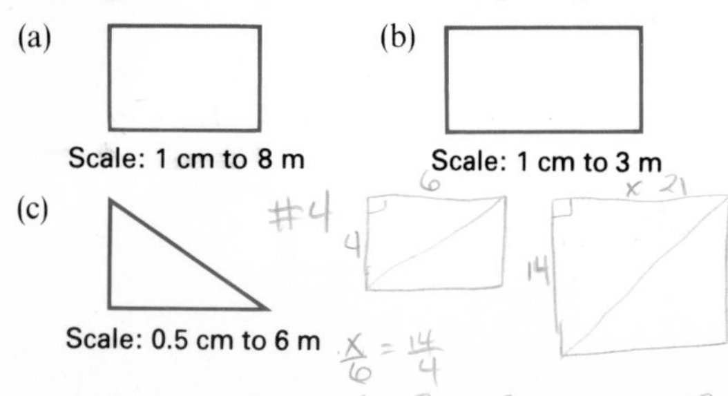

T 10. Suppose that $\triangle ABC \sim \triangle DEF$, where $\angle A \cong \angle D$, $\angle B \cong \angle E$, and $\angle C \cong \angle F$. Let $m(\overline{AC}) = 7$ m, $m(\overline{DF}) = 14$ m, and $m(\overline{DE}) = 22$ m. If the perimeter of $\triangle ABC$ is 27 m, what is the perimeter of $\triangle DEF$?

11. Judy has a picture of her boyfriend that is 15 cm wide and 25 cm long. She wants a smaller, wallet-size picture made and an enlargement made for her wall. Find the scale for the reduction and enlargement if the widths are to be 6 cm and 25 cm, respectively.

12. Merry Christmas is looking for a Christmas tree. The tree, however, can be no more than 4 m tall. Merry finds a tree that casts a shadow of 2 m, whereas Merry (120 cm tall) casts a shadow of 0.8 m. Will the tree fit in Merry's room?

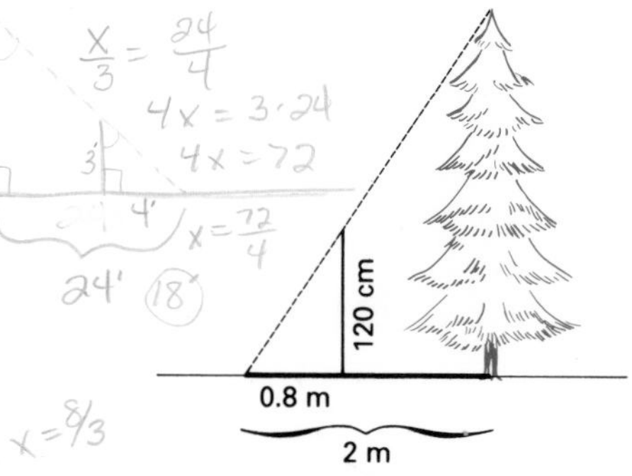

13. Prepare a scale drawing to represent the following floor plan with the scale 1 cm:2 m.

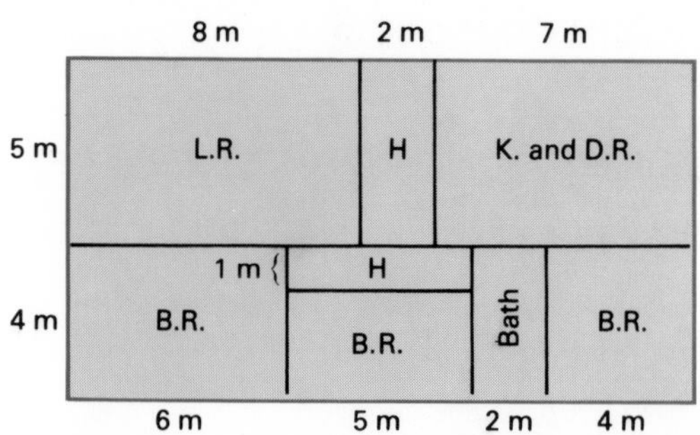

14. The idea of similarity of triangles can be expanded to include perimeters. The perimeters of two similar triangles are in proportion to any two corresponding sides.
   (a) Triangle A has a perimeter of 24. If one side measures 6 and it corresponds to a side of a similar $\triangle B$ measuring 2, then what is the perimeter of $\triangle B$?

(b) Triangle A has a perimeter of 12, and $\triangle B$ has a perimeter of 36. If two of the sides of $\triangle A$ measure 2 and 6, what is the measure of *each* side of $\triangle B$?

15. An observer on the shore saw a ship anchored off the coast. To find the distance to the ship, he made the measurements shown in the picture. How far is it from the shoreline to the ship?

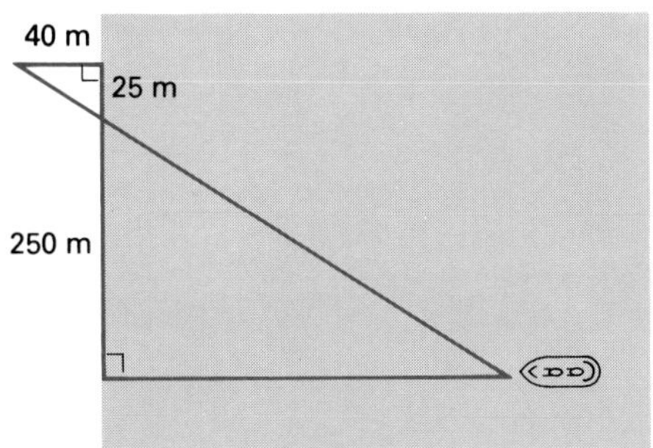

16. A farmer wishes to measure the distance across a river on his farm. Devise a plan to help the farmer find this distance.

17. At summer camp, a swimming course runs the length of a small lake. In order to determine the length of the swimming course, the counselors measure the two "dry" legs of a right triangle. What is the length of the swimming course? (*Hint:* Do not use similar triangles to make the indirect measurement.)

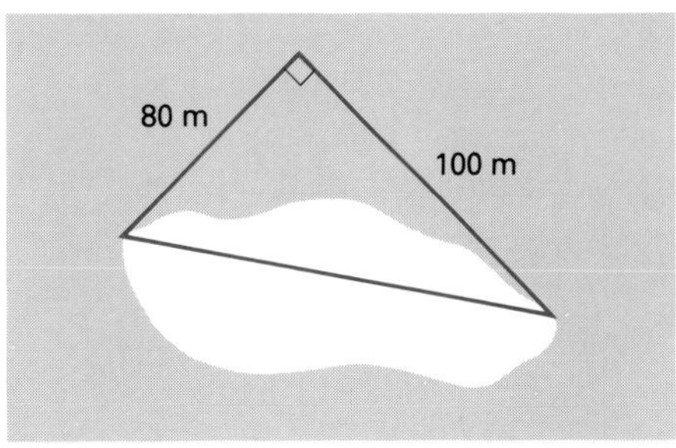

18. Using the fact that corresponding angles of two similar figures are congruent, construct a triangle similar to $\triangle ABC$ and make the side corresponding to $\overline{AC}$ 6 cm long.

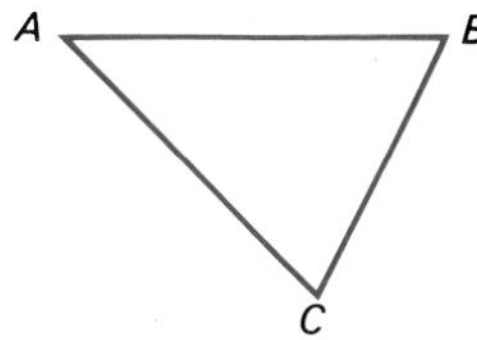

Solution to introductory problem

Understand the problem. This printing is a mirror image or a reflection.

Make a plan. The technique to use is another reflection. You can either hold the message up to a mirror upside down or hold it in front of a light upside down.

Carry out the plan. YOU TURKEY, HAVEN'T YOU GOT ANYTHING BETTER TO DO THAN MESS WITH OLD BOTTLES?

Summary and review

1. A transformation of a plane is a one-to-one correspondence between two sets of points on a plane.
2. A reflection r_m over line m is a transformation mapping each point P onto point P' as follows.
   (a) If P is on m, then $P = P'$.
   (b) If P is not on m, then m is the perpendicular bisector of $\overline{PP'}$.

3. A composite of transformations results from successive applications of transformations. For two reflections, the composite (r_m, r_n) is obtained by applying r_m and then r_n.
4. A transformation is a translation (T) if and only if $T =$ composite (r_m, r_n), where r_m is a reflection over line m, r_n is a reflection over line n, and m is parallel to n.
5. A transformation is a rotation (R) about point P if and only if $R =$ composite (r_m, r_n), where m and n intersect at P.
6. Two figures A and B are said to be congruent if and only if there is a composite of reflections, translations, or rotations such that the image of A is B.
7. Congruent figures have the same size and shape.
8. A similarity is a transformation of a plane that changes distances by a constant factor k, the scale factor of similarity. If A' and B' are the images of A and B, then $A'B' = k(AB)$.
9. Corresponding angles of similar figures are congruent.

Review exercise set 7

R 1. Show the image of $\triangle ABC$ for each transformation. In each case A', the image of A is given.

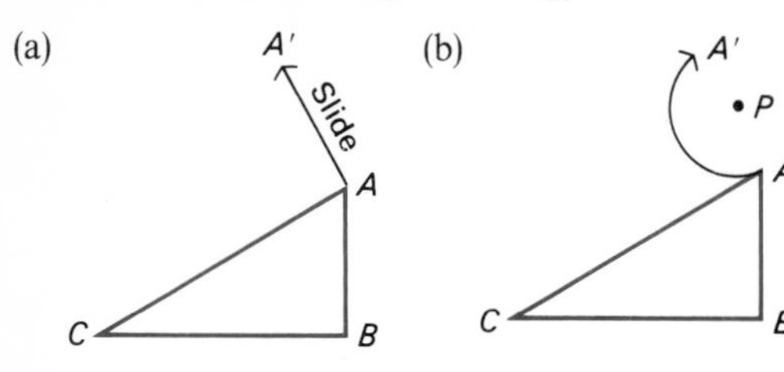

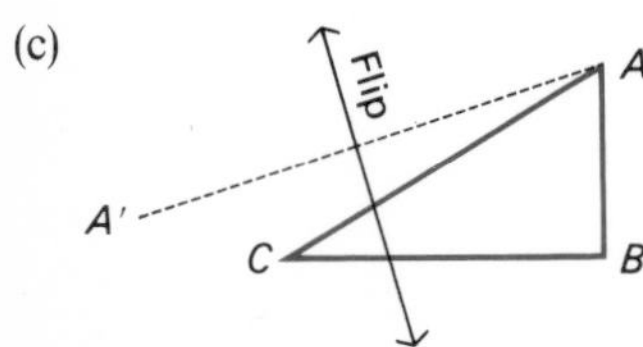

2. Jack has to measure the height of a tree in his backyard for a science experiment. He decides to use his knowledge of similar triangles to help him in his task. Jack notices that the tree's shadow is 2 m long, and his shadow is 0.9 m long. If Jack is 180 cm tall, how tall is the tree (in meters)?
3. On a certain map, 1 cm represents 20 km. If the distance between two cities is 160 km, how far apart should they be on the map?
4. Indicate whether a translation, a reflection, or a rotation has occurred in producing the shaded figure in each pair of figures.

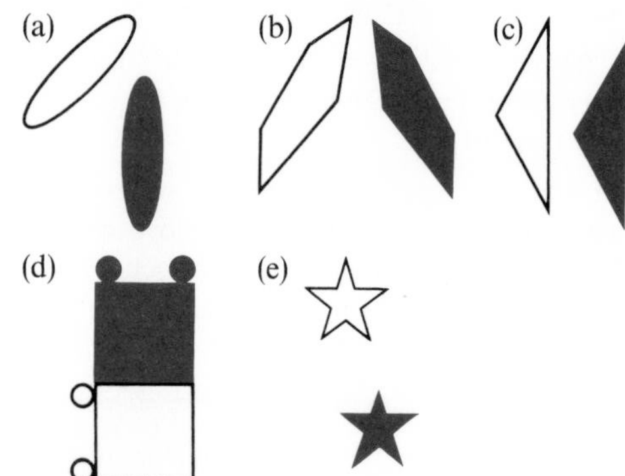

5. Given that these two triangles are similar, find x.

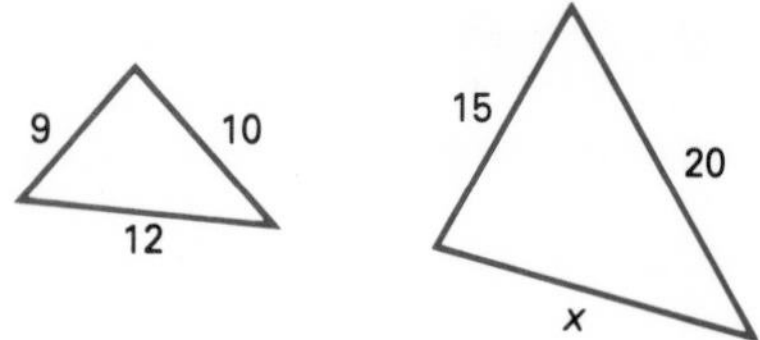

6. Determine whether the following statements are true or false.
   (a) All transformations are translations.
   (b) A given point may be the image of two different points under a transformation.
   (c) A transformation may transform a given point into two different points.
   (d) The image of a triangle under a reflection is a triangle.

(e) A translation is a motion transforming point P about point R onto point P' such that $\angle PRP'$ is a given angle measure and $\overline{RP} \cong \overline{RP'}$.

(f) A 60° rotation in a clockwise direction is the same as a 60° rotation in a counterclockwise direction.

(g) Different transformations may preserve different properties of figures.

(h) The composite of two reflections is always a translation.

(i) Every composite of transformations produces congruent figures.

(j) If a figure has rotational symmetry, then it has line symmetry.

7. Complete: A transformation that preserves shape but not size is ______.

8. Complete: Two figures are congruent if and only if ______.

9. Name four properties that are preserved by all transformations that are composites of reflections.

T 10. Sketch figures with the following qualities

(a) Exactly one line of symmetry
(b) Exactly two lines of symmetry
(c) Exactly three lines of symmetry
(d) More than three lines of symmetry
(e) Rotational but not reflectional symmetry
(f) Reflectional but not rotational symmetry

11. Perform each of the following motions on the given figure.

(a) Reflect it about line $\overleftrightarrow{DC}$; then rotate it 15° counterclockwise about point A.

(b) Translate it the length of $\overline{AD}$ in the direction of $\overrightarrow{AD}$; then reflect it about line $\overleftrightarrow{BC}$.

(c) Rotate it 90° clockwise about point B; next, translate it to the left the length of $\overline{AB}$; then reflect it about $\overleftrightarrow{AD}$.

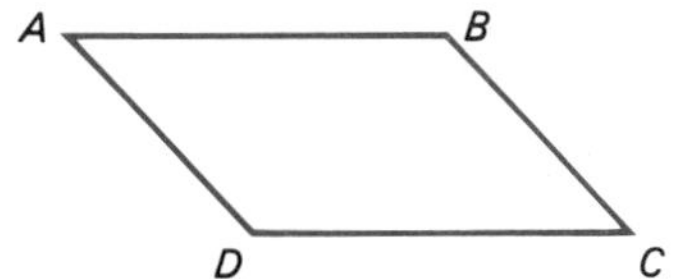

12. Sally's room is 4 m wide and 5 m long. She is making a scale drawing of it for a drafting class. If on the drawing her room is 2 cm wide, how long is it?

Exercises 13 through 18 refer to the following figure.

A	B	C	D	E
F	G	H	I	J
K	L	M	N	O
P	Q	R	S	T

C 13. A translation maps K onto H. Name the image of (a) R; (b) $\overline{LM}$; (c) $FGLK$

14. A reflection maps K onto Q. Find the image of (a) G; (b) $\overline{HI}$; (c) $FGLK$

15. What is the image of $\triangle GCH$ after a 90° clockwise rotation about L?

16. State the measure of the rotation about K that maps G onto Q.

17. Find a translation and a rotation about N that maps $PQLK$ onto $IHMN$.

18. Find a reflection that maps $BCHG$ onto $TONS$.

Bibliography

1. Bentley, W. A. & W. J. Humphrey, *Snow Crystals.* New York: McGraw-Hill, 1931.

2. Bidwell, James K. "Using Reflections to Find Symmetric and Asymmetric Patterns." *AT* (March, 1987), 10–15.

3. Coxford, Arthur F., & Zalman P. Usiskin. *Geometry, a Transformation Approach.* River Forest, Ill. Laidlaw.

4. Dana, Marcia, & Mary M. Lindquist. "The Surprising Circle." *AT* (January 1978), 4–10.

5. Hoffer, Alan R. *Geometry, a Model of the Universe.* Reading, Mass.: Addison-Wesley, 1979.

6. Horak, Virginia M., & Willis J. Horak. "Let's Do It Using Geometry Tiles as a Manipulative for Developing Basic Concepts." *AT* (April 1983), 8–15.

7. Johnson, Martin L., "Generating Patterns from Transformations." *AT* (March 1977), 191–195.

8. Kidder, F. Richard, "Euclidean Transformations: Elementary School Spaceometry," *AT* (March 1977), 201–207.

9. Lockwood, E. H. "Symmetry in Wallpaper Patterns." *Mathematics in Schools* (July 1973).

10. Maletsky, E. "Designs with Tessellations," *MT* (April 1974). 335–338.

11. Mold, J. *Tessellations.* London: Cambridge University Press, 1971.

12. Morris, Janet P. "Investigating Symmetry in the Primary Grades." *AT* (March 1977), 181–186.

13. Nichols, Eugene D., Mervine L. Edwards, E. Henry Garland, Sylvia A. Hoffman, Albert Mamary, & William F. Palmer. *Holt Geometry.* New York: Holt, Rinehart, and Winston, 1978.

14. Rahim, Medhat H. "Pythagorean Theorem and Transformation Geometry." *MT* (October 1979), 512–515.

15. Sanok, Gloria. "Living in a World of Transformations." *AT* (April 1978), 36–40.

16. Sawada, Daiyo. "Symmetry and Tessellations from Rotational Transformations on Transparencies." *AT* (December 1985), 12–13.

17. Swadener, Marc. "Pictures, Graphs, and Transformations: A Distorted View of Plane Figures for Middle Grades." *AT* (May 1974), 383–389.

18. Troccolo, Joseph A. "A Strip of Wallpaper." *MT* (January 1977), 55–58.

19. Van de Walle, John, & Charles S. Thompson. Concepts, Art, and Fun from Simple Tiling Patterns." *AT* (November 1980), 4–8.

20. Van de Walle, John, and Charles S. Thompson. "Cut and Paste for Geometric Thinking." *AT* (September 1984), 8–13.

21. Weyl, H. *Symmetry.* Princeton, N.J.: Princeton University Press, 1952.

22. Willcutt, Bob. "Triangular Tiles for Your Patio." *AT* (May, 1987), 43–45.

23. Zurstadt, Betty K. "Tessellations and the Art of M. C. Escher." *AT* (January 1984), 54–56.

CHAPTER 10

Additional Topics of Geometry

Can you trace the figures given above by beginning at one point and not raising your pencil until you have covered every point, being certain not to trace any part (except possibly an isolated point) more than once? Are there any that cannot be done? Make up more complicated figures and try to trace them. What is the secret?

In Chapter 9, we learned that two geometric figures were congruent if there existed a composite of reflections, translations, and rotations that would map one onto the other. Intuitively, congruent figures have the same size and shape. In this chapter, we use the ideas of congruence to determine congruence relationships for triangles. We then use these concepts to develop area formulas.

We conclude this chapter with two concepts that may be new to you: topological transformations and graph theory. You will undoubtedly find our brief and elementary encounter with these new concepts to be both interesting and accessible. The introductory problem will be very easy for you after a study of graph theory. However, you may wish to work the problem now before studying graph theory.

1 Congruence and triangles

■ You have 12 sticks with the following lengths: 3 are 2 units long, 3 are 3 units long, 3 are 4 units long, and 3 are 5 units long. What is the largest number of triangles you can make? What type of triangle did you make? ■

In the figure below, locate the three sticks of length 5 units, 3 of length 4 units, 3 of length 3 units, 3 of length 2 units, and 3 of length one unit. How many equilateral triangles do you locate?

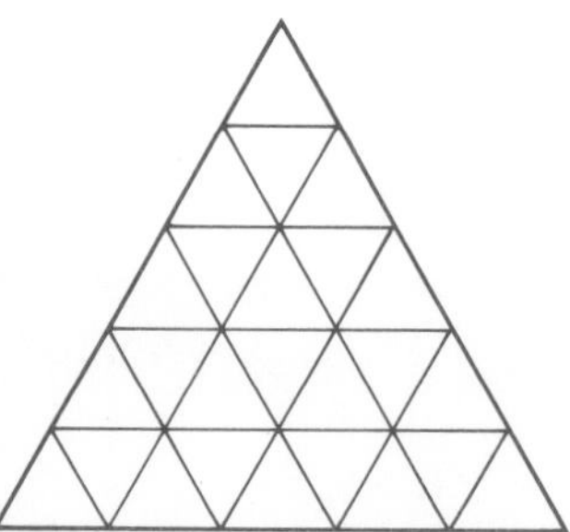

One of the advantages of the transformation approach to geometry is the understanding that accompanies the proving of geometric relationships such as the congruence of triangles. Euclid's approach will be replaced in this section by demonstrating a composite of reflections, translations and rotations that will map one congruent triangle on another.

The angles of a triangle are of special interest because they provide much insight into the properties of triangles. One very basic fact in geometry is that the sum of the measures of the three angles of any triangle is 180°. This statement can be verified approximately by using a protractor to measure the angles of a triangle. It can also be demonstrated by paper folding. See if you can fold a triangular region of paper such that the sum of the three angles makes approximately a straight line.

Congruent angles play an important role in determining the congruence of triangles.

Congruent triangles

Two **triangles,** $\triangle ABC$ and $\triangle DEF$, are **congruent** if and only if the vertices A, B, C and D, E, F can be paired so that corresponding angles and corresponding sides are congruent. If A is paired with D, B with E, and C with F, then $\overline{AB} \cong \overline{DE}$, $\overline{BC} \cong \overline{EF}$, $\overline{CA} \cong \overline{FD}$, $\angle ABC \cong \angle DEF$, $\angle BCA \cong \angle EFD$, and $\angle CAB \cong \angle FDE$.

Fortunately, it is unnecessary to check all six congruence relations in order to say that two triangles are congruent. Let's see how these requirements can be reduced by considering several constructions. We start with the case where two sides and the included angle of one triangle are congruent to two sides and the included angle of another triangle.

Construction 1 • Given: $\triangle ABC$ in Figure 1(a). We construct a second triangle such that $\angle A'B'C' \cong \angle ABC$, $B'A' = BA$, and $B'C' = BC$.

Figure 1

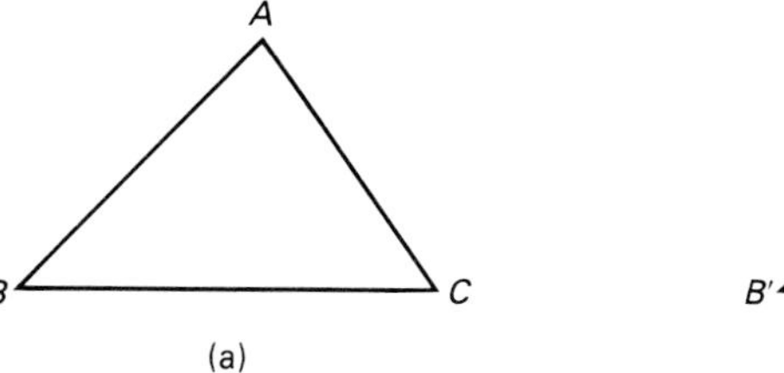

(a)

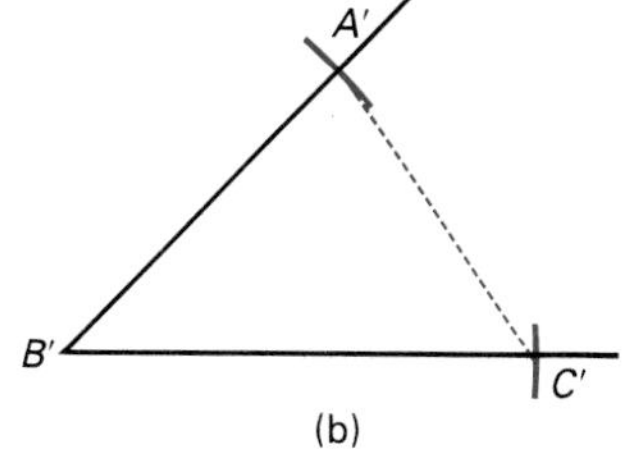

(b)

(a) As you learned in Chapter 9, construct $\angle A'B'C'$ so that it is congruent to $\angle ABC$. (See Figure 1(b).)
(b) Set your compass to measure $\overline{BA}$ in Figure 1(a) and, with this compass setting, draw in (b) an arc with B' as center intersecting the side of the angle at point A' ($B'A' = BA$).
(c) Set your compass to measure $\overline{BC}$ in (a) and, with B' as center, draw an arc intersecting the side of the angle not containing A' at point C' ($B'C' = BC$).
(d) Draw a line segment connecting A' and C'.
(e) If you make a trace of $\triangle A'B'C'$, you will note that it can be made to coincide with $\triangle ABC$. Thus, $\triangle A'B'C' \cong \triangle ABC$. •

The preceding discussion provides an intuitive understanding of what we now verify using transformation geometry.

Side, angle, side: SAS

> If two sides and the included angle (the angle formed by the rays containing the two sides) of one triangle are congruent to two sides and the included angle of a second triangle, then the two triangles are congruent. This statement is abbreviated SAS for "side, angle, side."

Consider the two triangles in Figure 2, where $\angle B \cong \angle E$, $AB = DE$, and $BC = EF$. Note that for this case, $\triangle ABC$ and $\triangle DEF$ have the same order or orientation. First translate $\triangle ABC$ a distance BE in a direction $\overrightarrow{BE}$ so that B coincides

Figure 2

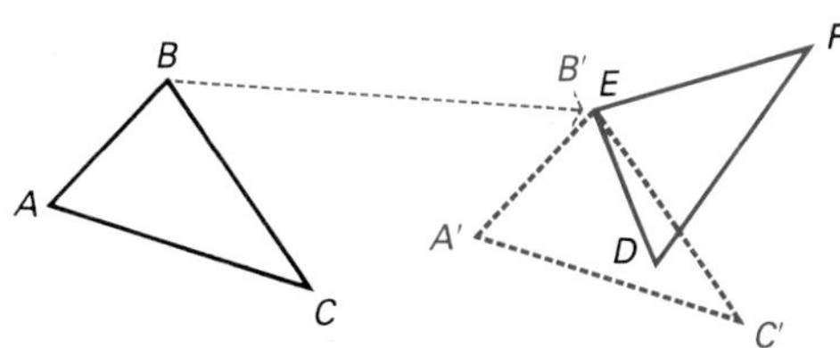

with E. Why is $\triangle A'B'C' \cong \triangle ABC$? Now rotate $\triangle A'B'C'$ about E until $\angle A'B'C'$ coincides with $\angle DEF$. ($\angle A'B'C' \cong \angle ABC \cong \angle DEF$) Since $\overline{AB} \cong \overline{A'B'}$, then $\overline{A'B'}$ coincides with $\overline{DE}$ (since $AB = DE$). Likewise $\overline{B'C'}$ coincides with $\overline{EF}$, since $BC = EF$. Obviously, A' coincides with D and C' coincides with F. Thus, $\overline{DF} \cong \overline{A'C'}$, $\angle EFD \cong \angle B'C'A'$, and $\angle EDF \cong \angle B'A'C'$. Therefore, since each part of $\triangle ABC$ is congruent to the corresponding part of $\triangle A'B'C'$, then $\triangle ABC \cong \triangle DEF$.

Suppose $\triangle ABC$ and $\triangle DEF$ have different orientations, as in Figure 3. Choose any line m and reflect $\triangle ABC$ over m to reverse the orientation of the image, and then translate as in the first part of this proof. $\triangle ABC \cong \triangle A''B''C''$, since both reflection and translation preserve size and shape. Thus, by the reasoning in the first part, $\triangle ABC \cong \triangle DEF$.

Figure 3

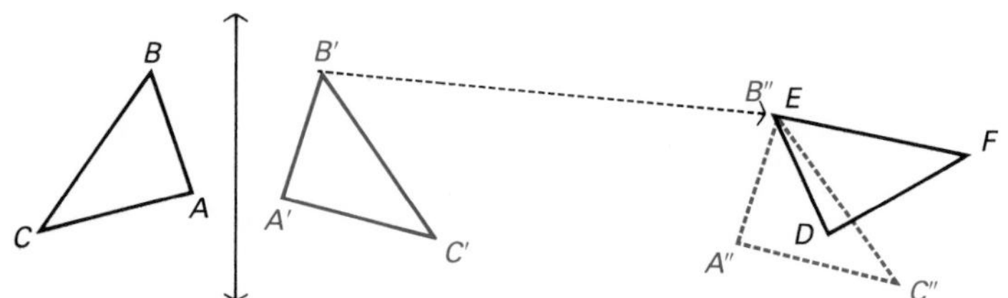

The next discussion of sufficiency requirements for congruent triangles can be developed from the SAS requirement. This will be assigned in the exercise set. In addition, we will show the feasibility of the requirement by a construction and validate the result with transformations.

Side, side, side: SSS

> Two triangles are congruent if three sides of one triangle are congruent to three sides of another triangle.

Construction 2 • Given: $\triangle ABC$ in Figure 4(a). Construct a triangle $A'B'C'$ such that $AB = A'B'$, $BC = B'C'$, and $CA = C'A'$.

Figure 4

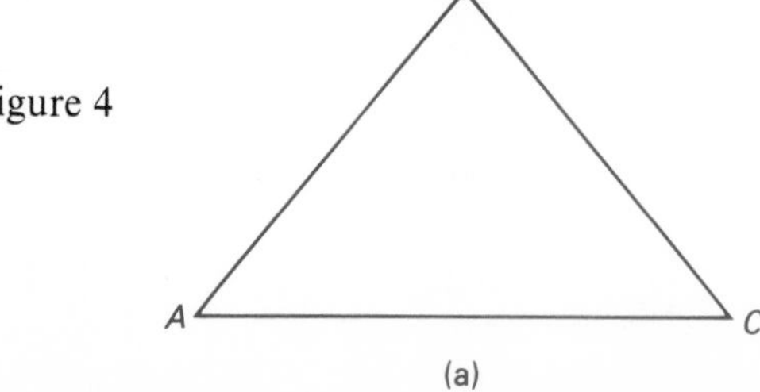

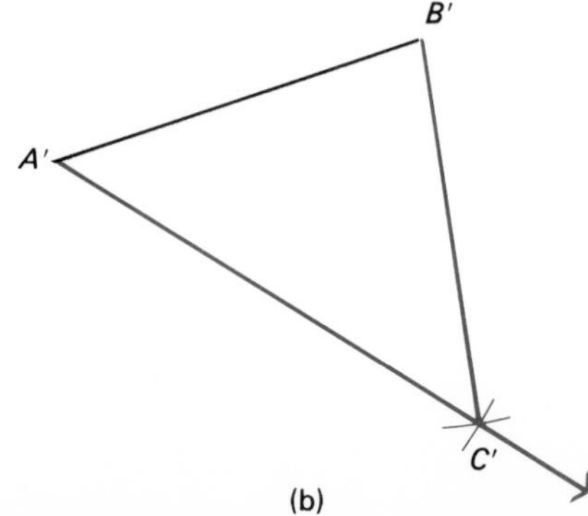

(a) Set your compass to measure $\overline{AB}$ and, with this compass setting, mark two points A' and B' and connect with a straightedge to get $\overline{A'B'}$ in Figure 4(b). Note that $AB = A'B'$.
(b) Set your compass to measure $\overline{BC}$ and, with B' as the center and the same compass setting, draw an arc on one side of $\overleftrightarrow{A'B'}$. For any point C' on this arc, $B'C' = BC$.
(c) Set your compass to measure $\overline{AC}$ and, with A' as center, draw an arc intersecting the arc in Figure 4(b) at point C'. Note that $A'C' = AC$. Why?
(d) Draw $\overline{A'C'}$ and $\overline{B'C'}$ to form $\triangle A'B'C'$.
(e) It appears that $\triangle A'B'C'$ as constructed is congruent to $\triangle ABC$. In fact, a trace of $\triangle A'B'C'$ fits $\triangle ABC$ exactly. Since we performed the construction by constructing the three sides of $\triangle A'B'C'$ to be congruent to the sides of $\triangle ABC$, it seems that congruence of all three sides is sufficient to ensure congruence of triangles. •

Suppose we are given in Figure 5 $\triangle ABC$ and $\triangle DEF$, with corresponding sides congruent. We know that a composite of reflections, translations, and rotations exists such that $\overline{AC}$ can be mapped onto $\overline{DF}$, with A' coinciding with D and C' with F to form quadrilateral $DB'FE$. From what is given and properties of a composite of reflections, translations and rotations, $EF = C'B'$ and $DE = A'B'$. Draw $\overline{EB'}$. Now $\overleftrightarrow{DF}$ is the perpendicular bisector of $\overline{EB'}$, because points equidistant from the ends of a segment are on the perpendicular bisector of a segment. (See Chapter 9.) Therefore, if $\triangle A'B'C'$ is reflected across $\overleftrightarrow{DF}$, B' coincides with E. Thus, $\triangle A'B'C' \cong \triangle DEF$, and hence $\triangle ABC \cong \triangle DEF$.

Figure 5

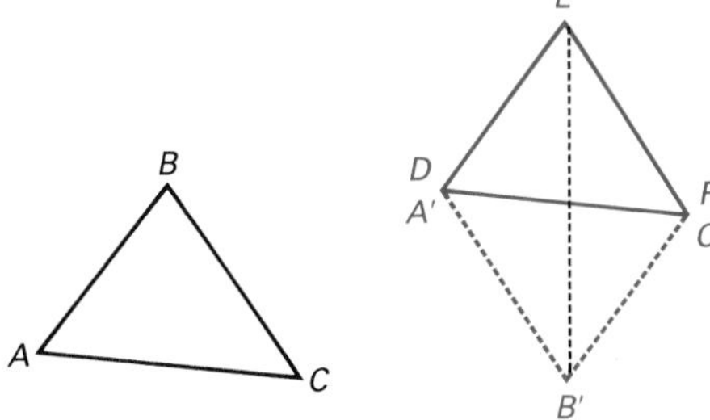

A third requirement for congruence of triangles can be given as follows.

Angle, side, angle: ASA

> Two triangles are congruent if two angles and the included side of one triangle are congruent to two angles and the included side of another triangle.

We validate this result in two different ways in the exercise set.

Example 1 • In Figure 6, congruent angles are denoted by the same number of arcs and sides by the same number of slashes. Are the pairs of triangles in Figure 6 congruent? If so, why?

Figure 6

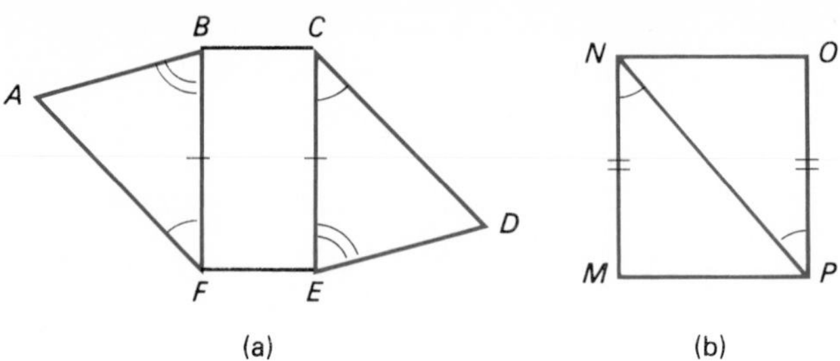

Solution (a) The marks on the sides and the arcs on the angles indicate that

$$\overline{BF} \cong \overline{EC}$$

$$\angle ABF \cong \angle DEC$$

$$\angle AFB \cong \angle DCE$$

$\overline{BF}$ is the side between $\angle ABF$ and $\angle AFB$.
$\overline{EC}$ is the side between $\angle DEC$ and $\angle DCE$.
Therefore, $\triangle ABF \cong \triangle DEC$ by ASA.

(b) Marks on the drawing indicate

$$\overline{MN} \cong \overline{OP}$$

$$\overline{NP} \cong \overline{PN} \qquad \text{Why?}$$

$$\angle MNP \cong \angle OPN$$

$\angle MNP$ is included between $\overline{MN}$ and $\overline{NP}$. $\angle OPN$ is included between $\overline{OP}$ and $\overline{PN}$. By SAS, $\triangle MNP \cong \triangle OPN$. •

The following properties can be proved using the congruence of triangles.

Isosceles and Equilateral Triangles

1. If a **triangle** is **isosceles,** the angles opposite congruent sides are congruent.
2. If two angles of a triangle are congruent, then the triangle is an **isosceles triangle,** where the sides opposite congruent angles are congruent.
3. If a **triangle** is **equilateral, all three angles are congruent.**
4. If three angles of a triangle are congruent, then the triangle is an **equilateral triangle.**

We conclude this section with an explanation of three terms that will be used in the problems of the exercise set.

Parts of a triangle

(a) A median of a triangle is a segment from a vertex of the triangle to the midpoint of the opposite side of the triangle.
(b) An altitude of a triangle is a segment from a vertex of the triangle perpendicular to the line containing the opposite side of the triangle.

Just for fun

Water was precious in the desert town of Dry Gulch. When Water Willy brought two 40-liter barrels of water to town, two townspeople hurried out with their buckets. But Water Willy had forgotten his measuring device. If the buckets held 4 L and 5 L, respectively, and each person wanted 2 L of water, how did Willy give the correct amounts?

Exercise set 1

R 1. Are there congruent triangles in the figures shown? If so, tell why. Marks on line segments that are alike indicate congruent line segments. Congruent angles are indicated by ∡ or ∡.

(a)

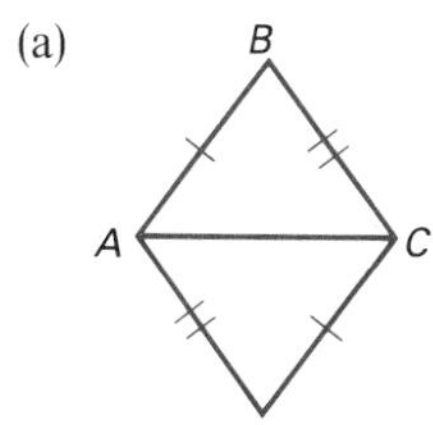

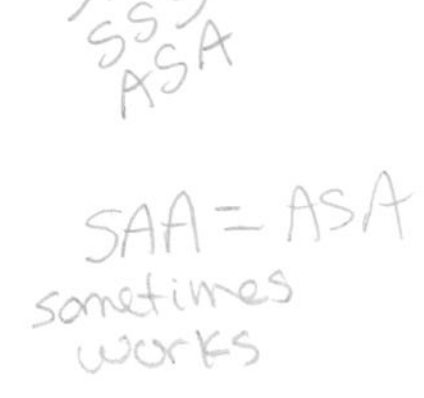

(b)

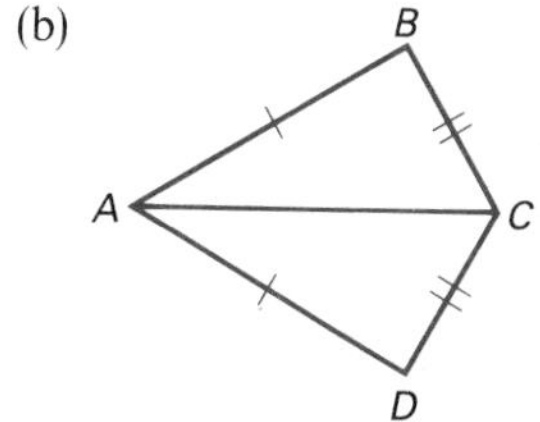

(c)

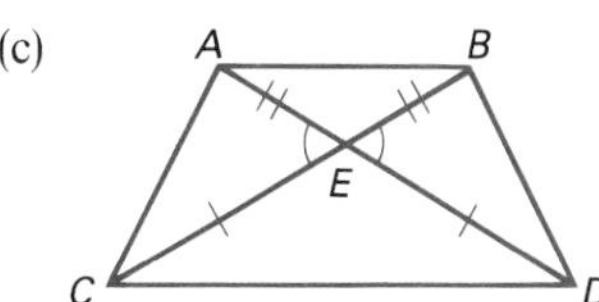

(d)

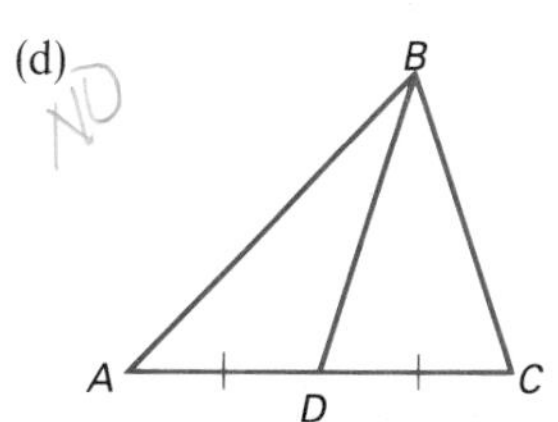

(e)

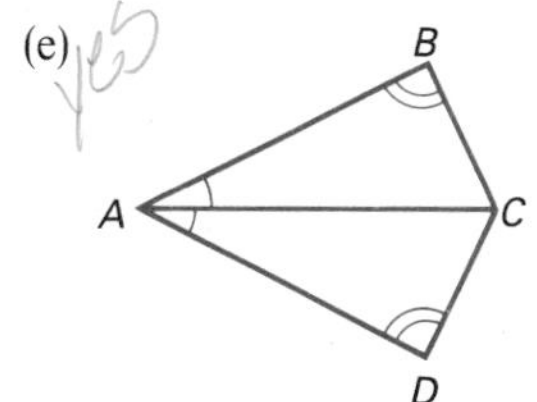

(f)

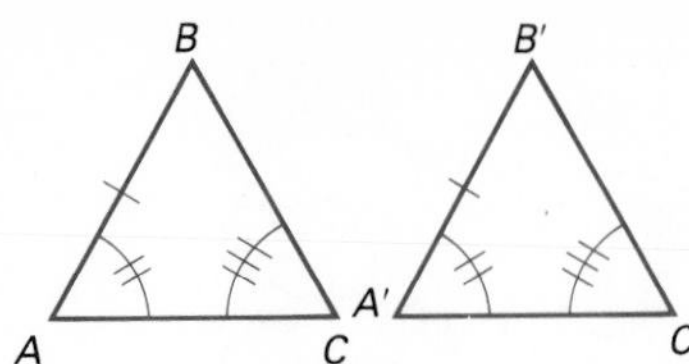

(g)

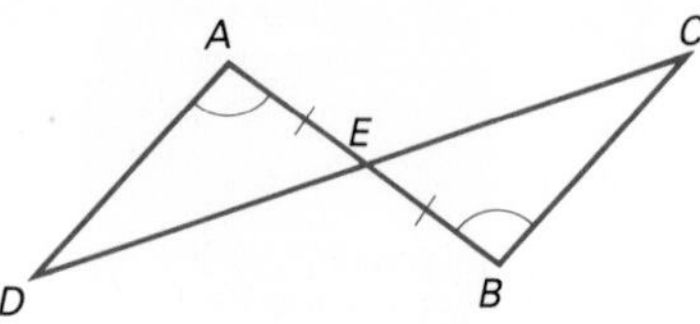

(h)

(i)

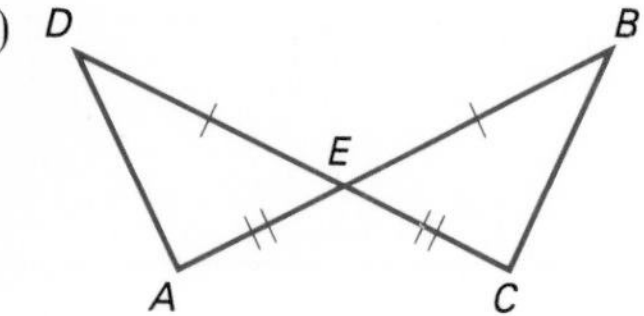

(j)

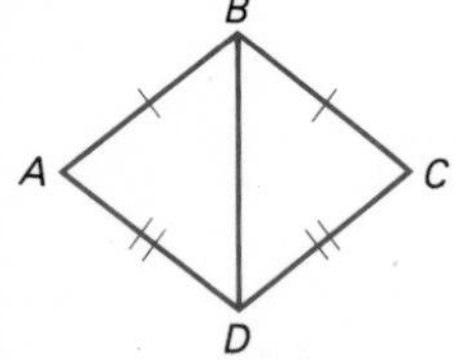

2. Classify as true or false. Explain those that are false.

(a)

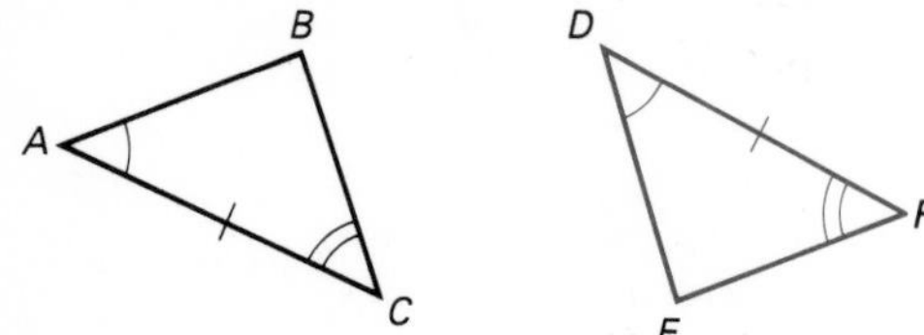

There is a composite of reflections, translations, and rotations that will map $\triangle ABC$ onto $\triangle DEF$.

(b) In (a) the composite can be a composite of reflections.

(c) $\triangle ABC$ can be mapped onto $\triangle DEF$ by a composite of translations and rotations.

3. Can an isosceles triangle have
   (a) three acute angles?
   (b) a right angle?
   (c) an obtuse angle?

4. A translation slides $\triangle ABC$ onto $\triangle A'B'C'$ where $A \leftrightarrow A'$, $B \leftrightarrow B'$, and $C \leftrightarrow C'$. If $m(\angle A) = 70°$, $m(\angle B = 80°)$, find $m(\angle C')$.

5. For which of the pairs of triangles given is the information inadequate to determine whether they are congruent?

(a)

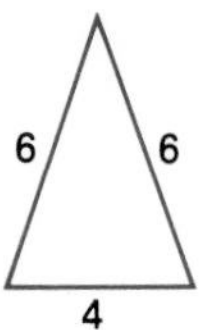

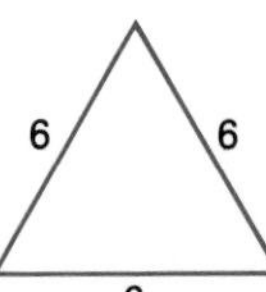

(b)

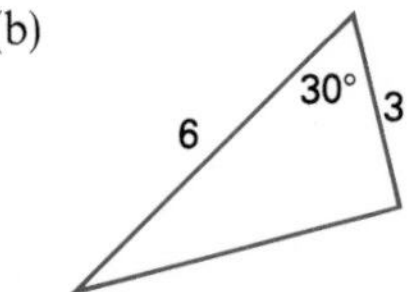

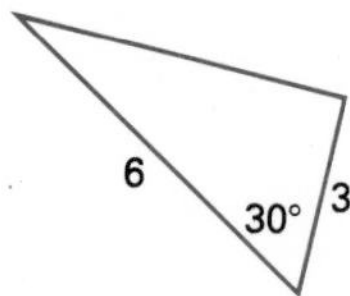

(c)

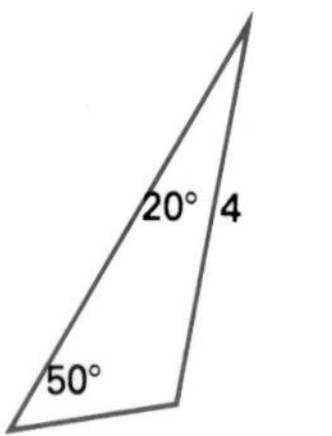

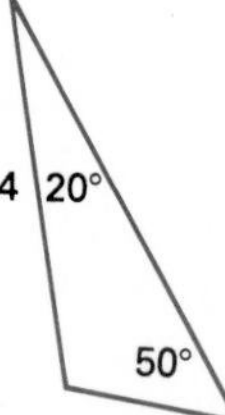

(d)

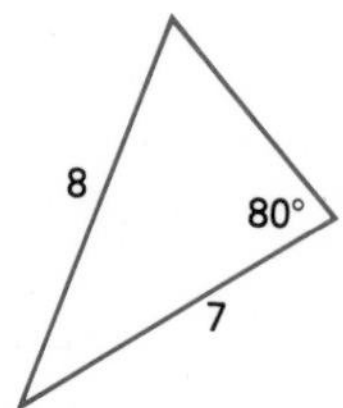

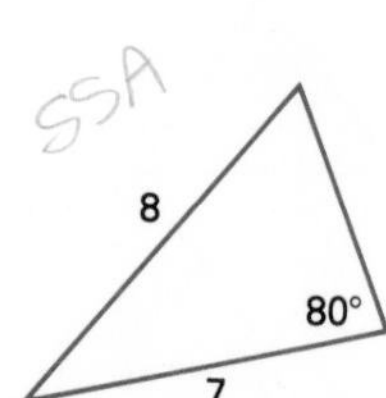

6. Using the theorems discussed in this section, prove that the following triangles are congruent.

(a)

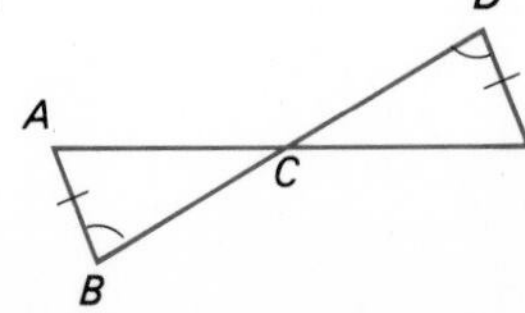

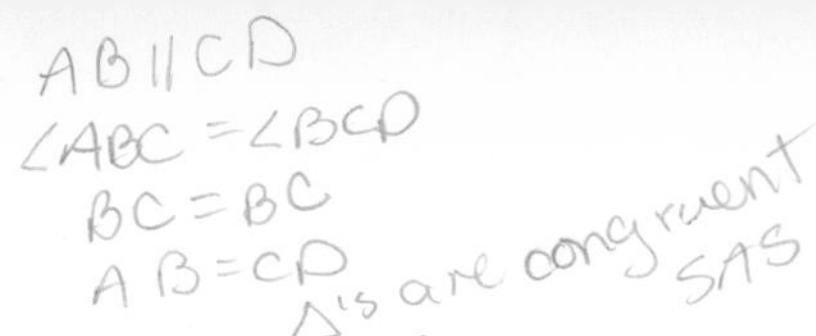

(b)

(c)

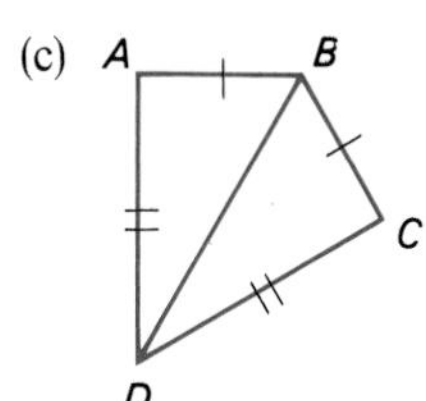

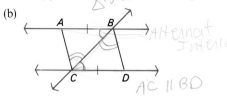

7. Discuss whether or not the following pairs of triangles are congruent. Give reasons for your answers.

(a)

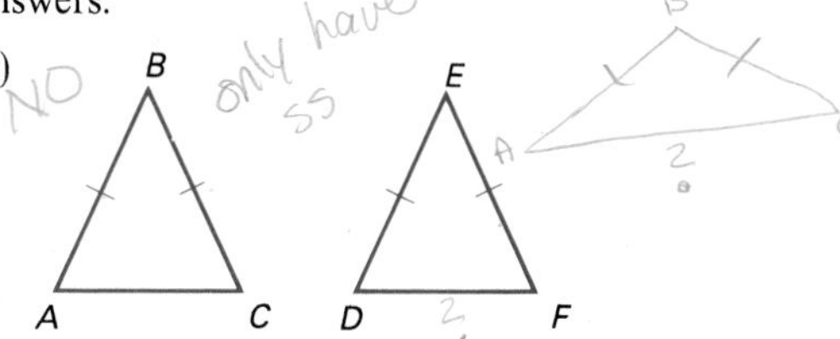

(b)

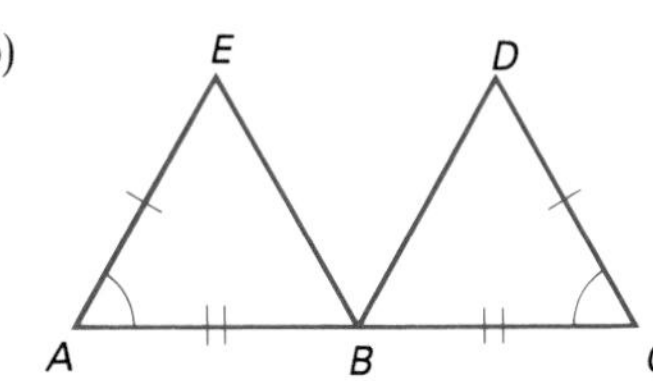

(c)

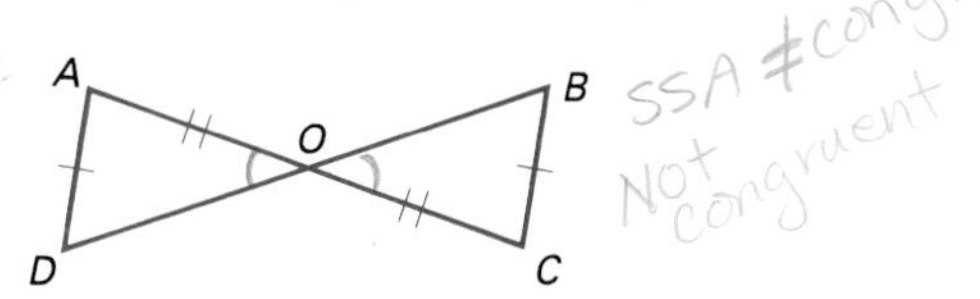

(d)

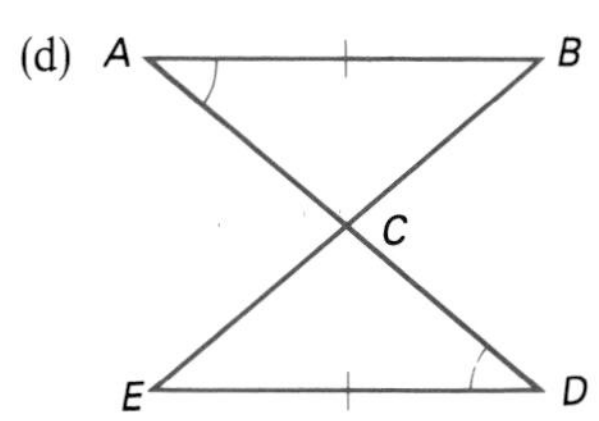

(e)

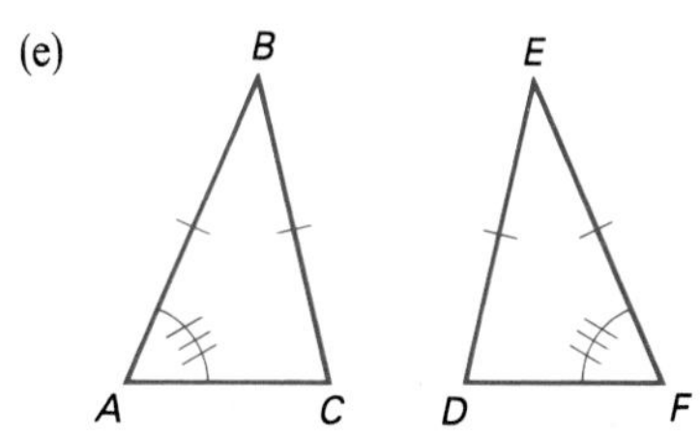

T 8. Construct a triangle with three sides as given.

9. Draw a triangle with an acute angle at A. Construct each of the following.
   (a) The altitude from vertex A.
   (b) The median from vertex A.

10. The sum of the measures of the angles of a triangle is 180°. Use this fact to find the sum of the measures of the angles of the following convex polygons.
    (a) Quadrilateral (b) Pentagon
    (c) Hexagon (d) n-sided polygon
    Hint: Form triangles by drawing diagonals from one vertex.

C 11. Why is there no SSA congruence relationship?

12. Describe a composite of transformations that will map $\triangle ABC$ onto $\triangle A'B'C'$.

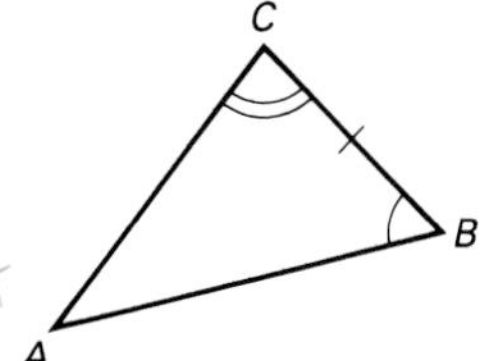

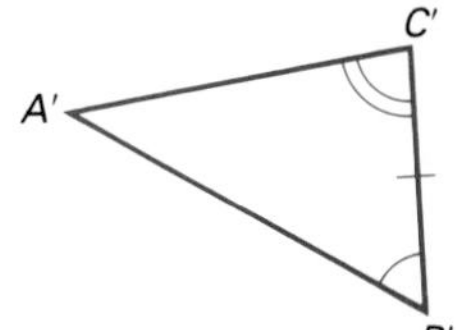

13. Use "If two sides and the included angle of one triangle are congruent to two sides and the included angle of a second triangle, then the two triangles are congruent" to prove the following properties.
    (a) If two angles and the included side of one triangle are congruent to two angles and the included side of another triangle, then the two triangles are congruent.
    (b) If three sides of one triangle are congruent to three sides of another triangle, then the two triangles are congruent.
    (c) If a triangle is isosceles, the angles opposite congruent sides are congruent.
    (d) If two angles of a triangle are congruent, then the triangle is an isosceles triangle.

14. Outline transformations that will map one triangle onto another satisfying ASA, if A, B, C are clockwise and A', B', C' are counter-clockwise.

2 Characteristics of right triangles

■ Leroy is helping his dad install siding on the end of their house. Leroy's dad said, "Cut the siding to fit snug against the gable." Leroy was in luck because he had just studied congruent right triangles in geometry. Here is his solution.

He needs to trim each piece of siding so that $\angle 1 \cong \angle 2$. This of course occurs when $\angle 3 \cong \angle 4$. Thus, Leroy cuts from each piece of siding congruent right triangles with $\angle 3 \cong \angle 4$.

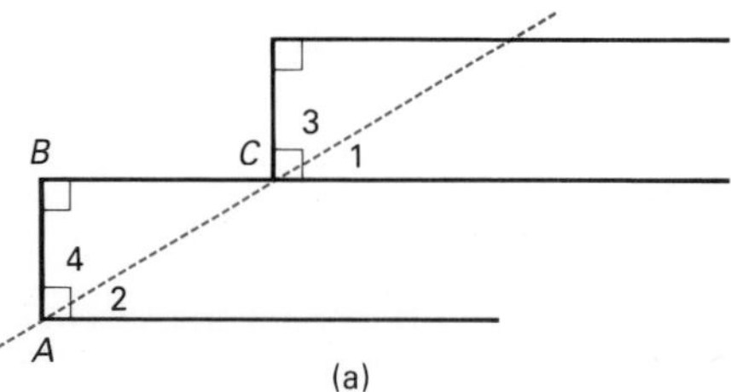

(a)

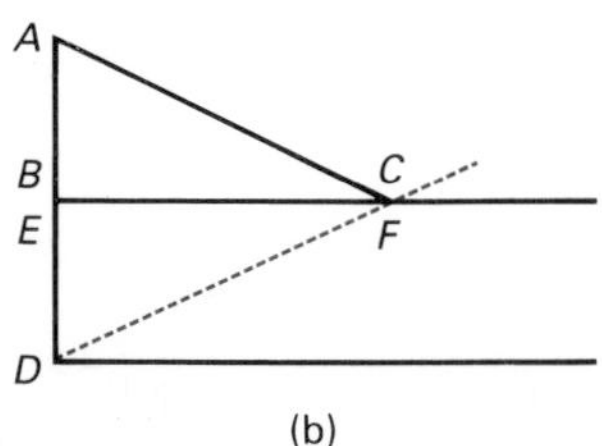

(b)

Here is the way he accomplishes this task. First he places a piece of siding against the gable, marks point C, and cuts off right triangle ABC. Then he places this against the next piece of siding, marks point F, and cuts off right triangle DEF. These two right triangles, $\triangle ABC$ and $\triangle DEF$, are congruent. Why? That these two triangles are congruent is one of the things we will learn in this chapter, as we study the characteristics of right triangles. ■

In the preceding section, we demonstrated that two triangles are congruent by the following congruence relationship.

side–side–side (SSS)
side–angle–side (SAS)
angle–side–angle (ASA)

We use these relationships to determine conditions for the congruence of two right triangles. In addition we show one relationship for congruence of right triangles by using transformations.

First we note that the sides of a right triangle have special names. The side opposite the right angle is called the **hypotenuse** and the other two sides are called **legs.** (See Figure 7.)

Figure 7

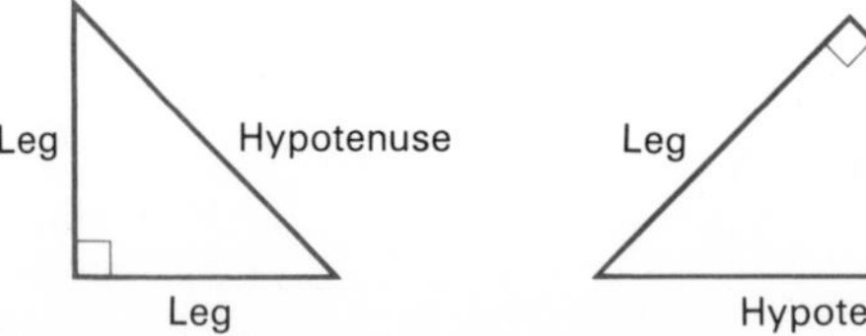

In Figure 8, suppose $\overline{AC} \cong \overline{A'C'}$ and $\overline{AB} \cong \overline{A'B'}$ (that is, corresponding legs are congruent). Since the included angle between $\overline{AC}$ and $\overline{AB}$ is a right angle and the angle between $\overline{A'C'}$ and $\overline{A'B'}$ is a right angle, $\triangle ABC \cong \triangle A'B'C'$ by the SAS requirement.

Figure 8

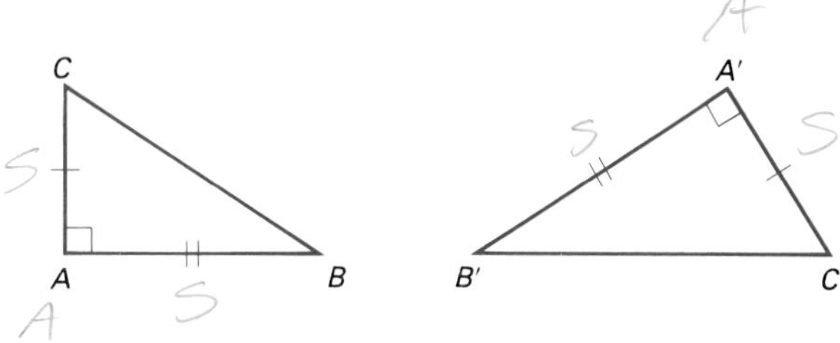

Right triangle: LL

> If two legs of one right triangle are congruent to the corresponding legs of another right triangle, the right triangles are congruent.

A similar relationship exists when a leg and an angle (other than the right angle) of one right triangle are congruent to the corresponding parts of another. Since the sum of the three angles of a right triangle is 180°, and since one of the angles is a right angle, then $m(\angle A) = 90° - m(\angle B)$ and $m(\angle A') = 90° - m(\angle B')$. If $m(\angle A) = m(\angle A')$, then

$$90° - m(\angle B) = 90° - m(\angle B')$$

$$m(\angle B) = m(\angle B')$$

Thus, if one acute angle of a right triangle is congruent to an acute angle of the second right triangle, then the other pair of corresponding acute angles are congruent. The leg (regardless of which leg) lies between angles (including right angles) that are congruent to the corresponding angles of the second right triangle. Thus, by the ASA condition for the congruence of triangles in general, the two right triangles are congruent.

Right triangle: LA

> If a leg and an acute angle of one right triangle are congruent to the corresponding parts of another right triangle, the triangles are congruent.

Example 2 • In Figure 9, right $\triangle ABC \cong$ right $\triangle A'B'C'$ because $\overline{AC} \cong \overline{A'C'}$ and $\angle C \cong \angle C'$.

Figure 9

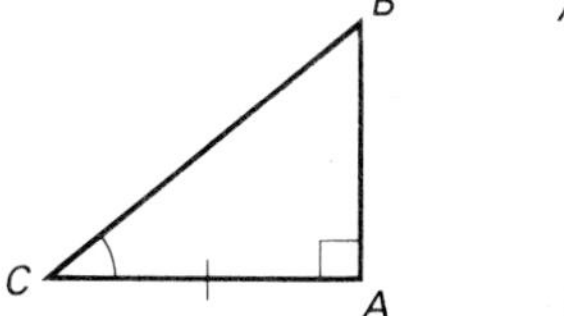

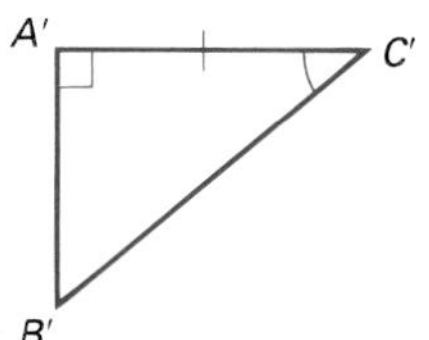

•

A similar proof can be given for the hypotenuse-angle correspondence.

Right triangle: HA

If the hypotenuse and an acute angle of one right triangle are congruent to the corresponding parts of another right triangle, the triangles are congruent.

Suppose for right triangles ABC and $A'B'C'$ in Figure 10 that $AB = A'B'$ and $AC = A'C'$ (that is, the hypotenuse and leg of one right triangle are congruent to the hypotenuse and leg of another.) Using a composite of reflections, translations, and rotations, we will show that right triangle $ABC \cong$ right triangle $A'B'C'$. Since $AC = A'C'$, there exists a composite that will map $\overline{AC}$ on $\overline{A'C'}$ so that A'' coincides with A', and C'' with C', and $\triangle ABC \cong \triangle A''B''C''$.

$$m(\angle 1) + m(\angle 2) = 180°.$$

So B'', C', and B' are collinear, producing $\triangle A'B'B''$. This is an isosceles triangle, since $A'B' = AB = A''B''$. Why? Hence, $\angle B' \cong \angle B''$. Right triangle $A'B'C' \cong$ right triangle $A''B''C''$ by LA. Thus, $\triangle ABC \cong \triangle A''B''C'' \cong \triangle A'B'C'$.

Figure 10

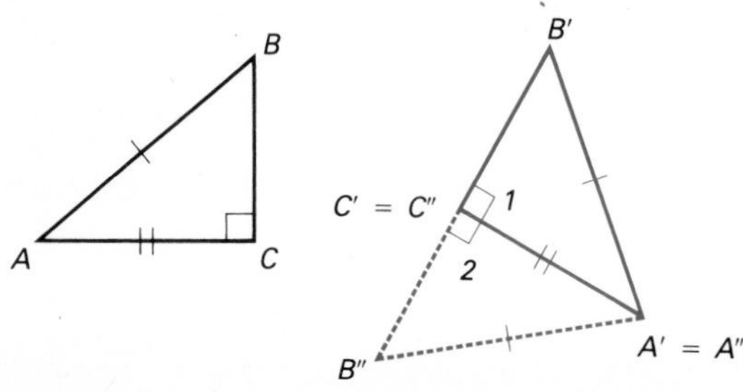

Right triangle: HL

If the hypotenuse and the leg of one right triangle are congruent to the corresponding parts of another right triangle, the triangles are congruent.

Example 3 • Why are the pairs of triangles in each part of Figure 11 congruent?

Figure 11

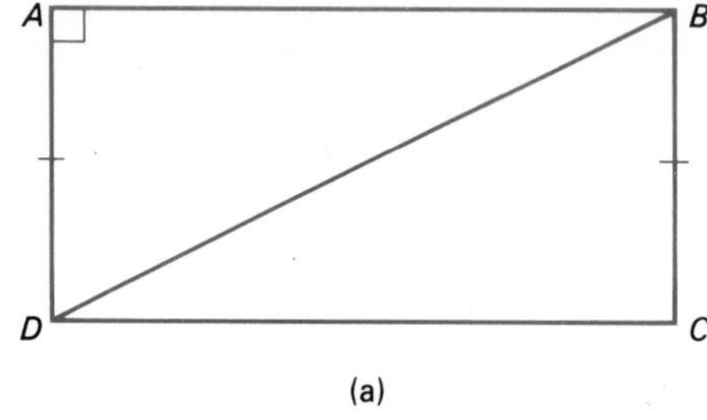

(a)

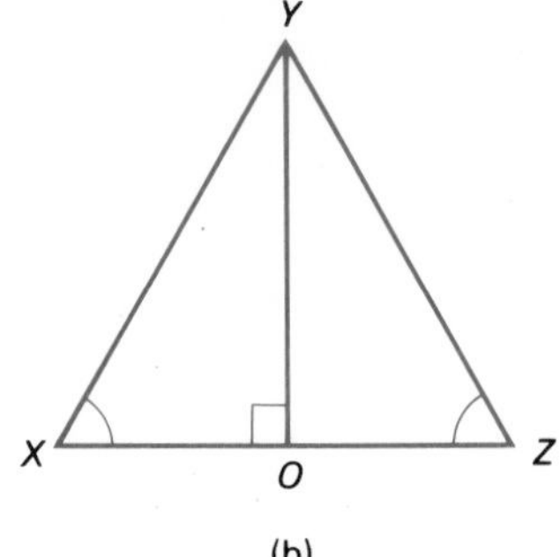

(b)

Solution In Figure 11(a), right $\triangle ABD \cong \triangle CDB$ by HL because $\overline{AD} \cong \overline{CB}$ and $\overline{BD} \cong \overline{DB}$. In Figure 11(b), right $\triangle XOY \cong \triangle ZOY$ by LA because the marks indicate that $\angle YXO \cong \angle YZO$ and legs YO of each triangle are congruent. •

The Babylonians learned long ago that right triangles have a very special property. They learned that the square of the length of the side opposite the right angle is equal to the sum of the squares of the other two sides. For example, if the legs are 3 and 4, then the length of the hypotenuse must be 5, since $5^2 = 3^2 + 4^2$. This relationship can be visually verified by counting the number of squares on the legs and then on the hypotenuse in Figure 12.

Figure 12

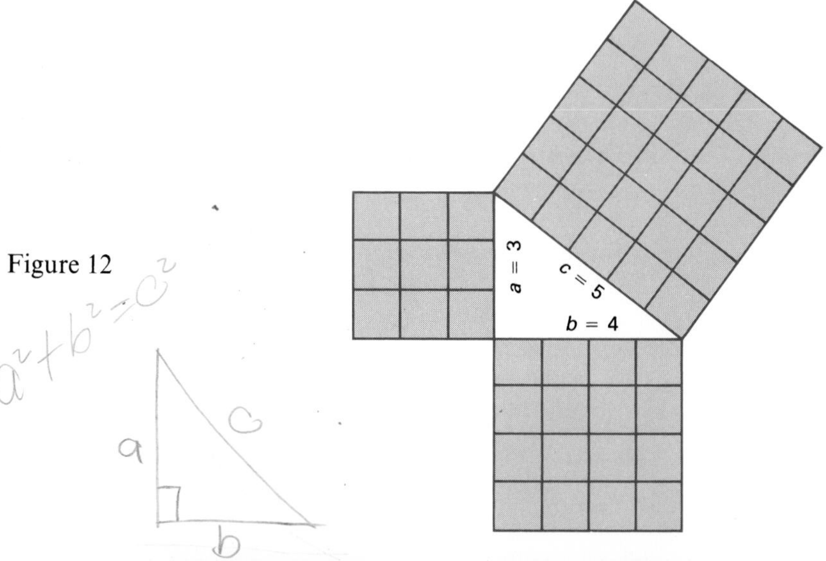

Pythagorean Theorem

> In a right triangle, the square of the measure of the hypotenuse is equal to the sum of the squares of the measures of the legs: $c^2 = a^2 + b^2$. (See Figure 13(a).)

Example 4 • In Figure 13(a), if the right triangle has legs of 2 cm and 5 cm, what is the length of the hypotenuse?

Figure 13

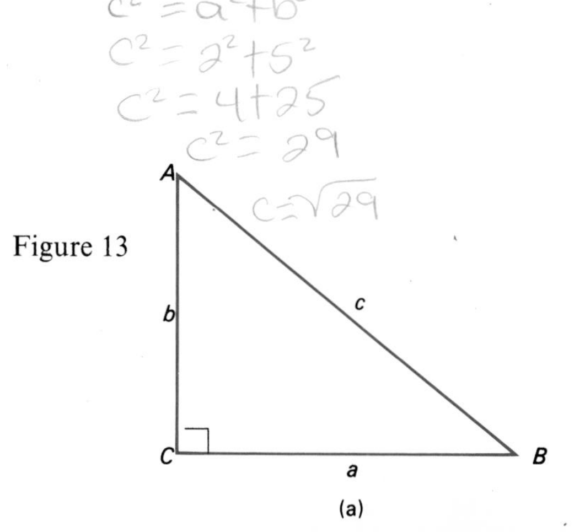

(a)

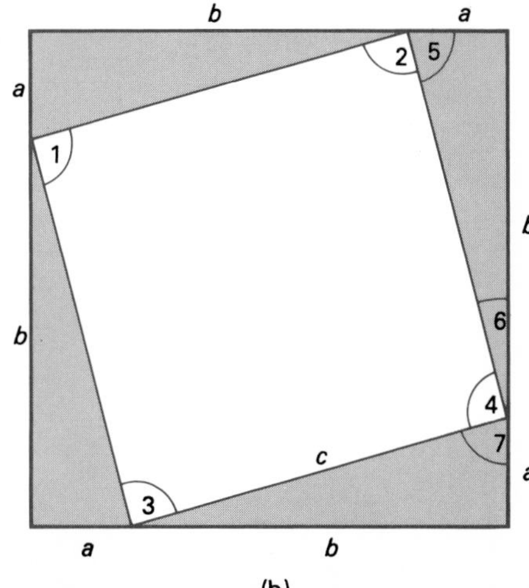

(b)

Solution Since $a^2 + b^2 = c^2$, then $(2)^2 + (5)^2 = c^2$; $4 + 25 = c^2$; $29 = c^2$; and the length of the hypotenuse is $\sqrt{29}$ cm. •

The converse of the Pythagorean Theorem is also true.

Converse of the Pythagorean Theorem

> If the square of the measure of one side of a triangle is equal to the sum of the squares of the measures of the other two sides, the triangle is a right triangle.

Over the years, many proofs of the Pythagorean Theorem have been presented. In fact, the theorem was known to the Babylonians about a thousand years before Pythagoras, but the first general proof of the theorem is believed to have been given by Pythagoras about 525 B.C. There have been many conjectures as to the type of proof Pythagoras might have given; it is generally believed today that his proof was probably a dissection-type proof, such as shown in Figure 13(b). Note that the properties of congruence figure prominently in the proof.

Figure 13(b) is a square with sides that have measures of length $a + b$. In this square, we draw four right triangles with legs of measure a and b. The four right triangles are congruent. Why? Thus, the four hypotenuses are congruent.

$$m(\angle 6) + m(\angle 4) + m(\angle 7) = 180^\circ \qquad \text{(See Exercise 10, p. 457.)}$$

$$m(\angle 5) + m(\angle 6) = 90^\circ \qquad \text{Why?}$$

Since the triangles are congruent,

$$m(\angle 7) = m(\angle 5) \qquad \text{Why?}$$

Hence,

$$m(\angle 7) + m(\angle 6) = 90^\circ \quad \text{and so} \quad m(\angle 4) + 90^\circ = 180^\circ$$

which means that $m(\angle 4) = 90^\circ$. In like manner, it can be shown that

$$m(\angle 1) = m(\angle 2) = m(\angle 3) = 90^\circ$$

Thus, the figure bounded by the four hypotenuses is a square.

The area of the large square is equal to the area of the smaller square plus the area of the four congruent right triangles. The area of the large square is

$$(a + b)^2 = a^2 + 2ab + b^2$$

The area of the small square is c^2. Since the area of a right triangle is one-half the product of the legs, the total area of the four right triangles is

$$\frac{4(ab)}{2} = 2ab$$

Thus

$$a^2 + 2ab + b^2 = c^2 + 2ab$$

Subtract $2ab$ from each side of this equation to obtain $a^2 + b^2 = c^2$, thus completing the proof.

Just for fun

Henry Perigal, a London stockholder and amateur astronomer, discovered a paper-and-scissors "proof" of the Pythagorean Theorem. Can you?

Hint: In the large square (2) on the leg of the right triangle, draw a line through the center of the square perpendicular to the hypotenuse. Then draw a line through the center of square (2) perpendicular to the first line. Cut out the pieces of square (2) and use them to form the square on the hypotenuse.

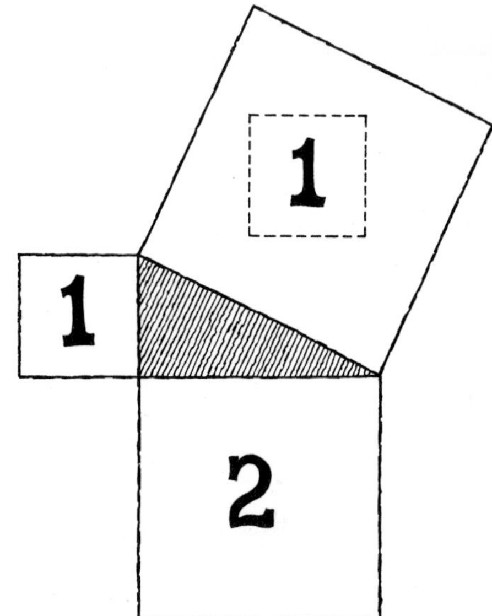

Exercise set 2

R 1. Are there congruent right triangles in the figures shown? If so, tell why. Right angles are marked with ⌝.

(a)

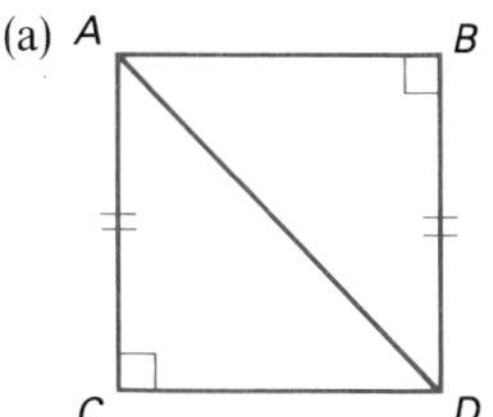

(b)

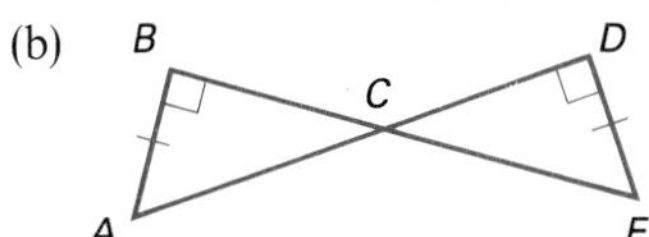

(c)

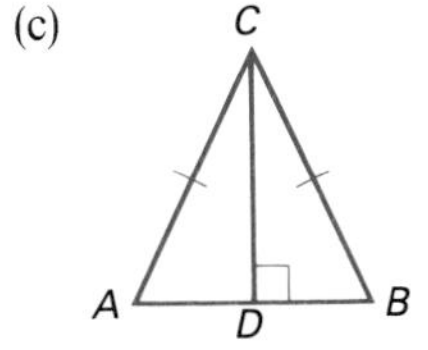

(d)

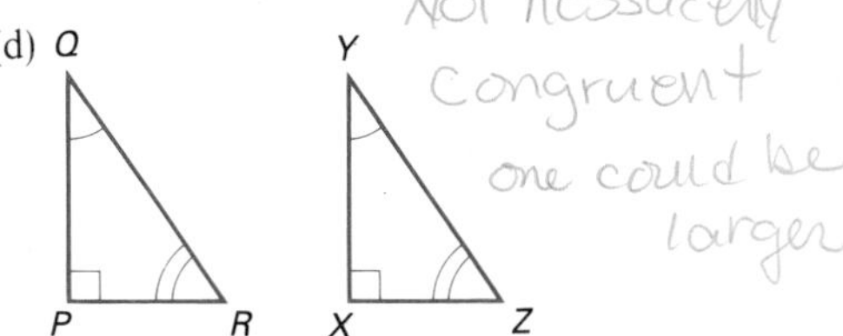

(e)

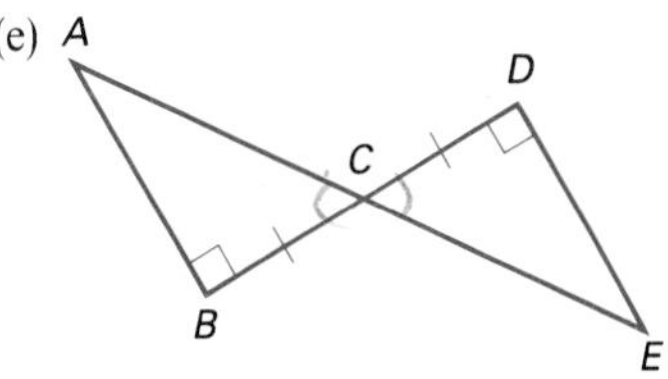

2. Find the length of the missing side on the following triangles.

(a)

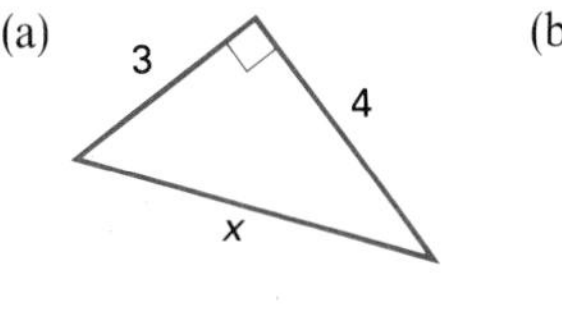

(b)

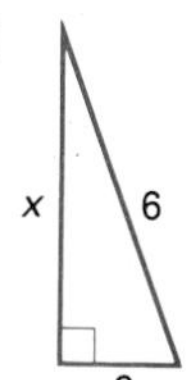

(c)

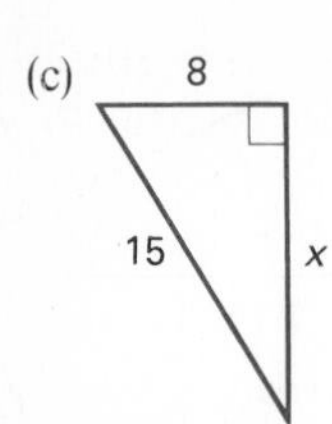

(d)

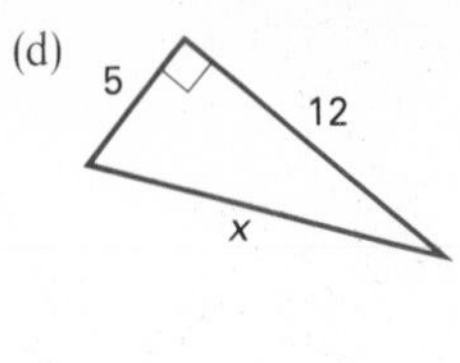

(e)

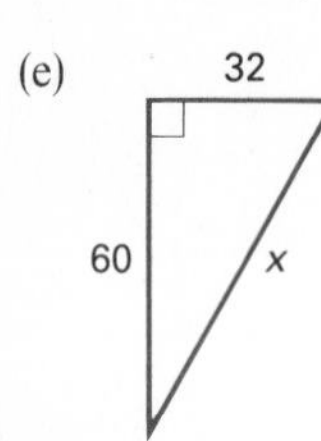

(f)

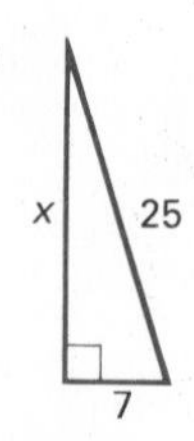

3. Which of the following triplets of numbers are the lengths of the sides of a right triangle?

(a) 5, 12, 13 (b) 20, 23, 31 (c) 2, 36, 42
(d) 4, 5, 6 (e) 4, $\sqrt{5}$, $\sqrt{21}$ (f) 18, 24, 30

4. If m is the axis of reflection and B and C are reflection images, what kind of a triangle is $\triangle ABC$?

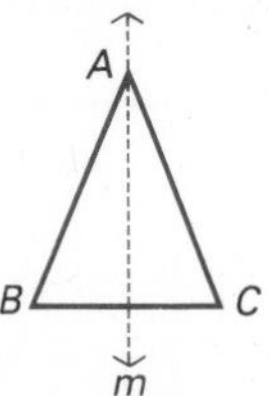

5. Prove that $\triangle ABC \cong \triangle DCB$.

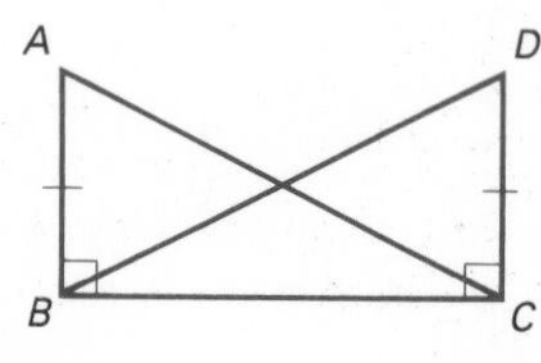

6. Show that $a^2 - y^2 = b^2 - x^2$.

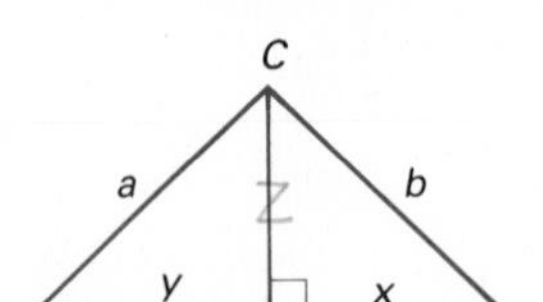

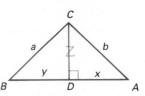

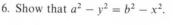

7. Show that $PQ = PR$.

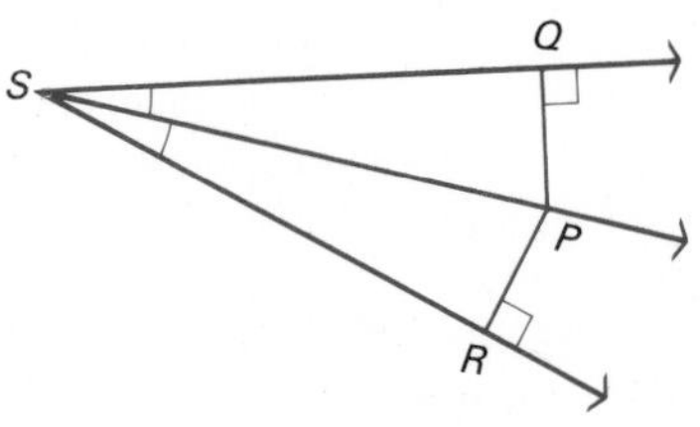

T 8. Show that $(BD)^2 + (AC)^2 = (AB)^2 + (DC)^2$.

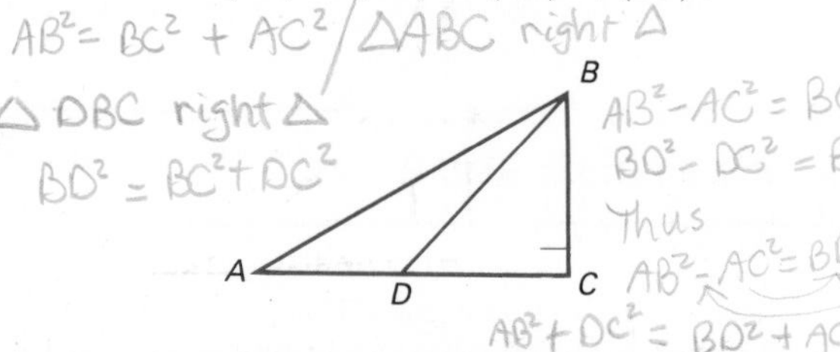

9. Find the length of a rectangle with a width of 8 m and a diagonal of length 17 m.

C 10. $BD = 6$, $AD = 4$. Find BC.

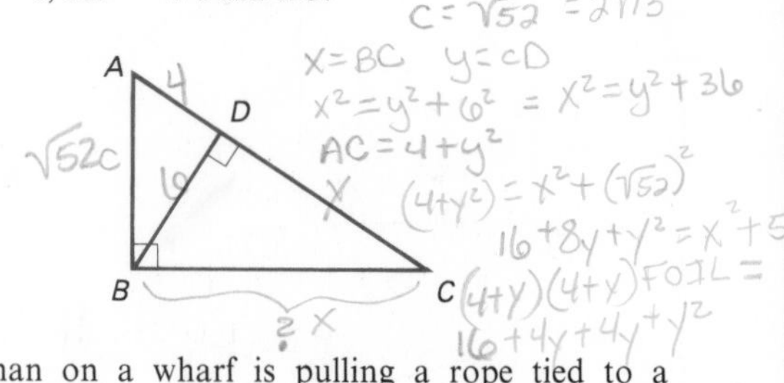

11. A man on a wharf is pulling a rope tied to a boat.

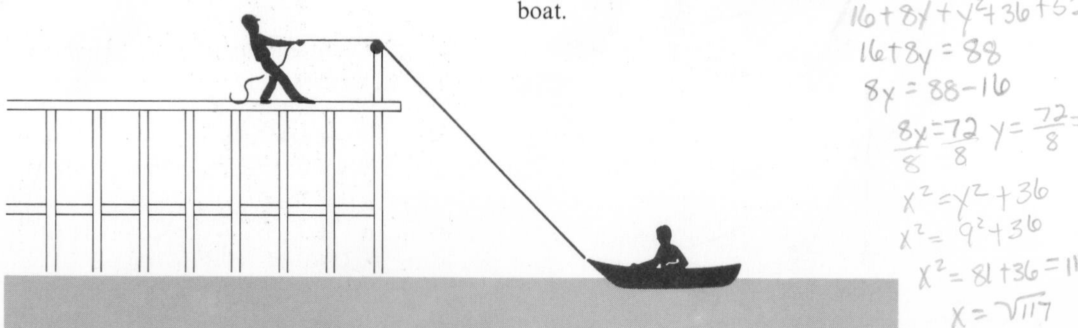

He pulls the rope x meters. Intuitively, will the boat move

(a) fewer than x meters?
(b) exactly x meters?
(c) more than x meters?

12. Joyce said that the distance from her house straight across the lake to Jim's house is 5 km. Jim knows that when he goes from his home to Joyce's by the road, he travels 3 km due south and then 4 km due west. Is Joyce correct about the distance across the lake?

13. John has a ladder that is 15′ long. If he sets the base 2′ from the house, will the ladder reach the gutter, which is 13′ from the ground?

14. The minute hand of Big Ben is 8 ft. long, and the hour hand is 6 ft. long. What is the distance between the tip of the hands at 3:00 P.M.?

15. Prove that $(XJ)^2 + (HB)^2 = (XH)^2 + (JB)^2$.

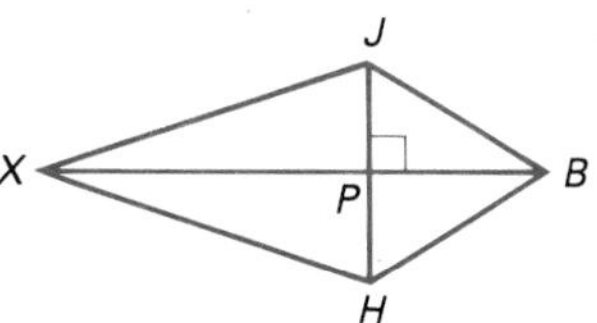

16. You want to take an 8-ft.-wide piece of plywood through a 36-inch-wide door. How tall must the door be in order to handle the plywood?

17. If you know the length of x, show how you would construct a segment of length

(a) $\sqrt{2x}$ (b) $\sqrt{5x}$
(c) $\sqrt{10x}$ (d) $\sqrt{13x}$

18. If $a = 2mn$, $b = n^2 - m^2$, and $c = n^2 + m^2$, show that

$$c^2 = a^2 + b^2$$

Thus, a, b, and c satisfy the Pythagorean Theorem. Values may be substituted for m and n to secure what are called *Pythagorean triplets*. Find 5 triplets.

3 Area and perimeter

■ The sixth-grade mathematics class painted three figures with integral lengths: an isosceles triangle with two sides 10 ft. long and the third side 8 ft.; a square with one side 8 ft.; and a rectangle whose longest side is of the same length as the side of the square. Which figure has the biggest perimeter? Of course, the answer is the ________. You will learn to find all such perimeters in this section. ■

The standard formulas for the area of regions bounded by well-known figures are reviewed in this section. These are the formulas you will need to remember if you teach elementary school mathematics. These will also provide valuable tools for the everyday experiences of life. To help you remember these formulas, we show how they can be developed. This understanding should assist you in remembering how to use each one.

Recall that pairs of opposite sides of a parallelogram are parallel. So a perpendicular to one side will be perpendicular to the opposite side. Any one of the sides of a parallelogram may be designated as its **base.** Once a base is selected, a line segment perpendicular to the base and having its endpoints on the base and on the side parallel to the base is called the **altitude** of the parallelogram. These two terms are used in the formula for the area bounded by a parallelogram.

Consider parallelogram $ABCD$ in Figure 14. Draw altitudes intersecting the base at points E and F. Now $\angle D \cong \angle C$. Why? $AE = BF$. Why? Then $\triangle AED \cong \triangle BFC$, by LA. Therefore, we can translate $\triangle AED$ until it coincides with $\triangle BFC$.

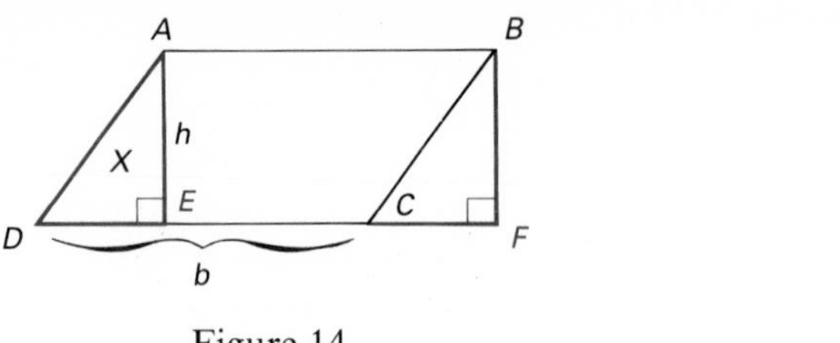

Figure 14

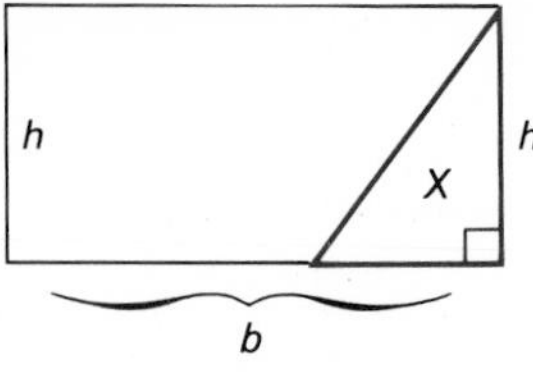

Figure 15

Can you prove that the new figure is a rectangle? The areas of the region bounded by the parallelogram and the region bounded by the rectangle are the same. But the area of the region bounded by the rectangle is *hb*. *Therefore, the area of the region bounded by the* **parallelogram** *ABCD is hb, where h is the measure of the altitude and b is the measure of the base.* (See Figure 15.)

In Figure 16(a), the measure of the altitude (from vertex $A \perp$ to base $\overline{CB}$) of $\triangle ABC$ is given as *h*. Through *B* construct a line parallel to $\overleftrightarrow{AC}$, and through *A* construct a line parallel to $\overleftrightarrow{CB}$, forming parallelogram *ADBC*. Can you prove that $\triangle ABC \cong \triangle BAD$ by ASA?

$$\text{Area } \triangle ABC = \frac{1}{2} \text{ area of parallelogram } ADCB$$

$$= \frac{1}{2} hb$$

Figure 16

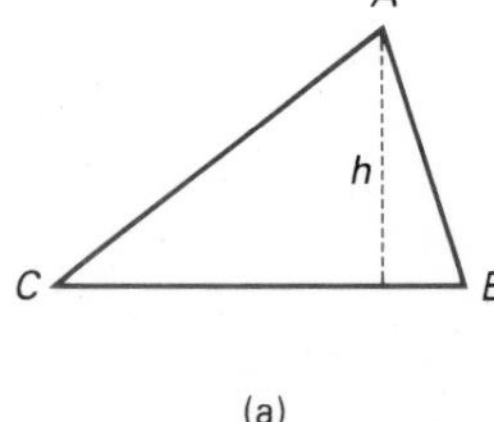

(a)

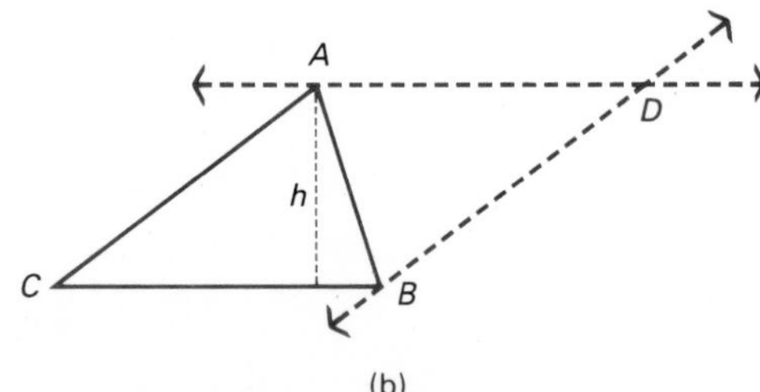

(b)

Let's examine a second approach to showing that the area of a triangle is one-half the base times the altitude. In Figure 17(a), through the vertices at the end of the base, construct lines that are parallel to the altitude. Then through the other vertex of the triangle construct a line parallel to the base. Can you prove that you have constructed rectangle *ADEB*? Now can you prove that the area of triangle *ABC* is one-half the area of the rectangle *ADEB*? But the area of the rectangle is *hb*. Therefore, the area of the triangle *ABC* is $\frac{1}{2}hb$.

Figure 17

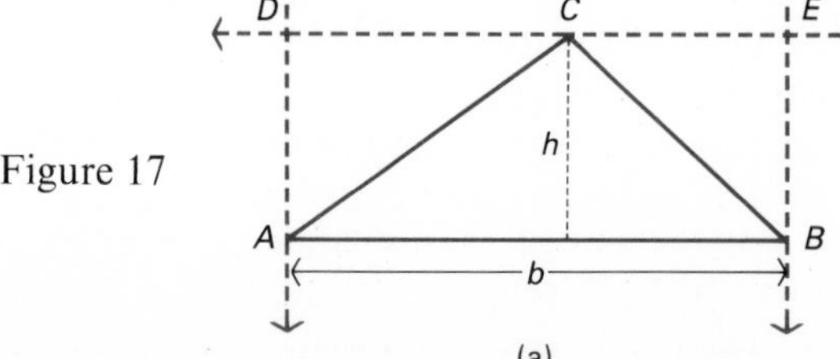

(a)

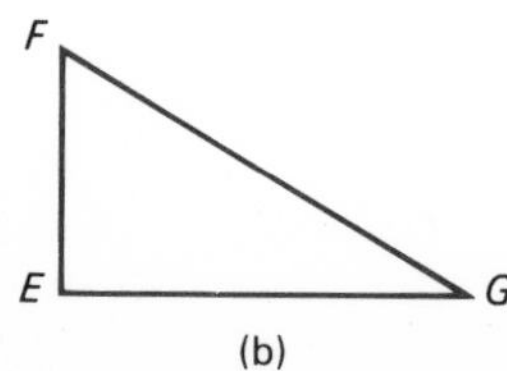

(b)

Of course, if the triangle is a right triangle, such as Figure 17(b), the altitude can be one leg and the base the other leg of the right triangle. Consequently, the area of the region bounded by a right triangle is one-half the product of the legs.

The area of the region bounded by a trapezoid may be considered as the sum of the areas of regions bounded by a rectangle and one or two triangles. For instance, consider trapezoid $ABCD$ in Figure 18. The area of the region bounded by

$$\begin{aligned} ABCD &= (\text{area of the region bounded by rectangle } BCEF) \\ &\quad + (\text{area of the region bounded by } \triangle ABF) \\ &\quad + (\text{area of the region bounded by } \triangle ECD) \\ &= ch + \frac{dh}{2} + \frac{eh}{2} = \left(c + \frac{1}{2}d + \frac{1}{2}e\right)h = \left(\frac{2c + d + e}{2}\right)h \\ &= \left[\frac{c + (c + d + e)}{2}\right]h = \left(\frac{c + b}{2}\right)h \end{aligned}$$

where $b = c + d + e$. *Thus the area of the region bounded by a* **trapezoid** *is one-half the sum of the measures of the opposite parallel sides times the measure of the altitude—that is,* $\frac{1}{2}(c + b)h$.

Figure 18

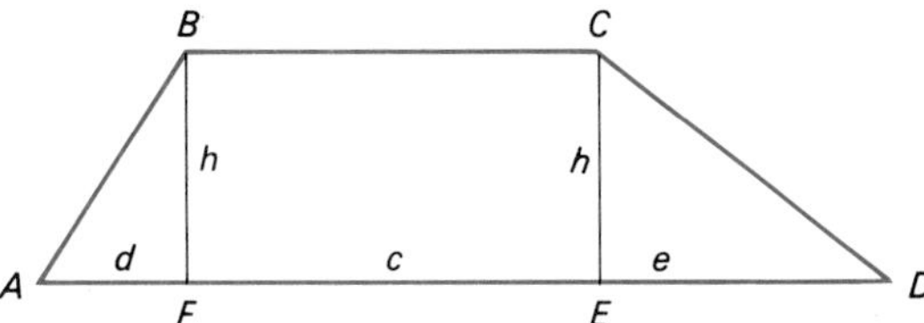

The formulas for the areas of several regions bounded by polygons are given as follows.

1. The **area** of the region bounded by a **parallelogram** is found by multiplying the measure of the base b and the measure of the altitude, or height, h.

$$A = bh$$

2. The area of the region bounded by a **triangle** is one-half the product of the measure of the base b and the measure of the altitude, or height, h.

$$A = \frac{1}{2}bh$$

3. The area of the region bounded by a **trapezoid** is the product of one-half of the sum of the measures of the opposite parallel sides b and c and the measure of the altitude, or height, h.

$$A = \frac{1}{2}(b + c)h$$

$(b + c)/2$ is sometimes called the **average base.**

Example 5 • What is the area of a region bounded by a parallelogram with a base of 5 cm and a height of 3 cm?

Solution

$$A = bh$$

$$A = 3 \cdot 5 = 15 \text{ cm}^2 \quad \bullet$$

Example 6 • Find the area of a triangular region with a base of 6 cm and a height of 3 cm.

Solution

$$A = \frac{1}{2}bh = \frac{1}{2}(6)(3) = 9$$

The area is 9 cm^2. •

Example 7 • Find the area of a region bounded by a trapezoid where the parallel sides measure 6 cm and 10 cm and the height measures 5 cm.

Solution

$$A = \frac{1}{2}(b + c)h$$

$$= \frac{1}{2}(6 + 10)5$$

$$= 40$$

The area is 40 cm^2. •

Example 8 • A residential property developer discovers what he considers to be a perfect tract of land, a rectangle measuring 1 hm by 2 hm. How many hectares are contained in this tract?

Solution

$$A \text{ (of rectangle)} = l \cdot w = 1 \cdot 2 = 2 \text{ hm}^2 = 2 \text{ ha} \quad \bullet$$

Another useful measure associated with a polygon is perimeter. Perimeter is used to find the amount of weatherstripping to go around a window or the length of fence for your backyard.

The **perimeter,** P, of a simple closed curve is the distance around the curve. It can be called the measure of the simple closed curve. The perimeter of a polygon is the sum of the measures of the sides of the polygon. For example, the perimeter of the polygon in Figure 19(a) is

$$P = 8 + 8 + 8 + 6 + 4 = 34$$

It is easy to develop formulas for the perimeter of certain much-used polygons.

Example 9 • Develop formulas for the perimeters of rectangles and squares.

Solution For the rectangle in Figure 19(b),

$$P = l + w + l + w = 2l + 2w = 2(l + w)$$

Figure 19

(a) (b) (c)

For the square in Figure 19(c),

$$P = s + s + s + s = 4s \quad \bullet$$

Example 10 • The perimeter of a regular pentagon is 65 cm. What is the length of each side?

Solution Since a regular pentagon has five sides of equal measure,

$$5s = 65$$
$$s = 13 \quad \bullet$$

Consider two special triangles, an isosceles triangle in Figure 20(a) and an equilateral triangle in Figure 20(b). For the isosceles triangle, denote by s the congruent sides and by $2b$ the base. Draw a perpendicular from vertex B to the base. By finding congruent triangles, you can easily show that this perpendicular bisects the base. Next, consider the right triangle BDC, with legs h and b and hypotenuse s. $h^2 + b^2 = s^2$, or $b = \sqrt{s^2 - h^2}$. The area of the isosceles triangle is

$$\begin{aligned} A &= \frac{1}{2}h(2b) \\ &= hb \\ &= h\sqrt{s^2 - h^2} \end{aligned}$$

In a like manner, draw a perpendicular in Figure 20(b) from vertex B to $\overline{AC}$. This segment bisects side $\overline{AC}$. Then consider the right triangle BDC.

Figure 20

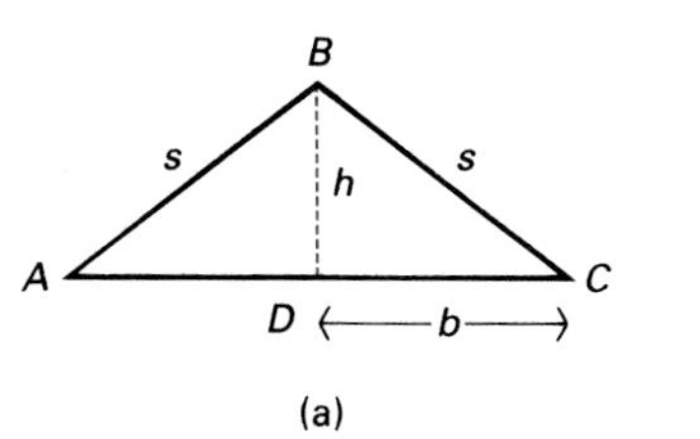

(a)

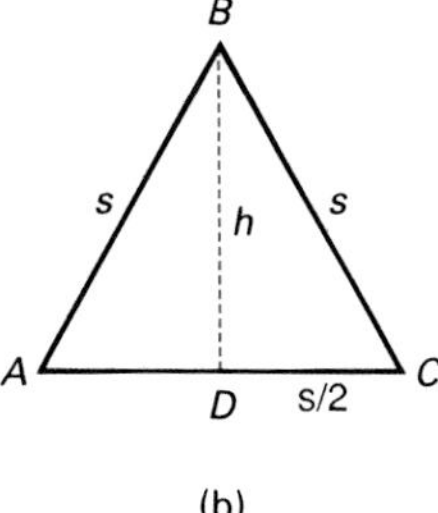

(b)

$$h^2 + \left(\frac{s}{2}\right)^2 = s^2$$

$$h^2 = s^2 - \frac{s^2}{4}$$

$$h^2 = \frac{3s^2}{4}$$

$$h = \frac{\sqrt{3}s}{2}$$

Therefore, the area of the equilateral triangle is

$$A = \frac{1}{2}hs$$

$$= \frac{\sqrt{3}}{4}s^2$$

Areas of isosceles and equilateral triangles

(a) The area of the isosceles triangle with congruent sides of measure s and altitude to the third side (h) is

$$A = h\sqrt{s^2 - h^2}$$

(b) The area of the equilateral triangle with side s is

$$A = \frac{\sqrt{3}s^2}{4}$$

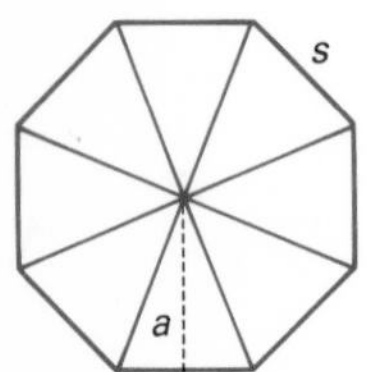

Figure 21

For a regular polygon with n sides, called an ***n*-gon,** draw lines from the center of the polygon to each vertex. (See the octagon in Figure 21). These lines from the center to each vertex, along with the sides of the polygon, give n isosceles triangles. Since the polygon is regular, these n triangles are congruent.

The area of the n-gon is the sum of the areas of the n congruent isosceles triangles. Thus the area of the polygon is n times the area of an isosceles triangle. Let the length of each side of the polygon be s and the length of the perpendicular be a. (The distance a from the center of a regular polygon to each of the sides is called the **apothem** of the polygon.) The area of each isosceles triangle is $\frac{1}{2}sa$. The area of the polygon is

$$A = n(\text{area of each triangle})$$

$$= n\frac{1}{2}sa$$

$$= \frac{1}{2}(ns)(a)$$

But ns is the perimeter (p) of the n-gon. Thus, the area of the polygon is $\frac{1}{2}pa$.

Just for fun

A famous puzzle about a water lily was introduced by Henry Longfellow in his novel Kavanagh. When the stem of a water lily is vertical, the blossom is 22 cm above the surface of the water. When moved by the gentle breeze, keeping the stem straight, the blossom touches the water at a spot 45 cm from where the stem formerly cut the surface. How deep is the water?

Area of an n-gon

The area of a regular n-gon with sides s is given by

$$A = \frac{pa}{2}$$

where $p = ns$ is the perimeter of the n-gon and a is the apothem.

Exercise set 3

R 1. The perimeter of a rectangle is 50 m. Its length is 12 m. What is its width?

2. Find the perimeter of a regular hexagon when each side is 3.2 cm.

3. Find the areas bounded by the polygons described here.

Rectangles:
(a) 3 m by 5 m
(b) 21.4 cm by 170 dm (express in terms of dm^2)

Triangles:
(c) $b = 34$, $h = 52$
(d) $b = 23$ cm, $h = 497$ mm (express in mm^2)

Parallelograms:
(e) $b = 24$, $h = 12$
(f) $b = 53$ m, $h = 539$ dm (express in m^2)

Trapezoids:
(g) $h = 3$, $b = 3$, $c = 2$
(h) $h = 3$ cm, $b = 124$ mm, $c = 5$ cm (express in cm^2)

4. Find the area of each figure.

(a)

7
4
8
12
6

(b)

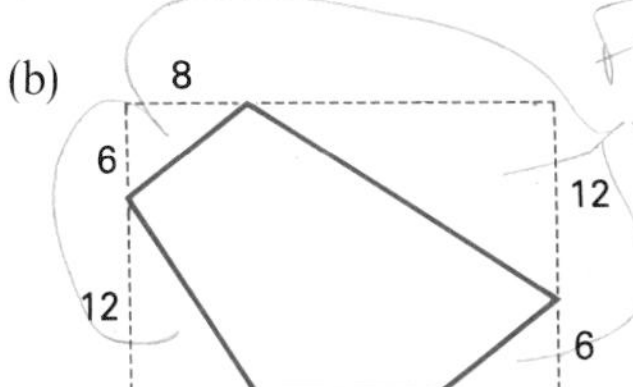

5. Find the area of the 12-sided regular polygon with side 8 and apothem $4\sqrt{3}$.

6. Find the measurement of the regions bounded by each figure shown, using the dimensions given.

(a)

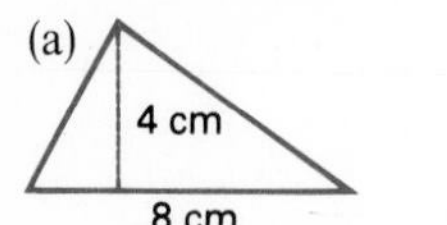

(b)

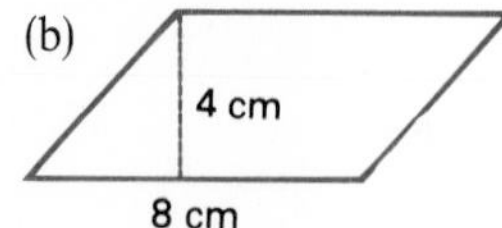

7. Find the area of the equilateral triangle with side 6 cm.

8. Find the area of the regular polygon with perimeter 24 and apothem 4.

9. Find the area of the region bounded by a trapezoid with altitude a and bases b and c if

   (a) $a = 6$, $b = 10$, $c = 8$.
   (b) $a = 4\frac{1}{2}$, $b = 6\frac{1}{2}$, $c = 7\frac{1}{2}$

10. Find the area of the equilateral triangle with sides of length 10.

T 11. In trapezoid $ABCD$, $\overline{AB}$ is parallel to $\overline{CD}$, and $\overline{AD}$ is perpendicular to $\overline{CD}$. If $AB = 6$, $AD = 4$, and $DC = 8$, find the area of the trapezoidal region.

12. If a triangle and a parallelogram has equal areas and equal bases, what can you say about their altitudes?

13. A rancher decided to give each of four children an equal portion of land. The rancher, however, made sure that he himself kept certain favorite parts of the land. The land to be divided among the children looked similar to the figure shown. How much land would each child receive?

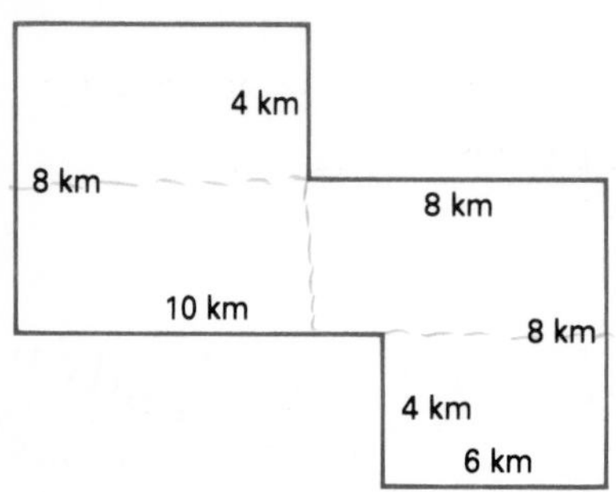

14. For triangular region ABC with altitude to base $\overline{AB}$ denoted by $\overline{CD}$, what happens to the area of the triangular region if

   (a) AB is doubled?
   (b) AB and CD are both doubled?
   (c) CD is tripled?
   (d) AB and CD are both tripled?

15. Discuss the truth of the statement that two polygons with equal areas are congruent.

16. Find the length of each side of an equilateral triangle with area $5\sqrt{3}$.

17. The Joneses plan to remodel parts of their home: the hall, living room, dining room, kitchen, and den.

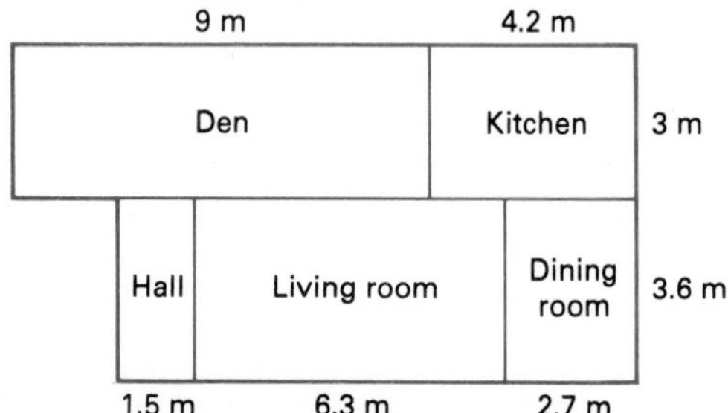

   (a) If the hall is to be carpeted in one pattern and the living and dining rooms in another, how many square meters of each carpet would Mrs. Jones need to purchase?
   (b) How much carpet would be needed if all three rooms in part (a) were carpeted in the same pattern?
   (c) Mr. Jones wants to tile the den and kitchen himself. How much area must Mr. Jones tile in the kitchen? In the den?
   (d) Suppose that Mr. Jones plans to buy 15-cm square tiles. If the kitchen is one pattern and the den another, how many tiles of each pattern should he buy? How much tile should he buy if he uses the same pattern for both rooms?

18. Find the perimeter of a regular polygon if its area is 100 units^2 and its apothem is 10.

19. The roof of a stadium has the shape of a regular hexagon with side 60 m. Find the area of the roof.

Prove or disprove the statements in Exercises 20 through 23.

C 20. The ratio of the areas of two similar regular polygons is equal to the ratio of their perimeters.

21. The ratio of the areas of two similar regular polygons is equal to the ratio of their apothems.

22. The ratio of the apothems of two similar regular polygons is equal to the ratios of the lengths of two corresponding sides.

23. The ratio of the areas of two similar regular polygons is equal to the square of the ratio of their apothems.

4 Properties of circles

■ For the set of gears shown, if the left gear is turning clockwise in which direction will the right gear turn? ■

The above application of gears emphasizes the importance of circles. In this section, we define several geometric terms associated with circles, learn the relationship between central angles and arcs, discuss inscribing and circumscribing circles with polygons, and finally compute the circumference and area of circles.

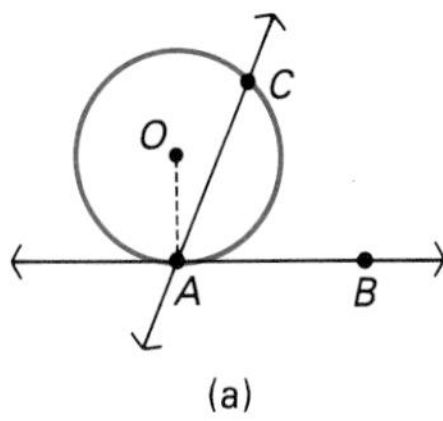

(b)

Figure 22

Let's review the definition and terms associated with a circle before we introduce the area bounded by a circle. A circle is a set of coplanar points equidistant from a fixed point called the center. A **radius** of a circle is a segment joining the center to a point on the circle. We also use radius to mean the measure of such a segment. For example, we say a circle of radius 4. A **chord** of a circle is a segment that joins any two points on the circle. The **diameter** of a circle is a chord containing the center of the circle.

The **interior** of a circle is the set of points whose distance from the center of the circle is less than the radius. We will be interested in the area of the interior of a circle. The **exterior** of a circle is the set of points whose distance from the center is greater than the radius.

In general, a line intersects a circle in two points. Such a line is called a **secant line** ($\overleftrightarrow{AC}$ in Figure 22(a)). A line that intersects a circle in only one point is called a **tangent line** or **tangent** $\overleftrightarrow{AB}$ in Figure 22(a)).

Three interesting properties of tangents to circles are summarized as follows.

Properties of tangents

(a) If a line is tangent to a circle, then it is perpendicular to the radius drawn to the point of tangency.
(b) If a radius is perpendicular to a line at a point on the circle, then the line is tangent to the circle.
(c) Tangent circles have a common tangent line at a point of intersection (Figure 22(b)).

The **central angle** of a circle is an angle whose vertex is at the center of a circle. Such angles are very useful in drawing a regular polygon **inscribed** in a circle (the circle contains each vertex of the polygon). The number of degrees in the central angle of the polygon (also central angle of the circle) is $360°/n$, where n is the number of sides of the polygon.

Construction 3 • To construct a ten-sided regular polygon (a decagon), each central angle will be $360° \div 10 = 36°$, as shown in Figure 23. •

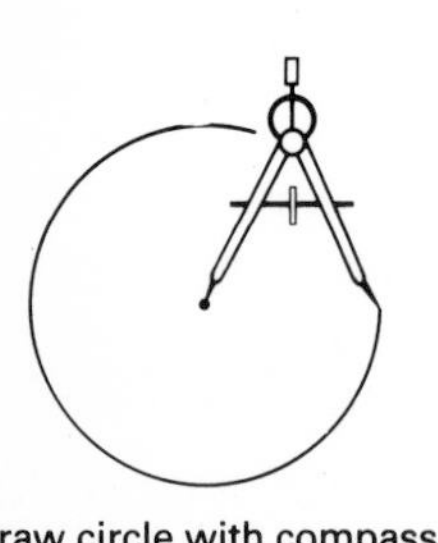
Draw circle with compass

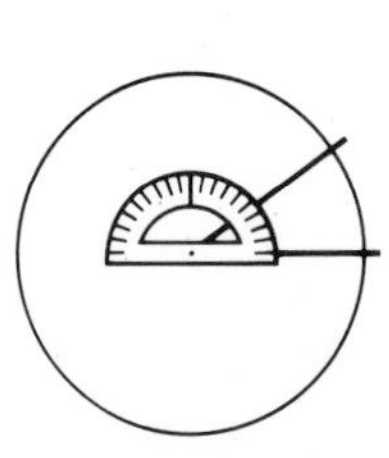
Measure off 36° angle

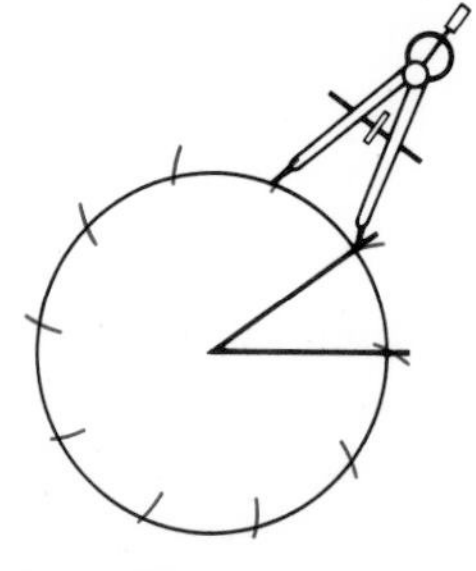
"Pace off" 10 equal arcs

Connect the 10 points

Figure 23

A polygon is **circumscribed** about a circle if the circle is inscribed in the polygon (each side of the polygon is tangent to the circle, as in Figure 24).

Figure 24

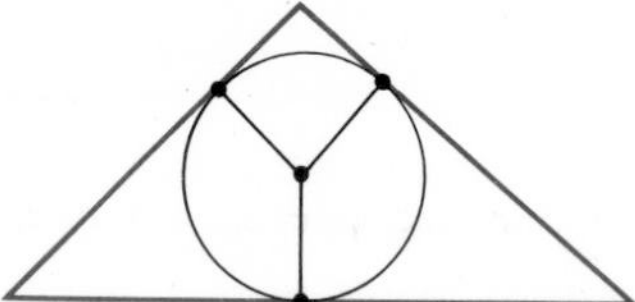

An arc of a circle is a set of points on a circle that are in the interior of a central angle. That is, the central angle actually defines the arc. In this discussion, the measure of an arc such as in Figure 25 is defined in terms of the degree measure of the central angle.

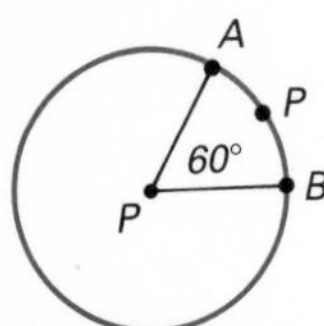

Figure 25

$$m(\widehat{AB}) = 60°$$

The perimeter of a circle is called the *circumference of the circle* and is, of course, the total length of the simple closed curve. Intuitively, one could stretch a string along the circle and then measure the string. Thus the measurement of a circumference is given in linear units.

The ancient Greeks found by experimentation that the circumference of a circle divided by the diameter seemed to give the same answer regardless of the size of the circle. At first they used 3 and then 3.14. Later, it was proved that the quotient of the circumference of the circle and the diameter is a real number that is called *pi* and denoted by the lowercase Greek letter π. Thus,

$$C = \pi d$$

Pi (π) is an irrational number that can be approximated by a rational number to as many decimal places as desired; accurate to eight places, $\pi = 3.14159265\ldots$. The rational number $\frac{355}{113}$ approximates π to six decimal places. Perhaps the most widely used approximation for π is $\frac{22}{7}$, which agrees with the decimal representation of π to only two decimal places.

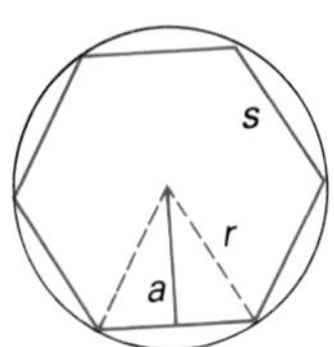

Figure 26

Suppose a regular n-gon is inscribed in a circle (see the hexagon in Figure 26). Let the perimeter of the polygon be p and the apothem a. From the preceding section, the area of the polygon is

$$\frac{ap}{2}$$

As the number of sides n increases, a approaches r and p approaches the circumference of the circle. Thus, for very large values of n, $ap/2$ is approximately

$$\frac{rC}{2} = \frac{r}{2}(2\pi r) = \pi r^2$$

So the area of the polygon as the number of sides increases is approaching the area of the circle and is getting close to πr^2. It can be shown that πr^2 is actually the area of a circle.

Circle

There is a real number π in terms of which the area (A) and circumference (C) of a circle are computed.

1. $C = 2\pi r = \pi d$
2. $A = \pi r^2$

Example 11 • Find the circumference of a circle with radius of measure 20 cm. Approximate π by $\frac{22}{7}$.

Solution

$$\begin{aligned} C &= 2\pi r \\ &= 2\pi(20) \text{ cm} \\ &\approx 40 \cdot \frac{22}{7} \text{ cm} \\ &\approx 126 \text{ cm} \end{aligned}$$ •

Example 12 • Find the area of the circular region where the radius measures 8 cm. Approximate π by $\frac{22}{7}$.

Solution

$$\begin{aligned} A &= \pi r^2 \\ &\approx \left(\frac{22}{7}\right) \cdot (8)^2 \text{ cm}^2 \\ &\approx 201 \text{ cm}^2 \end{aligned}$$ •

Example 13 • If a bicycle wheel has a radius of 50 cm, how many times does it turn in going 1 km? Approximate π by 3.14.

Solution

$$\text{Circumference} = 2\pi(50) = 100\pi \text{ cm}$$
$$\approx 100(3.14) \text{ cm}$$
$$\approx 314 \text{ cm}$$
$$1 \text{ km} = 100{,}000 \text{ cm}$$
$$\frac{100{,}000}{314} \approx 318 \text{ revolutions} \quad \bullet$$

Exercise set 4

R 1. Given that O is the center of the circle, let C and A represent the circumference and the area, respectively.

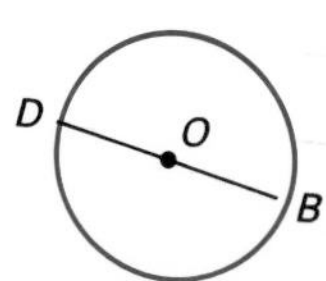

(a) If $DB = 6$, find C and A.
(b) If $C = 8\pi$, find OB and A.
(c) If $A = 16\pi$, find OB and C.
(d) If $OB = \pi$, find C and A.

2. Complete the following table, using the information on each line. Leave answers in terms of π.

Diameter	*Radius*	*Area*	*Circumference*
____	6	____	____
____	____	16π	____
____	____	____	22π
18	____	____	____

3. Find the measurement of the area of each circular region for which the measurement of the radius is given. This time, approximate π by 3.1416. Assume all measurements are accurate to 5 significant digits.

(a) 4 ft. (b) 10 m (c) 12 cm (d) 9 yd.

4. The given figure is a semicircle; the area of the interior of this closed curve is 24π. Find the length around the closed curve.

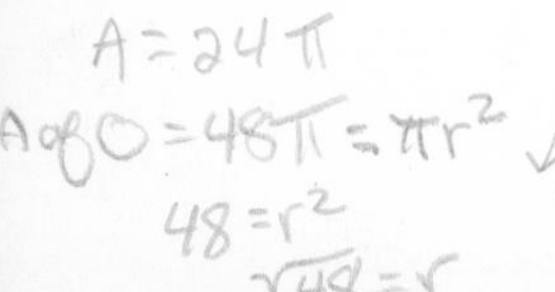

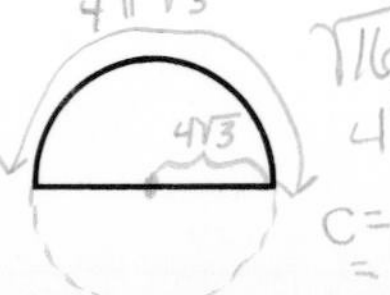

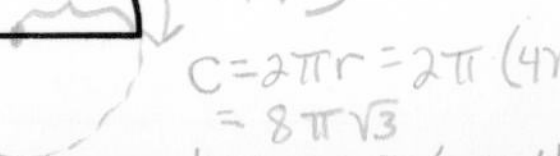

5. A circle has a circumference 120π cm. What is its area?

Find the circumference of the circle circumscribed about the following.

T 6. A square with side 10.

7. A right triangle whose legs are 6 and 8.

8. An equilateral triangle with side 10.

9. A regular hexagon with side 10.

Find the area of each shaded region of 10 through 14.

10. 8, 4

11. 8, 4

12. 13.

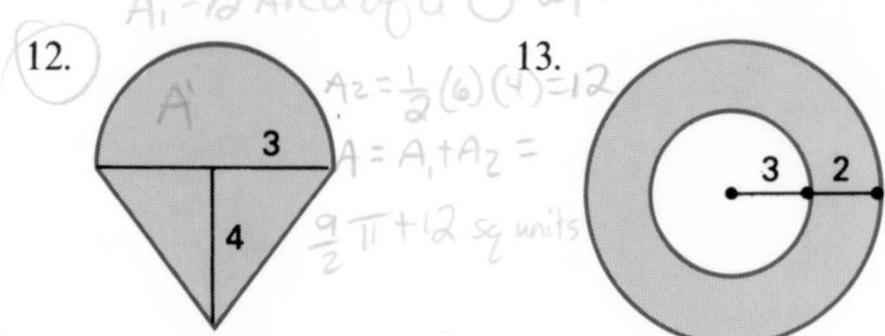

14.

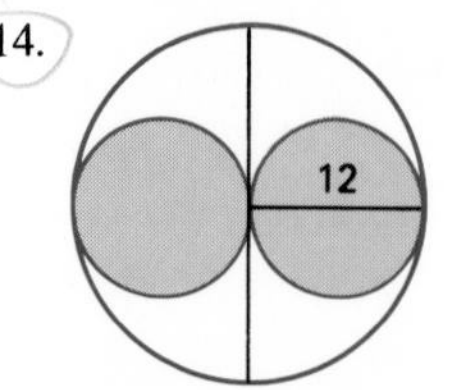

15. If the radii of two circles are in ratio of 1 to 3, what is the ratio of their circumferences? Their areas?

C 16. The famous Yang-Yin symbol shown below has been constructed from a circle of radius r. Compute the length of the arc that divides the colored and white regions.

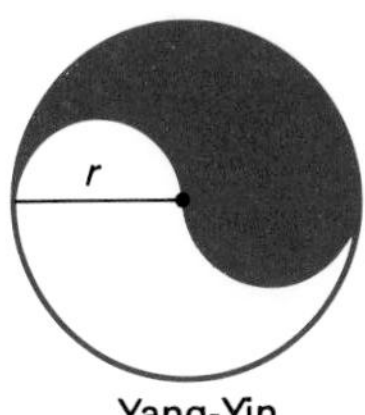

Yang-Yin

17. A regular polygon with sides of length 8 and apothem 10 is inscribed in a circle. Find the area of the circle.

18. The formula for the area of a circle, $A = \pi r^2$, is given in terms of the radius. Express the formula for the area of a circle in terms of the circumference.

19. If a point of a circular wheel traverses the circumference of the wheel as the wheel is rotated, the wheel is said to have been rotated through one revolution. How many revolutions are made in 1 km by a car wheel with a diameter of 81 cm?

20. If a car goes 80 km per hour, how many revolutions does a wheel 77 cm in diameter turn in a half hour?

21. Find the ratio of the area of an inscribed square to a circumscribed square for a circle.

Prove or disprove the following statements for two circles.

22. The ratio of the areas equals the ratio of the radii.

23. The ratio of the areas equals the ratio of the squares of circumference.

*5 Topological motions

No discussion of geometry would be complete without a brief introduction to topological motions.

■ In his book *The Mathematics of Distortion,* D. Bergamini classifies topology as the "mathematics of distortion." Topological geometry is a special kind of geometry that studies figures and surfaces when they are distorted from one shape to another. Such figures can be bent, twisted, enlarged, or shrunk. A figure may be changed so much by a topological motion that one would not readily recognize that one figure is the topological image of the other. ■

In this section, we shall give only a brief introduction to give you some idea of the subject. A more thorough presentation would be too complicated.

Let's begin this discussion with a definition.

Topological mapping

A mapping of one geometric figure to another is a topological mapping if

1. the mapping is a one-to-one correspondence between the two sets.
2. the mapping is continuous.

The word *continuous* used in this definition has not been defined. For the purpose of our discussion, we shall use the intuitive notion that, for a continuous mapping, the image of an unbroken curve must be unbroken. Thus in a topological mapping, you cannot deform a curve by cutting it or by punching a hole in the region it bounds.

Topologically equivalent

> If there is a topological mapping from one set to another, the two sets are said to be *topologically equivalent.*

In a plane, topology is sometimes called "rubber-sheet" geometry, because topological mappings in a plane can generally be considered as distorting the rubber sheet. Figure 27 shows a topological mapping, and Figure 28 does not. Why not?

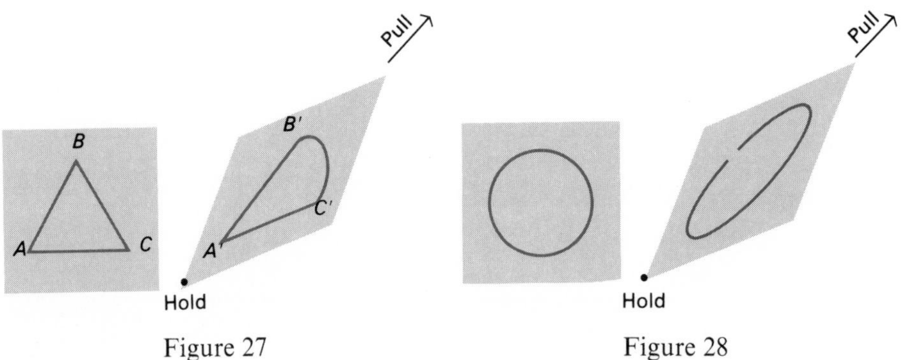

Figure 27

Figure 28

Example 14 • All simple closed curves such as those pictured in Figure 29 are topologically equivalent.

Figure 29

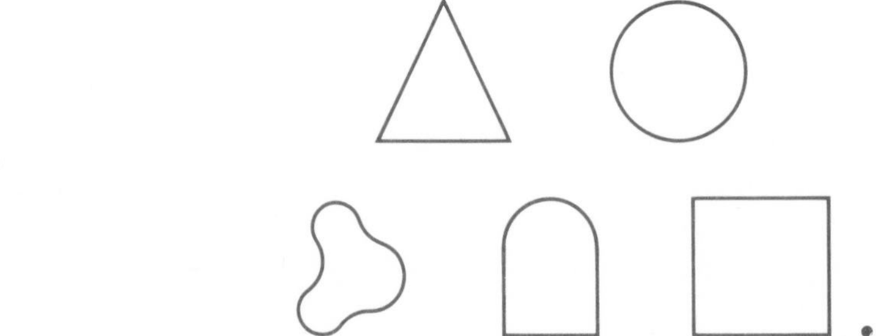

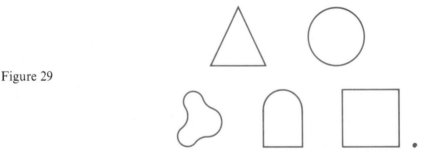

Did you note in the preceding example that topologically, every circle is a square, every triangle is a circle, etc.?

Example 15 • All simple nonclosed curves such as those in Figure 30 are topologically equivalent. (Some like to think of these as a broken rubber band that can be stretched into the different shapes.)

Figure 30

Just for fun *Tie a piece of cord to each wrist. Then tie a cord to the wrists of a second person and through your cord so that you are tied together. Now see if you can separate yourselves without cutting the cord.*

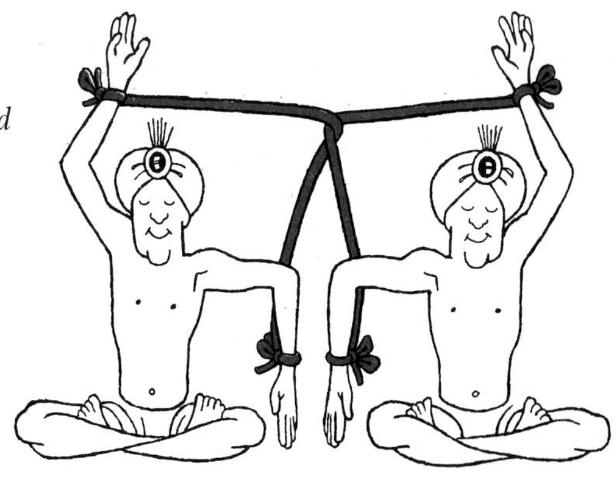

Example 16 • All bounded convex three-dimensional figures such as spheres, cylinders (with ends), and polyhedra are topologically equivalent. (See Figure 31.)

Figure 31

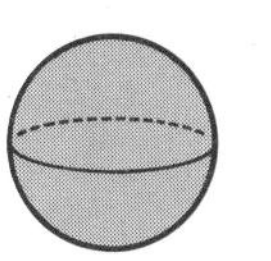
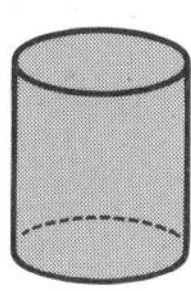
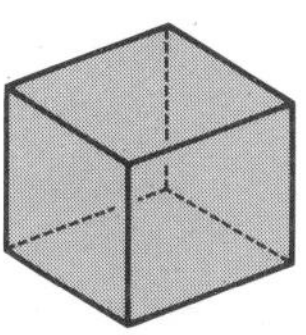

•

Exercise set 5

R 1. Group the objects in (a) to (r) into classes so that all the elements within each class are topologically equivalent and no elements from different classes are topologically equivalent.

(a) A glass (b) A ring
(c) A straw (d) A bowling ball
(e) A circular saw blade
(f) A record
(g) A sheet of two-ring-binder paper
(h) A section of pipe
(i) A sheet of typing paper
(j) A pitcher with two handles
(k) A cup (l) A doughnut
(m) A ruler (n) A sewing needle
(o) A funnel with a handle
(p) A brick
(q) A banana (r) A pencil

2. Group (a) through (o) into classes so that all the elements within each class are topologically equivalent and no elements from different classes are topologically equivalent.

(a) 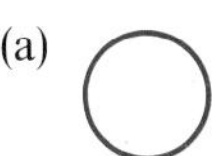
(b) 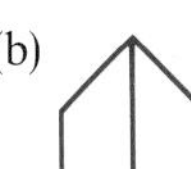
(c) 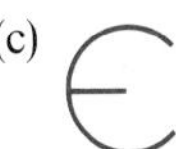
(d) 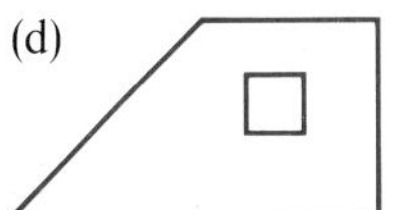
(e) 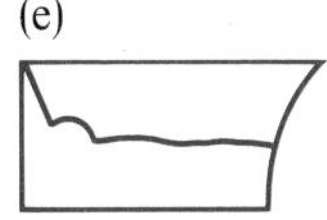
(f) 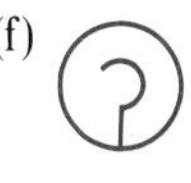
(g) 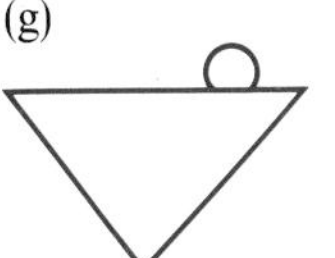
(h) 
(i) 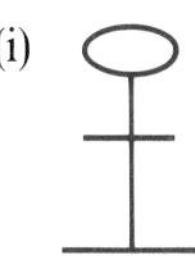
(j) 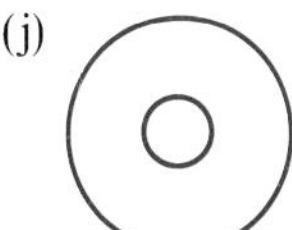
(k) 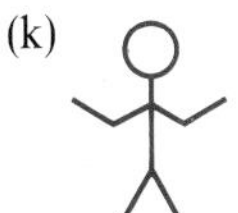
(l) 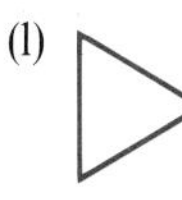

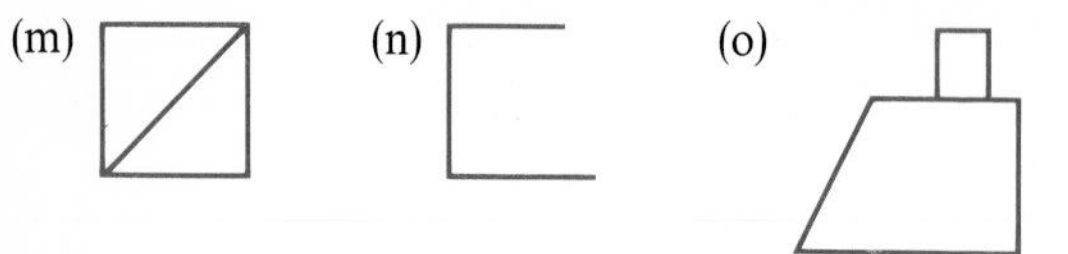

(e)

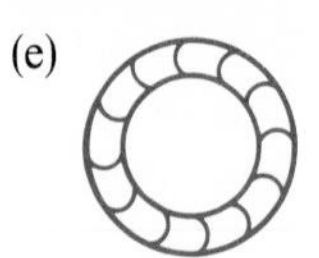

3. Classify the following printed letters into sets in which all the letters in each set are topologically equivalent.

ABCDEFGHIJKLM
NOPQRSTUVWXYZ

4. The following sets are listed as topologically equivalent. Indicate as true or as false.

5. Which pairs of digits are topologically equivalent?

0, 1, 2, 3, 4, 5, 6, 7, 8, 9

T 6. In each part shown, complete the figure on the right so that it is topologically equivalent to the figure on the left.

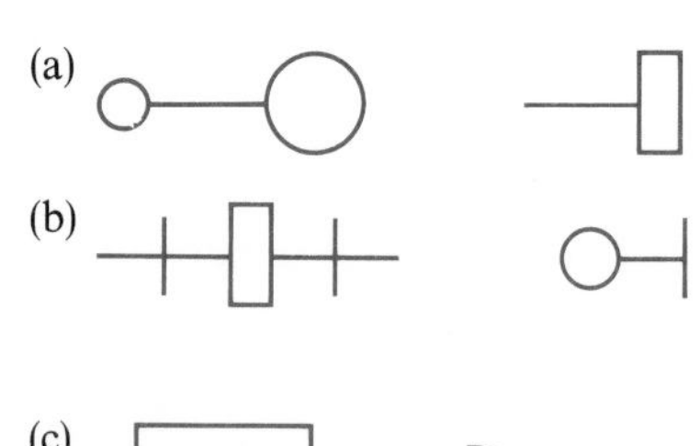

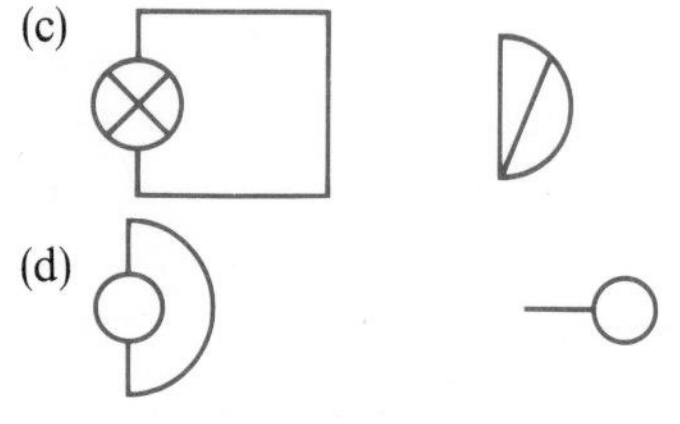

C 7. Using three matchsticks, how many topologically different patterns can be constructed if the matches meet only at endpoints?

8. Repeat Exercise 7 for six matches.

6 Traversing a graph or a network

■ We now consider a branch of mathematics that can be applied to mail routes, scheduling, modeling, and puzzles and that is even useful to the garbage man. It is called **graph theory** or **network theory.** A family tree is a graph or a network; so are road maps and electric circuits. Thus, these graphs or networks are much different from those you may have studied previously. You can easily construct a network. Just put some points on a paper and connect some of them with lines. You have a network. Isn't this easy? ■

Tradition tells that network theory originated in an attempt to solve a famous 18th-century problem about the city of Königsberg in Germany (now Kaliningrad in Russia) located on the banks and on two islands of the Pregel River. The various regions of the city were connected by seven bridges, as shown in Figure 32. The famous problem is as follows: Is it possible to take a walk around the city in such a way that you cross each bridge exactly once?

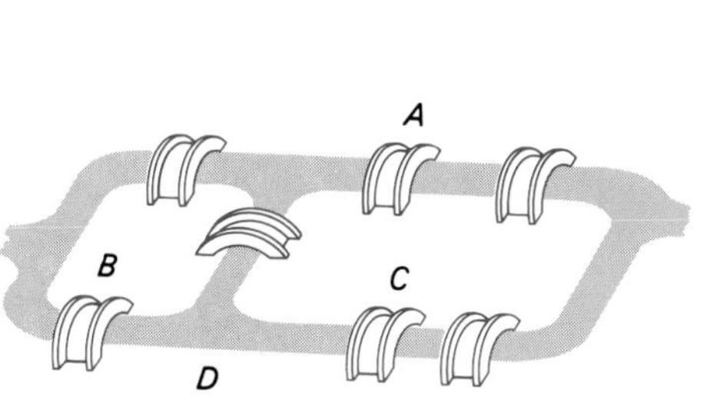

Figure 32

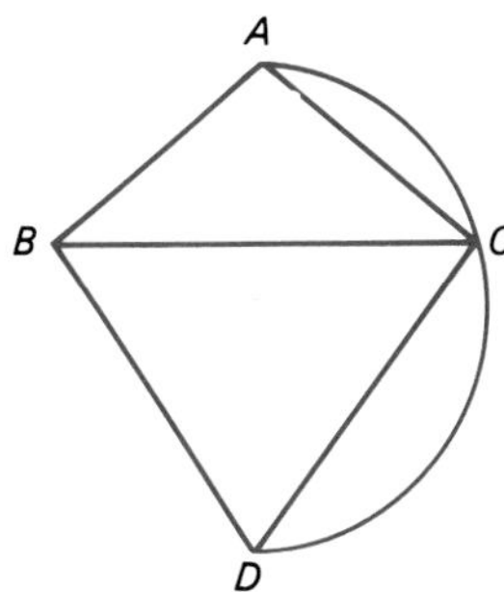

Figure 33

A discussion of this problem was given by the Swis mathematician Leonhard Euler. He represented the four land areas A, B, C, and D by four points (Figure 33) and the seven bridges by seven line segments or curves joining the points. This type of diagram is called a **network.**

Network

> A network consists of a collection of points, called **vertices,** and line segments or curves, called **arcs,** that join all, some, or none of the vertices.

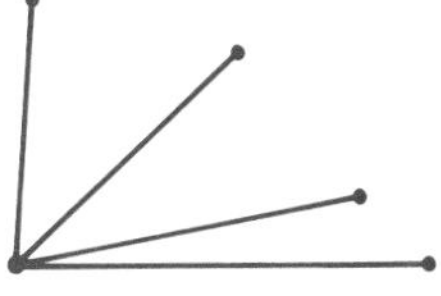

Figure 34

Figure 33 is a network. So is Figure 34. In designing a model for a given situation, objects or events are vertices, and arcs are used to show relationships between two vertices. In the Königsberg problem, note again that the land areas are represented by vertices and the bridges by arcs. Our problem is to determine if there is a path following arcs so that one can start at one vertex and travel along each arc one and only one time. If this happens, we say we have **traversed** the network.

Traversable

> A network is traversable if it is possible to trace a path passing from vertex to vertex on arcs so that the path traces each arc exactly once.

We now introduce the theory that will be used to determine whether or not the Königsberg graph is traversable. The theory leading to this solution involves the order of a vertex.

Order of a vertex

> The order of a vertex is the number of arcs that terminate at a vertex. A vertex with an odd number of arcs is called an *odd vertex.* Likewise, a vertex with an even number of arcs is an *even vertex.* A vertex with no arcs is said to be *null.*

Note for the Königsberg problem that all vertices are odd. What does this say about a possible solution? Let's look at some possible situations. Since every arc that has a beginning must have an end, the sum of the orders of the vertices of a graph is an even number. Therefore, for any graph, there will be 0, 2, 4, 6, . . . odd vertices.

Now as we traverse through a vertex, we use two arcs—one arriving and one leaving. What does this mean? The only way we can have an odd vertex on a traversable graph is for that vertex to be either the beginning or the ending (but not both) of a path.

The preceding discussion should provide intuitive insight into the discovery made many years ago by Leonhard Euler.

Traversable networks

A network with no null vertices is traversable if and only if it satisfies one of the following descriptions.

1. All vertices are of even order. In this case, the path can begin and end at the same vertex. Such a path is called an **Euler circuit.**
2. Two vertices are of odd order, and all the rest are of even order. In this case, the path begins at one odd vertex and ends at the other. Such a path is called an **Euler path.**

Since the Königsberg graph has four odd vertices, it is obviously not traversable.

Example 17 • Is the network in Figure 35 traversable? If so, describe the path.

Figure 35

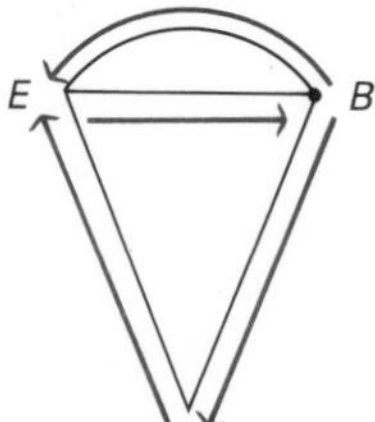

Figure 36

Solution There are two odd vertices and one even vertex. Therefore, the path is traversable. With two odd vertices, the path is an Euler path and must start at one of the two odd vertices. One possible path is given in Figure 36. •

An interesting problem using this theory may be stated as follows.

Example 18 • Is it possible to take a trip through the house described in Figure 37 and pass through each door once and only once?

Solution We use a network theory model as follows. In Figure 38, we let each room be a vertex and each door an arc. Label the rooms A, B, C, D, and E. Label the area outside of the house as F. Now a door connects the outside and A; so draw an

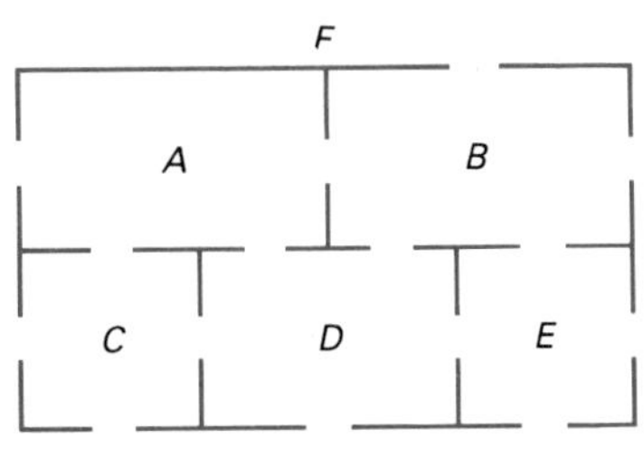

Figure 37

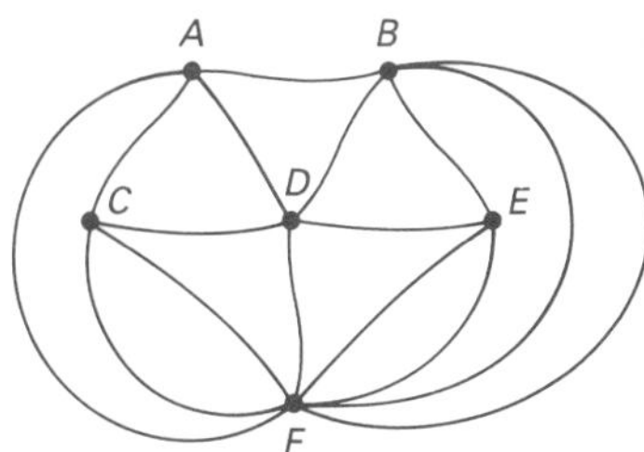

Figure 38

arc between *A* and *F*; likewise, between *D* and *F*. *C*, *E*, and *B* are connected with *F* twice. *A* is connected to *B*, *C*, *D*, and *F*; *C* is connected to *A*, *D*, and *F*; *D* is connected to *A*, *C*, *B*, *E*, and *F*; *E* is connected to *D*, *B*, and *F*; and *B* is connected to *A*, *D*, *E*, and *F*. *B* and *D* are odd vertices, and the other vertices are even. Thus, an Euler path is possible; it must start with either *B* or *D*. See Figure 39. •

Figure 39

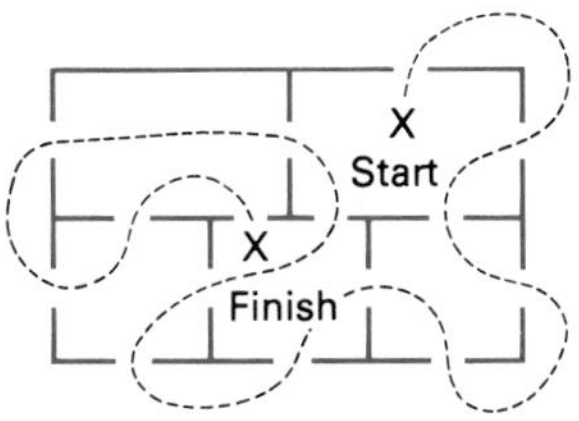

Let's apply the theory of networks to coloring problems. Actually, this type of problem encompasses much more than the coloring of graphs. For example, this study is applicable to scheduling problems of various kinds.

One of the best known problems in mathematics has been in existence for over a century. Adjoining states or countries of a map should be painted different colors. How many colors are needed? Through the years, it has been conjectured that only four colors are needed to color the most complicated map (proved using a computer in 1976). For example, two colors are all that are needed in Figure 40, but three colors are needed to color the map in Figure 41.

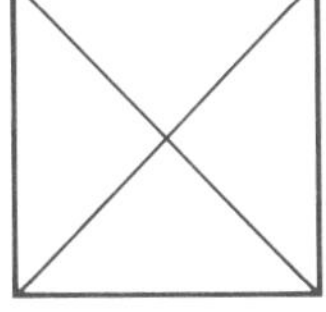
Figure 40

Figure 41

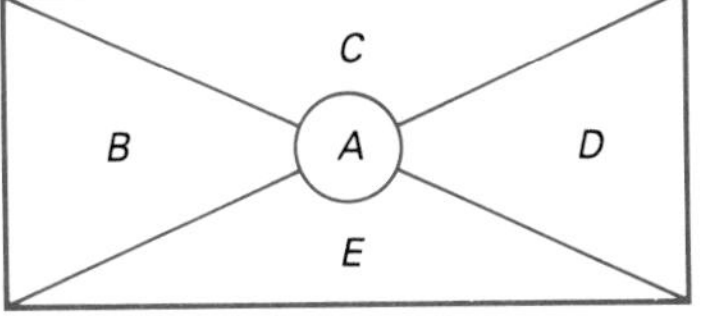

As we study this problem with network theory, we let a vertex represent a region to be colored and arcs between vertices to represent common boundaries. Figure 41 would then become the network in Figure 42. Vertex *A* is of the fourth order, and all other vertices are of the third order. To color this map, we shall use what is known as the **Largest First Algorithm.** (Please note that this algorithm does not always give the fewest possible colors.) Use Color 1 at *A* since this vertex has the highest order. The next vertices all have the same order. If one were higher than the others, it would be selected. So let's select Vertex *B*. Which color do we

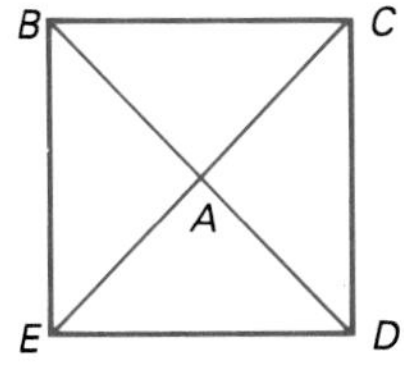

Figure 42

Just for fun *Can you traverse this path?*

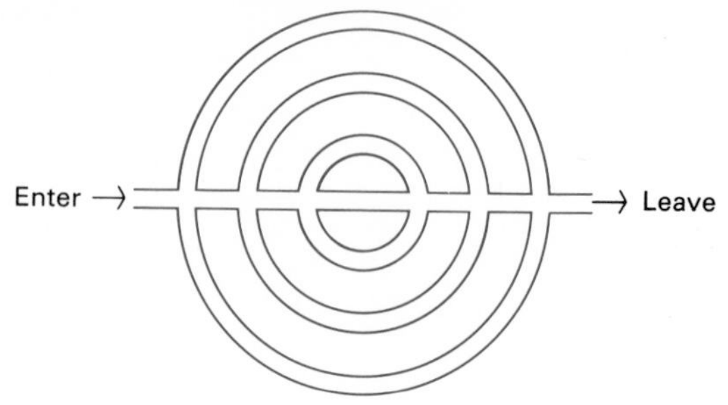

use? If all vertices adjacent to it are uncolored, use Color 1. If some adjacent vertices are colored, then use the lowest color number that is not adjacent. A is colored, so we use Color 2 at B. Repeat this process on the other vertices. Since Color 2 was used at B, Color 3 will have to be used at C. Color 2 can be used at D, and Color 3 at E.

Example 19 • Color the map for the network in Figure 43, using the Largest First Algorithm.

Figure 43

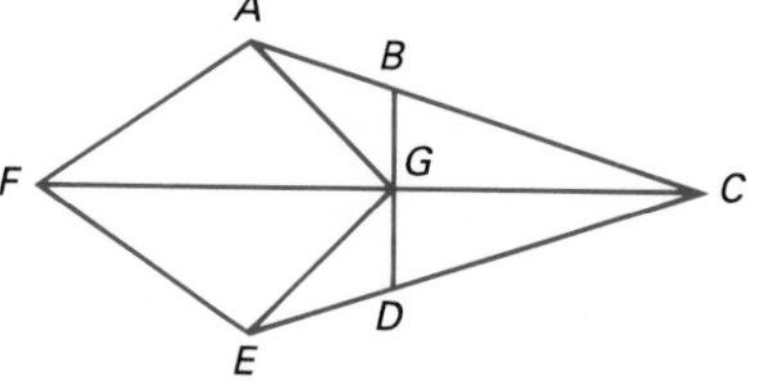

Solution Since G is of order 6, use Color 1. All the other vertices are of order 3. Starting with A, use Color 2; at B, you must use Color 3; at C, Color 2 can be used; at E, Color 2; at F, Color 3; D requires Color 3. •

Exercise set 6

1. Complete the table for networks (a) through (i).

Network	*Number of even vertices*	*Number of odd vertices*	*Traversable (yes or no)*
(a)			
(b)			
(c)			
(d)			
(e)			
(f)			
(g)			
(h)			
(i)			

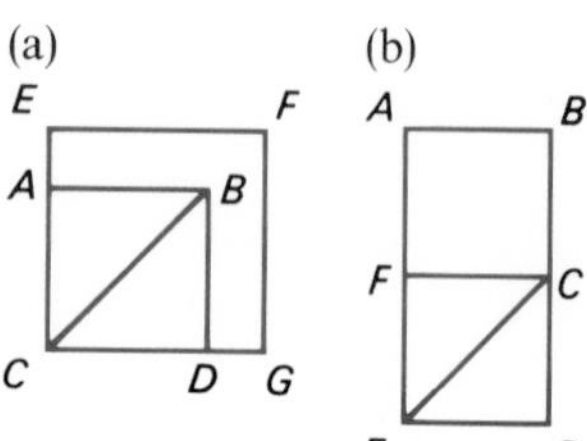

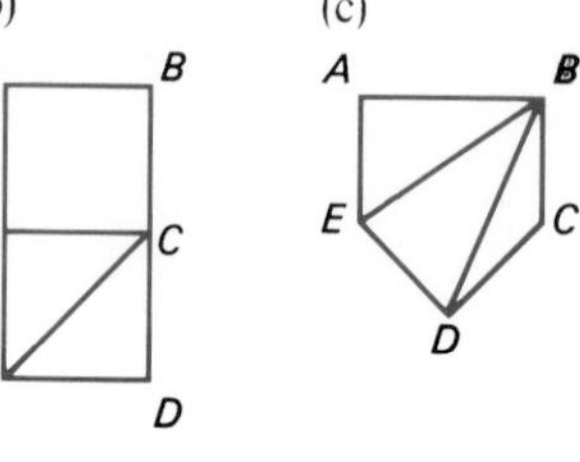

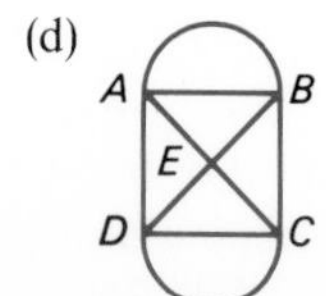

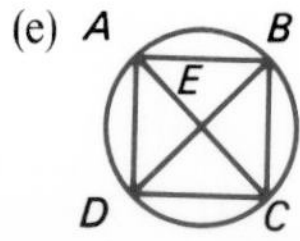

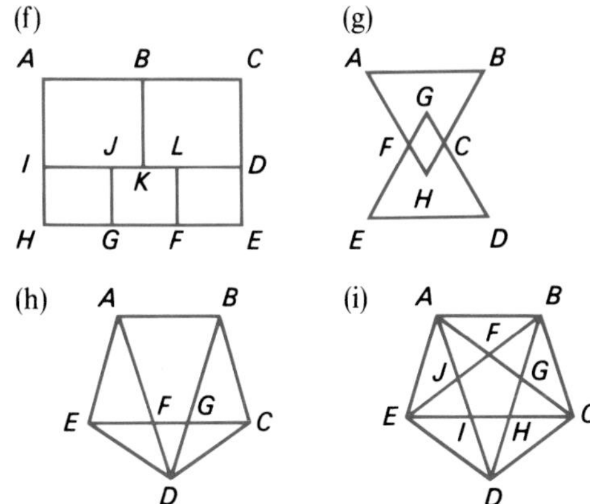

2. For each traversable network of Exercise 1, classify the path as an Euler circuit or an Euler path.

3. In each traversable network of Exercise 1, find a path.

4. Is it possible to take an entire trip through the houses whose floor plans are shown here and pass through each door once and only once?

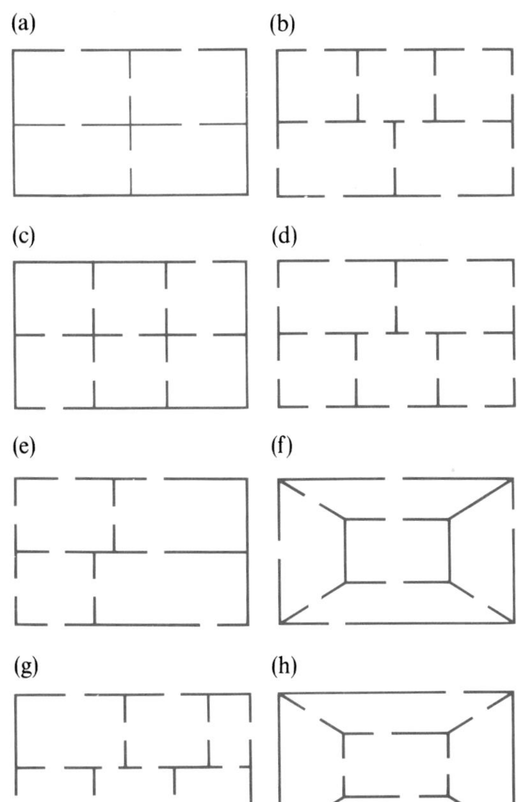

5. Many times, an arrangement of rooms is given with the understanding that there is a door on each wall. Then a path must cross each wall of a room one and exactly one time. Are the following rooms traversable?

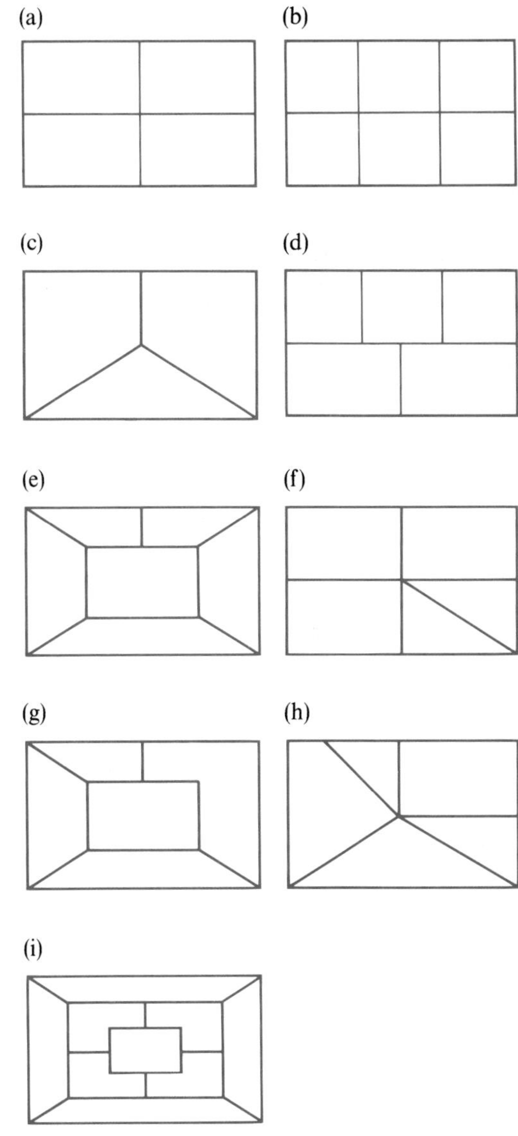

T 6. Find a path for each traversable drawing in Exercise 5.

7. Use each drawing in Exercise 5 as a map. Use the Largest First Algorithm to color the map.

8. Draw a map with seven regions that can be colored with three colors so that neighboring regions are of different colors.

C 9. The vertices and edges of a polyhedron are a three-dimensional network. Which ones of the following are traversable?

(a) Tetrahedron (b) Cube
(c) Octahedron (d) Dodecahedron
(e) Icosahedron

10. After many years, an eighth bridge was built in Königsberg connecting regions B and D (see Figure 32). Is this new network traversable?

Solution to introductory problem

Understand the problem. Before a study of graph theory, you probably used guess procedures to discover if the graphs were traversable. After graph theory, you know the following.

All vertices even, Euler circuit

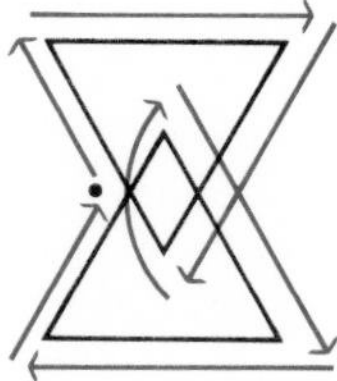

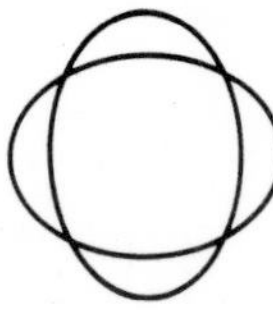

All vertices even, Euler circuit

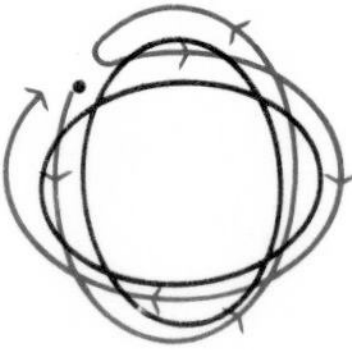

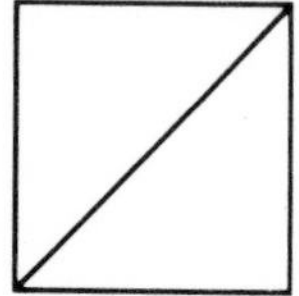

Two odd vertices, Euler path

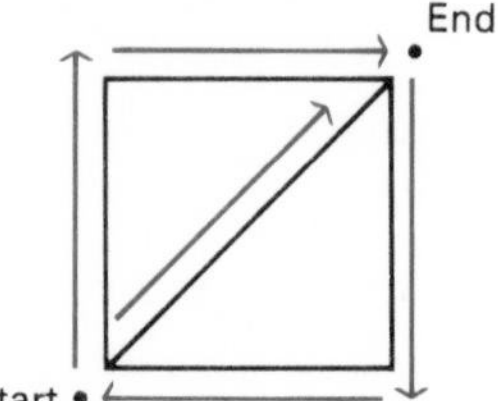

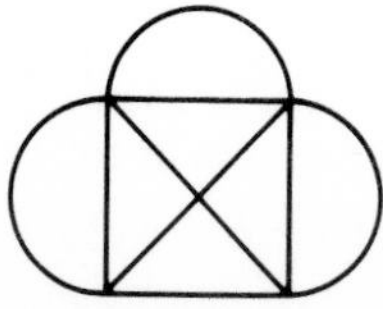

Two odd vertices, Euler path

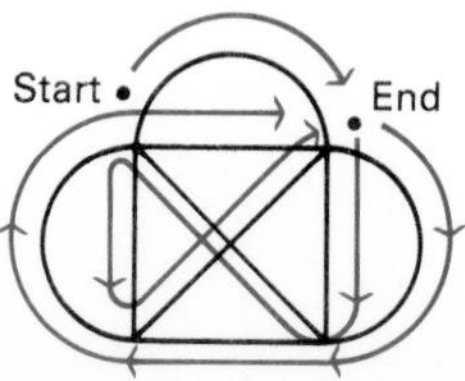

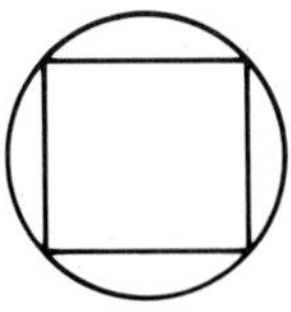

All vertices even, Euler circuit

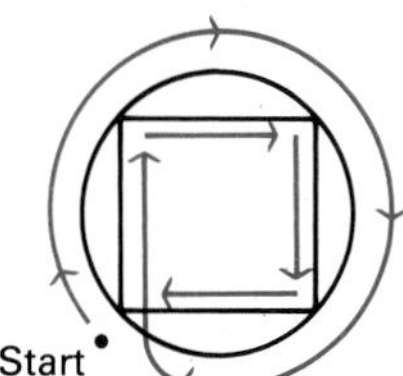

Four odd vertices, not traversable

Summary and review

1. In general, for two figures to be congruent, one must be the image of the other under a composite of reflections, translations, and rotations.
2. Two triangles are congruent if the following corresponding parts are congruent:
   (a) Three sides (SSS)
   (b) Two sides and the included angle (SAS)
   (c) Two angles and a side (ASA)
3. Two right triangles are congruent if the following corresponding parts are congruent:
   (a) Hypotenuse and a leg (HL)
   (b) Hypotenuse and an angle (HA)
   (c) Leg and an angle (LA)
   (d) Two legs (LL)
4. Some important formulas for area include:
   (a) Parallelogram: $A = bh$
   (b) Rectangle: $A = bh$
   (c) Triangle: $A = \dfrac{bh}{2}$
   (d) Isosceles triangle: $A = h\sqrt{s^2 - h^2}$
   (e) Equilateral triangle: $A = \dfrac{\sqrt{3}s^2}{4}$
   (f) Trapezoid: $A = \frac{1}{2}(b + c)h$
   (g) Regular n-gon: $A = \dfrac{pa}{2}$
   (h) Circle: $A = \pi r^2$
5. Networks
   (a) The order of a vertex is given by the number of arcs that terminate at the vertex.
   (b) A network with no null vertices is traversable if and only if it satisfies one of the following descriptions.
      i. All vertices are of even order. In this case, the path can begin at a vertex and will end at same vertex.
      ii. Two vertices are of odd order and all the rest are of even order. In this case, the path starts at one odd vertex and ends at another.

Review exercise set 7

R *Find the area of each of the following figures.*

1. (a)

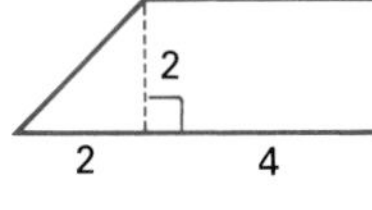

(b)

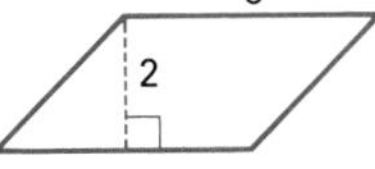

2. (a)

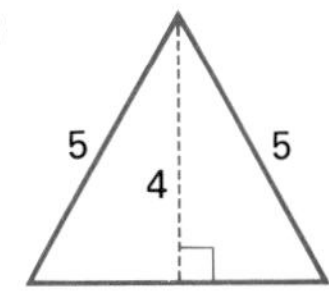

(b)

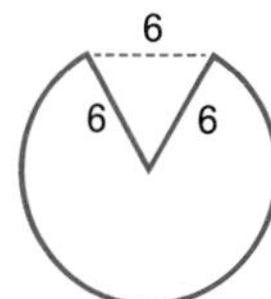

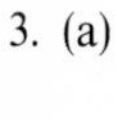

3. (a)

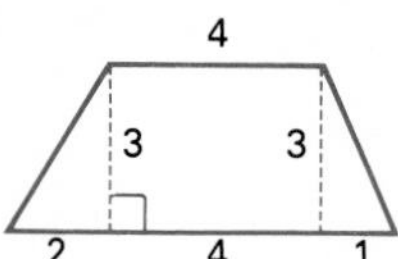

(b)

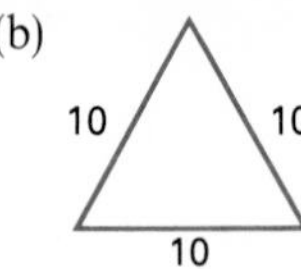

4. Indicate whether the networks in (a) and (b) are traversable. If the network is traversable, show how.

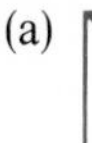

(a)

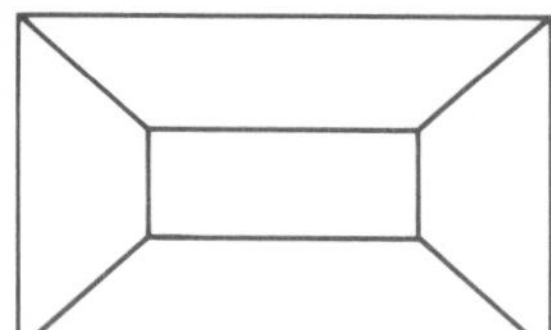

(b)

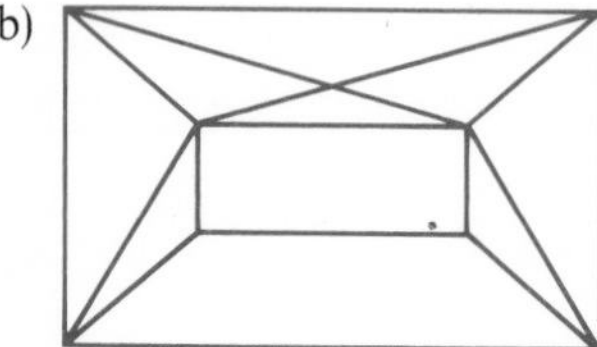

5. Which of the following are not topologically equivalent to S?

6. Using the figures in Exercise 4 as a map, color the map using the algorithm given in this chapter. Can the map be colored with fewer colors?

7. Find x.

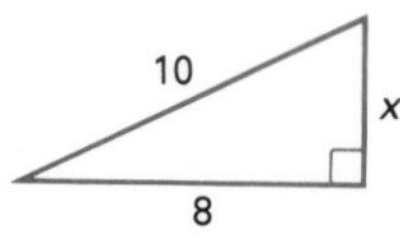

8. Which of the following can be the sides of a right triangle?

(a) $1, 5, \sqrt{26}$ (b) 15, 20, 25
(c) $2, 4\sqrt{2}, 6$ (d) 5, 12, 13

9. If the perimeter of a square is equal to the circumference of a circle, compare the areas of the bounded regions.

10. The area of a circular region is equal to the circumference of the circle. Find the measure of the radius of the circle.

T 11. Find the area of the shaded region between the circles of radius 4 that touch as shown.

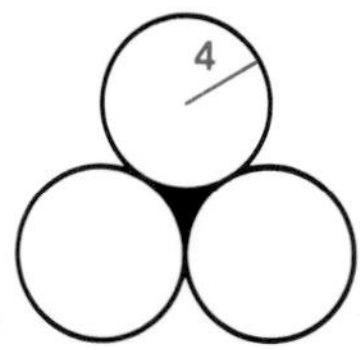

C 12. If an 80-cm-diameter wheel on a car traveling 50 km per hour turns 225 revolutions, how long has it been traveling?

13. For each floor plan, indicate whether you can devise a path that will pass through each door only once.

(a)

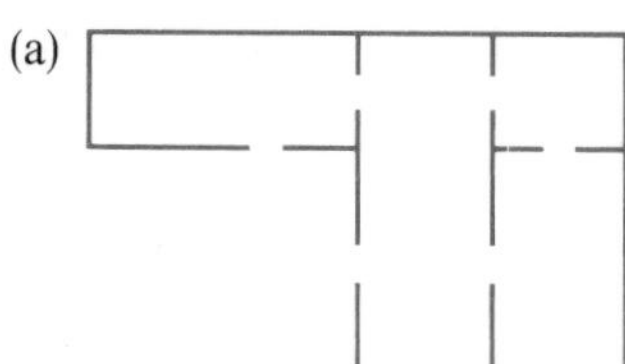

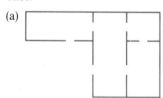

(b)

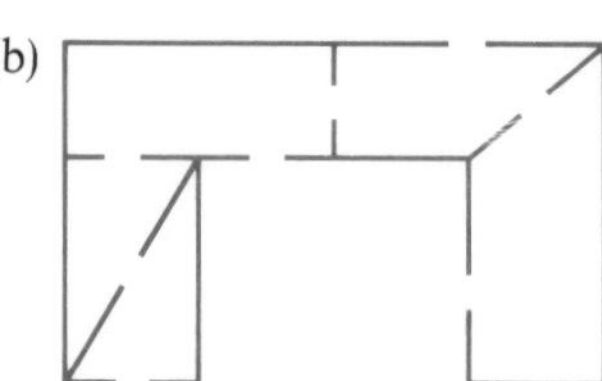

14. Sally's room is 4 m wide and 5 m long. She is making a scale drawing of it for a drafting class. If on the drawing her room is 2 cm wide, how long is it?

Bibliography

1. Blum, Hamilton S. "Introducing the Metric System." *AT* (March 1975), 214–216.

2. Bradley, Mark E. "Isosceles Triangle with the Same Perimeter and Area." *MT* (April 1980), 264–266.

3. Gardner, Martin. *Aha!* New York, N.Y.: W. H. Freeman, 1978.

4. Gardner, M. "Four Mathematical Diversions Involving Concepts of Topology." *Scientific American* (April 1958), 124–129.

5. Hawkins, Vincent J. "The Pythagorean Theorem Revisited: Weighing the Results." *AT* (December 1984), 36–37.

6. Hopkins, R. A. *The International (SI) Metric System and How It Works.* Tarzana, Calif.: Polymetric Services, 1973.

7. Leutzinger, Larry P., & Glenn Nelson. "Meaningful Measurements." *AT* (March 1980), 6–11.

8. Litwiller, Bonnie H., & David R. Duncan. "Areas of Polygons on Isometric Dot Paper: Pick's Formula Revised." *AT* (April 1983), 38–40.

9. Moon, Leland, Jr. "Laboratory Experiences with Perimeter, Area, and Volume." *AT* (April 1975), 281–286.

10. National Council of Teachers of Mathematics, *Measurement in School Mathematics—1976 Yearbook.* Reston, Va.: NCTM.

11. Nelson, Rebecca S. and Donald R. Whitaker. "Another Use for Geoboards." *AT* (April 1983), 34–37.

12. Poggi, Jeanlee. "An Invitation to Topology." *AT* (December 1985), 8–11.

13. Russell, Dorothy S., & Elaine M. Bologna. "Teaching Geometry with Tangrams." *AT* (October 1982), 34–38.

14. Spitler, Gail. "The Shear Joy of Area." *AT* (April 1982), 36–38.

15. Vervoort, Gerardus. "Inching Our Way Towards the Metric System." *AT* (April 1973), 275–279.

16. Zaslavsky, Claudia. "Network—New York Subways, a Piece of String and African Tradition." *AT* (October 1981), 42–47.

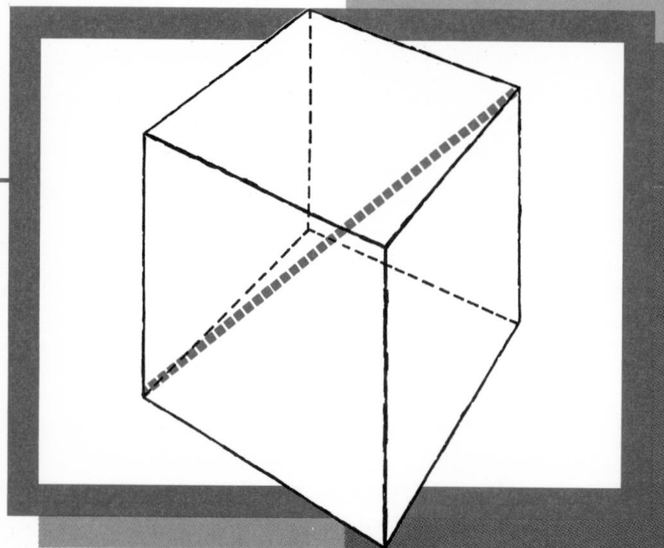

CHAPTER 11
Three-Dimensional Geometry

Devise a procedure to measure with a ruler the diameter of a cube [Gardner, 1978].

The world around us is three-dimensional, so the study of this chapter should assume added significance. In this chapter, volume and surface area are defined. The lateral area, the total surface area, and volume are discussed for prisms, cylinders, pyramids, cones, and spheres.

We emphasize the relationship between formulas for prisms and cylinders, pyramids, and cones. However, since the derivation of most formulas is tedious, much of our discussion will be of an intuitive nature.

1 Simple closed surfaces

■ Can you find a relationship between the sides, the points (or vertices), and the edges of the following figures? ■

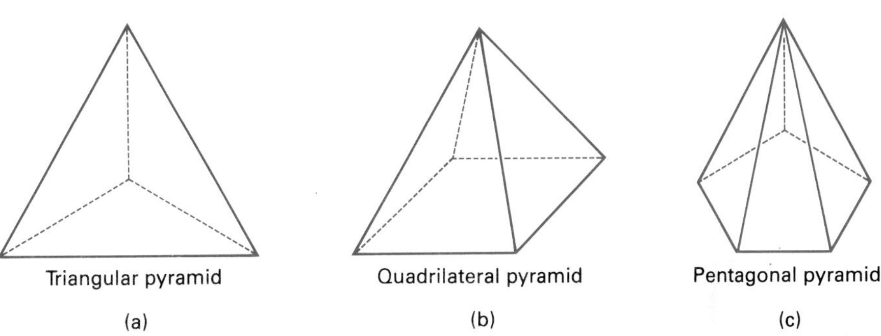
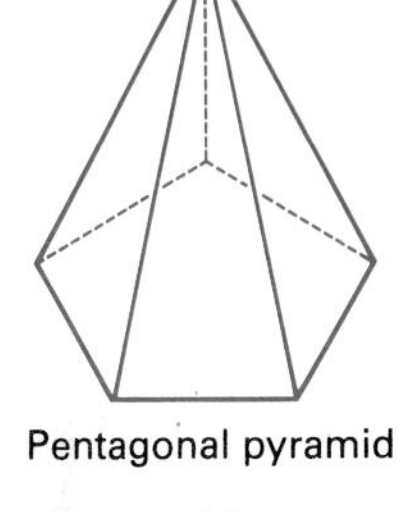

Triangular pyramid (a)

Quadrilateral pyramid (b)

Pentagonal pyramid (c)

An unusual relationship is the answer to the preceding question. If you do not discover this relationship, it will be given to you in this section.

Space figures are no more difficult to understand than plane figures, but you may encounter some difficulty in drawing space figures and in visualizing them. Simple closed surfaces (sometimes called *solids* or *three-dimensional figures*) are analogous to simple closed curves in a plane. In this section, you will learn to recognize common three-dimensional figures such as spheres, cubes, pyramids, and cylinders. Actually, a knowledge of these figures preceded most of modern mathematics. Over 5000 years ago, the civilization that occupies what we now know as Iran and Iraq used solid figures as parts of contracts in business transactions. These figures may not have had the same names, but archaeology has validated the existence of these solids made from clay. A knowledge of this section will be of much value to you as you learn to find volumes and surface areas of space figures later in this chapter.

A simple closed surface in space is analogous to the concept of a simple closed curve in a plane. Although we shall not attempt to give a rigorous definition, note that a simple closed surface cannot have holes through it and that it must separate the points of space into three disjoint sets of points: the set of points on the surface, the set of points interior to the surface, and the set of points exterior to the surface. Hence the surface in Figure 1(a) is and in (b) is not a simple closed surface. We

Figure 1

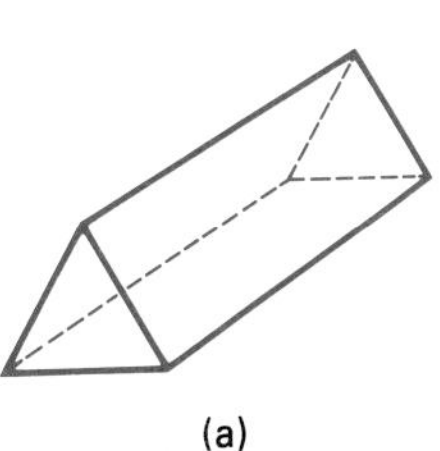

(a)

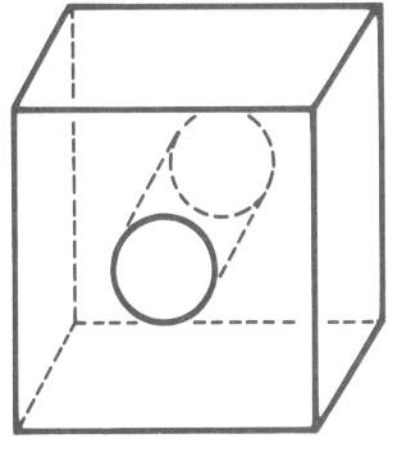

(b)

shall consider the union of the points on the surface and the points interior to the surface as the *solid* bounded by the closed surface or as a **space region.**

Let's begin with the types of space surfaces formed by polygonal regions.

Polyhedron

> A *polyhedron* is a simple closed surface in space whose boundary is composed of polygonal regions.

Notice the different types of polyhedrons in Figure 2. In (a), a square bounds each polygonal region. In (b), the polygonal regions are bounded by triangles. In (c), the regions are bounded by pentagons.

Figure 2

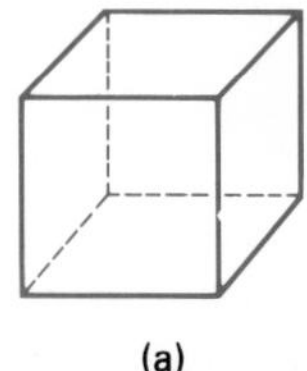

(a)

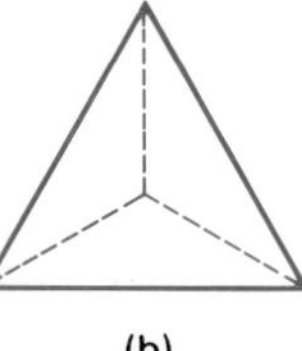

(b)

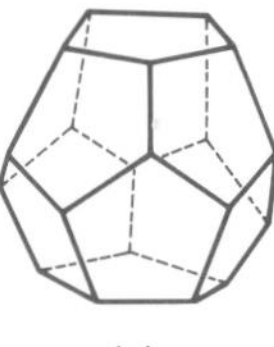

(c)

Consider next a special kind of polyhedron.

Prism

> A *prism* is a polyhedron formed by two congruent polygonal regions in parallel planes, along with three or more regions bounded by parallelograms joining the two polygonal regions so as to form a closed space figure.

The polygonal regions in parallel planes are called **bases** of the prism. The parallel edges joining the two bases are called **lateral edges,** and the regions bounded by parallelograms are called **lateral faces.** If the lateral faces are rectangular regions, and the planes containing the lateral faces are perpendicular to the planes containing the bases of the prisms, then the prism is called a *right prism.*

A prism is named in terms of its bases. The **triangular prism** (Figure 3(a)) resembles a wedge or a trough. The **quadrilateral prism** (in (b)) could be a closed shoe box or a room. Sometimes chalk is shaped in the form of a **hexagonal prism** (in (d)).

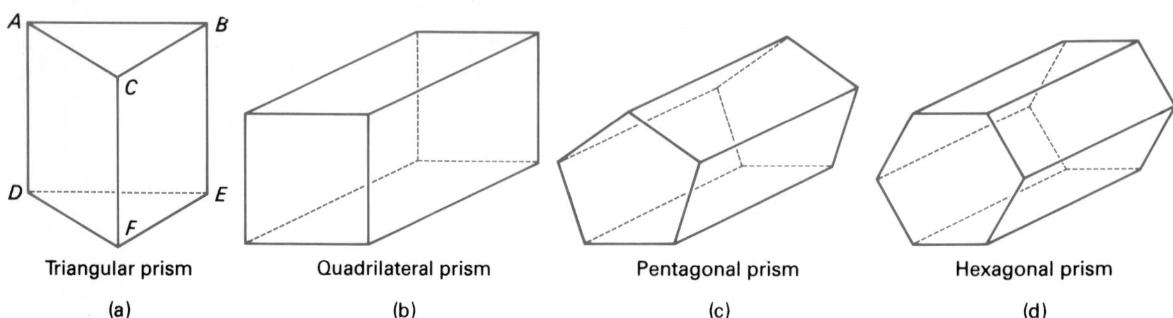

Triangular prism (a) — Quadrilateral prism (b) — Pentagonal prism (c) — Hexagonal prism (d)

Figure 3

In the triangular prism in Figure 3(a), $\triangle ABC$ and $\triangle DEF$ are congruent and are in parallel planes. The two triangular regions $\triangle ABC$ and $\triangle DEF$, called *bases*, are joined by $ACFD$, $ABED$, and $BCFE$, which are called **lateral faces** of the prism. The **sides** include both bases and lateral faces. The segments that form the boundaries of the sides of a prism, such as $\overline{BE}$, $\overline{CF}$, and $\overline{AD}$ in Figure 3, are called **lateral edges,** and the corners—points A, B, C, D, E, and F—are called the **vertices** of the prism.

Another polyhedron of interest is the pyramid.

Pyramid

A *pyramid* is a polyhedron formed by a simple closed polygonal region (called the **base**), a point (called the **vertex**) not in the plane of the region, and the triangular regions joining the point and the edges of the polygonal region. The triangular regions are congruent in a regular pyramid.

A **regular pyramid** has a base that is a regular polygon, and the line joining the vertex and the center of the base is perpendicular to the base. A pyramid is classified according to the polygonal region forming the base, as in Figure 4. When we think of pyramids, we think of the fabulous quadrilateral pyramids of Egypt.

Figure 4

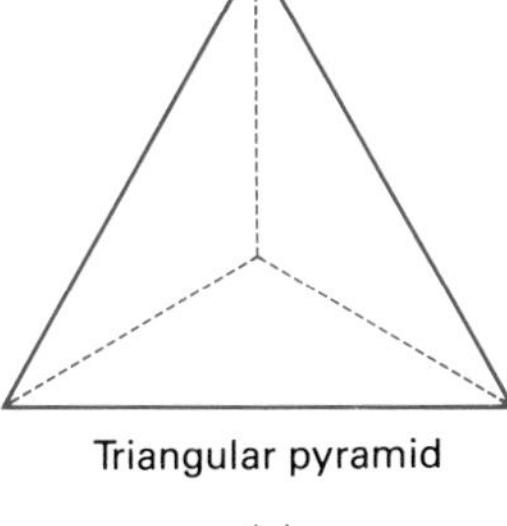

Triangular pyramid

(a)

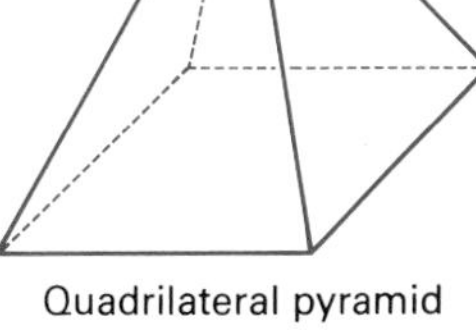

Quadrilateral pyramid

(b)

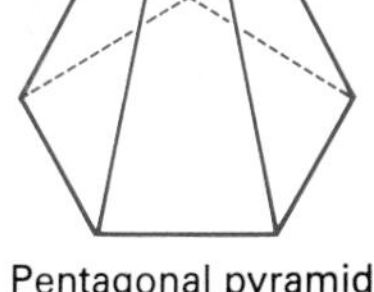

Pentagonal pyramid

(c)

A somewhat unexpected relationship exists between the number of vertices, edges, and sides of polyhedra.

Euler's formula

Let V represent the number of vertices, E the number of edges, and S the number of sides of a polyhedron. Then

$$V + S - E = 2$$

Consider the triangular prism in Figure 3. Count the number of vertices, edges, and sides. Did you get 6 vertices, 9 edges, and 5 sides? Since $6 + 5 - 9 = 2$, Euler's formula is satisfied. Count the number of sides, edges, and vertices in the quadrilateral prism and verify that the formula is satisfied here also.

Just as polygons are named according to the number of sides, so regular polyhedra are named according to the number of faces (see Table 1).

Table 1

Polyhedra	
Name	*Number of faces*
Tetrahedron	4
Hexahedron	6
Octahedron	8
Dodecahedron	12
Icosahedron	20

Regular polyhedron

If the faces of a polyhedron are congruent regular polygonal regions, and if each vertex is the intersection of the same number of edges, then the polyhedron is called a *regular polyhedron.*

The drawings in Figure 5 are examples of regular solids. A tetrahedron is formed by 4 congruent triangular regions, a cube is formed by 6 congruent quadrilateral regions, an octahedron by 8 congruent triangular regions, a dodecahedron by 12 congruent pentagonal regions, and an icosahedron by 20 congruent triangular regions. Greek mathematicians (such as Thaetetus) discovered that these are the only possible regular solids. Because of the prominence of Plato in the Greek world, they are sometimes called the *Platonic solids.*

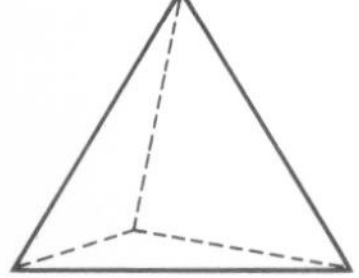
Regular tetrahedron

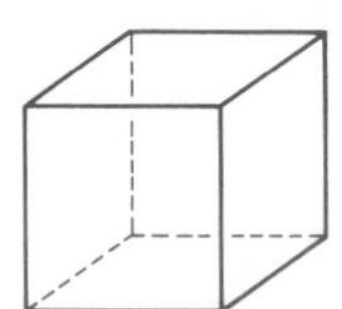
Regular hexahedron (cube)

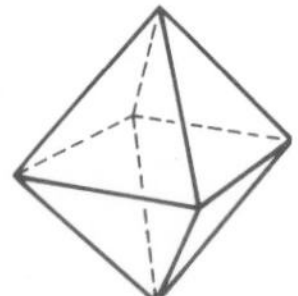
Regular octahedron

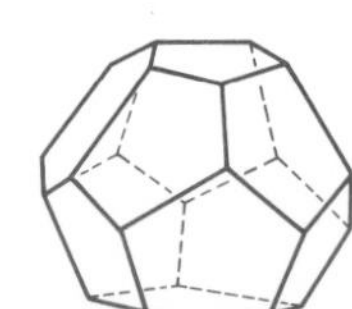
Regular dodecahedron

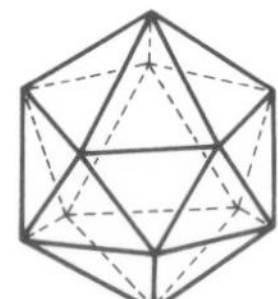
Regular icosahedron

Figure 5

Nature also utilizes the Platonic solids. In Figure 6, note the different regular solids found in the enlarged pictures of skeletons of minute marine animals called *Radiolaria.*

You may have observed that some of the angles in polyhedra are different from those studied previously. Whereas plane angles are formed by intersecting lines, dihedral angles are formed by intersecting planes. Thus two intersecting planes form four dihedral angles in the same manner that two intersecting lines form four plane angles.

Dihedral angle

A *dihedral angle* is the union of two distinct intersecting, noncoplanar half planes and their line of intersection (common edge).

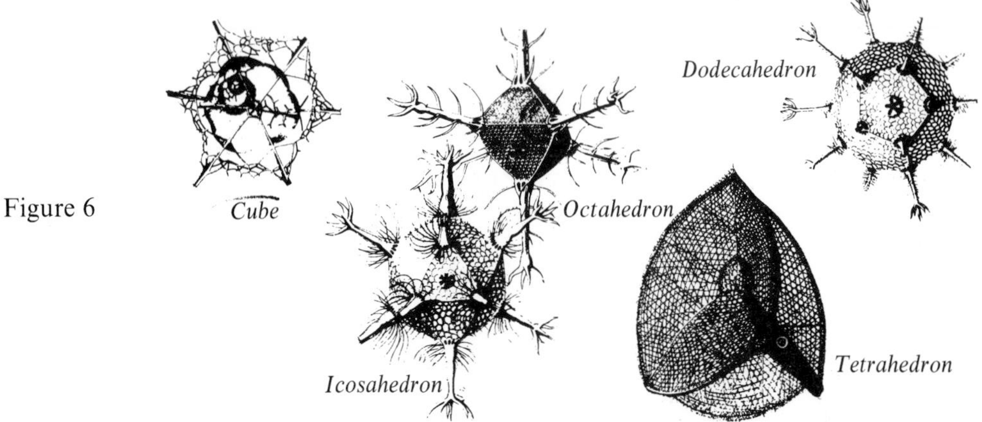

Figure 6

The two half planes are called the **faces** of the dihedral angle, and the common line is called the **edge** of the dihedral angle. Figure 7 illustrates a dihedral angle.

A plane angle formed by two rays, one in each face of a dihedral angle—each ray having its endpoint on the edge of the dihedral angle and each ray perpendicular to the edge—is called a *plane edge of the dihedral angle.* The measure of a dihedral angle is the measure of the plane angle of the dihedral angle. *Thus a right dihedral angle has a right plane angle.*

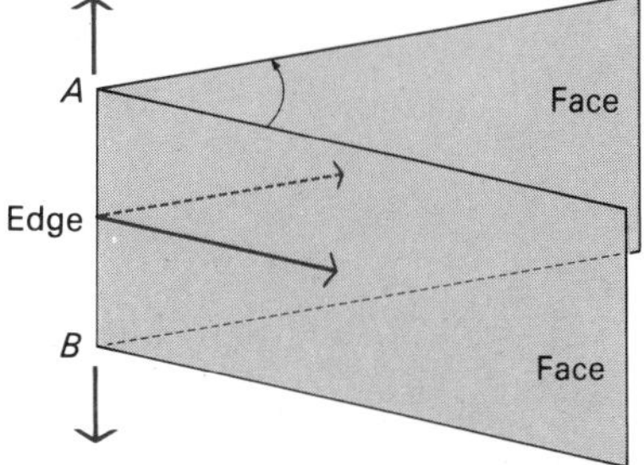

Figure 7

Can you visualize a set of points in space equidistant from a fixed point? This set of points can be considered as the union of infinitely many congruent circles, all having the same center. Such a space surface is called a *sphere.* (See Figure 8(a).)

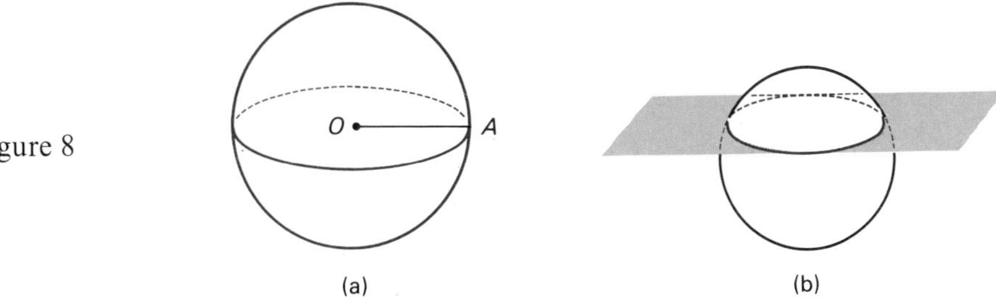

Figure 8

Sphere

A set of all points in space of equal distance from a fixed point is called a *sphere.*

The fixed point of the sphere is called the **center** of the sphere, and any line segment from the center to a point on the sphere is called a **radius** of the sphere.

The *interior of a sphere* is the set of all points whose distance from the center of the sphere is less than the measure of the radius. The set of points whose distance from the center of the sphere is greater than the measure of the radius of the sphere is called the *exterior of the sphere.*

Consider the intersection of a sphere and a plane. If the intersection is not empty, it will be either one point or a circle. (See Figure 8(b).) If the plane is *tangent to the sphere,* the intersection is a single point. Such a situation is intuitively understood by visualizing a ball resting on a flat surface.

Consider a circle and a fixed line that is not in the plane of the circle and is not parallel to the plane of the circle. The union of *all* lines through the circle that are parallel to the fixed line is called a **circular cylinder.** A more general definition of a cylinder replaces the circle with a simple closed curve. (See Section 2.)

In this book, we shall not use the general definition of a cylinder but shall instead consider a cylinder as a simple closed surface bounded on two ends by circular bases. (See Figure 9.) A circular cylinder resembles a prism except that its bases are circular regions. In Figure 9(a), the cylinder is called a **right circular cylinder.** In such a cylinder, the line segment joining the centers of the bases is perpendicular to the bases. The cylinders in (b) and (c) are not right circular cylinders.

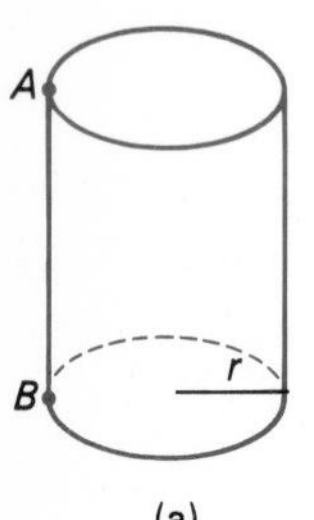

(a)

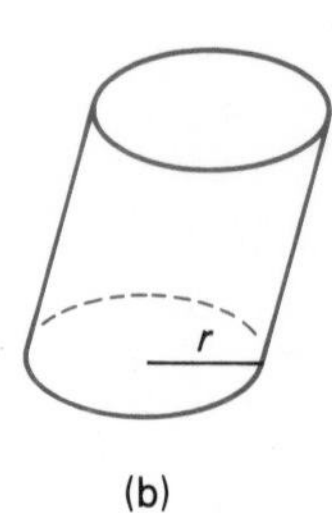

(b)

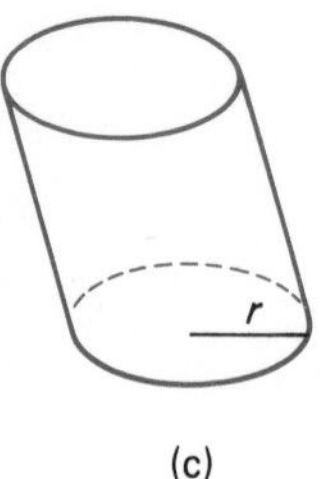

(c)

Figure 9

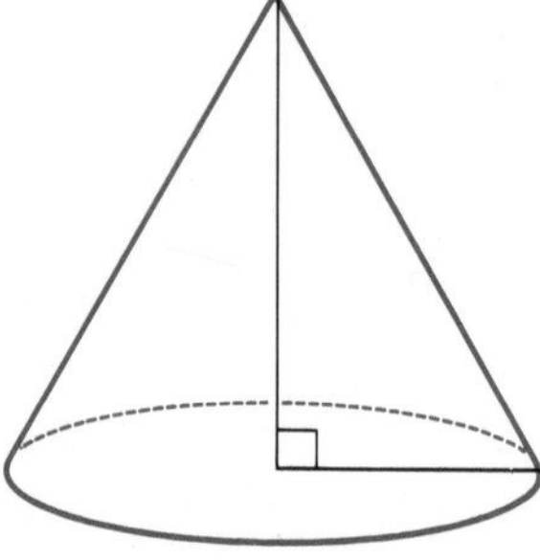

Figure 10

Another useful simple closed surface is the cone. A circular cone (Figure 10) consists of a circular base and segments joining a fixed point (vertex) to points on the base. In general, a cone can have any simple closed curve bounding the base. With this definition, pyramids are cones with polygonal bases. Circular cones have circular bases, and it is the circular cones that we usually call *cones.* This is the terminology that will be used in this chapter.

Just for fun *These are pictures of the same ABC block. What letter is opposite C? Is opposite E? Is opposite A?*

A circular cone in which the line segment from the vertex of the cone to the center of the circular base is perpendicular to the base is called a **right circular cone.** (See Figure 10.)

Exercise set 1

R 1. Identify the types of polyhedra given here. (Answers may not be unique.)

(a)

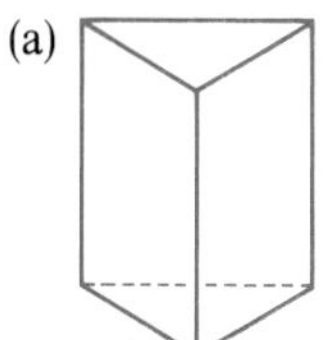

(b)

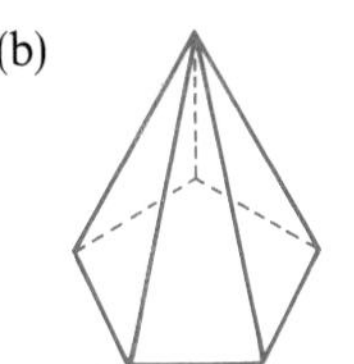

(c)

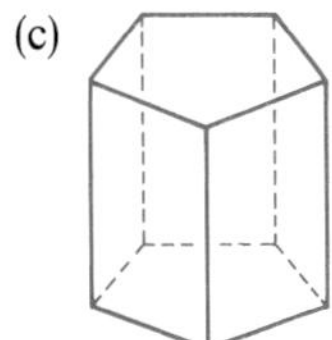

(d)

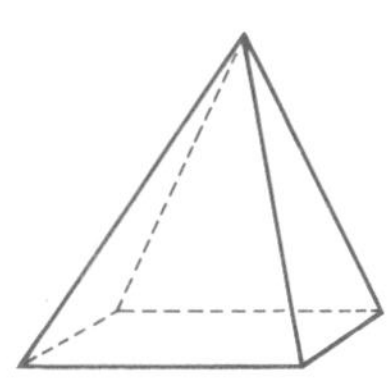

2. Verify Euler's formula for the figures in Exercise 1.

3. Answer true or false.
   (a) A pyramid is a polyhedron.
   (b) A pyramid is a prism.
   (c) Every triangular prism is a right prism.
   (d) Some polyhedra are prisms.
   (e) The bases of a prism lie in perpendicular planes.
   (f) The faces of a pyramid are parallelograms.
   (g) A pyramid with seven sides has a hexagon as the boundary of its base.
   (h) Euler's formula does not hold for pyramids.

4. Give the name for each space figure shown. List the segments that are the lateral edges and the points that are vertices.

(a)

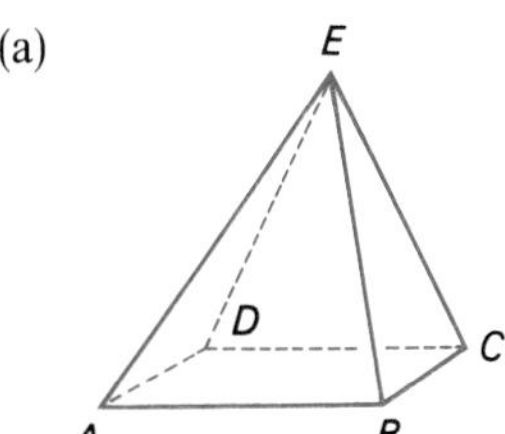

(b)

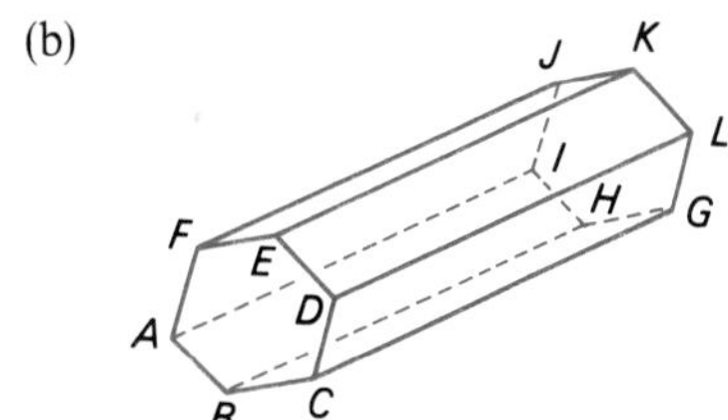

5. Draw a pyramid and a prism having each of the following as a base.
   (a) Triangle
   (b) Quadrilateral
   (c) Pentagon
   (d) Hexagon

6. Verify Euler's formula for the number of sides, vertices, and edges of each space figure named.
   (a) A pentagonal pyramid
   (b) A hexagonal prism
   (c) A triangular prism
   (d) A quadrilateral pyramid
   (e) A hexagonal pyramid
   (f) A pentagonal prism
   (g) A quadrilateral prism
   (h) A triangular pyramid
7. Trace and cut out the given figure. Enlarge and then glue. What do you have?

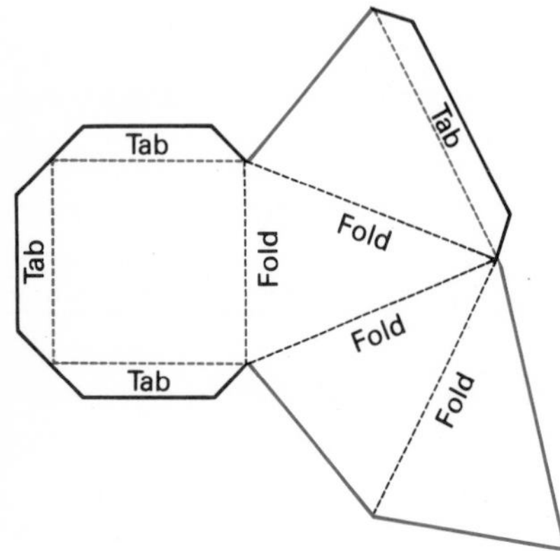

8. If you enlarge the patterns shown and form regular solids, you will have a better understanding of this section. Name the solids.

(a) (b)

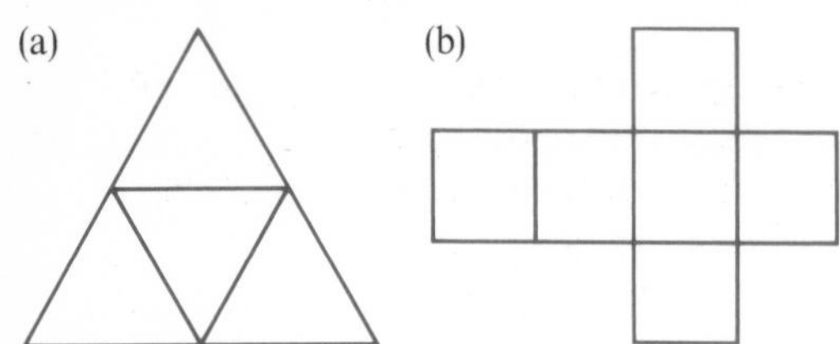

(c)

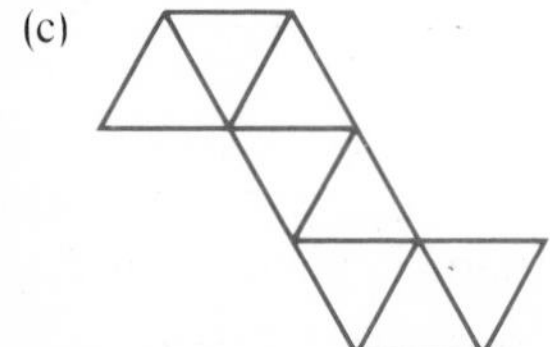

(d)

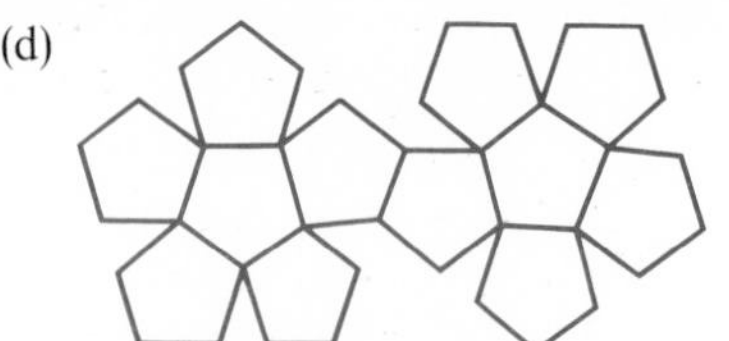

Pattern for constructing a dodecahedron

(e)

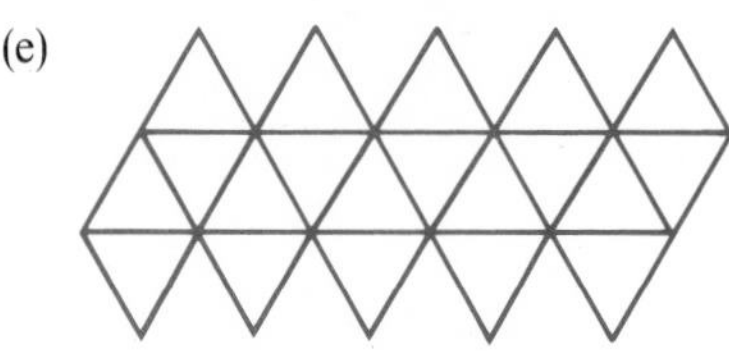

Pattern for constructing an icosahedron

9. Complete the table, indicating the number of vertices, edges, faces, and Euler's formula for each of the following regular solids.

Surface	V	E	S	$V + S - E$
Tetrahedron				
Cube				
Octahedron				
Dodecahedron				
Icosahedron				

10. Indicate whether the following are true for a pyramid or a prism or both.
    (a) Its lateral faces are parallelograms.
    (b) It has the same number of faces as vertices.
    (c) It has one base.
    (d) It always has an even number of vertices.
    (e) It always has an even number of lateral faces.
    (f) It can have as few as four faces.
    (g) It has two bases.
    (h) Its lateral faces are triangles.
11. Consider a prism with each of the following bases. How many lateral faces does it have? How many vertcies? How many edges?
    (a) Triangle (b) Quadrilateral
    (c) Pentagon (d) Hexagon
12. Do Exercise 11 for a pyramid.

T 13. Count the number of vertices, edges, and sides on what is left of a quadrilateral prism after one corner is cut off. Does Euler's formula hold?

14. A quadrilateral prism has a tunnel cut through it in the form of a pentagonal prism. Find V, E, S, and $V + S - E$.

15. Space figures have planes of symmetry for reflection and lines of symmetry of rotation. List the planes of symmetry and lines of rotational symmetry for the following.

    (a) A right circular cylinder
    (b) A right triangular prism
    (c) A cube
    (d) A right quadrilateral pyramid
    (e) A right circular cone

C 16. Every convex polyhedron has at least two more vertices than half the number of sides, or $V \geq 2 + \frac{1}{2}S$, where V represents the number of vertices and S represents the number of sides. Examine the truth of this statement for the following figures.

    (a) Pentagonal pyramid
    (b) Quadrilateral prism
    (c) Hexagonal prism
    (d) Quadrilateral pyramid

17. Generalize the answers for Exercise 11 for an n-sided base.

18. Generalize the answers for Exercise 12 for n-sided base.

2 Surface area

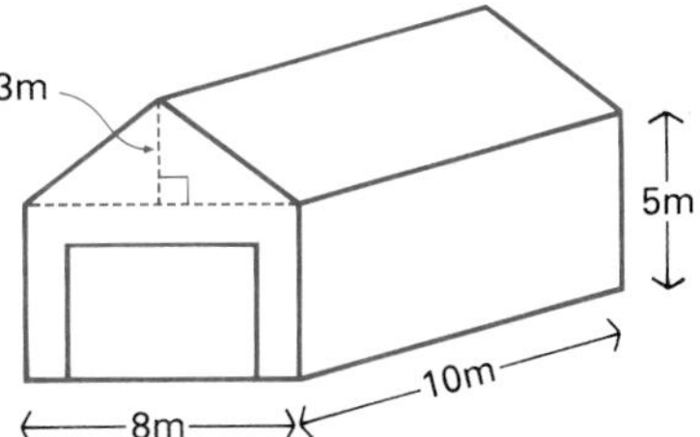

■ One gallon of paint covers 60 m of area. If paint costs \$12 per gallon, how much does it cost to paint the outside of the garage pictured above? Paint the garage door, but do not paint the roof. ■

In this section, we discuss area in space called **surface area.** The same units of measure will be used for this discussion as were used for area in the preceding chapter. We generalize our discussion by defining a cylindrical surface and defining other surfaces as subsets of this one.

In the preceding chapter, we considered the concept of area for certain plane figures. In this section, we extend the concept of area to surface area of space solids. The easiest way to introduce this concept is by looking at patterns that could be used to form these solids.

Example 1 • Consider the right prism in Figure 11. Imagine a pattern cut out of cardboard that could be glued to form the solid. This pattern is pictured in Figure 11(b). The area of the rectangle in the middle is called the **lateral area** of the solid. The area of the two bases (B) and the lateral area is called the **total surface area** (T.S.A.) of the solid.

Figure 11

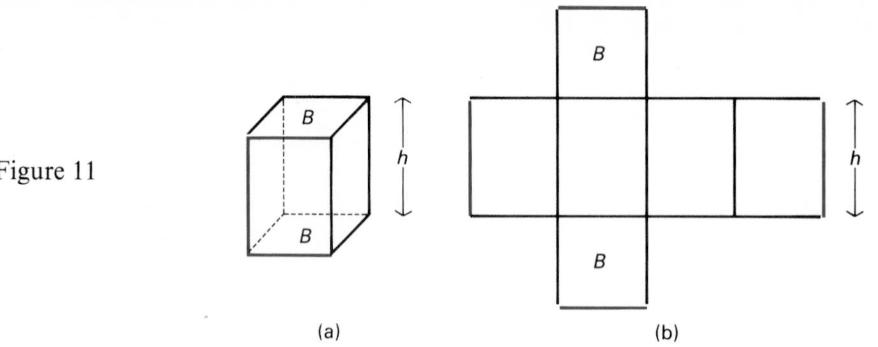

Example 2 • Look at the pattern for the right pentagonal prism in Figure 12. Again, the lateral surface area is given by the rectangle in the middle, and the total surface area is obtained by adding the area of the two bases.

Figure 12

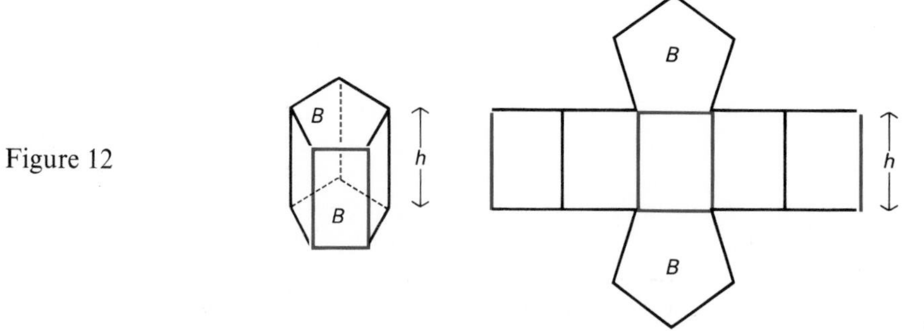

Example 3 • Consider the right cylinder in Figure 13. The lateral surface area is obtained from the rectangle in the middle, of length $2\pi r$ (circumference of the circle). To get the total surface area, we add the area of the two bases.

Figure 13

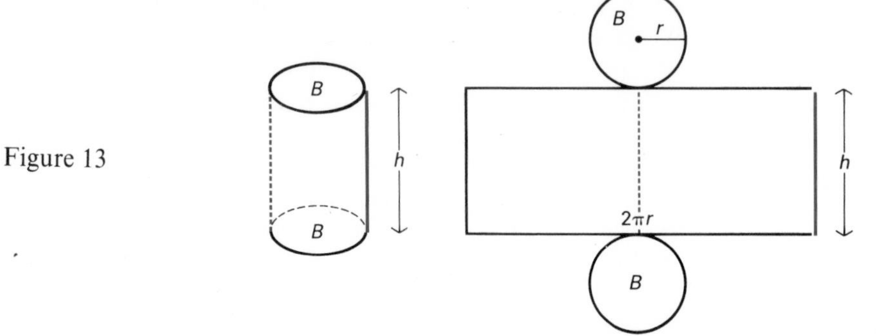

The preceding examples suggest the following definition, which can be used to determine the total surface area of many solids. The **lateral** area of a surface is the sum of the areas of all the faces or sides (other than bases) of the space

Just for fun

House A faces left and house B faces right. Can you move one line in house A to make it face right?

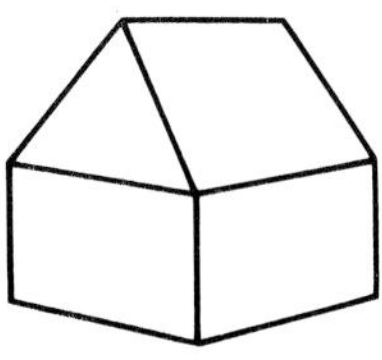
House A

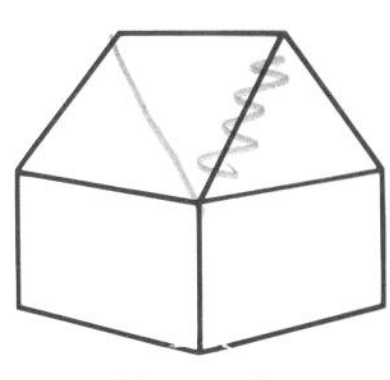
House B

figure. The **total** surface area is the lateral area added to the area of the bases of the space figure.

Example 4 • Find the total surface area of a right rectangular prism, where $l = 2$, $w = 4$, and $h = 5$.

Solution Total surface area $= 2(2 \cdot 4) + 2(2 \cdot 5) + 2(4 \cdot 5) = 76$ •

Often, a more general concept of surface area is useful. To introduce this concept, we give a very general definition of a cylinder, of which a prism will be a subset. Let C be any curve in a plane α. In this book, we will consider only closed curves. Let m be a line not in the plane α but intersecting α in a point on C (see Figure 14). The union of m and all lines parallel to m and passing through C is defined to be a cylindrical surface. The surface may be thought of as that generated by a line that moves parallel too itself and always intersects curve C. Each of the lines is called an **element** of the surface. Note that a prism is a special case of a cylindrical surface where curve C is a polygon.

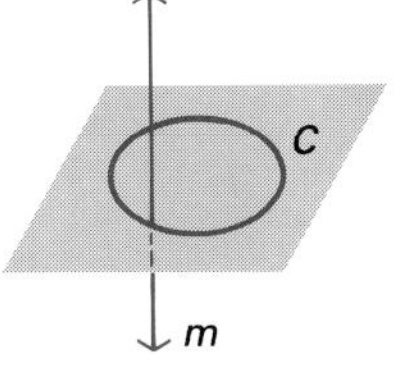

Figure 14

In this book, we usually limit a cylinder to that part of a cylindrical surface between two parallel planes that intersect all the elements of the surface. The portion of the surface that lies in one of the parallel planes is called a **base** of the cylinder. That portion of the surface between the parallel planes is called the **lateral surface** of the cylinder. A segment perpendicular to the planes of the bases and joining a point of one of these planes to a point of the other plane is called an **altitude** of the cylinder.

Since a prism is a special case of a cylindrical surface, all of the above definitions hold for prisms. The elements that are edges of the prism are called **lateral edges.** A right prism has faces that are rectangles.

Now let's consider the prism in Figure 15. Although it is drawn with five faces, it could have any number. Let's look at the surface area of one face, with base p_1 and edge e. The surface area of this face is ep_1. The same reasoning could be used on all faces. Thus

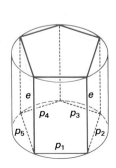

$$\text{S.A.} = ep_1 + ep_2 + \cdots + ep_n = eP$$

Figure 15

where P is the perimeter. If we think of the number of sides of a prism inscribed inside a cylinder as increasing in an unlimited fashion, then the surface area of a cylinder is the perimeter of the cylinder times an element. If the cylinder is a circular cylinder, the perimeter becomes the circumference of the circle ($2\pi r$). So the surface area is

$$\begin{aligned}(\text{Cylinder})\ \text{S.A.} &= 2\pi re \\ &= Ce\end{aligned}$$

Exercise set 2

R 1. Find the total surface area of each figure.

(a) A cylinder with diameter measuring 8 and altitude measuring 10

(b) A right rectangular prism with $l = 4$, $w = 6$, $h = 5$

(c) A right rectangular prism with $l = \frac{3}{4}$, $w = \frac{7}{8}$, $h = 3\frac{1}{2}$

In Exercises 2 through 5, find the lateral area of the following figures.

2.

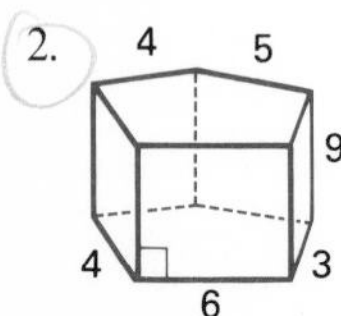

3. A prism with a perimeter of right section (perpendicular to edges) of 30 cm and a lateral edge of 10 cm

4. A right prism whose bases are equilateral triangles with sides 6 cm and with an altitude of 10 cm

5. A right prism whose base is a right triangle with legs 3 ft. and 4 ft. and whose altitude is 8 ft.

In Exercises 6 through 10, find the lateral surface area of each.

6. Right hexagonal prism with altitude 10, base perimeter 50, and base area 35

7.

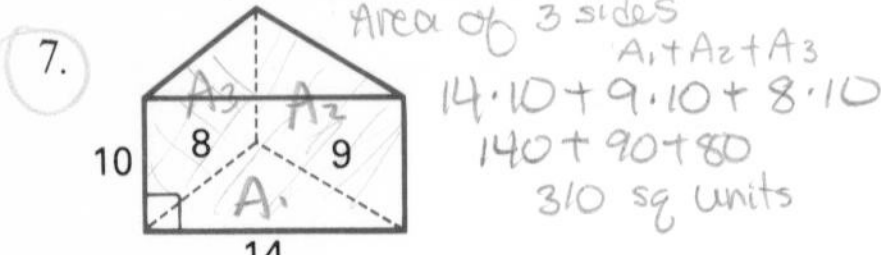

8.

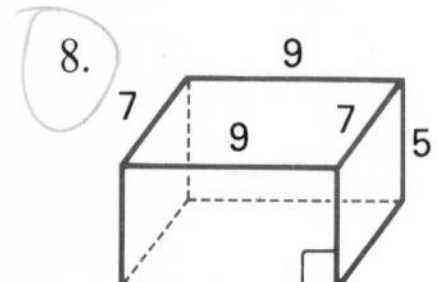

9.

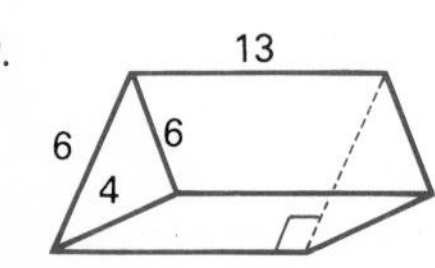

10. Right pentagonal prism with altitude 6, base area 40, and perimeter 60.

Find the total surface area in Exercises 11 and 12.

T 11. Area of $B = 24\sqrt{3}$

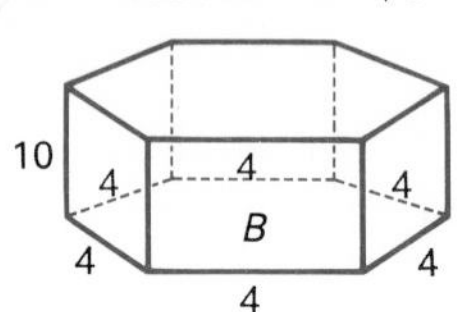

12. Right cylinder whose altitude is 10 m and radius of base 6 m.

13. Find the total surface area of each figure.

(a)

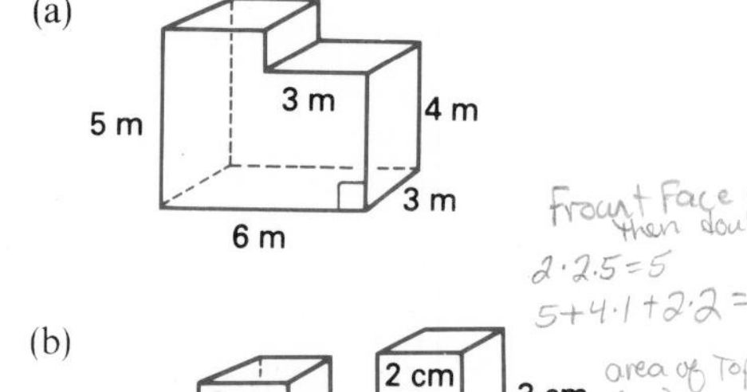

(b)

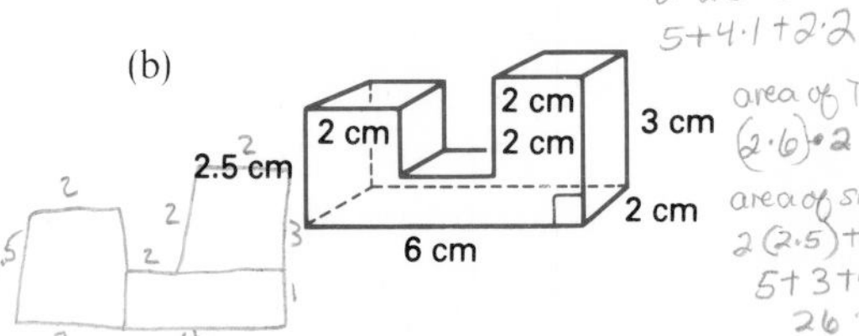

14. How much does it cost to paint the garage described in the introduction to this section?

15. How many liters of paint will be needed to paint the bottom and sides of a swimming pool that is 5 m by 10 m by 3 m? (The pool is 3 m deep all the way across.) Assume that 1 L of paint covers 24 m^2 and that the paint is purchased in 1-L cans.

16. How many liters of paint (to be purchased in 1-L cans) are needed to paint a cylindrical tank (including the top and bottom) if the radius of the tank is 2.8 m and the altitude of the tank is 3.7 m? Again, assume that 1 L of paint covers 24 m^2.

17. Compare the total surface areas of the two right cylinders.

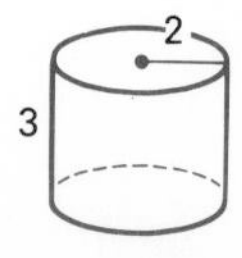

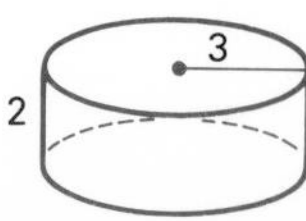

C 18. Consider two cylinders that are similar, with radii 4 and 5 and altitudes 8 and 10.

(a) Find the lateral surface area of each cylinder.
(b) What is the ratio of the two surface areas?
(c) What is the similarity ratio of the two cylinders?
(d) Make a conjecture about the ratio of surface areas of similar cylinders.

19. Consider two cubes that are similar, with sides 6 and 8.

(a) What is the ratio of surface areas?
(b) What is the similarity ratio of the two cubes?
(c) Make a statement about the ratio of surface areas of similar cubes.

20. The perimeter of the base of a cube is 32 inches. Find the total surface area.

21. (a) Find the total surface area of a rectangular parallelepiped with edges 4, 4, and 10.
(b) Inscribe a cylinder with altitude 10 in the parallelepiped. Find the surface area of the cylinder.

22. What is the edge of a cube that has a total surface area of 288 square inches?

3 Volume and measurements

■ What is the ratio of the number of cubes that touch the floor to the number of cubes that do not touch the floor in the following stack of cubes? ■

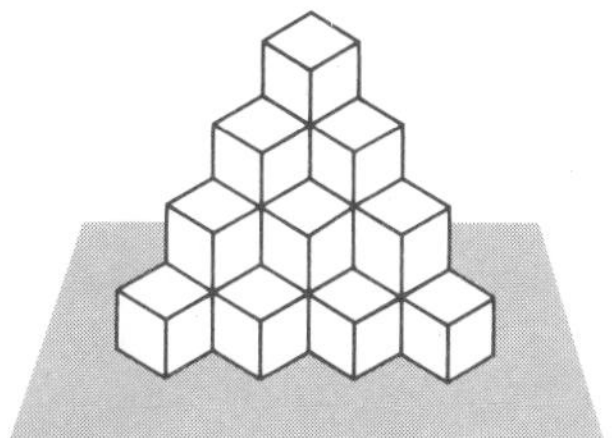

As you might expect, we turn our attention next to the volume of space regions. We will consider both English and metric units of measurement. In addition, we will develop formulas for obtaining volumes of space regions.

Just as length and area associate numbers, volume measure associates a unique number with each closed-space region. Although this number is not always obtainable using a physical process, there is exactly one number for each closed-space region using a given unit of measurement. The unit of measurement of volume is ordinarily the unit cube. In a unit cube, each edge of each square face measures one unit. As an illustration of the use of a unit cube, consider the volume bounded by a right rectangular prism (sometimes called a *rectangular parallelepiped*). A right rectangular prism is similar to a cube except that its faces are rectangles that are not necessarily squares. To find the volume of the region bounded by a rectangular prism, determine how many unit cubes can be fitted into the prism. Consider the rectangular prism in Figure 16(a). Here, 6 cubes can be placed on one edge (the length), 4 on a second edge (the width), and 3 on the third edge (the height). When finding the total number of cubic units, we find the total number of cubes on the base ($4 \cdot 6 = 24$) and recognize that this number of cubes is used on three levels ($24 \cdot 3 = 72$); so the total volume is 72 cubic units.

Figure 16

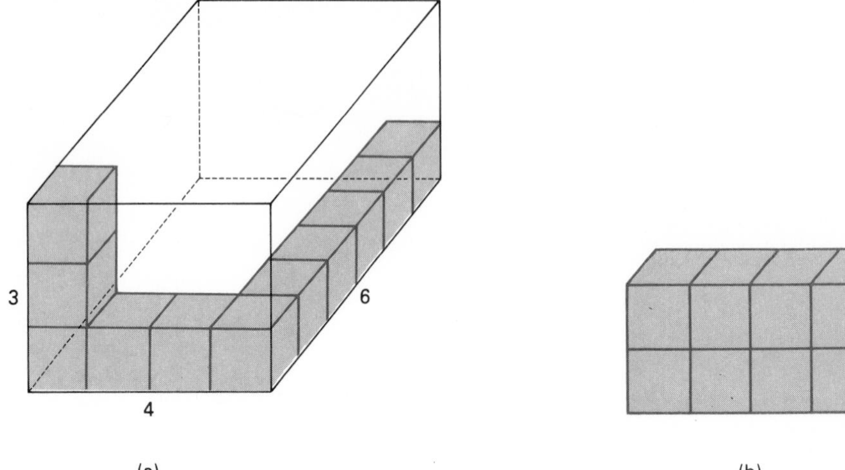

(a) (b)

To obtain an understanding of volume quickly, students are often asked to place unit cubes in a space region or to count the number of unit cubes in a region.

Example 5 • With the given unit cube, find the volume of the space region in Figure 16(b).

Solution Since there is only one layer of cubes, you can easily see that the volume is 8. •

Example 6 • By counting, find the volume of the space regions in (a) and (b) of Figure 17.

Figure 17

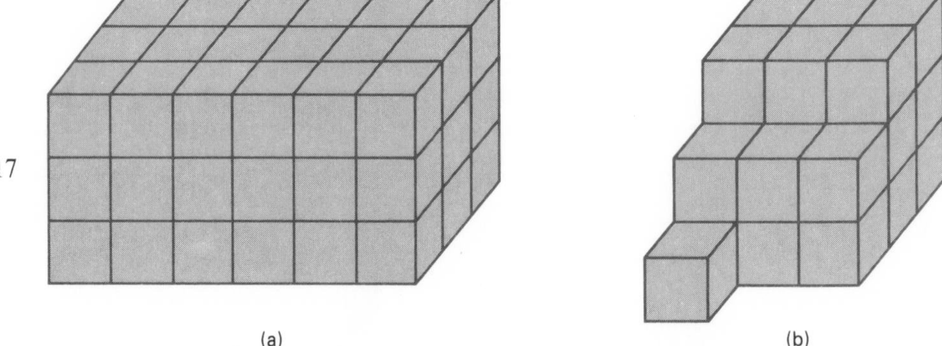

(a) (b)

Solution (a) In Figure 17(a), there are 3 layers of cubes. Each layer contains $6 \cdot 3$ cubes. So the volume is

$$(6 \cdot 3) \cdot 3 = 54 \text{ unit cubes}$$

(b) In Figure 17(b), the bottom layer has $3 \cdot 3 + 1 = 10$ cubes. The second layer has $3 \cdot 3 = 9$ cubes. The third layer has $2 \cdot 3 = 6$ cubes.

$$V = 10 + 9 + 6 = 25 \text{ cubes}$$ •

The preceding discussion should have suggested the following definition of the volume of a prism.

Volume of a prism

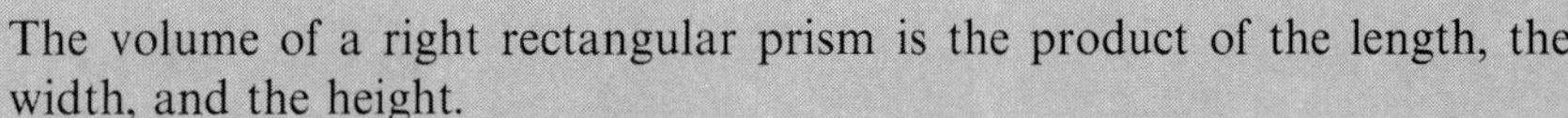

The volume of a right rectangular prism is the product of the length, the width, and the height.

$$V = lwh$$

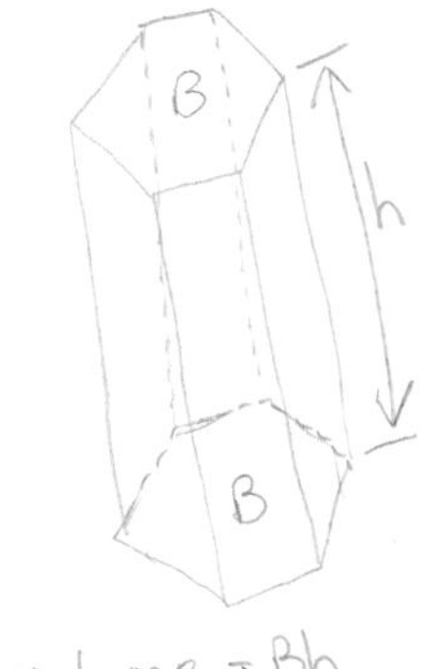

Congruence in three dimensions may be defined by considering space transformations. First we define a reflection over a plane instead of a line. Then a translation can be shown to be a composite of two reflections over parallel planes. Likewise, a rotation about a line can be shown to be the composite of reflections about intersecting planes. Of course, each of these transformations preserves size and shape. Therefore, two solids are congruent if one can be obtained from the other using a composite of space reflections, translations, and rotations. This definition of congruence helps us to discuss the properties of volume of space regions.

In the previous discussion, you have already used some of the properties of volume of space regions. Can you recognize which ones you have used?

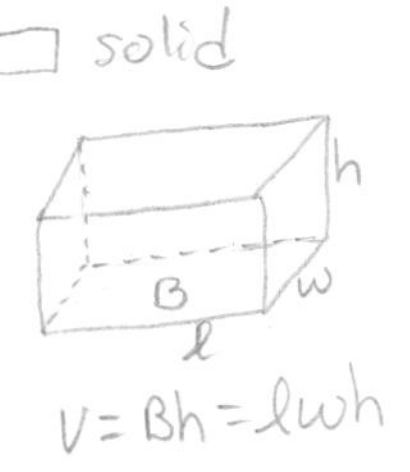

Properties of Volume

1. Volume is additive; that is, the volume of the whole is equal to the sum of the volumes of its nonoverlapping parts. If R and S are nonoverlapping space regions (possibly with surfaces in common), then the volume of $R \cup S$ is equal to the volume of R plus the volume of S.
2. If space region $R \cong$ space region S, then the volume of space region R equals the volume of space region S.
3. If a space region is cut into parts and reassembled in a nonoverlapping manner to form another space region, then the two space regions have the same volume.

You are familiar with the basic units for volume in the English system.

$$1728 \text{ in.}^3 \text{ (cubic inches)} = 1 \text{ ft.}^3 \text{ (cubic feet)}$$

$$27 \text{ ft.}^3 \text{ (cubic feet)} = 1 \text{ yd.}^3 \text{ (cubic yard)}$$

Under the SI system, volume measure is based on the cubic meter (m^3). For small volumes cubic centimeter (cm^3) or cubic millimeter (mm^3) is used for the unit of volume.

Figure 18

The volume of the figure depicted in Figure 18 illustrates a cubic centimeter (cm^3), which is about the size of a sugar cube. Table 2 summarizes the relationships

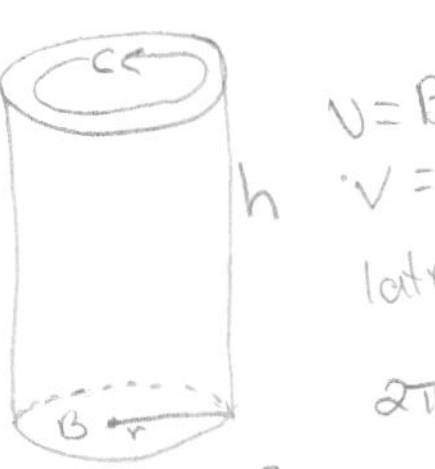

Table 2

$1 \text{ km}^3 = 10^9 \text{ m}^3$, since $1 \text{ km} = 10^3 \text{ m}$
$1 \text{ hm}^3 = 10^6 \text{ m}^3$, since $1 \text{ hm} = 10^2 \text{ m}$
$1 \text{ dam}^3 = 10^3 \text{ m}^3$, since $1 \text{ dam} = 10 \text{ m}$
$1 \text{ dm}^3 = 10^{-3} \text{ m}^3$ (called a *liter*, symbolized by L)
$1 \text{ cm}^3 = 10^{-6} \text{ m}^3$ (called a *milliliter*, symbolized by mL)
$1 \text{ mm}^3 = 10^{-9} \text{ m}^3$, since $1 \text{ mm} = 10^{-3} \text{ m}$
$1\ \mu\text{m}^3 = 10^{-18} \text{ m}^3$, since $1\ \mu\text{m} = 10^{-6} \text{ m}$

among measures of volume in the SI system. Note that since

$$1 \text{ km} = 10^3 \text{ m}$$

then

$$\begin{aligned} 1 \text{ km}^3 &= (10^3 \text{ m})(10^3 \text{ m})(10^3 \text{ m}) \\ &= 10^9 \text{ m}^3 \end{aligned}$$

The cubic meter, the cubic centimeter, and the cubic millimeter are the units of volume used in ordinary transactions.

One can use either Table 3 or a metric scale for cubic units, realizing that between consecutive cubic units one must move the decimal point three places.

Example 7 • Convert 321 m^3 to cubic centimeters.

Solution

$$1 \text{ m}^3 = 1{,}000{,}000 \text{ cm}^3$$

$$321 \text{ m}^3 = 321(1{,}000{,}000) = 321{,}000{,}000 \text{ cm}^3 \quad \bullet$$

As we continue our study of volume, we examine a principle first stated by Bonaventra Cavalieri (1598–1647).

Cavalieri's Principle

Given: a plane and any two solids. If every plane parallel to the given plane intersects the two solids in cross sections that have the same area, then the solids have the same volume.

In Figure 19, for example, if the area of the cross section of the circle in (a) is equal to the area of the triangle in (b), then the triangular prism and the cylinder have the same volume.

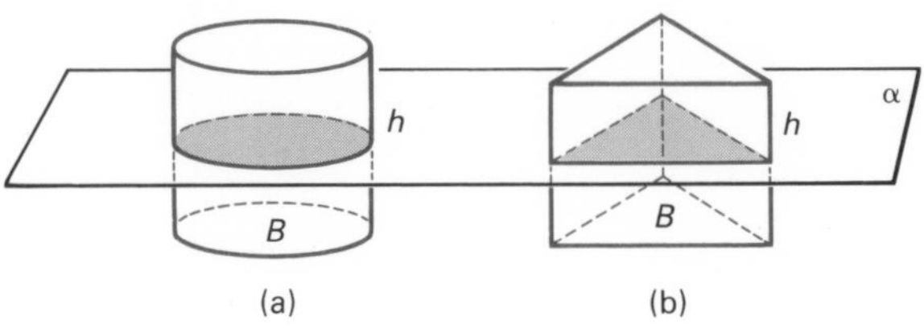

Figure 19

Cavalieri pictured a solid as being made up of infinitely many thin slices (Figure 20).

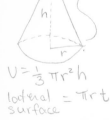

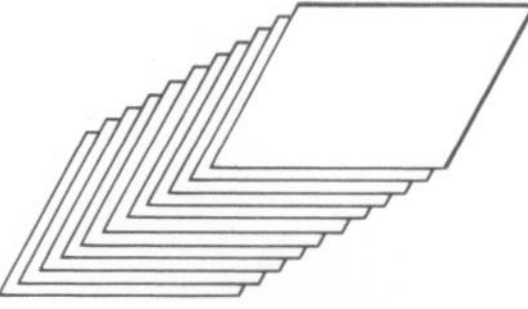

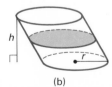

Figure 20

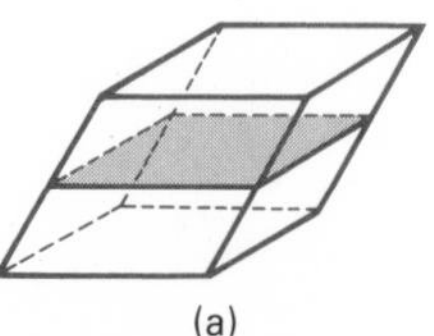

Figure 21

Just for fun

A farmer had four cylindrical silos 4 m in radius and 10 m tall. While pumping grain into the silos, he discovered that one had a hole and that the grain spilled over to fill the space between the silos. How many liters of grain did the farmer store?

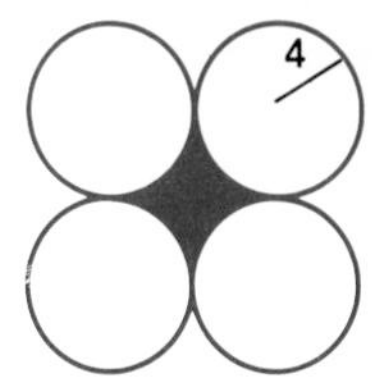

Thus, an oblique prism and an oblique cylinder (Figure 21) would utilize the same formula for volume as a right prism or right cylinder, namely the area of the cross section multiplied by the altitude.

Volume of prism or cylinder

> The volume of any prism or cylinder is the area (B) of a cross section cut by a plane parallel to the base multiplied by the altitude (h).
>
> $$V = Bh$$

Sometimes it is more convenient to find the volume of a prism in terms of a lateral edge. Then, instead of using the area of the base, we use the area of a right section of the prism.

Volume of a prism

> Volume = (area of a right section) · (lateral edge of prism)

Example 8 • Consider a right triangular prism with an altitude of 10. The base and altitude of the triangular bases are 6 and 8, respectively. Find the volume of the space region bounded by the prism.

Solution

$$\text{Area of base} = \frac{1}{2}(6)(8) = 24$$

$$\text{Volume} = 24 \cdot 10 = 240 \quad \bullet$$

Example 9 • The area of a right section of a rectangular prism is 140 cm^2. The measure of the lateral edge of the prism is 20 cm. Find the volume bounded by the prism.

Solution

$$V = 140 \cdot 20 = 2800 \text{ cm}^3 \quad \bullet$$

Example 10 • A tin can has a diameter of 10 and a height of 25. Find the volume and the total surface area of the can.

Solution

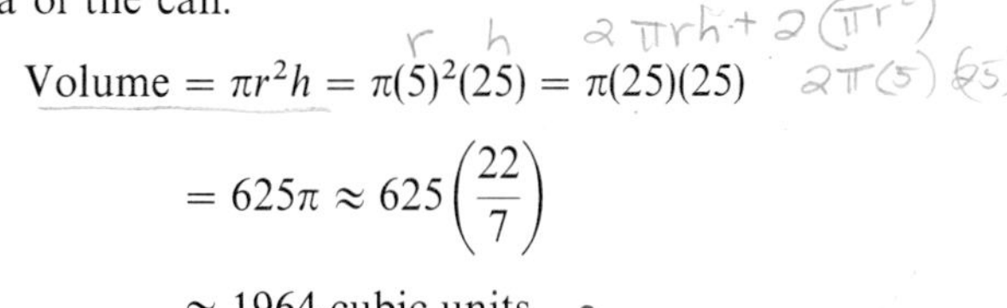

$$\text{Volume} = \pi r^2 h = \pi(5)^2(25) = \pi(25)(25)$$

$$= 625\pi \approx 625\left(\frac{22}{7}\right)$$

$$\approx 1964 \text{ cubic units} \quad \bullet$$

Exercise set 3

R 1. Find the volume of each figure.

(a) A cylinder with diameter measuring 8 and altitude measuring 10

(b) A right rectangular prism with $l = 4$, $w = 6$, $h = 5$

(c) A right rectangular prism with $l = \frac{3}{4}$, $w = \frac{7}{8}$, $h = 3\frac{1}{2}$

2. Find the volume of the following space regions (all prisms are right prisms).

(a)

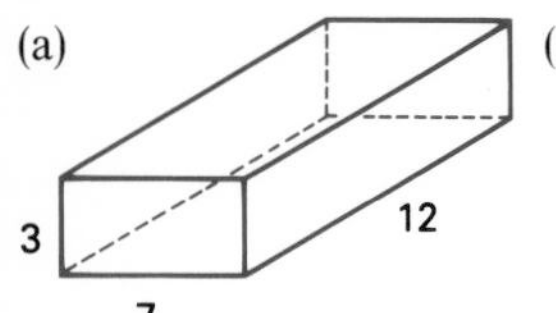

(b)

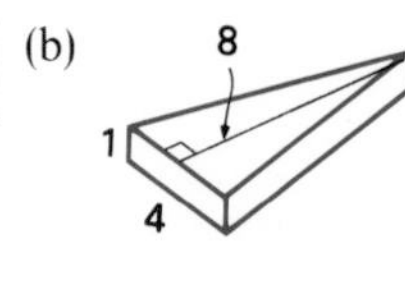

(c)

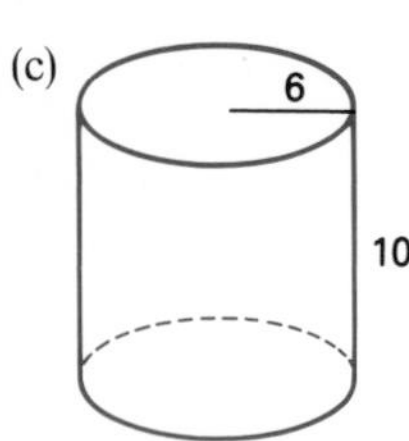

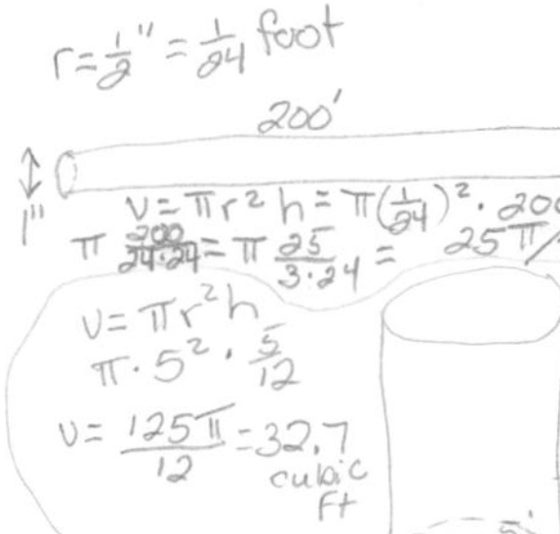

3. Convert each of the following to cubic centimeters.

(a) $34\ m^3$ (b) $328\ km^3$ (c) $3100\ mm^3$
(d) $41\ km^3$ (e) $42\ mm^3$ (f) $863\ km^3$

4. Which of the following boxes is larger if each has the given capacity?

(a) One holding $13\ m^3$ or one holding $132{,}000\ cm^3$

(b) One holding $4\ km^3$ or one holding $40{,}000\ m^3$

(c) One holding $9\ m^3$ or one holding $9(10)^6\ mm^3$

(d) One holding $5\ km^3$ or one holding $500{,}000\ m^3$

(e) One holding $123{,}456\ mm^3$ or one holding $20\ cm^3$

5. Find the volume of each figure.

(a)

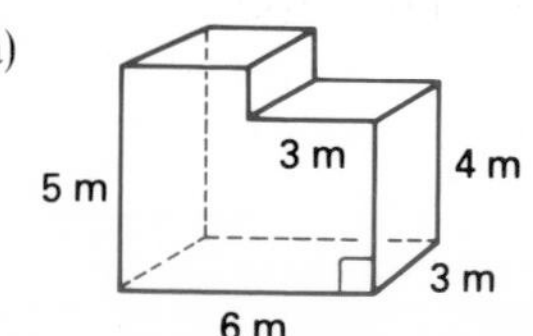

(b)

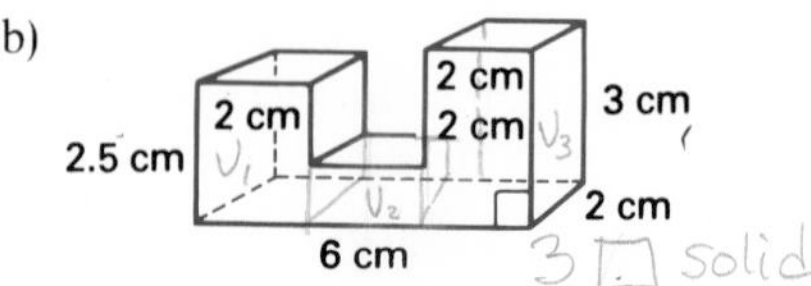

6. One cubic centimeter of aluminum weighs about 2.7 grams. An aluminum bar is 30 cm long, 12 cm wide, and 15 cm thick. What is the weight of the bar?

T 7. A cement box without a top is being poured, with outside measurements 1.2 m wide, 4 m long and 0.8 m thick. Inside dimensions are 1 m, 3.8 m, and 0.7 m. How many cubic meters of cement are needed?

8. Ideally, each classroom should allow an average of $8\ m^3$ of air space per student. How many students could be placed in a classroom 10 m by 8 m by 4 m?

9. A pipe 200 feet long with an inside diameter of 1 inch will hold how many cubic feet of water?

10. A cylindrical tank has a diameter of 10 feet. As the tank is being filled how many cubic feet of water are needed to raise the water level in the tank by 5 inches?

11. Estimate the volume of the following in cubic meters.

(a) An average-sized bedroom
(b) A refrigerator
(c) A shoe box
(d) An average-sized house
(e) An oil drum
(f) A filing cabinet

12. Find the measurement of the volume of the walls of a pipe if the measurement of the inner radius is 6 cm, the outer radius is 7 cm, and the length is 135 cm.

13. Which has the greatest volume and by how much: a square prism with 40 as the perimeter of the base, or a cylinder with circumference 40, if both have altitude 20?

C 14. A cylinder is inscribed in a prism with a square base. If the dimensions of the prism are $4 \times 4 \times 10$, what is the volume between the cylinder and the prism?

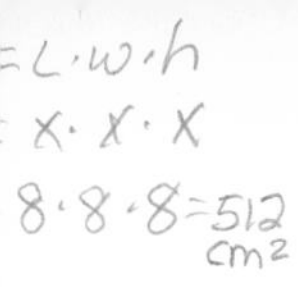

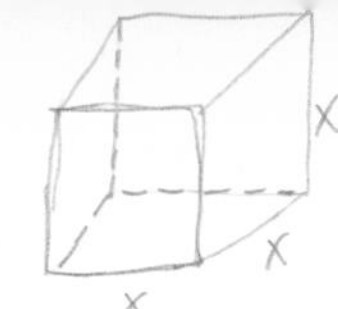

15. If the total surface area of a cube is 384 cm^2, what is the volume of the cube?

16. A cylinder is inscribed in a cube. What is the ratio of the volume of the cylinder to the volume of the cube?

17. A cylinder is circumscribed about a cube. What is the ratio of the volume of the cube to the volume of the cylinder?

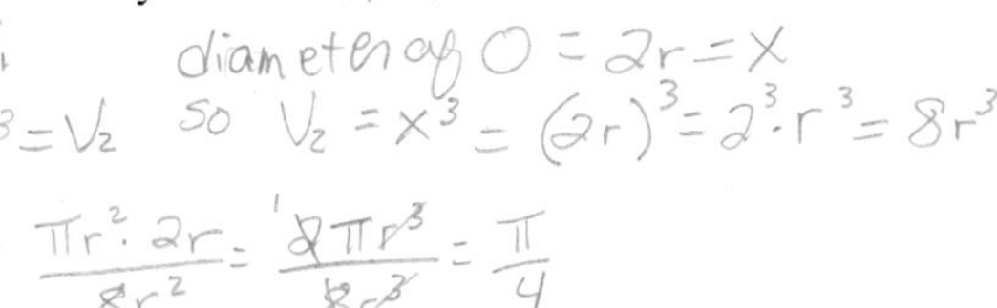

*4 Pyramids, cones, and spheres

■ The largest of all man-made geometric wonders is the Great Pyramid in Egypt. This pyramid is only one of more than 80 such structures built by the ancient Egyptians. It is comparable in height to a 40-story building, and its base covers more than 13 acres. Its structure contains more than 2 million stones weighing from 2 to 100 tons each. It contains more than 91,000,000 cubic feet of rock. How can such a volume be determined? We will develop formulas not only for the volume of a pyramid but also for a cone and a sphere in this section. ■

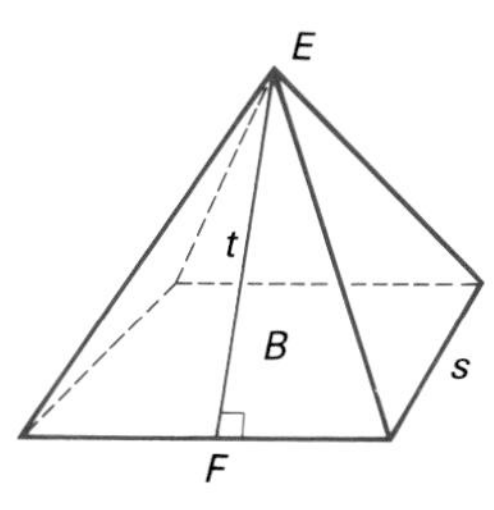

Figure 22

A pyramid is said to be regular if and only if its lateral faces are isosceles triangles. This is true regardless of the number of edges of the polygonal base.

When the lateral faces of a pyramid are isosceles triangles, the base of the pyramid is a regular polygon, and the altitude of the pyramid from the vertex to the base intersects the base at its center. A perpendicular from the vertex of the pyramid to the base of one of the isosceles triangles is perpendicular to the base of the triangle and is sometimes called the *slant height* of the pyramid. The area of each of these isosceles triangles will be one-half the slant height times an edge of the polygonal base, or $ts/2$ where t represents the slant height and s the side of the polygonal base. (See Figure 22.)

Now if there are n edges of the polygonal base, the surface area of the n triangles would be

$$\frac{n(ts)}{2} = \frac{t(ns)}{2}$$

But ns is the perimeter of the base. Therefore, the lateral surface area of a pyramid can be expressed as

$$\frac{tP}{2}$$

where t is the slant height and P is the perimeter of the base.

Surface area of a pyramid

For a regular pyramid with slant height t, perimeter of the base P, and area of the base B,

$$\text{Lateral surface area} = \frac{tP}{2}$$

$$\text{Total surface area} = \frac{tP}{2} + B$$

Example 11 • Find the total surface area of a regular pyramid with square base of side 10 cm and slant height of 8 cm.

Solution Figure 23 presents a visual interpretation. The total area would be the area of 4 triangles plus the area of the square, or

$$4\left(\frac{8 \cdot 10}{2}\right) + 10^2 = 260$$

Figure 23

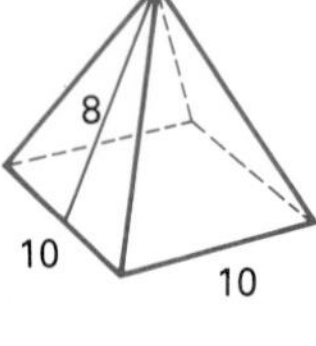

(a)

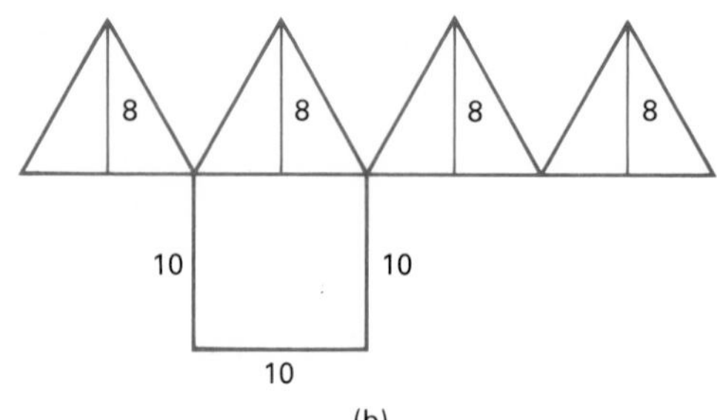

(b)

If we consider a pyramid with n faces inscribed in a right circular cone and let n increase without bound, then the surface area of the pyramid approaches the surface area of the cone. The perimeter of the pyramid approaches the circumference of the circular base of the cone and the lateral surface area of the cone is

$$\frac{tC}{2}$$

where C is the circumference of the base.

Surface area of a cone

Let t be the slant height of a cone and $C = 2\pi r$ the circumference.

$$\text{Lateral surface area} = \frac{Ct}{2} = \pi rt$$

$$\text{Total surface area} = \pi rt + \pi r^2$$

Consider in Figure 24 the triangular prism $ABCDEF$ with base of area B and altitude h. It is possible to cut the prism into three pyramids that do not overlap. It can be shown that these pyramids have equal volumes. Therefore, each has one-third the volume of the prism, which is Bh.

Figure 24

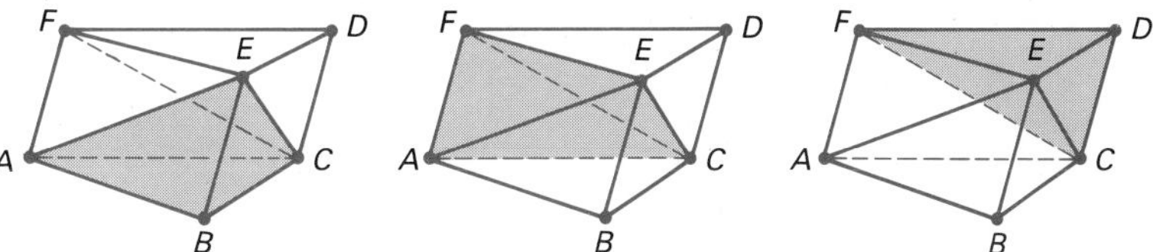

For any pyramid, there is a triangular pyramid with the same altitude and with a base of the same area. Therefore, by Cavalieri's Principle, the two pyramids have the same volume.

Volume of a pyramid

The volume of any pyramid is

$$\frac{Bh}{3}$$

where B is the area of the base and h is the altitude.

For any cone, we know that there exists a pyramid with the same altitude and with a base of the same area. Therefore, by Cavalieri's Principle, the following is true.

Volume of a cone

The volume of any cone is

$$\frac{Bh}{3}$$

where B is the area of the base and h is the height.

Example 12 • What is the volume of a water tank whose bottom is a cylinder of radius 3 m and altitude 4 m and whose top is a cone of altitude 6 m?

Solution

$$V = \pi r^2 h_1 + \frac{1}{3}\pi r^2 h_2$$

$$= \pi \cdot 3^2 \cdot 4 + \frac{1}{3}\pi \cdot 3^2 \cdot 6$$

$$= 54\pi \text{ m}^2 \quad \bullet$$

The formula for the volume of a sphere, obtained from Cavalieri's Principle using a cone inscribed in a cylinder, is given by:

Volume of a sphere

The volume of a sphere of radius r is

$$V = \frac{4}{3}\pi r^3$$

It can be shown that the surface area of a sphere is $4\pi r^2$.

Exercise set 4

R 1. (a) Two regular pyramids with bases with same area and with equal altitudes have ________ volumes.
(b) A cylinder and a cone have the same base and equal altitudes. The volume of the cone is ________ the volume of the cylinder.
(c) Suppose a right prism and a pyramid have equal altitudes and congruent bases. The volume of the pyramid is ________ the volume of the prism.

Find the total surface area and the volume of each solid in Exercises 2 through 4.

2.

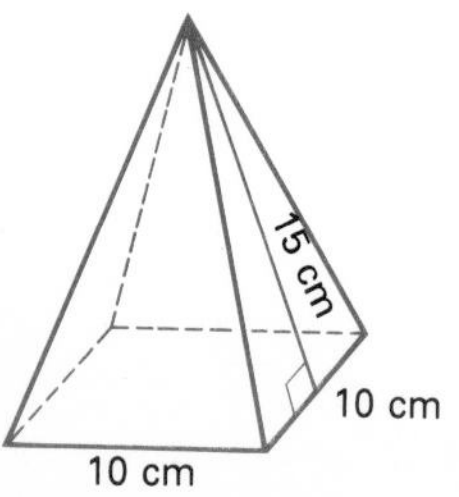

3.

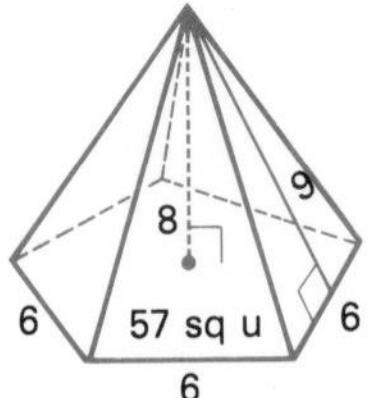

4.

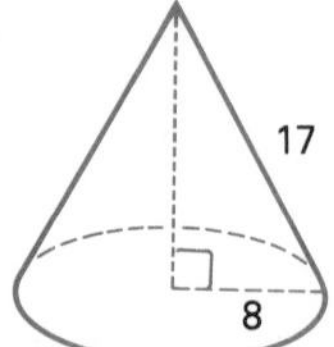

5. If the surface area of a sphere is 400π units, find the volume.

6. A pyramid has a rectangular base 10 cm by 8 cm. The altitude is 14 cm. Find the volume of the pyramid.

T 7. Find the volume of the figure shown.

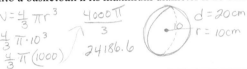

8. Suppose two cones are similar, with ratio k.
   (a) Find the ratio of the lateral area of one to the lateral area of the other.
   (b) Find the ratio of the volume of one to the volume of the other.

9. The volume of a circular cone is 320π cubic inches, and the altitude is 20 inches. What is the radius of the base?

10. How many cubic centimeters of air can be pumped into a basketball if its maximum diameter is 20 cm?

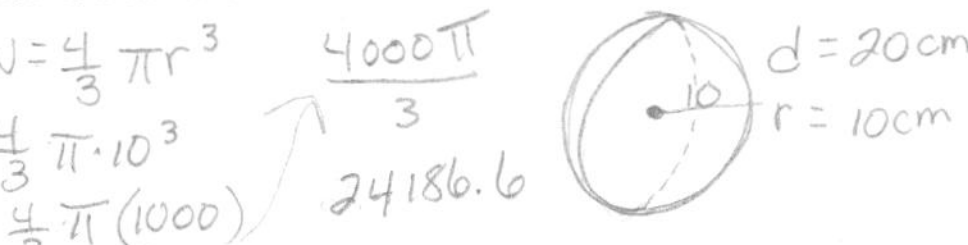

11. What happens to the surface area of a sphere if its radius is doubled?

12. What happens to the volume of sphere if its radius is doubled?

13. Suppose two spheres are similar, with constant of similarity k. What is the ratio of their volumes? Their surface areas?

C 14. Suppose a sphere of radius r is inscribed inside a cylinder. Compare their surface areas.

15. (a) A cone-shaped cup is filled with water to a depth equal to half of the altitude. What part is this of a full cup of water?
    (b) What should the water level be in order for the cup to be half full (that is, contain one-half the volume of a full cup)?

16. A sphere is inscribed inside a cylinder. Find the ratio of the volume of the sphere to the volume of the cylinder.

*5 Other SI units of measure

■ At a SPL fraternity party, Larry proclaimed that his 12-ounce can of soft drink did not weigh 12 ounces. Immediately there was disagreement and bets on both sides. Could Larry be right? ■

Larry's fraternity friends were surprised to learn that the 12 referred to fluid ounces, not weight. A fluid quart and a dry quart are different. You will find the metric system to be much less complicated.

A person, an apple, and a tree have *mass*. To understand the concept of mass, let us compare it to weight, a measure with which you are more familiar. How do you measure your weight? A scale could give your weight, but this weight would vary, depending on gravitational pull. For instance, a rock on the earth might show a weight of 30, whereas the weight might be only 5 on the moon.

Mass, on the other hand, remains the same whether it is on the moon or on the earth. Sometimes described as the property of an object that resists acceleration, mass is technically a measure of the inertia of an object.

In this section, you will review the SI units for mass and capacity.

In addition to the ease with which it is possible to convert among the units of length, the units of area, and the units of volume in the SI system, there are useful relationships among volume, capacity, and mass in this system. At certain specified conditions of temperature and atmosphere (4°C and 1 atm pressure), a **cube of water one centimeter on a side occupies a milliliter of space (capacity) and weighs one gram (weight).**

A **liter** is the basic unit for measuring capacity. From preceding discussion,

$$1 \text{ liter (L)} = 1 \text{ dm}^3$$

Table 3 summarizes these measurements, the most common of which are the liter and milliliter.

Table 3

1 liter	= 1000 milliliters (mL)
	= 100 centiliters (cL)
	= 10 deciliters (dL)
	= 1/10 dekaliter (daL)
	= $1/10^2$ hectoliter (hL)
	= $1/10^3$ kiloliter (kL)
	= $1/10^6$ megaliter (ML)

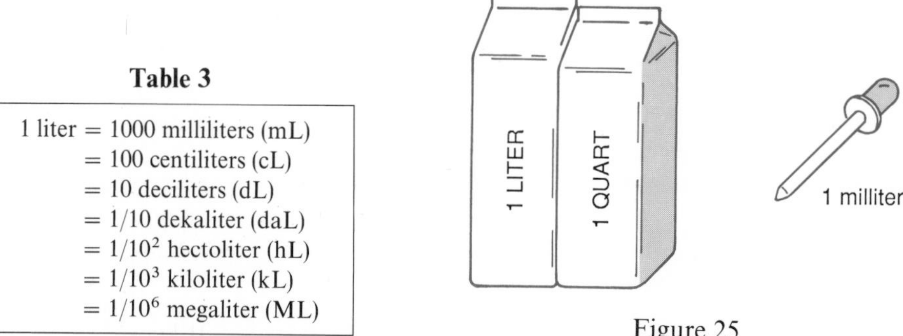

Figure 25

Remember that a liter (L) is a little more than a quart (Figure 25) and that a milliliter (mL) is approximately the amount of liquid in an eyedropper. The metric scale in Figure 26 is useful in comparing metric units.

Figure 26

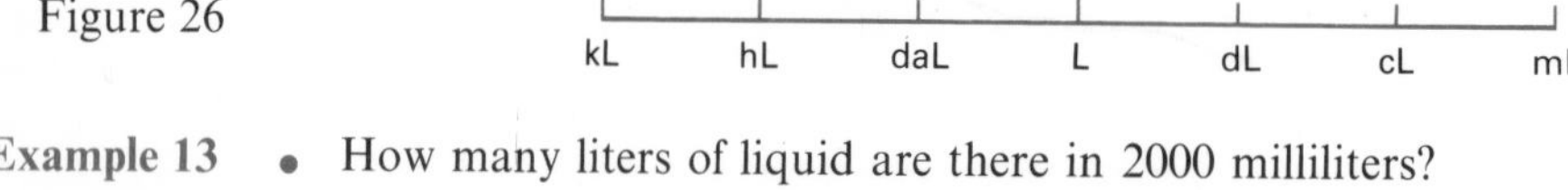

Example 13 • How many liters of liquid are there in 2000 milliliters?

Solution

$$1 \text{ mL} = 0.001 \text{ L}$$

$$2000 \text{ mL} = 2000(0.001) = 2 \text{ L} \quad \bullet$$

As indicated in the introduction, scientists distinguish between mass and weight. However, the word *weight* is used by the average person instead of mass. Since it has such widespread acceptance, *weight* will be used in the problems in this book as a synonym for mass.

Table 4

1 kilogram (kg)	= 10^3 grams (g)
1 hectogram (hg)	= 10^2 grams
1 dekagram (dag)	= 10 grams
1 decigram (dg)	= 1/10 gram
1 centigram (cg)	= 1/100 gram
1 milligram (mg)	= 1/1000 gram

The basic metric unit of mass is the **kilogram.** A *gram* ($\frac{1}{1000}$ of a kilogram) is rather small. A paper clip has a mass of approximately 1 gram and a nickel a mass of approximately 5 grams. Table 4 gives the different metric measures of mass, and Figure 27 shows these units on a metric scale. The kilogram, the gram, and the milligram are the units in practical use today.

Figure 27

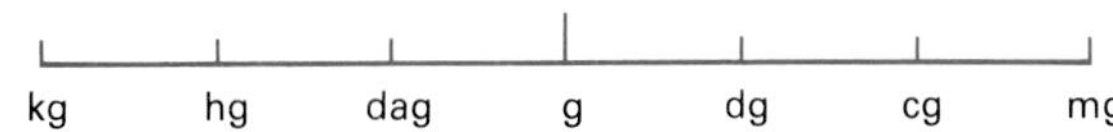

Example 14 • (a) Change 20 kg to grams.
(b) Change 36,000 g to kilograms.

Solution (a) $1 \text{ kg} = 1000 \text{ g}$
$20 \text{ kg} = 20(1000) = 20{,}000 \text{ g}$

(b) $1 \text{ g} = 0.001 \text{ kg}$
$36{,}000 \text{ g} = 36{,}000(0.001) = 36 \text{ kg}$ •

Example 15 • Jan weighs 52 kg. The bag of groceries weighs 5000 g. What is their combined weight?

Solution

$$1 \text{ g} = 0.001 \text{ kg}$$

$$5000 \text{ g} = 5000(0.001) = 5 \text{ kg}$$

The combined weight is $52 + 5 = 57$ kg. •

Example 16 • A detergent is packaged in super size at 2 kg for \$2.10 and in regular size at 720 g for 60¢. Which is actually the money-saving box?

Solution The super size costs ($2.10 \div 2000$), or about \$.001 per gram. The regular size costs ($0.60 \div 720$), or about \$.0008 per gram. The regular size is the better buy. •

Two units of mass used extensively in daily business transactions are the **metric ton** and what is often called the **kilo.** The metric ton (t) is 100 kilograms. Kilo is frequently used in place of kilogram. "I ordered 4 kilos of potatoes" is the same as "I ordered 4 kilograms of potatoes." Since 1 kg ≈ 2.20 lb., a metric ton is about 2200 lb.

The unit of temperature in the metric system is the kelvin or K (no degree sign), which is especially suited for scientific calculations. A unit derived from the kelvin unit and in common usage is the degree Celsius or C°. This unit is named for Anders Celsius, a Swedish astronomer, who in 1712 devised a scale with a 0° reading for freezing water and 100° for boiling water. Figure 28 gives comparisons between temperature in degrees Celsius and degrees Fahrenheit, and the formula

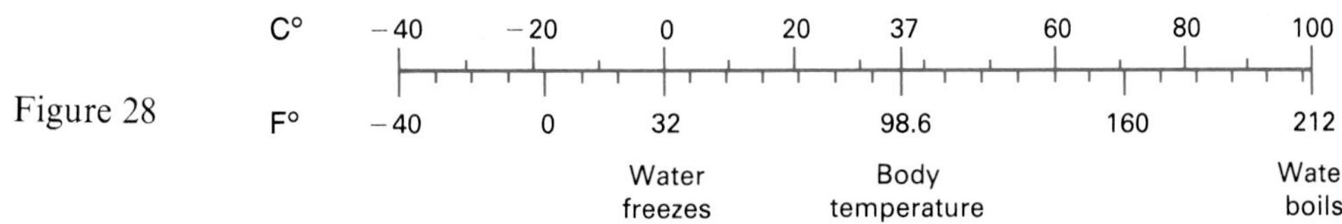

Figure 28

Just for fun

A man with a 5-L and a 3-L bucket went to the well to get exactly 4L of water. How was he able to get the 4L of water by using only the two buckets?

relating the two is

$$F^\circ = 32^\circ + \frac{9}{5}C^\circ$$

Also

$$C^\circ = \frac{5}{9}(F^\circ - 32^\circ)$$

Exercise set 5

R 1. Convert to liters.

(a) 125,000 mL (b) 16 mL
(c) 1230 kL (d) 14 kL
(e) 236 cL (f) 135,000 cL

2. Perform the following conversions.

(a) 1689 kg = ______ g (b) 346 kg = ______ cg
(c) 367 cg = ______ kg (d) 6152 mg = ______ g
(e) 14 cg = ______ mg (f) 179 kg = ______ g

3. Choose the more reasonable measurement for weight.

(a) Pair of tennis shoes (1 g or 1 kg)
(b) Nickel (5 g or 5 kg)
(c) Tennis ball (57 g or 57 kg)
(d) Large dog (35 g or 35 kg)

4. Complete the table.

Article	mg	g	kg
Vitamin C tablet	250	______	______
Adult	______	______	80
Chicken	______	610	______

5. Choose the better measurement.

(a) A full bathtub (210 mL or 210 L)
(b) An eyedropper (1 mL or 1 L)
(c) A bucket (8 mL or 8 L)
(d) A gas tank in a car (80 mL or 80 L)

6. Choose the more sensible temperature.

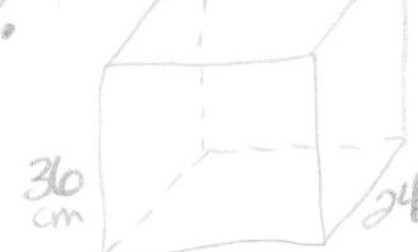

(a) A warm day (5°C or 30°C)
(b) Cold milk (4°C or 40°C)
(c) Hot soup (20°C or 80°C)
(d) A snowy day (−8°C or 18°C)
(e) A warm shower (50°C or 110°C)
(f) Baseball weather (25°C or 75°C)

7. A right rectangular prism has edges that measure 16 cm, 24 cm, and 36 cm. How many liters will the container hold?

8. A container in the shape of a right prism is 110 cm high and contains 1 L. How many square centimeters are in the base?

9. A conical drinking cup is 8 cm high and has a diameter of 6 cm. How many of the cups can be filled from a 2-liter bottle of cola?

10. One cubic centimeter of iron ore weighs about 0.008 kg. What is the weight of an iron bar 50 cm by 20 cm by 5 cm?

11. Convert the following temperatures from degrees Fahrenheit to degrees Celsius.

    (a) 50° (b) 68° (c) 32° (d) −2°

12. Convert the following temperatures from degrees Celsius to degrees Fahrenheit.

    (a) 20° (b) 10° (c) 30° (d) 0°

13. The normal body temperature is 98.6°F. Convert this to degrees Celsius.

14. An aquarium holds 35 L of water. How much does the water weigh?

15. A car is driven 180 km and uses 12 L of gasoline. What is gasoline consumption in kilometers per liter?

T 16. Estimate the capacity of each item in liters or milliliters.

(a) A half-gallon carton of milk
(b) A cup of water
(c) A gallon can of gasoline
(d) A can of cola
(e) A quart jar of jelly
(f) A teaspoon of medicine

17. Estimate in grams or kilograms the weight or mass of each of these items.

(a) A glass of water (b) A quarter
(c) This math book (d) A can of cola
(e) A ball-point pen
(f) A regular-sized automobile
(g) A sack of fertilizer (h) A piece of chalk
(i) A basketball (j) A tennis racket
(k) A compact automobile
(l) A bicycle

18. A distributor intends to sell a metric ton of sugar in 657-gram containers at 22¢ each.

    (a) How many containers will be obtained from the metric ton?
    (b) For how much money will this metric ton sell?

19. Estimate these temperatures in degrees Celsius.

    (a) Comfortable room temperature
    (b) Hot soup
    (c) Cold milk
    (d) Boiling water

20. Formulate rules for performing the following conversions.

    (a) Kilogram to gram (b) Gram to milligram

C 21. Susan must take 5 mL of medicine twice a day. If her bottle holds 4 dm^3, how many days will her medicine last?

22. A car is advertised as getting 15 km/L. Is this gasoline mileage "good"?

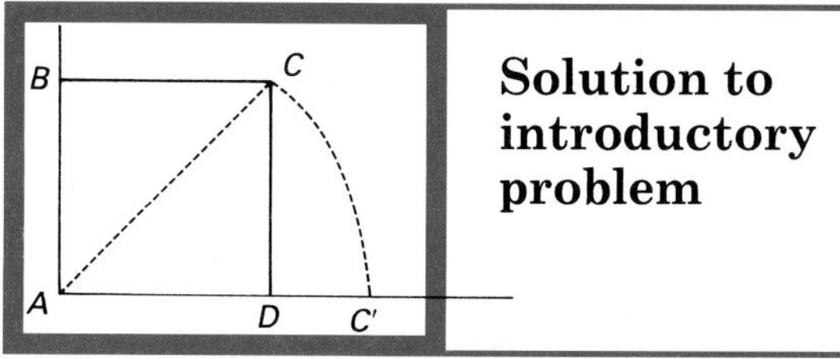

Solution to introductory problem

On the corner of a table place the cube with bottom $ABCD$ as shown. On the table mark the point B. "Rotate the bottom of the cube" about A until $\overline{AC}$ coincides with the edge of the table. Now mark the position occupied by C as C'. Remove the cube and measure $\overline{BC'}$. This is the length of the diagonal shown. Why?

Summary and review

1. A polyhedron is a simple closed surface in space whose boundary is composed of polygonal regions.
   (a) A prism is a polyhedron formed by two congruent polygonal regions in parallel planes, along with three or more regions bounded by parallelograms joining the two polygonal regions so as to form a closed-space figure.
   (b) A pyramid is a polyhedron formed by a simple closed polygonal region and a point not in the plane of the region and the triangular regions joining the point and the edges of the polygonal region. The triangular regions are congruent in a regular pyramid.
2. The plantonic solids consist of
   (a) Regular tetrahedron
   (b) Regular hexagon
   (c) Regular octagon
   (d) Regular dodecahedron
   (e) Regular icosahedron
3. Let V represent the number of vertices, E the number of edges, and S the number of sides of a polyhedron. Then Euler's formula states that $V + S - E = 2$.
4. In general, a cylinder is the surface generated by movement of parallel lines, each of which intersects a closed figure in a plane. Usually we consider the case where the closed figure is a circle, and the bases bound the ends.
5. All points in space equidistant from a fixed point make up a sphere.
6. A cone consists of segments joining a fixed point not in a given plane to points on a closed curve in the given plane. Usually the closed curve is a circle.
7. Cavalieri's Principle. Given a plane and any two solids. If every plane parallel to the given plane intersects the two solids in cross sections that have the same area, then the solids have the same volume.
8. Some important formulas for total surface area include the following, where B is the area of the base, P is the perimeter of the base, e is a lateral edge, and t is the slant height. Make sure you can associate these formulas with appropriate solids.

$$\text{T.S.A.} = eP + 2B$$

$$\text{T.S.A.} = Ce + 2B$$

$$\text{T.S.A.} = \frac{tp}{2} + B$$

$$\text{T.S.A.} = \pi rt + \pi r^2$$

9. Some formulas for volume include the following:

$$V = lwh$$

$$V = Bh$$

$$V = (\text{area of right section}) \cdot (\text{lateral edge})$$

$$V = \frac{Bh}{3}$$

$$V = \frac{4\pi r^3}{3}$$

Review exercise set 6

R 1. Perform these conversions.
   (a) 5001 cm = ______ km
   (b) 300 mL = ______ L
   (c) 81 km = ______ m
   (d) 17 ha = ______ m^2
   (e) 4738 cg = ______ kg
   (f) 1022 cL = ______ mL

2. The road gang was told to dig a trench 2 km long, 180 cm wide, and 1 m deep. How many cubic meters of dirt will be removed?

3. Find the volume and surface area of the following figures.

(a) A milk carton with height 30 cm and a square base with each side measuring 8 cm (assume that the top is flat and closed)
(b) A tin can with circumference of 6π cm and a height of 11 cm

4. Which unit—meter, square centimeter, liter, gram, milliliter, square meter, cubic centimeter—would most likely be used to measure the following?
   (a) The weight of a book
   (b) The capacity of a cone-shaped paper cup
   (c) The length of a swimming pool
   (d) A dosage of liquid medicine
   (e) The area of a wall to be painted
   (f) The volume of a shoe box
   (g) The area of a sheet of notebook paper
   (h) The capacity of a milk bottle

5. Find the lateral surface areas and the volume of each of the solids.
   (a)

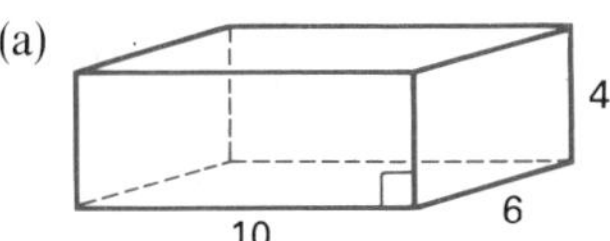

   (b)

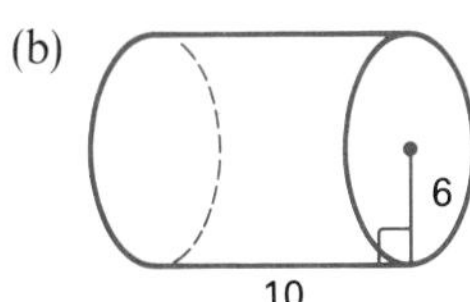

   (c)

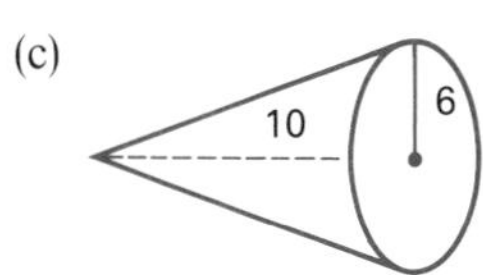

T 6. What happens to the volume of the interior of a right circular cylinder if both the circumference of the base and the altitude are doubled?

7. If the perimeter of a square is equal to the circumference of a circle, compare the areas of the bounded regions.

8. The area of a circular region is equal to the circumference of the circle. Find the measure of the radius of the circle.

C 9. Suppose that a room 2 m by 3 m by 5 m is used by a chemical company for storage. If the chemists have containers, each of which contains the following equivalents, what is an upper limit on the number of each type of container that they could put in the room?
   (a) 213 cm^3 of paper
   (b) 4 m^3 of powdered chemicals
   (c) 25 L of acid

10. If an 80-cm diameter wheel on a car traveling 50 km per hour turns 225 revolutions, how long has it been traveling?

11. Classify as true or false the following statements concerning a portion of the triangular pyramid when viewed from above.
   (a) Edges $\overline{AD}$ and $\overline{CF}$ intersect.
   (b) $\overleftrightarrow{DF}$ and $\overleftrightarrow{BE}$ intersect.
   (c) $\overrightarrow{BE}$ and $\overrightarrow{CF}$ intersect.

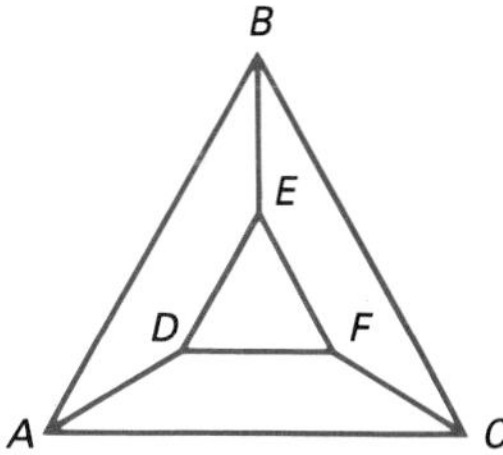

12. Verify Euler's formula for the following.
   (a)

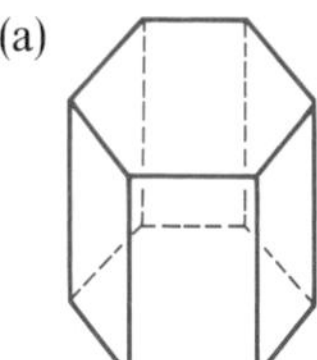

   (b)

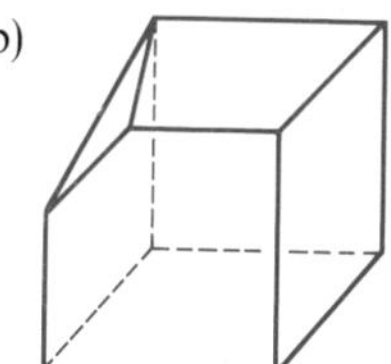

   (c) The figure in Exercise 11

13. A cylindrical fuel tank has a height of 5 m and a radius of 1.3 m. How many liters of fuel will it hold?

14. James' water hose is 16 m long with an inner radius of 5 cm. Disregarding any thickness of the wall of the hose, how many liters of water could be within the hose at one time?

Bibliography

1. Boag, Tom, Charles Boberg, & Lyn Hughes. "On Archimedean Solids." *MT* (May 1979), 371–376.
2. Campbell, Patricia F. "Cardboard, Rubber Bands, and Polyhedron Models." *AT* (October 1983), 48–52.
3. Clements, Douglas C. and Michael Ballista. "Geometry and Geometric Measurement." *AT* (February 1986), 29–32.
4. Gardner, Martin. *Aha!* New York, N.Y.: W. H. Freeman, 1978.
5. Hirsch, Christian R., Mary Ann Roberts, Dwight O. Coblentz, Andrew J. Samine, and Harold L. Schoen. *Geometry*, Glenview, Illinois: Scott Foresman and Company, 1969.
6. "Ideas," *AT* (April 87), 26–33.
7. Malcom, Paul S. "Braided Polyhedra." *AT* (May 1976), 386–388.
8. Nelson, Norman N., & Forest N. Fisch. "The Classical Cake Problem." *MT* (November 1973), 659–661.
9. Nichols, Eugene D., Mervine L. Edwards, E. Henry Garland, Sylvia A. Hoffman, Albert Mamary, and Will F. Palmer. *Holt Geometry*. New York: Holt, Rinehart and Winston, 1978.
10. Reynolds, Jean. "Build a City." *AT* (September 1985), 12–15.
11. Shumway, Richard J., & Larry Sachs. "Don't Just Think Metric—Live Metric." *AT* (February 1975), 103–110.
12. Young, Jerry L. "Improving Spatial Abilities with Geometric Activities." *AT* (September 1982), 38–43.

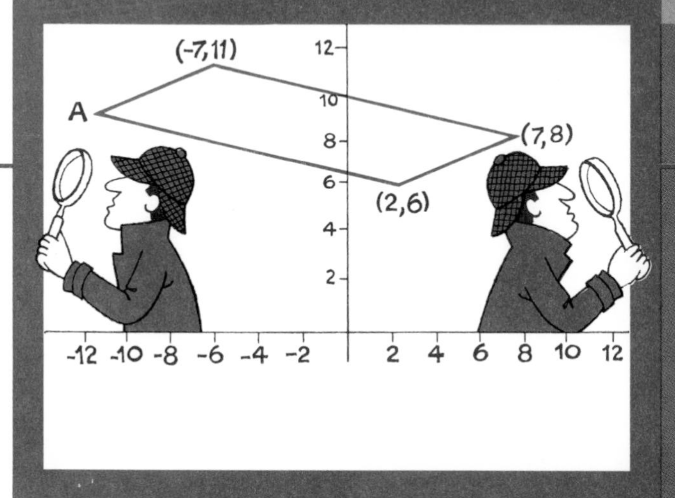

The coordinates of three verticles of a parallelogram are given above. Find the coordinates of vertex *A*, in order for the figure to be a parallelogram. (Note that if the figure were not given, there could be three possible locations for the fourth vertex.)

CHAPTER 12
Coordinate Geometry and Transformations

One morning, while meditating in bed, as was his custom, a Frenchman named René Descartes had a simple but profound idea. He thought that in order to describe the relationships between points in the plane, it would be helpful to have a frame of reference on which to base the discussion. Out of this inspiration grew the rectangular coordinate system (or coordinate plane, sometimes called *Cartesian* plane after Descartes) and the notion of coordinate graphing as a useful tool in understanding mathematical relationships.

After studying this chapter, you will be able to solve the introductory problem. In fact, this problem uses three concepts you will soon master. In this chapter, we set up the machinery not only to look at transformations in a different way but also to introduce basic concepts of analytic geometry: distance formulas, slopes, and equations of lines.

1 Coordinate geometry

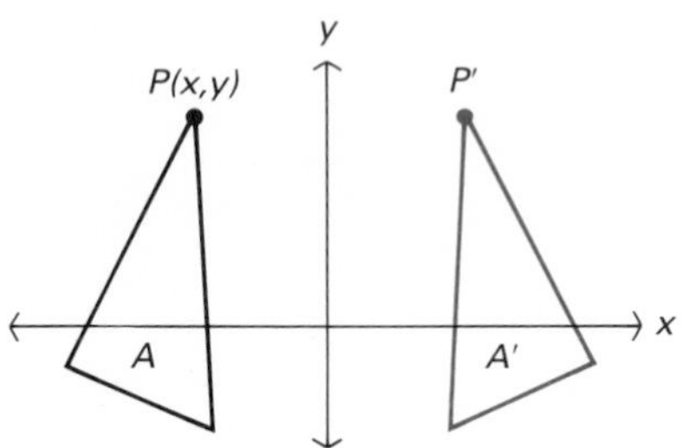

■ Triangle A has been reflected across the y axis to get triangle A'. If P has coordinates (x, y), what are the coordinates of P'? ■

As noted in the preceding problem we are beginning a study of coordinate geometry. One-dimensional number lines or real number lines serve as an excellent introduction to coordinate geometry in two dimensions. In this section, you will review the plotting of points on a rectangular Cartesian coordinate system. Then you will use coordinate geometry to discuss translations and reflections. We conclude the section with a distance formula and a midpoint property that enable us to prove geometric properties algebraically.

In Chapter 8, we considered a real number line with one-dimensional coordinates. In this section, we use number lines in two directions to represent an ordered pair of numbers. These numbers are called **coordinates** (in two dimensions), and the coordinate system is called the **Cartesian coordinate system.**

Coordinate system

The x axis is a horizontal number line that extends infinitely to the left and right; the y axis is a vertical number line that extends infinitely up and down. The two axes cross at right angles, and their point of intersection is the point 0 on the vertical number line and 0 on the horizontal number line, which is called the *origin.*

Since the two axes are number lines, units are marked off on each axis using positive integers going up and to the right and negative integers going down and to the left as seen in Figure 1.

To locate a pair of coordinates in the Cartesian coordinate system, find the x coordinate on the x axis, and mentally draw a vertical line through this point. Then find the y coordinate on the y axis, and mentally draw a horizontal line through this point. The point of intersection of these two lines is the point with the desired coordinates. Every pair of coordinates (x, y) corresponds to a point in the coordinate plane, and every point on the plane has coordinates that can be written in the form (x, y). Locating points on the plane is called **plotting** the points.

BIOGRAPHY

René Descartes
1596–1650

René Descartes, the founder of analytic geometry, considered himself more a philosopher than a mathematician. Indeed, today Descartes is known primarily as a philosopher. Yet his mathematical discoveries provided the stimulus for the growth of mathematics for 200 years. John Stuart Mill called Descartes' invention of the Cartesian plane "the greatest single step ever made in the progress of the exact sciences."

Descartes, born in 1596 near Tours into the lesser French nobility, was a youth of delicate health. To protect his health, René developed a lifelong habit of lying in bed until late in the morning. These hours of meditation seem to have been his most productive.

Example 1 • Plot the point with the coordinates (3, 5).

Solution Move three places to the right of 0 on the x axis. Mentally draw a vertical line through that point. Then move five places up on the y axis. Mentally draw a horizontal line through that point. The intersection of these two lines, seen in Figure 1, is the point (3, 5). •

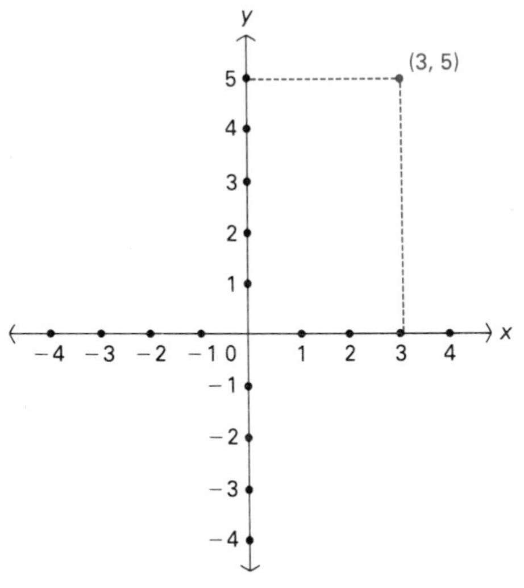

Figure 1

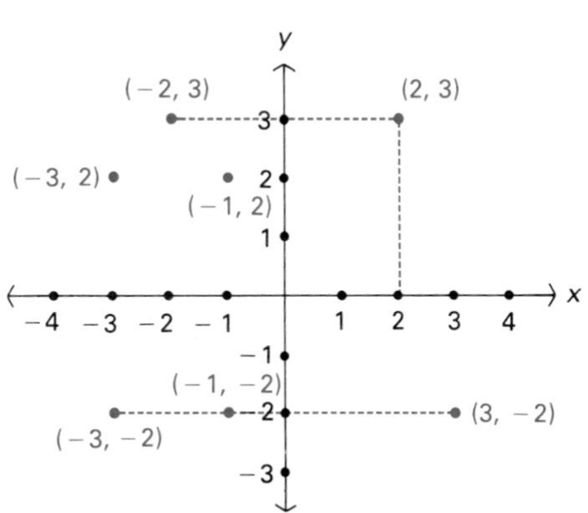

Figure 2

Reflect the point (2, 3) about the y axis in Figure 2. Do you see that the reflection image is $(-2, 3)$? Likewise, the reflection image of $(-3, -2)$ is $(3, -2)$.

Reflection about y axis

> The reflection image of (x, y) about the y axis is the point $(-x, y)$.

Example 2 • Find the coordinates of $\triangle A'B'C'$, the reflection image of $\triangle ABC$ about the y axis, where $\triangle ABC$ is defined by $A(2, 1)$, $B(3, -2)$, and $C(3, 4)$.

Solution The images of A, B, and C are $A'(-2, 1)$, $B'(-3, -2)$, and $C'(-3, 4)$ •

In Figure 2, the reflection image of $(-1, 2)$ about the x axis is $(-1, -2)$, and the reflection image of $(-3, -2)$ about the x axis is $(-3, 2)$.

Reflections about x axis

> The reflection image of (x, y) about the x-axis is the point $(x, -y)$.

Example 3 • Find the coordinates of $\triangle A'B'C'$, the reflection image of $\triangle ABC$ about the x axis, where $\triangle ABC$ is defined by $A(2, 1)$, $B(3, -2)$, and $C(3, 4)$.

Solution The images of A, B, and C are $A'(2, -1)$, $B'(3, 2)$, and $C'(3, -4)$. •

In Figure 3, suppose that $P(2, 3)$ is translated 3 units to the right and 1 unit up. P' will have coordinates $(2 + 3, 3 + 1)$ or $(5, 4)$. This leads to the following property.

Translation

> If $P(x, y)$ is translated r units in an x direction and s units in a y direction, the image of P is
>
> $$P'(x + r, y + s)$$

Example 4 • Find the coordinates of $\triangle A'B'C'$, the image of $\triangle ABC$ translated 3 units to the right and down 4 units, where $\triangle ABC$ is defined by $A(2, 1)$, $B(3, -2)$, and $C(3, 4)$.

Solution The images of A, B, and C are $A'(5, -3)$, $B'(6, -6)$ and $C'(6, 0)$. •

We learned in Chapter 8 how to find the length of a line segment when the segment is parallel to one of the axes.

Example 5 • Find the length of the segment $\overline{AB}$ where
(a) A is given by $(2, 3)$ and B by $(2, -1)$.
(b) A is given by $(2, 5)$ and B by $(7, 5)$.

Solution (a) The change in y is from 3 to -1 or

$$AB = |3 - (-1)| = 4$$

(b) The change in x is from 2 to 7 or

$$AB = |7 - 2| = 5 \quad \bullet$$

Figure 3

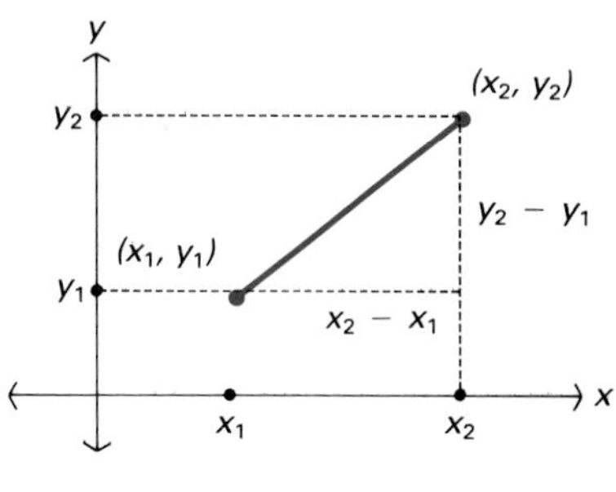

Figure 4

The Pythagorean Theorem can be used to find the length of a line segment not parallel to an axis. For example, consider the segment from $A(x_1, y_1)$ to $B(x_2, y_2)$ in Figure 4. The change in y is given by $|y_2 - y_1|$ and the change in x by $|x_2 - x_1|$. By the Pythagorean Theorem,

$$AB = \sqrt{(x_2 - x_1)^2 + (y_2 - y_1)^2}$$

Distance formula

> The distance AB between any two points A and B with coordinates (x_1, y_1) and (x_2, y_2) is given by
>
> $$AB = \sqrt{(x_2 - x_1)^2 + (y_2 - y_1)^2}$$

Another useful formula in looking at geometric properties is the formula for the midpoint M of a segment. In Figure 5(a), consider the midpoint of $\overline{AB}$ where A is given by (x_1, y_1) and B by (x_1, y_2). $AB = y_2 - y_1$. So one-half of $AB = (y_2 - y_1)/2$. The y coordinate of the midpoint is

$$y_1 + \frac{AB}{2} = y_1 + \frac{y_2 - y_1}{2}$$

$$= \frac{y_2 + y_1}{2}$$

The coordinates of M are $(x_1, (y_2 + y_1)/2)$. Similarly, the coordinates of the midpoint M of AB where A is (x_1, y_1) and B is (x_2, y_1) is $(x_1 + x_2)/2, y_1)$.

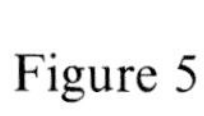

Figure 5

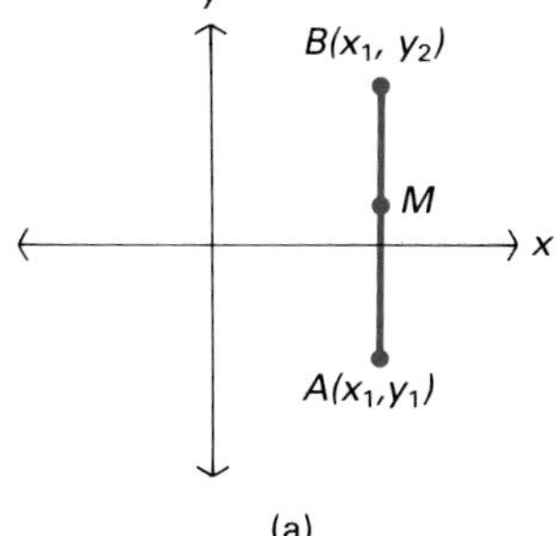

(a)

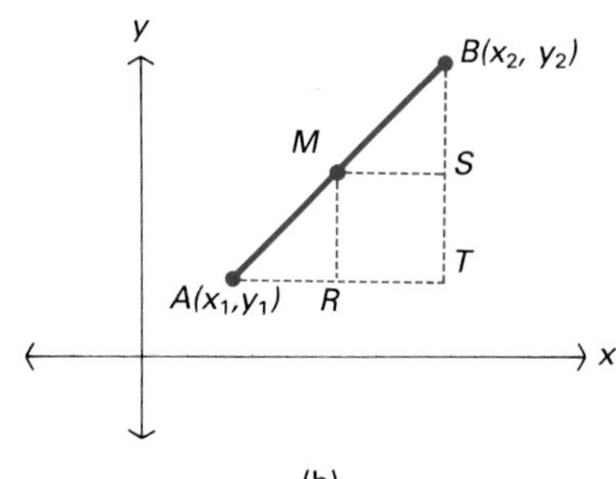

(b)

Just for fun *Jodi and Aaron walk to school each day on the routes indicated. Who walks the shortest distance?*

To find the midpoint M of $\overline{AB}$ in Figure 5(b), where A is (x_1, y_1) and B is (x_2, y_2), draw vertical and horizontal segments forming right $\triangle ATB$, right $\triangle MSB$, and right $\triangle ARM$. Can you show that right $\triangle MSB \cong$ right $\triangle ARM$ and that $RTSM$ is a rectangle? Then

$$AR = MS = RT \qquad \text{and} \qquad BS = MR = ST$$

Thus, R is the midpoint of $\overline{AT}$ and S is the midpoint of $\overline{BT}$. The coordinates of S are $(x_2, (y_2 + y_1)/2)$ and of R are $((x_1 + x_1)/2, y_1)$. Since by construction the midpoint M has the x coordinate of R and the y coordinate of S, we have the following midpoint formula.

Midpoint formula

Let A and B have coordinates (x_1, y_1) and (x_2, y_2) respectively, then the midpoint of $\overline{AB}$ has coordinates

$$M = \left(\frac{x_1 + x_2}{2}, \frac{y_1 + y_2}{2}\right)$$

Exercise set 1

R 1. Plot the following points on a coordinate system.

(a) (2, 3) (b) (−1, −2)
(c) (−1, 4) (d) (0, 5)
(e) (5, 0) (f) (−2, 3)

Plot the given points, locate the distance between them, and find the coordinates of each midpoint.

2. $A(-1, 3)$ and $B(7, -7)$
3. $A(2, 4)$ and $B(-2, -3)$
4. $A(-4, -1)$ and $B(3, -5)$
5. $A(-4, -3)$ and $B(-2, -7)$

Graph $\angle ABC$. Does the angle appear to be acute, right, or obtuse?

6. $A(5, -1)$, $B(2, -1)$, $C(2, 3)$

7. $A(3, -2)$, $B(-2, 4)$, $C(7, 0)$
8. $A(2, 5)$, $B(3, 5)$, $C(2, -3)$
9. $A(2, 3)$, $B(0, 4)$, $C(-5, -2)$

T 10. Show that $\triangle ABC$ is a right triangle: $A(5, -1)$, $B(2, -1)$, $C(2, 3)$.

11. Show that $\triangle ABC$ is a right triangle: $A(6, 1)$, $B(2, -3)$, $C(-4, 3)$.
12. Show that $\triangle ABC$ is an isosceles triangle: $A(1, 3)$, $B(5, 0)$, $C(-2, -1)$.
13. Show that the diagonals of $ABCD$ are of equal length if $A(3, 2)$, $B(3, -1)$, $C(7, -1)$, and $D(7, 2)$.

Find the coordinates of the image reflected about the y axis; about the x axis. Draw these figures. Does your work look accurate?

14. $A(1, 1)$, $B(-2, 3)$, $C(2, 4)$
15. $A(1, 1)$, $B(2, 3)$, $C(-1, 2)$, $D(-2, -1)$
16. Find the coordinates of the image of the figure in Exercise 14 by translating the figure four units to the right and three units down. Draw the new figure.
17. Work Exercise 16 for data given in Exercise 15.
18. Determine whether or not the triangle with vertices $(6, -2)$, $(-1, -1)$, and $(4, 4)$ is equilateral.

C 19. Discuss a procedure for algebraically determining reflection about $y = x$.

20. Work Exercise 19 for $y = -x$.
21. Find the image of $(-2, 3)$ after a reflection about $y = x$.
22. Find the image of $(-2, -3)$ after a reflection about $y = -x$.

2 Functions and their graphs

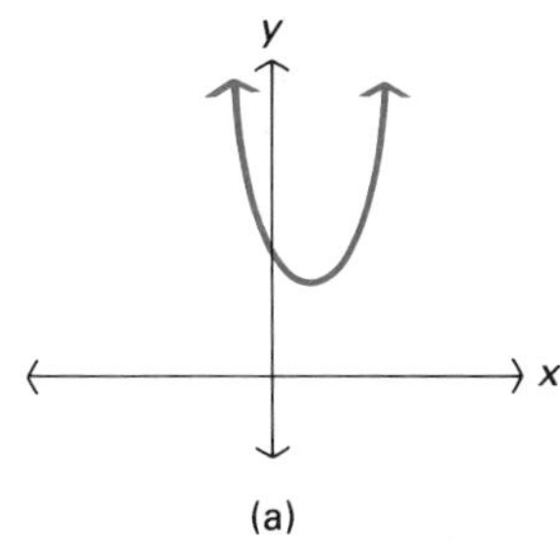

(a)

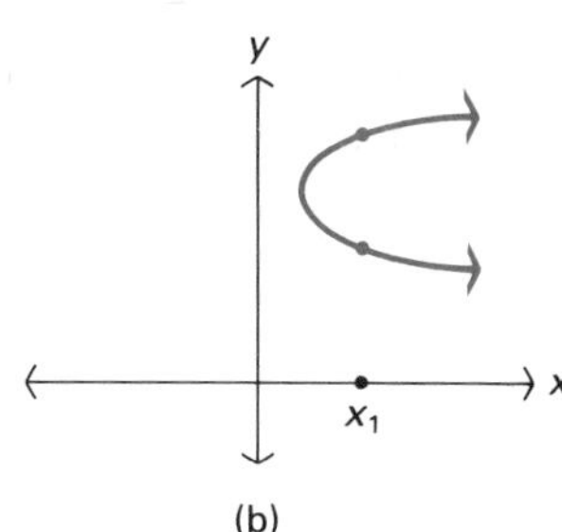

(b)

■ One of the above graphs represents a function; the other one does not. In this chapter, we will learn to recognize graphs that represent functions. ■

An important concept that influences the study of mathematics from elementary grades through graduate work is the concept of a function. Its study is often introduced to students through a "function machine"; that is, a function is regarded as a machine that produces a unique output number for a given input number. This concept, along with numerous simple examples, will be used to help you to understand the meaning of a function.

Functions are of great importance because most attempts to describe our everyday world produce a function. That is, a function can be used to describe a relationship between two varying quantities. The speed of a falling rock depends on the length of time since it was dropped; home use of electricity depends on the severity of the weather; and the profits enjoyed by a manufacturer are related to the price set for the product. The list of functional relationships is endless.

Let's first think of a function as an operation that transforms elements of a first set A into elements of a second set B. This way of thinking of a function can be demonstrated by what is often called a "function machine." The first set can be considered as *input* for the machine and the second set, or some subset of it, as *output*.

Example 6 • Notice in Figure 6 that we place an *input* into the machine, operate with the rule, and obtain *output*. Suppose that the rule, or function, is "Add a 4 to the input." This rule can be expressed as $x + 4$ when the input is x. If the input is 3, the rule operates to give $3 + 4$, and the output is 7.

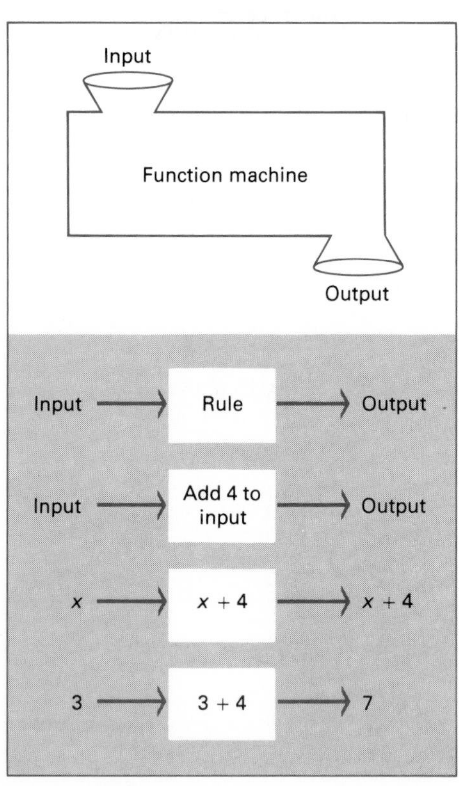

Figure 6

In addition, notice that this function $x + 4$ is a rule or procedure that associates with each element of set A (the input) one and only one element of set B (the output). •

Example 7 • As a more complex example, consider the function machine described in Figure 7, in which the machine consists of a multiplier and an adder. Examining this concept closely, we see that each element of A is paired with exactly one element of B. This property defines a function. •

A function

> A *function* is a rule that assigns to each element of the nonempty input set A (the domain) exactly one element of the output set B.

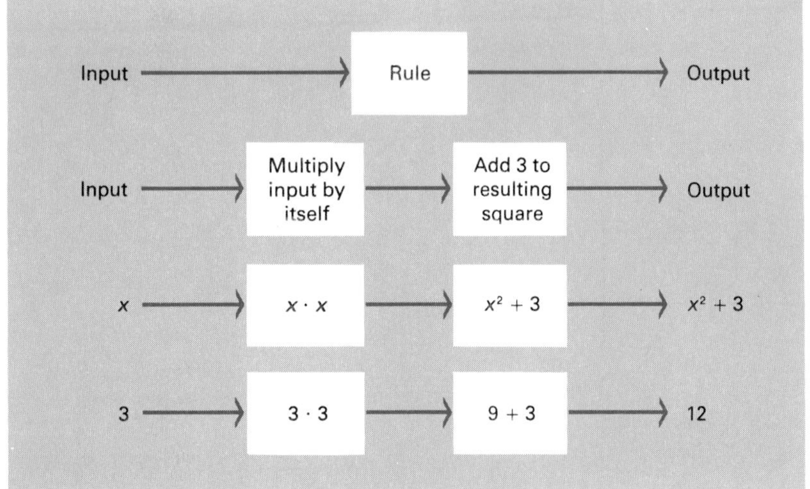

Figure 7

If we write the rule "Add 4 to x," or $x + 4$, as an equation

$$y = x + 4$$

we have another procedure for expressing a function. Using this equation, let x assume the value 2. Then $y = 2 + 4 = 6$. Also

$$x = 3, \qquad y = 7$$

$$x = 4, \qquad y = 8$$

$$x = 8, \qquad y = 12$$

The ordered pairs (2, 6), (3, 7), (4, 8), (8, 12), . . . also describe this function. Using ordered pairs to describe a function leads to the following concept. A function from set A to set B is a relation from A to B with the property that each element of A is paired with exactly one element of B. This result may be formally stated as follows.

A function

A function is a relation in which no two of the ordered pairs have the same first element.

That is, for each element in A (called the **domain**) there can be only one element in B (called the **range**).

The relation in Figure 8(a) is a function; the relation in Figure 8(b) is not. Why not?

Example 8 • Consider the following relations from A to B, where $A = \{2, 3, 4\}$ and $B = \{3, 4, 5, 6\}$. Which of these relations are functions?

(a) $R = \{(2, 3), (3, 4), (4, 5)\}$
(b) $R = \{(3, 3), (3, 4)\}$
(c) $R = \{(2, 3), (3, 4), (4, 5), (2, 6)\}$
(d) $R = \{(2, 5), (3, 5), (4, 5)\}$

Figure 8

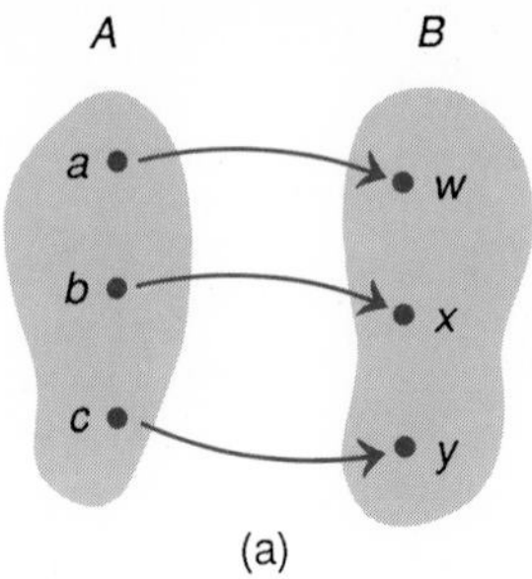

(a)

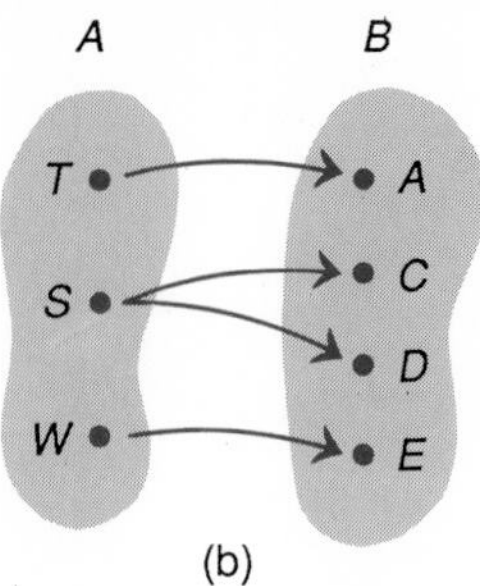

(b)

Solution (a) and (d) are functions. (c) is not a function, because 2 is paired with both 3 and 6. (b) is not a function, because 3 is paired with both 3 and 4. •

Consider the function given by the following rule. Given a number, square it and add 4. The results of this rule are shown by the ordered pairs in Table 1 when the numbers 1, 2, 3, and 4 are used as input values for the function.

Table 1

Input value	*Output value*
1	$1^2 + 4 = 5$
2	$2^2 + 4 = 8$
3	$3^2 + 4 = 13$
4	$4^2 + 4 = 20$

However, a much simpler expression for the function can be obtained if we use a letter to represent the elements of the input set (the domain). When x is used to represent an arbitrary element of the domain, x is called the *independent variable* (or *input variable*).

When the number x is used as the independent variable, the output value is often written as $f(x)$. Thus, the function discussed earlier can be written as the mathematical formula $f(x) = x^2 + 4$. This notation can be explained as follows.

$$\overbrace{f(x)}^{\text{names the output}} = \underbrace{x^2 + 4}_{\text{describes the rule for computing the output for a given input}}$$

(x names the input)

Example 9 • $f(x) = 3x - 1$ defines a function on the real numbers such that when

(a) $x = 1$, $f(1) = 3 \cdot 1 - 1 = 2$

(b) $x = 2$, $f(2) = 3 \cdot 2 - 1 = 5$

(c) $x = \frac{1}{3}$, $f(\frac{1}{3}) = 3 \cdot (\frac{1}{3}) - 1 = 0$

(d) $x = t$, $f(t) = 3t - 1$ •

Example 10 • For $f(x) = 3x^2 + 2$, find the value of $f(x)$ when $x = 4$.

Solution Using $x = 4$ as input gives

$$\begin{aligned} f(4) &= 3(4)^2 + 2 \\ &= 3(16) + 2 \\ &= 48 + 2 \\ &= 50 \quad \bullet \end{aligned}$$

It is often quite useful to represent a function relationship with a graph. In this circumstance, we use y in place of $f(x)$. We compute ordered pairs that satisfy the formula $y = f(x)$ and plot the pairs on a coordinate system.

Example 11 • Evaluate the function $y = f(x) = 2x - 3$ at several points. Then sketch the graph of the function.

$$\text{When } x = 0,\ y = f(0) = 2(0) - 3 = -3$$

$$\text{When } x = 2,\ y = f(2) = 2(2) - 3 = 1$$

$$\text{When } x = 4,\ y = f(4) = 2(4) - 3 = 5$$

As we will discuss, $y = f(x) = 2x - 3$ is the equation of a straight line. Hence, three points will be sufficient for the graph of this equation with one point serving as a check. (See Figure 9.)

Figure 9

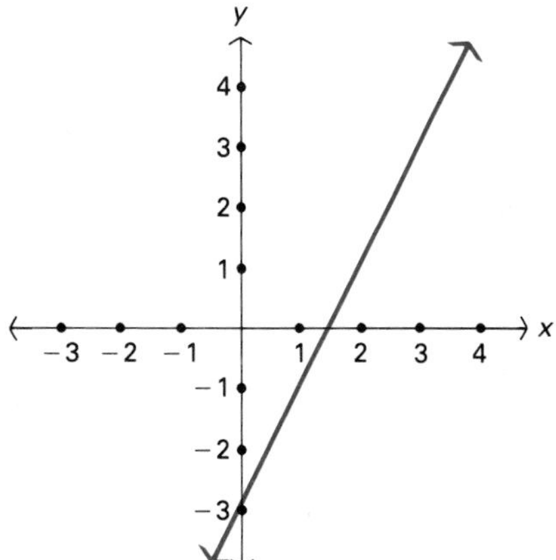

If the graph of a function is a straight line, the function is called a **linear function.** Hence, $f(x) = 2x - 3$ is a linear function. •

We now examine equations of the form $ax + by + c = 0$. If every solution of such an equation were found and then plotted, the solutions would form a line. Thus, an equation of the form $\boldsymbol{ax + by + c = 0}$**, where** $\boldsymbol{a}$ **and** $\boldsymbol{b}$ **are not both 0, is called a linear equation.**

Writing down every solution for a linear equation is impossible. However, every solution can be visually represented on the Cartesian coordinate system. This representation is a line, and a line is determined by two points. Hence, to graph a linear equation, determine at least two solutions of the equation; plot these points; and join them by a straight line. Every point on this line is a solution to the equation. As a check, it is always wise to plot three or four points. If you have made no error. all the points will be on the same straight line.

Example 12 • Graph $x + y = 4$.

Solution (a) A number of solutions of this equation are listed in Table 2. The following discussion shows how two of these solutions are obtained.

Table 2

x	y
1	3
2	2
3	1
4	0

If $x = 1$,
$1 + y = 4$
$y = 3$

If $x = 2$,
$2 + y = 4$
$y = 2$

(b) Now plot each of the four points (Figure 10). If the work is correct, these four points can be joined by a straight line. This line represents the equation $x + y = 4$. Any point on that line is a solution to the equation. •

Figure 10

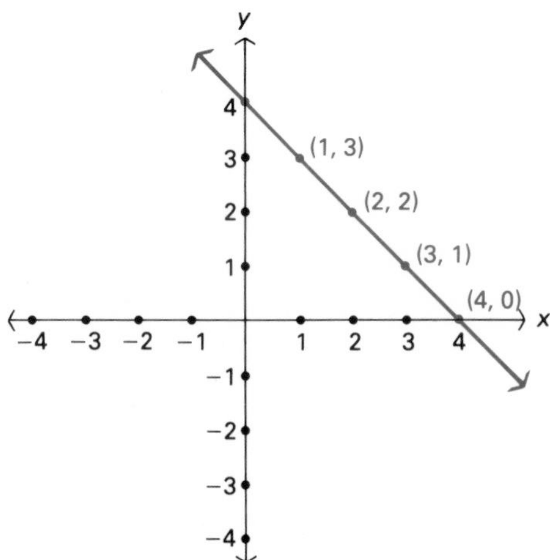

Linear inequalities are graphed in much the same way as equations. The only difference exists in the representation of the inequality on the coordinate system. The following steps indicate the modifications that must be made in the technique.

1. Treat the inequality as an equation and plot points as if to graph the line.
2. If the relationship in the inequality is $<$ or $>$, draw the line with dashes since the points on the line are not solutions of the inequality but, rather, are on the

boundary. If the relationship is $\leq$ or $\geq$, draw a solid line since the points on the line are solutions to the inequality.

3. By testing a point on each side of the line in the inequality, determine which side of the line represents the set of solution points.
4. Shade the side of the coordinate system that was found to contain the solution points.

Follow these four steps in the following example.

Example 13 • Graph $3x + 2y < 5$.

Solution 1. Four points on the line represented by $3x + 2y = 5$ are given in Table 3.

Table 3

x	y
0	$\frac{5}{2}$
1	1
−1	4
3	−2

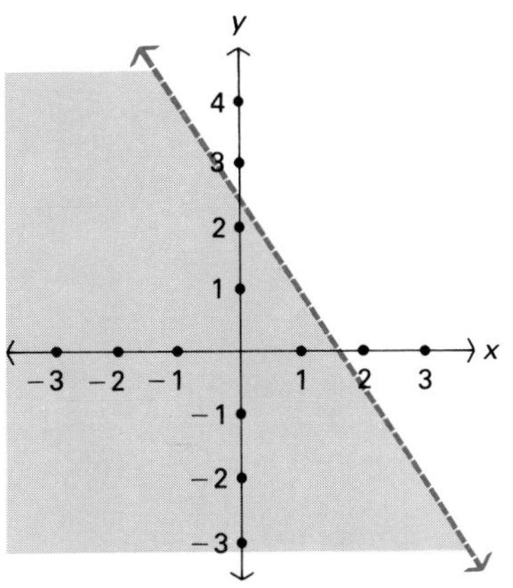

Figure 11

2. Sketch the line. Since no point on the line itself is a solution of $3x + 2y < 5$, draw the line with dashes.
3. Determine which side of the line consists of points (x, y) such that $3x + 2y < 5$. Select any point, say (3, 3), that is to the right of the line. Does the inequality become a true statement if this point is substituted for the x and y values? Test it. Is $3(3) + 2(3)$ less than 5?

$$3(3) + 2(3) = 9 + 6 = 15$$

$$15 \nless 5$$

Therefore, points to the right of the line are not solutions of the inequality. Now test any point to the left of the line. Substitute the point $(1, -1)$ to see if it gives a true statement. Is $3(1) + 2(-1)$ less than 5?

$$3(1) + 2(-1) = 3 + (-2) = 1$$

$$1 < 5$$

This point is one solution of the inequality.

4. Shade the area to the left side of the broken line, forming the graph in Figure 11. •

Just for fun

The four toothpicks in the illustration form a goalpost, with a football between the posts. By moving only two toothpicks, form another goalpost that does not contain the football.

Example 14 • Graph the solution of $y \geq x + 1$.

Solution **1.** The points in Table 4 help to determine the line representing $y = x + 1$.

Table 4

x	y
0	1
1	2
2	3
−1	0

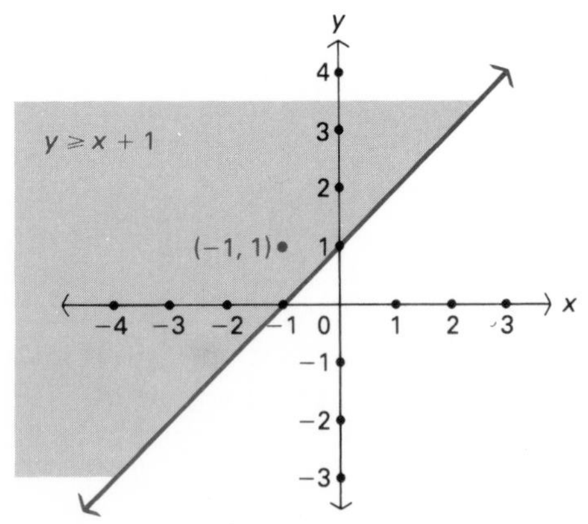

Figure 12

2. Since the relationship is $\leq$, the line $y = x + 1$ is sketched as a solid line in Figure 12.

3. Testing the points (0, 4) and (0, 0), we see that $4 \geq 0 + 1$ and $0 \ngeq 0 + 1$.

4. Shade the solution in Figure 12. •

Example 15 • Determine which of the graphs in Figure 13 is the graph of a function $y = f(x)$.

Solution In the graph of Figure 13(a), y is a function of x. However, in Figure 13(b), the graph does not represent a function as most x values have been assigned to two different y values, as indicated by the colored dots at $x = x_1$.

Figure 13

(a)

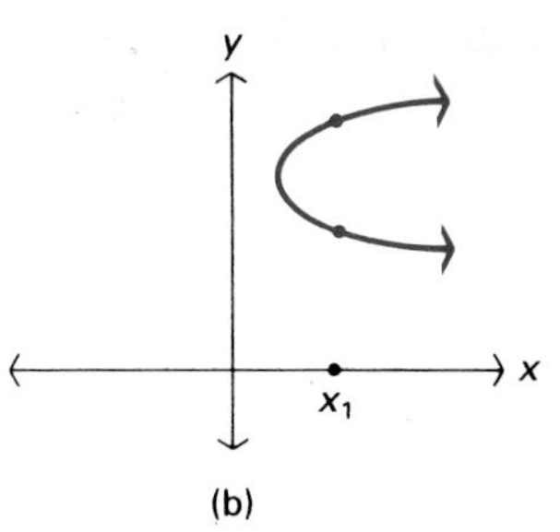

(b)

Exercise set 2

R 1. Given the function rules, find $f(1)$ and $f(5)$.

(a) $f(x) = x + 5$ (b) $f(x) = 2x + 3$
(c) $f(x) = 3x + 1$ (d) $f(x) = x^2$
(e) $f(x) = x^2 + 2x + 5$

2. Consider the following functions given by the rule $y = f(x)$. Compute $f(2)$, $f(3)$, and $f(4)$.

(a) $f(x) = x - 1$ (b) $f(x) = 2x - 1$
(c) $f(x) = x^2$ (d) $f(x) = x^2 + 2$

3. Find y for each equation when $x = -1, 0, 1$.

(a) $2x + 5y = 3$ (b) $\frac{x}{3} - 5y = -4$

4. Find five solutions for each of the following equations and inequalities.

(a) $2x + 3y = 6$ (b) $y < x + 2$
(c) $15 - 3y = 5x$ (d) $5x + 3y = 7$
(e) $\frac{3}{4}x + y = 2y - 1$ (f) $x + y + 3 < 0$

5. Graph the following equations on a coordinate system.

(a) $4y - x = 2$ (b) $y + 1 = 2x$
(c) $\frac{x}{3} + \frac{y}{3} = 3$ (d) $3x - 4y = 8$
(e) $x + y = 1$ (f) $x = 5 + y$

6. Which of the following relations are functions?

(a) $R = \{(3, 4), (5, 10), (6, 4), (7, 1)\}$
(b) $R = \{(1, 5), (1, 6), (2, 5), (3, 10)\}$
(c) $R = \{(3, 7), (7, 3), (8, 3)\}$
(d) $R = \{(4, 6), (5, 6)\}$
(e) $R = \{(5, 3), (5, 4)\}$

7. Which of the following do not define functions?

(a) 1  0
2
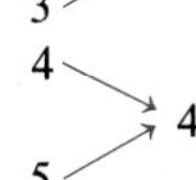 2
3
4
4
5

(b) 1 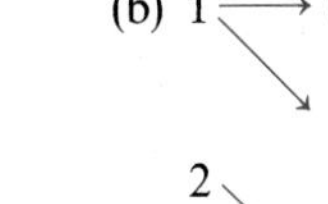 0
2
2
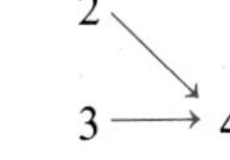
3 → 4

(c)

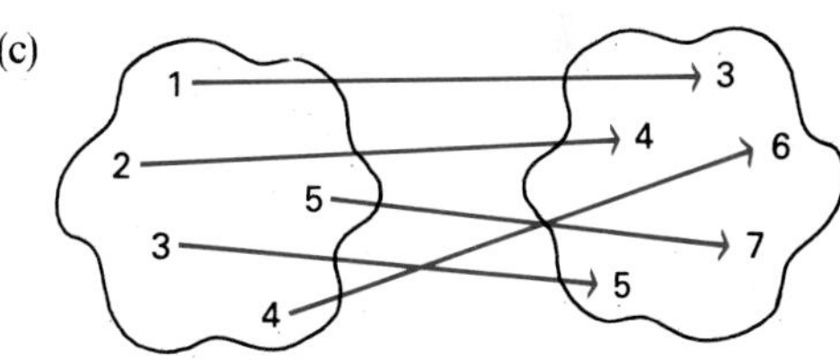

8. Graph each linear inequality on a coordinate system.

(a) $4x - 12 \geq 0$ (b) $4 - x < 0$
(c) $2x - 6 > 0$ (d) $2 + 3x \leq 0$
(e) $3x - 4 < 0$ (f) $4 + 3x \leq 0$

9. Construct a graph for each equation or inequality.

(a) $\{(x, y)|y = 2x - 3\}$ (b) $\{(x\ y)|y > x + 2\}$
(c) $\{(x, y)|x - y = 2\}$ (d) $\{(x, y)|y \leq x\}$
(e) $\{(x, y)|2x + 3y < 6\}$ (f) $\{(x, y)|5x + 3y = 7\}$
(g) $\{(x, y)|2x + y = 1\}$ (h) $\{(x, y)|x + 1 = 0$
(i) $\{(x, y)|y < -2x + 2\}$ (j) $\{(x, y)|y > 0\}$

T 10. If $A = \{1, 2, 3, 4\}$ and $B = \{2, 5\}$, tabulate tne elements (x, y) of $A \times B$ so that (a) x is less then y; (b) x is unequal to y; (c) x is equal to y; (d) x is greater than or equal to y. Which of these are functions?

11. Which of the following are graphs of functions?

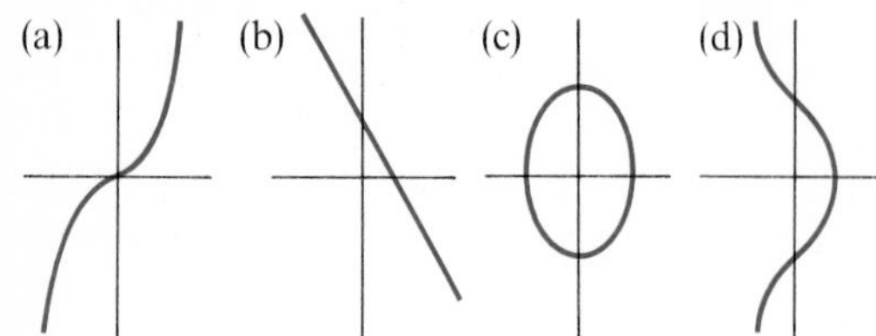

12. The charges of a taxicab company are computed by the following rule, $y = 30x + 50$, where y is given in cents and x in miles. Draw a graph of this relationship.
13. (a) Sketch the line $x = 3$.
    (b) What do lines of the form $x = b$ look like?
    (c) Sketch the line $y = 7$.
    (d) What do lines of the form $y = d$ look like?
14. Give the equation of each of the axes in the coordinate plane.

C 15. Consider the following functions given as subsets of $A \times B$, where $A = \{1, 2, 3, 4, 5\}$ and $B = \{1, 2, 3, 4, \ldots, 30\}$. In each case, use your skill with patterns to find a rule or equation that describes this function
    (a) $\{(1, 3), (2, 4), (3, 5), (4, 6), (5, 7)\}$
    (b) $\{(1, 1), (2, 4), (3, 9), (4, 16), (5, 25)\}$
    (c) $\{(1, 2), (2, 5), (3, 10), (4, 17), (5, 26)\}$
    (d) $\{(1, 3), (2, 6), (3, 9), (4, 12), (5, 15)\}$

16. Given $f(x) = (3x^2 - 6)/(x - 1)$, find the following
    (a) $f(0.015)$ (b) $f(-7.612)$
    (c) $f(\frac{4}{3})$ (d) $f(2t)$
    (e) $f(2t + 1)$ (f) $f(0.1t + 0.005)$
17. In Exercise 1(c), find
    (a) $f(t)$. (b) $f(t + 3)$.
    (c) $f(2t)$. (d) $f(2t + 1)$.
18. Can you determine a formula for the function denoted by the following table?

x	1	2	3	4	0	-1	-2
y	8	11	16	23	7	8	11

3 Slopes and linear equations

■ Suppose that the enrollment at a college with \$3000 tuition per student was 3400, enrollment at the same college with \$4000 tuition was 3000, and enrollment at the college with \$5000 tuition was 2600. Given this trend, what will be the enrollment at the college if the tuition cost is \$6000? \$7000? ■

It was possible to find the answers in this example without expressing the relationship between x and y in the form of an equation, but what if it is necessary to find a projected enrollment for a tuition of \$2670 or of \$5230? In this case, an equation should be found. In this section we develop equations of lines from data representing real-life situations. We define the slope of a line and use this concept to find the graph of the line and also to find the equation of the line. In addition, we learn to find the equation of a line through two points.

From Chapter 8, we know that two points determine a straight line (Figure 14). Let us investigate the relationship between points on a line and the equation of a line. Any two points in a plane can be considered the endpoints of some line segment. Let $P_1(x_1, y_1)$ and $P_2(x_2, y_2)$ be endpoints of a line segment, as in Figure 14. If we now construct through P_2 a line parallel to the y axis and through P_1 a line parallel to the x axis, the lines will meet at P_3. Note that the x coordinate of P_3 is the same as the x coordinate of P_2 and that the y coordinate of P_3 is the

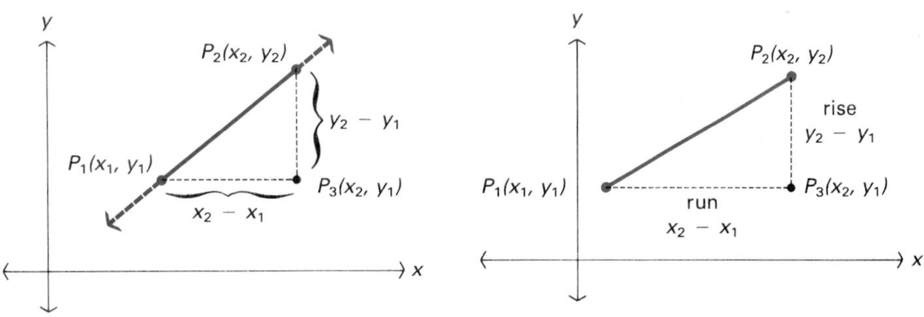

Figure 14

Figure 15

same as the y coordinate of P_1. The distance from P_3 to P_2 is $y_2 - y_1$, and the distance from P_1 to P_3 is $x_2 - x_1$.

These concepts are important in discussing the *inclination* of a line, which is measured by comparing the rise $(y_2 - y_1)$ to the run $(x_2 - x_1)$, as shown in Figure 15.

Slope of a line segment

The ratio of the rise to the run of a line segment is called the **slope** and is designated by the letter m. Thus, the slope of the line segment from $P_1(x_1, y_1)$ to $P_2(x_2, y_2)$ is

$$m = \frac{y_2 - y_1}{x_2 - x_1}$$

If P_2 is to the right of P_1, $x_2 - x_1$ will necessarily be positive, and the slope will be positive or negative as $y_2 - y_1$ is positive or negative. Thus, positive slope indicates that a line rises to the right; negative slope indicates that it falls to the right. Since

$$\frac{y_2 - y_1}{x_2 - x_1} = \frac{-(y_1 - y_2)}{-(x_1 - x_2)} = \frac{y_1 - y_2}{x_1 - x_2}$$

the restriction that P_2 be to the right of P_1 is not necessary, and the order in which the points are considered is immaterial in determining the slope.

If a line segment is parallel to the x axis, then $y_2 - y_1 = 0$, and the line has slope 0; but if it is parallel to the y axis, then $x_2 - x_1 = 0$, and its slope is not defined. These two special cases are shown in Figure 16.

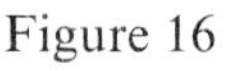
Figure 16

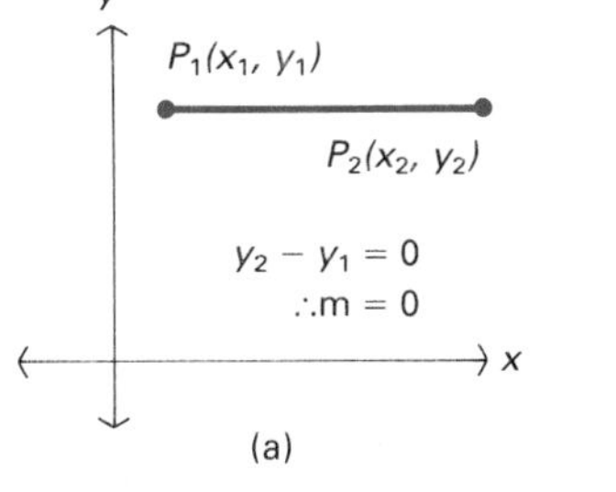

(a)

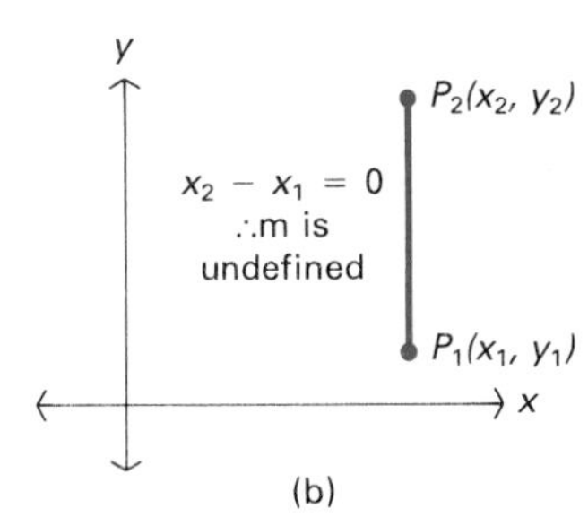

(b)

Example 16 • The slope of the line in Figure 17 is given by

$$m = \frac{y_2 - y_1}{x_2 - x_1} = \frac{5 - 2}{2 - 1} = \frac{3}{1} = 3 \quad \bullet$$

Example 17 • The slope of the previous tuition-enrollment example may be found by using any two of the known points.

$$m = \frac{3400 - 3000}{3000 - 4000} = -0.4 \quad \bullet$$

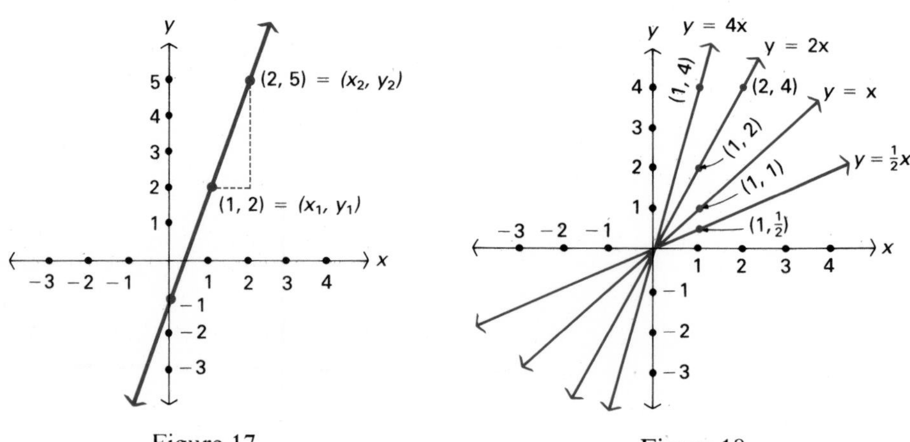

Figure 17

Figure 18

Let's now look at slopes in relation to linear equations of the form $ax + dy + c = 0$, where a and d are not both zero.

In the preceding section, we noted that graphs of $y = 2x - 3$, and $x + y = 4$ were straight lines. In fact, the graph of any linear equation with no more than two unknowns is a straight line. In fact, any line may be described by an equation of the form $ax + dy + c = 0$. So a given graph is a line if and only if it has an equation of the form $ax + dy + c = 0$, where a and d are not both zero.

A linear equation may be solved for y, if $d \neq 0$, to obtain

$$y = \frac{-a}{d}x + \frac{-c}{d}$$

By letting $m = -a/d$ and $b = -c/d$, the expression becomes

$$y = mx + b$$

Now consider the graph of $y = mx$ ($b = 0$ for this example) when m has values $\frac{1}{2}$, 1, 2, and 4 (see Figure 18). Since (0, 0) and $(1, \frac{1}{2})$ are two points on the line $y = \frac{1}{2}x$, we note that the y value increases by $\frac{1}{2}$ as x increases by 1. Thus, the slope is $\frac{1}{2}$.

Note also that $m = \frac{1}{2}$. Since (0, 0) and (1, 2) are points on $y = 2x$, the slope is $(2 - 0)/(1 - 0) = 2$, and the coefficient of x in $y = 2x$ is likewise 2. Intuitively, you may have already decided that in $y = mx + b$, m is the slope of the line.

Slope of a line

> For the line $ax + dy + c = 0$, when expressed as
>
> $$y = \frac{-a}{d}x + \frac{-c}{d} \qquad \text{where } d \neq 0$$
>
> the slope is
>
> $$m = \frac{-a}{d}$$
>
> The y intercept is given by $b = -c/d$.

If (x_1, y_1) is a fixed point on a given straight line, and (x, y) is any point on the line, then the slope from (x_1, y_1) to (x, y) is

$$m = \frac{y - y_1}{x - x_1} \qquad \text{or} \qquad y - y_1 = m(x - x_1)$$

Since the coordinates x and y are variables denoting any point on the line, the equation $y - y_1 = m(x - x_1)$ represents the relationship between x and y. Thus, the equation of the line with slope m passing through the fixed point (x_1, y_1) is $y - y_1 = m(x - x_1)$. A linear equation written in this form is said to be in **point-slope form.** If the slope and one point of the line are known, then the equation of the line can be easily obtained.

Point-slope form

> If a line has slope m and passes through the point (x_1, y_1), then the equation of the line is given by
>
> $$y - y_1 = m(x - x_1)$$

Example 18 • Find the equation of the line through (2, 1) with a slope of 3.

Solution

$$y - 1 = 3(x - 2)$$
$$y = 1 + 3x - 6$$
$$y = 3x - 5 \quad \bullet$$

Example 19 • Suppose that we wish to find the equation of the line in the tuition-enrollment example, given the following facts. When the tuition is \$3000, the enrollment is 3400 and the slope is -0.4. Thus we have a point, (3000, 3400), and the slope of the line.

Solution

$$y - 3400 = -0.4(x - 3000)$$
$$y - 3400 = -0.4x + 1200$$
$$0.4x + y - 4600 = 0 \quad \bullet$$

As a special case, the fixed point may be chosen to be the point where the line crosses the y axis. The coordinates of this point are usually written as $(0, b)$. Then the equation of the line becomes $y = mx + b$, the equation discussed earlier. The b in this equation is the value of y when $x = 0$, or the **y intercept.**

Slope-intercept form

If a line has a slope of m and a y intercept of b, then the equation of the line is given by

$$y = mx + b$$

Example 20 • If the slope of a line is 3 and the y intercept is 2, what is the equation of the line?

Solution Since $m = 3$ and $b = 2$, the equation is $y = mx + b$ or $y = 3x + 2$. •

Example 21 • Find the equation of the line that crosses the y axis at $(0, -5)$ and has a slope of 2.

Solution

$$y = 2x - 5 \quad \bullet$$

Sometimes, instead of being given a point and the slope, you are given only two points along the line. The *two-point method* of finding the equation of a line consists of using these two points to determine the slope of the line and then using the point-slope formula to establish the equation.

Example 22 • Find the equation of the line containing the points $(2, 3)$ and $(-1, 4)$.

Solution The two points can be used in either order to find the slope of the line; one way is

$$m = \frac{4 - 3}{-1 - 2} = -\frac{1}{3}$$

The slope along with either of the fixed points can now be used to determine the equation of the line; that is,

$$y - 3 = \left(-\frac{1}{3}\right)(x - 2) \qquad \text{or} \qquad y - 4 = \left(-\frac{1}{3}\right)(x + 1)$$

Both of these may be simplified to give $y = -\frac{1}{3}x + \frac{11}{3}$ as the equation of the line in slope-intercept form. •

As a special case, the graph of a linear equation may be a horizontal line or a vertical line. The equation of a **horizontal line** is of the form $y = b$ and has a slope of 0. The equation $y = 1$ is the equation of a horizontal line, as seen in Figure 19. By choosing two points, say, $(2, 1)$ and $(4, 1)$, the slope is seen to be 0. That is,

$$m = \frac{1 - 1}{4 - 2} = \frac{0}{2} = 0$$

Just for fun

Given three houses A, B, and C and the utility centers of electricity (E), gas (G), and water (W), connect each of the three houses to each of the three utilities without the lines crossing.

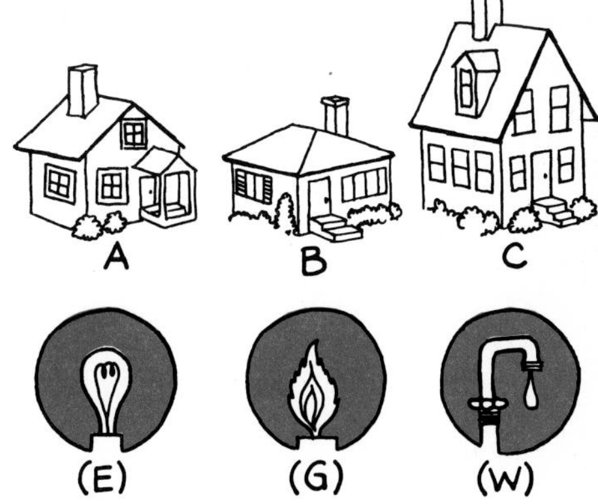

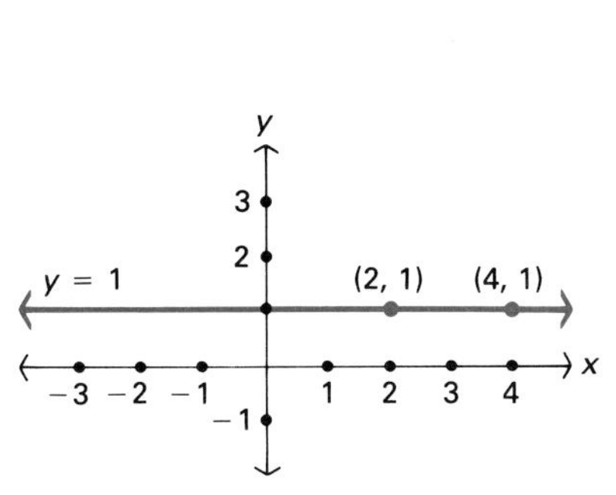

Figure 19

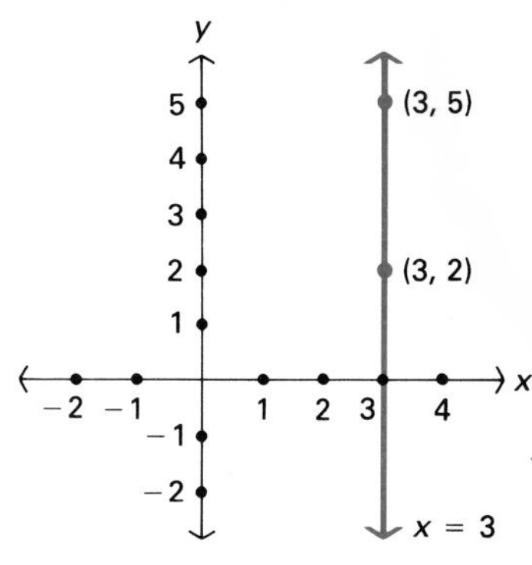

Figure 20

The equation of a **vertical line** is of the form $x = h$, and the slope is undefined. The equation of the line shown in Figure 20 is $x = 3$. By choosing two points on the line, say, (3, 5) and (3, 2), we can see that the slope is undefined. That is,

$$m = \frac{5-2}{3-3} = \frac{3}{0}$$

which does not exist.

The box below contains a summary of the types of equations introduced in this section.

$y - y_1 = m(x - x_1)$	**Point-slope form:** slope is m; line passes through (x_1, y_1)
$y = mx + b$	**Slope-intercept form:** slope is m; y intercept is b
$y = b$	**Horizontal line:** y intercept is b; line has slope of 0
$x = h$	**Vertical line:** x intercept is h; slope of the line is undefined

Exercise set 3

R 1. Compute the slope or indicate that the slope is not defined for the line through each pair of points.

(a) (3, 6), (4, 1) (b) (0, 1), (2, 3)
(c) $(-3, -5)$, (4, 2) (d) $(7, -1)$, $(-3, 1)$
(e) (0, 4), (4, 0) (f) $(-1, -7)$, $(-6, -5)$
(g) (4, 3), $(4, -1)$ (h) $(7, -1)$, (7, 4)
(i) (3, 1), (7, 1) (j) $(-1, 2)$, (7, 2)

2. Find an equation for and graph the line that has:

(a) a slope of 4 and goes through the point (2, 3)
(b) a slope of -2 and goes through the point $(4, -1)$
(c) a slope of $\frac{1}{2}$ and goes through the point $(-1, 1)$
(d) a slope of $7/-2$ and goes through the point (3, 4)

3. In each of the following linear equations, what is the slope? What is the y intercept of each? Graph each equation.

(a) $y = 3x + 2$ (b) $y + 2x - 1 = 0$
(c) $y = 3x - 1$ (d) $2y = 1 - x$
(e) $y = \dfrac{x-4}{2}$ (f) $x = \dfrac{y-1}{3}$
(g) $4x + 3y - 7 = 0$ (h) $3x - 2y = 5$

4. Classify the following statements as either true or false.

(a) The slope of the y axis is 0.
(b) The line segment joining (a, b) and (c, b) is horizontal.
(c) The line with a negative slope rises to the right.
(d) The line that is almost vertical has a slope close to 0.

5. Find an equation of the line ℓ with slope m containing point Z.

(a) $m = \frac{1}{2}$, $Z = (1, 3)$ (b) $m = 1$, $Z = (0, 2)$
(c) $m = \dfrac{-1}{3}$, $Z = (-1, -2)$
(d) $m = 0$, $Z = (-3, 1)$

T 6. Find an equation of the line through each of the following pairs of points.

(a) (1, 1), (2, 5) (b) $(-1, 1)$, (2, 5)
(c) (1, 3), $(1, -2)$ (d) (2, 4), (4, 2)

7. Find an equation of each line with the following characteristics:

(a) The line contains the two points $(1, -3)$ and (4, 5).
(b) The line has a slope of -3 and goes through the point (7, 1).
(c) The line has a slope of 1 and goes through the point $(-7, 1)$.
(d) The line contains the two points (0, 1) and (4, 3).
(e) The line has a y intercept of 4 and a slope of 5.
(f) The line has a y intercept of 6 and a slope of -3.

8. (a) Find an equation of the horizontal line through $(-4, -6)$.
(b) Find an equation of the vertical line through $(-5, 4)$.
(c) Write the equation of the x axis.
(d) Write the equation of the y axis.

9. Suppose the equation of a line is written in the form

$$\frac{x}{a} + \frac{y}{b} = 1$$

What is the x intercept? The y intercept?

10. Use the intercept form of the equation of a line (Exercise 9) to find equations for lines with the following intercepts:

(a) $x = 2$, and $y = -3$ (b) $x = 3$, and $y = 5$

C 11. Find the y intercept of the line that passes through the point $(3, -2)$ with a slope of 2.

12. What is the slope of a line with a y intercept of -3 that passes through the point $(-4, 1)$?

If two lines are parallel, they have the same slope. Lines that are perpendicular to a given line whose slope is m have slopes of $-1/m$, $m \neq 0$. Use this information in the next two problems.

13. For each part of Exercise 5, find the equation of the line parallel to ℓ through the point (2, 3).

14. For each part of Exercise 5, find the equation of the line perpendicular to ℓ through the point (2, 3).

15. The population of a small town seems to be increasing linearly. In 1970 the population was 15,000. In 1980 the population was 20,000. Find the linear equation that represents this change. Estimate the population in 1988 and in 1990.

16. For each part of Exercise 2, find the equation of the line perpendicular to the line and through the point (2, 3).

17. A store chain specializes in selling first-quality winter coats. Last year they sold 3400 coats for \$250 each, 3000 coats for \$300 each, and 2600 coats for \$350 each. If the selling trend continued in this manner, how many coats priced at \$375 should they have sold? How many coats should they have sold at \$530 each?

18. In a certain industrial city, it is believed that the pollution count increases linearly from 8:00 A.M. until 3:00 P.M. At 10:00 A.M. the count is 80; at 1:00 P.M. the count is 155. What is your prediction of the pollution count at each of the following times?

(a) 8:00 A.M. (b) 12:00 Noon
(c) 3:00 P.M. (d) 9:30 A.M.

4 Systems of linear equations

■ Suppose that a candy store has 40 pounds of candy selling for \$1.40 a pound. Since this candy has not been selling well lately, the owner believes that the price may be a little too high, but he does not want to lose any of his profit. How much candy that sells for \$1 a pound should he mix with the 40 pounds of \$1.40-a-pound candy in order to make a mixture that would sell at \$1.25 a pound? ■

Having learned to find the equations for lines, we are ready to find the solution set of two linear equations in two variables. The solution set of a pair of equations is the intersection of the solution sets, where the solution set of each equation is the set of ordered pairs satisfying the equation.

Many times, a system of linear equations represents an application problem better than one equation. Consequently, you will learn to apply systems of equations to a variety of application problems.

Representing a linear equation by means of a straight line permits us to solve the system of two linear equations graphically. The graph of the solution set of each equation is a straight line. Hence, the graph of the solution (x, y) of the system of two equations is the intersection of the two lines. From a geometric point of view, three possibilities may occur: (1) the two lines may intersect in exactly one point, (2) the two lines may coincide, or (3) the two lines may be parallel. Each of these possibilities will be examined in the examples of this section.

Example 23 • Solve the following system graphically.

$$3x + y = 3$$

$$x + 2y = -4$$

Solution The first equation is satisfied by infinitely many ordered pairs, three of which are $(0, 3)$, $(1, 0)$, $(3, -6)$. Similarly, some number pairs that satisfy $x + 2y = -4$ are $(-4, 0)$, $(0, -2)$, $(2, -3)$. The graphs of these two equations are given in Figure

21. The intersection of the two lines in the figure seems to be the point $(2, -3)$. We can prove that this number pair is a solution of the system of equations by substituting it into each equation.

Figure 21

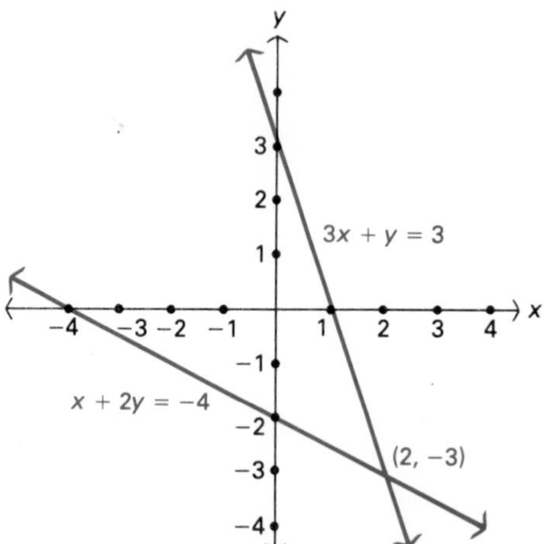

Although solving a system of two linear equations in two variables by graphs gives an excellent picture of the relationship between the two variables, the method is time-consuming and may not be accurate if the numbers that form the ordered pairs in the solution set are not integers. Consequently, an algebraic method of solution, usually called the *elimination method,* is more common. This method seeks to combine the equations in such a way that one of the unknowns appears with a zero coefficient. The derived equation in one unknown is then solved. To eliminate one unknown from a system of two equations in two unknowns, multiply one or both of the equations by numbers that will make the coefficients (numbers multiplied by the variables) of one of the unknowns the same. Then subtract like terms (terms with same variables) of the equations to obtain an equation with only one unknown. To illustrate the algebraic solution, the solution set will now be obtained for the same system that was solved graphically in the preceding example.

Example 24 • Find the solution set of

$$(1) \quad 3x + y = 3$$

$$(2) \quad x + 2y = -4$$

Solution To make the coefficients of y the same in both equations, multiply each term of the first equation by 2 to obtain

$$6x + 2y = 6$$

Subtract corresponding terms of the second equation from the first,

$$\begin{aligned} 6x + 2y &= 6 \\ x + 2y &= -4 \qquad 2(\text{equation } 1) - 1(\text{equation } 2) \\ \hline 5x &= 10 \end{aligned}$$

to get

Divide both sides by 5 to obtain

$$x = 2$$

Substituting $x = 2$ into the first equation (actually, in either equation) gives

$$3(2) + y = 3$$

Therefore

$$y = 3 - 6 = -3$$

The solution set is $\{(2, -3)\}$. We can prove that this point lies on both lines by substituting its coordinates into both equations and showing that they satisfy both equations.

Substituting $x = 2$ and $y = -3$ into the first equation gives

$$3(2) + (-3) = 6 - 3 = 3$$

Substituting $x = 2$ and $y = -3$ into the second equation gives

$$(2) + 2(-3) = 2 - 6 = -4 \quad \bullet$$

Example 25 • Solve the system of equations.

$$(1) \quad 3x + 2y = 7$$

$$(2) \quad 2x + 5y = 12$$

Solution To eliminate x in this pair of equations, it is convenient to multiply the first equation by 2 and the second equation by 3 to get

$$6x + 4y = 14 \qquad 2(\text{equation 1}) - 3(\text{equation 2})$$

$$6x + 15y = 36$$

Subtracting the second equation from the first equation gives $-11y = -22$, or $y = 2$. Substituting this value into the first equation gives $x = 1$. The solution is $(1, 2)$. •

Example 26 • Find the solution set of

$$(1) \quad 3x + y = 3$$

$$(2) \quad 6x + 2y = 6$$

Solution We shall first eliminate the x. Multiply both sides of the first equation by 2 to obtain

$$6x + 2y = 6$$

Subtract the second equation

$$6x + 2y = 6 \qquad 2(\text{equation 1}) - 1(\text{equation 2})$$

term by term to obtain

$$0 = 0$$

Not only did we eliminate x, we also eliminated y. The fact that 0 is always equal to 0 indicates that all number pairs such as (0, 3), (1, 0), (2, −3) that satisfy $3x + y = 3$ also satisfy $6x + 2y = 6$. Hence, these lines have every point in common; that is, the lines coincide. (See Figure 22.) Every ordered pair that is a solution of one equation is a solution of the other. •

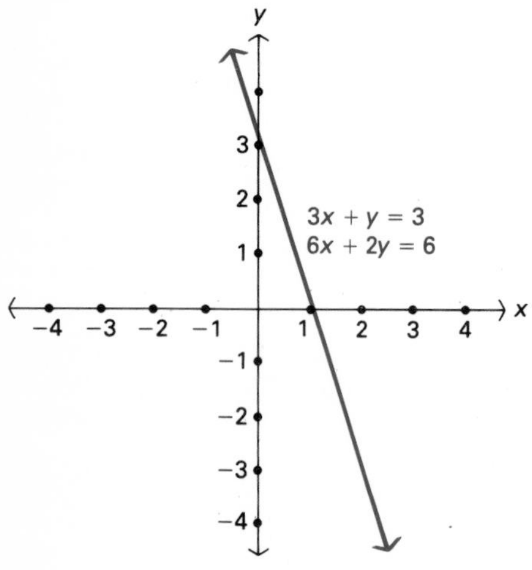

Figure 22

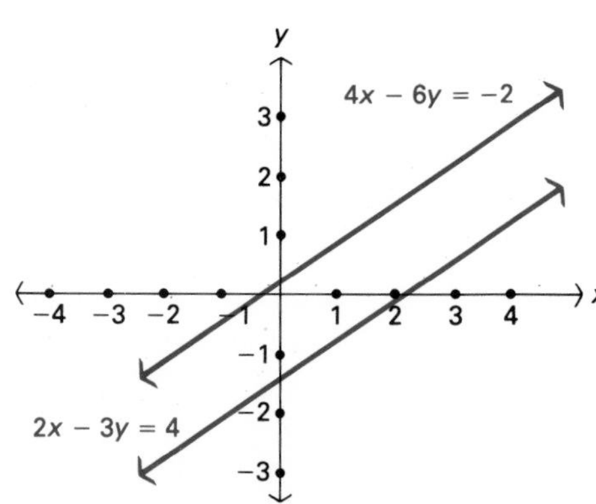

Figure 23

Example 27 • Solve (if possible) graphically.

$$4x - 6y = -2$$

$$2x - 3y = 4$$

Solution The graph for each of these equations is shown in Figure 23. This figure indicates that the two lines are parallel and will never intersect. •

Example 28 • Algebraically, find the solution set of

$$(1)\quad 4x - 6y = -2$$

$$(2)\quad 2x - 3y = 4$$

Solution Our strategy is to eliminate y. Multiplying the second equation by 2 gives

$$4x - 6y = 8$$

Subtracting this equation termwise from the first equation yields zero on the left side of the equation and −10 on the right side of the equation. Since $0 \neq -10$, no numbers x and y can satisfy both equations. Consequently, the graphs of these equations must be parallel lines. (See Figure 23.) •

The relationship between two linear equations can be summarized by the following discussion. If the two lines intersect, the two equations have a unique simultaneous solution; if the lines are parallel, the system has no solution. If two lines, such as $2x - y = 4$ and $4x - 2y = 8$, coincide, there is an infinite number of solutions, since every ordered pair that satisfies one equation also satisfies the other equation.

Example 29 • (Introductory problem) Suppose that a candy store has 40 pounds of candy selling for \$1.40 a pound. How much candy that sells for \$1 a pound should be mixed with the 40 pounds of \$1.40-a-pound candy in order to make a mixture that would sell at \$1.25 a pound?

Solution This problem can be solved as follows. As experienced problem solvers, we first identify the unknowns.

x = number of pounds of \$1-a-pound candy to be mixed with the \$1.40-a-pound candy

y = number of pounds of \$1.25-a-pound candy

When we decompose the problem into its essential facts, we learn

$$\begin{pmatrix}\text{amount of money}\\ \text{received from}\\ \text{\$1.40-a-pound candy}\\ \text{before mixing}\end{pmatrix} + \begin{pmatrix}\text{amount of money}\\ \text{received from}\\ \text{\$1-a-pound candy}\\ \text{before mixing}\end{pmatrix} = \begin{pmatrix}\text{amount of money}\\ \text{received from mixture}\\ \text{at \$1.25 a pound}\end{pmatrix}$$

Translating this information into algebra gives

$$(\$1.40)(40) + \$1x = \$1.25y \quad \text{or} \quad 56 + x = 1.25y$$

Similarly,

$$\begin{pmatrix}\text{number of pounds}\\ \text{of \$1.40 candy}\end{pmatrix} + \begin{pmatrix}\text{number of pounds}\\ \text{of \$1 candy}\end{pmatrix} = \begin{pmatrix}\text{number of pounds}\\ \text{of mixture}\end{pmatrix}$$

In algebraic terms, we write

$$40 + x = y$$

Now we need the solution set of the following system of two equations in two unknowns:

$$56 + x = 1.25y$$
$$40 + x = y$$

Solving these equations by the methods of this section, we find

$$x = 24 \qquad y = 64$$

Therefore, 24 pounds of \$1.00-a-pound candy should be mixed with 40 pounds of the \$1.40-a-pound candy to make a 64-pound mixture of \$1.25-a-pound candy. •

Example 30 • Find two numbers whose sum is 21 and whose difference is 3.

Solution Denoting the larger of the two numbers by x and the smaller by y, the two numbers must satisfy conditions

$$x + y = 21$$
$$x - y = 3$$

Solving by the procedures of this section yields

$$x = 12 \qquad \text{and} \qquad y = 9$$ •

Just for fun

A wife said to her husband, "When you are 15 times the age I was when you were half as old as I am, you will be one-half again as old as I would be were I as much older than you are as you are older than I am." Said the husband to the wife, "When you will be as much older than you are as you are younger than I was year before last, our combined ages will be 50." What are their ages? [Reference 8]

Exercise set 4

R 1. Solve the following systems of equations graphically.

(a) $2x + y = 3$; $x - y = 3$ (b) $3x - 4y = -2$; $-6x + 8y = 4$

(c) $x - 2y = 0$; $-2x + 4y = 6$ (d) $2x + 2y = 2$; $3x + 5y = 7$

(e) $3x + y = 6$; $3x - 2y = 3$ (f) $5x - 2y = 1$; $3x + y = 5$

2. Solve the systems in Exercise 1 algebraically.

3. Use the technique of eliminating a variable to find solutions for the following systems of equations.

(a) $x + y = 7$; $x - y = 4$ (b) $2x + y = 3$; $2x + 2y = 6$

(c) $2x + y = 4$; $6x + 3y = 12$ (d) $2x + 3y = 5$; $4x - 2y = 6$

(e) $x - y = 4$; $y = 3$ (f) $2x + y = 6$; $12x + 6y = 2$

(g) $3x + y = 4$; $x + 1 = 2y$ (h) $y - 4 = 0$; $x - 3y = -2$

(i) $x - 6 = 3y$; $2x - 6 = y$ (j) $3y = 12 + 6x$; $2x + 4 = y$

(k) $6.2x - 0.3y = 1.7$; $3.1x + 2.1y = 7.1$ (l) $\frac{1}{2}x - \frac{1}{4}y = 1$; $\frac{1}{3}x + \frac{1}{2}y = 6$

T 4. Mr. Black invests $4000, part at 10% annual interest and part at 8% annual interest. How much does he invest at each of these rates if he earns $344 in interest in 1 year?

5. A coin collector has $45 in quarters and dimes. How many coins of each kind does he have if the total number of his coins is 240?

6. A chemist has a 90% solution of sulfuric acid and a 70% solution of the same acid. How many gallons of each must be mixed in order to make 200 gallons of 75% solution?

7. A candy store proprietor wishes to mix caramels that sell for $2.40 a kilogram with chocolates that sell for $3.00 a kilogram to make a mixture that would sell for $2.80 a kilogram. How many kilograms of each kind of candy should be used if he wishes to make 160 kilograms of the $2.80 candy?

8. An investor with $10,000 wishes to invest part of it at 11% simple interest and the rest at 12% simple interest. How much should he invest at 11% if he is to receive a total of $1115 in interest from both investments that year?

C 9. A 6-minute phone call between two cities costs $4.40, whereas a 10-minute call costs $7.60. Find the basic charge for the first 3-minute period and the rate for each additional minute.

*5 Quadratic functions and the parabola

The paths of balls, bullets, or other objects that are thrown or shot in the air are parabolas. In fact, the parabola can be found in many places. Radar dishes, reflectors on spotlights, components of microphones, cables of suspension in bridges are all in the shape of parabolas.

The function whose graph is a parabola is called a *quadratic function.* As you can imagine, such functions have many important applications. In this section, we shall examine the quadratic function, its graph, and a very important equation related to the quadratic function: the quadratic equation.

A function that can be written in the form

$$f(x) = ax^2 + bx + c$$

where $a \neq 0$ is called a **quadratic function.**

Example 31 • Each of the following is a quadratic function. Find a, b, and c in each case.

(a) $f(x) = x^2 + 1$
(b) $f(x) = 2x^2 + x + 2$
(c) $f(x) = x^2 - 3x - 2$

Solution (a) $f(x) = x^2 + 1 = (1)x^2 + 0(x) + 1$; so $a = 1$, $b = 0$, $c = 1$.
(b) $f(x) = 2x^2 + (1)x + 2$; so $a = 2$, $b = 1$, $c = 2$.
(c) $f(x) = (1)x^2 + (-3)x + (-2)$; so $a = 1$, $b = -3$, $c = -2$. •

The graph of a quadratic function always has a distinct shape; the graph is called a **parabola.**

Example 32 • Find several solutions of the equation $y = f(x) = x^2 + 2x - 3$, plot these points, and draw the graph of the function.

Solution Assign to x the values $\{-5, -4, -3, -2, -1, 0, 1, 2, 3\}$ to obtain the following set of ordered pairs that satisfy $y = x^2 + 2x - 3$.

$$\{(-5, 12), (-4, 5), (-3, 0), (-2, -3), (-1, -4), (0, -3), (1, 0), (2, 5), (3, 12)\}$$

These points are then plotted on the coordinate system in Figure 24, and a smooth curve is drawn through the points, illustrating the shape of a parabola. Notice that the curve seems to be symmetric about the dotted line (called the **line of symmetry** or **axis of symmetry**) shown in the figure, and the lowest point on the graph (called the **minimum** or **vertex**) seems to be attained on this line. •

The graph of the function in this example opens upward. After you draw a number of graphs of quadratic functions, you will notice that **a parabola opens upward when the coefficient of x^2 is positive and downward when the coefficient of**

Figure 24

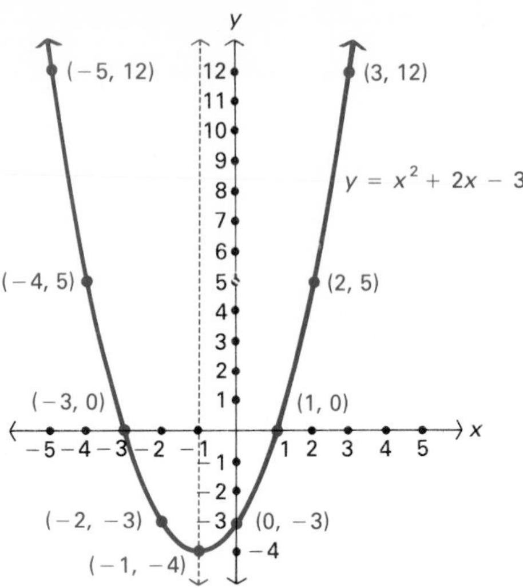

x^2 is negative. When the graph of the quadratic function turns downward, the function has a maximum value at the vertex.

Graph of a quadratic function

> The graph of a **quadratic function** turns **upward** and has a **minimum value** when a in
>
> $$f(x) = ax^2 + bx + c$$
>
> is a positive number. When a is negative, the graph turns **downward** and has a **maximum value.**

It is quite easy to ascertain where the minimum or maximum of the graph of a quadratic equation occurs.

Line of symmetry

> The maximum (or minimum) of the quadratic function
>
> $$f(x) = ax^2 + bx + c$$
>
> occurs at $x = -b/2a$. The line $x = -b/2a$ is called the *line of symmetry* of the function.

Example 33 • Find the line of symmetry, the coordinates of the maximum or minimum, and several ordered pairs that satisfy the equation, and sketch the graph of

$$y = f(x) = x^2 - 4x + 3$$

Solution For this quadratic function, $a = 1$, $b = -4$, and $c = 3$. Thus, the axis of symmetry is $x = -(-4)/2(1)$, or $x = 2$. Since $a > 0$, this function has a minimum when $x = 2$. Hence, to find that minimum, we substitute $x = 2$ to get $y = 2^2 - 4(2) + 3 = -1$. The coordinates of the minimum are $(2, -1)$. •

Assign to x the values $\{-1, 0, 1, 2, 3, 4, 5\}$ (note that some are larger than 2 and some are smaller than 2) to obtain the following set of ordered pairs.

$$\{(-1, 8), (0, 3), (1, 0), (2, -1), (3, 0), (4, 3), (5, 8)\}$$

Then sketch the curve. (See Figure 25.)

Figure 25

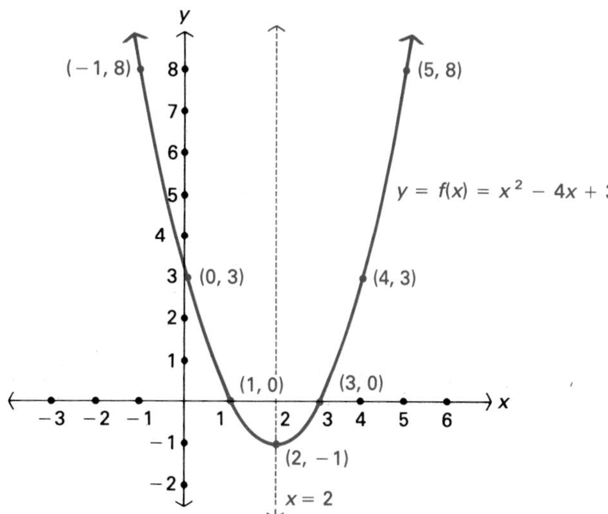

A common problem in applications of the quadratic function is, "Find the value of x for which $f(x) = 0$," which is equivalent to the problem, "Solve $ax^2 + bx + c = 0$."

Quadratic equations

> Equations of the form
>
> $$ax^2 + bx + c = 0$$
>
> are called *quadratic equations.*

One can solve quadratic equations by looking at the graph of the function $y = ax^2 + bx + c$ and determining the values of x where y equals 0.

Example 34 • Solve the quadratic equation $x^2 + 2x - 3 = 0$ graphically.

Solution In Figure 24, we have a sketch of $y = x^2 + 2x - 3$. To find the solution of $x^2 + 2x - 3 = 0$, we look at the points where the parabola crosses the x axis (where $y = 0$). One solution is at $x = 1$ and the other at $x = -3$. •

The difficulty with a graphical solution is that it is fairly long, and often we cannot get precise answers. It can be shown using algebra that a formula can be found that always gives the solutions to a quadratic equation if those solutions exist. We shall obtain this formula by a procedure called **completing the square.** The result of this procedure will be used as the quadratic formula.

$$ax^2 + bx + c = 0 \qquad a \neq 0$$

$$x^2 + \frac{b}{a}x + \frac{c}{a} = 0 \qquad \text{Divide by } a.$$

$$x^2 + \frac{b}{a}x = -\frac{c}{a} \qquad \text{Subtract } \frac{c}{a}.$$

To get a perfect square on the left side, add the square of one-half of the coefficient of x to both sides of the equation.

$$x^2 + \frac{b}{a}x + \left(\frac{b}{2a}\right)^2 = -\frac{c}{a} + \left(\frac{b}{2a}\right)^2 = \frac{b^2 - 4ac}{4a^2}$$

Multiply

$$\left(x + \frac{b}{2a}\right) \cdot \left(x + \frac{b}{2a}\right) = x^2 + \frac{b}{a}x + \left(\frac{b}{2a}\right)^2$$

$$\left(x + \frac{b}{2a}\right)^2 = \frac{b^2 - 4ac}{4a^2}$$

$$x + \frac{b}{2a} = \pm\sqrt{\frac{b^2 - 4ac}{4a^2}}$$

$$x = \frac{-b \pm \sqrt{b^2 - 4ac}}{2a}$$

This result is called the **quadratic formula.**

Quadratic formula

If a, b, and c are real numbers, then the solutions of $ax^2 + bx + c = 0$ (if they exist) are

$$x = \frac{-b + \sqrt{b^2 - 4ac}}{2a} \quad \text{and} \quad x = \frac{-b - \sqrt{b^2 - 4ac}}{2a}$$

Example 35 • Solve $x^2 + 2x - 3 = 0$, using the quadratic formula.

Solution

$$x = \frac{-b \pm \sqrt{b^2 - 4ac}}{2a} = \frac{-2 \pm \sqrt{(2)^2 - 4(1)(-3)}}{2(1)} = \frac{-2 \pm \sqrt{4 + 12}}{2}$$

$$= \frac{-2 \pm \sqrt{16}}{2} = \frac{-2 \pm 4}{2}$$

$$x = \frac{-2 + 4}{2} = \frac{2}{2} = 1 \qquad x = \frac{-2 - 4}{2} = \frac{-6}{2} = -3$$

The solutions for $x^2 + 2x - 3 = 0$ are 1 and -3. •

Just for fun *Jodi is three times as old as Brooke was when Jodi was as old as Brooke is now. In 24 years, Brooke will be twice as old as Jodi was when Brooke was half as old as Jodi is now. Find their present ages.*

Exercise set 5

R 1. For the following functions, find the equations of the lines of symmetry of their graphs.

(a) $f(x) = x^2 + x - 6$
(b) $f(x) = 2x^2 - 9x - 5$
(c) $f(x) = -x^2 - 2x + 15$
(d) $f(x) = -2x^2 + 7x - 3$
(e) $f(x) = 3x^2 + 18x + 24$
(f) $f(x) = 2x^2 + 6x - 30$
(g) $f(x) = -6x^2 + 12x + 18$

2. Find the point at which a maximum or minimum value occurs for each function in Exercise 1.

3. Without drawing the graphs, state whether the graphs of the functions in Exercise 1 turn upward or downward.

4. Make a table of ordered pairs for each part of Exercise 1.

5. Draw the graphs of each part of Exercise 1, indicating with a dotted line the line of symmetry and plotting the maximum or minimum.

6. (a) Solve the quadratic equation $x^2 - x - 6 = 0$ graphically by sketching $y = x^2 - x - 6$.
(b) Solve the quadratic equation $x^2 - x - 6 = 0$ using the quadratic formula.

7. (a) Solve the equation $x^2 + 4x + 4 = 0$ using the quadratic formula. How many solutions did you get?
(b) Draw the graph of $y = x^2 + 4x + 4$. How does this graph relate to (a)?

8. Solve the following quadratic equations using the quadratic formula.

(a) $x^2 - 2x = 3$
(b) $3x^2 + 9x - 12 = 0$
(c) $3x^2 + 11x = 4$
(d) $6x^2 - 7x - 3 = 0$

T 9. (a) Solve $5x^2 - 12x + 4 = 0$ graphically.
(b) Solve $5x^2 - 12x + 4 = 0$ using the quadratic formula. Can you see why graphing yields only approximate solutions?

10. A sociologist studying in the town of Mud Flat determined that she can relate the number of responses received per month (y) to the number of invitations offered (x) to fund raising social occasions by the quadratic function $y = -x^2 + 30x + 4$.

(a) How many invitations per month should a resident of Mud Flat offer in order to receive the maximum number of responses?
(b) How many responses would be received?

11. Industry analysts have determined that if the Greedo Corporation produces x widgets each month, the profit will be

$$P(x) = 10 + 600x - 2x^2$$

(a) How many widgets should Greedo make to produce maximum profits?
(b) What would that maximum profit be?

C 12. Suppose $(x - 2)(x - 3) = 0$. Then either $(x - 2) = 0$ or $(x - 3) = 0$. Hence, $x = 2$ and $x = 3$ are solutions of the equation. If $x^2 + bx + c$ can be written as a product of two factors such as $(x - d)(x - e)$, the equation can be solved quite easily.

EXAMPLE. $x^2 - 2x = 0$ can be rewritten $x(x - 2) = 0$. Then either $x = 0$ or $x - 2 = 0$. Solutions are $x = 0$ or $x = 2$.

Solve:

(a) $x(x - 4) = 0$ (b) $(x - 1)(x + 1) = 0$
(c) $x^2 + 7x = 0$ (d) $x^2 + 5x + 6 = 0$

13. Certain kinds of quadratic equations can be solved very easily. If an equation can be written as $x^2 = d$ where $d \geq 0$, one can compute x by taking the square root of both sides of the equation. Observe that each positive number has two square roots $[(-2)^2 = 4$ and $2^2 = 4$, so $\sqrt{4} = 2$ and $-2]$.

EXAMPLE. $x^2 - 6 = 0$

$$x^2 = 6$$

$$x = \sqrt{6} \text{ and } -\sqrt{6}$$

Solve:

(a) $x^2 - 16 = 0$ (b) $x^2 - 22 = 0$
(c) $3x^2 - 147 = 0$ (d) $x^2 - 4 = 32$

14. Farmer Jones has 300 m of fence with which he wishes to enclose a rectangular field. One side of the field will be along a straight river bank (and hence will need no fence). What is the area of the largest field he can enclose?

15. If $a < 0$, then a has no real square root. For example, $\sqrt{-1}$ does not exist since there is no real number b such that $b \cdot b = -1$. Hence, when using the quadratic formula, if one encounters the square root of a negative number, the quadratic equation has no solution among the real numbers.

(a) Solve $x^2 - x + 3 = 0$ using the quadratic formula.

(b) Graph $y = x^2 - x + 3$. Use this graph to show why there is no solution to the equation $x^2 - x + 3 = 0$.

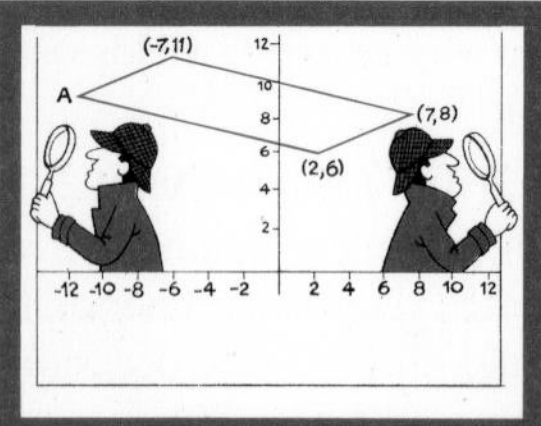

Solution to introductory problem

Understand the problem. Let's remember that for the figure to be a parallelogram, opposite sides must be parallel. If opposite sides are parallel, they have the same slopes.

Devise a plan. We let (x, y) be the coordinates of the unknown point. Then set equal the slopes of parallel lines. One of the slopes each time will be in terms of x and y. Solve the two linear equations in x and y, getting the coordinates (x, y).

Carry out the plan. $m_1 = \dfrac{8-6}{7-2} = \dfrac{2}{5}$ $m_1 = \dfrac{11-y}{-7-x}$

$$\frac{2}{5} = \frac{11-y}{-7-x} \quad \text{or} \quad 5y - 2x = 69$$

$$m_2 = \frac{8-11}{7-(-7)} = \frac{-3}{14} \qquad m_2 = \frac{6-y}{2-x}$$

$$\frac{-3}{14} = \frac{6-y}{2-x} \quad \text{or} \quad 14y + 3x = 90$$

Solving

$$5y - 2x = 69$$
$$14y + 3x = 90$$

yields

$$x = -12 \qquad y = 9$$

Looking back. Could this be done by a counting process? From (7, 8) to (−7, 11), you travel 14 left and 3 up. If you do the same from (2, 6) where do you stop? Did you get (−12, 9)?

Summary and review

1. *Coordinate system.* Make certain you can plot points, transform coordinates, and obtain a midpoint and the distance between two points.

2. *Functions*

   (a) Define a function using ordered pairs, relations, graphs, and equations.
   (b) Make sure you can graph both equations and inequalities.

3. *Slope*

$$m = \frac{y_2 - y_1}{x_2 - x_1}$$

4. Equation of a line

   (a) **Point-slope form:** $y - y_1 = m(x - x_1)$
   slope is m; line passes through (x_1, y_1)
   (b) **Slope-intercept form:** $y = mx + b$
   slope is m; y-intercept is b
   (c) **Horizontal line:** $y = b$
   y-intercept is b; line has slope of 0
   (d) **Vertical line:** $x = h$
   x-intercept is h; slope of the line is undefined

5. *Solve systems of linear equations*

   (a) By graphs
   (b) By addition or subtraction

6. *Quadratic functions*

   (a) In $f(x) = ax^2 + bx + c$, the graph turns downward and has a maximum value when a is negative; it turns upward and has minimum when a is positive.
   (b) Line of symmetry $x = -b/2a$

7. *Quadratic equations*

   $ax^2 + bx + c = 0$ has solutions (if they exist)

$$x = \frac{-b + \sqrt{b^2 - 4ac}}{2a} \qquad x = \frac{-b - \sqrt{b^2 - 4ac}}{2a}$$

Review exercise set 6

R 1. Solve the systems of equations and check by graphs.

   (a) $3x + 2y = 7$
   $x - 4y = -7$
   (b) $6x - 4y = -2$
   $4x + y = -5$

2. Draw a graph and determine the equations of the lines with properties given. Also find the points of intersection.

   (a)

x	0	-4
y	2	4

   and

x	1	-1
y	5	2

   (b)

x	-2	3
y	-3	1

   and

x	1	3
y	0	-2

   (c) Point $(-2, 4)$, $m = \frac{3}{2}$; and point $(4, -2)$, $m = \frac{2}{3}$
   (d) Point $(0, -5)$, $m = \frac{4}{5}$; and point $(-5, 0)$, $m = \frac{4}{5}$

3. Graph.

   (a) $x + 7y - 4 < 2$
   (b) $2x + y - 5 < 3$
   (c) $6x + 4y < 5$
   (d) $3x + y - 4 < 0$

4. If the domain for the function $f(x) = 2x^2 - x + 4$ is the set of reals, find the following.

   (a) $f(-1)$
   (b) $f(2)$
   (c) $f(0)$
   (d) $f(\frac{1}{2})$
   (e) $f(0.1)$
   (f) $f(100)$

5. Construct a graph for each of the following.

   (a) $\{(x, y) | y < 3x + 2\}$
   (b) $\{(x, y) | x = 2y - 7\}$
   (c) $\{(x, y) | 3x + 2y = 1\}$
   (d) $\{(x, y) | 3y + 2x > 1\}$
   (e) $\{(x, y) | x + y < 5\}$
   (f) $\{(x, y) | 2x + y < 6\}$

6. Find the equation of the line through $(-1, 4)$ and $(-1, 6)$.

7. Find the equation of the line through the point $(-1, 7)$ parallel to the line $3y + 2x = 7$.

8. Find the equation of the line through the points $(1, -2)$ and $(6, 4)$, and find the value of y when $x = 5$.

T 9. What is the image of $A(-2, 3)$, $B(1, 4)$, and $C(0, 6)$ under a reflection over the x axis?

10. What is the image of the triangle in Exercise 9 under a reflection over the y axis?
11. What is the image of the triangle in Exercise 9 under a reflection over the line $y = x$?
12. Find the image in Exercise 9 if the triangle is translated four units to the right and three units down.
13. Find the distance between and the midpoint of the segment determined by $(1, -2)$ and $(2, 3)$.
14. Are the following lines parallel, coincident, or neither?

    (a) $2y + 3x - 5 = 0$ and $4y + 6x - 7 = 0$
    (b) $4y + 6x - 9 = 0$ and $4y = 6x + 9$
    (c) $y = \frac{4x}{3} + 2$ and $3y - 4x - 6 = 0$
15. Find the equations of the lines of symmetry for the following quadratic functions.

    (a) $y = x^2 - x - 6$ (b) $y = x^2 + 5x + 6$
    (c) $y = x^2 - 5x - 6$ (d) $y = x^2 - 7x + 12$
16. Find the point at which a maximum or minimum value occurs for each function in Exercise 15.
17. Without drawing the graphs, state whether the graphs of the functions in Exercise 15 turn upward or downward.
18. Make a table of ordered pairs for each function given in Exercise 15.

C 19. Draw the graph of each function in Exercise 15.

20. Solve by using the quadratic formula.

    (a) $x^2 = 7x - 12$ (b) $x^2 - 8x + 2 = 0$
    (c) $3x^2 - 9x - 5 = 0$
21. Find the image in Exercise 9 if the triangle is rotated 90° clockwise.
22. Find the image of the triangle in Exercise 9 under a rotation of 90° clockwise about the origin, a reflection about $y = x$, a rotation of 270° counterclockwise about the origin, and finally a reflection about the y axis.

Bibliography

1. Bernstein, Barbara E. "'Equation' Means 'Equal'." *AT* (December 1974), 697–698.
2. Bestgen, Barbara J. "Making and Interpreting Graphs and Tables: Results and Implications from National Assessment." *AT* (December 1980), 26–29.
3. Brieske, T. "Functions, Mappings, and Mapping Diagrams." *MT* (May 1973), 463–468.
4. Bruni, James V., & Helene Silverman. "Graphing as a Communication Skill." *AT* (May 1975), 354–366.
5. Cetorelli, Nancy D. "Teaching Function Notation." *MT* (November 1979), 590–591.
6. Coxford, Arthur F., & Zalman P. Usiskin. *Geometry, a Transformation Approach.* River Forest, Ill.: Laidlaw Brothers, 1971.
7. Erckmann, Ruth. "What We Can Do with $a \times b = c$." *AT* (March 1976), 181–183.
8. Jones, S. I. *Mathematical Nuts.* Nashville, Tenn.: S. I. Jones, 1932.
9. Pereira-Mendoza, Lionel. "Graphing and Prediction in the Elementary School." *AT* (February 1977), 112–113.
10. Ranucci, Ernest R. "Space Filling in Two Dimensions." *MT* (November 1971), 587–593.
11. Reys, Robert E. "Functioning with a Sticky Model." *AT* (September 1981), 18–23.
12. Sullivan, Delia, & Mary Ann O'Neill. "This Is Us! Great Graphs for Kids." *AT* (September 1980), 14–18.
13. Terc, Michael. "Coordinate Geometry—Art and Mathematics." *AT* (October 1985), 22–24.
14. Tucker, Benny F., & Edna F. Bazik. "IDEAS." *AT* (February 1983), 31–42.
15. Van de Walle, John, & Charles S. Thompson. "A Poster-Board Balance Helps Write Equations." *AT* (May 1981), 4–8.

CHAPTER 13
An Introduction to Probability Theory

A couple wants to have 3 or 4 children, and they want exactly 2 girls. Is it more likely that they will get their wish with 3 children or with 4?

Archaeological artifacts indicate that many of the early peoples played some version of dice either for recreation or to determine the will of the tribal deity. As more elaborate games were developed, the players began to observe certain patterns in the results, but they did not have the language of probability with which to describe and to analyze them.

One of the first people who attempted to apply mathematical analysis to games of chance was a 16th-century Italian gambler, Girolamo Cardano. Probability was firmly established as a legitimate field of inquiry by the work of Blaise Pascal and Pierre Fermat, who attempted to answer questions concerning a dice game. Pascal corresponded with Fermat on this problem, and they each arrived at a solution by different methods. Their interest in these questions generated widespread interest in probability. As time passed, it became clear that probability was much more than just a plaything for gamblers. In the 19th and 20th centuries, probability has been used in many different applications. Physicists use probability in studying the various gas and heat laws as well as in the theory of atomic physics. Biologists apply the techniques of probability in genetics, the theory of natural selection, and learning theory. Managers in government and industry use probabilistic techniques in decision-making processes. Furthermore, probability is the theoretical basis of statistics, a discipline that permeates modern thinking.

1 The language of probability

One of the significant characteristics of our increasingly complex society is that we must deal with questions for which there is no known answer but instead one or more probable (or improbable) answers. Statements like, "I am 90% confident that the mean weight is between 76.2 grams and 78.6 grams"; "This cancer treatment has a 0.6 chance of leading to complete remission"; and "The candidate has a 20.4% chance of carrying the state of Ohio" are common in conversations of the day. A necessary skill for our times is the ability to measure the degree of uncertainty in an undetermined situation. In this section, we discuss probability—the language of uncertainty. As the language of the undetermined and the uncertain, probability is an important tool in many phases of our uncertain modern life.

In the study of probability, a number between zero and one is assigned to an uncertain outcome. The number indicates how likely or unlikely the outcome may be.

One or zero probability

1. If an outcome is impossible, it has a probability of 0.
2. If an outcome will surely happen, it has a probability of 1.

In this section, we discuss procedures for assigning probabilities, rules that govern these probabilities, and what can be deduced from a probability once it is assigned. Since probability is a language of uncertainty, any discussion of probability presupposes a process of observation or measurement in which the outcomes are not certain. Such a process is called an **experiment.** Any possible result of an experiment is called an **outcome.**

Example 1 • *Experiment:* A coin is tossed.
Possible outcomes: Head (H) or tail (T). •

Example 2 • *Experiment:* A die (one member of a pair of dice) is tossed.
Possible outcomes: The top side of the die shows 1, 2, 3, 4, 5, or 6. •

The outcomes in the two previous examples are **equally likely,** because any one outcome has the same chance of occurring as another.

Example 3 • Suppose we have a spinner that is as likely to stop at one place as another. First of all, consider a spinner divided into four equal sections colored red, black, white, and green. (See Figure 1.) Let an experiment consist of spinning and observing the color of the region where the needle stops. (If the needle stops on a line separating two regions, we agree to record the color of the region into which the needle would move if rotated clockwise.) The set of possible outcomes of this experiment consists not of numbers but colors:

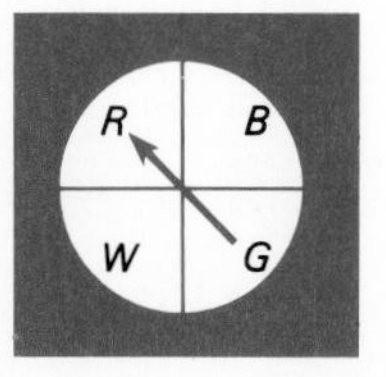

Figure 1

$$\{\text{red, black, white, green}\}$$ •

BIOGRAPHY

Girolamo Cardano

1501–1576

The first inquiry into the science of probability was made by Girolamo Cardano, one of the real rogues in the history of mathematics. One of the lesser of his vices was compulsive gambling. As a result, he investigated several interesting probability problems and wrote a handbook for gamblers (with information on how to cheat).

Cardano predicted the date of his own death, and since he was healthy at the end of the day, he poisoned himself to make the prediction come true.

Before we can analyze an experiment, we must decide upon a **sample space** (defined shortly), since different sample spaces can result from the same experiment, depending on how the observer chooses to record the outcomes.

Example 4 • Consider the experiment of drawing one card from six cards marked by numbers 1 through 6 and observing the number on the card. The set of possible outcomes is

$$\{1, 2, 3, 4, 5, 6\} \quad \bullet$$

Example 5 • In another experiment with the six cards, we might observe whether the number is even or odd. Thus, the possible outcomes of this experiment are

$$\{\text{even, odd}\} \quad \bullet$$

Consider again the two examples of drawing a card. Notice that associated with the experiment of drawing a card are several sets that classify the outcomes (Table 1).

Although the classifying sets are different, notice that they share certain properties. In each example, the set of outcomes classifies completely or exhausts the possibilities of what can happen if the experiment is performed. The number on the card is either even or odd; one of the numbers 1, 2, 3, 4, 5, 6 must appear on the card; and the number on the card either is less than or equal to 3 or is greater

Table 1

Question	*Set of possible classifications of outcomes*
Is the number on the card even or odd?	{even, odd}
What is the number on the card?	{1, 2, 3, 4, 5, 6}
Is the number on the card greater than 3?	{less than or equal to 3, greater than 3}

than 3. Moreover, the members of a set of outcomes for an experiment are distinct; that is, they do not overlap. In other words, each possible result of the experiment is classified by exactly one member of the set. This discussion suggests the following definition.

Sample space

A *sample space* (denoted by S) is a list of the outcomes of an experiment constructed in such a way that

1. the categories do not overlap.
2. no result is classified more than once.
3. the list is complete (exhausts all the possibilities).

Example 6 • A coin is flipped two times in succession.

Sample space A: One complete listing of the outcomes is $S = \{(H, H), (H, T), (T, H), (T, T)\}$. The letter listed first in each pair indicates the result of the first flip, and the letter listed second gives the result of the second flip.

Sample space B: An alternate way to list the outcomes is to ignore the order in which the heads and tails occur and to record only how many of each appear.

$$S = \{(2H), (1H \text{ and } 1T), (2T)\} \quad \bullet$$

Example 7 • A sack contains 5 chocolate candies, 3 butterscotch candies, and 1 peppermint candy. One candy is drawn from the sack and eaten. Then a second candy is drawn and eaten. What is a sample space for the experiment?

Solution Notice that the peppermint cannot be chosen on the second draw if the only peppermint was eaten on the first draw.

$$S = \{(C, C), (C, B), (C, P), (B, C), (B, B), (B, P), (P, C), (P, B)\} \quad \bullet$$

Once we have tabulated the sample space of an experiment, we must turn to the task of determining the probabilities of the various outcomes. The techniques of assigning probabilities to outcomes are heavily dependent on the following interpretation of probability.

Relative frequency interpretation of probability

Suppose an experiment is performed N times where N is a very large number. The probability of an outcome A should be approximately equal to the following ratio.

$$P(A) = \frac{\text{Number of times } A \text{ occurs}}{N}$$

In other words, the probability of an outcome should represent the *long-range relative frequency* of the outcome.

Example 8 • A fair die is rolled 10,000 times. Table 2 itemizes the number of times a 1 has occurred at various stages of the process. Notice that as N becomes larger, the relative frequency stabilizes in the neighborhood of $0.166 \approx \frac{1}{6}$. •

Table 2

Number of rolls (N)	*Number of 1's occurring (m)*	*Relative frequency (m/N)*
10	4	0.4
100	20	0.2
1000	175	0.175
3000	499	0.166333 . . .
5000	840	0.168
7000	1150	0.164285714 . . .
10,000	1657	0.1657

The probability describes the fraction of times one would expect the outcome E to occur if the experiment were performed a large number of times.

Suppose a thumbtack lands with "point up" 1000 times out of 10,000 trials. The relative frequency is $1000/10{,}000 = \frac{1}{10}$. If we continue the experiment 10,000 more times and find that the ratio is still approximately $\frac{1}{10}$, we are willing to assign this number as a measure of our degree of belief that it will land point up on the next toss. The ratio $\frac{1}{10}$ is assigned as the probability that a thumbtack will fall "point up" if dropped on the table.

Example 9 • If a student were selected at random from Lamor University (see Table 3), the probability that the student is enrolled in school A is $\frac{176}{600}$. (To say that a student is selected *at random* means that each student is equally likely to be selected.) This statement is true, since the relative frequency of those enrolled in school A is $\frac{176}{600}$, or $\frac{22}{75}$. The probability that a student is in school B is $\frac{244}{600}$ or $\frac{61}{150}$ and in school C is $\frac{180}{600}$, or $\frac{3}{10}$. •

Table 3

Number of students at Lamor University	
School A	176
School B	244
School C	180
Total	600

Properties of probability

A **probability rule on a sample space** must satisfy two properties.

1. If A is an outcome, then the probability of A, $P(A)$, is between 0 and 1 $[0 \le P(A) \le 1]$.
2. The sum of the probabilities of all outcomes in the experiment equals 1.

Example 10 • In the repeated tossing of a coin, it is noted that $P(T) = \frac{1}{2}$ and $P(H) = \frac{1}{2}$. Notice that these probabilities satisfy the properties.

Just for fun

Try to place the pennies in four equal stacks subject to the following rules:
1. Each penny can be moved only once.
2. A penny can be jumped over at most two other pennies or stacks.
3. If an empty space occurs, ignore it.

Good luck.

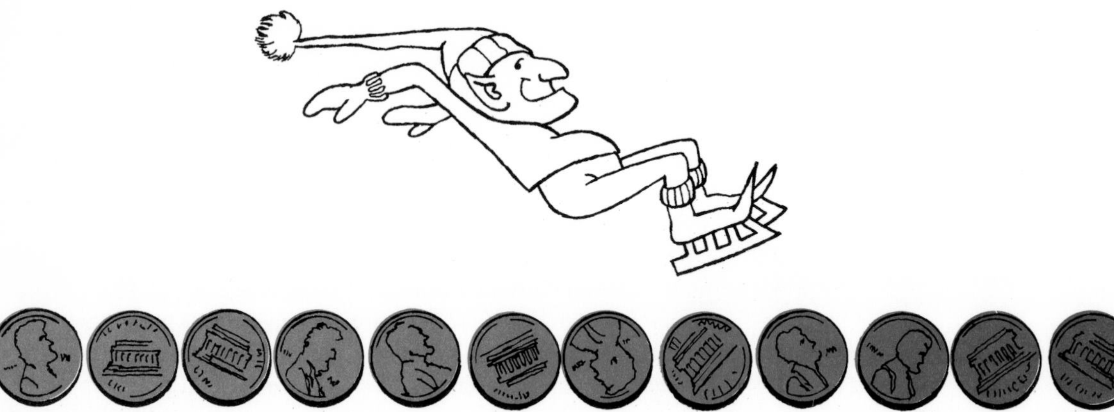

(a) $0 \leq P(T) \leq 1$ and $0 \leq P(H) \leq 1$

(b) $P(H) + P(T) = \dfrac{1}{2} + \dfrac{1}{2} = 1$ •

There is one whole class of sample spaces whose probability assignments are particularly easy to determine.

Uniform sample space

> If each outcome of the sample space is equally likely to occur, the sample space is called a *uniform sample space.*

Suppose a uniform sample space consists of n outcomes. Since each of the outcomes is equally likely, it seems reasonable to assign to each outcome the same probability, denoted by $P(A)$.

Since the sum of the probabilities of the n individual outcomes must be 1, we note that

$$n \cdot P(A) = 1 \qquad \text{or} \qquad P(A) = \frac{1}{n}$$

Thus, each of the n outcomes has probability $1/n$.

Equal probabilities

> In a **uniform sample space** with n outcomes, each outcome has probability $1/n$.

Example 11 • Eight identical balls numbered 1 to 8 are placed in an urn. Find a sample space and a probability rule describing the experiment of randomly drawing one of them from the urn.

Solution A suitable sample space is $\{1, 2, 3, 4, 5, 6, 7, 8\}$, each number representing one of the eight balls. Since each ball is equally likely to be drawn, we assign a probability of $\frac{1}{8}$ to each outcome.

$$P(1) = \frac{1}{8}, \quad P(2) = \frac{1}{8}, \quad \ldots, \quad P(8) = \frac{1}{8} \quad \bullet$$

Exercise set 1

R 1. Give the sample space for each of the four spinners.

(a) 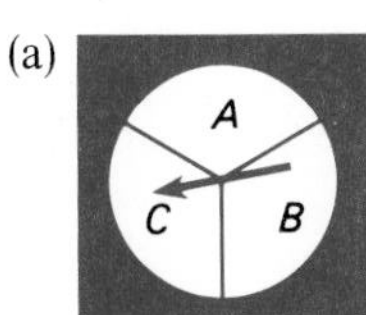

(b) 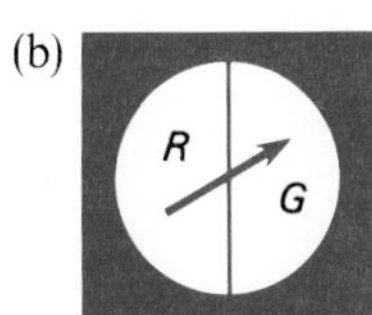

(c) 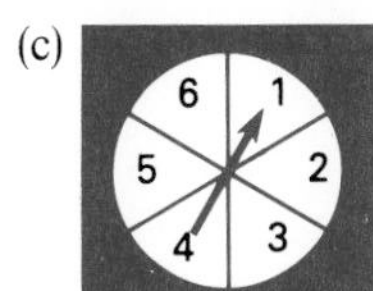

(d) 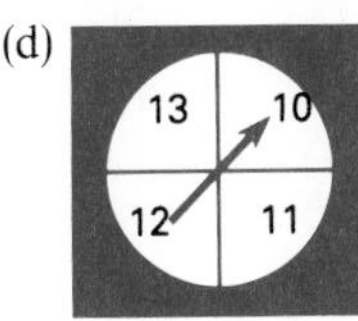

2. Assign a probability to each outcome in the sample spaces of Exercise 1.

3. A fair die is to be rolled. Find a sample space for this experiment and assign probabilities to each outcome.

4. Doug, Tom, Ben, Fred, and Louis place their names in a hat. One name is to be drawn to determine who will confess to breaking a window. Tabulate a sample space and a probability rule.

5. Give a sample space and assign a probability tc each outcome.

(a) 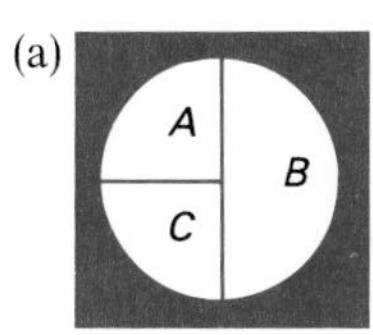

(b) 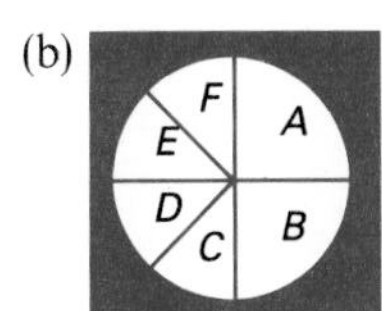

6. Consider the given sales record indicating the number of days on which a certain number of sales were made.

Units sold	*Days*
0	20
1	80
2	120
3	250
4	260
5	190
6	80

(a) Assign a probability to the event of selling 3 units per day.
(b) Assign a probability of selling 4 or 5 units per day.
(c) Assign a probability of selling no units per day.
(d) Assign a probability of selling less than 3 units per day.

7. What is the probability of getting heads when a two-headed coin is tossed?

8. In Exercise 7, what is the probability of getting tails?

9. Which of the following could not be a probability? Why?

(a) $\frac{-1}{2}$ (b) $\frac{17}{16}$ (c) 0.001 (d) 0

(e) 1.03 (f) $-.01$ (g) $\frac{5}{4}$ (h) 1

10. Dr. Paul indicates he has found a cure for a certain blood disease. Out of 80,000 patients, 74,000 recovered after using his medication. Assign a probability that a person suffering from the blood disease will recover using Dr. Paul's medicine.

T 11. An experiment consists of spinning two of the spinners in Exercise 1. List a sample space if the two spinners are

(a) (a) and (b). (b) (c) and (a).
(c) (a) and (d). (d) (b) and (c).
(e) (d) and (b). (f) (c) and (d).

12. List the elements in a sample space for the simultaneous tossing of a coin and drawing of a card from a set of six numbered 1 through 6.

13. An experiment consists of an ordinary coin being tossed three times. What is a sample space?

14. A box contains three red balls and four black balls. Let R represent a red ball and B a black ball. Tabulate a sample space if

(a) one ball is drawn at a time.
(b) two balls are drawn at a time.
(c) three balls are drawn at a time.
(d) Are these sample spaces uniform?

15. Ed cannot decide which car to purchase. He writes the names of seven cars {Ford, Chevrolet, Pontiac, Oldsmobile, Buick, Plymouth, Dodge} on slips of paper and puts them in a box. Ed is planning to buy the car named on the slip of paper drawn from the box.

(a) What are the possible choices?
(b) If Ed decided to purchase two makes of automobile, tabulate a set of possible outcomes.

16. For the given pair of spinners, list the pairs of possible outcomes of spinning both spinners.

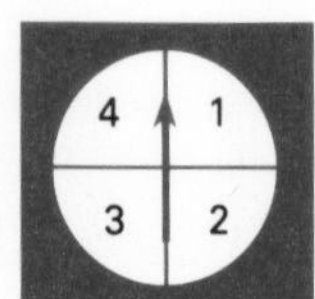

17. Three coins are tossed, and the number of heads is recorded. Which of the following sets is a sample space for this experiment? Why do the other sets fail to qualify as sample spaces?

(a) {1, 2, 3} (b) {0, 1, 2}
(c) {0, 1, 2, 3, 4} (d) {0, 1, 2, 3}
(e) $\{x \mid x$ is a whole number less than 2 and greater than 1$\}$

18. An urn contains four balls numbered 1 to 4. Record a sample space for the following experiments.

(a) A ball is drawn and the number is recorded. The ball is returned and a second ball is drawn and recorded.
(b) A ball is drawn and recorded. Without replacing the first ball, a second is drawn and recorded.

C 19. A bag contains 4 red marbles, 3 blue marbles, and 3 green marbles. A single marble is drawn.

(a) Tabulate a sample space with 3 outcomes.
(b) Tabulate a sample space with 10 outcomes.
(c) Which of these sample spaces is uniform?

20. A sample of the employment status of the residents in a certain town is given in the following table.

	Employed	*Unemployed*
Male	1000	40
Female	800	160

Assign a probability that each of the following is true.

(a) An unemployed person is female.
(b) An unemployed person is male.
(c) A male is unemployed.
(d) A female is employed.

21. Let a sample space S be represented by $S = \{1, 2, 3, 4\}$. This set can be considered as the sample space of the outcomes of tossing a tetrahedron—that is, a solid with four faces—considering as an outcome the face on which it lands. Tabulate the following subsets of S.

(a) A, the occurrence of an even number
(b) B, the occurrence of an odd number
(c) C, the occurrence of a number greater than 3
(d) $B \cup C$

22. Three coins are tossed, and the number of heads is recorded. Which of the following sets are sample spaces for this experiment? If a set fails to qualify as a sample space, give the reason.

(a) {0, 2, an odd number}
(b) $\{x \mid x$ is a whole number and $x < 4\}$

2 The probability of events

You were introduced to the assigning of probabilities in Section 1. A poor means of assigning probabilities to outcomes in a sample space is subjective judgment; yet subjective judgment is often the only tool accessible when very limited data are available. Hence a sales manager might project that "We have a probability of $\frac{2}{3}$ of getting the XYZ contract" and the ubiquitous man on the street might prophesy that "The probability of Ron Rigon being elected is $\frac{40}{100}$." Generally, such assignments are merely measures of the strength of the person's belief. However, if the person making the projection has much experience in the area and a keen sense of the undefinable "intuition," the probability model might still be of use. This is evidenced daily in the decision-making centers of government, education, and industry.

In this section, we shall study rules for assigning probabilities to events. The rules are correct, but decisions made using them can be disastrous if the experiment in question does not meet the assumptions under which the rule was devised. For example, if one assumes a coin is fair and assigns $P(H) = \frac{1}{2}$ and $P(T) = \frac{1}{2}$, and the coin has in fact been modified to be unbalanced, the probability rule is not useful. If one bases a decision about production in the Nuts and Bolts factory on the evidence from last month's output, and this month a new machine is installed, the decisions might be very inappropriate. Certainly any probability rule built on subjective judgment is suspect. One of the greatest sources of error in practical application of probability (and its sister science, statistics) is the use of an inappropriate probability rule to make decisions.

In general, we are interested not in the probability of a single outcome in a sample space but in the probability of an event. For instance, in drawing a card from a set of six, we may be interested in the outcomes that are even numbers: 2, 4, 6. We note that these numbers are elements of a subset of the sample space $\{1, 2, 3, 4, 5, 6\}$. This observation leads to the definition of an event.

Event

An *event* is a subset of a sample space.

The occurrence of any collection of outcomes of an experiment can be considered as an event. Consider the experiment of drawing a card from our set of six cards numbered 1 through 6 and observing the number that appears on the card.

Some possible events are	The subsets are
1. observing a 1.	$\{1\}$.
2. observing a 3.	$\{3\}$.
3. observing a 6.	$\{6\}$.
4. observing an even number.	$\{2, 4, 6\}$.
5. observing an odd number.	$\{1, 3, 5\}$.
6. observing a number divisible by 3.	$\{3, 6\}$.

7. observing a number less than 4. $\{1, 2, 3\}$.
8. observing a number greater than 4. $\{5, 6\}$.
9. observing a digit. $\{1, 2, 3, 4, 5, 6\}$.
10. observing a number greater than 17. $\varnothing$.

Of course, these are not all the events that could be listed relative to the experiment. Certain characteristics of the events listed are apparent. Events 1, 2, and 3 differ from the remaining events listed in that each contains only one element of the whole sample space, whereas events 4 through 9 all involve more than one element, and 10 is empty.

When an event comprises only one outcome of the sample space, it is called a simple event; a compound event involves more than one sample outcome. There are only six simple events associated with the preceding experiment of drawing a card. Can you name them? In like manner, there are many compound events associated with this experiment. Can you name some of those not already listed?

Example 12 • Tabulate the sample space and the event of getting at least one head when two coins are tossed.

Solution The sample space is tabulated as

$$\{HH, HT, TH, TT\}$$

The event of getting at least one head is $\{HH, HT, TH\}$. •

Example 13 • A nickel, a dime, and a penny are tossed. List the sample space of possible outcomes and the event of getting at least one head.

Solution The outcomes of this experiment are tabulated in Figure 2. The event of getting at least two heads is contained in the closed curve. •

Now let's summarize, for event A, the properties of probability.

Probability of event A, denoted by $P(A)$

1. If A is the empty set, then $P(A) = 0$.
2. If A is the universe (consists of all outcomes), then $P(A) = 1$.
3. If A is an event with one outcome from a uniform sample space with $n(S)$ outcomes, then $P(A) = \dfrac{1}{n(S)}$.
4. If A is an event that consists of two outcomes from a uniform sample space, then

$$P(A) = \frac{1}{n(S)} + \frac{1}{n(S)} = \frac{2}{n(S)}$$

5. Let S be a uniform sample space (each outcome in S is equally likely). Suppose further that event A is a subset of S. Then $P(A)$, the probability of event A, is given by

$$P(A) = \frac{\text{number of equally likely outcomes in event } A}{\text{number of possible equally likely outcomes}} = \frac{n(A)}{n(S)}$$

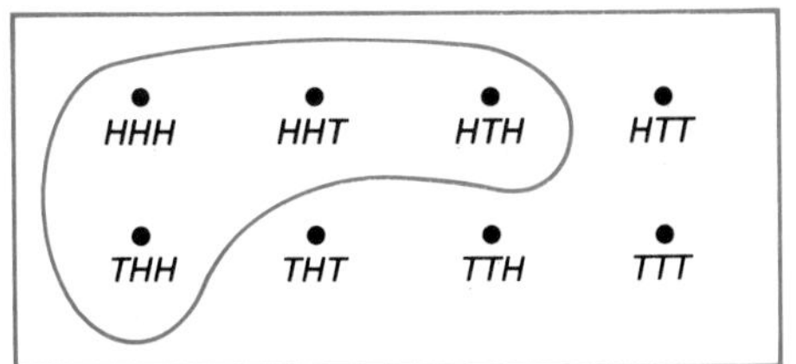

Figure 2

This rule is the classical definition of probability. Suppose that there are N possible equally likely outcomes of an experiment. If r of these outcomes have a particular characteristic, so that they can be classified as a success, then the probability of a success is defined to be r/N.

Example 14 • Suppose that a card is drawn from a set of six (numbered 1 through 6). There are six equally likely possible outcomes of the experiment. Let two of these, a 3 and a 6, represent a success, E. Then

$$P(E) = \frac{n(E)}{n(S)} = \frac{2}{6} = \frac{1}{3} \quad \bullet$$

Example 15 • A card is dealt from a shuffled deck (52 cards with four suits: clubs, diamonds, hearts, and spades). What is the probability that the card is an ace?

Solution There are 52 cards in the deck; so $n(S) = 52$. Only 4 cards are aces; therefore $n(A) = 4$.

$$P(\text{ace}) = P(A) = \frac{n(A)}{n(S)} = \frac{4}{52} = \frac{1}{13} \quad \bullet$$

Example 16 • Consider the experiment of tossing two coins. What is the probability of tossing two heads? At this juncture, we must be careful. We have discussed two sample spaces for this experiment: $\{HH, HT, TT\}$, in which we do not observe which coin has a head and which coin has a tail, and $\{HH, HT, TH, TT\}$, in which we classify each toss with an ordered pair that takes into consideration which coin has a head and which coin has a tail. Note that we cannot apply the previous definition in the first case, because it is not a uniform sample space; however, the definition is applicable to the second sample space. Now let A be the event of tossing two heads. Using the second sample space,

$$P(A) = \frac{n(A)}{n(S)} = \frac{1}{4} \quad \bullet$$

Example 17 • Consider the two spinners in Figure 3. If both spinners are activated, find the probability of spinning a total of five points.

Figure 3

Just for fun *Move only one glass so that empty glasses alternate with full glasses.*

Solution Spinning both spinners results in 24 equally likely outcomes. Four of them,

$$\{(1, 4), (2, 3), (3, 2), (4, 1)\}$$

give a total of five points. Thus,

$$P(\text{five points}) = \frac{4}{24} = \frac{1}{6}$$ •

Exercise set 2

R 1. Suppose that there is an equally likely probability that the spinner will stop at any one of the six sections for the given spinner.

(a) What is the probability of stopping on an even number?
(b) What is the probability of stopping on a multiple of 3?
(c) What is the probability of stopping on an even number or a multiple of 3?
(d) What is the probability of stopping on an even number or an odd number?
(e) What is the probability of stopping on a number less than 3?

2. A card is drawn from an ordinary deck. What is the probability of getting

(a) a heart? (b) an ace?
(c) the jack of spades? (d) a red card?
(e) a red ace?

3. A multiple-choice question has five possible answers. You haven't studied and hence have no idea which answer is correct, so you randomly choose one of the five answers. What is the probability that you select the correct answer? An incorrect answer?

4. A survey course in history contains 20 freshmen, 8 sophomores, 6 juniors, and 1 senior. A student is chosen at random from the class roll.

(a) Describe in words a uniform sample space for this experiment.
(b) What is the probability that the student is a freshman?
(c) What is the probability that the student is a senior?
(d) What is the probability that the student is a junior or a senior?
(e) What is the probability that the student is a freshman or a sophomore or a junior or a senior?

5. A box contains four black balls, seven white balls, and three red balls. If a ball is drawn, what is the probability of getting the following colors?

(a) Black (b) Red
(c) White (d) Red or white
(e) Black or white
(f) Red or white or black

6. A professor has decided to award the following grades in his class.

5 A's
8 B's
16 C's
7 D's
4 F's

The 40 grades are placed on cards and put in a box. You randomly select a card. What is the probability that you select

(a) a C?
(b) an F?
(c) a B or C?
(d) a grade higher than C?
(e) a grade lower than C?

7. One card is drawn from each of two sets of four cards (each set numbered 1 through 4). List a uniform sample space to show that the two cards can be drawn in 16 different ways. What is the probability that the sum of the numbers is

(a) greater than 4? (b) equal to 9?
(c) equal to 7? (d) greater than 7?

8. A spinner with three equally likely sections is spun along with the spinner in Exercise 1. What is the probability that the sum of the numbers is

(a) greater than 7? (b) equal to 8?
(c) equal to 3? (d) less than 5?

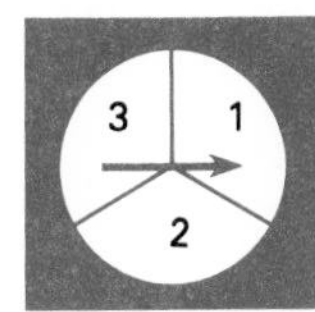

9. Two coins are tossed. What is the probability of getting

(a) two heads? (b) exactly one head?
(c) at most one head?

T 10. Discuss the following statements.

(a) Since there are 50 states, the probability that a person was born in Rhode Island is $\frac{1}{50}$.
(b) A's chances of winning an election are $\frac{6}{10}$, B's chances are $\frac{3}{10}$, and C's chances are $\frac{1}{10}$. If C withdraws from the race, A believes his chances of winning will decrease.
(c) Twenty tosses of a coin yield 18 heads and 2 tails.
(d) A candidate is running for two offices A and B. He assigns a probability of .4 of being elected to A, a probability of .3 of being elected to B. and a probability of .6 of being elected to both A and B.

11. A medical survey of the cause of death among a group of 190 females was categorized according to the cause of death and the age of the subject at the time of death.

Cause of death	Age at death: Below 40	40–60	Over 60
Heart disease	1	6	10
Stroke	2	5	12
Cancer	7	12	16
Pneumonia	0	4	5
Diabetes	4	0	1
Tuberculosis	0	2	0
Other	34	28	41

If one of these subjects is chosen at random, what is the probability that she

(a) died of cancer?
(b) was over 60 years old when she died?
(c) died of diabetes?
(d) died of a stroke or of heart disease?
(e) was over 60 and died of diabetes?
(f) was under 60 and died of stroke or heart disease?

12. If you flipped a fair coin 15 times and got 15 heads, what would be the probability of getting a head on the 16th toss?

13. In drawing a card from a standard deck of cards, you reason that you can get a spade (S) or cannot get a spade. Therefore, there are two outcomes. Hence, $P(S) = \frac{1}{2}$. Is this reasoning correct?

C 14. You have forgotten the last digit of a telephone number. You randomly try a digit. What is the probability that you select the correct number?

15. A pair of fair tetrahedral (four-faced) dice with the numbers 1 through 4 on the faces is rolled. Let A = the sum is 4 and B = the sum exceeds 4

(a) Tabulate a uniform sample space for this experiment.
(b) Compute $P(A)$.
(c) Compute $P(B)$.

16. A sample space is $\{HH, HT, TH, TT\}$. List all possible events and describe each in words.

17. The experiment consists of rolling a regular dodecahedron with the numbers 1 through 12 on its faces.

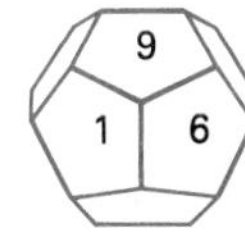

What probability would you assign to each outcome? The following events are described in words. Tabulate each event and find its probability.

(a) An odd number
(b) A prime number
(c) A divisor of 12
(d) A multiple of 1
(e) A multiple of 16
(f) The square of 4
(g) A multiple of 2
(h) A multiple of 3
(i) A multiple of 2 or 3
(j) A multiple of both 2 and 3
(k) A prime or a composite number
(l) A prime and a composite number

3 Tree diagrams and the Fundamental Principle of Counting

■ The Hard College Bulldogs are purchasing uniforms. Members can purchase red or white shorts. They can choose red, white, or striped shirts. How many possible choices are there for the uniforms? ■

In this section, we learn a procedure for counting such possibilities. If one can count the number of outcomes in a uniform sample space and in an event, one can always assign a probability to an event. However, many times the counting of outcomes becomes tedious. In this section, we shall introduce two tools that may be of help—tree diagrams and the Fundamental Principle of Counting.

The college chorale is planning a concert tour with performances in Dallas, St. Louis, and New Orleans. In how many ways can they arrange their itinerary? If there is no restriction on the order of the performances, any one of the three cities can be chosen as the first stop. After the first city is selected, either of the other cities can be second, and the remaining city can be the last stop. A **tree diagram** can aid the chorale in determining their possible tour schedule. Starting at the campus, draw lines to each of the three cities. Then draw lines from these cities to each of the remaining choices. The result is a diagram resembling the branches of a tree. (See Figure 4.) A quick glance at the tree diagram in Figure 4 reveals that there are six ways the chorale can arrange its tour.

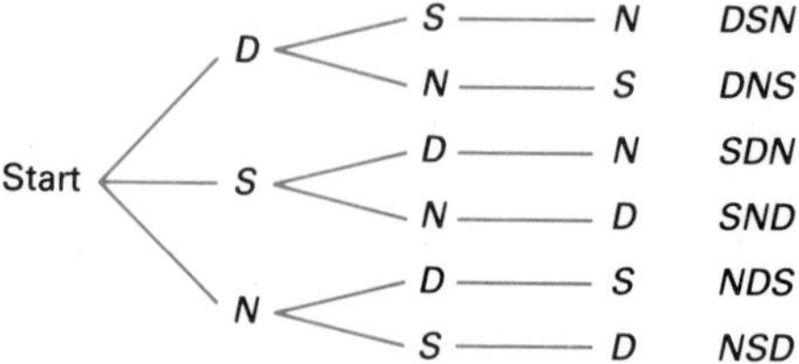

Figure 4

The following procedure was used in constructing the tree diagram. Draw at the position called *start* a number of *branches* corresponding to the number of ways the first step can be performed. At the end of each branch in this first step, draw branches corresponding to the ways the second step can be performed. Any sequence of branches that moves from *start* to the end of the last branch is called a **path.** Count the number of paths to obtain the number of ways an experiment can be performed.

Whenever a task can be done in two or more stages and each stage can be done in a number of ways, a tree diagram provides a good illustration of the choices involved and serves as an aid in determining the number of ways the whole task can be accomplished.

Example 18 • In how many ways can a two-toned car be painted with red, white, and black paint?

Solution As shown in Figure 5, there are six paths, so with three paints, six two-toned color combinations are possible.

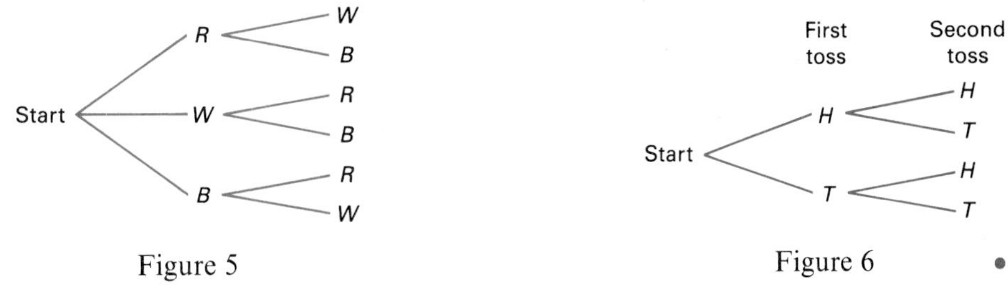

Figure 5

Figure 6 •

Example 19 • A coin is tossed twice. Draw a tree diagram to illustrate the possible outcomes.

Solution The four branches of the diagram in Figure 6 correspond to the outcomes *HH*, *HT*, *TH*, and *TT*. •

Now let's return to the first example of this section.

Example 20 • The members of the chorale decided to sing first in New Orleans, next in Dallas, and then close in St. Louis. Now, they must decide on their modes of transportation. They can travel from the campus to New Orleans by bus or plane and from New Orleans to Dallas by bus, plane, or train and from Dallas to St. Louis by bus or train. The tree diagram in Figure 7 indicates the choices of the chorale.

Figure 7

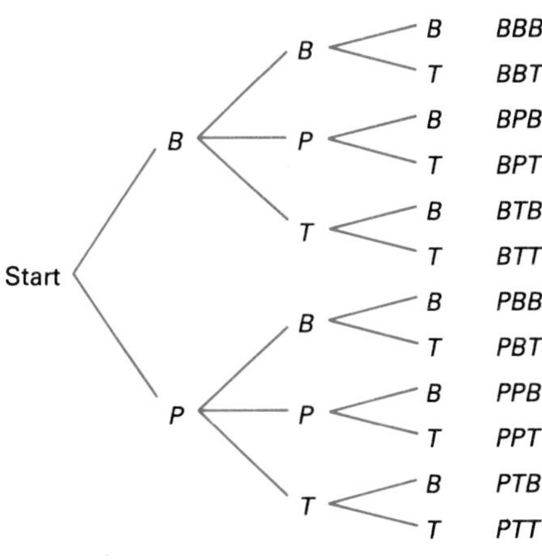

The first part of the trip can be made in 2 ways, the second part in 3 ways, and the last part in 2 ways. Notice that the number of ways the transportation can be chosen is $2 \cdot 3 \cdot 2 = 12$ ways. •

This example introduces the following principle.

The Fundamental Principle of Counting

1. If two experiments are performed in order with n_1 possible outcomes of the first experiment and n_2 possible outcomes from the second experiment, then there are

$$n_1 \cdot n_2$$

combined outcomes of the first experiment followed by the second.
2. In general, if k experiments are performed in order with possible number of outcomes $n_1, n_2, n_3, \ldots, n_k$, respectively, then there are

$$n_1 \cdot n_2 \cdot n_3 \cdot \cdots \cdot n_k$$

possible outcomes of the experiments performed in order.

For example, if the first stage of a task can be accomplished in m ways, the second stage in n ways, and so forth, and the last stage can be done in z ways, then the whole task can be performed in $m \cdot n \cdot \cdots \cdot z$ ways. The Fundamental Principle of Counting is helpful in solving problems such as the following.

Example 21 • In many states, automobile license plates have on them a combination of three letters and three digits. If all letters and digits may be used repeatedly, how many combinations are available to each of these states?

Solution There are 26 letters to choose from for each of the 3 letter places, and there are 10 digits to choose from for the digit places. By the Fundamental Principle of Counting, the number of combinations is

$$26 \cdot 26 \cdot 26 \cdot 10 \cdot 10 \cdot 10 = 17{,}576{,}000 \quad \bullet$$

The Fundamental Principle of Counting is helpful in solving problems such as the following.

Example 22 • An urn contains five red balls and seven white balls. A ball is drawn, its color is noted, and a second ball is drawn. What is the probability of drawing a red ball followed by a white ball?

Solution By the counting principle, there are 12 ways of drawing the first ball and 11 ways of drawing the second. Therefore, there are

$$12 \cdot 11 = 132$$

ways of drawing the two balls. At the same time, there are 5 ways of drawing the red ball on the first draw and 7 ways of drawing the white ball on the second draw.

Just for fun

Take all the different cards of one suit in a standard deck of cards and count the number of letters in each name:

Ace 3
Two 3
Three 5
Four 4
⋮

Now find the total of the numbers. What is surprising about your answer?

Therefore, the number of ways of drawing a red ball and then a white ball is

$$5 \cdot 7 = 35$$

Thus,

$$P(R \text{ followed by } W) = \frac{35}{132} \quad \bullet$$

Example 23 • If a couple plans to have 3 children, what is the probability that exactly 2 will be boys? (Assume that it is equally likely for a child to be a boy or a girl.)

Solution The tree diagram in Figure 8 indicates 8 paths or ways that the couple can have 3 children. By counting, one notes that 3 of these have exactly 2 boys. Therefore,

$$P(\text{exactly 2 boys}) = \frac{3}{8} \quad \bullet$$

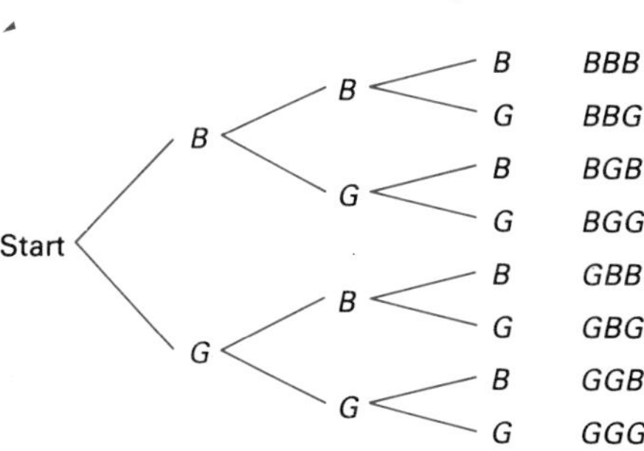

Figure 8

Example 24 • A card is drawn from a deck of cards. Then the card is replaced, the deck is reshuffled, and a second card is drawn. What is the probability of an ace on the first draw and a king on the second?

Solution Since the first card is returned, there are $52 \cdot 52$ ways of drawing the 2 cards. There are 4 ways of drawing an ace on the first draw and 4 ways of drawing a king on the second draw, or $4 \cdot 4$ ways of drawing both. Thus,

$$P(\text{ace followed by a king}) = \frac{4 \cdot 4}{52 \cdot 52} = \frac{1}{169} \approx 0.0059 \quad \bullet$$

Exercise set 3

R 1. A student plans a trip from Atlanta to Boston to London. From Atlanta to Boston, he can travel by bus, train, or airplane. However, from Boston to London, he can travel only by ship or airplane.

(a) In how many ways can the trip be made?
(b) Verify your answer by drawing an appropriate tree diagram and counting the routes.

2. There are six roads from A to B and four roads between B and C.

(a) In how many ways can Joy drive from A to C by way of B?
(b) In how many ways can Joy drive round trip from A to B to C and return to A through B?

3. A sociology quiz contains a true/false question and two multiple-choice questions with possible answers (a), (b), (c), and (d).

(a) In how many possible ways can the test be answered?
(b) Draw a tree diagram and count the options to check (a).
(c) If Jodi guesses on each problem, what is the probability she will get all three correct?

4. Joyce desires to purchase an automobile. She has a choice of two body styles (standard and sports model) and four colors (green, red, black, blue). What is the probability she selects a red sports model?

T 5. In a family of exactly two children, what is the probability that

(a) both are girls?
(b) the first is a boy and the second a girl?
(c) at least one is a boy?
(d) none is a girl?

6. Suppose that in a local election only two parties are represented, D and R. Draw a tree diagram illustrating four consecutive elections and determine how many possibilities result in

(a) at least one party change.
(b) no party changes.
(c) exactly one partly change.
(d) more than two party changes.

7. In an interview, people are classified according to sex (M or F), political affiliation (D, I, R), and age (Y, M, O). Construct a tree diagram showing all possible classifications. Does the number of branches agree with the Fundamental Principle of Counting?

C 8. A box contains six different-colored balls: red, white, blue, black, green, and yellow. If two balls are drawn at random, one at a time, and replaced, what is the probability of getting

(a) a yellow ball followed by a red ball?
(b) a red ball followed by a blue ball?
(c) a yellow ball and a red ball? (*Hint:* This can happen in more than one way.)

9. Assume that a family wishes to have four children. What is the probability that

(a) exactly four are girls?
(b) all the children are the same sex?
(c) at least three are boys?
(d) at most three are boys?

10. Work Exercise 9 for a family with five children.

11. A restaurant offers the following menu.

Main course	*Vegetables*	*Beverage*
Beef	Potatoes	Milk
Ham	Green beans	Coffee
Fried chicken	Green peas	Tea
Shrimp	Asparagus	

If you choose one main course, two vegetables, and one beverage, in how many ways could your order a meal?

12. Two cards are drawn from a standard deck of playing cards. What is the probability that a king is drawn followed by an ace

(a) if the first card is replaced before the second is drawn?
(b) if the first card is not replaced before the second is drawn?

*4 Properties of probability

■ Of the freshmen who entered Hard College last year, 12% failed freshman English, 8% failed history, and 4% failed both English and history. An admissions counselor would like to know what percent failed English or history. ■

The material in this section should enable you to assist the admissions counselor in finding the answer.

Consider now some combinations of events. Three of these relationships are of such importance that we list them as special events. These events are defined using the operators of set theory: union, intersection, and complementation.

And, or, and complement

1. The event $A \cup B$ is the collection of all outcomes that are **in A or in B or in both A and B.**
2. The event $A \cap B$ is the set of all outcomes that are in **both A and B.**
3. The **complement** of an event A, denoted $\bar{A}$, is the collection of all outcomes that are in the sample space and are not in A.

We illustrate these concepts with examples involving the roll of a die.

Example 25 • In the rolling of a fair die, what is the probability of an odd number or a 4?

Solution We let O represent an odd number and F represent a 4 and seek $P(O \cup F)$. In Figure 9(b) we see that

$$P(O \cup F) = \frac{4}{6} = \frac{2}{3}$$

Figure 9

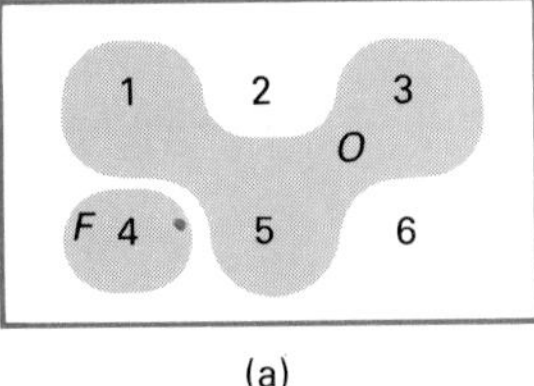

(a)

1 2 3
4 5 6

(b) $O \cup F$

Note in Figure 9(a) that

$$P(O) = \frac{3}{6} \qquad \text{and} \qquad P(F) = \frac{1}{6}$$

Thus,

$$P(O \cup F) = P(O) + P(F)$$

as

$$\frac{4}{6} = \frac{3}{6} + \frac{1}{6} \quad \bullet$$

Example 26 • In the rolling of the same fair die, what is the probability of an even number or a 4?

Solution Let E represent an even number and F represent a 4. We seek $P(E \cup F)$. In Figure 10(b) we see that

$$P(E \cup F) = \frac{3}{6}$$

Figure 10

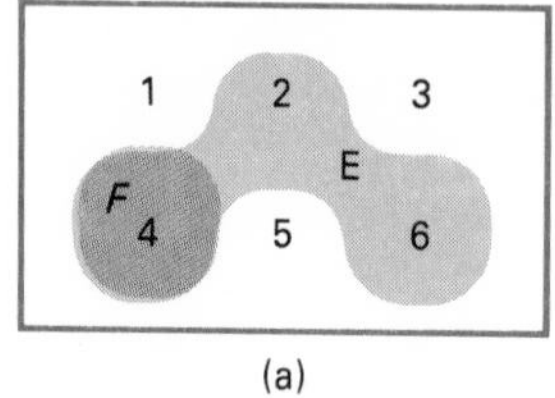

(a)

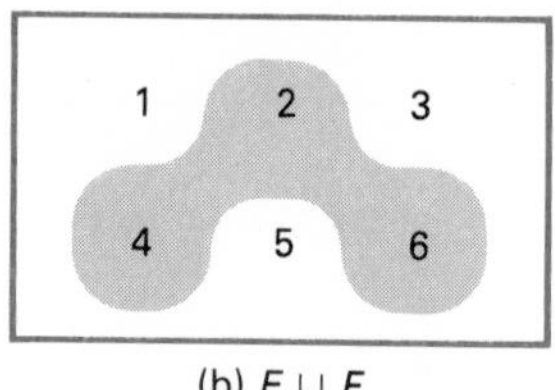

(b) $E \cup F$

Note in Figure 10(a) that

$$P(E) = \frac{3}{6} \qquad \text{and} \qquad P(F) = \frac{1}{6}$$

$$P(E \cup F) \neq P(E) + P(F)$$

$$\frac{3}{6} \neq \frac{3}{6} + \frac{1}{6} \quad \bullet$$

What is the difference in the problems in the two previous examples? For $P(O \cup F)$, F and O had no points in common. For $P(E \cup F)$, E and F overlapped. This discussion suggests the following definition and property of probability:

Mutually exclusive events

1. Events A and B are **mutually exclusive** if they have no outcomes in common.
2. If events A and B are mutually exclusive,

$$P(A \cup B) = P(A) + P(B)$$

Example 27 • From a standard deck of cards, we draw one card. What is the probability of our getting a spade or a red card?

Solution Verify that

$$P(S) = \frac{13}{52} \qquad \text{and} \qquad P(R) = \frac{26}{52}$$

Since getting a spade and getting a red card are mutually exclusive,

$$P(S \cup R) = P(S) + P(R)$$
$$= \frac{13}{52} + \frac{26}{52} = \frac{39}{52} = \frac{3}{4} \quad \bullet$$

Now let's return to the preceding example, where we noted that $P(E \cup F) = \frac{3}{6}$, $P(E) = \frac{3}{6}$, and $P(F) = \frac{1}{6}$. Why is $P(E \cup F) \neq P(E) + P(F)$? The outcome 4 is in both

E and F and is thus counted twice in $P(E) + P(F)$. (See Figure 10(a)). The probability that 4 is in both E and F is

$$P(E \cap F) = \frac{1}{6}$$

Since $E \cap F$ is included twice in $P(E) + P(F)$ we subtract one of these and note that

$$P(E \cup F) = P(E) + P(F) - P(E \cap F)$$
$$\frac{3}{6} = \frac{3}{6} + \frac{1}{6} - \frac{1}{6}$$

We can generalize this concept by realizing that in set theory the number of outcomes in event A *or* in event B is the number in A plus the number in B less the number in $A \cap B$, which has been counted in both A and B. (See Figure 11.) Thus,

$$n(A \cup B) = n(A) + n(B) - n(A \cap B)$$

Figure 11

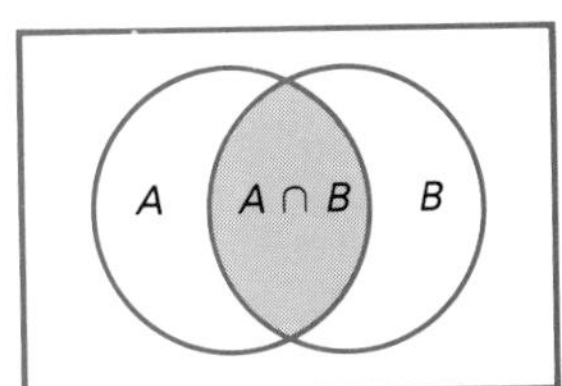

Divide both sides of the equation by N, the number of elements in a sample space, to obtain

$$\frac{n(A \cup B)}{N} = \frac{n(A)}{N} + \frac{n(B)}{N} - \frac{n(A \cap B)}{N}$$

or

$$P(A \cup B) = P(A) + P(B) - P(A \cap B)$$

Probability of *A* or *B*

For any two events A and B, the probability of A **or** B is given by

$$P(A \cup B) = P(A) + P(B) - P(A \cap B)$$

Example 28 • (Introductory problem) Of the freshmen at Hard College last year, 12% failed English, 8% failed history, and 4% failed both English and history. What percent failed English or history?

Solution

$$P(E) = 0.12$$
$$P(H) = 0.08$$
$$P(E \cap H) = 0.04$$
$$P(E \cup H) = P(E) + P(H) - P(E \cap H)$$
$$= 0.12 + 0.08 - 0.04$$
$$= 0.16$$

16% failed English or history. •

Example 29 • In drawing a card from eight cards numbered 1 through 8, what is the probability of getting an even number or a number less than 5?

Solution Let A represent the event of getting a number less than 5 and B the event of getting an even number.

$$P(A) = \frac{4}{8} = \frac{1}{2}$$

$$P(B) = \frac{4}{8} = \frac{1}{2}$$

But 2 and 4 are both even and less than 5; consequently,

$$P(A \cap B) = \frac{2}{8} = \frac{1}{4}$$

Thus,

$$\begin{aligned} P(A \cup B) &= P(A) + P(B) - P(A \cap B) \\ &= \frac{1}{2} + \frac{1}{2} - \frac{1}{4} = \frac{3}{4} \end{aligned}$$ •

Let us now make an observation concerning the probability that an event does **not** occur. The probability of getting a 6 on the toss of a die is $\frac{1}{6}$. What is the probability of not getting a 6? There are five equally likely ways of not getting a 6—namely, getting a 1, 2, 3, 4, or 5. Thus, the probability of *not getting* a 6 is $\frac{5}{6}$. Note that $\frac{5}{6} = 1 - \frac{1}{6}$.

In general, let's divide all the events in the sample space into two mutually exclusive sets A and $\bar{A}$, as shown in Figure 12. The set $\bar{A}$ is called the **complement** of A in relation to the sample space S.

Figure 12

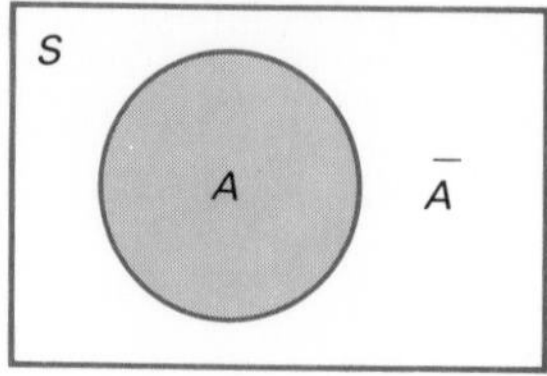

Note that

$$A \cap \bar{A} = \varnothing \qquad \text{and} \qquad A \cup \bar{A} = S$$

Thus

$$\begin{aligned} P(A \cup \bar{A}) &= P(A) + P(\bar{A}) - P(A \cap \bar{A}) \\ 1 &= P(A) + P(\bar{A}) - 0 \\ P(\bar{A}) &= 1 - P(A) \end{aligned}$$

Just for fun

The "Birthday Problem" gives a result that defies intuition. Suppose that 35 people are in a room. What is the probability that at least 2 of them have their birthdays on the same day of the year? Since there are 365 days in the year and only 35 people, it seems that the probability should be fairly small. In fact, it is quite large, namely 0.814.

The Probability P That Among n People at Least Two Will Have the Same Birthday

n	P
5	0.027
10	0.117
15	0.253
20	0.411
23	0.500
25	0.569
30	0.706
35	0.814
40	0.891
45	0.941
50	0.970
55	0.986
60	0.994
65	0.998
70	0.999
75	0.999 +

Probability of a complement

If A is any event in the sample space S, and if $\bar{A}$ denotes the *complement of A*, then

$$P(\bar{A}) = 1 - P(A)$$

Example 30 • What is the probability of not getting an ace in the drawing of a card from a standard deck of cards?

Solution

$$P(\text{no ace}) = 1 - P(\text{ace})$$
$$= 1 - \frac{4}{52} = \frac{12}{13} \quad \bullet$$

Exercise set 4

R 1. If A and B are events with $P(A) = .6$, $P(B) = .3$, and $P(A \cap B) = .2$, find $P(A \cup B)$.

2. In Brooks College, 30% of the freshmen failed mathematics, 20% failed English, and 15% failed

both mathematics and English. What is the probability that a freshman failed mathematics or English?

3. An experiment consists of tossing a coin seven times. Describe in words the complement of each of the following.

   (a) Getting at least two heads
   (b) Getting three, four, or five tails
   (c) Getting one tail
   (d) Getting no heads

4. A number x is selected at random from the set of numbers $\{1, 2, 3, \ldots, 8\}$. What is the probability that

   (a) x is less than 5?
   (b) x is even?
   (c) x is less than 5 and even?
   (d) x is less than 5 or a 7?

5. A single card is drawn from a 52-card deck. What is the probability that it is

   (a) either a heart or a club?
   (b) either a heart or a king?
   (c) not a jack?
   (d) either red or black?

6. From a bag containing six red balls, four black balls, and three green balls, one ball is drawn. What is the probability that it is

   (a) red or black?
   (b) red or black or green?
   (c) not black?
   (d) not red or not black?

T 7. If A and B are events in a sample space such that $P(A) = .6$, $P(B) = .2$, and $P(A \cap B) = .1$, compute each of the following.

   (a) $P(\bar{A})$ (b) $P(\bar{B})$
   (c) $P(A \cup B)$ (d) $P(\overline{A \cap B})$

8. If A and B are events with $P(A \cup B) = \frac{5}{8}$, $P(A \cap B) = \frac{1}{3}$, and $P(\bar{A}) = \frac{1}{2}$, compute the following.

   (a) $P(A)$ (b) $P(B)$
   (c) $P(\bar{B})$ (d) $P(\bar{A} \cup B)$

9. In a survey, families with children were classified as C, and those without children as $\bar{C}$. At the same time, families were classified according to D, husband and wife divorced, and $\bar{D}$, not divorced. Of 200 families surveyed, the given results were obtained.

	C	$\bar{C}$	*Total*
D	60	20	80
$\bar{D}$	90	30	120
Total	150	50	200

What is the probability that a family selected at random

   (a) has children?
   (b) has husband and wife who are not divorced?
   (c) has children or husband and wife who are divorced?
   (d) has no divorce or no children?

10. A recent survey found that 60% of the people in a given community drink Lola Cola and 40% drink other soft drinks; 15% of the people interviewed indicated that they drink both Lola Cola and other soft drinks. What percent of the people drink either Lola Cola or other soft drinks?

11. A fair coin is flipped 4 times.

   (a) Tabulate the uniform sample space of the set of all possible outcomes involving H and T.
   (b) Tabulate the event $A = \{\text{exactly one head appears}\}$.
   (c) Tabulate the event $B = \{\text{at least one tail occurs}\}$.
   (d) Tabulate $\{A \text{ or } B\}$.
   (e) Tabulate $\{A \text{ and } B\}$.
   (f) Tabulate $\bar{A}$.

12. For the event A from Exercise 11, compute $P(A)$ and $P(\bar{A})$, and verify that $P(A) = 1 - P(\bar{A})$.

An extension of the additive formula states that

$$P(A \cup B \cup C) = P(A) + P(B) + P(C) - P(A \cap B) - P(A \cap C) - P(B \cap C) + P(A \cap B \cap C)$$

Use the formula to work the following exercises.

C 13. In a survey of 100 families of a school district in 1980, each family was asked the following questions.

   A. Do you have children in public school?
   B. Do you object to the modern approach of teaching mathematics?
   C. Do you object to placing students in classes according to IQ tests?

The *yes* answers to these questions were tabulated as follows:

$$n(A) = 70 \qquad n(A \cap B) = 15$$
$$n(B) = 30 \qquad n(A \cap C) = 6$$
$$n(C) = 10 \qquad n(B \cap C) = 8$$
$$n(A \cap B \cap C) = 5$$

If a family is selected at random, what is the probability that

(a) the answer from this family is *yes* on either A or B?
(b) the answer from this family is *yes* to one of the three questions A or B or C?
(c) it does not object to modern mathematics or does not object to placing students in classes by IQ tests?

14. In a sample of 50 people, it was found that 28 planned to vote for the Democratic candidates for mayor and assistant to the mayor, 10 planned to vote for the Republican candidates for mayor and assistant to the mayor, 5 planned to vote for a Republican mayor and a Democratic assistant to the mayor, and 7 planned to vote for a Democratic mayor and Republican assistant to the mayor. What is the probability that a randomly selected person plans to vote
   (a) for a Republican mayor?
   (b) for at least one Republican?
   (c) for a Republican mayor and a Democratic assistant mayor?
   (d) for a Republican mayor or a Democratic assistant mayor?

15. Three-digit numbers range all the way from 100 to 999. A three-digit number is selected at random. Give the probability that it
   (a) is even.
   (b) is made up of only even digits.
   (c) is not comprised of only even digits.
   (d) is a square.
   (e) has all prime digits.
   (f) has all composite digits.

*5 Expected value and the probability of compound events

■ When dealing with an uncertain situation, we might expect that as more information is obtained, the probabilities would change. Alternatively, we might say that as more information is available, the sample space is modified. Suppose, for instance, that the top executives for the P. G. and Y. Corporation are evaluating their chances of obtaining a large fabrication contract. They feel that P. G. and Y., the Go-Slo Corporation, and Leary Enterprises are equally likely to win the bidding. Hence, in their minds, the probability is $\frac{1}{3}$ that they will win the contract. Then comes information that Go-Slo has withdrawn from the bidding. Excitement reigns at P. G. and Y., because in this modified sample space, the probability of success for P. G. and Y. is reevaluated at $\frac{1}{2}$. Is this true? ■

Probabilities by themselves do not always supply all the information that is useful when making a decision in an uncertain situation. Equally important is information about what the decision-maker stands to gain or lose in the transaction. The tool that is used to discuss such expected gain or loss is **expected value.**

Conditional probability is the mathematical term used to describe probability with additional information. The symbol $P(A|B)$ denotes the conditional probability that event A will occur, given the information or condition that event B has occurred; $P(A|B)$ is read "the probability of A, given B." We consider in this section two procedures for computing conditional probability. In the first procedure, information is used to obtain a new sample space reflecting the fact that event B has occurred. This procedure is demonstrated by the next four examples.

Example 31 • In a sample of 120 students, suppose 80 are enrolled in English, 60 in mathematics, and 20 in both English and mathematics (see Figure 13). What is the probability that a student selected at random is enrolled in English? What is the probability that a student selected at random is enrolled in English, given that the same student is enrolled in mathematics?

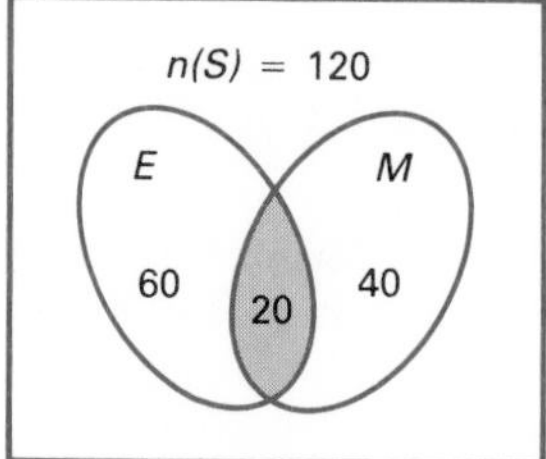

Figure 13

Solution The probability that a student is enrolled in English is

$$P(E) = \frac{80}{120} = \frac{2}{3} \quad \bullet$$

The given condition of being enrolled in mathematics reduces the number of possibilities to 60, of which 20 are enrolled in English; thus,

$$P(E|M) = \frac{20}{60} = \frac{1}{3}$$

Example 32 • A poll is taken to determine whether 700 hourly employees of a company favor a strike. The 700 employees are divided into three groups, X, Y, and Z, according to salary (see Table 4). Suppose an hourly employee is selected at random. The probability that he is in favor of a strike is $\frac{380}{700}$. Now suppose an hourly employee is selected at random from group X. What is the probability that she is in favor of a strike?

Table 4

	In favor of a strike	*Not in favor of a strike*	*No opinion*	*Total*
Group X	150	50	10	210
Group Y	100	80	8	188
Group Z	130	170	2	302
Total	380	300	20	700

Solution This conditional probability is denoted by

$$P(\text{in favor of a strike} \mid \text{group X}) = \frac{150}{210}$$

Other conditional probabilities are

$$P(\text{not in favor of a strike} \mid \text{group Z}) = \frac{170}{302}$$

and

$$P(\text{group Y} \mid \text{in favor of a strike}) = \frac{100}{380} \quad \bullet$$

From our assignment of probabilities in this example, we note an interesting relationship among $P(A \mid B)$, $P(A \cap B)$, $P(A)$, and $P(B)$.

Probability of A and B

If A and B are **any events**, $P(A) \neq 0$, and $P(B) \neq 0$, then

$$P(A \mid B) = \frac{P(A \cap B)}{P(B)} \quad \text{or} \quad P(B \mid A) = \frac{P(A \cap B)}{P(A)}$$

If we multiply by the denominators, we get

$$P(A \cap B) = P(B) \cdot P(A \mid B) \quad \text{and} \quad P(A \cap B) = P(A) \cdot P(B \mid A)$$

Multiplication rule

The probability that both of two events will occur is equal to the probability that the first event will occur multiplied by the conditional probability that the second event will occur when it is known that the first event has occurred:

$$P(A \cap B) = P(A) \cdot P(B \mid A)$$

$$P(A \cap B) = P(B) \cdot P(A \mid B)$$

We quickly observe that this rule gives a procedure for computing the probability of A and B, something that has been missing from our repertoire of skills. However, it should be noted that this relationship is helpful only if one of the relevant conditional probabilities is known or can be computed.

Example 33 • A basket contains two red balls and two white balls. A ball is drawn and its color is noted. Then a second ball is drawn. What is the probability that both balls are red?

Solution Let R_1 be the event of drawing a red ball on the first draw. Then

$$P(R_1) = \frac{2}{4}$$

To find $P(R_2 \mid R_1)$, where R_2 represents a red ball on the second draw, we consider only the outcomes after R_1 has occurred. Since the red ball has not been replaced, there are three balls in the basket, and one of these is red. Thus,

$$P(R_2 \mid R_1) = \frac{1}{3}$$

Substituting these values in the appropriate multiplication rule gives

$$P(R_1 \cap R_2) = P(R_1) \cdot P(R_2 | R_1) = \frac{2}{4} \cdot \frac{1}{3} = \frac{1}{6} \quad \bullet$$

If $P(A|B) = P(A)$, it indicates that knowing that B has occurred does not yield any additional information about the occurrence or nonoccurrence of A. If $P(A|B) = P(A)$, we say event A is **independent** of event B. In this instance, $P(A \cap B) = P(B) \cdot P(A|B) = P(B) \cdot P(A)$. Thus, in this very special case, the probability of the intersection of the events is equal to the product of the probabilities of the events. We can use the relationship $P(A \cap B) = P(A) \cdot P(B)$ in situations in which two events are performed and it is clear that what happens on the first trial has no influence on what occurs on the second trial.

Independent events

If two events, A and B, are **independent**, then

$$P(A \cap B) = P(A) \cdot P(B)$$

Example 34 • A card is drawn from a deck of cards. Then the card is replaced, the deck is reshuffled, and a second card is drawn. What is the probability of an ace on the first draw and a king on the second?

Solution $P(A_1 \cap K_2) = P(A_1) \cdot P(K_2|A_1)$. However, knowing that an ace is drawn on the first draw yields no information about what occurs on the second draw, since the first card is replaced and the deck is reshuffled. Hence, $P(K_2|A_1) = P(K_2)$.

$$\begin{aligned} P(A_1 \cap K_2) &= P(A_1) \cdot P(K_2) \\ &= \left(\frac{4}{52}\right) \cdot \left(\frac{4}{52}\right) \\ &\approx .0059 \quad \bullet \end{aligned}$$

Example 35 • A basket contains two red balls and two white balls. A ball is drawn, inspected, and returned to the box. Then a second ball is randomly drawn. What is the probability of drawing two red balls?

Solution Let R_1 represent getting a red ball on the first draw and R_2 represent getting a red ball on the second draw. Since the ball was returned after the first draw, R_1 and R_2 are independent events.

$$P(R_1 \cap R_2) = P(R_1) \cdot P(R_2) = \frac{2}{4} \cdot \frac{2}{4} = \frac{1}{4} \quad \bullet$$

An important property associated with probability is that of **expectation** or **expected value.** If we toss a coin 100 times we would expect to get a head

$$100\left(\frac{1}{2}\right) = 50 \text{ times}$$

If we spin a spinner with 10 equal sectors 1000 times we would expect the spinner to stop on any given sector

$$1000\left(\frac{1}{10}\right) = 100 \text{ times}$$

The concept is perhaps most easily explored in the analysis of a simple game of chance. Suppose some poor benighted soul were persuaded to play the following game. A fair coin is flipped. If a head appears we receive \$5; if a tail appears our demented opponent receives \$2.

We investigate what happens if we play the game 100 times. In 100 flips of the coin we can expect approximately 50 heads and 50 tails. Hence we can expect a payoff of approximately 50(\$5) from the heads and a payoff of (50)(−\$2) from the tails. Our net profit would thus be (50)(\$5) + 50(−\$2) = \$150. Since the game was played 100 times, our average profit per game would be $\frac{150}{100}$ = \$1.50.

It is very important to observe that there is an alternate way to compute this average gain per game.

$$P(\text{winning } \$5) = P(H) = \frac{1}{2}$$

$$P(\text{winning } -\$2) = P(T) = \frac{1}{2}$$

$$\begin{aligned}(\$5)\cdot P(\text{winning } \$5) + (-\$2)P(\text{winning } -\$2) &= \frac{1}{2}\cdot \$5 + \frac{1}{2}\cdot -\$2 \\ &= \$2.50 - \$1 = \$1.50\end{aligned}$$

This second set of computations motivates the following definition of expected value.

Expected value

Suppose that there are n payoff values in a given experiment: $A_1, A_2, A_3, \ldots, A_n$. The expected value for that experiment is

$$A_1 \cdot P(A_1) + A_2 \cdot P(A_2) + \cdots + A_n \cdot P(A_n)$$

Example 36 • Let x be a variable representing the number of tails that can appear in the toss of three coins. Of course, x can assume values 0, 1, 2, or 3. Tabulating the results as in Table 5 assists in computing the expected value of x. The expected number of tails is $\frac{3}{2}$.

Table 5

x	$P(x)$	$xP(x)$
0	$\frac{1}{8}$	0
1	$\frac{3}{8}$	$\frac{3}{8}$
2	$\frac{3}{8}$	$\frac{6}{8}$
3	$\frac{1}{8}$	$\frac{3}{8}$
Total	$E(x) = \frac{12}{8} = \frac{3}{2}$	

•

Just for fun

Arrange two nickels and two dimes as shown in the figure. Can you move the coins, one at a time, by moving one space or by jumping over one coin (no two coins may occupy the same space at the same time) so that the nickels are on the right and the dimes are on the left? What is the minimum number of moves? Can you do the same if there is only one space between nickels and dimes?

One interpretation of the expected value is that it is the average payoff per experiment when the experiment is performed a large number of times.

Example 37 • A nationwide promotion promises a first prize of \$25,000, two second prizes of \$5000, and four third prizes of \$1000. A total of 950,000 persons enter the lottery.

(a) What is the expected value if the lottery costs nothing to enter?
(b) Is it worth the stamp required to mail the lottery form?

Solution (a) Since

$$P(\$25{,}000) = \frac{1}{950{,}000} \qquad P(\$5000) = \frac{2}{949{,}999}$$

and

$$P(\$1000) = \frac{4}{949{,}997}$$

the expected value is

$$(\$25{,}000)\frac{1}{950{,}000} + (\$5000)\frac{2}{949{,}999} + (\$1000)\frac{4}{947{,}997} = 4.1¢$$

(b) Hardly! •

Note again that expected value is not something to be expected, in the ordinary sense of the word. It is a long-run average of repeated experimentation.

Exercise set 5

R 1. A single card is drawn at random from a standard deck. Let $B = \{\text{the card is black}\}$, $H = \{\text{the card is a heart}\}$, and $C = \{\text{the card is a club}\}$.

(a) Describe in words a sample space for the experiment.
(b) How is the sample space changed if we have the additional information that a black card is drawn?
(c) Compute $P(H|B)$.
(d) Compute $P(C|B)$.
(e) Compute $P(B|C)$.

2. Roll a single fair die. Let $A = \{\text{the die shows less than 4}\}$ and $B = \{\text{the die shows an odd number}\}$. Compute

(a) $P(A|B)$ (b) $P(B|A)$

3. If $P(A) = .6$, $P(B|A) = .7$, and $P(B) = .6$, compute
   (a) $P(A \cap B)$ (b) $P(A|B)$
   (c) $P(\bar{B})$ (d) $P(A \cup B)$

T 4. Given the following table, compute the probabilities requested.

	C	*D*	*E*	*Total*
A	.20	.10	.05	.35
B	.30	.20	.15	.65
Total	.50	.30	.20	1

   (a) $P(A)$ (b) $P(E)$ (c) $P(B \cap C)$
   (d) $P(A \cap E)$ (e) $P(\bar{E})$ (f) $P(\bar{B})$
   (g) $P(C|A)$ (h) $P(A|C)$ (i) $P(B|D)$

5. A new low-flying missile has a probability of .9 of penetrating the enemy defenses and a probability of .7 of hitting the target if it penetrates the defenses. What is the probability that the missile will penetrate the defenses and hit the target?

6. A card is drawn from a standard deck of cards. What is the probability that it is a jack, given that it is a face card (that is, a king, queen, or jack)?

7. From an urn containing five red balls and three white balls, two balls are drawn successively at random, without replacement. What is the probability that the first is white and the second is red?

8. Assume that two cards are drawn from a standard deck of playing cards. What is the probability that a king is drawn, followed by an ace
   (a) if the first card is replaced before the second is drawn?
   (b) if the first card is not replaced before the second is drawn?

9. An ABC block is rolled on the floor 360 times. How many times would you expect the A to be on the top?

10. In a lottery, 200 tickets are sold for \$1 each. There are four prizes, worth \$50, \$25, \$10, and \$5. What is the expected value for someone who purchases one ticket?

11. A highway engineer knows that her crew can repair 4 miles of road a day in dry weather and 2 miles of road a day in wet weather. If the weather in her region is rainy 25% of the time, what is the average number of miles of repairs per day that she can expect?

C 12. Several students decide to play the following game for points. A single die is rolled. If it shows an even number, the student receives points equal to twice the number of dots showing. If it shows an odd number, the student loses points equal to 3 times the number of dots showing. What is the expected value of the game?

13. A box contains three red balls and four white balls. What is the probability of drawing two white balls
   (a) if the first is replaced before the second one is drawn?
   (b) if the first ball is not replaced?

14. In a certain college, 30% of the students failed mathematics, 20% failed English, and 15% failed both mathematics and English.
   (a) If a student failed English, what is the probability that she failed mathematics?
   (b) If a student failed mathematics, what is the probability that he failed English?
   (c) What is the probability that a student failed mathematics or English?
   (d) If a student did not fail mathematics, what is the probability that he failed English?
   (e) If a student did not fail English, what is the probability that she did not fail mathematics?

15. An urn contains the following balls: five colored red and white, three black and white, four green and white, six red and black, four red and green, and five black and green.
   (a) Given that you have drawn a ball that is partly green, what is the probability that it is partly white?
   (b) Given that the ball you have drawn is partly white, what is the probability that it is partly red?

16. A man who rides a bus to work each day determines that the probability that the bus will be on time is $\frac{7}{16}$. The probability that the bus will be 5 minutes late is $\frac{3}{16}$; 10 minutes late, $\frac{1}{4}$; and 15 minutes late, $\frac{1}{8}$. What is his expected waiting time if the man arrives at the bus stop at the scheduled time?

17. Two coins, each biased (or weighted) $\frac{1}{3}$ heads and $\frac{2}{3}$ tails, are tossed. The payoff is \$5 for matching heads. \$3 for matching tails, and $-\$2$ if they don't match. What is the expected value of the game?

Solution to introductory problem

Understand the problem. A couple plans to have either 3 or 4 children. They want exactly 2 girls. We want to find the probability of exactly 2 girls with 3 children and the probability of exactly 2 girls with 4 children.

Devise a plan. List all possible combinations in order of birth with 3 children and then with 4 children. Let B represent boy and G represent girl. Circle all those with exactly 2 girls.

Carry out the plan.

Three children		*Four children*		
BBB	(BGG)	BBBB	(BGGB)	(GBBG)
BBG	(GBG)	BBBG	(BBGG)	GBBB
BGB	(GGB)	BBGB	(BGBG)	GBGG
GBB	GGG	BGBB	(GBGB)	GGBG
		GBBB	(GGBB)	GGGB
				GGGG

$$P\text{ (exactly 2 girls from 3 children)} = \frac{3}{8}$$

$$P\text{ (exactly 2 girls from 4 children)} = \frac{6}{16} = \frac{3}{8}$$

Surprised?

Summary and review

Review to ensure that you understand the following terms as applied to probability:

- Experiment
- Outcome
- Sample space
- Relative frequency interpretation of probability
- Probability rule on a sample space
- Uniform sample space
- Event
- Simple event
- Compound event
- Probability of an event
- A priori probability
- $A \cup B$ (A or B)
- $A \cap B$ (A and B)
- Complement
- Mutually exclusive events
- Conditional probability
- Dependent events
- Independent events

Review the following concepts used in counting:

- Tree diagrams
- Fundamental Principle of Counting

Success in solving probability problems depends on the ability to use the correct formula. Do you know the conditions that allow you to use each of the following formulas?

$$P(A) = \frac{\text{number of times } A \text{ occurs}}{N}$$

$$0 \leq P(A) \leq 1$$

$$P(A) = \frac{1}{n(S)}$$

$$P(A) = \frac{n(A)}{n(S)}$$

$$E = A_1P(A_1) + A_2P(A_2) + \cdots + A_nP(A_n)$$

$$P(A \cup B) = P(A) + P(B)$$

$$P(A \cup B) = P(A) + P(B) - P(A \cap B)$$

$$P(\bar{A}) = 1 - P(A)$$

$$P(A|B) = \frac{P(A \cap B)}{P(B)}$$

$$P(A \cap B) = P(A) \cdot P(B|A)$$

$$P(A \cap B) = P(B) \cdot P(A|B)$$

$$P(A \cap B) = P(A) \cdot P(B)$$

Review exercise set 6

R 1. A number x is selected at random from the set of numbers

$$\{1, 2, 3, \ldots, 10\}$$

What is the probability that

(a) x is less than 5? (b) x is divisible by 3?
(c) x is even? (d) $x \geq 2$?
(e) x is less than 5 and even?
(f) $x < 7$?
(g) x is less than 5 or even?
(h) $x^2 > 3$?

2. A bag contains 6 red balls, 4 black balls, and 3 green balls. Tabulate a sample space for the following experiments.

(a) A single ball is drawn (a sample space with three outcomes).
(b) A single ball is drawn (a uniform sample space).
(c) A ball is drawn and pocketed. A second ball is drawn. (A tree diagram might be helpful.)
(d) A ball is drawn, its color is recorded, and it is replaced. A second ball is drawn. (Try a tree diagram.)
(e) A ball is drawn and pocketed; a second ball is drawn and pocketed; a third ball is drawn and pocketed. (Try a tree diagram.)

3. A ball is drawn from the bag in Exercise 2. What is the probability that it is

(a) red or black?
(b) blue?
(c) red or black or green?
(d) not red and not green?
(e) not black?
(f) green?
(g) not red or not black?
(h) red and black?
(i) not red or not green?
(j) not green?

T 4. A ball is drawn from the bag in Exercise 2, its color is recorded, and the ball is replaced. A second ball is drawn. What is the probability of

(a) two red balls?
(b) a red ball followed by a green ball?
(c) two black balls?

5. At the Fresh Air Farm Charity Bazaar, one booth sells 60 identically wrapped surprise packages, where 8 of the packages contain tape players, 12 of the packages contain electric razors, 16 of the packages contain rubber galoshes, and 24 contain a box of Kleenex.

(a) What is the probability of getting a tape player or an electric razor?
(b) What is the probability of getting something other than the Kleenex?

6. Suppose $P(A) = 0.35$, $P(B) = 0.51$, and $P(A \cap B) = 0.17$. Compute

(a) $P(\bar{A})$. (b) $P(A \cup B)$.

7. As the semester draws to a close, several of Professor Wheeler's students become desperate. They offer to play the following game for points. A single die is rolled. If it shows an even number, the student receives points equal to twice the number of dots showing. If the die shows an odd number, the student loses points equal to three times the number of dots showing. What is the expected value of the game for the student?

C 8. If two cards are drawn without replacement from a deck of cards, what is the probability that

(a) both cards are red?
(b) the two cards are the king of spades and then the queen of hearts?
(c) both cards will be aces?

9. The letters f, o, u, and r are written on four cards and placed in a hat. They are drawn one at a time and arranged according to the order drawn.

(a) Count the number of possible arrangements (without tabulating them).
(b) Tabulate all possible arrangements of these letters.
(c) What is the probability that the arrangement drawn will have the two vowels adjacent to one another?

10. A 13-card bridge hand is drawn from a standard deck. What is the probability that it consists of

(a) all hearts?
(b) first 6 red cards and then 7 black cards?
(c) all face cards (jack, king, queen)?

(d) first 4 hearts, then 6 diamonds, and then 3 other cards?

11. In Exercise 8, find the probability that one card is red and one is black.

12. In Exercise 4, what is the probability that one ball is red and one is black?

13. A shipment contains 96 good items and 4 defective items. Three items are drawn, one at a time, from the shipment.
    (a) What is the probability of 3 defective items?
    (b) What is the probability of no defective items?
    (c) What is the probability of 2 good items and 1 defective item?
    (d) What is the probability in part (a) if the items are replaced after each drawing?
    (e) What is the probability in part (b) if the items are replaced after each drawing?

Bibliography

1. Billstein, Rick. "A Fun Way to Introduce Probability." *AT* (January 1977), 39–42.
2. Booth, Ada. "Two-thirds of the Most Successful . . . ." *MT* (November 1973), 593–597.
3. Bruni, James V., & Helene J. Silverman. "Developing Concepts in Probability and Statistics—and Much More." *AT* (February 1986), 34–37.
4. Burns, Marilyn. "Put Some Probability in Your Classroom." *AT* (March 1983) 21–22.
5. Choate, Stuart. "Activities in Applying Probability Ideas." *AT* (February 1979), 40–42.
6. Enman, Virginia. "Probability in the Intermediate Grades." *AT* (February 1979), 38 –39.
7. Fennell, Francis. "Ya Gotta Play to Win: A Probability and Statistics Unit for the Middle Grades." *AT* (March 1984), 26–30.
8. Fey, J. T. "Probability, Integers, and Pi." *MT* (April 1971), 329–331.
9. Hoffman, Nathan. "Pascal's Triangle." *AT* (March 1974), 190–198.
10. Horak, Virginia M., & Willis J. Horak. "Take a Chance." *AT* (May 1983), 8–15.
11. Jones, Graham. "A Case for Probability." *AT* (February 1979), 37, 57.
12. Jordan, Arthur E. "Thank You, Mr. Pascal." *AT* (December 1979), 32–34.
13. Kasner, E., & J. Newman. "Buffon's Needle Problem." *Mathematics and the Imagination.* New York: Simon & Schuster, 1940.
14. Lai, Theodore. "Bingo and the Law of Equal Ignorance." *AT* (January 1977), 83–84.
15. Litwiller, Bonnie H., & David R. Durcan. "The Probability of Winning Dice Games." *MT* (September 1979), 458–461.
16. National Council of Teachers of Mathematics. "Chapter 5 (Probability)." *Twenty-Fourth Yearbook.*
17. National Council of Teachers of Mathematics "Chapter 8 (Probability in the Elementary School)." *Twenty-Seventh Yearbook.*
18. Niman, John, & Robert D. Postman. "Probability on the Geoboard." *AT* (March 1973), 167–170.
19. Reeves, Charles. "Volleyball and Probability." *MT* (October 1978), 595–596.
20. Starr, Norton. "A Paradox in Probability Theory." *MT* (February 1973), 166–168.
21. Rudd, D. "A Problem in Probability." *MT* (February 1974), 180–181.
22. Stone, Janine S. "Place Value and Probability (with Promptings from Pascal)." *AT* (March 1980), 47–49.
23. Tkaczyk, John. "A Problem in Probability." *MT* (November 1979), 595–597.

CHAPTER 14
The Uses and Misuses of Statistics

The management of Acme Manufacturing published an average annual salary for 50 employees of $17,720. The Trainsters Union, trying to organize the Acme plant, published an average annual salary of $12,000. Some of the employees made an informal survey and published an average annual salary of $16,000. The interesting point is that all were correct. How can this be?

When Aunt Jane asserts that smoking is not harmful to the health because Uncle Joe lived to be 88 years of age and smoked two packs of cigarettes every day of his adult life, Aunt Jane is using statistical thinking; that is, she has organized the data of her experience (Uncle Joe) and then made a statement on the basis of her data. However, she lacks an understanding of how much data are needed, how the data should be organized, and what conclusions are appropriate or inappropriate relative to the data. This chapter should help you to avoid making the types of mistakes that Aunt Jane made.

In addition to presenting a basic introduction to the field of statistics, we demonstrate some misuses of statistics and discuss some basic concepts that will aid in determining when statistical information is accurate.

If people on the street were asked for their interpretations of statistics, they might answer "Statistics is a hodgepodge of numbers." Or they might comment that statistical statements are used in an attempt to "improve" a statement that may be only partly true. These reactions are normal. We have been bombarded with so much misuse of statistics that many people have doubts about the field. Yet when we ourselves lack adequate information to defend our position, we often present our opponents with a mumbo jumbo of statistical facts. These tactics would not be possible if everyone had an adequate knowledge of statistics.

Statistics can be divided into two subdivisions: *descriptive statistics* and *inferential statistics*. **Descriptive statistics** includes those techniques used to summarize and describe the main characteristics of a set of data. Most of the material in this chapter can be classified as descriptive statistics.

In **inferential statistics,** a small sample is selected from a large population. Then the information from the sample is used to draw inferences about the population. For the purposes of this discussion, a **population** is any predetermined set of elements or items upon which we make observations or measurements. Since most populations are too large for the examination of every element, we examine a part of the population, called a *sample.* A **sample** is a group of measurements, observations, or objects selected from a larger group called the *population.* In much of statistics, a sample is used in which each element of a population has equally likely probability of being in the sample. Such a sample is called a **random sample.**

1 Summarizing data with frequency distributions

We are immersed daily in a torrent of numbers flowing in bubbling splendor from our televisions, our radios, our newspapers, our barber, and our favorite Uncle Al. Although data are a part of our daily life, it is evident that we often do not know how to organize, interpret, or understand the message being conveyed. In this section, we shall learn to organize and summarize data for better understanding.

Sets of data that are numerical in nature are called **quantitative data.** Throughout your studies and in everyday experience you will encounter quantitative data: price-to-earnings stock ratios, grade-point averages, lifespans in different civilizations, weights of members of Obesity Anonymous, salaries of employees, and so on.

The first objective of a would-be statistician is to make sense out of a large amount of data. Suppose you have collected the numbers in Table 1, which shows the number of colds experienced during a winter by a group of 30 elementary school children.

Table 1

7	1	1	0	3	4	5	5	3	2	3	3	6	6	2
4	2	1	0	0	3	4	5	6	3	1	4	1	3	4

A quick glance at this array of numbers tells us very little about what is implied about the group of people represented in the data. Closer observation indicates that the largest number of colds experienced was 7, and the smallest number experienced was 0. *The difference in the largest and smallest entry in the data is called the* **range** *of the data.* In this case, the range is $7 - 0 = 7$. To understand better the significance of this list of numbers, we might summarize it in a **frequency distribution.** To do this, we record the number of students who reported each number of colds.

Example 1 • Make a frequency distribution for the data in Table 1.

Solution From the summary in Table 2, it is easily seen that 3 colds was the number most often reported. The summary also shows how the number of colds was distributed through the 30 students. •

Table 2

Number of colds	*Tally*	*Frequency*
0	///	3
1	~~////~~	5
2	~~///~~	3
3	//// //	7
4	~~////~~	5
5	///	3
6	///	3
7	/	1
	Total	30

In a grouped frequency distribution, we cover the range of data by intervals of equal length and record the number of pieces of data that fall in each interval. There are three main steps in the construction of a grouped frequency distribution.

Steps in Making a Grouped Frequency Distribution

1. Choose the number and size of the classes into which the information will be grouped.
2. Make a record of each bit of data in a given class; that is, data is tabulated into the selected classes.
3. Count the data in each class. This number is the **frequency** of the class.

Example 2 • Table 3 gives the lengths of engagements (by the number of months) of 30 newly married students. Construct a grouped frequency distribution.

Solution The range of the data is

$$18 - 1 = 17$$

We arbitrarily select six classes for our grouping. Since 17 ÷ 6 is 2.833, the length of the classes (if the classes are of equal length) must be more than 2.833 in order to include all the data in six classes. Whenever feasible, it is desirable to have classes of equal integral length. Thus, we arbitrarily select the following class limits: 1–3, 4–6, 7–9, 10–12, and so on. The grouped frequency distribution is found in Table 4. •

Table 3

Lengths of engagements				
10	2	9	6	11
17	4	10	7	3
1	4	11	6	3
8	15	12	9	12
8	18	12	6	10
8	18	12	6	9

Table 4

Class intervals	*Tallies*	*Frequency*
1–3	////	4
4–6	~~////~~ /	6
7–9	~~////~~ //	7
10–12	~~////~~ ////	9
13–15	/	1
16–18	///	3

In Table 4, the class interval 7–9 includes all measurements for 6.5–9.5. If there were an engagement of 6.7 months, this entry would be placed in the class 7–9. However, an engagement of 9.6 months would go in the class 10–12. The numbers that are halfway between class limits (such as 6.5–9.5, 9.5–12.5, etc.) are called **class boundaries.**

The *length* of a class interval can be found by taking the difference in the boundaries of a class. The middle value of a class interval is called the **class mark.** It can also be computed as the sum of a lower class boundary and one-half of the length of the class interval. The class marks for the lengths of engagements are 2, 5, 8, 11, 14, and 17.

Difficulty arises when class limits are chosen so that they overlap. In this case, some decision must be made as to which class will contain the endpoint. For example, consider the following set of intervals.

200–300
300–400
400–500
500–600

In which interval would you place 500? This problem can be solved by writing the intervals

200–299
300–399
400–499
500–599

One of the ways by which erroneous conclusions can be obtained from grouped frequency distributions is the use of unequal intervals. Table 5 is a tabula-

Just for fun

Make a guess as to which digit in this calendar appears most often. List the digits 0 through 9 and tabulate the frequencies. Did you correctly guess the answer?

November						
S	*M*	*T*	*W*	*T*	*F*	*S*
				1	2	3
4	5	6	7	8	9	10
11	12	13	14	15	16	17
18	19	20	21	22	23	24
25	26	27	28	29	30	

tion of the data in Table 3. It seems that the most common length of engagement is in the interval 6–9. This is misleading. Why?

Table 5

Class	*Tallies*	*Frequencies*
1–3	////	4
4–5	//	2
6–9	~~////~~ ~~////~~ /	11
10–12	~~////~~ ////	9
13–15	/	1
16–18	///	3

Exercise set 1

R 1. In a transportation survey, bus riders on the Friday evening run were asked how many times they had ridden the bus that week. Summarize the data in a frequency distribution.

4	8	6	4
7	2	2	8
2	5	8	1
7	9	8	3
8	2	4	8
10	3	3	9

2. Park officials sought to understand the use of a municipal park. One evening 36 people were interviewed and their ages recorded.

(a) Summarize the resulting data in a grouped frequency distribution with seven intervals. Let the first interval be 5 to 14.

(b) What trend can you see in the data?

7	18	35	73	18	28
15	19	41	61	16	24
51	65	12	65	61	26
16	62	14	73	72	48
17	59	16	62	43	68
21	16	17	19	32	72

3. The following is a tabulation of the ages of mothers of the first babies born in Morningside Hospital in 1987.

Class	*Tallies*	*Frequency*
15–19	////	4
20–24	~~////~~ //	7
25–29	~~////~~	5
30–34	//	2
35–39	//	2

(a) Find class boundaries.
(b) Find class marks.
(c) Can you determine the number of mothers in the tabulation?
(d) Find the number of mothers younger than 29.5.
(e) Find the number of mothers who were at least 19.5 years of age.
(f) Find the number of mothers whose ages were between 19.5 and 34.5.

T 4. The following is a tabulation of salaries at the Acme Publishing House.

Salary	*Frequency*
0–$9999.99	4
10,000–19,999.99	16
20,000–29,999.99	32
30,000–39,999.99	10
40,000 and up	8

(a) How many salaries were tabulated?
(b) Can you find the class marks?
(c) How many salaries are less than $30,000?
(d) How many salaries are $20,000 or more?
(e) In what interval would you find a $20,000 salary?
(f) Can you find the range?
(g) What would you use for class boundaries?

5. The grades of 66 students in a mathematics course were recorded as follows. Make a frequency distribution, starting the first interval at 44 and using 7 intervals of minimum integral length.

97	73	44	78	75	74	88	82	92	80	73
79	73	83	82	88	94	96	54	65	67	87
84	72	59	98	54	75	62	69	68	64	62
56	82	63	88	77	75	72	66	94	72	47
75	72	78	84	86	85	95	57	64	66	95
76	49	90	85	76	77	76	62	92	91	81

For Exercises 6 through 10, group the data in the manner outlined in each exercise. Then use the tabulated data to complete parts (a) through (d).

(a) *Make a grouped frequency distribution.*
(b) *What is the range?*
(c) *Find class boundaries.*
(d) *Find class marks.*

C 6. The 25 scores below were achieved by a group of high school seniors on a mathematics placement test. Tabulate this information into five groups of minimum integral length, the first beginning with the first class at 450.

477	485	527	483	582
567	513	609	596	525
566	540	451	519	530
576	656	525	621	603
648	555	535	528	546

7. The amounts (rounded to the nearest $1) that a sample of 50 freshmen spent on textbooks per class during a fall semester are listed here.

33	41	35	53	42	47	41	31	38	37
30	38	37	33	41	35	42	50	41	38
39	42	41	40	40	38	37	41	45	48
35	36	35	38	33	39	40	40	47	38
37	38	37	34	35	44	44	46	40	39

Using six intervals (of minimum integral length), make a frequency distribution, with the first class beginning at 30.

8. The grades of 60 students in a mathematics course were recorded as follows.

96	71	43	77	74	73	87	81	91	79
78	72	82	81	87	93	95	53	64	66
83	71	58	97	53	74	61	68	67	63
55	81	62	87	76	74	71	65	93	71
74	71	77	83	85	84	94	56	63	65
75	48	89	84	75	76	75	61	91	90

Make a frequency distribution, starting the first class at 43 and using seven intervals of minimum integral length.

2 Summarizing data with graphs

■ Line graphs are often used to make predictions about the future. Based on the trend from 1980 to 1985, estimate world population in 2000. ■

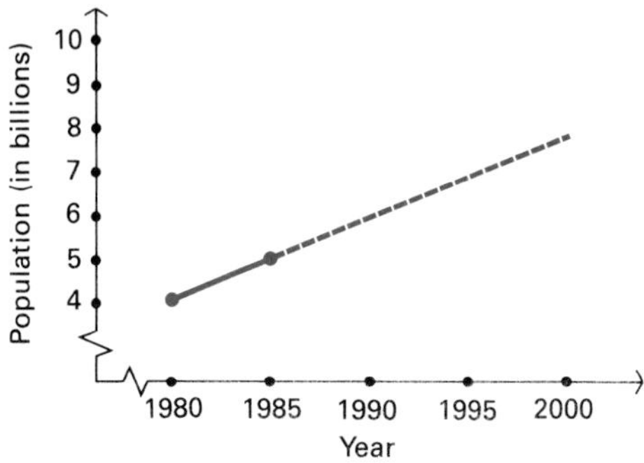

Tables and distributions are useful in summarizing data, but often they are not as useful as graphs. In fact, one of the best ways to summarize data for public consumption is through the use of visual displays or graphs.

In this section, we shall consider bar charts (or histograms), line graphs, circle charts, and frequency polygons. These graphs will enable you to present most data visually.

We have seen throughout this text that if we can draw a graph, picture, or diagram, much is gained in terms of understanding. There are several ways to represent graphically a conglomeration of data. One such representation is a **bar graph.** To construct a bar graph, first construct a frequency distribution or a grouped frequency distribution, whichever is appropriate. Then plot the frequencies on the y axis and the data values or intervals on the x axis. A bar is then drawn to show the relationship between the values and the frequencies.

Example 3 • Draw a bar graph representing the number of graduates as shown in Table 6.

Table 6

Year	*1975*	*1976*	*1977*	*1978*	*1979*	*1980*	*1981*	*1982*	*1983*	*1984*	*1985*	*1986*
Number of graduates	152	163	197	185	201	196	210	189	195	205	200	180

Solution Figure 1 is a bar graph of the data in Table 6. Notice that the number of graduates is measured on the vertical axis and that the years are given on the horizontal axis. The break in the vertical axis, denoted by ⦚, indicates that the scale is not accurate from 0 to 150. The height of each bar represents the number of students who graduated in a given year. To determine from the bar graph the number of students who graduated in 1978, locate the bar labeled 1978 and draw a horizontal line from the top of the bar to the vertical axis. The point where this horizontal line meets the vertical axis gives the number of graduating students. We see that about 185 students graduated in 1978. •

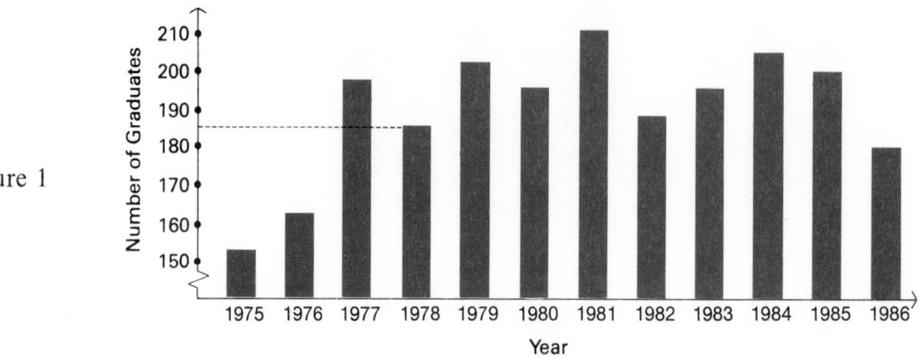

Figure 1

Bar graphs can also be used to create false impressions. By changing scales on graphs, it is possible to give different first impressions. For instance, although Figure 2(a) and (b) represent exactly the same data, (b) is drawn to give the impression that there are greater frequencies in each class.

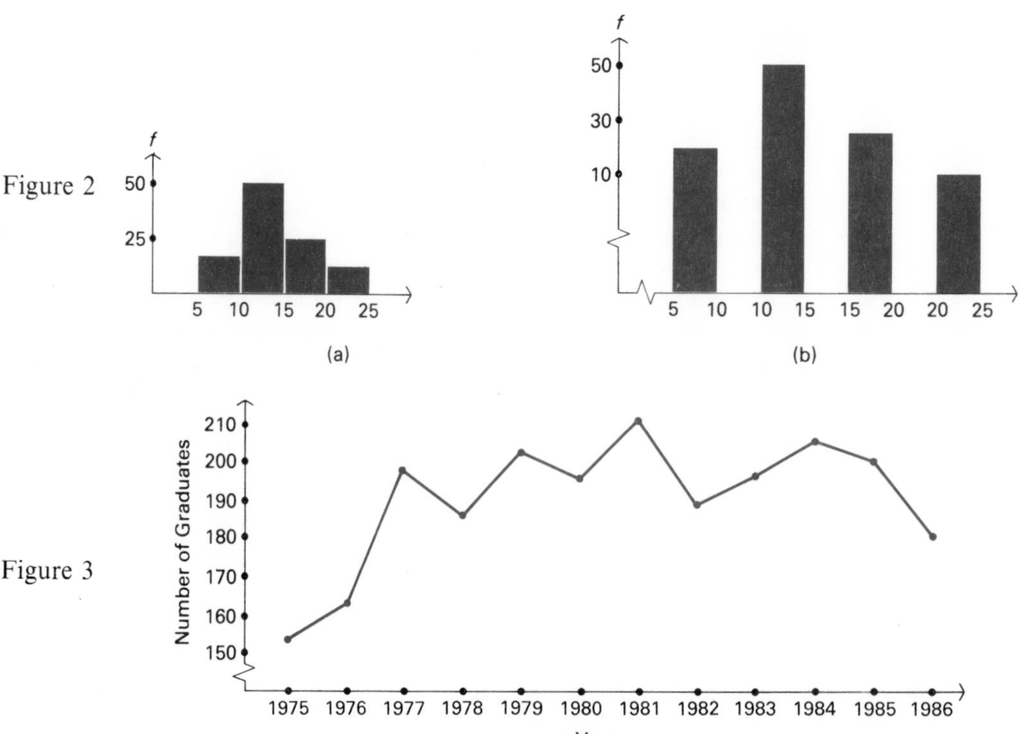

Figure 2

Figure 3

A line graph shows the fluctuations and emphasizes the changes that have taken place better than the bar graph. The line graph in Figure 3 represents the data in Table 6. Looking at this graph, we can readily see the changes that occur from year to year in the number of graduates.

There are several ways to manipulate a line graph so that it creates a false impression. Compare the graph shown in Figure 4(a) with the graphs shown in (b) through (d). In (b), the vertical axis has been stretched, making the graph appear steeper. In (c), the horizontal axis has been stretched, and the graph appears flatter. The effect on the graph is especially dramatic if one of the axes is stretched while the other is compressed. Notice the horizontal scale in (d). The same interval has been used to represent units of length 1, 2, and 4.

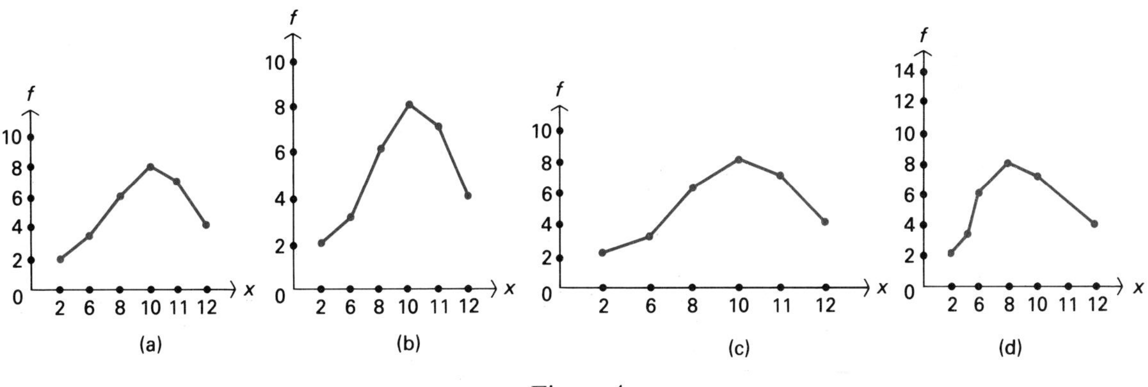

Figure 4

Sometimes, data are so erratic that a line graph would have little meaning. In this case, a *scattergram* improves the understanding of the data. A scattergram consists of data points and a line drawn to visually "fit" the data. Figure 5 shows data relating ACT scores in mathematics and students' averages in a class at the end of a semester. The scattering of the points about a trend line shows how the data correlate. From Figure 5, there is a **positive correlation:** as the ACT mathematics score increases, there seems to be a corresponding increase in the class

Figure 5

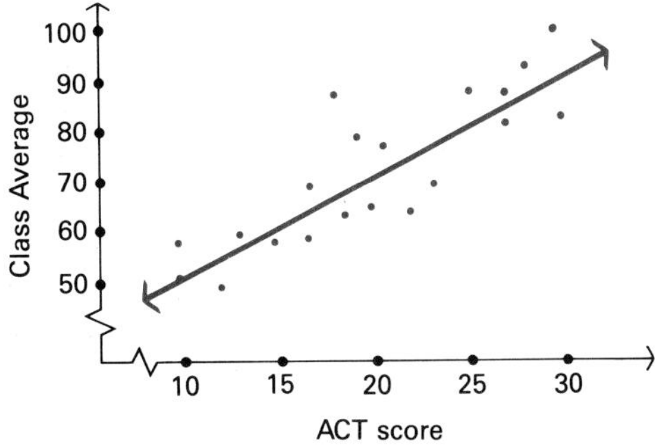

average of a student. If the slope of the line were negative, there would be a **negative correlation.** For perfect correlation, all points would be on the same line; that is, there would be no scattering about the line. When the points are so scattered that it is nearly impossible to draw a trend line, then we say that there is very little correlation.

A bar graph representing a grouped frequency distribution is called a **histogram.** To construct a histogram, first construct a grouped frequency distribution. Then represent each interval by marking its upper and lower class boundaries on the x axis. The sides of the rectangles (or bars) will be drawn at the class boundaries; the height of a rectangle will represent the frequency of an interval.

Example 4 • Draw a histogram of the data in Table 7.

Table 7

Class	*Tallies*	*Frequency*
15–19	////	4
20–24	~~////~~ //	7
25–29	~~////~~	5
30–34	//	2
35–39	//	2

Solution Figure 6 is the histogram that corresponds to the data in Table 7. •

A line graph used to represent a grouped frequency distribution is sometimes called a **frequency polygon.** To draw a frequency polygon, plot the midpoints (class marks) of the intervals versus the frequency of the intervals, and then connect the resulting points by straight-line segments. Finally, connect the first and last points to points on the horizontal axis located one-half of an interval from the end class boundary.

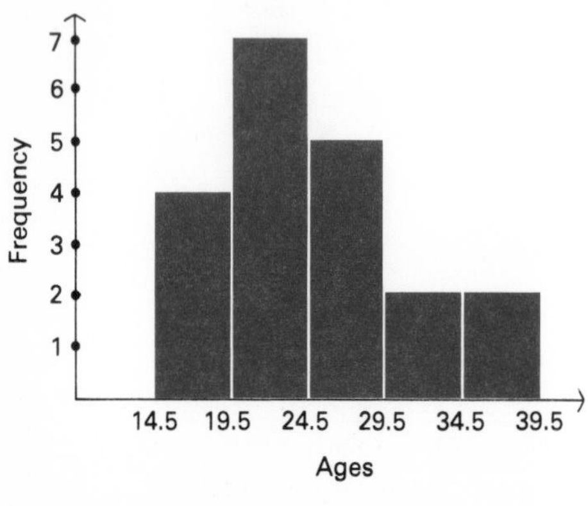

Figure 6

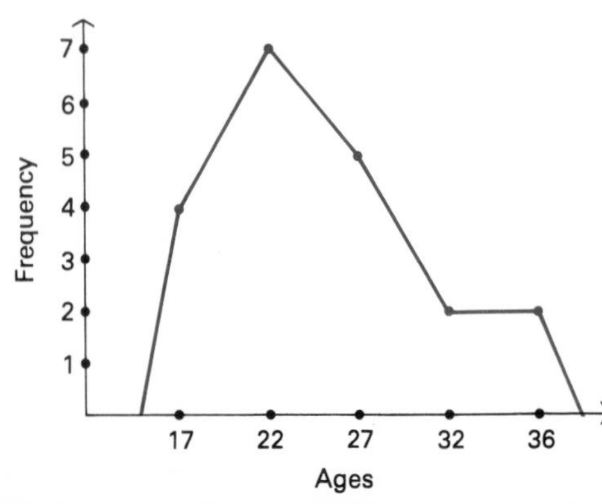

Figure 7

Just for fun

The scout left a trail marker of ten stones arranged as shown. However, his enemy secretly changed the marker to point in the opposite direction by moving three stones. What stones did he move and where did he move them?

Example 5 • Draw a frequency polygon for the data in Table 7.

Solution Figure 7 presents a frequency polygon for the grouped frequency distribution of Table 7. •

One of the simplest types of graphs is the **circle graph,** sometimes called a **pie chart.** It consists of a circle partitioned into sections, each sector of which represents a percentage of the whole.

Example 6 • Table 8 records examination grades in a class. Represent the data by a circle graph.

Solution The circle graph in Figure 8 is a visual representation of the data. We see that the largest percentage of the class made C's. In fact, more than half of the class made C's.

Table 8

Final-examination grades	*Frequency*
A	4
B	15
C	36
D	3
F	2
Total	60

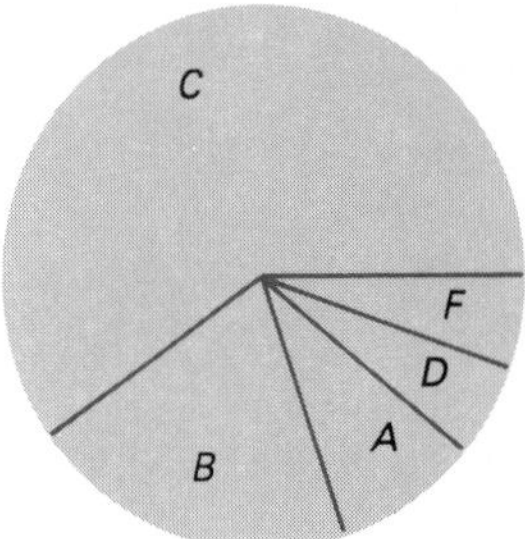

Final-examination grades

Figure 8

In constructing the circle graph shown in Figure 8, we use a protractor to obtain angle measurement. Since 36 out of 60 or $\frac{36}{60}$ or 60% of the students made C's, the sector representing C's has an angle of $0.60(360°) = 216°$. Similarly, $\frac{1}{15}$ of the grades were A's; so the section representing A's encompasses $(\frac{1}{15}) \cdot (360°) = 24°$. The remaining sections are constructed similarly. •

Exercise set 2

R 1. In the given pie chart or circle graph,

(a) what percent are professionals?
(b) what percent are craftsmen?
(c) what percent are managers or clerical?
(d) what percent are neither managers nor professionals?

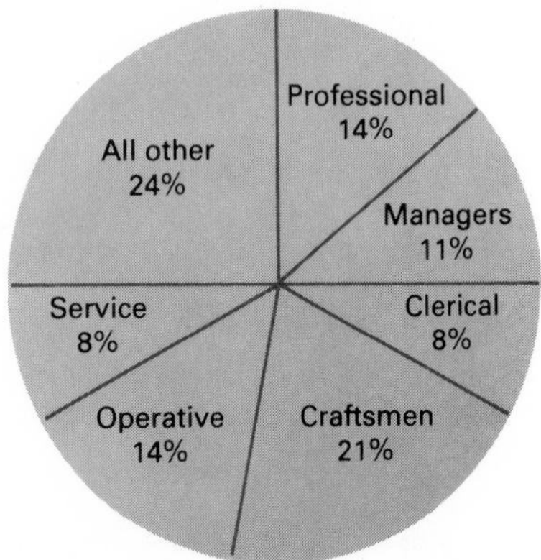

2. The following table gives the number of students who had a specific number of absences in a given semester.

Number of absences	*Frequency*
0	25
1	18
2	20
3	31
4	34
5	14
6	13
7	12
8	8
9	3
10	1

(a) Display these data with a bar graph.
(b) Represent these data with a line graph.

3. Consider a grouped frequency distribution defined by the given table.

(a) Find the class marks.
(b) Construct a histogram.
(c) Construct a frequency polygon.

Class	*Frequency*
20–29	1
30–39	2
40–49	4
50–59	5
60–69	9
70–79	6
80–89	3

4. Alabaster College has 1426 students. The given table presents a tabulation of their ages.

(a) Present the data as a frequency polygon.
(b) Present the data as a histogram.

Age	*Number of students*
15–19	562
20–24	450
25–29	350
30–34	58
35–39	6

5. (a) Indicate on a bar graph a comparison of the number of students in freshman mathematics courses majoring in the following academic fields.

Academic fields	*Number of students*
Business administration	110
Social sciences	100
Life sciences	60
Humanities	30
Physical sciences	60
Elementary education	140

(b) Display the data given in part (a) with a circle graph.
(c) Draw a line graph for these data.

6. The new vice president of the Seashore Oil Company claims that production has doubled during the first 12 months of his tenure. To present this fact to the board of directors, he has the following graph prepared.

What is misleading about this graph? (*Hint:* The viewer mentally compares volumes. What happens to the volume of a cylinder if you double its height and its radius?)

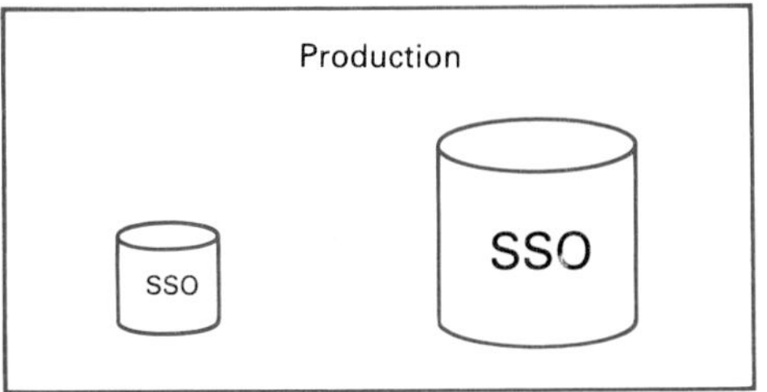

7. These circle graphs record the contributions to federal candidates for office in the late 1970's.

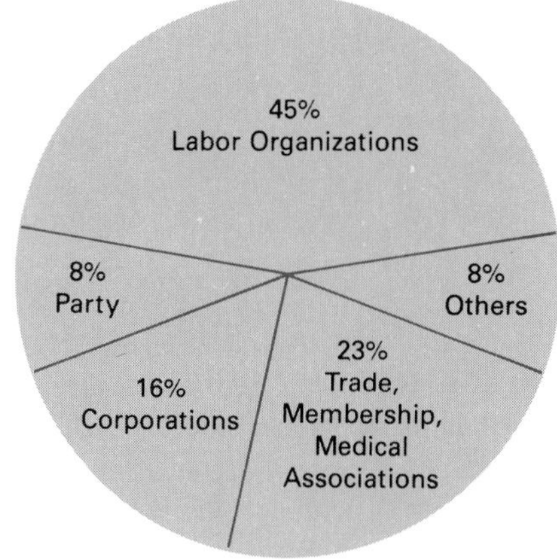

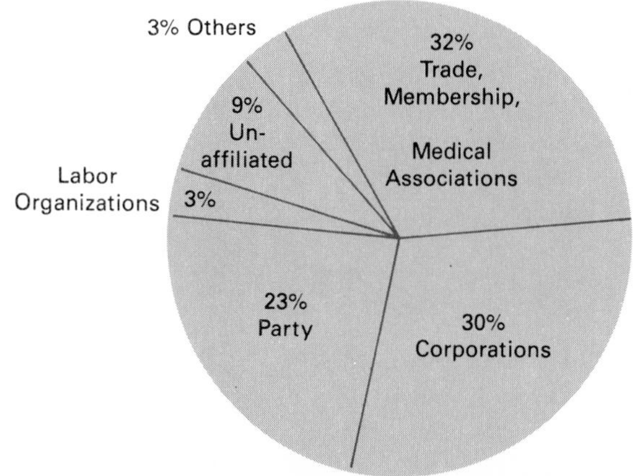

(a) Compare the percent of contributions to Democratic and Republican candidates that came from labor organizations; from corporations.
(b) What dollar amount of support for Democratic candidates came from the party?
(c) What dollar amount of support for Republican candidates came from corporations?

In the next three exercises, use the data given in Exercise Set 1 of this chapter to (a) draw a histogram and (b) draw a frequency polygon.

T 8. Exercise 8

9. Exercise 7

10. Exercise 9

11. Use a protractor to construct a pie chart showing the percentage of women who work in the various occupations. These percentages are given below.

Professional	16%
Managers	4%
Clerical	35%
Craftsmen	2%
Operative	14%
Service	17%
All other	12%

C 12. Given the histogram shown, tabulate a frequency distribution and find the class marks.

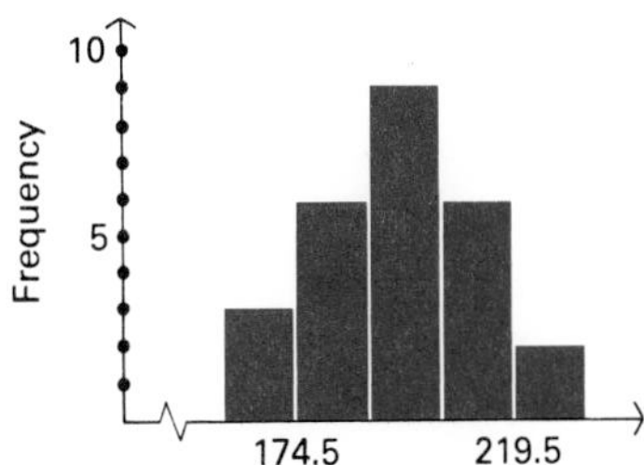

13. Given the frequency polygon shown, tabulate a grouped frequency distribution.

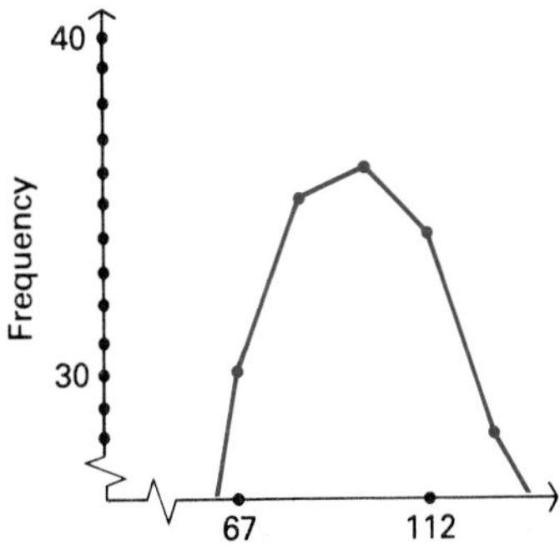

14. The following table gives a sampling of attendance at the Redshirts' baseball games, along with the average temperature (°F) during the game.

Temperature (F)	*Attendance*
40°	15,000
50°	48,000
62°	50,000
48°	18,000
60°	24,000
100°	60,000
72°	80,000
70°	66,000
84°	78,000
65°	40,000

(a) Plot the given points.
(b) Draw a trend line.
(c) Does there appear to be a correlation between temperature and attendance?
(d) In (c), what kind of correlation?

3 What is average?

■ Sandra receives 69, 71, 78, 82, and 73 on her five tests in Math 102. She gives her average grade as 74.6, but her friend Sam claims her average is 73. Which average is correct? In this section, we learn that both averages are correct. Sandra found the mean and Sam the median. ■

One common use of statistics is comparing sets of data by means of averages (or, more accurately, **measures of central tendency**). Three measures are in general use—the **arithmetic mean,** the **median,** and the **mode.** The fact that there are these three (as well as others) often leads to misuses of statistics. One measure may be quoted, and the reader automatically thinks of another. When a measure is quoted, immediately ask the question, "Which one?"

In the preceding section, we discussed grouping data and the graphical techniques for describing and analyzing data. However, there are occasions when even these tools are too cumbersome. Often all that is needed is a single number that estimates the location of the center of a set of data. Such a number is called a **measure of central tendency** (or an *average*). In this section, we discuss the **mean,** the **median,** and the **mode,** three measures of central tendency.

point of interest

Consider a seesaw consisting of a plank and a support called the *fulcrum.* The mean of 4, 6, 7, 11, and 12 is 8. If we place unit weights at 4, 6, 7, 11, and 12, the seesaw will balance if the fulcrum is at 8.

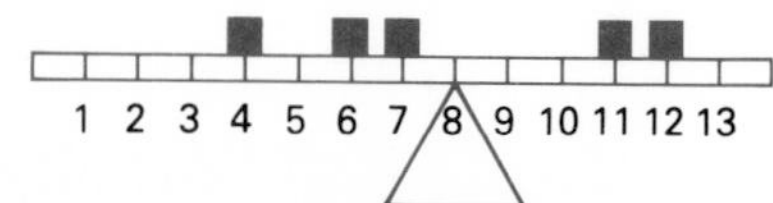

The most widely used measure of central tendency is the *arithmetic mean* (sometimes called *arithmetic average*). The arithmetic mean of a set of n measurements is the sum of the measurements divided by n.

Arithmetic mean

> Consider n measurements $x_1, x_2, x_3, \ldots, x_n$. The formula for the arithmetic mean, denoted by $\bar{x}$, is
>
> $$\bar{x} = \frac{x_1 + x_2 + x_3 + \cdots + x_n}{n}$$

Example 7 • Find the arithmetic mean of 8, 16, 4, 12, and 10.

Solution

$$\bar{x} = \frac{8 + 16 + 4 + 12 + 10}{5} = 10 \quad \bullet$$

Example 8 • Find the arithmetic mean of 25, 25, 25, 25, 30, 30, 30, 40, 40, 40, 40, 50.

Solution

$$\bar{x} = \frac{25 + 25 + 25 + 25 + 30 + 30 + 30 + 40 + 40 + 40 + 40 + 50}{12}$$

$$= \frac{25(4) + 30(3) + 40(4) + 50(1)}{4 + 3 + 4 + 1}$$

$$= \frac{400}{12} = 33\tfrac{1}{3} \quad \bullet$$

In the preceding example, observe that the 4, 3, 4, and 1 are the frequencies of the 25, 30, 40, and 50, respectively. Note that the mean is obtained by multiplying each value by the frequency of occurrence of the value and by dividing the sum of these products by the sum of the frequencies. Let us now generalize the formula for finding the arithmetic mean to include frequencies of the observations.

Arithmetic mean

> Let $x_1, x_2, \ldots, x_m$ be different measurements.
>
> $$\bar{x} = \frac{x_1 f_1 + x_2 f_2 + x_3 f_3 + \cdots + x_m f_m}{f_1 + f_2 + f_3 + \cdots + f_m}$$
>
> where f_i is the frequency of x_i for $i = 1, 2, 3, \ldots, m$.

Example 9 • Find the arithmetic mean of the following set of data.

x	4	14	24	34
f	2	8	20	10

Solution

$$\bar{x} = \frac{4(2) + 14(8) + 24(20) + 34(10)}{2 + 8 + 20 + 10} = \frac{940}{40} = 23.5 \quad \bullet$$

Sometimes, information for finding the arithmetic mean is placed in tabular form, as seen in the following example.

Example 10 • Find the arithmetic mean of the data in Table 9.

Solution

$$\bar{x} = \frac{495}{20} = 24.75 \quad \bullet$$

Table 9

x	f	xf
17	4	68
22	7	154
27	5	135
32	2	64
37	2	74
Total	20	495

A small company has four employees, with annual salaries of \$5500, \$6053, \$7144, and \$7553. The president of the company has an annual salary of \$35,000. The mean of the five salaries is \$12,250. This is a true average as given by the arithmetic mean, but most people would not accept it as a meaningful measure of central tendency. The median salary is more representative, because of one large salary.

The median of a set of observations is the middle number when the observations are ranked according to size.

Median

> If $x_1, x_2, x_3, \ldots, x_n$ is a set of data placed in increasing or decreasing order, the median is the middle entry, if n is odd. If n is even, the median is the mean of the two middle entries.

Example 11 • Consider the set of five measurements 7, 1, 2, 1, 3. Arranged in increasing order, they may be written as 1, 1, 2, 3, 7. Hence, the median is 2. •

Example 12 • The array

25, 2, 5, 6, 5, 23, 7, 10, 22, 15, 21, 23

can be arranged in decreasing order as

25, 23, 23, 22, 21, 15, 10, 7, 6, 5, 5, 2

so the median is

$$\frac{15 + 10}{2} = 12.5 \quad \bullet$$

The third measure of central tendency is called the *mode*, the measurement that appears most often in a given set of data.

Mode

> The mode of a set of measurements is the observation that occurs most often. If every measurement occurs only once, then there is no mode. If two measurements occur with the same frequency, the set of data is *bimodal*. It may be the case that there are three or more modes.

Example 13 • Baseball caps with the following head sizes were sold in a week by the Glo-Slo Sporting Goods Store: 7, $7\frac{1}{2}$, 8, 6, $7\frac{1}{2}$, 7, $6\frac{1}{2}$, $8\frac{1}{2}$, $7\frac{1}{2}$, 8, $7\frac{1}{2}$. Find the mode head size.

Solution The mode is $7\frac{1}{2}$—it occurs four times, more times than any other size. •

Example 14 • In one series of games against the Dodgers, the Reds won six of seven games by the following scores (see Table 10). Find the mean, median, and mode of these scores.

Table 10

Dodgers	2	6	1	15	4	2	2
Reds	4	7	2	1	5	3	3

Solution When the mean scores are computed, the following results are obtained.

$$\text{Dodgers' mean score} = 4.57$$

$$\text{Reds' mean score} = 3.57$$

Although the Reds won the series impressively, the Dodgers' mean score was substantially higher. In this case, the mean is not a good average to use because the Dodgers had one game whose extraordinarily high score biased the mean. In such cases, it is often better to use the median.

Reds scores (placed in order): 1 2 3 ③ 4 5 7

↑ median ↓

Dodgers' scores (placed in order); 1 2 2 ② 4 6 15

In this case, then, the median offers a better measure with which to make a comparison of the scores. The mode score for the Reds is 3, and the mode score for the Dodgers is 2. •

The decision as to which measure of central tendency to use in a given situation is not always easy to make. When extraordinarily large or small values are included in the data set, the median is usually better than the mean. However, the mean is the average most often used, since it gives equal weight to the value of each measurement. The mode is used when the "most common" measurement is desired.

Just for fun

The average salary of six office workers in the XYZ Corporation is \$14,000. John remembers five of the six salaries: \$10,000, \$13,000, \$19,000, \$16,000, and \$12,000. If the average is the median, describe for John the missing salary. Do the same if the missing salary is the mean.

Unfortunately, people use the average that is suitable to the objectives they hope to accomplish. A factory might report an average income of \$15,000, while the union representative might report only \$12,000. Don't be misled. Determine whether the statistic is the mean, the median, or the mode.

Exercise set 3

R 1. Compute the arithmetic mean, the median, and the mode (or modes if any) for the given sets of data.

(a) 3, 4, 5, 8, 10
(b) 4, 6, 6, 8, 9, 12
(c) 3, 6, 2, 6, 5, 6, 4, 1, 1
(d) 7, 1, 3, 1, 4, 6, 5, 2
(e) 21, 13, 12, 6, 23, 23, 20, 19
(f) 18, 13, 12, 14, 12, 11, 16, 15, 21

2. An elevator has a capacity of 15 people and a load limit of 2250 lb. What is the mean weight of the passengers if the elevator is loaded to capacity with people and weight?

3. Find the mean of the given distribution.

x	*Frequency*
10	2
20	6
30	8
40	4

4. Make up a set of data with four or more measurements, not all of which are equal, with each of the following characteristics.

(a) The mean and median are equal.
(b) The mean and mode are equal.
(c) The mean, median, and mode are equal.
(d) The mean and median have values of 8.
(e) The mean and mode have values of 6.
(f) The mean, median and mode have values of 10.

5. The mean score of a set of eight scores is 65. What is the sum of the eight scores?

6. The mean score of nine of ten scores is 81. The tenth score is 100. What is the mean of the ten scores?

7. Which of the three averages should be used for the following data?

(a) The average salary of four salesmen and the owner of a small store
(b) The average height of all male students in We-Fail High School
(c) The average dress size sold at Acme Apparel

8. At the initial meeting of Obesity Anonymous, the weights of the members were found to be 220, 275, 199, 246, 302, 333, 401, 190, 286, 254, 302, 323, 221.

   (a) Compute the mean, median, and mode of the data.
   (b) Which measure is most representative of the data?

T 9. The weights in kilograms of the smallest members of the Lo-Ho College football squad are as follows: 75, 60, 62, 94, 78, 80, 72, 74, 76, 89, 95, 98, 97, 80, 98, 91, 96, 90, 84, 73, 80, 92, 94, 96, 99, 84, 60, 68, 74, 80, 92, 96, 88, 74, 84, 94, 72, 76, 64, 80.

   (a) What is the mean weight of the "small" football squad?
   (b) What is the median weight?
   (c) What is the modal weight?
   (d) If you were a sportswriter assigned to do a story on this squad, how would you describe the (average) weight?

10. The given grades were recorded for a test on Chapter 9. Find the arithmetic mean.

Score	*Frequency*
100	3
90	5
80	7
70	15
60	14
50	3
40	3

11. The table shows the distribution of scores on a test administered to freshmen at Laneville College. Find the arithmetic mean.

Score	*Frequency*
140–149	3
130–139	4
120–129	8
110–119	13
100–109	4
90–99	2
80–89	0
70–79	1

12. The President of We-Work Factory draws a salary of \$100,000 per year. Four foremen have salaries of \$20,000 each. Twenty laborers have salaries of \$10,000 each. Discuss each of the following.

   (a) The President says the average salary is \$15,200.
   (b) The union says the average salary is \$10,000.
   (c) Which average is more representative of the factory salaries?

13. If 99 people have a mean income of \$4000, how much is the mean income increased by the addition of a man with an income of \$150,000?

14. A student has an average of 85 on nine tests. What will she need to make on the tenth test to have an average of 90?

15. Using the class mark as the variable, find the mean for the following distributions in Exercise set 2.

   (a) Exercise 3. (b) Exercise 4.

16. (a) What is the median of the data given by the histogram below?
   (b) What is the mean?

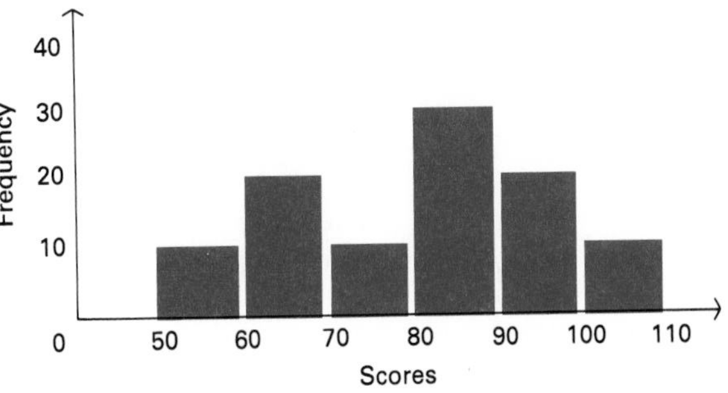

C 17. Let the letter x represent the values in a set of data. If $y = (x - a)/b$ (or $x = by + a$) where a and b are any numbers ($b \neq 0$), then the mean of the x's is b times the mean of the y's, plus a; that is $\bar{x} = b\bar{y} + a$.

   (a) In Exercise 11, obtain a new set of data y from the old data x using $y = (x - 114.5)/10$. Compute $\bar{y}$, and then $\bar{x}$. Compare with your answer in Exercise 11.
   (b) In Exercise 10, obtain new y data using $y = (x - 70)/10$. Compute $\bar{y}$ and then $\bar{x}$. Compare with your answer in Exercise 10.

18. Write a BASIC program to input a list of data and output the mean.

19. Modify Exercise 18 to include a frequency distribution.

20. (a) Run the program in Exercise 18 for the data in Exercise 9.
   (b) Run the program in Exercise 19 for the data in Exercise 10.

4 How to measure scattering

■ Misuses of statistics frequently involve a disregard for the *dispersion* or *scattering* of the observations. The fact that the average salary of the Do-Good Company exceeds the average salary of the We-Fix-Um Company does not mean that the salaries of the Do-Good Company are superior to those of the We-Fix-Um Company. Compare four monthly salaries: \$3000, \$3100, \$3100, and \$20,000, with a mean of \$7300; and four salaries: \$6800, \$6900, \$7000, and \$7200, with a mean of \$6975. The mean salary of the first company exceeds the mean salary of the second company, but the lowest salary of the second company is better than all the salaries except the largest salary of the first company.

Joe Smith had a score of 450 on an intelligence test. This score is considered exceptional because it is well above the mean of 400. Is it? The answer is *no* because, on this test, 45% of the scores are between 450 and 550. To be exceptional on this test, one must score above 500.

The average depth of the Cahaba River is 1 ft. This river should be a nice river in which to go wading. Wait a minute! There are many shallow areas, but there are also a number of potholes 15 to 16 ft. deep. ■

The preceding misuses of statistics emphasize the need for a measure for dispersion or scattering of data.

There are several ways to measure dispersion of observations. The measure of scattering that is easiest to calculate is the range, introduced in Section 1.

Example 15 • For the set of data, 7, 3, 1, 15, 41, 74, and 35, the range is 73 since $74 - 1 = 73$. •

Although the range is easy to obtain, it is not always a good measure of dispersion because it is so radically affected by a single extreme value. For example, suppose the 74 in the set of observations listed previously was miscopied and listed as 24 instead. Note that the range changes from 73 to 40.

Since the range is affected significantly by extreme values, other measures of dispersion are preferable. In this section, we consider **variance,** denoted by s_x^2, and **standard deviation,** s_x, since these measures of scattering are satisfactory for inferential statistics. (*Note:* Some books use σ_x^2 for population variance.)

Variance

Variance for a set of data can be obtained in three easy steps.

1. Compute the difference of each observation from the arithmetic mean.
2. Square each difference.
3. Divide the sum of the differences squared by n, where n is the number of observations.

Example 16 • Compute the variance of 112, 108, 114, 100, and 116.

Solution (a) Compute the mean of the data,

$$\bar{x} = \frac{x_1 + x_2 + \cdots + x_n}{n}$$

$$\frac{112 + 108 + 114 + 100 + 116}{5} = 110$$

(b) Determine how much each piece of data x deviates from the mean; that is, compute $x - \bar{x}$.

Data	*Deviation from the mean*	*Square of the deviation*
112	$112 - 110 = 2$	4
108	$108 - 110 = -2$	4
114	$114 - 110 = 4$	16
100	$100 - 110 = -10$	100
116	$116 - 110 = 6$	36

(c) Square the deviation of each data value; that is, compute $(x - \bar{x})^2$.

(d) Find the arithmetic mean of the squares of the deviations (divide by n).

$$s_x^2 = \frac{(x_1 - \bar{x})^2 + (x_2 - \bar{x})^2 + \cdots + (x_n - \bar{x})^2}{n}$$

$$s_x^2 = \frac{4 + 4 + 16 + 100 + 36}{5} = 32 \quad \bullet$$

These computations are indicated by the following formula for variance.

Formula for variance

Variance, denoted by s_x^2, is

$$s_x^2 = \frac{(x_1 - \bar{x})^2 + (x_2 - \bar{x})^2 + (x_3 - \bar{x})^2 + \cdots + (x_n - \bar{x})^2}{n}$$

where $\bar{x}$ is the mean of the observations.

Another measure of dispersion or variation is the *standard deviation.* The standard deviation is the square root of the variance and has the advantage that it is expressed in the same units as the original data.

Standard deviation

The standard deviation of the data, $x_1, x_2, \ldots, x_n$, with the mean $\bar{x}$, is given by

$$s_x = \sqrt{\frac{(x_1 - \bar{x})^2 + (x_2 + \bar{x})^2 + \cdots + (x_n - \bar{x})^2}{n}}$$

Standard deviation, denoted by s_x, is the positive square root of the variance.

Example 17 • Find the standard deviation for the data 5, 7, 1, 2, 3, and 6, using Table 11.

Table 11

	x	$x - \bar{x}$	$(x - \bar{x})^2$
	5	1	1
	7	3	9
	1	−3	9
	2	−2	4
	3	−1	1
	6	2	4
Total	24		28

Solution

$$\bar{x} = \frac{24}{6} = 4$$

$$s_x^2 = \frac{28}{6} = 4.67$$

The standard deviation is $\sqrt{4.67}$, which is approximately 2.16. •

Again, a more general formula can be used when the observations are repeated with given frequencies.

Variance

Suppose $x_1, x_2, \ldots, x_m$ have respective frequencies $f_1, f_2, \ldots, f_m$.

$$s_x^2 = \frac{(x_1 - \bar{x})^2 f_1 + (x_2 - \bar{x})^2 f_2 + \cdots + (x_m - \bar{x})^2 f_m}{n}$$

where $n = f_1 + f_2 + \cdots + f_m$.

Example 18 • Find the variance and standard deviation of the distribution as tabulated in Table 12.

Table 12

x	f	xf	$x - \bar{x}$	$(x - \bar{x})^2$	$(x - \bar{x})^2 f$
2	3	6	−4	16	48
4	4	16	−2	4	16
6	6	36	0	0	0
8	4	32	2	4	16
10	3	30	4	16	48
	20	120			128

$$\bar{x} = \frac{120}{20} = 6$$

$$s_x^2 = \frac{128}{20} = 6.4$$

$$s_x = \sqrt{6.4} \approx 2.53 \quad \bullet$$

One of the most common misuses of statistics is the making of inappropriate comparisons.

Example 19 • Juan made scores of 90, 82, 70, 61, and 94 on five tests. Of course, Juan did his best work relative to the rest of the class on the last test.

Or did he? What do we know about the rest of the class? Maybe everyone in the class made a higher score than 90 on the last test. •

The preceding example illustrates a need for what are called **z scores.**

z scores

> A score or measurement, denoted by x, from a population with mean $\bar{x}$ and standard deviation s_x has a corresponding **z score** given by
>
> $$z = \frac{x - \bar{x}}{s_x}$$
>
> the number of standard deviations from the mean.

The z score is a measurement in **standard units** or without units. For example, if x and $\bar{x}$ are in feet, then s_x is in feet, and the division eliminates the units. Consequently, z scores are of value in comparing two sets of data with different units. In many comparisons, the mean and standard deviation of the population are not known, and the z score is approximated by using the mean and standard deviation of a sample.

Example 20 • Teresa scores a 76 on the entrance test at school X and an 82 at school Y. At which school did she have the best score?

Solution To answer this question, we need to know that the mean score at school X was 70 with a standard deviation of 12 and the mean score at school Y was 76 with a standard deviation of 16. The z scores are then as follows:

$$\text{School X:} \quad z = \frac{76 - 70}{12} = 0.5$$

$$\text{School Y:} \quad z = \frac{82 - 76}{16} = 0.375$$

Just for fun *In a class of six students, each guesses the number of pennies in a jar. The six guesses are 52, 59, 62, 65, 49, and 42. One guess is 12 away, and the others are 1, 4, 6, 9, and 11 away. How can you use a statistic to help you find the number of pennies in the jar?*

Since 0.5 is greater than 0.375, Teresa's score at school X was superior in comparison with others who took the test. •

Exercise set 4

R 1. For the given sets of observations, find the range, the variance, and the standard deviation.

(a) 6, 8, 8, 14
(b) 10, 12, 13, 14, 16
(c) 1, 4, 5, 7, 13
(d) 80, 75, 80, 70, 80
(e) 15, 17, 19, 23, 26
(f) 16, 14, 12, 13, 15, 18, 24, 8, 10, 4
(g) 18, 47, 64, 32, 41, 92, 84, 27, 14, 12

2. Compute s_x^2.

x	10	14	18	22
f	4	6	8	2

3. Compute the variance for this frequency distribution.

x	11	13	18	21
f	5	6	7	2

4. The following data show the miles per gallon reported by owners of five small automobiles from different manufacturers.

	Manufacturer				
	A	*B*	*C*	*D*	*E*
Miles per gallon	18	18	24	21	18
	19	18	16	18	18
	20	20	18	19	19
	21	21	20	18	27
	22	24	22	20	18
	22	19	24	21	18

(a) Which sample suggests the best gasoline mileage?
(b) Which sample has the lowest standard deviation?
(c) Which manufacturer has the least variability in gasoline mileage?
(d) Is a large or small variability desirable?

5. Find the standard deviation of the given data.

x	*Frequency*
1	10
4	20
7	30
10	40

6. The following table shows the distribution of scores on a test administered to freshmen at Hard College. Find the standard deviation.

Score	*Frequency*
144	3
134	4
124	8
114	13
104	4
94	2
84	0
74	1

7. A pollster tabulated the ages of 30 users of a vitamin pill designed to make one feel young. The results are shown in the table. Find the standard deviation.

Age	*Frequency*
25	1
35	2
45	4
55	5
65	9
75	6
85	3

8. Find the standard deviation for the distributions in Exercise 4, p. 614.

T 9. You have just given an examination to your students. What do you know about your students' performance on the test if the distribution of scores has

(a) a large range but a small standard deviation?
(b) a mean higher than the median?
(c) a mean lower than the median?
(d) a small range but a large standard deviation?
(e) a standard deviation of 0?

10. When John entered his profession in 1975, the average salary in the profession was $11,500 with a standard deviation of $1000. In 1985, the average salary in the profession was $21,000 with a standard deviation of $3000. John made $11,000 in 1975 and $20,000 in 1985. In which year did he do better in comparison to the rest of the profession?

11. (a) What effect does adding the same amount to each observation have on the standard deviation? Test your conjecture by adding 5 to each member of the set 10, 12, 13, 14, 16.
(b) What effect does dividing each entry by the same number have on the standard deviation? Check your conjecture by dividing 80, 85, 80, 70, 80 by 5.
(c) What effect does subtracting a number from each observation and then dividing each result by a number have on the standard deviation? Check your conjecture by subtracting 75 from each entry and then dividing by 5 for 80, 75, 80, 70, 80.

C 12. If $y = (x - a)/b$ (or $x = by + a$) where a and b are any numbers with $b \neq 0$, then the standard deviation of the x's equals b times the standard deviation of the y's; that is, $s_x = bs_y$.

(a) In Exercise 6, let $y = (x - 114)/10$ where x represents the data of Exercise 6. Compute s_y and then s_x. Compare with the answer for s_x in Exercise 6.
(b) In Exercise 7, let $y = (x - 45)/10$. Compute s_y and then s_x. Compare with the answer for s_x in Exercise 7.

13. It can be shown with a lot of algebra that the standard deviation of n data values can be computed as follows. Let A be the sum of the data values and B the sum of the squares of the data values.

$$s = \frac{1}{n}\sqrt{nB - A^2}$$

Verify that the formula works for Exercises 1(a) through (d).

14. (a) Write a BASIC program to find the standard deviation of a set of data. (*Hint:* See Exercise 13.)
(b) Modify the program in (a) to find the standard deviation of a frequency distribution.

15. Use the BASIC programs to find the standard deviation in

(a) Exercise 6 (b) Exercise 1(g).

*5 The normal distribution

■ The grades on a certain standardized test are normally distributed with a mean of 70 and a standard deviation of 10. What is the probability that a randomly selected student who took the test scored between 70 and 85? The key to answering this question is an understanding of the phrase *normally distributed.* ■

One of the fortunate surprises in statistics is that many line graphs and bar graphs are approximately bell-shaped. In fact, if we modify some line graphs to indicate probability rather than frequency, the resulting graphs will approximate closely a smooth, bell-shaped curve called the *normal probability curve. If this is true, the data involved are said to be normally distributed.*

We now plot a bar graph and a line graph for the data given in Table 13 using probability rather than frequency. (See Figure 9.)

The height of each rectangle represents the probability that the variable falls in an interval. For example, the probability that the variable falls between 19.5 and 24.5 is .30. However, for a probability curve, we want the area of the rectangle instead of the height to represent the probability. Since the width of each rectangle is 5, the area of each rectangle is five times the probability. We make the area of each rectangle equal to the probability by dividing the height of each rectangle

Table 13

Class	*Frequency*	*Relative frequency*
5–9	1	$\frac{1}{20} = .05$
10–14	2	$\frac{2}{20} = .10$
15–19	4	$\frac{4}{20} = .20$
20–24	6	$\frac{6}{20} = .30$
25–29	4	$\frac{4}{20} = .20$
30–34	2	$\frac{2}{20} = .10$
35–39	1	$\frac{1}{20} = .05$
	20	

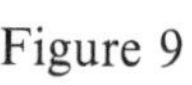
Figure 9

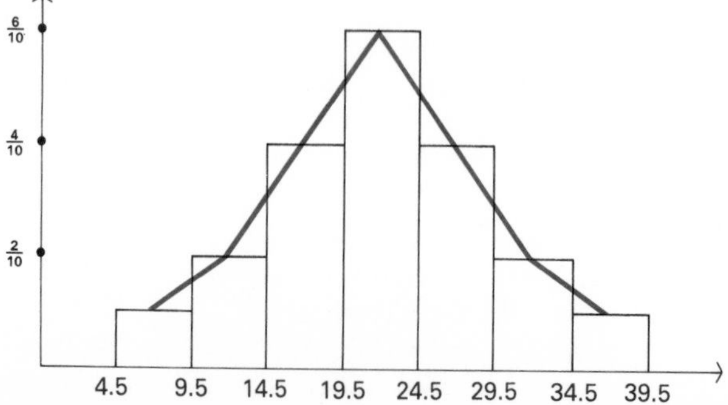

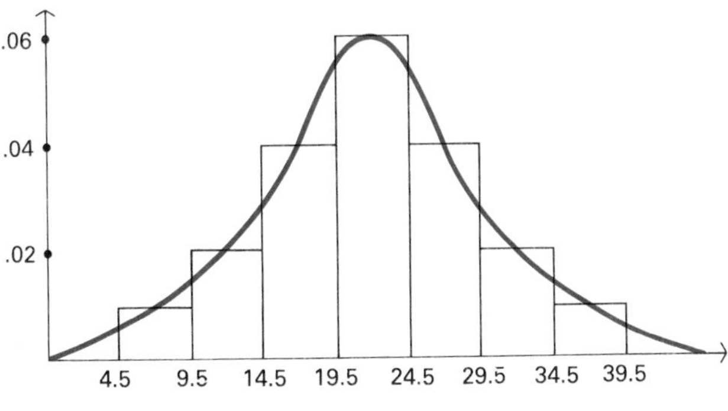

Figure 10

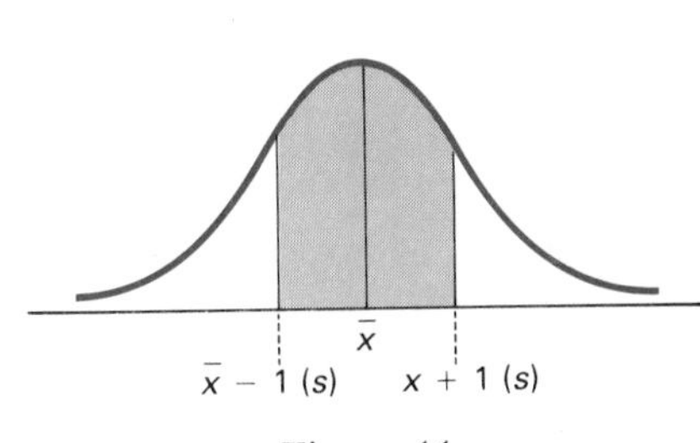

Figure 11

by 5. The resulting graph is drawn in Figure 10. This smooth curve is an example of a normal curve. Note some interesting properties of the normal probability curve by studying the approximating bar graph in Figure 10.

The sum of the areas of all the rectangles is 1. Therefore, it is reasonable to assume that the area under the smooth, approximating curve is 1. The mean of the data is 22, and the bar graph is symmetric about a vertical line through the mean. Similarly, the smooth curve is symmetric about a vertical line through the mean. The standard deviation of the data is approximately 7.5. The interval about the mean that extends for one standard deviation on either side of the mean is the interval from $(22 - 7.5)$ to $(22 + 7.5)$. This interval from 14.5 to 29.5 contains $\frac{14}{20}$ or 70% of the data (see Figure 11). The interval about the mean that extends for two standard deviations on either side of the mean is the interval from $(22 - 15)$ to $(22 + 15)$, or from 7 to 37. Assume that half of the frequency in the first and last class intervals belong in this range from 7 to 37. Then the interval (mean ± 2 standard deviations) contains $\frac{19}{20}$ or 95% of the data. Note also that the probability that x lies between 19.5 and 29.5 is .50, which is the area of two rectangles. The area of these two rectangles is approximately the area under the curve between 19.5 and 29.5.

These properties for the bar graph and the approximating smooth curve demonstrate the following well-known properties of a normal curve.

Properties of a normal curve

1. The area under a normal curve is 1.
2. The normal curve is symmetric about a vertical line through the mean of the set of data.
3. The interval extending from 2 standard deviations to the left of the mean to 2 standard deviations to the right of the mean contains approximately 95% of the area; the corresponding interval extending 1 standard deviation on each side of the mean contains approximately 68% of the area; the corresponding interval extending 3 standard deviations on each side of the mean contains 99% of the area.
4. If x is a data value from a set of data that is normally distributed, then the probability that x is greater than a and less than b is the area under the normal curve between a and b.

In the previous section, we discussed the process of converting data to standard units. Recall that to express x in standard units, we subtract the mean, then divide by the standard deviation,

$$z = \frac{x - \bar{x}}{s}$$

When a data value from a normal distribution is standardized, the resulting data value lies in the standard normal distribution. The standard normal distribution has a mean of 0 and standard deviation of 1.

The curve in Figure 12 is the standard normal distribution. We use z to represent the standard normal variable and y to represent the frequency. The maximum value of the curve is attained at $z = 0$. The standard normal curve has perfect symmetry. Because of this characteristic, the mean and median of the distribution have the same value—namely, 0. The range is not defined, because there are values occurring as far out as you wish to go—that is, the curve never intersects the axis.

Figure 12

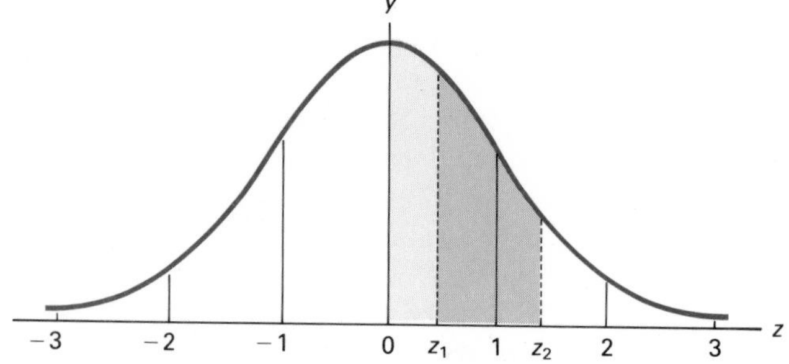

The area under the standard normal curve is 1. Thus, to find the probability that z is between z_1 and z_2, we obtain the area under the curve between z_1 and z_2 (the region in Figure 12). Table 14 gives the area under the normal curve less than or equal to $z = z_1$, and greater than or equal to $z = 0$. That is, the area indicated by the light shading in Figure 12 is given in Table 14 at $z = z_1$. The area from $z = 0$ to $z = z_1$ is the same as the probability that z is less than or equal to z_1 and greater than or equal to 0. The table stops at $z = 3.09$, because the area under the curve beyond $z = 3.09$ is negligible.

The fact that the standard normal curve is symmetric about $z = 0$ means that the area under the curve on either side of 0 is 0.5. This symmetry allows us to compute the probabilities that do not specifically occur in the table.

Example 21 • Find $P(z \leq 1.84)$.

Solution Since the area under either half of the curve in Figure 13 is 0.5000,

$$\begin{aligned} P(z \leq 1.84) &= .5000 + P(0 \leq z \leq 1.84) \\ &= .5000 + .4671 \\ &= .9671 \quad \bullet \end{aligned}$$

Table 14

z	*.00*	*.01*	*.02*	*.03*	*.04*	*.05*	*.06*	*.07*	*.08*	*.09*
0.0	.0000	.0040	.0080	.0120	.0160	.0199	.0239	.0279	.0319	.0359
0.1	.0398	.0438	.0478	.0517	.0557	.0596	.0636	.0675	.0714	.0753
0.2	.0793	.0832	.0871	.0910	.0948	.0987	.1026	.1064	.1103	.1141
0.3	.1179	.1217	.1255	.1293	.1331	.1368	.1406	.1443	.1480	.1517
0.4	.1554	.1591	.1628	.1664	.1700	.1736	.1772	.1808	.1844	.1879
0.5	.1915	.1950	.1985	.2019	.2054	.2088	.2123	.2157	.2190	.2224
0.6	.2257	.2291	.2324	.2357	.2389	.2422	.2454	.2486	.2517	.2549
0.7	.2580	.2611	.2642	.2673	.2704	.2734	.2764	.2794	.2823	.2852
0.8	.2881	.2910	.2939	.2967	.2995	.3023	.3051	.3078	.3106	.3133
0.9	.3159	.3186	.3212	.3238	.3264	.3289	.3315	.3340	.3365	.3389
1.0	.3413	.3438	.3461	.3485	.3508	.3531	.3554	.3577	.3599	.3621
1.1	.3643	.3665	.3686	.3708	.3729	.3749	.3770	.3790	.3810	.3830
1.2	.3849	.3869	.3888	.3907	.3925	.3944	.3962	.3980	.3997	.4015
1.3	.4032	.4049	.4066	.4082	.4099	.4115	.4131	.4147	.4162	.4177
1.4	.4192	.4207	.4222	.4236	.4251	.4265	.4279	.4292	.4306	.4319
1.5	.4332	.4345	.4357	.4370	.4382	.4394	.4406	.4418	.4429	.4441
1.6	.4452	.4463	.4474	.4484	.4495	.4505	.4515	.4525	.4535	.4545
1.7	.4554	.4564	.4573	.4582	.4591	.4599	.4608	.4616	.4625	.4633
1.8	.4641	.4649	.4656	.4664	.4671	.4678	.4686	.4693	.4699	.4706
1.9	.4713	.4719	.4726	.4732	.4738	.4744	.4750	.4756	.4761	.4767
2.0	.4772	.4778	.4783	.4788	.4793	.4798	.4803	.4808	.4812	.4817
2.1	.4821	.4826	.4830	.4834	.4838	.4842	.4846	.4850	.4854	.4857
2.2	.4861	.4864	.4868	.4871	.4875	.4878	.4881	.4884	.4887	.4890
2.3	.4893	.4896	.4898	.4901	.4904	.4906	,4909	.4911	.4913	.4916
2.4	.4918	.4920	.4922	.4925	.4927	.4929	.4931	.4932	.4934	.4936
2.5	.4938	.4940	.4941	.4943	.4945	.4946	.4948	.4949	.4951	.4952
2.6	.4953	.4955	.4956	.4957	.4959	.4960	.4961	.4962	.4963	.4964
2.7	.4965	.4966	.4967	.4968	.4969	.4970	.4971	.4972	.4973	.4974
2.8	.4974	.4975	.4976	.4977	.4977	.4978	.4979	.4979	.4980	.4981
2.9	.4981	.4982	.4982	.4983	.4984	.4984	.4985	.4985	.4986	.4986
3.0	.4987	.4987	.4987	.4988	.4988	.4989	.4989	.4989	.4990	.4990

Figure 13

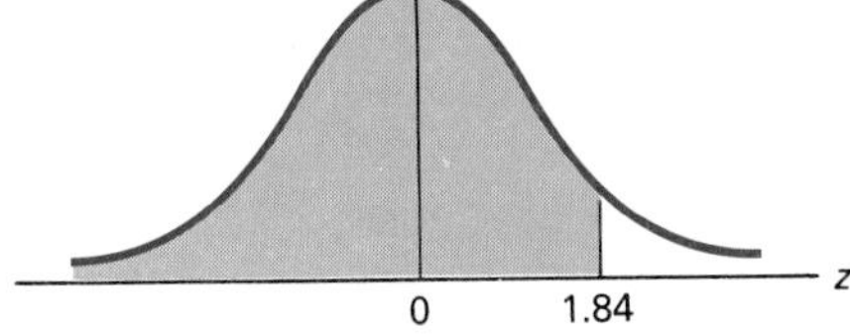

For the normal curve, $P(z \leq a) = P(z < a)$, and $P(z \geq a) = P(z > a)$, for all a. In general, for distributions (called *continuous distributions*) like the standard normal curve, the probability that z is less than or equal to a number is the same as the probability that z is less than the number.

Example 22 • Find $P(z \leq 1.2)$.

Solution From Table 14, we read that the value for $z = 1.2$ is .3849. Thus,

$$P(z \leq 1.2) = .3849 + .5000 = .8849 \quad \bullet$$

Recall that the normal curve is symmetrical about $z = 0$. This fact is important as we discuss areas under the curve. The fact that the standard normal curve is symmetric about the origin means that the area under the curve on either side of 0 is the same. For example,

$$\begin{aligned} P(-1.05 \leq z \leq 1.05) &= 2P(0 \leq x \leq 1.05) \\ &= 2(.3531) \\ &= .7062 \end{aligned}$$

Example 23 • Use Table 14 to verify that $P(-2 \leq z \leq 2) = .9544$. Note that in the table,

$$\begin{aligned} P(-2 \leq z \leq 2) &= 2P(0 \leq z \leq 2) \\ &= 2(.4772) \\ &= .9544 \quad \bullet \end{aligned}$$

Since the total area under the curve is 1, the area to the right of $z = 1.66$ is 1 minus the area to the left of 1.66. That is,

$$\begin{aligned} P(z \geq 1.66) &= 1 - P(z \leq 1.66) \\ &= 1 - [.5000 + P(0 \leq z \leq 1.66)] \\ &= 1 - [.5000 + .4515] \\ &= 1 - .9515 \\ &= .0485 \end{aligned}$$

Sometimes we need to compute the probability that z is in a certain range—say, between 0.4 and 1.4 (see Figure 14). This probability is indicated by $P(0.4 < z < 1.4)$. It can be obtained by finding the probability that z is less than 1.4 and subtracting from this the probability that z is less than 0.4.

$$\begin{aligned} P(0.4 < z < 1.4) &= P(z < 1.4) - P(z < 0.4) \\ &= .9192 - .6554 \\ &= .2638 \end{aligned}$$

Figure 14

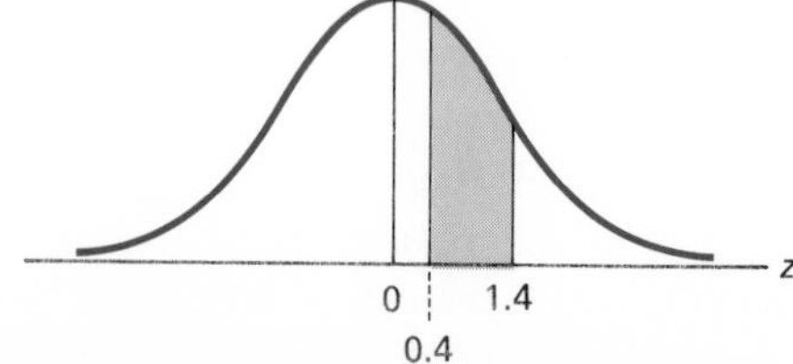

Just for fun *The combination of a lock is given by the following distribution of numbers: 14, 5, 6, 11, 13, 8, 3 and 4. Turn left the mean of the distribution; turn right the value of the median; turn left the largest positive integer in the standard deviation; and turn right one-half the range. Can you open the lock?*

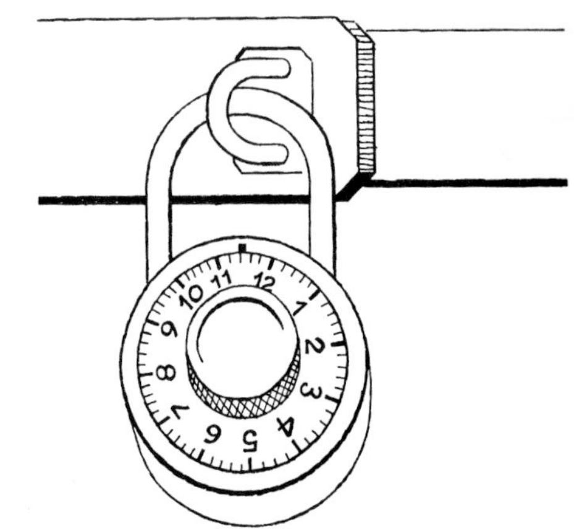

Example 24 • Find the probability that the normal variable x, with mean 175 and standard deviation 20, is less than or equal to 215.

Solution

$$z = \frac{215 - 175}{20} = 2$$

$$P(x \leq 215) = P(z \leq 2) = .9772 \quad \bullet$$

Example 25 • The grades on a certain test are known to be normally distributed, with mean 74 and standard deviation 8. What is the probability that a student will make less than 58 on this test?

Solution

$$z = \frac{58 - 74}{8} = -2$$

$$\begin{aligned} P(x < 58) &= P(z < -2) \\ &= P(z > 2) \\ &= 1 - P(z \leq 2) \\ &= 1 - .9772 \\ &= .0228 \end{aligned}$$

Thus, 2.28% of the students will make less than 58 on the test. Equivalently, the probability that a student chosen at random will make less than 58 is .0228. •

Exercise set 5

R 1. Find the area under the standard normal curve that lies between the following pairs of values of z.

(a) $z = 0$ to $z = 2.40$
(b) $z = 0$ to $z = 0.41$
(c) $z = 0$ to $z = 1.67$
(d) $z = -0.36$ to $z = 0.36$

2. Assuming these sketches represent the standard normal curve, compute the shaded areas.

(a)

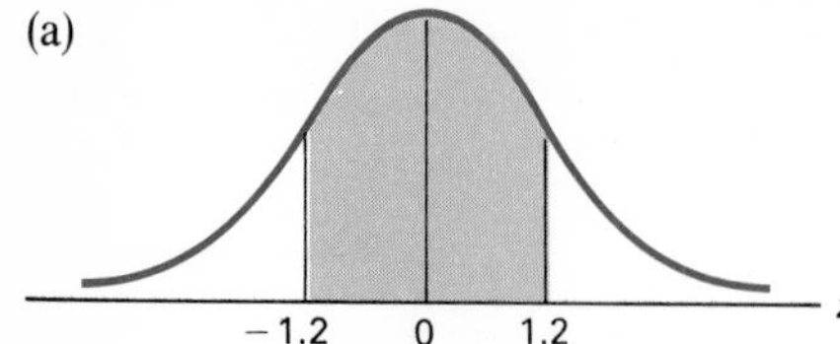

(b)

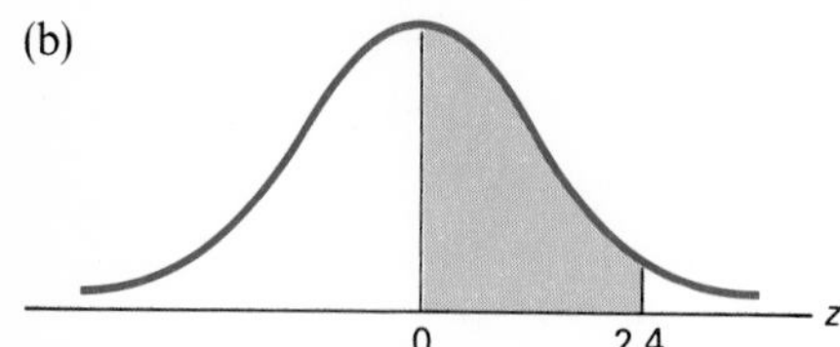

(c)

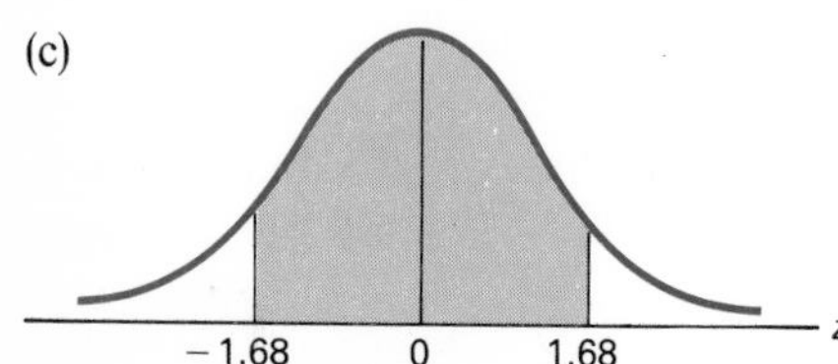

(d)

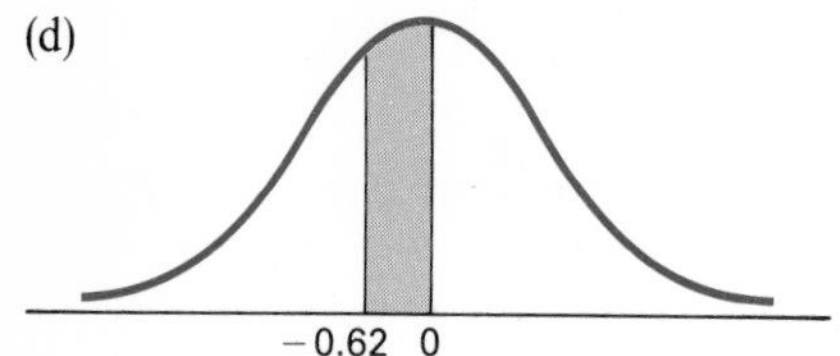

3. Find the following probabilities from Table 14.

(a) $P(z \leq -2.1)$ (b) $P(z \geq -1.4)$
(c) $P(z \leq 0.1)$ (d) $P(z < -1.6)$
(e) $P(z > 1.5)$ (f) $P(z > 2.4)$
(g) $P(z > -2.1)$ (h) $P(z > -1.8)$
(i) $P(1.3 \leq z \leq 2.4)$ (j) $P(2.1 < z < 2.8)$
(k) $P(-1.2 \leq z \leq 0.3)$ (l) $P(-2.6 \leq z \leq 1.4)$

4. If x is a variable having a normal distribution with $\bar{x} = 12$ and $s_x = 4$, find the probability that x assumes the following values.

(a) $x \leq 16$ (b) $x \geq 10$
(c) $10 \leq x \leq 14$ (d) $8 \leq x \leq 16$
(e) $x \geq -4$

5. Given that x is normally distributed with mean 50 and standard deviation 10, find the following probabilities.

(a) $P(x \geq 50)$ (b) $P(x \leq 50)$
(c) $P(x \geq 60)$ (d) $P(x \leq 70)$
(e) $P(x \leq 40)$ (f) $P(x \geq 46)$
(g) $P(38 \leq x \leq 54)$ (h) $P(32 \leq x \leq 61)$

T 6. Given that x is normally distributed with mean 6 and standard deviation 1.4, find the following probabilities.

(a) $P(6 \leq x \leq 7)$ (b) $P(6 \leq x \leq 8)$
(c) $P(6 \leq x \leq 9)$ (d) $P(6 \leq x \leq 10)$
(e) $P(4.6 \leq x \leq 7.4)$ (f) $P(5.2 \leq x \leq 6.8)$

7. The mean of a normal distribution is 35 with a standard deviation of 4. What percentage of the data should be in the following intervals?

(a) 31.5 to 38.5 (b) 24.5 to 45.5
(c) 17.5 to 52.5 (d) 35 to 38.5
(e) 35 to 45.5 (f) 35 to 52.5

C 8. In Exercise 7, compute the percent of the data that should be in the following intervals.

(a) 38.5 to 45.5 (b) 45.5 to 52.5
(c) 17.5 to 24.5 (d) 24.5 to 31.5

In each of the following problems, assume that the data are normally distributed.

9. It is known from experience that the number of telephone orders made daily to a company approximates a normal curve with mean 350 and standard deviation 20. What percentage of the time will there be more than 400 telephone orders per day?

10. Radar is used to check the speed of traffic on Interstate 75 north of Atlanta. If the mean speed of the traffic is 60 miles per hour with a standard deviation of 4 miles per hour, what percentage of the cars are exceeding the legal speed of 55 miles per hour?

11. Young rabbits placed on a certain high-protein diet for a month show a weight gain with a mean of 120 grams and a standard deviation of 12 grams.
(a) What is the probability that a given rabbit will gain at least 100 grams in weight?
(b) If 15,000 rabbits are placed on this diet, how many can be expected to gain at least 140 grams?

12. The heights of men in a certain army regiment are normally distributed, with mean 177 centimeters and standard deviation 4 centimeters.
(a) What percentage of the men are between 173 and 181 centimeters in height?
(b) What percentage of the men are between 169 and 185 centimeters in height?

13. The scores on the entrance exam for the Kentucky Police Academy are normally distributed with a mean of 60 and a standard deviation of 6. What is

the probability that a randomly selected test score will lie between 60 and 75?

14. The grades in a certain class are normally distributed with mean 76 and standard deviation 6. The lowest D is 61, the lowest C is 70, the lowest B is 82, and the lowest A is 91. What percentage of the class will make A's? What percentage will make B's? C's? D's? F's?

6 Misuses of statistics

We conclude this short chapter on an introduction to statistics with some examples of the misuse of statistics. You probably have heard the statement "There are lairs, there are damn liars, and there are statisticians." This statement emphasizes the point that some people quote statistics purposely to give results that are misleading. Some of these abuses are indicated by Darrell Huff in his book, *How to Lie with Statistics.* Consider the following examples of the misuse of statistics.

One misuse of statistics involves the shifting of the definition to suit your purpose. For example, in 1949 a Russian assertion stated that there were 14 million unemployed in the United States. This number far exceeded the number of unemployed as given by the United States Bureau of Labor. What was the discrepancy? The Russian figure involved everyone who worked less than 40 hours a week, while the American figure represented those who were not employed. Thus, by changing a definition one can manipulate a statistic to prove an argument. In accurate statistical statements, all terms should be completely defined.

A second misuse of statistics is due to inaccurate measurement or tabulation of data. Many opinion-polling organizations have difficulty in securing personnel to take opinions and record data accurately.

A third misuse is due to the method of selecting the group from which the information is to be obtained. For example, each child in a school system was asked how many children are in his or her family. From these data, the mean number of children in a family was obtained. The answer was much too large. Why? The answer was too large because a family with seven in school occurred seven times in the survey, whereas a family of one child with only one in school occurred only once. The method of taking the survey was weighted in favor of large families.

Undoubtedly, the most prevalent misuse of statistics involves inappropriate comparisons. A toothpaste commercial states that a group using Droat toothpaste has 25% fewer cavities than a group using other toothpastes. Who knows the composition of the other toothpaste preparations? Maybe they include ingredients that will cause cavities. Or maybe the people using Droat toothpaste have fewer teeth, hence fewer cavities.

Another misuse is due to the failure to include representative groups within a sample. Ronald Clodhopper polled his neighbors and found the average salary was $60,000. We certainly would not conclude that the average salary in the United States was $60,000—especially since Ronald lived in an exclusive suburb of a prosperous metropolitan area.

Of course, we have already learned that an average without some indication of dispersion can be misleading. The mean temperature in a particular city is 76°F. Do you want to move there? It could be exceedingly hot in summer and exceedingly cold in winter. Don't be misled by averages!

If you get nothing else out of this chapter, we hope you take time to analyze statements involving statistics. Such analysis is the answer of an educated person to the misuse of statistics.

(Exercises will be found in Review Exercise Set 6.)

Solution to introductory problem

Understand the problem. Of course, the key word in understanding the problem is "average." What do you mean by average? Which average?

Solution to the problem. A tabulation of the salaries at Acme Manufacturing is given in the following table.

Salary	*Frequency*
\$150,000	1
80,000	1
60,000	1
16,000	20
12,000	10
10,000	10
8,000	7
Total	50

The management used the arithmetic mean as average and you can verify this to be \$17,720. The union used the median of \$12,000. The employees talked to those with the most common salary, \$16,000, so they used the mode. All were correct; they simply used different averages.

Summary and review

As you read the following review list, be sure you understand the meaning (or definition) of each of the terms. If you are not completely satisfied with your knowledge of the term, find the relevant discussion in the chapter and review it.

Frequency	Line graph
Frequency distribution	Frequency polygon
Grouped frequency distribution	Circle graph
Class boundaries	Mean
Class marks	Median
Bar graph	Mode
Histogram	Expected value

No explanation is given for the following formulas. If you do not immediately recognize a formula, then you need to look it up and review how to use it.

$$\bar{x} = \frac{x_1 + x_2 + x_3 + \cdots + x_n}{n}$$

$$\bar{x} = \frac{x_1 f_1 + x_2 f_2 + \cdots + x_m f_m}{f_1 + f_2 + \cdots + f_m}$$

$$s_x^2 = \frac{(x_1 - \bar{x})^2 + (x_2 - \bar{x})^2 + \cdots + (x_n - \bar{x})^2}{n}$$

$$s_x^2 = \frac{(x_1 - \bar{x})^2 f_1 + (x_2 - \bar{x})^2 f_2 + \cdots + (x_m - \bar{x})^2 f_m}{n}$$

$$s_x = \sqrt{\text{variance}}$$

$$z = \frac{x - \bar{x}}{s_x}$$

Normal curve

68% of data lie within one standard deviations of the mean.

95% of data lie within two standard deviations of the mean.

99% of data lie within three standard deviations of the mean.

Review exercise set 6

R 1. Consider the set of scores 1, 8, 16, 18, 20, 20, 21, 23, 24, 29. Find the following.

(a) Mean
(b) Median
(c) Mode
(d) Range
(e) Standard deviation

2. The mean salary of all employees at the Brown Corporation is $30,000. Make up an example to show how this statistic may be misleading.

3. Find the mean, median, and standard deviation for the following frequency distribution.

x	10	20	30	40	50
f	4	6	8	4	3

4. The following test scores were received by 24 students.

63, 71, 85, 96, 94, 90, 75, 72, 77, 71, 62, 84,
81, 76, 61, 54, 87, 94, 32, 81, 94, 77, 63, 60

Find the following.

(a) Mean
(b) Median
(c) Mode
(d) Standard deviation

5. Group the test scores in Exercise 4 into classes 95–99, 90–94, and so on. Then compute each part of Exercise 4.

6. For the data in Exercise 5, construct a

(a) histogram
(b) frequency polygon

7. On a test, the grades were distributed as shown below. Show these grades on a pie chart.

A	20%
B	25%
C	35%
D	10%
F	10%

T 8. A set of measurements is approximately normally distributed, with mean of 100 and standard deviation of 10. Find the percentage of measurements in the interval from 80 to 110.

9. Comment on the following "misuses" of statistics.

(a) The average number of people in a family is 3.5. Therefore, government policy should be to construct public housing units to accommodate 4 people.
(b) More people died in accidents in the United States last year than died in fighting in World War II. Therefore, it is more dangerous to live in the United States today than it was to fight in World War II.
(c) More people die in hospitals than at home. There are severe deficiencies in our medical care system.
(d) One candidate says that more people are employed presently than at any time in our history. A second candidate bemoans that we suffer from the highest unemployment rate in history. Can both be right? How?
(e) Senator Gladrag brags that two years before he came to office, workers earned only $4500; now they earn $9000 each year.

10. The following are the IQ scores of 30 first-grade students in one classroom.

128	133	100	115	82	99
107	142	98	112	152	100
105	78	114	84	86	110
96	93	101	94	86	124
120	100	102	107	94	128

What is the median IQ score?

C 11. In Exercise 3, subtract 30 from each data value x and divide the result by 10, getting a new data value y. For the y's, find the

(a) mean
(b) standard deviation
(c) How do these relate to the answers in Exercise 3?

Bibliography

1. American Statistical Association and NCTM. *Statistics: A Guide to the Unknown.* San Francisco, Calif.: Holden-Day, 1974, 40–51, 92–100, 102–111.
2. Ball, J. "Finding Averages with Bar Graphs." *AT* (October 1969), 487–489.
3. Beckenbach, Edwin F. "Baseball Statistics." *MT* (May 1979), 351–352.
4. Bruni, James V., & Helene J. Silverman "Developing Concepts in Probability and Statistics—and Much More." *AT* (February 1986), 34–37.
5. Feltman, James. "Cryptics and Statistics." *MT* (March 1979), 189—191.
6. Grass, B. A. "Statistics Made Simple." *AT* (March 1965), 196–198.
7. Hyatt, D. "M and M's Candy: A Statistical Approach." *AT* (January, 1977), 34.
8. Joiner, B., & C. Campbell. "Some Interesting Examples for Teaching Statistics." *MT* (May 1975), 364–369.
9. Klitz, R., & J. Hofmeister. "Statistics in the Middle School." *AT* (February 1979), 35–36.
10. National Council of Teachers of Mathematics. *More Topics in Mathematics for Elementary Teachers.* In Thirteenth Yearbook of NCTM, Booklet 16, Washington, D.C.: National Council of Teachers of Mathematics, 1969.
11. Noether, G. "The Nonparametric Approach in Elementary Statistics." *MT* (February 1974), 123–126.
12. Pincus, M., & F. Morgenstern. "Graphs in the Primary Grades." *AT* (October 1970), 499–501.
13. Shaw, Jean M. "Dealing with Data." *AT* (May 1984), 9–15.
14. Shulte Albert. "A Case for Statistics." *AT* (February 1979), 24.

Appendix

(Continuation of Logic)

A.1 Logically true and quantifiers

In this section, we continue to use truth tables to prove that assertions and conjectures are true or not true. Then we define what is meant by a statement being **logically true.**

If two statements are logically equivalent, then an implication connecting the two is logically true. A definition of *logically true* is given as follows.

Tautology

A statement having other statements as components is said to be **logically true** if it is always true regardless of the truth values of the statements that compose it. Such a statement is called a *tautology*.

It is easy to determine when statements are logically true by means of truth tables. If a statement is logically true, the column for this statement in a truth table will contain all T's.

Example 1 • The statement $p \to (p \vee q)$ is logically true, as seen in Table 1, since the last column contains all T's.

Table 1

p	q	$p \vee q$	$p \to (p \vee q)$
T	T	T	T
T	F	T	T
F	T	T	T
F	F	F	T

•

Example 2 • Show that $\sim[p \wedge (\sim p)]$ is a tautology.

Solution This is verified in Table 2.

Table 2

p	$\sim p$	$p \wedge (\sim p)$	$\sim[p \wedge (\sim p)]$
T	F	F Always false	T Always true
F	T	F	T

•

The fact that $\sim[p \wedge (\sim p)]$ is always true is sometimes called the **law of the excluded middle.**

Example 3 • Show that $\sim(p \vee q) \rightarrow \sim p$ is a tautology.

Solution This fact is verified in Table 3.

Table 3

p	q	$\sim p$	$p \vee q$	$\sim(p \vee q)$	$\sim(p \vee q) \rightarrow \sim p$
T	T	F	T	F	T Always true
T	F	F	T	F	T
F	T	T	T	F	T
F	F	T	F	T	T

•

Contradiction

A compound statement is said to be **logically false** if the statement is always false, regardless of the truth values of the simple statements that comprise it. A logically false statement is called a **contradiction.**

Note in Table 2 that "p and not p," denoted $p \wedge (\sim p)$, is always false. The fact that $p \wedge (\sim p)$ is always false is sometimes called the **law of contradiction.** Can you show by means of a truth table that $\sim[p \vee (\sim p)]$ is a contradiction?

Two statements r and s are said to be *logically equivalent* if and only if the biconditional statement $r \leftrightarrow s$ is logically true.

Thus, r and s are logically equivalent if they have the same truth values.

We now introduce terms that play a significant role in determining whether a statement is true or false.

Quantified statement

A statement that indicates something about "how many" is said to be **quantified.**

Some examples of quantified statements include the following:

Some women have red hair.
All bananas are yellow.
No professors are bald-headed.
Some students do not work hard.

The quantifiers (words that convey the idea of quantity) in the preceding statements are *some, all,* and *no.*

Certain statements are intended to be *universally* true for those factors under consideration. Such statements usually contain the words *all, every,* or *for each,* which are called **universal quantifiers.**

All prime numbers greater than 2 are odd.
Every automobile pollutes the atmosphere.
For each counting number, $x + 3 = 3 + x$.
All men have black hair.

Expressions involving *all* may be expressed as conditionals in the following manner. "All illegal acts are immoral" can be written as "If an act is illegal, then it is immoral." Similarly, the expression "No men use hair curlers" can be written "If a person is a man, then he does not use hair curlers."

Other statements are intended to indicate that there exists at least one case in which the statement is true. Such statements generally involve the **existential quantifiers** *some, there exist,* or *there exists at least one.*

There exist students who do not work hard.
There exists at least one student who does not work hard.
Some men have black hair.

In saying "Some men have black hair," we mean that there is at least one man with black hair.

Quantifiers are most useful in rewriting assertions that cannot be classified as true or false. Consider the assertion "Men are tall." By adding *all* to obtain "All men are tall," we know that the statement is false. However, by adding *some* to obtain "Some men are tall," we know that the statement is true.

In addition to this use, quantifiers are extremely helpful in expressing sentences so that they can be identified as true or false. The statement "$x + 4 = 6$" cannot be classified either as true or as false. However, the statement "There exists a counting number x such that $x + 4 = 6$" is true, because $x = 2$ is such a counting number. "For all counting numbers, $x + 4 = 6$" is false, because there exists a counting number 5 such that $5 + 4$ is not equal to 6. However, the statement "For all x, $3 + x = x + 3$" is true.

Sometimes confusion occurs in the use of statements like "all are not" and "not all are." Consider the following statements, which say the same thing.

Not all dogs have tails.
Some dogs do not have tails.

As the preceding examples demonstrate, it is sometimes difficult to state the negations of statements involving quantifiers. In such instances, the following rule is helpful. **In order to negate a quantified statement involving a single quantifier, change the quantifier as illustrated below and negate the statement it quantifies.**

Statement	*Negation*
All p are q.	Some p are not q.
Some p are q.	No p is q.
Some p are not q.	All p are q.
No p is q.	Some p are q.

Example 5 • Write the negations of the statements on the left.

Statement	*Negation*
Some women have red hair. (At least one woman has red hair.)	No woman has red hair.
All bananas are yellow. (Every banana is yellow.)	Some bananas are not yellow. (At least one banana is not yellow.)
No professor is bald-headed.	Some professors are bald-headed. (At least one professor is bald-headed.)
Some students do not work hard. (At least one student does not work hard.)	All students work hard. •

Notice that in order to show that a universally quantified statement is false, it is only necessary to find one case for which the statement is false; that is, only one **counterexample** is needed. However, in showing that an existentially quantified statement is false, we must show that it is false for all possibilities. Similarly, an existentially quantified statement is true if we can find one case in which it is true, but a universally quantified statement is true only if it is true for all cases.

The negations of mathematical expressions are also difficult to express. Let $p(x)$ be some mathematical expression, and consider the statement "For all x, $p(x)$ holds." The negation of this statement is "It is not true for all x that $p(x)$ holds." The negation of "There exists an x such that $p(x)$ holds" is "There does not exist an x such that $p(x)$ holds." However, this negation can also be expressed as "There is no x such that $p(x)$ holds." In a similar way, the negation of "For all x, $p(x)$ holds" can be written as "There exists an x for which $p(x)$ does not hold."

Statement	*Negation*
There exists an x such that $x + 4 = 6$.	There is no x such that $x + 4 = 6$.
For all x, $(x - 2)(x) = x^2 - 2x$.	There exists an x such that $(x - 2)(x) \neq x^2 - 2x$.

Exercise set A.1

1. Using truth tables, determine which of the statements below are logically true.
   (a) $p \to p$
   (b) $[p \wedge (p \to q)] \to \sim q$
   (c) $(p \to q) \to p$
   (d) $[(\sim p \vee q) \wedge p] \to q$
   (e) $(p \to q) \to (p \vee q)$
   (f) $(p \wedge q) \to q$
   (g) $(\sim p \vee q) \to (p \to q)$
   (h) $(p \to q) \to q$
2. Assign a value for x so as to make each statement true.
   (a) $x + 4 = 7$
   (b) $x - 7 = 4$
   (c) $x^2 = 16$
   (d) $x + 3 = 3 + x$
3. If possible, assign a value for x that makes each statement in Exercise 2 false.
4. Use a quantifier to make the statements in Exercise 2 true.
5. Use a quantifier to make the statements in Exercise 2 false.
6. Write the negation of each of the following statements without using the expression "It is not true that."
   (a) All athletes over 7 feet tall play basketball.
   (b) Some students work hard at their studies.
   (c) All men use Bob-Bob hair oil.
   (d) Some professors are not intelligent.
   (e) No man weighs more than 500 pounds.
   (f) Some rabbits have white tails.
   (g) Some bunnies are not rabbits.
   (h) No prime numbers are perfect numbers.
   (i) All prime numbers greater than 2 are odd.
7. Write the negation of the statements below. Do not use "It is not true that."
   (a) There exists a counting number greater than 50.
   (b) Not all counting numbers are greater than 5.
   (c) There exists an x such that $x + 2 = 7$.
   (d) For all x, $(x + 3) + 2 = 3 + (2 + x)$.
   (e) For all x, $x(x + 2) = x^2 + 2x$.
   (f) For every a, $a(a + 1) = a^2 + a$.
8. Use truth tables to verify the following statements, where $=$ means logically equivalent, c is a contradiction, and t is a tautology.
   (a) $p \wedge q = q \wedge p$ and $p \vee q = q \vee p$ — Commutative properties
   (b) $p \wedge (q \wedge r) = (p \wedge q) \wedge r$ and
   (c) $p \vee (q \vee r) = (p \vee q) \vee r$ — Associative properties
   (d) $p \vee (q \wedge r) = (p \vee q) \wedge (p \vee r)$ and
   (e) $p \wedge (q \vee r) = (p \wedge q) \vee (p \wedge r)$ — Distributive properties
   (f) $p \vee p = p$ and $p \wedge p = p$ — Idempotent properties
   (g) $p \wedge \sim p = c$ and $p \vee \sim p = t$ — Complement properties
   (h) $p \vee c = p$ and $p \wedge t = p$ — Identity properties
   (i) $\sim(p \wedge q) = \sim p \vee \sim q$ and
   (j) $\sim(p \vee q) = \sim p \wedge \sim q$ — De Morgan's properties
9. Write the negations of the statements shown. Do not use "It is not true that."
   (a) For all x, $p(x)$ and $q(x)$.
   (b) For all x, $p(x)$ or $q(x)$.
   (c) For all x, if $p(x)$, then $q(x)$.
   (d) There exists an x such that "$p(x)$ and $q(x)$" is true.
   (e) There exists an x such that "$p(x)$ or $q(x)$" is true.
   (f) There exists an x such that if $p(x)$, then $q(x)$.
10. Prove for each of the following that the statements on each side of $\leftrightarrow$ are logically equivalent, or prove that the complete statement is logically true.
    (a) $[r \to s] \leftrightarrow [\sim r \vee s]$
    (b) $[(r \wedge \sim s) \to s] \leftrightarrow [r \to s]$
    (c) $[s \to t] \leftrightarrow [(t \to s) \to (s \to t)]$
    (d) $[r \wedge (s \vee t)] \leftrightarrow [(r \wedge s) \vee (r \wedge t)]$
    (e) $[r \vee (s \wedge t)] \leftrightarrow [(r \vee s) \wedge (r \vee t)]$
11. In the remainder of this appendix we will be using the fact that the following statements are logically true. Verify that each is logically true by using a truth table.
    (a) $[(p \to q) \wedge p] \to q$
    (b) $[(p \to q) \wedge \sim q] \to \sim p$
    (c) $[(p \to q) \wedge (q \to r)] \to (p \to r)$
12. Verify, using truth tables, that each statement below is or is not a tautology.
    (a) $[(p \to q) \wedge (q \to r) \wedge p] \to r$
    (b) $[(r \to s) \wedge (\sim r \to \sim t) \wedge \sim s] \to \sim t$
13. Simplify the statements below by using the properties of the algebra of statements given in exercise 8.
    (a) $\sim p \vee (\sim p \vee q)$
    (b) $p \vee (\sim p \wedge q)$
    (c) $\sim(p \vee q) \wedge (p \vee \sim q)$
    (d) $p \wedge (\sim p \wedge q)$
    (e) $\sim(p \wedge q) \vee (p \wedge \sim q)$
    (f) $\sim(p \wedge q) \vee (\sim p \wedge q)$
    (g) $(q \wedge p) \vee (\sim q \vee \sim p)$
    (h) $(q \vee p) \wedge (q \vee \sim q)$

A.2 Making use of deductive logic

To keep from making truth tables each time we need to test the validity of a situation, two well-known procedures, the **Rule of Detachment** and the **Chain Rule,** are validated using truth tables and then used in other situations. In this section, we introduce **deductive** reasoning to determine whether arguments are valid or invalid.

We consider an argument to be an assertion consisting of two parts. In an argument, a given set of statements called **premises** yields another statement called the **conclusion.** An argument is **valid** if the truth of the **conjunction** of the premises implies the truth of the conclusion. If the truth of the premises does not imply the truth of the conclusion, then the argument is said to be **invalid.** Invalid arguments are sometimes called **fallacies.** If we assume that the statements constituting the premises are all true for a valid argument, then the conclusion must be true.

Note that *the validity of an argument is independent of the truth values of its premises and its conclusion.*

Arguments are often written symbolically in the following manner:

If inflation occurs, the price of the automobiles increases.	$p \to q$
Inflation occurs.	p
Therefore, the price of automobiles increases.	$\therefore q$

The three dots $\therefore$ are read "therefore."

When verbal statements are changed to symbolic form, we note that the same conditional statement often occurs. For example, the conditional $[(p \to q) \wedge p] \to q$ occurs frequently in testing the validity of arguments. Consequently, we state this implication as a theorem so that it can be used without having to construct a truth table each time.

Rule of Detachment

The fact that $p \to q$ is true and p is true implies that q is true; that is,

$$[(p \to q) \wedge p] \to q$$

is logically true.

Example 6 • Use a truth table (Table 4) to show that $[(p \to q) \wedge p] \to q$ is logically true.

Solution

Table 4

p	q	$p \to q$	$(p \to q) \wedge p$	$[(p \to q) \wedge p] \to q$
T	T	T	T	T
T	F	F	F	T
F	T	T	F	T
F	F	T	F	T

Since for each possible combination of truth values for p and q, $[(p \to q) \wedge p] \to q$ is true, then $[(p \to q) \wedge p] \to q$ is logically true. •

Example 7 •

If he is a thief, he is a lawbreaker.	$p \to q$
He is a thief.	p
Therefore he is a lawbreaker.	Therefore q

This argument is valid because of the Rule of Detachment. •

Another often used argument, which can be stated as a theorem and proved by means of a truth table, is

$$\begin{array}{r} p \to q \\ q \to r \\ \hline \therefore\ p \to r \end{array}$$

Chain Rule

This rule asserts that if the conditionals "$p \to q$" and "$q \to r$" are both true, then the conditional "$p \to r$" is true; that is,

$$[(p \to q) \wedge (q \to r)] \to (p \to r)$$

is logically true.

Written in terms of connectives, the theorem may be stated as "$p \to r$ is true when $(p \to q) \wedge (q \to r)$ is true." Table 5 verifies the fact that $[(p \to q) \wedge (q \to r)] \to (p \to r)$ is always true.

Table 5

p	q	r	$p \to q$	$q \to r$	$(p \to q) \wedge (q \to r)$	$p \to r$	$[(p \to q) \wedge (q \to r)] \to (p \to r)$
T	T	T	T	T	T	T	T
T	T	F	T	F	F	F	T
T	F	T	F	T	F	T	T
T	F	F	F	T	F	F	T
F	T	T	T	T	T	T	T
F	T	F	T	F	F	T	T
F	F	T	T	T	T	T	T
F	F	F	T	T	T	T	T

Example 8 • Friends of Joyce know that the two conditionals

If Joyce is smiling, then she is happy.
If Joyce is happy, then she is polite.

are both true statements. So they know, by the Chain Rule, that "If Joyce is smiling, then she is polite." •

In the next example, we make use of our knowledge of equivalent statements in order to test the validity of the argument.

Example 9 • Show that the following argument is valid.

If Earl is not guilty, then John is telling the truth.	$\sim p \to q$
John is not telling the truth.	$\sim q$
Therefore, Earl is guilty.	$\therefore p$

Solution Since a conditional and its contrapositive are equivalent, $\sim p \to q$ may be replaced by $\sim q \to \sim(\sim p)$. But $\sim(\sim p)$ is equivalent to p. Thus, the first premise may be replaced by $\sim q \to p$, and the argument becomes $\sim q \to p$; $\sim q$; therefore, p. This argument is valid by the Rule of Detachment. •

Example 10 • Show that the following argument is not valid.

If someone is a college graduate, then he or she is intelligent.	$p \to q$
Juan is intelligent.	q
Therefore, Juan is a college graduate.	$\therefore p$

Solution Note that the q in "Juan is intelligent" is merely an instance of the q in "Someone is intelligent." Of course, this is not a valid argument, because it would imply that the truth of the conclusion of a true conditional statement leads to the truth of the hypothesis, but a statement and its converse are not logically equivalent. •

An extension of the chain rule gives

$$[(p \to q) \wedge (q \to r) \wedge (r \to s) \wedge \cdots \wedge (w \to z)] \to (p \to z)$$

Example 11 • Show that the following argument is valid.

If a person is a good speaker, she or he is a good teacher.	$p \to q$
If a person is a good teacher, she or he is friendly.	$q \to r$
If a person is friendly, she or he is polite.	$r \to s$
If a person is polite, she or he is well liked.	$s \to t$
Therefore, if a person is a good speaker, he is well liked.	$\therefore p \to t$

Solution This argument is valid by the chain rule. •

Example 12 • Show that the following argument is valid.

If I go bowling, I will not study.	$p \to \sim q$
If I do not study, I will take a nap.	$\sim q \to r$
I will not take a nap.	$\sim r$
Therefore, I will not go bowling.	$\therefore \sim p$

Solution Now $p \to \sim q$ and $\sim q \to r$ can be replaced by $p \to r$, by the chain rule. Likewise, $p \to r$ can be replaced by its contrapositive, $\sim r \to \sim p$. Thus, the argument can be written as $\sim r \to \sim p$ and $\sim r$, which leads to the conclusion $\sim p$, by the Rule of Detachment. •

Exercise set A.2

1. Determine whether the following arguments are valid and specify the rules of logic that you use to arrive at a valid conclusion.

   (a) $p \to q$; q; $\therefore p$

   (b) $a \to b$; $b \to c$; $\therefore c$

   (c) If two sides of a triangle are equal, the angles opposite these sides are equal. Side BC equals side AB in triangle ABC.
   Therefore, the angles opposite BC and AB are equal.

   (d) When the movie is over, we must go home.
   The movie is not over.
   Therefore, we must go home.

   (e) If R, then $\sim S$.
   If $\sim S$, then T.
   Therefore, if R, then T.

2. Test the validity of the following arguments, either by appealing to the principles given in this section or by constructing truth tables.

   (a) ($p \to q$ and q) implies p
   (b) [(p or q) and not p] implies q
   (c) [($p \to q$) and ($q \to t$) and p] implies t
   (d) [p and ($q \to \sim r$) and $p \to r$] implies $\sim q$

3. Determine whether the following arguments are valid.

   (a) $p \to q$; $p \to r$; $\therefore q \to r$

   (b) $p \vee q$; p; $\therefore \sim q$

   (c) $p \to \sim q$; $q \vee r$; p; $\therefore r$

   (d) $p \vee \sim q$; $\sim p \vee r$; $\therefore q \to r$

   (e) $q \to \sim p$; $q \wedge r$; $\therefore \sim p \to \sim r$

   (f) $q \to r$; $\sim p \vee q$; p; $\therefore r$

4. Demonstrate the truth of the following arguments. If an argument is not logically true, use a truth table to illustrate this fact.

   (a) If a man is a good speaker, he is a good teacher.
   Mr. Faulkner is a good teacher.
   Therefore he must be a good speaker.

   (b) If I can't go to town, I will go to the shopping center.
   I can go to town.
   Therefore I will not go to the shopping center.

   (c) When the movie is over, we must go home.
   The movie is over.
   Therefore we must go home.

   (d) You will fail this course if you do not study enough.
   You are not studying enough.
   Therefore you will fail this course.

   (e) If R, then not S.
   If not S, then T.
   Therefore if R, then T.

   (f) If a quadrilateral is a parallelogram, then opposite sides are parallel. Opposite sides of quadrilateral $ABCD$ are parallel.
   Therefore $ABCD$ is a parallelogram.

   (g) If q, then not p.
   q and r are true.
   Therefore if not p, then not r.

   (h) If a's are b's, then c's are d's.
   a's are not b's.
   Therefore c's are not d's.

   (i) If c's are d's, then a's are e's.
   a's are not e's.
   Therefore c's are not d's.

5. Make a valid conclusion from each set of statements given below and then explain why your conclusion is valid.

   (a) If Susan is a freshman, then Susan takes mathematics. Susan is a freshman.
   (b) You will fail this test if you do not study. You do not fail the test.
   (c) You cry if you are sad. You are sad.
   (d) You will receive your allowance if you cut the grass. You do not receive your allowance.

6. Arrange the statements below in logical order to prove that Joan teaches seventh grade, given that Joan is tall.

   (a) If Joan has red hair, she teaches seventh grade.
   (b) If Joan is tall, she wears contacts.
   (c) Therefore if Joan is tall, she teaches seventh grade.
   (d) If Joan wears contacts, she has red hair.

7. Margaret said "If I have hamburgers tonight, it will be raining. If I do not have hamburgers, I will have

chicken. If I have chicken, I will have tossed salad." Margaret did not have tossed salad. How was the weather?

8. Demonstrate the truth or falsity of the following arguments.
   (a) If Paul burns the hot dog, then Ed will get indigestion. If Ed does not get indigestion, then either Paul didn't burn the hot dog or Ed threw it away. Ed didn't throw it away. Therefore, he will get indigestion.
   (b) If the wheel needs greasing, then it squeaks. If it squeaks, then Bob gets the chills. The wheel is not squeaking. Therefore, Bob does not have the chills.
   (c) If you like mathematics, then you will like this course. You do not like mathematics. Therefore, you will not like this course.

9. In each of the following, two premises are given. Find a conclusion that is a logical result of the premises.
   (a) All college students are clever.
   Diana is a college student.
   (b) All right angles are equal.
   Angles A and B are not equal.
   (c) All college women are beautiful.
   Kay is not beautiful.

Table A.1 Annual Percentage Rate Table for Monthly Payment Plans[a]
See Instructions for Use of Tables on page 314.

	Annual Percentage Rate															
Number of payments	*10.00 %*	*10.25 %*	*10.50 %*	*10.75 %*	*11.00 %*	*11.25 %*	*11.50 %*	*11.75 %*	*12.00 %*	*12.25 %*	*12.50 %*	*12.75 %*	*13.00 %*	*13.25 %*	*13.50 %*	*13.75 %*
	(Finance Charge per $100 of Amount Financed)															
1	0.83	0.85	0.87	0.90	0.92	0.94	0.96	0.98	1.00	1.02	1.04	1.06	1.08	1.10	1.12	1.15
2	1.25	1.28	1.31	1.35	1.38	1.41	1.44	1.47	1.50	1.53	1.57	1.60	1.63	1.66	1.69	1.72
3	1.67	1.71	1.76	1.80	1.84	1.88	1.92	1.96	2.01	2.05	2.09	2.13	2.17	2.22	2.26	2.30
4	2.09	2.14	2.20	2.25	2.30	2.35	2.41	2.46	2.51	2.57	2.62	2.67	2.72	2.78	2.83	2.88
5	2.51	2.58	2.64	2.70	2.77	2.83	2.89	2.96	3.02	3.08	3.15	3.21	3.27	3.34	3.40	3.46
6	2.94	3.01	3.08	3.16	3.23	3.31	3.38	3.45	3.53	3.60	3.68	3.75	3.83	3.90	3.97	4.05
7	3.36	3.45	3.53	3.62	3.70	3.78	3.87	3.95	4.04	4.12	4.21	4.29	4.38	4.47	4.55	4.64
8	3.79	3.88	3.98	4.07	4.17	4.26	4.36	4.46	4.55	4.65	4.74	4.84	4.94	5.03	5.13	5.22
9	4.21	4.32	4.43	4.53	4.64	4.75	4.85	4.96	5.07	5.17	5.28	5.39	5.49	5.60	5.71	5.82
10	4.64	4.76	4.88	4.99	5.11	5.23	5.35	5.46	5.58	5.70	5.82	5.94	6.05	6.17	6.29	6.41
11	5.07	5.20	5.33	5.45	5.58	5.71	5.84	5.97	6.10	6.23	6.36	6.49	6.62	6.75	6.88	7.01
12	5.50	5.64	5.78	5.92	6.06	6.20	6.34	6.48	6.62	6.76	6.90	7.04	7.18	7.32	7.46	7.60
13	5.93	6.08	6.23	6.38	6.53	6.68	6.84	6.99	7.14	7.29	7.44	7.59	7.75	7.90	8.05	8.20
14	6.36	6.52	6.69	6.85	7.01	7.17	7.34	7.50	7.66	7.82	7.99	8.15	8.31	8.48	8.64	8.81
15	6.80	6.97	7.14	7.32	7.49	7.66	7.84	8.01	8.19	8.36	8.53	8.71	8.88	9.06	9.23	9.41
16	7.23	7.41	7.60	7.78	7.97	8.15	8.34	8.53	8.71	8.90	9.08	9.27	9.46	9.64	9.83	10.02
17	7.67	7.86	8.06	8.25	8.45	8.65	8.84	9.04	9.24	9.44	9.63	9.83	10.03	10.23	10.43	10.63
18	8.10	8.31	8.52	8.73	8.93	9.14	9.35	9.56	9.77	9.98	10.19	10.40	10.61	10.82	11.03	11.24
19	8.54	8.76	8.98	9.20	9.42	9.64	9.86	10.08	10.30	10.52	10.74	10.96	11.18	11.41	11.63	11.85
20	8.98	9.21	9.44	9.67	9.90	10.13	10.37	10.60	10.83	11.06	11.30	11.53	11.76	12.00	12.23	12.46
21	9.42	9.66	9.90	10.15	10.39	10.63	10.88	11.12	11.36	11.61	11.85	12.10	12.34	12.59	12.84	13.08
22	9.86	10.12	10.37	10.62	10.88	11.13	11.39	11.64	11.90	12.16	12.41	12.67	12.93	13.19	13.44	13.70
23	10.30	10.57	10.84	11.10	11.37	11.63	11.90	12.17	12.44	12.71	12.97	13.24	13.51	13.78	14.05	14.32
24	10.75	11.02	11.30	11.58	11.86	12.14	12.42	12.70	12.98	13.26	13.54	13.82	14.10	14.38	14.66	14.95
25	11.19	11.48	11.77	12.06	12.35	12.64	12.93	13.22	13.52	13.81	14.10	14.40	14.69	14.98	15.28	15.57
26	11.64	11.94	12.24	12.54	12.85	13.15	13.45	13.75	14.06	14.36	14.67	14.97	15.28	15.59	15.89	16.20
27	12.09	12.40	12.71	13.03	13.34	13.66	13.97	14.29	14.60	14.92	15.24	15.56	15.87	16.19	16.51	16.83
28	12.53	12.86	13.18	13.51	13.84	14.16	14.49	14.82	15.15	15.48	15.81	16.14	16.47	16.80	17.13	17.46
29	12.98	13.32	13.66	14.00	14.33	14.67	15.01	15.35	15.70	16.04	16.38	16.72	17.07	17.41	17.75	18.10
30	13.43	13.78	14.13	14.48	14.83	15.19	15.54	15.89	16.24	16.60	16.95	17.31	17.66	18.02	18.38	18.74
31	13.89	14.25	14.61	14.97	15.33	15.70	16.06	16.43	16.79	17.16	17.53	17.90	18.27	18.63	19.00	19.38
32	14.34	14.71	15.09	15.46	15.84	16.21	16.59	16.97	17.35	17.73	18.11	18.49	18.87	19.25	19.63	20.02
33	14.79	15.18	15.57	15.95	16.34	16.73	17.12	17.51	17.90	18.29	18.65	19.08	19.47	19.87	20.26	20.66
34	15.25	15.65	16.05	16.44	16.85	17.25	17.65	18.05	18.46	18.86	19.27	19.67	20.08	20.49	20.90	21.31
35	15.70	16.11	16.53	16.94	17.35	17.77	18.18	18.60	19.01	19.43	19.85	20.27	20.69	21.11	21.53	21.95
36	16.16	16.58	17.01	17.43	17.86	18.29	18.71	19.14	19.57	20.00	20.43	20.87	21.30	21.73	22.17	22.60
37	16.62	17.06	17.49	17.93	18.37	18.81	19.25	19.69	20.13	20.58	21.02	21.46	21.91	22.36	22.81	23.25
38	17.08	17.53	17.98	18.43	18.88	19.33	19.78	20.24	20.69	21.15	21.61	22.07	22.52	22.99	23.45	23.91
39	17.54	18.00	18.46	18.93	19.39	19.86	20.32	20.79	21.26	21.73	22.20	22.67	23.14	23.61	24.09	24.56
40	18.00	18.48	18.95	19.43	19.90	20.38	20.86	21.34	21.82	22.30	22.79	23.27	23.76	24.25	24.73	25.22
41	18.47	18.95	19.44	19.93	20.42	20.91	21.40	21.89	22.39	22.88	23.38	23.88	24.38	24.88	25.38	25.88
42	18.93	19.43	19.93	20.43	20.93	21.44	21.94	22.45	22.96	23.47	23.98	24.49	25.00	25.51	26.03	26.55
43	19.40	19.91	20.42	20.94	21.45	21.97	22.49	23.01	23.53	24.05	24.57	25.10	25.62	26.15	26.68	27.21
44	19.86	20.39	20.91	21.44	21.97	22.50	23.03	23.57	24.10	24.64	25.17	25.71	26.25	26.79	27.33	27.88
45	20.33	20.87	21.41	21.95	22.49	23.03	23.58	24.12	24.67	25.22	25.77	26.32	26.88	27.43	27.99	28.55
46	20.80	21.35	21.90	22.46	23.01	23.57	24.13	24.69	25.25	25.81	26.37	26.94	27.51	28.08	28.65	29.22
47	21.27	21.83	22.40	22.97	23.53	24.10	24.68	25.25	25.82	26.40	26.98	27.56	28.14	28.72	29.31	29.89
48	21.74	22.32	22.90	23.48	24.06	24.64	25.23	25.81	26.40	26.99	27.58	28.18	28.77	29.37	29.97	30.57
49	22.21	22.80	23.39	23.99	24.58	25.18	25.78	26.38	26.98	27.59	28.19	28.80	29.41	30.02	30.63	31.24
50	22.69	23.29	23.89	24.50	25.11	25.72	26.33	26.95	27.56	28.18	28.90	29.42	30.04	30.67	31.29	31.92
51	23.16	23.78	24.40	25.02	25.64	26.26	26.89	27.52	28.15	28.78	29.41	30.05	30.68	31.32	31.96	32.60
52	23.64	24.27	24.90	25.53	26.17	26.81	27.45	28.09	28.73	29.38	30.02	30.67	31.32	31.98	32.63	33.29
53	24.11	24.76	25.40	26.05	26.70	27.35	28.00	28.66	29.32	29.98	30.64	31.30	31.97	32.63	33.30	33.97
54	24.59	25.25	25.91	26.57	27.23	27.90	28.56	29.23	29.91	30.58	31.25	31.93	32.61	33.29	33.98	34.66
55	25.07	25.74	26.41	27.09	27.77	28.44	29.13	29.81	30.50	31.18	31.87	32.56	33.26	33.95	34.65	35.35
56	25.55	26.23	26.92	27.61	28.30	28.99	26.69	30.39	31.09	31.79	32.49	33.20	33.91	34.62	35.33	36.04
57	26.03	26.73	27.43	28.13	28.84	29.54	30.25	30.97	31.68	32.39	33.11	33.83	34.56	35.28	36.01	36.74
58	26.51	27.23	27.94	28.66	29.37	30.10	30.82	31.55	32.27	33.00	33.74	34.47	35.21	35.95	36.69	37.43
59	27.00	27.72	28.45	29.18	29.91	30.65	31.39	32.13	32.87	33.61	34.36	35.11	35.86	36.62	37.37	38.13
60	27.48	28.22	28.96	29.71	30.45	31.20	31.96	32.71	33.47	34.23	34.99	35.75	36.52	37.29	38.06	38.83

[a] Courtesy of the Federal Reserve System, the Truth-in-Lending, Regulation Z, Annual Percentage Rate Tables.

ANSWERS

CHAPTER 1

Exercise Set 1, Page 7

1. (a) ⓜathematics (c) ⓓeadly
(e) fⓡankly
3. Northwest **5.** (d)
7. (a) $7 - 3$ (c) $8 \div 4$ (e) $3 \cdot 17$
(g) $16 = 2 \cdot 8$
8. (a) 3 (c) 5 (e) 9
9. (a) 4 (c) 0
10. (a) $x + 4$ (c) $x - 15$ (e) $9x + 6$
(g) $10x - 16$ (i) $5x - 20$ (k) $6 + x = 15$
(m) $x + 2 > 8$
11. (a) $xy - (x + y)$ (c) $(4 \div 6) \cdot x = 8$
(e) $x \div 20$ (g) $2x - 3$
12. (a) 5, 6, 7 (c) No solution exists. (e) 1, 2, 3
13. Friday (Suggestion: Make a table.)
14. (a) $x + 7 = 10$; 3 (c) $x - 5 = 12$; 17
(e) $(3x \div 2) = x + 6$; 12 (g) $2x + 6 = 20$; 7

Exercise Set 2, Page 17

1. (b)
2. (a) △, □ (c) b, a (e) M, O
3. (a) D, S, R (c) V, W, U
5. *HIJEFG*, *GHIJEF*
6. (a) N (c) S **7.** (a) No (c) Yes
9. bludo wasca **10.** (a) 1,234,321
11.

```
. . . . .     . . . . . .     . . . . . . .
. . . . .     . . . . . .     . . . . . . .
. . . . .     . . . . . .     . . . . . . .
. . . . .     . . . . . .     . . . . . . .
. . . . .     . . . . . .     . . . . . . .
   25         . . . . . .     . . . . . . .
                  36          . . . . . . .
                                   49
```

13.

```
                    •
          •        • •
  •      • •      • • •
-------------------------
 • •    • • •    • • • •
 • •    • • •    • • • •
        • • •    • • • •
                 • • • •
```

14. (a) 2, 5, 10, 17, 26 (c) 3, 7, 11, 15, 19
15. (a) Starting with 7, add 5 to each term to get the next term.
(c) Starting with an exponent of 1, increase the exponent of 3 by 1 to obtain successive terms; or starting with 3, multiply each term by 3 to get the next term.
16. (a) Starting with 2, each term is 3 times the preceding term: 486, 1458, 4374.
(c) Starting with 3, add successive even numbers (beginning with 2) to each term to obtain the following terms: 33, 45, 59.
(e) The odd-numbered terms start with 4 and increase each time by 1. The even-numbered terms start with 1 and increase each time by 1:3, 7, 4.
17. 17, 22 **18.** (a) 81; 243
19. (a) 11 (c) 15 **21.** (a) 156
23. 8 pairs; 13 pairs; 21 pairs
24. (a) Number of ancestors in each generation is equal to the corresponding number in the Fibonacci sequence where the first 1 represents the present generation.
(c) Number of female ancestors in each generation is equal to the corresponding number of the Fibonacci sequence.
25. (a) Square the counting numbers; $4^2 = 16$; the fifth term would be 25.
27. 55 **28.** (a) $5n + 2$ (c) 3^n
29. First four terms: 7 First five terms: 12
First six terms: 20
30. (a) 1,111,110 (c) 1456
31. \$48,828.00

Exercise Set 3, Page 25

1. Identify what you are trying to find. A man could never marry his widow's sister because if his wife is a widow, then he is dead.
3. Determine the information available. The number of doughnuts doubles each minute, so at 10:29 A.M. there are half as many doughnuts as at 10:30. At 10:30 the container is full, so at 10:29 it is half full.
5. Use only pertinent information; that is, eliminate unjustified assumptions and irrelevant facts. It took the same amount of time for the return trip. The amount of time was stated in a different manner. One hour and 20 minutes is the same time as 80 minutes.
7. Use only pertinent information; that is, eliminate unjustified assumptions and irrelevant facts. There is no dirt in a hole.
9. Use only pertinent information; that is, eliminate unjustified assumptions and irrelevant facts. Of course, when they are together, they are the same distance from New York.
11. Use only pertinent information; that is, eliminate unjustified assumptions and irrelevant facts. Don't assume that the two brothers for each daughter are different brothers. There could be only 8 children in the Brown family.
13. Can you solve the problem with the coin of greatest value being a dime?
    Can you solve the problem with one quarter? Two quarters?
    Solution: five dimes, one nickel, and two pennies, or one quarter, one dime, four nickels, and two pennies.
15. How many boxes are involved? What is the total weight of all boxes? What is the relationship between the lightest box and the next to lightest? Between the second lightest and third lightest? Between the third box and the heaviest?
    Solution: 4, 8, 16, 32 lb.
17. What is the cost of gasoline per gallon? How many gallons of gasoline does the tank hold?
    Solution: $12
19. Dick is Tom's grandfather.
21. 2 hours **23.** 55¢
25. 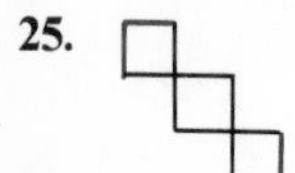

27. The same

29. 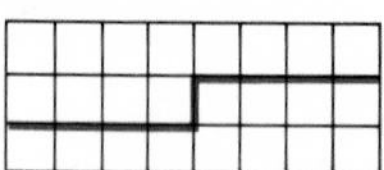 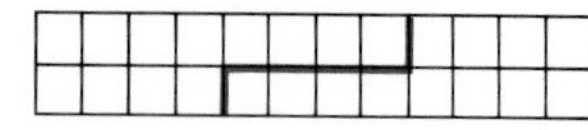

31. Fill a 4-mL container and pour into a 7-mL container. Repeat the process and pour until the 7-mL container is full. 1 mL remains in the 4-mL container.

Exercise Set 4, Page 32

1. (a) Start with a number of smaller squares and look for a pattern.
   (c) Guess, test, and revise
   (e) Restate the problem
   (g) Guess, test, and revise **3.** 55
5. Use 6 coins of the 9 to see if 3 will balance 3. If they balance, use 2 of the remaining 3 (not part of the 6 selected) to see if they balance. If they balance, the remaining coin is the counterfeit. If they do not, select the lighter coin from the 2 that do not balance. If in the original weighing the 3 coins do not balance 3 coins, select 2 from the lighter 3. If they balance, the remaining coin is the counterfeit. If they do not balance, the lighter coin is counterfeit.
7. 

8. (a)

9. 10 pigs and 3 chickens **11.** $1.19 **13.** $7
15. (a) 1; 6 5; 2 4 3 (c) No
17. 242 eggs **19.** 4; 7
21. (a) Separate the coins into two stacks and weigh to determine the lighter stack.
    (b) Weigh 4 coins of the lighter stack with 2 in each group.
       (i) If they balance, the remaining coin is counterfeit.
       (ii) If they do not balance, take the lighter stack and break it up to determine the lighter coin.
23. Four trucks A, B, C, and D start across the desert with five days of gasoline each. At the end of one day, D supplies gasoline for one day to A, B, and

C, and has one day of gasoline to return. At this point A, B, and C have gasoline for five days each. At the end of the second day C gives one day of gasoline to A and B, has used one, and has two days of gasoline to return. At the end of the third day B shares one day of gasoline with A, has used one day, and has gasoline for three days to return. A now has five days of gasoline, so this truck can cross the desert.

25.

26. (a) 8 (c) 6

27. Remove a ball from the box with incorrect label of 1 red and 1 black. If the ball is red, you know 2 red balls are in this box. Therefore, the box with the incorrect label of 2 black balls must contain 1 red ball and 1 black ball since we know it cannot contain 2 red balls. Finally the box with the incorrect label of 2 red balls must contain 2 black balls. The same reasoning would be used if a black ball had been drawn from the box with the incorrect label of 1 red and 1 black.

29. 1, 3, 9, and 27

31. 435

Exercise Set 5, Page 39

1. (a)

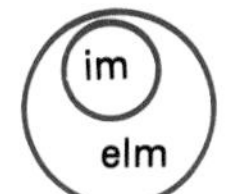

(c)

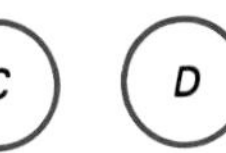

2. (a)

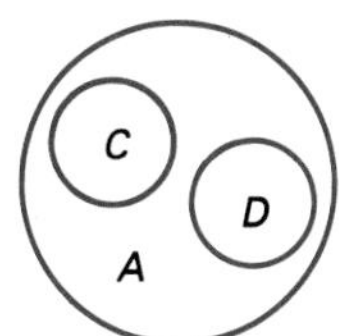

(c)

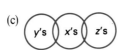

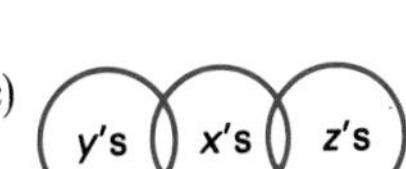

(one possibility)

(e)

x's
z's
y's

(one possibility)

3. (a) Is a meat product from a
(c) Is used to wash
(e) Grows at the end of a
(g) Will result in one being

5. (d) **7.** (a)

9. (a) Multiplied by 5 is (c) Is smaller than

11.

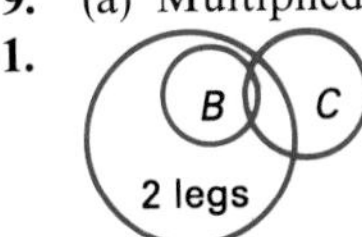

The diagram above shows that some cows have 2 legs, and the given conditions can be satisfied without all birds being cows. Also the diagram shows that it is not necessary for all cows to have 2 legs.

13.

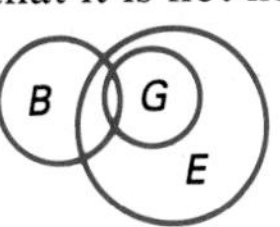

The diagram shows that there are beans that are not edible. The same conditions are satisfied by

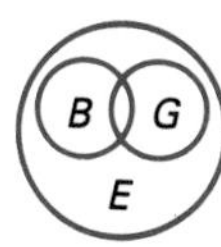

From this diagram we cannot be certain that some beans are not edible.

15.

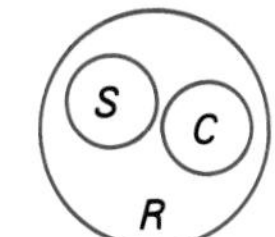

False

17.

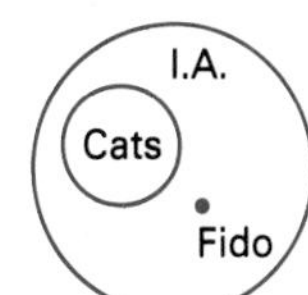

False

19.

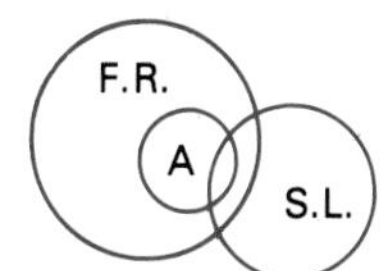

False

21. Consumption must come from production.

23. (e)

25. (b)

27. (c)

28. (a) Dr. X is not dull, not boring, and not a lady professor.
(c) Some R's are S's; however, no R is a T.

29.

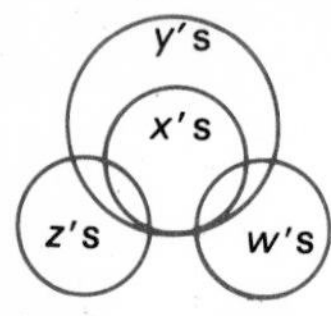

(a) Incorrect (c) Correct

Exercise Set 6, Page 46

1. (a) Statement, false (c) Statement, true
   (e) Not a statement
   (g) Not a statement unless x is known
2. (a) F (c) F (e) T
3. (a) F (c) F (e) T
4. (a) F (c) T (e) T
5. (a) November has 30 days and Thanksgiving is always on November 25. (False)
   November has 30 days or Thanksgiving is always on November 25. (True)
   (c) $2 + 3 = 4 + 1$ or $8 \cdot 6 = 4 \cdot 12$ (True)
   $2 + 3 = 4 + 1$ and $8 \cdot 6 = 4 \cdot 12$ (True)
6. (a) $A \wedge D$ (c) $C \wedge (\sim B)$
7. (a) It is snowing and the roofs are not white.
   (c) It is snowing and the roofs are white or the streets are not slick.
   (e) It is not true that it is snowing and the trees are not beautiful.
8. (a) T (c) F (e) T
9. (a)

p	q	$\sim q$	$p \vee \sim q$
T	T	F	T
T	F	T	T
F	T	F	F
F	F	T	T

(c)

p	$\sim p$	$p \wedge \sim p$
T	F	F
F	T	F

(e)

p	q	r	$\sim q$	$p \vee \sim q$	$(p \vee \sim q) \wedge r$
T	T	T	F	T	T
T	T	F	F	T	F
T	F	T	T	T	T
T	F	F	T	T	F
F	T	T	F	F	F
F	T	F	F	F	F
F	F	T	T	T	T
F	F	F	T	T	F

10. (a) $\sim q \wedge p$ (c) $\sim q \vee \sim p$
11. (a) Two (c) Eight
12. (a) The stock market is not bullish, and the Dow average is increasing.
    (c) The stock market is bullish or the price of utilities is decreasing, and the Dow average is increasing.
13. (a) $S \vee A$ (c) $\sim G \wedge \sim A$
14. (a) $p \wedge q$ (c) $\sim p \wedge q$
15. (a) False (c) False

Exercise Set 7, Page 52

1. (a) $A \to D$ (c) $\sim C \to \sim A$
2. (a) True (c) False
3. (a) *Converse:* If one angle of a triangle has a measure of 90 degrees, then the triangle is a right triangle.
   Inverse: If a triangle is not a right triangle, then no angle of the triangle has a measure of 90 degrees.
   Contrapositive: If no angle of a triangle has a measure of 90 degrees, then the triangle is not a right triangle.
   (c) *Converse:* If two alternate interior angles are equal, then two lines are parallel.
   Inverse: If two lines are not parallel, then alternate interior angles are not equal.
   Contrapositive: If alternate interior angles are not equal, then two lines are not parallel.
   (e) *Converse:* If x is divisible by 5, then x is divisible by 10.
   Inverse: If x is not divisible by 10, then x is not divisible by 5.
   Contrapositive: If x is not divisible by 5, then x is not divisible by 10.
4. (a) If a polygon is a triangle, then it is not a square.
   (c) If she is an honest politician, then she does not accept bribes.
5. (a) *Converse:* If a polygon is not a square, then it is a triangle.
   Contrapositive: If a polygon is a square, then it is not a triangle.
   (c) *Converse:* If she does not accept bribes, she is an honest politician.
   Contrapositive: If she accepts bribes, she is not an honest politician.
6. (a) If it is not snowing, then the roofs are white.
   (c) If the roofs are not white and the streets are not slick, then it is snowing.
   (e) It is not true that if it is snowing, then the roofs are not white.

7. (a) True (c) True (e) False

9. (a) *Converse:* If you have fewer cavities, then you brush your teeth with White-as-Snow.
Inverse: If you do not brush your teeth with White-as-Snow, then you do not have fewer cavities.
Contrapositive: If you do not have fewer cavities, then you do not brush your teeth with White-as-Snow.
Only the contrapositive has the same truth values as the original expression.
(c) *Converse:* If you eat Barlies for breakfast, then you want to be strong.
Inverse: If you do not want to be strong, then you do not eat Barlies for breakfast.
Contrapositive: If you do not eat Barlies for breakfast, then you do not want to be strong.
Only the contrapositive has the same truth values.
(e) *Converse:* If you use Velvet after-shave lotion, then you are tall, dark, and sexy.
Inverse: If you are not tall, dark, and sexy, then you do not use Velvet after-shave lotion.
Contrapositive: If you do not use Velvet after-shave lotion, then you are not tall, dark, and sexy.
Only the contrapositive has the same truth values.

10. (a) If I do not use leaded gasoline, then I do not pollute the atmosphere.

11. (a)

p	q	$\sim p$	$\sim q$	$\sim p \to \sim q$
T	T	F	F	T
T	F	F	T	T
F	T	T	F	F
F	F	T	T	T

(c)

p	q	r	$q \wedge r$	$p \to (q \wedge r)$
T	T	T	T	T
T	T	F	F	F
T	F	T	F	F
T	F	F	F	F
F	T	T	T	T
F	T	F	F	T
F	F	T	F	T
F	F	F	F	T

(e)

p	q	r	$p \vee q$	$r \wedge p$	$(p \vee q) \to (r \wedge p)$
T	T	T	T	T	T
T	T	F	T	F	F
T	F	T	T	T	T
T	F	F	T	F	F
F	T	T	T	F	F
F	T	F	T	F	F
F	F	T	F	F	T
F	F	F	F	F	T

(g)

p	q	$p \vee q$	$q \vee p$	$(p \vee q) \leftrightarrow (q \vee p)$
T	T	T	T	T
T	F	T	T	T
F	T	T	T	T
F	F	F	F	T

12. (a) False (c) False (e) True

13.

p	q	$p \to q$	$p \wedge \sim q$	$\sim(p \wedge \sim q)$
T	T	T	F	T
T	F	F	T	F
F	T	T	F	T
F	F	T	F	T

(columns $p \to q$ and $\sim(p \wedge \sim q)$: same)

14. (a) $\sim(\sim p)$; p

p	$\sim p$	$\sim(\sim p)$
T	F	T
F	T	F

Since the first and last columns are identical, the two statements are equivalent.

(c)

p	q	$p \wedge q$	$\sim(p \wedge q)$	$\sim p$	$\sim q$	$\sim p \wedge \sim q$
T	T	T	F	F	F	F
T	F	T	F	F	T	F
F	T	T	F	T	F	F
F	F	F	T	T	T	T

Since the fourth and seventh columns are identical, the statements are equivalent.

(e)

p	q	$p \to q$	$\sim p$	$\sim p \vee q$
T	T	T	F	T
T	F	F	F	F
F	T	T	T	T
F	F	T	T	T

Since the third and fifth columns are identical, the statements are equivalent.

(g)

p	q	$\sim p$	$\sim q$	$\sim p \vee q$	$\sim(\sim p \vee q)$	$p \vee \sim q$
T	T	F	F	T	F	T
T	F	F	T	F	T	T
F	T	T	F	T	F	F
F	F	T	T	T	F	T

The statements are not logically equivalent because the last two columns are not identical.

(i)

r	s	t	$s \wedge t$	$r \vee s$	$r \vee t$	$r \vee (s \wedge t)$	$(r \vee s) \wedge (r \vee t)$
T	T	T	T	T	T	T	T
T	T	F	F	T	T	T	T
T	F	T	F	T	T	T	T
T	F	F	F	T	T	T	T
F	T	T	T	T	T	T	T
F	T	F	F	T	F	F	F
F	F	T	F	F	T	F	F
F	F	F	F	F	F	F	F

The statements are logically equivalent, since the last two columns are identical.

Review Exercise Set 8, page 55

1. (a) Arithmetic is useful implies that logic is easy and algebra is hard.
   (c) Arithmetic is not useful and algebra is not hard.
2. (a) No
3. Animal
5. (a) $3x - 2$ (c) $4x - 5 = 7$
6. (a) 4 (c) 3 (e) 3
7. (a) *Converse:* If we play tennis, then it does not rain.
   Contrapositive: If we do not play tennis, then it rains.
9. Mary, 1; Jean; 7

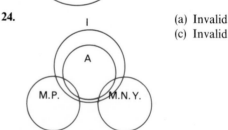

11.
```
                  1
               1     1
            1     2     1
         1     3     3     1
      1     4     6     4     1
   1     5    10    10     5     1
1     6    15    20    15     6     1
```
13. (a) *T, I, R* (c) 17, 21, 25
14. (a) Yes
15. (a) F (c) F (e) T
17. Start timing with both hourglasses at the same time. When 7 minutes have elapsed turn the 7-minute hourglass over. When the 11 minutes have elapsed, turn the 7-minute hourglass over again. It should have 4 more minutes, making a total of 15 minutes.
19. All triangles are plane figures.
21. Henry may or may not be a radical and may or may not be a Republican.
23. (a) Invalid
    (c) Invalid 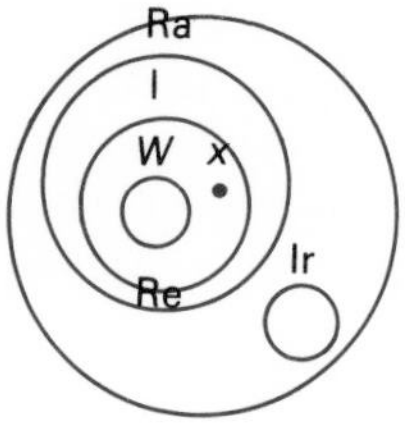
24. (a) Invalid
    (c) Invalid

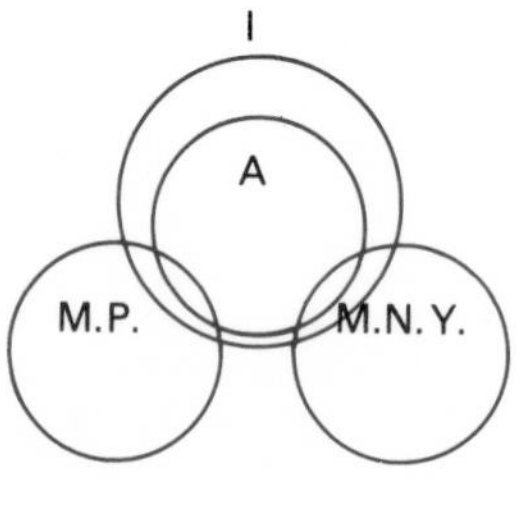

CHAPTER 2

Exercise Set 1, Page 67

1. (a) T (c) F (A set is not an element.) (e) T
2. (a) $\{1, 2, 3, \ldots, 16\}$ (c) $\varnothing$
3. (a) $\{x|x$ is a counting number less than or equal to $16\}$
   (c) $\{x|x$ is a woman president of the United States$\}$
4. (a) (a), (e) (c) (a), (c), (e)
5. (a) T (c) T (e) F
6. (a) No (c) Yes
7. (a) $\in$ (c) $\subseteq$

8. (a) $R \cup T = \{5, 10, 15, 20\}$
$R \cap T = \{15\}$
(c) $A \cup B = \{0, 10, 100, 1000\}$
$A \cap B = \{10, 100\}$
(e) $A \cup B = \{x, y, z, r, s, t\}$
$A \cap B = \{x, y\}$

9. (a) The elements of set A that are not in B
(c) The elements of set B that are not in A

10. (a) $\{a, c, e, g\}$ (c) $\{b, f, h\}$ (e) $\{c\}$

11. (a) $A \cap D$ (c) $B \cap \bar{C} \cap \bar{D}$

12. (a) U A B C

(c)

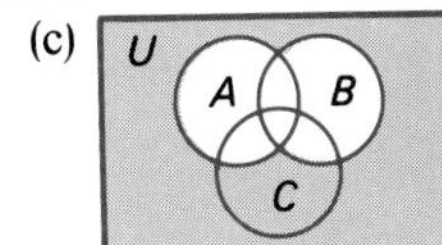

(e)

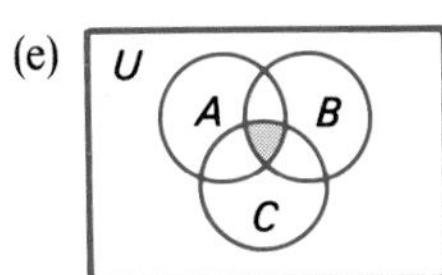

13. (a)

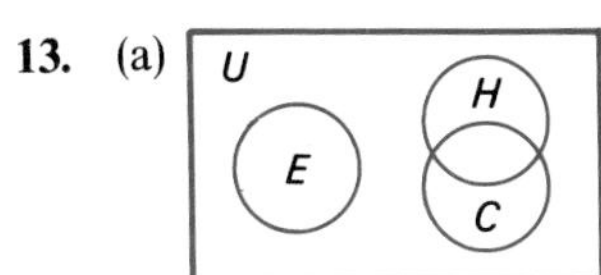

14. (a) (c) ∪ (f) ∪ (g) ∪ (h) (c) (e)
(e) (a) ∪ (b) ∪ (c) ∪ (d) ∪ (e) ∪ (f) ∪ (g)
(g) (g) ∪ (h)

15. (a) $\varnothing, \{a\}, \{b\}, \{a, b\}$
(c) $\varnothing, \{a\}, \{b\}, \{d\}, \{a, b\}, \{a, d\}, \{b, d\}$

16. (a) $B - A = \{7, 8\}$

17. (a) 8 (c) 32

18. (a) $A \cap \bar{B} \cap \bar{C}$ (c) $B \cap \bar{A} \cap \bar{C}$
(e) $A \cap B \cap C$ (g) $C \cap \bar{A} \cap \bar{B}$

19. (a)

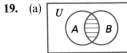

U A B

$A \cap B$

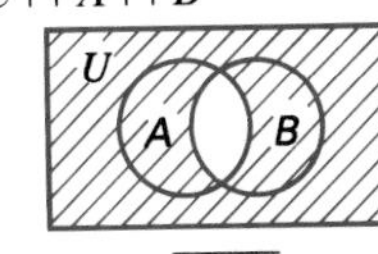

$\overline{A \cap B}$

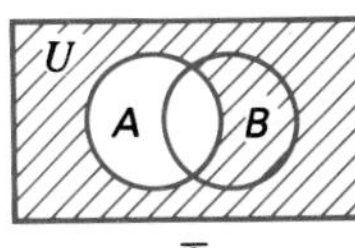

$\bar{A}$

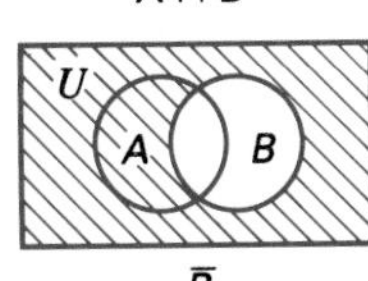

$\bar{B}$

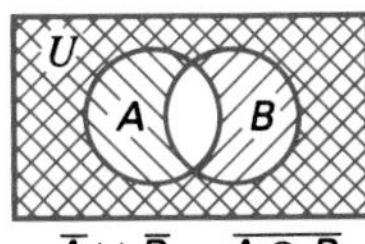

$\bar{A} \cup \bar{B} = \overline{A \cap B}$

21. (a)

$x \in A$	$x \in B$	$x \in C$	$x \in B \cup C$	$x \in A \cup (B \cup C)$	$x \in A \cup B$	$x \in (A \cup B) \cup C$
T	T	T	T	T	T	T
T	T	F	T	T	T	T
T	F	T	T	T	T	T
T	F	F	F	T	T	T
F	T	T	T	T	T	T
F	T	F	T	T	T	T
F	F	T	T	T	F	T
F	F	F	F	F	F	F

(columns $x \in A \cup (B \cup C)$ and $x \in (A \cup B) \cup C$: same)

22. (a)

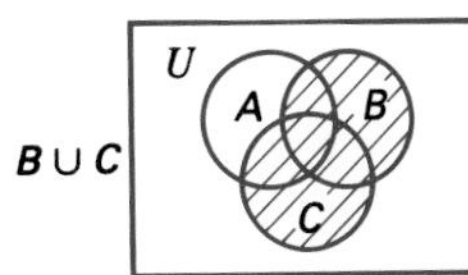

$B \cup C$

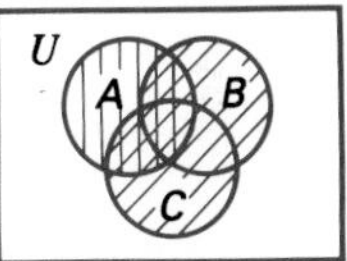

$A \cup (B \cup C)$

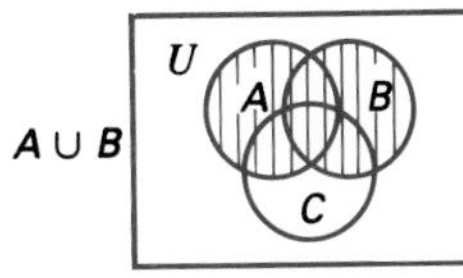

$A \cup B$

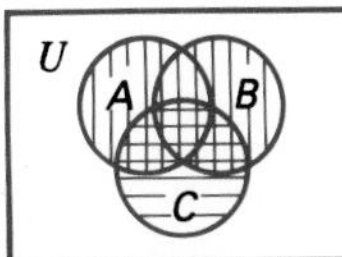

$(A \cup B) \cup C$

$(A \cup B) \cup C = A \cup (B \cup C)$

Associative property of union

23. $\{A, B, C\}, \{A, B, D\}, \{A, B, E\}, \{A, C, D\}, \{A, C, E\}, \{A, D, E\}, \{B, C, D\}, \{B, C, E\}, \{B, D, E\}, \{C, D, E\}, \{A, B, C, D\}, \{A, B, C, E\}, \{A, B, D, E\}, \{A, C, D, E\}, \{B, C, D, E\}, \{A, B, C, D, E\}$

24. (a) $\{AB\}$ $\{AC\}$ $\{BC\}$ $\{ABC\}$ $\{ABD\}$ $\{ABE\}$ $\{ACD\}$ $\{ACE\}$ $\{BCD\}$ $\{BCE\}$ $\{ABCD\}$ $\{ABCE\}$ $\{ABDE\}$ $\{ACDE\}$ $\{BCDE\}$ $\{ABCDE\}$
(c) No

Exercise Set 2, Page 77

1. (a) y is a manufacturer of product x.

2. (a) • A is heavier than • B is heavier than • C

Transitive is heavier than

• A lives in the same city as A Reflexive

• A lives in the same city as B • Symmetric

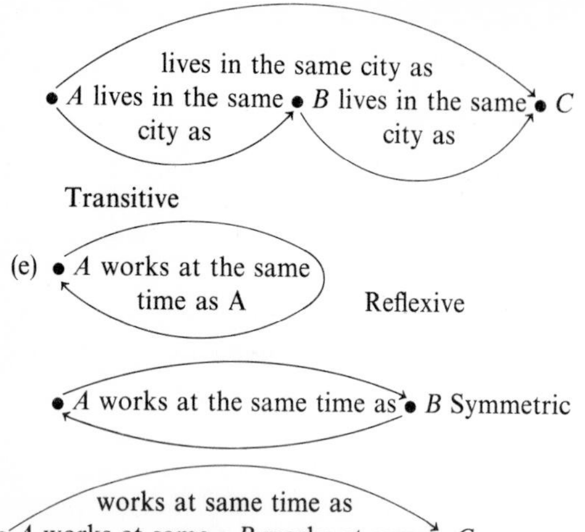

(g) None

3. (a) Symmetric (c) Transitive
(e) Transitive (g) An equivalence relation
(i) Symmetric
4. (a) An equivalence relation
(c) Reflexive, transitive
5. (a) $A \times B = \{(a, r), (a, s), (a, t), (b, r), (b, s), (b, t), (c, r), (c, s), (c, t)\}$
(c) $B \times A = \{(r, a), (r, b), (r, c), (s, a), (s, b), (s, c), (t, a), (t, b), (t, c)\}$
6. (a) $A \times (B \cap C) = \{(a, c), (b, c), (c, c)\}$
(c) $(A \cap B) \times C = \{(c, a), (c, c), (c, x)\}$
7. (a) $B \times C = \{(3, 0)\}$
8. (a) $B = \{1, 4\}$, $C = \{1, 2, 3\}$
(c) $B = \{6\}$, $C = \{6, 7, 8\}$
9. $P \times S = \{$(Mardis, Janet), (Mardis, Judy), (Mardis, Marion), (Boyd, Janet), (Boyd, Judy), (Boyd, Marion), (Kiefer, Janet), (Kiefer, Judy), (Kiefer, Marion)$\}$
10. (a) 9; 9 (c) r^2; rs; s^2
11. (a) $\{(1, 2), (1, 5), (2, 5), (3, 5), (4, 5)\}$ (c) $\{(2, 2)\}$
12. (a) No (c) Yes
13. (a) $B \times (C \times A) = \{[3, (4, 1)], [3, (4, 2)], [3, (5, 1)], [3, (5, 2)], [4, (4, 1)], [4, (4, 2)], [4, (5, 1)], [4, (5, 2)], [5, (4, 1)], [5, (4, 2)], [5, (5, 1)], [5, (5, 2)]\}$
15. (a) $R = \{(1, 3), (3, 5), (1, 5)\}$
(c) $R = \{(1, 1), (1, 3), (3, 3), (3, 5), (5, 5), (1, 5)\}$
16. (a) {Asa, Jim, Ben, Bob, Joe, Amy}
{Allen, Betty, Alisa}
{John, Jack, Alto, Bill, Jill}

17.

$x \in A$	$x \in B$	$y \in C$	$x \in (A \cup B)$	$(x, y) \in (A \cup B) \times C$	$(x, y) \in A \times C$	$(x, y) \in B \times C$	$(x, y) \in [(A \times C) \cup (B \times C)]$
T	T	T	T	T	T	T	T
T	T	F	T	F	F	F	F
T	F	T	T	T	T	F	T
T	F	F	T	F	F	F	F
F	T	T	T	T	F	T	T
F	T	F	T	F	F	F	F
F	F	T	F	F	F	F	F
F	F	F	F	F	F	F	F

Since the truth values are the same, $(A \cup B) \times C = (A \times C) \cup (B \times C)$.

Exercise Set 3, Page 84

1. (a) C (c) O (e) C (g) C (i) C
(k) C
2. (a) Yes (c) Yes (e) No
3. (a) $\{b\}$ (c) $\{a, b, c, d, e\}$
$\updownarrow$ $\updownarrow\ \updownarrow\ \updownarrow\ \updownarrow\ \updownarrow$
$\{1\}$ $\{1, 2, 3, 4, 5\}$
(e) $\{10, 40, 30, 50, 70\}$
$\updownarrow\ \updownarrow\ \updownarrow\ \updownarrow\ \updownarrow$
$\{1, 2, 3, 4, 5\}$
4. (a) 17 (c) 5 (e) 30 (g) 47 (i) 42
5. (a) True (c) True (e) False (g) True
(i) True (k) True
6. (a) A and R are not equal sets because the elements are not pairwise equal. In fact, none of the elements are equal.
7. (a) True (c) True (e) True
8. (a) $A \cup B = \{2, 3, 4, \ldots, 10\}$ so $n(A \cup B) = 9$, $n(A) = 5$, and $n(B) = 5$.

$$A \cap B = \{6\} \text{ so } n(A \cap B) = 1.$$

Therefore $n(A \cup B) = n(A) + n(B) - n(A \cap B)$, since $9 = 5 + 5 - 1$.

(c) $A \times A = \{(2, 2), (2, 3), (2, 4), (2, 5), (2, 6), (3, 2), (3, 3), (3, 4), (3, 5), (3, 6), (4, 2), (4, 3), (4, 4), (4, 5), (4, 6), (5, 2), (5, 3), (5, 4), (5, 5), (5, 6), (6, 2), (6, 3), (6, 4), (6, 5), (6, 6)\}$
$n(A \times A) = 25$ and $n(A) = 5$
Therefore $n(A \times A) = n(A) \cdot n(A)$, since $25 = 5 \cdot 5$

9. (a) False: Let $A = \{a, b, c, d, e\}$ and $B = \{e, f, g, h, i\}$. $n(A \cup B) = 9$, $n(A) = 5$, and $n(B) = 5$ $n(A \cup B) \neq n(A) + n(B)$, since $5 + 5 \neq 9$.

(c) False: Let $A = \{x, y, z, r, w\}$ and $B = \{\square, \triangle, \bigcirc, 3, 4\}$. $n(A) = n(B) = 5$, but $A \neq B$.

(e) True

(g) False: $U = \{1, 2, \ldots, 10\}$; $A = \{1, 2, \ldots, 5\}$; $B = \{6, 7, \ldots, 10\}$. $n(\bar{A} \cup \bar{B}) = 10$; $n(\overline{A \cup B}) = 0$; $10 \neq 0$

11. (a)

U; S; O; 650; 150; 150; 50

(c) 850
(e) 150

12. (a)

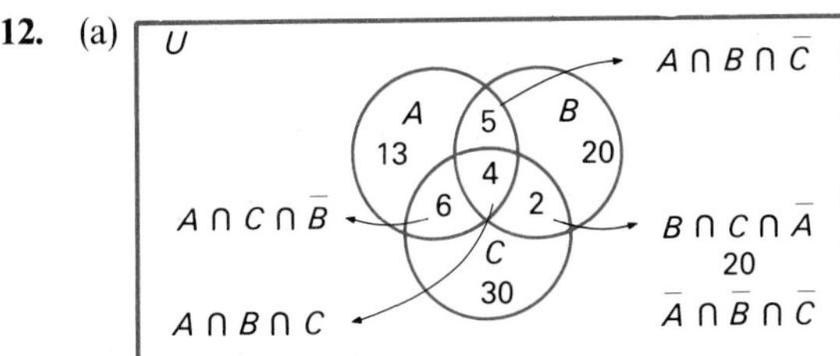

A = car pools
B = buses
C = drove alone (c) 13

13. No, only 49

15. The poll was not accurate if all used one of the shampoos. This may be seen by the diagram drawn from the given information.

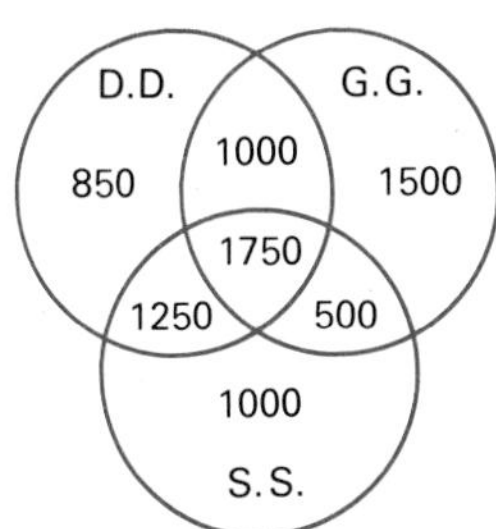

The total number of people is 7850, not 7900 as it should have been.

Exercise Set 4, Page 92

1. (a) $n(R) = 4, n(S) = 3, n(R \cup S) = 7, 4 + 3 = 7$; yes
(c) $n(R) = 4, n(S) = 4, n(R \cup S) = 7, 4 + 4 \neq 7$; no
(e) $n(R) = 0, n(S) = 1, n(R \cup S) = 1, 0 + 1 = 1$; yes

2. (a) Commutative (c) Additive identity
(e) Associative (g) Commutative
(i) Commutative

3. (c) $(56 + 8) + 70 = 134$

4. (a) 7 (c) 9

5. (a)

8; 3; 5; 0 1 2 3 4 5 6 7 8 9

(c) The commutative property of addition

6. (a) $\boxed{3} + (\triangle\!\!\!2 + ①) = (\boxed{3} + \triangle\!\!\!2) + ①$
$3 + 3 = 5 + 1$
$6 = 6$
The associative property of addition

(c) $\boxed{3} + ⑤ = ⑤ + \boxed{3}$
$8 = 8$
The commutative property of addition

7. (a) Yes (c) No (e) Yes (g) Yes

8. (a) The sum of the sets does not equal 5; the sum of the cardinal numbers of the sets equals 5.
(c) $3 + 2 \neq 4$ (Sets are not disjoint.)
(e) It is not correct to speak of the union of two numbers. Union is defined for sets.
(g) It is not correct to speak of the union of a number and the null set.
(i) It is not correct to add sets; the sum of a number and a set has no meaning.

9. (a) Numbers symmetric about the diagonal from upper left to lower right are equal.
(c) Yes, 0. The first column and row are identical to the number added to zero.
(e) Closure property for addition

10. (a) No (c) No, $a + 0 = a$ but $0 + a = 2a$.

11. (a) Commutative and associative properties hold; no additive identity.
(e) Commutative and associative properties hold; no additive identity.
(g) Commutative and associative properties hold; additive identity is 0.

13. (a) $\{5\}$ (c) $\{3\}$ (e) $\{\ \}$

15. x = distance Xan hiked
$x + 6$ = distance John hiked
$x + (x + 6) = 22$
$x = 8$; $x + 6 = 14$

17. (a) $(2 + 4) + 3$
$= 3 + (2 + 4)$ Commutative property of addition
$= (3 + 2) + 4$ Associative property of addition
$(2 + 4) + 3$
$= (3 + 2) + 4$ Transitive property of equalities
(c) $8 + (5 + 2)$
$= (8 + 5) + 2$ Associative property of addition
$= 2 + (8 + 5)$ Commutative property of addition
$8 + (5 + 2)$
$= 2 + (8 + 5)$ Transitive property of equalities
(e) $(a + b) + (c + d)$
$= (a + b) + (d + c)$ Commutative property of addition
$= [(a + b) + d] + c$ Associative property of addition
$= [a + (b + d)] + c$ Associative property of addition
$= [(b + d) + a] + c$ Commutative property of addition
$= (b + d) + (a + c)$ Associative property of addition
$(a + b) + (c + d)$
$= (b + d) + (a + c)$ Transitive property of equalities

18. (a) $a + (c + b)$
$= a + (b + c)$ Commutative property of addition
(c) $(b + c) + a$
$= a + (b + c)$ Commutative property of addition
(e) $c + (a + b)$
$= (a + b) + c$ Commutative property of addition
$= a + (b + c)$ Associative property of addition
$c + (a + b)$
$= a + (b + c)$ Transitive property of equalities

19. (a) Yes (c) Yes

21. (a) Let $A = \{l, m, n, o, p, q\}$ $B = \{r, s, t, u\}$
Then $A \cup B = \{l, m, n, o, p, q, r, s, t, u\} = B \cup A$
Thus, $n(A \cup B) = n(B \cup A)$
Therefore, $n(A) + n(B) = n(B) + n(A)$
or $6 + 4 = 4 + 6$

Exercise Set 5, Page 102

1. (a) False (c) False (e) False
2. (a) Commutative property of multiplication
(c) Multiplicative property of 0
(e) Associative property of multiplication
3. (a) $18 \cdot 5 = 9 \cdot (2 \cdot 5) = 9 \cdot 10$
(c) $8 \cdot 36 = (8 \cdot 4) \cdot 9 = 32 \cdot 9$
4. (a) $(2 \cdot 3) + (2 \cdot 4)$ (c) $4(2 + 3)$
(e) $2x(2ay + z)$ (g) $(a + 5)x$
(i) $(2 + 3)(x + 4)$
5. (a) Commutative property of addition
(c) Commutative property of multiplication
(e) Additive identity
(g) Multiplicative identity and commutative property of addition
(i) Distributive property of multiplication over addition
(k) Commutative property of addition
(m) Commutative property of multiplication
7. (a) $5(3 + 7 + 5 + 6) = 5(21) = 105$
$5(3 + 7 + 5 + 6) = 5(3) + 5(7) + 5(5) + 5(6) = 15 + 35 + 25 + 30 = 105$
8. (a) $\{2\}$ (c) $\{0\}$
9. Let x = total number of points scored.
$$3 \cdot 8 = x \quad \text{and} \quad x = 24$$
11. Let x = number of Smackos purchased.
$$20x = 80 \quad \text{and} \quad x = 4$$
13. (a) Numbers symmetric about the diagonal from upper left to lower right are equal.
(c) Yes (1); the repetition of the numbers under the "1" column and in the "1" row
(e) Closure property for multiplication of whole numbers
(g) For multiplication by 2, adding 2 to the previous product gives next product. All products are even numbers. For multiplication by 5, adding 5 to the previous product gives the next product. All products are multiples of 5.
14. (a) No; $2 + (3 + 5) \neq (2 + 3) + (2 + 5)$ as $10 \neq 5 + 7$
(c) No; $2(3 \cdot 4) \neq (2 \cdot 3)(2 \cdot 4)$ as $24 \neq 6 \cdot 8$
15. (a) Not closed, identity 1
(c) Closed, no identity
(e) Closed, no identity (g) Closed, identity 1
(i) Closed, just one element, not 1
16. (a) $4x + 12$
(c) $x^2 + ax$ $(x \cdot x = x^2)$
(e) $4y^2 + 4yx + 8yz$
17. (a) $3 \cdot 10 + 3 \cdot 2 = 36$

(c) $3(100 + 30 + 2) = 3(100) + 3(30) + 3(2)$
$= 396$

18. (a) 248

19. (a) $x(x + 2) + 3(x + 2) = x^2 + 2x + 3x + 6 = x^2 + 5x + 6$
(c) $x(x + y) + y(x + y) = x^2 + xy + yx + y^2 = x^2 + 2xy + y^2$

20. (a) $8(x + 2)$ (c) $6x(x + 3)$
(e) $(x + 2)(3 + x)$

21. (a)

$(2 \cdot 3)4 = 4(2 \cdot 3)$	Commutative property of multiplication
$= (4 \cdot 2)3$	Associative property of multiplication
$(2 \cdot 3)4 = (4 \cdot 2)3$	Transitive property of equalities

(c)

$(6 + 1)5$	
$= (6 \cdot 5) + (1 \cdot 5)$	Distributive property of multiplication over addition
$= (1 \cdot 5) + (6 \cdot 5)$	Commutative property of addition
$= (1 \cdot 5) + (5 \cdot 6)$	Commutative property of multiplication
$(6 + 1)5$ $= (1 \cdot 5) + (5 \cdot 6)$	Transitive property of equalities

(e)

$2(3 \cdot 1) = (2 \cdot 3)1$	Associative property of multiplication
$= 1(2 \cdot 3)$	Commutative property of multiplication
$= (1 \cdot 2)3$	Associative property of multiplication
$2(3 \cdot 1) = (1 \cdot 2)3$	Transitive property of equalities

22. (a) Yes (c) Yes

23. (a) Yes (c) Yes

24. (a) 6 (c) 110 (e) 37

Exercise Set 6, Page 109

1. (a) 2 (c) No answer (e) 3

2. (a)

2 −5
7
0 1 2 3 4 5 6 7

(c)

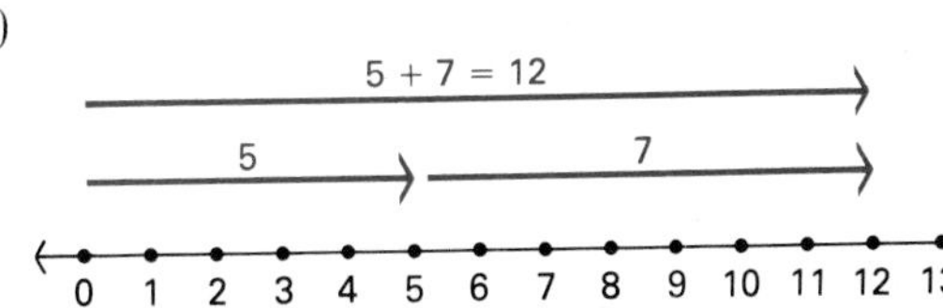

(e)

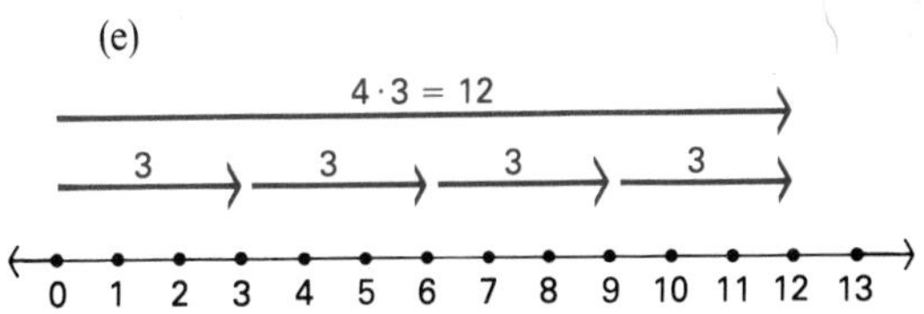

(g)

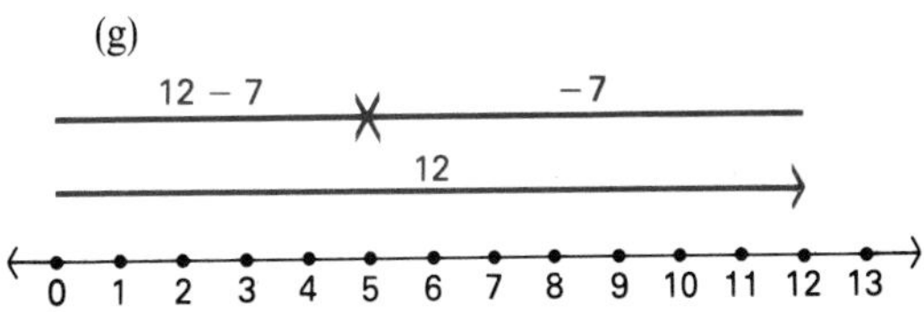

3. (a) True (c) False (e) True (g) True

4. (a) $21 - 16 = 5$; $21 - 5 = 16$
(c) $f - a = c$; $f - c = a$

5. (a) $4 - 6 = ?$ (no whole number)
(c) $(5 - 3) - 2 = 5 - (3 - 2)$
$2 - 2 = 5 - 1$
$0 \neq 4$

6. (a) 8 (c) 0 (e) 6

7. 3

8. (a)

36
−9
27
−9
18
−9
9
−9
0

→ 4

9. (a) $6(8 - 5) = 6(3) = 18$
$6(8 - 5) = 6(8) - 6(5) = 18$
(c) $17(10 - 4) = 17(6) = 102$
$17(10 - 4) = 17(10) - 17(4) = 102$

10. (a) False (c) False

11. No; $3 \not< 3$ (not reflexive).
If $3 < 5$, then $5 \not< 3$ (not symmetric).
If $3 < 5$, and $5 < 7$, then $3 < 7$ (transitive).

12. (a) $\{0, 1, 2, 3\}$ (c) $\{0, 1, 2\}$

13. (a) Let $x = 4$, $y = 3$
$4 - 3 \neq 3 - 4$
Let $x = 4$, $y = 4$
$4 - 4 = 4 - 4$
(c) Let $x = 2$
$2 - 0 \neq 0 - 2$
Let $x = 0$
$0 - 0 = 0 - 0 = 0$

15. Let x = number of football tickets purchased.
$4x < 11$
$x = \{0, 1, 2\}$
He purchased 2 tickets; he has \$3 left.
17. Let x = number of baskets
$8x = 60 \cdot 6$
$x = 45$ baskets
18. (a) $(8 + 4) \div 2 = (8 \div 2) + (4 \div 2)$
$12 \div 2 = 4 + 2$
$6 = 6$
19. (a) $a = b \neq 0$ (c) 3
20. (a) $a \geq b$ (c) $a \geq c$ and $(a - c) \geq b$
21. (a) $\{3\}$ (c) $\{1\}$ (e) $\{0, 1\}$
22. (a) 40 (c) 385
23. (a) all whole numbers (c) $x > 7$ and a whole number (e) $\{4, 2\}$

Review Exercise Set 7, Page 113

1. (a) 0 (c) Not 0 (e) 0 (g) Not 0
2. (a) True (c) True (e) False (g) True
3. (a) $\{x \mid x \in A \text{ and } x \in B\}$
(c) $\{(x, y) \mid x \in A \text{ and } y \in B\}$
4. (a) $\{1, 2, \ldots, 10\}$ (c) $\{4, 5, 6, 7, 8\}$
5. (a) False; (b, c) and (c, d) are elements of the relation but (b, d) is not.
(c) False: the set has 1,000,000 elements.
(e) False: $A \times B \neq B \times A$ and $(A \times B) \times C \neq A \times (B \times C)$
(g) True
6. (a) $A \cap \varnothing = \varnothing$ (c) $A \cap U = A$
(e) $A \cap \bar{A} = \varnothing$ (g) $U \cap \varnothing = \varnothing$
7. (a) $A \cap B = \{b, c\} = B \cap A$
(c) $(A \cap B) \cap C = \{c\} = A \cap (B \cap C)$
(e) $A \cap (B \cup C) = \{b, c\} = (A \cap B) \cup (A \cap C)$
8. (a) Associative property of addition
(c) Commutative property of multiplication
9. $7 + 3 = 10$
$5 \cdot 6 = 30$
$7 = 5 + 2$ and 2 is a natural number
$3 + 11 = 14$ and 11 is a natural number
10. (a) $3 + 3 + 3$
11. (a) 0
12. (a) $A \cap (B \cup C) = \{4, 5\}$
(c) $C \cup (B \cup A) = \{1, 2, 3, 4, 5, 6, 7, 8, 9, 10, 11, 12, 13, 14, 15\}$
13. (a) True (c) True
14. (a) Not reflexive; not symmetric; transitive; not an equivalence relation
15. (a) $x \boxdot y = y \boxdot x$
(c) $(x \boxdot y) \boxdot z = x \boxdot (y \boxdot z)$
(e) $x \boxdot (y \oplus z) = (x \boxdot y) \oplus (x \boxdot z)$
16. (a) $\{2\}$ (c) $x > 4$ and a whole number
(e) $\{0, 1, 2\}$
17. (a)

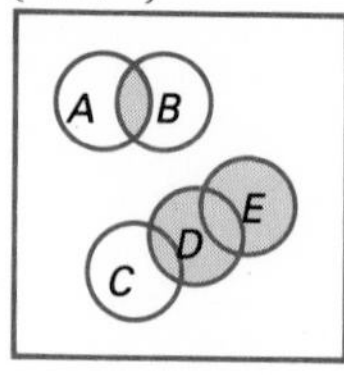

18. (a) $\overline{(A \cup B)} \cap C$ (c) $B \cup \bar{A}$
19. (a) $(110)(9) - (9)(6)$
$= (9)(110) - (9)(6)$ Commutative property of multiplication
$= 9(110 - 6)$ Distributive property of multiplication over subtraction
$(110)(9) - (9)(6) = 9(110 - 6)$ Transitive property of equalities

(c) $4(23 \cdot 25) = (4 \cdot 23)25$ Associative property of multiplication
$= 25(4 \cdot 23)$ Commutative property of multiplication
$= 25(23 \cdot 4)$ Commutative property of multiplication
$4(23 \cdot 25) = 25(23 \cdot 4)$ Transitive property of equalities

(e) $c(d \cdot e) = (c \cdot d)e$ Associative property of multiplication
$= e(c \cdot d)$ Commutative property of multiplication
$= e(d \cdot c)$ Commutative property of multiplication
$c(d \cdot e) = e(d \cdot c)$ Transitive property of equalities
21. (a) 20 (c) 20
22. (a) 2 (c) 2 (e) 43

CHAPTER 3

Exercise Set 1, Page 122

1. (a) $43{,}736 = 4(10{,}000) + 3(1000) + 7(100) + 3(10) + 6$
(c) $14{,}136 = 1(10{,}000) + 4(1000) + 1(100) + 3(10) + 6$
(e) $3{,}111{,}005 = 3(1{,}000{,}000) + 1(100{,}000) + 1(10{,}000) + 1(1000) + 0(100) + 0(10) + 5$
2. (a) 34,786 (c) 740
3. (a) Thousands (c) Ones

4. (a) Thousands (c) Ten thousands
5. (a) The number represented by 2
6. (a) LXXVI (c) CLXXXIX (e) CXLVIII
7. (a) 753
9. (a) 43,210
10. (a) 29 (c) 1776 (e) 10,000,649
11. (a) 4 (c) 34 (e) 200,320
12. (a) 2 (c) 11 (e) 4862
13. (a) 142 (c) 3410
14. (a) 6 (c) 7 (e) 9 (g) 843
15. (a) ∩|||| (c) 9∩∩∩∩∩∩∩∩||||
16. (a) ‹ ▼▼▼▼ (c) ▼▼▼ ▼▼▼▼
17. (a) 一
十
四 (c) 一
百
八
十
四
18. (a) ••••
═ (c) ••••
—
••••
19. (a) 200,320; 36,601; 142
20. (a) MLXXXVIII; MXC
(c) ∩|∩|∩|∩|∩|∩|∩|∩|∩
9
21. (a) Sum = 78; difference = 50
(c) Sum = 874; difference = 490
23. (a) 9∩∩∩∩∩, 9∩∩∩∩∩∩∩|||||

Exercise Set 2, Page 133

1.

10^9	1,000,000,000	billion
10^8	100,000,000	hundred million
10^7	10,000,000	ten million
10^6	1,000,000	million
10^5	100,000	hundred thousand
10^4	10,000	ten thousand
10^3	1000	thousand
10^2	100	hundred
10^1	10	ten
10^0	1	one

2. (a) 2^5 (c) $7^3 \cdot 5^2$ (e) 3^7
3. (a) 81 (c) 1
4. (a) $12 \cdot 10^7$
5. (a) 4578 (c) 8035 (e) 405 (g) 46,307
6. (a) $3(10)^3 + 4(10)^2 + 3(1)$
(c) $3(10)^7 + 4(10)^6 + 1(10)^3 + 6(10)^2 + 2(10)^1$
(e) $4(10)^6$
7. (a) 4,560,000
(c) When a number is multiplied by 10, the units digit becomes a tens digit, the tens digit becomes a hundreds digit, the hundreds digit becomes a thousands digit, and so on. Repeat for additional 10's.
8. (a) (i) $\begin{array}{r} 364 \\ +423 \\ \hline \end{array} = \begin{array}{l} 3(100) + 6(10) + 4(1) \\ 4(100) + 2(10) + 3(1) \\ \hline 7(100) + 8(10) + 7(1) = 787 \end{array}$

(ii) $\begin{array}{r} 364 \\ +423 \\ \hline \end{array} = \begin{array}{l} 3(10)^2 + 6(10) + 4(1) \\ 4(10)^2 + 2(10) + 3(1) \\ \hline 7(10)^2 + 8(10) + 7(1) = 787 \end{array}$

(c) (i) $\begin{array}{r} 426 \\ -\ 14 \\ \hline \end{array} = \begin{array}{r} 4(100) + 2(10) + 6(1) \\ -[1(10) + 4(1)] \\ \hline 4(100) + 1(10) + 2(1) \end{array} = 412$

(ii) $\begin{array}{r} 426 \\ -\ 14 \\ \hline \end{array} = \begin{array}{r} 4(10)^2 + 2(10) + 6(1) \\ -[1(10) + 4(1)] \\ \hline 4(10)^2 + 1(10) + 2(1) \end{array} = 412$

(e) (i) $\begin{array}{r} 1894 \\ -\ 562 \\ \hline \end{array} = \begin{array}{r} 1(1000) + 8(100) + 9(10) + 4(1) \\ -[5(100) + 6(10) + 2(1)] \\ \hline 1(1000) + 3(100) + 3(10) + 2(1) \end{array}$
$= 1332$

(ii) $\begin{array}{r} 1894 \\ -\ 562 \\ \hline \end{array} = \begin{array}{r} 1(10)^3 + 8(10)^2 + 9(10) + 4(1)] \\ -[5(10)^2 + 6(10) + 2(1)] \\ \hline 1(10)^3 + 3(10)^2 + 3(10) + 2(1) \end{array}$
$= 1332$

9. (a) $5(10)^2 + 13(10)$
$= 5(10)^2 + (10 + 3)10$ Renaming 13
$= 5(10)^2 + 10(10) + 3(10)$ Distributive property of multiplication over addition
$= [5(10)^2 + 1(10)^2] + 3(10)$ Associative property and renaming 100
$= (5 + 1)(10)^2 + 3(10)$ Distributive property of multiplication over addition
$= 6(10)^2 + 3(10)$ Addition

(c) $16(10) + 17$

$= 16(10) + 1(10) + 7$ Renaming 17

$= (16 + 1)(10) + 7$ Associative property of addition and distributive property of multiplication over addition

$= 17(10) + 7$ Addition

$= (10 + 7)(10) + 7$ Renaming 17

$= 10(10) + 7(10) + 7$ Distributive property of multiplication over addition

$= 1(10)^2 + 7(10) + 7$ Renaming 100

10. (a)
$$\begin{array}{r} 700 + 60 + 8 \\ 500 + 70 + 4 \\ \hline 1200 + 130 + 12 = 1342 \end{array}$$

11. (a)
$$\begin{array}{r} 40 + 6 \\ 70 + 5 \\ \hline 110 + 11 = 121 \end{array}$$

(c)
$$\begin{array}{r} 60 + 0 \\ -[30 + 7] \\ \hline \end{array} \quad \text{or} \quad \begin{array}{r} 50 + 10 \\ -[30 + 7] \\ \hline 20 + 3 \end{array} = 23$$

13. (a)
$$\begin{array}{rrrr} 8 & 6 & 4 & 2 \\ 7 & 5 & 3 & 1 \\ \hline 16 & 1 & 7 & 3 \end{array}$$

14. (a)
$$\begin{array}{rrrr} 8 & 7 & 6 & 5 \\ 1 & 2 & 3 & 4 \\ \hline 7 & 5 & 3 & 1 \end{array}$$

15. (a) $4^0, 10^2, 7^3, 2^{10}, 3^7, 9^4, 4^9$

16. (a) 2^{15} (c) 2^{10} (e) 2^{26}

17. (a) 4 (c) 2 (e) 3 (g) 1 (i) 2

19. (a)
$$\begin{array}{r} 30 + 4 \\ 20 + 7 \\ \hline 50 + 11 \end{array} \qquad \begin{array}{r} 34 \\ 27 \\ \hline 11 \\ 50 \\ \hline 61 \end{array} \qquad \begin{array}{r} \scriptstyle 1 \\ 34 \\ 27 \\ \hline 61 \end{array}$$

(c)
$$\begin{array}{r} 100 + 60 + 7 \\ 200 + 40 + 5 \\ \hline 300 \quad 100 + 12 \end{array} \qquad \begin{array}{r} 167 \\ 245 \\ \hline 12 \\ 100 \\ 300 \\ \hline 412 \end{array} \qquad \begin{array}{r} \scriptstyle 11 \\ 167 \\ 245 \\ \hline 412 \end{array}$$

20. (a) 32 (c) 1,048,576

21. (a) Renaming 28
(c) Definition of addition
(e) Associative property of addition
(g) Definition of addition

22. (a) Steps (a) through (e)

23. (a) In the first example, $7 + 3 = 10$, which is written down. The $9(10) + 2(10) = 11(10)$, which is written as 11 in the "tens" place. Then $6(10)^2 + 3(10)^2 = 9(10)^2$, which is written as 9 in the "hundreds" place. Then the three "presums" are added.

(c)
$$\begin{array}{r} 745 \\ 628 \\ 211 \\ \hline 14 \\ 7 \\ 15 \\ \hline 1584 \end{array}$$

24. (a) $(5000 + 700 + 6) - (3000 + 400 + 7)$
$(5000 + 700 + 16) - (3000 + 400 + 10 + 7)$
$(5000 + 700 + 100 + 16) - (3000 + 500 + 10 + 7)$
$(2000 + 200 + 90 + 9) = 2299$

(c) $(1000 + 600 + 30 + 4) - (900 + 80 + 5)$
$(1000 + 600 + 30 + 14) - (900 + 90 + 5)$
$(1000 + 600 + 130 + 14) - (1000 + 90 + 5)$
$(600 + 40 + 9) = 649$

25. (a)
$$\begin{array}{r} 6299 \\ -4000 \\ \hline 2299 \end{array}$$
(c)
$$\begin{array}{r} 1649 \\ -1000 \\ \hline 649 \end{array}$$

26. (a)
$$\begin{array}{r} 5706 \\ 6592 \\ \hline 12298 \\ -11 \\ \hline 2299 \end{array}$$
(c)
$$\begin{array}{r} 1634 \\ 14 \\ \hline 1648 \\ -11 \\ \hline 649 \end{array}$$

27. (a)
$$\begin{array}{r} \scriptstyle 16 \\ 5706 \\ -3407 \\ \hline 2\not{3}\not{1}\not{0}9 \\ 29 \\ 2299 \end{array}$$
(c)
$$\begin{array}{r} \scriptstyle 13\,14 \\ 163\,4 \\ -98\,5 \\ \hline \not{7}\not{5}\,9 \\ 64 \\ 64\,9 \end{array}$$

Exercise Set 3, Page 142

1. (a)

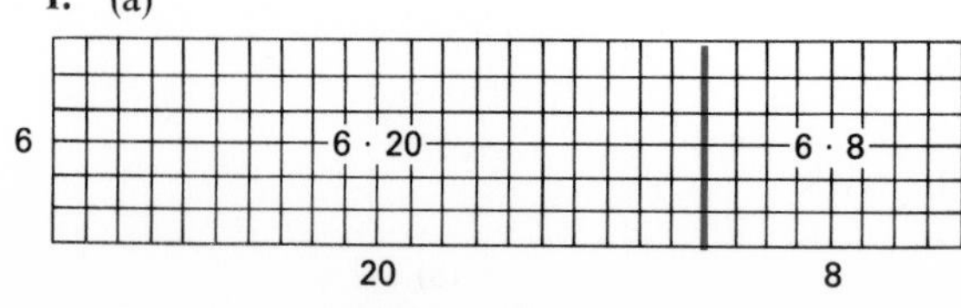

(c)

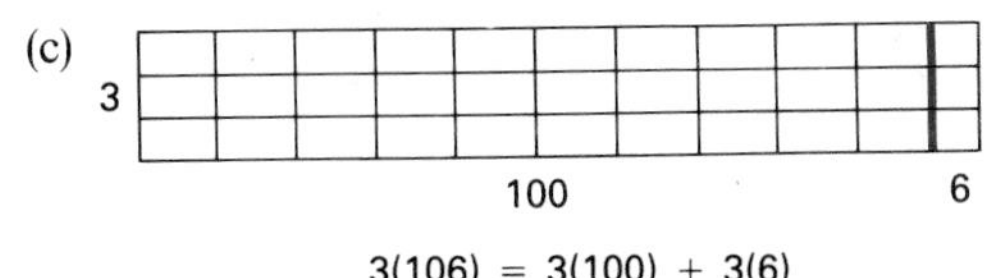

3(106) = 3(100) + 3(6)

2. (a) (i) $6 \cdot 28 = 6(20 + 8) = 120 + 48 = 168$

(ii)
```
  28
 · 6
  48   (6 · 8)
 120   (6 · 20)
 168
```
(iii)
```
  4
 28
· 6
168
```

(c) (i) $3 \cdot 106 = 3(100 + 6) = 300 + 18 = 318$

(ii)
```
 106
  ·3
  18   (3 · 6)
 300   (3 · 100)
 318
```
(iii)
```
  1
 106
  ·3
 318
```

3. See Exercise 2.

4. (a) $q = 6, r = 6$ (c) $q = 0, r = 11$
(e) $q = 3, r = 9$ (g) $q = 6, r = 0$

5. (a) (i)
```
6 ⟌ 166
    120    20 (6)
     46
     42     7 (6)
      4    27
```
(ii)
```
     27
6 ⟌ 166
    120
     46
     42
      4
```
(iii)
```
     27
6 ⟌ 166
    12
     46
     42
      4
```

(c) (i)
```
8 ⟌ 1425
     800    100 (8)
     625
     560     70 (8)
      65
      64      8 (8)
       1    178
```
(ii)
```
      178
8 ⟌ 1425
     800
     625
     560
      65
      64
       1
```
(iii)
```
     178
8 ⟌ 1425
    8
     62
     56
      65
      64
       1
```

6. (See Exercise 5.)

7. (a)
```
   47
   84
  188
 376
3948
```
(c)
```
        32
21 ⟌ 678
     63
      48
      42
       6
```

8. (a) (i) $32(20 + 6) = 32 \cdot 20 + 32 \cdot 6$
$= (30 + 2)20 + (30 + 2)6$
$= 600 + 40 + 180 + 12 = 832$

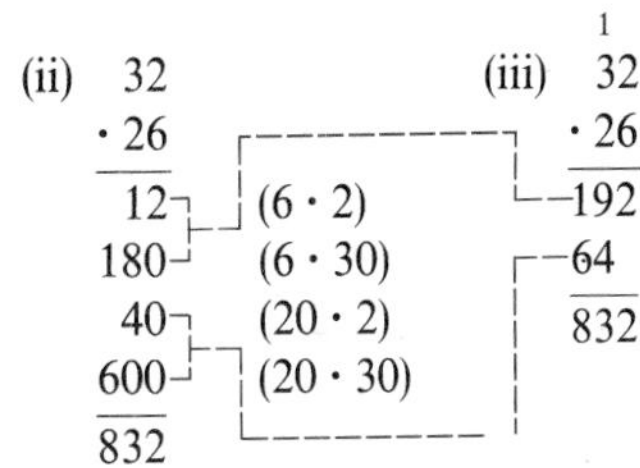

(c) (i) $92(70 + 4) = 92 \cdot 70 + 92 \cdot 4$
$= (90 + 2)70 + (90 + 2)4$
$= 90 \cdot 70 + 2 \cdot 70 + 90 \cdot 4 + 2 \cdot 4$
$= 6300 + 140 + 360 + 8$
$= 6808$

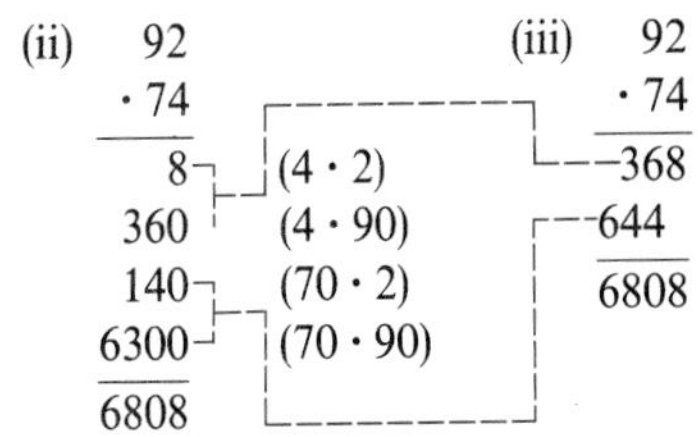

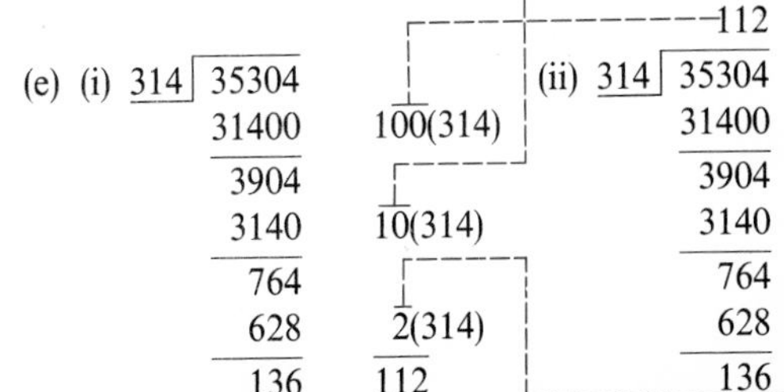

(iii)

$$\begin{array}{r} 112 \\ 314\overline{)35304} \\ 314 \\ \hline 390 \\ 314 \\ \hline 764 \\ 628 \\ \hline 136 \end{array}$$

9. (See Exercise 8.)

10. (a)

$$\begin{array}{r|l} 745 & \\ 74 & \\ \hline 5215 & 7 \cdot 745 \\ 2980 & 4 \cdot 745 \\ \hline 55{,}130 & \end{array}$$

(c)

$$\begin{array}{r|l} 1724 & \\ 848 & \\ \hline 13792 & 8 \cdot 1724 \\ 6896 & 4 \cdot 1724 \\ 13792 & 8 \cdot 1724 \\ \hline 1461952 & \end{array}$$

11. (a) This 4 represents 4(10). This is the largest whole number multiple of a power of 10 multiplied by 33 that gives an answer less than 1605.
(c) This process is the same as subtracting the zero, since zero is the identity element, and adding or subtracting the identity element does not change the sum. It is a place-holder.

12. (a) There are numerals after the decimal point. 764 R.4

13. 107 R.12 **15.** 123,454,321

16. (a)

Halving	*Doubling*
~~14~~	~~36~~
7	72
3	144
1	288
	504

(c)

Halving	*Doubling*
23	102
11	204
5	408
~~2~~	~~816~~
1	1632
	2346

17. (a)

34(176) = 5984

(c)

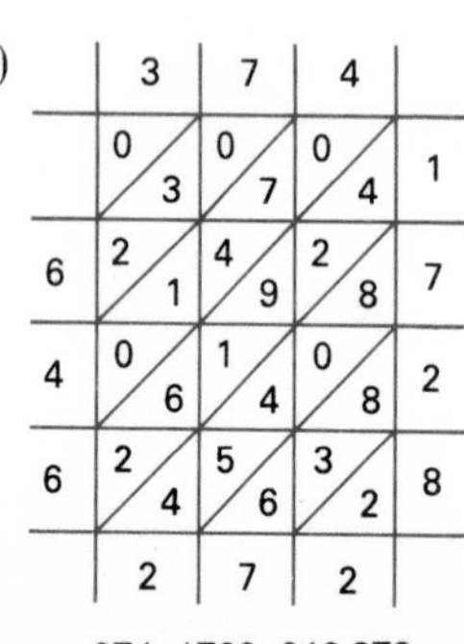

374 · 1728=646,272

18. (a) 100: $q = 4$, $r = 38$
10: $q = 43$, $r = 8$
1: $q = 438$, $r = 0$
(c) When dividing by a power of 10, such as 10^k, the last k digits of a comprise the r and the remaining digits of a comprise the q.

Exercise Set 4, Page 148

1. (a) 1, 2, 3, 4, 5, 6, 10, 11, 12, 13, 14, 15, 16, 20, 21, 22, 23, 24, 25, 26 (base seven)

2. (a)

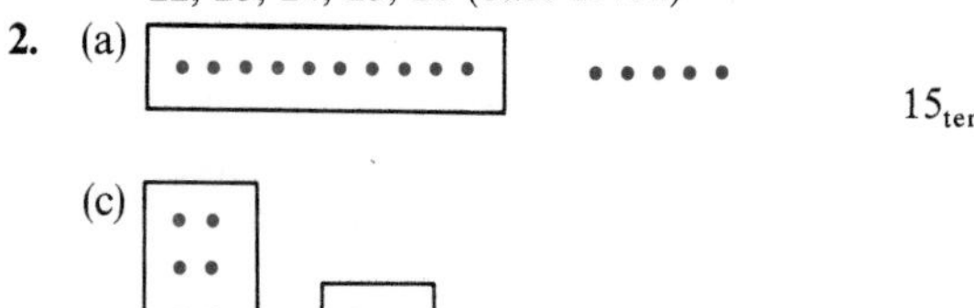

15_{ten}

(c)

1111_{two}

3. (a)

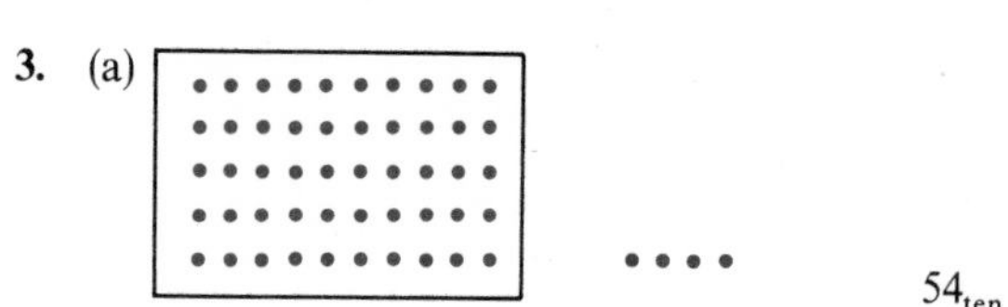

54_{ten}

(c)

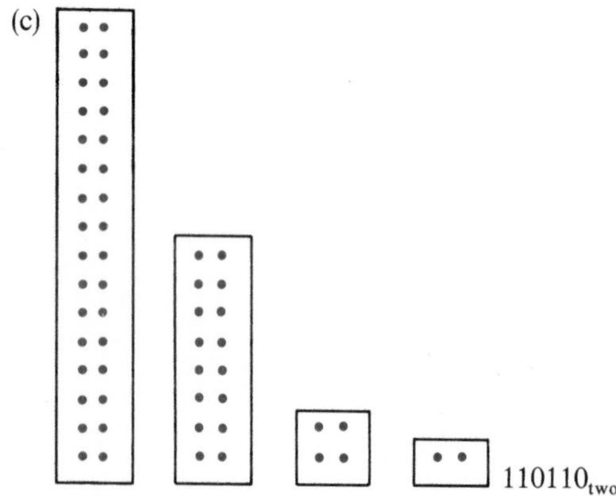

110110_{two}

4. (a) 15_{seven}, 20_{seven} (c) 110_{two}, 1000_{two}

5. (a) 295 (c) 14

6. (a) 12_{seven} (c) 106_{seven} (e) 2626_{seven}

7. (a) 1001_{two} (c) 110111_{two}
(e) 1111101000_{two}

9. (a) 0 (c) 5

11. 1, 2, 3, 4, 10, 11, 12, 13, 14, 20, 21, 22, 23, 24, 30, 31, 32, 33, 34, 40, 41, 42, 43, 44, 100, 101, 102, 103, 104, 110, 111, 112, 113, 114, 120, 121, 122, 123, 124, 130, 131, 132, 133, 134, 140, 141, 142, 143, 144, 200 (base five)

12. (a)

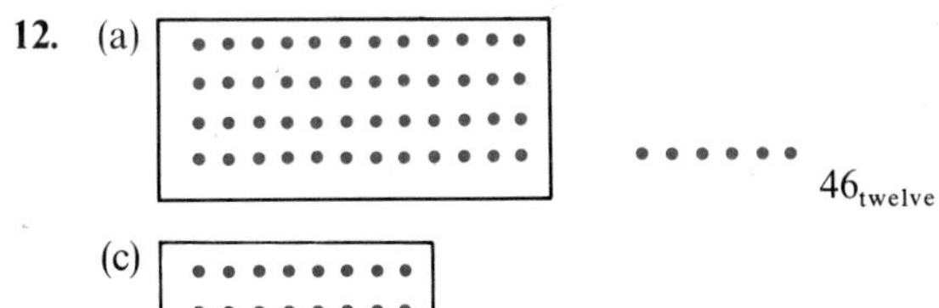

(c)

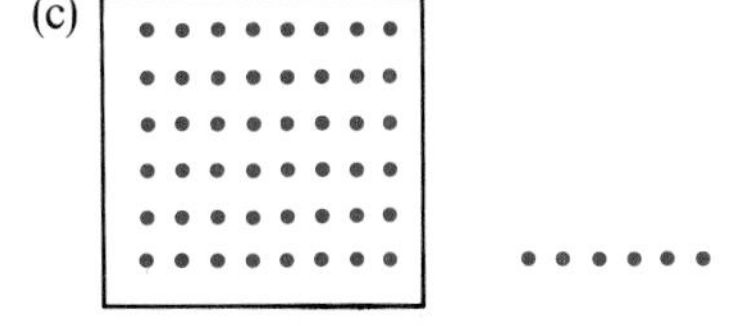

13. (a) EET_{twelve}; 1000_{twelve} (c) 221_{three}; 1000_{three}

14. (a) 133 (c) 184 (e) 280

15. (a) 9_{twelve} (c) 47_{twelve} (e) $6E4_{\text{twelve}}$

16. (a) 14_{five} (c) 210_{five} (e) 13000_{five}

17. (a) 101_{three} (c) ET_{twelve}

18. (a) Base eleven (c) Base twelve

19. 1241

21. (a) 333_{four}; 888_{nine}; EEE_{twelve}
(c) 63; 728; 1727; 64, 729; 1728

22. (a) 39_{twelve}

23. Base seven

24. (a) 12 (c) 9 (e) 59

25. Yes; no; yes; no

27. $2(5)^2 + 4(5) + 2 = 242_{\text{five}}$

Exercise Set 5, Page 155

1. (a) 1003_{seven} (c) 10100_{two}

2. (a) 1001_{two} (c) 223_{seven}

3. (a) 100111_{two} (c) 130366_{seven}

4. (a)

+	0	1	2
0	0	1	2
1	1	2	10
2	2	10	11

(c)

+	0	1	2	3	4	5	6	7	8	9	T	E
0	0	1	2	3	4	5	6	7	8	9	T	E
1	1	2	3	4	5	6	7	8	9	T	E	10
2	2	3	4	5	6	7	8	9	T	E	10	11
3	3	4	5	6	7	8	9	T	E	10	11	12
4	4	5	6	7	8	9	T	E	10	11	12	13
5	5	6	7	8	9	T	E	10	11	12	13	14
6	6	7	8	9	T	E	10	11	12	13	14	15
7	7	8	9	T	E	10	11	12	13	14	15	16
8	8	9	T	E	10	11	12	13	14	15	16	17
9	9	T	E	10	11	12	13	14	15	16	17	18
T	T	E	10	11	12	13	14	15	16	17	18	19
E	E	10	11	12	13	14	15	16	17	18	19	1T

5. (a) $4T_{\text{twelve}}$ (c) 5_{twelve} (e) $E51_{\text{twelve}}$
(g) 214_{twelve} (i) 222_{five}

6. (a)

·	0	1	2
0	0	0	0
1	0	1	2
2	0	2	11

(c)

·	0	1	2	3	4	5	6	7	8	9	T	E
0	0	0	0	0	0	0	0	0	0	0	0	0
1	0	1	2	3	4	5	6	7	8	9	T	E
2	0	2	4	6	8	T	10	12	14	16	18	1T
3	0	3	6	9	10	13	16	19	20	23	26	29
4	0	4	8	10	14	18	20	24	28	30	34	38
5	0	5	T	13	18	21	26	2E	34	39	42	47
6	0	6	10	16	20	26	30	36	40	46	50	56
7	0	7	12	19	24	2E	36	41	48	53	5T	65
8	0	8	14	20	28	34	40	48	54	60	68	74
9	0	9	16	23	30	39	46	53	60	69	76	83
T	0	T	18	26	34	42	50	5T	68	76	84	92
E	0	E	1T	29	38	47	56	65	74	83	92	T1

7. (a) 341_{five} (c) 31044_{five} (e) 120022_{three}

8. (a) 444_{five}; 100_{five}; 344_{five}
(c) 222_{three}; 100_{three}; 122_{three}

9. (a) 10 hours, 32 minutes, 8 seconds; 2 hours, 48 minutes, 54 seconds

10. (a) 1001 R10 (c) 16 R31

11. (a) Base eight (c) Base five (e) Base seven

13. (a) Base eight (c) Base eleven

15. (a) In base two there are no numerals 8 or 9; 10110011_{two}.
(c) Units digit should be 4 instead of 6; or 604_{twelve}.

16. (a) 301_{seven} (c) 10000100_{two}

Review Exercise Set 6, Page 157

1. (a) 729 (c) 4 (e) 128

2. (a) 10111_{two}; 23 (c) 500_{seven}; 245

3. (a) Units digit

4. (a) 30 (c) 100

5. (a) 65_{seven}; 101111_{two}
(c) 1141_{seven}; 110100101_{two}

6. (a) 276 (c) 13

7. (a) (i)

$$\begin{array}{l} 400 + 30 + 6 \\ \underline{200 + 40 + 3} \\ 600 + 70 + 9 = 679 \end{array}$$

(ii)

$$\begin{array}{l} 4(10)^2 + 3(10) + 6 \\ \underline{2(10)^2 + 4(10) + 3} \\ 6(10)^2 + 7(10) + 9 = 679 \end{array}$$

(c) (i)
$$\begin{array}{r} 400 + 70 + 6 \\ -[200 + 40 + 3] \\ \hline 200 + 30 + 3 \end{array} = 233$$

(ii)
$$\begin{array}{r} 4(10)^2 + 7(10) + 6 \\ -[2(10)^2 + 4(10) + 3] \\ \hline 2(10)^2 + 3(10) + 3 \end{array} = 233$$

(c) (i)
$$\begin{array}{r|l} 37 \,\overline{)\, 23{,}410} & \\ 22{,}200 & 600(37) \\ \hline 1210 & \\ 1110 & 30(37) \\ \hline 100 & \\ 74 & 2(37) \\ \hline 26 & 632 \end{array}$$

8. (a)

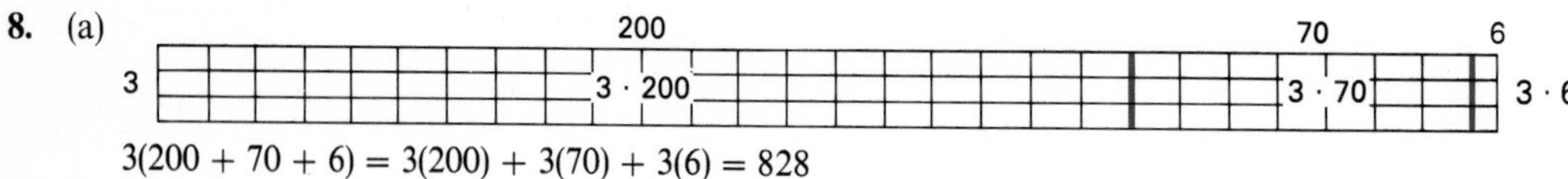

$3(200 + 70 + 6) = 3(200) + 3(70) + 3(6) = 828$

9. (a) $k = 3$ (c) $k = 6$ (e) $k = 4$

10. (a) (i)
$$\begin{array}{r} 700 + 50 + 4 \\ 600 + 40 + 7 \\ \hline 1300 + 90 + 11 \end{array} = 1401$$

(ii)
$$\begin{array}{r} {\scriptstyle 1\,1} \\ 754 \\ 647 \\ \hline 1401 \end{array}$$

(c) (i)
$$\begin{array}{rrrr} & 13(100) & 12(10) & 11 \\ \cancel{1000} + & \cancel{4(100)} + & \cancel{3(10)} + & \cancel{1} \\ -\,[& 5(100) + & 7(10) + & 2] \\ \hline & 8(100) + & 5(10) + & 9 \end{array} = 859$$

(ii)
$$\begin{array}{r} 1431 \\ -572 \\ \hline 859 \end{array}$$

11. (a)
$$\begin{array}{r} 1(10)^2 \quad 1(10) \quad\; \\ \cancel{7(10)^2} + \cancel{5(10)} + \cancel{4} \\ 6(10)^2 + 4(10) + 7 \\ \hline 1(10)^3 + 4(10)^2 + 0(10) + 1 \end{array} = 1401$$

(c)
$$\begin{array}{rrrr} & 13(10)^2 & 12(10) & 11 \\ \cancel{1(10)^3} + & \cancel{4(10)^2} + & \cancel{3(10)} + & \cancel{1} \\ -\,[& 5(10)^2 + & 7(10) + & 2] \\ \hline & 8(10)^2 + & 5(10) + & 9 \end{array} = 859$$

12. (a) (i)
$$\begin{array}{r|l} 32 \,\overline{)\, 1728} & \\ 1600 & 50(32) \\ \hline 128 & \\ 128 & 4(32) \\ \hline & 54 \end{array}$$

(ii)
$$\begin{array}{r} 54 \\ 32 \,\overline{)\, 1728} \\ 1600 \\ \hline 128 \\ 128 \\ \hline \end{array}$$

(iii)
$$\begin{array}{r} 54 \\ 32 \,\overline{)\, 1728} \\ 160 \\ \hline 128 \\ 128 \\ \hline \end{array}$$

(ii)
$$\begin{array}{r} 632 \\ 37 \,\overline{)\, 23{,}410} \\ 22{,}200 \\ \hline 1210 \\ 1110 \\ \hline 100 \\ 74 \\ \hline 26 \end{array}$$

(iii)
$$\begin{array}{r} 632 \\ 37 \,\overline{)\, 23{,}410} \\ 222 \\ \hline 121 \\ 111 \\ \hline 100 \\ 74 \\ \hline 26 \end{array}$$

13. (a) 810 (c) 718

14. (a) 10442_{five} (c) 4974_{twelve} (e) 21132_{six}

15. (a) 𝟗||||||||| (c) 一百九

16. (a) $(457 \text{ with a remainder of } 4)_{\text{twelve}}$

17. (a) Base twelve (c) Base seven (e) Base twelve

CHAPTER 4

Exercise Set 1, Page 167

1. (a) 1, 5, 7 (c) 1, 7, 11
2. (a) False, $9 \cdot 5 = 45$ (c) False; $8 = 2 \cdot 4$ (e) False; $8 \mid 24$ (g) False, $6/12$ (i) True
3. (a) Divisible by 4 but not by 8 (c) Divisible by 4 but not by 8
4. (a) Yes (c) Yes
5. (a) Yes (c) Yes
6. (a) By 5 and 10 (c) By neither
7. (a) False; $3 \mid (5 + 4)$, but $3 \nmid 5$ and $3 \nmid 4$
   (c) False; $6 \nmid 4$ and $6 \nmid 3$, but $6 \mid (4 \cdot 3)$
   (e) False; $6 \mid 12$ or $6 \mid (4 \cdot 3)$, but $6 \nmid 4$ and $6 \nmid 3$
   (g) False; $2 \nmid 5$ and $2 \nmid 3$, but $2 \mid (5 + 3)$
   (i) True
   (k) False; $4 \nmid 1$ where $a = 4$

8. (a) $a|a \quad a \neq 0$
(c) $a|b$, $a \nmid c$ but $a|(bc)$
(e) $a|b$, $a|c$, $a|d$, $a|e$, then $a|(b + c + d + e)$

9. (a) Divisible by 8 and 7 (c) Divisible by 5
(e) None (g) None
(i) Divisible by 8 (k) Divisible by 11

10. (a) Yes; $2^3|2^4$, $3^2|3^9$, and $5^7|5^7$
(c) Yes; $7^2|7^5$, $3^5|3^6$

11. No, 53 is not divisible by any number except 1.

12. (a) 17,595 is one of many possibilities.
(c) 175,616 is one of many possibilities.

13. (a) Leap year (c) Leap year
(e) Leap year (g) Not a leap year

14. (a) 15 rows

15. (a) Yes (c) Yes

17. (a) 362 (c) Impossible

19. $726{,}664 = 7(100{,}000) + 2(10{,}000) + 6(1000) + 664$
Now $8|7(100{,}000)$, since $8|100{,}000$
$8|2(10{,}000)$, since $8|10{,}000$
$8|6(1000)$, since $8|1000$
If $8|664$, then $8|726{,}664$.

20. (a) $c = 6 \cdot 5 \cdot 4 \cdot 3 \cdot 2 \cdot 1 = 720$
(c) $c = 5 \cdot 4 \cdot 1 = 20$

21. (a) $N = a_k 10^k + \cdots + a_3 10^3 + a_2 10^2 + a_1 10 + a_0$
$4|10^2$, $4|10^3$, and $4|10^k$ if $k \geq 2$. So $4|a_k 10^k + \cdots + a_3 10^3 + a_2 10^2$. Therefore, $4|N$ if $4|(a_1 10 + a_0)$.
(c) $N = a_k 10^k + \cdots + a_3 10^3 + a_2 10^2 + a_1 10 + a_0$
$8|10^3$, and $8|10^k$ if $k \geq 3$. So $8|(a_k 10^k + \cdots + a_3 10^3)$. Therefore $8|N$, if $8|(a_2 10^2 + a_1 10 + a_0)$.
(e) Proof for k odd: $N = a_k 10^k + \cdots + a_4 10^4 + a_3 10^3 + a_2 10^2 + a_1 10 + a_0 = a_k(10^k + 1) + \cdots + a_4(10^4 - 1) + a_3(10^3 + 1) + a_2(10^2 - 1) + a_1(10 + 1) + (-a_k + \cdots + a_4 - a_3 + a_2 - a_1 + a_0)$. Now $11|(10 + 1)$, $11|(10^2 - 1)$, $11|(10^3 + 1)$, and $11|(10^k + 1)$, k odd. Therefore, $11|N$ if $11|(-a_k + \cdots + a_4 - a_3 + a_2 - a_1 + a_0)$. A similar proof holds when k is even.

22. (a) True, $5|5!$ If $5|(5! + 1)$, then $5|1$. This is not true, so $5 \nmid (5! + 1)$.

23. (a) $232{,}323 \div 3 = 77{,}441$ no remainder
$565{,}656 \div 3 = 188{,}552$ no remainder
$787{,}878 \div 3 = 262{,}626$ no remainder
$484{,}848 \div 3 = 161{,}616$ no remainder
xyx, yxy is divisible by 3.
(c) $232{,}323 \div 13 = 17{,}871$ no remainder
$565{,}656 \div 13 = 43{,}512$ no remainder
$787{,}878 \div 13 = 60{,}606$ no remainder
$484{,}848 \div 13 = 37{,}296$ no remainder
xyx, yxy is divisible by 13.

24. (a) $12{,}341{,}234 \div 73 = 169{,}058$
(c) $wx{,}yzw{,}xyz = w{,}xyz(10{,}001) = w{,}xyz(137)(73)$
$wx{,}yzw{,}xyz \div 73 = w{,}xyz(137)$

25. $9 \cdot 8 \cdot 7 \cdot 5 = 2520$

27. (a) 7254_{eight} is divisible by 3 because $(7 + 5) - (2 + 4) = 6$, which is divisible by 3.

28. (a) A number in base seven is divisible by 2 if the sum of the digits is divisible by 2. Likewise, the number is divisible by 3 if the sum of the digits is divisible by 3.

Exercise Set 2, Page 174

1. (a) $144 = 2 \cdot 2 \cdot 2 \cdot 2 \cdot 3 \cdot 3$
(c) $612 = 2 \cdot 2 \cdot 3 \cdot 3 \cdot 17$

2. (a) Prime (c) Not a prime
(e) Not a prime (g) Not a prime

3. (a) 9 (c) 22 (e) 13

4. (a) $54 = 2 \cdot 3 \cdot 3 \cdot 3$ (c) $120 = 2 \cdot 2 \cdot 2 \cdot 3 \cdot 5$
(e) $141 = 3 \cdot 47$ (g) $144 = 2 \cdot 2 \cdot 2 \cdot 2 \cdot 3 \cdot 3$
(i) $256 = 2 \cdot 2 \cdot 2 \cdot 2 \cdot 2 \cdot 2 \cdot 2 \cdot 2$

5.

1	2	3	4	5	6	7	8	9	10
11	12	13	14	15	16	17	18	19	20
21	22	23	24	25	26	27	28	29	30
31	32	33	34	35	36	37	38	39	40
41	42	43	44	45	46	47	48	49	50
51	52	53	54	55	56	57	58	59	60
61	62	63	64	65	66	67	68	69	70
71	72	73	74	75	76	77	78	79	80
81	82	83	84	85	86	87	88	89	90
91	92	93	94	95	96	97	98	99	100
101	102	103	104	105	106	107	108	109	110
111	112	113	114	115	116	117	118	119	120
121	122	123	124	125	126	127	128	129	130
131	132	133	134	135	136	137	138	139	140
141	142	143	144	145	146	147	148	149	150
151	152	153	154	155	156	157	158	159	160
161	162	163	164	165	166	167	168	169	170
171	172	173	174	175	176	177	178	179	180
181	182	183	184	185	186	187	188	189	190
191	192	193	194	195	196	197	198	199	200

201 202 203 204 205 206 207 208 209 210
(211) 212 213 214 215 216 217 218 219 220
221 222 (223) 224 225 226 (227) 228 (229) 230
231 232 (233) 234 235 236 237 238 (239) 240
(241) 242 243 244 245 246 247 248 249 250

7. (a) 15 prime numbers less than 50
25 prime numbers less than 100
35 prime numbers less than 150

8. (a) $1512 = 2 \cdot 2 \cdot 2 \cdot 3 \cdot 3 \cdot 3 \cdot 7$

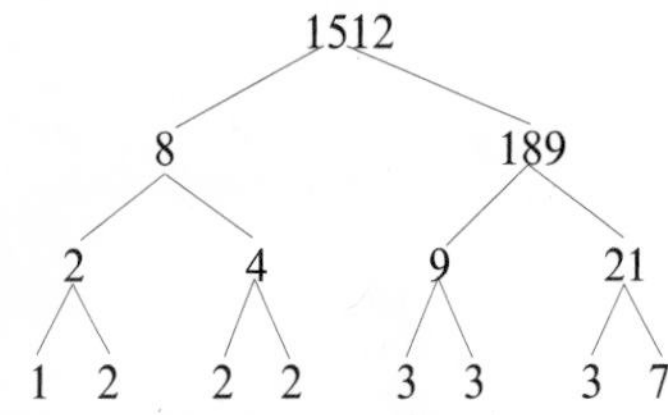

(c) $1836 = 2 \cdot 2 \cdot 3 \cdot 3 \cdot 3 \cdot 17$

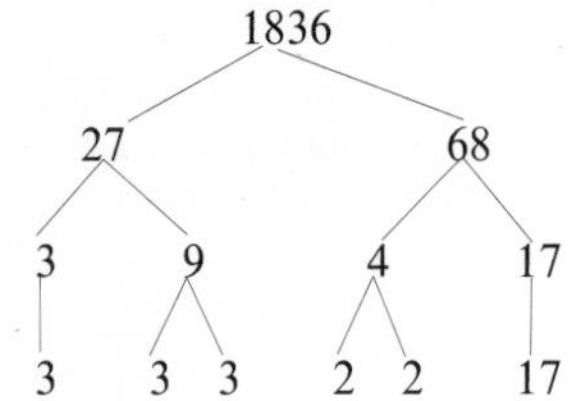

9. (a)
2) 1512
2) 756
2) 378
3) 189
3) 63
3) 21
7) 7
1

(c)
2) 1836
2) 918
3) 459
3) 153
3) 51
17) 17
1

10. (a) 23

11. (a) 6 (c) 210
(e) 30,030 (used smallest primes in increasing order)

12. (a) $3^3 \cdot 3^3$; $3^5 \cdot 3$; $3^2 \cdot 3^4$
(c) $10 \cdot 10$; $20 \cdot 5$; $4 \cdot 25$

13. (a) 1, 3, 3^2, 3^3, 3^4

15. (a) It is obvious that they are consecutive whole numbers because they differ by 1. Now $2|(5!+2)$, $3|(5!+3)$, $4|(5!+4)$, and $5|(5!+5)$. Therefore, they are composite numbers.

16. (a) $11!+2, 11!+3, 11!+4, 11!+5, \ldots, 11!+11$

17. Yes. $n|(n!+n)$ for $n = (2, 3, \ldots, n)$

18. (a) {1, 4, 7, 10, 13, 16, 19, 22, 25, 28, 31, 34, 37, 40, 43, 46, 49, 52, 55, 58}
(c) Yes

19. (a) A prime on this set of numbers is any number whose only divisors are 1 and itself: 3, 5, 7, 11, 13, 17, 19, 23, 29, 31.

Exercise Set 3, Page 182

1. (a) 15; 210 (c) 6; 990 (e) 2; 336
(g) 1; 90

2. (a) 6; 840 (c) 4; 7,341,600 (e) 1; 2730

3. (a) $3 \cdot 5 \cdot 2$ (c) $3^3 \cdot 5^2 \cdot 2^3$

4. (a) When a and b are relatively prime
(c) When a and b are both 1

5. (a) g.c.d. = 4
l.c.m. = $(92 \cdot 44)/4 = \frac{4048}{4} = 1012$

2	44	92
2	22	46
	11	23

$2 \cdot 2 \cdot 11 \cdot 23 = 1012 = \text{l.c.m.}$

(c) g.c.d. = 2; l.c.m. = $(146 \cdot 124)/2 = 9052$

2	146	124
2	73	62
	73	31

$2 \cdot 2 \cdot 73 \cdot 31 = 9052 = \text{l.c.m.}$

(e) g.c.d. = $2^3 \cdot 3 \cdot 5 = 120$
l.c.m. = $(840 \cdot 1800)/120 = 1{,}512{,}000/120$
$= 12{,}600$

5	1800	840
3	360	168
2	120	56
2	60	28
2	30	14
3	15	7
	5	7

$2^3 \cdot 3^2 \cdot 5^2 \cdot 7 = 12{,}600 = \text{l.c.m.}$

6. (a) 3 (c) 22 (e) 5 (g) 24

7. (a) g.c.d. = 1; l.c.m. = ab
(c) g.c.d. $(a, a) = a$; l.c.m. $(a, a) = a$
(e) $b|a$ (g) ab

9. 20 minutes

11. 7

13. 7:30 P.M.

15. 48

17. (a) Yes (c) Yes (e) No
18. (a) Yes (c) Yes (e) No
19. (a) Yes
20. (a) Let $d =$ g.c.d. (a, b). Then $xd|xa$ and $xd|xb$. So xd is a divisor of xa and xb. By definition, g.c.d. (xa, xb) must contain x as a factor. Let g.c.d. (xa, xb) be some $xk > xd$. $k|a$ and $k|b$. Also $k > d$. This is a contradiction. Therefore,

$$\text{g.c.d. } (xa, xb) = x \cdot \text{g.c.d. } (a,b)$$

Exercise Set 4, Page 187

1. (a) A (c) P (e) D (g) D (i) P
2. (a) The set of primes, excluding 2
3. (a) True (c) True
5. (a) 1 is the only proper divisor of a prime. Therefore, the sum of the proper divisors are less than p, and every prime is deficient.
6. (a) $n = 1$ (41), $n = 2$ (43), $n = 3$ (47), $n = 4$ (53), $n = 5$ (61), $n = 6$ (71), $n = 7$ (83), $n = 8$ (97), $n = 9$ (113), $n = 10$ (131)
   (c) No
7. (a) 41, 43 (c) 71, 73
8. (a) Yes (c) Yes (e) No
9. $1184 = 1 + 2 + 5 + 10 + 11 + 22 + 55 + 110 + 121 + 242 + 605$
   $1210 = 1 + 2 + 4 + 8 + 16 + 32 + 37 + 74 + 148 + 296 + 592$
10. (a) $4(2) + 1 = 9$, a composite number.
    (c) $1^2 + 1 = 2$, a prime.
    (e) $8 \cdot 2 + 1 = 17$, a prime.
11. (a) Odd numbers: $2m + 1$ and $2n + 1$
    Sum: $(2m + 1) + (2n + 1) = 2[(m + n) + 1]$
    This is an even number and not an odd number. The odd numbers are not closed under addition.
12. (a) One gets primes for all values of n except $n = 17$.
13. 101, 103; 107, 109; 137, 139; 149, 151; 179, 181; 191, 193; 197, 199
14. (a) $a = 6 \quad p = 5 \qquad \dfrac{a^p - a}{p} = \dfrac{6^5 - 6}{5} = 1554$
    (c) $\dfrac{11^3 - 11}{3} = \dfrac{1331 - 11}{3} = 440$
15. (a) $1 \cdot 2 \cdot 3 \cdot 4 + 1 = 25$; $5|25$
    (c) $1 \cdot 2 \cdot 3 \cdot 4 \cdot 5 \cdot 6 \cdot 7 \cdot 8 \cdot 9 \cdot 10 + 1 = 3{,}628{,}801$; $11|3{,}628{,}801$
16. (a) Yes (c) Yes
17. (a) $24 = 11 + 13 = 5 + 19$ (c) $44 = 7 + 37$
18. (a) $25 = 5 + 7 + 13$ (c) $43 = 7 + 13 + 23$
19. Every odd number greater than 7 can be written as an even number greater than 4 plus 3. Since every even number can be written as the sum of two odd primes (the first of Goldbach's Conjectures), two odd primes + 3 consists of three odd primes.
20. (a) The proof is given in Exercise 11.
    (c) Even number: $2m$
    Odd number: $2n + 1$
    Product: $2m(2n + 1) = 4mn + 2m = 2(2mn + m)$
    Therefore the product is an even number.
21. No, there are no other sets of prime triplets since 3 will divide every third odd number.
23. The sum of the proper divisors of A is greater than A, because A is abundant. All the divisors of A divide nA for any natural number n. Likewise, n times any divisor of A divides nA. The sum of n times any divisor of A is equal to n(sum of the divisors of A), which is greater than nA.
    Since there are infinitely many natural numbers, there are infinitely many natural numbers times 12.
25. $1 + 2 + 4 + 8 + 16 + 32 + 64 + 127 + 254 + 508 + 1016 + 2032 + 4064 = 8128$. Therefore $2^6(2^7 - 1)$ is perfect.

Exercise Set 5, Page 196

1. (a) 1 (c) 8 (e) 0 (g) Not possible
2. (a) 1 (c) 2 (e) 2 (g) 3
3. (a) 4:00 (c) 1:00 P.M. the next day
4. (a) 3 (c) 10 (e) 3
5. (a) 15 (c) 22 (e) 3
7. 2 years; next year
9. (a) Yes, $(4 + 7) \div 8$ has a remainder of 3. $(15 + 4) \div 8$ has a remainder of 3.
   (c) (mod 6)
10. (a) False (c) False (e) True (g) True
11. (a) 1 (c) 2 (e) 4 (g) 1
    (i) 0 (k) 8 (m) 4
12. (a) 2 (c) 2 (e) 2 (g) 6
    (i) 1, −1, 2, −2
13. (a) Off (c) Brightest (e) Brightest
14. (a) Yes
15. (a) Tuesday (c) Monday
16. (a) An Abelian group (c) An Abelian group

17.

⊕	*0*	*1*	*2*	*3*	*4*
0	0	1	2	3	4
1	1	2	3	4	0
2	2	3	4	0	1
3	3	4	0	1	2
4	4	0	1	2	3

⊙	*0*	*1*	*2*	*3*	*4*
0	0	0	0	0	0
1	0	1	2	3	4
2	0	2	4	1	3
3	0	3	1	4	2
4	0	4	3	2	1

(a) Yes (c) Addition

18.

⊕	*0*	*1*	*2*	*3*	*4*	*5*	*6*
0	0	1	2	3	4	5	6
1	1	2	3	4	5	6	0
2	2	3	4	5	6	0	1
3	3	4	5	6	0	1	2
4	4	5	6	0	1	2	3
5	5	6	0	1	2	3	4
6	6	0	1	2	3	4	5

⊙	*0*	*1*	*2*	*3*	*4*	*5*	*6*
0	0	0	0	0	0	0	0
1	0	1	2	3	4	5	6
2	0	2	4	6	1	3	5
3	0	3	6	2	5	1	4
4	0	4	1	5	2	6	3
5	0	5	3	1	6	4	2
6	0	6	5	4	3	2	1

(a) Yes (c) Addition

Exercise Set 6, Page 207

1. (a) $^{-}5$ (c) 0 (e) ^{-}a

2. (a) $^{-}2$

$^{-}3$ $^{-}2$ $^{-}1$ 0

(c) $^{-}3$

$^{-}4$ $^{-}3$ $^{-}2$ $^{-}1$ 0

(e) 6

0 1 2 3 4 5 6 7

3. (a) $^{-}2$

$^{-}5$ $^{-}4$ $^{-}3$ $^{-}2$ $^{-}1$ 0

(c) $^{-}3$

$^{-}6$ $^{-}5$ $^{-}4$ $^{-}3$ $^{-}2$ $^{-}1$ 0

(e) 6

$^{-}3$ $^{-}2$ $^{-}1$ 0 1 2 3 4

4. (a) Yes (c) No (e) Yes (g) No

5. (a) $^{-}3$ (c) $^{-}8$ (e) $^{-}13$ (g) $^{-}11$
(i) $^{-}16$ (k) 4 (m) $^{-}2$ (o) 0 (q) a

6. (a)

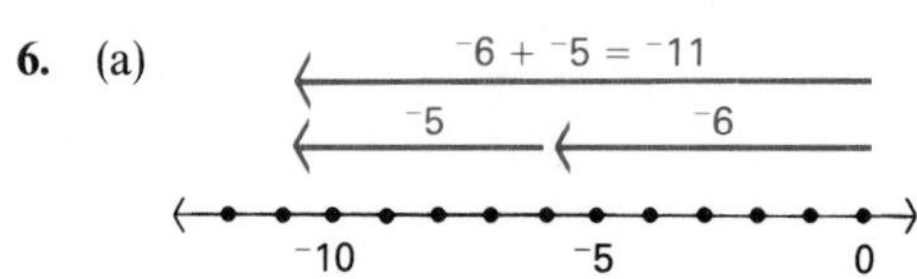

(c) $^{-}5 + {}^{-}3 = {}^{-}8$

$^{-}3$ $^{-}5$

$^{-}8$ $^{-}6$ $^{-}4$ $^{-}2$ 0

(e)

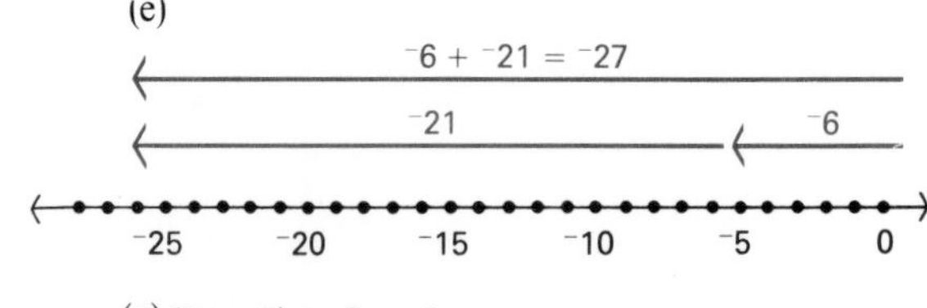

(g) $(8 + {}^{-}2) + {}^{-}7 = {}^{-}1$

$^{-}7$ $^{-}2$

8

$^{-}2$ 0 2 4 6 8

(i)

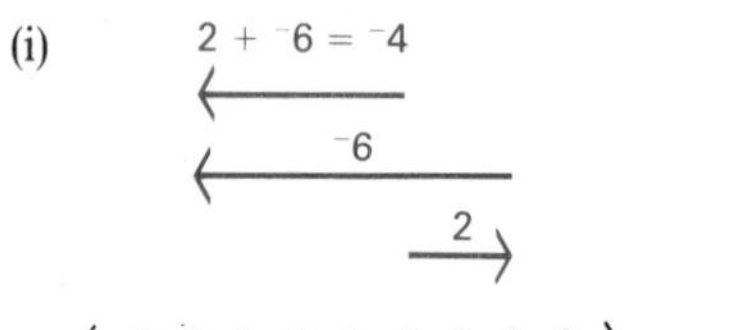

7. (a) No; $(8 - 2) - 3 = 6 - 3 = 3$
$8 - (2 - 3) = 8 - ({}^{-}1) = 9$, and $3 \neq 9$

8. (a) $x = 5$

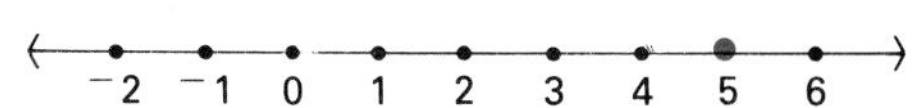

(c) $x < 3$ and an integer

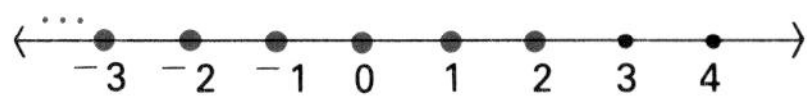

9. (a) $11 + {}^-17 = {}^-6$
11. $^-8°C$ 13. $198
14. (a) Yes, 5 extra yards (c) Yes, 2 extra yards
15. (a) $^-10{,}500$ feet (c) 3500 feet
16. (a) False (c) True (e) True
17. (a) $^-150$ meters (c) $^-12°C$
18. (a) 7 (c) 12 (e) 5
19. (a) 26; 26 (c) c; $|c|$ (e) 5; 5
20. (a) 0 (c) Unlimited (e) $^-1$
21. (a) ^-10x (c) $^-11y^2$
22. (a) $\varnothing$ (c) All integers except zero
 (e) $\{1, 2, 3, 4, \ldots\}$
23. (a) $(8 + 4) + {}^-2$
 $= {}^-2 + (8 + 4)$ Commutative property of addition
 $= ({}^-2 + 8) + 4$ Associative property of addition
 $= (8 + {}^-2) + 4$ Commutative property of addition
 $(8 + 4) + {}^-2$
 $= (8 + {}^-2) + 4$ Transitive property of equalities
 (c) $({}^-4 + 6) + {}^-2$
 $= {}^-2 + ({}^-4 + 6)$ Commutative property of addition
 $= {}^-2 + (6 + {}^-4)$ Commutative property of addition
 $= ({}^-2 + 6) + {}^-4$ Associative property of addition
 $({}^-4 + 6) + {}^-2$
 $= ({}^-2 + 6) + {}^-4$ Transitive property of equalities
24. (a) $^-3 < x < 4$ and an integer

(c) $x < 2$ and an integer

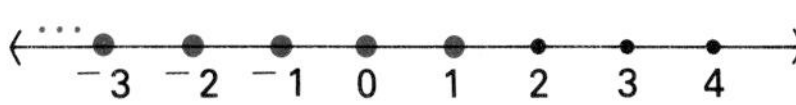

Exercise Set 7, Page 210

1. (a) $^-45$ (c) 0 (e) $^-322$ (g) xy
 (i) 2 (k) $^-135$
2. (a) 72 (c) $^-384$ (e) $^-3$ (g) 8
 (i) 24 (k) ^-xyz
3. (a) $^-1(3 + {}^-5) = {}^-1({}^-2) = 2$
 $^-1(3 + {}^-5) = {}^-1(3) + {}^-1({}^-5) = {}^-3 + 5 = 2$
4. (a) $^-5a + {}^-5b + {}^-5c$ (c) $^-3y + 2yx + {}^-4yz$
5. (a) Both factors are positive or both are negative.
 (c) One factor must be positive and one must be negative.
6. (a) a must be equal to b; it does not matter what values a assumes except 0.
 (c) 3
7. (a) $x = {}^-2$ (c) $x = {}^-4$
9. Let x be difference in the altitude of the plane and a mountain of height 2000 m.
 $10{,}000 - 800(10) + x = 2000$
 $x = 0$
 Same altitude
11. Let x be the temperature five days ago.
 $x - 4 \cdot 5 = {}^-45$
 $x = {}^-25°C$
12. (a) Let x be the temperature two days ago.
 $x - 4 \cdot 2 = {}^-45$
 $x = {}^-37°C$
13. $^-40° - 6°t$
15. Let t be the time in days until the temperature is 88°F.
 $94 - 2 \cdot t = 88$
 $t = 3$ days
16. (a) W, M, I (c) W, I
17. $x = {}^-y$ when x and y are additive inverses.
19. (a) $2 \cdot 3 - 2 \cdot 5 = {}^-4$
20. (a) 66
21. (a) $x^2 - 4$ (c) $4x^2 - y^2$
22. (a) $(x - 3)(x + 3)$ (c) $(3x + y)(3x - y)$
23. (a) $x < {}^-3$ and an integer
 (c) $x > {}^-1$ (integer)

Review Exercise Set 8, Page 214

1. (a) No (c) Yes
2. (a) $2 \cdot 3 \cdot 3 \cdot 7$ (c) $3 \cdot 3 \cdot 5 \cdot 5 \cdot 11$
3. (a) $3 \cdot 7 = 21$ (c) 21
4. (a) 2 (c) 1
5. (a) 3 (c) 4

7. 2; 36,875,354

9. They would be equal or identical.

10. (a) False (c) False (e) True (g) False
(i) True (k) False (m) False (o) False
(q) False (s) False (u) True

11. (a) 25 (c) $^{-}17$ (e) 0 (g) $^{-}41$
(i) $xz - xy$ (k) 40

12. (a)

Add	+	−
+	+	?
−	?	−

(c)

·	+	−
+	+	−
−	−	+

13. (a) $4(^{-}3) + (^{-}4)(2)$
$= 4(^{-}3) + [(^{-}1)(4)](2)$ Renaming -4
$= 4(^{-}3) + [(4)(^{-}1)](2)$ Commutative property of multiplication
$= 4(^{-}3) + 4[(^{-}1)(2)]$ Associative property of multiplication
$= 4(^{-}3) + 4(^{-}2)$ Multiplication
$= 4(^{-}3 + {}^{-}2)$ Distributive property of multiplication over addition
$4(^{-}3) + (^{-}4)(2)$
$= 4(^{-}3 + {}^{-}2)$ Transitive property of equalities

(c) $^{-}4(^{-}8 \cdot 2)$
$= (^{-}8 \cdot 2) \cdot {}^{-}4$ Commutative property of multiplication
$= [(^{-}1 \cdot 8) \cdot 2] \cdot {}^{-}4$ Rewriting $^{-}8$
$= [(8 \cdot {}^{-}1) \cdot 2] \cdot {}^{-}4$ Commutative property of multiplication
$= [8(^{-}1 \cdot 2)] \cdot {}^{-}4$ Associative property of multiplication
$= (8 \cdot {}^{-}2) \cdot {}^{-}4$ Definition of multiplication
$^{-}4(^{-}8 \cdot 2)$
$= (8 \cdot {}^{-}2) \cdot {}^{-}4$ Transitive property of equalities

14. (a)

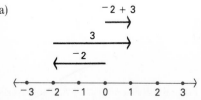

(c)

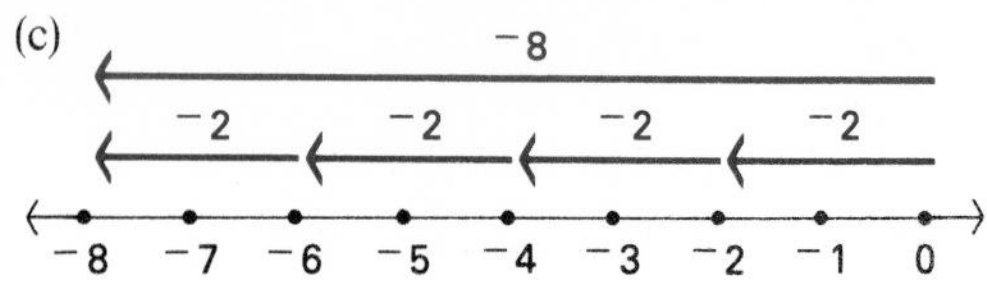

(e)

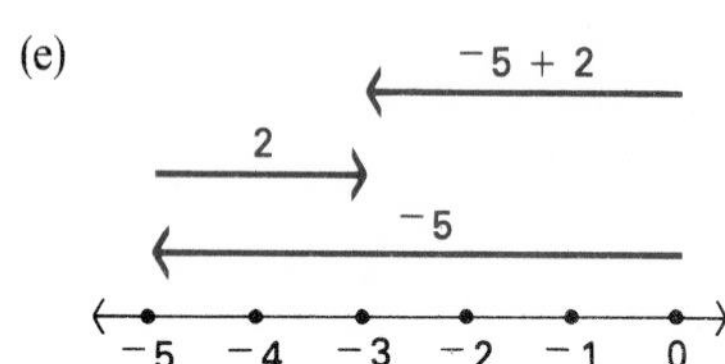

15. (a)

⊗	*1*	*2*	*3*	*4*	*5*
1	1	1	1	1	1
2	1	2	2	2	2
3	1	2	3	3	3
4	1	2	3	4	4
5	1	2	3	4	5

(c) Yes (e) No

16. (a) False, $5|10$ but $5 \nmid 7$ and $5 \nmid 3$.
(c) False, $5|10$ but $5 \nmid 7$ and $5 \nmid 3$.
(e) True (g) True
(i) False, $6|150$ but $6 \nmid 10$ or $6 \nmid 15$

17. Thursday; Friday

19. (a) 24, 25, 26, 27, 28 (c) 24

21. No, the seventh number after a prime (not 2) will be even, not a prime.

23. (a) $x = 2$ (c) $x = {}^{-}1$
(e) $x < 2$ and an integer

24. (a) $n^3 - n = n(n^2 - 1)$; $n^3 - n = n(n-1)(n+1)$. So $n(n-1)(n+1)$ is a product of three consecutive whole numbers. Of any three consecutive whole numbers, one will be a multiple of 3. Thus, $3|(n^3 - n)$.
(c) No, if $n = 2$, $9 \nmid (2^9 - 2)$ as $9 \nmid 510$.

25. $2^4 \cdot 3^2 \cdot 5 \cdot 7 \cdot 11 \cdot 13 \cdot 17 \cdot 19$

26. (a) It is not a group.

CHAPTER 5

Exercise Set 1, Page 223

1. (a) $\frac{3}{8}$ (c) $\frac{1}{2}$ (e) $\frac{7}{8}$
2. (a) $\frac{6}{8}, \frac{12}{16}, \frac{18}{24}, \frac{24}{32}, \frac{36}{48}, \ldots$ (c) $\frac{0}{1}, \frac{0}{2}, \frac{0}{3}, \frac{0}{100}, \ldots$
3. (a) Yes (c) Yes (e) No
4. (a) 3 (c) 3
5. (a) 3 (c) 0
6. (a) $\frac{1}{2}$ (c) $\frac{1}{4}$

7. (a) $\frac{81}{44}$ (c) $\frac{14}{19}$ (e) x^2/z^2 (g) x/yz^3
   (i) $-\frac{13}{16}$
8. (a) $\frac{49}{21}$ (c) Impossible (e) $\frac{30}{40}$
   (g) Impossible (i) $\frac{10}{24}$ (k) $\frac{6}{10}$
9. (a) $\frac{2}{3}, \frac{3}{2}, \frac{4}{1}, \frac{1}{4}, \frac{0}{5}$ (c) $\frac{2}{1}, \frac{3}{2}, \frac{4}{3}, \frac{4}{4}, \frac{5}{4}$
   (e) $\frac{1}{1}, \frac{2}{2}, \frac{3}{3}, \frac{4}{4}, \frac{5}{5}$
10. (a) $\left\{\frac{3k}{5k} \middle| k = 1, 2, 3, \ldots\right\}$

    (c) $\left\{\frac{xk}{yk} \middle| k = 1, 2, 3, \ldots\right\}$
11. (a) Yes (c) No (e) Yes
13. (a) $\frac{3}{8} = \frac{3}{8}$
14. (a) 15 (c) 45
15. (a) True (c) True (e) True
16. (a) $ad = cd$ if $a = c$ by substitution.
    (c) c must be equal to a.
17. (a) $(3 \cdot 5^2 \cdot 7)/(2 \cdot 11)$ (c) $x^2/2y^2$, $x \neq 0$, $y \neq 0$

    (e) $\dfrac{a}{b^2}$, $b \neq 0$, $x + y \neq 0$ (g) $\dfrac{a}{2a + b}$, $b \neq -2a$

    (i) $\dfrac{ab(a + b)}{a - b}$, $a \neq b$
18. (a) $\frac{1}{4}, \frac{4}{16}, \frac{2}{8}$
19. (a) Yes, $18{,}000 = 18{,}000$ (c) No, $6237 \neq 2666$
21. (a) $a/b = a/b$ (Reflexive)
    $a/b = c/d$, then $c/d = a/b$ (Symmetric)
    $a/b = c/d$ and $c/d = e/f$, then
    $a/b = e/f$ (Transitive)
22. (a) $\frac{1}{2}$

Exercise Set 2, Page 231

1. (a) $\frac{3}{4}$ (c) $\frac{29}{36}$ (e) $13/2x$
2. (a) $\frac{3}{4}$ (c) $\frac{11}{36}$ (e) $3/2x$
3. (a) 8; $\frac{19}{8}$ (c) 54; $\frac{10}{9}$ (e) 385; $\frac{1187}{385}$
4. (a) $19\frac{1}{3}$ (c) $6\frac{1}{8}$
5. (a) $\frac{23}{6}$ (c) $\frac{49}{3}$
6. (a) $\frac{1}{8} + \frac{3}{16}$ (c) $\frac{15}{4} + \frac{19}{4}$
7. (a) $10\frac{13}{21}, 5\frac{16}{21}$ (c) $31\frac{19}{36}, 15\frac{35}{36}$
8. (a) $2\frac{1}{24}$ (c) $\frac{17}{24}$
   (e) Associative property of addition
9. (a) $11\frac{7}{12}$ (c) $11\frac{13}{18}$ (e) $7\frac{49}{60}$ (g) $11\frac{17}{40}$
   (i) $99\frac{11}{12}$
10. (a) $\left(\frac{3}{4} + \frac{2}{5}\right) \stackrel{?}{=} \left(\frac{2}{5} + \frac{3}{4}\right)$ (c) $\frac{3}{4} + \frac{0}{1} \stackrel{?}{=} \frac{3}{4}$

    $\frac{15 + 8}{20} \stackrel{?}{=} \frac{8 + 15}{20}$ $\frac{3 + 0(4)}{4} \stackrel{?}{=} \frac{3}{4}$

    $\frac{23}{20} = \frac{23}{20}$ $\frac{3 + 0}{4} = \frac{3}{4}$

11. (a)

    $\frac{3}{5}$

    $\frac{2}{5}$ $\frac{1}{5}$

    0 $\frac{1}{5}$ $\frac{2}{5}$ $\frac{3}{5}$ $\frac{4}{5}$ 1

    (c)

    $\frac{7}{12}$

    $\frac{1}{3}$ $\frac{1}{4}$

    0 $\frac{1}{12}$ $\frac{2}{12}$ $\frac{3}{12}$ $\frac{4}{12}$ $\frac{5}{12}$ $\frac{6}{12}$ $\frac{7}{12}$
12. (a) Yes
    (c) No

    $$\frac{2}{5} - \left(\frac{4}{5} - \frac{1}{5}\right) \neq \left(\frac{2}{5} - \frac{4}{5}\right) - \frac{1}{5}$$

    $$-\frac{1}{5} \neq -\frac{3}{5}$$

    (e) Yes, $\frac{2}{5} - 0 = \frac{2}{5}$
13. (a) $x = \frac{11}{12}$ (c) $x = \frac{10}{9}$
14. (a) $-\frac{59}{40}$ (c) $-\frac{19}{45}$
15. (a) $S = \dfrac{55}{273}$ $d = \dfrac{43}{273}$

    (c) $S = \dfrac{17}{2 \cdot 3^4}$ $d = \dfrac{29}{2^2 \cdot 3^4}$

    (e) $S = 58\frac{49}{253}$ $d = 25\frac{72}{253}$
16. (a) $\dfrac{3 + 3y^2z}{xy^3z}; \dfrac{3 - 3y^2z}{xy^3z}$ (c) $\dfrac{3 + 5b}{ab}; \dfrac{3 - 5b}{ab}$

    (e) $\dfrac{132 + n^3x}{12mn^2}; \dfrac{132 - n^3x}{12mn^2}$
17. (a) Let x = total weight.
    $x = \frac{1}{2} + 6\frac{1}{4} + 2\frac{3}{4} + 12\frac{1}{8} + 1$
    (c) Let x be the fractional part of shares purchased by Brooke.

    $$\frac{1}{3} + \frac{1}{3} + x = \frac{3}{4}$$

    (e) Let x = time of trip.
    $x = 4\frac{5}{8} + 1\frac{1}{3} + 2\frac{1}{4}$
18. (a) $x = 22\frac{5}{8}$ pounds (c) $x = \frac{1}{12}$
    (e) $x = 8\frac{5}{24}$ hours
19. (a) $\dfrac{2 + 3}{2} = \dfrac{2}{2} + \dfrac{3}{2}$, not $\dfrac{2}{2} + 3$, because a common denominator is needed to add fractions.

(c) With a common denominator, you add only the numerators.

20. (a) Let $x = 1$, $y = 2$, $w = 3$, and $z = 4$.

$$\frac{1}{2} + \frac{3}{4} = \frac{\frac{4}{2} + 3}{4} = \frac{2 + 3}{4} = \frac{5}{4}$$

21. (a) Yes (c) Yes (e) No

22. (a) $(\frac{1}{4} + \frac{3}{5})_{\text{seven}} = [(5 + 15)/26]_{\text{seven}} = (\frac{23}{26})_{\text{seven}}$

(c) $[43 + (12 + 13)/20]_{\text{six}} = (43 + \frac{25}{20})_{\text{six}}$
$= (44\frac{5}{20})_{\text{six}}$

Exercise Set 3, Page 240

1. (a)

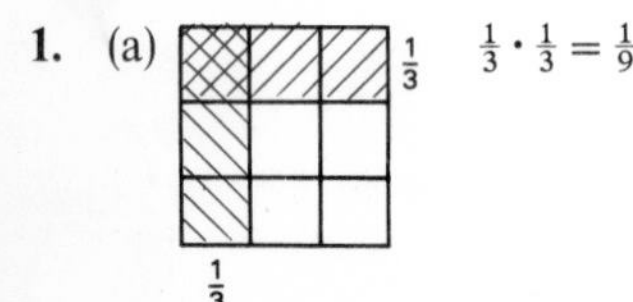

$\frac{1}{3} \cdot \frac{1}{3} = \frac{1}{9}$

2. (a) $\frac{1}{7}; \frac{7}{1} \cdot \frac{1}{7} = \frac{7}{7} = 1$ (c) Impossible
(e) $z/w;\ w/z \cdot z/w = wz/wz = 1$

3. (a) $\frac{1}{4}$ (c) $\frac{5}{21}$ (e) 36 (g) $\frac{26}{21}$
(i) $\frac{64}{63}$ (k) $\frac{2}{27}$ (m) $\frac{2}{45}$

4. (a) a (c) $ab(ad + be)/cde$ (e) x^2/y^2

5. (a) $44\frac{17}{32}$ (c) $17\frac{1}{2}$ (e) $1\frac{19}{26}$ (g) $2\frac{11}{14}$

6. (a) $\frac{3}{2}$ (c) $\frac{22}{7}$ or $3\frac{1}{7}$ (e) $3\frac{15}{91}$
(g) $\frac{29}{7}$ (i) $(z + y)/z^3$

7. (a) $6 \div 12 = \frac{1}{2}$; $12 \div 6 = 2$ $\frac{1}{2} \neq 2$
$10 \div (5 \div 2) = 4$; $(10 \div 5) \div 2 = 1$; $4 \neq 1$
(c) Yes, $(\frac{3}{5} + \frac{6}{7}) \div 3 = \frac{51}{35} \div 3 = \frac{17}{35}$;
$\frac{3}{5} \div 3 + \frac{6}{7} \div 3 = \frac{1}{5} + \frac{2}{7} = \frac{17}{35}$

8. (a) $x = \frac{5}{2}$ (c) $x = \frac{5}{28}$ (e) $x = 3$

9. (a) Let x = number of hours.
$x = 2(\frac{5}{4})$

(c) Let x equal the number of empty parking spaces.

$$x + \frac{7}{8}(64) = 64$$

(e) Let t be time for desalinating.

$$12\frac{1}{3}(t) = 100$$

(g) Let x be those who like all three.
$x = \frac{1}{4} \cdot \frac{1}{2} \cdot \frac{2}{3}$

(i) Let x equal number of pieces left.

$$\frac{1}{12}x = \frac{2}{3}$$

(k) Let x be the number of pounds collected by girl scouts.

$$\left(2\frac{1}{2}\right)(x + 50) + x + 50 = 490$$

(m) Let n equal the number of bows.

$$n\left(\frac{3}{8}\right) = \left(2\frac{1}{3}\right)(3)$$

10. (a) $x = 2\frac{1}{2}$ hours
(c) $x = 8$ spaces
(e) $t = 8\frac{4}{37}$ hours
(g) $\frac{1}{12}$
(i) 8 slices (No)
(k) The girl scouts collected 90 pounds; boy scouts 350 pounds
(m) 18; $\frac{2}{3}$ left over

11. (a) $8\frac{1}{3}$ kilometers

12. (a) $\frac{1}{4}$ pound

13. (a) 1300 mL flour 12 eggs
24 mL salt 240 mL buter
180 mL sugar 1500 mL milk
32 mL baking soda

15. (a) $\dfrac{adf + cbf + ebd}{bdf}$ (c) $\dfrac{(ad + bc)fh}{bd(eh + fg)}$

16. (a) $\left(\dfrac{1}{11}\right)_{\text{seven}}$

Exercise Set 4, Page 251

1. (a) $\frac{9}{10} > \frac{7}{11}$ (c) $\frac{5}{7} > \frac{2}{5}$ (e) $\frac{17}{51} < \frac{6}{7}$

2. (a) $\frac{71}{100} < \frac{23}{30} < \frac{3}{2}$
(c) $\frac{16}{27} < \frac{11}{18} < \frac{2}{3} < \frac{67}{100}$
(e) $\frac{156}{50} < \frac{22}{7} < \frac{10}{3} < \frac{7}{2}$

3. (a) $\{x \mid x \text{ is an integer } < 2\}$

4. (a) False (c) False

5. (a) Q, I, N (c) Q, I
(e) W, F, Q, I, N (g) Q, I

6. (a) If $p/q < r/s$, then $p/q + t/v < r/s + t/v$.
(c) If $p/q < r/s$ and $t/v > 0$, then $p/q \cdot t/v < r/s \cdot t/v$.
(e) If $p/q < r/s$ and $t/v < 0$, then $p/q \cdot t/v > r/s \cdot t/v$.
(g) Commutative property of multiplication of rational numbers

7. (a) Yes (c) Yes (e) No

9. Let a/b and c/d be fractions such that b and d are positive and $a/b < c/d$. Then $\frac{a+c}{b+d} < \frac{c}{d}$ if $ad + cd < bc + cd$ or $ad < bc$. But $ad < bc$, since $\frac{a}{b} < \frac{c}{d}$. Thus, $\frac{a+c}{b+d} < \frac{c}{d}$. Likewise, $\frac{a}{b} < \frac{a+c}{b+d}$ if $ab + ad < ab + bc$ or $ad < bc$. This is true, so $\frac{a}{b} < \frac{a+c}{b+d}$.

10. (a) $\{7\}$ (c) $\{z \mid z$ is a rational number $\geq 4\}$
(e) $\{2\}$ (g) $\{x \mid x$ is a rational number $\geq 1\}$
(i) $\{-2\}$ (k) $\{-5\}$
(m) $\{x \mid x$ is a rational number $> -4\}$

11. (a) $x + 8 = 15$ Given
$(x + 8) + (-8) = 15 + (-8)$ If $a = b$, then $a + c = b + c$.
$(x + 8) + -8 = 7$ Addition
$x + [8 + (-8)] = 7$ Associative property of addition
$x + 0 = 7$ Additive inverse
$x = 7$ Additive identity

(c) $2x = 12$ Given
$\frac{1}{2}(2x) = \frac{1}{2}(12)$ If $a/b = c/d$, then
$$\frac{a}{b} \cdot \frac{e}{f} = \frac{c}{d} \cdot \frac{e}{f}$$
$x = 6$ Associative property of multiplication and multiplication

(e) $y + 7 = 8$ Given
$(y + 7) + (-7) = 8 + (-7)$ If $a = b$, then $a + c = b + c$.
$(y + 7) + (-7) = 1$ Addition
$y + (7 + -7) = 1$ Associative property of addition
$y + 0 = 1$ Additive inverses
$y = 1$ Additive identity

(g) $5x + 7 = 12$ Given
$(5x + 7) + (-7) = 12 + (-7)$ If $a = b$, then $a + c = b + c$.
$(5x + 7) + (-7) = 5$ Addition
$5x + (7 + -7) = 5$ Associative property of addition
$5x + 0 = 5$ Additive inverses
$5x = 5$ Additive identity
$5x(\frac{1}{5}) = 5(\frac{1}{5})$ If $a = b$, then $a \cdot (1/c) = b \cdot (1/c)$.
$x = 1$ Associative and commutative properties of multiplication and multiplication

12. (a) $x + 2 < 6$ Given
$(x + 2) + (-2) < 6 + (-2)$ If $a < b$, then $a + c < b + c$.
$(x + 2) + (-2) < 4$ Addition
$x + (2 + -2) < 4$ Associative property of addition
$x + 0 < 4$ Additive inverses
$x < 4$ Additive identity
$\{x \mid x$ is a rational number $< 4\}$

(c) $-2x - 4 < -6$ Given
$(-2x - 4) + 4 < -6 + 4$ If $a < b$, then $a + c < b + c$.
$(-2x - 4) + 4 < -2$ Addition
$-2x + (-4 + 4) < -2$ Associative property of addition and definition of subtraction
$-2x + 0 < -2$ Additive inverses
$-2x < -2$ Additive identity
$-2x\left(\frac{-1}{2}\right) > -2 \cdot \left(\frac{-1}{2}\right)$ If $a/b < c/d$ and $e/f < 0$, then $\frac{a}{b} \cdot \frac{e}{f} > \frac{c}{d} \cdot \frac{e}{f}$
$x > 1$ Commutative and associative properties of multiplication
$\{x \mid x$ is a rational number $> 1\}$

13. (a) $\{x \mid x$ is a rational number $< 3\}$
(c) $\{x \mid x$ is a rational number $< -\frac{11}{24}\}$
(e) $\{x \mid x$ is a rational number $< 8\}$
(g) $\{x \mid x$ is a rational number $< -9\}$
(i) $\{x \mid x$ is a rational number $\leq \frac{5}{8}\}$

15. Let x be Mrs. Jones' bill.
$6x + x = 84$
$x = \$12$ (Mrs. Jones)
$6x = \$72$ (Mrs. Smith)

17. Let x be Ralph's weight.
$3x + 54 = 300$
$x = 82$ kg. (Ralph's weight)

19. (a) Additive identity
(c) Additive inverse for each element

21. (a) Yes

22. (a) No (c) Yes (e) No

23. (a) $R = A + C - P$ (c) $t = (A - P)/Pr$
(e) $d = (l - a)/(n - 1)$ (g) $x = (y - b)/m$

24. (a) Let p/q and r/s be expressed with positive denominators.
$\frac{p}{q} < \frac{r}{s}$ Given
$ps < qr$ Definition of less than for rational numbers

$psv^2 < qrv^2$ Positive multiplication property of less than for integers

$psv^2 + qvst <$ $qrv^2 + qvst$ Additive property of less than for integers

$(pv)(sv) + (qt)(sv) <$ $(rv)(qv) + (st)(qv)$ Commutative and associative properties of multiplication of integers

$(pv + qt)(sv) <$ $(rv + st)(qv)$ Distributive property of multiplication over addition

$\frac{pv + qt}{qv} < \frac{rv + st}{sv}$ Definition of less than for rational numbers

$\frac{p}{q} + \frac{t}{v} < \frac{r}{s} + \frac{t}{v}$ Addition of rational numbers

(c) Let p/q and r/s be expressed with positive denominators.

$\frac{p}{q} < \frac{r}{s}$ Given

$ps < qr$ Definition of less than for rational numbers

$\frac{t}{v} < 0$ Given

Let t be such that v is positive. If $t/v < 0$ and $v > 0$, then $t < 0$. So $tv < 0$.

$(ps)(tv) > (qr)(tv)$ Since $tv < 0$, the inequality is reversed.

$(pt)(sv) > (qv)(rt)$ Associative and commutative properties of multiplication of integers

$\frac{pt}{qv} > \frac{rt}{sv}$ Definition of less than for rational numbers

$\frac{p}{q} \cdot \frac{t}{v} > \frac{r}{s} \cdot \frac{t}{v}$ Multiplication of rational numbers

25. (a) $\frac{p}{q} + \frac{t}{v} < \frac{r}{s} + \frac{t}{v}$ Given

$\left(\frac{p}{q} + \frac{t}{v}\right) + -\frac{t}{v} <$ $\left(\frac{r}{s} + \frac{t}{v}\right) + -\frac{t}{v}$ Additive property of inequalities

$\frac{p}{q} + \left(\frac{t}{v} + -\frac{t}{v}\right) <$ $\frac{r}{s} + \left(\frac{t}{v} + -\frac{t}{v}\right)$ Associative property of addition

$\frac{p}{q} + 0 < \frac{r}{s} + 0$ Additive inverses

$\frac{p}{q} < \frac{r}{s}$ Additive identity

Exercise Set 5, Page 260

1. (a) $1(10) + 4(1) + 7(1/10) + 2(1/10)^2$
(c) $1(\frac{1}{10}) + 4(\frac{1}{10})^2 + 6(\frac{1}{10})^3$

2. (a) 43.26 (c) 7.0042

3. (a) 85/1000 (c) 1274/100 (e) 275/100

4. (a) 31.6 (c) 1.2 (e) 5.425 (g) 0.375 (i) 1.2

5. (a) Yes (c) No (e) No

6. (a) 2.4

7. (a) 35.841 (c) 5.30311 (e) 315.87 (g) 683.174

8. (a) $28\frac{32}{100} + 7\frac{521}{1000} = 35\frac{841}{1000} = 35.841$

(c) $\frac{3}{10} + 5\frac{311}{100{,}000} = 5\frac{30311}{100{,}000} = 5.30311$

(e) $354\frac{51}{100} - 38\frac{64}{100} = 315\frac{87}{100} = 315.87$

(g) $691\frac{7}{10} - 8\frac{526}{1000} = 683\frac{174}{1000} = 683.174$

9. (a)
$$\begin{array}{r} 2(10) + 8 + 3(10)^{-1} + 2(10)^{-2} \phantom{+ 1(10)^{-3}} \\ 7 + 5(10)^{-1} + 2(10)^{-2} + 1(10)^{-3} \\ \hline 2(10) + 15 + 8(10)^{-1} + 4(10)^{-2} + 1(10)^{-3} \\ 3(10) + 5 + 8(10)^{-1} + 4(10)^{-2} + 1(10)^{-3} \end{array}$$
$= 35.841$

(c)
$$\begin{array}{r} 3(10)^{-1} \phantom{+ 0(10)^{-2} + 3(10)^{-3} + 1(10)^{-4} + 1(10)^{-5}} \\ 5 + 0(10)^{-1} + 0(10)^{-2} + 3(10)^{-3} + 1(10)^{-4} + 1(10)^{-5} \\ \hline 5 + 3(10)^{-1} + 0(10)^{-2} + 3(10)^{-3} + 1(10)^{-4} + 1(10)^{-5} \end{array}$$
$= 5.30311$

(e)
$$\begin{array}{r} 3(10)^2 + 5(10) + 4 + 5(10)^{-1} + 1(10)^{-2} \\ -\,[3(10) + 8 + 6(10)^{-1} + 4(10)^{-2}] \\ \hline 3(10)^2 + 1(10) + 5 + 8(10)^{-1} + 7(10)^{-2} \end{array}$$
$= 315.87$

(g) $6(10)^2 + 9(10) + 1 + 7(10)^{-1}$
$$\frac{\quad - [8 + 5(10)^{-1} + 2(10)^{-2} + 6(10)^{-3}]}{6(10)^2 + 8(10) + 3 + 1(10)^{-1} + 7(10)^{-2} + 4(10)^{-3}}$$
$= 683.174$

10. (a) 417.6; 4176; 41,760
(c) 0.0021; 0.021; 0.21

11. (a) 10 (c) 1

13. Let x be money in Chang's account.

$$x + 2(25) + 13.42 = 142.36$$
$$x = \$78.94$$

15. Let x be cost of long distance calls.

$$x + 13.75 = 30$$
$$x = \$16.25$$

17. (a) $(\frac{1}{3})_{ten} = (\frac{1}{10})_{three} = 0.1_{three}$
(c) $(\frac{1}{3})_{ten} = (\frac{3}{9})_{ten} = 3(\frac{1}{10})_{nine} = 0.3_{nine}$

18. (a) $\frac{2}{7}$ is 0.2 in base seven.

19. (a) $(1 \cdot 2^2 + 1 \cdot 2^0 + 1 \cdot 2^{-1} + 1 \cdot 2^{-2})_{ten}$
(c) $(6 \cdot 12 + 10 \cdot 12^0 + 7 \cdot 12^{-1} + 1 \cdot 12^{-2})_{ten}$

20. (a) 100.111_{two} (c) 2.211_{three}

Exercise Set 6, Page 267

1. (a) 0.16002 (c) 2.64

2. (a) $\frac{16002}{100,000}$ (c) $\frac{264}{100}$

3. (a) 16.3 (c) 17.1

4. Part (a) is the same as parts (b), (e), and (f). I know these quotients are identical by moving the decimal point.

5. (a) 0.04230 (c) 6.63

6. (a) $>$ (c) $>$ (e) $<$

7. (a) 16; 16.259; 16.2591; 16.3
(c) 0.036; 0.063; 0.306; 0.603

8. (a) $0.\overline{142857}$ (c) $0.\overline{230769}$

9. (a) $\frac{173}{500}$ (c) $\frac{2}{11}$ (e) $\frac{823}{50}$ (g) $\frac{69}{55}$
(i) $\frac{62}{9}$ (k) $\frac{595}{99}$ (m) $\frac{1}{110}$

10. (a) $0.\overline{6}$ (c) They are equal.

11. (a) Let

$$N = 7.9\overline{9}$$
$$10N = 79.\overline{9}$$
$$10N - N = 72$$
$$9N = 72$$
$$N = 8$$

(c) $3(\frac{10}{3}) = 3(3.333\ldots)$
$10 = 9.999\ldots$

13. Let x be the weight of the satchel.

$$x + 2(1.03) + 3(.72) = 5.46$$
$$x = 1.24 \text{ kg}$$

15. Let a represent average speed.

$$6(a) = 300$$
$$a = 50 \text{ mi/hr}$$

16. (a) \$38.36

17. (a) 17.62 kg (c) \$43.20

19. (a) $x = 0.81$

20. (a) 24.02_{five} (c) 1654.134_{seven} (c) 10.0011_{two}

21. (a) $x/9$ (c) $xyz/999$

Exercise Set 7, Page 273

1. 56,000; 62,619.48

3. 1600; 1804.406

5. 4050; 4070.79

7. 0.083; 0.08365

9. 453,000; 453,039

11. 3400; 3398.68

13. 0.0002; 0.00021622

15. 5,000,000; 4,717,741.9

17. (a) Correct (c) Incorrect (e) Incorrect
(g) Correct

18. (a) 60; 66.444 (c) 2; 2.120878
(e) 1900; 1877.8952

19. (a) 28; 30.3366 (c) 300; 300

20. (a) $1.1(10)^7$ (c) $3.3(10)^{-6}$ (e) $3.6754(10)^{-2}$

21. (a) 4 significant digits
0.005, greatest possible error
(c) 3 significant digits
5, greatest possible error
(e) 3 significant digits
500, greatest possible error

22. (a) 0.46 (c) 0.000006 (e) 28,900,000,000

23. (a) 91.48, 91.5, 91.5 (c) 0.85, 0.9, 0.854
(e) 273.87, 273.9, 274

24. (a) $4499.\overline{9}$; 3500 (c) $2.14\overline{9}$; 2.05

25. (a) 57.76 (c) 0.00595 (e) 943.

26. (a) $1.2(10)^2$ (c) 1(10)

27. (a) 4.24 (c) 17.616

28. (a) 3.7 kg (c) 4.5 qt.

29. (a) 1.035 (c) $0.5\overline{37}$

Review Exercise Set 8 Page 277

1. $\frac{6}{16}, \frac{9}{24}, \frac{12}{32}$
2. (a) $\frac{2}{3}$ (c) $\frac{5}{18}$
3. (a) $\frac{34}{35}, \frac{6}{35}, \frac{8}{35}, \frac{10}{7}$
   (c) $\frac{33}{20}, \frac{3}{20}, \frac{27}{40}, \frac{6}{5}$
4. (a) $\frac{9}{32}$ (c) $\frac{1363}{48}$ (e) 2.3601
5. (a) $\frac{7}{5}$ (c) Does not exist
7. (a) $\frac{66}{119}$
8. (a) 0.375 (c) 0.625
9. (a) $\frac{1409}{99}$ (c) $\frac{9904}{330}$ or $\frac{4952}{165}$
   (e) $\frac{89}{4950}$ (g) $\frac{13}{2}$ (i) $\frac{1}{4}$
10. (a) 5.2 (c) 12,000
11. (a) $x = -3$
12. (a) $x < -2$ and a rational number
13. (a) $\frac{3}{4}$ (c) $(a+1)/12$ (e) 0
14. (a) 5 (c) 1 (e) 3
15. (a) $\{x \mid x = -1\}$ (c) $\{x \mid x = -16\}$
    (e) $\{x \mid x = 888\frac{8}{9}\}$ (g) $\{x \mid x = 19/6\}$
17. 1 quarter, 2 nickels, and 2 dimes
19. 3 hours upstream
    1 hour downstream
21. (a) $1\frac{5}{16}$ (c) $3\frac{1}{8}$
22. (a) $0.\overline{3}_{seven}$ (c) $0.\overline{15}_{seven}$
23. (a) When the denominator can be expressed in the form 2^r5^s.
    (c) Must be of the form 2^s.

CHAPTER 6

Exercise Set 1, Page 285

1. (a) Irrational (c) Irrational
   (e) Irrational (g) Irrational (i) Rational
   (k) Irrational (m) Rational (o) Rational
2. (a) False (c) False (e) True (g) False
   (i) False (k) True (m) True
   (o) False (q) True
3.

Number	Whole number	Integer	Rational number	Irrational number	Real number
-5	No	Yes	Yes	No	Yes
$2.\overline{71}$	No	No	Yes	No	Yes
0.717117111 . . .	No	No	No	Yes	Yes
0	Yes	Yes	Yes	No	Yes
$5\sqrt{2}$	No	No	No	Yes	Yes
3π	No	No	No	Yes	Yes
$0.\overline{45}$	No	No	Yes	No	Yes
4	Yes	Yes	Yes	No	Yes

4. (a) Yes, 7 (c) No (e) No
5. No
6. (a) $\sqrt{2}\cdot\sqrt{3} = \sqrt{6}$, irrational;
   $\sqrt{2}\cdot\sqrt{2} = 2$, rational
   (c) $3\sqrt{2} - 2\sqrt{2} = \sqrt{2}$, irrational;
   $6\sqrt{2} - 6\sqrt{2} = 0$, rational
7. $(3) + (\sqrt{5}) = 3 + \sqrt{5}$
9. (a) 4 (c) An infinite number; $1/\sqrt{2}, \sqrt{2}, \sqrt{3}$
10. (a) 9 (c) Infinite number
    (e) Infinite number
11. (a) False (c) False
13. $\sqrt{0.216}$ and 0.46614114111 . . .
14. (a) 0.56521
15. (a) Third (c) Seventh (e) Fourth
17. $1/x < 1/y$ since multiplying both sides by xy gives $y < x$.
19. (a) Assume $3\sqrt{2}$ is rational $= a/b$. Then $\sqrt{2} = a/3b$, which is rational. But $\sqrt{2}$ is irrational. Contradiction. Therefore, $3\sqrt{2}$ is irrational.
    (c) Assume $3/\sqrt{2}$ is rational. $3/\sqrt{2} = a/b$ or $3b/a = \sqrt{2}$. $3b/a$ is rational. Thus, $\sqrt{2}$ is rational. However, $\sqrt{2}$ is irrational. Contradiction. Therefore, $3/\sqrt{2}$ is irrational.
    (e) Assume $3 - 2\sqrt{2}$ is rational. $3 - 2\sqrt{2} = a/b$ or $(3b - a)/b = 2\sqrt{2}$. $(3b - a)/b$ is rational, so $2\sqrt{2}$ is rational. But $2\sqrt{2}$ is irrational. Contradiction. Thus, $3 - 2\sqrt{2}$ is irrational.
    (g) Assume $2 + \sqrt{3}$ is rational. Then $2 + \sqrt{3} = a/b$ or $\sqrt{3} = a/b - 2$, which is rational. Thus $\sqrt{3}$ is rational. This is a contradiction. Thus $2 + \sqrt{3}$ is irrational.
21.

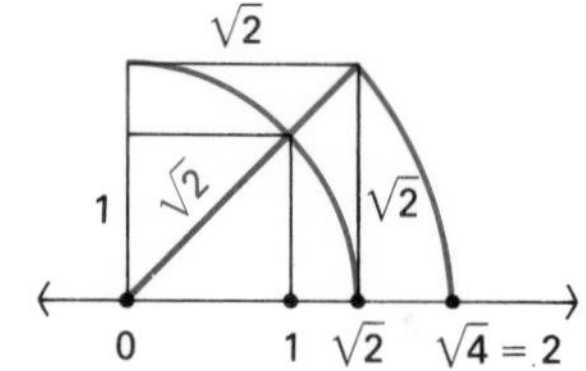

Exercise Set 2, Page 291

1. (a) 7 (c) 3 (e) -9
2. (a) $4\sqrt{2}$ (c) $5\sqrt{5}$
3. (a) $3\sqrt{5}$ (c) $4\sqrt{2}$ (e) $20\sqrt{3}$
4. (a) $5\sqrt{2}$ (c) $(\frac{7}{6})\sqrt{3}$
5. (a) 4 (c) 2
6. (a) $\sqrt{10}$ (c) $6\sqrt{6}$ (e) $\sqrt{3}$ (g) $2\sqrt{10}$

7. (a) $\frac{5\sqrt{2}}{2}$ (c) $\frac{\sqrt{14}}{2}$ (e) $\frac{\sqrt{105}}{7}$ (g) $\frac{\sqrt{6}}{4}$
8. (a) $\frac{1}{160}$ (c) $\frac{1}{4}$
9. (a) $-3\sqrt[3]{3}$ (c) $-2\sqrt[3]{3}$ (e) 2
10. (a) $x = 2 - \frac{\sqrt{2}}{4}$ (c) $y = -\frac{3}{2} + 2\sqrt{3}$
11. 0.1001000100001 . . .
0.1221222122221 . . .
0.1331333133331 . . .
0.1441444144441 . . .
13. (a) 6.08276 (c) 2.2804 (e) 15.65247

Exercise Set 3, Page 297

1. (a) $\frac{5}{3}$ (c) $\frac{2}{7}$ (e) $\frac{1}{230}$
2. (a) 1:9 (c) 34:45
3. (a) $x = 8$ (c) $x = 15\frac{6}{23}$ (e) $x = 40$
4. (a) 2:3 (c) 2:5
5. (a) $\frac{43}{2}$ miles per gallon
(c) $\frac{3}{17}$ administrators to employees
6. (a) $x = \frac{6}{5}$ (c) $x = 5\frac{1}{7}$
7. (a) 7:3 (c) 12
8. (a)

Price	49¢	46¢
Grams	900	850
Ratio	$\frac{49}{900} = 0.0544$	$\frac{46}{850} = 0.0541$

The 850-g can for 46¢ is better buy.
9. (a) 29-oz can at \$1.69¢ per oz.
(c) 3-lb ham for \$1.83 per lb.
(e) 8 oz of spaghetti for \$.0375 per oz.
11. (a) vanilla $\frac{1}{2}$ tsp.
flour $1\frac{1}{2}$ cups
sugar $\frac{7}{8}$ cup
(c) eggs 4
vanilla 2 tsp
flour 6 cups
13. $1\frac{7}{8}$ g
15. 10 L of green
10 L of yellow
40 L of white
17. The number of boys is larger in the second class.
19. 54 gallons
21. 28.44 g
22. (a) Let $a = 2, b = 4, c = 3, d = 6$. For $a/b = c/d$ if and only if $b/a = d/c$, $\frac{2}{4} = \frac{3}{6}$ if and only if $\frac{4}{2} = \frac{6}{3}$, which is true. For $a/b = c/d$ if and only if $a/c = b/d$, $\frac{2}{4} = \frac{3}{6}$ if and only if $\frac{2}{3} = \frac{4}{6}$, which is true. (One of many examples)

Exercise Set 4, Page 304

1. (a) 0.055 (c) 0.0031 (e) 0.436
2. (a) 14.7% (c) 0.5% (e) 861%
3. (a) 120% (c) 40% (e) 27.5%
4. (a) 34 (c) 123.45 (e) 0.66
5. (a) 34 (c) 123.45 (e) 0.66
7. 0.36 (\$1.56)
0.72 (\$3.12)
0.51 (\$2.21)
1.08 (\$4.68)
8. (a) 25% (c) 32 (e) 12.48 (g) 300%
9.

	Cost	% Markup based on cost	Selling price
(a)	\$1400	24	\$1736
(c)	\$70	20	\$84

10. (a) 19.35% (c) $16\frac{2}{3}$%
11. $16\frac{2}{3}$%
13. \$988; 20.24%
15. 20 hours
17. \$20.01
19. 31.82%
21. 200 g
23. 314.3%
24. (a) \$94.08
25. \$898
27. 57.97%
29. \$19.99
30. (a) \$17.66
31. \$21,767.10
33. 91%

Exercise Set 5, Page 311

1. (a) \$80 (c) \$250
2. (a) \$580 (c) \$750
3. Interest is \$960; amount is \$3960.
5. \$1545
7. (a) Compound interest = \$2024.64
Compound amount = \$7024.64
(c) Compound interest = \$2128.80
Compound amount = \$7128.80
8. (a) Compound interest = \$928.20
Compound amount = \$2928.20
(c) Compound interest = \$969.01
Compound amount = \$2969.01
9. (a) Compound interest = \$1255.56
Compound amount = \$4255.56
(c) Compound interest = \$1288.51
Compound amount = \$4288.51

10. (a) \$4880 (c) \$124
11. (a) $r = 9\%$ (c) $r = 16\%$
13. \$4053.39
15. \$5330.74
17. \$29.33
19. Compounded quarterly for \$1; \$1.43. Simple interest on \$1; \$1.36. The 12% compounded quarterly requires a larger payment per dollar borrowed.
20. (a) \$1559.02
21. 12%
23. \$735.23
25. (a) \$71,633.91 (c) \$6.27
27. (a) \$103,749.70 (c) \$9.08

Exercise Set 6, Page 315

1. (a) \$140 (c) \$825
2. (a) \$785 (c) \$148
3. (a) \$199.38
5. \$193
7. (a) Option 2. He pays \$1400 finance charge, as compared to \$1150 in option 1.
8. (a) \$50
9. (a) \$54.94
10. (a) \$552 (c) 11.25%
11. 12.50%
13. (c) 11.00%

Review Exercise Set 7, Page 317

1. (a) $x = \frac{21}{4}$
2. (a) \$3.30 (c) 112.5%
3. \$25
5. (a) 26.97% (c) \$35
6. (a) False. It is a terminating decimal. (c) True
   (e) False ($\sqrt{2} \cdot \sqrt{2} = 2$ rational)
   (g) False. If a number can be represented by a repeating decimal, it is a rational number.
   (i) False. If $b = 0$, then a could equal any rational number and $a \cdot b = 0$
   (k) False. The largest rational number less than 3 cannot be defined.
   (m) True (o) True
   (q) False. $-\frac{1}{2}$ is larger than -1.
7. (a) $8\sqrt{3}$ (c) $\frac{6}{5}\sqrt{2}$ (e) 3 (g) 4
9. Compound interest = \$1067.08
   Compound amount = \$5067.08
11. \$300
13. \$9.60; \$2.00
15. 72.75% based on cost
    42.11% based on selling price

CHAPTER 7

Exercise Set 1, page 328

1. (a) 9 (c) 1.6 (e) 12 (g) 2.2
   (i) 0.1428571429 (k) 2 (m) 17 (o) 9
   (q) 16 (s) 900 (u) 20,700
2. (a) `6 * 5^3` (c) `16^10` (e) `(16.1 + 17.3)^4`
   (g) `(9.4 - 3.1)^4/(3 + 5.6)`
3. (a) `4 * X + 7` (c) `2 * X^2 + 7 * X + 4`
   (e) `2/X + 5 * (X + 7)`
   (g) `(5 * X^2 + 1) * (2 * X + 5)/(4 * X - 1)^2`
4. (a) $4x^2 - 3x + 2$ (c) $(x - 5)^2 + 4(x - 5)^3$
   (e) $(15x^2 + 7/5)^2/3$
5. (a) 10 (c) THE SOLUTION IS 1.25.
   (e) 11
   30, 121
6. (a) Correct
   (c) Incorrect; need a single variable on the left of the inequality and need a line number. Probably, "`10 A = B + C`."
   (e) Needs a line number
   `20 PRINT 2 * A`
7. ROW ROW ROW YOUR BOAT
   GENTLY DOWN THE STREAM
8. (a) 0.000763 (c) 715.6
9. (a) 4.1762E + 03 (c) 1.3E − 05
   (e) 1E + 06
10. (a) Commands are backward. Should read
    ```
    10 C = 10
    20 B = 20
    ```
11. (a) 46.24 (c) 4.8365385 (e) 18.396
    (g) 22.773647
12. (a) 8.3022E − 01 (c) 4.3E − 16
13. (a) The purpose of this program is to find the length of the hypotenuse of a right triangle. *Note:* SQR(C) is a function that finds the square root of c.

Exercise Set 2, Page 333

1. (a) 3, 4, 81
   (c) ENTER THE RATE AND BASE
2. (a)
    ```
    10 INPUT P, R, T
    20 A = P + P * R * T
    30 PRINT A
    40 END
    ```
   (c)
    ```
    10 INPUT R, B
    20 P = R * B
    30 IF P < 100 THEN 50
    40 PRINT "TOO LARGE"
    50 PRINT "TOO SMALL"
    60 END
    ```

3. (a) There are no line numbers.
(c) The last part of an IF . . . THEN statement should be a line number.

4. (a) 87.2 cm
(c)
```
10  INPUT M
20  IF M < 1 THEN 60
30  H = M/100
40  PRINT H
50  GOTO 80
60  C = 100 * M
70  PRINT C
80  END
```

5.
```
10  PRINT "ENTER A CENTIGRADE ";
15  PRINT "TEMPERATURE"
20  INPUT C
30  F = (9/5) * C + 32
40  PRINT "THE FAHRENHEIT TEMPER";
45  PRINT "ATURE IS", F
50  END
```
(a) 86 (c) 32

7.
```
10  PRINT "INPUT PRINCIPAL, INTEREST   ";
20  PRINT "AND NUMBER OF PERIODS"
30  INPUT P, I, N
40  A = P * (1 + I)^N
50  PRINT "THE COMPOUND AMT IS" A
60  END
```
(a) THE COMPOUND AMT IS 112.4864

8.
```
10  PRINT "INPUT AMOUNT, INTEREST ";
20  PRINT "AND NUMBER OF PERIODS"
30  INPUT A, I, N
40  P = A/(1 + I)^N
50  PRINT "THE PRESENT VALUE IS   " P
60  END
```
(a) THE PRESENT VALUE IS 771.680349

9.
```
10  INPUT H
20  P = 4 * H
30  IF H < 40 THEN 50
40  P = P + 1.50 * (H - 40)
50  PRINT P
60  END
```

Exercise Set 3, Page 338

1. (a)
```
  1
  8
 27
 64
125
216
```
(c)
```
1   3
2   8
3  15
4  24
```
(e)
```
X = 0   X^2 = 0
X = 2   X^2 = 4
X = 4   X^2 = 16
X = 6   X^2 = 36
X = 8   X^2 = 64
```

2. (a) Incorrect: should have a statement number.
(c) Incorrect: you cannot use END after GOTO.
(e) Incorrect: a number is missing after "IF X> <____."

3.
```
10  N = 1
20  CORRECT
30  IF N > 10 THEN 60
40  S = S + 2 * N
45  CORRECT
50  GOTO 30
60  CORRECT
70  CORRECT
```

5. (a)
```
10  N = 1
20  S = 0
30  S = S + N
40  IF N = 500 THEN 70
50  N = N + 1
60  GOTO 30
70  PRINT S
80  END
```
(c)

N	1	2	3	4	5	6	7
S	1	3	6	10	15	21	28

7. (a)
```
10  N = 1
15  T = 1
20  IF N > 10 THEN 70
30  PRINT T
40  T = T + 3
50  N = N + 1
60  GOTO 20
70  END
```

9.
```
 10  PRINT "ENTER A MEASUREMENT IN CM"
 20  INPUT X
 30  IF X > 100 THEN 70
 40  IF X < 1 THEN 100
 50  PRINT X "CM"
 60  GOTO 120
 70  M = X/100
 80  PRINT X "CM EQUALS   " M "M"
 90  GOTO 120
100  M = 10 * X
110  PRINT X "CM EQUALS   " M "MM"
120  END
```

11. (a)
```
10  N = 1
20  S = 0
30  S = S + N^2
40  IF N = 100 THEN 70
50  N = N + 1
60  GOTO 30
70  PRINT S
80  END
```

13.
```
10  N = N
20  A = N^2
30  B = N^3
40  C = SQR (N)
15  PRINT "NUMBER", "SQUARE",
16  PRINT "CUBE", "SQUARE ROOT"
70  PRINT N, A, B, C
80  IF N = 25 THEN 92
90  LET N = N + 1
91  GO TO 20
92  END
```

15.
```
10  N = 1
11  D = 30
20  D = D - 3
21  IF D = 0 THEN 30
22  D = D + 2
23  N = N + 1
24  GOTO 20
30  PRINT N "DAYS ARE REQUIRED"
31  END
```

Exercise Set 4, Page 344

1. (a) 18 (c) X^2 − Y^2 = 15

(e)

BASE	HEIGHT	AREA
4	3	6
2	1	1
3	6	9

2. (a) Incorrect. Should be "15 B = 4".

(c) Correct

(e) Incorrect. Should be "40 C = 2 * A".

3. Should be
```
10  READ A, B
20  DATA 5, 7, 6, 4, 10, 11, 13, 15, -1, -1
30  PRINT A^2 - B
35  IF B < 0 THEN 50
40  GOTO 10
50  END
```

4. (a)
```
10  READ A, B
20  DATA 13, 24
30  S = A + B
40  P = A * B
50  PRINT "THE SUM IS   " S
60  PRINT "THE PRODUCT IS   " P
70  END
```

(c)
```
10  READ A, B
20  DATA 13, 24
30  S = A + B
40  P = A * B
50  PRINT "SUM", "PRODUCT"
60  PRINT
70  PRINT S, P
80  END
```

5.
```
10  PRINT "ENTER THE NUMBER"
20  INPUT N
30  IF INT (N/17) = N/17 THEN 60
40  PRINT N "IS NOT DIVISIBLE BY 17"
50  GOTO 70
60  PRINT N "IS DIVISIBLE BY 17"
70  END
```

7.
```
10  INPUT N
20  S = 0
30  FOR I = 1 TO N
40  E = 2 * I
50  S = S + E
60  NEXT I
70  PRINT "THE SUM IS" S
80  END
```

9. (a)
```
10  PRINT "KILOMETERS", "METERS"
20  READ K
30  DATA (Data to be supplied)
40  IF K < 0 THEN 80
50  M = 1000 * K
60  PRINT K, M
70  GOTO 20
80  END
```

11. (a)
```
10  INPUT C
20  PRINT C "CM IS EQUIV TO"
30  PRINT C * 10 "MM"
40  PRINT C/10 "DM"
50  PRINT C/100 "M"
60  PRINT C/1000 "DAM"
70  PRINT C/10000 "HM"
80  PRINT C/100000 "KM"
90  END
```

Exercise Set 5, Page 348

1. (a) (c)

2. (a) CS LT 90 FD 50 RT 90 FD 70 RT 90 FD 50

(c) CS FD 20 RT 90 FD 20 LT 90 FD 20 RT 90 FD 20 LT 90 FD 20

3. (a) N (c) F

4. (a) CS LT 90 FD 60 RT 90 FD 20 RT 90 FD 60 RT 90 FD 20

5. (a)

6. (a) CS FD 60 RT 45 FD 60 PU BK 60 LT 90 PD FD 60 HT
   (c) CS FD 60 RT 90 FD 10 LT 90 FD 5 LT 90 FD 25 LT 90 FD 5 LT 90 FD 10 RT 90 FD 60 RT 90 FD 10 LT 90 FD 5 LT 90 FD 25 LT 90 FD 5 LT 90 FD 10

7. (a) CS FD 10 RT 90 FD 20 RT 90 FD 20 RT 90 FD 30 RT 90 FD 30 RT 90 FD 40 RT 90 FD 40 RT 90 FD 50

8. (a)

Exercise Set 6, Page 351

1. (a) REPEAT 3 [FD 30 RT 120]
2. (a)
3. (a) Need one space between FD and 1ØØ
   (c) No number indicating number of times to repeat
4. (a) REPEAT 4 [FD 30 BK 30 RT 90]
5. REPEAT 3 [FD 30 RT 120]
   REPEAT 6 [FD 30 RT 60]
6. (a) REPEAT 8 [FD 40 RT 360/8]
   (c) REPEAT 11 [FD 40 RT 360/11]
7. REPEAT 36 [FD 10 RT 360/36]

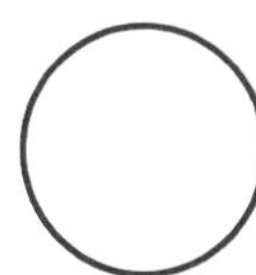

8. (a) RT 90 REPEAT 4 [LT 90 FD 30 RT 90 FD 15 RT 90 FD 30]
   (c) REPEAT 6 [REPEAT 4 [FD 30 R 90] RT 60]
9. (a) REPEAT 6 [REPEAT 3 [FD 20 RT 120] RT 60]
   (c) REPEAT 2 [REPEAT 6 [REPEAT 3 [FD 30 RT 120] RT 60] LT 30]
10. (a) REPEAT 6 [REPEAT 6 [FD 30 RT 360/6] RT 360/6]
    (c) REPEAT 7 [REPEAT 7 [FD 30 RT 360/7] RT 360/7]

Exercise Set 7, Page 355

1. (a) TO RECTANGLE
   REPEAT 2 [LINE 30 RT 90 LINE 40 RT 90]
   END
   (c) TO HEXAGON :LENGTH
   REPEAT 6 [LINE :LENGTH RT 360/6]
   END
2. (a) Need END
   (c) Need colon after FD
3. (a)
   (c)

4. (a) REPEAT 3 [SQUARE RT 90 PU FD 60 PD LT 90]
   (c) LT 90 REPEAT 4 [PU FD 100 RT 90 PD SQUARE]
5. (a) TO L
   FD 40 RT 180 PU FD 40 LT 90
   FD 20 PU FD 10 LT 90 PD HT
   END
   (c) TO G
   FD 40 RT 90 FD 30 RT 90
   FD 4 RT 180 PU FD 4 LT 90
   FD 30 LT 90 FD 40 LT 90
   PD FD 30 LT 90 FD 8
   RT 90 FD 4 RT 180 PU
   FD 4 PD FD 4 PU RT 180
   FD 4 RT 90 FD 8 LT 90
   FD 10 PD LT 90 HT
   END
7. (a) Left off RT 6Ø] at end

Review Exercise Set 8, Page 357

1. (a) 6 * (7 + 3) (c) 2/3 + 8
(e) 4^3 − (2 * 3^2)^4

2. (a)

5	12	60
6	13	78
7	14	98

3. (a) Incorrect; the variable to which the value is assigned must be on the left. A = 3.
(c) Incorrect; each letter that is printed must have quotation marks around it. 40 PRINT "I LOVE YOU"
(e) Incorrect; there should be a comma between each of the variables in a PRINT statement. 20 PRINT A, B
(g) Incorrect; there must be a statement number for each line. 50 PRINT 2 * A
(i) Incorrect; cannot go to 8Ø. IF X > 10 THEN 90 (the 9Ø can be anything other than 8Ø).
(k) Incorrect; following the word THEN, there should always be a statement number. 70 IF X = 4 THEN 90 (any line number other than 7Ø).

4. (a) (4 * X^2 + 7)/(X − 1)
(c) (2 * X^2 + 7 * X + 4)^3

5. (a) 21

7. (a)

8. (a) FD 40 RT 90 FD 20 RT 90 FD 20 RT 90 FD 20 LT 90 PU FD 20 LT 90 FD 30 PD
(c) LT 90 FD 30 LT 90 FD 20 LT 90 FD 20 LT 180 PU FD 20 PD LT 90 FD 20 LT 90 FD 30 PU FD 10 PD ST

9.
```
10 DATA . . . , −1
20 PRINT "FEET", "INCHES"
30 READ F
40 I = 12 * F
50 PRINT F, I
60 IF F < 0 THEN 80
70 GOTO 30
80 END
```

11.
```
10 INPUT Y
20 Z = 3 * Y^3 − 17 * Y^2 + 4 * Y + 3
30 IF Z > 0 THEN 50
40 GOTO 60
50 PRINT Z
60 END
```

13.
```
 10 PRINT "PRINCIPAL", "RATE"
 12 PRINT "TIME", "AMOUNT"
 15 PRINT
 20 R = 0.06
 30 T = 3
 40 DATA 100, 200, 300, 400, 0
 50 READ P
 60 IF P = 0 THEN 100
 70 A = P * (1 + R * T)
 80 PRINT P, R, T, A
 90 GOTO 50
100 END
```

15. (a)
```
TO STEP
FD 10 RT 90 FD 10 LT 90
END
```
(c)
```
TO MANYSTEP
REPEAT 4 [FOURSTEP RT 90]
END
```

16. (a)
```
TO LINE
FD 40 PU RT 180 FD 40 LT 90
FD 20 LT 90 PD ST
END
```

17. (c)
```
  5 DATA (TO BE SUPPLIED)
  6 READ K
  7 N = 2
 10 READ X
 20 S = X
 30 READ X
 40 IF X < S THEN 70
 50 IF N = K THEN 90
 55 N = N + 1
 60 GOTO 30
 70 S = X
 80 GOTO 50
 90 PRINT S "IS THE SMALLEST VALUE"
100 END
```

19.
```
 5 DATA . . .
10 READ T, S, D
20 IF D = 1 THEN 70
30 IF D = 2 THEN 81
40 IF D = 3 THEN 83
50 IF D = 4 THEN 86
60 GOTO 87
70 F = 5 * (D − 35)
80 GOTO 82
81 F = 6 * (D − 35)
82 PRINT T, F
83 GOTO 87
84 PRINT "NOT GUILTY"
85 GOTO 87
86 PRINT "JAILED"
87 GOTO 10
88 END
```

20. (a) PARALLEL LT 90 PU FD 20 PD
PARALLEL

CHAPTER 8

Exercise Set 1, Page 365

1. (a) Line segment (c) Plane (e) Point
(g) Intersecting lines through a point
2. (a) (c) (e)
3. 6
4. (a) $\overline{AB}$ (c) $\overrightarrow{DB}$, $\overrightarrow{DA}$ (e) $\overleftrightarrow{AB}$, $\overleftrightarrow{AC}$
5. (a) Parallel (c) Intersecting (e) Skew
6. (a) True
(c) False. A ray has one endpoint.
(e) False. $\overline{AB}$ is the same set of points as $\overline{BA}$. But $\overrightarrow{AB} \neq \overrightarrow{BA}$
(g) False. $\overline{AB} \subset \overrightarrow{AB}$
(i) False. Two lines can be skew lines.
(k) False. The endpoints of $\overline{BA}$ are B and A.
(m) False. A line has no endpoints.
(o) True
7. (a) No (c) Yes (e) Yes
9. (a) $\overline{QR}$ (c) $\overline{RS}$ (e) $\overline{RS}$ (g) $\{R\}$ (i) $\overline{QT}$
10. (a)
11. (a) (c) Impossible, 4 lines
(e) Impossible, 3 lines

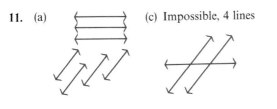

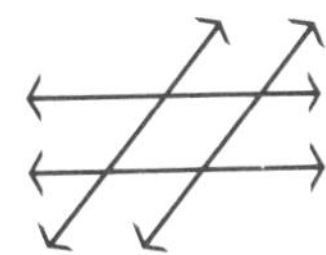

12. (a) $\overrightarrow{AB}$, $\overrightarrow{CD}$; $\overrightarrow{AC}$, $\overrightarrow{CD}$; and so on
(c) $\overrightarrow{AC}$, $\overrightarrow{DB}$; $\overrightarrow{AB}$, $\overrightarrow{DC}$; and so on
13. (a) $\overline{AC}$, $\overline{BC}$; $\overline{AD}$, $\overline{CD}$; and so on
(c) $\overline{AC}$, $\overline{BD}$; $\overline{AD}$, $\overline{BC}$; and so on

14. (a) Yes, $\overrightarrow{BC} \subset \overrightarrow{AC}$ A B C
15. (a) $\overline{AB}$ (c) $\{C\}$ (e) $\{D\}$ (g) $\overleftrightarrow{AD}$
(i) $\overrightarrow{AF}$
16. (a) 1 (c) 6 (e) 15
17. (a) 4 (c) 8
19. (a) 1 (c) 6 (e) 15
20. (a) Half line (c) Segment

21.

Number of lines	4	5	. . .	10
Number of regions	11	16		56

Exercise Set 2, Page 372

1. The intersection of the three planes with the fold on the tabletop is a line. The intersection of the three planes with the folded paper on its end is a point.
2. (a) Sometimes true (c) Always true
(e) Always true
3. (a) No, if the lines are skew.
(c) Yes, as two parallel lines determine a plane.
4. (a) DEF and CAB
(c) ABC, $AEFB$, $BFDC$ intersect in point B. (There are many other examples.)
(e) $\overleftrightarrow{AE}$ intersects EDF in one point. (There are many other examples.)
5. Yes
7. No, if the point is located on the line
8. (a) They are the same plane.
9. Only one
11. Four
13. (a) At least one (c) At least one
(e) At least one
15.

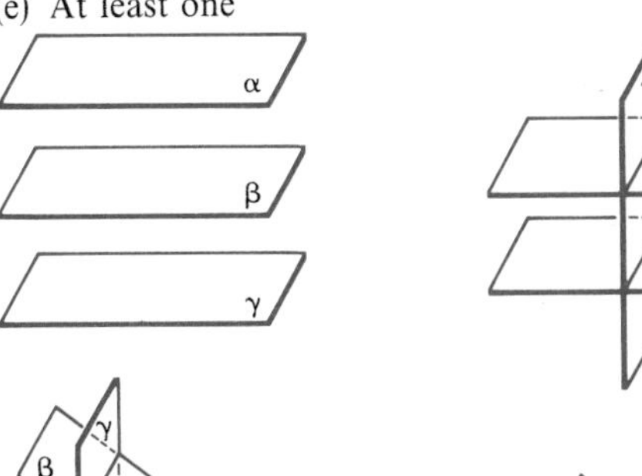

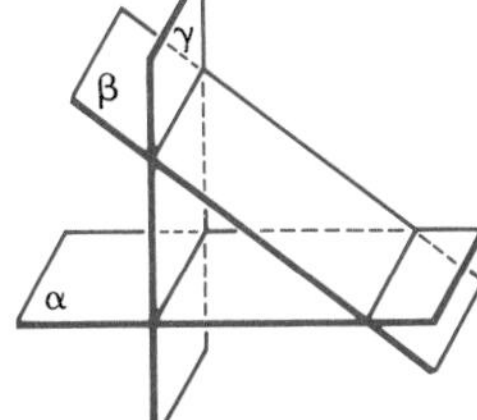

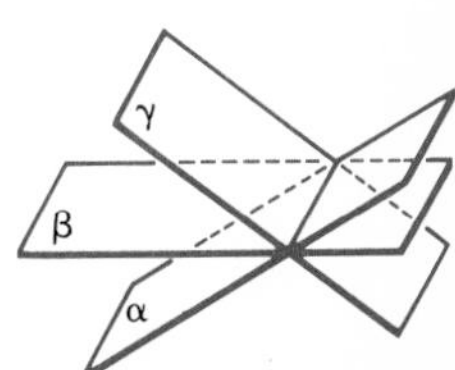

17. (a) Given: A line and a point not on that line.
Prove: A line and a point not on that line determine a plane.

Statement	*Reason*
1. A line has two points, A and B.	Two points determine a line.
2. There is a point C that is not on $\overleftrightarrow{AB}$.	Given
3. $\overleftrightarrow{AB}$ and point C determine a plane.	Three noncollinear points determine a plane (points A, B, and C.

(c) Given: Two parallel lines
Prove: Two parallel lines determine a plane.

Statement	*Reason*
1. $\overleftrightarrow{AB}$ is parallel to $\overleftrightarrow{CD}$.	Given
2. $\overleftrightarrow{AB}$ contains two points A, B. $\overleftrightarrow{CD}$ contains two points C, D.	Two points determine a line.
3. Two parallel lines determine a plane.	Three noncollinear points determine a plane (points A, B, and C).

Exercise Set 3, Page 378

1. $\angle BAC$, $\angle CAB$, $\angle \theta$; 40°
2. (a) False (c) True (e) True
3. (a) Yes (c) No; no common vertex
4. (a) $\angle z$ and $\angle w$ are adjacent and supplementary angles. $\angle x$ and $\angle y$ are vertical angles.
5. (a)

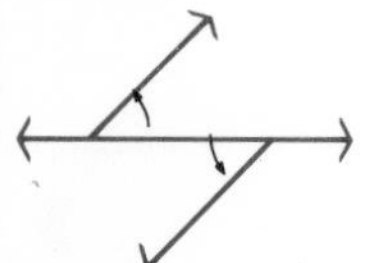

(c)

(e)

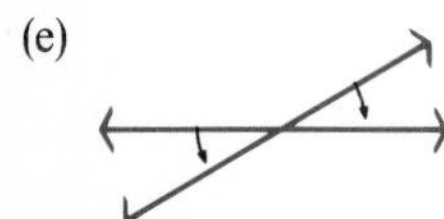

6. (a) $\angle AOF$, $\angle COD$; $\angle DOE$
(c) $\angle BOC$, $\angle DOF$; $\angle AOE$
7. (a) $90° - x$ (c) measure
9. (a) Acute (c) Right (e) Obtuse
10. (a) 140° (c) 140°
11. (a) 145° (c) 145° (e) 35° (g) 55°
12. (a) $\angle AFB$, $\angle EFD$, $\angle AFC$, $\angle EFC$
(c) adjacent; supplementary
(e) 40° (g) 90°
13. (a) 12: $\angle EOC$, $\angle EOB$, $\angle COB$, $\angle COF$, $\angle BOF$, $\angle EOA$, $\angle EOD$, $\angle AOF$, $\angle DOF$, $\angle AOD$, $\angle AOC$, $\angle BOD$
(c) $\angle EOB$ and $\angle AOF$; $\angle EOC$ and $\angle DOF$; $\angle COB$ and $\angle AOD$; $\angle AOC$ and $\angle BOD$; $\angle COF$ and $\angle EOD$; $\angle BOF$ and $\angle EOA$
15. (a) 6 lines, 48 angles
16. (a) 3 (c) 3; 180°

Exercise Set 4, Page 386

1. (a) 370 m (c) 12 m (e) 35,470,000 m
2. (a) 37,890 mm (c) 1,284,000 mm
(e) 35,478,000,000 mm (g) 160 mm
3. (a) 0.017 km (c) 1.6 km
5. (a) 0.032 m (c) $3(10)^6$ mm (e) 134 cm
(g) 48,000,000 μm
6. (a) $m(\overline{AB}) = 1\frac{1}{2}$ (c) $m(\overline{DB}) = 2$
(e) $m(\overline{GA}) = 6$
7. (a) m (c) km (e) cm (g) mm
(i) cm (k) km
8. (a) 100 cm, 10 m, 100,000 mm
(c) 0.1 cm, 0.01 m, 0.001 km
9. (a) 21.55, 21.65 (c) 1.5, 2.5
10. (a) 1075 cm or 10.75 m (c) 457 cm or 4.57 m
11. (a) 185 cm (c) 50 mm (e) 100 cm
12. (a) To change meters to kilometers, divide the number of meters by 1000.
13. (a) 214 (c) 5.6 km
14. (a) $x = 3\frac{1}{2}$ or $\frac{1}{2}$ (c) $x = 7$ or -1
15. (a) $B = 1\frac{1}{2}$ or $B = -\frac{1}{2}$ (c) $B = -3$ or $B = -5$
(e) $B = 0$ or $B = -2$
17. (a) 68 cm (c) 4 m (e) 4 mm
18. (a) 60 cm (c) 8 cm (e) 3 mm (g) 24 m
(i) 210 cm
19. 1375 m

Exercise Set 5, Page 393

1. (a) Simple plane curve (c) Simple plane curve
(e) Simple plane curve
(g) Not a simple plane curve
(i) Not a simple plane curve
2. (a) Not closed (c) Closed (e) Closed
3. (a) Not a polygon (c) Not a simple polygon
(e) Simple polygon

4. (a) Convex (c) Not convex
(e) Not convex

5. (a) (c)

(e) Not possible

6. (a) (c)

7. (a) (c)

8. (a) (c) (e)

9. (a) False (c) True (e) False (g) True
(i) True (k) False

10. (a) ii, iv, v, vi (c) i, iv, v, vi (e) i, iii, iv, v, vi

11. (a) First and third (c) Yes

13. Impossible

14. (a) None (c) 20

15. (a) False (c) False (e) False (g) False

16. (a) Let $\overleftrightarrow{AB}$ divide plane α into half planes α_1 and α_2. α_2 is indicated by horizontal shading in the figure. Let $\overleftrightarrow{BC}$ divide plane β into half planes β_1 and β_2, with β_2 being indicated in the figure by diagonal lines. Let $\overleftrightarrow{AC}$ divide plane θ into half planes θ_1 and θ_2, with θ_2 being indicated by vertical lines. The interior of $\triangle ABC$ is $\alpha_2 \cap \beta_2 \cap \theta_2$.

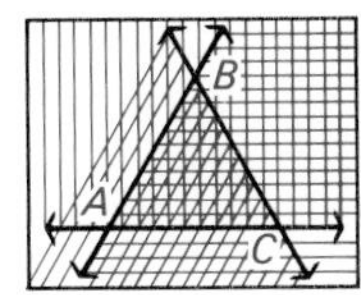

17. (a) $\{C\}$ (c) $\overline{CE}$ without the endpoints
(e) Interior of $\triangle DBF$

19. (a) 180° (c) 540° (e) 900°
(g) $(n-2)180°$

20. (a) This is false; $m(\angle B) + m(\angle D) \neq 180°$.

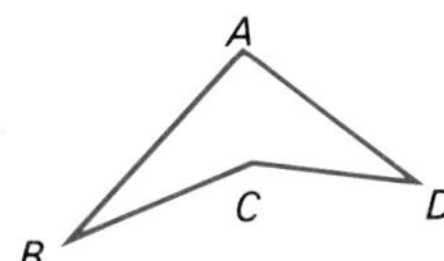

(c) This is false because some diagonals do not intersect.

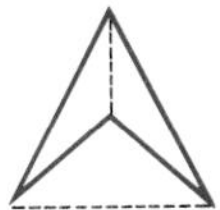

(e) True

Exercise Set 6, Page 399

1. 15, yes

3. 9, 12, 15, 18, 21, 24, 27, 30
They differ by the measure of the width.

4. (a) 4 (c) 9

5. (a) $10 < A < 17$ (c) $4 < A < 17$

6. (a) $11.5 < A < 14.5$ (c) $6 < A < 11$

7. (a) 4 (c) 2

8. (a) 8 (c) 8 (e) 16 (g) 15

9. Yes, $A = ba$

11. Yes, $A = (\frac{1}{2})ba$

13. 1 ha

14. (a) 34 km^2 (c) 456,000 m^2

15. (a) $34(10)^4$ cm^2 (c) 31 cm^2
(e) $42(10)^{10}$ cm^2

16. (a) Divide square meters by 10,000 to get hectares.

17. (a) 400 cm^2

18. (a) 48 (c) 16

Review Exercise Set 7, Page 403

1. (a) False; a plane separates space.
(c) False
A B
(e) False
A B D
(g) True
(i) False; isosceles triangle has two congruent angles, but it is not required to have three congruent angles.
(k) True
(m) False; a polygon is the union of three or more line segments.
(o) False
(q) False; some closed curves have more than three regions, which would nullify the definition of a simple closed curve.
(s) False; a polygon has three or more sides.
(u) True

3. Planes must be parallel or must intersect in a line parallel to l and to m.

4. (a) A (c) S (e) S (g) A (i) S

5. (a) (c)

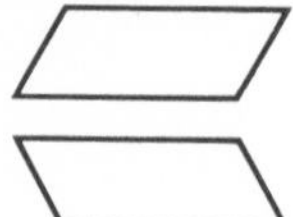

(e) (g) Impossible

7. (a) (c)

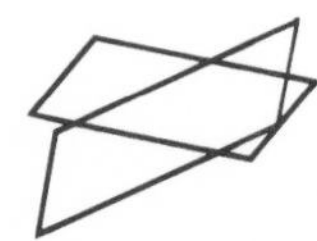

(e)

8. (a) (c)

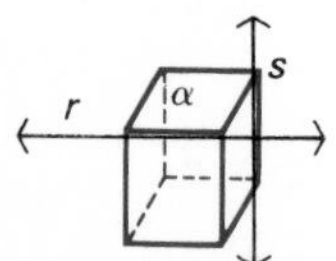

9. (a) 0.05001 km (c) 81,000 m
(e) 261,000 cm

10. (a) cm (c) mm (e) m

11. (a) $\overline{AE}$ (c) $\overline{BE}$ excluding B and E
(e) $\overline{BA} \cup \overline{AE}$ (g) $\{D\}$

13. Three noncollinear points
A line and a point not on the line
Two intersecting lines
Two parallel lines

16. (a)

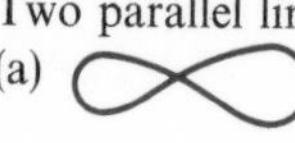

(c)

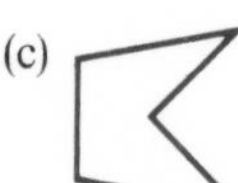

18. (a)

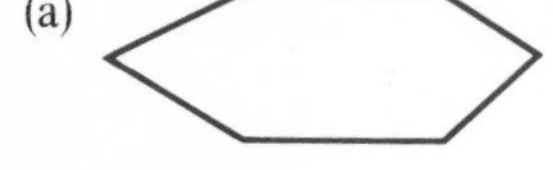

(c) Impossible
(e) Impossible

19. (a)

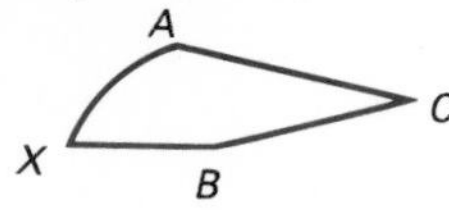

(c)

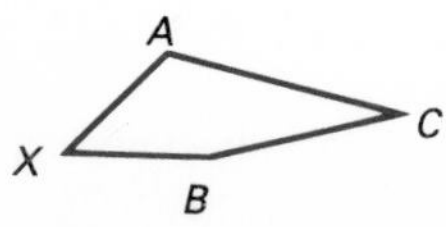

20. (a) False (c) True (e) True

21. No

23. 40°, 60°, 80°

24. (a) When points A, B, C, and D are collinear

25. Interior

CHAPTER 9

Exercise Set 1, Page 412

1.

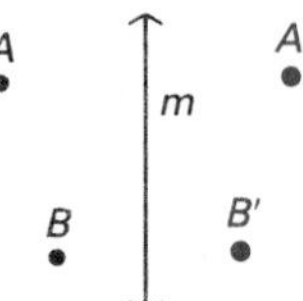

3.

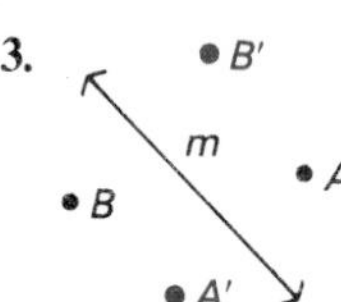

5.

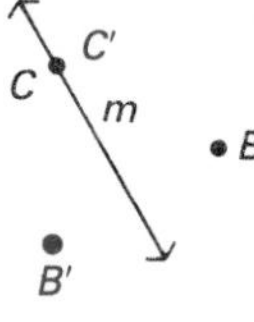

7. (a) The image of A is B, and the image of B is A.
(c) Collinear

9. (a)

(c)

10. (a) False (c) True (e) True (g) False

11.

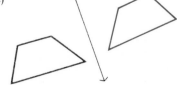

13.

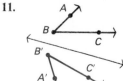

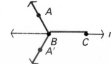

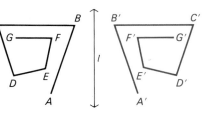

15.

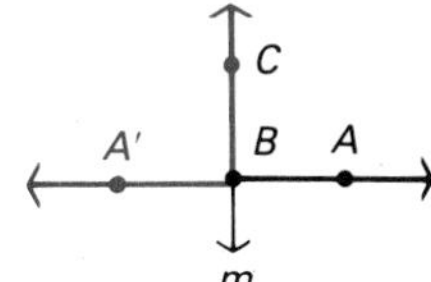

17. (a)

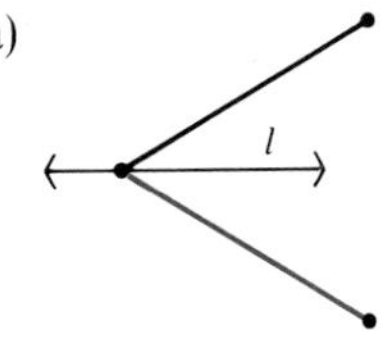

(c) • •P
•
l
•
• •P′

(e)

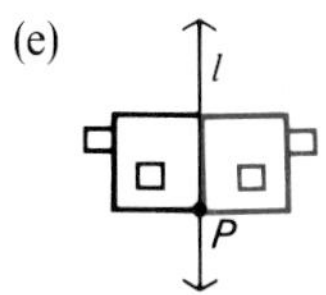

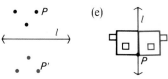

18. (a)

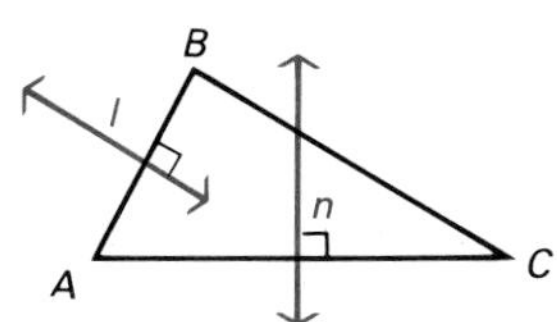

19.

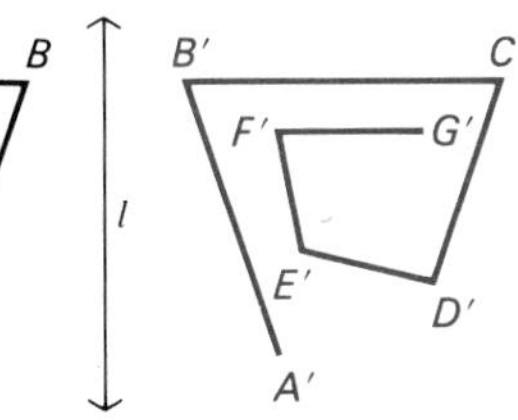

21. Distance, collinearity of points, betweenness of points, and angle measure.

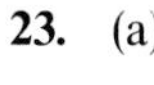

23. (a)

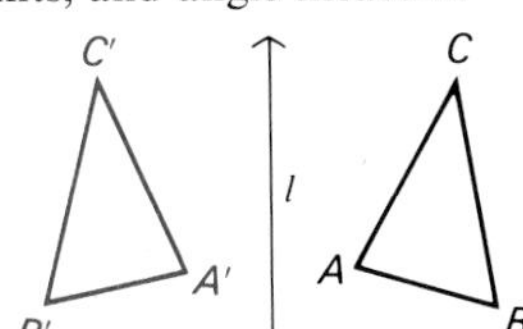

(c)

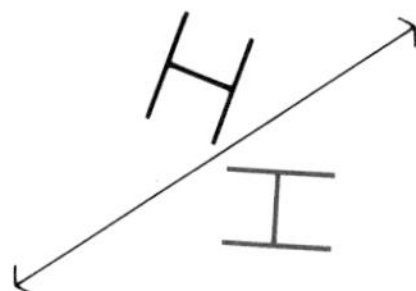

25. (a)

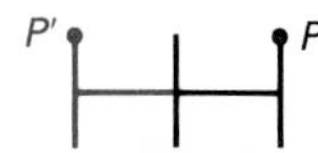

26. (c) Using points P and Q as your pivot points for the compass, intersect arcs $\overset{\frown}{PC}$ and $\overset{\frown}{QC}$ on the opposite side of l to construct C'. Similarly, intersect arcs $\overset{\frown}{PD}$ and $\overset{\frown}{QD}$ to get D' and $\overset{\frown}{PE}$ and $\overset{\frown}{QE}$ to get E'.

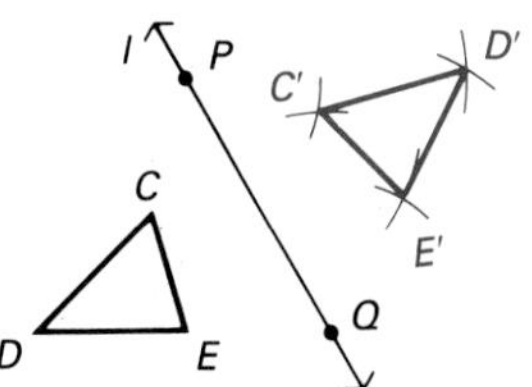

27. Yes, as seen below

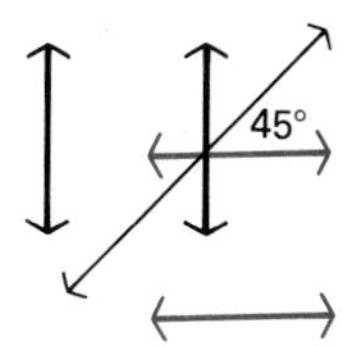

Exercise Set 2, Page 418

1. (a) Line n
(c) Distance, collinearity of points, betweenness of points, and angle measure
(e) Its image
(g) When the specified distance is zero

2. (a) (c)

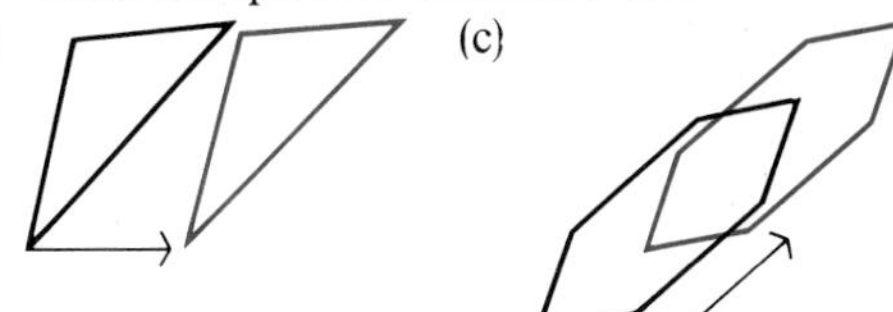

3. (a)

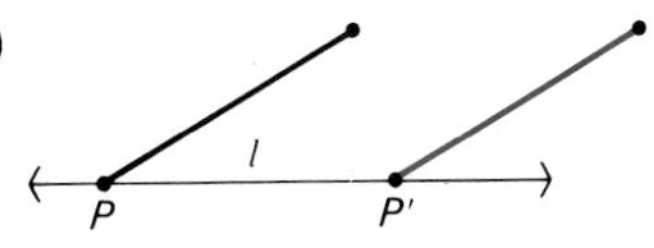

(c)

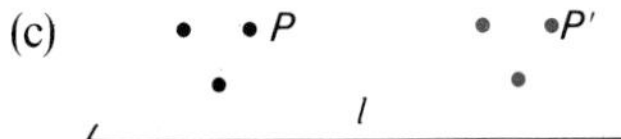

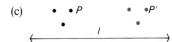

(e)

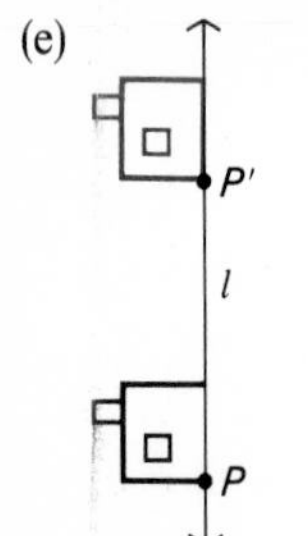

4. (a)

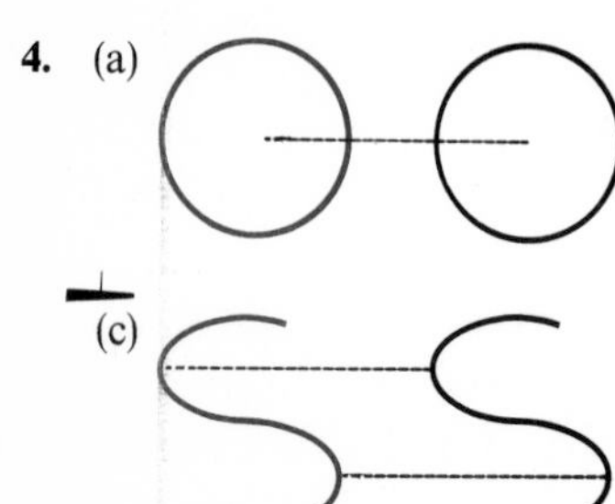

(c)

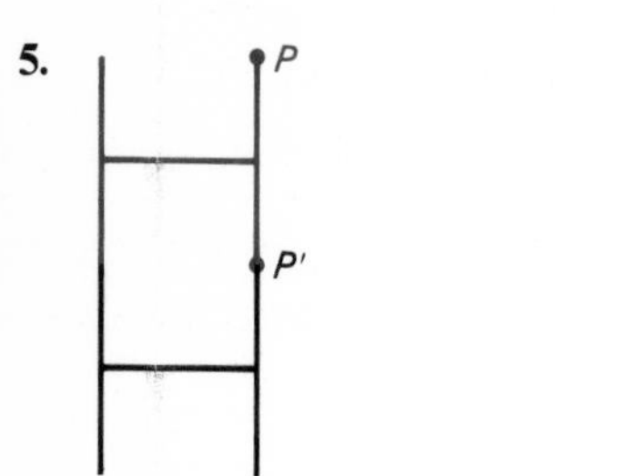

5.

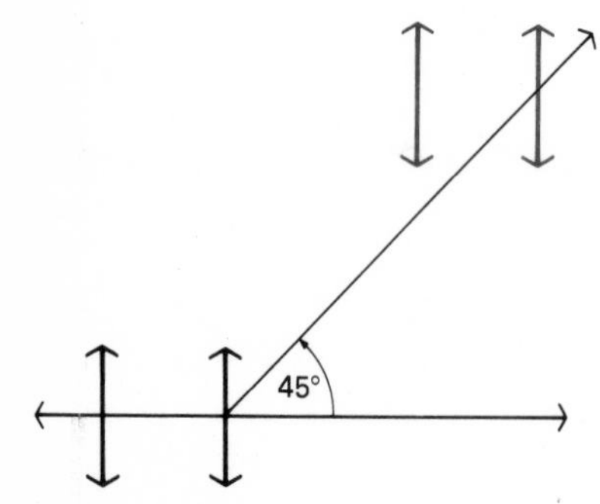

7.

45°

Yes; yes

8. (a) G (c) S (e) $\angle GTS$

9. (a) K; V (c) U; R (e) $\angle KVU$; $\angle VSR$

11. Simply measure the distance between corresponding points. Yes, if corresponding points are known and the distances are equal.

12. (a) $(-1, 2)$ (c) $(1, -3)$

13. (a) $(0, 4)$ (c) $(2, -1)$

15.

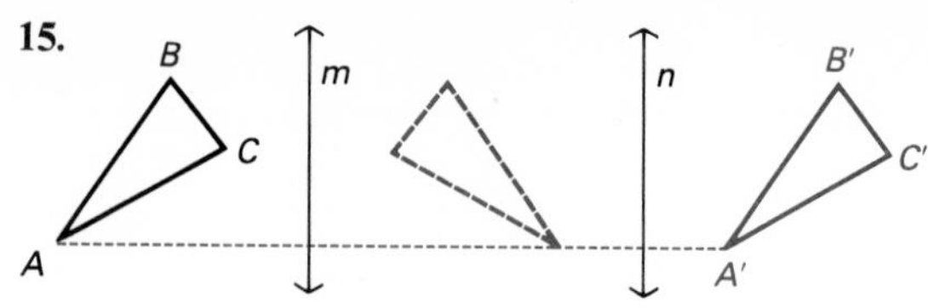

Exercise Set 3, Page 424

1. (a) (c)

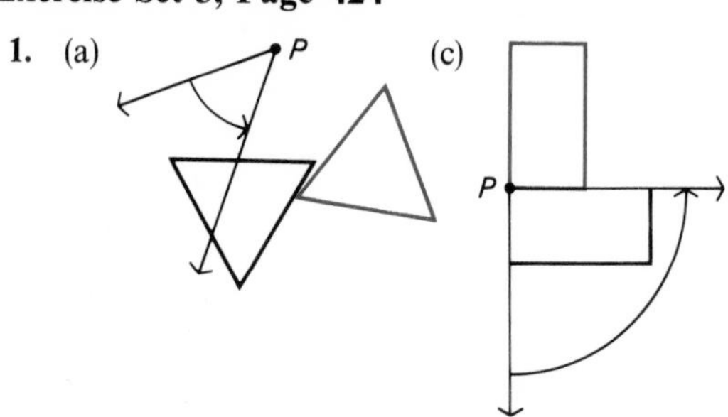

2. (a)

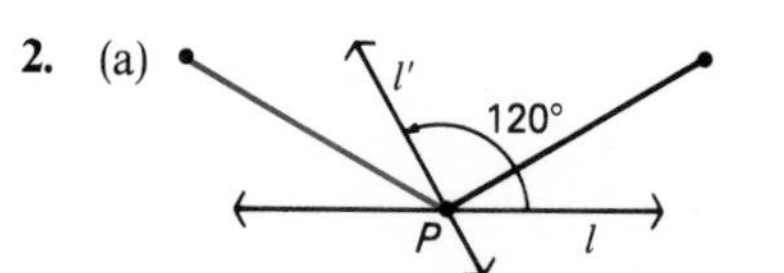

(c) (e)

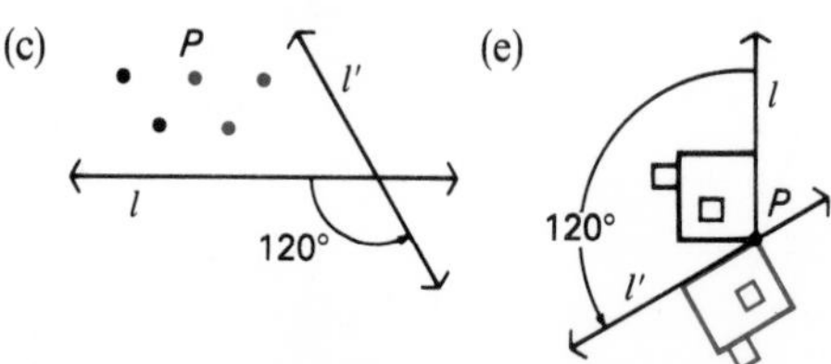

3.

4. (a)

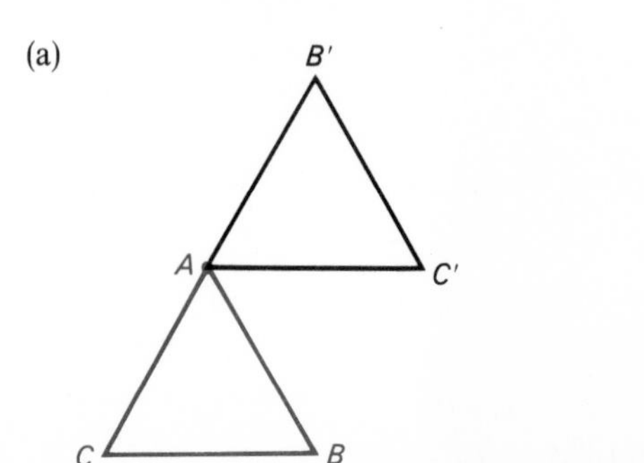

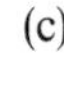

(c)

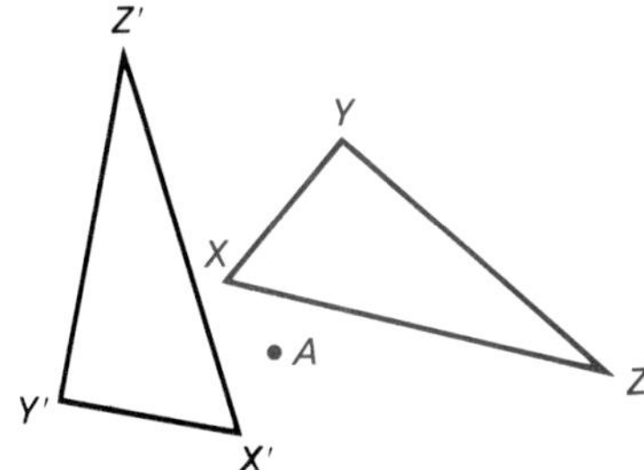

5.

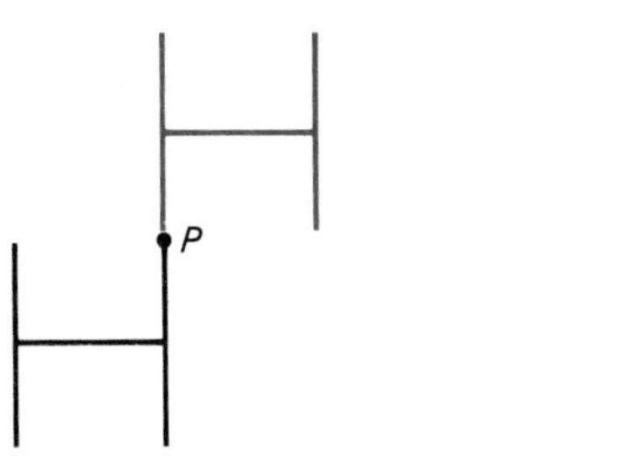

6. (a)

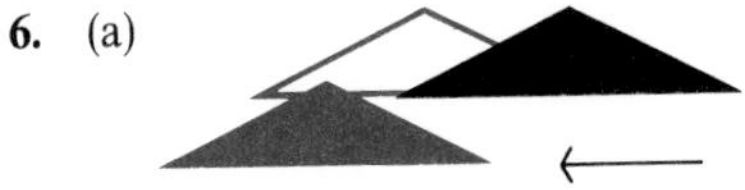

(c)

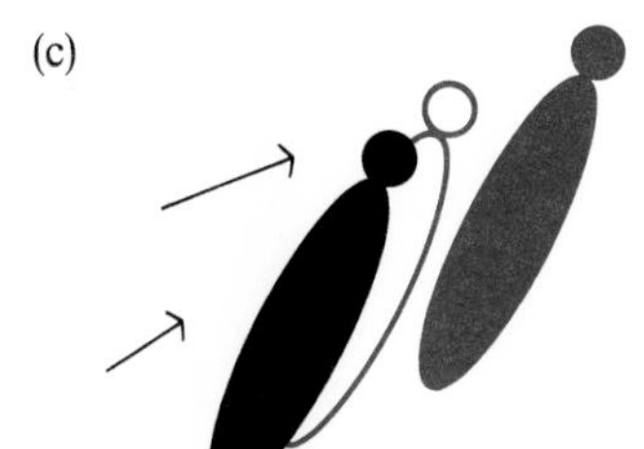

(Answers demonstrate that the order makes no difference.)

7. (a) (c)

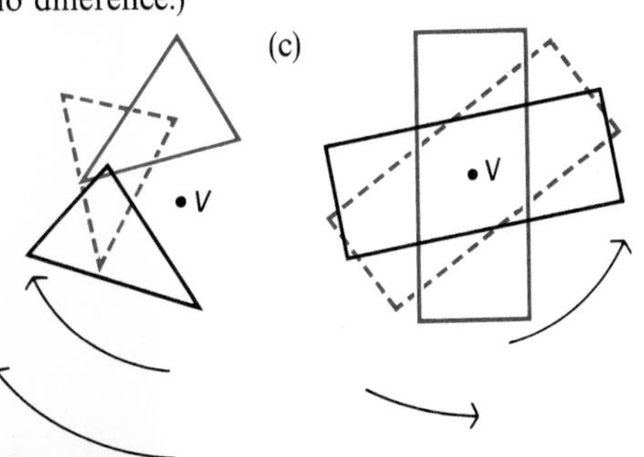

8. (a)

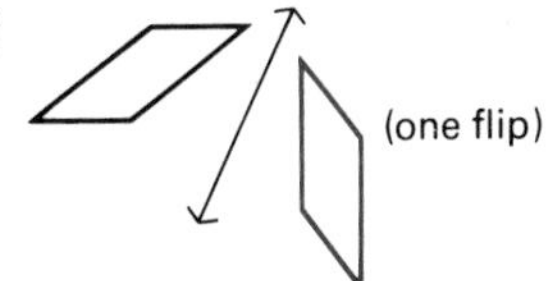

(c)

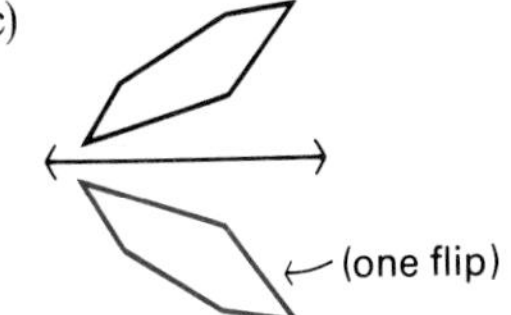

9. (a) Reflection (c) Reflection
10. (a) Reflection (c) Rotation
11. (a) Rotate 90° clockwise about lower right corner; reflect about the vertical line through the tip.

13.

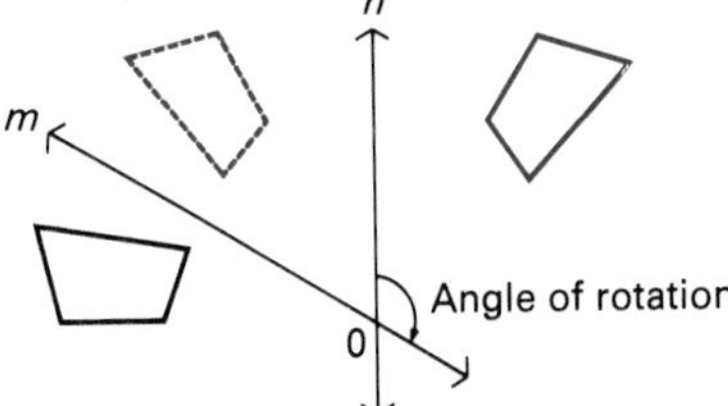

15.

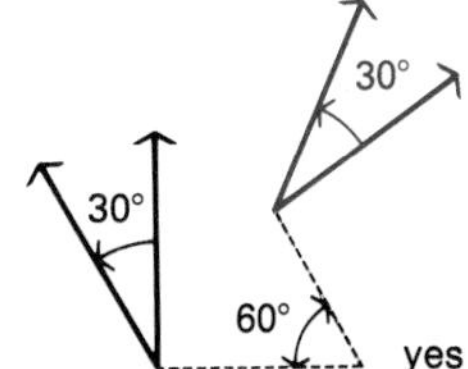

Exercise Set 4, Page 432

1. (a) Line of symmetry
   (c) Not a line of symmetry
2. (a) (c)

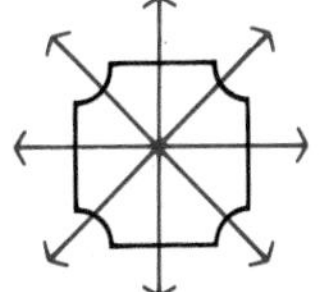

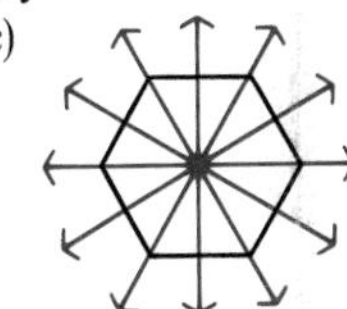

3. (a) Rotational symmetry about Q (multiples of 90°)
   (c) Rotational symmetry about P (multiples of 60°)
4. (a) 3, 120° (c) 5, 72° (e) 7, 51.43°
5. (a) H, O, X (c) None

6. (a) Tessellates the plane

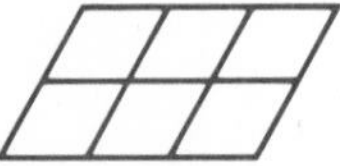

(c) Tessellates the plane

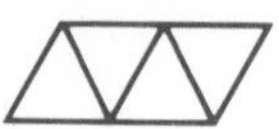

(e) Does not tessellate plane

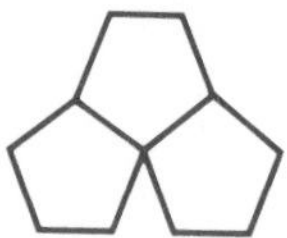

(g) Tessellates the plane

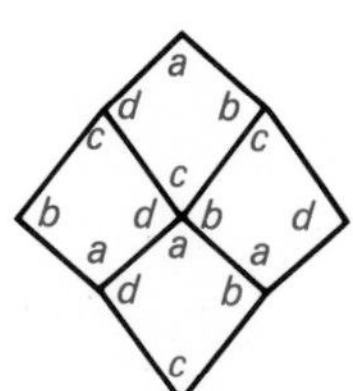

(i) Tessellates the plane

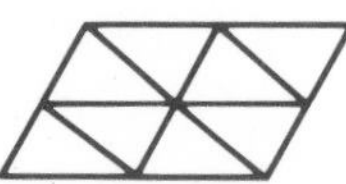

(k) Does not tessellate the plane

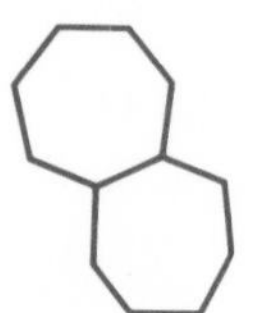

7. (a) Square (c) Equilateral triangle
8. (a) Reptile (c) Reptile (e) Reptile
9. (a) No lines of symmetry; 45° rotational symmetry
   (c) 4 lines of symmetry; 90° rotational symmetry
11. (a) (c) (e)

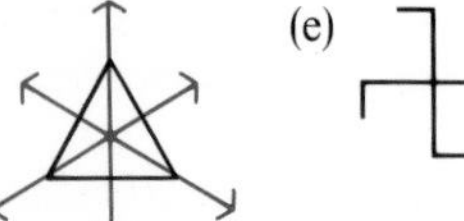

12. (a) Yes (c) No (e) No
13. (a) Yes (c) No (e) No (g) No
    (i) No

Exercise Set 5, Page 436

1. Yes, yes
3. (a) Yes (c) Yes
4. (a) False (c) False (e) True
5. (a) $\overline{HE}$ (c) $\triangle FGH$
6. (a) Measures are equal, not congruent.
   (c) Notation acceptable
7. (a) The equilateral triangles can be obtained by reflections about sides or as rotations.
   (c) The equilateral triangles with hexagons at each vertex can be obtained by 60° rotations about vertices (centers of hexagons).
9. (a) Translation (c) Translation
   (e) Rotation about C
10. (a)

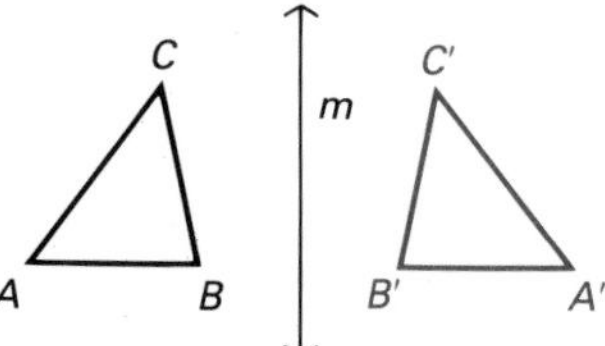

(c)

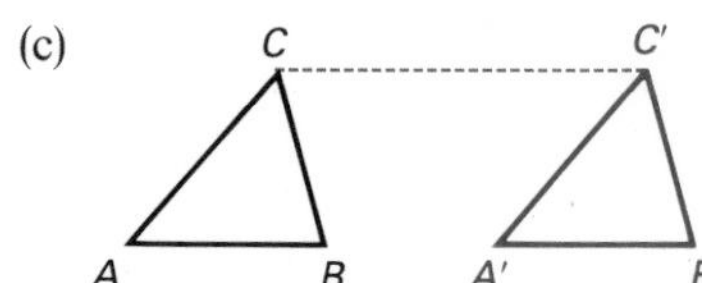

11. (a)

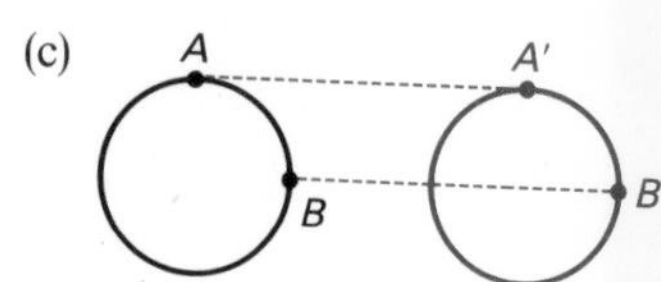

(c)

13. $\triangle ACB \cong \triangle ECD$ because one can be obtained from the other by a rotation about C.

Exercise Set 6, Page 443

1. (a) $\frac{40}{7}$ (c) 9 (e) 12
2. (a) $x = \frac{16}{3}$ (c) $z = 27$, $w = \frac{27}{2}$
3. (a) $m(\angle C) = 70°$; $m(\angle F) = 70°$
   (c) $ST = 8$, $WX = 4$, $m(\angle WVX) = 60°$
5. 18 ft.
7. (a) $\frac{8}{3}$ (c) 4

9. (a) 8 m by 12 m
   (c) Base = 18 m, height = 12 m, hypotenuse ≈ 21.6 m
11. For reduction, change 1 cm to $\frac{2}{5}$ cm
    For enlargement, change 1 cm to $\frac{5}{3}$ cm
14. (a) 8
15. 400 m from the shoreline to the ship
17. 128.06 m

Review Exercise Set 7, Page 447

1. (a)

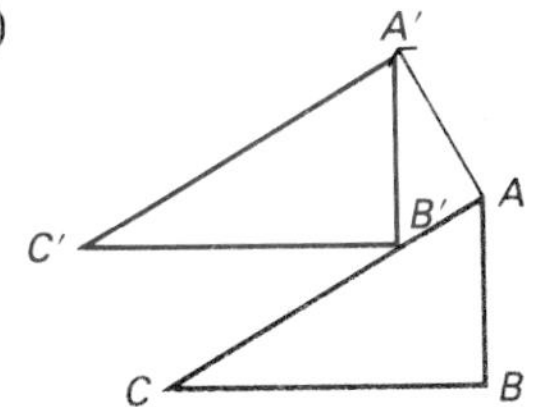

(c)

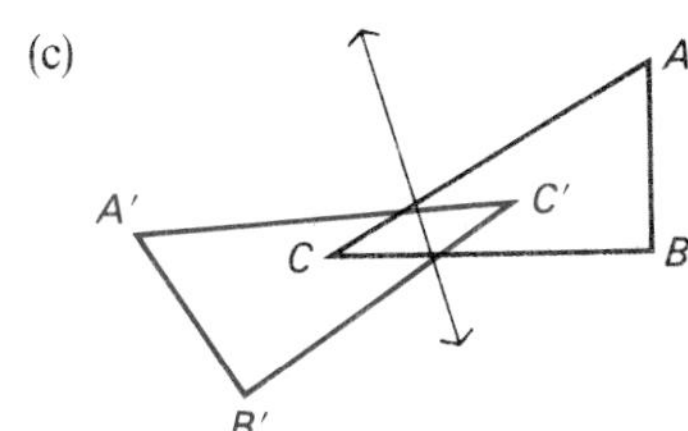

3. 8 cm
4. (a) Rotation (c) Translation
   (e) Translation
5. $x = 16\frac{2}{3}$
6. (a) False (c) False (e) False (g) True
   (i) False
7. A similarity transformation
9. Distance, collinearity of points, betweenness of points, angle measure
10. (a)

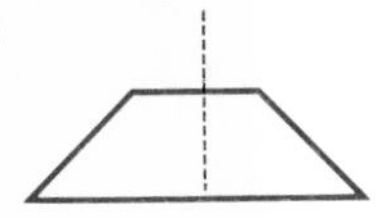

(c)

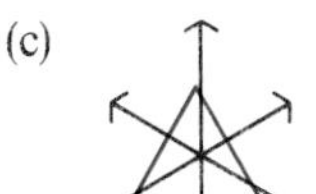

(e)

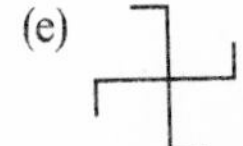

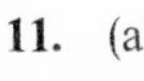

11. (a)

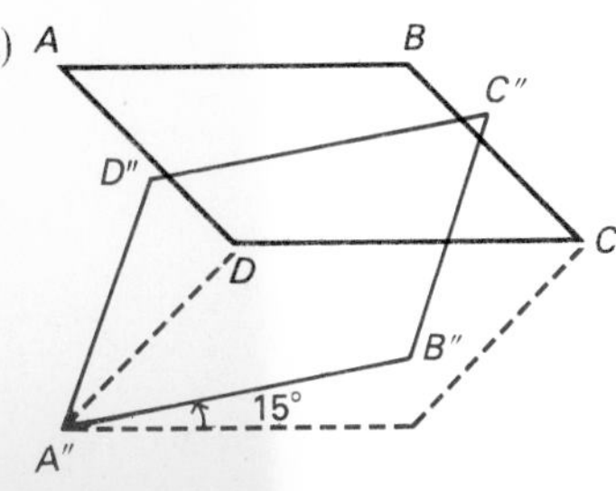

(c)

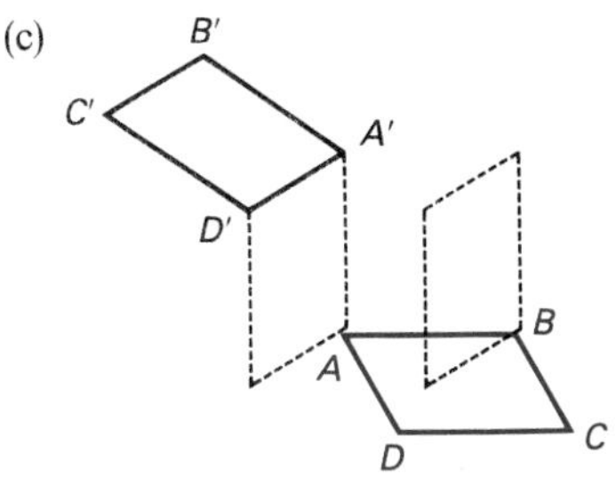

13. (a) O (c) $CDIH$
14. (a) M (c) $RMLQ$
15. $\triangle MSR$
17. A translation that maps L on M, then a 180° degree rotation counterclockwise about N

CHAPTER 10

Exercise Set 1, Page 455

1. (a) $\triangle ABC \cong \triangle CDA$ by SSS
   (c) $\triangle CEA \cong \triangle DEB$ by SAS
   (e) $\triangle ABC \cong \triangle ADC$ by ASA
   (g) $\triangle AED \cong \triangle BEC$ by ASA
   (i) $\triangle EAD \cong \triangle ECB$ by SAS
2. (a) True
   (c) False; the points A, B, and C in $\triangle ABC$ are listed clockwise. The points D, E, and F in $\triangle DEF$ are listed counterclockwise. A translation and a rotation do not reverse the order of the points.
3. (a) Yes (c) Yes
5. (a) Adequate information; not congruent
   (c) Adequate information; congruent
6. (a) $\angle B \cong \angle D$ Given
   $\angle ACB \cong \angle ECD$ They are vertical angles.
   $m(\angle A) = 180° - m(\angle B) - m(\angle ACB)$
   $= 180° - m(\angle D) - m(\angle ECD)$
   $= m(\angle E)$
   Since $AB = DE$, $\triangle ABC \cong \triangle EDC$ by ASA.
   (c) $\triangle ABD \cong \triangle CBD$ by SSS.
7. (a) Not necessarily; $\overline{AC}$ may not be congruent to $\overline{DF}$.
   (c) Not necessarily; $\angle OAD$ may not be congruent to $\angle OCB$. Then OD is not equal to OB.
   (e) Yes. $\angle A \cong \angle C$ and $\angle D \cong \angle F$, since they are angles of isosceles triangles. Thus, $\angle C \cong \angle D$. Hence, $\angle B \cong \angle E$. $\triangle ABC \cong \triangle FED$ by SAS.

9. (a)

10. (a) 360° (c) 720°
11. Because SSA does not fix a triangle.

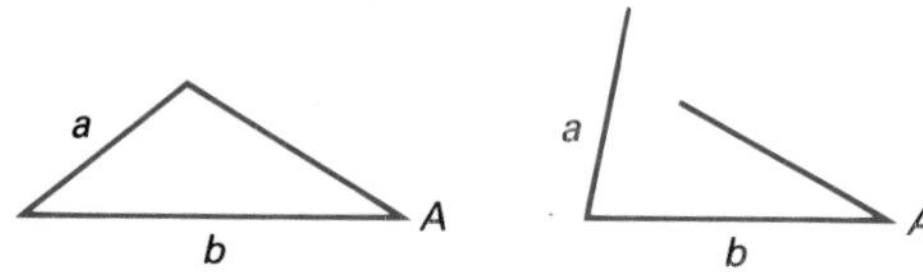

13. (a)

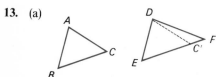

Given: $\angle ABC \cong \angle DEF$

$\angle BAC \cong \angle EDF \quad \overline{BA} \cong \overline{ED}$

There is a point C' on $\overline{EF}$ such that $\overline{BC} \cong \overline{EC'}$. Then $\triangle BAC \cong \triangle EDC'$ since $\overline{BC} \cong \overline{EC'}$, $\overline{BA} \cong \overline{ED}$ and $\angle ABC \cong \angle DEC'$ (by SAS). Thus, $\angle BAC \cong \angle EDC'$. But $\angle BAC \cong \angle EDF$. So $\angle EDC' \cong \angle EDF$. This means that $\overleftrightarrow{DC'}$ and $\overleftrightarrow{DF}$ coincide. Thus $\triangle BAC \cong \triangle EDF$.

(c)

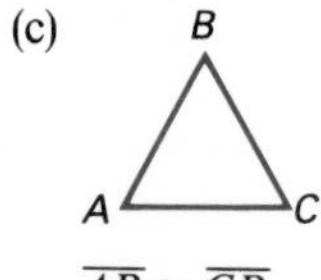

$\overline{AB} \cong \overline{CB}$

$\overline{BC} \cong \overline{BA}$

$\angle B \cong \angle B$

$\triangle ABC \cong \triangle CBA$ by SAS. Thus, $\angle A \cong \angle C$

Exercise Set 2, Page 463

1. (a) $\triangle ACD \cong \triangle DBA$ by HL
   (c) $\triangle ADC \cong \triangle BDC$ by HL
   (e) $\triangle CBA \cong \triangle CDE$ by LA
2. (a) 5 (c) $\sqrt{161}$ (e) 68
3. (a) Right triangle (c) Not a right triangle
   (e) Right triangle
5. $\triangle ABC \cong \triangle DCB$ by LL
7. $\triangle SQP \cong \triangle SRP$ by HA. Therefore, $PQ = PR$
9. 15 m
11. Fewer than x meters

13. John's ladder will reach the gutter and have some of the length of the ladder left over.
15. $(XJ)^2 + (HB)^2$
   $= (XP)^2 + (JP)^2 + (PH)^2 + (PB)^2$
   $= [(XP)^2 + (PH)^2] + [(JP)^2 + (PB)^2]$
   $= (XH)^2 + (JB)^2$
17. (a) (c)

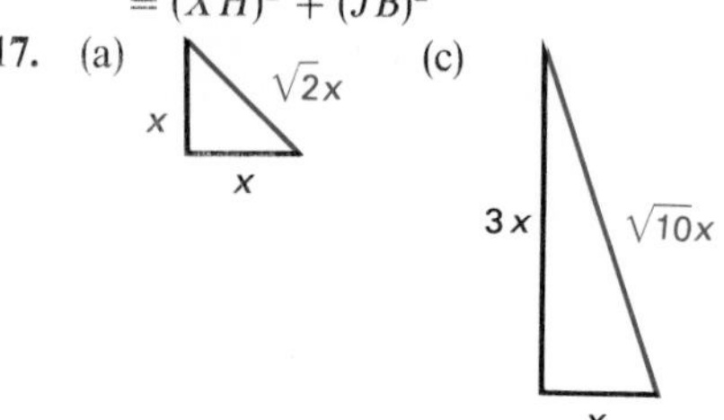

Exercise Set 3, Page 471

1. 13 m
3. (a) 15 m² (c) 884 (e) 288 (g) $\frac{15}{2}$
4. (a) 158
5. $192\sqrt{3}$
6. (a) 16 cm²
7. $9\sqrt{3}$
9. (a) 54
11. 28
13. 30 km²
14. (a) Area is doubled. (c) Area is tripled.
15. False. Their shapes could be completely different.
17. (a) 5.4 m²; 32.4 m² (c) 12.6 m²; 27 m²
19. $5400\sqrt{3}$
21. False; $\dfrac{A_1}{A_2} = k^2$ where k = ratio of similarity.
   $a_1 = ka_2$ and $p_1 = kp_2$
23. Yes; see Exercise 21

Exercise Set 4, Page 476

1. (a) $C = 6\pi$; $A = 9\pi$ (c) $OB = 4$; $C = 8\pi$
3. (a) 50.266 ft.² (c) 452.39 cm²
5. 3600π
7. 10π
9. 20π
11. $\frac{136}{7}$ square units
13. (c) $\frac{352}{7}$ square units
15. 1 to 3; 1 to 9
17. 116π
19. Nearly 393 revolutions
21. 1 to 2
23. Yes $A_1/A_2 = C_1^2/C_2^2$

Exercise Set 5, Page 479

1. (i) Bowling ball, a brick, a banana, a pencil, a glass, a sheet of typing paper, a ruler

(ii) A ring, a straw, a cup, a sewing needle, a record, a section of pipe, a doughnut, a circular saw blade
(iii) A sheet of two-ring binder paper, a funnel with a handle, and a pitcher with 2 handles

2. (i) (a) (h) (l)
(ii) (b) (e) (g) (m) (o)
(iii) (c)
(iv) (d) (j)
(v) (f)
(vi) (k) (i)
(vii) (n)

3. (i) A, R
(ii) B
(iii) C, I, J, L, M, N, S, U, V, W, Z
(iv) D, O
(v) E, F, G, Y, T
(vi) H, K
(vii) X
(viii) P
(ix) Q

4. (a) True (c) True (e) True

5. 0
1 3
2 5 7
4
6 9
8

6. (a) (c)

7. 3

Exercise Set 6, Page 484

1.

Network	Number of even vertices	Number of odd vertices	Traversable Yes or No
(a)	3	4	No
(c)	3	2	Yes
(e)	1	4	No
(g)	8	0	Yes
(i)	10	0	Yes

2. (c) Euler path (g) Euler circuit
(i) Euler circuit

3. (a) Not traversable (c)

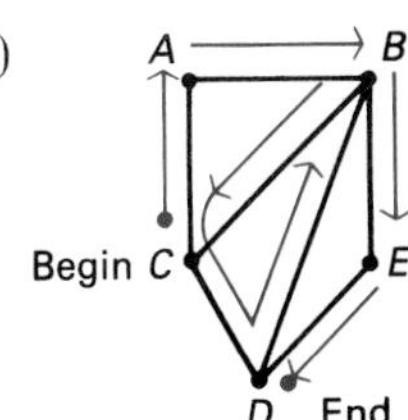

(e) Not traversable
(g)

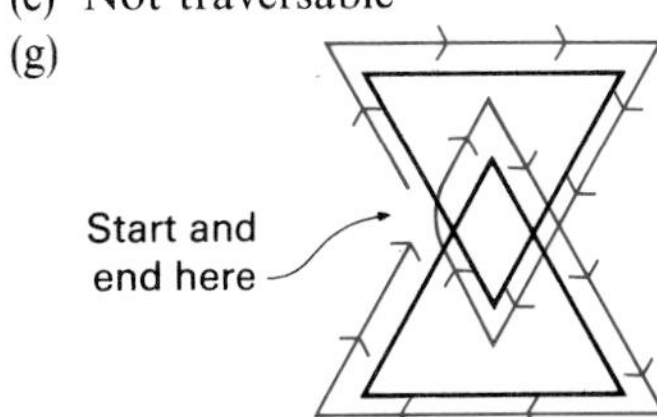

(i)

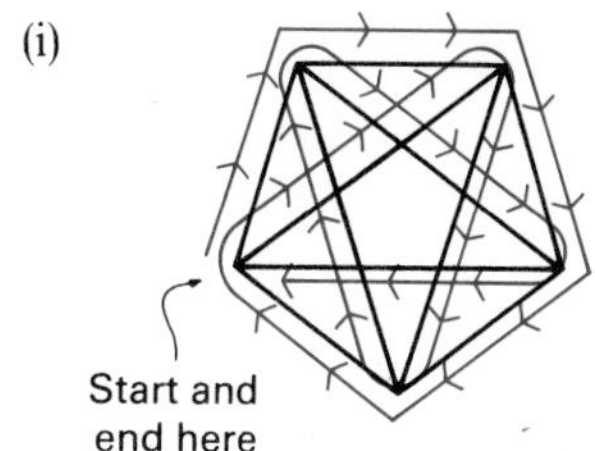

4. (a) (c) No

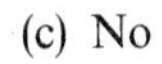

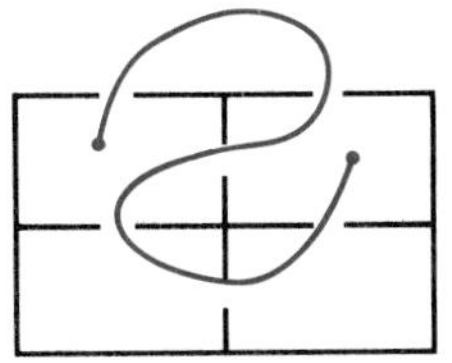

(e) (g)

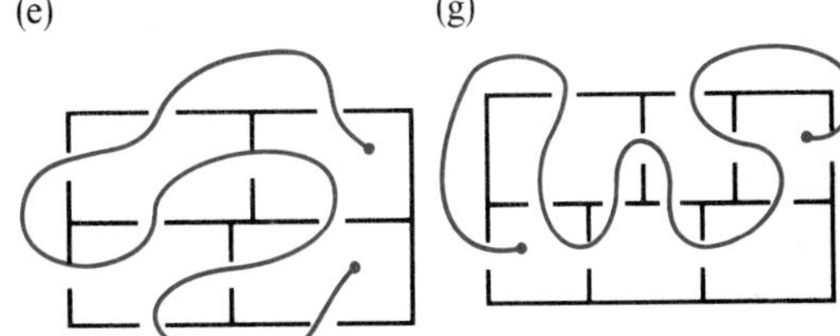

5. (a) Yes (c) Yes (e) Yes (g) Yes
(i) No

6. (a)

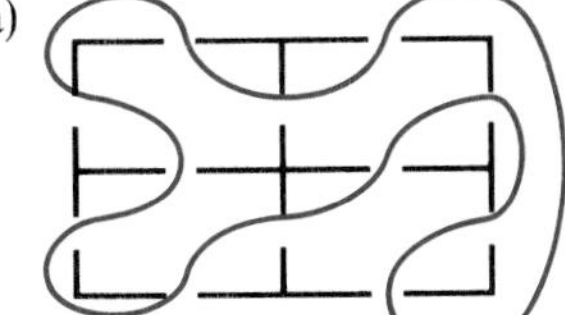

(c)

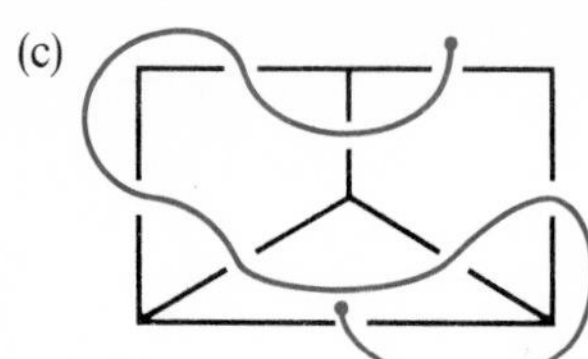

(f)

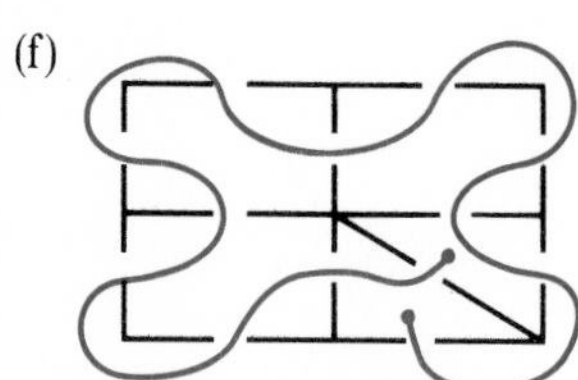

7. (a) 2 (c) 3

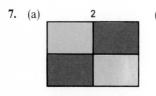

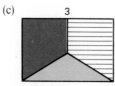

(e) 4 (g) 3

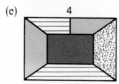

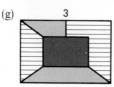

(i) 4

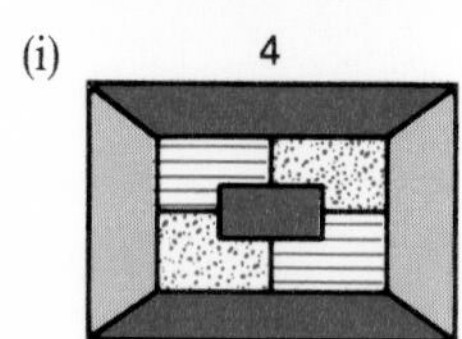

9. (a) Nontraversable (c) Traversable
   (e) Nontraversable

Review Exercise Set 7, Page 487

1. (a) 10
2. (a) 12
3. (a) $16\frac{1}{2}$
4. (a) Nontraversable
5. T , , +
6. (a)

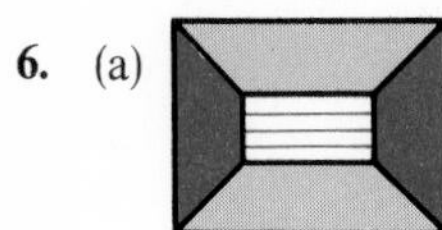

7. $x = 6$
8. (a) Yes (c) Yes

9. Area of the square = $\pi/4$ times the area of the circle.
11. $16\sqrt{3} - 8\pi$
13. (a)

CHAPTER 11

Exercise Set 1, Page 497

1. (a) Triangular prism (c) Pentagonal prism
2. (a) $V + S - E = 2$ $6 + 5 - 9 = 2$
   (c) $V + S - E = 2$ $10 + 7 - 15 = 17 - 15 = 2$
3. (a) True (c) False (e) False (g) True
4. (a) Quadrilateral pyramid,
   vertices = $\{A, B, C, D, E\}$; edges, $\overline{AE}, \overline{BE}, \overline{CE}, \overline{DE}, \overline{AB}, \overline{BC}, \overline{CD}, \overline{DA}$
5. (a)

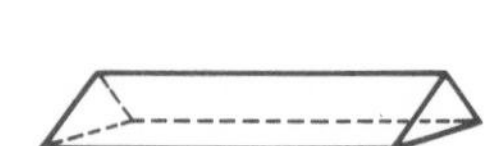

(c)

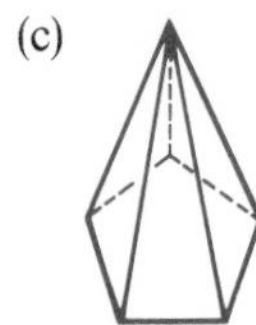

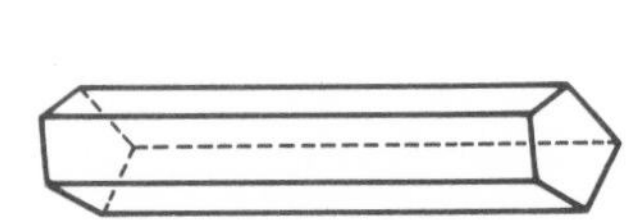

6. (a) $6 + 6 - 10 = 2$ (c) $6 + 5 - 9 = 2$
   (e) $7 + 7 - 12 = 2$ (g) $8 + 6 - 12 = 2$
7. Quadrilateral pyramid
8. (a) Tetrahedron (c) Octahedron
   (e) Icosahedron
9.

Surface	V	E	S	$V + S - E$
Tetrahedron	4	6	4	2
Hexahedron	8	12	6	2
Octahedron	6	12	8	2
Dodecahedron	20	30	12	2
Icosahedron	12	30	20	2

10. (a) Is true for prisms only
    (c) Is true for pyramids only

(e) Not true for either
(g) Is true for prisms only

11. (a) 5; 6; 9 (c) 7; 10; 15

12. (a) 4, 4, 6 (c) 6, 6, 10

13.

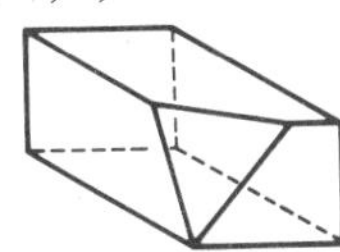

Vertices		Edges		Sides	
9	−	14	+	7	= 2

Euler's formula does hold.

For a smaller cut, 10 − 15 + 7 = 2.

15. (a) Line of rotational symmetry is through the center of the cylinder. Planes of symmetry are any planes which go through the line of rotational symmetry. Another plane of symmetry is perpendicular to this line of symmetry at the center of the cylinder. Any line in this plane through the center of the cylinder is a line for 180° rotational symmetry.

(c) Lines of rotational symmetry are l_1, l_2, and l_3. Planes of symmetry are any planes that contain two lines of rotational symmetry. Other planes of symmetry are formed by intersecting diagonals.

(e) l is a line of rotational symmetry. Planes of symmetry are any planes that contain the line of symmetry.

(a)

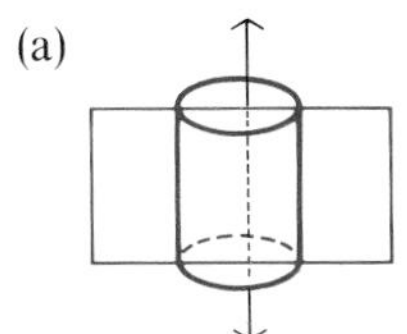

(c)

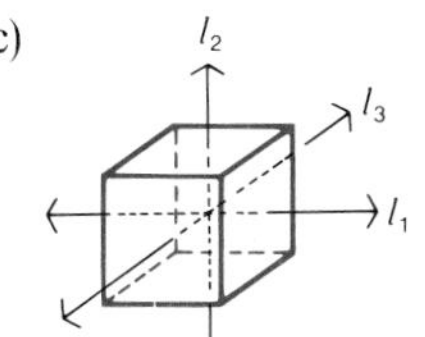

(e)

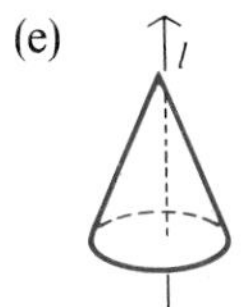

16. (a) True; $6 \geq 2 + \frac{1}{2}(6)$ (c) True; $12 \geq 2 + \frac{1}{2}(8)$

17. $n + 2$; $2n$; $3n$

Exercise Set 2, Page 502

1. (a) 112π, total surface area
(c) 203/16 total surface area

3. 300 cm^2

5. 96 ft.^{2}

7. 310

9. 208

11. $240 + 48\sqrt{3}$

13. (a) Total surface area, 120 m^2

15. 6 liters of paint

17. The shorter has 10π more surface area than the other.

18. (a) 64π; 100π (c) 4:5

19. (a) 216:384 or 9:16
(c) The ratio of surface area of similar cubes is the square of the ratio of the sides of the cubes.

21. (a) 192

Exercise Set 3, Page 508

1. (a) 160π (c) $\frac{147}{64}$

2. (a) 252 (c) 360π

3. (a) 34,000,000 cm^3 (c) 3.100 cm^3
(e) 0.042 cm^3

4. (a) 13 m^3 (c) 9 m^3 (e) 123,456 mm^3

5. (a) 81 m^3

7. 1.18 m^3

9. $\dfrac{50\pi}{144}$ ft.^{3}

11. (a) 58 (c) $6(10)^{-3}$ (e) $2(10)^{-1}$

13. Cylinder by

$$\frac{8000}{\pi} - 2000$$

15. 512 cm^3

17. 1 to $\pi/2$ or 2 to π

Exercise Set 4, Page 512

1. (a) equal (c) $\frac{1}{3}$

3. S.A. 192; volume 152

5. $4000\pi/3$

7. 56π

8. (a) 1 to k^2

9. $4\sqrt{3}$ in.

11. Increased by 4 times

13. 1 to k^3; 1 to k^2

15. (a) $\frac{1}{8}$

Exercise Set 5, Page 516

1. (a) 125 L (c) 1,230,000 L (e) 2.36 L

2. (a) 1,689,000 g (c) 0.00367 kg (e) 140 mg

3. (a) 1 kg (c) 57 g

5. (a) 210 L (c) 8 L

6. (a) 30°C (c) 80°C (e) 50°C

7. 13.824 L

9. 26
11. (a) 10°C (c) 0°C
12. (a) 68°F (c) 86°F
13. 37°C
15. 15 km/L
16. (a) 2 L (c) 4 L (e) 1 L
17. (a) about 280 g (c) about 1 kg
(e) about 56 g (g) about 23 kg
(i) about 3 kg (k) about 900 kg
18. (a) 1522 containers
19. (a) 23°C (c) 7°C
20. (a) Multiply kilograms by 1000 to get grams.
21. 400 days

Review Exercise Set 6, Page 518

1. (a) 0.05001 km (c) 81,000 m
(e) 0.04738 kg
3. (a) 1088 cm^2, total surface area
1920 cm^3, volume
4. (a) gram (c) meter (e) square meter
(g) square centimeter
5. (a) 128; 240 (volume)
(c) 60π; 96π
7. Area of the square = $\pi/4$ times the area of the circle.
9. (a) 140, 845 (c) 1200
11. (a) False (c) True
12. (a) $V + S - E = 12 + 8 - 18 = 2$
(c) $V + S - E = 6 + 5 - 9 = 2$
13. 8450π L

CHAPTER 12

Exercise Set 1, Page 526

1. (a)

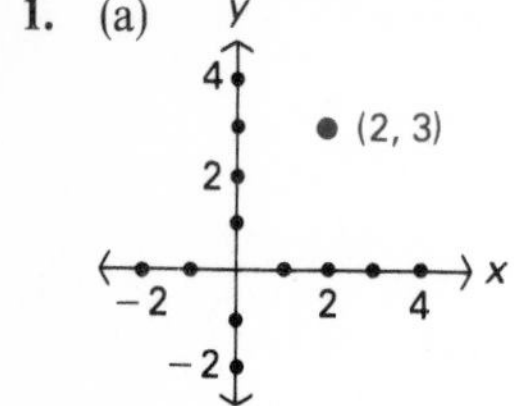

(c)

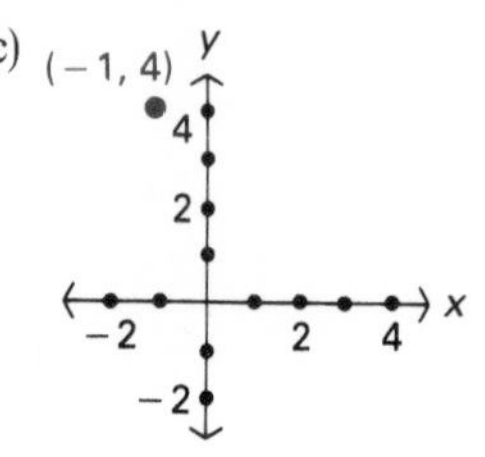

(e)

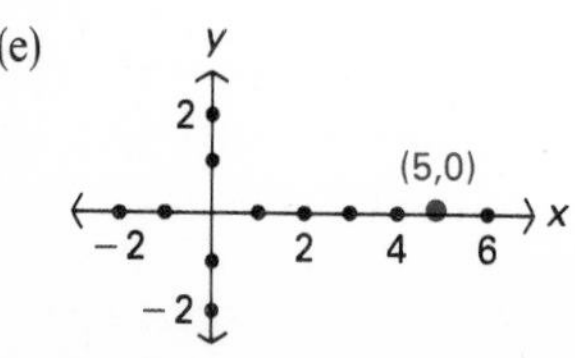

3. $\sqrt{65}$; $(0, \frac{1}{2})$

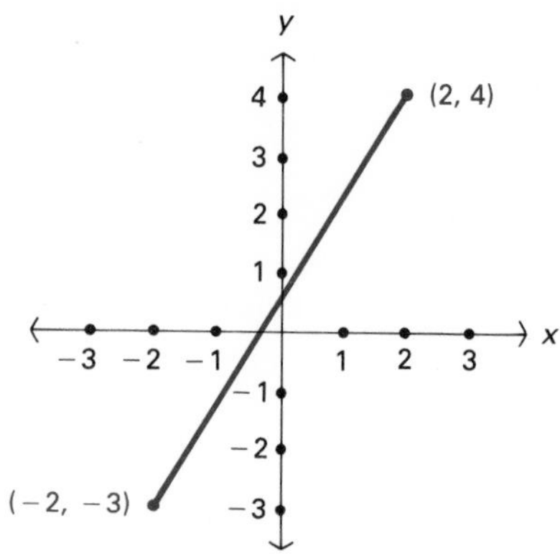

5. $2\sqrt{5}$; $(-3, -5)$

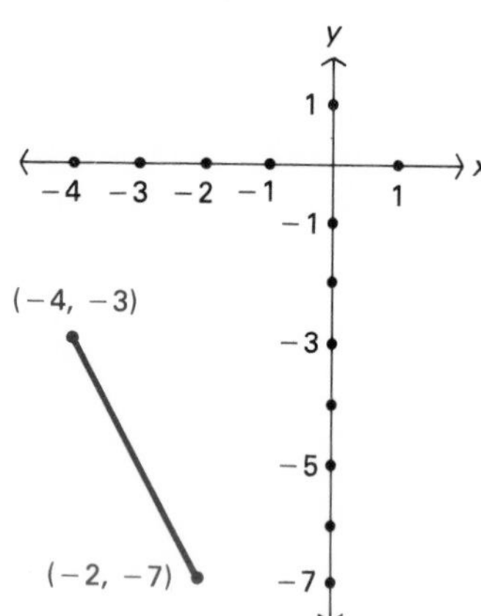

7. Acute

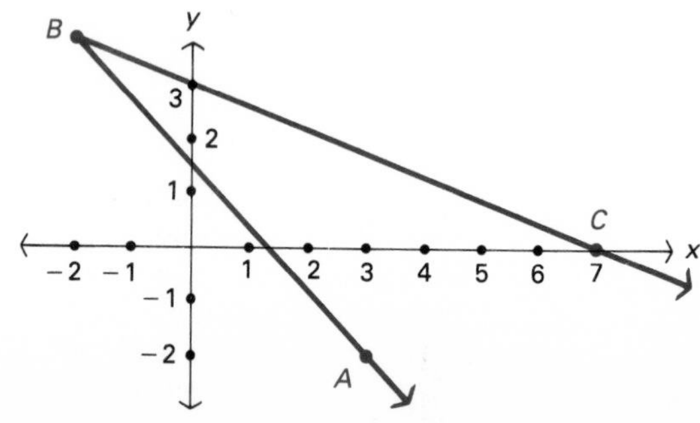

9. Obtuse

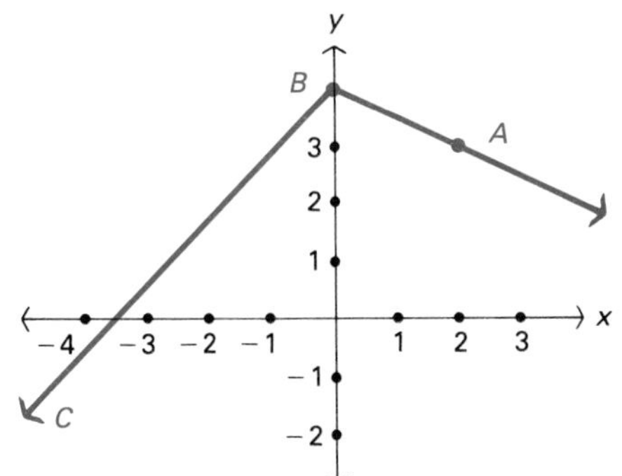

11. $AB = \sqrt{32}$; $BC = \sqrt{72}$; $AC = \sqrt{104}$. Thus, $(AB)^2 + (BC)^2 = (AC)^2$.

13. $AC = 5$ and $BD = 5$

15. $A'(-1, 1)$, $B'(-2, 3)$, $C'(1, 2)$, $D'(2, -1)$

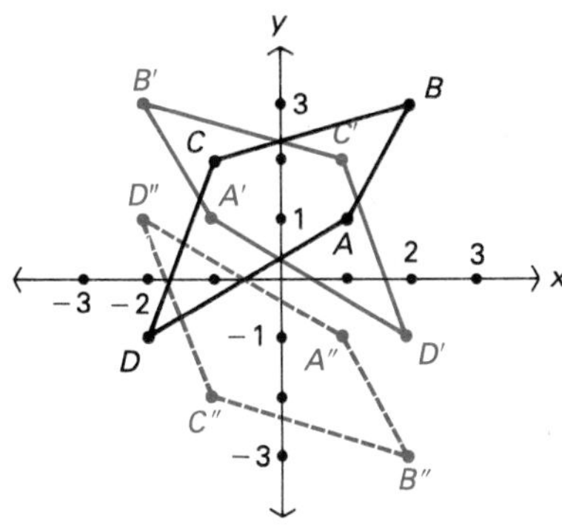

$A''(1, -1)$, $B''(2, -3)$, $C''(-1, -2)$, $D''(-2, 1)$

17. $A'(5, -2)$, $B'(6, 0)$, $C'(3, -1)$, $D'(2, -4)$

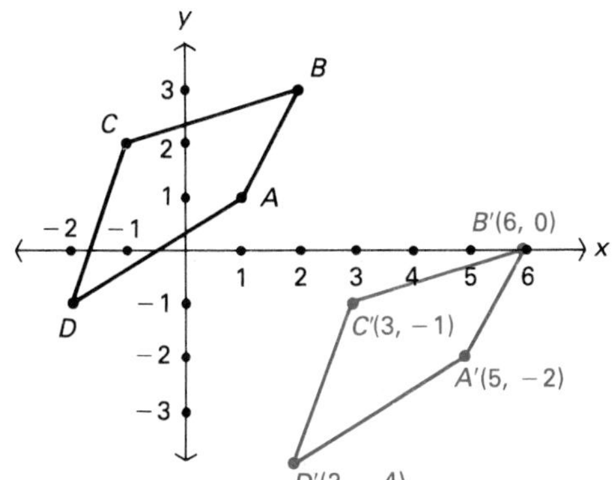

19. Interchange the x and y in (x, y)

21. $(3, -2)$

Exercise Set 2, Page 535

1. (a) $f(1) = 6$, $f(5) = 10$
(c) $f(1) = 4$, $f(5) = 16$
(e) $f(1) = 8$, $f(5) = 40$

2. (a) $f(2) = 1$, $f(3) = 2$, $f(4) = 3$ (c) $f(2) = 4$, $f(3) = 9$, $f(4) = 16$

3. (a) $1; \frac{3}{5}; \frac{1}{5}$

4. (a) $(0, 2)$, $(4, -\frac{2}{3})$, $(1, \frac{4}{3})$, $(-3, 4)$, $(2, \frac{2}{3})$
(c) $(-1, \frac{20}{3})$, $(1, \frac{10}{3})$, $(0, 5)$, $(4, -\frac{5}{3})$, $(5, -\frac{10}{3})$
(e) $(0, 1)$, $(-1, \frac{1}{4})$, $(3, \frac{13}{4})$, $(-3, -\frac{5}{4})$, $(1, \frac{7}{4})$

5. (a)

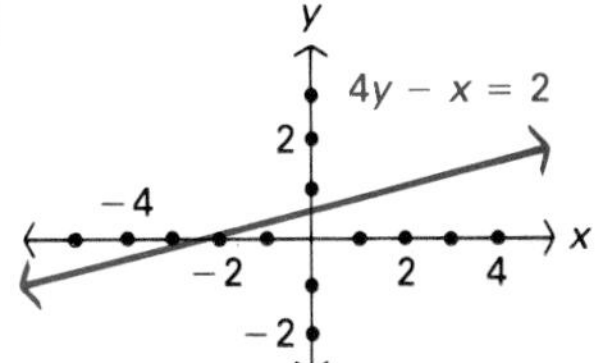

(c)

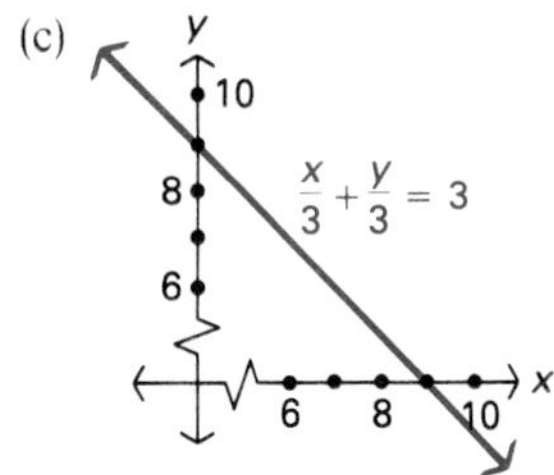

(e)

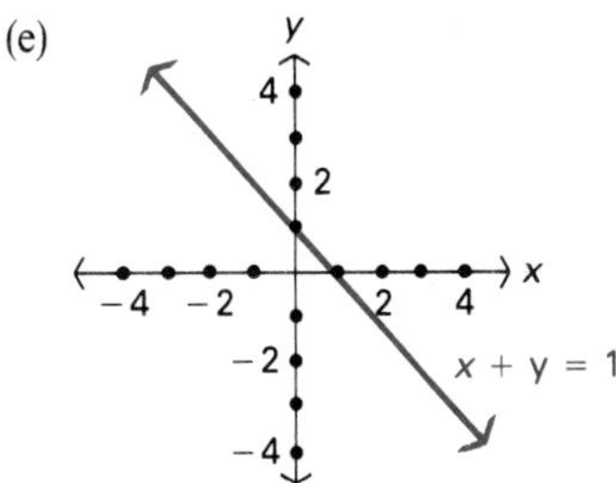

6. (a) is a function. (c) is a function.
(e) is not a function.

7. (a) is a function. (c) is a function.

8. (a)

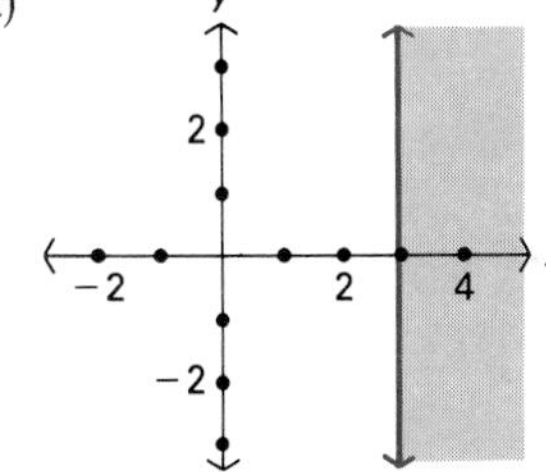

(c)

(e)

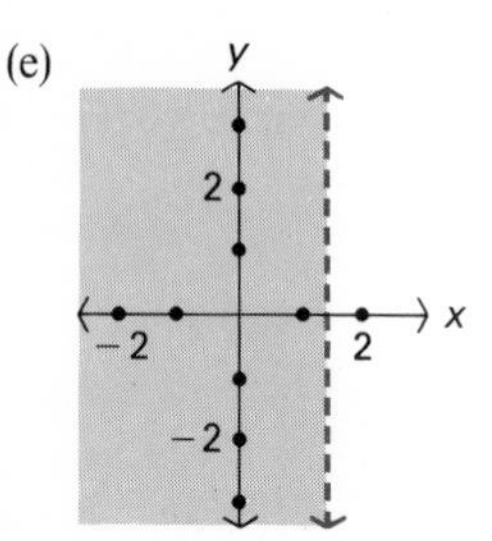

9. (a)

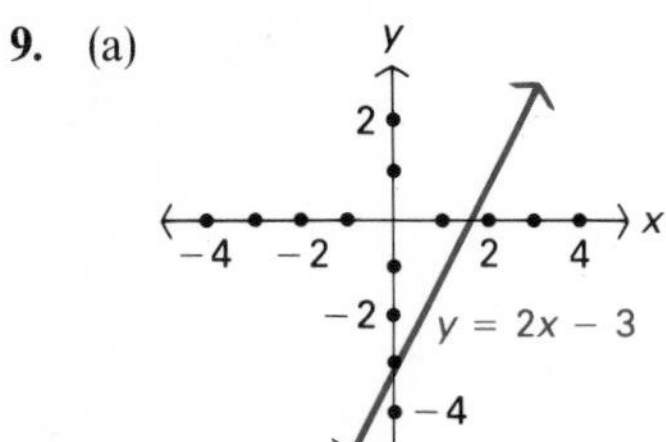

(c)

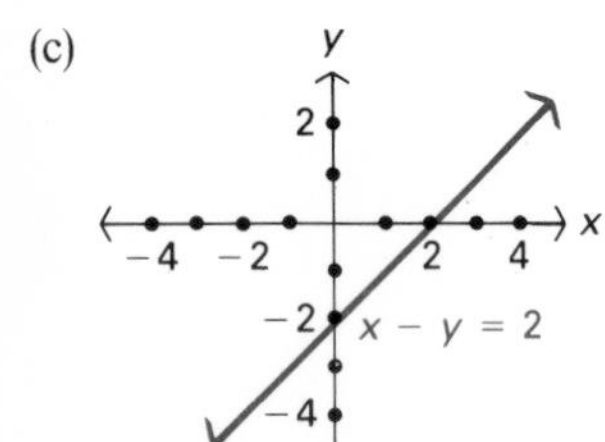

(e)

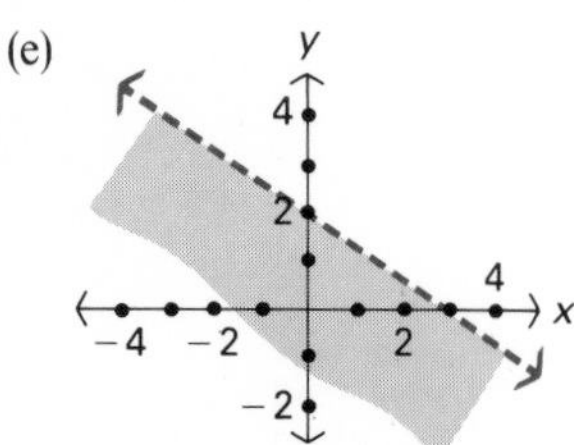

(g)

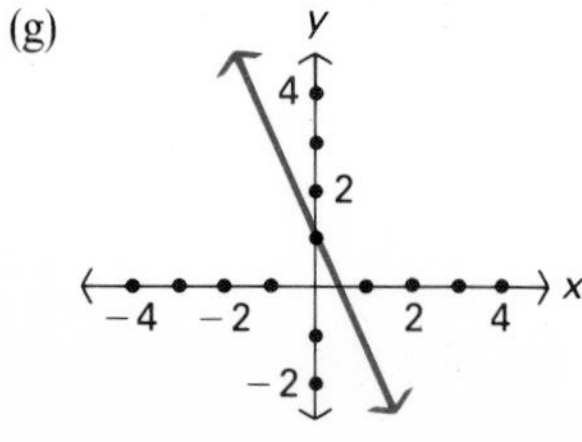

(i)

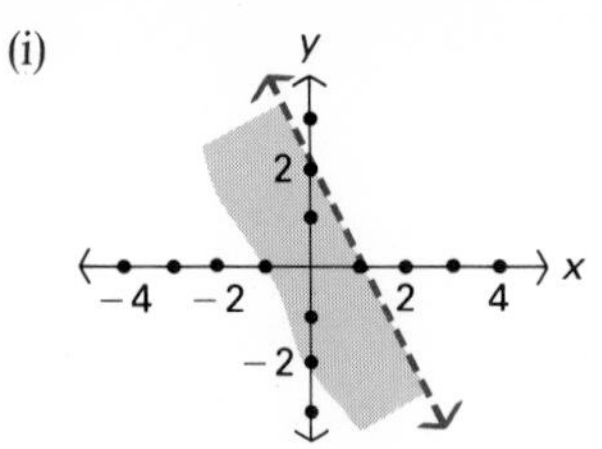

10. (a) $\{(1, 2), (1, 5), (2, 5), (3, 5), (4, 5)\}$; no
    (c) $\{(2, 2)\}$; yes

11. (a) Yes (c) No

13. (a)

(c)

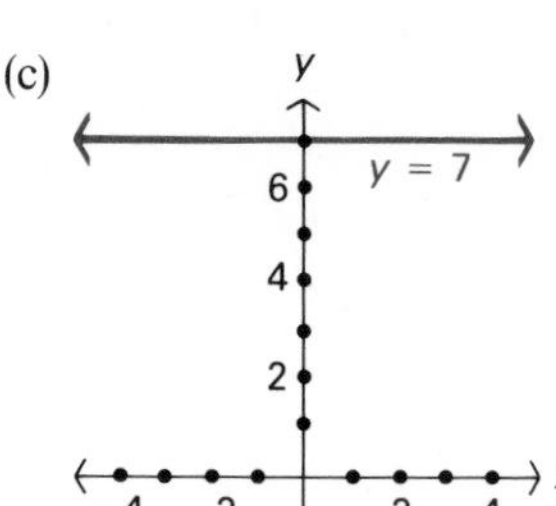

15. (a) $y = x + 2$ (c) $y = x^2 + 1$

16. (a) 6.0906853 (c) -2 (e) $6t + 6 - 3/2t$

17. (a) $3t + 1$ (c) $6t + 1$

Exercise Set 3, Page 542

1. (a) -5 (c) 1 (e) -1
   (g) The slope is not defined. (i) 0

2. (a)

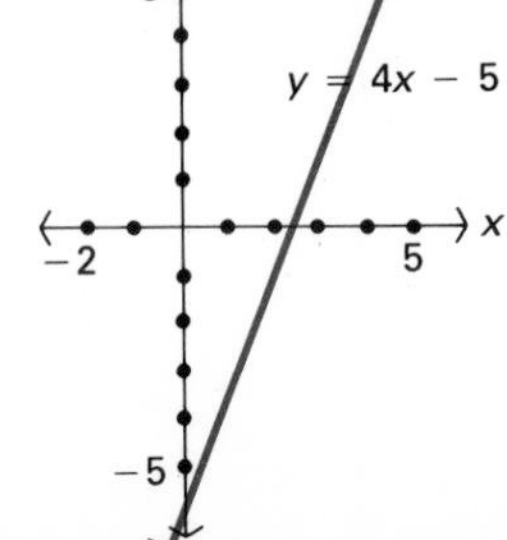

(c) $y = \frac{x}{2} + \frac{3}{2}$

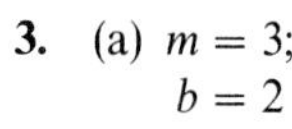
3. (a) $m = 3$; $b = 2$

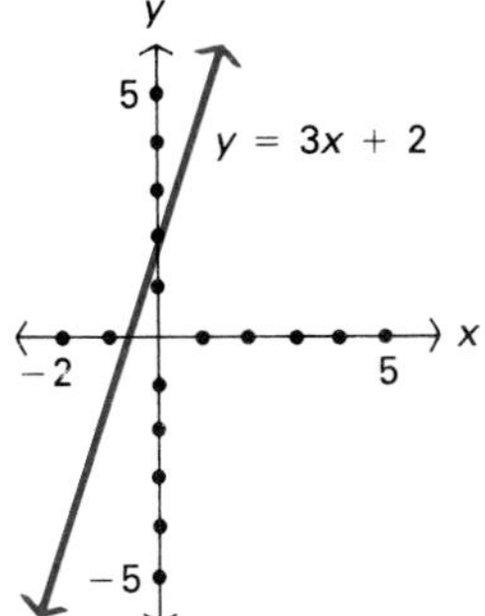

(c) $m = 3$; $b = -1$

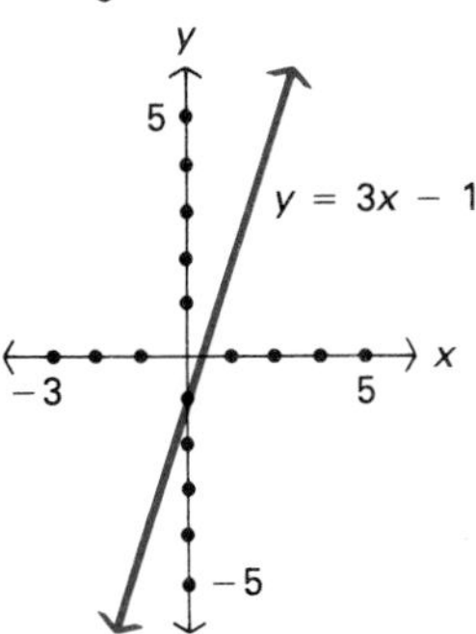

(e) $m = \frac{1}{2}$; $b = -2$

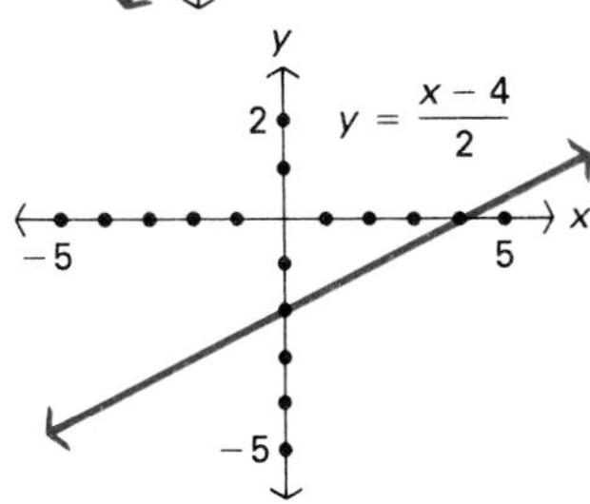

(g) $m = -\frac{4}{3}$; $b = \frac{7}{3}$

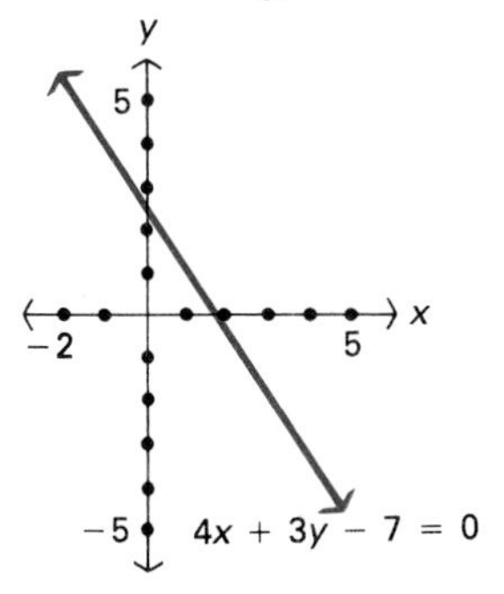

4. (a) False (c) False
5. (a) $x - 2y = -5$ (c) $x + 3y = -7$
6. (a) $4x - y = 3$ (c) $x = 1$
7. (a) $8x - 3y = 17$ (c) $x - y = -8$
   (e) $5x - y = -4$
8. (a) $y = -6$ (c) $y = 0$
9. a; b
10. (a) $\frac{x}{2} - \frac{y}{3} = 1$
11. $b = -8$
13. (a) $x - 2y = -4$ (c) $x + 3y = 11$
14. (a) $2x + y = 7$ (c) $3x - y = 3$
15. $y = 500x + 15{,}000$ ($x = 0$ at 1970)
    1988: 24,000 1990; 25,000
16. (a) $x + 4y = 14$ (c) $y + 2x = 7$
17. 2400 at \$375 1160 at \$530
18. (a) 30 at 8 A.M. (c) 205 at 3 P.M.

Exercise Set 4, Page 548

1. (a)

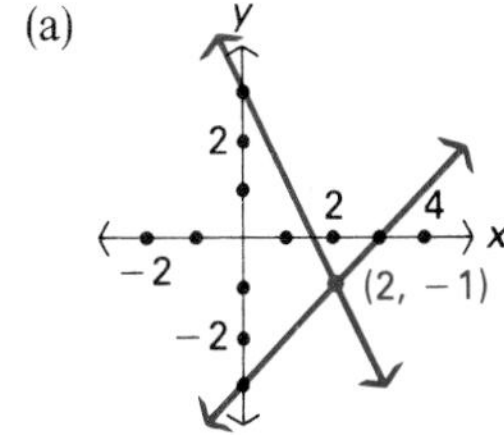

   (c) No solution

   (e)

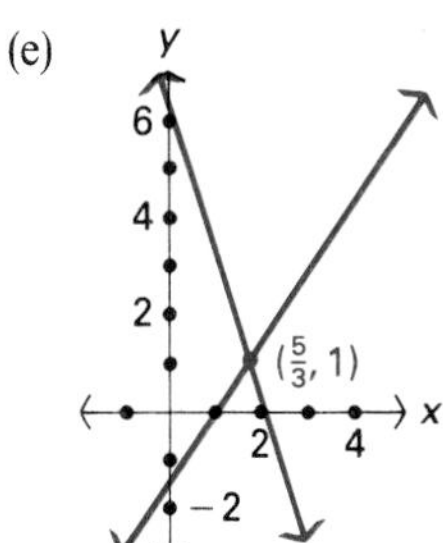

2. (a) $(2, -1)$ (c) No solution (e) $(\frac{5}{3}, 1)$
3. (a) $(\frac{11}{2}, \frac{3}{2})$
   (c) Infinite number of solutions
   (e) $(7, 3)$ (g) $(1, 1)$ (i) $(\frac{12}{5}, -\frac{6}{5})$
   (k) $(\frac{38}{93}, \frac{25}{9})$
5. 140 quarters and 100 dimes
7. $53\frac{1}{3}$ lb. of caramels; $106\frac{2}{3}$ lb. of chocolates
9. First 3 minutes cost \$2.00. Each additional minute costs \$0.80.

Exercise Set 5, Page 553

1. (a) $x = -\frac{1}{2}$ (c) $x = -1$ (e) $x = -3$
   (g) $x = 1$
2. (a) min. $= (-\frac{1}{2}, -\frac{25}{4})$ (c) max. $= (-1, 16)$
   (e) min. $= (-3, -3)$ (g) max. $= (1, 24)$
3. (a) Upward (c) Downward (e) Upward
   (g) Downward

4. (a)

x	-2	-1	0	1	2
y	-4	-6	-6	-4	0

(c)

x	-2	-1	0	1	2
y	15	16	15	12	7

(e)

x	-4	-3	-2	-1	0
y	0	-3	0	9	24

(g)

x	-1	0	1	2	3
y	0	18	24	18	0

5. (a)

(c)

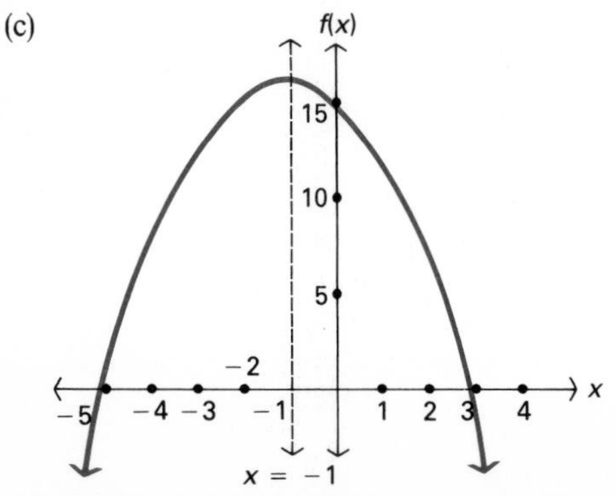

(e)

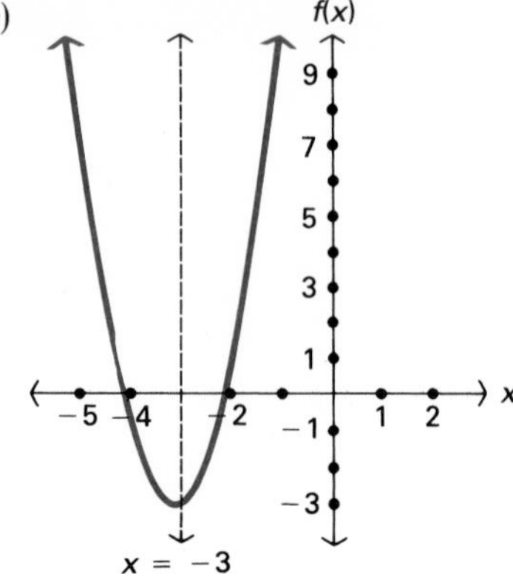

(g)

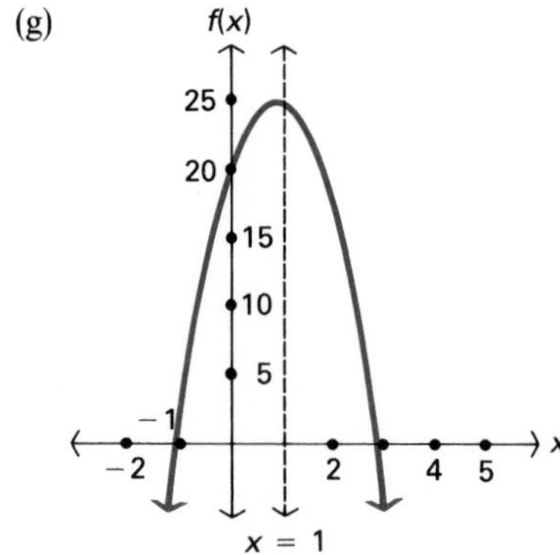

7. (a) $x = -2$; only one
8. (a) $x = 3, x = -1$ (c) $x = \frac{1}{3}, x = -4$
9. (a) $x = 2$ and $x = \frac{2}{5}$
10. (a) 15
11. (a) 150
12. (a) $x = 0, x = 4$ (c) $x = 0, x = -7$
13. (a) $x = 4, x = -4$ (c) $x = 7, x = -7$
15. (a) $x = \dfrac{1 \pm \sqrt{-11}}{2}$; no real solutions

Review Exercise Set 6, Page 555

1. (a) $x = 1, y = 2$
2. (a) $2y = 4 - x$ and $2y = 3x + 7$ intersect at $(-\frac{3}{4}, \frac{19}{8})$.

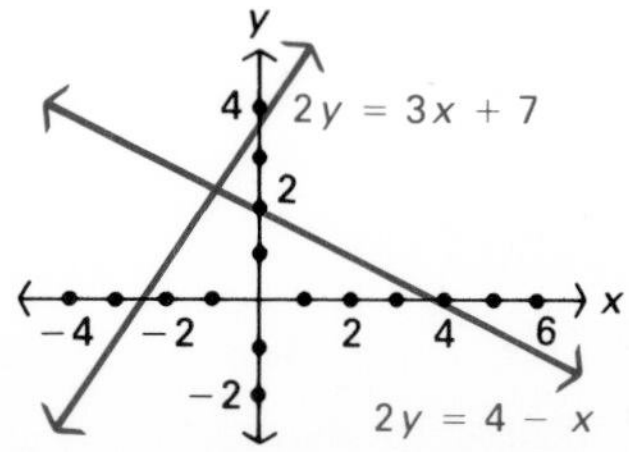

(c) $2y = 3x + 14$ and $3y = 2x - 14$ intersect at $(-14, -14)$.

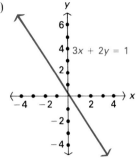

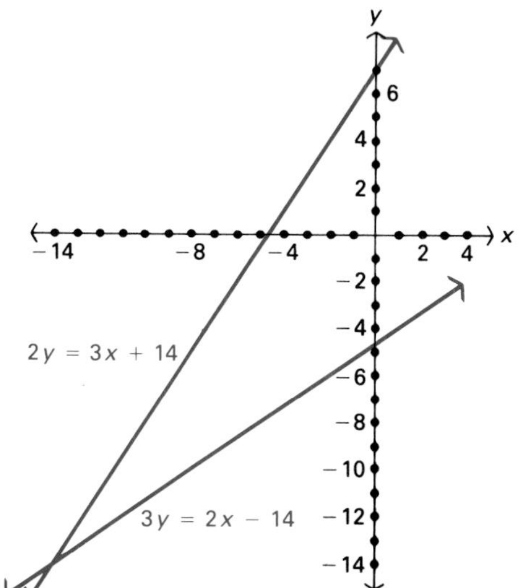

3. (a)

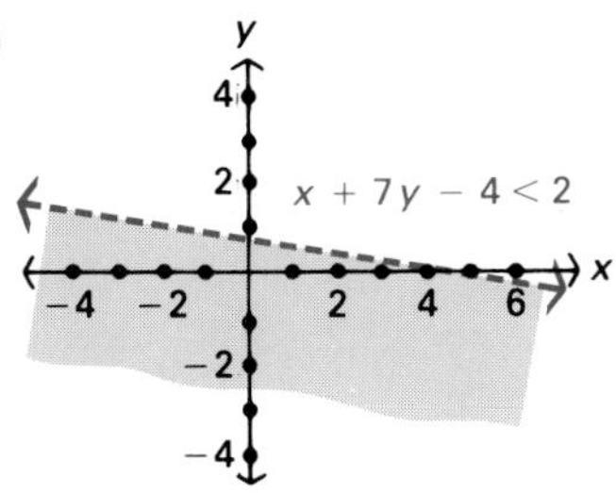

(c) $6x + 4y < 5$

4. (a) 7 (c) 4 (e) 3.92

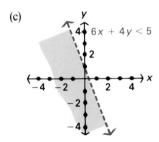

5. (a) $y < 3x + 2$

(c)

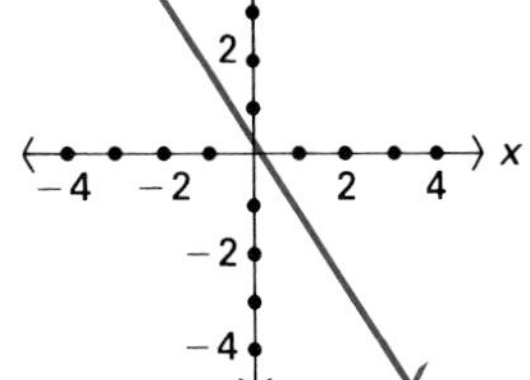

(e) $x + y < 5$

7. $2x + 3y = 19$

9. Triangle $A'(-2, -3)$, $B'(1, -4)$, $C'(0, -6)$

11. Triangle $A'(3, -2)$, $B'(4, 1)$, $C'(6, 0)$

13. $\sqrt{26}$; $(\frac{3}{2}, \frac{1}{2})$

14. (a) Parallel (c) Coincident

15. (a) $x = \frac{1}{2}$ (c) $x = \frac{5}{2}$

16. (a) $(\frac{1}{2}, -\frac{25}{4})$ (c) $(\frac{5}{2}, -\frac{49}{4})$

17. (a) Upward (c) Upward

18. (a)

x	−2	−1	0	1	2
y	0	−4	−6	−6	−4

(c)

x	−1	0	1	2	3
y	0	−6	−10	−12	−12

19. (a)

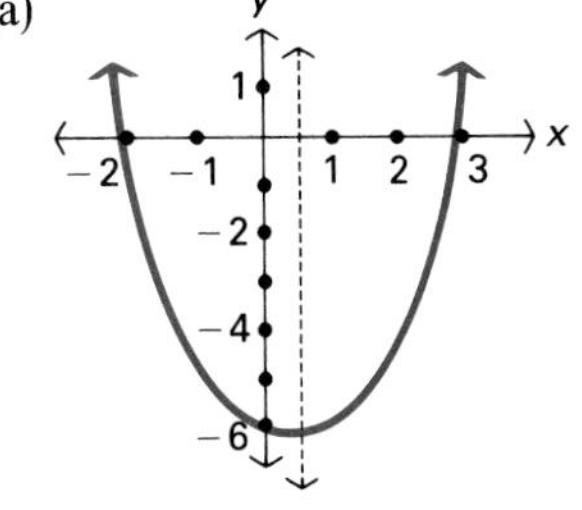

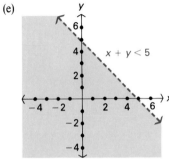

(c)

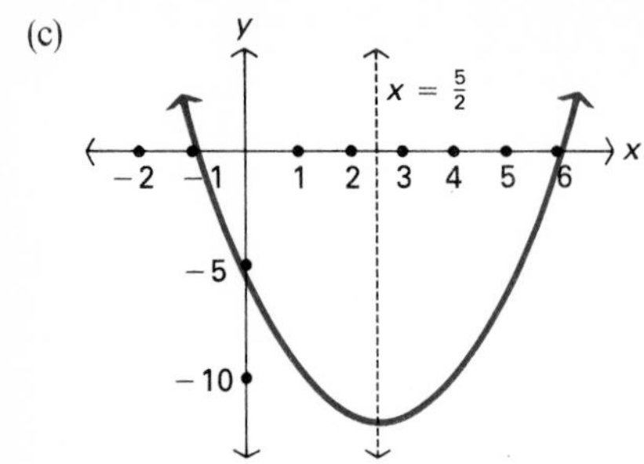

20. (a) $x = 3$ or 4
(c) $x = \dfrac{9 \pm \sqrt{141}}{6}$

21. $A'(3, 2)$, $B'(4, -1)$, $C'(6, 0)$

CHAPTER 13

Exercise Set 1, Page 563

1. (a) $\{A, B, C\}$ (c) $\{1, 2, 3, 4, 5, 6\}$
2. (a) $P(A) = P(B) = P(C) = \frac{1}{3}$
(c) $P(1) = P(2) = P(3) = P(4) = P(5) = P(6) = \frac{1}{6}$
3. $\{1, 2, 3, 4, 5, 6\}$
$P(1) = P(2) = P(3) = P(4) = P(5) = P(6) = \frac{1}{6}$
5. (a) $\{A, B, C\}$ $P(A) = P(C) = \frac{1}{4}$, $P(B) = \frac{1}{2}$
6. (a) $\frac{250}{1000} = \frac{1}{4}$ (c) $\frac{20}{1000} = \frac{1}{50}$
7. 1
9. (a) Probability cannot be negative.
(c) 0.001 can be a probability.
(e) Probability cannot exceed 1.
(g) Probability cannot exceed 1.
11. (a) $\{(A, R), (A, G), (B, R), (B, G), (C, R), (C, G)\}$
(c) $\{(A, 10), (A, 11), (A, 12), (A, 13), (B, 10), (B, 11), (B, 12), (B, 13), (C, 10), (C, 11), (C, 12), (C, 13)\}$
(e) $\{(10, R), (11, R), (12, R), (13, R), (10, G), (11, G), (12, G), (13, G)\}$
13. $\{(H, H, H), (H, H, T), (H, T, H), (T, H, H), (T, H, T), (T, T, H), (H, T, T), (T, T, T)\}$
14. (a) $\{R, B\}$
(c) $\{RRR, RRB, RBR, BRR, RBB, BRB, BBR, BBB\}$
15. (a) {Chevrolet, Ford, Pontiac, Oldsmobile, Buick, Plymouth, Dodge}
17. (a) Does not qualify for lack of the possibility of getting three tails or zero heads
(c) Does not qualify because it contains the possibility of four heads and there are only three coins
(e) Does not qualify. The set is empty.
18. (a) $\{(1, 1), (1, 2), (1, 3), (1, 4), (2, 1), (2, 2), (2, 3), (2, 4), (3, 1), (3, 2), (3, 3), (3, 4), (4, 1), (4, 2), (4, 3), (4, 4)\}$
19. (a) {red, blue, green} (c) The (b) part
20. (a) $\frac{4}{5}$ (c) $\frac{1}{26}$
21. (a) $A = \{2, 4\}$ (c) $C = \{4\}$
22. (a) This is a sample space.

Exercise Set 2, Page 568

1. (a) $\frac{1}{2}$ (c) $\frac{2}{3}$ (e) $\frac{1}{3}$
2. (a) $\frac{1}{4}$ (c) $\frac{1}{52}$ (e) $\frac{1}{26}$
3. $\frac{1}{5}$; $\frac{4}{5}$
4. (a) Any 1 of the 35 students would be an equally likely outcome.
(c) $\frac{1}{35}$ (e) 1
5. (a) $\frac{2}{7}$ (c) $\frac{1}{2}$ (e) $\frac{11}{14}$
6. (a) $\frac{2}{5}$ (c) $\frac{3}{5}$ (e) $\frac{11}{40}$
7. (a) $\frac{5}{8}$ (c) $\frac{1}{8}$
8. (a) $\frac{1}{6}$ (c) $\frac{1}{9}$
9. (a) $\frac{1}{4}$ (c) $\frac{3}{4}$
10. (a) False, since being from a given state is not equally likely, Rhode Island has fewer people than most of the other states.
(c) In probability, such is possible. However, one would guess this to be a loaded coin.
11. (a) $\frac{7}{38}$ (c) $\frac{1}{38}$ (e) $\frac{1}{190}$
13. No, because there are more ways *not* to get a spade than to get one.
15. (a) $\{(1, 1), (1, 2), (1, 3), (1, 4), (2, 1), (2, 2), (2, 3), (2, 4), (3, 1), (3, 2), (3, 3), (3, 4), (4, 1), (4, 2), (4, 3), (4, 4)\}$
(c) $\frac{5}{8}$
17. $\frac{1}{12}$
(a) $\{1, 3, 5, 7, 9, 11\}$, $\frac{1}{2}$ (c) $\{1, 2, 3, 4, 6, 12\}$, $\frac{1}{2}$
(e) $\{\ \}$, 0 (g) $\{2, 4, 6, 8, 10, 12\}$, $\frac{1}{2}$
(i) $\{2, 3, 4, 6, 8, 9, 10, 12\}$, $\frac{2}{3}$
(k) $\{2, 3, 4, 5, 6, 7, 8, 9, 10, 11, 12\}$, $\frac{11}{12}$

Exercise Set 3, Page 574

1. (a) 6
2. (a) 24
3. (a) 32 (c) $\frac{1}{32}$
5. (a) $\frac{1}{4}$ (c) $\frac{3}{4}$
6. (a) 14 (c) 6

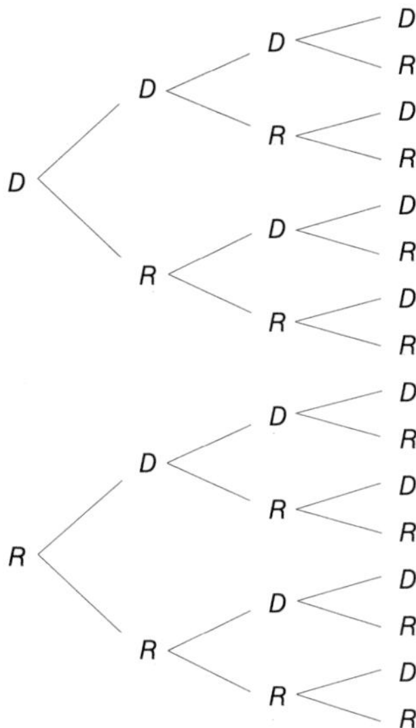

7. Yes, $2 \cdot 3 \cdot 3 = 18$ branches.

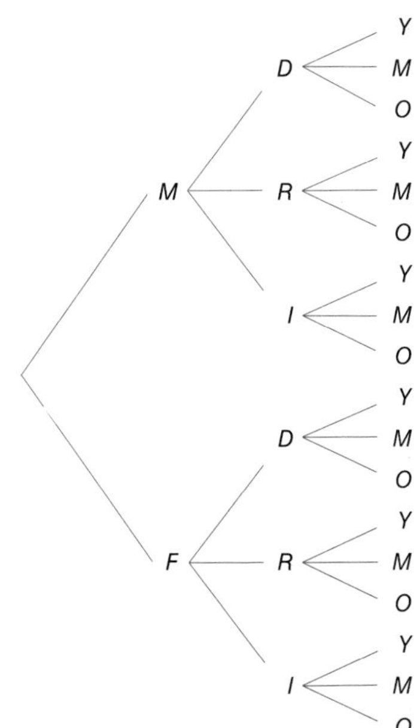

8. (a) $\frac{1}{36}$ (c) $\frac{1}{18}$
9. (a) $\frac{1}{16}$ (c) $\frac{5}{16}$
10. (a) $\frac{5}{32}$ (c) $\frac{1}{2}$
11. 72
12. (a) $\frac{1}{169}$

Exercise Set 4, Page 579

1. 0.7
3. (a) Getting no heads or one head
   (c) Getting less than six heads or getting seven heads
4. (a) $\frac{1}{2}$ (c) $\frac{1}{4}$
5. (a) $\frac{1}{2}$ (c) $\frac{12}{13}$
6. (a) $\frac{10}{13}$ (c) $\frac{9}{13}$
7. (a) 0.4 (c) 0.7
8. (a) $\frac{1}{2}$ (c) $\frac{13}{24}$
9. (a) $\frac{3}{4}$ (c) $\frac{17}{20}$
11. (a) $A = \{(HHHH), (HHHT), (HHTH), (HHTT), (HTHH), (HTHT), (HTTH), (HTTT), (THHH), (THHT), (THTH), (THTT), (TTHH), (TTHT), (TTTH), (TTTT)\}$
    (c) $B = \{(HHHT), (HHTH), (HHTT), (HTHH), (HTHT), (HTTH), (HTTT), (THHH), (THHT), (THTH), (THTT), (TTHH), (TTHT), (TTTH), (TTTT)\}$
    (e) $\{A \text{ and } B\} = \{(HTTT), (THTT), (TTHT), (TTTH)\}$
13. (a) $\frac{17}{20}$ (c) $\frac{23}{25}$
14. (a) $\frac{3}{10}$ (c) $\frac{1}{10}$
15. (a) $\frac{1}{2}$ (c) $\frac{8}{9}$ (e) $\frac{16}{225}$

Exercise Set 5, Page 586

1. (a) $S = \{$all the cards in the deck$\}$
   (c) 0 (e) 1
2. (a) $\frac{2}{3}$
3. (a) 0.42 (c) 0.4
4. (a) 0.35 (c) 0.3 (e) 0.8 (g) $\frac{4}{7}$ (i) $\frac{2}{3}$
5. $\frac{63}{100}$
7. $\frac{15}{36}$
9. 60
11. 3.5 miles
13. (a) $\frac{16}{49}$
14. (a) $\frac{3}{4}$ (c) $\frac{7}{20}$ (e) $\frac{13}{16}$
15. (a) $\frac{4}{13}$
17. 1

Review Exercise Set 6, Page 589

1. (a) $\frac{2}{5}$ (c) $\frac{1}{2}$ (e) $\frac{1}{5}$ (g) $\frac{7}{10}$
2. (a) {red, black, green}

(c) {(red, red), (red, black), (red, green), (black, red), (black, black), (black, green), (green, red), (green, black), (green, green)}

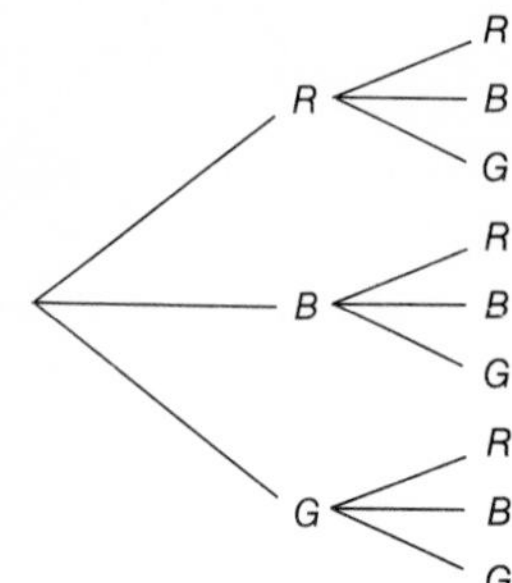

3. (a) $\frac{10}{13}$ (c) 1 (e) $\frac{9}{13}$ (g) 1 (i) 1
4. (a) $\frac{36}{169}$ (c) $\frac{16}{169}$
5. (a) $\frac{1}{3}$
6. (a) 0.65
7. $-\frac{1}{2}$ point
8. (a) $\frac{25}{102}$ (c) $\frac{1}{221}$
9. (a) 24 (c) $\frac{1}{2}$
10. (a) $1/6.35(10)^{11}$ (c) 0
11. $\frac{26}{51}$
12. $\frac{48}{169}$
13. (a) 1/40,425 (c) $\frac{304}{2695}$ (e) 13,824/15,625

CHAPTER 14

Exercise Set 1, Page 595

1.

Number of rides	*Tally*	*Frequency*
1	|	1
2	||||	4
3	|||	3
4	|||	3
5	|	1
6	|	1
7	||	2
8	𝍸 |	6
9	||	2
10	|	1
	Total	24

2. (a)

Class (age)	*Tally*	*Frequency*
5–14	|||	3
15–24	𝍸 𝍸 |||	13
25–34	|||	3
35–44	|||	3
45–54	||	2
55–64	𝍸	5
65–74	𝍸 ||	7
	Total	36

3. (a) 14.5–19.5 (c) 20 (e) 16
19.5–24.5
24.5–29.5
29.5–34.5
34.5–39.5

4. (a) 70 (c) 52 (e) \$20,000–\$29,999.99
(g) 0; 9,999.995; 19,999.995; 29,999.995; 39,999.995

5.

Class (grades)	*Tally*	*Frequency*
44–51	|||	3
52–59	𝍸	5
60–67	𝍸 𝍸	10
68–75	𝍸 𝍸 ||||	14
76–83	𝍸 𝍸 ||||	14
84–91	𝍸 𝍸 |	11
92–99	𝍸 ||||	9
	Total	66

6. (a)

Class	*Frequency*
450–491	4
492–533	7
534–575	6
576–617	5
618–659	3

(c) 449.5–491.5
491.5–533.5
533.5–575.5
575.5–617.5
617.5–659.5

7. (a)

Class	Frequency
30–33	5
34–37	12
38–41	21
42–45	6
46–49	4
50–53	2

(c) 29.5–33.5
33.5–37.5
37.5–41.5
41.5–45.5
45.5–49.5
49.5–53.5

8. (a)

Class	Frequency
43–50	2
51–58	5
59–66	9
67–74	13
75–82	13
83–90	10
91–98	8

(c) 42.5–50.5
50.5–58.5
58.5–66.5
66.5–74.5
74.5–82.5
82.5–90.5
90.5–98.5

Exercise Set 2, Page 602

1. (a) 14% (c) 19%
2. (a)

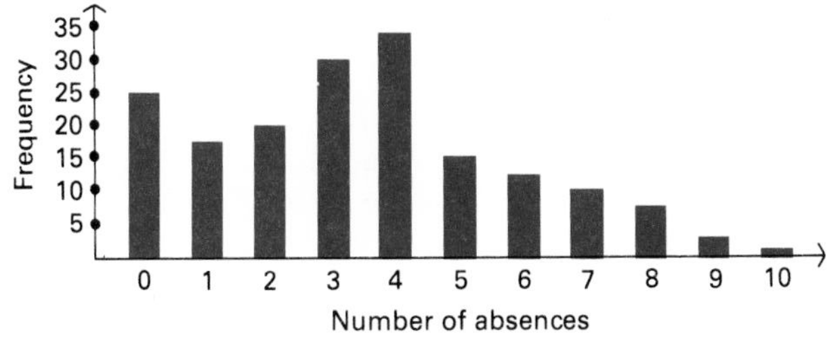

3. (a) 24.5, 34.5, 44.5, 54.5, 64.5, 74.5, 84.5

(c)

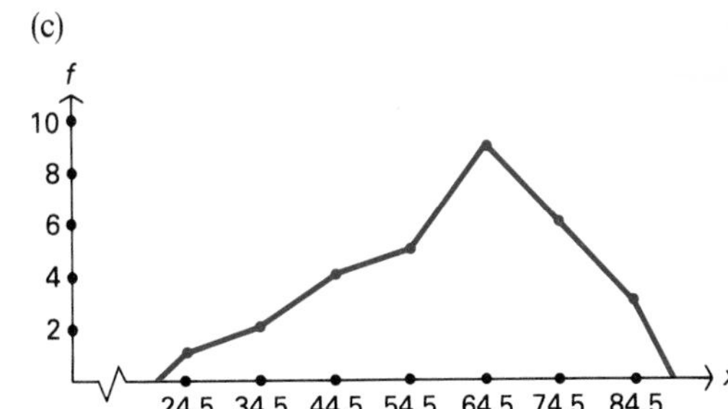

4. (a)

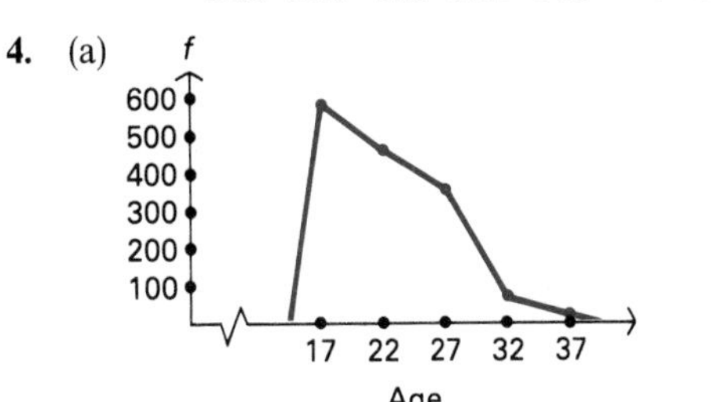

5. (a)

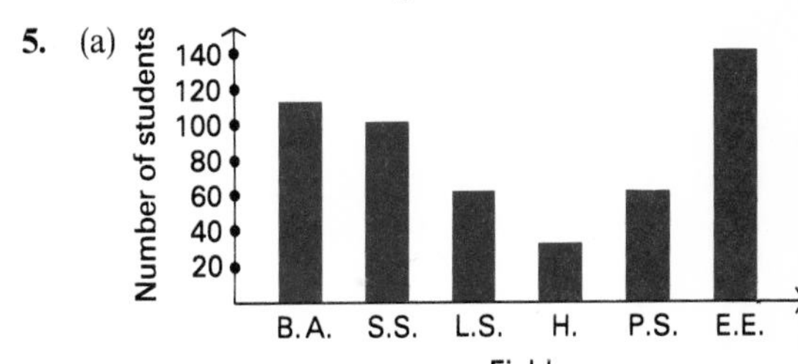

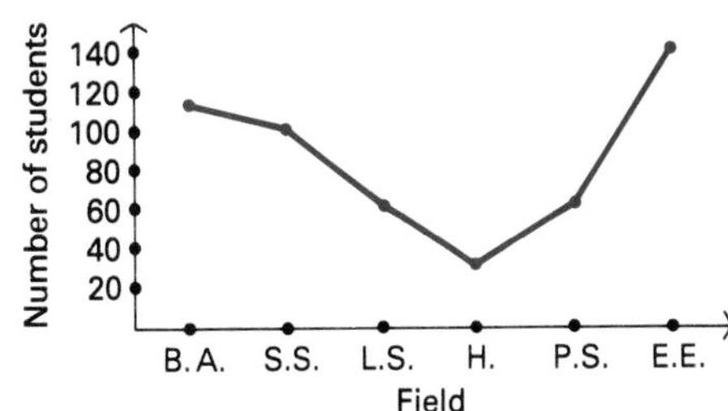

6. Increases eightfold
7. (a) 45% for Democratic candidates and 3% for Republican candidates
(c) $5.94 million
9. (a)

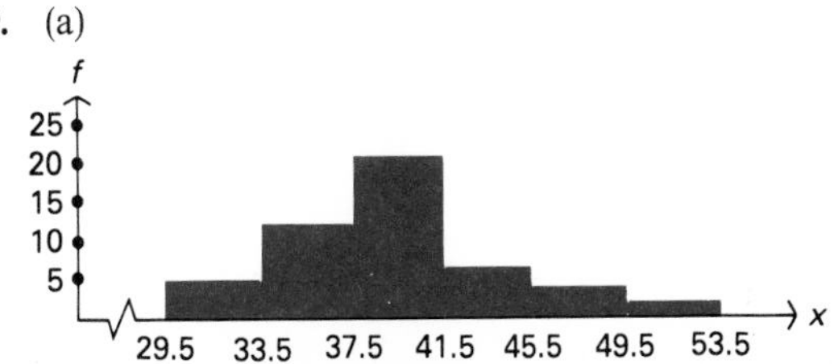

11.

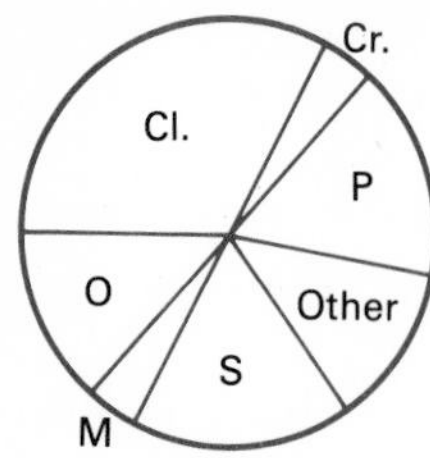

13.

Class	*Frequency*
60–74	30
75–89	35
90–104	36
105–119	34
120–134	28

Exercise Set 3, Page 608

1. (a) Mean = 6, median = 5, no mode
(c) Mean = $\frac{34}{9}$, median = 4, mode = 6
(e) Mean = $\frac{137}{8}$, median = 19.5, mode = 23

3. Mean = 27

4. (a) 5, 6, 7, 8 (c) 3, 4, 4, 5 (e) 3, 6, 6, 9

5. 520

7. (a) Median (c) Mode

8. (a) Mean = 273.23 lb. Median = 275 lb.
Mode = 302 lb.

9. (a) Mean = 82.725 (c) Mode = 80

11. Mean = 118.21

13. Mean becomes \$5460 instead of \$4000.

15. (a) 60.83

16. (a) 83.33

17. (a) $\bar{y} = 0.371$ $\bar{x} = 114.5 + 10(0.371) = 118.21$

19. (b)

```
10  LET N = 0
20  LET S = 0
30  DATA     (to be supplied; last pair of
             values should be 1000000, 1)
40  READ P, F
50  IF P = 10^6 THEN 90
60  LET S = S + P * F
70  LET N = N + F
80  GOTO 40
90  LET M = S/N
100 PRINT "THE MEAN IS   " M
110 END
```

20. (a) 82.725

Exercise Set 4, Page 614

1. (a) Range = 8
$s_x^2 = 9$ $s_x = 3$
(c) Range = 12
$s_x^2 = 16$ $s_x = 4$
(e) Range = 11
$s_x^2 = 16$ $s_x = 4$
(g) Range = 80
$s_x^2 = 734.69$ $s_x = 27.11$

3. 11.95

4. (a) C (c) D

5. 3

7. 15.16

9. (a) Most students scored near the average, but a small number had either extremely low or extremely high scores.
(c) More than half of the students scored above the mean.
(e) All students had the same score.

11. (a) No effect
(c) Reduces the standard deviation to $\frac{1}{5}$ its original size

12. (a) $s_y = 1.456$ $s_x = 14.56$ $s_x = 10s_y$

13. (a) 3 (c) 4 (e) 4 (g) 27.11
(i) 27.13

14. (a)

```
10  LET I = 0
20  LET A = 0
30  LET B = 0
40  READ N
50  DATA (to be supplied)
60  READ X
65  LET I = I + 1
70  LET A = A + X
80  LET B = B + X^2
90  IF I < N THEN 60
100 LET S = (1/N) * SQR(N * B - A^2)
110 PRINT S
120 END
```

15. (a) 14.56

Exercise Set 5, Page 621

1. (a) 0.4918 (c) 0.4525

2. (a) 0.7698 (c) 0.9070

3. (a) .0179 (c) .5398 (e) .0668 (g) .9821
(i) .0886 (k) .5028

4. (a) .8413 (c) .3830 (e) Approximately 1

5. (a) .5000 (c) .1587 (e) .1587 (g) .5403

6. (a) .2611 (c) .4838 (e) .6826

7. (a) 62.12% (c) 100% (e) 49.57%

8. (a) 18.51% (c) 0.43%

9. 0.62%

11. (a) .9525

12. (a) 68.26%

13. 49.38%

Review Exercise Set 6, Page 625

1. (a) 18 (c) 20 (e) 7.69

3. Mean = 28.4; Median = 28.125; Standard deviation = 12.22

4. (a) 75 (c) 94

5.

Interval	*Frequency*
30–34	1
35–39	0
40–44	0
45–49	0
50–54	1
55–59	0
60–64	5
65–69	0
70–74	3
75–79	4
80–84	3
85–89	2
90–94	4
95–99	1

(a) 75.125
(b) 60–64 modal class

6. (a)

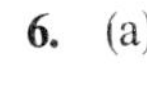

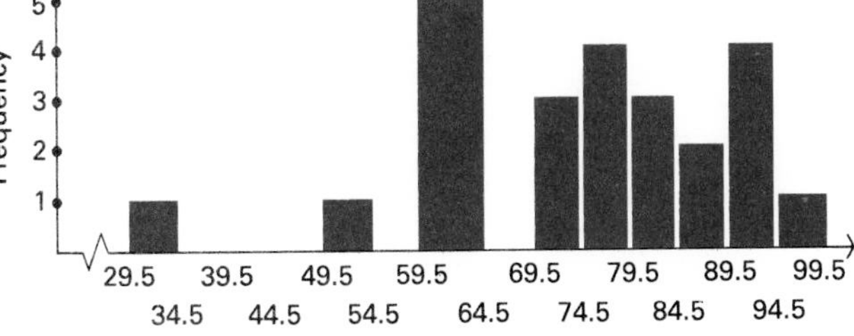

7.

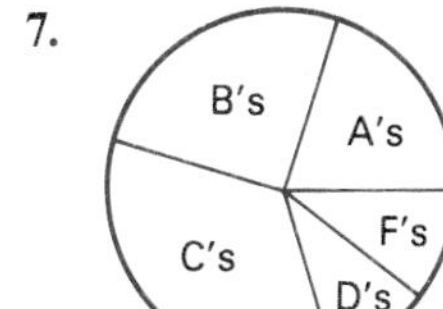

9. (a) Probably 50% of the families have fewer than 3.5 people. Therefore, housing to accommodate 4 people would be too large for these.
(c) False. The people with serious illness are more prone to die and these people are found at hospitals.
(e) One needs to know how long he has been in office and the inflation rate.

11. (a) -0.16 (c) $-0.16 = \dfrac{(28.4 - 30)}{10}$

APPENDIX

Exercise Set A.1 Page 631

1. (a)

p	$p \to p$
T	T
F	T

Logically true

(c)

p	q	$p \to q$	$(p \to q) \to p$
T	T	T	T
T	F	F	T
F	T	T	F
F	F	T	F

Not logically true

(e)

p	q	$p \to q$	$p \vee q$	$(p \to q) \to (p \vee q)$
T	T	T	T	T
T	F	F	T	T
F	T	T	T	T
F	F	T	F	F

Not logically true

(g)

p	q	$\sim p$	$\sim p \vee q$	$p \to q$	$(\sim p \vee q) \to (p \to q)$
T	T	F	T	T	T
T	F	F	F	F	T
F	T	T	T	T	T
F	F	T	T	T	T

Logically true

2. (a) 3 (c) 4 or -4

3. (a) $x = 2$ (There are many possibilities.)
(c) $x = 3$ (There are many possibilities.)

4. (a) There exists an x for which $x + 4 = 7$.
(c) There exist x's for which $x^2 = 16$.

5. (a) There is no x such that $x + 4 = 7$.
(c) There is no x such that $x^2 = 16$.

6. (a) Some athletes over 7 feet tall do not play basketball.
(c) Some men do not use Bob-Bob hair oil.
(e) Some man weighs more than 500 pounds.
(g) All bunnies are rabbits.
(i) Some prime numbers greater than 2 are not odd.

7. (a) There is no counting number greater than 50.
(c) There is no x such that $x + 2 = 7$.
(e) There exists an x such that $x(x + 2) \neq x^2 + 2x$.

8. (a)

p	q	$p \wedge q$	$q \wedge p$	$p \vee q$	$q \vee p$
T	T	T	T	T	T
T	F	F	F	T	T
F	T	F	F	T	T
F	F	F	F	F	F

Since the third and fourth columns are identical, $p \wedge q = q \wedge p$.
Since the fifth and sixth columns are identical, $p \vee q = q \vee p$.

(c)

p	q	r	$p \vee q$	$(p \vee q) \vee r$	$q \vee r$	$p \vee (q \vee r)$
T	T	T	T	T	T	T
T	T	F	T	T	T	T
T	F	T	T	T	T	T
T	F	F	T	T	F	T
F	T	T	T	T	T	T
F	T	F	T	T	T	T
F	F	T	F	T	T	T
F	F	F	F	F	F	F

Since the seventh and fifth columns are identical, $p \vee (q \vee r) = (p \vee q) \vee r$.

(e)

p	q	r	$q \vee r$	$p \wedge (q \vee r)$	$p \wedge q$	$p \wedge r$	$(p \wedge q) \vee (p \wedge r)$
T	T	T	T	T	T	T	T
T	T	F	T	T	T	F	T
T	F	T	T	T	F	T	T
T	F	F	F	F	F	F	F
F	T	T	T	F	F	F	F
F	T	F	T	F	F	F	F
F	F	T	T	F	F	F	F
F	F	F	F	F	F	F	F

Since the fifth and eighth columns are identical, $p \wedge (q \vee r)$ and $(p \wedge q) \vee (p \wedge r)$ are logically equivalent.

(g)

p	$\sim p$	$p \wedge \sim p$	$p \vee \sim p$
T	F	F	T
T	T	F	T

Since the third column contains all F's, $p \wedge \sim p$ is a contradiction.

Since the fourth column contains all T's, $p \vee \sim p$ is a tautology.

(i)

p	q	$p \wedge q$	$\sim(p \wedge q)$	$\sim p$	$\sim q$	$\sim p \vee \sim q$
T	T	T	F	F	F	F
T	F	F	T	F	T	T
F	T	F	T	T	F	T
F	F	F	T	T	T	T

Since the fourth and seventh columns are identical, $\sim(p \wedge q)$ and $\sim p \vee \sim q$ are logically equivalent.

9. (a) There exists an x for which "$p(x)$ and $q(x)$" is not true.
(c) There exists an x for which "if $p(x)$, then $q(x)$" is not true.
(e) There does not exist an x such that "$p(x)$ or $q(x)$" is true.

10. (a)

r	s	$\sim r$	$r \to s$	$\sim r \vee s$
T	T	F	T	T
T	F	F	F	F
F	T	T	T	T
F	F	T	T	T

Since the last two columns are identical, the two statements are equivalent.

(c)

s	t	$s \to t$	$t \to s$	$(t \to s) \to (s \to t)$
T	T	T	T	T
T	F	F	T	F
F	T	T	F	T
F	F	T	T	T

These two statements are equivalent, because the third and fifth columns are identical.

(e)

r	s	t	$s \wedge t$	$r \vee s$	$r \vee t$	$r \vee (s \wedge t)$	$(r \vee s) \wedge (r \vee t)$
T	T	T	T	T	T	T	T
T	T	F	F	T	T	T	T
T	F	T	F	T	T	T	T
T	F	F	F	T	T	T	T
F	T	T	T	T	T	T	T
F	T	F	F	T	F	F	F
F	F	T	F	F	T	F	F
F	F	F	F	F	F	F	F

Since the last two columns are identical, the statements are equivalent.

11. (a)

q	p	$p \to q$	$(p \to q) \wedge p$	$[(p \to q) \wedge p] \to q$
T	T	T	T	T
T	F	F	F	T
F	T	T	F	T
F	F	T	F	T

(c)

p	q	r	$p \to q$	$q \to r$	$p \to r$	$(p \to q) \wedge (q \to r)$	$[(p \to q) \wedge (q \to r)] \to (p \to r)$
T	T	T	T	T	T	T	T
T	T	F	T	F	F	F	T
T	F	T	F	T	T	F	T
T	F	F	F	T	F	F	T
F	T	T	T	T	T	T	T
F	T	F	T	F	T	F	T
F	F	T	T	T	T	T	T
F	F	F	T	T	T	T	T

p	q	r	$p \to q$	$q \to r$	$(p \to q) \wedge (q \to r) \wedge p$	$[(p \to q) \wedge (q \to r) \wedge p] \to r$
T	T	T	T	T	T	T
T	T	F	T	F	F	T
T	F	T	F	T	F	T
T	F	F	F	T	F	T
F	T	T	T	T	F	T
F	T	F	T	F	F	T
F	F	T	T	T	F	T
F	F	F	T	T	F	T

13. (a) $\sim p \vee q$ (c) $\sim p \vee \sim q$ (e) $\sim q \vee \sim p$
(g) tautology

Exercise Set A.2, Page 635

1. (a) Invalid (c) Valid; Rule of Detachment
(e) Valid; Chain Rule

2. (a) Invalid (c) Valid

3. (a) Invalid (c) Valid (e) Invalid

4. (a) Invalid. Let p represent "A man is a good speaker" and q represent "A man is a good teacher." Does $(p \to q) \wedge q$ imply p?

p	q	$p \to q$	$(p \to q) \wedge q$	$[(p \to q) \wedge q] \to p$
T	T	T	T	T
T	F	F	F	T
F	T	T	T	F
F	F	T	F	T

Invalid, since $[(p \to q) \wedge q] \to p$ is not logically true

(c) Valid, Rule of Detachment
(e) Valid, Chain Rule
(g)

p	q	r	$\sim p$	$q \to \sim p$	$q \wedge r$	$(q \to \sim p) \wedge (q \wedge r)$	$\sim r$	$\sim p \to \sim r$	$[(q \to \sim p) \wedge (q \wedge r)] \to (\sim p \to \sim r)$
T	T	T	F	F	T	F	F	T	T
T	T	F	F	F	F	F	T	T	T
T	F	T	F	T	F	F	F	T	T
T	F	F	F	T	F	F	T	T	T
F	T	T	T	T	T	T	F	F	F
F	T	F	T	T	F	F	T	T	T
F	F	T	T	T	F	F	F	F	T
F	F	F	T	T	F	F	T	T	T

Since the last column does not contain T for every case, it is not logically true.

(i) Valid, contrapositive, and Rule of Detachment

5. (a) Therefore, Susan takes mathematics (by the Rule of Detachment).
(c) Therefore, you cry (by the Rule of Detachment).

6. (b) If Joan is tall, she wears contacts.
(d) If Joan wears contacts, she has red hair.
(a) If Joan has red hair, she teaches seventh grade.
(c) Therefore, if Joan is tall, she teaches seventh grade.

7. It was raining.

8. (a) Let p be "Paul burns the hot dog," q be "Ed will get indigestion," and r be "Ed threw the hot dog away."

p	q	r	$p \to q$	$\sim p$	$\sim p \vee r$	$\sim q$	$\sim q \to (\sim p \vee r)$	$(p \to q) \wedge [\sim q \to (\sim p \vee r)]$	$\sim r$
T	T	T	T	F	T	F	T	T	F
T	T	F	T	F	F	F	T	T	T
T	F	T	F	F	T	T	T	F	F
T	F	F	F	F	F	T	F	F	T
F	T	T	T	T	T	F	T	T	F
F	T	F	T	T	T	F	T	T	T
F	F	T	T	T	T	T	T	T	F
F	F	F	T	T	T	T	T	T	T

(continued below)

$[(p \to q) \wedge (\sim q \to (\sim p \vee r))] \wedge \sim r$	$[((p \to q) \wedge (\sim q \to (\sim p \vee r))) \wedge \sim r] \to q$
F	T
T	T
F	T
F	T
F	T
T	T
F	T
T	F

Invalid, because $[((p \to q) \wedge (\sim q \to (\sim p \vee r))) \wedge \sim r] \to q$ is not logically true

(c) Invalid

9. (a) Therefore, Diana is clever.
   (c) Kay is not a college girl.

ANSWERS TO JUST FOR FUN

(Problem-solving techniques for Just for Fun Problems are outlined in the *Activities Manual.*)

Chapter 1, Page 6

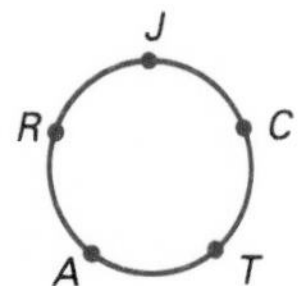

Chapter 1, Page 16

\$1.40

Chapter 1, Page 25

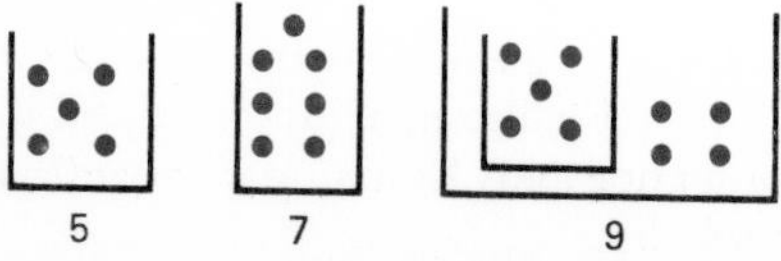

Chapter 1, Page 31

Blue ribbon (see *Activities Manual* for explanation).

Chapter 1, Page 39

Consider couples AA', BB', CC', where A', B', and C' are women. AA' cross the river, A returns; $B'C'$ cross the river; C' returns; AB cross the river; AA' returns; AC cross the river; B' returns; $A'C'$ cross the river; C' returns; $B'C'$ cross the river.

Chapter 1, Page 45

Bushy threw the rock. (see *Activities Manual.*)

Chapter 1, Page 51

Godchild was the murderer.
(See *Activities Manual.*)

Chapter 2, Page 66

2^n (see *Activities Manual* for discussion).

Chapter 2, Page 76

Carry out the plan:
Reds vs. Angels
Blues vs. Jets
Bulldogs vs. Panthers
Tigers vs. Rockets

Chapter 2, Page 85

3

Chapter 2, Page 91

	8	3
		7

Finish the problem.

Chapter 2, Page 101

$[(6 \cdot 6) - 6] \div 6 = 5$

Chapter 2, Page 109

$7 = 4 + 4 - (4 \div 4)$
$8 = (4 \cdot 4 \div 4) + 4$
$9 = 4 + 4 + (4 \div 4)$
(See *Activities Manual* for others.)

Chapter 3, Page 122

Multiplication table for 9.

Chapter 3, Page 132

. . . fiefotfot, foefotfot, fumfotfot, fotfotfot, feefotfotfot, fiefotfotfot, foefotfotfot, fumfotfotfot, fotfotfotfot.

Chapter 3, Page 141

```
      173          586
215 ) 37195         88
      215         4688
      1569        4688
      1505       51568
       645
       645
```

Chapter 3, Page 148

Each card represents a place value in binary notation. *C* is 1; *A* is 2; *R* is 4; and *D* is 8.

Chapter 3, Page 154

The message is "key."

Chapter 4, Page 167

333333331 is divisible by 17.

Chapter 4, Page 173

5	3	7
7	5	3
3	7	5

Chapter 4, Page 182

Dogs barked at once (see *Activities Manual* for solution).

Chapter 4, Page 187

524

Chapter 4, Page 195

A	B	C	D	E	F	G	H	I	J	K	L
L	K	J	I	H	G						
G	L	K	J	I	H						
H	G	L	K	J	I						
I	H	G	L	K	J						
J	I	H	G	L	K						
K	J	I	H	G	L						
D	E	F				J	K	L			
E	F	D				K	L	J			
B		E	F			H		K	L		
F	C		E			I	J			L	
C	D			F		L	I		K		

Chapter 4, Page 205

The girl (see *Activities Manual* for explanation).

Chapter 4, Page 210

$1 - [2 + 3 - (4 + 5 - 6)] = -1$

Chapter 5, Page 223

$\frac{1}{3}$

Chapter 5, Page 231

$\frac{5}{6}$	$3\frac{3}{4}$	$1\frac{2}{3}$
$2\frac{11}{12}$	$2\frac{1}{12}$	$1\frac{1}{4}$
$2\frac{1}{2}$	$\frac{5}{12}$	$3\frac{1}{3}$

Chapter 5, Page 239

According to Ahab's will, the following camels should have been received:

Oldest child: $\frac{1}{2}(17) = 8\frac{1}{2}$

Second child: $\frac{1}{3}(17) = 5\frac{2}{3}$

Third child: $\frac{1}{9}(17) = 1\frac{8}{9}$

However, $\frac{1}{2} + \frac{1}{3} + \frac{1}{9} \neq 1$, so Ahab did not give away all his estate. If the $8\frac{1}{2}$, $5\frac{2}{3}$, and $1\frac{8}{9}$ are rounded to 9, 6, 2, all the estate is used since $9 + 6 + 2 = 17$. The smart lawyer recognized this so he pulled the borrowing technique where $\frac{1}{2}(18) = 9$, $\frac{1}{3}(18) = 6$, and $\frac{1}{9}(18) = 2$.

Chapter 5, Page 250

240

Chapter 5, Page 260

.06	.03	.12
.09	.15	.21
.18	.27	.24

Chapter 5, Page 267

$(44.4 - 4.4) \div 4$

Chapter 5, Page 273

Note the answer will always be 5 regardless of the number selected.

Chapter 6, Page 284

$$100\left(1 + 1 + \frac{1}{2} + \frac{1}{2^2} + \cdots + \frac{1}{2^{n-2}}\right)$$

where n is the nth time it hits the ground

Chapter 6, Page 291

Yes (see *Activities Manual.*)

Chapter 6, Page 297

11/20

Chapter 6, Page 304

120 (see *Activities Manual.*)

Chapter 6, Page 310

$8.75

Chapter 6, Page 315

No, because the sum of the balance is not necessarily the total deposited

Chapter 7, Page 327

Yes; yes. (See *Activities Manual.*)

Chapter 7, Page 333

Write each line as a PRINT statement. (See *Activities Manual.*)

Chapter 7, Page 338

Output: ROW ROW ROW YOUR BOAT GENTLY DOWN THE STREAM

Chapter 7, Page 343

Three degrees below zero; (the B.S. degree, the M.S. degree, and the Ph.D. degree)

Chapter 7, Page 347

```
FD 40 REPEAT 4[RT 144 FD 40]
```

Chapter 7, Page 351

```
RT 80 REPEAT 36[LT 80 FD 15 RT 160 FD 15 LT 80
RT 360/36]
```

Chapter 8, Page 365

$1 + 2 + 3 + \cdots + 11 = 66$

Chapter 8, Page 371

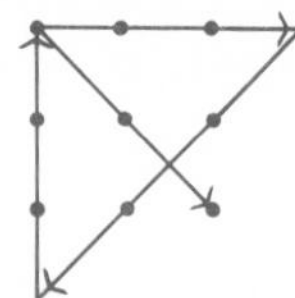

Chapter 8, Page 378

At 4 o'clock, the measure of the angle is 120°. At 3 o'clock, the measure of the angle is 90°.

Chapter 8, Page 386

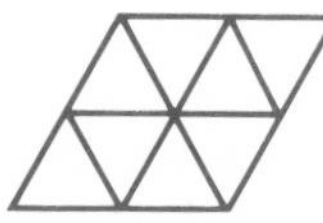

Chapter 8, Page 392

Trace the curve with a pencil and you will see that it is a simple curve.

Chapter 8, Page 399

Move the dotted lines to form the new lines.

Chapter 9, Page 412

The same

Chapter 9, Page 417

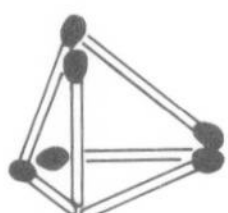

Chapter 9, Page 423

Yes, 60° rotation

Chapter 9, Page 430

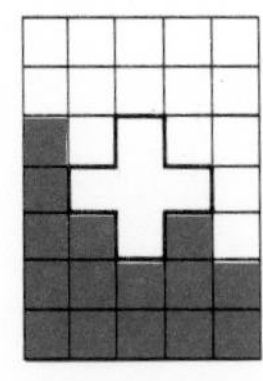

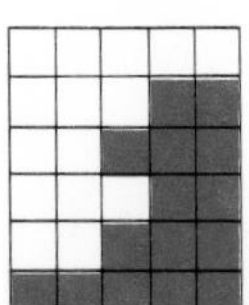

Chapter 9, Page 436

10

Chapter 9, Page 443

All are of the same length.

Chapter 10, Page 455

From barrel A, pour 5 L into the 5-liter bucket. Pour the water in the 5-liter bucket into the 4-liter bucket, leaving 1 L in the 5-liter bucket. Pour the water in the 4-liter bucket back into A. Now A holds 39 liters. Pour the 1 L in the 5-liter bucket into the 4-liter bucket. Fill the 5-liter bucket from A. Fill the 4-liter bucket from the 5-liter bucket. Now there are 2 L in the 5-liter bucket Empty the 4-liter bucket into barrel A, which now contains 38 L. Fill the 4-liter bucket from B. Pour the 4-liter bucket into A unitl A is full, leaving 2 L in the 4-liter bucket. Now each bucket contains 2 L.

Chapter 10, Page 463

The pieces a, b, c, and d will fit as shown around the square e.

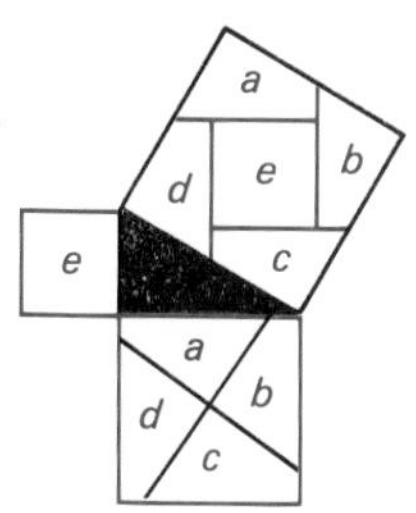

Chapter 10, Page 471

$35\frac{1}{44}$ cm

Chapter 10, Page 479

Have the one whose rope is on top hold his arm up so that the other person can bend over and crawl under his arm. Move until the arms and cord of one person are around the other. Then move arms and cord over the head of the second person.

Chapter 10, Page 480

Yes; there are several possibilities. (see *Activities Manual.*)

Chapter 11, Page 497

F; *B*; *D*

Chapter 11, Page 501

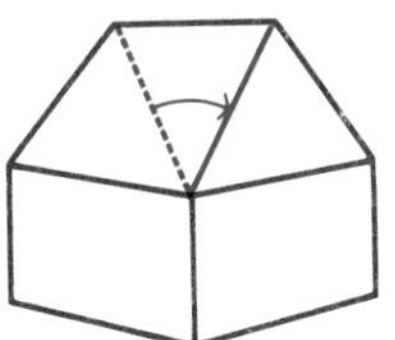

Chapter 11, Page 507

$.640 - .160\pi$ L

Chapter 11, Page 516

The plan is to get either 1 L of water in the 5-L container or 2 liters of water in the 3-L container. Why would these situations lead to a solution? One of the situations is impossible. The other leads to a solution. Find it. Now work backward to find the complete solution.

Chapter 12, Page 526

Same distance.

Chapter 12, Page 534

Move as indicated.

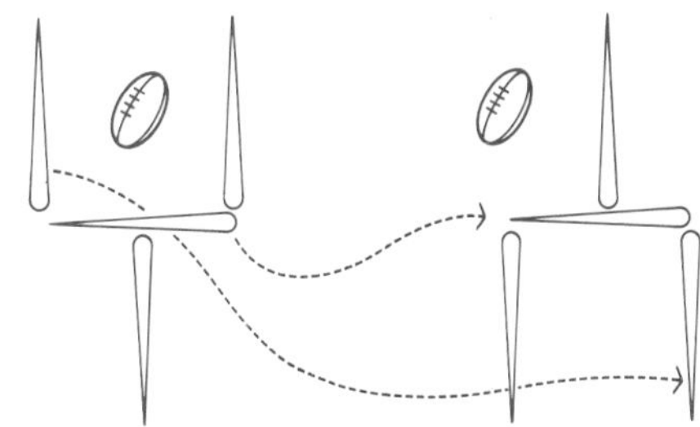

Chapter 12, Page 541

When two houses are connected to all utilities, there exists at least one simple closed curve with the remaining house and one of the utilities in the interior and the exterior of the simple closed curve. Thus, the solution does not exist.

Chapter 12, Page 548

Wife's age 18; husband's age 24

Chapter 12, Page 553

Jodi is 72.
Brooke is 48.

Chapter 13, Page 562

5 2 3 6
4 10 11 12
1 2 3 4 5 6 7 8 9 10 11 12

Jump penny 4 over pennies 2 and 3 and put 4 on top of 1. Jump 5 over 2 and 3 and put 5 on top of 4. Move 12 over 10 and 11 and put it on top of 9. Move 11 over 10 and 9 and put it on top of 8. Move 10 over 8 and 9 on to the top of 7. Move 2 over 3 and 6 and put it on top of the stack on 7. Move 3 over 6 and the stack on 7 to the top of 8. Move 6 over the stack on 7 and the stack on 8 to the top of 9.

Chapter 13, page 568

Move the second full glass and pour into the next-to-the-last empty glass, and return the glass to its original position.

Chapter 13, Page 573

Devise a plane: Write out all the names. Count the number of letters and total. Did you get 52? This is significant because 52 is the number of cards in a standard deck.

Chapter 13, Page 586

Label the boxes and enumerate each step in the procedure.

1	2	3	4	5	6
N	N			D	D

Move the dime in box 6 to box 4 by jumping over the dime in box 5. Now box 6 is empty. Move the nickel in box 2 to box 3. Move the dime in box 4 by jumping it into box 2 over the nickel in box 3. Move the nickel in box 3 to box 4. Move the nickel in box 1 by jumping it into box 3 over the dime in box 2. Move the nickel in box 4 by jumping it into box 6 over the dime in box 5. Move the dime in box 5 to box 4. Move the nickel in box 3 to box 5 by jumping over the dime in box 4. Move the dime in box 4 to box 3. Move the dime in box 3 to box 1 by jumping over the dime in box 2. Now the dimes are on the left and the nickels are on the right. The minimum number of moves is 10.

Now see if you can work the problem with just one vacant space.

Chapter 14, Page 595

2

Chapter 14, Page 601

He moved the stones as follows.

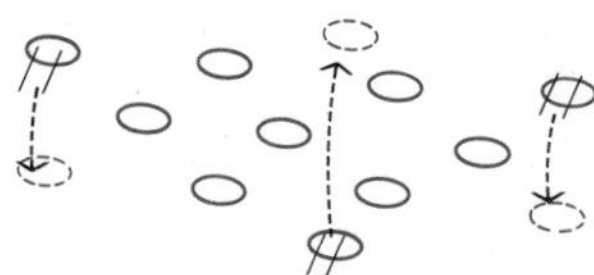

Now the marker points in the opposite direction

Chapter 14, Page 608

$15,000; $14,000

Chapter 14, Page 614

One answer is 62 pennies in the jar. Find the mean of the guesses (54.83) and the mean of the differences (7.17). If the differences are all positive, the total will give the number of pennies in the jar.

Chapter 14, Page 621

L8 R7 L3 R7

INDEX